Fernaroli's Handbook of Flavor Ingredients

CRC

FENAROLI'S HANDBOOK *of* FLAVOR INGREDIENTS

Second Edition

Volume 2

Edited, Translated, and Revised
by
THOMAS E. FURIA and NICOLÓ BELLANCA
Dynapol
Palo Alto, California

Adapted from the Italian language works of
PROF. DR. GIOVANNI FENAROLI
Director, Center for Studies of Aromatic Substances
University of Milano, Milano, Italy

Published by

CRC PRESS

CRC PRESS, Inc.
18901 Cranwood Parkway · Cleveland, Ohio 44128

Library of Congress Cataloging in Publication Data
Fenaroli, Giovanni.
Fenaroli's Handbook of flavor ingredients.

At head of title: CRC.
Includes bibliographies and index.
1. Flavoring essences--Handbooks, manuals, etc.
I. Chemical Rubber Company, Cleveland. II. Title.
III. Title: Handbook of flavor ingredients.
TX589.F4613 1975 664'.5 72-152143
ISBN O-87819-533-5

Comments, criticisms, and suggestions regarding the format and selection of subject matter are invited. Any errors or omissions in the data that appear in the handbook should be brought to the attention of the editors.

Library of Congress Catalog Card Number 72-152143

International Standard Book Number (ISBN)
Set 0-87819-533-5
Volume 2 0-87819-532-7

Direct all inquires and correspondence to
CRC Press, Inc., 18901 Cranwood Parkway, Cleveland, Ohio 44128

Printed in the United States of America

EDITORIAL FOREWORD, FIRST EDITION

The subject of flavor and flavor ingredients, much like its close allies fragrance and perfumery, is still viewed by many through a veil of mystery. To contemporary practitioners of the "pure sciences," the subject of flavors is grossly misunderstood, suggestive of alchemy, and at times subject to condescending criticism. No such indictments are further from reality; few areas of research are more serious or embellished more by sound, scientific thought, and sophisticated techniques. The giants of chemistry have often resorted to problems in flavors and fragrances as a platform for hallmark contributions. It is true that the final assessment of flavor is subjective in character, but, prior to reaching that level of development, most, if not all, of the more objective measuring parameters are *fully* utilized. TLC, GLC, Mass Spec, NMR, UV, and IR, to mention a few, are *routinely* employed not only for resolving flavor components and elucidating structures but also as product control measures. The interdisciplinary nature of the flavor and fragrance industry is characterized by numerous scientific specialties, including analytical, organic, physical, and natural products, chemistry, agronomy, botany, genetics, and taxonomy; food technology; engineering; and finally the artistic efforts of the *flavorist.*

The chemical armament of the flavor industry contains the most numerous constituents of all food ingredients. In the United States it consists of no less than 1,200 ingredients of which approximately 200 are well characterized products of natural origin, while approximately 1,000 are precisely defined synthetics. The task of selecting and blending those ingredients capable of imparting a desired flavor characteristic to complex substrates (often having flavor backgrounds of their own) and at the same time being mindful of toxicological, regulatory, and economic considerations is indeed formidable.

While detailed treatises on various aspects of flavor technology have appeared from time to time, first-source reference texts on the subject as a whole are not readily available in the English language. Those seriously interested in flavor technology have had to wade through numerous journals to locate pertinent information, while the novice or cursory inquirer had only one or two general sources. The scope of this volume is therefore to present a current, authoritative, first-source description of natural and synthetic flavor ingredients, their detailed characteristics, and their application in food. It is primarily intended for those using flavors rather than for the accomplished *flavorist.* Furthermore, it is our intent to revise and up-date this handbook periodically in order to keep pace with advancing technology.

We especially wish to thank Prof. Dr. Giovanni Fenaroli, who arranged to be with us late in 1968 for editorial sessions, and Ulrico Hoepli and L. di G. Pirola, publishers of Prof. Dr. Fenaroli's Italian language books (*Sostanze Aromatiche*, Vols. I, II, and III, and *Aromatizzazione*), for their kind permission to translate and adapt the information for this work. We are grateful to those members of the flavor and fragrance industry who supplied valuable information; outstanding among these is Mr. Frank Fischetti, Jr., of Fritzsche Dodge & Olcott, Inc. We also wish to thank Naarden, Inc., for their permission to reproduce the fold-out schematic flow sheet (Figure 1) in Part I, Volume 1.

Ardsley, New York
November, 1970

T.E.F./N.B.

PREFACE, FIRST EDITION

Flavor is undoubtedly a basic characteristic stimulating the desirability of food; together with color, texture, and nutrition, it forms the cornerstone of contemporary food industries throughout the world. The economic importance of flavor ingredients (essential oils, synthetics, compounded formulations, etc.) is evident from current estimates placing their value at about two-thirds the total sales of all products offered by the flavor and fragrance industry. It is widespread opinion that within a few years an even greater shift in the value of flavors as opposed to fragrances will occur not only from normal increases in productivity but mainly as the result of demands for new food products by both affluent societies and those now in the process of increasing their living standards. Within the decade, however, sharpest increases in the demand for flavor ingredients will arise from new requirements for finished foods geared toward solving the critical starvation problem in most underdeveloped nations. It is therefore evident that while present achievements in flavor technology are indeed remarkable, considering the wide range of products currently offered by the food industry, the tempo will increase sharply. New techniques and more sophisticated flavor formulations designed to respond more precisely to the exacting specifications of future food manufacturers will be required. An example of the quickening pace and exacting specifications that we speak of is the problem of countering flavor modification and alteration in high-protein foods ("protein breakthrough").

With these basic thoughts in mind, it is my primary goal to summarize for the reader my personal experiences lived in this climate of continuing evolution and to rapidly update current efforts in flavor technology. I am convinced that the present state of the art furnishes a sufficiently broad and dynamic base to permit subsequent advances; it is therefore appropriate to describe and examine it in some detail. Certain flavor secrets employed by nature and their exact function are not as yet disclosed, but the investigational means are at our disposal; this makes us confident of the future success in identifying extremely complex systems. The more difficult problems concerning biological factors linked in the formation of naturally occurring flavors, as well as the assessment of the physiological reactions connected with their perception and assimilation by humans and animals, will also be resolved in due course. In the present work I wish to define flavor ingredients as to what they are, how they are manufactured, and how they are used.

In attempting to organize this handbook into a logical sequence capable of future expansion, my primary concern was to present as broad an information base as possible. This was accomplished by presenting the information in four interrelated parts. Part I treats definitions, nomenclature, assessment, and general methods of preparing flavor ingredients; in this section the viewpoint is generalized. (Those who argue that flavor manufacturing methods deserve detailed attention are probably correct, and we shall endeavor to do so in subsequent editions.) Parts II and III consist of detailed descriptions of the natural and synthetic flavor ingredients including the FDA regulatory status and FEMA proposals. Part IV deals exclusively with formulation technology and the uses of compounded flavors of natural and synthetic origin to achieve the most effective flavoring of foods and kindred products. From a technical viewpoint this section presents the greatest challenge, since I felt obliged to discuss not only the organoleptic effects of flavor ingredients and their combined effects in certain food substrates but also auxiliary flavor components, such as non-nutritive sweeteners and flavor potentiators. The flavor formulations presented in Part IV under various substrate categories are typical; while faithfully reproduced and tested, they are intended as guidelines only. In this sense I have attempted to explain the study and evolution of a formulation by stressing those general concepts that must be kept in focus (together with personal taste and experience) during the compounding of a flavor complex or flavoring a finished food product. While of paramount importance, the examples must be considered in light of current regulatory statutes, since it is nearly impossible to predict future restrictions aimed toward safeguarding public health. No doubt, future editions of the handbook will reflect significant changes of this section to keep pace with the application of flavors to food manufacturing problems.

The information presented in this text is partially derived from other works that I have written. Thanks to the Italian publishers Ulrico Hoepli (Milano) and L. di G. Pirola, I was able to summarize and enlarge herein the more basic parts concerning flavor ingredients appearing in *Sostanze Aromatiche Isolate e Sintetiche*, Vol. II, Tomo II, U. Hoepli, Milano, 1968, and *Aromatizzazione*, L. di G. Pirola, Milano, 1969. A personal note of thanks to Dr. U. Hoepli and Messrs. L. A. Bosisio and D. Caremoli of Curt Georgi Imes for their kind cooperation and assistance.

Finally, if I am at all successful in fulfilling my goal, it will be through the invaluable cooperative efforts of T. Furia and N. Bellanca, who not only translated and edited the original manuscript but also interpreted it in its present fashion so as to stress those areas of importance, especially with regard to U.S. regulatory statutes.

Longmeadow, Massachusetts
December 15, 1968

GIOVANNI FENAROLI

PREFACE TO THE SECOND EDITION

The scientific and technological progress which has characterized numerous sectors of industry during the latter half of the 20th century has likewise been extremely apparent in the area of flavor ingredients. Since the publication of the first edition in 1971 (certainly since its inception in 1968), the flavor industry has traveled a rather precipitous road at what must often appear to peripheral observers as hurtling speeds. However, during this period several bench marks have been established:

1. The need for flavors has been recognized at all levels of food processing.
2. The concept that flavor ingredients should be considered food ingredients rather than isolated components of the flavor subunit has been accepted as rational from a technological viewpoint and punctuated by regulatory agencies concerned with safety.
3. Possibly most important, a more valid scientific notion of "flavoring" has been formulated with respect to the preparation of flavors and their uses.

In addition, the food industry is constantly presenting flavor manufacturers with new and increasingly difficult problems. To meet such needs, the number of new flavor ingredients has been expanded, and many await review by regulatory agencies.

Towards this end, the preparation of a revised and expanded second edition, rather than a second printing, was initiated.

As in the first edition, the same four-part format has been maintained, with corrections, additions, and the contributions of specific authors inserted in what was felt to be appropriate sections. The new material presented includes the following:

1. Data on new synthetic flavor ingredients appearing in FEMA lists 4 through 8 (1970–74).
2. Updating of the natural occurrence of flavor ingredients.
3. Addition of references, augmenting many of the topics already discussed while inserting new ones.
4. Through the cooperation of CRC Press, utilization of new, comprehensive reviews on significant flavor topics.

In expressing my thanks to all those who have contributed to the publication of this edition, I would like to highlight the efforts of Thomas E. Furia and Nicolo Bellanca of Dynapol (Palo Alto, California), who edited, translated, and revised the overall publication. Further, during the intervening years since publication of the first edition, they have maintained a close watch on developments in the flavor industry, reminding us, among other things of regulatory issues and gathering new material for review.

Milano
April 1, 1975

Giovanni Fenaroli

CONTRIBUTORS

Basant K. Dwivedi
Associate Research Director
Foster D. Snell, Inc.
Florham Park, New Jersey
The Role of Enzymes in Food Flavor
Meat Flavors

Joseph A. Maga
Department of Food Science and Nutrition
Colorado State University
Fort Collins, Colorado
Bread Flavors

INTRODUCTION AND REFERENCE NOTATIONS FOR PARTS II*AND III

Introduction

The naturally derived and synthetic flavor ingredients listed in Parts II and III represent only those currently employed and approved for use in the United States under provisions of the Federal Food, Drug and Cosmetic Act (November, 1966) as administered by the FDA. While recent developments and changes in the regulations governing the use of flavor ingredients in the United States have been updated and bear the appropriate *date notice*, the reader is cautioned not to use this listing as an "official" statement of current legal status, since it is impossible in a work of this type to post the latest changes. Accordingly, the reader is advised to consult the *Federal Register* for latest official notices applicable under U.S. law.

Regulatory Status Citations

The following general citations are used in Parts II and III with respect to regulatory status:

GRAS Indicates products *generally recognized as safe* when used:
I. in foods as the spice, herb, or botanical and/or
II. in foods when used as the essential oil, oleoresin (solvent-free), and natural extractives, including distillates. All GRAS substances receive this status when prepared and used in accordance with "good manufacturing practice." The reader is advised to consult with FDA for opinions regarding "good manufacturing practices."

§121.1163 Refers to that section of the Federal Food, Drug and Cosmetic Act stating that the indicated natural flavoring substances and natural adjuvants may be used safely in food in accordance with the following conditions:
1. used in minimum quantity required to produce their intended effect and in accordance with all principles of "good manufacturing practice", and
2. they consist of one or more appropriate forms (e.g., plant parts, fluid and solid extracts, concretes, absolute, oils, gums, balsams, resins, oleoresins, waxes, and distillates) used alone or in combination with flavoring substances and adjuvants generally recognized as safe in food, previously sanctioned for use in food, or regulated in any section. The reader is advised to consult the full text of §121.1163 for an official statement of the provision.

§121.1164 Refers to that section of the Federal Food, Drug and Cosmetic Act stating that the indicated synthetic flavoring substances and adjuvants may be safely used in food in accordance with the following conditions:
1. used in minimum quantity required to produce their intended effect and in accordance with all principles of "good manufacturing practice", and
2. they consist of one or more of the appropriate forms used alone or in combination with flavoring substances and adjuvants generally recognized as safe in food, previously sanctioned for use in food or regulated in any section. The reader is advised, as in the case of natural flavoring substances and adjuvants, to consult the full text of §121.1164 for an official statement of the provision.

*For Part II, see volume 1.

REFERENCES FOR PART III

The following references have been used in Part III with respect to physical/chemical constants, methods, and use:

1. **Fenaroli, G.,** *Sostanze Aromatiche Naturali,* Vol. 1, Hoepli, Milan, Italy, 1963.
2. *EOA Specification and Standards,* Essential Oil Association of U.S.A., Inc., 60 East 42nd Street, New York, 10017.
3. FEMA (Flavor and Extract Manufacturers' Association) list of GRAS substances as published in various issues of *Food Technology,* starting with Vol. 19, 1965.
4. *Official Methods of Analysis of the Association of Official Agricultural Chemists,* Section 0.091, 10th edition, 1965.
5. **Fenaroli, G.,** *Sostanze Aromatiche Isolate e Sintetiche,* Vol. 2, Parts 1 and 2, Hoepli, Milan, Italy, 1968.

BIBLIOGRAPHY FOR PART III

Information on natural and synthetic flavoring ingredients is quite plentiful, but scattered throughout several texts, often in limited availability. Nonetheless, good central technical libraries should have these available as reference material. The following texts should be consulted for more detailed information:

Arctander, S., *Perfume and Flavor Chemicals,* Steffan Arctander, Montclair, N.J., 1969.
Arctander, S., *Perfume and Flavor Materials of Natural Origin,* Steffan Arctander, Elizabeth, N.J., 1960.
Bedoukian, P. Z., *Perfumery and Flavoring Synthetics,* 2nd ed., American Elsevier Publishing Co., 1967.
Chemicals Used in Food Processing, Publication 1274, National Academy of Sciences/National Research Council, 1965.
Clair, C., *Of Herbs and Spices,* Abelard-Schuman Ltd., 1961.
Gildemeister, E. and Hoffmann, F. R., *Die Aetherischen Oele,* Vols. 3a–3c, Akademie Verlag, 1960–1963.
Givaudan Index, 2nd ed., Givaudan-Delawanna, Inc., 1961.
Guenther, E., *The Essential Oils,* Vols. 1–6, D. Van Nostrand Co., 1948.
Merory, J. M., *Food Flavorings,* 2nd ed., Avi Publishing Co., 1968.

TABLE OF CONTENTS

PART III. SYNTHETIC FLAVORS

PART IV. USE OF FLAVOR INGREDIENTS

PART III

Synthetic Flavors

	Acetal	Acetaldehyde
Other names	Acetaldehyde diethyl acetal; 1,1-diethoxyethane; diethyl acetal	Ethanal
Empirical formula	$C_6H_{14}O_2$	C_2H_4O
Structure	$CH_3-CH(-O-CH_2-CH_3)(-O-CH_2-CH_3)$	$H_3C-C(=O)H$
Physical/chemical characteristics[5]		
Appearance	Colorless, mobile, flammable liquid; tends to polymerize on aging	Colorless, volatile, flammable liquid
Assay	97% min; free of dissolved metals, chlorides, sulfates, and water	
Molecular weight	118.18	44.05
Melting point		−120 C (−123.3 C)
Boiling point	97–112 C (20 C at 20 mm Hg)	20.8°C
Flash point		−30 C
Specific gravity	0.826–0.830 (0.8264) at 20 C	0.8009
Refractive index	1.3805 at 20 C	1.3392 at 18 C
Solubility	Slightly soluble in water (1:20); completely miscible with ethyl alcohol and ether; soluble in various aliphatic alcohols and other solvents	Water, ethyl alcohol, and ether; commercially available as a 50% aqueous solution or 99% pure in small steel cylinders
Organoleptic characteristics	Pleasant odor with a taste similar to nut	Characteristic pungent and penetrating odor
Synthesis	From ethyl alcohol and acetaldehyde in the presence of anhydrous calcium chloride[6] or small amounts of mineral acids (HCl)[7]	a. By oxidation of ethyl alcohol with potassium dichromate or manganese dioxide in the presence of sulfuric acid b. By addition of water to acetylene c. Forming during the natural alcoholic fermentation process. Recovery is effected by suitable fractionation, subsequent preparation of the acetaldehyde ammonia, and final treatment of the addition compound with diluted sulfuric acid.
Natural occurrence	Present in some liquors (e.g., sake and whisky); also detected and quantitatively assessed in rum[8]	In oak and tobacco leaves; in the fruital aromas of: pear,[9] apple,[10] raspberry,[11] strawberry,[11] and pineapple;[11] in the distillation waters of *Monarda punctata*,[12] orris, cumin, chenopodium; in the essential oils of: *Litsea cubeba*,[13] *Magnolia grandiflora*,[14] *Artemisia brevifolia*, rosemary, balm, clary sage, *Mentha arvensis*, daffodil, bitter orange, camphor, angelica, fennel, mustard, etc.; in rum[8]
Reported uses[3]	Non-alcoholic beverages, 7.3 ppm; ice cream, ices, etc., 52 ppm; candy, 39 ppm; baked goods, 60–120 ppm	Non-alcoholic beverages, 3.9 ppm; ice cream, ices, etc., 25 ppm; candy, 22 ppm; baked goods, 12 ppm; gelatins and puddings, 6.8 ppm; chewing gum, 20–270 ppm
Regulatory status	FDA 121.1164; FEMA No. 2002	FDA GRAS

REFERENCES

For References 1–5, see end of Part III.

6. **Adkins,** *J. Am. Chem. Soc.*, 44, 2750, 1922; *Org. Synth.*, 3, 1, 1933.
7. **de Melle,** German Patent, 566, 1927; U.S. Patent 1,850,836, 1927.
8. **Fenaroli et al.,** *Riv. Ital. EPPOS*, 9, 484, 1965; 2, 75, 1966.
9. **Winter et al.,** *Perfum. Essent. Oil Rec.*, 49, 250, 1958.
10. **Huelin,** *J. Sci. Res. Aust.*, 5, 328, 1952.
11. **Curtius and Franzen,** *Justus Liebigs Ann. Chem.*, 390, 89, 1912.
12. **Harwood,** *J. Am. Pharm. Assoc.*, 20, 631, 1931.
13. **Fürst and Fuestel,** *Z. Chem.*, 1, 22, 1961.
14. **Peyron,** *Fr. Ses Parfums*, 2, 41, 1959; *Soap Perfum. Cosmet.*, 33, 373,

For further information on acetaldehyde see the following:

Karrer, *Konstitution und Vorkommen der Organischen Pflanzenstoffe*, Birkhäuser, 1958, 144, 344.

Gildemeister and Hoffmann, *Die Aetherischen Oele*, Vol. 3c, Akademie Verlag, 1963, 8.

Guenther, *The Essential Oils*, Vol. 2, D. Van Nostrand Co., 1949.

	Acetaldehyde, butyl phenethyl acetal	Acetaldehyde phenethyl propyl acetal
Other names	2-Butoxy-2-phenylethoxy-ethane	Acetal R Pepital 1-Phenethoxy-1-propoxy-ethane Propyl phenethyl acetal
Empirical formula	$C_{14}H_{22}O_2$	$C_{13}H_{20}O_2$
Structure	$CH_3-CH(O-CH_2-CH_2-C_6H_5)(O-CH_2-CH_2-CH_2-CH_3)$	$H_3C-CH(O-CH_2-CH_2-C_6H_5)(O-CH_2-CH_2-CH_3)$
Physical/chemical characteristics[5]		
Appearance		Colorless, stable liquid
Molecular weight	222.32	208.30
Flash point		95°C
Specific gravity		0.944–0.950 at 25°/25°C
Refractive index		1.4770–1.4820 at 20°C
Solubility		1:7 in 70% ethanol; 1:2 in 80% ethanol
Organoleptic characteristics		Very strong odor of green leaves
Synthesis		From acetaldehyde with a mixture of propyl and β-phenyl ethyl alcohols
Natural occurrence		Not reported found in nature
Reported uses[3]	Non-alcoholic beverages 5.0 ppm Ice cream, ices, etc. 5.0 ppm Candy 5.0 ppm Gelatins and puddings 5.0 ppm Chewing gum 5.0 ppm	Candy 2.5 ppm Baked goods 2.5 ppm
Regulatory status	FEMA No. 3125	FDA 121.1164; FEMA No. 2004

REFERENCES

For References 1–5, see end of Part III.

	Acetanisole	Acetic acid
Other names	*p*-Acetyl anisole 4′-Methoxyacetophenone *p*-Methoxyacetophenone Navatone	Ethanoic acid
Empirical formula	$C_9H_{10}O_2$	$C_2H_4O_2$
Structure	$H_3C-C(=O)-C_6H_4-O-CH_3$	$CH_3—COOH$
Physical/chemical characteristics[5]		
Appearance	Yellowish-white crystals at room temperature	Colorless liquid
Assay	98% min.[2] (as a ketone)	
Molecular weight	150.18	60.05
Melting point	38°C (36°C)	17°C (16.6–16.7°C)
Boiling point	158/256°C	118°C (118.2–118.5°C)
Freezing point	34.7°C (not less than 36°C)[2]	
Specific gravity	1.0959 at 25°C; 1.0818 at 41.1°C	1.049 at 20°C
Refractive index	1.55489 at 25°C; 1.547 at 41.1°C	1.3718 at 20°C
Halogens	Free[2]	
Optical rotation	±0°	
Solubility	Slightly soluble in water; from 1:4–1:5 in 50% ethanol; soluble in most organic solvents	Completely miscible with water, ethanol, ether, and carbon tetrachloride; insoluble in carbon disulfide
Organoleptic characteristics	Odor similar to that of *p*-methylacetophenone; suggestive of hawthorn and floral note of heliotrope; bitter, unpleasant taste	Strong, pungent, characteristic odor
Synthesis	From anisole and acetyl chloride in the presence of aluminum chloride and carbon disulfide; from anisole and acetic acid in the presence of boron trifluoride	From the destructive distillation of wood; from acetylene and water; via acetaldehyde by subsequent oxidation with air
Natural occurrence	In anise seed	Identified among the constituents of petitgrain lemon oil,[6] in bitter orange oil,[7] and in strawberry aroma[8]
Reported uses[3]	Non-alcoholic beverages 2.3 ppm Ice cream, ices, etc. 2.5 ppm Candy 4.6 ppm Baked goods 5.8 ppm Chewing gum 840 ppm	Non-alcoholic beverages 39 ppm Ice cream, ices, etc. 32 ppm Candy 52 ppm Baked goods 38 ppm Gelatins and puddings 15 ppm Chewing gum 60 ppm Condiments 5,900 ppm
Regulatory status	FDA 121.1164; FEMA No. 2005	FDA GRAS; FEMA No. 2006

REFERENCES

For References 1–5, see end of Part III.

6. **Peyron,** *Riv. Ital. EPPOS,* 8, 413, 1965.
7. **Mehlitz and Minas,** *Ind. Obst. Gemueseverwert.,* 50, 861, 1965.
8. **Willhalm et al.,** *Helv. Chim. Acta,* 49, 66, 1966.

	Acetone	Acetoin
Other names	2-Propanone Dimethyl ketone β-Ketopropane Pyroacetic ether	Acetyl methyl carbinol 2,3-Butanolone Dimethylketol 3-Hydroxy-2-butanone γ-Hydroxy-β-oxobutane
Empirical formula	C_3H_6O	$C_4H_8O_2$
Structure	$CH_3-C(=O)-CH_3$	$CH_3-C(=O)-CH(OH)-CH_3$
Physical/chemical characteristics[5]		
Appearance	Colorless, mobile, highly flammable liquid	Yellowish liquid; forms dimer on standing that reconverts to monomer by melting
Molecular weight	58.08	88.10
Melting point	–95.35°C	15°C
Boiling point	57°C	148°C (140/142°C)
Flash point	–20°C	
Specific gravity	0.788 at 25°C 0.8186 at 0°C	0.9972 at 17°C
Refractive index	1.3588 at 20°C	1.4190 at 17.3°C
Optical rotation		–39.4° (pure) and –105° (aqueous solution) at 20°C; *also see* Synthesis (following)
Solubility	Soluble in water in all proportions; soluble in alcohol, ether, benzene, and chloroform	Miscible with water; soluble in ethanol; slightly soluble in ether
Organoleptic characteristics	Characteristic aromatic odor; pungent, somewhat sweet taste	Pleasant, butter-like odor and flavor
Synthesis	By fermentation, or by chemical synthesis from isopropanol, cumene, or propane[9]	From diacetyl by partial reduction with zinc and acid,[6] also a product of fermentation Acetoin is an optically active compound. The *d* (–) acetyl methyl carbinol is obtained from fermentation and, in mixture with other products, from the catalytic oxidation of 2,3-butanediol.[7] The *l* (+) acetyl methyl carbinol is obtained from fermentation. The optically pure form has not been isolated; the optically inactive form is prepared synthetically.
Natural occurrence	In apple, pear, grape, pineapple, strawberry, raspberry, tomato, black currant, citrus, onion, and potato; also reported found in cocoa leaves, in Mexican goosefoot, and in the oils of coriander and lavender; in trace amounts it has been reportedly identified in the oil of bitter orange, in distilled wine, and in coffee aroma[10]	Present in corn, wine, vinegar (from grape or other sources), honey, cocoa, butter, and roasted coffee; identified also in currant and strawberry aromas[8]
Reported uses	Non-alcoholic beverages 5.0 ppm Ice cream, ices, etc. 5.0 ppm Candy 8.0 ppm Baked goods 8.0 ppm Gelatins and puddings 5.0 ppm Preserves and spreads 5.0 ppm	Non-alcoholic beverages 7.4 ppm Ice cream, ices, etc. 3.3 ppm Candy 18 ppm Baked goods 32 ppm Gelatins and puddings 0.60–21 ppm Shortening 8.0 ppm Margarine 0.80–50 ppm Cheese (cottage) 7.0 ppm
Regulatory status	FEMA No. 3326	FDA GRAS; FEMA No. 2008

REFERENCES

For References 1–5, see end of Part III.

6. **Diels and Stephan,** *Berichte,* 40, 4338, 1907.
7. **Blom,** *J. Am. Chem. Soc.,* 67, 484, 1945.
8. **Coppens and Hoejenbons,** *Chem. Zentralbl.,* 2, 2339, 1939.
9. **Weiss,** *Chem. Eng. News,* 36, 79, 1958.
10. **Stoll et al.,** *Helv. Chim. Acta,* 50, 628, 1967.

	Acetophenone	3-Acetyl-2,5-dimethyl furan
Other names	Acetyl benzene Hypnone Methyl phenyl ketone	2,5-Dimethyl-3-acetyl furan
Empirical formula	C_8H_8O	$C_8H_{10}O_2$
Structure	O ‖ C—CH_3	$COCH_3$ H_3C CH_3 O
Physical/chemical characteristics[5]		
Appearance	Almost colorless liquid at room temperature	
Assay	98% min.	
Molecular weight	120.14	138
Melting point	19.5–20.5°C	
Boiling point	202°C	
Freezing point	19.5°C	
Flash point	76°C	
Specific gravity	1.0329 at 15°C (1.025–1.028 at 25°/25°C)[2]	
Refractive index	1.53631 at 15°C (1.533–1.5350 at 20°C)[2]	
Solubility	1:5 in 50% ethanol; soluble in essential oils and most organic solvents; insoluble in water	
Organoleptic characteristics	Characteristic sweet, pungent odor (acacia) and bitter, aromatic flavor	
Synthesis	From benzene and acetylchloride in the presence of aluminum chloride or by catalytic oxidation of ethyl benzene; also prepared by fractional distillation and crystallization from the essential oil of *Stirlingia latifolia*	
Natural occurrence	In the oils of labdanum, *Stirlingia latifolia*, *Urtica dioica*,[6] *Elsholtzia argyi* var. *nipponica*,[7] *Elsholtzia ciliata*,[8] in various species of *Orthodon*[9] (*O. citraliferum*, *O. linalooliferum* var. *laerolinalooliferum*, *O. linaloiferum*, *O. sabinoliferum* var. *taiwanense*), and in castoreum absolute[10]	
Reported uses[3]	Non-alcoholic beverages 0.98 ppm Ice cream, ices, etc. 2.8 ppm Candy 3.6 ppm Baked goods 5.6 ppm Gelatins and puddings 7.0 ppm Chewing gum 0.60–20 ppm	Non-alcoholic beverages 1.0 ppm Ice cream, ices, etc. 0.6 ppm Candy 1.5 ppm Baked goods 2.0 ppm Gelatins and puddings 1.5 ppm Meat, meat sauces, soups 1.0 ppm Milk, dairy products 0.6 ppm
Regulatory status	FDA 121.1164; FEMA No. 2009	FEMA No. 3391

REFERENCES

For References 1–5, see end of Part III.

6. **Naves and Ardizio,** *Perfum. Essent. Oil Rec.*, 46, 79, 1955.
7. **Fujita and Ueda,** *J. Chem. Soc. Jap. Pure Chem. Sect.*, 79, 1067, 1958.
8. **Fujita and Ueda,** *J. Chem. Soc. Jap. Pure Chem. Sect.*, 80, 1495, 1959.
9. **Fujita and Ueda,** *Miltizer Ber.*, 91, 1959; *J. Chem. Soc. Jap. Pure Chem. Sect.*, 1957, 1958, 1959.
10. **Lederer,** *Schimmel Ber.*, 25, 1950.

	2-Acetyl-3,5 (and 6)-dimethylpyrazine	2-Acetyl-3-ethylpyrazine
Other names	(1) 2-Acetyl-3,5-dimethylpyrazine (2) 3-Acetyl-2,5-dimethylpyrazine	2-Acetyl-3-ethyl-1,4-diazine 3-Ethyl-2-pyrazinyl methyl ketone
Empirical formula	$C_8H_{10}N_2O$	$C_8H_{10}N_2O$
Structure	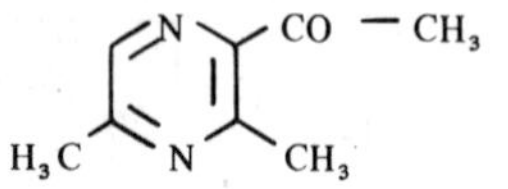 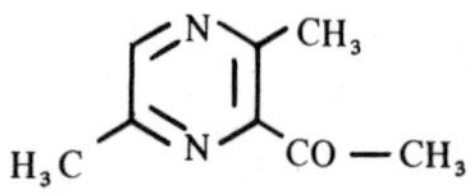	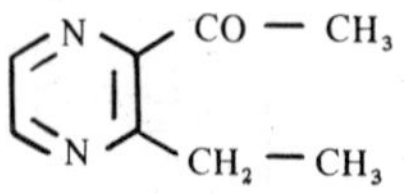
Physical/chemical characteristics		
Assay	C: 63.98%; H: 6.71%; N: 18.65%; O: 10.65%	C: 63.98%; H: 6.71%; N: 18.65%; O: 10.65%
Molecular weight	150.18 (calculated = 150.079312)	150.18 (calculated = 150.079312)
Boiling point		55°C at 1.1 mm Hg
Mass spectra[6]		150, 43, 107, 52, 79, 27
IR spectra[6]		3.3, 3.39, 3.41, 5.5, 5.9, 6.05, 6.45, 6.55, 6.85, 7.09, 7.14, 7.4, 7.6, 7.85, 8.02, 8.45, 8.6, 8.7, 9.2, 9.3, 9.45, 9.72, 10.3, 10.5, 11.65μ
NMR spectra[6]		($CDCl_3$) δ 1.28 (t, 3, $-CH_2-CH_3$), 2.69 (s, 3, $-C(=0)-$**CH_3**), 3.15 (q, 2, **$-CH_2$** **$-CH_3$**) 8.44 (d-1, ring –H), 8.61 (d,1, ring –H)
Synthesis	From the corresponding dimethylethylpyrazine by bromination, followed by oxidation to the ketone[6]	From 2-ethyl-3-methylpyrazine by bromination, followed by oxidation to the ketone[7]
Reported uses[3]	Ice cream, ices, etc. 1.0 ppm Candy 5.0 ppm Baked goods 5.0 ppm Gelatins and puddings 1.0 ppm Meat, meat sauces 5.0 ppm Soups 5.0 ppm Cereals 5.0 ppm	Non-alcoholic beverages 10.0 ppm Ice cream, ices, etc. 10.0 ppm Candy 10.0 ppm Baked goods 10.0 ppm Gelatins and puddings 10.0 ppm Meat, meat sauces, soups 10.0 ppm Milk, dairy products 10.0 ppm Cereals 10.0 ppm
Regulatory status	FEMA No. 3327	FEMA No. 3250

REFERENCES

For References 1–5, see end of Part III.

6. **Mookherjee and Klaiber,** *J. Org. Chem.*, 37(3), 511, 1972.
7. **Mookherjee and Klaiber,** *J. Org. Chem.*, 37, 511, 1972.

	2-Acetylfuran	Acetylpyrazine
Other names	2-Furyl methyl ketone 2-Acetofuran	Methylpyrazinyl ketone 2-Acetylpyrazine
Empirical formula	$C_6H_6O_2$	$C_6H_6ON_2$
Structure	(furan ring)—CO — CH_3	(pyrazine ring)—CO — CH_3
Physical/chemical characteristics[5]		
Appearance	Colorless crystals	
Assay		C: 59.01%, H: 4.95%; O: 13.10%; N: 22.94%
Molecular weight	110.11	122.13
Melting point	33°C	79–80°C
Boiling point	173°C 67°C at 10 mm Hg 60°C at 12 mm Hg	
Mass spectra[13]		43 (100), 52 (44), 53 (34), 79 (27), 80 (45), 122 (37)
IR spectra[14]		1683, 1570, 1282, 1163, 1102, 1043, 1020, 950, 858 cm^{-1}
UV spectra[14]		λ_{max}^{ETOH} 269 nm
NMR spectra[14]		T = 0.82 (1, db, ring – H), 1.28 (1, db, ring – H), 1.37 (1, qd, ring – H), 7.28 (3, S, – CH_3)
Specific gravity	1.098 at 20°C	
Refractive index	1.5017 at 20°C	
Solubility	Insoluble in water; soluble in alcohol, ether, and propylene glycol	
Organoleptic characteristics	Powerful, balsamic-sweet odor	Flavor reminiscent of popcorn
Synthesis	From furan and acetyl chloride by a Friedel-Kraft condensation[6]	By the ester condensation of EtO_2C-pyrazine;[10] by dehydrating pyrazynamide with $POCl_3$ and then reacting the resulting 2-cyanopyrazine with methyl magnesium bromide[11]
Natural occurrence	Reported found as a constituent in coffee aroma;[7] also isolated from beech and oak wood tar oils; an aroma component of yellow passion fruit (*Passiflora edulis* f. flavicarpa);[8] also a volatile flavor component of potato chips[9]	Reported present in peanuts, filberts, popcorn, and sesame oil[12]
Reported uses[3]	Baked goods 20.0 ppm Condiments 20.0 ppm Meats, meat sauces 20.0 ppm Soups 20.0 ppm	Non-alcoholic beverages 5.0 ppm Ice cream, ices, etc. 5.0 ppm Baked goods 5.0 ppm
Regulatory status	FEMA No. 3163	FEMA No. 3126

REFERENCES

For References 1–5, see end of Part III.

6. Reichstein, *Helv. Chim. Acta,* 13, 356, 1930.
7. Gianturco, Gianmarino and Friedel, *Nature,* 210, 1358, 1966.
8. Winter and Klöti, *Helv. Chim. Acta,* 55(6), 1916, 1972.
9. Deck, Pokorny, and Chang, *J. Food Sci.,* 38(2), 345, 1973.
10. Shindo, *Chem. Pharm. Bull.* (Tokyo), 8, 33, 1960.
11. British Patent 1,164,050, assigned to R. J. Reynolds Tobacco Co.
12. Maga and Sizer, *Crit. Rev. Food Technol.,* 4(1), 39, 1973.
13. Walradt et al., *J. Agric. Food Chem.,* 18, 926, 1970.
14. Roberts, U.S. Patent 3,402,051, 1968.

	2-Acetylpyridine		3-Acetylpyridine	
Other names	Methyl-2-pyridyl ketone 2-Acetopyridine		β-Acetylpyridine Methyl-3-pyridyl ketone	
Empirical formula	C_7H_7ON		C_7H_7ON	
Structure	[pyridine ring with COCH$_3$ at 2-position]		[pyridine ring with CO — CH$_3$ at 3-position]	
Physical/chemical characteristics[5]				
Appearance	Colorless liquid; tends to yellow on exposure to air		Colorless liquid; turns yellow in air	
Molecular weight	121.14		121.13	
Boiling point	192°C 78°C at 12 mm Hg		220°C 106°C at 12 mm Hg	
Refractive index	1.5203 at 20°C			
Solubility	Soluble in alcohol, ether, and acids		Soluble in water, ethanol, ether, and acids	
Organoleptic characteristics	Tobacco-like aroma			
Synthesis	From ethyl picolinate[6]		Dry distillation of calcium nicotinate with calcium acetate[7]	
Natural occurrence			In roasted filberts[8]	
Reported uses[3]	Candy	3.0 ppm	Non-alcoholic beverages	2.0 ppm
	Baked goods	5.0 ppm	Ice cream, ices, etc.	2.0 ppm
	Meats, meat sauces	3.0 ppm	Candy	3.0 ppm
	Soups	3.0 ppm	Baked goods	3.0 ppm
	Milk, dairy products	3.0 ppm	Gelatins and puddings	2.0 ppm
	Cereals	3.0 ppm		
Regulatory status	FEMA No. 3251		FEMA No. 3424	

REFERENCES

For References 1–5, see end of Part III.

6. **Kolloff and Hunter,** *J. Am. Chem. Soc.*, 63, 490, 1941.
7. **Engler et al.,** *Chem. Ber.*, 22, 597, 1889.
8. **Kinlin et al.,** *J. Agric. Food Chem.*, 20, 1021, 1972.

	2-Acetyl thiazole	Aconitic acid
Other names	Methyl-2-thiazolyl ketone 5-Acetyl thiazole Methyl-5-thiazolyl ketone	Achilleic acid Citridic acid Equisetic acid 1-Propene-1,2,3-tricarboxylic acid
Empirical formula	C_5H_5NOS	$C_6H_6O_6$
Structure	(thiazole ring with N and S)–CO–CH_3	CH–COOH ‖ C–COOH | CH_2–COOH
Physical/chemical characteristics		
Appearance	Oily liquid	White or yellowish crystalline solid
Molecular weight	127.16	174.11
Melting point	64.5–65.5°C	194–195°C (with decomposition)
Boiling point	95–105°C at 15 mm Hg 74°C at 11 mm Hg	
Refractive index	n_D^{20} 1.5480	
Solubility		Soluble in water and alcohol; slightly soluble in ether
Synthesis	By oxidation of the corresponding carbinol using dichromate[6]	By dehydration of citric acid with concentrated H_2SO_4.[7] Aconitic acid prepared by this or other methods has the *trans* configuration. The *cis*-isomer is little known.
Natural occurrence	In beef broth	In beet root, sugar cane, the leaves of *Aconitum napellus*, and other natural products (*Pelargonium* species)
Reported uses[3]	Non-alcoholic beverages 0.2 ppm Ice cream, ices, etc. 0.9 ppm Candy 1.4 ppm Chewing gum 0.6 ppm Protein foods 0.7 ppm	Non-alcoholic beverages 0.20–2.0 ppm Alcoholic beverages 20 ppm Ice cream, ices, etc. 0.60 ppm Candy 0.60–30 ppm Baked goods 0.60–15 ppm Chewing gum 28 ppm
Regulatory status	FEMA No. 3328	FDA GRAS; FEMA No. 2010

REFERENCES

For References 1–5, see end of Part III.

6. **Berand and Metzger,** *Bull. Soc. Chim. Fr.*, p. 2072, 1962.
7. **Bruce,** *Org. Synth.*, 2, 12, 1943.

	Adipic acid	β-Alanine
Other names	1,4-Butanedicarboxylic acid Hexanedioic acid	β-Aminopropionic acid 3-Aminopropanoic acid 3-Aminopropionic acid
Empirical formula	$C_6H_{10}O_4$	$C_3H_7O_2N$
Structure	$HOOC-(CH_2)_4-COOH$	$H_2N-CH_2-CH_2-COOH$
Physical/chemical characteristics[5]		
Appearance	White, crystalline solid	Needles (from water)
Molecular weight	146.14	89.10
Melting point	153°C	207°C (decomposes)
Boiling point	216°C at 15 mm Hg	
Specific gravity	1.360 at 25°C	1.437 at 19°C
Solubility	Slightly soluble in cold water; soluble in hot water; soluble in alcohol; insoluble in benzene	Soluble in water; slightly soluble in alcohol; insoluble in ether and acetone
Organoleptic characteristics		Slightly sweet taste
Synthesis	By oxidation of cyclohexanol with concentrated nitric acid; by catalytic oxidation of cyclohexanone with air[6,7]	By heating acrylic acid with concentrated aqueous ammonia under pressure;[8] by addition of acrylonitrile to phthalimide or to ammonia;[9] from β-aminopropionitrile;[10] from succinimide by the Hofmann degradation;[11] patented process[12]
Natural occurrence	In beet juice	Reported to occur as a component in amino acids: carnosine, auserine, pantothenic acid
Reported uses[3]	Non-alcoholic beverages 40 ppm Gelatins and puddings 5,000 ppm It can usually replace tartaric acid in baking powders because of its lower hygroscopicity	Non-alcoholic beverages 10.0 ppm Baked goods 10.0 ppm Condiments, pickles 10.0 ppm Meat, meat sauce 10.0 ppm Soups 10.0 ppm Milk, dairy products 10.0 ppm Cereals 10.0 ppm
Regulatory status	FDA 121.1164; FEMA No. 2011	FEMA No. 3252

REFERENCES

For References 1–5, see end of Part III.

6. **Bouveault and Locquin,** *Bull. Soc. Chim. Fr.*, 3, 438, 1908.
7. **Thorpe and Kon,** *Org. Synth.*, 5, 9, 1925.
8. **Baker and Ollis,** *J. Chem. Soc.*, p. 345, 1949.
9. **Horning,** *Organic Syntheses,* Coll. Vol. 3, Wiley & Sons, New York, 1955, 34.
10. **Blatt,** *Organic Syntheses,* Coll. Vol. 2, Wiley & Sons, New York, 1943, 19.
11. **Clarke and Behr,** *Org. Synth.*, 16, 1, 1936.
12. U.S. Patents 2,734,081 and 2,956,080.

	Allyl anthranilate	Allyl butyrate
Other names	Allyl 2-aminobenzoate Allyl *o*-aminobenzoate	
Empirical formula	$C_{10}H_{11}NO_2$	$C_7H_{12}O_2$
Structure	$C_6H_4(NH_2)-C(=O)-O-CH_2-CH=CH_2$	$CH_3-(CH_2)_2-C(=O)-O-CH_2-CH=CH_2$
Physical/chemical characteristics[5]		
Appearance		Liquid
Molecular weight	177.21	128.17
Boiling point		44–45°C at 15 mm Hg
Flash point		
Specific gravity		0.9017 at 20°C
Refractive index		1.4158 at 20°C
Synthesis		From allyl alcohol and butyric acid in the presence of concentrated H_2SO_4 or *p*-toluenesulfonic acid in benzene[6–8]
Natural occurrence	Not reported found in nature	
Reported uses[3]	Non-alcoholic beverages 1.1 ppm Ice cream, ices, etc. 0.67 ppm Candy 2.0 ppm Baked goods 0.02–1.0 ppm Gelatins and puddings 2.0 ppm	Non-alcoholic beverages 1.2 ppm Ice cream, ices, etc. 0.50–1.0 ppm Candy 1.3 ppm Baked goods 0.50–3.0 ppm Gelatins and puddings 1.0 ppm
Regulatory status	FDA 121.1164; FEMA No. 2020	FDA 121.1164; FEMA No. 2021

REFERENCES

For References 1–5, see end of Part III.

6. Golendejew, *Zh. Obshch, Khim.*, 6, 1844, 1936.
7. Jeffery and Vogel, *J. Chem. Soc.*, p. 663, 1948.
8. Salmi and Leimu, *Suom. Kemistil. B*, 20, 47, 1947.

	Allyl cinnamate	Allyl cyclohexaneacetate
Other names		
Empirical formula	$C_{12}H_{12}O_2$	$C_{11}H_{18}O_2$
Structure	$C_6H_5-CH=CH-C(=O)-O-CH_2-CH=CH_2$	$(S)-CH_2-C(=O)-O-CH_2-CH=CH_2$
Physical/chemical characteristics[5]		
Appearance	Almost colorless liquid	Liquid
Molecular weight	188.22	182.26
Boiling point	150–152°C	66°C at 1 mm Hg
Specific gravity	1.048 at 23°C	
Refractive index	1.5661 (1.530) at 23°C	1.4574 at 20°C
Solubility	Allyl 3-phenylpropenoate; allyl β-phenylacrylate	
Organoleptic characteristics	Peach or apricot-like aroma	Intense fruital (pineapple, peach, apricot) aroma
Synthesis	By esterification of cinnamic acid with allyl alcohol in the presence of concentrated H_2SO_4 [6]	By esterification of cyclohexane acetic acid and allyl alcohol in the presence of benzene
Natural occurrence	Not reported found in nature	Not reported found in nature
Reported uses[3]	Non-alcoholic beverages 1.0 ppm Ice cream, ices, etc. 1.4 ppm Candy 1.8 ppm Baked goods 2.6 ppm	Non-alcoholic beverages 1.1 ppm Ice cream, ices, etc. 1.6 ppm Candy 3.5 ppm Baked goods 4.0 ppm
Regulatory status	FDA 121.1164; FEMA No. 2022	FDA 121.1164; FEMA No. 2023

REFERENCES

For References 1–5, see end of Part III.

6. Seeligmann, Dissertation, Technical University of Karlsruhe, Germany, 1906, 17.

	Allyl cyclohexane butyrate	Allyl cyclohexanehexanoate
Other names		Allyl cyclohexylcaproate Allyl cyclohexylcapronate
Empirical formula	$C_{13}H_{22}O_2$	$C_{15}H_{26}O_2$
Structure	(S)$-(CH_2)_3-\overset{O}{\overset{\Vert}{C}}-O-CH_2-CH{=}CH_2$	(S)$-(CH_2)_5-\overset{O}{\overset{\Vert}{C}}-O-CH_2-CH{=}CH_2$
Physical/chemical characteristics[5]		
Appearance	Liquid	
Molecular weight	210.31	238.37
Boiling point	104°C at 1 mm Hg	
Refractive index	1.4608 at 20.5°C	
Organoleptic characteristics	Pineapple aroma	
Synthesis	By direct esterification in the presence of benzene	By direct esterification of the acid with allyl alcohol
Natural occurrence	Not reported found in nature	Not reported found in nature
Reported uses[3]	Non-alcoholic beverages 1.0 ppm Ice cream, ices, etc. 1.4 ppm Candy 3.3 ppm Baked goods 3.8 ppm	Non-alcoholic beverages 1.4 ppm Ice cream, ices, etc. 3.3 ppm Candy 8.0 ppm Baked goods 8.5 ppm
Regulatory status	FDA 121.1164; FEMA No. 2024	FDA 121.1164; FEMA No. 2025

REFERENCES

For References 1–5, see end of Part III.

	Allyl cyclohexane propionate	Allyl cyclohexanevalerate
Other names	Allyl-3-cyclohexylpropionate Allyl-β-cyclohexylpropionate	Allyl cyclohexylpentanoate
Empirical formula	$C_{12}H_{20}O_2$	$C_{14}H_{24}O_2$
Structure	$\langle S \rangle-(CH_2)_2-\overset{O}{\overset{\Vert}{C}}-O-CH_2-CH{=}CH_2$	$\langle S \rangle-(CH_2)_4-\overset{O}{\overset{\Vert}{C}}-O-CH_2-CH{=}CH$
Physical/chemical characteristics[5]		
Appearance	Colorless liquid	Liquid
Molecular weight	98% min.[2] (ester content)	224.34
Melting point	196.29	
Boiling point	91°C at 1 mm Hg	119°C at 1 mm Hg
Specific gravity	0.945–0.950[2] at 25°/25°C	
Refractive index	1.4595 (1.4570–1.4620)[2] at 22°C	1.4605 at 22°C
Acid value	Not more than 5.0	
Solubility	1:4 in 80% ethanol[2]	
Organoleptic characteristics	Pineapple aroma	Characteristic fruital (peach, apricot, apple) aroma
Synthesis	By direct esterification of cyclohexylpropionic acid with allyl alcohol in the presence of benzene	By direct esterification in the presence of benzene
Natural occurrence	Not reported found in nature	Not reported found in nature
Reported uses[3]	Non-alcoholic beverages 3.7 ppm Ice cream, ices, etc. 3.1 ppm Candy 13 ppm Baked goods 7.1 ppm Gelatins and puddings 7.7 ppm Chewing gum 30 ppm Icings 0.20 ppm	Non-alcoholic beverages 1.2 ppm Ice cream, ices, etc. 2.3 ppm Candy 4.4 ppm Baked goods 4.8 ppm
Regulatory status	FDA 121.1164; FEMA No. 2026	FDA 121.1164; FEMA No. 2027

REFERENCES

For References 1–5, see end of Part III.

	Allyl disulfide	Allyl 2-ethylbutyrate
Other names	Diallyl disulfide	
Empirical formula	$C_6H_{10}S_2$	$C_9H_{16}O_2$
Structure	$CH_2{=}CH{-}CH_2{-}S{-}S{-}CH_2{-}CH{=}CH_2$	$CH_3{-}CH_2{-}CH(CH_2{-}CH_3){-}C({=}O){-}O{-}CH_2{-}CH{=}CH_2$
Physical/chemical characteristics[5]		
Appearance	Liquid	Liquid
Molecular weight	146.26	156.23
Boiling point	138/139°C (79°C at 16 mm Hg)	165–167°C; 58–60°C at 6 mm Hg
Specific gravity	1.010 at 15°C	
Refractive index	1.541 at 20°C	1.4240 at 15°C
Solubility	Insoluble in water; soluble in most common organic solvents	
Organoleptic characteristics	Characteristic garlic odor	Ethereal aroma with a slightly fruit-like note
Synthesis	By oxidation of allyl mercaptan with iodine in the presence of pyridine and ethanol;[6] from sodium allyl thiosulfate with potassium hydroxide[7]	By prolonged heating of 2-ethylbutyric acid (sodium salt) with allyl bromide in xylene;[9] by direct esterification of the acid with allyl alcohol[10]
Natural occurrence	Main constituent of *Allium sativum* essential oil[8]	Not reported found in nature
Reported uses[3]	Condiments 6.5 ppm Meats 7.0 ppm	Non-alcoholic beverages 0.50–1.0 ppm Candy 2.0 ppm Gelatins and puddings 1.0 ppm
Regulatory status	FDA 121.1164; FEMA No. 2028	FDA 121.1164; FEMA No. 2029

REFERENCES

For References 1–5, see end of Part III.
6. **Smoll, Bailey and Cavallito,** *J. Am. Chem. Soc.,* 69, 1711, 1947.
7. **Challenger and Greenwood,** *Biochem. J.,* 44, 90, 1949.
8. **Semmler,** *Arch. Pharm.* (Wernheim), 230, 438, 1892.
9. **Arnold and Liggett,** *J. Am. Chem. Soc.,* 64, 2877, 1942.
10. **Seldner,** *Am. Perfum.,* 54(3), 296, 1949.

	Allyl heptylate	Allyl hexanoate
Other names	Allyl heptanoate Allyl heptoate Allyl oenanthate	Allyl caproate Allyl capronate 2-Propenyl hexanoate
Empirical formula	$C_{10}H_{18}O_2$	$C_9H_{16}O_2$
Structure	$H_3C-(CH_2)_5-C(=O)-O-CH_2-CH=CH_2$	$H_3C-(CH_2)_4-C(=O)-O-CH_2-CH=CH_2$
Physical/chemical characteristics[5]		
Appearance	Liquid	Colorless to light-yellow, mobile liquid; stable
Assay		98% min
Molecular weight	170.25	156.23
Boiling point	210°C	75/76°C at 15 mm Hg
Flash point		68°C
Specific gravity	0.890 at 15.5°C	0.8869 at 20°C; 0.885–0.890 at 25°C
Refractive index	1.4290 at 20°C	1.4243 (1.422–1.426) at 20°C
Acid value		1 max
Solubility		Insoluble in water; soluble in most organic solvents; 1:2 in 80% ethanol
Organoleptic characteristics	Characteristic wine odor with a slight banana note and banana-like flavor	Fruit-like aroma (pineapple)
Synthesis		By esterification of *n*-capronic acid with allyl alcohol in the presence of concentrated H_2SO_4[6] or of naphthalene-β-sulfonic acid in benzene under a nitrogen blanket[7]
Natural occurrence	Not reported found in nature	Not reported found in nature
Reported uses[3]	Non-alcoholic beverages 1.3 ppm Ice cream, ices, etc. 2.7 ppm Candy 6.4 ppm Baked goods 6.4 ppm Gelatins and puddings 2.9 ppm Chewing gum 86 ppm	Non-alcoholic beverages 7.0 ppm Ice cream, ices, etc. 11 ppm Candy 32 ppm Baked goods 25 ppm Gelatins and puddings 22 ppm Chewing gum 210 ppm
Regulatory status	FDA GRAS; FEMA No. 2031	FDA 121.1164; FEMA No. 2032

REFERENCES

For References 1–5, see end of Part III.

6. **Golendejew,** *Zh. Obshch. Khim.*, 10, 1411, 1940; *Chem. Abstr.*, p. 3067, 1941.
7. **Swern and Jordan,** *J. Am. Chem. Soc.*, 70, 2336, 1948.

	Allyl α-ionone	Allyl isothiocyanate
Other names	Butenyl α-cyclocitrylidenemethyl ketone Cetone V α-Cyclocitrylidenemethyl butenyl ketone 1-(2,6,6-Trimethyl-2-cyclohexen-1-yl)-1,6-heptadiene-3-one	Mustard oil (mustard oil volatile, synthetic)
Empirical formula	$C_{16}H_{24}O$	C_4H_5NS
Structure	H_3C CH_3 ring $-CH=CH-C(=O)-(CH_2)_2-CH=CH_2$; CH_3	$CH_2=CH-CH_2-N=C=S$
Physical/chemical characteristics[5]		
Appearance	Yellow liquid	Colorless liquid; tends to darken on aging
Assay	88% min	
Molecular weight	232.35	99.19
Melting point		−102.5°C
Boiling point	102/104°C at 0.15 mm Hg	152°C
Flash point	> 100°C	
Specific gravity	0.9289 at 20°C; 0.928–0.935 at 25°/25°C	1.02356 at 15°C
Refractive index	1.50402 (1.5030–1.5070) at 20°C	1.52481 at 15°C
Solubility	1:8 in 70% ethanol	1:8 in 80% ethanol; slightly soluble in water; completely miscible with ether, chloroform, and benzene
Organoleptic characteristics	Strong fruital aroma reminiscent of pineapple	Strong, pungent odor; lachrymatory
Synthesis	By condensation of citral and allyl acetone in the presence of sodium methoxide and subsequent ring closure in the presence of boron trifluoride	By distillation of sodium thiocyanate and allyl chloride
Natural occurrence	Not reported found in nature	In the essential oil from seeds of *Brassica nigra* Koch.,[6] *Brassica juncea* Hook. and Thoms.,[7] and *Thlaspi arvense*;[8] in the essential oil from roots of *Cochlearia armoracia*;[9] in the seeds and roots of *Alliaria officinalis*;[10,11] in onion juice;[12] and in the seeds of various *Cruciferae*[13-16]
Reported uses[3]	Non-alcoholic beverages 0.50 ppm Ice cream, ices, etc. 1.4 ppm Candy 2.6 ppm Baked goods 3.1 ppm Gelatins and puddings 1.0 ppm Jellies 2.0 ppm	Non-alcoholic beverages 0.02–0.50 ppm Ice cream, ices, etc. 0.50 ppm Candy 0.50 ppm Baked goods 5.2 ppm Condiments 52 ppm Meats 87 ppm Syrups 10–88 ppm
Regulatory status	FDA 121.1164; FEMA No. 2033	FDA 121.1164; FEMA No. 2034

REFERENCES

For References 1–5, see end of Part III.

6. **Will,** *Ann. Chem.*, 52, 1, 1840–44.
7. **Gildemeister and Hoffmann,** *Die Aetherischen Oele,* 3rd ed., Vol. 1, Akademie Verlag, 1963, 686.
8. **Pless,** *Ann. Chem.*, 58, 38, 1845.
9. **Hubatka,** *Ann. Chem.*, 47, 153, 1840.
10. **Pless,** *Ann. Chem.*, 58, 38, 1845.
11. **Wertheim,** *Ann. Chem.*, 52, 52, 1840.
12. **Kooper,** *Chem. Zentralbl.*, 2, 331, 1910.
13. **Raquet,** *Chem. Zentralbl.*, 2, 457, 1912.
14. **Grimme,** *Chem. Zentralbl.*, 2, 613, 1912.
15. **Carles,** *Chem. Zentralbl.*, 2, 297, 1913.
16. **Imbert and Juillet,** *Chem. Zentralbl.*, 2, 1170, 1913.

	Allyl isovalerate	Allyl mercaptan
Other names		Allyl sulfhydrate Allylthiol 2-Propene-1-thiol
Empirical formula	$C_8H_{14}O_2$	C_3H_6S
Structure	$(H_3C)_2CH-CH_2-C(=O)-O-CH_2-CH=CH_2$	$CH_2=CH-CH_2-SH$
Physical/chemical characteristics[5]		
Appearance	Liquid	Colorless liquid; tends to darken on aging
Molecular weight	142.19	74.14
Boiling point	89–90°C	67–68°C
Refractive index	1.4162 at 21°C	0.925 at 23°C
Solubility		Insoluble in water; completely miscible with alcohol and ether
Organoleptic characteristics	Fruit-like (apple, cherry) aroma	Strong, garlic odor
Synthesis	By direct esterification in the presence of benzene	
Natural occurrence	Not reported found in nature	In onion and garlic
Reported uses[3]	Non-alcoholic beverages 8.6 ppm Ice cream, ices, etc. 18 ppm Candy 22 ppm Baked goods 15–48 ppm Gelatins and puddings 1.0 ppm	Non-alcoholic beverages 0.25 ppm Ice cream, ices, etc. 0.50–2.0 ppm Candy 0.50 ppm Baked goods 0.50–2.0 ppm Chewing gum 2.0–3.0 ppm Condiments 0.50 ppm
Regulatory status	FDA 121.1164; FEMA No. 2045	FDA 121.1164; FEMA No. 2035

REFERENCES

For References 1–5, see end of Part III.

	Allyl methyl disulfide		Allyl methyl trisulfide	
Other names			Methyl allyl trisulfide	
Empirical formula	$C_4H_8S_2$		$C_4H_8S_3$	
Structure	$CH_2=CH-CH_2-S-S-CH_3$		$CH_2=CH-CH_2-S-S-S-CH_3$	
Physical/chemical characteristics				
Molecular weight	120.23		152.30	
Boiling point	30–33°C at 20 mm Hg		47°C at 0.8 mm Hg	
Refractive index	n_D^{25} 1.5453		n_D^{21} 1.5930	
Synthesis	By treating a mixture of corresponding aliphatic thiosulfates with an alkali metal sulfide[6]			
Natural occurrence	It occurs in *Allium sativum* (garlic), *Allium chinense,* and *Allium ampeloprasum;*[7] it is the main component of *Allium victorialis*[8]		It is a main volatile component of flavor of caucas (*Allium rictorialis*)	
Reported uses[3]	Baked goods	1.0 ppm	Baked goods	2.0 ppm
	Condiments, pickles	1.0 ppm	Condiments, pickles	2.0 ppm
	Meat, meat sauces, soups	1.0 ppm	Meat, meat sauces, soups	2.0 ppm
Regulatory status	FEMA No. 3127		FEMA No. 3253	

REFERENCES

For References 1–5, see end of Part III.
6. German Patent 2,020,051; *Chem. Abstr.,* 74, 86571z, 1971.
7. **Jacobsen, Bernhard and Mann.,** *Arch. Biochem. Biophys.,* 104(3), 473, 1964.
8. **Nishimura et al.,** *J. Agric. Food Chem.,* 19, 992, 1971.

	Allyl nonanoate	Allyl octanoate
Other names	Allyl pelargonate	Allyl caprylate Allyl octylate
Empirical formula	$C_{12}H_{22}O_2$	$C_{11}H_{20}O_2$
Structure	$CH_3-(CH_2)_7-C(=O)-O-CH_2-CH=CH_2$	$CH_3-(CH_2)_6-C(=O)-O-CH_2-CH=CH_2$
Physical/chemical characteristics[5]		
Appearance	Mobile liquid	Light, mobile liquid
Molecular weight	198.31	184.28
Boiling point	151°C at 50 mm Hg	87–88°C at 5.5 mm Hg
Specific gravity	0.8702 at 30°C	0.8729 at 30°C
Refractive index	1.4302 at 30°C	1.4271 at 30°C
Solubility		
Organoleptic characteristics	Characteristic fruital aroma (cognac, pineapple)	Fruit-like aroma
Synthesis	By esterification of nonanoic acid with allyl alcohol in benzene solution in the presence of naphthalene-β-sulfonic acid[6]	By esterification of caprylic acid with allyl alcohol in benzene in the presence of naphthalene-β-sulfonic acid[7]
Natural occurrence	Not reported found in nature	
Reported uses[3]	Non-alcoholic beverages 0.70 ppm Ice cream, ices, etc. 0.50–3.0 ppm Candy 5.0 ppm Baked goods 3.0–5.0 ppm Meats 1.0 ppm	Non-alcoholic beverages 1.7 ppm Ice cream, ices, etc. 3.3 ppm Candy 5.1 ppm Baked goods 4.0 ppm Gelatins and puddings 0.10 ppm
Regulatory status	FDA 121.1164; FEMA No. 2036	FDA 121.1164; FEMA No. 2037

REFERENCES

For References 1–5, see end of Part III.

6. **Swern and Jordan,** *J. Am. Chem. Soc.,* 70, 2337, 1948.
7. **Swern and Jordan,** *J. Am. Chem. Soc.,* 70, 2336, 1948.

	Allyl phenoxyacetate	Allyl phenylacetate
Other names	Acetate PA	Allyl α-toluate
Empirical formula	$C_{11}H_{12}O_3$	$C_{11}H_{12}O_2$
Structure	$C_6H_5-O-CH_2-C(=O)-O-CH_2-CH=CH_2$	$C_6H_5-CH_2-C(=O)-O-CH_2-CH=CH_2$
Physical/chemical characteristics[5]		
Appearance	Liquid	Liquid
Molecular weight	192.22	176.22
Boiling point	100–102°C at 1 mm Hg	89–93°C at 3 mm Hg
Specific gravity		
Refractive index	1.5131 at 25.5°C	1.5122 at 13.5°C
Solubility		
Organoleptic characteristics	Honey and pineapple-like aroma	Honey-like aroma
Synthesis	By direct esterification in benzene solution	By direct esterification in benzene solution
Natural occurrence	Not reported found in nature	Not reported found in nature
Reported uses[3]	Non-alcoholic beverages 0.82 ppm Ice cream, ices, etc. 0.004–40 ppm Candy 2.3 ppm Baked goods 0.02–1.0 ppm Gelatins and puddings 3.0 ppm	Non-alcoholic beverages 0.06–3.0 ppm Ice cream, ices, etc. 8.0 ppm Candy 14 ppm Baked goods 40 ppm
Regulatory status	FDA 121.1164; FEMA No. 2038	FDA 121.1164; FEMA No. 2039

REFERENCES

For References 1–5, see end of Part III.

	Allyl propionate	Allyl sorbate
Other names		Allyl-2,4-hexadienoate
Empirical formula	$C_6H_{10}O_2$	$C_9H_{12}O_2$
Structure	$CH_3-CH_2-C(=O)-O-CH_2-CH=CH_2$	$CH_3-CH=CH-CH=CH-C(=O)-O-CH_2-CH=CH_2$
Physical/chemical characteristics[5]		
Appearance	Liquid	
Molecular weight	114.15	152.19
Boiling point	122–123°C (121°C)	
Specific gravity	0.9140 at 20°C	
Refractive index	1.4105 at 20°C; 1.4142 at 14°C	
Solubility		
Organoleptic characteristics	Fruital (apricot, apple) aroma	
Synthesis	By direct esterification of the acid with allyl alcohol in benzene solution in the presence of concentrated H_2SO_4[6] or of *p*-toluene sulfonic acid[7]	
Natural occurrence	Not reported found in nature	Not reported found in nature
Reported uses[3]	Non-alcoholic beverages 0.06–3.0 ppm Ice cream, ices, etc. 16 ppm Candy 6.5 ppm Baked goods 10 ppm	Non-alcoholic beverages 0.86 ppm Ice cream, ices, etc. 0.50 ppm Candy 0.50–5.0 ppm Baked goods 1.0 ppm Gelatins and puddings 2.0 ppm
Regulatory status	FDA 121.1164; FEMA No. 2040	FDA 121.1164; FEMA No. 2041

REFERENCES

For References 1–5, see end of Part III.

6. **Jeffery and Vogel,** *J. Chem. Soc.,* p. 663, 1948.
7. **Salmi and Leimu,** *Chem. Abstr.,* 42, 4031, 1948.

	Allyl sulfide		Allyl thiopropionate	
Other names	Diallyl sulfide Thioallyl ether		Propionic acid, thioacrylic ester Thiopropionic acid, allyl ester	
Empirical formula	$C_6H_{10}S$		$C_6H_{10}OS$	
Structure	$CH_2{=}CH{-}CH_2{-}S{-}CH_2{-}CH{=}CH_2$		$CH_3{-}CH_2{-}C({=}S){-}O{-}CH_2{-}CH{=}CH_2$	
Physical/chemical characteristics				
Appearance	Colorless liquid			
Molecular weight	114.20		130.21	
Boiling point	139°C			
Specific gravity	0.888 at 27°C			
Refractive index	1.4877 at 27°C			
Solubility	Insoluble in water; miscible with alcohol, ether, and other organic solvents			
Organoleptic characteristics	Characteristic garlic odor			
Synthesis	See *Beilstein* B1[2], 478, and also reference 6			
Natural occurrences	Identified in garlic and garlic extracts;[6] also identified in the leaves of *Diplotazis tenuifolia* and the roots of *Armoracia lapathifolia*. Allyl disulfide appears to be a secondary rather than primary natural product.			
Reported uses[3]	Non-alcoholic beverages	0.04 ppm	Baked goods	0.25 ppm
	Ice cream, ices, etc.	0.06 ppm	Condiments, pickles	0.25 ppm
	Candy	0.07 ppm	Meat, meat sauces, soups	0.25 ppm
	Baked goods	0.05 ppm		
	Condiments	13 ppm		
	Meats	3.7 ppm		
Regulatory status	FDA 121.1164; FEMA No. 2042		FEMA No. 3329	

REFERENCES

For References 1–5, see end of Part III.

6. Wertheim, *Annalen,* 51, 289, 1844; 55, 297, 1845.

	Allyl tiglate		Allyl 10-undecenoate	
Other names	Allyl-*trans*-2,3-dimethylacrylate Allyl-*trans*-2-methyl-2-butenoate		Allyl hendecenoate Allyl undecylenate Allyl undecylenoate	
Empirical formula	$C_8H_{12}O_2$		$C_{14}H_{24}O_2$	
Structure	$CH_3-CH=C(CH_3)-C(=O)-O-CH_2-CH=CH_2$		$CH_2=CH-(CH_2)_8-C(=O)-O-CH_2-CH=CH$	
Physical/chemical characteristics[5]				
Appearance			Liquid	
Molecular weight	140.18		224.34	
Boiling point			180°C at 30 mm Hg; 114–118°C at 1 mm Hg	
Specific gravity			0.8802 at 30°C; 0.935 at 15.5°C	
Refractive index			1.4448 at 30°C; 1.4479 at 23°C	
Solubility			Insoluble in water; miscible with most organic solvents	
Organoleptic characteristics			Fruital aroma suggestive of pineapple	
Synthesis			By azeotropic distillation of a benzene solution of the corresponding acid and allyl alcohol in the presence of naphthalene-β-sulfonic acid[6]	
Natural occurrence	Not reported found in nature		Not reported found in nature	
Reported uses[3]	Non-alcoholic beverages	0.28 ppm	Non-alcoholic beverages	0.25–1.0 ppm
	Ice cream, ices, etc.	0.50 ppm	Ice cream, ices, etc.	0.50 ppm
	Candy	0.50–3.0 ppm	Candy	0.50 ppm
	Baked goods	0.50–3.0 ppm	Baked goods	0.50 ppm
Regulatory status	FDA 121.1164; FEMA No. 2043		FDA 121.1164; FEMA No. 2044	

REFERENCES

For References 1–5, see end of Part III.

6. Jordan and Swern, *J. Am. Chem. Soc.*, 71, 2378, 1949.

	Ammonium isovalerate	Ammonium sulfide
Other names	Isovaleric acid, ammonium salt	Ammonium monosulfide
Empirical formula	$C_5H_{13}O_2N$	H_8N_2S
Structure	$(H_3C)_2CH-CH_2-C(=O)-O^- \ NH_4^+$	$H_4N-S-NH_4$
Physical/chemical characteristics[5]		
Appearance	Deliquescent crystals	Liquid at room temperature (>−18°C, yellow crystals); tends to decompose at higher temperatures
Molecular weight	109	68.15
Solubility	Soluble in alcohol and in water	Very soluble in ammonia and in cold water; decomposes in hot water; soluble in alcohol
Organoleptic characteristics	Odor reminiscent of valeric acid; sharp, sweet taste; used in butter, nut, and cheese flavors	
Synthesis		
Natural occurrence		
Reported uses[3]	Baked goods 58 ppm Syrups 0.20 ppm	Baked goods 5.0 ppm Condiments 5.0 ppm
Regulatory status	FDA 121.1164; FEMA No. 2054	FDA 121.1164; FEMA No. 2053

REFERENCES

For References 1–5, see end of Part III.

	Amyl alcohol	Amyl butyrate
Other names	1-Pentanol Pentyl alcohol	Pentyl butyrate
Empirical formula	$C_5H_{12}O$	$C_9H_{18}O_2$
Structure	$CH_3-CH_2-CH_2-CH_2-CH_2OH$	$CH_3-(CH_2)_2-\overset{\overset{\displaystyle O}{\|}}{C}-O-(CH_2)_4-CH_3$
Physical/chemical characteristics[5]		
Appearance	Colorless liquid	Colorless liquid
Molecular weight	88.15	158.24
Melting point	−73.85°C (−78.5°C)	
Boiling point	138.06°C (137.8–138°C)	185–186°C
Flash point	39°C	
Specific gravity	0.8144 at 20°C	0.8659 at 20°C
Refractive index	1.4113 at 20°C	1.4123 at 20°C
Solubility	1:5 in water at 30°C; miscible with most organic solvents	
Organoleptic characteristics	Characteristic fusel-like odor and burning taste; somewhat more toxic than ethyl alcohol	Strong, penetrating odor; sweet taste
Synthesis	By hydrogenation of valeric aldehyde with sodium amalgam; from amyl chloride	From *n*-amyl alcohol and butyric acid in the presence of H_2SO_4[7]
Natural occurrence	In the essential oils of Brazilian and American peppermint, Spanish "origanum", *Thymus marshallianus*, *Artemisia herba-alba*, *Eucalyptus cornuta*, *Eucalyptus aggregata*, oreganum, and cottonseed oil. It has been identified also among the constituents of bitter orange oil[6]	Found in several natural products
Reported uses[3]	Non-alcoholic beverages 18 ppm Ice cream, ices, etc. 15 ppm Candy 35 ppm Baked goods 24 ppm Gelatins and puddings 7.7–50 ppm Chewing gum 150–340 ppm	Non-alcoholic beverages 19 ppm Ice cream, ices, etc. 32 ppm Candy 76 ppm Baked goods 43 ppm Gelatins and puddings 0.50–1.4 ppm Chewing gum 760 ppm Syrups 58 ppm
Regulatory status	FDA 121.1164; FEMA No. 2056	FDA 121.1164; FEMA No. 2059

REFERENCES

For References 1–5, see end of Part III.

6. **Mehlitz and Minas,** *Ind. Obst. Gemueseverwert.,* 50, 861, 1965.
7. **Vogel,** *J. Chem. Soc.,* p. 629, 1948.

	α-Amylcinnamaldehyde	α-Amylcinnamaldehyde dimethyl acetal
Other names	Amyl cinnamic aldehyde α-Amyl-β-phenylacrolein 2-Benzylidene heptanal α-Pentylcinnamaldehyde	α-*n*-Amyl-β-phenylacroleindimethylacetal 1,1-Dimethoxy-2-amyl-3-phenyl-2-propene 1,1-Dimethoxy-2-benzylidene-heptane
Empirical formula	$C_{14}H_{18}O$	$C_{16}H_{24}O_2$
Structure	$C_6H_5-CH=C(CHO)-CH_2-(CH_2)_3-CH_3$	$C_6H_5-CH=C(CH(O-CH_3)_2)-(CH_2)_4-CH_3$
Physical/chemical characteristics[5]		
Appearance	Yellowish, clear liquid	Yellowish liquid; stable
Assay	97% min	
Molecular weight	202.29	248.37
Melting point		
Boiling point	153–154°C at 10 mm Hg	
Flash point		92°C
Specific gravity	0.9715 at 15°C; 0.963–0.968 at 25°C	0.953–0.958 at 25°/25°C; 0.935 at 15.5°C
Refractive index	1.5552 (1.5540–1.5590) at 20°C	1.5050–1.5080 at 20°C; 1.5245 at 20°C
Acid value		1 max.
Solubility	1:6 in 80% ethanol; miscible with most organic solvents	1:9 in 80% alcohol; 1:2 in 90% alcohol
Organoleptic characteristics	Distinct floral (jasmine, lily) note	Floral note reminiscent of jasmine
Synthesis	By condensation of *n*-amyl aldehyde with cinnamic aldehyde. This method of condensation of aromatic aldehydes with aliphatic aldehydes has the maximum yield in α-amylcinnamic aldehyde with little formation of the inferior homologs. The methyl, ethyl, and propyl amylcinnamic aldehyde analogs exhibit a characteristic scent[6]	
Natural occurrence	Not reported found in nature	Not reported found in nature
Reported uses[3]	Non-alcoholic beverages 1.3 ppm Ice cream, ices, etc. 1.5 ppm Candy 4.0 ppm Baked goods 4.5 ppm Gelatins and puddings 0.03–0.05 ppm Chewing gum 15 ppm	Non-alcoholic beverages 0.80 ppm Ice cream, ices, etc. 1.5–2.0 ppm Candy 2.0 ppm Baked goods 2.6 ppm
Regulatory status	FDA 121.1164; FEMA No. 2061	FDA 121.1164; FEMA No. 2062

REFERENCES

For References 1–5, see end of Part III.

6. **Bedoukian,** *Perfumery Synthetics and Isolates,* D. Van Nostrand, 1951, 43.

	α-Amylcinnamyl acetate	**α-Amylcinnamyl alcohol**
Other names	α-*n*-Amyl-β-phenylacryl acetate Floxin acetate α-Pentylcinnamyl acetate	*n*-Amyl cinnamic alcohol 2-Amyl-3-phenyl-2-propen-1-ol 2-Benzylidene-heptanol α-Pentylcinnamyl alcohol
Empirical formula	$C_{16}H_{22}O_2$	$C_{14}H_{20}O$
Structure	$C_6H_5-CH=C(-(CH_2)_4-CH_3)-CH_2-O-C(=O)-CH_3$	$C_6H_5-CH=C(-(CH_2)_4-CH_3)-CH_2OH$
Physical/chemical characteristics[5]		
Appearance	Colorless, oily liquid	Yellowish liquid
Assay		95% min. (97%)
Molecular weight	246.36	204.31
Flash point		>100°C
Specific gravity	>1.0	0.954–0.962 at 25°C; 0.9675–0.9705 at 15°C
Refractive index		1.5330–1.5400 (1.5365–1.5385) at 20°C
Solubility	Soluble in alcohol	1:3 in 70% ethanol; miscible with water and ether
Organoleptic characteristics	Mild, fruity, green odor with a balsamic and slightly floral undernote; somewhat fruity flavor	Light, floral note
Synthesis	From amylcinnamic alcohol and acetic acid	
Natural occurrence	Not reported found in nature	Not reported found in nature
Reported uses[3]	Non-alcoholic beverages 0.92 ppm Ice cream, ices, etc. 3.5 ppm Candy 3.5 ppm Baked goods 3.0 ppm Chewing gum 3.0 ppm	Non-alcoholic beverages 0.47 ppm Ice cream, ices, etc. 1.5 ppm Candy 1.6 ppm Baked goods 1.5 ppm Chewing gum 2.0 ppm
Regulatory status	FDA 121.1164; FEMA No. 2064	FDA 121.1164; FEMA No. 2065

REFERENCES

For References 1–5, see end of Part III.

	α-Amylcinnamyl formate	α-Amylcinnamyl isovalerate
Other names	α-*n*-Amyl-β-phenylacryl formate α-Pentylcinnamyl formate	α-*n*-Amyl-β-phenylacryl isovalerate Floxin isovalerate α-Pentylcinnamyl isovalerate
Empirical formula	$C_{15}H_{20}O_2$	$C_{19}H_{28}O_2$
Structure	$C_6H_5-CH=C(-(CH_2)_4-CH_3)-CH_2-O-C(=O)H$	$C_6H_5-CH=C(-CH_2-CH_2-CH_2-CH_2-CH_3)-CH_2-O-C(=O)-CH_2-CH(CH_3)_2$
Physical/chemical characteristics[5]		
Appearance	Colorless liquid	Colorless liquid
Molecular weight	232.33	288.43
Solubility	Soluble in alcohol	Soluble in alcohol
Organoleptic characteristics	Sweet, oily, herbaceous, and somewhat green odor	Mild, fruity, somewhat tobacco-like odor; oily, fruity, hay-like, somewhat spicy flavor
Synthesis	By formylation of the alcohol	From amylcinnamic alcohol and isovaleric acid
Natural occurrence	Not reported found in nature	Not reported found in nature
Reported uses[3]	Non-alcoholic beverages 0.17 ppm Ice cream, ices, etc. 0.93 ppm Candy 1.5 ppm Baked goods 1.5 ppm Chewing gum 1.0 ppm	Non-alcoholic beverages 0.36 ppm Ice cream, ices, etc. 1.2 ppm Candy 1.3 ppm Baked goods 1.7 ppm Chewing gum 1.0 ppm
Regulatory status	FDA 121.1164; FEMA No. 2066	FDA 121.1164; FEMA No. 2067

REFERENCES

For References 1–5, see end of Part III.

	Amyl formate	Amyl heptanoate
Other names	Pentyl formate	Amyl heptoate Amyl heptylate Pentyl heptanoate
Empirical formula	$C_6H_{12}O_2$	$C_{12}H_{24}O_2$
Structure	$CH_3-(CH_2)_4-O-\overset{\overset{O}{\Vert}}{C}-H$	$CH_3-(CH_2)_5-\overset{\overset{O}{\Vert}}{C}-O-(CH_2)_4-CH_3$
Physical/chemical characteristics[5]		
Appearance	Colorless liquid	Colorless liquid
Molecular weight	116.16	200.32
Melting point	−73.5°C	−49.5°C (from pentane)
Boiling point	130.4°C	245.4°C (240°C)
Specific gravity	0.8846 at 20°C, 1.8926 at 15°C	0.85427 at 20°C; 0.882 at 15.5°C
Refractive index	1.3997 at 20°C; 1.3951 at 11.5°C	1.42627 (1.4240) at 20°C
Solubility	Very slightly soluble in water; miscible with alcohol, ether, and other organic solvents	Soluble in most organic solvents
Organoleptic characteristics	Fruit-like aroma	Characteristic, strong, ethereal, fruital (unripe banana) odor and corresponding taste
Synthesis	From *n*-amyl alcohol and formic acid in the presence of H_2SO_4[6]	From the corresponding heptanoic acid and *n*-amyl alcohol in the presence of mineral acids[7] or in benzene solution in the presence of *p*-toluene sulfonic acid[8]
Natural occurrence	Identified as a constituent of *Pyrus malus*	Not reported found in nature
Reported uses[3]	Non-alcoholic beverages 13 ppm Ice cream, ices, etc. 11 ppm Candy 31 ppm Baked goods 8.0 ppm Chewing gum 170 ppm	Non-alcoholic beverages 7.0 ppm Ice cream, ices, etc. 3.8 ppm Candy 7.5 ppm Baked goods 3.0 ppm Gelatins and puddings 3.5 ppm Chewing gum 53 ppm
Regulatory status	FDA 121.1164; FEMA No. 2068	FDA 121.1164; FEMA No. 2073

REFERENCES

For References 1–5, see end of Part III.

6. **Kohlraush and Pongratz,** *Z. Phys. Chem. Abt. B.*, 22, 379, 1933.
7. **Gartenmeister,** *Justus Liebigs Ann. Chem.*, 233, 284, 1886; Bilterys and Gisseleire, *Bull. Soc. Chem. Belg.*, 44, 569, 572, 1935.
8. **Hoback, Parsons and Bartlett,** *J. Am. Chem. Soc.*, 65, 1606, 1943.

	Amyl hexanoate	Amyl octanoate
Other names	Amyl caproate Amyl hexylate Pentyl hexanoate	Amyl caprylate Amyl octylate Pentyl octanoate
Empirical formula	$C_{11}H_{22}O_2$	$C_{13}H_{26}O_2$
Structure	$CH_3-(CH_2)_4-\overset{\overset{\displaystyle O}{\|}}{C}-O-(CH_2)_4-CH_3$	$CH_3-(CH_2)_6-\overset{\overset{\displaystyle O}{\|}}{C}-O-(CH_2)_4-CH_3$
Physical/chemical characteristics[5]		
Appearance	Colorless liquid	Liquid
Molecular weight	186.30	214.35
Melting point	−47°C	−34 to −35°C
Boiling point	226°C (116°C at 20 mm Hg)	260°C (124–216°C at 12 mm Hg)
Specific gravity	0.8612 at 25°C	0.8562 at 25°C
Refractive index	1.4202 at 25°C	1.4262 at 25°C
Solubility		
Organoleptic characteristics	Characteristic fruit-like (banana, pineapple) aroma	Odor suggestive of orris
Synthesis	By esterification of hexanoic acid with *n*-amyl alcohol in benzene solution in the presence of *p*-toluenesulfonic acid[6]	By esterification of octanoic acid with *n*-amyl alcohol in benzene solution in the presence of *p*-toluenesulfonic acid[6]
Natural occurrence	Occurs in several natural products	In strawberry and apple
Reported uses[3]	Non-alcoholic beverages 5.3 ppm Ice cream, ices, etc. 16 ppm Candy 22 ppm Baked goods 8.3 ppm Gelatins and puddings 0.30–3.7 ppm Chewing gum 110 ppm	Non-alcoholic beverages 5.0 ppm Ice cream, ices, etc. 3.5 ppm Candy 6.0 ppm Baked goods 3.5 ppm Gelatins and puddings 2.1 ppm
Regulatory status	FDA 121.1164; FEMA No. 2074	FDA 121.1164; FEMA No. 2079

REFERENCES

For References 1–5, see end of Part III.

6. **Hoback, Parsons and Bartlett,** *J. Am. Chem. Soc.,* 65, 1606, 1943.

	Anethole	α-Angelica lactone
Other names	Isoestragole 1-Methoxy-4-propenyl benzene *p*-Methoxypropenyl benzene *p*-Propenyl anisole *p*-Propenylphenyl methyl ether	4-Hydroxy-3-pentenoic acid lactone β, γ-Angelica lactone γ-Angelica lactone 5-Methyl-2 (3H)-furanone γ-Methyl-β-butenolide
Empirical formula	$C_{10}H_{12}O$	$C_5H_6O_2$
Structure	$H_3C-CH=CH-C_6H_4-OCH_3$	H_3C–(furanone ring)=O
Physical/chemical characteristics[5]		
Appearance	White, crystalline solid; melts at room temperature	Needle-like crystals; readily isomerizes to the beta form
Molecular weight	148.20	98.10
Melting point	20–23°C	18°C
Boiling point	234–237°C	167–170°C 56°C at 12 mm Hg
Specific gravity	0.983–0.987 at 25°C	1.084 at 20°C
Refractive index	1.558–1.561 at 25°C	1.4476 at 20°C
Solubility	1:8 in 80% alcohol; 1:1 in 90% ethanol; almost water insoluble; miscible with chloroform and ether	Slightly soluble in water; soluble in alcohol, ether, and carbon disulfide
Organoleptic characteristics	Characteristic anise odor and corresponding sweet taste	Sweet, herbaceous odor reminiscent of tobacco
Synthesis	By esterification of *p*-cresol with methyl alcohol and subsequent condensation with acetaldehyde (Perkins). The most common method of preparation is from pine oil. By fractional distillation of the essential oils of anise, star anise, and fennel. The anise essences contain an average of 85% anethole; fennel, from 60 to 70%.	By dry distillation of levulinic acid[6]
Natural occurrence	In addition to the mentioned oils, it is also present in the oil of *Illicium anisatum* L., in the oil from leaves of *Piper peltatum* Sur., in the roots of *Osmorrhiza longistylis* Eberh., in the oils of common fennel and basil, and in the oils from leaves of *Clausena anisata* Hook., *Backhousia anisata*, and *Magnolia salicifolia*.	
Reported uses[3]	Non-alcoholic beverages 11 ppm Alcoholic beverages 1,400 ppm Ice cream, ices, etc. 26 ppm Candy 340 ppm Baked goods 150 ppm Chewing gum 1,500 ppm	Ice cream, ices, etc. 0.2 ppm Candy 2.0 ppm Baked goods 4.0 ppm Gelatins and puddings 2.0 ppm Meats 2.0 ppm Milk, dairy products 0.2 ppm Cereals 2.0 ppm
Regulatory status	FDA GRAS; FEMA No. 2086	FEMA No. 3293

REFERENCES

For References 1–5, see end of Part III.
6. **Wolff,** *Annalen,* 229, 250, 1885.

	Anisole	Anisyl acetate
Other names	Methoxybenzene Methyl phenyl ether	*p*-Methoxybenzyl acetate
Empirical formula	C_7H_8O	$C_{10}H_{12}O_3$
Structure	$C_6H_5-OCH_3$	$CH_3O-C_6H_4-CH_2-O-C(=O)-CH_3$
Physical/chemical characteristics[5]		
Appearance	Colorless liquid	Colorless to slightly yellow liquid
Assay		97% min.[2]
Molecular weight	108.13	180.21
Melting point	37–38°C	270°C
Boiling point	155°C	
Flash point	42°C	
Specific gravity	0.999 (0.9980–1.0010) at 15°C	1.104–1.107[2] at 25°/25°C; 1.014 at 20°C
Refractive index	1.5179 (1.5165–1.5175) at 20°C	1.5110–1.5160[2] at 20°C; 1.515 at 15°C
Acid value		Not more than 1.0[2]
Solubility	Insoluble in water; soluble in alcohol and ether	1:6 in 60% alcohol; insoluble in water
Organoleptic characteristics	Characteristic, pleasant aromatic odor	Floral fruit-like odor; a slight pungent, sweet taste
Synthesis	By reacting phenol and dimethyl sulfate in the presence of aqueous NaOH; by passing methyl chloride into a suspension of sodium phenolate in liquid ammonia	May be prepared by the reaction of anisic alcohol with acetic anhydride
Natural occurrence	Identified in the oil of *Artemisia dracunculus* var. *turkestanica*	In currants
Reported uses[3]	Non-alcoholic beverages 9.0 ppm Ice cream, ices, etc. 16 ppm Candy 51 ppm Baked goods 34 ppm	Non-alcoholic beverages 6.3 ppm Ice cream, ices, etc. 8.0 ppm Candy 15 ppm Baked goods 12 ppm Gelatins and puddings 11 ppm Chewing gum 30 ppm
Regulatory status	FDA 121.1164; FEMA No. 2097	FDA 121.1164; FEMA No. 2098

REFERENCES

For References 1–5, see end of Part III.

	Anisyl alcohol	Anisyl butyrate
Other names	Anisic alcohol *p*-Methoxybenzyl alcohol	*p*-Methoxybenzyl butyrate
Empirical formula	$C_8H_{10}O_2$	$C_{12}H_{16}O_3$
Structure	CH_2OH (para) OCH_3 on benzene ring	$H_3C-(CH_2)_2-C(=O)-O-CH_2-C_6H_4-OCH_3$
Physical/chemical characteristics[5]		
Appearance	Colorless to slightly yellow liquid	Colorless liquid
Assay	97% min.[2]	
Molecular weight	138.16	208.26
Melting point	23.8°C (25–27°C)	Approx. 270°C
Boiling point	259°C (145°C at 18 mm Hg)	
Freezing point	>23.5°C	
Specific gravity	1.110–1.115[2] (1.110–1.125) at 25°/25°C	
Refractive index	1.5430–1.5450[2] (1.5412–1.5430) at 25°C	
Acid value	Not more than 1.0[2]	
Aldehyde content	Not more than 1.0%[2]	
Solubility	1:1 in 50% alcohol; 1:13 in 30% alcohol; soluble in most organic solvents	Insoluble in water; soluble in alcohol
Organoleptic characteristics	Floral odor; sweet, fruity (peach) taste	Weak, floral, intensely sweet, plum-like odor
Synthesis	By reduction from anisic aldehyde	From anisyl alcohol and *n*-butyric acid
Natural occurrence	Reported found in vanilla pods (Tahitian production) and anise seed oil[6]	Not reported found in nature
Reported uses[3]	Non-alcoholic beverages 7.4 ppm Ice cream, ices, etc. 8.0 ppm Candy 11 ppm Baked goods 12 ppm Gelatins and puddings 1.9 ppm	Non-alcoholic beverages 3.1 ppm Ice cream, ices, etc. 5.7 ppm Candy 10 ppm Baked goods 13 ppm
Regulatory status	FDA 121.1164; FEMA No. 2099	FDA 121.1164; FEMA No. 2100

REFERENCES

For References 1–5, see end of Part III.

6. Monod et al., *Chem. Abstr.*, 45, 3124, 1951.

	Anisyl formate	Anisyl phenylacetate
Other names	*p*-Methoxybenzyl formate	Anisyl α-toluate
Empirical formula	$C_9H_{10}O_3$	$C_{16}H_{16}O_3$
Structure	$CH_3O-C_6H_4-CH_2-O-CH(=O)$	$C_6H_5-CH_2-C(=O)-O-CH_2-C_6H_4-OCH_3$
Physical/chemical characteristics[5]		
Appearance	Colorless liquid	Colorless, oily liquid
Assay	90% min	
Molecular weight	166.18	256.30
Boiling point	100°C	370°C
Specific gravity	0.138–0.142 at 25°C	
Refractive index	1.5220–1.5240 at 20°C	
Acid Value	3 max	
Solubility	Insoluble in water; soluble in most organic solvents; 1:3 in 70% alcohol	Soluble in alcohol
Organoleptic characteristics	Sweet, floral odor; fruity (strawberry) taste. The floral note varies, depending on whether synthetic or natural anisic alcohol has been used (probably because of the purity of the starting alcohol).	Anise and honey-like odor
Synthesis	By direct esterification of anisic alcohol with formic acid	From anisyl alcohol and phenylacetic acid by direct esterification
Natural occurrence	Reported in *Vanilla fragrans* and *Ribes* species	
Reported uses[3]	Non-alcoholic beverages 3.2 ppm Ice cream, ices, etc. 3.9 ppm Candy 7.9 ppm Baked goods 14 ppm Gelatins and puddings 0.20 ppm	
Regulatory status	FDA 121.1164; FEMA No. 2101	FDA 121.1164

REFERENCES

For References 1–5, see end of Part III.

	Anisyl propionate	Arabinogalactan
Other names	*p*-Methoxybenzyl propionate	Larch gum
Empirical formula	$C_{11}H_{14}O_3$	$[(C_5H_8O_4)(C_6H_{10}O_5)_6]_n$
Structure	$CH_3-CH_2-C(=O)-O-CH_2-C_6H_4-OCH_3$	$[(C_5H_8O_4)(C_6H_{10}O_5)_6]_n$
Physical/chemical characteristics[5]		
Appearance		Complex polysaccharide containing *d*-galactose and *l*-arabinose in the approximate ratio of 6:1
Molecular weight	194.23	
Boiling point	277°C (100–103°C at 0.5 mm Hg)	
Specific gravity	1.078 at 15.5°C	
Refractive index	1.5490 at 20°C; 1.5112 at 15°C	
Solubility		Soluble in water
Organoleptic characteristics	Herbaceous odor; fruity (cherry, peach) taste	
Synthesis	By esterification of anisic alcohol with propionic anhydride	From larch wood, white pine, Monterrey pine, and Jack pine[6]
Natural occurrence	Not reported found in nature	
Reported uses[3]	Non-alcoholic beverages 5.6 ppm Ice cream, ices, etc. 6.1 ppm Candy 16 ppm Baked goods 20 ppm Gelatins and puddings 0.25 ppm	Candy, 85%
Regulatory status	FDA 121.1164; FEMA No. 2102	FEMA No. 3254

REFERENCES

For References 1–5, see end of Part III.

6. **Whistler,** in *Methods in Carbohydrate Chemistry,* Vol. 5, Whistler, R. L. and Wolfrom, M. L., Eds., Academic Press, New York, 1965, 75.

	L-Arabinose		Beechwood creosote
Other names	Pectin sugar		Wood creosote
Empirical formula	$C_5H_{10}O_5$		
Structure	[structure diagram: pyranose ring with H, OH, O, HOH, H, OH substituents]		A mixture of phenols (chiefly guaiacol, creosol, and *p*-cresol)
Physical/chemical characteristics[5]			
Appearance	Rhombic prisms		Colorless or yellowish oily liquid
Molecular weight	150.13		
Melting point	159.5°C		
Boiling point			200–220°C
Congealing point			Approx. −20°C
Specific gravity	α-, 1.585 at 20°C β-, 1.625 at 20°C		1.076–1.09 at 25°C
Optical rotation	+105° at 20°C (c = 3 in water)		
Solubility	Very soluble in water; slightly soluble in alcohol; insoluble in ether.		Insoluble in water; soluble in glycerol, glacial acetic acid, and alkali hydroxide solutions; miscible with alcohol, chloroform, and ether
Organoleptic characteristics			Characteristic smoky odor; caustic, burning taste
Synthesis	By partial hydrolysis of mesquite gum[6]		Starting materials are the heavy oils from wood tar. The material first is treated with NaOH to precipitate all saponifiable oils and subsequently with H_2SO_4. The raw creosote finally is purified by repeated fractional distillation.[7]
Reported uses[3]	Non-alcoholic beverages	450.0 ppm	
	Candy	450.0 ppm	
	Baked goods	450.0 ppm	
	Condiments, pickles	450.0 ppm	
	Preserves and spreads	450.0 ppm	
Regulatory status	FEMA No. 3255		FDA 121.1164

REFERENCES

For References 1–5, see end of Part III.

6. **White,** in *Methods in Carbohydrate Chemistry,* Vol. 1, Whistler, R. L. and Wolfrom, M. L., Eds., Academic Press, 1962, 76.
7. *Ullmanns Enzyklopädie der Technischen Chemie,* Vol. 3c, Urban and Schwarzenberg, 1957, 590.

	Benzaldehyde	Benzaldehyde dimethyl acetal
Other names	Benzenecarbonal Benzenemethylal Benzoic aldehyde	Dimethoxy-(phenyl)-methane α,α-Dimethoxy toluene
Empirical formula	C_7H_6O	$C_9H_{12}O_2$
Structure	C_6H_5–CHO	C_6H_5–HC(OCH_3)$_2$
Physical/chemical characteristics[5]		
Appearance	Colorless liquid	Liquid
Assay	98% min	
Molecular weight	106.12	152.19
Melting point	−26°C	
Boiling point	179°C	198°C (207°C)
Flash point	62°C	
Specific gravity	1.0500 at 15°C; 1.041–1.046 at 25°C	1.025 at 15.5°C
Refractive index	1.5450 (1.5440–1.5465) at 20°C	1.4950 at 20°C
Optical rotation	±0°	
Solubility	Miscible with alcohol and ether; 1:1–1:1.5 in 70% alcohol; 1:2.5 in 60% alcohol; 1:8 in 50% alcohol; 1:300 in water	
Organoleptic characteristics	Characteristic odor and aromatic taste similar to bitter almond	Floral note; also reminiscent of almond
Synthesis	By extraction and subsequent fractional distillation from botanical sources; synthetically, from benzyl chloride and lime or by oxidation of toluene	From benzaldehyde and methanol in the presence of calcium chloride and HCl;[7] from benzaldehyde and tetramethylorthosilicate in the presence of anhydrous HCl in methanol[8]
Natural occurrence	Present as cyanuric glucoside (amygdalin) in bitter almond, peach, apricot kernel, and other *Prunus* species. Amygdalin is also present in various parts of the following plants: *Sambucus nigra, Chrysophyllum arten, Anacyclus officinarum, Anacyclus pedunculatus, Davallia brasiliensis, Lacuma deliciosa, Lacuma multiflora,* and others. Free benzaldehyde has been reported found in several essential oils: hyacinth, citronella, orris, cinnamon, sassafras, labdanum, and patchouli.[6]	In potato
Reported uses[3]	Non-alcoholic beverages 36 ppm Alcoholic beverages 50–60 ppm Ice cream, ices, etc. 42 ppm Candy 120 ppm Baked goods 110 ppm Gelatins and puddings 160 ppm Chewing gum 840 ppm	Non-alcoholic beverages 26 ppm Alcoholic beverages 60 ppm Ice cream, ices, etc. 22 ppm Candy 56 ppm Baked goods 45 ppm Gelatins and puddings 50 ppm
Regulatory status	FDA GRAS; FEMA No. 2127	FDA 121.1164; FEMA No. 2128

REFERENCES

For References 1–5, see end of Part III.

6. **Gildemeister and Hoffmann,** *Die Aetherischen Oele,* Vol. 3c, Akademie Verlag, 1963, 128.
7. **Adams and Adkins,** *J. Am. Chem. Soc.,* 47, 1365, 1925.
8. **Helferich and Hansen,** *Ber. Dtsch. Chem. Ges.,* 57, 797, 1924.

	Benzaldehyde glyceryl acetal	**Benzaldehyde propylene glycol acetal**
Other names	Phenyl-*m*-dioxan-5-ol (α,α') 2-Phenyl-1,3-dioxan-5-ol (α,α')	4-Methyl-2-phenyl-*m*-dioxolane
Empirical formula	$C_{10}H_{12}O_3$	$C_{10}H_{12}O_2$
Structure	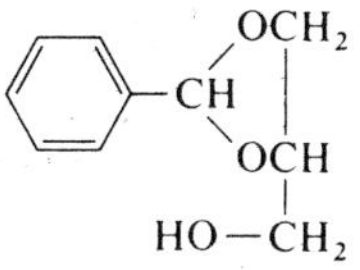 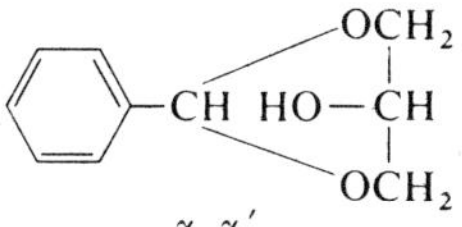 C_6H_5–CH(–OCH₂–CH(OCH)–) HO–CH₂ (α,β); C_6H_5–CH(OCH₂)₂CH–OH (α,α')	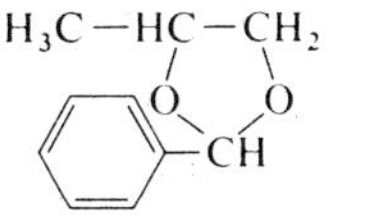 $H_3C-HC-CH_2$, O, O, CH–C_6H_5
Physical/chemical characteristics[5]		
Appearance	α,α', solid α,β, oily liquid	
Molecular weight	180.21	164.21
Melting point	α,α'(*cis*), 63.5–64.5°C α,α' (*trans*), 83–84°C	
Boiling point	α,α', 185°C at 20 mm Hg α,β, 143–144°C at 2 mm Hg	
Specific gravity	α,β, 1.1916 at 17°C	
Refractive index	α,β, 1.5389 at 17°C	
Solubility	α,α', slightly soluble in a ligroin-benzene mixture at room temperature; α,β, soluble in a ligroin-benzene mixture at room temperature	
Organoleptic characteristics		
Synthesis	The α,α'- and α,β-isomers are obtained in mixture by heating glycerol and benzaldehyde to 145–170°C under a stream of CO_2.[6,7] The isomers are subsequently isolated, exploiting the solubility differences. The α,α'-isomer is readily converted to the α,β-form by heating in the presence of HCl.	
Natural occurrence	Not reported found in nature	Not reported found in nature
Reported uses[3]	Non-alcoholic beverages 21 ppm Ice cream, ices, etc. 24 ppm Candy 110 ppm Baked goods 73 ppm Gelatins and puddings 100 ppm Chewing gum 840 ppm	Non-alcoholic beverages 34 ppm Ice cream, ices, etc. 27 ppm Candy 110 ppm Baked goods 96 ppm Gelatins and puddings 50 ppm
Regulatory status	FDA 121.1164; FEMA No. 2129	FDA 121.1164; FEMA No. 2130

REFERENCES

For References 1–5, see end of Part III.

6. **Hill, Whelen and Hibbert,** *J. Am. Chem. Soc.,* 50, 2239, 1928.
7. **Verkade and Van Roon,** *Recl. Trav. Chim. Pays-Bas,* 61, 834, 1942.

	Benzenethiol	Benzoic acid
Other names	Phenyl mercaptan Thiophenol	Benzenecarboxylic acid Dracylic acid Phenylformic acid
Empirical formula	C_6H_6S	$C_7H_6O_2$
Structure	(benzene ring with SH)	(benzene ring with COOH)
Physical/chemical characteristics[5]		
Appearance	Prism-like crystals from ligroin	Flexible, shiny plates or monoclinic needles
Molecular weight	110.04	122.12
Melting point	70°C	121–123°C (U.S.P.)
Boiling point	169.5°C	249°C 249–250°C
Specific gravity	1.0728 at 25°C	1.265 at 15.5°C
Refractive index	1.5931 at 14°C	
Solubility	Insoluble in water; slightly soluble in alcohol, ether, and benzene	1:380 in cold water; 1:15 in hot water; 1:2.5 in 90% alcohol; 1:10 in glycerol
Organoleptic characteristics	Repulsive, penetrating, garlic-like odor	Almost odorless or exhibiting a faint balsamic odor and a sweet-sour to acrid taste
Synthesis	By reduction of benzenesulfonyl chloride with zinc dust in sulfuric acid[6]	By oxidation of toluene with nitric acid or sodium bichromate or from benzonitrile
Natural occurrence	Not reported found in nature	Reported as being a constituent of various oils, resins, and flower absolutes: hyacinth, tuberose, neroli bigarade, Chinese cinnamon, cinnamon leaves, anise, vetiver, ylang-ylang, Tolu balsam, and clove. It is contained in fairly sizable amounts in gum benzoin, from which benzoic acid is extracted by sublimation.
Reported uses[3]		Non-alcoholic beverages 7.5 ppm Ice cream, ices, etc. 4.8 ppm Candy 8.9 ppm Baked goods 40 ppm Chewing gum 20–32 ppm Icings 250 ppm Used chiefly as a preservative at 0.1%; as a flavor ingredient it is used at ppm levels
Regulatory status	FDA 121.1164	FDA GRAS (0.1% preservative); FEMA No. 2131

REFERENCES

For References 1–5, see end of Part III.

6. Adams and Marvel, *Org. Synth.*, 1, 71, 1921.

	Benzoin	Benzophenone
Other names	Benzoyl phenylcarbinol 2-Hydroxy-2-phenylacetophenone	Benzoyl benzene Diphenyl ketone
Empirical formula	$C_{14}H_{12}O_2$	$C_{13}H_{10}O$
Structure	$C_6H_5-CH(OH)-C(=O)-C_6H_5$	$C_6H_5-C(=O)-C_6H_5$
Physical/chemical characteristics[5]		
Appearance	Monoclinic prisms or needles	White, crystalline material
Molecular weight	212.22	182.21
Melting point	132–137°C	48.5°C
Boiling point		170°C at 15 mm Hg
Congealing point		>47.0°C[2]
Specific gravity		1.0496 at 95°C
Refractive index		1.5893 at 45.2°C
Solubility	Soluble in acetone and boiling alcohol	1:10 in 80% alcohol;[2] insoluble in water; soluble in alcohol, ether, and acetic acid
Organoleptic characteristics		Delicate, persistent, rose-like odor
Synthesis	By treating an alcoholic solution of benzaldehyde with an alkali cyanide[6,7]	By Friedel-Craft condensation of benzene with benzoyl chloride in the presence of aluminum chloride[8]
Natural occurrence	Not reported found in nature	In grapes
Reported uses[3]	Non-alcoholic beverages 4.5 ppm Ice cream, ices, etc. 0.54 ppm Candy 2.0 ppm Baked goods 1.4 ppm Gelatins and puddings 0.10 ppm	Non-alcoholic beverages 0.50 ppm Ice cream, ices, etc. 0.61 ppm Candy 1.7 ppm Baked goods 2.4 ppm
Regulatory status	FDA 121.1164; FEMA No. 2132	FDA 121.1164; FEMA No. 2134

REFERENCES

For References 1–5, see end of Part III.

6. **Adams and Marvel,** *Org. Synth.,* 1, 33, 1921.
7. **Arnold and Fuson,** *J. Am. Chem. Soc.,* 58, 1295, 1936.
8. **Marvel and Sperry,** *Org. Synth.,* 8, 26, 1928.

	Benzothiazole	Benzyl acetate
Empirical formula	C_7H_5NS	$C_9H_{10}O_2$
Structure		$C_6H_5-CH_2-O-C(=O)-CH_3$
Physical/chemical characteristics[5]		
Appearance	Liquid	Colorless liquid
Assay		98% min[2] (as benzyl acetate)
Molecular weight	135.19	150.18
Melting point	2°C	
Boiling point	223–225°C (234°C) 131°C at 34 mm Hg	215.5–216.0°C (71°C at 3 mm Hg)
Flash point		93°C
Specific gravity	1.2460 at 20°C	1.052–1.056 at 25°/25°C[2]
Refractive index	1.6379 at 20°C	1.5010–1.5030 at 20°C[2]
Optical rotation		±0°
Acid value		Not more than 1.0
Solubility	Soluble in alcohol, acetone, and carbon disulfide; very slightly soluble in water	1:5 in 60% alcohol; 1:200 in 30% alcohol; slightly soluble in water; soluble in most organic solvents
Organoleptic characteristics	Odor similar to that of quinoline	Characteristic flowery (jasmine) odor; bitter, pungent taste
Synthesis	By refluxing a mixture of zinc *o*-aminophenylsulfide and formic acid, followed by steam distillation of the alkalized reaction mixture;[6] by heating formanilid or dimethylaniline with sulfur;[7] by oxidation of 2-mercaptobenzothiazole or of the corresponding disulfide[8]	By interaction of benzyl chloride and sodium acetate;[9] by acetylation of benzyl alcohol; or from benzaldehyde and acetic acid with zinc dust[10]
Natural occurrence		Present as a main constituent in several oils and flower absolutes: ylang-ylang, cananga, neroli, jasmine, hyacinth, gardenia, tuberose. It has been isolated in the essential oil from flowers of *Loiseleuria procumbens* Desv. (azalea).[11] Benzyl acetate can be isolated from natural sources (in which it may occur at levels up to 65%), but the commercial product is of synthetic origin.
Reported uses[3]	Non-alcoholic beverages 0.5 ppm Candy 0.5 ppm Baked goods 0.5 ppm Meat, meat sauces, soups 0.5 ppm Milk, dairy products 0.5 ppm	Non-alcoholic beverages 7.8 ppm Ice cream, ices, etc. 14 ppm Candy 34 ppm Baked goods 22 ppm Gelatins and puddings 23 ppm Chewing gum 760 ppm
Regulatory status	FEMA No. 3256	FDA 121.1164; FEMA No. 2135

REFERENCES

For References 1–5, see end of Part III.

6. **Kiprianov, Snitnik, and Grigoreva,** *Chem. Abstr.*, 30, 4859, 1936.
7. **Knowles and Watt,** *J. Org. Chem.*, 7, 56, 1942.
8. **Čech,** *Collect. Czech. Chem. Commun.*, 14, 55, 1949; *Chem. Abstr.*, 44, 5873, 1950.
9. **Seelig,** *J. Prakt. Chem.*, 39, 162, 1899.
10. **Tiemann,** *Berl. Ber.*, 19, 355, 1866.
11. **Salgues,** *C. R. Séances Acad. Sci.* (Paris), 240, 1136, 1955.

	Benzyl acetoacetate	**Benzyl alcohol**
Other names	Benzyl acetyl acetate Benzyl β-ketobutyrate Benzyl 3-oxobutanoate	α-Hydroxy toluene Phenyl carbinol Phenyl methanol
Empirical formula	$C_{11}H_{12}O_3$	C_7H_8O
Structure	$CH_3-C(=O)-CH_2-C(=O)-O-CH_2-C_6H_5$	$C_6H_5-CH_2OH$
Physical/chemical characteristics[5]		
Appearance	Oily liquid	Colorless liquid
Molecular weight	192.22	108.13
Melting point	Approx. 270°C (162–164°C at 16 mm Hg)	−15.3°C
Boiling point		205.45°C (202.5–206.5°C)
Specific gravity		1.04927 at 15°C; 1.041–1.046 at 25°C
Refractive index		1.54259 at 15°C; 1.5385–1.5405 at 20°C
Solubility	Soluble in alkali solutions at room temperature	Soluble 1:25 in water and most organic solvents; 1:8–9 in 30% alcohol; 1:1.5 in 50% alcohol
Organoleptic characteristics	Odor similar to that of ethyl acetate	Characteristic pleasant, fruity odor; slightly pungent, sweet taste. The note tends to become similar to that of benzyl aldehyde on aging.
Synthesis	By heating ethyl acetoacetate and benzyl alcohol to 160°C	By saponification of the ester present in Tolu and Peru balsams. Synthetically, it is obtained from benzyl chloride by the action of sodium or potassium carbonate.
Natural occurrence		The free alcohol is often present in several essential oils and extracts of: jasmine, tobacco, tea, neroli, copaiba, *Acacia farnesiana* Willd., *Acacia cavenia* Hook. and Arn., *Robinia pseudacacia*, ylang-ylang, *Pandanus odoratissimus*, *Michelia champaca*, *Prunus laurocerasus*, tuberose, orris, castoreum, violet leaves, clove buds, and others.
Reported uses[3]	Non-alcoholic beverages 2.7 ppm Ice cream, ices, etc. 6.0 ppm Candy 13 ppm Baked goods 13 ppm Gelatins and puddings 0.50–10 ppm Chewing gum 50 ppm	Non-alcoholic beverages 15 ppm Ice cream, ices, etc. 160 ppm Candy 47 ppm Baked goods 220 ppm Gelatins and puddings 21–45 ppm Chewing gum 1,200 ppm
Regulatory status	FDA 121.1164; FEMA No. 2136	FDA 121.1164; FEMA No. 2137

REFERENCES

For References 1–5, see end of Part III.

	Benzyl benzoate	Benzyl butyl ether
Other names	Benzyl benzene carboxylate Benzyl phenylformate	Butyl benzyl ether
Empirical formula	$C_{14}H_{12}O_2$	$C_{11}H_{16}O$
Structure	$C_6H_5-C(=O)-O-CH_2-C_6H_5$	$C_6H_5-CH_2-O-CH_2-CH_2-CH_2-CH_3$
Physical/chemical characteristics[5]		
Appearance	Leaflets or oily liquid	Colorless liquid
Assay	98% min	
Molecular weight	212.24	164.25
Melting point	18–19°C	
Boiling point	324°C (170–171°C at 11 mm Hg)	220–221°C at 744 mm Hg
Congealing point	18°C	
Flash point	148°C	
Specific gravity	1.116–1.120 (1.1121) at 25°/25°C	0.931 at 10°C
Refractive index	1.568–1.570 at 20°C; 1.5685 at 21.5°C	
Solubility	Insoluble in water or glycerol; soluble in alcohol, chloroform, and ether; 1:2 in 90% alcohol	Insoluble in water; miscible with alcohol and ether
Organoleptic characteristics	Light, balsamic odor reminiscent of almond; sharp, pungent taste	
Synthesis	By the dry esterification of sodium benzoate and benzoyl chloride in the presence of triethylamine[6] or by reaction of sodium benzylate on benzaldehyde[7]	Obtained in mixture by heating benzyl alcohol and butyl alcohol in the presence of sulfuric acid or sodium bisulfate[15,16]
Natural occurrence	Contained in Peru balsam[8] and in the concrete and absolute of tuberose flowers,[9] hyacinth,[10] *Narcissus jonquilla* L.,[11] and *Dianthus caryophillus* L.;[12] also in the oil of ylang-ylang and in Tolu balsam[13,14]	Not reported found in nature
Reported uses[3]	Non-alcoholic beverages 4.5 ppm Ice cream, ices, etc. 12 ppm Candy 39 ppm Baked goods 33 ppm Chewing gum 280 ppm	Non-alcoholic beverages 0.50–2.0 ppm Ice cream, ices, etc. 3.5 ppm Candy 8.0 ppm Baked goods 2.0–8.0 ppm Gelatins and puddings 2.0 ppm
Regulatory status	FDA 121.1164; FEMA No. 2138	FDA 121.1164; FEMA No. 2139

REFERENCES

For References 1–5, see end of Part III.

6. **Thorp and Nottorf,** *Ind. Eng. Chem.*, 39, 1300, 1947.
7. **Kamm,** *Org. Synth.*, 2, 5, 1922.
8. **Krant,** *Ann. Chem.*, 152, 131, 1869.
9. **Hesse,** *Ber. Dtsch. Chem. Ges.*, 36, 1467, 1903.
10. **Euklaar,** *Chem. Weekbl.*, 7, 11, 1910.
11. **Soden,** *J. Prakt. Chem.*, 110, 227, 1925.
12. **Treff and Wittrish,** *J. Prakt. Chem.*, 122, 139, 1929.
13. **Van Itallie and Harsma,** *Pharm. Weekbl.*, 62, 898, 1926.
14. **Van Itallie and Harsma,** *Chem. Zentralbl.*, 1, 167, 1926.
15. **Senderens,** *Compt. Rend.*, 178, 1415, 1924.
16. **Senderens,** *Compt. Rend.*, 188, 1075, 1929.

	Benzyl butyrate	Benzyl cinnamate
Other names	Benzyl butanoate	Benzyl β-phenylacrylate Benzyl-3-phenylpropenoate Cinnamein
Empirical formula	$C_{11}H_{14}O_2$	$C_{16}H_{14}O_2$
Structure	$CH_3-CH_2-CH_2-C(=O)-O-CH_2-C_6H_5$	$C_6H_5-CH=CH-C(=O)-O-CH_2-C_6H_5$
Physical/chemical characteristics[5]		
Appearance	Colorless liquid	White to pale-yellow, fused, crystalline solid, melting at room temperature to a yellowish liquid
Assay	98% min	99% min[2]
Molecular weight	178.23	238.29
Melting point		*trans*, 35–36°C;[8] *cis*, 30°C[9]
Boiling point	235–242°C (130–132°C at 15 mm Hg)	228–230°C at 22 mm Hg
Congealing point		33.0–34.5°C[2]
Flash point	>100°C	>100°C
Specific gravity	1.006–1.009 at 25°/25°C;[2] 1.014 at 19°C	1.109 at 15°C
Refractive index	1.4920–1.4960 at 20°C[2]	
Acid value	Not more than 1.0[2]	Not more than 1.0[2]
Solubility	Insoluble in water; miscible with alcohol and ether	1:8 in 90% alcohol
Organoleptic characteristics	Characteristic fruity (apricot) odor; sweet, pear-like taste	Sweet, balsamic odor; honey-like taste
Synthesis	By heating benzyl chloride and sodium butyrate in water[6] or butyric acid and benzyl chloride under pressure[7]	Usually prepared by esterification of benzyl alcohol and cinnamic acid[2] By heating benzyl chloride and excess sodium cinnamate in water to 110–115°C;[10] by heating sodium cinnamate with an excess of benzyl chloride in the presence of diethylamine[11]
Natural occurrence	In passion fruit	Reported found in Peru[12] and Tolu[13] balsam, in Sumatra and Penang benzoin,[14] and as the main constituent of copaiba balsam[15]
Reported uses[3]	Non-alcoholic beverages 4.5 ppm Ice cream, ices, etc. 6.9 ppm Candy 7.7 ppm Baked goods 9.9 ppm Gelatins and puddings 3.0 ppm Chewing gum 310 ppm	Non-alcoholic beverages 1.4 ppm Ice cream, ices, etc. 2.5 ppm Candy 6.7 ppm Baked goods 6.6 ppm Gelatins and puddings 3.0–5.0 ppm Chewing gum 5.3–120 ppm
Regulatory status	FDA 121.1164; FEMA No. 2140	FDA 121.1164; FEMA No. 2142

REFERENCES

For References 1–5, see end of Part III.
6. **Gomberg and Buchler,** *J. Am. Chem. Soc.*, 42, 2065, 1920.
7. **Thompson and Leuck,** *J. Am. Chem. Soc.*, 44, 2895, 1922.
8. **Manta,** *Bull. Soc. Chim. Fr.*, 53, 1285, 1933.
9. **Vorländer and Walter,** *Z. Phys. Chem. Frankfurt am Main*, 118, 13, 1925.
10. **Gomberg and Buchler,** *J. Am. Chem. Soc.*, 42, 2065, 1920.
11. **Volwiler and Vliet,** *J. Am. Chem. Soc.*, 43, 1674, 1921.
12. **Tschirch and Trog,** *Arch. Pharm.*, 232, 91, 1894.
13. **Tschirch and Oberländer,** *Arch. Pharm.*, 232, 559, 1894.
14. **Tschirch and Lüdy,** *Arch. Pharm.*, 231, 43, 1893.
15. **Machado,** *Rev. Quim. Ind. Rio de Janeiro*, 10, 1941; *Chem. Abstr.*, 36, 1735, 1942.

	Benzyl 2,3-dimethylcrotonate	Benzyl disulfide
Other names	Benzyl 2,3-dimethyl-2-butenoate Benzyl methyl tiglate	Dibenzyl disulfide
Empirical formula	$C_{13}H_{16}O_2$	$C_{14}H_{14}S_2$
Structure	$C_6H_5-CH_2-O-C(=O)-C(CH_3)=C(CH_3)-CH_3$	$C_6H_5-CH_2-S-S-CH_2-C_6H_5$
Physical/chemical characteristics[5]		
Appearance		Pale-yellow leaflets (from alcohol)
Molecular weight	204.27	246.40
Melting point		71–72°C (74°C)
Boiling point		>270°C (with decomposition)
Solubility		Almost insoluble in water; soluble in hot alcohol and ether
Organoleptic characteristics		Powerful, burnt-caramel odor; irritating when concentrated
Synthesis		From benzyl chloride and Na_2S_2; also from benzyl mercaptan via oxidation
Natural occurrence	Not reported found in nature	Not reported found in nature
Reported uses[3]	Non-alcoholic beverages 0.75 ppm Ice cream, ices, etc. 2.8 ppm Candy 1.8 ppm Baked goods 1.5 ppm	
Regulatory status	FDA 121.1164; FEMA No. 2143	FDA 121.1164

REFERENCES

For References 1–5, see end of Part III.

	Benzyl ethyl ether	Benzyl formate
Other names	Ethyl benzyl ether	Formic acid benzyl ester
Empirical formula	$C_9H_{12}O$	$C_8H_8O_2$
Structure	$C_6H_5-CH_2-O-CH_2-CH_3$	$C_6H_5-CH_2-O-CH(=O)$
Physical/chemical characteristics[5]		
Appearance	Oily liquid	Colorless liquid
Assay		95% min
Molecular weight	136.19	136.15
Boiling point	186°C	202–203°C
Flash point		83°C
Specific gravity	0.9490 at 20°C	1.0930–1.0965 at 15°C
Refractive index	1.4955 (1.4970) at 20°C	1.5100–1.5120 at 20°C
Acid value		0.2 max
Solubility	Insoluble in water; miscible with alcohol and ether	1:3–4 in 70% alcohol
Organoleptic characteristics	Pleasant, fruity (pineapple) odor	Intense, pleasant, floral-fruity odor; sweet taste reminiscent of apricot and pineapple
Synthesis	From benzyl chloride and sodium ethylate in alcoholic solution[6]	By heating a mixture of formic/acetic anhydride and benzyl alcohol to 50°C;[7] by passing a mixture of formic acid and excess benzyl alcohol over a catalyst (ThO_2 or TiO_2) at high temperature[8]
Natural occurrence	In cocoa	It has been isolated from the oil of *Rosa rugosa*.[9]
Reported uses[3]	Non-alcoholic beverages 0.50–1.0 ppm Ice cream, ices, etc. 2.5 ppm Candy 7.5 ppm Baked goods 7.5 ppm	Non-alcoholic beverages 2.4 ppm Ice cream, ices, etc. 8.0 ppm Candy 12 ppm Baked goods 8.6 ppm Chewing gum 3.2 ppm
Regulatory status	FDA 121.1164; FEMA No. 2144	FDA 121.1164; FEMA No. 2145

REFERENCES

For References 1–5, see end of Part III.

6. **Franzen,** *J. Prakt. Chem.,* 97, 82, 1918.
7. **Béhal,** *Ann. Chim. Paris,* 20, 421, 1900.
8. **Sabatier and Maihle,** *Compt. Rend.,* 152, 1045, 1911.
9. **Tsumotu et al.,** *Nippon Kagaku Zasshi,* 83, 745, 1962.

	3-Benzyl-4-heptanone	Benzyl isobutyrate
Other names	Benzyl dipropyl ketone Morellone	Benzyl 2-methyl propanoate
Empirical formula	$C_{14}H_{20}O$	$C_{11}H_{14}O_2$
Structure	$CH_3-CH_2-CH(CH_2C_6H_5)-C(=O)-CH_2-CH_2-CH_3$	$(H_3C)_2CH-C(=O)-O-CH_2-C_6H_5$
Physical/chemical characteristics[5]		
Appearance		Colorless liquid
Assay		97% min[2]
Molecular weight	204.31	178.23
Boiling point		105–108°C at 4 mm Hg
Flash point		100°C
Specific gravity		1.001–1.005 at 25°/25°C[2]
Refractive index		1.4890–1.4920 at 20°C[2]
Acid value		Not more than 1.0
Solubility		1:6 in 70% alcohol
Organoleptic characteristics		Fruity, somewhat jasmine-like odor; sweet, strawberry-like taste
Synthesis		May be prepared from benzyl alcohol and isobutyric acid;[2] by the action of aluminum ethylate on a mixture of benzyl and butyric aldehydes[6]
Natural occurrence	Not reported found in nature	In mint
Reported uses[3]	Non-alcoholic beverages 1.2 ppm Ice cream, ices, etc. 4.6 ppm Candy 11 ppm Baked goods 11 ppm	Non-alcoholic beverages 5.2 ppm Ice cream, ices, etc. 12 ppm Candy 12 ppm Baked goods 25 ppm
Regulatory status	FDA 121.1164; FEMA No. 2146	FDA 121.1164; FEMA No. 2141

REFERENCES

For References 1–5, see end of Part III.

6. Orlow, *Bull. Soc. Chim. Fr.*, 35, 362, 1924.

	Benzyl isovalerate	Benzyl mercaptan
Other names	Benzyl 3-methyl butyrate	Benzyl hydrosulfide Thiobenzyl alcohol α-Toluenethiol
Empirical formula	$C_{12}H_{16}O_2$	C_7H_8S
Structure	$(H_3C)_2CH-CH_2-C(=O)-O-CH_2-C_6H_5$	$C_6H_5-CH_2-SH$
Physical/chemical characteristics[5]		
Appearance	Colorless liquid	Colorless liquid
Assay	98% min[2]	
Molecular weight	192.26	124.19
Boiling point	245°C	194–195°C
Flash point	>100°C	
Specific gravity	0.985–0.991 at 25°/25°C;[2] 0.9932–0.9983 at 15°C	1.058 at 20°C
Refractive index	1.4860–1.4900 at 20°C[2]	
Acid value	Not more than 1.0[2]	
Solubility	1:2 in 80% ethanol; 1:1 in 90% ethanol	
Organoleptic characteristics	Fruity odor reminiscent of apple or pineapple	Repulsive, garlic-like odor; oxidizes in air to dibenzyl disulfide
Synthesis	May be prepared by esterification of isovaleric acid with benzyl alcohol[6]	From benzyl chloride and potassium hydrosulfide[7,8]
Natural occurrence	Not reported found in nature	In coffee
Reported uses[3]	Non-alcoholic beverages 2.2 ppm Ice cream, ices, etc. 3.4 ppm Candy 16 ppm Baked goods 9.4 ppm Gelatins and puddings 56 ppm Chewing gum 200 ppm	Non-alcoholic beverages 0.15–0.25 ppm Ice cream, ices, etc. 0.15–0.50 ppm Candy 0.50–0.75 ppm Baked goods 0.50–0.75 ppm
Regulatory status	FDA 121.1164; FEMA No. 2152	FDA 121.1164; FEMA No. 2147

REFERENCES

For References 1–5, see end of Part III.

6. **Thompson and Leuk,** *J. Am. Chem. Soc.*, 44, 2895, 1922.
7. **Scheibler and Voss,** *Ber. Dtsch. Chem. Ges.*, 53, 382, 1920.
8. **Hoffman and Reid,** *J. Am. Chem. Soc.*, 45, 1883, 1923.

	Benzyl methoxyethyl acetal	Benzyl phenylacetate
Other names	Acetaldehyde benzyl β-methoxyethyl acetal 1-Benzoxy-1-(2-methoxyethoxy)-ethane 1-Benzyloxy-1-(β-methoxy)-ethoxyethane	Benzyl α-toluate
Empirical formula	$C_{12}H_{18}O_3$	$C_{15}H_{14}O_2$
Structure	$H_3C-HC(O-CH_2-C_6H_5)(O-CH_2-CH_2-OCH_3)$	$C_6H_5-CH_2-C(=O)-O-CH_2-C_6H_5$
Physical/chemical characteristics[5]		
Appearance	Colorless liquid	Colorless liquid
Assay		98% min
Molecular weight	210.27	226.28
Boiling point		320°C
Flash point		$>$100°C
Specific gravity		1.095–1.099 at 25°C
Refractive index		1.5530–1.5580 at 20°C
Acid value		Not more than 1.0
Solubility		1:3 in 90% ethanol
Organoleptic characteristics	Mild, sweet, green, fruital odor	Sweet, floral (jasmine-rose) odor; a slight honey-like taste
Synthesis		By direct esterification of benzyl alcohol with phenylacetic acid
Natural occurrence	Not reported found in nature	Not reported found in nature. Phenylacetic acid, on the other hand, has been reported in several essential oils.
Reported uses[3]	Non-alcoholic beverages 0.50 ppm Ice cream, ices, etc. 1.0 ppm Candy 1.0 ppm Baked goods 1.0 ppm	Non-alcoholic beverages 1.3 ppm Ice cream, ices, etc. 2.6 ppm Candy 6.6 ppm Baked goods 4.3 ppm Toppings 5.0 ppm
Regulatory status	FDA 121.1164; FEMA No. 2148	FDA 121.1164; FEMA No. 2149

REFERENCES

For References 1–5, see end of Part III.

	Benzyl propionate	Benzyl salicylate
Other names	Benzyl propanoate	Benzyl *o*-hydroxybenzoate
Empirical formula	$C_{10}H_{12}O_2$	$C_{14}H_{12}O_3$
Structure	$C_6H_5-CH_2-O-C(=O)-CH_2-CH_3$	o-$HO-C_6H_4-C(=O)-O-CH_2-C_6H_5$
Physical/chemical characteristics[5]		
Appearance	Colorless liquid	Colorless, oily liquid
Assay	98% min[2]	98% min[2]
Molecular weight	164.21	228.25
Boiling point	219–220°C (227°C)	208°C at 26 mm Hg; 170–175°C at 7 mm Hg
Congealing point		>23.5°C[2]
Flash point	100°C	>100°C
Specific gravity	1.028–1.032 at 25°/25°C;[2] 1.040 at 15.5°C	1.176–1.180 at 25°/25°C;[2] 1.1799 at 20°C
Refractive index	1.496–1.500[2] (1.5100) at 20°C	1.5790–1.5820[2] (1.5805) at 20°C
Acid value	1.0 max[2]	Not more than 1.0[2]
Solubility	1:3 in 70% alcohol	1:9 in 90% ethanol
Organoleptic characteristics	Sweet, fruity odor; peach-, apricot-like taste	Faint, sweet odor; a sweet, currant-like taste
Synthesis	By esterification of benzyl alcohol with propionic acid[2]	By esterification of salicylic acid with benzyl alcohol[2]
Natural occurrence	In strawberry	It has been reported in small amounts in carnation oil (*Dianthus caryophyllus* L.)[6] and in larger amounts in the oil of *Primula auricula*.[7]
Reported uses[3]	Non-alcoholic beverages 4.1 ppm Ice cream, ices, etc. 5.8 ppm Candy 19 ppm Baked goods 17 ppm Chewing gum 19–150 ppm Icings 40 ppm	Non-alcoholic beverages 1.4 ppm Ice cream, ices, etc. 0.89 ppm Candy 1.8 ppm Baked goods 0.01–2.2 ppm
Regulatory status	FDA 121.1164; FEMA No. 2150	FDA 121.1164; FEMA No. 2151

REFERENCES

For References 1–5, see end of Part III.

6. **Treff and Wittrisch,** *J. Prakt. Chem.,* 122, 339, 1939.
7. *Ullmanns Enzyklopädie der Technischen Chemie,* Vol. 5, Urban and Schwarzenberg, 1953, 313.

	Benzyl tiglate		Biphenyl	
Other names	Benzyl *trans*-2-methyl 2-butenoate Benzyl *trans*-2,3-dimethyl acrylate Benzyl *trans*-2-methyl crotonate		Phenylbenzene Diphenyl	
Empirical formula	$C_{12}H_{14}O_2$		$C_{12}H_{10}$	
Structure	$CH_3-CH=C(CH_3)-C(=O)-O-CH_2-C_6H_5$		$C_6H_5-C_6H_5$	
Physical/chemical characteristics[5]				
Appearance	Colorless liquid		White to slightly yellow crystals	
Molecular weight	190.24		154.21	
Melting point			71°C	
Boiling point			254–255°C 227°C at 400 mm Hg 217°C at 300 mm Hg 145°C at 22 mm Hg 70–78°C at 0.2 mm Hg	
Specific gravity			0.8660 at 20°C	
Refractive index			1.475 at 20°C	
Solubility	Insoluble in water; soluble in alcohol and oils		Insoluble in water; soluble in alcohol and ether; very soluble in benzene, methanol, carbon tetrachloride, and carbon disulfide	
Organoleptic characteristics	Odor reminiscent of mushroom, with a rosy undertone		Pungent odor, reminiscent of rose on dilution	
Synthesis	By direct esterification of tiglic acid with benzyl alcohol under azeotropic conditions		By thermal dehydrogenation of benzene[6]	
Natural occurrence			Reported found in coal tar	
Reported uses[3]	Baked goods	8.0 ppm	Non-alcoholic beverages	2.0 ppm
	Condiments	8.0 ppm	Ice cream, ices, etc.	2.0 ppm
			Candy	2.0 ppm
			Baked goods	2.0 ppm
			Gelatins and puddings	2.0 ppm
Regulatory status	FEMA No. 3330		FEMA No. 3129	

REFERENCES

For References 1–5, see end of Part III.

6. **Kirk and Othmer,** *Encyclopedia of Chemical Technology,* 2nd Ed., Vol. 7, Wiley & Sons, 1965, 191.

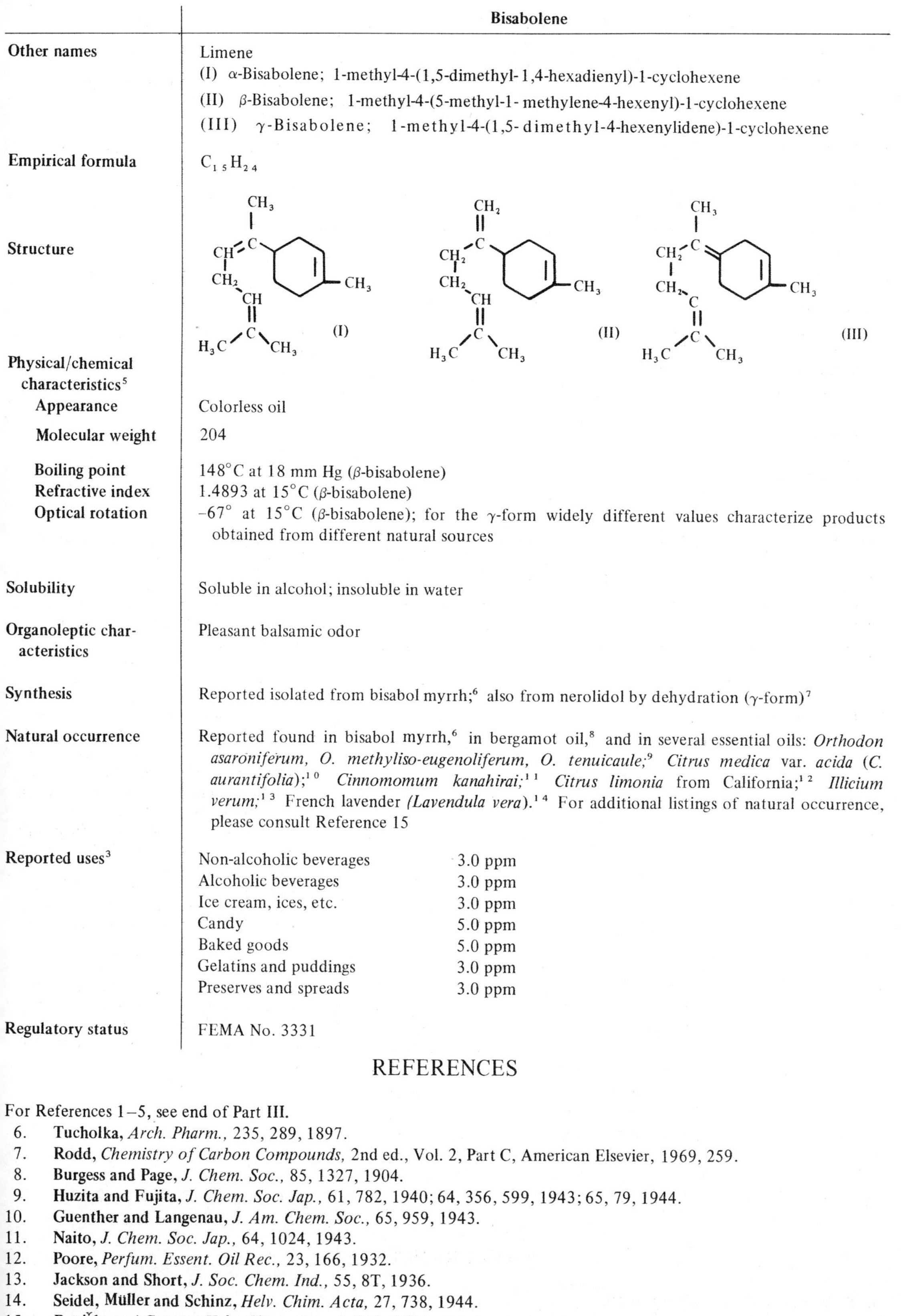

	Bisabolene
Other names	Limene (I) α-Bisabolene; 1-methyl-4-(1,5-dimethyl-1,4-hexadienyl)-1-cyclohexene (II) β-Bisabolene; 1-methyl-4-(5-methyl-1-methylene-4-hexenyl)-1-cyclohexene (III) γ-Bisabolene; 1-methyl-4-(1,5-dimethyl-4-hexenylidene)-1-cyclohexene
Empirical formula	$C_{15}H_{24}$
Structure	(I) (II) (III)
Physical/chemical characteristics[5]	
Appearance	Colorless oil
Molecular weight	204
Boiling point	148°C at 18 mm Hg (β-bisabolene)
Refractive index	1.4893 at 15°C (β-bisabolene)
Optical rotation	-67° at 15°C (β-bisabolene); for the γ-form widely different values characterize products obtained from different natural sources
Solubility	Soluble in alcohol; insoluble in water
Organoleptic characteristics	Pleasant balsamic odor
Synthesis	Reported isolated from bisabol myrrh;[6] also from nerolidol by dehydration (γ-form)[7]
Natural occurrence	Reported found in bisabol myrrh,[6] in bergamot oil,[8] and in several essential oils: *Orthodon asaroniferum, O. methyliso-eugenoliferum, O. tenuicaule;*[9] *Citrus medica* var. *acida* (*C. aurantifolia*);[10] *Cinnomomum kanahirai;*[11] *Citrus limonia* from California;[12] *Illicium verum;*[13] French lavender (*Lavendula vera*).[14] For additional listings of natural occurrence, please consult Reference 15
Reported uses[3]	Non-alcoholic beverages 3.0 ppm Alcoholic beverages 3.0 ppm Ice cream, ices, etc. 3.0 ppm Candy 5.0 ppm Baked goods 5.0 ppm Gelatins and puddings 3.0 ppm Preserves and spreads 3.0 ppm
Regulatory status	FEMA No. 3331

REFERENCES

For References 1–5, see end of Part III.

6. **Tucholka,** *Arch. Pharm.*, 235, 289, 1897.
7. **Rodd,** *Chemistry of Carbon Compounds*, 2nd ed., Vol. 2, Part C, American Elsevier, 1969, 259.
8. **Burgess and Page,** *J. Chem. Soc.*, 85, 1327, 1904.
9. **Huzita and Fujita,** *J. Chem. Soc. Jap.*, 61, 782, 1940; 64, 356, 599, 1943; 65, 79, 1944.
10. **Guenther and Langenau,** *J. Am. Chem. Soc.*, 65, 959, 1943.
11. **Naito,** *J. Chem. Soc. Jap.*, 64, 1024, 1943.
12. **Poore,** *Perfum. Essent. Oil Rec.*, 23, 166, 1932.
13. **Jackson and Short,** *J. Soc. Chem. Ind.*, 55, 8T, 1936.
14. **Seidel, Müller and Schinz,** *Helv. Chim. Acta*, 27, 738, 1944.
15. **Ružička and Capato,** *Helv. Chim. Acta*, 8, 263, 1925.

	Bis (2-furfuryl) disulfide	Borneol
Other names	Difurfuryl disulfide Furfuryl disulfide	2-Bornanol; borneo camphor; bornyl alcohol; 2-camphanol
Empirical formula	$C_{10}H_{10}O_2S_2$	$C_{10}H_{18}O$
Structure	(furan ring)—CH_2—S—S—CH_2—(furan ring)	(bicyclic structure with CH_3, OH, H, H_3C, CH_3)
Physical/chemical characteristics[5]		
Appearance	Pale-yellow oily liquid	Colorless plates
Molecular weight	226.32	154.24
Melting point	10°C	208°C (*dl*-borneol)
Boiling point	112–113°C at 0.5 mm Hg	212°C
Flash point		150°C
Specific gravity		1.011 at 20°C
Optical rotation		+37°44′ at 20°C
Solubility	Slightly soluble in water; very soluble in alcohol and organic solvents	Insoluble in water; soluble in alcohol, chloroform, and ether
Organoleptic characteristics	Powerful, repulsive sulfide odor	Pungent, camphor-like odor and burning taste somewhat reminiscent of mint
Synthesis	From furfural and sodium hydrogen sulfide[6]	Racemic borneol is prepared synthetically by reduction of camphor or from pinene.
Natural occurrence		Unlike isoborneol free or esterified borneol has been identified in more than 250 distillates from plants, herbs, leaves, or bark: *Compositae*, *Ericaceae*, *Lauraceae*, *Labiatae*, *Rutaceae*. Natural borneol can be *d* or *l* rotatory but very seldom also racemic. Most frequently encountered is the *l*-borneol characteristic of *Compositae*, *Graminaceae*, and almost all *Pinaceae*. *d*-Borneol is characteristic of *Cupressaceae*, *Zingiberaceae*, lavender, lavandin, and spike oils.
Reported uses[3]	Non-alcoholic beverages 3.0 ppm Ice cream, ices, etc. 3.0 ppm Candy 3.0 ppm Baked goods 3.0 ppm Gelatins and puddings 3.0 ppm Chewing gum 3.0 ppm Condiments, pickles 3.0 ppm Meat, meat sauces, soups 3.0 ppm Milk, dairy products 3.0 ppm	Non-alcoholic beverages, 0.25–1.4 ppm; ice cream, ices, etc., 1.4 ppm; candy, 3.7 ppm; baked goods, 5.1 ppm; chewing gum, 0.30 ppm; syrups, 0.30 ppm
Regulatory status	FEMA No. 3257	FDA 121.1164; FEMA No. 2157

REFERENCES

For References 1–5, see end of Part III.

6. Gilman and Hewlett, *J. Am. Chem. Soc.*, 52, 2141, 1930.

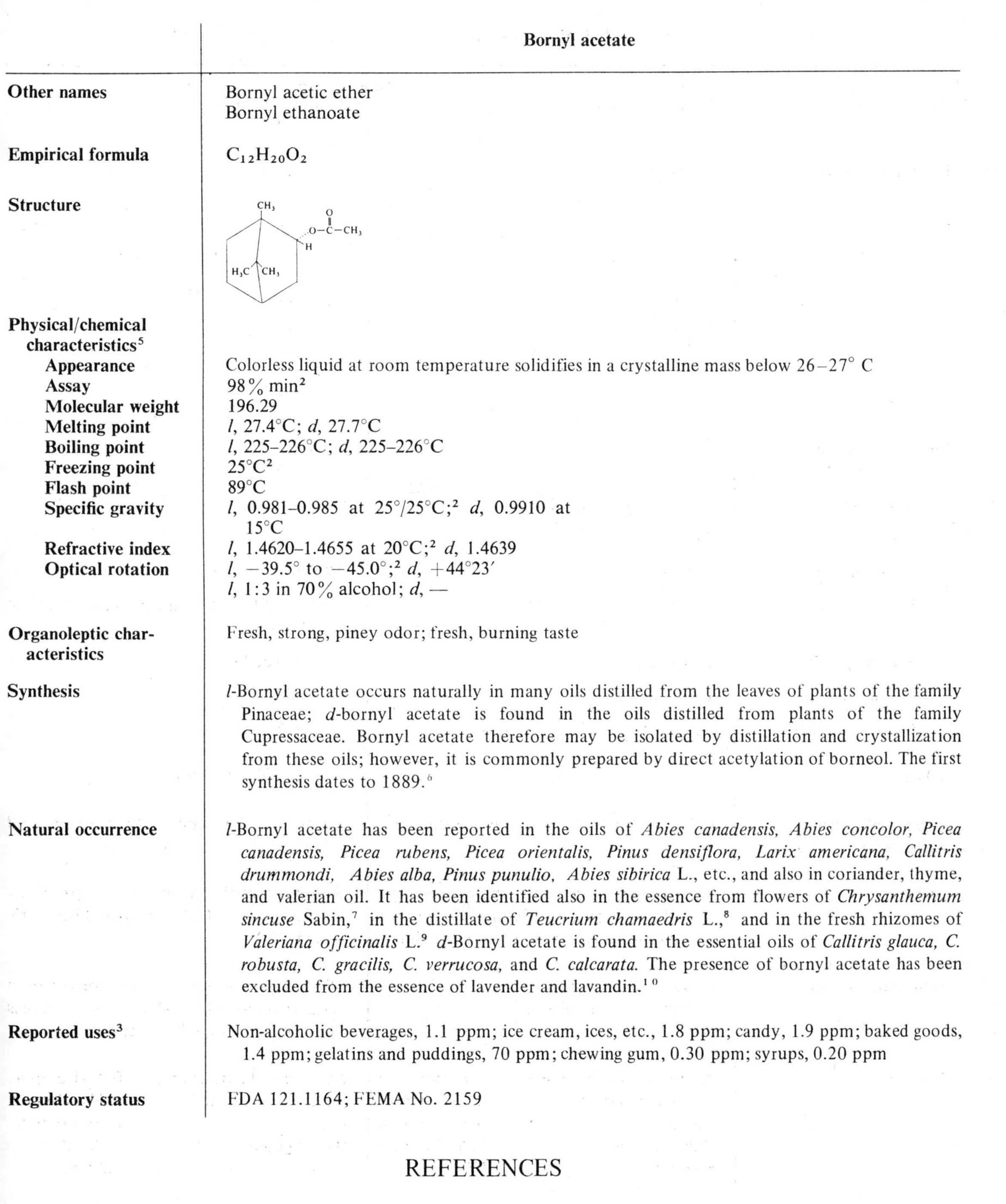

	Bornyl acetate
Other names	Bornyl acetic ether Bornyl ethanoate
Empirical formula	$C_{12}H_{20}O_2$
Structure	CH_3, O, O–C–CH_3, H, H_3C, CH_3
Physical/chemical characteristics[5]	
Appearance	Colorless liquid at room temperature solidifies in a crystalline mass below 26–27° C
Assay	98% min[2]
Molecular weight	196.29
Melting point	*l*, 27.4°C; *d*, 27.7°C
Boiling point	*l*, 225–226°C; *d*, 225–226°C
Freezing point	25°C[2]
Flash point	89°C
Specific gravity	*l*, 0.981–0.985 at 25°/25°C;[2] *d*, 0.9910 at 15°C
Refractive index	*l*, 1.4620–1.4655 at 20°C;[2] *d*, 1.4639
Optical rotation	*l*, −39.5° to −45.0°;[2] *d*, +44°23′ *l*, 1:3 in 70% alcohol; *d*, —
Organoleptic characteristics	Fresh, strong, piney odor; fresh, burning taste
Synthesis	*l*-Bornyl acetate occurs naturally in many oils distilled from the leaves of plants of the family Pinaceae; *d*-bornyl acetate is found in the oils distilled from plants of the family Cupressaceae. Bornyl acetate therefore may be isolated by distillation and crystallization from these oils; however, it is commonly prepared by direct acetylation of borneol. The first synthesis dates to 1889.[6]
Natural occurrence	*l*-Bornyl acetate has been reported in the oils of *Abies canadensis, Abies concolor, Picea canadensis, Picea rubens, Picea orientalis, Pinus densiflora, Larix americana, Callitris drummondi, Abies alba, Pinus punulio, Abies sibirica* L., etc., and also in coriander, thyme, and valerian oil. It has been identified also in the essence from flowers of *Chrysanthemum sincuse* Sabin,[7] in the distillate of *Teucrium chamaedris* L.,[8] and in the fresh rhizomes of *Valeriana officinalis* L.[9] *d*-Bornyl acetate is found in the essential oils of *Callitris glauca, C. robusta, C. gracilis, C. verrucosa,* and *C. calcarata.* The presence of bornyl acetate has been excluded from the essence of lavender and lavandin.[10]
Reported uses[3]	Non-alcoholic beverages, 1.1 ppm; ice cream, ices, etc., 1.8 ppm; candy, 1.9 ppm; baked goods, 1.4 ppm; gelatins and puddings, 70 ppm; chewing gum, 0.30 ppm; syrups, 0.20 ppm
Regulatory status	FDA 121.1164; FEMA No. 2159

REFERENCES

For References 1–5, see end of Part III.

6. **Haller,** *C. R. Séances Acad. Sci.* (Paris), 109, 29, 1889.
7. **Kotake and Nonaka,** *Miltizer Ber.,* p. 26, 1958.
8. **Rovesti,** *Riv. Ital. EPPOS,* 40, 163, 1958.
9. **Stoll et al.,** *Helv. Chim. Acta,* 40, 1205, 1957.
10. **Naves,** *Helv. Chim. Acta,* 42, 2744, 1959.

	Bornyl formate	Bornyl isovalerate
Other names	Bornyl methanoate	Bornyl-3-methylbutanoate Bornyval
Empirical formula	$C_{11}H_{18}O_2$	$C_{15}H_{26}O_2$
Structure	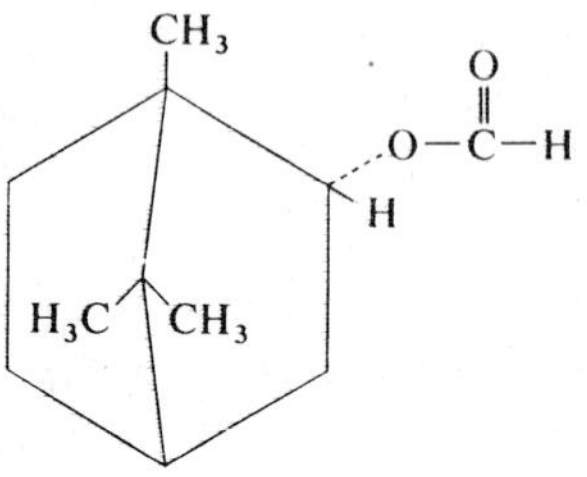	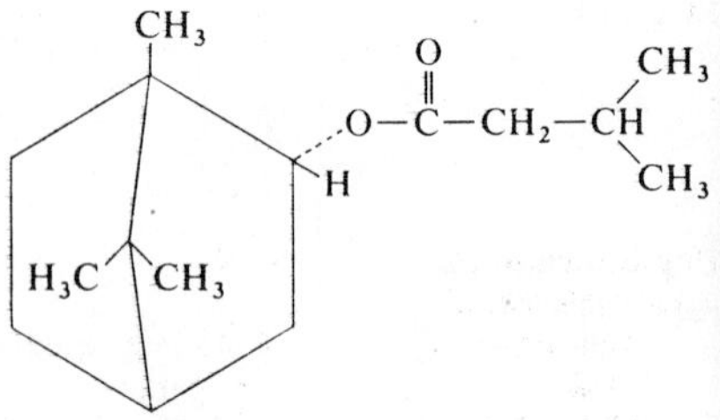
Physical/chemical characteristics[5]		
Appearance	Colorless, oily liquid	Colorless oil
Assay	95% min	
Molecular weight	182.66	238.37
Boiling point	106–108°C at 21 mm Hg	*d*, 151–152°C at 26 mm Hg; *l*, 255–260°C at 760 mm Hg
Specific gravity	1.010 at 20°C	*d*, 0.9486 at 18°C
Refractive index	1.4689 at 20°C	*d*, 1.4605 at 18°C
Optical rotation	At 20°C: *l*, −48.6°; *d*, +48.45°	*d*, +36.7°
Solubility		*d*, 1:4 in 80% alcohol
Organoleptic characteristics	Odor similar to that of bornyl acetate	Aromatic odor similar to valerian and borneol
Synthesis	The corresponding ester is prepared by heating to 50°C a mixture of formic/acetic anhydride with *d*-bornyl alcohol[6]	*d*-Bornyl isovalerate can be prepared by heating to 140°C *d*-borneol and isovaleric acid.[9]
Natural occurrence	It has been reported in the essential oil from roots of valerian.[7,8]	*l*-Bornyl isovalerate has been reported in valerian root oil.[7,8] *d*-Bornyl isovalerate is not reported found in nature.
Reported uses[3]	Non-alcoholic beverages 3.7 ppm Ice cream, ices, etc. 0.30–3.0 ppm Candy 0.80–2.0 ppm Baked goods 0.80–2.0 ppm Syrups 0.04 ppm	Non-alcoholic beverages 0.06–1.0 ppm Ice cream, ices, etc. 0.40–1.0 ppm Candy 0.90–2.0 ppm Baked goods 0.90–2.0 ppm Syrups 1.2 ppm
Regulatory status	FDA 121.1164; FEMA No. 2161	FDA 121.1164; FEMA No. 2165

REFERENCES

For References 1–5, see end of Part III.

6. **Béhal,** *Ann. Chim. Paris,* 20, 421, 1900.
7. **Bruylants,** *Ber. Dtsch. Chem. Ges.,* 11, 455, 1878.
8. **Haller,** *Ann. Chim. Paris,* 27, 396, 1892.
9. **Vavon and Peignier,** *Bull. Soc. Chim. Fr.,* 39, 936, 1926.

	2-Butanone	Butan-3-one-2-yl butanoate
Other names	Ethyl methyl ketone Methyl ethyl ketone	Acetoyl butyrate
Empirical formula	C_4H_8O	$C_8H_{14}O_3$
Structure	$CH_3-C(=O)-CH_2-CH_3$	$CH_3-CH_2-CH_2-C(=O)-O-CH(CH_3)-C(=O)-CH_3$
Physical/chemical characteristics[5]		
Appearance	Colorless, flammable liquid	
Molecular weight	72.10	158.19
Melting point	−85.9°C (−86.4°C)	
Boiling point	78.6–79.57°C	
Specific gravity	0.8255 at 4°C; 0.8054 at 20°C	
Refractive index	1.3785 (1.3790) at 20°C	
Solubility	Soluble in water, alcohol, and ether; miscible with oils	
Organoleptic characteristics	Acetone-like odor	
Synthesis	By catalytic dehydrogenation of secondary butyl alcohol; by dehydration of butane-2,3-diol by refluxing with 25% aqueous H_2SO_4[6,7] Industrially, it is also prepared by controlled oxidation of butane, by dry distillation of calcium acetate and calcium propionate, or by refluxing methyl acetoacetate and diluted H_2SO_4.	Reported not found in nature
Natural occurrence	It has been reported found as an impurity among the products from the dry distillation of wood and in the oil (extracted with ether) of black tea.[8] It is also present in coffee, cheese, bread, some citrus oils, and some other natural products (grape, raspberry).	
Reported uses[3]	Non-alcoholic beverages 70 ppm Ice cream, ices, etc. 270 ppm Candy 100 ppm Baked goods 100 ppm	Non-alcoholic beverages 5.0 ppm Alcoholic beverages 25.0 ppm Ice cream, ices, etc. 25.0 ppm Candy 50.0 ppm Baked goods 75.0 ppm Gelatins and puddings 30.0 ppm
Regulatory status	FDA 121.1164; FEMA No. 2170	FEMA No. 3332

REFERENCES

For References 1–5, see end of Part III.

6. **Akabori,** *J. Chem. Soc. Jap.*, 59, 1132, 1938.
7. **Neish et al.,** *Can. J. Res.*, 23, 281, 1945.
8. **Takel et al.,** *Bull. Inst. Phys. Chem. Res. Abstr. Tokyo,* 10, 45, 1937.

	Butter acids	Butter esters
Description	Naturally occurring fatty acids in butter	Ethyl ester of the mixed acids in butter
Physical/chemical characteristics[5]		
Appearance	Waxy solid	Waxy solid
Assay	Oleic/palmitic acids, 50–70%; stearic/myristic acids, about 20%; butyric acid, 3–4%; linoleic, lauric, palmitic, caproic, capric, and caprylic acids, <1.0% each	See Butter acids
Melting point	Low melting, variable	Low melting, variable
Solubility	Poorly soluble in alcohol; soluble in oils	Insoluble in water; soluble in alcohol
Organoleptic characteristics	Typical odor and flavor of fatty acids	Fatty, oily, waxy odor and flavor
Synthesis	Isolated from butter; mixtures of the acids from other sources	By esterification of the mixed butter acids with ethanol
Natural occurrence	In butter	
Reported uses[3]	Non-alcoholic beverages 2.0 ppm Ice cream, ices, etc. 3.0 ppm Candy 2,800 ppm Baked goods 8.3 ppm	Ice cream, ices, etc. 24 ppm Candy 78 ppm Baked goods 86 ppm Toppings 2.0 ppm Popcorn oil 1,200 ppm
Regulatory status	FDA 121.1164; FEMA No. 2171	FDA 121.1164; FEMA No. 2172

REFERENCES

For References 1–5, see end of Part III.

	Butter starter distillate	Butyl acetate
Other names	Flavor compounds distilled from cultured reconstituted skim milk	
Empirical formula		$C_6H_{12}O_2$
Structure		$CH_3-CH_2-CH_2-CH_2-O-\overset{O}{\overset{\parallel}{C}}-CH_3$
Physical/chemical characteristics[5]		
Appearance	Slightly viscous, clear liquid	Colorless liquid
Assay	Main component diacetyl; variable from 1–15 mg diacetyl/ml	95% min
Molecular weight		116.16
Melting point		−76.8°C
Boiling point		126.5°C
Flash point		38°C (22°C)
Specific gravity		0.882 at 20°C; 0.8830–0.8870 at 15°C
Refractive index		1.3942 (1.3930–1.3970) at 20°C
Acid value		0.2 max
Solubility	Soluble in water; partially soluble in oil	Very slightly soluble in water; soluble in most common organic solvents
Organoleptic characteristics		Strong, fruity odor; burning and then sweet taste reminiscent of pineapple
Synthesis	By steam-water distillation at 100°C of specially cultured skim milk	By esterification of *n*-butyl alcohol with acetic acid
Natural occurrence		Reported present in rum ether[6] and other natural products
Reported uses[3]	Ice cream, ices, etc. 20–40 ppm Candy 420 ppm Baked goods 720 ppm Shortenings 750–12,000 ppm	Non-alcoholic beverages 11 ppm Ice cream, ices, etc. 16 ppm Candy 32 ppm Baked goods 32 ppm Gelatins and puddings 13 ppm Chewing gum 220 ppm
Regulatory status	FDA 121.1164; FEMA No. 2173	FDA 121.1164; FEMA No. 2174

REFERENCES

For References 1–5, see end of Part III.

6. Fenaroli et al., *Riv. Ital. EPPOS*, 9, 484, 1965; 2, 75, 1966.

	Butyl acetoacetate	Butyl alcohol
Other names	Butyl-β-ketobutyrate Butyl 3-oxobutanoate	1-Butanol *n*-Butyl alcohol
Empirical formula	$C_8H_{14}O_3$	$C_4H_{10}O$
Structure	$CH_3-C(=O)-CH_2-C(=O)-O-(CH_2)_3-CH_3$	$CH_3-CH_2-CH_2-CH_2OH$
Physical/chemical characteristics[5]		
Appearance	Liquid	Colorless liquid
Molecular weight	158.20	74.12
Boiling point	100–103°C at 16 mm Hg	117–118°C
Freezing point		−90°C
Flash point		35–36°C
Specific gravity	0.9761 at 25°C	0.8096–0.8099 at 20°C
Refractive index	1.4245 at 25°C	1.39949 at 20°C
Solubility		1:11–1:12 in water; soluble in alcohol and ether
Organoleptic characteristics		Odor similar to amyl alcohol; dry, burning taste
Synthesis	By heating butyl acetate and sodium or potassium butylate; from ethyl acetoacetate and *n*-butyl alcohol; by reacting diketene and butyl alcohol in the presence of acetic acid and pyridine	*n*-Butyl alcohol is obtained by fermentation of glycerol, mannite, starches, and sugars in general, using *Bacillus butylicus* sometimes synergized by the presence of *Clostridium acetobutyricum*; synthetically, from acetylene
Natural occurrence		Reported present in peppermint oil from Brazil,[6] *Achillea ageratum*, tea, and in the apple aroma
Reported uses[3]	Non-alcoholic beverages 4.2 ppm Ice cream, ices, etc. 7.3 ppm Candy 26 ppm Baked goods 26 ppm	Non-alcoholic beverages 12 ppm Alcoholic beverages 1.0 ppm Ice cream, ices, etc. 7.0 ppm Candy 34 ppm Baked goods 32 ppm Cream 4.0 ppm
Regulatory status	FDA 121.1164; FEMA No. 2176	FDA 121.1164; FEMA No. 2178

REFERENCES

For References 1–5, see end of Part III.

6. Garnero, Bénezet and Igolen, *Ind. Parfum.*, 3, 353, 1958.

	Butylamine		Butyl anthranilate	
Other names	1-Aminobutane *m*-Butylamine		Butyl-2-aminobenzoate Butyl *o*-aminobenzoate	
Empirical formula	$C_4H_{11}N$		$C_{11}H_{15}NO_2$	
Structure	$CH_3—CH_2—CH_2—CH_2—NH_2$		(2-NH_2-phenyl)$-C(=O)-O-CH_2-CH_2-CH_2-CH_3$	
Physical/chemical characteristics				
Appearance	Colorless liquid		Liquid at room temperature	
Assay				
Molecular weight	73.14		193.25	
Melting point	-50°C		Approx 0°C	
Boiling point	78°C		303°C	
Flash point	45°C			
Specific gravity	0.7327 at 25/4°C		1.060 at 15.5°C	
Refractive index			1.5420 at 20°C	
Solubility	Miscible with water, alcohol, and ether			
Organoleptic characteristics	Ammoniacal odor		Sweet, faint, fruity (plum, petitgrain) note	
Synthesis	Catalytic alkylation of ammonia with butyl alcohol		Prepared by transesterification of methyl anthranilate (methyl 2-aminobenzoate) with *n*-butyl alcohol in the presence of HCl	
Natural occurrence	In mulberry leaves		Not reported found in nature	
Reported uses[3]	Alcoholic beverages	0.1 ppm	Non-alcoholic beverages	1.3 ppm
	Ice cream, ices, etc.	0.1 ppm	Ice cream, ices, etc.	2.6 ppm
	Candy	0.1 ppm	Candy	9.0 ppm
	Baked goods	0.1 ppm	Baked goods	6.7 ppm
	Gelatins and puddings	0.1 ppm		
Regulatory status	FEMA No. 3130		FDA 121.1164; FEMA No. 2181	

REFERENCES

For References 1–5, see end of Part III.

	2-Butyl-2-butenal	Butyl butyrate
Other names	2-Ethylidene hexanal	
Empirical formula	$C_8H_{14}O$	$C_8H_{16}O_2$
Structure	$CH_3—CH=C(—(CH_2)_3—CH_3)—CHO$	$CH_3—(CH_2)_2—C(=O)—O—(CH_2)_3—CH_3$
Physical/chemical characteristics[5]		
Appearance		Liquid
Molecular weight	126.19	144.21
Boiling point		165°C (162°C)
Specific gravity	1.4540 at 20°C	0.8692 at 20°C
Refractive index		1.4064 at 20°C
Solubility		Insoluble in water; miscible with alcohol and ether
Organoleptic characteristics		Fruity (pear-pineapple-like) odor
Synthesis	By isomerization of 2-butyl-3-butene-1,2-diol in acid media[6]	By passing vapors of *n*-butyl alcohol over MnO_2 or ZnO at 400°C;[6] also by passing vapors of *n*-butyl alcohol over CuO-VO at 180–200°C[7,8]
Natural occurrence		Reported found in several natural products
Reported uses[3]	Non-alcoholic beverages 0.5 ppm Ice cream, ices, etc. 0.6 ppm Candy 1.5 ppm Baked goods 2.0 ppm Gelatins and puddings 1.5 ppm Milk, dairy products 0.6 ppm	Non-alcoholic beverages 8.6 ppm Ice cream, ices, etc. 22 ppm Candy 24 ppm Baked goods 22 ppm Gelatins and puddings 14 ppm Chewing gum 150–1,500 ppm
Regulatory status	FEMA No. 3392	FDA 121.1164; FEMA No. 2186

REFERENCES

For References 1–5, see end of Part III.

6. **Gelin et al.,** *Bull Soc. Chim. Fr.*, p. 720, 1972.
7. E. I. du Pont de Nemours & Co., U.S. Patent 1,857,921, 1928.
8. **Iwannikow and Gawrilowa,** *Zh. Prikl. Khim.*, 9, 491, 1936.
9. **Iwannikow and Gawrilowa,** *Zh. Obshch. Khim.*, 17, 1103, 1947.

	Butyl butyryllactate	α-Butylcinnamaldehyde
Other names	Lactic acid, butyl ester, butyrate	2-Benzylidene hexanal Butyl cinnamic aldehyde α-Butyl-β-phenylacrolein
Empirical formula	$C_{11}H_{20}O_4$	$C_{13}H_{16}O$
Structure	CH_3 (on HC), O (double-bonded on C) $HC-O-C(=O)-CH_2-CH_2-CH_3$ $O=C-O-CH_2-CH_2-CH_2-CH_3$ (on HC)	$C_6H_5-CH=C(C_4H_9)-CHO$
Physical/chemical characteristics[5]		
Appearance	Liquid	Liquid
Molecular weight	216.28	188.27
Boiling point	90°C at 2 mm Hg	265°C
Specific gravity	0.9731 at 20°C	0.825 at 15.5°C
Refractive index	1.4215 at 20°C	
Solubility		
Organoleptic characteristics		Characteristic floral, jasmine, lily-like odor
Synthesis	By direct acetylation of the butyl ester of lactic acid with butyric anhydride in the presence of H_2SO_4[6,7]	
Natural occurrence	Not reported found in nature	Not reported found in nature
Reported uses[3]	Non-alcoholic beverages 13 ppm Ice cream, ices, etc. 9.0 ppm Candy 44 ppm Baked goods 58 ppm	Non-alcoholic beverages 0.50–1.0 ppm Ice cream, ices, etc. 1.0–2.8 ppm Candy 2.0–8.0 ppm Baked goods 2.0–8.0 ppm
Regulatory status	FDA 121.1164; FEMA No. 2190	FDA 121.1164; FEMA No. 2191

REFERENCES

For References 1–5, see end of Part III.
6. Fein and Fisher, *J. Am. Chem. Soc.*, 70, 53, 1948.
7. Fein and Fisher, *Ind. Eng. Chem.*, 36, 235, 1944.

	Butyl cinnamate	2-*sec*-Butylcyclohexanone
Other names	Butyl β-phenyl acrylate Butyl 3-phenyl propenoate	2-(1-Methylpropyl)-cyclohexanone
Empirical formula	$C_{13}H_{16}O_2$	$C_{10}H_{18}O$
Structure	$C_6H_5-CH=CH-C(=O)-O-(CH_2)_3-CH_3$	2-oxocyclohexyl $-CH(CH_3)-CH_2-CH_3$
Physical/chemical characteristics[5]		
Appearance	Mobile liquid	Colorless viscous liquid; solidifies to an opaque mass
Molecular weight	204.27	154.25
Boiling point	145°C at 13 mm Hg (295°C at 760 mm Hg)	76–78°C at 8 mm Hg
Specific gravity	1.012 at 18°C; 1.015 at 15.5°C	0.9152 at 20°C
Refractive index	1.5470 at 20°C	1.4586 at 20°C
Solubility	Soluble in 95% alcohol, chloroform, benzene, and ether; almost insoluble in water	Insoluble in water; soluble in alcohol; miscible with most oils
Organoleptic characteristics	Pleasant, balsamic, slightly cocoa-like odor	Woody, camphoraceous, somewhat dry and musty odor
Synthesis		By hydrogenation of *o-sec*-butyl phenol in the presence of palladium catalyst;[6] by hydrogenation of 2-*sec*-butylidene cyclohexanone[7]
Natural occurrence	Not reported found in nature	Reported found in yellow passion fruit[8]
Reported uses[3]	Non-alcoholic beverages 0.83 ppm Alcoholic beverages 2.0 ppm Ice cream, ices, etc. 2.6 ppm Candy 1.0–15 ppm Baked goods 1.0–15 ppm	Non-alcoholic beverages 25.0 ppm Ice cream, ices, etc. 25.0 ppm Candy 150.0 ppm Baked goods 100.0 ppm Chewing gum 1,000.0 ppm
Regulatory status	FDA 121.1164; FEMA No. 2192	FEMA No. 3261

REFERENCES

For References 1–5, see end of Part III.

6. **Dankert and Permoda,** U.S. Patent 3,129,614, 1964.
7. **Barkhash, Smironova, Prudchenko and Machinskaya,** *Zh. Obshch. Khim.*, 33(7), 2202, 1963.
8. **Winter and Klöti,** *Helv. Chim. Acta,* 55(6), 1916, 1972.

	Butyl 2-decenoate	1,3-Butylene glycol
Other names	*n*-Butyl decylenate	1,3-Butanediol β-Butyleneglycol 1,3-Dihydroxy butane
Empirical formula	$C_{14}H_{26}O_2$	$C_4H_{10}O_2$
Structure	$CH_3-(CH_2)_6-CH=CH-\overset{\overset{\displaystyle O}{\|}}{C}-O-(CH_2)_3-CH_3$	$\underset{OH}{CH_2}-CH_2-\underset{OH}{CH}-CH_3$
Physical/chemical characteristics[5]		
Appearance	Colorless liquid	Viscous liquid
Molecular weight	226.36	90.12
Boiling point		207°C
Specific gravity		1.00
Solubility	Insoluble in water; soluble in alcohol	Soluble in water and alcohol; insoluble in ether
Organoleptic characteristics	Rich, oily, fruity odor of peach and apricot; fruity, green taste	Sweet flavor with bitter aftertaste; odorless when pure
Synthesis	By esterification of *n*-butanol with 2-decenoic acid	From formaldehyde and propylene via pressure and a catalyst
Natural occurrence	Not reported found in nature	
Reported uses[3]	Non-alcoholic beverages 8.0 ppm Candy 1.5–22 ppm Baked goods 30 ppm Chewing gum 2000 ppm	The compound is used as a solvent for flavors.
Regulatory status	FDA 121.1164; FEMA No. 2194	FDA 121.1176

REFERENCES

For References 1–5, see end of Part III.

	sec-Butyl ethyl ether		Butyl ethyl malonate	
Other names			Ethyl butyl malonate	
Empirical formula	$C_6H_{14}O$		$C_9H_{16}O_4$	
Structure	$CH_3-CH_2-CH(CH_3)-O-CH_2-CH_3$		$CH_3-CH_2-O-C(=O)-CH_2-C(=O)-O-(CH_2)_3-CH_3$	
Physical/chemical characteristics[5]				
Appearance	Liquid			
Molecular weight	102.18		188.122	
Boiling point	81°C		222°C	
Specific gravity	0.7503 at 20°C		1.0257 at 17°C	
Refractive index	1.3802 at 20°C			
Solubility	Insoluble in water; very soluble in alcohol and ether			
Synthesis	From *sec*-butyl alcohol and ethyl bromide or ethyl sulfate in the presence of sodium metal[6,7]		It can be prepared by passing vapors of ethyl butyl oxalacetate over coke or pumice at 350°C.[8]	
Reported uses[3]	Non-alcoholic beverages	1.0 ppm	Non-alcoholic beverages	3.0 ppm
	Ice cream, ices, etc.	1.0 ppm	Candy	0.13 ppm
	Candy	1.0 ppm		
	Baked goods	1.0 ppm		
Regulatory status	FEMA No. 3131		FDA 121.1164; FEMA No. 2195	

REFERENCES

For References 1–5, see end of Part III.

6. Norris and Rigby, *J. Am. Chem. Soc.*, 54, 2097, 1932.
7. Marks, Lipkin and Bestman, *J. Am. Chem. Soc.*, 59, 947, 1937.
8. D.R.P., German Patent, 427,856.

	Butyl formate	Butyl heptanoate
Other names		Butyl heptoate Butyl heptylate
Empirical formula	$C_5H_{10}O_2$	$C_{11}H_{22}O_2$
Structure	$CH_3-CH_2-CH_2-CH_2-O-C(=O)H$	$CH_3-(CH_2)_5-C(=O)-O-(CH_2)_3-CH_3$
Physical/chemical characteristics[5]		
Appearance	Colorless liquid	Colorless liquid
Molecular weight	102.13	186.30
Melting point	−90°C	−68.4°C (from pentane)
Boiling point	106.8°C	226.2°C
Flash point	64°C	
Specific gravity	0.8917 at 20°C; 0.9108 at 0°C	0.85553 at 20°C
Refractive index	1.3890–1.3891 at 20°C	1.42280 at 20°C
Solubility	Slightly soluble in water; miscible with alcohol, ether, and other organic solvents	Soluble in most organic solvents
Organoleptic characteristics	Fruity, plum-like odor and corresponding taste	Characteristic herbaceous, slightly fruity odor and corresponding taste
Synthesis	By azeotropic distillation of formic acid and *n*-butyl alcohol with isopropyl formate,[6] by boiling *n*-butyl alcohol and formamide in the presence of ZnCl, $ZnSO_4$, or $HgCl_2$ [7,8]	By direct esterification of the acid with *n*-butyl alcohol in the presence of mineral acids[6,7] or in benzene solution in the presence of *p*-toluene sulfonic acid.[11] It has been prepared together with other products from heptyl aldehyde and aluminum butylate in butyl alcohol at 25–30°C.[12]
Natural occurrence	Reported found in several natural products	Not reported found in nature. The acid has been reported present in various plants.
Reported uses[3]	Non-alcoholic beverages 2.9 ppm Ice cream, ices, etc. 3.2 ppm Candy 11 ppm Baked goods 9.1 ppm Gelatins and puddings 5.0 ppm	Non-alcoholic beverages 0.50–1.0 ppm Ice cream, ices, etc. 2.0–10 ppm Candy 2.0–25 ppm Baked goods 2.0–25 ppm
Regulatory status	FDA 121.1164; FEMA No. 2196	FDA 121.1164; FEMA No. 2199

REFERENCES

For References 1–5, see end of Part III.

6. Consort. Elektrochem. Ind., German Patent 721,300, 1939.
7. Imp. Chem. Ind., U.S. Patent 1,877,847, 1928.
8. Imp. Chem. Ind., German Patent 523,189, 1928.
9. Gartenmeister, *Justus Liebigs Ann. Chem.*, 233, 284, 1886.
10. Bilterys and Gisseleire, *Bull. Soc. Chim. Belg.*, 44, 569, 1935.
11. Hoback, Parsons and Bartlett, *J. Am. Chem. Soc.*, 65, 1606, 1943.
12. Kay-Fries, U.S. Patent 2,412,469, 1943.

	Butyl hexanoate	3-Butylidenephthalide
Other names	Butyl caproate Butyl capronate Butyl hexylate	Ligusticum lactone
Empirical formula	$C_{10}H_{20}O_2$	$C_{12}H_{12}O_2$
Structure	$CH_3-(CH_2)_4-C(=O)-O-(CH_2)_3-CH_3$	(phthalide ring: benzene fused to C(=O)–O–C(=CH–CH₂–CH₂–CH₃))
Physical/chemical characteristics[5]		
Appearance	Liquid	Yellowish oily liquid
Assay		
Molecular weight	172.27	182.22
Melting point	−63°C to −64°C	82–83°C
Boiling point	208°C (99°C at 20 mm Hg)	134°C at 1.5 mm Hg (natural source) 141°C at 2.4 mm Hg (synthetic)
Specific gravity	0.8623 at 25°C; 0.865 at 15.5°C	1.0966 at 20°C (natural source) 1.1028 at 20°C (synthetic)
Refractive index	1.4153 at 25°C; 1.4170 at 20°C	1.5759 at 20°C (natural source) 1.5780 at 20°C (synthetic)
Solubility		Very slightly soluble in water; soluble in alcohol and oils
Organoleptic characteristics	Characteristic pineapple-like odor	Herbaceous odor reminiscent of lovage and celery
Synthesis	By esterification of hexanoic acid with *n*-butyl alcohol in benzene solution in the presence of *p*-toluene sulfonic acid.[6] It is also formed in the fermentation of carbohydrates yielding *n*-butyl alcohol and acetone.[7]	From sodium valerate, valeric anhydride, and phthalic anhydride[8]
Natural occurrence		Reported as occurring naturally in the fruits and roots of *Ligusticum acutilobum*
Reported uses[3]	Non-alcoholic beverages 1.7 ppm Ice cream, ices, etc. 3.9 ppm Candy 7.6 ppm Baked goods 10 ppm	Baked goods 8 ppm Condiments 5 ppm Meats 5 ppm
Regulatory status	FDA 121.1164; FEMA No. 2201	FEMA NO. 3334

REFERENCES

For References 1–5, see end of Part III.

6. **Hoback, Parsons and Bartlett,** *J. Am. Chem. Soc.*, 65, 1606, 1943.
7. **Fujita,** *J. Chem. Soc. Jap.*, 64, 1008, 1943.
8. **Berlingozzi and Lupo,** *Gazz. Chim. Ital.*, 57, 259, 1927.

	Butyl isobutyrate	Butyl isovalerate
Other names	*n*-Butyl 2-methyl propanoate	
Empirical formula	$C_8H_{16}O_2$	$C_9H_{18}O_2$
Structure	$(H_3C)_2CH{-}C({=}O){-}O{-}(CH_2)_3{-}CH_3$	$(H_3C)_2CH{-}CH_2{-}C({=}O){-}O{-}(CH_2)_3{-}CH_3$
Physical/chemical characteristics[5]		
Appearance	Colorless liquid	Liquid
Assay	97% min[2]	
Molecular weight	144.21	158.24
Boiling point	155–156°C	175–176°C
Specific gravity	0.859–0.864[2] at 25°/25°C; 0.8618 at 20°C	0.8608 at 20°C
Refractive index	1.4010–1.4040[2] at 20°C; 1.4025 at 20°C	1.4087 at 20°C
Acid value	Not more than 1.0[2]	
Solubility	Insoluble in water; 1:7 in 60% alcohol[2]	Insoluble in water; soluble in most organic solvents
Organoleptic characteristics	Strong, fruity odor and sweet, pineapple-like taste	Odor reminiscent of the isobutyl ester with a sweet, apple-like taste
Synthesis	Prepared from the corresponding acid and *n*-butyl alcohol by esterification in the presence of concentrated H_2SO_4 with or without solvent (boiling benzene)[7]	By prolonged boiling of the acid with *n*-butyl alcohol in benzene in the presence of concentrated H_2SO_4[7]
Natural occurrence	Reported present in Roman camomile essential oil.	Reported found in the oil from leaves of *Eriostemon coxii* and *Phebalium dentatum*
Reported uses[3]	Non-alcoholic beverages 8.7 ppm Ice cream, ices, etc. 4.0–5.0 ppm Candy 19 ppm Baked goods 39 ppm Chewing gum 2,000 ppm	Non-alcoholic beverages 4.6 ppm Ice cream, ices, etc. 12 ppm Candy 13 ppm Baked goods 15 ppm Gelatins and puddings 50 ppm
Regulatory status	FDA 121.1164; FEMA No. 2188	FDA 121.1164; FEMA No. 2218

REFERENCES

For References 1–5, see end of Part III.

6. Salmi and Leimu, *Suom. Kemistil.*, 208, 47, 1947.
7. Vogel, *J. Chem. Soc.*, p. 630, 1948.

	Butyl lactate			Butyl laurate
Other names	Butyl 2-hydroxypropanoate Butyl α-hydroxypropionate			Butyl dodecanoate Butyl dodecylate
Empirical formula	$C_7H_{14}O_3$			$C_{16}H_{32}O_2$
Structure	$CH_3-CH(OH)-C(=O)-O-(CH_2)_3-CH_3$			$CH_3-(CH_2)_{10}-C(=O)-O-(CH_2)_3-CH_3$
Physical/chemical characteristics[5]	*d-form*	*l-form*	*dl-form*	
Appearance	Liquid	Liquid	Liquid	Liquid
Molecular weight	146.16	146.16	146.16	256.44
Melting point				−7°C
Boiling point	77°C at 10 mm Hg	70–73°C at 10–11 mm Hg	185°C	154°C at 4.5 mm Hg
Specific gravity	0.9744 at 27.6°C	—	0.973 at 20°C	0.8603 at 20°C
Refractive index	—	—	1.4210 at 20°C	1.4362 at 20°C
Optical rotation	+13.63° at 27.3°C	—	—	
Solubility	Slightly soluble in water			Insoluble in water; soluble in most organic solvents
Organoleptic characteristics	Faintly sweet, pleasant odor			Characteristic fruity, peanut-like odor
Synthesis	Two optically active forms and one racemic form are known. The *d*-form is prepared by reacting zinc ammonium *l*-lactate with *n*-butyl alcohol in the presence of concentrated H_2SO_4.[6] The *l*-form is prepared by reacting zinc ammonium *d*-lactate with *n*-butyl alcohol in the presence of HCl.[7] The racemic form is prepared by several methods, one being from calcium or sodium lactate and *n*-butyl alcohol in benzene in the presence of H_2SO_4, with subsequent azeotropic distillation of the mixture.[8]			By esterification of lauric acid with *n*-butyl alcohol in the presence of H_2SO_4[9] or by conducting the esterification using gaseous HCl as a catalyst
Natural occurrence	Not reported found in nature			Not reported found in nature
Reported uses[3]	Non-alcoholic beverages 0.66 ppm Ice cream, ices, etc. 2.8 ppm Candy 6.5 ppm Baked goods 7.7 ppm			Non-alcoholic beverages 0.40–3.0 ppm Ice cream, ices, etc. 0.60 ppm Candy 17 ppm Baked goods 1.0–40 ppm
Regulatory status	FDA 121.1164; FEMA No. 2205			FDA 121.1164; FEMA No. 2206

REFERENCES

For References 1–5, see end of Part III.

6. Wood, Such and Scarf, *J. Chem. Soc.*, 123, 610, 1923
7. Wassmer and Guye, *Chem. Zentralbl.*, 2, 1419, 1903.
8. Smith and Claborn, *Ind. Eng. Chem.*, 32, 693, 1940.
9. Rheinboldt, König and Otten, *Ann. Chem.*, 473, 258, 1929.

	Butyl levulinate	*n*-Butyl-2-methylbutyrate
Other names	Butyl γ-butyrolactone Butyl 4-oxopentanoate	
Empirical formula	$C_9H_{16}O_3$	$C_9H_{18}O_2$
Structure	$CH_3-C(=O)-(CH_2)_2-C(=O)-O-(CH_2)_3-CH_3$	H_3C, CH_3, O, CH_3, O
Physical/chemical characteristics[5]		
Appearance	Liquid	
Molecular weight	172.23	158.23
Boiling point	238°C	
Specific gravity	0.9735 (0.9745) at 20°C	
Refractive index	1.4283 (1.4290) at 20°C	
Solubility	Slightly soluble in water; soluble in ether, alcohol, chloroform, and other organic solvents	
Organoleptic characteristics	Characteristic burning, bitter taste	
Synthesis	By esterification of levulinic acid with *n*-butyl alcohol in the presence of HCl[6–8]	H_3C, O, OH + HO, CH_3 → CH_3
Natural occurrence	Not reported found in nature	Reported found in strawberry[9,10]
Reported uses[3]	Non-alcoholic beverages 0.20–1.0 ppm Ice cream, ices, etc. 2.1 ppm Candy 4.6 ppm Baked goods 4.6 ppm	Non-alcoholic beverages 1.0 ppm Ice cream, ices, etc. 3.0 ppm Candy 5.0 ppm Baked goods 5.0 ppm Gelatins and puddings 3.0 ppm Jellies, preserves, spreads 2.0 ppm
Regulatory status	FDA 121.1164; FEMA No. 2207	FEMA No. 3393

REFERENCES

For References 1–5, see end of Part III.

6. Sah and Ma, *J. Am. Chem. Soc.*, 52, 4881, 1930.
7. Cox and Doods, *J. Am. Chem. Soc.*, 55, 3392, 1933.
8. Niacet Chemical Corp., U.S. Patent 2,029,412, 1934.
9. McFadden et al., *J. Chromatogr.*, 18, 10, 1965.
10. Nursten and Williams, *Chem. Ind.* (London), p. 486, 1967.

	Butyl phenylacetate	3-N-Butylphthalide
Other names	Butyl α-toluate	
Empirical formula	$C_{12}H_{16}O_2$	$C_{12}H_{14}O_2$
Structure	$C_6H_5-CH_2-C(=O)-O-(CH_2)_3-CH_3$	(phthalide ring: C(=O)–O–CH) $CH_2-CH_2-CH_2-CH_3$
Physical/chemical characteristics[5]		
Appearance	Colorless liquid	Almost colorless oily liquid
Assay	98% min	
Molecular weight	192.26	190.24
Melting point		
Boiling point	260°C	177–178°C at 15 mm Hg 106–108°C at 0.1 mm Hg
Congealing point		
Flash point	74°C	
Specific gravity	0.991–0.994 at 25°C	1.0672 at 20°C
Refractive index	1.4880–1.4910 at 20°C	1.5260 at 20°C
Optical rotation		–57° (c = 1.96 in chloroform)
Acid value	1.0 max	
Solubility	1:2 in 80% alcohol	Slightly soluble in water; soluble in alcohol and oils
Organoleptic characteristics	Pleasant rose and honey-like odor	Herbaceous celery-like odor reminiscent of concentrated vegetable soup
Synthesis	By direct esterification of the acid with *n*-butyl alcohol in the presence of gaseous HCl	From celery oil;[6] from phthalide via bromination and subsequent reaction of the resulting phthaldeidic acid with *n*-butyl magnesium bromide[7]
Natural occurrence	Not reported found in nature	Reported found in *Levisticum officinale* and *L. acutilobum*; also in celery oil
Reported uses[3]	Non-alcoholic beverages 0.50 ppm Ice cream, ices, etc. 2.1 ppm Candy 4.5 ppm Baked goods 4.6 ppm Gelatins and puddings 5.0 ppm	Alcoholic beverages 0.5 ppm Gelatins and puddings 2.0 ppm Condiments, pickles 2.0 ppm Meats, meat sauces 1.0 ppm Soups 1.0 ppm Preserves and spreads 2.0 ppm
Regulatory status	FDA 121.1164; FEMA No. 2209	FEMA No. 3334

REFERENCES

For References 1–5, see end of Part III.

6. Barton and de Vries, *J. Chem. Soc.*, 1916, 1963.
7. Naves, *Helv. Chim. Acta*, 26, 1281, 1943.

	Butyl propionate	Butyl stearate
Other names		Butyl octadecanoate Butyl octadecylate
Empirical formula	$C_7H_{14}O_2$	$C_{22}H_{44}O_2$
Structure	$CH_3{-}CH_2{-}C({=}O){-}O{-}(CH_2)_3{-}CH_3$	$CH_3{-}(CH_2)_{16}{-}C({=}O){-}O{-}(CH_2)_3{-}CH_3$
Physical/chemical characteristics[5]		
Appearance	Colorless liquid	Waxy or oily (>20°C), colorless material
Assay	96% min	
Molecular weight	130.18	340.35
Melting point		19–20°C
Boiling point	145–146°C	220–225°C at 25 mm Hg
Flash point	32°C	Approx. 172°C
Specific gravity	0.8795–0.8825 at 15°C; 0.8750 at 20°C	0.855–0.860 at 20°C
Refractive index	1.4000–1.4020 at 20°C; 1.4010 at 20°C	1.4430 at 20°C
Acid value	0.2 max	
Solubility	Slightly soluble in water; miscible with alcohol and ether	Insoluble in water; soluble in alcohol, ether, and mineral or vegetable oils
Organoleptic characteristics	Characteristic earthy, faintly sweet odor and apricot-like taste	Odorless or faintly fatty odor
Synthesis	By esterification of propionic acid with *n*-butyl alcohol in the presence of concentrated H_2SO_4[6] or *p*-toluene sulfonic acid[7]	By reacting silver stearate with *n*-butyl iodide at 100°C according to Whitly; by transesterification of glyceryl tristearate (tristearin) with *n*-butyl alcohol
Natural occurrence	Reported found in several natural products	Not reported found in nature
Reported uses[3]	Non-alcoholic beverages 4.0 ppm Ice cream, ices, etc. 5.2 ppm Candy 25 ppm Baked goods 27 ppm	Non-alcoholic beverages 1.0 ppm Alcoholic beverages 5.0 ppm Ice cream, ices, etc. 2.0 ppm Candy 190 ppm Baked goods 340 ppm Chewing gum 330 ppm
Regulatory status	FDA 121.1164; FEMA No. 2211	FDA 121.1164; FEMA No. 2214

REFERENCES

For References 1–5, see end of Part III.

6. Vogel, *J. Chem. Soc.*, p. 629, 1948.
7. Salmi and Leimu, *Chem. Abstr.*, 42, 4031, 1948.

	Butyl sulfide	Butyl 10-undecenoate
Other names	Dibutyl sulfide	Butyl 10-hendecenoate Butyl undecylenoate
Empirical formula	$C_8H_{18}S$	$C_{15}H_{28}O_2$
Structure	$CH_3-(CH_2)_3-S-(CH_2)_3-CH_3$	$CH_2{=}CH-(CH_2)_8-C(=O)-O-(CH_2)_3-CH_3$
Physical/chemical characteristics[5]		
Appearance	Liquid	Liquid
Molecular weight	146.30	240.39
Boiling point	182°C or 190–230°C (with decomposition)	125–128°C (252°C) at 3 mm Hg
Specific gravity	0.8386 at 20°C	0.8751 at 20°C
Refractive index	1.4530 at 20°C	1.4426 at 20°C
Solubility	Readily soluble in ether	
Organoleptic characteristics		Slightly fatty, buttery, wine-like, unpleasant odor
Synthesis	From butyl bromide and sodium sulfide in boiling ethanol.[6] According to some authors two forms exist, exhibiting different boiling points but identical solubilities in various solvents. Both forms are insoluble in water.[7]	By esterification of the acid with *n*-butyl alcohol at the boil in the presence of concentrated H_2SO_4[8] or in benzene solution[9]
Natural occurrence	Reported found in several natural products	Not reported found in nature
Reported uses[3]	Non-alcoholic beverages 0.02–1.0 ppm Ice cream, ices, etc. 0.01–1.0 ppm Candy 0.03–1.0 ppm Baked goods 0.03–1.0 ppm	Non-alcoholic beverages 0.90 ppm Alcoholic beverages 5.0 ppm Ice cream, ices, etc. 2.0 ppm Candy 6.6 ppm Baked goods 7.8 ppm Chewing gum 0.40–60 ppm Icings 5.0 ppm
Regulatory status	FDA 121.1164; FEMA No. 2215	FDA 121.1164; FEMA No. 2216

REFERENCES

For References 1–5, see end of Part III.

6. Man and Purdie, *J. Chem. Soc.*, p. 1556, 1935.
7. Hinsberg, *Ber. Dtsch. Chem. Ges.*, 62, 2167, 1929.
8. Dankowa and Cèll, *Zh. Obshch. Khim.*, 15, 191, 1945.
9. Jeffery and Vogel, *J. Chem. Soc.*, p. 662, 1948.

	n-Butyl valerate	Butyraldehyde
Other names	*n*-Butylpentanoate *n*-Butyl-*n*-valerianate	*n*-Butanal Butyl aldehyde Butyric aldehyde
Empirical formula	$C_9H_{18}O_2$	C_4H_8O
Structure	$C_4H_9-O-C(=O)-(CH_2)_3-CH_3$	$CH_3-CH_2-CH_2-CHO$
Physical/chemical characteristics[5]		
Appearance	Liquid	Colorless liquid
Molecular weight	158.24	72.10
Melting point		−99°C
Boiling point	186.5°C (175°C)	75–76°C
Specific gravity	0.8680 at 20°C; 0.865 at 15.5°C	0.817 at 20°C
Refractive index	1.4118 at 20°C	1.3843 at 20°C
Solubility	Slightly soluble in water; soluble in propylene glycol	1:27 in water
Organoleptic characteristics	Characteristic fruity (apple-raspberry) odor and corresponding sweet taste	Characteristic pungent odor
Synthesis	By esterification of the acid with *n*-butyl alcohol in the presence of concentrated H_2SO_4 at the boil[6]	By dry distillation of calcium butyrate and calcium formate[7]
Natural occurrence	Reported found in several natural products	Reported found in the essential oils from flowers, fruits, leaves, or bark of: *Monarda fistulosa*,[8] *Litsea cubeba*,[9] Bulgarian clary sage,[10] cajeput, *Eucalyptus cinerea*,[11] *E. globulus*,[12] and others, as well as in apple and strawberry aromas[13,14]
Reported uses[3]	Non-alcoholic beverages 3.0 ppm Ice cream, ices, etc. 2.6 ppm Candy 8.0 ppm Baked goods 6.8 ppm	Non-alcoholic beverages 0.71 ppm Alcoholic beverages 0.50 ppm Ice cream, ices, etc. 4.8 ppm Candy 2.9 ppm Baked goods 5.4 ppm Icings 0.25 ppm
Regulatory status	FDA 121.1164; FEMA No. 2217	FDA 121.1164; FEMA No. 2219

REFERENCES

For References 1–5, see end of Part III.

6. **Vogel,** *J. Chem. Soc.*, p. 630, 1948.
7. **Lieben and Rossi,** *Justus Liebigs Ann. Chem.*, 158, 137, 1871.
8. **Miller et al.,** *Chem. Zentralbl.*, 2, 453, 1920.
9. **Fürst and Fenstel,** *Z. Chem.*, 1, 22, 1961.
10. *Perfum. Essent. Oil Rec.*, 49, 233, 1958.
11. **Baker and Smith,** *Research Eucalyptus E.O.*, Sydney, 1920.
12. *Schimmel Ber.*, 4, 47, 1904.
13. **Meigh,** *J. Sci. Food Agric.*, 7, 396, 1956.
14. **McGlamphy,** *Perfum. Essent. Oil Rec.*, 44, 352, 1953.

	***n*-Butyric acid**
Other names	Butanoic acid Ethylacetic acid
Empirical formula	$C_4H_8O_2$
Structure	$CH_3{-}CH_2{-}CH_2{-}COOH$
Physical/chemical characteristics[5]	
Appearance	Colorless liquid
Molecular weight	88.10
Melting point	−5.55°C
Boiling point	163.55°C
Flash point	76°C
Specific gravity	0.960 at 15°C
Refractive index	1.3983–1.3984 at 20°C
Solubility	Soluble in water, alcohol, and ether
Organoleptic characteristics	Persistent, penetrating, rancid, butter-like odor and burning, acid taste
Synthesis	Obtained by fermentation of starches and molasses with selective enzymes (*Granulo saccharobutyricum*). It is subsequently isolated as the calcium salt.
Natural occurrence	Normally occurs in butter as a glyceride. It has been reported found in the essential oils of: citronella Ceylon, *Eucalyptus globulus*, *Araucaria cunninghamii*, *Lippia scaberrima*, *Monarda fistulosa*, cajeput, *Heracleum giganteum*, lavender, *Hedeoma pulegioides*, valerian, nutmeg, hops, *Pastinaca sativa*, Spanish anise, and others. It has been identified in strawberry aroma.[6]
Reported uses[3]	Non-alcoholic beverages 5.5 ppm Ice cream, ices, etc. 6.5 ppm Candy 82 ppm Baked goods 32 ppm Gelatins and puddings 0.19–45 ppm Chewing gum 60–270 ppm Margarine 18 ppm
Regulatory status	FDA GRAS; FEMA No. 2221

REFERENCES

For References 1–5, see end of Part III.
6. Willhalm et al., *Helv. Chim. Acta*, 49, 66, 1966.

	γ-Butyrolactone
Other names	4-Hydroxybutanoic acid, lactone 3 (or 4)-Hydroxybutyric acid, lactone 1,2-Butanolide 1,4-Butanolide
Empirical formula	$C_4H_6O_2$
Structure	H_2C—O—C=O ; H_2C—CH_2 (ring)
Physical/chemical characteristics[5]	
Appearance	Colorless to pale yellow, oily liquid
Molecular weight	86.09
Melting point	-44°C
Boiling point	206°C 89°C at 12 mm Hg
Specific gravity	1.1286 at 15°C 1.1441 at 0°C
Refractive index	1.4341 at 20°C
Solubility	Soluble in water in all proportions; very soluble in alcohol, ether, acetone, and benzene
Organoleptic characteristics	Faint, sweet, aromatic slightly buttery odor
Synthesis	From acetylene and formaldehyde;[6] also from a number of alternative sources: ethylene chlorohydrin, glutamic acid, γ-hydroxybutyric acid solutions, tetrahydrofuran, vinylacetic acid[7]
Natural occurrence	Reported found as a constituent in coffee aroma;[8,9] also a volatile flavor component in roasted filberts[10]
Reported uses[3]	Non-alcoholic beverages 10.0 ppm Ice cream, ices, etc. 10.0 ppm Candy 10.0 ppm Baked goods 20.0 ppm Meat, meat sauces, soups 10.0 ppm Milk, dairy products 10.0 ppm Cereals 20.0 ppm
Regulatory status	FEMA No. 3291

REFERENCES

For References 1–5, see end of Part III.

6. **Reppe,** *Chem. Ing. Tech.* 365, 1950; *Chem. Eng.*, 58(6), 176, 1951.
7. **Whitmore,** *Organic Chemistry,* 2nd ed., Dover Publ., New York, 1951.
8. **Stoll et al.,** *Helv. Chim. Acta,* 50, 628, 1967.
9. **Viani et al.,** *Helv. Chim. Acta,* 48, 1809, 1965.
10. **Sheldon** et al., *J. Food Sci.*, 37(2), 313, 1972.

Cadinene

Empirical formula $C_{15}H_{24}$

Structure

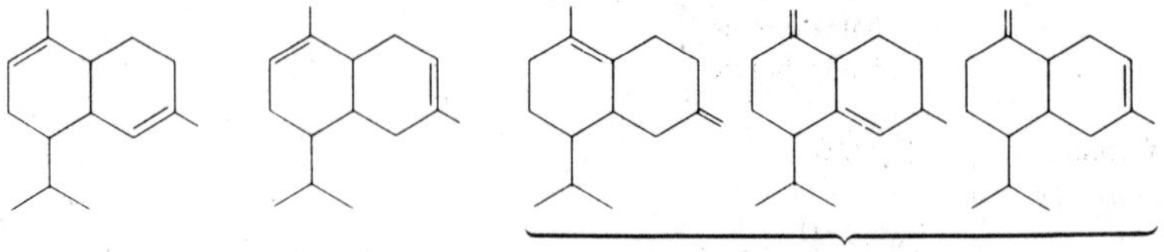

α β γ

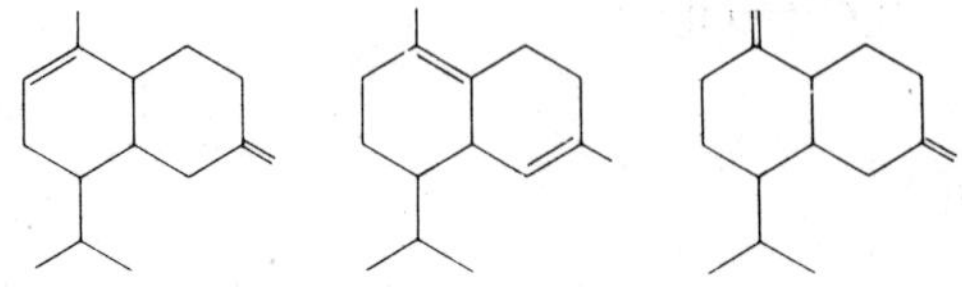

γ₁ δ ε

Physical/chemical characteristics[5]	α and β	γ	γ_1	δ	ε
Molecular weight	204.34	204.34	204.34	204.34	204.34
Boiling point at 9 mm Hg			120–121°C		
Flash point					
Specific gravity at 20°C	0.9239	0.9125[a] 0.9146[b] 0.9087[c]		0.9175[a] 0.9146[e] 0.9162[b]	0.9107
Refractive index at 20°C	1.5059	1.5075[a] 1.5080[b] 1.5056[c]	1.5152[d]	1.5086[a] 1.5085[e] 1.5090[b]	1.5090
Optical rotation at 20°C	−251°	+148°[a] +3°4′[b] +56°7′[c]	−18°9′	+94°[a] +77°[e] +84°6′[b]	47°2′

Organoleptic characteristics

Synthesis The relative position of the two double bonds characterizes the stereoisomers of cadinene; these have been identified.[6]

Natural occurrence Wallach confirmed the presence of cadinene in several essential oils:[7] α- and β-cadinene in the oil of *Dacrydium colensoi*; γ-cadinene in the oils of Java citronella, ylang-ylang, and *Populus balsamifera*; γ_1-cadinene in the oil of lemongrass Malabar; δ-cadinene in the oils of citronella, calamus, and ylang-ylang; ε-cadinene in the oil of ylang-ylang. See reference 8 for a tabular summary of essential oils containing cadinene and the corresponding plant and botanical family derivation.

[a]Citronella.
[b]Ylang-ylang.
[c]*Populus balsamifera.*
[d]At 26.5°C.
[e]Calamus.

REFERENCES

For References 1–5, see end of Part III.
6. **Campbell and Soffer,** *J. Am. Chem. Soc.*, 64, 417, 1942.
7. **Wallach,** *Annalen,* 238, 78, 1887.

	Caffeine	Camphene
Other names	Coffeine Guaramine Methyltheobromine Theine 1,3,7-Trimethyl-2,6-dioxopurine 1,3,7-Trimethylxanthine	3,3-Dimethyl-2-methylene norcamphane 2,2-Dimethyl-3-methylene norbornane
Empirical formula	$C_8H_{10}N_4O_2$	$C_{10}H_{16}$
Structure	[structure: 1,3,7-trimethylxanthine ring with H_3C, O, CH_3, N labels]	[structure: bicyclic with H_2C, CH_2, $C{=}CH_2$, $C-(CH_3)_2$, H labels]
Physical/chemical characteristics[5]		
Appearance	Prism-like crystals	Crystalline solid
Assay		Commercial product may contain approximately 20% tricyclene.
Molecular weight	194.19	136.23
Melting point	238°C; sublimes at 178°C	*d*, 45–46°C; *l*, 51–52°C; sublimes when heated
Boiling point		159°C
Optical rotation		*d*, +104°; *l*, −85°
Solubility	One gm soluble in 46 ml water, 66 ml alcohol, 22 ml of 60% alcohol, 50 ml acetone. Solubility in water increases with increasing temperature or by adding alkali benzoates, cinnamates, citrates, or salicylates.	Insoluble in water; moderately soluble in alcohol; soluble in ether and chloroform
Organoleptic characteristics		Terpene, camphoraceous taste
Synthesis	Usually obtained from tea dust in which it is present up to 1.5–3.5% or as a by-product from the manufacture of caffeine-free coffee. Synthetically prepared starting with dimethylurea and malonic acid[6]	From pinene by catalytic isomerization or from bornyl chloride by heating with alkali in the presence of abietenesulfonic acid[7,8]
Natural occurrence	Occurs in tea, coffee, and maté leaves; also in guarana paste and cola nuts	The *l*-form was isolated in 1888 by Gobolow in the oil *Abies sibirica*; its structure was described by G. Wagner.[9] It also occurs in the oils of: *Tsuga canadensis*, *Thuja occidentalis*, *Artemisia herba alba*, and others. The *d*-form is present in several essential oils: orange flowers, camphor, lavender, calamus, *Curcuma aromàtica*, and others.
Reported uses[3]	Non-alcoholic beverages 120 ppm	Non-alcoholic beverages 40–90 ppm Ice cream, ices, etc. 20 ppm Candy 160 ppm Baked goods 27 ppm
Regulatory status	FDA GRAS: 0.02% in cola-type beverages; FEMA No. 2224	FDA 121.1164; FEMA No. 2229

REFERENCES

For References 1–5, see end of Part III.

6. **Fischer and Ach,** *Berichte,* 28, 2473, 3135, 1895.
7. **Dupont, Dulou and Thuet,** *Bull. Soc. Chim. Fr.,* 8, 891, 1941.
8. **Tischchenko and Rudakov,** *Chem. Abstr.,* p. 4052, 1934.
9. **Wagner,** *Ber. Dtsch. Chem. Ges.,* 23, 2311, 1890.

	d-Camphor	Caramel
Other names	*d*-2-Bornanone; *d*-2-Camphanone	
Empirical formula	$C_{10}H_{16}O$	
Structure	(structure: CH_3, O, H_3C, CH_3)	
Physical/chemical characteristics[5]	(The following values refer to synthetic camphor.)	
Appearance	Crystalline, white solid	Dark-brown, thick liquid
Molecular weight	152.23	
Melting point	>170°C	
Boiling point	209°C	
Flash point	Vapor mixtures of camphor and air are explosive	
Specific gravity	0.990–0.995 at 15°C	Approx. 1.35
Optical rotation	+2° to +5° at 20°C (This measurement is utilized to differentiate between the synthetic and the natural product.)	
Solubility	Readily soluble in alcohol (1:1), ether, chloroform, acetone, acetic acid, essential oils, and benzene; insoluble in alkali and glycerol. A solution of synthetic camphor in pet ether must be absolutely clear with a residue after evaporation of less than 1.0%.	Soluble in water and dilute alcohol
Organoleptic characteristics	Characteristic penetrating odor; burning, bitter, fresh taste	Odor of burnt sugar and pleasant, bitter taste
Synthesis	Natural camphor is obtained by distillation from the plants of *Cinnamomum* or *Laurus camphora* from China and Japan, together with the corresponding essential oils. The raw camphor contains several impurities. It is separated from the water and the essential oil by pressure or by centrifugation and subsequently purified by sublimation or crystallization. Synthetic camphor is prepared from pinene isolated by fractional distillation of turpentine oil. Pinene is reacted to bornyl chloride with gaseous HCl under pressure and then to camphene. The distilled and purified camphene is then oxidized to camphor with Na or K bichromate in the presence of H_2SO_4.	Made by heating sugar or glucose, adding small amounts of alkali, alkaline carbonate, or a trace of mineral acid during the heating
Natural occurrence	Frequently occurring in nature as the *d*- or *l*-form; the optically inactive form is seldom encountered. The *d*-form has been reported found in *Cinnamomum camphora* Nees. (*Laurus camphora* L.) from China, Japan, and the East Indies; in *Sassafras officinale*, *Lavandula spica*, and in other *Labiatae*. The *l*-form is reported found in the essential oils of *Salvia grandiflora*, *Matricaria parthenium*, *Artemisia herba alba*. The optically inactive form is found in *Chrysanthemum japonicum sinense*. For additional information see reference 6.	
Reported uses[3]	Ice cream, ices, etc., 0.10 ppm; candy, 1.1–25 ppm; baked goods, 11 ppm; condiments, 20 ppm	Non-alcoholic beverages, 2,200 ppm; ice cream, ices, etc., 590 ppm; candy, 180 ppm; baked goods, 220 ppm; meats, 2,100 ppm; syrups, 2,800 ppm
Regulatory status	FDA 121.1164; FEMA No. 2230	FDA GRAS; FEMA No. 2235

REFERENCES

For References 1–5, see end of Part III.

6. **Gildemeister and Hoffmann,** *Die Aetherischen Oele,* Vol. 3c, Akademie Verlag, 1963, 291.

For further information on *d*-camphor see the following:

Gilman, H. (Ed.), *Organic Chemistry,* Vol. 4, John Wiley & Sons, 1953, 2907 ff.

Simonsen, *The Terpenes,* Vol. 2, Cambridge University Press, 1947, 373 ff.

Carvacrol

Other names	2-*p*-Cymenol 2-Hydroxy-*p*-cymene Isopropyl-*o*-cresol Isothymol 2-Methyl-5-isopropyl phenol
Empirical formula	$C_{10}H_{14}O$
Structure	CH_3, OH, CH, H_3C, CH_3
Physical/chemical characteristics[5]	
Appearance	Colorless or yellowish, thick liquid
Molecular weight	150.22
Melting point	0–1°C
Boiling point	236–237°C
Flash point	100°C
Specific gravity	0.9743 at 21°C; 0.974–0.979 at 25°C
Refractive index	1.5209 at 21°C; 1.5210–1.5260 at 20°C
Solubility	Insoluble in water; soluble in alcohol and ether
Organoleptic characteristics	Characteristic pungent, warm odor. The commercial product consists of a mixture of isomers differing because of the position of the isopropyl radical. The technical and pure quality are different because of the odor characteristics.
Synthesis	From *p*-cymene by sulfonation and subsequent alkali fusion[6,7]
Natural occurrence	Occurs in the bark oil of *Thuja articulata*,[8] camphor oil,[9] in the oils of *Monarda citriodora*[10] and *Monarda fistulosa*,[11] in some species of *Satureja*,[12–14] in the oil of various species of *Origanum*,[15–17] *Thymus*,[18–20] *Thymbra spicata* L.,[21] in the seed oil of *Ptychotis ajowan* D.C.,[22] and others
Reported uses[3]	Non-alcoholic beverages 26 ppm Ice cream, ices, etc. 34 ppm Candy 92 ppm Baked goods 120 ppm Condiments 37 ppm
Regulatory status	FDA 121.1164; FEMA No. 2245

REFERENCES

For References 1–5, see end of Part III.

6. **Hixson and McKee,** *Chem. Zentralbl.*, 2, 851, 1919.
7. **Gibbs and Phillips,** *J. Ind. Eng. Chem.*, 12, 145, 1920.
8. **Grimal,** *Compt. Rend.*, 139, 927, 1904.
9. *Schimmel Ber.*, October 1902.
10. **Prandel,** *Pharm. Rev.*, 22, 153, 1904.
11. **Kremers,** *Chem. Zentralbl.*, 2, 41, 1897.
12. **Jahns,** *Chem. Ber.*, 15, 816, 1882.
13. **Haller,** *Compt, Rend.*, 94, 132, 1882.
14. Schimmel and Co., *Chem. Zentralbl.*, 2, 1804, 1911.
15. **Jahns,** *Arch. Pharm.*, 215, 1, 1870.
16. **Umney and Bennett,** *Chem. Zentralbl.*, 1, 360, 1906.
17. **Pickles,** *J. Chem. Soc.*, 93, 866, 1908.
18. **Jahns,** *Arch. Pharm.*, 216, 277, 1880.
19. **Palazzo and Lutri,** *Ann. Chim.* (Rome), 14, 105, 1924.
20. **Puxeddu,** *Ann. Chim.* (Rome), 16, 329, 1926.
21. Schimmel and Co., *Chem. Zentralbl.*, (81), 1755, 1910.
22. **Sobt and Singh,** *Perfum. Essent. Oil Rec.*, 14, 399, 1923.

	Carvacryl ethyl ether	**Carveol**	
Other names	2-Ethoxy-*p*-cymene Ethyl carvacrol Ethyl carvacryl ether	*p*-Mentha-6,8-dien-2-ol 1-Methyl-4-isopropenyl-6-cyclohexen-2-ol	
Empirical formula	$C_{12}H_{18}O$	$C_{10}H_{16}O$	
Structure	$O-CH_2-CH_3$; H_3C, CH, CH_3, H_3C	CH_3, OH, C, H_3C, CH_2	
Physical/chemical characteristics[5]		*cis*-**Carveol**[7]	*trans*-**Carveol**[7]
Appearance	Oily, light liquid		
Molecular weight	178.27		
Boiling point	235°C	101°C (*dl*- at 10 mm Hg) 108°C (*dl*- at 16 mm Hg)	102.2–102.4°C (*d*- at 10 mm Hg) 108°C (*dl*- at 16 mm Hg)
Specific gravity		*dl*-, 0.9521 at 25°C	*d*-, 0.9484; *dl*-, 0.9510 at 18°C
Refractive index		*dl*-, 1.4959 at 25°C *dl*-, 1.4972 at 18°C	*d*-, 1.4942; *dl*-, 1.4956 at 19°C
Optical rotation		*d*-, +23°9′ at 25°C; *l*-, −23°9′	*d*-, +213°1′
Organoleptic characteristics	Odor similar to carrot		
Synthesis	By passing vapors of ethyl alcohol and carvacrol over a catalyst (ThO_2) at 400–500°C[6]	The various *d*-, *l*-, and *dl*- (*cis*- and *trans*-, respectively) forms have been prepared synthetically and isolated by means of the dinitrobenzoates according to Amvers's law. They are prepared by oxidation of limonene or, better, from carvone.	
Natural occurrence	Not reported found in nature	In small amounts, sometimes esterified, it has been reported present in nature (caraway, spearmint).	
Reported uses[3]	Non-alcoholic beverages 2.0–13 ppm Ice cream, ices, etc. 10 ppm Candy 21 ppm Baked goods 3.0–39 ppm	Non-alcoholic beverages 1.5–13 ppm Ice cream, ices, etc. 3.0 ppm Candy 3.0–39 ppm Baked goods 3.0–5.0 ppm	
Regulatory status	FDA 121.1164; FEMA No. 2246	FDA 121.1164; FEMA No. 2247	

REFERENCES

For References 1–5, see end of Part III.

6. **Sabatier and Maibhe,** *Compt. Rend.*, 158, 611, 1914.
7. **Johnston and Read,** *J. Chem. Soc.* (London), p. 233, 1934.

4-Carvomenthenol

Other names

1-*p*-Menthen-4-ol
1-Methyl-4-isopropyl-1-cyclohexen-4-ol
Origanol
4-Terpinenol

Empirical formula

$C_{10}H_{18}O$

Structure

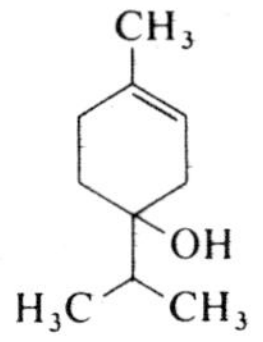

Physical/chemical characteristics[5]

	dextro	*levo*	*racemic*
Molecular weight	154.24	154.24	154.24
Boiling point	208–212°C	70–71°C (at 3 mm Hg)	92–97°C (at 14 mm Hg)
Specific gravity	0.9265 at 19°C	0.9266 at 20°C	0.9210 at 24°C
Refractive index	1.4785 at 19°C	1.47604 at 20°C	1.4778 at 24°C
Optical rotation	Approx +25°	−24°32′ at 20°C	—

Synthesis

One of several terpinenol isomers, depending on the position of the double bond and that of the hydroxyl group. This terpene, whose structure has been defined by Wallach, can be isolated by fractional distillation. It exists in nature as the *dextro*, *levo*, and racemic isomer. The synthetic product is always optically inactive. The 1-terpinenol or 1-methyl-4-isopropyl-3-cyclohexen-1-ol has been prepared by Wallach.[6]

Natural occurrence

4-Carvomenthenol (*dextro*) has been reported present in the oil of *Cupressus macrocarpa*, lavender, Spanish origanum, *Ledum palustre*, *Eucalyptus australiana* var. *A.*, *Thuja occidentalis*, etc. The *levo* form is present in the oil of *Eucalyptus dives* and in some other essences such as *Xanthoxylum rhetsa*, together with the racemic form. The racemic form is found in camphor oil.

Reported uses[3]

Non-alcoholic beverages	1.0–21 ppm
Ice cream, ices, etc.	1.0–84 ppm
Candy	7.0–63 ppm
Baked goods	7.0 ppm

Regulatory status

FDA 121.1164; FEMA No. 2248

REFERENCES

For References 1–5, see end of Part III.
6. **Wallach,** *Annalen,* 356, 218, 1907; 362, 269, 1908.

	Carvomenthone
Other names	*p*-Menthan-2-one Tetrahydrocarvone 1-Methyl-4-isopropylcyclohexan-2-one 5-Isopropyl-2-methylcyclohexanone
Empirical formula	$C_{10}H_{18}O$
Structure	CH_3 / cyclohexanone ring (=O) / CH / H_3C CH_3
Physical/chemical characteristics[5]	
Molecular weight	154.24
Melting point	
Boiling point	80–100°C at 12 mm Hg *l*-, *trans*-, 218–219°C at 705 mm Hg *dl*-, 220–221°C
Specific gravity	*l*-, *trans*-, 0.9001 at 30°C *dl*-, 0.900 at 20°C *l*-iso-0.9102 at 20°C *d*-, 0.9075 at 15°C
Refractive index	*l*-, *trans*- 1.4531 at 30°C *dl*-, 1.4554 at 20°C *l*-iso-1.4558 at 20°C *d*-, 1.4544 at 20°C
Optical rotation	*l*-, *trans*-, –9.33 at 30°C *l*-iso-–56.5 at 20°C *d*-, +17.15 at 21°C
Solubility	Very soluble in alcohol; soluble in benzene and chloroform
Synthesis	By oxidation of carvomenthol;[6,7] from α-methyl-β′-isopropyl pimelic acid, dimethyl ester[8]
Natural occurence	Reported found in the oil of *Blumea malcomii* and *Blumea eriantha;*[9,10] a flavor component in cognac[11]
Reported uses[3]	Non-alcoholic beverages 10.0 ppm Ice cream, ices, etc. 10.0 ppm Candy 10.0 ppm
Regulatory status	FEMA No. 3176

REFERENCES

For References 1–5, see end of Part III.

6. **Bayer,** *Berichte,* 26, 882, 1893.
7. **Wallach,** *Annalen,* 226, 133, 1893.
8. **Ruzička and Trebler,** *Helv. Chim. Acta,* 3, 779, 1920.
9. **Simonsen and Rao,** *J. Chem. Soc.,* 121, 876, 1922.
10. *Schimmel Ber.,* 8, 1937.
11. **Schaefer and Timmer,** *J. Food Sci.,* 35(1), 10, 1970.

	Carvone		Carvyl acetate
Other names	Carvol 6,8(9)-*p*-Menthadien-2-one 1-Menthyl-4-isopropenyl-6-cyclohexen-2-one		
Empirical formula	$C_{10}H_{14}O$		$C_{12}H_{18}O_2$
Structure	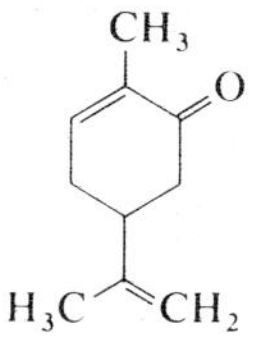		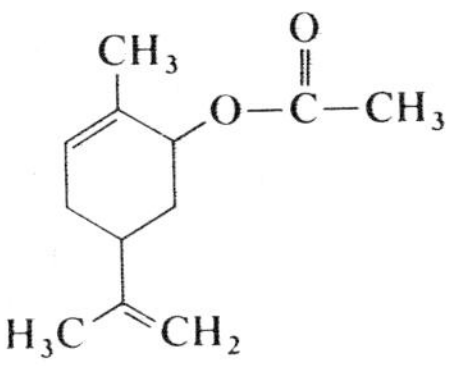
Physical/chemical characteristics[5]	***l*-Carvone**	***d*-Carvone**	
Appearance	Colorless to pale straw-colored liquid[2]	Colorless to light yellow liquid	Colorless liquid
Assay	97% min[2] (as ketone content)	95% min[2]	
Molecular weight	150.21	150.21	194.27
Boiling point	230°C at 755 mm Hg	230–231°C at 763 mm Hg	115–116°C
Specific gravity	0.956–0.960 at 25°C[2]	0.956–0.960 at 25°C[2]	0.9755 at 15°C
Refractive index	1.4950–1.4990 at 20°C[2]	1.4965–1.4990 at 20°C[2]	
Optical rotation	−57.0° to −62.0°[2]	+50° to +60°[2]	
Solubility	1:2 in 70% alcohol[2]	1:5 in 60% alcohol[2]	
Organoleptic characteristics	*l*-Carvone exhibits odor of spearmint;[2] *d*-carvone exhibits odor reminiscent of caraway[2]		Characteristic odor reminiscent of spearmint
Synthesis	Carvone occurs in the *dextro*, *levo*, and racemic form. *l*-Carvone can be isolated from the essential oil of spearmint or is commercially synthesized from *d*-limonene.[2] *d*-Carvone is usually prepared by fractional distillation of oil caraway; also from dillseed and dillweed oils, but this type differs in odor and flavor.[2]		By boiling carveol with acetic anhydride and sodium acetate[6]
Natural occurrence	The optically active and inactive forms have been reported among the constituents of about 70 essential oils. The *dextro* form is present in carvi, *Anethum graveolens*, *Anethum sowa*, *Lippia carviodora*, *Mentha arvensis*, etc. The *levo* form is present in *Mentha viridis* var. *crispa*, *Mentha longifolia* from South Africa, *Eucalyptus globulus*, and several mint species. The racemic form is present in ginger grass, *Litsea guatemaleusis*, lavender, and *Artemisia ferganensis*.		In *Mentha crispa* and *Mentha aquatica*
Reported uses[3]	Non-alcoholic beverages 850 ppm Alcoholic beverages 130 ppm Ice cream, ices, etc. 120 ppm Candy 180 ppm Baked goods 110 ppm		Non-alcoholic beverages 1.5–11 ppm Ice cream, ices, etc. 3.0–44 ppm Candy 20 ppm Baked goods 20 ppm
Regulatory status	FDA GRAS; FEMA No. 2249		FDA 121.1164; FEMA No. 2250

REFERENCES

For References 1–5, see end of Part III.

6. **Blumann and Zeitschel,** *Chem. Ber.*, 47, 2627, 1914.

	l-Carvyl propionate	β-Caryophyllene
Other names	*l*-*p*-Mentha-6,8-dien-2-yl propionate	Commercially known as caryophyllene
Empirical formula	$C_{13}H_{20}O_2$	$C_{15}H_{24}$
Structure		
Physical/chemical characteristics[5]		
Appearance	Colorless liquid	Colorless oil
Molecular weight	208.30	204.34
Boiling point	239°C	254–257°C[6]
Specific gravity	0.95	0.909–0.910 at 16°C[6]
Refractive index		1.5004–1.5027[6]
Optical rotation		−22°C[6]
Solubility	Insoluble in water; soluble in alcohol	Soluble in alcohol; insoluble in water
Organoleptic characteristics	Sweet, warm, minty, and fruity odor; sweet, fruity, minty taste	Terpene odor, midway between that of cloves and turpentine
Synthesis	By esterification of *l*-carveol with propionic anhydride; also from α-terpinyl propionate	Isolated from oil of clove stems and separated from eugenol by treating the oil with a 7% sodium carbonate solution, extracting with ether, repeating the carbonate treatment on the concentrated extracts, and finally steam distilling[7]
Natural occurrence	Not reported found in nature	Three isomers (α-, β-, and γ-caryophyllene) are found in nature. The β-isomer is the most frequently encountered and most abundant. This sesquiterpene hydrocarbon naturally occurs in approximately 60 essential oils, mainly in that of cloves, from which it was originally isolated.[8] The chemical structure has been thoroughly studied.[9–11] Other studies have been conducted on the isolation and the dipolar moment,[12,13] as well as on its oxide.[14]
Reported uses[3]	Non-alcoholic beverages 1.0 ppm Ice cream, ices, etc. 2.0 ppm Candy 2.0–24 ppm Baked goods 2.0–24 ppm	Non-alcoholic beverages 14 ppm Ice cream, ices, etc. 2.0 ppm Candy 34 ppm Baked goods 27 ppm Chewing gum 200 ppm Condiments 50 ppm
Regulatory status	FDA 121.1164; FEMA No. 2251	FDA 121.1164; FEMA No. 2252

REFERENCES

For References 1–5, see end of Part III.

6. **Schmidt,** *Ber. Deutsch. Chem. Ges.*, 80, 538, 1947.
7. **Schreiner and Kremers,** *Pharm. Arch.*, 2, 273, 1899.
8. **Gildemeister and Hoffmann,** *Die Aetherischen Oele,* Vol. 3a, Akademie Verlag, 1960, 287.
9. **Ružička et al.,** *Helv. Chim. Acta,* 14, 423, 1931.
10. **Simonsen,** *J. Chem. Soc.* (London), p. 1806, 1934.
11. **Rydon,** *J. Chem. Soc.* (London), p. 593, 1936; p. 1339, 1937.
12. **Naves,** *Helv. Chim. Acta,* 31, 378, 1948.
13. **Naves,** *Helv. Chim. Acta,* 24, 796, 1941.
14. **Treibs,** *Ber. Dtsch. Chem. Ges.*, 80, 56, 1947.

	Caryophyllene alcohol	Caryophyllene alcohol acetate
Other names		Caryophyllene acetate
Empirical formula	$C_{15}H_{26}O$	$C_{17}H_{28}O_2$
Structure	CH_3, H_3C, OH	CH_3, H_3C, $O-C(=O)-CH_3$
Physical/chemical characteristics[5]		
Appearance	Isomers are white crystalline solids	Crystalline solid
Molecular weight	222.36	264.41
Melting point	α-, 117°C; β-, 94–96°C	40°C
Boiling point	α-, 143–152°C at 10 mm Hg; β-, 287–297°C	149–152°C at 10 mm Hg
Specific gravity	α-, 0.9860 at 17°C	1.003 at 17°C
Refractive index	α-, 1.5010 at 17°C	1.4919 at 17°C
Optical rotation	α-, none; β-, −5.8° (alcohol)	
Solubility	Soluble in alcohol; insoluble in water	Rather soluble in alcohol and pet ether; insoluble in water
Organoleptic characteristics	Warm, moss-like, spicy odor; slightly minty, earthy flavor	Mild fruity and woody odor
Synthesis	Both isomers are obtained from caryophyllene. An ether solution of caryophyllene is treated with an ether solution of sulfuric acid monohydrate under cooling to prevent the temperature from raising above 10°C. After alkalinization of the solution, the β-caryophyllene alcohol is distilled off; from the residue where it is present as an ester, the α-caryophyllene alcohol is distilled after the ether solution has been made acid once again.	The β-caryophyllene alcohol acetate is prepared from chlordihydrocaryophyllene by boiling with acetic acid and sodium acetate[11]
Natural occurrence	The β-form occurs in the high boiling fraction of the oils of *Mentha arvensis* and *Mentha piperita*. The chemical structure of these compounds has been the subject of various investigations.[6–8] The structures of the α- and β-form are reported in references 9 and 10, respectively.	
Regulatory status	FDA 121.1164	FDA 121.1164

REFERENCES

For References 1–5, see end of Part III.

6. **Asahina and Tsukamoto,** *J. Pharm. Soc. Jap.*, 42, 484, 1922.
7. **Treibs,** *Ber. Dtsch. Chem. Ges.*, 80, 56, 1947.
8. **Gildemeister and Hoffmann,** *Die Aetherischen Oele*, Vol. 3b, Akademia Verlag, 1962, 337.
9. *J. Am. Chem. Soc.*, 86, 1437, 1652, 1964.
10. **Robertson and Todd,** *Chem. Ind.*, p. 437, 1953.
11. **Henderson, Robertson and Kerr,** *J. Chem. Soc.*, p. 64, 1926.

	Cedarwood oil alcohols	Cedarwood oil terpenes	
Other names	Cedarwood camphor Cedrenol* Cedrol	Cedrene (indicates the terpene fraction obtained by distillation from the wood of various cedar species)	
Empirical formula	$C_{15}H_{26}O$	$C_{15}H_{24}$	
Structure	(structure: H₃C, CH₃, OH, CH₃, CH₃)		
Physical/chemical characteristics[5]	*Note*: The physical constants refer to the purest grade of cedrol obtained by recrystallization; they do not cover the many grades obtained by distillation or fractional solidification.	**Natural**	**Synthetic**
Appearance	White crystals		
Molecular weight	222.36[2]	204.34	204.34
Melting point	>80°C[2] (86°C)		
Boiling point		262–263°C	—
Specific gravity	0.98	0.9320–0.9390 at 20°C	—
Refractive index		1.4989–1.5021 at 20°C	1.4983–1.4989
Optical rotation		−52°8′ to 71°3′	−77°67′ to 85°35′
Solubility	Soluble in benzyl benzoate; slightly soluble in glycol and mineral oil[2]	Soluble in alcohol; insoluble in water	
Organoleptic characteristics	Mild odor of cedarwood		
Synthesis	Usually obtained from cedarwood oil by fractional distillation, followed by recrystallization from suitable solvents of the appropriate solid fractions[2]	See reference 7. The natural product is also prepared from distillation of cedarwood.	
Natural occurrence	Cedrol is found in the wood of several conifers, particularly cypresses and cedars: *Cedrus atlantica*, *Cupressus sempervirens*, *Juniperus virginiana*, and others. It exhibits a positive optical rotation.[6]	A constituent of the oil of *Juniperus virginiana* and other junipers. It exists in nature in the α- and β- forms. Also found in the oils of clary sage, costus, lavender, and others. For ample description of these types of hydrocarbons, see reference 8.	
Reported uses[3]			
Regulatory status	FDA 121.1164	FDA 121.1164	

REFERENCES

For References 1–5, see end of Part III.

6. **Gildemeister and Hoffmann,** *Die Aetherischen Oele,* Vol. 3b, Akademie Verlag, 1962, 327.
7. **Naves et al.,** *Helv. Chim. Acta,* 26, 302, 1943.
8. **Gildemeister and Hoffmann,** *Die Aetherischen Oele,* Vol. 3a, Akademie Verlag, 1960, 311.

	1,4-Cineole	Cinnamaldehyde
Other names	1,4-Epoxy-*p*-menthane	Cinnamal Cinnamic aldehyde β-Phenylacrolein 3-Phenylpropenal
Empirical formula	$C_{10}H_{18}O$	C_9H_8O
Structure	CH_3–C(bridging O)–$(CH_2)_2$ / $(CH_2)_2$–CH–C(H_3C)(CH_3)	C_6H_5–CH=CH–CHO
Physical/chemical characteristics[5]		
Appearance	Colorless liquid	Greenish-yellow liquid
Assay		98% min[2]
Molecular weight	154.24	132.15
Melting point		−7.5°C
Boiling point	109–112°C (173°C) at 100 mm Hg	252°C
Flash point		120°C
Specific gravity	0.90	1.046–1.050 at 25°/25°C;[2] 1.050–1.058 at 15°C
Refractive index		1.6190–1.6230 at 20°C;[2] 1.61949 at 20°C
Acid value		5 max[2]
Solubility	Almost insoluble in water; soluble in alcohol	1:5 in 60% alcohol;[2] 1:25 in 50% alcohol; 1:2.5 in 70% alcohol; almost insoluble in water
Organoleptic characteristics	Light and mild camphoraceous odor; cool, mildly spice-like flavor	Pungent, spicy note and burning taste
Synthesis	Its structure has been confirmed through synthesis.[6] For its description see reference 8.	By isolation from natural sources. Synthetically, by condensation of benzaldehyde with acetaldehyde in the presence of sodium or calcium hydroxide[9]
Natural occurrence	Reported present in *Piper cubeba*.[7] It is also present in various essential oils: *Boldea fragrans* Juss., *Xanthoxylum rhetsa* DC, and *Ormenis multicaulis*.	It has been identified in the essential oils of: Ceylon and Madagascar cinnamon leaves, Ceylon, Seychelles, and Japanese (*Cinnamomum laureirii*) cinnamon bark, and in other cinnamon species in varying amounts (0.1–76%); also in the essential oils of: hyacinth, myrrh, Bulgarian rose, patchouli, and others.
Reported uses[3]		Non-alcoholic beverages 9.0 ppm Ice cream, ices, etc. 7.7 ppm Candy 700 ppm Baked goods 180 ppm Chewing gum 4,900 ppm Condiments 20 ppm Meats 60 ppm
Regulatory status	FDA 121.1164	FDA GRAS; FEMA No. 2286

REFERENCES

For References 1–5, see end of Part III.

6. **Wallach,** *Ann. Dtsch. Chem. Ges.*, 356, 204, 1907; 392, 62, 1912.
7. **Rao, Shintre and Simonsen,** *J. Soc. Chem. Ind.* (London), 47, 92, 1928.
8. **Wallach,** *Ann. Dtsch. Chem. Ges.*, 3366, 204, 1907; 392, 62, 1912.
9. **Peine,** *Berichte,* 17, 2117, 1884.

	Cinnamaldehyde ethylene glycol acetal	**Cinnamic acid**
Other names	Cinnamic aldehyde ethylene glycol acetal "Cinncloval" 2-Styryl-1,3-dioxolane 2-Styryl-*m*-dioxolane	β-Phenylacrylic acid 3-Phenylpropenoic acid
Empirical formula	$C_{11}H_{12}O_2$	$C_9H_8O_2$
Structure		*trans*-Cinnamic acid
Physical/chemical characteristics[5]		
Appearance	Colorless, oily liquid	White, colorless, crystalline powder
Molecular weight	176.22	148.16
Melting point		136°C; *cis*-, 57°C
Boiling point	265°C (approx.)	300°C
Specific gravity		1.2450 at 15°C; 1.285 at 15.5°C
Solubility	Insoluble in water; soluble in alcohol	Almost insoluble in water; slightly soluble in alcohol; soluble in benzene, chloroform, acetone, and essential oils
Organoleptic characteristics	Soft, warm, spicy odor reminiscent of cinnamon; sweet, spicy cinnamon-allspice flavor	Almost odorless with a burning taste, then turning sweet and reminiscent of apricot
Synthesis	From cinnamic aldehyde and ethylene glycol	Two isomers, *trans*- and *cis*-, exist. The *trans*-isomer is of interest for use in flavoring. In addition to the extraction from natural sources (storax), it can be prepared as follows: 1. From benzaldehyde, anhydrous sodium acetate, and acetic anhydride in the presence of pyridine (Perkin reaction) 2. From benzaldehyde and ethyl acetate (Claisen condensation) 3. From benzaldehyde and acetyl chloride 4. By oxidation of benzylidene acetone with sodium hypochlorite
Natural occurrence	Not reported found in nature	The *trans*-form has been found among the constituents of the essential oils of: basil, Chinese cinnamon, *Melaleuca bracteata*, *Alpinia galanga*, and in Peru balsam, Asian and American storax, cocoa leaves, etc. The *cis*-form is present in the oil of *Alpinia malacensis*.
Reported uses[3]	Candy 0.06–2.0 ppm Baked goods 2.0 ppm Condiments 5.0 ppm	Non-alcoholic beverages 31 ppm Ice cream, ices, etc. 40 ppm Candy 30 ppm Baked goods 36 ppm Chewing gum 10 ppm
Regulatory status	FDA 121.1164; FEMA No. 2287	FDA 121.1164; FEMA No. 2288

REFERENCES

For References 1–5, see end of Part III.

	Cinnamyl acetate	Cinnamyl alcohol
Other names		Cinnamic alcohol γ-Phenylallyl alcohol 3-Phenyl-2-propen-1-ol Styryl carbinol
Empirical formula	$C_{11}H_{12}O_2$	$C_9H_{10}O$
Structure	$C_6H_5-CH=CH-CH_2-O-C(=O)-CH_3$	$C_6H_5-CH=CH-CH_2OH$
Physical/chemical characteristics[5]		
Appearance	Colorless to yellowish liquid	White to yellowish solid
Assay	98% min	98% min[2]
Molecular weight	176.22	134.17
Melting point		33°C
Boiling point	265°C (114°C at 3 mm Hg)	257.5°C
Freezing point		>31°C[2]
Flash point	118°C	126°C
Specific gravity	1.047–1.051 at 25°C; 1.050 at 15.5°C	1.0440 at 20°C
Refractive index	1.5400–1.5430 at 20°C; 1.5300 at 20°C	1.5819 at 20°C
Aldehyde content		Not more than 1.5%[2] (as cinnamic aldehyde)
Solubility	Insoluble in water; 1:5 in 70% alcohol; soluble in most organic solvents	Somewhat soluble in water; 1:2 in 60% alcohol; soluble in most organic solvents
Organoleptic characteristics	Characteristic balsamic-floral odor and burning, sweet taste reminiscent of pineapple. The ester obtained from natural cinnamyl alcohol exhibits a more delicate (hyacinth, jasmine-like) note.	Pleasant, floral odor and bitter taste
Synthesis	By acetylation of cinnamyl alcohol with acetic acid[6]	Obtained originally by saponification of cinnamyl cinnamate (styracin). This method of extraction from storax is still usable. Synthetically, by reduction of cinnamaldehyde with sodium or potassium hydroxide
Natural occurrence	Reported in the essential oil of mimosa (black wattle)	Occurring as an ester or in the free state in several natural products: cinnamon leaves, hyacinth, *Aristolochia clematis*, and *Xanthorrhoea hastilis*; also in the essence of daffodil flowers[7]
Reported uses[3]	Non-alcoholic beverages 2.7 ppm Ice cream, ices, etc. 6.5 ppm Candy 16 ppm Baked goods 11 ppm Chewing gum 8.7 ppm Condiments 2.0 ppm	Non-alcoholic beverages 8.8 ppm Alcoholic beverages 5.0 ppm Ice cream, ices, etc. 8.7 ppm Candy 17 ppm Baked goods 33 ppm Gelatins and puddings 22 ppm Chewing gum 720 ppm
Regulatory status	FDA 121.1164; FEMA No. 2293	FDA 121.1164; FEMA No. 2294

REFERENCES

For References 1–5, see end of Part III.

6. U.S. Patent 3,077,495, 1963.
7. **Igolen,** *Chem. Abstr.*, 43, 2373, 1949.

	Cinnamyl anthranilate	Cinnamyl benzoate
Other names	Cinnamyl 2-aminobenzoate Cinnamyl *o*-aminobenzoate	
Empirical formula	$C_{16}H_{15}NO_2$	$C_{16}H_{14}O_2$
Structure	NH_2 (ortho) C_6H_4–C(=O)–O–CH_2–CH=CH–C_6H_5	C_6H_5–C(=O)–O–CH_2–CH=CH–C_6H_5
Physical/chemical characteristics[5]		
Appearance	Brownish powder	White, crystalline powder
Assay	96% min	
Molecular weight	253.30	238.29
Melting point	>60°C	31°C
Boiling point	332°C	209°C (335°C) at 13 mm Hg
Specific gravity	1.180 at 15.5°C	1.04 (liquid)
Refractive index		
Solubility	1 gm dissolves in 20 ml of 95% alcohol	Soluble in alcohol; insoluble in water
Organoleptic characteristics	Balsamic, fruity odor somewhat reminiscent of grape	Characteristic balsamic, aromatic, spicy odor
Synthesis	By esterification of anthranilic acid with cinnamyl alcohol	From cinnamyl alcohol and benzoyl chloride in pyridine[6]
Natural occurrence	Not reported found in nature	In Siam benzoin[7,8]
Reported uses[3]	Non-alcoholic beverages 6.8 ppm Ice cream, ices, etc. 1.7 ppm Candy 4.3 ppm Baked goods 5.3 ppm Gelatins and puddings 28 ppm Chewing gum 46–730 ppm	
Regulatory status	FDA 121.1164; FEMA No. 2295	FDA 121.1164

REFERENCES

For References 1–5, see end of Part III.

6. **Rupe and Müller,** *Helv. Chim. Acta,* 4, 845, 1921.
7. **Lüdy,** *Arch. Pharm.* (Wernheim), 231, 465, 1893.
8. **Reinitzer,** *Arch. Pharm.* (Wernheim), 264, 132, 1926.

	Cinnamyl butyrate	Cinnamyl cinnamate
Other names	Phenyl propenyl-*n*-butyrate	Cinnamyl β-phenyl acrylate Cinnamyl 3-phenyl propenoate Phenylallyl cinnamate Styracin
Empirical formula	$C_{13}H_{16}O_2$	$C_{18}H_{16}O_2$
Structure	$C_6H_5-CH{=}CH-CH_2-O-C({=}O)-(CH_2)_2-CH_3$	$C_6H_5-CH{=}CH-C({=}O)-O-CH_2-CH{=}CH-C_6H_5$
Physical/chemical characteristics[5]		
Appearance	Colorless to yellowish liquid	Low-melting, crystalline solid
Assay	98% min	
Molecular weight	204.27	264.31
Melting point		44°C
Boiling point	300°C	
Flash point	>100°C	
Specific gravity	1.010–1.015 (1.02) at 25°/25°C	1.1565 at 4°C
Refractive index	1.525–1.528 at 20°C	
Acid value	3 max	
Solubility	1:5 in 80% alcohol; insoluble in water	Insoluble in water; 1:20 in cold alcohol; 1:3 in boiling alcohol and in ether
Organoleptic characteristics	Fruity, slightly floral odor and sweet taste reminiscent of honey; used in orange and citrus flavors	Characteristic, faintly sweet, resinous odor
Synthesis	By esterification of cinnamic alcohol with *n*-butyric acid	It can be extracted from storax; synthetically, by prolonged reaction of aluminum ethylate with cinnamic aldehyde in absolute ether[6]
Natural occurrence	Not reported found in nature	One of the most important constituents of storax: *Liquidambar orientale*,[7–9] *Liquidambar styraciflumm* L.;[10,11] also in white Peru and Honduras balsams[12–14]
Reported uses[3]	Non-alcoholic beverages 1.6 ppm Ice cream, ices, etc. 8.5 ppm Candy 7.6 ppm Baked goods 11 ppm Gelatins and puddings 1.2 ppm	Non-alcoholic beverages 0.81 ppm Ice cream, ices, etc. 1.5 ppm Candy 10 ppm Baked goods 7.0 ppm
Regulatory status	FDA 121.1164; FEMA No. 2296	FDA 121.1164; FEMA No. 2298

REFERENCES

For References 1–5, see end of Part III.

6. **Endoh,** *Recl. Trav. Chim.*, 44, 871, 1925.
7. **Simon,** *Ann. Chem.*, 31, 272, 1835–40.
8. **Toel,** *Ann. Chem.*, 70, 2, 1845–49.
9. **Tschirch and Van Itallie,** *Arch. Pharm.*, 239, 514, 1901.
10. **Miller,** *Arch. Pharm.*, 220, 648, 1882.
11. **Tschirch and Van Itallie,** *Arch. Pharm.*, 239, 536, 1901.
12. **Thomas and Biltz,** *Chem. Zentralbl.*, 2, 1047, 1904.
13. **Hellström,** *Arch. Pharm.*, 243, 234, 1905.
14. **Tschirch and Werdmüller,** *Arch. Pharm.*, 248, 421, 1910.

	Cinnamyl formate	Cinnamyl isobutyrate
Other names		
Empirical formula	$C_{10}H_{10}O_2$	$C_{13}H_{16}O_2$
Structure	$C_6H_5-CH=CH-CH_2-O-C(=O)-H$	$C_6H_5-CH=CH-CH_2-O-C(=O)-CH(CH_3)_2$
Physical/chemical characteristics[5]		
Appearance	Colorless to yellowish liquid	Colorless to yellowish liquid
Assay	92% min	96% min
Molecular weight	162.18	204.27
Boiling point	250°C	254°C
Flash point	100°C	>100°C
Specific gravity	1.074–1.079 at 25°C; 1.040 at 15.5°C	1.008–1.014 (1.01) at 25°C
Refractive index	1.5500–1.5560 (1.5680) at 20°C	1.5230–1.5280 at 20°C
Acid value	3 max	3 max
Solubility	1:10 in 70% alcohol; 1:2 in 80% alcohol; insoluble in water; soluble in most organic solvents	1:2 in 70% alcohol; insoluble in water; poorly soluble in propylene glycol
Organoleptic characteristics	Balsamic, fruital-floral odor and bittersweet taste reminiscent of apple	Sweet, balsamic, fruital odor and sweet taste reminiscent of apple and banana
Synthesis	By esterification of cinnamyl alcohol with formic acid	By esterification of cinnamic alcohol with isobutyric acid
Natural occurrence	Not reported found in nature	Not reported found in nature
Reported uses[3]	Non-alcoholic beverages 1.3 ppm Ice cream, ices, etc. 9.1 ppm Candy 6.9 ppm Baked goods 8.0 ppm Chewing gum 0.60 ppm	Non-alcoholic beverages 1.5 ppm Ice cream, ices, etc. 5.0 ppm Candy 7.7 ppm Baked goods 8.5 ppm Gelatins and puddings 0.02–1.2 ppm Chewing gum 140 ppm Toppings 1.0 ppm
Regulatory status	FDA 121.1164; FEMA No. 2299	FDA 121.1164; FEMA No. 2297

REFERENCES

For References 1–5, see end of Part III.

	Cinnamyl isovalerate	Cinnamyl phenylacetate
Other names		Cinnamyl α-toluate
Empirical formula	$C_{14}H_{18}O_2$	$C_{17}H_{16}O_2$
Structure	C_6H_5–CH=CH–CH₂–O–C(=O)–CH₂–CH(CH₃)₂	C_6H_5–CH₂–C(=O)–O–CH₂–CH=CH–C_6H_5
Physical/chemical characteristics[5]		
Appearance	Colorless to yellowish liquid	Colorless liquid
Assay	95% min	
Molecular weight	218.29	252.32
Melting point		
Boiling point	313°C	333–335°C
Flash point	>100°C	
Specific gravity	0.991–0.995 (1.00) at 25°/25°C	1.086 (1.09) at 15.5°C
Refractive index	1.5180–1.5240 at 20°C	
Acid value	3 max	
Solubility	1:1 in 90% alcohol; insoluble in water; poorly soluble in propylene glycol	Insoluble in water; soluble in alcohol
Organoleptic characteristics	Characteristic rose-like odor; apple-like taste	Deep, rich, chrysanthemum-like odor; spicy, honey-like flavor
Synthesis	By esterification of cinnamic alcohol with isovaleric acid	From cinnamylchloride and sodium phenylacetate; also by esterification
Natural occurrence	Not reported found in nature	Not reported found in nature
Reported uses[3]	Non-alcoholic beverages 2.2 ppm Ice cream, ices, etc. 2.6 ppm Candy 4.1 ppm Baked goods 3.6 ppm Gelatins and puddings 11 ppm Chewing gum 19–30 ppm	Non-alcoholic beverages 2.7 ppm Ice cream, ices, etc. 0.25–2.0 ppm Candy 7.3 ppm Baked goods 7.3 ppm
Regulatory status	FDA 121.1164; FEMA No. 2302	FDA 121.1164; FEMA No. 2300

REFERENCES

For References 1–5, see end of Part III.

	Cinnamyl propionate
Other names	γ-Phenylallyl propionate 3-Phenyl-2-propenyl propanoate
Empirical formula	$C_{12}H_{14}O_2$
Structure	$C_6H_5-CH=CH-CH_2-O-C(=O)-CH_2-CH_3$
Physical/chemical characteristics[5]	
Appearance	Colorless to yellowish liquid
Assay	95% min (98%)
Molecular weight	190.24
Melting point	
Boiling point	289°C
Flash point	>100°C (95°C)
Specific gravity	0.991–0.995 at 25°/25°C; 1.0370–1.0410 at 15°C
Refractive index	1.5180–1.5240 (1.5330–1.5360) at 20°C
Acid value	3 max (0.9)
Solubility	1:1 in 90% alcohol; 1:2.5–3.5 in 80% alcohol; insoluble in water
Organoleptic characteristics	Spicy, fruital odor with a woody, balsamic undernote; sweet, warm, powerful, spicy taste
Synthesis	By esterification of cinnamic alcohol with propionic acid
Natural occurrence	Not reported found in nature
Reported uses[3]	Non-alcoholic beverages 1.0 ppm Ice cream, ices, etc. 4.3 ppm Candy 7.5 ppm Baked goods 8.8 ppm Gelatins and puddings 2.4–4.0 ppm Chewing gum 20–53 ppm
Regulatory status	FDA 121.1164; FEMA No. 2301

REFERENCES

For References 1–5, see end of Part III.

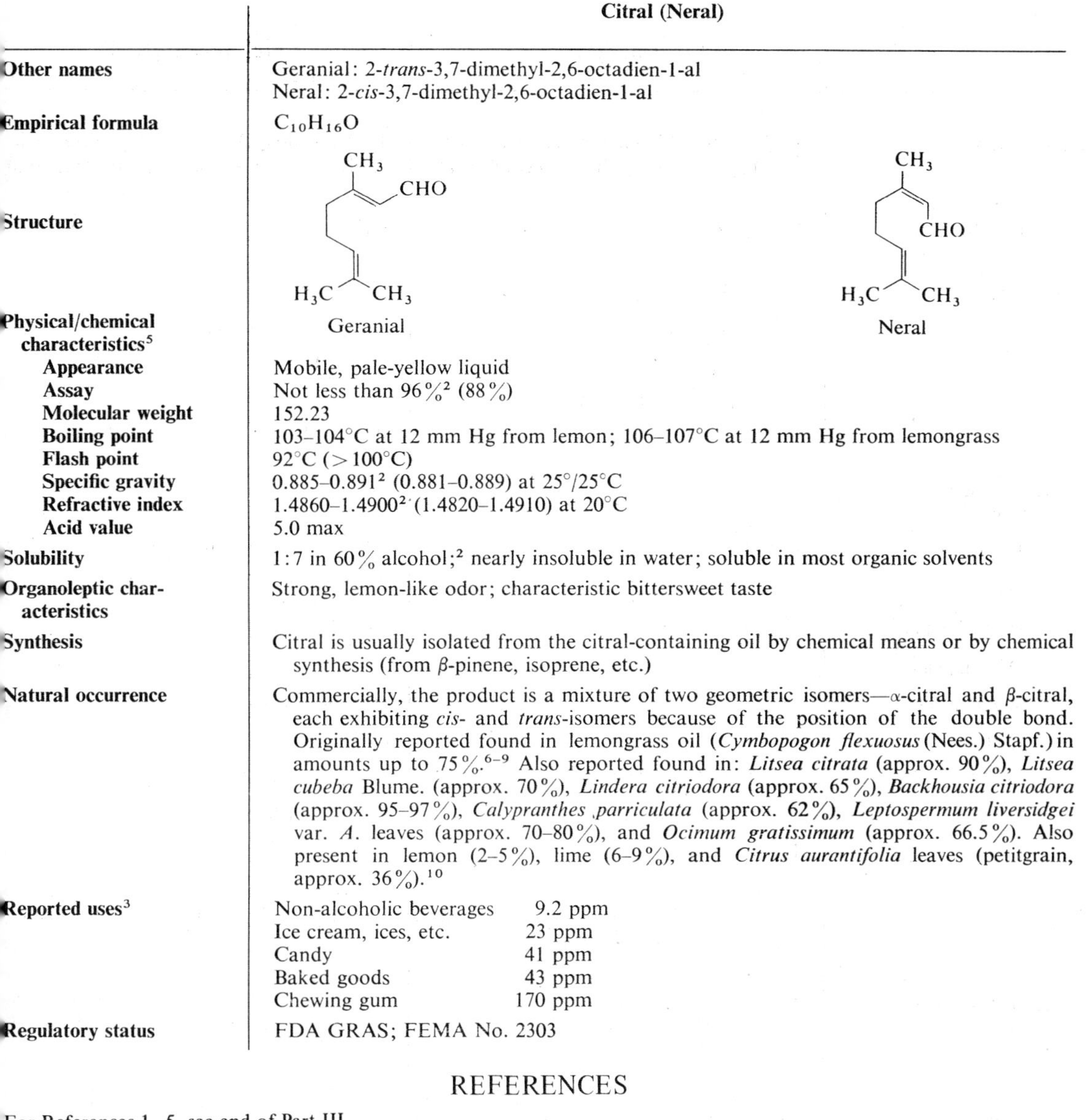

Citral (Neral)

Other names	Geranial: 2-*trans*-3,7-dimethyl-2,6-octadien-1-al Neral: 2-*cis*-3,7-dimethyl-2,6-octadien-1-al
Empirical formula	$C_{10}H_{16}O$
Structure	Geranial; Neral
Physical/chemical characteristics[5]	
Appearance	Mobile, pale-yellow liquid
Assay	Not less than 96%[2] (88%)
Molecular weight	152.23
Boiling point	103–104°C at 12 mm Hg from lemon; 106–107°C at 12 mm Hg from lemongrass
Flash point	92°C (>100°C)
Specific gravity	0.885–0.891[2] (0.881–0.889) at 25°/25°C
Refractive index	1.4860–1.4900[2] (1.4820–1.4910) at 20°C
Acid value	5.0 max
Solubility	1:7 in 60% alcohol;[2] nearly insoluble in water; soluble in most organic solvents
Organoleptic characteristics	Strong, lemon-like odor; characteristic bittersweet taste
Synthesis	Citral is usually isolated from the citral-containing oil by chemical means or by chemical synthesis (from β-pinene, isoprene, etc.)
Natural occurrence	Commercially, the product is a mixture of two geometric isomers—α-citral and β-citral, each exhibiting *cis*- and *trans*-isomers because of the position of the double bond. Originally reported found in lemongrass oil (*Cymbopogon flexuosus* (Nees.) Stapf.) in amounts up to 75%.[6–9] Also reported found in: *Litsea citrata* (approx. 90%), *Litsea cubeba* Blume. (approx. 70%), *Lindera citriodora* (approx. 65%), *Backhousia citriodora* (approx. 95–97%), *Calypranthes parriculata* (approx. 62%), *Leptospermum liversidgei* var. *A.* leaves (approx. 70–80%), and *Ocimum gratissimum* (approx. 66.5%). Also present in lemon (2–5%), lime (6–9%), and *Citrus aurantifolia* leaves (petitgrain, approx. 36%).[10]
Reported uses[3]	Non-alcoholic beverages 9.2 ppm Ice cream, ices, etc. 23 ppm Candy 41 ppm Baked goods 43 ppm Chewing gum 170 ppm
Regulatory status	FDA GRAS; FEMA No. 2303

REFERENCES

For References 1–5, see end of Part III.

6. **Bertram,** *Schimmel Ber.*, 10, 17, 1888.
7. **Dodge,** *J. Am. Chem. Soc.*, 12, 553, 1890.
8. **Barbier and Bouveault,** *C. R. Hébd. Séances Acad. Sci. Paris,* 131, 1159, 1895.
9. **Tiemann,** *Ber. Dtsch. Chem. Ges.*, 31, 2313, 1898; 32, 831, 1899.
10. **Payron,** *Fr. Ses Parfums,* 2(12), 41, 1959.

For further information on citral see the following:

Cookson, Hudec, Knight and Whitear, *Tetrahedron,* 19(12), 1995, 1963.
Hibbert and Cannon, *J. Am. Chem. Soc.*, 46, 119, 1924.
Yamashita and Matsumura, *J. Chem. Soc. Jap.*, 64, 506, 1943.
Lauchenauer and Schinz, *Helv. Chim. Acta,* 32, 1273, 1949.
Stoll and Commarmont, *Helv. Chim. Acta,* 32, 1354, 1949.
Wagner and Jauregg, *Annalen,* 488, 176, 1931; 496, 52, 1932.
Lennartz, *Ber. Dtsch. Chem. Ges.*, 76, 831, 1943.
Lebedewa et al., *J. Gen. Chem. USSR,* 28, 904, 1958.
Sancy et al., *Chimia,* 12, 326, 1958; *Helv. Chim. Acta,* 42, 1945, 1959.
Scamochwalow et al., *J. Gen. Chem. USSR,* 29, 2575, 1959.
Mitchell and Strauss, *Perfum. Essent. Oil Rec.*, 39, 312, 1948.
Tiemann, *Ber. Dtsch. Chem. Ges.*, 32, 117, 1899; 33, 880, 1900.
Naves, *Bull. Soc. Chim. Fr.*, p. 521, 1952.
Nadal, Chapel and Lecumberry, *Am. Chem. Perfum. Cosmet.,* 79(4), 43, 1964.

	Citral diethyl acetal	Citral dimethyl acetal
Other names	1,1-Diethoxy-3,7-dimethyl-2,6-octadiene 3,7-Dimethyl-2,6-octadienal diethylacetal	1,1-Dimethoxy-3,7-dimethyl-2,6-octadiene 3,7-Dimethyl-2,6-octadienal dimethylacetal
Empirical formula	$C_{14}H_{26}O_2$	$C_{12}H_{22}O_2$
Structure	CH_3; HC; $O-CH_2-CH_3$; $O-CH_2-CH_3$; H_3C; CH_3	CH_3; HC; $O-CH_3$; $O-CH_3$; H_3C; CH_3
Physical/chemical characteristics[5]		
Appearance	Colorless liquid	Colorless to yellowish liquid
Assay		93% min
Molecular weight	226.36	198.31
Melting point		
Boiling point	230°C	Approx 198°C
Flash point	79°C	70°C
Specific gravity	0.8745–0.8790 at 15°C	0.883–0.888 at 25°/25°C
Refractive index	1.4520–1.4545 at 20°C	1.4560–1.4630 at 20°C
Aldehyde content	11% max	
Solubility	1:2–2.6 in 80% alcohol; almost insoluble in water; soluble in propylene glycol	1:5 in 70% alcohol and 1:1 in 80% alcoho
Organoleptic characteristics	Mild, green, citrus odor; green, oily, citrus-peel flavor	Fresh, lemon-like odor. The commercial product consists of a mixture of *cis*- and *trans*-isomers.
Synthesis	From ethyl orthoformate and citral	
Natural occurrence	Not reported found in nature	Not reported found in nature
Reported uses[3]	Non-alcoholic beverages 0.03 ppm Candy 0.13 ppm Condiments 110 ppm	Non-alcoholic beverages 6.3 ppm Ice cream, ices, etc. 11 ppm Candy 60 ppm Baked goods 60 ppm Chewing gum 15 ppm
Regulatory status	FDA 121.1164; FEMA No. 2304	FDA 121.1164; FEMA No. 2305

REFERENCES

For References 1–5, see end of Part III.

	Citral propylene glycol acetal	Citric acid
Other names		2-Hydroxy-1,2,3-propanetricarboxylic acid
Empirical formula	$C_{13}H_{22}O_2$	$C_6H_8O_7$
Structure	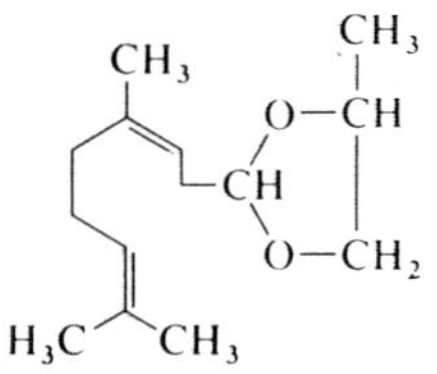	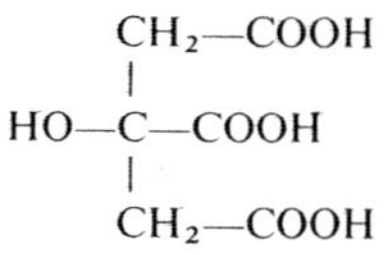 CH_2—COOH / HO—C—COOH / CH_2—COOH
Physical/chemical characteristics[5]		
Appearance	Colorless, oily liquid	White solid
Molecular weight	210.32	192.12
Melting point		153°C (anhydrous)
Solubility	Slightly soluble in water (with some decomposition); soluble in propylene glycol, alcohol, and oils	Extremely soluble in water and alcohol; slightly soluble in ether
Organoleptic characteristics	Sweet, citrus, lemon-orange odor; mild, lemony flavor changing to a lemon flavor in aqueous media	Pleasant acid taste; odorless
Synthesis	Condensation of citral with propylene glycol using a catalyst	By mycological fermentation using molasses and strains of *Aspergillus niger*; from citrus juices and pineapple wastes
Natural occurrence	Not reported found in nature	Extremely widespread in nature; also identified in the flowers of *Hibiscus subdariffa*[6]
Reported uses[3]		Non-alcoholic beverages 2,500 ppm; Ice cream, ices, etc. 1,600 ppm; Candy 4,300 ppm; Baked goods 1,200 ppm; Chewing gum 3,600 ppm
Regulatory status		FDA GRAS; FEMA No. 2306

REFERENCES

For References 1–5, see end of Part III.

6. **Indovina and Capotummino,** *Chem. Zentralbl.*, 2, 1295, 1939.

	Citronellal
Other names	3,7-Dimethyl-6-octenal 3,7-Dimethyl-6-octen-1-al Rhodinal (*dextro*-rotatory form)
Empirical formula	$C_{10}H_{18}O$
Structure	CH_3 / CHO / H_3C CH_3
Physical/chemical characteristics[5]	
Appearance	Colorless to slightly yellow liquid
Assay	Not less than 85%[2]
Molecular weight	154.24
Boiling point	*d*-, 69–71°C at 5 mm Hg *l*-, 88°C at 10.5 mm Hg *dl*-, 106–108°C at 15 mm Hg
Flash point	77°C
Specific gravity	0.850–0.860 at 25°/25°C[2]
Refractive index	1.4460–1.4560 at 20°C[2]
Optical rotation	−1° to +11°[2]
Acid value	<3.0[2]
Solubility	1:5 in 70% alcohol;[2] insoluble in water; soluble in most organic solvents
Organoleptic characteristics	Intense lemon-, citronella-, rose-type odor
Synthesis	Can be prepared by chemical synthesis or by fractional distillation of natural oils, such as citronella. Industrially prepared by hydrogenation of β-citronellol or by catalytic hydrogenation of citral;[6] also in the laboratory by dehydration of hydroxydihydrocitronellal.
Natural occurrence	The *d*-form of citronellal has been reported in the oil of citronella (Ceylon, Jammus, Kaschmis), in the oil from leaves of *Barosma pulchella*, in the oil from roots of *Phebalium nudum*, in the oils of *Eucalyptus citriodora*, *Leptospermum citratum*, and *Baeckea citriodora*. The *l*-form is present in the oils of: *Backhousia citriodora* var. *A*, *Eucalyptus citriodora*, *Litsea cubeba* (fruits), and lemongrass. Citronellal in general also is present in the oils of lemon, mandarin, *Lavandula delphinensis*, *Ocimum canum* f. *citrata*, and others.
Reported uses[3]	Non-alcoholic beverages 4.0 ppm Ice cream, ices, etc. 1.3 ppm Candy 4.5 ppm Baked goods 4.7 ppm Gelatins and puddings 0.60 ppm Chewing gum 0.30 ppm
Regulatory status	FDA 121.1164; FEMA No. 2307

REFERENCES

For References 1–5, see end of Part III.
6. **Paal,** German Patent 298,193.

Citronellic acid

Other names	3,7-Dimethyl-6-octenoic acid; Rhodinic acid; Rhodinolic acid
Empirical formula	$C_{10}H_{18}O_2$
Structure	CH_3–CH(CH_2COOH)–CH_2–CH_2–CH=C(CH_3)–CH_3
Physical/chemical characteristics[5]	
Appearance	Colorless liquid
Molecular weight	170.25
Boiling point	*d*-, 139–140°C at 10 mm Hg; *l*-, 117–119°C at 0.6 mm Hg; *dl*-, 126°C at 4 mm Hg
Specific gravity	*d*-, 0.9308 at 20°C; *l*-, 0.9274 at 25°C; *dl*-, 0.9557 at 15°C
Refractive index	*d*-, 1.4561 at 21.5°C; *l*-, 1.4563 at 24°C; *dl*-, 1.4606 at 20°C
Optical rotation	*d*-, +2.48; *l*-, −2.49
Solubility	Almost insoluble in water; soluble in alcohol and oils
Organoleptic characteristics	Green-grassy odor
Synthesis	By oxidation of citronellal; also by the conversion of pulegone[6]
Natural occurrence	The *d*-form is reportedly a constituent of Java citronella,[7] geranium, *Borosma pulchellum, Xanthoxylum piperitum,*[8] bitter-orange leaves,[9] lemon grass,[10] *Calytrix virgata,*[11] and *Calytrix tetragona*[12] leaves; the *l*-form has been reportedly identified in cypress oil (*Callitris glauca* R. Br.),[13] *Callitris intratropica,*[14] in the oil of *Chamaecyparis obtusa,*[15] and of *Thujopsis dolabrata;*[16] the *dl*-form is isolated from camphor oil[17]
Reported uses[3]	Non-alcoholic beverages 0.50 ppm Ice cream, ices, etc. 0.50 ppm Candy 0.50 ppm Baked goods 0.50 ppm Gelatins and puddings 0.50 ppm
Regulatory status	FEMA No. 3142

REFERENCES

For References 1–5, see Introduction to Part II.

6. **Plešck,** *Chem. Listy,* 50, 1854, 1956.
7. **Liu,** *cfr., Schimmel Ber.,* 32, 1955.
8. **Uchida,** *Chem. Zentralbl.,* 2, 2296, 1928.
9. **Uchida,** *Chem. Zentralbl.,* 2, 2197, 1928.
10. **Naves,** *Chem. Zentralbl.,* 2, 329, 1931.
11. **Penfold and Morrison,** *Chem. Zentralbl.,* 2, 2300, 1934.
12. **Penfold et al.,** *Chem. Zentralbl.,* 2, 770, 1935.
13. **Neuhaus and Reuter,** *Schimmel Ber.,* 12, 1951.
14. **Trikojus and White,** *Chem. Zentralbl.,* 2, 1456, 1935.
15. **Nozoe,** *Chem. Zentralbl.,* 2, 3142, 1936.
16. **Nozoe et al.,** *Chem. Abstr.,* 46, 11585, 1952.
17. **Rochussen,** *J. Prakt. Chem.,* 105(2), 124, 1922.

	Citronellol
Other names	*d*-Citronellol 3,7-Dimethyl-6-octen-1-ol For the *l*-form see Rhodinol
Empirical formula	$C_{10}H_{20}O$
Structure	CH_3 / CH_2OH / H_3C CH_3
Physical/chemical characteristics[5]	
Appearance	Colorless liquid[2]
Assay	90% min[2]
Molecular weight	156.26
Melting point	
Boiling point	*l*-, 103°C at 5 mm Hg *d*-, 105–105.5°C at 15 mm Hg
Flash point	102°C
Specific gravity	0.850–0.860 at 25°/25°C[2]
Refractive index	1.4540–1.4620 at 20°C[2]
Optical rotation	−1° to +5°[2]
Ester content	Not more than 1% (as citronellyl acetate)[2]
Aldehyde content	Not more than 1% (as citronellal)[2]
Solubility	1:2 in 70% alcohol;[2] very slightly soluble in water; soluble in most organic solvents
Organoleptic characteristics	Rose-like odor. Because odor plays such an important part in selecting this material, there may be special grades of citronellol that do not meet the E.O.A. specification. These limits have been broadened enough to include best qualities of commercial citronellol and chemically pure citronellol. *l*-Citronellol has a sweet, peach-like flavor; *d*-citronellol has a bitter taste.
Synthesis	It is generally accepted to distinguish rhodinol as the product isolated from geranium consisting of a mixture of *l*-citronellol and geraniol, whereas the name *l*-citronellol should be used to indicate the corresponding synthetic product with the highest level of purity. *dl*-Citronellol can be prepared by catalytic hydrogenation of geraniol[6] or by oxidation of allo-cymene.[7] *l*-Citronellol is prepared from (+) *d*-pinene via (+) *cis*-pinene to (+) 2,6-dimethyl-2,7-octadiene and finally isolating *l*-citronellol by hydrolysis of the aluminum-organo compound.
Natural occurrence	*l*-Citronellol has been found in nature in the plants of the *Rosaceae* family; *d*- and *dl*-citronellol have been identified in: *Verbenaceae*, *Labiatae*, *Rutaceae*, *Geraniaceae*, and others. Citronellol has been reported in about 70 essential oils and in the oil of *Rosa bourbonia*.[8] The Bulgarian rose oil has been reported to contain more than 50% *l*-citronellol, whereas East African geranium contains more than 80% of the *d*-isomer.[9] The natural product is always optically active. For identification see references 10, 11, 12, and 13.
Reported uses[5]	Non-alcoholic beverages 4.1 ppm Ice cream, ices, etc. 4.1 ppm Candy 16 ppm Baked goods 18 ppm Gelatins and puddings 5.8 ppm Chewing gum 29–52 ppm
Regulatory status	FDA 121.1164; FEMA No. 2309

REFERENCES

For References 1–5, see end of Part III.

6. *Schimmel Ber.*, 4, 62, 1898.
7. *Bull. Soc. Chim. Fr.*, 5, 761, 1956.
8. **Nigam, Gupta and Dhingra,** *Indian Perfum.*, 2(3), 81, 1959.
9. **Glichitch et al.,** *Parfums Fr.*, 5, 361, 1927.
10. **Sutherland,** *J. Am. Chem. Soc.*, 73, 2385, 1951; 74, 2688, 1952.
11. **Naves,** *C. R. Hébd. Séances Acad. Sci. Paris,* 237, 146, 1953.
12. **Carrol et al.,** *J. Chem. Soc.* (London), p. 3457, 1950.
13. **Nave,** *Perfum. Essent. Oil Rec.*, 49, 290, 1958.

	Citronellyl oxyacetaldehyde	Citronellyl acetate
Other names		3,7-Dimethyl-6-octen-yl acetate
Empirical formula	$C_{12}H_{22}O_2$	$C_{12}H_{22}O_2$
Structure	CH_3; $-O \cdot CH_2 \cdot CHO$; H_3C CH_3	CH_3; $CH_2-O-\overset{O}{\overset{\Vert}{C}}-CH_3$; H_3C CH_3
Physical/chemical characteristics[5]		
Appearance		Colorless liquid
Assay		92% min[2]
Molecular weight	198.28	198.31
Boiling point	130°C at 12 mm Hg	240°C; 119–121°C at 15 mm Hg
Specific gravity		0.883–0.890 at 25°/25°C[2]
Refractive index		1.4400–1.4500 at 20°C[2]
Optical rotation		−1° to +4°[2]
Acid value		Not more than 1[2]
Solubility		1:9 in 70% alcohol;[2] soluble in most organic solvents
Organoleptic characteristics	Strong, rose odor	Fresh, fruity odor reminiscent of rose; pungent taste at the beginning, turning to sweet, apricot-like taste afterwards
Synthesis	Interaction of bromoacetals with sodium or potassium alcoholates[6–10]	By direct acetylation of citronellol (natural or synthetic). Its physical-chemical characteristics vary, depending on the quality of the starting alcohol.
Natural occurrence	Not reported found in nature	Reported in the oil of citronella and, more recently, in *Chamaecyparis lawsoniana* Parl.
Reported uses[3]	Non-alcoholic beverages 0.005–1.0 ppm Ice cream, ices, etc. 1.4 ppm Candy 4.1 ppm Baked goods 4.3 ppm	Non-alcoholic beverages 3.4 ppm Ice cream, ices, etc. 4.2 ppm Candy 7.5 ppm Baked goods 9.7 ppm Gelatins and puddings 0.71–3.7 ppm Chewing gum 6.9–600 ppm
Regulatory status	FDA 121.1164; FEMA No. 2310	FDA 121.1164; FEMA No. 2311

REFERENCES

For References 1–5, see end of Part III.

6. **Pinner,** *Berichte,* 5, 159, 1872.
7. **Sabetay,** *Bull. Soc. Chim. Fr.,* 45, 1161, 1929.
8. **Sabetay et al.,** *Compt. Rend.,* 194, 617, 1932; 196, 1508, 1933.
9. **Rothart,** *Compt. Rend.,* 196, 2013, 1933; 197, 1225, 1933.
10. **Shoruigin and Korshak,** *Berichte,* 6B, 838, 1935; *Maslob. Zhir. Delo,* 11, 275, 1935; *Chem. Abstr.,* 29, 7941, 1935.

	Citronellyl butyrate	Citronellyl formate
Other names	3,7-Dimethyl-6-octen-l-yl butyrate	3,7-Dimethyl-6-octen-l-yl formate
Empirical formula	$C_{14}H_{26}O_2$	$C_{11}H_{20}O_2$
Structure	CH_3; $CH_2O{-}C(=O){-}CH_2{-}CH_2{-}CH_3$; H_3C CH_3	CH_3; $CH_2{-}O{-}C(=O)H$; H_3C CH_3
Physical/chemical characteristics[5]		
Appearance	Colorless liquid	Colorless, oily liquid
Assay	90% min (97% min)	86% min[2] (95% min)
Molecular weight	226.36	184.28
Boiling point	134–135°C at 12 mm Hg	97–98°C at 11 mm Hg; 235°C at 760 mm Hg
Flash point	>100°C (119°C)	92°C (93°C)
Specific gravity	0.873–0.883 at 25°/25°C; 0.880–0.886 at 15°C	0.890–0.903 at 25°/25°C;[2] 0.898–0.902 at 15°C
Refractive index	1.444–1.448 (1.4400–1.4440) at 20°C	1.4430–1.4490[2] (1.4400–1.4440) at 20°C
Optical rotation	+1°30′	±1°30′
Acid value	Not more than 1 (0.06)	Not more than 3 (0.2)
Solubility	1:6 to 1:9 in 80% alcohol (1:5 to 1:6.5)	1:2–1:3 in 80% alcohol;[2] insoluble in water
Organoleptic characteristics	Strong, fruity, and rose-like odor with sweet, plum-like taste	Strong, fruity, rose-like odor with a sweet, fruity taste
Synthesis		By direct esterification of citronellol with formic acid
Natural occurrence	Reported in Ceylon citronella oil	Reported present in geranium essential oil and in the oils of *Rosa borbonica*, *Cupressus lusitanica*, and others
Reported uses[3]	Non-alcoholic beverages 3.8 ppm Ice cream, ices, etc. 11 ppm Candy 13 ppm Baked goods 11 ppm Gelatins and puddings 3.1–4.2 ppm Chewing gum 2.3 ppm	Non-alcoholic beverages 14 ppm Ice cream, ices, etc. 13 ppm Candy 19 ppm Baked goods 32 ppm Chewing gum 63–100 ppm
Regulatory status	FDA 121.1164; FEMA No. 2312	FDA 121.1164; FEMA No. 2314

REFERENCES

For References 1–5, see end of Part III.

	Citronellyl isobutyrate	Citronellyl phenylacetate
Other names	3,7-Dimethyl-6-octen-l-yl isobutyrate	Citronellyl α-toluate 3,7-Dimethyl-6-octen-l-yl phenylacetate
Empirical formula	$C_{14}H_{26}O_2$	$C_{18}H_{26}O_2$
Structure	CH_3; $CH_2-O-C(=O)-CH(CH_3)_2$; H_3C CH_3	CH_3; $CH_2-O-C(=O)-CH_2-C_6H_5$; H_3C CH_3
Physical/chemical characteristics[5]		
Appearance	Colorless liquid	
Assay	92% min (96%)	
Molecular weight	226.36	274.41
Boiling point	249°C	342°C
Flash point	100°C (95°C)	
Specific gravity	0.870–0.880 at 25°/25°C; 0.8760–0.8830 at 15°C	0.992 at 15.5°C
Refractive index	1.4400–1.4480 (1.4375–1.4440) at 20°C	1.5100 at 20°C
Optical rotation	±1.30°	
Acid value	1 max (0.09)	
Solubility	1:6 in 80% alcohol (1:4.5–1:5.5); 1:1 in 90% alcohol; almost insoluble in water and propylene glycol	
Organoleptic characteristics	Sweet, fruity, and rose-like odor with a slightly sweet, apricot-like taste	Sweet, faintly honey, rose-like odor with a definite waxy undertone
Synthesis	By direct esterification of citronellol with isobutyric acid via azeotropic conditions or using isobutyric anhydride	
Natural occurrence	Not reported found in nature	Not reported found in nature
Reported uses[3]	Non-alcoholic beverages 2.3 ppm Ice cream, ices, etc. 1.7 ppm Candy 8.2 ppm Baked goods 12 ppm Gelatins and puddings 3.1 ppm	Non-alcoholic beverages 1.3 ppm Ice cream, ices, etc. 0.95 ppm Candy 2.4 ppm Baked goods 17 ppm
Regulatory status	FDA 121.1164; FEMA No. 2313	FDA 121.1164; FEMA No. 2315

REFERENCES

For References 1–5, see end of Part III.

	Citronellyl propionate	Citronellyl valerate
Other names	3,7-Dimethyl-6-octen-l-yl propionate	3,7-Dimethyl-6-octen-l-yl valerate
Empirical formula	$C_{13}H_{24}O_2$	$C_{15}H_{28}O_2$
Structure	$CH_2{-}O{-}C(=O){-}CH_2{-}CH_3$ (citronellyl skeleton with CH_3, H_3C, CH_3)	$CH_2O{-}C(=O){-}CH_2{-}(CH_2)_2{-}CH_3$ (citronellyl skeleton with CH_3, H_3C, CH_3)
Physical/chemical characteristics[5]		
Appearance	Colorless liquid	Liquid
Assay	90% min (97%)	
Molecular weight	212.33	240.39
Boiling point	120–124°C (242°C) at 14 mm Hg	237°C (195°C at 31 mm Hg)
Flash point	>100°C (99°C)	
Specific gravity	0.877–0.883 at 25°/25°C; 0.8810–0.8840 at 20°C	0.890 at 15.5°C
Refractive index	1.4430–1.4490 (1.4400–1.4425) at 20°C	1.4435 at 20°C
Optical rotation	±1.50°	
Acid value	1 max (0.2)	
Solubility	1:3 in 80% alcohol (1:2.5 to 1:4)	Characteristic rose, herb, honey-like odor
Organoleptic characteristics	Rose-like odor; bittersweet, plum-like taste	
Synthesis	By direct esterification of citronellol with propionic acid under azeotropic conditions on using propionic anhydride	
Natural occurrence	In tomato	Reported in the essence of *Boronia citriodora*
Reported uses[3]	Non-alcoholic beverages 3.1 ppm Ice cream, ices, etc. 9.0 ppm Candy 18 ppm Baked goods 19 ppm Chewing gum 0.80–15 ppm	Non-alcoholic beverages 1.0 ppm Ice cream, ices, etc. 2.5 ppm Candy 3.0 ppm Baked goods 7.7 ppm
Regulatory status	FDA 121.1164; FEMA No. 2316	FDA 121.1164; FEMA No. 2317

REFERENCES

For References 1–5, see end of Part III.

	Coumarin	*p*-Cresol
Other names	1,2-Benzopyrone; 5,6-Benzo-α-pyrone; *cis-o*-Coumaric acid anhydride; *o*-Hydroxycinnamic acid, lactone	4-Cresol; *p*-Hydroxytoluene; *p*-Methyl phenol
Empirical formula	$C_9H_6O_2$	C- H_8 OS
Structure	O, O	CH_3, OH
Physical/chemical characteristics[5]		
Appearance	White, crystalline solid (leaflets or rhomboids)	Low-melting, crystalline solid
Molecular weight	146.15	108.13
Melting point	68–70°C	35–36°C
Boiling point	297–299°C* (161–162°C at 14 mm Hg)	201–202°C
Congealing point	68.0°C min[2]	
Specific gravity		1.004 at 58°C
Refractive index		1.5395 at 20°C
Solubility	8% in alcohol; 0.3% in water; soluble in boiling water, alcohol, ether, and chloroform	Slightly soluble in water; soluble in alkali solutions and most organic solvents
Organoleptic characteristics	Sweet, fresh, hay-like, odor similar to vanilla seeds; burning taste with bitter undertone; nut-like flavour on dilution	Phenol-like odor
Synthesis	May be extracted from tonka beans.[6] From salicylaldehyde and acetic anhydride in the presence of sodium acetate.[7] Also from *o*-cresol and carbonyl chloride followed by chlorination of the carbonate and fusion with a mixture of alkali acetate, acetic anhydride, and a catalyst	It can be prepared by fractional distillation of coal tar, where it occurs together with the *ortho*- and *para*-isomers.
Natural occurrence	Reported in tonka bean (*Dipteryx odorata*) seed,[8,9] the flowers of *Melilotus officinalis*,[10–12] the leaves of *Melilotus albus*,[13] in *Liatris odoratissima*,[14] *Asperula odorosa*,[15] wild vanilla (*Achlys triphylla*),[16] lavender oil,[17] and several varieties of orchid[18]	Reported in *Acacia farnesiana*,[19] ylang-ylang oil (probably as *p*-cresyl acetate),[20] jasmine absolute,[21] orange oil from leaves,[22] the essence from flowers of *Lilium candidum*,[23] anise-seed oil,[24] the essence of *Artemisia santolinoflia*,[25] and some sea algae[26]
Reported uses[3]		Non-alcoholic beverages, 0.67 ppm; ice cream, ices, etc., 0.01–1.0 ppm; candy, 0.01–2.0 ppm; baked goods, 0.01–2.0 ppm
Regulatory status	Use in foods not permitted in USA	FDA 121.1164; FEMA No. 2337

* Sublimes; not volatile in steam except when superheated.

REFERENCES

For References 1–5, see end of Part III.

6. **Wöhler,** *Chem. Ann.*, 98, 66, 1856.
7. **Hickinbottom,** *Reactions of Organic Compounds,* Longmans, Green & Co., 1936, 146.
8. **Boullay and Boutron-Charlard,** *J. Pharm. Sci. Access.*, 11, 485, 1825.
9. **Vogel and Gilberts,** *Ann. Phys.* (Leipzig), 64, 163, 1820.
10. **Fontana,** *Pharm. Zentralbl.*, p. 648, 1833.
11. **Guillemette,** *J. Pharm. Sci. Access.*, 21, 172, 1835; *Ann. Chem.*, 14, 324, 1835.
12. **Zwenger and Bodenbender,** *Ann. Chem.*, 126, 257, 1863.
13. **Reinsch,** *Jahresber. Fortschr. Chem.*, p. 439, 1867.
14. *Schimmel Ber.*, p. 66, April 1900.
15. **Kosmann,** *Ann. Chem.*, 52, 387, 1844.
16. **Bradley,** *J. Am. Chem. Soc.*, 29, 606, 1907.
17. Schimmel and Co., *Chem. Zentralbl.*, 2, 969, 1900; 1, 1086, 1903.
18. **Herissey and Delaunay,** *Bull. Soc. Chim. Biol.*, 3, 573, 1921.
19. Schimmel and Co., *Chem. Zentralbl.*, 1, 1264, 1904.
20. **Darzens,** *Bull. Soc. Chim. Fr.*, 3(27), 83, 1902.
21. **Elze,** *Chem. Ztg.*, 34, 912, 1910.
22. **Igolen,** *Chem. Zentralbl.*, 2, 750, 1939.
23. **Igolen,** *C. R. Séances Acad. Sci. Paris,* 214, 772, 1942.
24. **Monod et al.,** *Chem. Abstr.*, 45, 3124, 1951.
25. **Gorayaev et al.,** *Chem. Abstr.*, 50, 9693, 1956.
26. **Kamayama,** *Chem. Abstr.*, 50, 13184, 1956.

	Cuminaldehyde	Cycloheptadeca-9-en-1-one
Other names	Cumaldehyde Cuminal Cuminic aldehyde *p*-Isopropyl benzaldehyde	Civettone
Empirical formula	$C_{10}H_{12}O$	$C_{17}H_{30}O$
Structure	CHO / benzene ring / CH(CH$_3$)... H_3C CH CH_3	CH — $[CH_2]_6$ — CH_2 ═ CH — $[CH_2]_6$ — CH_2 > C═O
Physical/chemical characteristics[5]		
Appearance	From colorless to yellowish liquid	Colorless crystals
Assay	95% min	
Molecular weight	148.20	250.4
Melting point		32.5°C
Boiling point	232°C	342°C at 742 mm Hg 158–160°C at 2 mm Hg
Specific gravity	0.976–0.980 at 25°C; 0.9818 at 15°C	0.917 at 33°C
Refractive index	1.5290–1.5340 at 20°C; 1.5301 at 20°C	1.4830 at 33.4°C
Solubility	1:4 in 70% alcohol; soluble in most organic solvents	
Organoleptic characteristics	Strong, pungent, cumin-like odor and similar taste	Disgustingly obnoxious animal-gland odor; pleasant at extreme dilutions
Synthesis	Prepared synthetically by heating *p*-isopropylbenzoyl chloride with an aqueous or alcoholic hexamethylenetetraamine solution	From aleuritic acid, the *trans*-form only;[6,7] the *cis*- and *trans*-forms from ethyl-9-keto hepta- decanedioate[8]
Natural occurrence	Reported in a large number of essential oils: cumin, *Acacia farnesiana*, cinnamon, bitter orange, Mexican lime, *Eucalyptus globulus*, rue, boldus, *Artemisia hausiliensis*, and others	In civet (*cis*-form).
Reported uses[3]	Non-alcoholic beverages 3.1 ppm Ice cream, ices, etc. 3.2 ppm Candy 4.0 ppm Baked goods 4.0 ppm Chewing gum 0.40–0.50 ppm Condiments 3.0 ppm	Non-alcoholic beverages 0.03 ppm Ice cream, ices, etc. 0.02 ppm Candy 0.03 ppm Baked goods 0.05 ppm Gelatins and puddings 0.02 ppm
Regulatory status	FDA 121.1164; FEMA No. 2341	FEMA No. 3425

REFERENCES

For References 1–5, see end of Part III.
6. **Hunsdiecker,** *Chem. Ber.,* 76, 142, 1943.
7. **Mathur et al.,** *J. Chem. Soc.,* 114, 1963.
8. **Stoll et al.,** *Helv. Chim. Acta,* 31, 543, 1948.

	Cyclohexaneacetic acid	Cyclohexaneethyl acetate
Other names	Cyclohexylacetic acid	Cyclohexylethyl acetate Hexahydrophenethyl acetate
Empirical formula	$C_8H_{14}O_2$	$C_{10}H_{18}O_2$
Structure	(S)$-CH_2-COOH$	(S)$-CH_2-CH_2-O-C(=O)-CH_3$
Physical/chemical characteristics[5]		
Appearance	Crystalline solid	Liquid
Molecular weight	142.19	170.25
Melting point	27–33°C	
Boiling point	244–246°C (116–117°C at 4 mm Hg)	104°C at 15 mm Hg
Specific gravity	1.007 at 20°C	
Refractive index	1.4558 at 30°C	
Solubility	Slightly soluble in water; soluble in most organic solvents	
Organoleptic characteristics	Waxy, fatty odor reminiscent of phenylacetic acid	
Synthesis	By catalytic reduction of phenylacetic acid	From the corresponding alcohol by acetylation with sodium acetate in acetic acid solution[6]
Natural occurrence	Not reported found in nature	Not reported found in nature
Reported uses[3]	Non-alcoholic beverages 1.0 ppm Ice cream, ices, etc. 2.0 ppm Candy 2.0 ppm Baked goods 2.0 ppm	Non-alcoholic beverages 2.0 ppm Ice cream, ices, etc. 3.0 ppm Candy 6.0 ppm Baked goods 20 ppm
Regulatory status	FDA 121.1164; FEMA No. 2347	FDA 121.1164; FEMA No. 2348

REFERENCES

For References 1–5, see end of Part III.

6. **Skita,** *Chem. Ber.,* 48, 1694, 1915.

	Cyclohexyl acetate	Cyclohexyl anthranilate
Other names		Cyclohexyl 2-aminobenzoate Cyclohexyl *o*-aminobenzoate
Empirical formula	$C_8H_{14}O_2$	$C_{13}H_{17}NO_2$
Structure	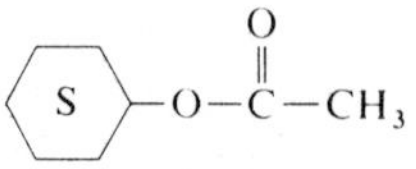	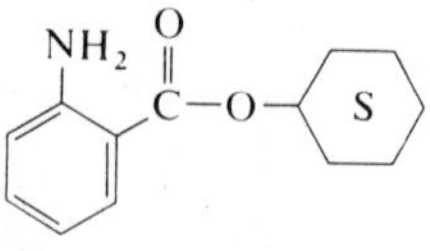
Physical/chemical characteristics[5]		
Appearance	Oily liquid	Pale-yellow liquid
Assay	99% min	
Molecular weight	142.19	219.28
Boiling point	175–177°C; 61–62°C at 12 mm Hg	318°C
Flash point	58–64°C	
Specific gravity	0.9750–0.9790 at 15°C; 0.981 at 12°C	1.018 (1.01) at 15.5°C
Refractive index	1.4400–1.4410 (1.4390) at 20°C	
Acid value	0.2 max	
Solubility	1:1.2–1:1.6 in 70% alcohol; insoluble in water; miscible with alcohol and ether	Insoluble in water; soluble in alcohol
Organoleptic characteristics	Odor reminiscent of amyl acetate	Faint, neroli-like, orange blossom odor; sweet, fruity, grape-like taste
Synthesis	By heating the corresponding alcohol with acetic anhydride or acetic acid in the presence of traces of sulfuric acid	From isatoic anhydride and cyclohexanol
Natural occurrence	Not reported found in nature	Not reported found in nature
Reported uses[3]	Non-alcoholic beverages 20 ppm Ice cream, ices, etc. 15 ppm Candy 100 ppm Baked goods 110 ppm	Non-alcoholic beverages 10 ppm Ice cream, ices, etc. 3.7 ppm Candy 3.7 ppm Baked goods 2.0–10 ppm Gelatins and puddings 1.0 ppm
Regulatory status	FDA 121.1164; FEMA No. 2349	FDA 121.1164; FEMA No. 2350

REFERENCES

For References 1–5, see end of Part III.

	Cyclohexyl butyrate	Cyclohexyl cinnamate
Other names		Cyclohexyl β-phenylacrylate Cyclohexyl 3-phenylpropenoate
Empirical formula	$C_{10}H_{18}O_2$	$C_{15}H_{18}O_2$
Structure	$CH_3-CH_2-CH_2-\overset{O}{\overset{\parallel}{C}}-O-$(cyclohexyl S)	(phenyl)$-CH=CH-\overset{O}{\overset{\parallel}{C}}-O-$(cyclohexyl S)
Physical/chemical characteristics[5]		
Appearance	Colorless liquid	Colorless, viscous liquid; solidifies when cold
Molecular weight	170.25	230.31
Melting point		28°C
Boiling point	212°C	195°C at 12 mm Hg
Specific gravity	0.957 at 15.5°C	1.054 at 14°C (supercooled liquid)
Refractive index	1.4490 at 20°C	
Solubility	Almost insoluble in water; soluble in alcohol	Insoluble in water; poorly soluble in propylene glycol; soluble in alcohol
Organoleptic characteristics	Fresh, floral odor reminiscent of benzyl butyrate; intense, sweet taste	Fruity, peach-, cherry-, and almond-like odor; similar flavor notes
Synthesis	By esterification of cyclohexanol with isobutyric acid	From cyclohexanol and cinnamic acid
Natural occurrence	Not reported found in nature	Not reported found in nature
Reported uses[3]	Non-alcoholic beverages 3.9 ppm Ice cream, ices, etc. 5.7 ppm Candy 9.2 ppm Baked goods 28 ppm Gelatins and puddings 0.54 ppm	Non-alcoholic beverages 2.0 ppm Ice cream, ices, etc. 1.0–5.0 ppm Candy 4.0–10 ppm Baked goods 4.0–20 ppm
Regulatory status	FDA 121.1164; FEMA No. 2351	FDA 121.1164; FEMA No. 2352

REFERENCES

For References 1–5, see end of Part III.

	Cyclohexyl formate	Cyclohexyl isovalerate
Other names		
Empirical formula	$C_7H_{12}O_2$	$C_{11}H_{20}O_2$
Structure	(S)−O−C(=O)−H	(S)−O−C(=O)−CH_2−CH(CH_3)(CH_3)
Physical/chemical characteristics[5]		
Appearance	Light, colorless liquid	Liquid
Molecular weight	128.17	184.28
Boiling point	162–163°C (158–160°C)	58–62°C at 0.5 mm Hg
Specific gravity	1.0057 at 0°C	
Refractive index	1.4417 at 24°C	1.4410 at 21°C
Solubility	Soluble in alcohol and acetic and formic acids; insoluble in water	
Organoleptic characteristics	Pleasant, cherry-like odor	Apple- and banana-like aroma
Synthesis	From cyclohexyl alcohol and concentrated formic acid in the presence of sulfuric acid;[6,7] by distilling the alcohol-acid mixture in the presence of excess acid	From cyclohexane and isovaleric acid[8] in the presence of perchloric acid[6]
Natural occurrence	Not reported found in nature	Not reported found in nature
Reported uses[3]	Non-alcoholic beverages 11 ppm Ice cream, ices, etc. 2.8 ppm Candy 8.0 ppm Baked goods 7.0–10 ppm	Non-alcoholic beverages 13 ppm Ice cream, ices, etc. 7.0–25 ppm Candy 9.3 ppm Baked goods 1.7–60 ppm
Regulatory status	FDA 121.1164; FEMA No. 2353	FDA 121.1164; FEMA No. 2355

REFERENCES

For References 1–5, see end of Part III.

6. **Senderens and Aboulenc,** *Compt. Rend.*, 155, 169, 1912.
7. **Senderens and Aboulenc,** *Ann. Chim.* (Paris), 9(18), 177, 1922.
8. **Dorofeenko and Dzhigirei,** *Ukr. Khim. Zh.*, 29, 616, 1963; *Chem. Abstr.*, 59, 11284b, 1963.

	Cyclohexyl propionate	Cyclopentanethiol
Other names		Cyclopentyl mercaptan
Empirical formula	$C_9H_{16}O_2$	$C_5H_{10}S$
Structure	S —O—C(=O)—CH_2—CH_3	SH
Physical/chemical characteristics		
Appearance	Liquid	
Molecular weight	156.23	102.19
Boiling point	193°C at 750 mm Hg; 46–50°C at 0.7 mm Hg	129–131°C at 745 mm Hg
Specific gravity	0.9718 at 0°C	0.9550 at 20°C
Refractive index	1.4430 at 20°C	1.4880 at 20°C
Organoleptic characteristics	Apple- and banana-like aroma	
Synthesis	By direct esterification of the alcohol with the acid in the presence of concentrated sulfuric acid[10,11] or in the presence of benzene	From cyclopentyl bromide and potassium hydrosulfide;[6] by treating cycloalkanone dimethyl dithioacetals with four equivalents of an alkaline metal;[7] from cyclopentyl naphthyl sulfide and an aluminosilicate catalyst[8]
Natural occurrence	Not reported found in nature	Reported found in the 111–150°C boiling fraction of petroleum[9]
Reported uses[3]	Non-alcoholic beverages 2.4 ppm Ice cream, ices, etc. 2.7 ppm Candy 2.0–3.0 ppm Baked goods 3.0 ppm Gelatins and puddings 5.0 ppm	Baked goods 0.1 ppm Condiments, pickles 0.1 ppm Meat, meat sauces, soups 0.1 ppm Milk, dairy products 0.1 ppm Cereals 0.1 ppm
Regulatory status	FDA 121.1164; FEMA No. 2354	FEMA No. 3262

REFERENCES

For References 1–5, see end of Part III.

6. Loevenich et al., *Berichte,* 62, 3090, 1929.
7. Brandsma, *Recl. Trav. Chim. Pays-Bas,* 89(6), 593, 1970.
8. *Chem. Abstr.,* 53, 8084h, 1959.
9. Coleman, Thompson, Hopkins and Rall, *J. Chem. Eng. Data,* 10(1), 80, 1965.
10. Senderens and Aboulenc, *Compt. Rend.,* 155, 170, 1912.
11. Senderens and Aboulenc, *Ann. Chim.* (Paris), 9(18), 180, 1922.

	p-Cymene
Other names	*p*-Isopropyltoluene 1-Methyl-4-isopropyl benzene
Empirical formula	$C_{10}H_{14}$
Structure	CH_3 (benzene ring) $CH(CH_3)_2$
Physical/chemical characteristics[5]	
Appearance	Colorless liquid
Molecular weight	134.21
Boiling point	175–176°C (179°C)
Specific gravity	0.857 at 20°C
Refractive index	1.4917 at 20°C
Solubility	Insoluble in water; soluble in alcohol
Organoleptic characteristics	Strong, characteristic odor reminiscent of carrot; tends to darken with aging
Synthesis	Obtained chiefly from the wash water of sulfite paper
Natural occurrence	Extensively widespread in nature, Reported in the oils of cypress,[6] *Artemisia cina* Bg.,[7] Ceylon cinnamon,[8,9] boldo leaf,[10] cascarilla,[11] *Cuminum cyminum* L.,[12,13] and lemon;[14] in the oils from seeds of coriander[15] and *Cicuta virosa;*[16] in the ether-extracted oils of *Monarda fisulosa*[17] and *Monarda punctata;*[18] in star anise and nutmeg essential oils.[19] Note that *p*-cymene reported in an essential oil may be formed from the conversion of a cyclic terpene. Its presence is indicative of aging in lemon essential oil.[20]
Reported uses[3]	Non-alcoholic beverages, 3.3 ppm; ice cream, ices, etc., 5.3 ppm; candy, 7.0 ppm; chewing gum, 250 ppm; condiments, 10–130 ppm
Regulatory status	FDA 121.1164; FEMA No. 2356

REFERENCES

For References 1–5, see Introduction to Part II.

6. *Schimmel Ber.*, p. 33, April 1904.
7. *Schimmel Ber.*, p. 13, April 1908.
8. **Walbaum and Hüthig,** *J. Prakt. Chem.*, 2(66), 50, 1902.
9. Schimmel and Co., German Patent 134,789; *Chem. Zentralbl.*, 2, 1486, 1902.
10. *Schimmel Ber.*, p. 16, October 1907.
11. **Fendler,** *Arch. Pharm.*, 238, 685, 1900.
12. **Gerhardt and Cahours,** *Ann. Chim.* (Paris), 3(1), 65, 102, 372, 1844.
13. *Schimmel Ber.*, p. 35, October 1909.
14. **Penfold,** *Chem. Zentralbl.*, 2, 753, 1927.
15. **Walbaum and Müller,** *Chem. Zentralbl.*, 2, 2160, 1909.
16. **Trapp,** *Chem. Ann.*, 108, 386, 1858.
17. **Kremers,** *Chem. Zentralbl.*, 2, 41, 1897.
18. **Schumann and Kremers,** *Chem. Zentralbl.*, 2, 42, 1897.
19. Schimmel and Co., *Chem. Zentralbl.*, 1, 1719, 1910.
20. **Rispoli, DiGiacomo and Tracuzzi,** *Riv. Ital, EPPOS*, 3, 118, 1965.

	L-Cysteine	2-*trans*, 4-*trans*-Decadienal
Other names	α-Amino-β-mercaptopropionic acid 2-Amino-3-mercaptopropanoic acid *l*-β-Mercaptoalanine α-Amino-β-thiolpropionic acid	Heptenyl acrolein
Empirical formula	$C_3H_7O_2NS$	$C_{10}H_{16}O$
Structure	$HS-CH_2-CH(NH_2)-COOH$	$CH_3-(CH_2)_4-CH=CH-CH=CH-CHO$
Physical/chemical characteristics[5]		
Appearance	Crystals; readily oxidizes to *l*-cystine	Colorless oily liquid
Molecular weight	121.16	152.23
Melting point	~ 240°C (decomposes)	
Boiling point		64–65°C at 0.1 mm Hg
Optical rotation	+6.5 at 25°C (c = 1 in 5*N* HCl) +9.8° at 30°C (c = 1.3 in water) +13° at 25°C (c = 1 in glacial acetic acid)	
Solubility	Very soluble in water, alcohol, and acetic acid; insoluble in ether, acetone, and benzene	Almost insoluble in water; soluble in alcohol; slightly soluble in propylene glycol; miscible with oils
Organoleptic characteristics		Powerful green odor; sweet, orange-like odor at high concentration; grapefruit/orange-like taste on dilution
Synthesis	By addition of a thiol compound to an unsaturated amino acid derivative;[6,7] by hydrolysis of proteins in the presence of carbon dioxide;[8] by treating an HCl hydrolysate of hair with CuO_2, followed by decomposition of the resulting copper-cysteine complex with hydrogen sulfide;[9] by the addition of thioacetic acid to α-acetamido acrylic acid;[10] by treatment of a keratine HCl hydrolysate with zinc to reduce the cystine present to cysteine;[11] by electrolytic reduction of cystine[12]	By autooxidation of methyl (*trans,trans*)-lineoleate hydroperoxide[13]
Natural occurrence	Reported found in protein hydrolysates, with the exception of gelatin, in which only traces of cysteine are present	Reported as occurring in peanut oil and, as a volatile component, in fried potatoes,[14] the *trans,trans*-form is reported found among the volatile flavor components of potato chips[15] and tomato[16]
Reported uses[3]	Non-alcoholic beverages 100.0 ppm Baked goods 100.0 ppm Condiments, pickles 100.0 ppm Meat, meat sauces, soups 100.0 ppm Milk, dairy products 100.0 ppm Cereals 100.0 ppm	Non-alcoholic beverages 10.0 ppm Ice cream, ices, etc. 10.0 ppm Candy 10.0 ppm Baked goods 10.0 ppm Gelatins and puddings 10.0 ppm Chewing gum 10.0 ppm Meat, meat sauces 10.0 ppm Cereals 10.0 ppm Vegetables 10.0 ppm
Regulatory status	FEMA No. 3263	FEMA No. 3135

REFERENCES

For References 1–5, see end of Part III.

6. **Farlow,** *J. Biol. Chem.*, 176, 71, 1948.
7. **Behringer,** *Chem. Ber.*, 81, 326, 1948.
8. **Okuda,** *Proc. Acad. Tokyo*, 2, 277, 1926; *Chem. Zentralb.*, 2, 2728, 1926.
9. **Rapline,** U.S. Patent 2,376, 186, 1945.
10. **Farlow,** U.S. Patent 2,406,362, 1946.
11. **Holloway and Young,** U.S. Patent 2,414,303, 1947.
12. **Koperina and Gavrilov,** *J. Gen. Chem. USSR*, 17, 1651, 1947.
13. **Hoffman and Keppler,** *Nature*, 185, 310, 1960.
14. **Lefort and Sorba,** *Bull. Soc. Chim. Fr.*, p. 69, 1956.
15. **Deck, Pokorny and Chang,** *J. Food Sci.*, 38(2), 345, 1973.
16. **Kazeniac and Hall,** *J. Food Sci.*, 35(5), 519, 1970.

	γ-Decalactone	δ-Decalactone
Other names	4-Hydroxydecanoic acid, γ-lactone	Decanolide-1,4 γ-*n*-Hexyl-γ-butyrolactone 5-Hydroxy-decanoic acid, δ-lactone
Empirical formula	$C_{10}H_{18}O_2$	$C_{10}H_{18}O_2$
Structure	$CH_3-(CH_2)_5-CH-CH_2-CH_2-C{=}O$ (ring closed between CH and C through O)	$CH_3-(CH_2)_4-CH-(CH_2)_3-C{=}O$ (ring closed between CH and C through O)
Physical/chemical characteristics[5]		
Appearance	Colorless liquid	Colorless, oily liquid
Molecular weight	170.25	170.25
Boiling point	153°C at 15 mm Hg; 114–116°C at 0.5 mm Hg	281°C
Specific gravity		0.95
Refractive index	1.4610 at 21.5°C	
Solubility	Very slightly soluble in water	Almost insoluble in water; soluble in alcohol
Organoleptic characteristics	Pleasant, fruity, peach-like odor	Oily, peach odor and taste
Synthesis	By heating γ-bromocapric acid in a sodium carbonate solution[7] by prolonged heating of 9-decen-1-oic acid with 80% H_2SO_4 at 90°C[8,9]	From hexylethylene oxide and sodium malonic ester; also from decanoic acid
Natural occurrence	Reported in peach and apricot aroma,[10] as well as in strawberry aroma[11]	Reported in coconut and raspberry[6]
Reported uses[3]	Non-alcoholic beverages, 2.0 ppm; ice cream, ices, etc., 4.5 ppm; candy, 5.7 ppm; baked goods, 7.1 ppm; gelatins and puddings, 0.08–8.0 ppm	Non-alcoholic beverages 5.0 ppm Ice cream, ices, etc. 10 ppm Candy 0.25–5.0 ppm Baked goods 0.25–8.0 ppm Toppings 50 ppm Margarine 10 ppm
Regulatory status	FDA 121.1164; FEMA No. 2360	FDA 121.1164; tolerance 10 ppm in margarine. FEMA No. 2361

REFERENCES

For References 1–5, see end of Part III.

6. **Ruys,** *Fr. Ses Parfums,* 47, 34, 1966.
7. **Fitting and Schneegans,** *Chem. Ann.,* 227, 92, 1885.
8. **Grün and Wirth,** *Chem. Ber.,* 55, 2117, 1922.
9. **Rochussen,** *Schimmel Ber.,* p. 307, 1929.
10. **Ruys,** *Fr. Ses Parfums,* 47, 34, 1966.
11. **Willhalm et al.,** *Helv. Chim. Acta,* 49, 65, 1966.

	Decanal	Decanal dimethyl acetal
Other names	Aldehyde C-10 Capric aldehyde Caprinaldehyde *n*-Decylaldehyde Decylic aldehyde	Aldehyde C-10 dimethyl acetal Capraldehyde dimethyl acetal Decylaldehyde dimethyl acetal 1,1-Dimethoxy decane 10,10-Dimethoxy decane
Empirical formula	$C_{10}H_{20}O$	$C_{12}H_{26}O_2$
Structure	$H_3C{-}(CH_2)_8{-}CHO$	$CH_3{-}(CH_2)_8{-}CH(OCH_3)_2$
Physical/chemical characteristics[5]		
Appearance	Colorless to light-yellow liquid	Colorless liquid
Assay	92% min[2]	
Molecular weight	156.26	202.34
Melting point	17–18°C	
Boiling point	207–210°C	218°C; 77–79°C at 1 mm Hg
Flash point	84°C	
Specific gravity	0.823–0.832 at 25°/25°C;[2] 0.8502 at 20°C	0.830 at 15.5°C
Refractive index	1.4260–1.4300[2] (1.4287) at 20°C	1.4244 at 24°C
Acid value	Not more than 10[2]	
Solubility	Soluble 1:1 in 80% alcohol; almost insoluble in water; soluble in most organic solvents	Almost insoluble in water; soluble in alcohol
Organoleptic characteristics	Pronounced, fatty odor that develops a floral character on dilution;[2] strong, orange-rose odor and fatty, citrus-like taste	Characteristic, herbaceous, citrus-floral odor; brandy, cognac flavor
Synthesis	Industrially prepared by oxidation of *n*-decanol or by reduction of the corresponding acid	From decanal and methyl alcohol
Natural occurrence	Among the aliphatic aldehydes, it has the largest natural occurrence in a variety of essential oils and extraction products: lemongrass, lavender, Taiwan citronella, sweet orange, mandarin, grapefruit, orris, coriander, *Acacia farnesiana* Willd., lemon (from different sources), bitter orange, petitgrain bergamot, petitgrain lime, lime, and Bulgarian clary sage	Not reported found in nature
Reported uses[3]	Non-alcoholic beverages 2.3 ppm Ice cream, ices, etc. 4.1 ppm Candy 5.7 ppm Baked goods 6.6 ppm Gelatins and puddings 3.0 ppm Chewing gum 0.60 ppm	Non-alcoholic beverages 1.0–2.0 ppm Alcoholic beverages 8.0 ppm Ice cream, ices, etc. 2.0 ppm Candy 8.0 ppm Baked goods 8.0 ppm Gelatins and puddings 3.0 ppm
Regulatory status	FDA GRAS; FEMA No. 2362	FDA 121.1164; FEMA No. 2363

REFERENCES

For References 1–5, see end of Part III.

	1-Decanol	4-Decenal
Other names	Alcohol C-10 Capric alcohol Decyl alcohol Nonyl carbinol	
Empirical formula	$C_{10}H_{22}O$	$C_{10}H_{18}O$
Structure	$H_3C{-}(CH_2)_8{-}CH_2OH$	$CH_3{-}(CH_2)_4{-}CH{=}CH{-}CH_2{-}CH_2{-}CHO$
Physical/chemical characteristics[5]		
Appearance	Viscous, colorless liquid	
Assay	98% min[2]	
Molecular weight	158.28	154.24
Melting point	6–7°C	
Boiling point	231°C	
Flash point	>5°C[2]	
Freezing point	>86°C	
Specific gravity	0.826–0.831 at 25°/25°C;[2] 0.8292 at 20°C	
Refractive index	1.4350–1.4390[2] (1.4378) at 20°C	
Acid value	Not more than 1[2]	
Solubility	1:3 in 60% alcohol;[2] almost insoluble in water; soluble in most organic solvents	
Organoleptic characteristics	Floral odor resembling orange flowers; slight, characteristic fatty taste	Strong orange-like odor
Synthesis	Synthetically prepared from coconut oil derivatives; by reduction of some capric esters, such as methyl caprate	From 1-octen-3-ol and vinyl ethyl ether in the presence of *p*-methylbenzene sulfonic acid[6]
Natural occurrence	Reported in the essential oils of ambrette seeds and almond flowers; also in citrus oils and fermented beverages	Reported formed in the autooxidation of lipids[7]
Reported uses[3]	Non-alcoholic beverages 2.1 ppm Ice cream, ices, etc. 4.6 ppm Candy 5.2 ppm Baked goods 5.2 ppm Chewing gum 3.0 ppm	Ice cream, ices, etc. 0.5 ppm Baked goods 1.0 ppm Condiments, pickles 0.5 ppm Meat, meat sauces, soups 0.5 ppm
Regulatory status	FDA 121.1164; FEMA No. 2365	FEMA No. 3264

REFERENCES

For References 1–5, see end of Part III.

6. **Lamparsky,** German Patent 2,018,898.
7. **Hrdlicka and Pokorny,** *Sb. Vys. Sk. Chem. Technol. Praze Oddil Fak. Potravin. Technol.*, 6(3), 161, 1962; cfr. *Chem. Abstr.*, 62, 2953d, 1965.

	3-Decen-2-one	Decyl acetate
Other names	Heptylidene acetone Oenanthylidene acetone	Acetate C-10 Decanyl acetate
Empirical formula	$C_{10}H_{18}O$	$C_{12}H_{24}O_2$
Structure	$CH_3{-}(CH_2)_5{-}CH{=}CH{-}\overset{\overset{\displaystyle O}{\|}}{C}{-}CH_3$	$CH_3{-}(CH_2)_8{-}CH_2{-}O{-}\overset{\overset{\displaystyle O}{\|}}{C}{-}CH_3$
Physical/chemical characteristics[5]		
Appearance	Needles (almost colorless, mobile liquid)	Colorless liquid
Assay		98% min
Molecular weight	154.25	200.32
Melting point	16–17°C	
Boiling point	125–126°C at 12 mm Hg	125.6–125.8°C at 15.5 mm Hg; 272°C at 760 mm Hg; 244°C
Specific gravity		0.862–0.866 at 250°C; 0.873 at 15.5°C
Refractive index		1.4250–1.4300 at 20°C
Solubility	Almost insoluble in water; soluble in alcohol and perfume oils	1:2 in 80% alcohol; insoluble in water
Organoleptic characteristics	Fruity-floral, jasmine-like odor with orris notes	Floral, orange-rose odor and characteristic flavor; at high dilution, sweet flavor with a pineapple-like undertone; traces used in apple, orange, and rum flavors
Synthesis	From acetone and heptyl alcohol in 1% sodium hydroxide solution;[6] by dehydration upon heating of 4-decanol-2-one over pumice or clay;[7] also by condensation of heptaldehyde and acetone	By direct acetylation of *n*-decanol with acetic acid via azeotropic conditions or using acetic anhydride
Natural occurrence	In *Boletus edulis*	In orange, lemon, melon, and apple
Reported uses[3]		Non-alcoholic beverages 3.4 ppm Ice cream, ices, etc. 2.7 ppm Candy 6.1 ppm Baked goods 10 ppm Gelatins and puddings 1.2 ppm Chewing gum 12 ppm
Regulatory status	FDA 121.1164	FDA 121.1164; FEMA No. 2367

REFERENCES

For References 1–5, see end of Part III.

6. **Rupe and Hinterlack,** *Ber. Dtsch. Chem. Ges.*, 40, 4767, 1907.
7. **Grignard and Chambret,** *Compt. Rend.*, 182, 300, 1926.

	Decyl butyrate	Decyl propionate
Other names		
Empirical formula	$C_{14}H_{28}O_2$	$C_{13}H_{26}O_2$
Structure	$CH_3-CH_2-CH_2-C(=O)-O-CH_2-(CH_2)_8-CH_3$	$CH_3-CH_2-C(=O)-O-CH_2-(CH_2)_8-CH_3$
Physical/chemical characteristics[5]		
Appearance	Oily liquid	Liquid
Assay		
Molecular weight	228.37	214.35
Melting point		
Boiling point	270°C; 135°C at 8 mm Hg	138/139°C at 15 mm Hg; 124°C at 8 mm Hg
Specific gravity	0.8617 at 20°C; 0.887 at 15.5°C	0.8639 at 20°C
Refractive index	1.43077 (1.4245) at 20°C	1.4291 at 20°C; 1.4280 at 22.5°
Solubility		
Organoleptic characteristics	Ethereal, apricot-like odor	Slightly fatty, aldehyde-like odor reminiscent of cognac
Synthesis	By esterification of the acid with the alcohol in the presence of concentrated H_2SO_4[6]	By direct esterification in the presence of benzene
Natural occurrence	Not reported found in nature	Not reported found in nature
Reported uses[3]	Non-alcoholic beverages 0.18 ppm Ice cream, ices, etc. 1.4 ppm Candy 5.9 ppm Baked goods 7.5 ppm	Non-alcoholic beverages 0.81 ppm Ice cream, ices, etc. 1.4 ppm Candy 5.9 ppm Baked goods 7.5 ppm
Regulatory status	FDA 121.1164; FEMA No. 2368	FDA 121.1164; FEMA No. 2369

REFERENCES

For References 1–5, see end of Part III.

6. **Komppa and Talvitie,** *J. Prakt. Chem.*, 2(135), 196, 1932.

	Diacetyl	Diallyl trisulfide
Other names	Biacetyl 2,3-Butanedione 2,3-Diketobutane Dimethylglyoxal "Dimethyl ketone" (incorrect)	Allyl trisulfide
Empirical formula	$C_4H_6O_2$	$C_6H_{10}S_3$
Structure	$H_3C{-}C({=}O){-}C({=}O){-}CH_3$	$CH_2{=}CH{-}CH_2{-}S{-}S{-}S{-}CH_2{-}CH{=}CH_2$
Physical/chemical characteristics[5]		
Appearance	Yellow to greenish, mobile liquid	Liquid
Molecular weight	86.09	178.34
Melting point	−3° to −4°C	
Boiling point	87–88°C	112–122°C at 16 mm Hg
Specific gravity	0.9734 at 22°C	1.0845 at 15°C
Refractive index	1.3933 at 18.5°C	
Solubility	Soluble in alcohol, carbitols, and many other organic solvents; miscible 1:4 with water	Insoluble in water and alcohol; miscible with ether
Organoleptic characteristics	Penetrating, very strong odor similar to butter on dilution; chlorine quinone-like odor when concentrated	Disagreeable odor
Synthesis	From methyl ethyl ketone by converting to the isonitroso compound and then decomposing to diacetyl by hydrolysis with HCl; by fermentation of glucose via methyl acetyl carbinol	
Natural occurrence	Reported in the oils of: Finnish pine, angelica, and lavender; in the flowers of *Polyalthia canangioides* Boerl. var. *angustifolia*[6] and *Fagroea racemosa* Jack. The following plants are also reported to contain diacetyl:[7] *Monodora grandiflora* Benth., *Magnolia tripetale* L., *Ximenia aegyptiaca* L., *Petasites fragrans* Presl., various narcissi, and tulips. It has been identified in certain types of wine,[8] the natural aromas of raspberry and strawberry, and the oils of lavender, lavandin, Reunion geranium, and Java citronella; finally, in *Cistus ladaniferus* L.[9] J. Pien has described its identification in butter.[10] Good quality butter should not contain less than 1.2 mg diacetyl, but a level up to 2.5 mg does not spoil the aroma (Holwerde and Van der Geest, 1948).	Reported found in garlic oil[11]
Reported uses[3]	Non-alcoholic beverages, 2.5 ppm; ice cream, ices, etc., 5.9 ppm; candy, 21 ppm; baked goods, 44 ppm; gelatins and puddings, 19 ppm; chewing gum, 35 ppm; shortening, 11 ppm	Baked goods 1.0 ppm Condiments, pickles 1.0 ppm Meat, meat sauces, soups 1.0 ppm
Regulatory status	FDA GRAS; FEMA No. 2370	FEMA No. 3265

REFERENCES

For References 1–5, see end of Part III.

6. **DeVries,** *Nature,* 1953.
7. **Maurer,** *Ind. Parfum.,* p. 9, 1954.
8. **Peynaud and Lafon,** *Bull. Soc. Chim. Fr.,* 44, 263, 1951.
9. **Fesan,** *Ind. Parfum.,* p. 5, 1950.
10. **Pien,** *Ann. Falsif. Fraudes,* p. 41, 1948.
11. **Semmler,** *Arch. Pharm.,* 231, 434, 1892.

	Di (butan-3-one-1-yl) sulfide	Dibenzyl ether
Other names		Benzyl ether
Empirical formula	$C_8H_{14}O_2S$	$C_{14}H_{14}O$
Structure	$CH_3-CO-CH_2-CH_2-S-CH_2-CH_2-CO-CH_3$	$C_6H_5-CH_2-O-CH_2-C_6H_5$
Physical/chemical characteristics		
Appearance		Colorless liquid
Molecular weight	174.25	198.25
Melting point		3–4°C
Boiling point		170°C at 16 mm Hg
Specific gravity		1.0428 at 20°C
Refractive index		1.561 at 20°C
Solubility		Insoluble in water; miscible with alcohol and ether
Organoleptic characteristics		Slightly earthy, mushroom-like odor with a rosy undertone
Synthesis		As a by-product in the preparation of benzyl alcohol by hydrolysis of benzyl chloride; by using a concentrated caustic instead of carbonate, yields can be improved to 50% or higher[6]
Natural occurrence		Not reported found in nature
Reported uses	Non-alcoholic beverages 1.0 ppm Ice cream, ices, etc. 1.0 ppm Candy 2.0 ppm Baked goods 2.0 ppm Gelatins and puddings 2.0 ppm Other uses 1.0 ppm	Non-alcoholic beverages, 8.3 ppm; ice cream, ices, etc., 5.6 ppm; candy, 23 ppm; baked goods, 25 ppm; chewing gum, 85–160 ppm
Regulatory status	FEMA No. 3335	FDA 121.1164; FEMA No. 2371

REFERENCES

For References 1–5, see end of Part III.

6. *Ullmanns Enzyklopädie der Technischen Chemie,* 3rd ed., Vol. 4, Urban and Schwarzenberg, 1960, 311.

	4,4-Dibutyl-γ-butyrolactone	Dibutyl sebacate
Other names	4-Butyl-4-hydroxyoctanoic acid, γ-lactone Dibutyl butyrolactone 4,4-Dibutyl-4-hydroxybutyric acid, γ-lactone	Butyl sebacate Dibutyl decanedioate Dibutyl 1,8-octanedicarboxylate
Empirical formula	$C_{12}H_{22}O_2$	$C_{18}H_{34}O_4$
Structure	$CH_3-(CH_2)_3$ and $CH_3-(CH_2)_3$ on C; $C-CH_2-CH_2-C{=}O$, ring closed through O	$CH_3-(CH_2)_3-O-C(=O)-(CH_2)_8-C(=O)-O-(CH_2)_3-CH_3$
Physical/chemical characteristics[5]		
Appearance	Colorless, oily liquid	Liquid
Molecular weight	198.31	314.47
Boiling point		349°C; 211–212°C at 11 mm Hg
Specific gravity		0.9405 at 15°C
Refractive index		1.4433 at 15°C
Solubility	Insoluble in water; poorly soluble in propylene glycol; soluble in alcohol	
Organoleptic characteristics	Oily, coconut-butter odor and creamy, coconut-like flavor	
Synthesis	From butyl pentanol and methylacrylate using a catalyst	By distillation of sebacic acid with butyl alcohol in the presence of concentrated HCl in benzene solution[6] or by reacting butyl alcohol and sebacyl chloride[7]
Natural occurrence	Not reported found in nature	Not reported found in nature
Reported uses[3]	Ice cream, ices, etc. 2.8–3.5 ppm Candy 4.4–15 ppm Baked goods 15 ppm	Non-alcoholic beverages 1.0–5.0 ppm Ice cream, ices, etc. 2.0–5.0 ppm Candy 15 ppm Baked goods 15 ppm
Regulatory status	FDA 121.1164; FEMA No. 2372	FDA 121.1164; FEMA No. 2373

REFERENCES

For References 1–5, see end of Part III.
6. **Kawaii et al.,** *Chem. Zentralbl.,* 2, 1694, 1931.
7. **Boissonnos,** *Helv. Chim. Acta,* 20, 776, 1937.

	Diethyl malate	Diethyl malonate
Other names	Diethylhydroxysuccinate Ethyl malate	Ethyl malonate Malonic ester
Empirical formula	$C_8H_{14}O_5$	$C_7H_{12}O_4$
Structure	$CH_3{-}CH_2{-}O{-}C(=O){-}CH_2{-}CH(OH){-}C(=O){-}O{-}CH_2{-}CH_3$	$CH_3{-}CH_2{-}O{-}C(=O){-}CH_2{-}C(=O){-}O{-}CH_2{-}$
Physical/chemical characteristics[5]		
Appearance	Liquid	Liquid
Molecular weight	190.20	160.17
Boiling point	91–92°C at 1 mm Hg	199–200°C; 82–83°C at 11 mm Hg
Specific gravity	1.4361 at 20°C	1.0551 at 20°C
Refractive index		1.4139 at 20°C
Optical rotation	−15.9° at 24°C (acetone; C=6)	
Solubility		Insoluble in water; miscible with alcohol and ether
Organoleptic characteristics		Faint, pleasant, aromatic odor
Synthesis	The *l*-isomer can be prepared by esterification of the *l*-isomer of the acid with alcohol in the presence of HCl.[6]	Reacting chloroacetic acid to cyanoacetic acid using sodium cyanide and subsequent saponification. Malonic acid is finally esterified by azeotropic distillation with ethanol in benzene.[8]
Natural occurrence	It exhibits two optically active isomers in addition to the optically inactive form; the most common is the *l*-isomer. Reported present in the cultures of certain fungi (*Rhizopus*).[7]	Reported found in several natural products
Reported uses[3]	Non-alcoholic beverages 5.5 ppm Ice cream, ices, etc. 6.5 ppm Candy 18 ppm Baked goods 44 ppm Gelatins and puddings 1.5 ppm	Non-alcoholic beverages 5.6 ppm Ice cream, ices, etc. 17 ppm Candy 20 ppm Baked goods 19 ppm Gelatins and puddings 20 ppm
Regulatory status	FDA 121.1164; FEMA No. 2374	FDA 121.1164; FEMA No. 2375

REFERENCES

For References 1–5, see end of Part III.

6. **D'Ianni and Adkins,** *J. Am. Chem. Soc.*, 61, 1680, 1939.
7. **Mirocha et al.,** *Phytopathology,* 51, 274, 1961.
8. **Kirk and Othmer,** *Encyclopedia of Chemical Technology,* Vol. 2, John Wiley & Sons, 1948, 301.

	2,3-Diethyl-5-methylpyrazine	2,3-Diethylpyrazine
Other names		
Empirical formula	$C_9H_{14}N_2$	$C_8H_{12}N_2$
Structure	(pyrazine ring with H_3C at 5-position and $CH_2—CH_3$ at 2- and 3-positions)	(pyrazine ring with C_2H_5 at 2- and 3-positions)
Physical/chemical characteristics		
Appearance		
Assay	C: 71.96%; H: 9.39%; N: 18.65%	C: 70.55%; H: 8.88%; N: 20.57%
Molecular weight	150.23	136.20
Mass spectra	39(37), 41(14), 54(10), 55(51), 66(9), 67(8), 79(4), 80(3), 93(3), 94(7), 106(1), 107(2), 121(23), 122(9), 135(62), 136(7), 149(64), 150(100)[8]	39 (16), 41 (12), 54 (12), 56 (13), 65 (4), 66(3), 80(22), 81 (18), 93 (3), 94 (2), 107 (17), 108 (11), 119 (7), 121 (83), 135 (46), 136 (100)[11]
IR spectra	S (3.4, 6.9, 8.5, 9.5) M (7.6, 7.9, 8.8, 9.9, 10.2, 10.3, 11.1) W (5.6, 6.4, 6.5, 8.1, 10.8, 11.6, 11.8, 11.9)[8]	7.07, 3.36, 11.71, 6.82, 8.59, 9.67, 3.28 μ[11]
UV spectra	278.5[8]	
NMR spectra[10]		τ = 1.84 (s, ring protons, 2H), 7.24 (q, CH_2 4H), 8.75 (t, chain CH_3 6H)[13]
Synthesis	By condensation of 3,4-hexanedione with propylene diamine[7]	By condensation of 3-4-hexanedione with ethylenediamine[12]
Natural occurrence	Reported to be present in coffee, filberts, and potato products[6]	
Reported uses[3]	Non-alcoholic beverages 0.1 ppm Ice cream, ices, etc. 0.2 ppm Candy 0.5 ppm Baked goods 1.0 ppm Gelatins and puddings 0.2 ppm Meat, meat sauces, soups 0.2 ppm Cereals 1.0 ppm	Non-alcoholic beverages 1.0 ppm Ice cream, ices, etc. 1.0 ppm Candy 1.0 ppm Gelatins and puddings 1.0 ppm Used to modify the flavor of natural substances[9] and as a flavor additive to chocolate fudge candy[10]
Regulatory status	FEMA No. 3336	FEMA No. 3136

REFERENCES

For References 1–5, see end of Part III.
6. **Maga and Sizer,** *Crit. Rev. Food Technol.,* 4(1), 39, 1973.
7. **Ishiguro and Matsumura,** *Yakugaku Zasshi,* 78, 229, 1958.
8. **Bondarovich et al.,** *J. Agric. Food Chem.,* 15, 1093, 1967.

	Diethyl sebacate	Diethyl succinate
Other names	Diethyl decanedioate Diethyl 1,8-octanedicarboxylate Ethyl sebacate	Diethyl butanedioate Diethyl ethanedicarboxylate Ethyl succinate
Empirical formula	$C_{14}H_{26}O_4$	$C_8H_{14}O_4$
Structure	$CH_3-CH_2-O-C(=O)-(CH_2)_8-C(=O)-O-CH_2-CH_3$	$CH_3-CH_2-O-C(=O)-(CH_2)_2-C(=O)-O-CH_2$
Physical/chemical characteristics[5]		
Appearance	Colorless to yellowish liquid	Colorless to yellowish, clear liquid
Assay		98% min[2]
Molecular weight	258.36	174.20
Melting point	1–3°C	
Boiling point	297°C; 151–153°C at 3 mm Hg	216°C; 105°C at 15 mm Hg
Specific gravity	0.9640 at 15°C	1.031–1.041[2] at 25°/25°C; 1.0416 at 20°C
Refractive index	1.4367 at 15°C	1.4190–1.4230[2] (1.4198) at 20°C
Acid value		Not more than 1.0[2]
Solubility	Partially water soluble; miscible with alcohol, ether, and other organic solvents	1:1 or more in 60% alcohol[2]
Organoleptic characteristics		Faint, pleasant odor
Synthesis	By heating sebacic acid and ethanol in the presence of concentrated H_2SO_4 or of other acid catalysts[6]	By direct esterification of the acid in th presence of concentrated H_2SO_4 in benzene solution;[7] from succinic anhydrid and ethyl alcohol[2]
Natural occurrence	Not reported found in nature	In apple and cocoa
Reported uses[3]	Non-alcoholic beverages 4.1 ppm Ice cream, ices, etc. 9.1 ppm Candy 21 ppm Baked goods 41 ppm Gelatins and puddings 3.2–19 ppm Chewing gum 2.7–450 ppm	Non-alcoholic beverages 7.3 ppm Ice cream, ices, etc. 11 ppm Candy 38 ppm Baked goods 45 ppm
Regulatory status	FDA 121.1164; FEMA No. 2376	FDA 121.1164; FEMA No. 2377

REFERENCES

For References 1–5, see end of Part III.

6. **Vogel,** *J. Chem. Soc.*, p. 632, 1948.
7. Commercial Solvents Corp., U.S. Patent 2,076,111, 1934.

	Diethyl tartrate	2,5-Diethyltetrahydrofuran
Other names	Diethyl 2,3-dihydroxybutanedioate Diethyl 2,3-dihydroxysuccinate Ethyl tartrate	
Empirical formula	$C_8H_{14}O_6$	$C_8H_{16}O$
Structure	$C_2H_5-O-C(=O)-CH(OH)-CH(OH)-C(=O)-O-C_2H_5$	$(C_2H_5)-HC$ (ring: $-O-$, $CH-C_2H_5$, H_2C-CH_2)
Physical/chemical characteristics[5]		
Appearance	Colorless, thick, oily liquid	Colorless liquid
Molecular weight	206.19	128.21
Boiling point	153°C at 12 mm Hg	116°C (approx)
Specific gravity	1.204 at 20°C	
Refractive index	1.4476 at 20°C	
Optical rotation	+7°5′ at 20°C	
Solubility	Insoluble in water; miscible with alcohol and ether	Slightly soluble in water; soluble in alcohol and propylene glycol
Organoleptic characteristics		Sweet, herbaceous, caramellic odor
Synthesis	By direct esterification of the acid in the presence of various catalysts, including Twitchell's reagent and HCl[6–8]	
Natural occurrence	The *d*-isomer has not been reported found in nature. The *l*-isomer and the racemic form are of little importance.	In mint
Reported uses[3]	Non-alcoholic beverages 50 ppm Ice cream, ices, etc. 200 ppm Candy 200 ppm Baked goods 200 ppm	The compound has been used in reconstructing mint flavors.
Regulatory status	FDA 121.1164; FEMA No. 2378	FDA 121.1164

REFERENCES

For References 1–5, see end of Part III.

6. **Zaganiaris and Varvoglis,** *Ber. Dtsch. Chem. Ges.*, p. 69, 1936.
7. **Badoche,** *Ann. Chim.* (Paris), 11(18), 146, 1943.
8. **Bretschneider,** *Monatsh. Chem.*, 80, 258, 1949.

	Difurfuryl ether	Difurfuryl sulfide
Other names	Furfuryl ether	Furfuryl sulfide Bis-(2-furfuryl)-sulfide
Empirical formula	$C_{10}H_{10}O_3$	$C_{10}H_{10}O_2S$
Structure	(furan-2-yl)$-CH_2-O-H_2C-$(furan-2-yl)	(furan-2-yl)$-CH_2-S-H_2C-$(furan-2-yl)
Physical/chemical characteristics[5]		
Appearance		Colorless liquid
Molecular weight	178.19	194.25
Melting point		31–32°C
Boiling point	101°C at 2 mm Hg 88–89°C at 1 mm Hg	135–143° at 14 mm Hg
Specific gravity	1.405 at 20°C	
Refractive index	1.5088 at 20°C	
Solubility	Insoluble in water	
Synthesis	By reacting furfuryl alcohol and furfuryl chloride[6] or furfuryl iodide[7] in boiling ether in the presence of KOH	By reacting furfuryl mercaptan and furfuryl bromide in ether solution in the presence of KOH[8]
Natural occurrence	Reported found as a constituent in coffee aroma[8]	Reported found as a constituent in coffee aroma[8]
Reported uses[3]	Baked goods 1.0 ppm Meat, meat sauces, soups 1.0 ppm	Non-alcoholic beverages 5.0 ppm Ice cream, ices, etc. 5.0 ppm Candy 5.0 ppm Baked goods 5.0 ppm Gelatins and puddings 5.0 ppm Chewing gum 5.0 ppm Condiments 5.0 ppm Meat, meat sauces, soups 5.0 ppm
Regulatory status	FEMA No. 3337	FEMA No. 3258

REFERENCES

For References 1–5, see end of Part III.

6. **Kirner,** *J. Am. Chem. Soc.,* 50, 1960, 1928.
7. **Zanetti,** *J. Am. Chem. Soc.,* 49, 1064, 1927.
8. **Stoll et al.,** *Helv. Chim. Acta,* 50, 628, 1967.

	Dihydrocarveol	Dihydrocarvone
Other names	8-*p*-Menthen-2-ol 6-Methyl-3-isopropenylcyclohexanol Tuberyl alcohol	*cis*-*l*-Dihydrocarvone 8-*p*-Menthen-2-one *cis*-*p*-Menthen-8(9)-one (2) 1-Methyl-4-isopropenyl cyclohexan-2-one
Empirical formula	$C_{10}H_{18}O$	$C_{10}H_{16}O$
Structure	CH_3, OH, H_3C, CH_2	CH_3, O, H_3C, CH_2
Physical/chemical characteristics[5]		
Appearance	Almost colorless, straw-colored liquid (somewhat viscous)	Colorless to straw-yellow liquid
Assay	Commercial products usually are a mixture of several stereoisomers; the *dextro* form is a more desirable product	
Molecular weight	154.24	152.23
Boiling point	*d*, 222.5–223°C at 749 mm Hg *l*, 107°C at 14 mm Hg	*l*, 220°C at 750 mm Hg
Specific gravity	*d*, 0.9204 at 20°C; *l*, 0.9202 at 23°C	*l*, 0.929 at 15°C
Refractive index	*d*, 1.47818 at 20°C; *l*, 1.4748 at 23°C	*l*, 1.47348 at 20°C
Optical rotation	*d*, +33.86°; *l*, −33.25° at 23°C	*l*, −15.6° (−18 to −19° at 20°C for synthetic product)
Solubility	Soluble in alcohol	Insoluble in water; soluble in alcohol
Organoleptic characteristics	Floral, woody odor; sweet, somewhat spicy flavor (peppery)	Warm, powerful, herb-like odor; spearmint-like flavor
Synthesis	By reducing carvone and separating the resulting isomers	By isomerization of limonene oxide; by oxidation of dihydrocarveol or reduction of carvone
Natural occurrence	For determination of structure see reference 6. Because of the presence of 3 asymmetric carbons, 8 optically active isomers are possible. The *d*-, *l*-, and *dl*-forms are known. Reported found (free or esterified) in the essential oils of: *Mentha longifolia*,[6] *Mentha verticillata*, *Artemisia juncea*, caraway, and others. Also identified in the essential oil of *Mentha viridis* var. *sativa* cultivated in Calabria.[7]	The following stereoisomers are known: *dl*- and *l*-dihydrocarvone and the *l*-isodihydrocarvone. The optically inactive form is usually prepared by synthesis. The *l*-forms are reported found among the constituents of the essential oils of: caraway, anethum, *Mentha spicata* var. *longifolia*, angelica seeds, and others. For the identification of chemical structure, see references 8, 9, 10, and 11.
Reported uses[3]	Non-alcoholic beverages 84 ppm Alcoholic beverages 500 ppm Ice cream, ices, etc. 300 ppm Candy 10–250 ppm Baked goods 10–250 ppm	
Regulatory status	FDA 121.1164; FEMA No. 2379	FDA 121.1164

REFERENCES

For References 1–5, see end of Part III.

6. **LaFace,** *Chem. Abstr.*, 48, 7261, 1954.
7. **Bonaccorsi,** *Chem. Zentralbl.*, 2, 1261, 1936.
8. **Wallach,** *Annalen*, 275, 114, 1893; *Ber. Dtsch. Chem. Ges.*, 28, 2704, 1895.
9. **Wallach and Schrader,** *Annalen*, 279, 377, 1894.
10. **Wagner,** *Ber. Dtsch. Chem. Ges.*, 27, 1652, 1894.

For further information on dihydrocarveol see the following:

Schmidt, *Chem. Ber.*, 83, 193, 1950; B: *Schimmel Ber.*, p. 136, 1951; C: *Miltizer Ber.*, 122, 1956.

	Dihydrocarvyl acetate	Dihydrocoumarin
Other names	8-*p*-Menthen-2-yl acetate *p*-Menth-8-(9)-en-2-yl acetate 6-Methyl-3-isopropenyl cyclohexyl acetate	Benzodihydropyrone 3,4-Dihydrocoumarin Hydrocoumarin
Empirical formula	$C_{12}H_{20}O_2$	$C_9H_8O_2$
Structure	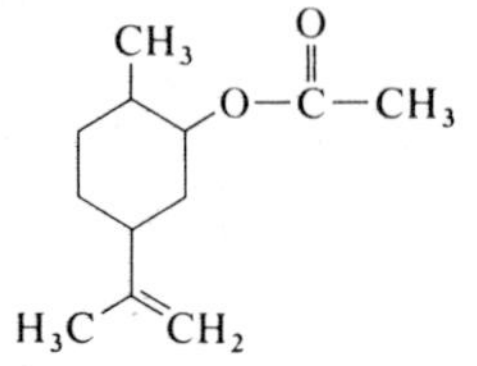	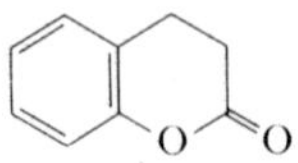
Physical/chemical characteristics[5]		
Appearance	Colorless liquid	Low-melting solid
Assay		98% min
Molecular weight	196.29	148.16
Melting point		25°C
Boiling point		272°C
Congealing point		23–23.6°C
Flash point		130°C
Specific gravity	0.96	1.955–1.980 at 15°C
Refractive index		1.5560–1.5575 at 20°C
Acid value		0.9
Solubility	Slightly soluble in water; soluble in alcohol	Slightly soluble in water; soluble in alcohol ether, and especially in chloroform 1:2.5–3.5 in 70% alcohol
Organoleptic characteristics	Sweet, floral, rose-like odor; slightly minty; used in imitation spearmint	Odor similar to coumarin at room temperature or reminiscent of nitrobenzene a higher temperature; burning taste
Synthesis	Acetylation of dihydrocarveol	By reduction of coumarin under pressure in the presence of nickel at 160–200°C[6] or in the presence of Pd-$BaSO_4$ in alcoholic solution[7]
Natural occurrence	In mint and celery	Reported found in *Melilotus officinalis*, from which it may be extracted by wate distillation
Reported uses[3]	Non-alcoholic beverages 2.0–5.0 ppm Ice cream, ices, etc. 20 ppm Candy 22 ppm Baked goods 22 ppm Condiments 10 ppm	Non-alcoholic beverages 7.8 ppm Ice cream, ices, etc. 21 ppm Candy 44 ppm Baked goods 28 ppm Gelatins and puddings 10 ppm Chewing gum 78 ppm
Regulatory status	FDA 121.1164; FEMA No. 2380	FDA status not fully defined; FEMA No. 2381

REFERENCES

For References 1–5, see end of Part III.

6. **Tetralin Gesellschaft,** German Patent 355,650.
7. **Paal and Schiedewitz,** *Ber. Dtsch. Chem. Ges.*, 63, 775, 1930.
8. **Phipson,** *Chem. News,* 32, 25, 1875.

	5,7-Dihydro-2-methylthieno(3,4d)pyrimidine		Dihydrosafrole
Other names	Thienyl pyrimidine; with the same number, FEMA also report *p*-tolylisovalevate		3,4-Methylenedioxy-propylbenzene
Empirical formula	$C_6H_{12}N_2S$		$C_{10}H_{12}O_2$
Structure	N, CH_3, N, S		$CH_2-CH_2-CH_3$, O, O—CH_2
Physical/chemical characteristics[5]			
Appearance			Colorless to yellowish liquid
Molecular weight	276.26		164.21
Boiling point			112°C at 15 mm Hg
Flash point			TCC 202°F
Specific gravity			1.063–1.070 at 25°C; 1.0759 at 15.6°C
Refractive index			1.5170–1.5200 at 20°C; 1.5228 at 15.6°C
Solubility			1:9 in 80% alcohol; 1:2 in 90% alcohol
Organoleptic characteristics			Odor reminiscent of sassafras
Synthesis	By reaction of acetamidine HCl with 3-carbomethoxytetrahydro-4-furanones[6]		From safrole or isosafrole by hydrogenation over nickel at 200°C;[6] by hydrogenation of either safrole in alcoholic solution in the presence of palladium or activated carbon[7] or of 3,4-methylenedioxy cinnamic aldehyde[8]
Natural occurrence			Not reported found in nature
Reported uses[3]	Non-alcoholic beverages Ice cream, ices, etc. Candy Baked goods Gelatins and puddings Chewing gum Condiments, pickles Meat, meat sauces, soups Cereals Milk, dairy products	1.0 ppm 1.0 ppm 1.0 ppm 1.0 ppm 1.0 ppm 1.0 ppm 1.0 ppm 1.0 ppm 1.0 ppm 1.0 ppm	
Regulatory status	FEMA No. 3338		Use in foods not permitted in USA

REFERENCES

For References 1–5, see end of Part III.

6. **Katz et al.**, (International Flavor and Fragrances), German Patent 2,141,916.
7. **Henrard,** *Chem. Zentralbl.*, 2, 1512, 1907.
8. **Spät and Quietensky,** *Ber. Dtsch. Chem. Ges.*, 60, 1887, 1927.
9. **Lohans,** *J. Prakt. Chem.*, 2(119), 259, 1928.

	4,5-Dihydro-3-(2H)thiophenone	*m*-Dimethoxybenzene
Other names	3-Tetrahydrothiophenone-3-thiophanone	1,3-Dimethoxybenzene Dimethylresorcinol Resorcinol dimethyl ether
Empirical formula	C_4H_6OS	$C_8H_{10}O_2$
Structure	[ring structure: five-membered ring with S, C=O]	[benzene ring with O—CH_3 at positions 1 and 3]
Physical/chemical characteristics[5]		
Appearance		Liquid
Molecular weight	102.15	138.16
Melting point		−52 to −55°C
Boiling point		215–217.5°C
Specific gravity		1.0803 at 0°C
Solubility		Slightly soluble in water; soluble in alcohol, ether, and benzene
Organoleptic characteristics		Acrid, fruity odor reminiscent of nerolin with a similar taste
Synthesis	From 4-methoxycarboxyl-3-oxothiophane treated with 10% sulfuric acid[6]	From resorcinol by methylation using dimethyl sulfate and alkali[7-9]
Natural occurrence		Not reported found in nature
Reported uses[3]	Non-alcoholic beverages 1.0 ppm Ice cream, ices, etc. 1.0 ppm Candy 1.0 ppm Gelatins and puddings 1.0 ppm Chewing gum 1.0 ppm Condiments, pickles 1.0 ppm Meat, meat sauces, soups 1.0 ppm Milk, dairy products 1.0 ppm Cereals 1.0 ppm	Non-alcoholic beverages 3.0 ppm Ice-cream, ices, etc. 5.0 ppm Candy 5.0 ppm Baked goods 8.0 ppm
Regulatory status	FEMA No. 3266	FDA 121.1164; FEMA No. 2385

REFERENCES

For References 1–5, see end of Part III.

6. **Arbuzor, Erastov and Remizor,** *Izv. Akad. Nauk SSSR Ser. Khim.*, (1), 180, 1969; *Chem. Abstr.*, 70, 160285g, 1969.
7. **Ullmanns,** *Chem. Ann.*, 327, 116, 1903.
8. **Ott and Nauen,** *Ber. Dtsch. Chem. Ges.*, 55, 928, 1922.
9. **Pfeiffer and Oberlin,** *Ber. Dtsch. Chem. Ges.*, 57, 209, 1924.

	p-Dimethoxybenzene	1,1-Dimethoxyethane
Other names	Dimethyl hydroquinone Hydroquinone dimethyl ether	Dimethylacetal Acetaldehyde dimethylacetal Ethylidene dimethyl ether
Empirical formula	$C_8H_{10}O_2$	$C_4H_{10}O_2$
Structure	$O-CH_3$ (benzene ring) $O-CH_3$	$CH_3-CH(OCH_3)(OCH_3)$
Physical/chemical characteristics		
Appearance	Low-melting solid	Mobile liquid
Molecular weight	138.16	90.12
Melting point	$>53°C$ (56°C)	
Boiling point	109°C at 20 mm Hg	64–65°C
Specific gravity	1.036 at 66°C	0.8501 at 20°C
Refractive index		1.3668 at 20°C
Solubility	1:10 in 95% alcohol	Miscible with water, alcohol, chloroform, and ether
Organoleptic characteristics	Bitter taste	
Synthesis	By methylation of hydroquinone using dimethyl sulfate and alkali[17]	From acetaldehyde and methanal[6,7]
Natural occurrence	Reported found in hyacinth (*Hyacinthus orientalis* L.) essential oil[18] and in *Rhodophyllus icterius*	In raspberry and blackberry,[8] strawberry,[9-11] hop oil,[12,13] coffee,[14,15] tea[16]
Reported uses[3]	Non-alcoholic beverages 8.1 ppm Ice cream, ices, etc. 5.0 ppm Candy 4.7 ppm Baked goods 5.8 ppm	Non-alcoholic beverages 3.0 ppm Ice cream, ices, etc. 3.0 ppm Candy 6.0 ppm Baked goods 6.0 ppm Gelatins and puddings 3.0 ppm Jellies, preserves, spreads 3.0 ppm
Regulatory status	FDA 121.1164; FEMA No. 2386	FEMA No. 3426

REFERENCES

For References 1–5, see end of Part III.

6. **Meadows et al.,** *Can. J. Chem.*, 30, 501, 1952.
7. **Frevel et al.,** U. S. Patent 2,691,684, assigned to Dow Chemical Co., 1954.
8. **Scanlan,** *J. Agric. Food Chem.*, 18, 744, 1970.
9. **Teranishi et al.,** *J. Food Sci.*, 28, 478, 1963.
10. **Nursten et al.,** *Chem. Ind. (London)*, p. 486, 1967.
11. **Gierschner et al.,** *Riechst. Aromen*, 18, 3, 37, 94, 134, 179, 220, 322, 1968.
12. **de Mets et al.,** *J. Inst. Brew.*, 74, 74, 1968.
13. **Drawert et al.,** *Brauwissenschaft*, 22, 169, 1969.
14. **Gianturco,** 4th Symposium on Foods, Oregon State University, 1965, Oregon State University Press, Corvallis, 431, 1967.
15. **Walter et al.,** *Z. Ernährungswiss.*, 9, 123, 1969.
16. **Bondarovich et al.,** *J. Agric. Food Chem.*, 15, 36, 1967.
17. **Dyson, George and Hunter,** *J. Chem. Soc.*, p. 440, 1927.
18. **Hoejenbons and Coppens,** *Recl. Trav. Chim. Pays-Bas*, 50, 708, 1931.

	2,6-Dimethoxyphenol		3,4-Dimethoxy-1-vinyl benzene	
Other names	Pirogallol, 1,3-dimethyl ether			
Empirical formula	$C_8H_{10}O_3$		$C_{10}H_{12}O_2$	
Structure				
Physical/chemical characteristics[5]				
Appearance	White or colorless crystals			
Molecular weight	154.17		164.20	
Melting point	56–57°C			
Boiling point	262°C			
Solubility	Slightly soluble in water; soluble in alcohol and aqueous alkalis; very soluble in ether			
Organoleptic characteristics	Woody, medicinal odor			
Synthesis	By reacting pyrogallol with methyl iodide in alkaline aqueous medium;[6] by demethylation of pyrogallol trimethyl ether in aqueous alkali or in alcohol[7]			
Natural occurrence	Reported found in beech tar creosote		Reported found in coffee[8-10]	
Reported uses[3]	Meat	3.0 ppm	Non-alcoholic beverages	2.0 ppm
	Soups	0.80 ppm	Ice cream, ices, etc.	2.0 ppm
	Seafood	2.0 ppm	Candy	2.0 ppm
			Gelatins and puddings	2.0 ppm
			Meat	2.0 ppm
			Soups	2.0 ppm
Regulatory status	FEMA No. 3137		FEMA No. 3138	

REFERENCES

For References 1–5, see end of Part III.

6. Graebl and Hess, *Ann.*, 340, 235, 1905.
7. German Patent 1,626,658.
8. Reymond et al., *J. Gas Chromatogr.*, 4, 28, 1966.
9. Gautschi et al., *J. Agric. Food Chem.*, 15, 15, 1967.
10. Walter et al., *Z. Ernährungswiss.*, 9, 123, 1969.

	2,4-Dimethylacetophenone		2,4-Dimethyl-5-acetylthiazole	
Other names	Methyl 2,4-dimethylphenyl ketone			
Empirical formula	$C_{10}H_{12}O$		C_7H_9NOS	
Structure	[structure: acetophenone ring with $C(=O)-CH_3$, CH_3 at 2- and 4-positions]		[structure: thiazole ring with H_3C, N, H_3COC, S, CH_3]	
Physical/chemical characteristics[5]				
Appearance	Oily, colorless to yellowish liquid			
Assay	97% min			
Molecular weight	148.20		155.22	
Boiling point	228°C; 110°C at 13 mm Hg			
Flash point	208°F			
Specific gravity	0.993–0.996 at 25°/25°C			
Refractive index	1.533–1.535 at 20°C			
Acid value	1 max			
Solubility	1:6 in 60% alcohol; 1:2 in 70% alcohol			
Organoleptic characteristics	Odor reminiscent of peppermint or floral, sweet odor			
Synthesis	By condensation of acetyl chloride and *m*-xylene in the presence of aluminum chloride[6] or in the presence of ferric chloride[7,8]			
Natural occurrences	Not reported found in nature			
Reported uses[3]	Non-alcoholic beverages	0.78 ppm	Ice cream, ices, etc.	10.0 ppm
	Alcoholic beverages	1.0 ppm	Candy	10.0 ppm
	Ice cream, ices, etc.	0.77 ppm	Baked goods	6.0 ppm
	Candy	3.9 ppm	Meat, meat sauces, soups	10.0 ppm
	Baked goods	2.7 ppm		
Regulatory status	FDA 121.1164; FEMA No. 2387		FEMA No. 3267	

REFERENCES

For References 1–5, see end of Part III.

6. **Claus,** *Ber. Dtsch. Chem. Ges.*, 19, 230, 1886.
7. **Meissel,** *Ber. Dtsch. Chem. Ges.*, 32, 2420, 1899.
8. **Perkin and Stone,** *J. Chem. Soc.*, 127, 2283, 1924.

	2,4-Dimethylbenzaldehyde		p, α-Dimethylbenzyl alcohol	
Other names	2,4-Xylylaldehyde 1-Formyl-2,4-dimethylbenzene		Methyl-*p*-tolyl carbinol 1-*p*-Tolyl-1-ethanol 4-(α-Hydroxyethyl)toluene 4-Methyl-α-phenethyl alcohol	
Empirical formula	$C_9H_{10}O$		$C_9H_{12}O$	
Structure	CHO, CH_3, CH_3		CH_3, H_3C, CH — OH	
Physical/chemical characteristics				
Appearance			Colorless, slightly oily liquid	
Molecular weight	134.17		136.19	
Melting point	-9°C			
Boiling point	212–215°C at 710 mm Hg		219°C	
Congealing point			134°C at 14 mm Hg	
Flash point			*dl*- 115–116°C at 11 mm Hg	
Specific gravity			0.9668 at 15.5°C	
Solubility			Sparingly soluble in water; miscible with alcohol and ether; soluble in isopropanol and liquid petrolatum	
Organoleptic characteristics			Odor reminiscent of menthol	
Synthesis	By passing HCl into a mixture of *o*-xylole, aluminum chloride, and sodium cyanide at 100°C[6]		From *p*-tolylmagnesium bromide and acetaldehyde in ether;[7,8] from 4-methylacetophenone with sodium metal in ethanol[9]	
Natural occurrence			Reported found as a constituent in the essential oil of *Curcuma longa* L. and related species (Zingiberaceae); the product obtained from natural sources is probably levorotatory	
Reported uses[3]	Non-alcoholic beverages	2.0 ppm	Non-alcoholic beverages	10.0 ppm
	Ice cream, ices, etc.	2.0 ppm	Ice cream, ices, etc.	10.0 ppm
	Candy	3.0 ppm	Candy	10.0 ppm
	Gelatins and puddings	2.0 ppm		
	Condiments	1.0 ppm		
	Jellies, preserves, spreads	1.0 ppm		
Regulatory status	FEMA No. 3427		FEMA No. 3139	

REFERENCES

For References 1–5, see end of Part III.

6. **Niedzielski,** *J. Org. Chem.*, 8, 147, 1943.
7. **Auwers and Kolligs,** *Berichte,* 55, 42, 1922.
8. **Eisenlohr,** *Berichte,* 57, 1816, 1924.
9. **Gastaldi and Cherchi,** *Gazz. Chim. Ital.*, 45(2), 274, 1915.

	α,α-Dimethylbenzyl isobutyrate	3,4-Dimethyl-1,2-cyclopentanedione
Other names	Phenyldimethyl carbinyl isobutyrate 2-Phenylpropan-2-yl isobutyrate	
Empirical formula	$C_{13}H_{18}O_2$	$C_7H_{10}O_2$
Structure	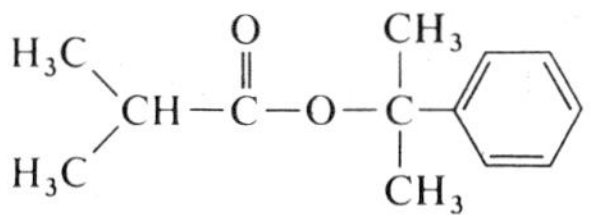	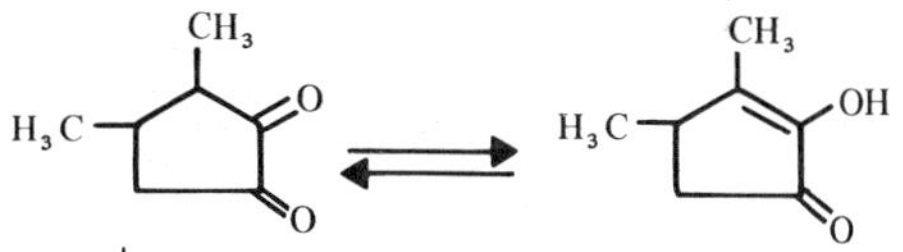
Physical/chemical characteristics[5]		
Appearance	Colorless liquid	
Molecular weight	206.28	126.15
Melting point	71–72°C	
Solubility	Insoluble in water; soluble in alcohol	
Organoleptic characteristics	Sweet, fruity, apricot, peach, and plum odor; green, dry, banana-, peach-, plum-like flavor	
Synthesis	By esterification of dimethylphenylcarbinol with isobutyric acid	By alkylation of 3.5-dicarbetoxy-4-methylcyclopentane-1,2-dione followed by hydrolysis and decarboxylation of the resulting 3,5-dicarbetoxy-3,4-dimethylcyclopentane-1,2-dione[6]
Natural occurrence	Not reported found in nature	Reported found as a constituent in coffee[7]
Reported uses[3]	Non-alcoholic beverages 5.0 ppm Ice cream, ices, etc. 40 ppm Candy 30 ppm Baked goods 20 ppm	Ice cream, ices, etc. 0.3 ppm Candy 4.0 ppm Chewing gum 1.9 ppm Protein foods 3.5 ppm
Regulatory status	FDA 121.1164; FEMA No. 2388	FEMA No. 3268

REFERENCES

For References 1–5, see end of Part III.

6. Gianturco, Gianmarino and Pitcher, *Tetrahedron Lett.*, 19, 2051, 1963.
7. Gianturco, Gianmarino and Friedel, *Nature*, 210(5043), 1358, 1966.

	3,5-Dimethyl-1,2-cyclopentadione	2,6-Dimethyl-4-heptanol
Other names	3,5-Dimethyl-1,2-cyclopentadione	Diisobutyl carbinol 4-Hydroxy-2,6-dimethyl heptane
Empirical formula	$C_7H_{10}O_2$	$C_9H_{20}O$
Structure	[structure: CH_3, H_3C, O, O ⇌ CH_3, OH, H_3C, O]	$(H_3C)_2CH-CH_2-CH(OH)-CH_2-CH(CH_3)_2$
Physical/chemical characteristics[5]		
Molecular weight	126.15	144.26
Melting point	91–92°C	
Boiling point		178°C 79°C at 15 mm Hg
Specific gravity		0.8097 at 20°C
Refractive index		1.4238 at 20°C
Solubility		Insoluble in water; soluble in alcohol and ether
Synthesis	By direct alkylation of the methyl ester of the commercially available 3-methylcyclo-pent-2-en-2-ol-1-one, followed by the cleavage of the ether function[6]	By catalytic hydrogenation of diisobutyl ketone[8,9]
Natural occurrence	Reported found as a volatile constituent in coffee aroma[7]	
Reported uses[3]	Non-alcoholic beverages 2.4 ppm Ice cream, ices, etc. 6.0 ppm Candy 5.0 ppm	Non-alcoholic beverages 20.0 ppm Ice cream, ices, etc. 20.0 ppm Candy 20.0 ppm Gelatins and puddings 20.0 ppm Chewing gum 20.0 ppm
Regulatory status	FEMA No. 3269	FEMA No. 3140

REFERENCES

For References 1–5, see end of Part III.

6. Gianturco, Gianmarino and Pitcher, *Tetrahedron Lett.*, 19, 2051, 1963.
7. Gianturco, Gianmarino and Friedel, *Nature*, 210 (5043), 1358, 1966.
8. Ipatieff and Haensel, *J. Org. Chem.*, 7, 195, 1942.
9. Cox and Adkins, *J. Am. Chem. Soc.*, 61, 3369, 1939.

	2,6-Dimethyl-5-heptenal	2,6-Dimethyl-10-methylene-2,6,11-dodecatrienal
Other names	Melonal®*	β-Sinensal (Note: the old name of α-sinensal was recently changed to β-sinensal in agreement with the structure of α- and β-farnesene; α-sinensal is now 2,6,10-trimethyl-2,6,9,11-dodecatetraenal[6-8])
Empirical formula	$C_9H_{16}O$	$C_{15}H_{22}O$
Structure	$(H_3C)_2C{=}CH{-}CH_2{-}CH_2{-}CH(CH_3){-}CHO$	CH_2, CH_3, CH_2, CH_3, CHO
Physical/chemical characteristics[5]		
Appearance	Yellow liquid	Colorless liquid
Assay	85% min[2]	
Molecular weight	140.23	218.34
Boiling point		>300°C 180°C at 1 mm Hg 100°C at 0.1 mm Hg
Flash point	144°F	
Specific gravity	0.845–0.855 at 25°/25°C[2]	
Refractive index	1.4410–1.4470 at 20°C[2]	1.5077 at 20°C
Acid value	Not more than 5.0[2]	
Organoleptic characteristics	Characteristic odor of melon and corresponding taste	
Synthesis		From myrcene by oxidation and other synthetic routes[7,8]
Natural occurrence	Not reported found in nature	Reported found in orange oil (*Citrus sinensis* L.)[9]
Reported uses[3]	Non-alcoholic beverages 2.8 ppm Ice cream, ices, etc. 1.7 ppm Candy 8.4 ppm Baked goods 19 ppm Gelatins and puddings 0.02–10 ppm Chewing gum 0.80 ppm	Non-alcoholic beverages 10.0 ppm Ice cream, ices, etc. 10.0 ppm Candy 10.0 ppm
Regulatory status	FDA 121.1164; FEMA No. 2389	FEMA No. 3141

*Trademark of Givaadan–Delawanna, Inc.

REFERENCES

For References 1–5, see end of Part III.

6. **Teranishi, Thomas, Schudel and Büchi,** *Chem. Commun.*, p. 928, 1968.
7. **Büchi and Wüst,** *Helv. Chim. Acta,* 50, 2440, 1967.
8. **Bertele and Schudel,** *Helv. Chim. Acta,* 50, 2445, 1967.
9. **Stevens, Lundin and Teranishi,** *J. Org. Chem.*, 30, 1690, 1965.

	3,7-Dimethylocta-2,6-dienyl-2-ethyl butanoate		2,6-Dimethyl octanal	
Other names	Geranyl 2-ethyl butyrate		2,6-Dimethyl octanoic aldehyde Isoaldehyde C-10 Isodecylaldehyde	
Empirical formula	$C_{16}H_{27}O_2$		$C_{10}H_{20}O$	
Structure	$CH_3-CH_2-C(CH_2-CH_3)-C(=O)-O-CH_2-CH=C(CH_3)-CH_2-CH_2-CH=C(CH_3)-CH_3$		$(H_3C)_2CH-(CH_2)_3-CH(CH_3)-CH_2-CH$	
Physical/chemical characteristics[5]				
Appearance			Colorless liquid	
Molecular weight	251.38		156.26	
Boiling point			*d*, 81°C at 12 mm Hg; *dl*, 82–84°C at 15 mm Hg	
Specific gravity			*d*, 0.8253 at 20°C; *d*, 0.823 at 22°C	
Refractive index			*d*, 1.4257 at 20°C; *d*, 1.4285 at 17°C	
Optical rotation			*d*, +9°	
Solubility			Insoluble in water; soluble in alcohol	
Organoleptic characteristics			Powerful, sweet odor of fruit; somewhat green flavor	
Synthesis			The *d*-isomer is prepared by heating *d*-dihydrocitronellol and isopropyl aluminate at 70–100°C and 14 mm Hg and subsequently adding cinnamaldehyde at 140–160°C;[6] or by hydrogenation of citronellal with palladium or strontium carbonate in methanol under pressure (3 atm).[7] The *dl*-form is prepared by hydrogenation of citral in ethanol under pressure in the presence of palladium and strontium carbonate[8] or of palladium and calcium carbonate.[9,10]	
Natural occurrence	Reported found in geranium		Not reported found in nature	
Reported uses[3]	Non-alcoholic beverages	10.0 ppm	Non-alcoholic beverages	0.44 ppm
	Alcoholic beverages	40.0 ppm	Ice cream, ices, etc.	3.2 ppm
	Ice cream, ices, etc.	40.0 ppm	Candy	1.9 ppm
	Candy	75.0 ppm	Baked goods	1.9 ppm
	Baked goods	100.0 ppm		
	Gelatins and puddings	50.0 ppm		
	Preserves and spreads	50.0 ppm		
Regulatory status	FEMA No. 3339		FDA 121.1164; FEMA No. 2390	

REFERENCES

For References 1–5, see end of Part III.

6. **Lauchenauer and Schinz,** *Helv. Chim. Acta,* 32, 273, 1949.
7. **Walker,** *J. Chem. Soc.,* p. 2002, 1949.
8. **Rydon,** *J. Chem. Soc.,* p. 1546, 1939.
9. **Caldwell and Jones,** *J. Chem. Soc.,* p. 600. 1946.
10. **Braum and Rudolph,** *Ber. Dtsch. Chem. Ges.,* 67, 1738, 1934.

	3,7-Dimethyl-1-octanol	2,4-Dimethyl-2-pentenoic acid
Other names	Dihydrocitronellol Tetrahydrogeraniol	
Empirical formula	$C_{10}H_{22}O$	$C_7H_{12}O_2$
Structure	$(H_3C)_2CH-(CH_2)_3-CH(CH_3)-CH_2-CH_2OH$	$CH_3-CH(CH_3)-CH=C(CH_3)-COOH$
Physical/chemical characteristics[5]		
Appearance	Colorless liquid	
Assay	90% min[2]	
Molecular weight	158.29	128.17
Boiling point	221–223°C; 81–82°C at 2.4 mm Hg	
Specific gravity	0.826–0.842 at 25°/25°C;[2] 0.8308 at 20°C	
Refractive index	1.4350–1.4450[2] (1.4360) at 20°C	
Acid value	Not more than 1[2]	
Solubility	1:3 and more in 70% alcohol;[2] insoluble in water	
Organoleptic characteristics	Sweet, rosy odor and bitter taste	
Synthesis	Usually prepared by hydrogenation of geraniol, citronellol, or citronellal	
Natural occurrence	Reported found in nature. Corresponds to the *dl*-form of dihydrocitronellol	
Reported uses[3]	Non-alcoholic beverages 4.3 ppm Ice cream, ices, etc. 2.0–44 ppm Candy 15 ppm Baked goods 19 ppm Chewing gum 2.9 ppm	Candy 1.0 ppm Baked goods 1.5 ppm
Regulatory status	FDA 121.1164; FEMA No. 2391	FEMA No. 3143

REFERENCES

For References 1–5, see end of Part III.

	α,α-Dimethylphenethyl acetate	α,α-Dimethylphenethyl alcohol
Other names	Benzyldimethylcarbinyl acetate Benzylpropyl acetate	Benzyl dimethyl carbinol Dimethyl benzyl carbinol 1,1-Dimethyl-2-phenylethanol 2-Methyl-l-phenyl-propanol-2
Empirical formula	$C_{12}H_{16}O_2$	$C_{10}H_{14}O$
Structure	$H_3C-C(=O)-O-C(CH_3)_2-CH_2-C_6H_5$	$C_6H_5-CH_2-C(CH_3)_2-OH$
Physical/chemical characteristics[5]		
Appearance	Watery, white liquid; solidifies at room temperature	Low-melting, white, crystalline solid. It may remain super-cooled as a white, oily liquid
Assay	98% min[2]	97%[2]
Molecular weight	192.26	150.22
Melting point	29.5°C	24°C
Boiling point		108°C at 11 mm Hg
Congealing point	>28.0°C[2]	>22°C[2]
Flash point	>100°C	92°C
Specific gravity	0.995–1.002 at 25°/25°C[2] (supercooled liquid)	0.972–0.977 at 25°/25°C;[2] 0.9790 at 16°C
Refractive index	1.4910–1.4950 at 20°C[2] (supercooled liquid)	1.5140–1.5170 at 20°C (supercooled liquid);[2] 1.5174 at 16°C
Acid value	Not more than 1.0[2]	Not more than 1.0[2]
Solubility	1:4 and more in 70% alcohol[2]	1:3 in 50% alcohol; 1:2 in 60% alcohol
Organoleptic characteristics	Floral, fruity odor reminiscent of pear[2]	Fresh, floral odor and bitter taste
Synthesis	By acetylation of dimethylbenzyl carbinol	From acetone and benzyl magnesium chloride[6] or benzyl magnesium bromide[7]
Natural occurrence	Not reported found in nature	Not reported found in nature
Reported uses[3]	Non-alcoholic beverages 2.8 ppm Ice cream, ices, etc. 8.0 ppm Candy 22 ppm Baked goods 19 ppm Chewing gum 2.9 ppm	Non-alcoholic beverages 3.3 ppm Ice cream, ices, etc. 3.2 ppm Candy 4.0 ppm Baked goods 4.9 ppm Gelatins and puddings 0.01 ppm Jellies 3.2 ppm
Regulatory status	FDA 121.1164; FEMA No. 2392	FDA 121.1164; FEMA No. 2393

REFERENCES

For References 1–5, see end of Part III.

6. **Tiffernau and Delange,** *Compt. Rend.,* 137, 575, 1903.
7. **Grignard,** *Compt. Rend.,* 130, 1924, 1900.

	α,α-Dimethylphenethyl butyrate	α,α-Dimethylphenethyl formate
Other names	Benzyl dimethylcarbinyl butyrate Dimethylbenzylcarbinyl butyrate D.M.B.C. butyrate	Benzyl dimethylcarbinyl formate Dimethylbenzylcarbinyl formate
Empirical formula	$C_{14}H_{20}O_2$	$C_{11}H_{14}O_2$
Structure	$C_6H_5-CH_2-C(CH_3)_2-O-C(=O)-(CH_2)_2-CH_3$	$C_6H_5-CH_2-C(CH_3)_2-O-C(=O)-H$
Physical/chemical characteristics[5]		
Appearance	Colorless liquid	Colorless liquid
Molecular weight	220.31	178.23
Solubility	Insoluble in water; soluble in alcohol	Almost insoluble in water; soluble in alcohol
Organoleptic characteristics	Mild, herbaceous, fruity (plum-prune) odor; apricot, peach, plum-like taste	Dry, herbaceous, green, lily-jasmine odor; spicy taste, quite warm and herbaceous
Synthesis	By esterification of dimethylbenzylcarbinol with *n*-butyric acid under catalysis or azeotropic conditions	From dimethylbenzylcarbinol and formic acid using acetic anhydride
Natural occurrence	Not reported found in nature	Not reported found in nature
Reported uses[3]	Non-alcoholic beverages 10 ppm Ice cream, ices, etc. 20 ppm Candy 20 ppm Gelatins and puddings 20 ppm	Non-alcoholic beverages 2.0 ppm Ice cream, ices, etc. 10 ppm Candy 10 ppm
Regulatory status	FDA 121.1164; FEMA No. 2394	FDA 121.1164; FEMA No. 2395

REFERENCES

For References 1–5, see end of Part III.

	2,3-Dimethylpyrazine
Other names	2,3-Dimethyl-1,4-diazine
Empirical formula	$C_6H_8N_2$
Structure	(pyrazine ring with N; $-CH_3$, $-CH_3$)
Physical/chemical characteristics	
Appearance	Liquid (volatile with steam)
Assay	C: 66.64%; H: 7.46%; N: 25.90%
Molecular weight	108.14
Boiling point	156°C
Congealing point	
Flash point	
Specific gravity	1.0218 at 0°C
Refractive index	
Mass spectra[8]	40 (20), 42 (23), 51 (6), 52 (11), 67 (100), 68 (6), 93 (4), 94 (1), 108 (97), 109 (6)
NMR spectra	
IR spectra[8]	S (6.9, 7.0, 7.1, 8.5, 10.1, 10.5, 11.8) M (3.3, 3.4, 7.3, 9.8, 11.3) W (5.2, 5.5, 6.5, 8.0, 8.3)
UV spectra[8]	(273.0, 269.5 nm)
Solubility	Soluble in water, alcohol, and ether
Organoleptic characteristics	Odor threshold: 2,500 ppb in water[9]
Synthesis	By heating the reaction product of diacetyl and ethlenediamine with potassium hydroxide in alcohol[6]
Natural occurrence	Reported to be present in the following foods: bakery products, roasted barley, cocoa products, coffee, dairy products, meat, peanuts, pecans, filberts, popcorn, potato products, rum and whisky, soy products[7]
Reported uses[3]	Non-alcoholic beverages 10.0 ppm Ice cream, ices, etc. 10.0 ppm Candy 10.0 ppm Baked goods 10.0 ppm Gelatins and puddings 10.0 ppm Meat, meat sauces, soups 10.0 ppm
Regulatory status	FEMA No. 3271

REFERENCES

For References 1–5, see end of Part III.

6. **Ishiguro and Matsumura,** *Yakugaku Zasshi,* 78, 229, 1958; *Chem. Abstr.,* 52, 11862a, 1958.
7. **Maga and Sizer,** *Crit. Rev. Food Technol.,* 4(1), 39, 1973.
8. **Bondarovich et al.,** *J. Agric. Food Chem.,* 15, 1093, 1967.
9. **Guadagni et al.,** *J. Sci. Food Agric.,* 23, 1435, 1972.

2,5-Dimethylpyrazine

Other names	2,5-Dimethyl-1,4-diazine Glycoline Ketine
Empirical formula	$C_6H_8N_2$
Structure	[structure: pyrazine ring with CH_3 and H_3C substituents]
Physical/chemical characteristics	
Appearance	Colorless liquid; solidifies in cold weather
Assay	C: 66.64%; H: 7.46%; N: 25.90%
Molecular weight	108.14
Melting point	15°C
Boiling point	155°C
Specific gravity	0.9887 at 20°C
Refractive index	1.4980 at 20°C
Mass spectra[8]	39(35), 42(87), 51(4), 52(6), 64(1), 66(2), 80(3), 81(18), 93(1), 108(100), 109(6)
IR spectra[9]	S (6.7, 7.2, 7.5, 8.6, 9.6)
UV spectra[9]	(277.0, 272.0 nm)
Solubility	Soluble in water, alcohol, and ether in all proportions; also soluble in acetone
Organoleptic characteristics	Characteristic flavor reminiscent of potato chips; odor threshold: 1,800 ppb in water,[10] 1,000 ppb in water[11]
Synthesis	By the interaction of acrolein and ammonia when heated in glycerol in the presence of ammonium salts;[6] by self-condensation of aminoacetone, followed by oxidation with mercury chloride[7]
Natural occurrence	Reported to be present in the following foods: bakery products, roasted barley, cocoa products, roasted coffee, dairy products, meat, peanuts, pecans, filberts, popcorn, potato products, rum and whisky, soy products;[8] a volatile component in roasted filberts[12] and in potato chips;[13] reportedly identified in the fat of cooked beef;[14] also reported found in toasted off-flavors[15]
Reported uses[3]	Non-alcoholic beverages 10.0 ppm Ice cream, ices, etc. 10.0 ppm Candy 10.0 ppm Baked goods 10.0 ppm Gelatins and puddings 10.0 ppm Meat 10.0 ppm Mil, dairy products 10.0 ppm Cereals 10.0 ppm
Regulatory status	FEMA No. 3272

REFERENCES

For References 1–5, see end of Part III.

6. **Stoehr,** *J. Fr. Chim.*, 51, 449, 1895.
7. **Cornforth,** *J. Chem. Soc.*, p. 1178, 1958.
8. **Maga and Sizer,** *Crit. Rev. Food Technol.*, 4(1), 39, 1973.
9. **Bondarovich et al.,** *J. Agric. Food Chem.*, 15, 1093, 1967.
10. **Guadagni et al.,** *J. Agric. Food Chem.*, 15, 1093, 1967.
11. **Deck and Chang,** *Chem. Ind.*, p. 1343, 1965.
12. **Sheldon et al.,** *J. Food Sci.*, 37(2), 313, 1972.
13. **Watanabe and Sato,** *Agric. Biol. Chem.*, 35, 756, 1971.
14. **Goldman et al.,** *Helv. Chim. Acta*, 50, 694, 1967.
15. **Sapers et al.,** *J. Food Sci.*, 36(1), 93, 1971.

	2,6-Dimethylpyrazine
Other names	2,6-Dimethyl-1,4-diazine
Empirical formula	$C_6H_8N_2$
Structure	H_3C–(pyrazine ring, N at 1,4)–CH_3
Physical/chemical characteristics	
Appearance	White crystals
Assay	C: 66.64%; H: 7.46%; N: 25.90%
Molecular weight	108.14
Melting point	47–48°C
Boiling point	155.6°C
Specific gravity	0.9647 at 50°C
Mass spectra[8]	40(43), 42(74), 52(4), 54(2), 66(7), 67(8), 80(1), 81(7), 92(3), 93(1), 108(100), 109(7)
IR spectra[8]	S (6.8, 7.1, 7.2, 8.0, 8.6, 9.8, 11.6) M (3.3, 3.4, 6.5, 7.8, 8.4, 10.7) W (5.1, 5.3, 5.6, 5.8, 7.6, 10.2)
UV spectra[8]	(275.6, 271.0 nm)
Solubility	Soluble in water, alcohol, ether, and essential oils
Organoleptic characteristics	Sweet, "fried" odor; odor threshold: 54,000 ppb in water,[9] 1,500 ppb in water[10]
Synthesis	By condensation of 1,2-diaminopropane, followed by column chromatography to separate the 2,6-methylpyrazine from the 2,5-dimethylpyrazine[6]
Natural occurrence	Reported to be present in the following foods: cocoa products, coffee, dairy products, meat, peanuts, pecans, filberts, potato products, rum and whisky, soy products;[7] reported found in grilled beef;[11] also a volatile flavor component of potato chips[12,13]
Reported uses[3]	Non-alcoholic beverages 10.0 ppm Ice cream, ices, etc. 10.0 ppm Candy 10.0 ppm Baked goods 10.0 ppm Gelatins and puddings 10.0 ppm Meat, meat sauces, soups 10.0 ppm Milk, dairy products 10.0 ppm Cereals 10.0 ppm
Regulatory status	FEMA No. 3273

REFERENCES

For References 1–5, see end of Part III.

6. **Peer and Heijden,** *Recueil,* 88, 1335, 1969.
7. **Maga and Sizer,** *Crit. Rev. Food Technol.,* 4(1), 39, 1973.
8. **Bondarovich et al.,** *J. Agric. Food Chem.,* 15, 1093, 1967.
9. **Koehler et al.,** *J. Food Sci.,* 36, 816, 1971.
10. **Guadagni et al.,** *J. Sci. Food Agric.,* 23, 1435, 1972.
11. **Flament and Ohloff,** *Helv. Chim. Acta,* 54(7), 1911, 1971.
12. **Deck, Pokorny and Chang,** *J. Food Sci.,* 38(2), 345, 1973.
13. **Watanabe and Sato,** *Agric. Biol. Chem.,* 35, 756, 1971.

	p,α-Dimethyl styrene	Dimethyl succinate
Other names	*p*-Isopropenyl toluene 1-Methyl-4-isopropenyl benzene 2-*p*-Tolyl propene	Dimethyl butanedioate Methyl succinate
Empirical formula	$C_{10}H_{12}$	$C_6H_{10}O_4$
Structure	$H_3C-C_6H_4-C(=CH_2)-CH_3$	$CH_3-O-C(=O)-(CH_2)_2-C(=O)-O-CH_3$
Physical/chemical characteristics[5]		
Appearance	Liquid; oxidizes and slowly polymerizes in the presence of air	Colorless liquid (at room temperature); solidifies when cold
Molecular weight	132.18	146.15
Melting point	-20°C (-28°C)	18–19°C
Boiling point	186–189°C 82°C at 21 mm Hg	195–196°C; 80°C at 11 mm Hg
Specific gravity	0.9038 at 15.5°C	1.1198 at 20°C
Refractive index	1.5329 at 20°C (1.5350 at 20°C)	1.4196 at 20°C
Solubility		Soluble at 1% in water; 3% in alcohol
Organoleptic characteristics		Pleasant, ethereal, winey odor; fruity, winey, and burning flavor
Synthesis	By passing a stream of *p*-cymene and water vapors over activated aluminum oxide at 730°C[6]	By direct esterification of the acid with the alcohol in benzene solution at the boil in the presence of concentrated H_2SO_4[8]
Natural occurrence	Reported present in Indian hemp *(Cannabis indica)*[7]	In filbert nuts
Reported uses[3]	Non-alcoholic beverages 0.15 ppm Candy 2.8 ppm Gelatins and puddings 1.0 ppm	Non-alcoholic beverages 1.0–100 ppm Ice cream, ices, etc. 5.0 ppm Candy 15 ppm Baked goods 15 ppm Chewing gum 5.0 ppm
Regulatory status	FEMA No. 3144	FDA 121.1164; FEMA No. 2396

REFERENCES

For References 1–5, see end of Part III.

6. U.S. Patent 2,441,095.
7. **Simonsen and Todd,** *J. Chem. Soc.* (London), p. 188, 1942.
8. **Vogel,** *J. Chem. Soc.*, p. 631, 1948.

	4,5-Dimethyl thiazole		Dimethyl trisulfide	
Other names			Methyl trisulfide	
Empirical formula	C_5H_7NS		$C_2H_6S_3$	
Structure	(structure: 4,5-dimethyl thiazole ring, two H_3C groups, N, S)		$CH_3-S-S-S-CH_3$	
Physical/chemical characteristics				
Appearance			Colorless to pale-yellow mobile, oily liquic	
Molecular weight	113.18		126.27	
Melting point	83–84°C			
Boiling point	158° 81–83°C at 59 mm Hg		165–170°C	
Refractive index			n_D^{20} 1.6010	
Solubility	Soluble in alcohol and ether		Slightly soluble in water; soluble in alcoho propylene glycol, and oils	
Organoleptic characteristics			Powerful, diffusive, penetrating odo reminiscent of fresh onion	
Synthesis	From methyl-α-bromoethyl ketone;[6] by oxidation of 2-mercaptothiazoles with hydrogen peroxide in aqueous solution;[7] from 2-mercapto-4,5-dimethyl thiazole[8]			
Natural occurrence			Reported as occurring in the volatile portio of fresh onion juice; the major arom component in cooked vegetables of th *Brassica* genus[9]	
Reported uses[3]	Ice Cream, ices, etc.	10.0 ppm	Baked goods	1.0 ppı
	Candy	10.0 ppm	Condiments, pickles	1.0 pp
	Baked goods	10.0 ppm	Meat, meat sauces, soups	1.0 ppı
	Gelatins and puddings	6.0 ppm		
	Condiments, pickles	20.0 ppm		
	Meat, meat sauces, soups	20.0 ppm		
Regulatory status	FEMA No. 3274		FEMA No. 3275	

REFERENCES

For References 1–5, see end of Part III.

6. **Kurkjy and Brown,** *J. Am. Chem. Soc.*, 74, 5778, 1952.
7. **Russell,** U.S. Patent 2,509,453; *Chem. Abstr.*, 44, 7885, 1950.
8. **Hoffman-La Roche, A. G.,** British Patent 492,637; *Chem. Abstr.*, 33, 1760, 1939.
9. **Maruyama,** *J. Food Sci.*, 35(5), 540, 1970.

	2,4-Dimethyl-5-vinylthiazole		Diphenyl disulfide	
Other names			Phenyl disulfide	
Empirical formula	C_7H_9NS		$C_{12}H_{10}S_2$	
Structure	H_3C, N, CH_3, $H_2C{=}HC$, S		S — S	
Physical/chemical characteristics				
Appearance			Needles (from alcohol)	
Molecular weight	139.21		218.34	
Melting point			61–62°C	
Boiling point			310°C 192°C at 15 mm Hg	
Specific gravity			1.353 at 20°C	
Solubility			Insoluble in water; soluble in alcohol, ether, benzene, and carbon disulfide	
Synthesis			By heating and passing a stream of air over an ammoniacal solution of thiophenol[7]	
Natural occurrence	In cooked beef[6]			
Reported uses[3]	Non-alcoholic beverages	0.10 ppm	Non-alcoholic beverages	1.0 ppm
	Ice cream, ices, etc.	0.50 ppm	Ice cream, ices, etc.	1.0 ppm
	Candy	0.50 ppm	Candy	1.0 ppm
	Baked goods	0.50 ppm	Gelatins and puddings	1.0 ppm
	Gelatins and puddings	0.50 ppm		
Regulatory status	FEMA No. 3145		FEMA No. 3225	

REFERENCES

For References 1–5, see end of Part III.

6. *J. Agric. Food Chem.*, 21(5), 874, 1973.
7. Wessely and Grill, *Monatsch. Chem.*, 77, 282, 1947.

	1,3-Diphenyl-2-propanone	Dipropyl trisulfide
Other names	Benzyl ketone Dibenzyl ketone	Propyl trisulfide
Empirical formula	$C_{15}H_{14}O$	$C_6H_{14}S_3$
Structure	$C_6H_5-CH_2-C(=O)-CH_2-C_6H_5$	$CH_3-CH_2-CH_2-S-S-S-CH_2-CH_2-C$
Physical/chemical characteristics[5]		
Appearance	Low-melting solid	Colorless to pale-yellow mobile liquid
Assay	98% min	
Molecular weight	210.28	182.38
Melting point	35°C	
Boiling point	331°C	68–69°C at 0.9 mm Hg
Refractive index		n_D^{20} 1.5424
Solubility	Soluble in ether; 1:2 in 95% alcohol; insoluble in water	Almost insoluble in water; soluble in alcoh and oils
Organoleptic characteristics	Fruity odor reminiscent of bitter almond	Very powerful, diffusive garlic-like odor
Synthesis	By heating α,α′-phenyl benzyl ethylene glycol or α,α′-phenyl benzyl ethylene oxide in the presence of diluted H_2SO_4 or $ZnCl_2$;[7,8] also by dry distillation of phenylacetate and magnesium chloride ($C_6H_5CH_2COOMgCl$)[9] or other salts of phenylacetic acid	By an adaptation of Westlake's procedure
Natural occurrence	Not reported found in nature	Reported found as a volatile constituent onion oil
Reported uses[3]	Non-alcoholic beverages 1.7 ppm Ice cream, ices, etc. 4.5 ppm Candy 9.5 ppm Baked goods 13 ppm	Baked goods 1.0 pp Condiments, pickles 1.0 pp Meat, meat sauces, soups 1.0 pp
Regulatory status	FDA 121.1164; FEMA No. 2397	FEMA No. 3276

REFERENCES

For References 1–5, see end of Part III.

6. **Westlake et al.,** *J. Am. Chem. Soc.,* 72, 436, 438, 1950; *Chem. Abstr.,* 44, 4739a, 1950.
7. **Lévy and Gombinska,** *Compt. Rend.,* 188, 713, 1929.
8. **McKenzie, Luis and Mitchell,** *Chem. Ber.,* 65, 800, 1932.
9. **Ivanow,** *Bull. Soc. Chim. Fr.,* 4(43), 445, 1928.

	Disodium succinate	spiro[2,4-Dithia-1-methyl-8-oxabicyclo [3.3.0] octane-3,3′-(1′-oxa-2′-methyl)cyclopentane] plus spiro [dithia-6-methyl-7-oxabicyclo[3.3.0] octane-3,3′-(1′-oxa-2-methyl)cyclopentane]
Other names	Sodium succinate Succinic acid, disodium salt	
Empirical formula	$C_4H_4Na_2O_4 \cdot 6H_2O$	$C_{10}H_{16}O_2S_2$
Structure	CH_2—COONa | · $6H_2O$ CH_2—COONa	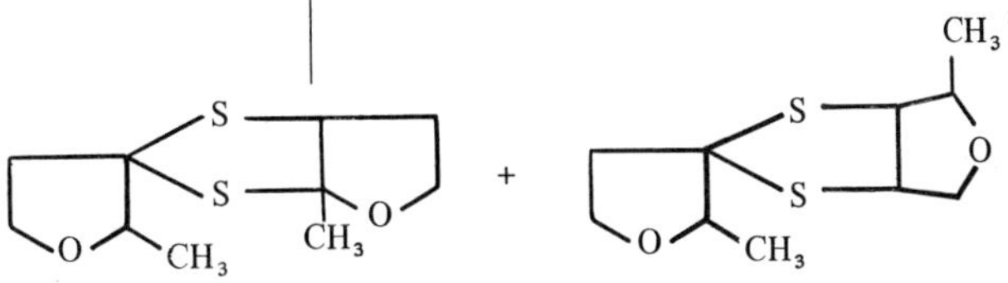 65% 35%
Physical/chemical characteristics[5]		
Appearance	Granules or crystalline powder; anhydrous when heated to 120°C	
Assay	Anhydrous, 59.99%; hydrate, 40.01%; as succinic acid, 43.71%	95% by VPC (65/35 mixture)
Molecular weight	270.16	232.2
Melting point		
Boiling point		135–140°C at 3 mm Hg
Refractive index		n_D^{20} 1.5619
Solubility	Soluble in about five parts of water; insoluble in alcohol	
Organoleptic characteristics		Meaty aroma
Synthesis	By a patented process[6]	
Natural occurrence	Reported found in condiments[7]	
Reported uses[3]	Non-alcoholic beverages 60.0 ppm Baked goods 60.0 ppm Meat, meat sauces, soups 60.0 ppm Milk, dairy products 60.0 ppm Cereals 60.0 ppm Extensively employed in the microencapsulation of flavoring oils[8]	Non-alcoholic beverages 0.045 ppm Ice cream, ices, etc. 0.25 ppm Candy 0.10 ppm Meat, meat sauces, soups 0.07 ppm Protein foods 0.40 ppm Mayonnaise 2.4 ppm
Regulatory status	FEMA No. 3277	FEMA No. 3270

REFERENCES

For References 1–5, see end of Part III.

6. VEB Filmfabrik Agfa, Wolfen, British Patent 927,450, 1963.
7. Kyowa Fermentation Industry Co. Ltd., French Patent 1,400,912, 1965.
8. Anonymous, *Fed. Regist.*, 33(232), 1968; *Chem. Abstr.*, 70, 27717w, 1969.

	3,3′-Dithio-bis- (2-methylforan)	γ-Dodecalactone
Other names	Bis(2-methyl-3-furyl) disulfide 2-Methyl-3-furyl disulfide	Dodecanolide-1,4 4-Hydroxydodecanoic acid, γ-lactone γ-Octyl-γ-butyrolactone γ-*n*-Octyl-γ-*n*-butyrolactone
Empirical formula	$C_{10}H_{10}O_2S_2$	$C_{12}H_{22}O_2$
Structure	(2-methyl-3-furyl)–S–S–(2-methyl-3-furyl): two furan rings, each bearing CH_3 / H_3C, joined through S—S	$CH_3-(CH_2)_7-CH-CH_2-CH_2-C{=}O$, with CH and C=O joined through O (ring)
Physical/chemical characteristics[5]		
Appearance		Colorless, oily liquid
Molecular weight	226.31	198.31
Boiling point		170–171°C (258°C) at 11 mm Hg 130°C at 5.1 mm Hg
Specific gravity		0.9382 at 15°C
Refractive index		1.4522 at 20°C
Solubility		Insoluble in water; poorly soluble in propylene glycol; soluble in alcohol
Organoleptic characteristics		Fatty, peachy odor; buttery, peach-like flavor
Synthesis		From 1-dodecen-12-oic acid with H_2SO_4 at 90°C;[6] from 4-hydroxydodecanoic acid by lactonization; also from methyl acrylate and octanol
Natural occurrence		Reported found in *Prunus persica*
Reported uses[3]	Baked goods 0.1 ppm Condiments, pickles 0.1 ppm Meat, meat sauces, soups 0.1 ppm	Non-alcoholic beverages 3.3 ppm Ice cream, ices, etc. 4.3 ppm Candy 13 ppm Baked goods 11 ppm Gelatins and puddings 0.15 ppm Jellies 0.01 ppm
Regulatory status	FEMA No. 3259	FDA 121.1164; FEMA No. 2400

REFERENCES

For References 1–5, see end of Part III.

6. **Chint et al.,** *Helv. Chim. Acta,* 10, 114, 1927.

	δ-Dodecalactone	2-Dodecenal
Other names	*n*-Heptyl-δ-valerolactone 5-Hydroxydodecanoic acid, δ-lactone	*n*-Dodecen-2-ol; *trans*-2-dodecen-l-ol; 3-nonyl acrolein
Empirical formula	$C_{12}H_{22}O_2$	$C_{12}H_{22}O$
Structure	$CH_3-(CH_2)_6-CH-(CH_2)_3-C{=}O$ (ring closed through O between CH and C=O)	$CH_3-(CH_2)_8-CH{=}CH-C(=O)H$
Physical/chemical characteristics[5]		
Appearance	Colorless to very pale straw-yellow, viscous liquid	Colorless, oily liquid
Molecular weight	198.31	182.31
Boiling point		108–109°C (272°C) at 1 mm Hg
Solubility	Insoluble in water; soluble in alcohol; poorly soluble in propylene glycol	Insoluble in water; soluble in alcohol
Organoleptic characteristics	Powerful, fresh-fruit, oily odor; at low levels has peach, pear, plum-like flavor	Powerful, citrus, mandarin orange-like odor at low levels; mandarin taste
Synthesis	By lactonization of 5-hydroxydodecanoic acid	See Synthesis by R. Delaby;[6] by condensation of acetaldehyde with decanal; also from α-bromolauric acid by way of the ethyl ester and alcohol
Natural occurrence	Reported found in several natural products	Its presence has been reported in the essential oils of: *Eryngium factidum*,[7] *Achasma walang* Val.,[8] Guinea orange,[9] and bitter orange[10]
Reported uses[3]	Baked goods 0.06 ppm Gelatins and puddings 0.06 ppm Toppings 10 ppm	Non-alcoholic beverages, 2.9 ppm; ice cream, ices, etc., 3.1 ppm; candy, 2.8 ppm; baked goods, 2.8 ppm
Regulatory status	FDA 121.1164; tolerance 10 ppm in margarine; FEMA No. 2401	FDA 121.1164; FEMA No. 2402

REFERENCES

For References 1–5, see end of Part III.

6. **Delaby,** *Bull. Soc. Chim. Fr.*, 5(3), 2375, 1936.
7. **Koolas,** *Chem. Zentralbl.*, 2, 630, 1932.
8. **Van Romburg,** *Chem. Zentralbl.*, 2, 2511, 1938.
9. **Naves,** *Perfum. Essent. Oil. Rec.*, 38, 237, 1947.
10. **Mehlitz and Minas,** *Ind. Obst Gemueseverwert.*, 50, 861, 1965.

	Estragole	*p*-Ethoxybenzaldehyde
Other names	*p*-Allyl anisole; chavicol methyl ether; methyl chavicol; *p*-methoxyallyl benzene	
Empirical formula	$C_{10}H_{12}O$	$C_9H_{10}O_2$
Structure	$O-CH_3$ (benzene ring) $CH_2-CH=CH_2$	$O-C_2H_5$ (benzene ring) CHO
Physical/chemical characteristics[5]		
Appearance	Colorless liquid	Colorless liquid; solidifies
Molecular weight	148.20	134.18
Melting point		14°C
Boiling point	214–216°C; 97–97.5°C at 12 mm Hg	140°C (249°C) at 20 mm Hg
Specific gravity	0.9600 at 25°C; 0.9645 at 20°C	
Refractive index	1.51372 at 22°C; 1.5230 at 17.5°C	
Solubility	Forms azeotropic mixtures with water (18%) and acetamide (18%); soluble in most organic solvents; 1:6 in 80% alcohol and 1:1 in 90% alcohol	Almost insoluble in water; soluble in alcohol
Organoleptic characteristics	Odor reminiscent of anise with a corresponding sweet taste (differing from anethole)	Sweet and floral odor; sweet, floral tas
Synthesis	Obtained by fractional distillation of the oil of turpentine or by treating a solution of the same oil in ether with an aqueous solution of mercuric acetate and subsequently heating the aqueous phase with zinc and sodium hyydroxide;[6] from allyl bromide and magnesium *p*-methoxy phenate in ether[7,8]	By ethylation of *p*-hydroxybenzaldehy using aluminum chloride catalyst
Natural occurrence	Isolated initially from the rind of *Persea gratissima* Gartn.[9] and subsequently from the oil of estragon.[10] It has been found in large amounts (as much as 60 – 90%) in the oils of: Chinese star anise, Russian anise, basil, fennel, turpentine, *Feronia elephatum* Corr., *Solidago odora*, *Agasnache rugosa*, and *Orthodon methylchavicoliferum*; obtained by steam distillation in the oil of *Fagara manishurica*[11]	Not reported found in nature
Reported uses[3]	Non-alcoholic beverages, 10 ppm; ice cream, ices, etc., 11 ppm; candy, 36 ppm; baked goods, 41 ppm; chewing gum, 50 ppm; condiments, 2.0 ppm	Non-alcoholic beverages 0.06–0.08 pp Ice cream, ices, etc. 0.36–0.50 pp Candy 1.0 ppm Baked goods 1.0 ppm
Regulatory status	FDA 121.1164; FEMA No. 2411	FDA 121.1164; FEMA No. 2413

REFERENCES

For References 1–5, see end of Part III.
6. **Hasselstroh and Hampton,** *J. Am. Chem. Soc.*, 60, 3086, 1938.
7. **Tiffeneau,** *Compt. Rend.*, 139, 482, 1904.
8. **Verley,** German Patent 154,654.
9. *Schimmel Ber.*, 1892.
10. **Grimand,** *C. R. Séances Acad. Sci. Paris*, 117, 1089, 1893.
11. **Goto,** *J. Pharm. Jap.*, 61, 91, 1941.

	2-Ethoxythiazole	Ethyl acetate
Other names	Ethyl-2-thiazosyl ether	Acetic ether Vinegar naphtha
Empirical formula	C_5H_7OSN	$C_4H_8O_2$
Structure	(thiazole ring with N, S; 2-OC_2H_5)	$CH_3-C(=O)-O-CH_2-CH_3$
Physical/chemical characteristics		
Appearance		Colorless, mobile liquid
Assay		99–100%
Molecular weight	129.18	88.10
Melting point		−83.4°C
Boiling point		75–76°C (77.11°C)
Flash point		3°C
Specific gravity		0.9000–0.9040 (0.9065–0.9070) at 15°C
Refractive index		1.3723 (1.3470) at 20°C
Acid value		Not more than 0.01%
Solubility		Soluble in most organic solvents; soluble in alcohol, ether, and approximately 1:10 in water
Organoleptic characteristics		Characteristic ether-like odor reminiscent of pineapple with a bittersweet, wine-like, burning taste
Synthesis	2-Bromothiazole is treated with sodium alkoxides to give 2-alkoxythiazole[6]	By reacting acetic acid and ethanol in the presence of sulfuric acid; by distillation of sodium, potassium, or lead acetate with ethanol in the presence of sulfuric acid; by polymerization of acetaldehyde in the presence of aluminum ethylate or aluminum acetate as catalysts
Natural occurrence		Although it has been reported present in some natural fruital aromas and in some distillates (rum, rum ether),[7] it has not been reported yet as a constituent of essential oils. It has been identified also in the petals of *Magnolia fuscata.*[8]
Reported uses[3]	Ice cream, ices, etc. 0.2 ppm Candy 1.0 ppm Baked goods 1.0 ppm Meat, meat sauces, soups 0.5 ppm Cereals 2.0 ppm	Non-alcoholic beverages 67 ppm Alcoholic beverages 50–65 ppm Ice cream, ices, etc. 99 ppm Candy 170 ppm Baked goods 170 ppm Gelatins and puddings 200 ppm Chewing gum 1,400 ppm
Regulatory status	FEMA No. 3340	FDA GRAS; FEMA No. 2414

REFERENCES

For References 1–5, see end of Part III.
6. Friedmann, *C. R. Acad. Sci. Ser. C,* 269(24), 1560, 1969; *Chem. Abstr.,* 72, 78934p, 1970.
7. Fenaroli et al., *Riv. Ital. EPPOS,* p. 9, 1965.
8. Göppert, *Justus Liebigs Ann. Chem.,* 111, 127, 1859.

	Ethyl acetoacetate	Ethyl 2-acetyl-3-phenylpropionate
Other names	Acetoacetic ester Ethyl β-ketobutyrate Ethyl-3-oxobutanoate	Ethyl α-acetylhydroxycinnamate Ethyl benzylacetoacetate Ethyl-3-oxo-2-benzylbutanoate
Empirical formula	$C_6H_{10}O_3$	$C_{13}H_{16}O_3$
Structure	$CH_3-C(=O)-CH_2-C(=O)-O-CH_2-CH_3$	$C_6H_5-CH_2-CH(C(=O)CH_3)-C(=O)-O-CH_2-CH_3$
Physical/chemical characteristics[5]		
Appearance	Colorless liquid	Colorless liquid
Molecular weight	130.14	220.26
Melting point	−45°C	
Boiling point	180–184°C; 78°C at 13 mm Hg	290°C (with decomposition); 164°C at 18 mm Hg
Flash point	84.5°C (85°C)	
Specific gravity	1.0358 at 17°C; 1.0315–1.0350 at 15°/15°C	1.061 at 25°C
Refractive index	1.41696 at 25°C; 1.4180–1.4195 at 20°C	
Solubility	Slightly soluble in water; soluble in most organic solvents	Insoluble in water; completely miscible with alcohol and ether
Organoleptic characteristics	Characteristic ether-like, fruity, pleasant, refreshing odor	
Synthesis	Ethyl acetoacetate is a mixture of two tautomer forms: the enolic and the ketonic. The liquid ester at equilibrium contains approximately 70% of the enolic form. It is prepared by Claisen condensation of ethyl acetate in the presence of sodium ethylate; also by reacting diketene with ethanol in the presence of sulfuric acid or triethylamine and sodium acetate, with or without solvent.	By reacting benzyl chloride over hot sodium acetoacetate
Natural occurrence	Naturally occurring in strawberry and bread	Not reported found in nature
Reported uses[3]	Non-alcoholic beverages 17 ppm Ice cream, ices, etc. 24 ppm Candy 110 ppm Baked goods 120 ppm Gelatins and puddings 93 ppm Chewing gum 530 ppm	Non-alcoholic beverages 0.10–5.0 ppm Ice cream, ices, etc. 2.0 ppm Candy 7.0 ppm
Regulatory status	FDA 121.1164; FEMA No. 2415	FDA 121.1164; FEMA No. 2416

REFERENCES

For References 1–5, see end of Part III.

	1-Ethyl-2-acetyl pyrrole	Ethyl aconitate, mixed esters
Other names		Ethyl-2-carboxyglutaconate Ethyl 1-propene-1,2,3-tricarboxylate A partially esterified product containing mono-, di-, and triethyl aconitate
Empirical formula	$C_8H_{11}NO$	$C_8H_{10}O_6$ $C_{10}H_{14}O_6$ $C_{12}H_{18}O_6$
Structure	(pyrrole ring with N–CH_2—CH_3 and —CO —CH_3)	CH—COO—(C_2H_5) (3) ‖ C—COO—(C_2H_5) (2) | CH_2—COO—(C_2H_5) (1)
Physical/chemical characteristics[5]		
Appearance		Oily, colorless liquid
Molecular weight	137.18	Mono- = 202.17; di- = 230.22; tri- = 258.28
Boiling point		172°C (260°C) at 18 mm Hg
Specific gravity		1.0961 at 25°C
Refractive index		1.45771 at 14.5°C
Solubility		Slightly soluble in water; soluble in alcohol
Organoleptic characteristics		Sweet, fruity, winey odor and flavor
Synthesis		By esterification of the acid with ethanol in the presence of acid catalysts; also by heating the citric acid triethyl ester Aconitic acid also yields the diethyl and the monoethyl esters upon esterification.
Natural occurrence	Reported found in roasted coffee[6-8]	Not reported found in nature
Reported uses[3]	Non-alcoholic beverages 5.0 ppm Ice cream, ices, etc. 5.0 ppm Candy 5.0 ppm Gelatins and puddings 5.0 ppm	Non-alcoholic beverages 3.6 ppm Ice cream, ices, etc. 12 ppm Candy 55 ppm Baked goods 66 ppm Gelatins and puddings 2.5 ppm
Regulatory status	FEMA No. 3147	FDA 121.1164; FEMA No. 2417

REFERENCES

'or References 1–5, see end of Part III.

6. **Stoffelsma et al.,** *Recl. Trav. Chim.*, 87, 241, 1968.
7. **Stoffelsma et al.,** *J. Agric. Food Chem.*, 16, 1000, 1968.
8. **Walter et al.,** *Z. Ernährungswiss.*, 9, 123, 1969.

	Ethyl acrylate	Ethyl *p*-anisate
Other names	Ethyl propenoate	Ethyl *p*-methoxybenzoate
Empirical formula	$C_5H_8O_2$	$C_{10}H_{12}O_3$
Structure	$CH_2{=}CH{-}C({=}O){-}O{-}CH_2{-}CH_3$	*p*-$CH_3O{-}C_6H_4{-}C({=}O){-}O{-}CH_2{-}CH_3$
Physical/chemical characteristics[5]		
Appearance	Liquid	Liquid (at room temperature)
Molecular weight	100.12	180.21
Melting point	−71 to −75°C	7–8°C
Boiling point	99–100°C	263°C; 136–137°C at 13 mm Hg
Specific gravity	0.9234 at 20°C	1.1090 at 14.4°C
Refractive index	1.4068 at 20°C	1.5245 at 18.5°C
Solubility	Slightly soluble in water	Insoluble in water; soluble in alcohol and ether
Organoleptic characteristics	Characteristic penetrating and persistent odor	Sweet, fruity, anise-like taste and similar odor
Synthesis	By esterification of acrylic acid; by heating acetylene with HCl in alcoholic solution in the presence of $Ni(CO)_4$; also from ethyl-3-chloropropionate passed over activated carbon at high temperature	By esterification of anisic acid with ethanol in the presence of an acid catalyst
Natural occurrence	Reported found in the fruits of *Ananas sativus* Lindl.[6]	In guava
Reported uses[3]	Non-alcoholic beverages 0.13–0.26 ppm Ice cream, ices, etc. 0.06–1.0 ppm Candy 1.1 ppm Baked goods 1.1 ppm Chewing gum 0.10 ppm	Non-alcoholic beverages 2.6 ppm Ice cream, ices, etc. 0.96 ppm Candy 8.8 ppm Baked goods 7.2 ppm
Regulatory status	FDA 121.1164; FEMA No. 2418	FDA 121.1164; FEMA No. 2420

REFERENCES

For References 1–5, see end of Part III.

6. **Haagen-Smit et al.,** *J. Am. Chem. Soc.*, 67, 1646, 1945.

	Ethyl anthranilate	Ethyl benzoate
Other names	Ethyl-2-aminobenzoate Ethyl *o*-aminobenzoate	Ethyl benzenecarboxylate
Empirical formula	$C_9H_{11}NO_2$	$C_9H_{10}O_2$
Structure	(benzene ring with NH_2 and $C(=O)-O-CH_2-CH_3$ in ortho positions)	(benzene ring with $C(=O)-O-CH_2-CH_3$)
Physical/chemical characteristics[5]		
Appearance	Liquid (at room temperature)	Colorless liquid
Assay		98% min[2]
Molecular weight	165.19	150.18
Melting point	13°C	−34 to −35°C
Boiling point	145–147°C at 15 mm Hg	213°C (94°C at 14 mm Hg)
Flash point		88°C
Specific gravity	1.1174 at 20°C	1.043–1.046 at 25°/25°C;[2] 1.0458 at 25°C
Refractive index	1.56455 at 20°C	1.5030–1.5060 at 20°C;[2] 1.5068 at 18°C
Acid value		Not more than 1[2]
Solubility	1:7 in 60% alcohol; 1:2 in 70% alcohol	1:6 in 60% alcohol;[2] 1:3 in 70% alcohol; almost insoluble in water
Organoleptic characteristics	Faint, orange-flowers odor and similar taste	Somewhat fruity odor similar to ylang-ylang but milder than methyl benzoate
Synthesis	By esterification of anthranilic acid with ethanol in the presence of acid catalysts; by reacting sodium hypochlorite with an alkaline alcoholic solution of phthalimide	By esterification of ethyl alcohol and benzoic acid in the presence of anhydrous aluminum sulfate and a trace of sulfuric acid;[6] by transesterification of methyl benzoate with ethanol in the presence of potassium ethylate[7]
Natural occurrence	In grapes	Reported found in volatiles from hard, mature peaches;[8] also found in pineapple and currant's
Reported uses[3]	Non-alcoholic beverages 5.9 ppm Ice cream, ices, etc. 7.6 ppm Candy 19 ppm Baked goods 23 ppm Gelatins and puddings 14 ppm Chewing gum 79 ppm	Non-alcoholic beverages 2.8 ppm Alcoholic beverages 0.50 ppm Ice cream, ices, etc. 2.8 ppm Candy 9.0 ppm Baked goods 10 ppm Gelatins and puddings 0.06 ppm Chewing gum 59 ppm
Regulatory status	FDA 121.1164; FEMA No. 2421	FDA 121.1164

REFERENCES

For References 1–5, see end of Part III.
6. **Kotake and Fujita,** *Chem. Zentralbl.*, 2, 1545, 1928.
7. **Reimer and Downes,** *J. Am. Chem. Soc.*, 43, 948, 1921.
8. **Do et al.,** *J. Food Sci.*, 34, 618, 1969.

	Ethyl benzoylacetate	α-Ethylbenzyl butyrate
Other names	Benzoyl acetic ester Ethyl β-keto-β-phenylpropionate Ethyl 3-phenyl-3-oxopropanoate	Ethyl phenyl carbinyl butyrate α-Phenylpropyl butyrate
Empirical formula	$C_{11}H_{12}O_3$	$C_{13}H_{18}O_2$
Structure	$C_6H_5-C(=O)-CH_2-C(=O)-O-C_2H_5$	$C_6H_5-(CH_2)_3-O-C(=O)-(CH_2)_2-CH_3$
Physical/chemical characteristics[5]		
Appearance	Colorless, oily liquid; turns yellow on storage and/or exposure to sunlight	Liquid
Assay		97% min
Molecular weight	192.22	206.28
Boiling point	147–149°C at 11 mm Hg (270°C)	282°C; 146–148°C at 15 mm Hg
Flash point		118°C
Specific gravity	1.12	0.9875–0.9905 at 15°C
Refractive index	1.5338 at 20°C	1.4875–1.4895 at 20°C
Acid value		0.06 max
Solubility	Insoluble in water; miscible with alcohol and ether	
Organoleptic characteristics	Brandy-like odor; bitter, whisky-like taste	Floral-fruity odor reminiscent of jasmine and apricot; sweet, plum-like taste
Synthesis	By condensation of ethyl benzoate with ethyl acetate (via Claisen condensation) using sodium ethoxide; another method also known	
Natural occurrence	Not reported found in nature	Not reported found in nature
Reported uses[3]	Non-alcoholic beverages 0.70 ppm Ice cream, ices, etc. 5.0 ppm Candy 10 ppm Baked goods 10 ppm	Non-alcoholic beverages 0.13–1.0 ppm Ice cream, ices, etc. 0.12–0.20 ppm Candy 1.0 ppm Baked goods 0.14 ppm
Regulatory status	FDA 121.1164; FEMA No. 2423	FDA 121.1164; FEMA No. 2424

REFERENCES

For References 1–5, see end of Part III.

	Ethyl brassylate	2-Ethylbutyl acetate
Other names	Cyclo-1,13-ethylenedioxytridecane-1,13-dione Ethylene brassylate Tridecanedioic acid, cyclic ethylene glycol diester	
Empirical formula	$C_{15}H_{26}O_4$	$C_8H_{16}O_2$
Structure	$CH_2-C(=O)-O-CH_2$ / $(CH_2)_9$ / $CH_2-C(=O)-O-CH_2$ (ring)	$CH_3-CH_2-CH(CH_2CH_3)-CH_2-O-C(=O)-CH_3$
Physical/chemical characteristics[5]		
Appearance	White to light-yellow liquid[2]	Liquid
Assay	97% min[2]	
Molecular weight	270.37	144.21
Melting point	332°C	
Congealing point	0–7°C	
Boiling point		160–163°C; 64–65°C at 20 mm Hg
Specific gravity	1.040–1.047[2] (1.05) at 25°/25°C	0.8784 at 20°C
Refractive index	1.4690–1.4730 at 20°C[2]	1.4109 at 20°C
Acid value	Not more than 1.0[2]	
Solubility	Soluble in alcohol and most organic solvents; insoluble in water; poorly soluble in propylene glycol	
Organoleptic characteristics	Musk-like character;[2] sweet odor	
Synthesis	Esterification of brassylic acid[2]	By reacting 2-ethylbutanol with acetic anhydride or acetic acid in the presence of sulfuric acid[6,7]
Natural occurrence	Not reported found in nature	Not reported found in nature
Reported uses[3]		Non-alcoholic beverages 5.0 ppm Ice cream, ices, etc. 2.0 ppm Candy 0.03–7.0 ppm
Regulatory status	FDA 121.1164	FDA 121.1164; FEMA No. 2425

REFERENCES

For References 1–5, see end of Part III.

6. **Fourneau and Matti,** *J. Pharm. Chim.*, 8(14), 513, 1931.
7. Carbide and Carbon, U.S. Patent 1,972,579, 1933.

	2-Ethylbutyraldehyde	Ethyl butyrate
Other names		
Empirical formula	$C_6H_{12}O$	$C_6H_{12}O_2$
Structure	$CH_3-CH_2-CH(CH_2CH_3)-CHO$	$CH_3-CH_2-CH_2-C(=O)-O-CH_2-CH_3$
Physical/chemical characteristics[5]		
Appearance	Colorless, flammable liquid	Colorless liquid
Assay		98% min[2] (97.5%)
Molecular weight	100.16	116.16
Melting point	−89°C	
Boiling point	116–117°C; 80–135°C for commercial product	121–122°C
Flash point		26°C
Specific gravity	0.814 (0.817–0.823) at 20°C	0.870–0.877 at 25°/25°C;[2] 0.8825–0.8860 at 15°C
Refractive index	1.40398	1.3910–1.3940[2] (1.3910–1.3935) at 20°C
Acid value	Not more than 2%	Not more than 1 (0.06)
Solubility	Slightly soluble in water; soluble in alcohol and ether	1:3 in 60% alcohol;[2] 1:1–1.5 in 70% alcohol
Organoleptic characteristics		Fruity odor with pineapple undernote and sweet, analogous taste
Synthesis	From diethyl carbinol and anhydrous oxalic acid according to Béhal and Sommelet or with sulfuric acid according to Stoernier. A more recent synthetic route (Zeisel-Neuwirth method) calls for the reduction of α-vinylcrotonaldehyde using iron dust and acetic acid.	By esterification of *n*-butyric acid with ethyl alcohol in the presence of Twitchells reagent[6] or $MgCl_2$;[7] also by heating *n*-butyl alcohol and ethanol over $CuO + UO_3$ catalyst at 270°C[8]
Natural occurrence	Not reported found in nature	Reported found in strawberry juice;[9] also identified by gas chromatography in olive oil and other vegetable oils[10]
Reported uses[3]	Non-alcoholic beverages 10 ppm Ice cream, ices, etc. 40 ppm Candy 0.12–25 ppm Baked goods 0.20–20 ppm	Non-alcoholic beverages 28 ppm Ice cream, ices, etc. 44 ppm Candy 98 ppm Baked goods 93 ppm Gelatins and puddings 54 ppm Chewing gum 1,400 ppm
Regulatory status	FDA 121.1164; FEMA No. 2426	FDA GRAS; FEMA No. 2427

REFERENCES

For References 1–5, see end of Part III.

6. **Zaganiaris and Varvoglis,** *Ber. Dtsch. Chem. Ges.*, 69, 2280, 1936.
7. **Petjumin,** *Zh. Obshch. Khim.*, 10, 37, 1940.
8. **Iwannikiow,** *Chem. Zentralbl.*, p. 7846, 1940.
9. **Coppens and Hoejembos,** *Recl. Trav. Chim. Pays-Bas*, 58, 683, 1939.
10. **Ratay,** *J. Assoc. Of. Agric. Chem.*, 4(47), 780, 1964.

	2-Ethylbutyric acid	Ethyl cinnamate
Other names	α-Ethylbutyric acid	Ethyl β-phenylacrylate Ethyl-3-phenylpropenoate
Empirical formula	$C_6H_{12}O_2$	$C_{11}H_{12}O_2$
Structure	$CH_3-CH_2-CH(CH_2CH_3)-C(=O)-OH$	$C_6H_5-CH=CH-C(=O)-O-CH_2-CH_3$
Physical/chemical characteristics[5]		
Appearance	Liquid	Oily, almost colorless liquid
Assay		98% min
Molecular weight	116.16	176.22
Melting point	−32°C	7°C;[7] 12°C[8]
Boiling point	193°C; 99–101°C at 18 mm Hg	271–272°C; 158–159°C at 24 mm Hg
Flash point		>100°C
Specific gravity	0.9243 at 20°C	1.045–1.048 at 25°/25°C
Refractive index	1.4132 at 20°C	1.5590–1.5610 at 20°C
Acid value		Not more than 1
Solubility		1:5 in 70% alcohol; insoluble in water; soluble in ether
Organoleptic characteristics	Faint, somewhat pleasant odor	Pleasant, cinnamon-like, balsamic, honey-like odor and sweet, peach/apricot-like taste
Synthesis	By catalytic oxidation of diethylacetaldehyde or by decarboxylation of diethylmalonic acid	By heating to 100°C cinnamic acid, alcohol, and sulfuric acid in the presence of aluminum sulfate;[6] also by Claisen condensation of benzaldehyde and ethyl acetate
Natural occurrence		Normally occurring in the *trans*-form; a *cis*-form also exists. Reported found in Oriental styrax,[9,10] in the oil of *Campheria galanga*,[11-13] and in the rhizomes of *Hedychium spicatum*[14]
Reported uses[3]	Non-alcoholic beverages 5.0 ppm Ice cream, ices, etc. 20 ppm Candy 20–35 ppm Baked goods 20 ppm	Non-alcoholic beverages 4.1 ppm Ice cream, ices, etc. 8.8 ppm Candy 9.5 ppm Baked goods 12 ppm Gelatins and puddings 2.4 ppm Chewing gum 11–40 ppm
Regulatory status	FDA 121.1164; FEMA No. 2429	FDA 121.1164; FEMA No. 2430

REFERENCES

For References 1–5, see end of Part III.

6. **Kotake and Fujita,** *Bull. Phys. Chem. Res. Tokyo,* 1, 65, 1927; *Chem. Zentralbl.,* 2, 1545, 1928.
7. **Van Duin,** *Recl. Trav. Chim. Pays-Bas,* 45, 347, 1926.
8. **Weger,** *Chem. Ann.,* 221, 75, 1883.
9. **Miller,** *Chem. Ann.,* 188, 203, 1877.
10. **Tschirch and Van Itallie,** *Arch. Pharm.,* 239, 506, 1901.
11. **Van Romburgh,** *K. Ned. Akad. Wet. Versl. Gewone Vargad. Afd. Natuurk.,* p. 627, 1902.
12. *Schimmel Ber.,* p. 38, April 1903.
13. *Chem. Zentralb.,* 1, 1086, 1903.
14. **Nakao and Shibuye,** *Chem. Zentralbl.,* 1, 947, 1925.

	Ethyl crotonate	Ethyl cyclohexanepropionate
Other names	*trans*-2-Butenoic acid ethyl ester Ethyl *trans*-2-butenoate Ethyl β-methylacrylate	Cyclohexane ethyl propionate Ethyl-3-cyclohexylpropanoate Ethyl cyclohexylpropionate Hexahydro phenylethyl propionate
Empirical formula	$C_6H_{10}O_2$	$C_{11}H_{20}O_2$
Structure	$CH_3-CH=CH-C(=O)-O-CH_2-CH_3$	(S)$-CH_2-CH_2-O-C(=O)-CH_2-CH_3$
Physical/chemical characteristics[5]		
Appearance	Liquid	Colorless, oily liquid
Molecular weight	114.15	184.28
Boiling point	136°C	
Specific gravity	0.9183 at 20°C	
Refractive index	1.4247 at 20°C	
Solubility		Insoluble in water; soluble in alcohol
Organoleptic characteristics		Powerful, fruity, sweet odor resembling pineapple; sweet, pineapple-like taste at high dilutions
Synthesis	By esterification of crotonic acid with ethyl alcohol in the presence of concentrated H_2SO_4[6-8]	By esterification of ethyl cyclohexanol with propionic acid or anhydride
Natural occurrence	Reported found in *Fragaria vesca*	Not reported found in nature
Reported uses[3]		Non-alcoholic beverages 9.0 ppm Candy 0.03–30 ppm Baked goods 24 ppm
Regulatory status	FDA 121.1164	FDA 121.1164; FEMA No. 2431

REFERENCES

For References 1–5, see end of Part III.

6. Michael, *Ber. Dtsch. Chem. Ges.*, 33, 3766, 1900.
7. Erdmann and Bedford, *Ber. Dtsch. Chem. Ges.*, 42, 1327, 1909.
8. Jeffery and Vogel, *J. Chem. Soc.*, p. 666, 1948.

	Ethyl-*trans*-2,*cis*-4-decadienoate	Ethyl decanoate
Other names		Ethyl caprate Ethyl decylate
Empirical formula	$C_{12}H_{20}O_2$	$C_{12}H_{24}O_2$
Structure	$CH_3-(CH_2)_4-CH=CH-CH=CH-C(=O)-O-CH_2-CH_3$	$CH_3-(CH_2)_8-C(=O)-O-CH_2-CH_3$
Physical/chemical characteristics[5]		
Appearance	Colorless oil	Colorless liquid
Assay		98% min[2]
Molecular weight	196.28	200.31
Melting point		−20°C
Boiling point	70–72°C at 0.05 mm Hg	241–242°C; 118–121°C at 15 mm Hg
Specific gravity		0.863–0.868 at 25°/25°C;[2] 0.865 at 15.5°C
Refractive index		1.4240–1.4270[2] (1.4260) at 20°C
Acid value		Not more than 1.0[2]
Solubility		1:4 in 80% alcohol[2]
Organoleptic characteristics	Characteristic pear-like flavor	Fruity odor reminiscent of grape (cognac)
Synthesis	Synthetically via the lithium vinyl cuprates[6]	By esterification of decanoic acid and ethyl alcohol in the presence of HCl[9] or H_2SO_4[10]
Natural occurrence	Reported found as a volatile component in Bartlett pears[7,8]	Reported present in cognac
Reported uses[3]	Non-alcoholic beverages 10.0 ppm Ice cream, ices, etc. 10.0 ppm Candy 10.0 ppm Gelatins and puddings 10.0 ppm	Non-alcoholic beverages 2.1 ppm Alcoholic beverages 3.0–10 ppm Ice cream, ices, etc. 4.5 ppm Candy 8.3 ppm Baked goods 23 ppm Gelatins and puddings 5.3 ppm
Regulatory status	FEMA No. 3148	FDA 121.1164; FEMA No. 2432

REFERENCES

For References 1–5, see end of Part III.

6. **Näf and Degen,** *Helv. Chim. Acta,* 54(7), 1939, 1971.
7. **Heinz and Jennings,** *J. Food Sci.,* 31(1), 69, 1966.
8. **Jennings,** *Int. Fruchtsaftunion Wiss. Tech. Komm. Ber.,* 6, 277, 1965.
9. **Deffet,** *Bull. Soc. Chim. Belg.,* 40, 388, 1931.
10. **Vogel,** *J. Chem. Soc.,* p. 631, 1948.

	2-Ethyl-3,5 or 6-dimethylpyrazine
Other names	(1) 3,5-Dimethyl-2-ethylpyrazine (2) 2,5-Dimethyl-3-ethylpyrazine
Empirical formula	$C_8H_{12}N_2$
Structure	(ring structure: N, CH_2-CH_3, H_3C, N, CH_3)
Physical/chemical characteristics	
Appearance	C: 70.55%; H: 8.88%; N: 20.57%
Assay	136.20
Melting point	64-66°C at 8 mm Hg (2-ethyl-3,5-dimethylpyrazine)
Mass spectra[9]	39(72), 42(53), 53(18), 56(63), 65(8), 66(13), 80(7), 81(3), 94(6), 95(4), 108(11), 109(20), 119(2), 120(10), 135(100), 136(15)
NMR spectra[8]	(CCl_4) δ 8.05 (S, 1, ring – H); 270 (q, J = 8,2, CH_2-CH_3) 2.42 (S, 3, ring – CH_3) 2.38 (S, 3, mg – CH_3) 1.24 (t, J = 8, 3, CH_2-CH_3)
IR spectra[9]	8.80, 7.17, 3.37, 8.52, 7.33, 9.78, 7.83 μm
Organoleptic characteristics	(1) Odor threshold: 15,000 ppb in water[8] (2) Odor threshold: 43,000 ppb in water[8]
Synthesis	The 2-ethyl-3,5-dimethylpyrazine is obtained by alkylation of dimethyl-2,6-pyrazine;[6] the 2,5-dimethyl-3-ethylpyrazine is obtained by alkylation of 2,5-dimethylpyrazine with ethyllithium;[10] a mixture of the two can be obtained by condensation of 2,3-pentanedione with propylenediamine; a mixture of the two is obtained from the side chain alkylation of trimethylpyrazine[10]
Natural occurrence	Reported to be present in the following foods: roasted barley, cocoa products, coffee, meat, peanuts, filberts, pecans, popcorn, rum and whisky, soy products[7]
Reported uses[3]	Non-alcoholic beverages 5.0 ppm Candy 5.0 ppm Baked goods 5.0 ppm Gelatins and puddings 2.0 ppm Condiments 2.0 ppm Meat, meat sauces 2.0 ppm Milk, dairy products 1.0 ppm Cereals 2.0 ppm
Regulatory status	FEMA No. 3149

REFERENCES

For references 1–5, see end of Part III.

6. **Klein and Spoerri,** *J. Am. Chem. Soc.*, 72, 1844, 1950.
7. **Maga and Sizer,** *Crit. Rev. Food Technol.*, 4(1), 39, 1973.
8. **Koehler et al.,** *J. Food Sci.*, 36, 816, 1971.
9. **Friedel et al.,** *J. Agric. Food Chem.*, 19, 530, 1971.
10. **Rizzi,** *J. Org. Chem.*, 33(4), 1333, 1968.

	3-Ethyl-2,6-dimethylpyrazine	Ethyl-2,4-dioxo hexanoate
Other names	2,6-Dimethyl-3-ethylpyrazine 2-Ethyl-3,5-dimethylpyrazine 3,5-Dimethyl-2-ethylpyrazine	Ethyl-2,4-diketocaproate Ethyl propionyl pyruvate
Empirical formula	$C_8H_{12}N_2$	$C_8H_{12}O_4$
Structure	(pyrazine ring with H_3C, CH_3, CH_2-CH_3 substituents)	$CH_3-CH_2-C(=O)-CH_2-C(=O)-COOC_2H_5$
Physical/chemical characteristics		
Assay	C: 70.55%; H: 8.88%; N: 20.57%	
Molecular weight	136.20	172.18
Boiling point		100–105°C at 6 mm Hg
Mass spectra[11]	39(72), 42(53), 53(18), 56(63), 65(8), 66(13), 80(7), 81(3), 94(6), 95(4), 108(11), 109(20), 119(2), 120(10), 135(100), 136(15)	
IR spectra[11]	6.20, 7.17, 3.37, 8.52, 7.33, 9.78, 7.83μm	
NMR spectra[12]	(CCl_4) δ 8.05 (S, 1, ring –H), 2.70 (q, J = 8, 2, CH_2 –CH_3), 2.42 (S, 3, ring –CH_3), 2.38 (S, 3, ring –CH_3), 1.24 (t, J = 8, 3, CH_2 –CH_3)	
Organoleptic characteristics	Odor threshold: 15,000 ppb in water[9]	
Synthesis	By alkylation of 2,6-dimethylpyrazine with ethyllithium[10]	From methyl ethyl ketone and ethyl oxalate[6]
Natural occurrence	Reported found in the basic fractions of whisky and rum,[6] in roasted peanuts,[7] in roasted barley,[8] in coffee, filberts, and potato and soy products[13]	
Reported uses[3]	Non-alcoholic beverages 5.0 ppm Ice cream, ices, etc. 5.0 ppm Candy 5.0 ppm Gelatins and puddings 5.0 ppm	Non-alcoholic beverages 15.0 ppm Candy 20.0 ppm Baked goods 120.0 ppm Syrups 60.0 ppm Jellies 30.0 ppm Milk, dairy products 10.0 ppm Hard candy 120.0 ppm
Regulatory status	FEMA No. 3150	FEMA No. 3278

REFERENCES

For References 1–5, see end of Part III.

6. **Wobben et al.,** *J. Food Sci.,* 36, 464, 1971.
7. **Waller et al.,** *J. Agric. Food Chem.,* 19, 1020, 1971.
8. **Collins,** *J. Agric. Food Chem.,* 19, 533, 1971.
9. **Koehler et al.,** *J. Food Sci.,* 36, 816, 1971.
10. **Rizzi,** *J. Org. Chem.,* 33(4), 1333, 1968.
11. **Friedel et al.,** *J. Agric. Food Chem.,* 19, 530, 1971.
12. **Rizzi,** *J. Agric Food Chem.,* 20, 1081, 1972.
13. **Maga and Sizer,** *Crit. Rev. Food Technol.,* 4(1) 39, 1973.
14. **Diels, Sielisch and Müller,** *Chem. Ber.,* 39, 1328, 1906.

	Ethyl 2-ethyl-3-phenyl propanoate	Ethyl formate
Other names	Ethyl 2-ethyl dihydrocinnamate	Formic ether
Empirical formula	$C_{13}H_{18}O_2$	$C_3H_6O_2$
Structure	$C_6H_5-CH_2-CH(CH_2-CH_3)-C(=O)-O-CH_2-CH_3$	$H-C(=O)-O-CH_2-CH_3$
Physical/chemical characteristics[5]		
Appearance		Colorless, mobile, flammable liquid
Assay		95% min[2]
Molecular weight	206.27	74.08
Melting point		−80.5°C
Boiling point		64.3°C
Flash point		−20°C
Specific gravity		0.918–0.921 at 25°/25°C;[2] 0.9250–0.9315 at 15°C
Refractive index		1.3590–1.3630[2] (1.3600–1.3615) at 20°C
Free acid content		0.1% max[2] (0.9%) (as formic acid)
Solubility		1:0.5 and more in 50% alcohol;[2] soluble in most organic solvents; slightly soluble in water
Organoleptic characteristics		Characteristic pungent odor similar to ethyl acetate and reminiscent of pineapple; slightly bitter taste
Synthesis		Usually prepared by esterification of ethyl alcohol and formic acid or by distillation of ethyl acetate and formic acid in the presence of concentrated H_2SO_4.[6]
Natural occurrence	Reported not found in nature	Reported in the oil of *Boronia dentigeroides*.[7] It has been identified in Florida orange juice,[8] several varieties of honey,[9] apple and pear,[10] and distilled liquors, such as rum.[11-13]
Reported uses[3]	Non-alcoholic beverages 5.0 ppm Alcoholic beverages 25.0 ppm Ice cream, ices, etc. 5.0 ppm Candy 50.0 ppm Preserves and spreads 50.0 ppm	Non-alcoholic beverages 9.4 ppm Alcoholic beverages 10 ppm Ice cream, ices, etc. 21 ppm Candy 50 ppm Baked goods 98 ppm Gelatins and puddings 0.35–11 ppm Chewing gum 430 ppm
Regulatory status	FEMA No. 3341	FDA 121.1164 and 121.1096; FEMA No. 2434

REFERENCES

For References 1–5, see end of Part III.

6. E. I. du Pont de Nemours & Co., U.S. Patent 1,822,803, 1928.
7. Penfold, *J. Proc. R. Soc. N.S.W.*, 62, 269, 1928.
8. Grimer and Riedmann, *Z. Naturforsch.*, 1(19), 76, 1964.
9. Wolford et al., *J. Food Sci.*, 28, 320, 1963.
10. Drawert et al., *Int. Fruchtsaftunion Wiss. Tech. Komm. Ber.*, 4, 235, 1962.
11. Fenaroli, Poy and Maroni, *Riv. Ital. EPPOS*, 9, 484, 1965.
12. Fenaroli, Poy and Maroni, *Riv. Ital. EPPOS*, 1, 593, 1965.
13. Fenaroli, Poy and Maroni, *Riv. Ital. EPPOS*, 2, 75, 1966.

	2-Ethylfuran	Ethyl 2-furanpropionate
Other names		Ethyl furfurylacetate Ethyl-3-(2-furyl)-propanoate Ethyl furylpropionate
Empirical formula	C_6H_8O	$C_9H_{12}O_3$
Structure	(furan ring)—C_2H_5	(furan ring)—$CH_2-CH_2-C(=O)-O-C_2H_5$
Physical/chemical characteristics[5]		
Appearance	Colorless liquid	Low-melting solid; turns yellow on exposure to air
Molecular weight	96.13	168.19
Melting point		24.5°C
Boiling point	93°C	260°C; 120–121°C at 17 mm Hg
Refractive index		1.54876 at 20°C
Solubility	Almost insoluble in water; soluble in alcohol	
Organoleptic characteristics	Powerful, sweet, burnt odor; when dilute, warm, sweet odor; coffee-like flavor (aroma)	Fruity odor reminiscent of camomile
Synthesis	By dehydration of furyl methyl carbinol followed by reduction	By direct esterification of the acid, which in turn is prepared by condensation of furfural with acetic anhydride and sodium acetate, followed by hydrogenation of the furacrylic acid[6,7]
Natural occurrence	In tomato, coffee, and *Mentha piperita* L.	Not reported found in nature
Reported uses[3]		Non-alcoholic beverages 1.6 ppm Ice cream, ices, etc. 1.6 ppm Candy 5.6 ppm Baked goods 7.5 ppm
Regulatory status	FDA 121.1164	FDA 121.1164; FEMA No. 2435

REFERENCES

For References 1–5, see end of Part III.

6. **Gillman and Blatt (Eds.),** *Organic Syntheses,* Coll. Vol. 1, John Wiley & Sons, 1941, 313.
7. *Ullmanns Enzyklopädie der Technischen Chemie,* 3rd ed., Vol. 14, Urban and Schwarzenberg, 1951–1962, 769.

	4-Ethylguaiacol	Ethyl heptanoate
Other names	4-Ethyl-2-methoxyphenol	Ethyl heptoate Ethyl heptylate
Empirical formula	$C_9H_{12}O_2$	$C_9H_{18}O_2$
Structure	(benzene ring with OH at 1, $O-CH_3$ at 2, CH_2-CH_3 at 4)	$CH_3-(CH_2)_5-C(=O)-O-CH_2-CH_3$
Physical/chemical characteristics[5]		
Appearance	Oily liquid	Colorless liquid
Assay		98% min[2] (97.5%)
Molecular weight	152.19	158.24
Melting point	229–230°C	−66.3°C
Boiling point		188.6°C (192°C)
Flash point		65°C
Specific gravity		0.867–0.872 at 25°/25°C;[2] 0.8725–0.8755 at 15°C
Refractive index		1.4110–1.4150[2] (1.4120–1.4140) at 20°C
Acid value		1.0 max[2] (0.2 max)
Solubility		1:3 in 70% alcohol;[2] also soluble in most organic solvents
Organoleptic characteristics		Fruity odor reminiscent of cognac with a corresponding taste
Synthesis	Reported found in several distillates of plant origin. It was identified in camphor oil,[9] and it had been reported previously in Tolu and Peru balsam, as well as in various wood distillates.	By esterification of heptoic acid; by reacting the silver salt of the acid with ethyliodide[6] or with ethyl alcohol in the presence of mineral acids[7,8]
Natural occurrence		Reported found in several natural products.
Reported uses[3]	Non-alcoholic beverages 0.05 ppm Ice cream, ices, etc. 1.1 ppm Gelatins and puddings 0.23 ppm	Non-alcoholic beverages 6.8 ppm Alcoholic beverages 8.5–20 ppm Ice cream, ices, etc. 7.5 ppm Candy 17 ppm Baked goods 24 ppm Gelatins and puddings 0.06–350 ppm Chewing gum 340 ppm
Regulatory status	FDA 121.1164; FEMA No. 2436	FDA 121.1164; FEMA No. 2437

REFERENCES

For References 1–5, see end of Part III.
6. **Mehlis,** *Justus Leibigs Ann. Chem.*, 185, 366, 1877.
7. **Gartenmeister,** *Justus Liebigs Ann. Chem.*, 233, 282, 1886.
8. **Deffet,** *Bull. Soc. Chim. Belg.*, 40, 388, 1931.
9. **Rochussen,** *Chem. Zentralbl.*, 1, 1540, 1923.

	2-Ethyl-2-heptenal	Ethyl hexanoate
Other names	2-Ethyl-3-butylacrolein	Ethyl caproate Ethyl capronate Ethyl hexylate
Empirical formula	$C_9H_{16}O$	$C_8H_{16}O_2$
Structure	$CH_3-(CH_2)_3-CH=C(CH_2CH_3)-C(=O)-H$	$CH_3-(CH_2)_4-C(=O)-O-CH_2-CH_3$
Physical/chemical characteristics[5]		
Appearance		Colorless to yellowish liquid
Assay		98% min[2]
Molecular weight	140.23	144.21
Melting point		−67 to −68°C
Boiling point		168°C (166°C)
Specific gravity		0.867–0.871 at 25°/25°C;[2] 0.8712 at 20°C
Refractive index		1.4060–1.4090[2] (1.4071) at 20°C
Acid value		Not more than 1[2]
Solubility		1:2 and more in 70% alcohol;[2] insoluble in water; soluble in most organic solvents
Organoleptic characteristics		Powerful, fruity odor with a pineapple-banana note
Synthesis		By esterification of caproic acid with ethyl alcohol in the presence of concentrated H_2SO_4 or HCl[6,7]
Natural occurrence	Not reported found in nature	Reported found in the fruits of *Ananas sativus*[8]
Reported uses[3]	Non-alcoholic beverages 0.40 ppm Candy 0.03–2.0 ppm	Non-alcoholic beverages 7.0 ppm Ice cream, ices, etc. 18 ppm Candy 12 ppm Baked goods 12 ppm Gelatins and puddings 10 ppm Chewing gum 32 ppm Jellies 1.3 ppm
Regulatory status	FDA 121.1164; FEMA No. 2438	FDA 121.1164; FEMA No. 2439

REFERENCES

For References 1–5, see end of Part III.

6. Vogel, *J. Chem. Soc.*, p. 630, 1948.
7. Whitmore and Williams, *J. Am. Chem. Soc.*, 55, 408, 1933.
8. Haagen-Smit et al., *J. Am. Chem. Soc.*, 67, 1647, 1945.

	2-Ethyl-1-hexanol	Ethyl 3-hexenoate
Other names	2-Ethyl hexyl alcohol	Hydrosorbic acid, ethyl ester
Empirical formula	$C_8H_{18}O$	$C_8H_{14}O_2$
Structure	$CH_3-CH_2-CH_2-CH_2-CH(CH_2-CH_3)-CH_2OH$	$CH_3-CH_2-CH=CH-CH_2-C(=O)-O-CH_2-CH_3$
Physical/chemical characteristics[5]		
Appearance	Colorless oily liquid	
Molecular weight	130.22	142.19
Boiling point	184–185°C	63–64°C at 12 mm Hg 63–65°C at 11 mm Hg
Flash point	81°C	
Specific gravity	0.8344 at 20°C	
Refractive index	1.4300 at 20°C	1.4233 at 24°C
Solubility	Insoluble in water; soluble in alcohol and oils	
Organoleptic characteristics	Mild, oily, sweet, slightly floral odor reminiscent of rose; sweet, fatty-floral taste with a fruital note	Green, fruity aroma
Synthesis	By hydrogenation of aldehydes obtained by the oxo process;[6] also synthesized from propylene;[7,8] by catalytic reduction of 2-ethyl-2-hexenal[9] and other similar patented processes	From 3-hexenoic acid and dicyclohexylcarbodiimide;[13] by fungal fermentation (20°C) of 3-hexenoic acid using *Saccharomyces cerevisiae;*[10] from trialkylboranes and ethyl-4-bromocrotonate in the presence of 2,6-di-*tert*-butylphenoxide;[11] by pyrolysis of acetates[12]
Natural occurrence		Reported found in pineapple;[13] a volatile flavor component in pineapple;[14] the *cis*-form has been reported found as a flavor component of passion fruit *(Passiflora edulis* f. *flavicarpa)*[15]
Reported uses[3]	Non-alcoholic beverages 10.0 ppm Ice cream, ices, etc. 10.0 ppm Candy 10.0 ppm Chewing gum 10.0 ppm	Non-alcoholic beverages 2.0 ppm Ice cream, ices, etc. 2.0 ppm Candy 4.0 ppm Baked goods 4.0 ppm Gelatins and puddings 2.0 ppm Preserves and spreads 2.0 ppm
Regulatory status	FEMA No. 3151	FEMA No. 3342

REFERENCES

For References 1–5, see end of Part III.

6. **Falbe and Tummes,** French Patent Appl. 2,011,758.
7. **Lemke and Duval,** French Patent, 1,563,043, 1969.
8. **Orlicek,** German Patent 1,913,198, 1970.
9. **Boettger et al.,** German Patent 1,949,296; *Chem. Abstr.,* 74, 140943n, 1971.
10. **Nordstrom,** *Nature,* 210(5031), 99, 1966.
11. **Brown,** *J. Am. Oil Chem. Soc.,* 92(6), 1761, 1970.
12. **Prochazka and Palecek,** *Collect. Czech. Chem. Commun.,* 35(5), 1399, 1970.
13. **Näf-Müller and Willhalm,** *Helv. Chim. Acta,* 54, 1880, 1971.
14. **Näf-Müller and Willham,** *Helv. Chim. Acta,* 54(7), 1880, 1971.
15. **Winter and Klöti,** *Helv. Chim. Acta,* 55(6), 1916, 1972.

	Ethyl-3-hydroxybutyrate		3-Ethyl-2-hydroxy-2-cyclopenten-1-one	
Other names	Ethyl-3-hydroxybutanoate		3-Ethyl-2-cyclopenten-2-ol-1-one	
Empirical formula	$C_6H_{12}O_3$		$C_7H_{10}O_2$	
Structure	$CH_3-CH(OH)-CH_2-COOC_2H_5$		[ring structure: O, OH, CH_2-CH_3]	
Physical/chemical characteristics				
Molecular weight	132.16		126.13	
Melting point			38–40°C	
Boiling point	81°C at 18 mm Hg		78–80°C at 4 mm Hg	
Specific gravity	1.0052 at 25°C		65–68°C at 1 mm Hg	
Refractive index	1.4200 at 25°C			
Organoleptic characteristics			Caramel-like flavor, often employed to impart a caramel flavor or coconut notes in food; a threshold value of 5 ppm, similar to the 3-methyl analog, has been determined by comparative sensory evaluation; the compound exhibits flavor-enhancing characteristics[21]	
Synthesis	By catalytic hydrogenation of acetoacetate[6-8]		From 5-methyl-3,5-dicarbethoxy-2-cyclopenten-2-ol-1-one and phosphoric acid;[17] by a patented process;[18] also from dimethyl adipate[19]	
Natural occurrence	In passion fruit,[9] rum,[10] sherry,[11] and wine[12-16]		Reported found in coffee[20]	
Reported uses[3]	Non alcoholic beverages	3.0 ppm	Non-alcoholic beverages	10.0 ppm
	Ice cream, ices, etc.	3.0 ppm	Ice cream, ices, etc.	10.0 ppm
	Candy	5.0 ppm	Candy	10.0 ppm
	Baked goods	5.0 ppm	Baked goods	10.0 ppm
	Gelatins and puddings	3.0 ppm	Gelatins and puddings	10.0 ppm
	Chewing gum	50.0 ppm	Condiments	10.0 ppm
	Milk, dairy products	3.0 ppm	Meat, meat sauces, soups	10.0 ppm
			Milk, dairy products	10.0 ppm
			Cereals	10.0 ppm
Regulatory status	FEMA No. 3428		FEMA No. 3152	

REFERENCES

For References 1–5, see end of Part III.

6. **Adkins et al.,** *J. Am. Chem. Soc.*, 53, 1093, 1931.
7. **Lease et al.,** *J. Am. Chem. Soc.*, 55, 807, 1933.
8. **Mozingo et al.,** *J. Am. Chem. Soc.*, 70, 228, 1948.
9. **Winter et al.,** *Helv. Chim. Acta*, 55, 1916, 1972.
10. **Maarse et al.,** *J. Food Sci.*, 31, 951, 1966.
11. **Webb et al.,** *Am. J. Enol. Viti.*, 18, 190, 1967.
12. **Webb,** *Biotechnol. Bioeng.*, 9, 305, 1967.
13. **van Wyk et al.,** *J. Food Sci.*, 32, 669, 1967.
14. **Drawert et al.,** *Chromatographia*, 1, 446, 1968.
15. **Anonymous,** *S. Afr. J. Agric. Sci.*, 11, 611, 1968.
16. **Kahn,** *J. Assoc. Off. Agric. Chem.*, 52, 1166, 1969.
17. **Tonari et al.,** *Nippon Nogei Kagaku Kaishi*, 44(1), 46, 1970.
18. **Stephens et al.,** German Patent 2,005,160, assigned to Chas. Pfizer and Company, 1970.
19. **Leir,** *J. Org. Chem.*, 35(10), 3203, 1970.
20. **Gianturco and Friedel,** *Tetrahedron*, 19(12), 2039, 1963.
21. **Stephens et al.,** U.S. Patent 3,628,970, assigned to Chas. Pfizer and Company.

	5-Ethyl-3-hydroxy-4-methyl-2(5H)-furanone	Ethyl isobutyrate
Other names	2,4-Dihydroxy-3-methyl-2-hexenoic acid, γ-lactone 2-Ethyl-3-methyl-4-hydroxydihydro-2,5-furan-5-one 2-Hydroxy-3-methyl-γ-2-hexenolactone	
Empirical formula	$C_7H_{10}O_3$	$C_6H_{12}O_2$
Structure	H_3C, OH; H_3C-H_2C; O, O (ring structure)	$(H_3C)_2CH-C(=O)-O-CH_2-CH_3$
Physical/chemical characteristics[5]		
Appearance		Colorless liquid
Molecular weight	142.15	116.16
Melting point		−88°C
Boiling point		112–113°C
Specific gravity		0.8693 at 20°C
Refractive index		1.3903 at 20°C
Solubility		Slightly soluble in water; completely miscible in most organic solvents
Organoleptic characteristics		Apple-like odor
Natural occurrence		Reported in the phlegma oils from molasses[6] and in some varieties of Russian champagne.[7] It is one of the volatile constituents of strawberry juice,[8] honey, and beer.
Reported uses[3]	Candy 1.0 ppm Baked goods 1.0 ppm Gelatins and puddings 1.0 ppm	Non-alcoholic beverages 10 ppm Ice cream, ices, etc. 25 ppm Candy 73 ppm Baked goods 200 ppm Gelatins and puddings 6.0 ppm Toppings 1.5 ppm
Regulatory status	FEMA No. 3153	FDA 121.1164; FEMA No. 2428

REFERENCES

For References 1–5, see end of Part III.

6. Dutt, *Proc. Natl. Acad. Sci. India,* 8, 105, 1938; *Chem. Zentralbl.*, 2, 753, 1939.
7. Rodopulo and Egorov, *Chem. Abstr.*, 61, 11291, 1964.
8. Teranishi et al., *J. Food Sci.*, 28, 478, 1963.

	Ethyl isovalerate	Ethyl lactate
Other names	Ethyl β-methylbutyrate	Ethyl α-hydroxy propionate
Empirical formula	$C_7H_{14}O_2$	$C_5H_{10}O_3$
Structure	$(H_3C)_2CH{-}CH_2{-}C({=}O){-}O{-}CH_2{-}CH_3$	$CH_3{-}CH(OH){-}C(O){-}O{-}CH_2{-}CH_3$
Physical/chemical characteristics[5]		
Appearance	Colorless liquid	
Assay	97% min[2] (98%)	
Molecular weight	130.18	119.12
Boiling point	134°C	154°C; *d*-, 58–59°C at 18 mm Hg; *l*-, 58°C at 20 mm Hg
Flash point	35°C	
Specific gravity	0.860–0.865 at 25°/25°C;[2] 0.8695–0.8725 at 15°C	1.0302 (*l*-, 1.033; *d*-, 1.0315) at 20°C
Refractive index	1.3950–1.3990[2] (1.3960–1.3970) at 20°C	1.4124 at 20°C (*l*-, 1.4125; *d*-, 1.4150 at 16°C)
Acid value	Not more than 1.0[2] (0.06)	
Optical rotation		At 16°C: *l*-, −5.7°; *d*-, +11.4° (undiluted)
Solubility	1:4 in 60% alcohol[2]	
Organoleptic characteristics	Fruity odor reminiscent of apple with a corresponding sweet taste	
Synthesis	By esterification of isovaleric acid with ethyl alcohol in the presence of concentrated H_2SO_4[6]	*d*-Ethyl lactate is obtained from *d*-lactic acid by azeotropic distillation with ethyl alcohol or benzene in the presence of concentrated H_2So_4.[9] The *l*-form is prepared in a similar fashion starting from *l*-lactic acid. The racemic product is prepared by boiling for 24 hours optically inactive lactic acid with ethyl alcohol in carbon tetrachloride,[10] or with an excess of ethyl alcohol in the presence of chlorosulfonic acid,[11] or in the presence of benzenesulfonic acid in benzene solution.[12]
Natural occurrence	Reported in pineapple fruits;[7] also in some varieties of Russian champagne	Both the optically active and the racemic forms are reported found in nature.
Reported uses[3]	Non-alcoholic beverages 4.9 ppm Ice cream, ices, etc. 7.5 ppm Candy 29 ppm Baked goods 27 ppm Gelatins and puddings 5.0 ppm Chewing gum 80–430 ppm Condiments 1.0 ppm	Non-alcoholic beverages 5.4 ppm Alcoholic beverages 1,000 ppm Ice cream, ices, etc. 17 ppm Candy 28 ppm Baked goods 71 ppm Gelatins and puddings 8.3 ppm Chewing gum 580–3,100 ppm Syrups 35 ppm
Regulatory status	FDA 121.1164; FEMA No. 2463	FDA 121.1164; FEMA No. 2440

REFERENCES

For References 1–5, see end of Part III.

6. **Vogel,** *J. Chem. Soc.,* p. 630, 1948.
7. **Haagen-Smit et al.,** *J. Am. Chem. Soc.,* 67, 1647, 1945.
8. **Rodopulo and Egorov,** *Chem. Abstr.,* 61, 11291, 1964.
9. **Gerrard, Kenyon and Phillips,** *J. Chem. Soc.,* p. 155, 1937.
10. **D'Ianni and Adkins,** *J. Am. Chem. Soc.,* 61, 1680, 1939.
11. **Gonzalez,** *Ciencia (Mexico),* 8, 175, 1947.
12. **Ciocca and Semproni,** *Ann. Chim. Appl.,* 25, 319, 1935.

	Ethyl laurate	Ethyl levulinate
Other names	Ethyl dodecanoate Ethyl dodecylate	Ethyl γ-ketovalerate Ethyl-4-oxopentanoate
Empirical formula	$C_{14}H_{28}O_2$	$C_7H_{12}O_3$
Structure	$CH_3-(CH_2)_{10}-C(=O)-O-CH_2-CH_3$	$CH_3-C(=O)-CH_2-CH_2-C(=O)-O-C_2H_5$
Physical/chemical characteristics[5]		
Appearance	Colorless, oily liquid	Liquid
Assay	98% min	
Molecular weight	228.37	144.17
Boiling point	272–273°C; 141–142°C at 8 mm Hg	206°C; 103–105°C at 24 mm Hg
Flash point	>100°C	
Specific gravity	0.858–0.862 at 25°C	1.0135 at 20°C
Refractive index	1.4300–1.4340 at 20°C	1.4218 at 20°C
Acid value	1 max	
Solubility	1:9 in 80% alcohol; 1:1 in 90% alcohol	Rather soluble in water; soluble in alcohol and most organic solvents
Organoleptic characteristics	Floral, fruity odor	
Synthesis	From lauroyl chloride and ethyl alcohol in the presence of Mg in ether solution,[6] or by transesterification of coconut oil with ethyl alcohol in the presence of HCl[7]	Usually obtained by direct esterification of levulinic acid with ethanol and H_2SO_4 in benzene[8] or in toluene;[9] also in the presence of HCl in ethanol.[10–12] Other methods starting from glucose, fructose, and others are of less importance.
Natural occurrence	Reported found in nature	Reported found in nature
Reported uses[3]	Non-alcoholic beverages 1.7 ppm Alcoholic beverages 3.0 ppm Ice cream, ices, etc. 3.7 ppm Candy 17 ppm Baked goods 17 ppm Gelatins and puddings 4.4 ppm Chewing gum 39 ppm	Non-alcoholic beverages 5.8 ppm Ice cream, ices, etc. 11 ppm Candy 12 ppm Baked goods 12 ppm
Regulatory status	FDA 121.1164; FEMA No. 2441	FDA 121.1164; FEMA No. 2442

REFERENCES

For References 1–5, see end of Part III.

6. **Papuot and Bouquet,** *Bull. Soc. Chim. Fr.*, p. 322, 1947.
7. **Sauer, Hain and Boutwell,** in *Organic Syntheses,* Coll. Vol. 3, Horning, E. C., Ed., John Wiley & Sons, 1955, 606.
8. **Frank et al.,** *J. Am. Chem. Soc.*, 66, 5, 1944.
9. **Mitchovitch,** *Bull. Soc. Chim. Fr.*, 5(4), 1667, 1937.
10. **Schiette and Cowley,** *J. Am. Chem. Soc.*, 53, 3485, 1931.
11. **Cox and Dodds,** *J. Am. Chem. Soc.*, 55, 3392, 1933.
12. Niacet Chemical Corp., U.S. Patent 2,029,412, 1934.

	Ethyl maltol		Ethyl-2-mercapto propionate	
Other names	3-Hydroxy-2-ethyl-4-pyrone 2-Ethyl pyromeconic acid 3-Ethyl-2-hydroxy-4H-pyran-4-one 2-Ethyl-3-ol-4H-pyran-4-one		2-Mercapto propionic acid, ethyl ester	
Empirical formula	$C_7H_8O_3$		$C_5H_{10}O_2S$	
Structure	O, OH, O, CH_2CH_3		$CH_3-CH\ (SH)\ COOCH_2-CH_3$	
Physical/chemical characteristics				
Appearance	White crystalline powder			
Assay	99% minimum			
Assay procedure	See References 6–8			
Molecular weight	140.14		134.19	
Melting point	89–93°C			
Boiling point	Considerable volatility at room temperature			
Solubility	In water: 1 g/65 ml at 15°C, 1 g/55 ml at 25°C; 12% in alcohol, 5% in phenylethyl alcohol, 5.5% in propylene glycol; slowly soluble in glycerine (1 g in 500 ml); 1 g/5 ml in chloroform; 1 g in 10 ml benzyl alcohol			
Organoleptic characteristics	Very sweet, caramel-like odor of immense tenacity; sweet, fruity taste with initial bitter-tart flavor; rapid loss of flavor *per se*; four to six times more potent than maltol			
Synthesis	From kojic acid[9-11]		By a patented process[12]	
Natural occurrence	Not reported found in nature			
Reported uses[3]	Non-alcoholic beverages	1.5–6 ppm	Non-alcoholic beverages	0.1 ppm
	Ice cream, ices, etc.	5–15 ppm	Ice cream, ices, etc.	0.1 ppm
	Candy	5–50 ppm	Candy	1.2 ppm
	Gelatins and puddings	5–50 ppm	Baked goods	1.5 ppm
	Chewing gum	5–50 ppm	Jellies	0.2 ppm
	Condiments (e.g., catsup)	5–15 ppm	Mint oils	10.0 ppm
	Toppings	5–50 ppm	Other	0.01 ppm
	Jellies	5–15 ppm		
	Soups (e.g., tomato)	5–15 ppm		
	Desserts	5–50 ppm		
Regulatory status	FDA 121.1164		FEMA No. 3279	

REFERENCES

For References 1–5, see end of Part III.

6. **Pazdera and McMullen**, Chromatography: Paper, in *Treatise on Analytical Chemistry,* Vol. 3, Part 1, Kolthoff, I. M. and Elving, P. J., Eds., Wiley & Sons, New York, 1961, 1593.
7. **Pazdera et al.**, *Anal. Chem.*, 29, 1649, 1957.
8. **Pfizer**, Chemical Division Data Sheet No. 635, 1970.
9. **Pfizer**, Belgian Patent 651,427, 1965.
10. **Pfizer**, U.S. Patent 3,468,916.
11. **Pfizer**, U.S. Patent 3,594,195.
12. **Dumesnil**, U.S. Patent 3,061,511, 1958.

	2-Ethyl (3 or 5 or 6)-methoxypyrazine (85%) plus 2-methyl (3 or 5 or 6)-methoxypyrazine (13%)
Other names	(1a) 2-Ethyl-3-methoxypyrazine (1b) 2-Ethyl-5-methoxypyrazine (1c) 2-Ethyl-6-methoxypyrazine (2a) 2-Methoxy-3-methylpyrazine (2b) 2-Methoxy-5-methylpyrazine (2c) 2-Methoxy-6-methylpyrazine
Empirical formula	$C_7H_{10}N_2O$ (ethyl substituent) $C_6H_8N_2O$ (methyl substituent)
Structure	(1a) (1b) (1c) (2a) (2b) (2c)
Physical/chemical characteristics	
Assay	(1) C: 60.85%; H: 7.30%; N: 20.27%; O: 11.58% (2) C: 58.05%; H: 6.50%; N: 22.57%; O: 12.89%
Molecular weight	138.17 (ethyl substituent) 124.14 (methyl substituent)
Mass spectra[8]	(1a): 41(27), 42(18), 52(23), 53(19), 54(20), 56(23), 68(25), 107(22), 123(65), 137(35), 138(100)
IR spectra[8]	(1a): S (6.5, 6.85, 6.9, 7.16, 7.2, 7.5, 8.6, 8.9, 9.9) M (7.7, 7.9, 8.4, 9.1, 9.2, 11.9) W (6.3, 9.4, 11.2, 13.7) VW (5.3, 5.6, 10.5, 10.8, 12.5)
Synthesis	2-methylchloropyrazine in methanol is added to a solution of sodium in methanol, and the mixture is refluxed to give a fraction (b.p. = 48–50°C at 15 mm) containing ∿75% 2-methyl-3-methoxypyrazine and ∿25% mixture of 2-methyl-5-methoxypyrazine and 2-methyl-6-methoxypyrazine; the ethyl substituent can be obtained in a similar manner[6]
Natural occurrence	The 2-ethyl-3-methoxypyrazine occurs in some brands of dehydrated and mashed potatoes[7]
Reported uses[3]	Non-alcoholic beverages 5.0 ppm Ice cream, ices, etc. 5.0 ppm Candy 5.0 ppm Baked goods 5.0 ppm Gelatins and puddings 5.0 ppm Meat, meat sauces, soups 5.0 ppm Milk, dairy products 5.0 ppm Cereals 5.0 ppm
Regulatory status	FEMA No. 3280

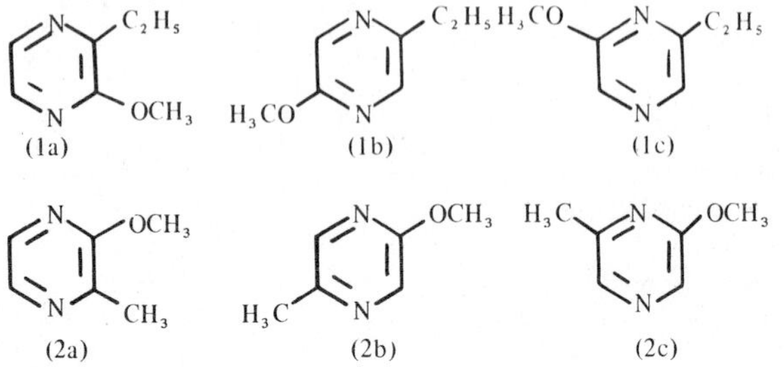

REFERENCES

For References 1–5, see end of Part III.

6. Firmenich & Co., French Patent 1,391,212, 1965.
7. **Guadagni et al.,** *J. Food Sci.,* 36, 373, 1971.
8. **Seifert et al.,** *J. Agric. Food Chem.,* 20(1), 137, 1972.

	Ethyl 2-methylbutyrate
Empirical formula	$C_7H_{14}O_2$
Structure	$CH_3-CH_2-CH(CH_3)-C(=O)-O-CH_2-CH_3$
Physical/chemical characteristics[5]	
Molecular weight	130.18
Boiling point	132–133°C
Specific gravity	0.8695 at 22°C; 0.8687 at 20°C
Refractive index	1.3970 at 20°C
Synthesis	The racemic form can be prepared catalytically by several methods: from butene and $Ni(CO)_4$ under nitrogen in ethyl alcohol/acetic acid solution,[6,7] or from ethylene and CO under pressure using HBF_4 and HF as catalysts[8,9]
Natural occurrence	Reported found in nature; the ethyl *l*-methylbutyrate has been identified in strawberry juice;[10] because of the presence of the asymmetric carbon, the compound should exhibit optically active forms as well as the racemic form; however, only the *d*-form and the racemic form are known
Reported uses[3]	Non-alcoholic beverages 0.50 ppm Ice cream, ices, etc. 3.0 ppm Candy 5.0 ppm
Regulatory status	FDA 121.1164; FEMA No. 2443

REFERENCES

For References 1–5, see end of Part III.
6. I. G. Farbenindustrie, German Patent 765,969, 1940.
7. **Reppe,** *Chem. Ann.,* 582, 53, 1953.
8. **Schering A.G.,** German Patent 629,086, 1963.
9. **Schering A.G.,** *Chem. Abstr.,* 60, 10553h, 1964.
10. **Willhalm, Pauly and Winter,** *Helv. Chim. Acta,* 49, 65, 1966.

	2-Ethyl-5-methylpyrazine
Other names	2-Methyl-5-ethylpyrazine
Empirical formula	$C_7H_{10}N_2$
Structure	H_3C–(pyrazine ring)–CH_2 — CH_3
Physical/chemical characteristics	
Assay	C: 68.82%; H: 8.25%; N: 22.93
Molecular weight	122.17
Boiling point	79–80°C at 50 mm Hg
Refractive index	1.4925 at 25°C
Mass spectra[8]	39(24), 42(10), 54(8), 56(16), 66(2), 67(2), 79(1), 80(3), 93(2), 94(13), 107(4), 121(100), 122(68)
IR spectra[8]	S (3.4, 6.7, 8.6, 9.6) M (3.2, 3.5, 6.8, 7.2, 7.4, 7.4, 8.4, 9.3, 11.2)
UV spectra[8]	W (5.5, 8.0, 8.2, 11.4, 11.7, 12.1) (272.0, 276.7)
NMR spectra[9]	(CCl_4) δ 8.18 (S,2, ring–H), 2.74 (g,2,–CH_2 –CH_3) 2.45(S,3, ring–CH_3), 1.28 (t,J = 8,3, CH_2 –CH_3)
Organoleptic characteristics	Odor threshold: 100 ppb in water[10]
Synthesis	By alkylation of dimethyl-2,5-pyrazine[6]
Natural occurrence	Reported found as a constituent in coffee aroma,[7] in roasted barley, cocoa, meats, dairy products, peanuts, pecans, filberts, popcorn, and potato and soy products;[11] a volatile flavor component in potato chips[12]
Reported uses[3]	Non-alcoholic beverages 5.0 ppm Candy 5.0 ppm Baked goods 5.0 ppm Gelatins and puddings 2.0 ppm Condiments 2.0 ppm Meat, meat sauces, soups 2.0 ppm Milk, dairy products 1.0 ppm Cereals 2.0 ppm
Regulatory status	FEMA No. 3154

REFERENCES

For References 1–5, see end of Part III.

6. **Kamal and Levine,** *J. Org. Chem.*, 27, 1355, 1962.
7. **Stoll et al.,** *Helv. Chim. Acta,* 50, 628, 1967.
8. **Bondarovich et al.,** *J. Agric. Food Chem.*, 15, 1093, 1967.
9. **Rizzi,** *J. Agric. Food Chem.*, 15, 549, 1967.
10. **Guadagni et al.,** *J. Sci. Food Agric.*, 23, 1435, 1972.
11. **Maga and Sizer,** *Crit. Rev. Food Technol.*, 4(1), 39, 1973.
12. **Deck, Pokorny and Chang,** *J. Food Sci.*, 38(2), 345, 1973.

	3-Ethyl-2-methylpyrazine	Ethyl-3-methylthiopropionate
Other names	2-Ethyl-3-methylpyrazine	Ethyl-β-methylthiopropionate
Empirical formula	$C_7H_{10}N_2$	$C_6H_{12}O_2S$
Structure	(pyrazine ring with $-CH_3$ and $-CH_2-CH_3$ substituents)	$CH_3-S-CH_2-CH_2-C(=O)-O-CH_2-CH_3$
Physical/chemical characteristics		
Assay	C: 68.82%; H: 8.25%; N: 22.93%	
Molecular weight	122.17	148.23
Boiling point	57°C at 10 mm Hg 69–70°C at 16 mm Hg	
Refractive index	1.5023 at 22°C	
Mass spectra[8]	39 (18), 43 (16), 52 (7), 53 (13), 65 (6), 67 (30), 80 (14), 81 (16), 93 (9), 94 (19), 107 (7), 121 (100), 122 (92)	
NMR spectra[10]	δ 1.88 (m, ring protons, 2H), 7.57, (S, ring–CH_3, 3H), 7.28 (8, CH_2, 2H), 8.77 (t, chain – CH_3, 3H)	
IR spectra[8]	S (3.4, 6.8, 7.1, 8.6, 8.8, 9.7, 11.7) M (3.3, 3.5, 7.3, 7.5, 9.4, 10.2) W (5.2, 5.5, 6.5, 8.1, 9.2, 11.5, 12.1)	
UV spectra[8]	(273.0, 269.2 nm)	
Organoleptic characteristics	Odor threshold: 130 ppb in water[9]	
Synthesis	By condensation of ethylenediamine with 2,3- pentanedione[6]	
Natural occurrence	Reported to be present in the following foods: bakery products, roasted barley, coffee, peanuts, filberts, potato products, soy products[7]	Flavor constituent in pineapple[11-13]
Reported uses[3]	Non-alcoholic beverages 3.0 ppm Ice cream, ices, etc. 3.0 ppm Candy 3.0 ppm Gelatins and puddings 3.0 ppm	Non-alcoholic beverages 1.0 ppm Ice cream, ices, etc. 1.0 ppm Candy 2.0 ppm Baked goods 2.0 ppm Gelatins and puddings 1.0 ppm Preserves and spreads 1.0 ppm
Regulatory status	FEMA No. 3155	FEMA No. 3343

REFERENCES

For References 1–5, see end of Part III.

6. **Flament and Stoll,** *Helv. Chim. Acta,* 50, 1754, 1967.
7. **Maga and Sizer,** *Crit. Rev. Food Technol.,* 4(1), 39, 1973.
8. **Bondarovich et al.,** *J. Agric. Food Chem.,* 15, 1093, 1967.
9. **Guadagni et al.,** *J. Sci. Food Agric.,* 23, 1435, 1972.
10. **Rizzi,** *J. Org. Chem.,* 33(4), 1336, 1968.
11. **Connell,** *Aust. J. Chem.,* 17(1), 130, 1964; *Chem. Abstr.,* 56, 13308f, 1962.
12. **Silverstein,** Syrup Foods, 4th Symposium on Foods, Oregon State University, 1965, Oregon State University Press, 1967, 456.
13. **Rodin, Coulson, Silverstein and Leeper,** *J. Food Sci.,* 31(5), 721, 1966; *Chem. Abstr.,* 65, 19225d, 1966.

	Ethyl myristate	Ethyl nitrite
Other names	Ethyl tetradecanoate	Nitrous ether Sweet spirit of nitre
Empirical formula	$C_{16}H_{32}O_2$	$C_2H_5O_2N$
Structure	$CH_3{-}(CH_2)_{12}{-}C({=}O){-}O{-}CH_2{-}CH_3$	$CH_3{-}CH_2{-}O{-}N{=}O$
Physical/chemical characteristics[5]		
Appearance	Colorless crystals	Very mobile, colorless liquid
Molecular weight	256.44	75.04
Melting point	10.5°C	
Boiling point	295°C	17.4°
Specific gravity	0.860 at 15.5°C	0.9009 at 15.5°C
Refractive index	1.4400 at 20°C	
Solubility	Very slightly soluble in alcohol and ether; slightly soluble in ligroin	Very slightly soluble in water; soluble in alcohol
Organoleptic characteristics	Odor reminiscent of orris	Characteristic ether-like odor
Synthesis	By esterification of the acid with ethyl alcohol in the presence of gaseous HCl	From sodium nitrite in aqueous solution by displacing the acid with H_2SO_4 in the presence of ethyl alcohol. It forms azeotropic mixtures with isopentane (85%), amyl bromide (40%), and carbon disulfide (96%).
Natural occurrence	Reported found in the residue of fusel oil from molasses	Not reported found in nature
Reported uses[3]	Non-alcoholic beverages 6.7 ppm Alcoholic beverages 30 ppm Ice cream, ices, etc. 8.0 ppm Candy 10 ppm Baked goods 14 ppm	Non-alcoholic beverages 3.0 ppm Ice cream, ices, etc. 4.5 ppm Candy 0.10–8.0 ppm Baked goods 0.10 ppm Chewing gum 3.9 ppm Syrups 52 ppm Icings 13 ppm
Regulatory status	FDA 121.1164; FEMA No. 2445	FDA 121.1164; FEMA No. 2446

REFERENCES

For References 1–5, see end of Part III.

	Ethyl nonanoate	Ethyl 2-nonynoate
Other names	Ethyl nonylate Ethyl pelargonate	Ethyl octyne carbonate
Empirical formula	$C_{11}H_{22}O_2$	$C_{11}H_{18}O_2$
Structure	$CH_3(CH_2)_7-\overset{\overset{O}{\parallel}}{C}-O-C_2H_5$	$CH_3-(CH_2)_5-C\equiv C-\overset{\overset{O}{\parallel}}{C}-O-CH_2-CH_3$
Physical/chemical characteristics[5]		
Appearance	Colorless liquid	Oily liquid
Assay	99% min	
Molecular weight	186.30	182.26
Melting point	−44.5°C (−36.7°C)	
Boiling point	228°C (220°C)	227°C (121–122°C at 13 mm Hg)
Flash point	85°C	
Specific gravity	0.8700–0.8725 at 15°C	0.9032 at 25°C; 0.935 at 15.5°C
Refractive index	1.4215–1.4230 at 20°C	1.4464 at 25°C; 1.4527 at 20°C
Acid value	0.9 max	
Solubility	Insoluble in water; soluble in water-ether mixtures; 1:1.5–2 in 80% alcohol	
Organoleptic characteristics	Odor reminiscent of cognac with a rosy-fruity note	Characteristic green, violet-like odor
Synthesis	By distillation of pelargonic acid and ethyl alcohol in toluene in the presence of small amounts of muriatic acid (HCl); also by hydrogenation of oenanthylidene acetate in the presence of Ni at 180°C	
Natural occurrence	In pineapple, banana, and apple	Not reported found in nature
Reported uses[3]	Non-alcoholic beverages 3.9 ppm Alcoholic beverages 20 ppm Ice cream, ices, etc. 4.0 ppm Candy 14 ppm Baked goods 15 ppm Gelatins and puddings 15 ppm Chewing gum 580 ppm Icings 39 ppm	Non-alcoholic beverages 0.56 ppm Ice cream, ices, etc. 0.55 ppm Candy 0.52 ppm Baked goods 1.2 ppm
Regulatory status	FDA 121.1164; FEMA No. 2447	FDA 121.1164; FEMA No. 2448

REFERENCES

For References 1–5, see end of Part III.

	Ethyl octanoate	Ethyl *cis*-4-octenoate
Other names	Ethyl caprylate Ethyl octylate	
Empirical formula	$C_{10}H_{20}O_2$	$C_{10}H_{18}O_2$
Structure	$CH_3-(CH_2)_6-C(=O)-O-CH_2-CH_3$	$CH_3-CH_2-CH_2-CH=CH-CH_2-CH_2-C(=O)-O-CH_2-CH_3$
Physical/chemical characteristics[5]		
Appearance	Colorless liquid	
Assay	98% min[2] (97.5%)	
Molecular weight	172.27	170.24
Melting point	−43°C	
Boiling point	208°C	105°C at 11 mm Hg
Flash point	85°C	
Specific gravity	0.865–0.869 at 25°/25°C;[2] 0.8710–0.8750 at 15°C	
Refractive index	1.4170–1.4190[2] at 20°C	
Acid value	Not more than 1[2] (1.9)	
Solubility	1:4 and more in 70% alcohol;[2] 1:1–2 in 90% alcohol; insoluble in water; soluble in most organic solvents	
Organoleptic characteristics	Pleasant, fruity, floral odor (wine-apricot note)	
Synthesis	Usually prepared by esterification of caprylic acid with ethyl alcohol and sulfuric acid as catalyst;[6] also by alcoholysis of coconut oil in the presence of HCl[7]	From *cis*-1-pentenyl bromide via the lithium vinyl cuprates[8]
Natural occurrence	Reported found in nature	Reported found as a volatile flavor component apple;[9] also reported found among the aroma components of yellow passion fruit[10]
Reported uses[3]	Non-alcoholic beverages 4.1 ppm Ice cream, ices, etc. 2.4 ppm Candy 9.0 ppm Baked goods 11 ppm Gelatins and puddings 0.10–2.7 ppm Chewing gum 4.0–60 ppm	Non-alcoholic beverages 3.0 ppm Ice cream, ices, etc. 3.0 ppm Candy 5.0 ppm Baked goods 5.0 ppm Gelatins and puddings 3.0 ppm Preserves and spreads 3.0 ppm
Regulatory status	FDA 121.1164; FEMA No. 2449	FEMA No. 3344

REFERENCES

For References 1–5, see end of Part III.

6. **Van Renesse,** *Chem. Ann.,* 171, 382, 1874.
7. **Sauer, Hain and Boutwall,** in *Organic Syntheses,* Coll. Vol. 3, Horning, E. C., Ed., John Wiley & Sons, 1955, 607.
8. **Naf and Degen,** *Helv. Chim. Acta,* 54(7), 1939, 1971.
9. **Näf-Müller and Willhalm,** *Helv. Chim. Acta,* 54(7), 1880, 1971.
10. **Winter and Klöti,** *Helv. Chim. Acta,* 55(6), 1916, 1972.

	Ethyl oleate	*p*-Ethylphenol
Other names	Ethyl 9-octadecenoate	1-Ethyl-4-hydroxybenzene 4-Ethylphenol
Empirical formula	$C_{20}H_{38}O_2$	$C_8H_{10}O$
Structure	$CH_3-(CH_2)_7-CH=CH-(CH_2)_7-C(=O)-O-CH_2-CH_3$	OH (ring) CH_2-CH_3
Physical/chemical characteristics[5]		
Appearance	Oily, slightly yellowish liquid	Colorless or white needles; tend to yellow on exposure to light
Assay	99% min	
Molecular weight	310.52	122.17
Melting point		47–48°C
Boiling point	207°C at 13 mm Hg; 205–208°C for commercial product	218–219°C 99.5°C at 10 mm Hg
Flash point	175°C	
Specific gravity	0.87196 at 20°C	1.011 at 20°C
Refractive index	1.4515 at 20°C	1.5239 at 25°C
Solubility	Insoluble in water; soluble in alcohol and ether	Slightly soluble in water; soluble in alcohol, ether, benzene, carbon disulfide, and acetone
Organoleptic characteristics	Faint, floral note	Powerful woody-phenolic, yet somewhat sweet odor
Synthesis	By direct esterification of oleic acid with ethyl alcohol in the presence of HCl at the boil;[6] in the presence of Twitchell reagent[7] or chlorosulfonic acid[8]	By catalytic reaction between phenol and ethylene or ethanol;[9] by heating 8-chloro-3-ethylbenzene with NaOH in the presence of copper powder[10]
Natural occurrence	Reported found in nature	Reported found in coal tar; also reported found in yellow passion fruit[11]
Reported uses[3]	Non-alcoholic beverages 0.10 ppm Ice cream, ices, etc. 0.10 ppm Candy 0.10–40 ppm Baked goods 0.10–55 ppm Gelatins and puddings 0.10 ppm	Baked goods 0.20 ppm Margarine 0.20 ppm
Regulatory status	FDA 121.1164; FEMA No. 2450	FEMA No. 3156

REFERENCES

For References 1–5, see end of Part III.

6. **Kailan and Kohberger,** *Monatsh. Chem.*, 59, 19, 1932.
7. **Zaganiaris and Varvoglis,** *Ber. Dtsch. Chem. Ges.*, 69, 2281, 1936.
8. **Gonzales,** *Ciencia (Mexico)*, p. 8, 1947.
9. *Beilstein's Encyclopedia of Organic Chemistry,* Vol. 6, Part 3, p. 1663, Springer Verlag.
10. **Meharge and Allen,** *J. Am. Chem. Soc.*, 54, 2920, 1932.
11. **Winter and Klöti,** *Helv. Chim. Acta,* 55(6), 1916, 1972.

	Ethyl phenylacetate	Ethyl-4-phenylbutyrate
Other names	Ethyl α-toluate	Ethyl phenylbutyrate Ethyl-γ-phenylbutyrate
Empirical formula	$C_{10}H_{12}O_2$	$C_{12}H_{16}O_2$
Structure	$C_6H_5-CH_2-C(=O)-O-CH_2-CH_3$	$C_6H_5-(CH_2)_3-C(=O)-O-CH_2-CH_3$
Physical/chemical characteristics[5]		
Appearance	Colorless or nearly colorless liquid	Colorless, somewhat oily liquid
Assay	98% min[2]	
Molecular weight	164.21	192.26
Boiling point	229°C (100°C at 10 mm Hg)	
Flash point	>100°C	
Specific gravity	1.027–1.032 at 25°/25°C[2]	
Refractive index	1.4960–1.5000[2] (1.4960–1.4980) at 20°C	
Acid value	1 max[2]	
Solubility	1:3 in 70% alcohol;[2] 1:10 in 60% alcohol; insoluble in water	Insoluble in water; soluble in alcohol
Organoleptic characteristics	Pleasant, strong, sweet odor suggestive of honey; bittersweet flavor	Sweet, fruity, plum-like odor and cooked plum-prune taste
Synthesis	By heating to the boil phenylacetonitrile and sulfuric acid in alcohol solution;[6,7] by esterification of the acid catalyzed by HCl[8,9] or H_2SO_4[10]	By esterification of ethanol with γ-phenylbutyric acid obtained by Grignard reaction from γ-bromopropylbenzene
Natural occurrence	Reported found in nature	Not reported found in nature
Reported uses[3]	Non-alcoholic beverages 2.4 ppm Ice cream, ices, etc. 5.2 ppm Candy 8.1 ppm Baked goods 6.0 ppm Syrups 24 ppm	Non-alcoholic beverages 0.06–1.0 ppm Ice cream, ices, etc. 0.06 ppm
Regulatory status	FDA 121.1164; FEMA No. 2452	FDA 121.1164; FEMA No. 2453

REFERENCES

For References 1–5, see end of Part III.

6. **Leonard,** *J. Am. Chem. Soc.*, 47, 1777, 1925.
7. **Gilman and Blatt (Eds.),** *Organic Syntheses*, Coll. Vol. 1, John Wiley & Sons, 1941, 270.
8. **Radriszewski,** *Ber. Dtsch. Chem. Ges.*, 2, 208, 1869.
9. **Bodroux,** *Compt. Rend.*, 157, 940, 1913.
10. **Senderens and Aboulenc,** *Compt. Rend.*, 152, 1856, 1911.

	Ethyl-3-phenylglycidate	Ethyl 3-phenylpropionate
Other names	Ethyl 3-phenyl-2,3-epoxypropionate	Ethyl hydrocinnamate
Empirical formula	$C_{11}H_{12}O_3$	$C_{11}H_{14}O_2$
Structure	$C_6H_5-CH-CH-C(=O)-O-CH_2-CH_3$ (with epoxide O bridging the two CH)	$C_6H_5-CH_2-CH_2-C(=O)-O-CH_2-CH_3$
Physical/chemical characteristics[5]		
Appearance	Colorless to pale-yellow liquid	Colorless liquid
Assay	99% min[2] (98%)	
Molecular weight	192.22	178.23
Boiling point		249°C (123°C at 16 mm Hg)
Flash point	>100°C	
Specific gravity	1.120–1.128[2] (1.120–1.125) at 25°/25°C	1.015 at 20°C
Refractive index	1.5190–1.5230[2] (1.516–1.521) at 20°C	1.4951 at 17.8°C
Acid value	Not more than 1.0[2]	
Solubility	1:6 in 70% alcohol;[2] 1:1 in 80% alcohol	Insoluble in water; soluble in most organic solvents
Organoleptic characteristics	Strong, fruity odor suggestive of strawberry with a corresponding sweet flavor	Floral odor
Synthesis	Usually prepared by the reaction of benzaldehyde and the ethyl ester of monochloracetic acid in the presence of an alkaline condensing agent; by reacting the silver salt of phenyl glycidic acid with ethyl iodide[6]	By hydrogenation of the corresponding ethyl cinnamate in the presence of nickel in alcohol solution[7,8]
Natural occurrence	Not reported found in nature	Not reported found in nature
Reported uses[3]	Non-alcoholic beverages 4.6 ppm Ice cream, ices, etc. 12 ppm Candy 18 ppm Baked goods 20 ppm Gelatins and puddings 10–70 ppm	Non-alcoholic beverages 1.8 ppm Ice cream, ices, etc. 1.0 ppm Candy 2.5 ppm Baked goods 0.50–30 ppm
Regulatory status	FDA 121.1164; FEMA No. 2454	FDA 121.1164; FEMA No. 2455

REFERENCES

For References 1–5, see end of Part III.

6. **Glaser,** *Ann. Chem.*, 147, 104, 1868.
7. **Tanaka,** *Chem. Ztg. Chem. Appar.*, 48, 25, 1924.
8. **Tanaka,** *Chem. Zentralbl.*, 1, 1878, 1924.

	Ethyl propionate	2-Ethylpyrazine
Other names		2-Ethyl-1,4-diazine Ethylpyrazine
Empirical formula	$C_5H_{10}O_2$	$C_6H_8N_2$
Structure	$CH_3-CH_2-C(=O)-O-CH_2-CH_3$	(pyrazine ring with $-CH_2-CH_3$)
Physical/chemical characteristics		
Appearance	Colorless, mobile liquid	C: 66.64%; H: 7.46%; N: 25.90%
Assay	98% min	108.14
Molecular weight	102.14	
Melting point	99°C	155°C
Boiling point		122–124°C at 90 mm Hg
Flash point	12°C (approx 15°C)	
Specific gravity	0.8945–0.8985 at 15°C; 0.8917 at 20°C	
Refractive index	1.3830–1.3850 at 20°C	
Acid value	Not more than 0.2	
Mass spectra[8]		39 (16), 40 (4), 52 (18), 53 (18), 64 (1), 66 (3), 80 (23), 81 (14), 92 (1), 93 (2), 107 (100), 108 (82)
IR spectra[8]		S (3.4, 6.8, 7.1, 8.6, 9.4, 9.8, 11.8) M (3.3, 3.5, 6.3, 6.5, 6.9, 7.7, 8.1, 10.3) W (5.2, 5.4, 5.6, 5.7, 7.5, 10.7)
UV spectra[8]		(266.2, 272.2 nm)
Solubility	Completely miscible in 95% alcohol; readily soluble in most organic solvents; 1:1–2 in 60% alcohol	
Organoleptic characteristics	Odor reminiscent of rum and pineapple	Odor threshold: 22,000 ppb in water, 6,000 ppb in water[10]
Synthesis	From propionic acid, ethyl alcohol, and concentrated H_2SO_4 in chloroform at the boil[11]	By alkylation of methylpyrazine with methyl iodide[6]
Natural occurrence	Reported found in several type of wine,[12] in white grape var. *Sauvignon*,[13,14] and in cocoa	Reported to be present in the following foods: bakery products, cocoa products, coffee, meat, peanuts, filberts, potato products[7]
Reported uses[3]	Non-alcoholic beverages 7.7 ppm Ice cream, ices, etc. 29 ppm Candy 78 ppm Baked goods 110 ppm Gelatins and puddings 10–15 ppm Chewing gum 1,100 ppm	Non-alcoholic beverages 10.0 ppm Ice cream, ices, etc. 10.0 ppm Candy 10.0 ppm Baked goods 10.0 ppm Gelatins and puddings 10.0 ppm Meat, meat sauces, soups 10.0 ppm Milk, dairy products 10.0 ppm Cereals 10.0 ppm
Regulatory status	FDA 121.1164; FEMA No. 2456	FEMA No. 3281

REFERENCES

For References 1–5, see end of Part III.
6. **Behun and Levine,** *J. Org. Chem.*, 26, 3379, 1961.
7. **Maga and Sizer,** *Crit. Rev. Food Technol.*, 4(1), 39, 1973.
8. **Bondarovich et al.,** *J. Agric. Food Chem.*, 15, 1093, 1967.
9. **Koehler et al.,** *J. Food Sci.*, 36, 816, 1971.
10. **Guadagni et al.,** *J. Sci. Food Agric.*, 23, 1435, 1972.
11. **Henecka,** *Houben-Weyl*, 8, 524, 1952.
12. **Pisarnitskii,** *Chem. Abstr.*, 62, 16919a, 1965.
13. **Chandhary et al.,** *Am. J. Enol. Vitic.*, 15, 190, 1964.
14. **Chandhary et al.,** *Chem. Abstr.*, 62, 9453g, 1965.

	3-Ethylpyridine		Ethyl pyruvate	
Other names	β-Ethylpyridine β-Lutidine		Ethyl α-ketopropionate Ethyl-2-oxopropanoate	
Empirical formula	C_7H_9N		$C_5H_8O_3$	
Structure	(pyridine ring, N at top)—CH_2—CH_3		CH_3—C(=O)—C(=O)—O—CH_2—CH_3	
Physical/chemical characteristics[5]				
Appearance	Colorless to brownish liquid		Colorless liquid	
Molecular weight	107.15		116.12	
Melting point			−50°C	
Boiling point	162–165°C at 762 mm Hg		155–156°C (66°C at 18 mm Hg)	
Specific gravity	0.940 at 23°C 0.9539 at 0°C		1.080 at 14°C	
Refractive index	1.5021 at 22°C		1.408 at 15.4°C	
Solubility	Slightly soluble in water; soluble in alcohol and ether		Slightly soluble in water; miscible with alcohol and ether	
Organoleptic characteristics	Tobacco flavor[7]			
Synthesis			By direct esterification of pyruvic acid with absolute ethyl alcohol at the boil and subsequent vacuum distillation; by esterification via oxidation of vapors of ethyl lactate in the presence of V_2O_5 at 155°C	
Natural occurrence	Reported found in coffee[6]		Reported found in nature	
Reported uses[3]	Non-alcoholic beverages	0.05 ppm	Non-alcoholic beverages	50 ppm
	Ice cream, ices, etc.	0.05 ppm	Ice cream, ices, etc.	20–150 ppm
	Candy	0.06 ppm	Candy	35 ppm
	Baked goods	0.06 ppm	Baked goods	40 ppm
	Gelatins and puddings	0.05 ppm		
	Meat, meat sauces, soups	0.05 ppm		
Regulatory status	FEMA No. 3394		FDA 121.1164; FEMA No. 2457	

REFERENCES

For References 1–5, see end of Part III.

6. **Stoll et al.,** *Helv. Chim. Acta,* 50, 628, 1967.
7. *Chem. Abstr.,* 74, 1083353j, 1971.

	Ethyl salicylate	Ethyl sorbate
Other names	Ethyl *o*-hydroxy benzoate	Ethyl-2,4-hexadienoate
Empirical formula	$C_9H_{10}O_3$	$C_8H_{12}O_2$
Structure	(OH)(C_6H_4)$-C(=O)-O-CH_2-CH_3$	$CH_3-CH=CH-CH=CH-C(=O)-O-CH_2-CH_3$
Physical/chemical characteristics[5]		
Appearance	Colorless liquid	
Assay	99% min[2]	
Molecular weight	166.18	140.18
Boiling point	232–235°C (107–108°C at 12 mm Hg)	81 C at 15 mm Hg
Specific gravity	1.126–1.130 at 25°/25°C;[2] 1.130–1.135 at 15°C	0.9560 at 20 C
Refractive index	1.520–1.523 at 20°C[2]	1.502 at 20 C
Acid value	Not more than 1[2]	
Solubility	1:4 in 80% alcohol;[2] slightly soluble in water	
Organoleptic characteristics	Characteristic aromatic odor similar to wintergreen; tends to darken on exposure to light and air	
Synthesis	By esterification of salicylic acid with ethyl alcohol and concentrated H_2SO_4 at 100°C in the presence of aluminum sulfate;[6] by heating to the boil an alkaline solution of salicylic acid and ethyl *p*-toluenesulfonate[7]	By reacting gaseous HCl with a solution of sorbic acid in ethyl alcohol,[8] or by direct esterification by reaction of sorbyl chloride with absolute ethyl alcohol
Natural occurrence	Reported found in nature (e.g., currant and strawberry)	Reported found in nature
Reported uses[3]	Non-alcoholic beverages 2.8 ppm Ice cream, ices, etc. 11 ppm Candy 10 ppm Baked goods 16 ppm Gelatins and puddings 0.04 ppm Chewing gum 16 ppm	Non-alcoholic beverages 5.5 ppm Ice cream, ices, etc. 14 ppm Candy 15 ppm Baked goods 18 ppm
Regulatory status	FDA 121.1164; FEMA No. 2458	FDA 121.1164; FEMA No. 2459

REFERENCES

For References 1–5, see end of Part III.

6. **Kotake and Fujita,** *Chem. Zentralbl.,* 2, 454, 1928.
7. **Finzi,** *Ann. Chim. Appl.,* 15, 49, 1925.
8. **Hoffmann,** *Annalen,* 110, 137, 1859.

	Ethyl thioacetate	Ethyl tiglate
Other names	Thioacetic acid, ethyl ester	Ethyl *trans*-2,3-dimethyl acrylate Ethyl *trans*-2-methyl-2-butenoate
Empirical formula	C_4H_8OS	$C_7H_{12}O_2$
Structure	$CH_3-C(=O)-S-CH_2-CH_3$	$CH_3-CH=C(CH_3)-C(=O)-O-CH_2-CH_3$
Physical/chemical characteristics		
Appearance		Liquid
Molecular weight	104.18	128.17
Boiling point	116.4°C	156 C (55–57 C at 15 mm Hg)
Specific gravity	0.9792 at 20°C	0.9239 at 16.8 C
Refractive index	1.4583 at 21°C	1.4347 at 16.8 C
Solubility	Insoluble in water; very soluble in alcohol and ether	Soluble in most organic solvents
Synthesis	From acetyl chloride and ethyl mercaptan[6]	By direct esterification of tiglic acid with ethyl alcohol in the presence of concentrated H_2SO_4[7] or by reaction of bromomethylethyl acetic acid ethyl ester with dimethylaniline[8]
Natural occurrence		Not reported found in nature
Reported uses[3]	Non-alcoholic beverages 1.0 ppm Ice cream, ices, etc. 1.0 ppm Candy 1.0 ppm Baked goods 1.0 ppm Gelatins and puddings 1.0 ppm Chewing gum 1.0 ppm Condiments, pickles 1.0 ppm Meat, meat sauces, soups 1.0 ppm Milk, dairy products 1.0 ppm Cereals 1.0 ppm	Non-alcoholic beverages 5.3 ppm Ice cream, ices, etc. 6.0 ppm Candy 20 ppm Baked goods 6.5 ppm
Regulatory status	FEMA No. 3282	FDA 121.1164; FEMA No. 2460

REFERENCES

For References 1–5, see end of Part III.

6. **Schaefgen,** *J. Am. Chem. Soc.*, 70, 1308, 1948.
7. **Auwers,** *Ann. Chem.*, 432, 71, 1923.
8. **Gardner and Rydon,** *J. Chem. Soc.*, p. 53, 1938.

	2-Ethylthiophenol		Ethyl (*p*-tolyloxy) acetate	
Other names	2-Ethylphenyl mercaptan-2-ethylbenzene-thiol		Ethyl *p*-cresoxy acetate	
Empirical formula	$C_8H_{10}S$		$C_{11}H_{14}O_2$	
Structure	SH; $-CH_2-CH_3$		$CH_2-C(=O)-C_2H_5$; O; CH_3	
Physical/chemical characteristics				
Molecular weight	138.23		178.22	
Natural occurrence			Not reported found in nature.	
Reported uses[3]	Non-alcoholic beverages	0.2 ppm	Non-alcoholic beverages	8.0 ppm
	Ice cream, ices, etc.	0.2 ppm	Candy	40.0 ppm
	Candy	0.3 ppm		
	Baked goods	0.3 ppm		
	Gelatins and puddings	0.2 ppm		
	Meat, meat sauces, soups	0.2 ppm		
Regulatory status	FEMA No. 3345		FEMA No. 3157	

REFERENCES

For References 1–5, see end of Part III.

	Ethyl undecanoate	Ethyl 10-undecenoate
Other names	Ethyl hendecanoate Ethyl undecylate	Ethyl 10-hendecenoate Ethyl undecylenoate
Empirical formula	$C_{13}H_{26}O_2$	$C_{13}H_{24}O_2$
Structure	$CH_3-(CH_2)_9-\overset{\overset{\displaystyle O}{\parallel}}{C}-O-CH_2-CH_3$	$CH_2=CH-(CH_2)_8-\overset{\overset{\displaystyle O}{\parallel}}{C}-O-CH_2-CH_3$
Physical/chemical characteristics[5]		
Appearance	Liquid	Liquid
Molecular weight	214.35	212.33
Boiling point	255 C (102–107 C at 1 mm Hg)	258–259 C (131.5 C at 14.5 mm Hg)
Specific gravity	0.870 at 15.5 C	0.8788 at 20 C
Refractive index	1.9325 at 20 C	1.4382 at 20 C
Solubility	Insoluble in water	
Organoleptic characteristics	Odor similar to the corresponding ethyl heptanoate and coconut-like aroma	Wine-like odor
Synthesis	By reduction of the *l*-undecen-ll-oic acid ethyl ester with hydrogen in the presence of Ni at 180 C,[6] or by direct esterification of *n*-undecanoic acid with ethyl alcohol under reflux	By heating the corresponding acid together with HCl in ethyl alcohol solution at 70 C[7] or with H_2SO_4 in alcoholic solution[8]
Natural occurrence	In rum, whisky, and wine	Not reported found in nature
Reported uses[3]		Non-alcoholic beverages 1.7 ppm Alcoholic beverages 5.0 ppm Ice cream, ices, etc. 8.7 ppm Candy 10 ppm Baked goods 11 ppm
Regulatory status	FDA 121.1164	FDA 121.1164; FEMA No. 2461

REFERENCES

For References 1–5, see end of Part III.

6. **Darzens,** *C.R. Hébd. Séances Acad. Sci. Paris,* 144, 330, 1907.
7. **Sörensen and Sörensen,** *Acta Chem. Scand.,* 2, 173, 1948.
8. **Jeffery and Vogel,** *J. Chem. Soc.,* p. 662, 1948.

	Ethyl valerate	Ethyl vanillin
Other names		3-Ethoxy-4-hydroxybenzaldehyde
Empirical formula	$C_7H_{14}O_2$	$C_9H_{10}O_3$
Structure	$CH_3-(CH_2)_3-C(=O)-O-CH_2-CH_3$	Benzene ring with CHO; $O-CH_2-CH_3$; OH
Physical/chemical characteristics[5]		
Appearance	Colorless liquid	Crystalline, white powder
Molecular weight	130.18	166.17
Melting point		77–78 C
Boiling point	144 C at 734 mm Hg; 72–74 C at 65 mm Hg	285 C
Specific gravity	0.8736 at 20 C	
Refractive index	1.4004 at 20 C; 1.3980 at 24 C	
Solubility	Slightly soluble in water; soluble in alcohol and ether	Slightly soluble in water (1.3% at 50°C); soluble in alcohol, ether, glycerin, and chloroform; 1 gm in 3 cc of 95% alcohol or more
Organoleptic characteristics	Fruity odor suggestive of apple	Intense vanilla odor and sweet taste. Its flavoring power is 2–4 times stronger than vanillin. One must be careful not to exceed certain use levels, as this might cause an unpleasant flavor in the product. The product is not stable; in contact with iron or alkali, it exhibits a red color and loses its flavoring power.
Synthesis	By refluxing valeric acid and ethyl alcohol in the presence of concentrated H_2SO_4[6]	From safrole by isomerization to isosafrole and subsequent oxidation to piperonal. The methylene linkage is then broken by heating piperonal in an alcoholic solution of KOH; finally the resulting protocatechualdehyde is reacted with ethyl alcohol. From guaethol by condensation with chloral to yield 3-ethoxy-4-hydroxyphenyl trichloromethyl carbinol. This is then boiled with an alcoholic solution of KOH or NaOH, acidified, and extracted with chloroform to yield ethyl vanillin.[7,8]
Natural occurrence	Reported found in the essential oil of rue	Not reported found in nature. It can be distinguished from vanillin because of the yellow color developed in the presence of concentrated H_2SO_4.
Reported uses[3]	Non-alcoholic beverages, 4.2 ppm; ice cream, ices, etc., 4.4 ppm; candy, 15 ppm; baked goods, 8.3 ppm; gelatins and puddings, 5.5 ppm; chewing gum, 260 ppm	Non-alcoholic beverages, 20 ppm; alcoholic beverages, 100 ppm; ice cream, ices, etc. 47 ppm; candy, 65 ppm; baked goods 63 ppm; gelatins and puddings, 74 ppm, chewing gum, 110 ppm; icings and toppings, 140–200 ppm; chocolate, 250 ppm; imitation vanilla extracts, 28,000 ppm
Regulatory status	FDA 121.1164; FEMA No. 2462	FDA GRAS; FEMA No. 2464

REFERENCES

For References 1–5, see end of Part III.

6. **Vogel,** *J. Chem. Soc.*, p. 630, 1948.
7. **Bedoukian,** *Perfumery: Synthetics and Isolates,* D. Van Nostrand Co., 1951, 418.
8. **Guglielmina,** *Choc. Confiserie Fr.*, 24, May 1965.

	Eucalyptol
Other names	Cajeputol Cineole (1,8-cineole) 1,8-Epoxy-*p*-menthane
Empirical formula	$C_{10}H_{18}O$
Structure	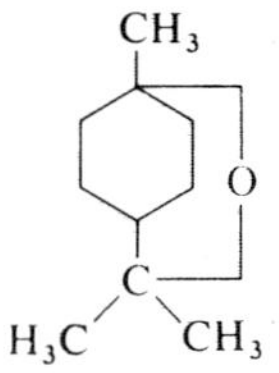 (structure: ring with CH_3 at top; O bridge; C bearing H_3C and CH_3)
Physical/chemical characteristics[5]	
Appearance	Colorless, mobile liquid
Molecular weight	154.24
Boiling point	176–177°C
Congealing point	0 to +1°C
Flash point	47–48°C
Specific gravity	0.9245 (0.928–0.930) at 20°C
Refractive index	1.4574 (1.454) at 20°C
Solubility	Almost insoluble in water; 1:2 in 70% alcohol; soluble in alcohol, ether, and most organic solvents
Organoleptic characteristics	Characteristic camphoraceous odor and fresh, pungent taste
Synthesis	By fractional distillation (170–180°C) from those essential oils containing high levels of eucalyptol, such as *Eucalyptus globulus* (approx 60%), and subsequent separation of the product by congealing the distillate
Natural occurrence	Its name is derived from its presence in the essential oils of *Eucalyptus globulus* and *Melaleuca leucadendron* L. (essential oil of cajeput). It was originally identified in the essential oil of *Artemisia maritima*[6] and subsequently in a large number (approx 270) of other essential oils: rosemary, laurel leaves, clary sage, myrrh, cardamom, star anise, camphor, lavender, peppermint, *Litsea guatemalensis*, *Luvunga scadens* Roxb., *Achillea micrantha*, and *Salvia triloba*. The essential oil of *Eucalyptus polibractea* has been reported to contain up to 91% eucalyptol.[7]
Reported uses[3]	Non-alcoholic beverages 0.13 ppm Ice cream, ices, etc. 0.50 ppm Candy 15 ppm Baked goods 0.50–4.0 ppm Chewing gum 190 ppm
Regulatory status	FDA 121.1164; FEMA No. 2465

REFERENCES

For References 1–5, see end of Part III.

6. **Völckel,** *Annalen,* 87, 312, 1853; 89, 358, 1854.
7. **Berry and Swanson,** *J. Proc. R. Soc. N.S.W.,* 75, 65, 1941.

	Eugenol	Eugenyl acetate
Other names	4-Allylguaiacol 4-Allyl-2-methoxyphenol 1-Hydroxy-2-methoxy-4-allyl benzene	Acetyl eugenol Eugenol acetate
Empirical formula	$C_{10}H_{12}O_2$	$C_{12}H_{14}O_3$
Structure	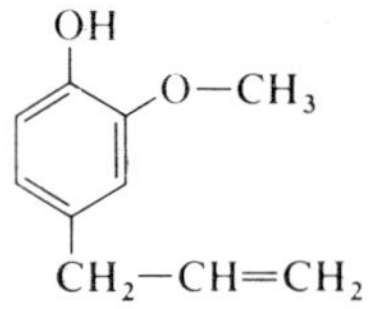 OH; $O-CH_3$; $CH_2-CH=CH_2$	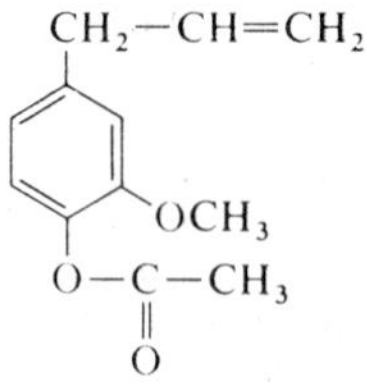 $CH_2-CH=CH_2$; OCH_3; $O-C(=O)-CH_3$
Physical/chemical characteristics[5]		
Appearance	Colorless to slightly yellow liquid; tends to darken on exposure to air	Semisolid mass at room temperature; yellowish liquid when heated
Assay	95–100%	98% min
Molecular weight	164.21	206.24
Melting point	10.3°C	29–30°C
Boiling point	253°C (121°C at 10 mm Hg)	281–282°C at 752 mm Hg
Congealing point		>25°C
Flash point	Approx 104°C	
Specific gravity	1.053–1.064 at 25°C; 1.0664 at 20°C	1.077–1.082 at 25°C; solid—1.0842 at 15°C
Refractive index	1.5380–1.5420 at 20°C; 1.5416 at 19.5°C	1.52069 at 20°C
Optical rotation	−1°30′	
Solubility	1:2 in 60% alcohol; almost insoluble in water; soluble in most organic solvents	1:5 in 70% alcohol; soluble in essential oils and most organic solvents
Organoleptic characteristics	Strong, clove-like odor and spicy, pungent taste	Characteristic odor reminiscent of clove oil, with burning, aromatic flavor
Synthesis	The oil containing eugenol is treated with a 3% aqueous solution of NaOH; the nonacid components are extracted with ether. The alkaline solution is acidified to isolate the phenols and subsequently is fractionally distilled under reduced pressure. To avoid the formation of emulsions, a pretreatment of the oil with tartaric acid is preferred.[6] Eugenol is the starting material in one of the syntheses for the preparation of vanillin.	By direct acetylation of eugenol using acetic anhydride
Natural occurrence	Reported found as a constituent in several volatile oils: clove oil, laurel, and cinnamon leaf oil. Smaller amounts of eugenol also are present in the oil of camphor, Java citronella, California laurel, and acacia flowers. Remarkable amounts of eugenol are found in *Ocimum sanctum* (70%) and *Ocimum gratissimum* (60%). Eugenol is also found in the oil from violet flowers (21%).[7] In some plants eugenol probably occurs as glucoside.[8]	Reported found in the essential oils of *Laurus nobilis* and clove buds and in clove leaves[6]
Reported uses[3]	Non-alcoholic beverages 1.4 ppm Ice cream, ices, etc. 3.1 ppm Candy 32 ppm Baked goods 33 ppm Gelatins and puddings 0.60 ppm Chewing gum 500 ppm Condiments 9.6–100 ppm Meats 40–2,000 ppm	Non-alcoholic beverages 0.43 ppm Ice cream, ices, etc. 3.3 ppm Candy 20 ppm Baked goods 10 ppm Chewing gum 25–100 ppm Condiments 2.0 ppm
Regulatory status	FDA GRAS; FEMA No. 2467	FDA 121.1164; FEMA No. 2469

REFERENCES

For References 1–5, see end of Part III.

6. **Rowaan and Insiger,** *Chem. Weekbl.*, 36, 642, 1939.
7. **Sabetay and Trabaud,** *Compt. Rend.*, 209, 843, 1939.
8. **Bourguelot and Herissey,** *Compt. Rend.*, 140, 870, 1905.
9. **Smith,** *Perfum. Essent. Oil Rec.*, 37, 143, 1946.

	Eugenyl benzoate	Eugenyl formate
Other names	4-Allyl-2-methoxyphenyl benzoate Benzoyl eugenol Eugenol benzoate	4-Allyl-2-methoxyphenyl formate Eugenol formate
Empirical formula	$C_{17}H_{16}O_3$	$C_{11}H_{12}O_3$
Structure	$C_6H_5-C(=O)-O-C_6H_3(OCH_3)-CH_2-CH=CH_2$	$CH_2-CH=CH_2$ on ring; $O-CH_3$; $O-C(=O)-H$
Physical/chemical characteristics[5]		
Appearance	Colorless, crystalline solid	
Assay		94% min
Molecular weight	268.32	192.22
Melting point	69–70 C	
Boiling point	360 C	270°C
Flash point		102°C
Specific gravity		1.1050–1.1090 at 15 C; 1.120 at 15.5 C
Refractive index		1.5240–1.5265 at 20 C
Acid value		0.9 max
Solubility	Insoluble in water; soluble in alcohol and ether	1:2–3.5 in 80% alcohol
Organoleptic characteristics	Balsamic odor with an undertone reminiscent of clove	Odor reminiscent of orris; often used in combination with methylionones to yield orris-like notes
Synthesis		
Natural occurrence	Not reported found in nature	Not reported found in nature
Reported uses[3]	Non-alcoholic beverages 0.03–0.13 ppm Ice cream, ices, etc. 0.25–2.0 ppm Candy 0.25–10 ppm Baked goods 0.13–10 ppm	Condiments 0.20 ppm
Regulatory status	FDA 121.1164; FEMA No. 2471	FDA 121.1164; FEMA No. 2473

REFERENCES

For References 1–5, see end of Part III.

	Eugenyl methyl ether
Other names	4-Allylveratrole 1,2-Dimethoxy-4-allylbenzene Methyl eugenol Methyl eugenyl ether
Empirical formula	$C_{11}H_{14}O_2$
Structure	O—CH_3, O—CH_3 (benzene ring), CH_2—CH=CH_2
Physical/chemical characteristics[5]	
Appearance	Colorless to pale-yellow liquid
Molecular weight	178.23
Boiling point	244–245°C
Specific gravity	1.032–1.036 at 25°/25°C;[2] 1.055 at 15°C
Refractive index	1.5320–1.5360 at 20°C[2]
Eugenol content	Not more than 1%[2]
Solubility	1:4 in 60% alcohol;[2] 1:2 in 70% alcohol; almost insoluble in water. Forms azeotropic mixtures with ethylene glycol (31.5%), eugenol (55%), benzoic acid (11%), and others.
Organoleptic characteristics	Delicate clove-carnation odor with a bitter, burning taste
Synthesis	Usually prepared by methylation of eugenol
Natural occurrence	Reported in the essential oils of *Myrtaceae* and *Lauraceae*. It was identified originally in the essential oil from roots of *Asarum europaeum* L. and *Asarum canadense* L.;[6] subsequently it was identified as the main constituent of the oil from wood of *Dacrydium franklinii* Hook (97.5%), in *Melaleuca bracteata* F.v.M. (leaves, 90–95%), in *Cinnamomum oliveri* Bail. (leaves, 90–95%), and as a minor constituent in the oils of: betel,[7] citronella,[8] Japanese calamus,[9] pimenta, hyacinth, rose, basil, bay, cajeput, and others.
Reported uses[3]	Non-alcoholic beverages 10 ppm Ice cream, ices, etc. 4.8 ppm Candy 11 ppm Baked goods 13 ppm Jellies 52 ppm
Regulatory status	FDA 121.1164; FEMA No. 2475

REFERENCES

For References 1–5, see end of Part III.

6. **Peterson,** *Ber. Dtsch. Chem. Ges.*, 21, 1057, 1888.
7. *Schimmel Ber.*, p. 15, 1907.
8. Schimmel and Co., *Chem. Zentralbl.*, 2, 879, 1889.
9. **Asahina,** *Apoth. Ztg.*, 21, 987, 1906.

Farnesol

Other names	2,6,10-Trimethyl-2,6,10-dodecatrien-12-ol 3,7,11-Trimethyl-2,6,10-dodecatrien-1-ol
Empirical formula	$C_{15}H_{26}O$
Structure	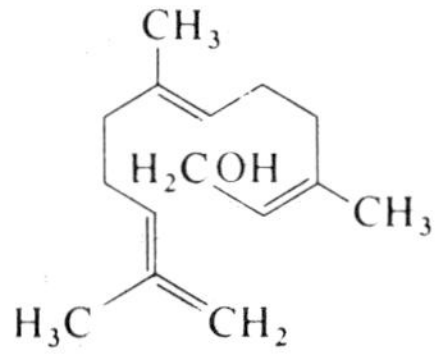
Physical/chemical characteristics[5]	The physical constants vary slightly, depending on the source and the method of preparation.
Appearance	Colorless liquid
Assay	97% min
Molecular weight	222.36
Boiling point	131–132°C at 3 mm Hg (from petitgrain) 124–125°C at 2.2 mm Hg (from Tolu balsam) 118–120°C at 2 mm Hg (from cabreuva) 118–120°C at 2 mm Hg (from geranyl acetone)
Specific gravity	0.887–0.889 at 25°C (0.8880 from petitgrain) (0.8883 at 20°C from Tolu balsam) (0.8886 from cabreuva) (0.8886 from geranyl acetone)
Refractive index	1.4890–1.4910 at 20°C (1.4891 from petitgrain) (1.4892 at 20°C from Tolu balsam) (1.4894 from cabreuva) (1.48906 from geranyl acetone)
Solubility	1:3 in 70% alcohol; almost insoluble in water; soluble in most organic solvents
Organoleptic characteristics	Characteristic flowery odor
Synthesis	One method uses cabreuva as the starting material (Swiss Patent 261,120—Givaudan & Co.), while a second method starts from ambrette seeds (German Patent 149,603—Haarmann & Reimer).
Natural occurrence	For structure see references 6, 7, and 8. The presence of this terpene alcohol in nature has been reported in more than 30 essential oils. The levels are generally low (0.5—1.0%) with the exception of cabreuva, which contains up to 2.5% farnesol,[9] and the distillate from flowers of *Oxystigma buccholtzii* Harms., which contains up to 18% farnesol.[10] Among the essential oils containing farnesol are lemon grass, Ceylon citronella, cananga, ambrette seeds, ylang-ylang, *Acacia farnesiana*, Peru balsam, palmarosa, tuberose, and others.
Reported uses[3]	Non-alcoholic beverages 0.76 ppm Ice cream, ices, etc. 0.40 ppm Candy 1.4 ppm Baked goods 1.7 ppm Gelatins and puddings 0.10 ppm
Regulatory status	FDA 121.1164; FEMA No. 2478

REFERENCES

For References 1–5, see end of Part III.

6. **Kirschbaum,** *Ber. Dtsch. Chem. Ges.*, 46, 1732, 1913.
7. **Ružička,** *Experientia* (Basel), 9, 357, 1953.
8. **Henfrickson,** *Tetrahedron*, 7, 82, 1959.
9. **Naves,** *Helv. Chim. Acta*, 30, 275, 1947.
10. **Salgues,** *Miltizer Ber.*, p. 112, 1957.

	d-Fenchone	Fenchyl alcohol
Other names	*d*-2-Fenchanone Fenchone 1,3,3,-Trimethylbicyclo-1,2,2-heptanone-2 *d*-1,3,3-Trimethyl-2-norbornanone	2-Fenchanol Fenchol 1,3,3-Trimethyl-2-norbornanol
Empirical formula	$C_{10}H_{16}O$	$C_{10}H_{18}O$
Structure		α-form; β-form
Physical/chemical characteristics[5]		
Appearance	Colorless, oily liquid; solidifies when cold	Colorless to straw-yellow crystals for the α-form; β-forms are liquids at room temperature
Molecular weight	152.23	154.24
Melting point	*dl*, > 0°C; *d,l*, 5–6°C	*d*, 48°C; *l*, 49°C
Boiling point	193°C	
Specific gravity	*d*, 0.948 at 18°C; *l*, 0.948 at 20°C; 0.9501 at 15°C	
Refractive index	*d*, 1.46355 at 18°C; *l*, 1.4636 at 20°C; 1.4702 at 20°C	
Optical rotation	*d*, +66°9′ at 20°C; *l*, −66°9′ (−58°57′) at 23°C from ethanol	
Solubility	Very soluble in alcohol and ether; insoluble in water	Very slightly soluble in water; soluble in alcohol
Organoleptic characteristics	Camphor-like odor, powerful and sweet; warm, somewhat bitter, burning taste	Camphor-like odor with citrus notes; bitter, lime-like flavor
Synthesis	Isolated from cedarleaf oil (thuja oil)	By reduction of fenchone or from pine terpenes
Natural occurrence	Reported found in several essential oils: *Thuja plicata*, *Thuja occidentalis*, *Thuja standishii*, Russian anise, fennel, a few *Artemisia* varieties (*A. frigida*, *A. verlotorum*, *A. santolinaefolia*), *Lavanda stoechas*, *Lavanda burmannii*, and others. The highest levels are found in fennel oil (12–19%). Two optically active forms have been isolated.[6] For structural studies see reference 7.	Both the α- and the β-fenchyl alcohols are reported found in nature. In mixture they are present in the essential oils of: *Picea alba*, *Pinus silvestris*, pine oil, *Artemisia santolinaefolia*, and others. For structural studies and additional information, see references 8 and 9. The isofenchyl alcohol (1,5,5-trimethyl-2-norbornanol) is not found in nature.
Reported uses[3]	Non-alcoholic beverages 0.13–0.80 ppm Alcoholic beverages 5.0 ppm Ice cream, ices, etc. 0.25 ppm Candy 0.25–30 ppm Baked goods 0.25 ppm	Non-alcoholic beverages 1.8 ppm Ice cream, ices, etc. 0.25 ppm Candy 4.7 ppm Baked goods 0.25 ppm
Regulatory status	FDA 121.1164; FEMA No. 2479	FDA 121.1164; FEMA No. 2480

REFERENCES

For References 1–5, see end of Part III.

6. **Wallach,** *Annalen*, 263, 129, 1891; 272, 102, 1892.
7. **Jacob, Owisson and Rassat,** *Bull. Soc. Chim. Fr.*, p. 1374, 1959.
8. **Treibs,** *Annalen*, 556, 10, 1944.
9. **Gildemeister and Hoffmann,** *Die Aetherischen Oele*, Vol. 3b, Akademie Verlag, 1962, 209.

	Fenchyl acetate
Other names	1,3,3-Trimethyl-2-norbornanyl acetate
Empirical formula	$C_{12}H_{20}O_2$
Structure	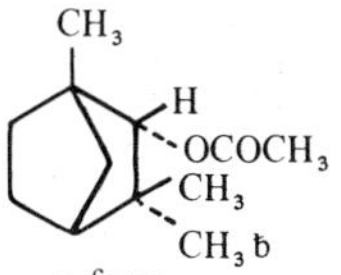 α-form 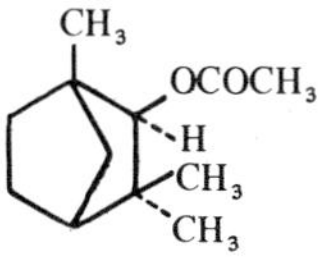 β-form
Physical/chemical characteristics[5]	
Appearance	Colorless, mobile liquid
Molecular weight	196.29
Boiling point	*d*-α, 87°C at 10 mm Hg *dl*-α, 79°C at 5 mm Hg
Specific gravity	*dl*-α, 09744 at 15°C *dl*-β, 0.9073 at 15°C
Refractive index	*dl*-α, 1.4560 at 20°C *dl*-β, 1.3577 at 20°C
Optical rotation	*d*-α, +69.5° at 20°C (undiluted) *l*-α, −68.7°C at 20°C *l*-β, +39.6° at 20°C
Solubility	Insoluble in water; soluble in alcohol and oils; slightly soluble in propylene glycol
Organoleptic characteristics	Mild, sweet odor reminiscent of fir needles oil
Synthesis	By esterification of fenchyl alcohol with acetic anhydride in pyridine[6-10]
Natural occurrence	Reported found in the oil from leaves and terminal branches of *Juniperus rigida*;[11] in *Seseli sibiricum*;[12,13] in rosemary oil;[14] in fennel oil;[15] in the oil of hinoki leaves[16]
Reported uses[3]	Non-alcoholic beverages 0.2 ppm Ice cream, ices, etc. 0.2 ppm Candy 0.3 ppm Baked goods 0.3 ppm Gelatins and puddings 0.2 ppm Preserves and spreads 0.2 ppm
Regulatory status	FEMA No. 3390

REFERENCES

For References 1–5, see end of Part III.

6. **Hirsjarvi,** *Ann. Acad. Sci. Fenn. Ser. A (II)*, 84, 10, 1957.
7. **Toivonen,** *Suom Kemistil.*, 24B, 62, 1951.
8. **Zeitschel and Todenhöfer,** *J. Prakt. Chem.*, 133, 374, 1932.
9. **Toivonen,** *Suom Kemistil.*, 24B, 62, 1951.
10. **Schmidt and Schultz,** *Schimmel Ber.*, p. 97, 1935.
11. **Nishimura and Hirose,** *Chem. Abstr.*, 59, 4971, 1963.
12. **Handa, Smith and Levi,** *Perfum. Rec.*, 53, 607, 1962.
13. *Riechst. Aromen,* 13, 181, 1963.
14. **Schwenker and Klöhn,** *Arch. Pharm.*, 296, 844, 1963.
15. **Dorronsoro,** *Mem. R. Acad. Cienc. Exactas Fis. Nat. Madr.*, p. 29, 1919.
16. **Thomas,** *Perfum. Rec.*, 56, 301, 1965.

	Formic acid	2-Formyl benzofuran
Other names		2-Benzofuran carboxaldehyde
Empirical formula	CH_2O_2	$C_9H_6O_2$
Structure	H—C(=O)—OH	(benzofuran ring)—CHO
Physical/chemical characteristics[5]		
Appearance	Colorless, fuming liquid	
Molecular weight	46	146.15
Melting point	8.3°C	
Boiling point	100–101°C	130–131°C at 13 mm Hg
Specific gravity	1.2196 at 20°C	
Refractive index	1.3714	
Solubility	Completely miscible with water, alcohol, ether, and glycerol	
Organoleptic characteristics	Pungent, penetrating odor	
Synthesis	From anhydrous sodium formate and concentrated H_2SO_4 at low temperature followed by distillation	By the Gattermann process;[12] also from coumaryloyl cyanide[13]
Natural occurrence	Widespread in a large variety of plants. Reported identified in *Cistus labdanum*[6] and the oil of *Artemisia transiliensis*;[7] also found among the constituents of petitgrain lemon[8] and bitter orange essential oil;[9] reported found in strawberry aroma.[10]	
Reported uses[3]	Non-alcoholic beverages 1.0 ppm Ice cream, ices, etc. 5.0 ppm Candy 5.0–18 ppm Baked goods 5.0–6.1 ppm	Ice cream, ices, etc. 10.0 ppm Candy 10.0 ppm Baked goods 20.0 ppm Gelatins and puddings 10.0 ppm
Regulatory status	FDA 121.1164; FEMA No. 2487	FEMA No. 3128

REFERENCES

For References 1–5, see end of Part III.

6. Sabetay and Trabaud, *Chem. Zentralbl.*, 1, 133, 1951.
7. Goryaev and Pugachev, *Chem. Abstr.*, 50, 1684, 1956.
8. Peyron, *Riv. Ital. EPPOS*, 8, 413, 1965.
9. Mehlitz and Minas, *Ind. Obst Gemueseverwert.*, 50, 861, 1965.
10. Willhalm, Pauly and Winter, *Helv. Chim. Acta*, 49, 66, 1966.
11. Wassermann, *Z. Phys. Chem.*, 146, 428, 1933.
12. Birch and Robertson, *J. Chem. Soc.*, p. 306, 1938.
13. Reichstein and Reichstein, *Helv. Chim. Acta*, 13, 1275, 1930.

	2-Formyl-6,6-dimethyl bicyclo [3.1.1] hept-2-ene
Other names	Myrtenal Benihinal 2-Formyl-6,6-dimethyl-2-norpinene 6,6-Dimethyl-2-norpinene-2-aldehyde
Empirical formula	$C_{10}H_{14}O$
Structure	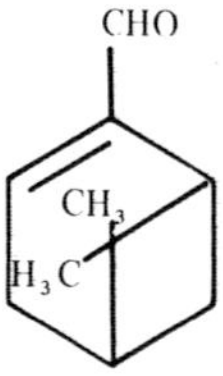 CHO, CH_3, H_3C
Physical/chemical characteristics[5]	
Appearance	Colorless liquid; readily polymerizes in the presence of air
Molecular weight	150.22
Boiling point	199°C 99–100°C at 15 mm Hg
Specific gravity	0.9872 at 20°C
Refractive index	1.5030 at 20°C
Optical rotation	+14.75° at 20°C
Solubility	Insoluble in water; soluble in alcohol and in oils
Organoleptic characteristics	Refreshing, spicy-herbaceous odor of bay leaf, reminiscent of cinnamon[14,15]
Synthesis	From myrtenol by oxidation with chromic acid; by isolation from higher-than-cineole boiling fractions in the distillation of eucalyptus oil[13] 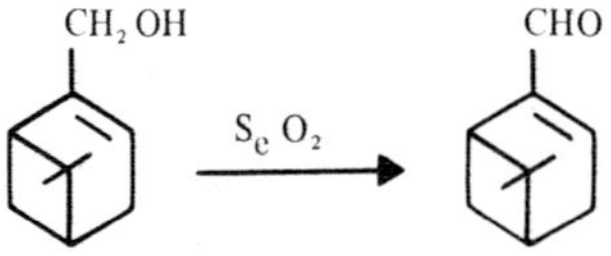 CH_2OH —S_eO_2→ CHO
Natural occurrence	Reported found in the oils from leaves of *Eucalyptus globulus*[6] and *Callytrix tetragona;*[7] in the oils of *Hernandia peltata*[8] and *Perovskia abrotanoides;*[9] in the oils from wood of *Chamaecyparis formonensis*[10] and *Chamaecyparis obtusa;*[11] also in cumin seeds and in mint[12]
Reported uses[3]	Non-alcoholic beverages 0.5 ppm Candy 2.0 ppm
Regulatory status	FEMA No. 3395

REFERENCES

For References 1–5, see end of Part III.

6. **Schmidt,** *Chem. Ber.,* 62, 2947, 1929; 80, 528, 1947.
7. **Gildemeister and Hoffmann,** *Die Aetherischen Oele,* Vol. 3, Akademie Verlag, 1960, 117.
8. **Semmler et al.,** *Chem. Ber.,* 44, 815, 1911.
9. **Goryaev and Sekebaeva,** *Izv. Akad. Nauk Kaz. S.S.R. Ser. Khim.,* (1), 107, 1961; *Chem. Abstr.,* 55, 22718f, 1961; **Goryaev and Sharipova,** *Izv. Akad. Nauk Kaz. S.S.R. Ser. Khim.,* (1), 112, 1961; *Chem. Abstr.,* 55, 22718e, 1961.
10. **Kafuku et al.,** *Bull. Chem. Soc. Jap.,* 8, 371, 1933.
11. **Katsura,** *J. Chem. Soc. Jap.,* 63, 1483, 1942.
12. *Weurman Report,* 3rd ed.
13. **Rodd,** *Chemistry of Carbon Compounds,* Vol. 2, Parts A and B, American Elsevier, New York, 1953, 569.
14. *Soap Perfum. Cosmet.,* 24, 806, 1951.
15. **Arctander,** *Perfume and Flavor Chemicals,* Montclair, N.J., 1969, No. 2293.

	2-Furanmethanethiol formate	Fumaric acid
Other names		Allomalenic acid Boletic acid *trans*-Butenedioic acid *trans*-1,2-Ethylenedicarboxylic acid
Empirical formula	$C_6H_6O_2S$	$C_4H_4O_4$
Structure	(furan ring, O)—CH_2—S—C(=O)—H	HO—C(=O)—CH=HC—C(=O)—OH
Physical/chemical characteristics		
Appearance		Colorless, crystalline solid; sublimes at 200°C
Molecular weight	142.17	116.07
Melting point		300–302°C[11] (287°C)
Specific gravity		1.625
Solubility		Slightly soluble in water and most organic solvents
Organoleptic characteristics		Used as a replacement for tartaric acid; odorless, tart, acidic-sour flavor
Synthesis		By the action of certain fungi (*Rhizopus nigricans*) on glucose; by oxidation of furfural with sodium chlorate in the presence of vanadium pentoxide
Natural occurrence		Reported found in several plants: *Fumaria officinalis* L., *Boletus scaber* Boll., and others
Reported uses[3]	Non-alcoholic beverages 1.0 ppm Ice cream, ices, etc. 1.0 ppm Candy 1.0 ppm	Non-alcoholic beverages 50 ppm Baked goods 1,300 ppm Gelatins and puddings 3,600 ppm
Regulatory status	FEMA No. 3158	FDA 121.1130; FEMA No. 2488

REFERENCES

For References 1–5, see end of Part III.

	Furfural	2-Furfuryl disulfide
Other names	2-Furaldehyde α-Furfuraldehyde Pyromucic aldehyde	2,2′-(Dithiodimethylene)difuran
Empirical formula	$C_5H_4O_2$	$C_{10}H_{10}O_2S_2$
Structure	(furan ring)–CHO	(furan ring)–CH_2 —S —S — CH_2–(furan ring)
Physical/chemical characteristics[5]		
Appearance	Colorless liquid when freshly prepared; tends to darken on aging	Pale-yellow, oily liquid
Molecular weight	96.08	226.32
Melting point	−36.5 C	10°C
Boiling point	161.7 C (72 C at 25 mm Hg)	112–113°C at 0.5 mm Hg
Flash point	65 C	
Specific gravity	1.1544 at 25 C; 1.1598 at 10 C	
Refractive index	1.52345 at 25 C; 1.5260 at 20 C	
Solubility	Rather soluble (1:1) in water; soluble in alcohol, ether, and most common organic solvents	Slightly soluble in water; very soluble in alcohol and in organic solvents
Organoleptic characteristics	Characteristic penetrating odor	Repulsive mercaptan-like odor; "roasted" at high dilution
Synthesis	Industrially prepared from pentosans that are contained in cereal straws and brans. These materials are previously digested with diluted H_2SO_4, and the formed furfural steam is distilled.	From furfural and hydrogen sulfide[13]
Natural occurrence	Reported found in several essential oils from plants of the *Pinaceae* family, in the essential oil from Cajenne linaloe,[6] in the oil from leaves of *Trifolium pratense*[7] and *Trifolium incarnatum*,[8] in the distillation waters of several essential oils, such as ambrette[9] and angelica seeds[10], in Ceylon cinnamon essential oil,[11] in petitgrain oil,[12] ylang-ylang, lavender, lemongrass, calamus, eucalyptus, neroli, sandalwood, tobacco leaves, and others. Also reported found in rum and toasted coffee.	
Reported uses[3]	Non-alcoholic beverages, 4.0 ppm; alcoholic beverages, 10 ppm; ice cream, ices, etc., 13 ppm; candy, 12 ppm; baked goods, 17 ppm; gelatins and puddings, 0.80 ppm; chewing gum, 45 ppm; syrups, 30 ppm	Non-alcoholic beverages 1.0 ppm Ice cream, ices, etc. 1.0 ppm Candy 1.0 ppm Baked goods 1.0 ppm Gelatins and puddings 1.0 ppm Chewing gum 1.0 ppm Condiments 1.0 ppm Meat, meat sauces, soups 1.0 ppm Milk, dairy products 1.0 ppm
Regulatory status	FDA status not fully defined; FEMA No. 2489	FEMA No. 3146

REFERENCES

For References 1–5, see end of Part III.

6. Schimmel and Co., *Chem. Zentralbl.*, 1, 1717, 1912.
7. **Power and Salway,** *J. Chem. Soc.* (London), 97, 232, 1910.
8. **Rogerson,** *J. Chem. Soc.* (London), 97, 1005, 1910.
9. Schimmel and Co., *Chem. Zentralbl.*, 2, 880, 1899.
10. Schimmel and Co., *Chem. Zentralbl.*, 1, 1972, 1913.
11. Schimmel and Co., *Chem. Zentralbl.*, 1, 1059, 1902; 2, 1486, 1902.
12. Schimmel and Co., *Chem. Zentralbl.*, 2, 1208, 1902.
13. **Gilman and Hewlett,** *J. Am. Chem. Soc.*, 52, 2141, 1930.

	Furfuryl isopropyl sulfide		Furfuryl isovalerate	
Other names			Furfuryl-2-methylbutanoate	
Empirical formula	$C_8H_{12}OS$		$C_{10}H_{14}O_3$	
Structure	(furan-2-yl)–CH₂–S–CH(CH₃)₂		(furan-2-yl)–CH₂–O–C(=O)–CH₂–CH(CH₃)₂	
Physical/chemical characteristics				
Molecular weight	156.25		182.21	
Boiling point	84°C at 16 mm Hg		97–98°C at 11 mm Hg	
Synthesis			By esterification of furfuryl alcohol with isovaleric chloride[6]	
Natural occurrence			Reported found as a constituent in coffee aroma[6]	
Reported uses[3]	Non-alcoholic beverages	0.5 ppm	Non-alcoholic beverages	5.0 ppm
	Ice cream, ices, etc.	0.5 ppm	Ice cream, ices, etc.	5.0 ppm
	Candy	0.5 ppm	Candy	10.0 ppm
	Baked goods	0.5 ppm	Gelatins and puddings	5.0 ppm
Regulatory status	FEMA No. 3161		FEMA No. 3283	

REFERENCES

For References 1–5, see end of Part III.
6. **Stoll et al.,** *Helv. Chim. Acta,* 50, 628, 1967.

	Furfuryl mercaptan	Furfuryl methyl ether
Other names	2-Furanmethanethiol 2-Furyl methanethiol	
Empirical formula	C_5H_6OS	$C_6H_8O_2$
Structure	(furan ring)–CH_2SH	(furan ring)–CH_2-O-CH_3
Physical/chemical characteristics[5]		
Appearance	Oily liquid	
Molecular weight	130.17	112.14
Boiling point	155°C (47°C at 12 mm Hg)	134–135°C 108–109° at 750 mm Hg
Specific gravity	1.1319 at 20°C	1.0163 at 20°C 1.046 at 25°C
Refractive index	1.5329 at 20°C	1.4570 at 20°C
Solubility	Insoluble in water; soluble in most organic solvents and diluted alkali solutions	Insoluble in water; soluble in alcohol; very soluble in ether
Organoleptic characteristics	Characteristic unpleasant odor	
Synthesis	Prepared by reacting thiourea and furfuryl chloride with subsequent hydrolysis of the reaction product;[6] also by reduction of difurfuryl disulfide in alcoholic solution using zinc dust and a small amount of acetic acid or using activated alumina[7]	From furfuryl chloride and potassium methoxide in ether solution[8]
Natural occurrence	Reported found in nature. Tends to polymerize when heated in the presence of mineral acids	Reported found as a constituent in the aroma of roasted coffee[9]
Reported uses[3]	Non-alcoholic beverages, 0.52 ppm; ice cream, ices, etc., 0.78 ppm; candy, 2.0 ppm; baked goods, 2.1 ppm; gelatins and puddings, 0.10 ppm; icings, 0.50 ppm	Non-alcoholic beverages 2.0 ppm Candy 2.0 ppm Gelatins and puddings 2.0 ppm Sauces 2.0 ppm
Regulatory status	FDA status not fully defined; FEMA No. 2493	FEMA No. 3159

REFERENCES

For References 1–5, see end of Part III.

6. **Kirner and Richter,** *J. Am. Chem. Soc.*, 51, 3134, 1929.
7. German Patent 484,244.
8. **Kirner,** *J. Am. Chem. Soc.*, 50, 1960, 1928.
9. **Stoll, Winter, Gautschi, Flament and Willhalm,** *Helv. Chim. Acta,* 50, 628, 1967.

	Furfuryl methyl sulfide	α-Furfuryl octanoate
Other names		α-Furfuryl caprylate
Empirical formula	C_6H_8OS	$C_{13}H_{20}O_3$
Structure	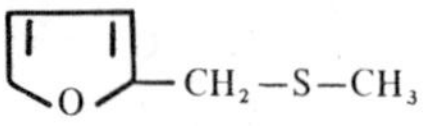 (furan-2-yl)–CH_2–S–CH_3	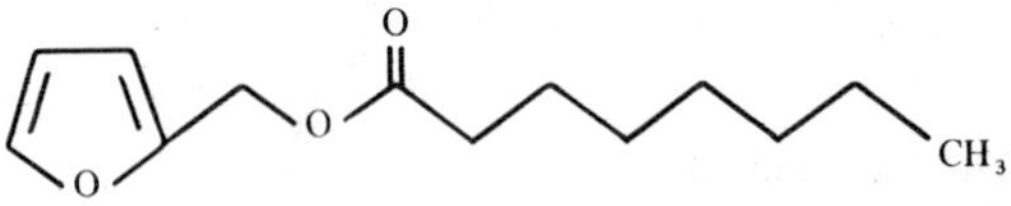 (furan-2-yl)–CH_2–O–C(=O)–$(CH_2)_6$–CH_3
Physical/chemical characteristics		
Molecular weight	128.11	224.29
Boiling point	62–63°C at 28 mm Hg	
Refractive index	1.5210 at 22.5°C	
Synthesis	From 2-furfuryl mercaptan and dimethyl sulfate[6]	Furfuryl alcohol (furan-2-yl–CH_2OH) + HO–C(=O)–$(CH_2)_6$–CH_3 →
Natural occurrence	Reported found as a minor constituent in coffee aroma	
Reported uses[3]	Non-alcoholic beverages 1.0 ppm Ice cream, ices, etc. 1.0 ppm Candy 1.0 ppm Gelatins and puddings 1.0 ppm	Baked goods 4.0 ppm Gelatins and puddings 2.0 ppm Meat, meat sauces, soups 2.0 ppm Margarine 2.0 ppm Snacks 10.0 ppm
Regulatory status	FEMA No. 3160	FEMA No. 3396

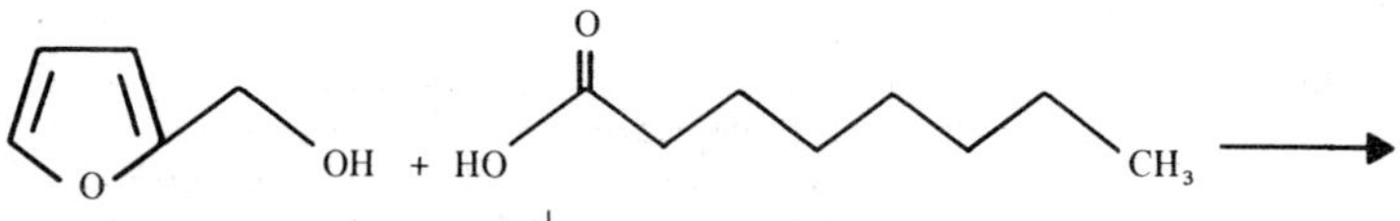

REFERENCES

For References 1–5, see end of Part III.

6. Gianturco et al., *Tetrahedron Lett.*, 20, 2951,1964.

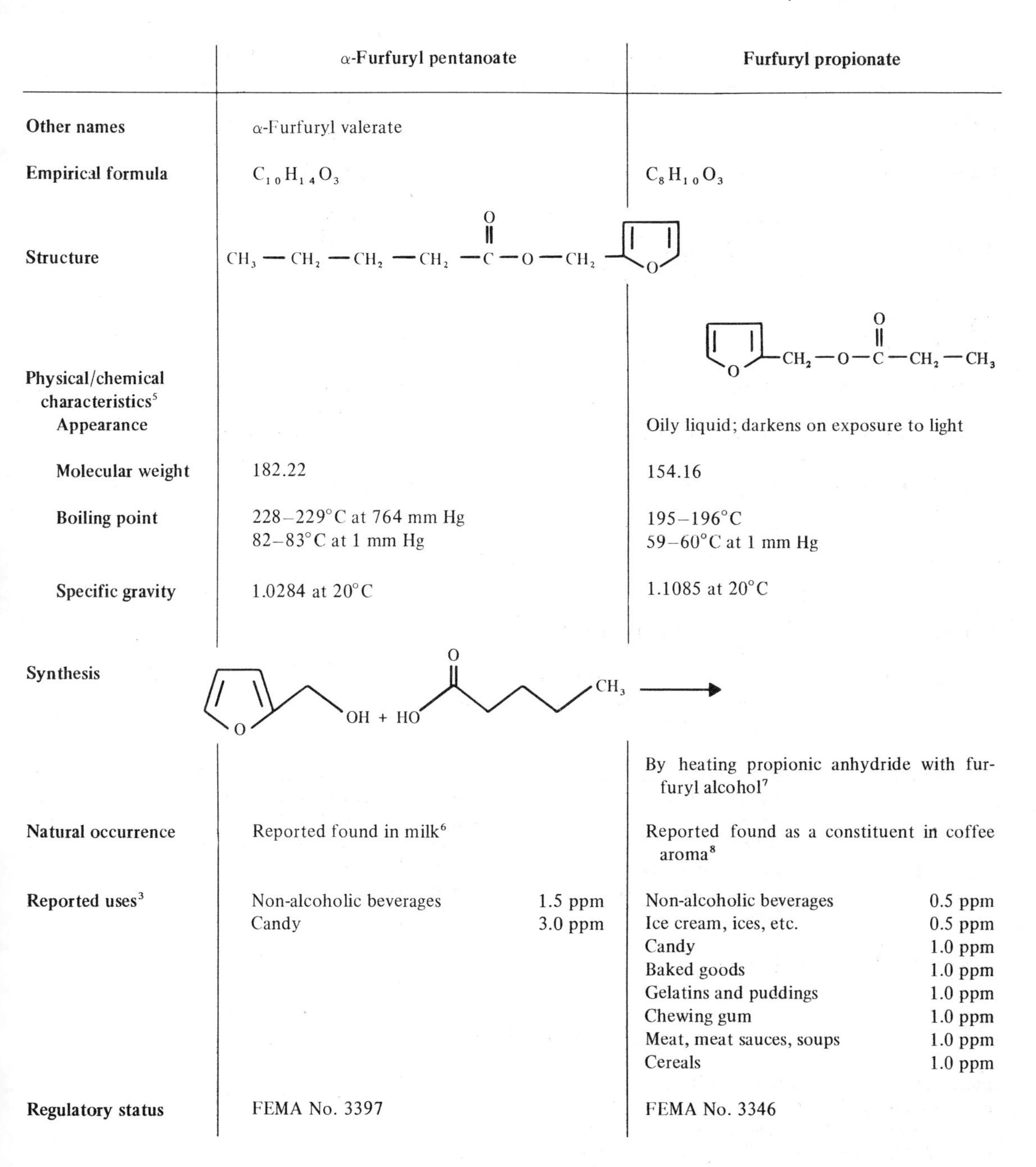

	α-Furfuryl pentanoate	Furfuryl propionate
Other names	α-Furfuryl valerate	
Empirical formula	$C_{10}H_{14}O_3$	$C_8H_{10}O_3$
Structure	$CH_3—CH_2—CH_2—CH_2—C(=O)—O—CH_2$–(2-furyl)	(2-furyl)–$CH_2—O—C(=O)—CH_2—CH_3$
Physical/chemical characteristics[5]		
Appearance		Oily liquid; darkens on exposure to light
Molecular weight	182.22	154.16
Boiling point	228–229°C at 764 mm Hg 82–83°C at 1 mm Hg	195–196°C 59–60°C at 1 mm Hg
Specific gravity	1.0284 at 20°C	1.1085 at 20°C
Synthesis	Furfuryl alcohol (OH) + HO–C(=O)–$CH_2CH_2CH_2CH_3$ →	By heating propionic anhydride with furfuryl alcohol[7]
Natural occurrence	Reported found in milk[6]	Reported found as a constituent in coffee aroma[8]
Reported uses[3]	Non-alcoholic beverages 1.5 ppm Candy 3.0 ppm	Non-alcoholic beverages 0.5 ppm Ice cream, ices, etc. 0.5 ppm Candy 1.0 ppm Baked goods 1.0 ppm Gelatins and puddings 1.0 ppm Chewing gum 1.0 ppm Meat, meat sauces, soups 1.0 ppm Cereals 1.0 ppm
Regulatory status	FEMA No. 3397	FEMA No. 3346

REFERENCES

For References 1–5, see end of Part III.

6. **Kinlin et al.,** *J. Agric. Food Chem.*, 20, 1021, 1972.
7. **Zanetti,** *J. Am. Chem. Soc.*, 47, 535, 1925.
8. **Stoll et al.,** *Helv. Chim. Acta*, 50, 628, 1967.

	N-Furfuryl pyrrole	Furfuryl thioacetate
Other names	1-(2-furfuryl)pyrrole	Furfuryl thiol acetate
Empirical formula	C_9H_9ON	$C_7H_8O_2S$
Structure	(furan ring)–CH_2–N(pyrrole ring)	(furan ring)–CH_2–S–OCH_3
Physical/chemical characteristics[5]		
Molecular weight	147.08	156.20
Boiling point	76–78°C at 1 mm Hg	90–92°C at 12 mm Hg
Refractive index	1.5317 at 21°C	
Solubility	Soluble in most common organic solvents	
Synthesis	From furfuryl amino mucate[6]	
Natural occurrence	Reported found as a constituent in coffee aroma;[7] a volatile flavor component in roasted filberts[8]	Reported found as a constituent in coffee[9]
Reported uses[3]	Non-alcoholic beverages 2.0 ppm Ice cream, ices, etc. 2.0 ppm Candy 2.0 ppm Baked goods 2.0 ppm Gelatins and puddings 2.0 ppm Milk, dairy products 2.0 ppm Cereals 2.0 ppm	Non-alcoholic beverages 0.20 ppm Ice cream, ices, etc. 1.0 ppm Candy 1.5 ppm Gelatins and puddings 1.0 ppm
Regulatory status	FEMA No. 3284	FEMA No. 3162

REFERENCES

For References 1–5, see end of Part III.

6. **Reichstein,** *Helv. Chim. Acta,* 13, 349, 1930.
7. **Reichstein and Standinger,** *Angew. Chem.,* 62, 212, 1950.
8. **Sheldon et al.,** *J. Food Sci.,* 37(2), 313, 1972.
9. **Stoll et al.,** *Helv. Chim. Acta,* 50(2), 628, 1967.

	Furfuryl thiopropionate		(2-Furyl)-2-propanone	
Other names			Furfuryl methyl ketone Furyl acetone Methyl furfuryl ketone	
Empirical formula	$C_8H_{10}O_2S$		$C_7H_8O_2$	
Structure	$CH_3-CH_2-C(=S)-O-CH_2-$(2-furyl)		(2-furyl)$-CH_2-C(=O)-CH_3$	
Physical/chemical characteristics				
Appearance			Liquid	
Molecular weight	170.23		124.14	
Boiling point	95–97°C at 10 mm Hg		179–180°C	
Solubility			Soluble in diluted HCl and most organic solvents	
Organoleptic characteristics			Odor and taste suggestive of radish	
Synthesis			By heating the α-methyl-β-(α-furyl) glycidic acid ethyl ester in aqueous NaOH[6]	
Natural occurrence			Reported found in roasted coffee	
Reported uses[3]	Non-alcoholic beverages	1.0 ppm	Ice cream, ices, etc.	5.0 ppm
	Ice cream, ices, etc.	1.0 ppm	Candy	3.8–20 ppm
	Candy	1.0 ppm	Baked goods	2.0–20 ppm
	Baked goods	1.0 ppm		
Regulatory status	FEMA No. 3347		FDA 121.1164; FEMA No. 2496	

REFERENCES

For References 1–5, see end of Part III.

6. **Darzens,** *Compt. Rend.,* 142, 215, 1906.

	Fusel oil, refined
Other names	Amyl alcohol, commercial (Not well defined)
Structure	A volatile, oily mixture consisting largely of amyl alcohols. Isoamyl alcohol and active amyl alcohol (2-methyl-1-butanol) are the chief constituents. Acids, esters, and aldehydes are also present.
Physical/chemical characteristics[5]	This specification refers to the rectified material.
Appearance	Water-white liquid
Molecular weight	
Boiling point	Distillation range: $<110°C$—none $<120°C$—$>15\%$ $<130°C$—$>60\%$ $>135°C$—none
Flash point	123°F
Specific gravity	0.811–0.815 at 20°/20°C
Solubility	Soluble in water, alcohol, and ether
Organoleptic characteristics	Disagreeable odor. An appreciable percentage of fusel oil in a liquor has an adverse effect, while trace amounts are necessary for the characteristic flavor.
Synthesis	As a by-product in the fermentation of starch- or sugar-containing materials (potatoes, grapes, beetroots, grain); purified by rectification
Reported uses[3]	Non-alcoholic beverages 21 ppm Alcoholic beverages 2.5 ppm Ice cream, ices, etc. 4.1 ppm Candy 30 ppm Baked goods 34 ppm Gelatins and puddings 4.0 ppm Chewing gum 270 ppm
Regulatory status	FDA 121.1164; FEMA No. 2497

REFERENCES

For References 1–5, see end of Part III.

Geraniol

	Geraniol
Other names	2-*trans*-3,7-Dimethyl-2,6-octadien-1-ol
Empirical formula	$C_{10}H_{18}O$
Structure	CH_3, CH_2OH, H_3C, CH_3
Physical/chemical characteristics[5]	The physical constants vary for the various commercial products, depending on the total geraniol content. Specific gravity and refractive index may be indicative of the purity of the product.
Appearance	Colorless liquid
Assay	88% min.[2]
Molecular weight	154.24
Boiling point	110–111°C at 10 mm Hg
Flash point	101°C
Specific gravity	0.870–0.885 at 25°/25°C;[2] 0.8793 at 20°C
Refractive index	1.4690–1.4780 at 20°C;[2] 1.47773 at 20°C
Optical rotation	−2° to +2°[2] (0°)
Ester content	Not more than 1%[2] (as geranyl acetate)
Aldehyde content	Not more than 1%[2] (as citronellol)
Solubility	1:2 in 70% alcohol;[2] almost insoluble in water; soluble in most common organic solvents
Organoleptic characteristics	Characteristic rose-like odor
Synthesis	By fractional distillation from those essential oils rich in geraniol, or synthetically from myrcene. Commercial geraniol cannot be classified according to its alcohol content, as most of the recurring impurities are alcoholic in nature (nerol, citronellol, tetrahydrogeraniol). Gas-chromatography techniques may be usefully employed to determine the geraniol content in a product.
Natural occurrence	The presence of geraniol in nature has been reported in more than 160 essential oils: ginger grass, lemongrass, Ceylon and Java citronella, tuberose, oak musk, orris, champaca, ylang-ylang, mace, nutmeg, sassafras, Cajenne bois-de-rose, rose, *Acacia farnesiana*, geranium, clary sage, spike, lavandin, lavender, jasmine, coriander, carrot, myrrh, eucalyptus, linaloe, lime, mandarin petitgrain, bergamot petitgrain, bergamot, lemon, orange, and others. The essential oils of palmarosa and *Cymbopogon winterianus* contain the highest levels of geraniol (approx 80–95%).[6]
Reported uses[3]	Non-alcoholic beverages 2.1 ppm Ice cream, ices, etc. 3.3 ppm Candy 10 ppm Baked goods 11 ppm Gelatins and puddings 2.0 ppm Chewing gum 0.80–2.9 ppm Toppings 1.0 ppm
Regulatory status	FDA GRAS; FEMA No. 2507

REFERENCES

For References 1–5, see end of Part III.
6. **Pietramellara,** *Riv. Ital. EPPOS,* 39, 60, 1957.

	Geranyl acetate	Geranyl acetoacetate
Other names	2,6-Dimethyl-2,6-octadiene-8-yl acetate Geraniol acetate	*trans*-3,7-Dimethyl-2,6-octadien-l-yl acetoacetate Geranyl β-ketobutyrate Geranyl 3-oxobutanoate
Empirical formula	$C_{12}H_{20}O_2$	$C_{14}H_{22}O_3$
Structure	CH_3 … $CH_2-O-C(=O)-CH_3$; H_3C CH_3	CH_3 … $CH_2-O-C(=O)-CH_2-C(=O)-CH_3$; H_3C CH_3
Physical/chemical characteristics[5]		
Appearance	Colorless liquid	Liquid
Assay	90% min;[2] 92% (85% tech.)	
Molecular weight	196.29	238.33
Boiling point	127–129°C (245°C) at 22 mm Hg	
Flash point	104°C	
Specific gravity	0.900–0.914 at 25°/25°C;[2] 0.900–0.910; 0.896–0.913 tech.	0.9625 at 25°C
Refractive index	1.4580–1.4640 at 20°C;[2] 1.4570–1.4620	1.4670 at 25°C
Optical rotation	−2.0° to +3.0°[2] (±1°)	
Solubility	1:8 in 70% alcohol;[2] 1:25–30 in 60% alcohol; soluble in most organic solvents	
Organoleptic characteristics	Pleasant, flowery odor reminiscent of rose lavender; burning taste, initially somewhat bitter and then sweet	
Synthesis	From geraniol by acetylation or by fractional distillation of essential oils in which it is present	From diketene and geraniol in the presence of the sodium salt of geraniol
Natural occurrence	Reported found in a large number of essential oils: Ceylon citronella, palmarosa, lemongrass, petitgrain, neroli bigarade, geranium, coriander, lavender, carrot, sassafras, in various *Callitris* species (*C. verrucosa*, *C. robusta*, and others) and *Eucalyptus* species (*E. acervula*, *E. urnigera*, etc.). A 60% level has been reported in the essential oil of *Eucalyptus macarthuri*[6] and up to 50% in the essential oil of *Orthodon citraliferum*.[7] Also identified in bitter orange essential oil.[8]	Not reported found in nature
Reported uses[3]	Non-alcoholic beverages 1.6 ppm Ice cream, ices, etc. 6.5 ppm Candy 15 ppm Baked goods 17 ppm Gelatins and puddings 6.8–7.5 ppm Chewing gum 0.30–1.2 ppm Syrups 1.0 ppm	Non-alcoholic beverages 0.50 ppm Ice cream, ices, etc. 1.0 ppm Candy 1.0–3.0 ppm Baked goods 1.0–10 ppm
Regulatory status	FDA GRAS; FEMA No. 2509	FDA 121.1164; FEMA No. 2510

REFERENCES

For References 1–5, see end of Part III.

6. **Arctander,** *Parfuem. Kosmet.,* 39, 7, 1958.
7. **Fujita and Minamino,** *Nippon Kagaku Zasshi,* 79, 1017, 1958.
8. **Mehlitz and Minas,** *Ind. Obst Gemueseverwert.,* 50, 861, 1965.

	Geranyl benzoate	Geranyl butyrate
Other names		
Empirical formula	$C_{17}H_{22}O_2$	$C_{14}H_{24}O_2$
Structure	(geranyl chain with CH_3, $CH_2-O-C(=O)-C_6H_5$; H_3C, CH_3)	(geranyl chain with CH_3, $CH_2-O-C(=O)-(CH_2)_2-CH_3$; H_3C, CH_3)
Physical/chemical characteristics[5]		
Appearance	Yellowish, oily liquid	Colorless liquid
Assay	92% min	92% min
Molecular weight	258.37	224.34
Boiling point	198–200°C at 15 mm Hg	151–153°C at 18 mm Hg; 142–143°C at 13 mm Hg
Flash point	>100°C	93°C
Specific gravity	0.978–0.984 at 25°/25°C	0.891–0.896 at 25°/25°C; 0.9008 at 17°C
Refractive index	1.513–1.518 at 20°C	1.456–1.462 at 20°C
Optical rotation	±1°	±1°
Acid value	1 max	1 max
Solubility	1:4 in 90% alcohol	1:7 in 80% alcohol; 1:1 in 90% alcohol; almost insoluble in water
Organoleptic characteristics	Sweet, light odor reminiscent of ylang-ylang	Characteristic fresh, fruity, rose-like odor and sweet apricot-like taste
Synthesis	From geraniol and benzoyl chloride in anhydrous pyridine;[6] also from geraniol and benzoyl chloride using the Schotten-Bauman reaction[7]	By heating geraniol and butyryl chloride in the presence of pyridine,[8] or from geraniol and butyric anhydride in the presence of camphor sulfonic acid[9]
Natural occurrence	Not reported found in nature	Reported found in the essential oil of *Darwinia grandiflora*. It has been identified in lavender oil[10] and in other essential oils.
Reported uses[3]	Non-alcoholic beverages 0.10–0.13 ppm Ice cream, ices, etc. 0.16–0.25 ppm Baked goods 0.50 ppm Gelatins and puddings 0.50 ppm	Non-alcoholic beverages 1.6 ppm Ice cream, ices, etc. 2.8 ppm Candy 10 ppm Baked goods 10 ppm Gelatins and puddings 5.3 ppm Chewing gum 0.30–1.5 ppm
Regulatory status	FDA 121.1164; FEMA No. 2511	FDA 121.1164; FEMA No. 2512

REFERENCES

For References 1–5, see end of Part III.

6. **Erdmann and Huth,** *J. Prakt. Chem.*, 2(56), 14, 1897.
7. **Bacon,** *Chem. Zentralbl.*, 2, 947, 1908.
8. **Erdmann,** *Ber. Dtsch. Chem. Ges.*, 31, 356, 1898.
9. **Reycher,** *Chem. Zentralbl.*, 1, 1042, 1908.
10. **Sfiras,** *Riv. Ital. EPPOS*, 7, 450, 1966.

	Geranyl formate	Geranyl hexanoate
Other names		Geranyl caproate Geranyl hexylate
Empirical formula	$C_{11}H_{18}O_2$	$C_{16}H_{28}O_2$
Structure	$(CH_3)_2C=CH-CH_2-CH_2-C(CH_3)=CH-CH_2-O-C(=O)-H$	$(CH_3)_2C=CH-CH_2-CH_2-C(CH_3)=CH-CH_2-O-C(=O)-(CH_2)_4-CH_3$
Physical/chemical characteristics[5]		
Appearance	Colorless to pale yellow liquid	
Assay	85% min[2]	
Molecular weight	182.26	252.40
Boiling point	113–114°C at 15 mm Hg; 92°C at 11 mm Hg	240°C
Specific gravity	0.906–0.920[2] (0.9886) at 25°/25°C	0.890 at 15.5°C
Refractive index	1.4580–1.4660 at 20°C[2]	1.4500 at 20°C
Optical rotation	±1°	
Acid value	Not more than 3.0[2]	
Solubility	1:3 in 80% alcohol;[2] insoluble in water	
Organoleptic characteristics	Fresh, green, leafy, rose odor with a bitter taste	Rose-geranium odor with a strong undertone reminiscent of pineapple and banana
Synthesis	Prepared by direct esterification of geraniol with formic acid	
Natural occurrence	Reported found in geranium oil and in the oil of *Ledum palustre*[6]	Reported found in the essential oil of palmarosa and in the leaves of *Phebalium dentatum*
Reported uses[3]	Non-alcoholic beverages 1.9 ppm Ice cream, ices, etc. 1.6 ppm Candy 7.5 ppm Baked goods 4.1 ppm Gelatins and puddings 3.4 ppm Chewing gum 0.80 ppm	Non-alcoholic beverages 1.3 ppm Ice cream, ices, etc. 0.90 ppm Candy 3.2 ppm Baked goods 2.0 ppm
Regulatory status		FDA 121.1164; FEMA No. 2515

REFERENCES

For References 1–5, see end of Part III.

6. Hasebe, *J. Chem. Soc. Jap.*, 64, 1041, 1943.

	Geranyl isobutyrate	Geranyl isovalerate
Other names		
Empirical formula	$C_{14}H_{24}O_2$	$C_{15}H_{26}O_2$
Structure	(geranyl)–CH_2–O–C(=O)–CH(CH_3)$_2$	(geranyl)–CH_2–O–C(=O)–CH_2–CH(CH_3)$_2$
Physical/chemical characteristics[5]		
Appearance	Liquid	Liquid
Molecular weight	224.34	238.37
Specific gravity	0.8997 at 15°C	
Refractive index	1.4576 at 20°C	1.4538 at 20°C
Solubility	Insoluble in water; soluble in most organic solvents	
Organoleptic characteristics	Light rose odor and sweet, apricot-like taste	Rose odor with an apple-pineapple undertone and sweet apple taste
Synthesis	By heating geraniol and acetic isobutyric aldehyde in the presence of sodium acetate in toluene and then distilling the formed acetic acid as an azeotrope[6]	
Natural occurrence	Reported found in the essential oils of Japanese hops[7] and valerian	In *Eucalyptus citriodora* Hook.
Reported uses[3]	Non-alcoholic beverages 1.0 ppm Ice cream, ices, etc. 0.80 ppm Candy 5.0 ppm Baked goods 4.9 ppm Gelatins and puddings 0.60 ppm Chewing gum 15 ppm	Non-alcoholic beverages 4.2 ppm Ice cream, ices, etc. 11 ppm Candy 10 ppm Baked goods 6.8 ppm
Regulatory status	FDA 121.1164; FEMA No. 2513	FDA 121.1164; FEMA No. 2518

REFERENCES

For References 1–5, see end of Part III.

6. Norda Essential Oil and Co., U.S. Patent 2,423,545, 1944.
7. **Shigematzu and Kitazawa,** *Chem. Abstr.*, 60, 8343d, 1964.

	Geranyl phenylacetate	Geranyl propionate
Other names	Geranyl α-toluate	2,6-Dimethyl octadien-6-yl-8-*n*-propionate
Empirical formula	$C_{18}H_{24}O_2$	$C_{13}H_{22}O_2$
Structure	$(CH_3)_2C{=}CH{-}CH_2{-}CH_2{-}C(CH_3){=}CH{-}CH_2{-}O{-}C(=O){-}CH_2{-}C_6H_5$	$(CH_3)_2C{=}CH{-}CH_2{-}CH_2{-}C(CH_3){=}CH{-}CH_2{-}O{-}C(=O){-}CH_2{-}CH_3$
Physical/chemical characteristics[5]		
Appearance	Yellowish liquid	Colorless liquid
Assay	97% min	92% min[2]
Molecular weight	272.39	210.31
Boiling point		253°C (128°C at 15 mm Hg)
Flash point	99°C	99°C
Specific gravity	0.971–0.978 at 25°/25°C	0.896–0.913 at 25°/25°C[2]
Refractive index	1.5070–1.5110 at 20°C	1.4570–1.4650 at 20°C[2]
Optical rotation	±1°	±1°
Acid value	2 max	Not more than 2; not more than 1[2]
Solubility	1:4 in 90% alcohol	1:4 or more in 80% alcohol[2]
Organoleptic characteristics	Honey- and rose-like odor	Fruity, somewhat flowery odor and bitter taste
Synthesis		By esterification of geraniol with propionic acid in the presence of a catalyst
Natural occurrence	Not reported found in nature	Reported found in *Fortunella margarita*
Reported uses[3]	Non-alcoholic beverages 1.1 ppm Ice cream, ices, etc. 3.1 ppm Candy 6.7 ppm Baked goods 4.7 ppm Chewing gum 11 ppm	Non-alcoholic beverages 1.5 ppm Ice cream, ices, etc. 1.3 ppm Candy 3.7 ppm Baked goods 4.9 ppm Gelatins and puddings 3.0 ppm Chewing gum 30–70 ppm
Regulatory status	FDA 121.1164; FEMA No. 2516	FDA 121.1164; FEMA No. 2517

REFERENCES

For References 1–5, see end of Part III.

	L-Glutamic acid
Other names	Aminoglutaric acid 2-Aminopentanedioic acid 1-Aminopropane-1,3-dicarboxylic acid
Empirical formula	$C_5H_9NO_4$
Structure	$HOOC-CH_2-CH_2-CH(NH_2)-COOH$
Physical/chemical characteristics[5]	
Appearance	White or colorless platelets
Molecular weight	147.13
Melting point	247–249°C (decomposes)
Boiling point	175°C at 10 mm Hg (sublimes)
Specific gravity	1.538 at 20°C (*in vacuo*)
Optical rotation	+31.4° at 22°C (c = 1 in 6*N* HCl)
Solubility	Slightly soluble in water and ether; insoluble in alcohol, acetone, benzene, methanol, and acetic acid
Organoleptic characteristics	Very faint odor reminiscent of yeast or freshly baked bread; mild, somewhat sweet, meat-like taste
Synthesis	By hydrolysis of gluten (wheat, corn, or other vegetable sources);[6] by fermentation from glucose-containing raw materials;[7] the racemic acid may be resolved into the *d*- and *l*-isomer by fractional crystallization;[8] from 2-cyclopentenylamine;[9] by microbial conversion of α-ketoglutaric acid;[10] or by an alternative method, using *Bacillus megatherium-cereus*;[11,12] from fumaric acid, using *B. pumilus*;[13] from starch[14]
Natural occurrence	Reported as occurring in many vegetable proteins,[15,16] in beef fibrin, in the chrysalis of silkworm, in the hydrolysate of crystalline insulin;[17] also present in other important peptides, such as glutathione, tyrocidin, folic acid, β-lactoglobulin, secretin, and bacitracin, and in a growth hormone[18,19]
Reported uses[3]	Non-alcoholic beverages 400.0 ppm Baked goods 400.0 ppm Condiments, pickles 400.0 ppm Meat, meat sauces, soups 400.0 ppm Milk, dairy products 400.0 ppm Cereals 400.0 ppm
Regulatory status	FEMA No. 3285

REFERENCES

For References 1–5, see end of Part III.

6–8. **Kirk and Othmer,** *Encyclopedia of Chemical Technology,* Vol. 2, Wiley & Sons, 1963, 198, 212.
9. **Norman,** U.S. Patent 2,900,391, to International Minerals & Chemicals, 1959.
10. **Good,** U.S. Patent 2,933,434, to International Minerals & Chemicals, 1960.
11. **Borel,** U.S. Patent 3,022,224, to Hercules Powder Co., 1962.
12. **Foster,** U.S. Patent 3,032,474, to Merck & Co., 1962.
13. **Ogawa et al.,** U.S. Patent 2,971,890, to Ajinomoto Co., 1961.
14. **Ogawa et al.,** U.S. Patent 3,042,585, to Ajinomoto Co. and Sauranku Distillers Co., 1962.
15. **Jones and Moeller,** *J. Biol. Chem.,* 79, 429, 1928.
16. **Furth,** *Sci. Pharm.,* 5, 21, 1934.
17. **Jensen and Wintersteiner,** *J. Biol. Chem.,* 97, 93, 1932.
18. **Frankin et al.,** *J. Biol. Chem.,* 169, 515, 1947.
19. **Synge,** *Q. Rev.,* 3, 245, 1949.

	Glucose pentaacetate	Glycerol tributyrate
Other names	Pentaacetyl glucose 1,2,3,4,6-Pentaacetyl-α-*d*-glucose 1,2,3,4,6-Pentaacetyl-β-*d*-glucose *Not* 2,3,4,5,6-Pentaacetyl-*d*-glucose	Butyrin Glyceryl tributyrate Tributyrin
Empirical formula	$C_{16}H_{22}O_{11}$	$C_{15}H_{26}O_6$
Structure	α-*d*: 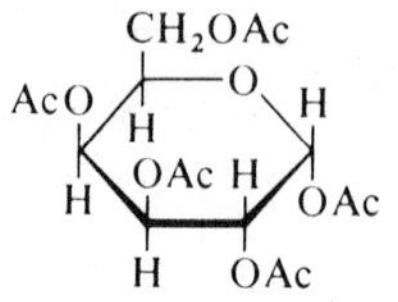 CH_2OAc, AcO, O, H, H, OAc H, H, OAc, H, OAc β-*d*: CH_2OAc, AcO, O, OAc, H, OAc H, H, H, H, OAc 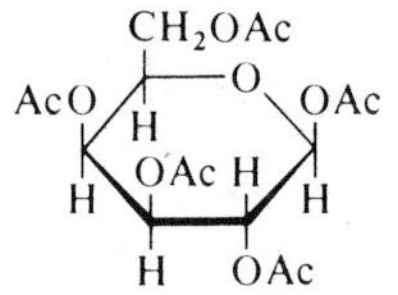	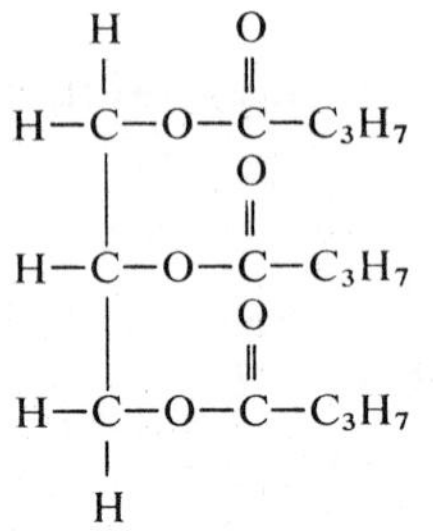 H–C–O–C(=O)–C_3H_7 H–C–O–C(=O)–C_3H_7 H–C–O–C(=O)–C_3H_7 H
Physical/chemical characteristics[5]		
Appearance	White, crystalline powder	Colorless, somewhat oily liquid
Assay	Commercial product probably a mixture of 90% β-*d* and 10% α-*d*	Not less than 99% of $C_{15}H_{26}O_6$
Molecular weight	390.34	302.37
Melting point	Pure α-*d*, 114°C; pure β-*d*, 135°C	−75°C
Boiling point		287–288°C; 190°C at 15 mm Hg; distills between 305–310°C
Specific gravity		1.034–1.037 at 20°C
Refractive index		1.4359 at 20°C
Optical rotation	At 20°C: α-d, + 101.6°; β-*d*, + 3.8°	
Solubility	Insoluble in water and pet ether; soluble in alcohol, benzene, and ether; α-*d*, 5 gm/100 ml $CHCl_3$; β-*d*, 7 gm/100 ml $CHCl_3$	Insoluble in water; soluble in benzene, acetone, and alcohol; very soluble in ether
Organoleptic characteristics	Odorless; bitter flavor	Characteristic odor and bitter taste
Synthesis	By acetylation of glucose using any number of techniques, including: $ZnCl_2$ and pyridine, sodium acetate, and acetic anhydride and pyridine. For an extensive review including methods of preparation, see reference 6.	See Reference 7.
Natural occurrence	Not reported found in nature	
Reported uses[3]	Non-alcoholic beverages 100 ppm Baked goods 1,500 ppm	
Regulatory status	FDA 121.1164; FEMA No. 2524	FDA GRAS

REFERENCES

For References 1–5, see end of Part III.

6. **Bates et al.**, Polarimetry, Saccharimetry and the Sugars, Circular C440, U.S. Department of Commerce, National Bureau of Standards, 1942, 485.
7. *Beilstein's Encyclopedia of Organic Chemistry,* B2[2], Springer Verlag, p. 249.

	Glyceryl tribenzoate	Glyceryl tripropanoate
Other names	1,2,3-Propanetriol tribenzoate Tribenzoin	Propionic acid, triglyceride Tripropionin
Empirical formula	$C_{24}H_{19}O_6$	$C_{12}H_{20}O_6$
Structure	$CH_2-O-C(=O)-$ $CH-O-C(=O)-$ $CH_2-O-C(=O)-$	$CH_2-O-C(=O)-CH_2-CH_3$ $CH_3-CH_2-C(=O)-O-CH$ $CH_2-O-C(=O)-CH_2-CH_3$
Physical/chemical characteristics		
Appearance		
Assay		
Molecular weight	404.43	260.29
Melting point	76°C	
Boiling point		175–176°C at 20 mm Hg 130–132°C at 3 mm Hg
Congealing point		
Flash point		
Specific gravity	1.228 at 12°C	1.100 at 20°C/18°C
Refractive index		1.4318 at 19°C
Solubility	Very soluble in ether, ethyl acetate, benzene, and chloroform; soluble in hot	Insoluble in water; soluble in alcohol and chloroform; very soluble in ether
Reported uses[3]	Non-alcoholic beverages 190 ppm	Non-alcoholic beverages 20.0 ppm Baked goods 50.0 ppm Desserts 20.0 ppm
Regulatory status	FEMA No. 3398	FEMA No. 3286

REFERENCES

For References 1–5, see end of Part III.

	Glycine
Other names	Glycocoll Aminoacetic acid Aminoethanoic acid
Empirical formula	$C_2H_5O_2N$
Structure	H_2N-CH_2-COOH
Physical/chemical characteristics[5]	
Appearance	Monoclinic or trigonal prisms (from diluted alcohol); exists in three different polymorphic forms (α, β, γ)[6]
Assay	
Molecular weight	75.07
Melting point	262°C (decomposes; turns brown at 228°C and is completely sintered at 290°C)
Refractive index	0.828 at 17° C
Solubility	Very soluble in water; insoluble in alcohol and ether; slightly soluble in acetone and pyridine
Organoleptic characteristics	Sweet taste
Synthesis	From chloroacetic acid and ammonia;[7] from protein sources, such as gelatin and silk fibroin;[8] from ammonium bicarbonate and sodium cyanide;[9] by catalytic cleavage of serine;[10] from hydrobromic acid and methyleneaminoacetonitrile[11]
Natural occurrence	Gelatin and silk fibroin are reportedly the best natural sources of this amino acid
Reported uses[3]	Non-alcoholic beverages 150.0 ppm Candy 150.0 ppm Baked goods 150.0 ppm Condiments, pickles 150.0 ppm Meat, meat sauces, soups 150.0 ppm Preserves and spreads 150.0 ppm
Regulatory status	FEMA No. 3287

REFERENCES

For References 1–5, see end of Part III.

6. **Iitaka,** *Nature,* 183, 390, 1959.
7. **Gilman and Blatt,** *Organic Syntheses,* Coll. Vol. 1, Wiley & Sons, New York, 1941, 300.
8. **Stein and Moore,** *Biochem. Prep.,* 1, 9, 1949.
9. **White and Wysong,** U.S. Patent 2,663,713, assigned to Dow Chemical Co., 1953.
10. **Metzler et al.,** *J. Am. Chem. Soc.,* 76, 639, 1954.
11. **Clarke and Taylor,** *Org. Synth.,* 4, 31, 1925.
12. **Greenberg,** *Amino Acids and Proteins,* C. C Thomas, Springfield, Ill., 1951, 232.

	Guaiacol	Guaiacyl acetate
Other names	*o*-Hydroxyanisole *o*-Methoxyphenol Methylcatechol Pyrocatechol monomethyl ether	Acetyl guaiacol *o*-Methoxyphenyl acetate
Empirical formula	$C_7H_8O_2$	$C_9H_{10}O_3$
Structure	OH, OCH_3	O, $O-C-CH_3$, $O-CH_3$
Physical/chemical characteristics[5]		
Appearance	Yellowish, low-melting crystals or liquid; tends to darken on exposure to light	Colorless liquid
Molecular weight	124.14	166.18
Melting point	27.9°C	
Boiling point	205°C; 107°C at 24 mm Hg	235–240°C; 123–124°C at 13 mm Hg
Specific gravity	1.1395 (1.112 liquid)	1.156 at 0°C
Refractive index	1.5341 at 34.8°C	
Solubility	Slightly soluble in water; soluble in alcohol, ether, and chloroform	Almost insoluble in water; miscible with alcohol and ether
Organoleptic characteristics	Characteristic sweet odor; slightly phenolic	Odor similar to that of guaiacol
Synthesis	Obtained from hardwood tar or synthetically from *o*-nitrophenol via *o*-anisidine[6]	From guaiacol and excess acetic anhydride in the presence of trace amounts of H_2SO_4
Natural occurrence	Discovered in the distillation product from guaiac resin. Guaiacol is found in castoreum oil,[7] in the essential oil from flowers of *Pandanus odoratissimus* L.,[8] in the distillation waters of orange leaves,[9] in the essential oil of *Ruta montana* L.,[10] in the essential oil from celery seeds,[11] and in the oil from tobacco leaves[12]	
Reported uses[3]	Non-alcoholic beverages 0.95 ppm Ice cream, ices, etc. 0.52 ppm Candy 0.96 ppm Baked goods 0.75 ppm	
Regulatory status	FDA 121.1164; FEMA No. 2532	FDA 121.1164

REFERENCES

For References 1–5, see end of Part III.

6. **Kalle,** German Patent 95,339, 1896.
7. **Lederer,** *Prog. Chim. Subst. Org. Nat.,* 6, 113, 1950.
8. **Janistyn,** *Chem. Zentralbl.,* 1, 2263, 1938.
9. **Igolen,** *Chem. Zentralbl.,* 2, 750, 1939.
10. **St. Pfau,** *Helv. Chim. Acta,* 22, 382, 1939.
11. **Small,** *Chem. Abstr.,* 42, 6149, 1948.
12. **Onishi et al.,** *Chem. Abstr.,* 50, 15028, 1956.

	Guaiacyl phenylacetate	Guaiene
Other names	Guaiacol phenylacetate *o*-Methoxyphenyl phenylacetate	1,4-Dimethyl-7-isopropenyl-Δ-9,10-octahydroazulene
Empirical formula	$C_{15}H_{14}O_3$	$C_{15}H_{24}$
Structure	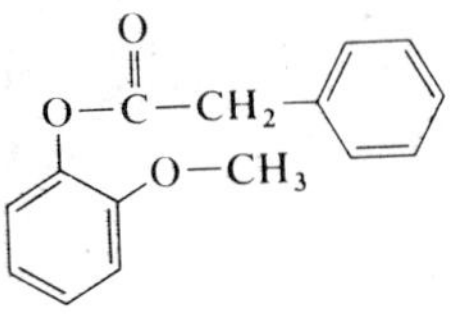	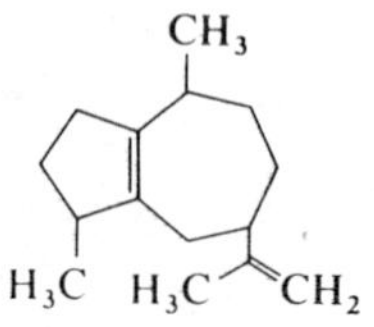
Physical/chemical characteristics[5]		
Appearance	Amber, viscous liquid	Mobile, greenish-yellow liquid
Molecular weight	242.28	
Boiling point		123°–124°C at 9 mm Hg; 128°–130°C at 12 mm Hg
Specific gravity		0.9115 at 19°C; 0.9085 at 20°C
Refractive index		1.5022 at 19°C; 1.5049 at 20°C
Optical rotation		−40°21′ (−16°48′) at 20°C
Solubility	Insoluble in water; soluble in alcohol	Soluble in alcohol
Organoleptic characteristics	Heavy, dry, woody, herbaceous odor; woody, spicy, smoky flavor	Very little odor; delicate, woody flavor
Synthesis	From phenylacetyl chloride and guaiacol in aqueous alkaline solutions or using pyridine	Usually prepared by dehydrating guai[illegible] with $KHSO_4$
Natural occurrence	Not reported found in nature	Guaiene is a bicyclic sesquiterpene four[illegible] naturally in patchouli oil and oth[illegible] essential oils in small quantities.
Reported uses[3]	Non-alcoholic beverages 0.38 ppm Ice cream, ices, etc. 1.0 ppm Candy 2.2 ppm Baked goods 3.2 ppm Toppings 1.0 ppm	
Regulatory status	FDA 121.1164; FEMA No. 2535	FDA 121.1164

REFERENCES

For References 1–5, see end of Part III.

	Guaiol acetate
Other names	1,4-Dimethyl-7-(α-hydroxy-isopropyl)-Δ-9,10-octahydroazulene acetate
Empirical formula	$C_{17}H_{28}O_2$
Structure	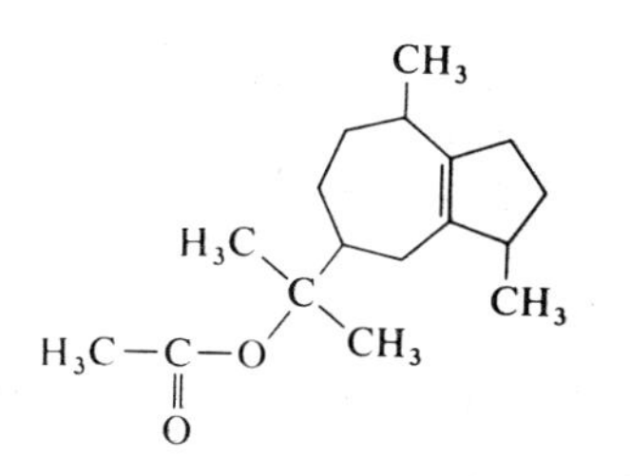
Physical/chemical characteristics[5]	
Appearance	Yellowish liquid
Assay	83% min (70% min)
Molecular weight	264.41
Flash point	>100°C
Specific gravity	0.975–0.985 (0.975–0.990) at 25°C
Refractive index	1.487–1.495 (1.4880–1.4980) at 20°C
Acid value	1 max
Solubility	1: 6 in 80% alcohol; 1:1 in 90% alcohol
Organoleptic characteristics	Faint but lasting rose odor; sweet taste suggestive of currant; not very stable
Synthesis	By acetylation of guaiol, which is one of the constituents of the essential oil from guaiac wood
Natural occurrence	Not reported found in nature
Regulatory status	FDA 121.1164

REFERENCES

For References 1–5, see end of Part III.

	2,4-Heptadienal
Empirical formula	$C_7H_{10}O$
Structure	$CH_3—CH{=}CH—CH{=}CH—CH_2—CHO$
Physical/chemical characteristics[5]	
Molecular weight	110.15
Boiling point	84.0–84.5°C at 20 mm Hg 58–60°C at 5 mm Hg
Specific gravity	0.881 at 25°C
Refractive index	1.5313 at 27°C
Solubility	
Organoleptic characteristics	In autooxidizing fats and oils, it retards or prevents the development of "off-flavors"[6]
Synthesis	By reduction with $LiAlH_4$ of the dienoic acid prepared by the Doebner synthesis, followed by oxidation with MnO_2 of the resulting dienol to the corresponding dienol;[7] by the method of Marshall and Whiting[8]
Natural occurrence	Reported found as a volatile component in boiled chicken,[9] in peas by enzymatic formation from lipids,[10] as the odorous component in *Vaccinium macrocarpon;*[11] additional products in which 2,4-heptadienal has been reportedly identified include soy bean oil,[12-14] fishy butterfat,[15] the oxidized flavor of skim milk,[16] homogenized milk,[17] salmon oil,[18] black-tea aroma (*trans-* and *cis-*form);[19] bilberry aroma (*Vaccinium myrtillus*);[20] frozen peas,[21] sunflower oil,[22] autooxidized lard,[23] salted salmon, sturgeon caviar, and salmon roe;[24] the *trans,trans*-form has been reported found among the volatile flavor components of potato chips[25] and tomato[6]
Reported uses[3]	Candy 1.0 ppm Baked goods 1.0 ppm Condiments 1.0 ppm Meat, meat sauces, soups 1.0 ppm
Regulatory status	FEMA No. 3164

REFERENCES

For References 1–5, see end of Part III.

6. **Unilever N. V.,** Netherlands Patent Appl. 6,900,214; *Chem. Abstr.,* 73, P75860w, 1970.
7. **Forss and Hancox,** *Aust. J. Chem.,* 9, 420, 1956.
8. *Chem. Abstr.,* 51, 4993c, 1957; 53, 6997i, 1959.
9. **Pippen, Nonaka, Jones and Stitt,** *Food Res.,* 23, 103, 1958.
10. **Anjon and von Sydow,** *Acta Chem. Scand.,* 21(8), 2076, 1967.
11. **Grosch,** *Z. Lebensm. Unters. Forsch.,* 135(2), 75, 1967.
12. **Hoffmann,** *J. Am. Oil Chem. Soc.,* 38, 30, 1961.
13. **Akiya,** *Yukagaku,* 14(7), 347, 1965; *Chem. Abstr.,* 59, 11962c, 1965.
14. **Smouse and Chang,** *J. Am. Oil Chem. Soc.,* 44(8), 509, 1967.
15. **Forss, Dunstone and Stark,** *J. Dairy Res.,* 27, 211, 1960.
16. **Forss, Pont and Stark,** *J. Dairy Res.,* 22, 345, 1955.
17. **Day, Lillard and Montgomery,** *J. Dairy Res.,* 46, 291, 1963.
18. **Wyatt and Day,** *J. Food Sci.,* 28, 305, 1963.
19. **Bricout et al.,** *Helv. Chim. Acta,* 50(6), 1517, 1967.
20. **von Sydow and Anjon,** *Lebensm. Wiss. Technol.,* 2(4), 78, 1969.
21. **Whitfield and Shipton,** *J. Food Sci.,* 31(3), 328, 1966.
22. **Swoboda and Lee,** *J. Sci. Food Agric.,* 16(11), 680, 1965.
23. **Gaddis, Ellis and Currie,** *J. Am. Oil Chem. Soc.,* 43(3), 147, 1966.
24. *J. Chromatogr.,* 61(1), 133, 1971.
25. **Sheldon et al.,** *J. Food Sci.,* 37(2), 313, 1972.
26. **Kazeniac and Hall,** *J. Food Sci.,* 35(5), 519, 1970.

	γ-Heptalactone	Heptanal
Other names	Heptanolide-(4,1) 4-Hydroxyheptanoic acid, γ-lactone γ-*n*-Propyl-γ-butyrolactone	Aldehyde C-7 Enanthaldehyde *n*-Heptaldehyde *n*-Heptyl aldehyde
Empirical formula	$C_7H_{12}O_2$	$C_7H_{14}O$
Structure	$CH_3{-}CH_2{-}CH_2{-}CH{-}CH_2{-}CH_2{-}C(=O){-}O$ (ring closed from CH to O)	$CH_3{-}(CH_2)_5{-}CHO$
Physical/chemical characteristics[5]		
Appearance	Colorless, slightly oil liquid	Colorless liquid
Molecular weight	128.17	114.18
Melting point		43.71°C
Boiling point		152.6°C
Specific gravity		0.8219 at 20°C
Refractive index		1.4133 at 20°C
Solubility	Almost insoluble in water; soluble in alcohol	Insoluble in water; 1:12 in 50% alcohol; 1:4 in 60% alcohol; 1:2 in 70% alcohol; soluble in most organic solvents
Organoleptic characteristics	Sweet, nut-like, caramel odor; malty, caramel, sweet, herbaceous taste	Very strong, fatty, harsh, pungent odor; unpleasant, fatty taste
Synthesis	Obtained in low yields by hydrogenation of ethyl β-furylacrylate;[6] by lactonization of heptenoic acid; also by condensation of methylacrylate butyl alcohol using a catalyst	Obtained by distilling castor oil, preferably under reduced pressure
Natural occurrence	Reported found in *Prunus persica*	Heptanal is a constituent of the essential oils of ylang-ylang, clary sage, California lemon, bitter orange,[7] rose, and hyacinth.
Reported uses[3]	Non-alcoholic beverages 18 ppm Ice cream, ices, etc. 40 ppm Candy 28 ppm Baked goods 26 ppm	Non-alcoholic beverages 4.9 ppm Alcoholic beverages 4.0 ppm Ice cream, ices, etc. 1.2 ppm Candy 2.0 ppm Baked goods 2.6 ppm
Regulatory status	FDA 121.1164; FEMA No. 2539	FDA 121.1164

REFERENCES

For References 1–5, see end of Part III.

6. Bel'skii et al., *Chem. Abstr.*, 58, 2365c, 1962.
7. Mehlitz and Minas, *Ind. Obst Gemueseverwert.*, 50, 861, 1965.

	Heptanal dimethyl acetal	Heptanal 1,2- and 1,3-glyceryl acetal
Other names	1,1-Dimethoxy heptane	A mixture of acetals in the 1,2 and 1,3 positions: A. 2-Hexyl-4-hydroxymethyl-1,3-dioxolan B. 2-Hexyl-4-hydroxy-1,3-dioxan
Empirical formula	$C_9H_{20}O_2$	$C_{10}H_{20}O_3$
Structure	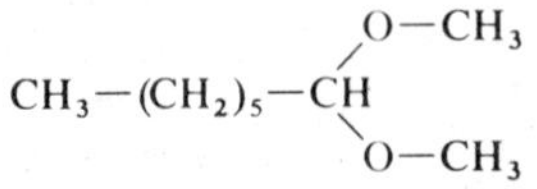 $CH_3-(CH_2)_5-CH(O-CH_3)(O-CH_3)$	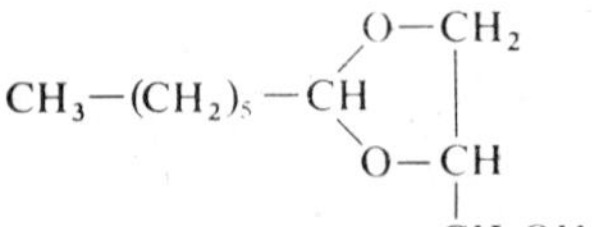 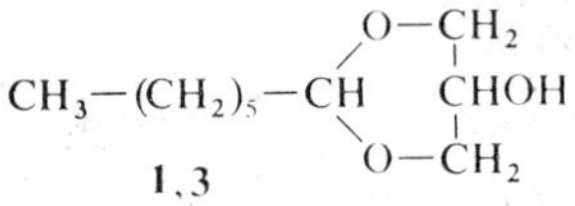 1,2: $CH_3-(CH_2)_5-CH$ ring with $O-CH_2$, $O-CH$, CH_2OH 1,3: $CH_3-(CH_2)_5-CH$ ring with $O-CH_2$, $CHOH$, $O-CH_2$
Physical/chemical characteristics[5]		
Appearance	Oily liquid	Colorless, viscous liquid
Molecular weight	160.26	188.27
Commercial product		Mixture of 1,2 and 1,3 acetals
Boiling point	164–165°C; 59–61°C at 1 mm Hg	
Specific gravity	0.849 at 20°/20°C	
Refractive index	1.4130 at 20°C; 1.4089 at 26°C	
Solubility		Very soluble in water; soluble in alcohol
Organoleptic characteristics	Pleasant odor reminiscent of walnut and cognac	Weak, fungus-like, sweet odor; mushroom (champignon) taste
Synthesis	From heptyl aldehyde and HCl in methanol solution;[6,7] from heptyl aldehyde and methanol in the presence of Twitchell reagent[8]	By condensing heptaldehyde with glycerin
Natural occurrence		Not reported found in nature
Reported uses[3]	Non-alcoholic beverages 0.10–0.13 ppm Ice cream, ices, etc. 0.25 ppm Candy 0.25 ppm Baked goods 0.25 ppm Condiments 1.0 ppm	Non-alcoholic beverages 5.0 ppm Ice cream, ices, etc. 10 ppm Candy 10 ppm Baked goods 10 ppm Condiments 100 ppm
Regulatory status	FDA 121.1164; FEMA No. 2541	FDA 121.1164; FEMA No. 2542

REFERENCES

For References 1–5, see end of Part III.

6. **Kirrmann,** *Bull. Soc. Chim. Fr.*, 4(41), 323, 1927.
7. **Kirrmann,** *Ann. Chim.*, 10(11), 261, 1929.
8. **Zaganiaris,** *Ber. Dtsch. Chem. Ges.*, 71, 2004, 1938.

	2,3-Heptanedione		Heptanoic acid	
Other names	Acetyl pentanoyl Acetyl valeryl Valeryl acetyl		Oenanthic acid Heptoic acid Oenanthylic acid Enanthic acid	
Empirical formula	$C_7H_{12}O_2$		$C_7H_{14}O_2$	
Structure	$CH_3-\overset{\overset{O}{\parallel}}{C}-\overset{\overset{O}{\parallel}}{C}-(CH_2)_3-CH_3$		$CH_3-(CH_2)_5-COOH$	
Physical/chemical characteristics[5]				
Appearance	Yellowish liquid		Colorless oily liquid; tends to solidify in cold weather	
Molecular weight	128.17		130.19	
Melting point			-7.5°C	
Boiling point			223°C 115–116°C at 11 mm Hg	
Congealing point			-6.26°C	
Specific gravity			0.9214 at 15°C 0.9200 at 20°C	
Refractive index			1.4289 at 15°C 1.4216 at 20°C	
Solubility	Slightly soluble in water; soluble in alcohol		Slightly soluble in water; soluble in alcohol, ether, acetone, and nitric acid; also soluble in dimethylformamide and dimethyl sulfoxide (DMSO)	
Organoleptic characteristics	Powerful, cheesy, pungent odor with a sweet, oily character; sweet, buttery taste		Disagreeable rancid odor; the spectroscopically pure acid exhibits a faint tallow-like odor	
Synthesis	From butyl acetoacetate, ethyl butyl ketone, or methyl-*n*-amyl ketone		By oxidation of heptaldehyde with potassium permanganate in diluted sulfuric acid[6]	
Natural occurrence	Not reported found in nature		Reported as occurring naturally in calamus, hops, *Acacia dealbata*, and Japanese peppermint and violet leaves; its presence in rancid oils has been observed; also reported found in yellow passion fruit[7]	
Reported uses[3]	Non-alcoholic beverages	0.96 ppm	Non-alcoholic beverages	2.0 ppm
	Ice cream, ices, etc.	3.1 ppm	Ice cream, ices, etc.	2.0 ppm
	Candy	8.2 ppm	Candy	6.0 ppm
	Baked goods	7.9 ppm	Baked goods	8.0 ppm
	Chewing gum	1.7 ppm	Gelatins and puddings	6.0 ppm
			Preserves and spreads	2.0 ppm
			Margarine	5.0 ppm
Regulatory uses	FDA 121.1164; FEMA No. 2543		FEMA No. 3348	

REFERENCES

For References 1–5, see end of Part III.

6. **Ruhoff,** in *Organic Syntheses,* Coll. Vol. 2, Blatt, A. H., Ed., Wiley & Sons, New York, 1943, 315.
7. **Winter and Klöt,** *Helv. Chim. Acta,* 55(6), 1916, 1972.

	2-Heptanol	3-Heptanol
Other names	2-Hydroxyheptane *n*-Amyl methyl carbinol *sec*-Heptyl alcohol	Butyl ethyl carbinol Ethyl butyl carbinol
Empirical formula	$C_7H_{16}O$	$C_7H_{16}O$
Structure	$CH_3-(CH_2)_4-CH(OH)-CH_3$	$CH_3-CH_2-CH(OH)-(CH_2)_3-CH_3$
Physical/chemical characteristics[5]		
Appearance	Colorless liquid	Colorless, oily liquid
Molecular weight	116.20	116.20
Boiling point	157–160°C *d*-, 73.5°C at 20 mm Hg *l*-, 74.5°C at 23 mm Hg	157°C at 750 mm Hg
Specific gravity	0.8344 at 0°C; 0.8193 at 20°C *d*-, 0.8190 at 20°C *l*-, 0.8184 at 20°C	
Refractive index	1.4213	
Optical rotation	+11.45 at 20°C (in ethanol) −10.48 at 20°C	
Solubility	Soluble in alcohol, ether, and benzene; slightly soluble in water	Insoluble in water; soluble in alcohol
Organoleptic characteristics	Brassy, herbaceous odor reminiscent of lemon; fruity, green, somewhat bitter taste	Powerful, herbaceous odor; pungent, slightly bitter taste
Synthesis	From amyl magnesium bromide and acetaldehyde;[6] by reduction of amyl methyl ketone with sodium metal in alcohol;[7] by the action of *Penicillium palitans* on peanut oil[8]	By catalytic hydrogenation of ethyl-*n*-butyl ketone
Natural occurrence	Reported found in bud oil[9] and, in small amounts, in rank coconut	In strawberry, mint, and banana
Reported uses[3]	Non-alcoholic beverages 0.5 ppm Ice cream, ices, etc. 0.5 ppm Candy 1.0 ppm Baked goods 15.0 ppm Gelatins and puddings 5.0 ppm Chewing gum 10.0 ppm Dips, spreads 12.0 ppm Salad dressing 80.0 ppm	
Regulatory status	FEMA No. 3288	FDA 121.1164

REFERENCES

For References 1–5, see end of Part III.

6. **Henry and DeWael,** *Rec. Trav. Chim.,* 28, 446, 1909.
7. **Whitemore and Ottenbacher,** *Org. Synth.,* 10, 60, 1930.
8. **Stokoe,** *Biochem. J.,* 22, 82, 1928.
9. **Masson,** *Compt. Rend.,* 149, 630, 1909.

	2-Heptanone	3-Heptanone
Other names	Amyl methyl ketone Ketone C-7 Methyl amyl ketone	Butyl ethyl ketone Ethyl butyl ketone Ethyl-*n*-butyl ketone
Empirical formula	$C_7H_{14}O$	$C_7H_{14}O$
Structure	$CH_3{-}(CH_2)_4{-}\overset{\overset{\large O}{\|}}{C}{-}CH_3$	$CH_3{-}(CH_2)_3{-}\overset{\overset{\large O}{\|}}{C}{-}CH_2{-}CH_3$
Physical/chemical characteristics[5]		
Appearance	Colorless liquid	Colorless liquid
Assay		95% min
Molecular weight	114.18	114.18
Melting point	−26.9°C	
Boiling point	150.6°C; 20°C at 2.6 mm Hg	148.5°C (143–148°C)
Flash point	Approx 49°C	
Specific gravity	0.8166 (0.8243) at 20°C	0.8191 (0.8183) at 20°/20°C
Refractive index	1.4043 at 30°C; 1.4110 at 20°C	1.4115 at 20°C
Acid value		0.034% (as acetic acid)
Solubility	Very slightly soluble in water; soluble in most organic solvents	Insoluble in water; completely miscible in most organic solvents
Organoleptic characteristics	Characteristic banana, slightly spicy odor	Powerful, green, fatty, fruity odor; melon, banana flavor
Synthesis	By oxidation of methyl *n*-amyl carbinol;[6] also from *n*-amyl propionic acid or from ethyl butyl acetate	From *n*-hept-2-one by hydration
Natural occurrence	Reported found in clove essential oil,[7] Ceylon cinnamon,[8] and rancid coconut oil; also identified in *Ruta montana;*[9] meat,[10] and cheese	Reported found in bread
Reported uses[3]	Non-alcoholic beverages 2.7 ppm Ice cream, ices, etc. 6.0 ppm Candy 6.4 ppm Baked goods 13 ppm Condiments 10–25 ppm	Non-alcoholic beverages 0.13–2.0 ppm Ice cream, ices, etc. 0.25–170 ppm Candy 67 ppm Baked goods 0.25–130 ppm
Regulatory status	FDA 121.1164; FEMA No. 2544	FDA 121.1164; FEMA No. 2545

REFERENCES

For References 1–5, see end of Part III.

6. **Schorlemmer,** *Annalen,* 161, 263, 1872.
7. **Schimmel and Co.,** *Chem. Zentralbl.,* 1, 1059, 1902.
8. **Schimmel and Co.,** *Chem. Zentralbl.,* 2, 1486, 1902.
9. **St. Pfau,** *Helv. Chim. Acta,* 22, 382, 1939.
10. **Anonymous,** *Flavour Ind.,* 2, 212, 1971; *Z. Naturforsch. Teil B,* 24, 781, 1969; *J. Agric. Food Chem.,* 19, 772, 1971.

	4-Heptanone	2-Heptenal
Other names	Butyrone Dipropyl ketone	β-Butylacrolein 2-Heptenic aldehyde 4-Propylcrotonaldehyde α,β-Heptenoic aldehyde
Empirical formula	$C_7H_{14}O$	$C_7H_{12}O$
Structure	$CH_3-CH_2-CH_2-C(=O)-CH_2-CH_2-CH_3$	$CH_3-CH_2-CH_2-CH_2-CH=CH-CH$[O]
Physical/chemical characteristics[5]		
Appearance	Colorless liquid	Colorless mobile liquid
Molecular weight	114.18	112.17
Boiling point	144°C	165–167°C 80–85°C at 14 mm Hg
Specific gravity	0.8145 at 20°C	0.864 at 17°C
Refractive index	1.4069 at 20°C	1.4314 at 20°C
Solubility	Insoluble in water; soluble in alcohol and ether	Almost insoluble in water; soluble in alcoh
Organoleptic characteristics	Penetrating odor and burning taste	Pungent green somewhat fatty odo pleasant only on extreme dilution
Synthesis	By passing butyric acid over wood coal at 425 C and then over cerium oxide at 500 C or over thorium oxide;[6–8] also over manganese oxide at 400–425 C[9,10]	By oxidation of butylallyl alcohol[12,13]
Natural occurrence	Reported found in tobacco smoke[11]	Reported found in soybean oil; a volat flavor component in potato chips[14]
Reported uses[3]	Non-alcoholic beverages 7.8 ppm Ice cream, ices, etc. 11 ppm Candy 19 ppm Baked goods 27 ppm Gelatins and puddings 0.60–8.0 ppm	Candy 1.0 pp Baked goods 1.0 pp Condiments 1.0 pp Meat, meat sauces, soups 1.0 pp
Regulatory status	FDA 121.1164; FEMA No. 2546	FEMA No. 3165

REFERENCES

For References 1–5, see end of Part III.

6. **Maison Camus Duchemin S. A.**, German Patent 630,909, 1930.
7. **Fischer,** *Kraftstoff,* 15, 5, 1939.
8. **Fischer,** *Chem Zentralbl.,* 2, 4413, 1939.
9. **Slobodin,** *Zh. Obshch. Khim.,* 8, 250, 1938.
10. **Cowan, Jeffery and Vogel,** *J. Chem. Soc.,* p. 173, 1940.
11. **Neuberg and Burkard,** *Biochem. Z.,* 243, 480, 1931.
12. **Delaby and Guillot-Allégre,** *Compt. Rend.,* 192, 1467, 1931.
13. **Martin, Schepartz and Daubert,** *J. Am. Chem. Soc.,* 70, 2601, 1948.
14. **Deck, Pokorny and Chang,** *J. Food Sci.,* 38(2), 345, 1973.

	4-Heptenal (*cis*- and *trans*-)	4-Heptenal diethyl acetal
Other names	4-Hepten-1-al *n*-Propylidene butyraldehyde	1,1-Diethoxy-4-heptene
Empirical formula	$C_7H_{12}O$	$C_{11}H_{22}O_2$
Structure	$CH_3-CH_2-CH=CH-CH_2-CH_2-CHO$	$CH_3-CH_2-CH=CH-CH_2-CH_2-CH(OC_2H_5)_2$
Physical/chemical characteristics[5]		
Appearance	Colorless oily liquid	Colorless liquid
Molecular weight	112.17	186.30
Boiling point	*trans*-, 112–113°C	
Solubility	Insoluble in water; soluble in alcohol and oils	Almost insoluble in water; soluble in alcohol and oils
Organoleptic characteristics	Odor reminiscent of heptaldehyde; suggestive of fried fats on dilution; used to impart fried, buttery flavor	Imparts cream or butter-like flavor to food products[10]
Synthesis	From *Hemiptera* bugs (*trans*-form);[7] from penten-1-en-3-ol converted to the corresponding vinyl ether, which undergoes a Claisen rearrangement to 4-heptenal (*trans*-form)[8,9]	
Natural occurrences	Reported found as a constituent in butter fat[6] and in black *Hemiptera* bugs[7]	
Reported uses[3]	Non-alcoholic beverages 1.0 ppm Candy 1.0 ppm Baked goods 1.0 ppm Meat, meat sauces, soups 1.0 ppm Milk, dairy products 1.0 ppm Cereals 1.0 ppm	Non-alcoholic beverages 1.0 ppm Candy 1.0 ppm Baked goods 1.0 ppm Meat, meat sauces, soups 1.0 ppm Milk, dairy products 1.0 ppm Cereals 1.0 ppm
Regulatory status	FEMA No. 3289	FEMA No. 3349

REFERENCES

For References 1–5, see Introduction to Part II.

6. Begemann and Koster, *Nature,* 202, 552, 1964.
7. Mukerji and Sharma, *Tetrahedron Lett.,* p. 2479, 1966.
8. Vig, Matta and Ray, *J. Indian Chem. Soc.,* 41, 752, 1964.
9. Burgstahler, *J. Am. Chem. Soc.,* 82, 4681, 1960.
0. Unilever Ltd., British Patent 1,068,712.

	2-Hepten-4-one		3-Hepten-2-one	
Other names	Propenyl propyl ketone Ethyl ethylidene acetone 1-Butyryl propylene		1-Acetyl-1-pentene Butylidene acetone Methyl pentenyl ketone	
Empirical formula	$C_7H_{12}O$		$C_7H_{12}O$	
Structure	$CH_3-CH_2-CH_2-C(=O)-CH=CH-CH_3$		$CH_3-CH_2-CH_2-CH=CH-C(=O)-CH_3$	
Physical/chemical characteristics[5]				
Appearance			Colorless oily liquid	
Molecular weight	112.17		112.17	
Boiling point	156–157°C 80.2°C at 51 mm Hg 74–75°C at 12 mm Hg		162°C *cis*-, 34–39°C at 11 mm Hg *trans*-, 51–52°C at 11 mm Hg	
Specific gravity			*cis*-, 0.8840 at 20°C *trans*-, 0.8496 at 20°C	
Refractive index	1.4430 at 20°C		*cis*-, 1.4325 at 20°C *trans*-, 1.4436 at 20°C	
Solubility			Almost insoluble in water; soluble in alcohol and oils	
Organoleptic characteristics			Powerful green-grassy, pungent odor[1,2]	
Synthesis	From 4-heptanone by a catalytic process[6,7]; also (2-pentanone + acetaldehyde (CHO), $B^{\ominus}$ →)		By reacting 1-pentyne with acetic anhydride to yield 3-heptyn-2-one, which is then converted to *cis*-3-hepten-2-one by partial catalytic hydrogenation;[9] the *trans*-form is probably obtained from *trans*-2-hexenic acid treated with methyl lithium;[10] also (butanal (CHO) + acetone, B^- →)	
Natural occurrence	Reported found in roasted filberts[8]		Reported found in roasted filberts[11]	
Reported uses[3]	Non-alcoholic beverages	1.0 ppm	Non-alcoholic beverages	1.0 ppm
	Ice cream, ices, etc.	1.0 ppm	Ice cream, ices, etc.	1.0 ppm
	Candy	2.0 ppm	Candy	2.0 ppm
	Baked goods	2.0 ppm	Baked goods	2.0 ppm
	Gelatins and puddings	2.0 ppm	Gelatins and puddings	2.0 ppm
	Milk, dairy products	1.0 ppm	Milk, dairy products	0.7 ppm
Regulatory status	FEMA No. 3399		FEMA No. 3400	

REFERENCES

For References 1–5, see Introduction to Part II.

6. British Patent 1,152,817, assigned to Mobil Oil Corp.; *Chem. Abstr.*, 71, 112504q, 1969.
7. **Baltz et al.**, German Patent 2,050,565; *Chem. Abstr.*, 75, 19800f, 1971.
8. **Kinlin et al.**, *J. Agric. Food Chem.*, 20, 1021, 1972.
9. **Theus et al.**, *Helv. Chim. Acta*, 38, 239, 1955.
10. **Theus et al.**, *Helv. Chim. Acta*, 38, 239, 1955.
11. **Kinlin et al.**, *J. Agric. Food Chem.*, 20, 1021, 1972.
12. **Arctander**, *Perfume and Flavor Chemicals*, Montclair, N.J., 1969, No. 472.

	3-Heptyl-5-methyl-2(3H)furanone	
Other names	The alternate name published by FEMA as α-*n*-heptyl-γ-valerolactone is incorrect	
Empirical formula	$C_{12}H_{22}O_2$	
Structure	$(CH_2)_6CH_3$; CH_3; O; O	
Physical/chemical characteristics		
Assay	99.0% by VPC	
Molecular weight	198.3	
Reported uses[3]	Baked goods	1.5 ppm
	Meat, meat sauces, soups	1.5 ppm
	Milk, dairy products	1.5 ppm
Regulatory status	FEMA No. 3350	

REFERENCES

For References 1–5, see end of Part III.

	Heptyl acetate	Heptyl alcohol
Other names		Alcohol C-7 Enanthic alcohol 1-Heptanol Hydroxy heptane
Empirical formula	$C_9H_{10}O_2$	$C_7H_{16}O$
Structure	$CH_3{-}(CH_2)_5{-}CH_2{-}O{-}C({=}O){-}CH_3$	$CH_3{-}(CH_2)_5{-}CH_2OH$
Physical/chemical characteristics[5]		
Appearance	Colorless liquid	Colorless liquid
Assay		97% min
Molecular weight	150.18	116.20
Melting point	−50°C	−34.6°C
Boiling point	46–47°C at 1 mm Hg; 192°C	175°C
Flash point		71°C
Specific gravity	0.87505 at 15°C	0.8219 at 20°C; 0.820–0.824 at 25°C
Refractive index	1.4150 (1.4200) at 20°C	1.42410 (1.4230–1.4270) at 20°C
Solubility	Insoluble in water; soluble in alcohol and ether	Insoluble in water; soluble in most organic solvents; 1:2.3 in 60% alcohol; 1:1 in 70% alcohol
Organoleptic characteristics	Pleasant odor suggestive of pear (rose); sweet, apricot-like taste	Faint, aromatic, fatty odor; pungent, spicy taste
Synthesis	From *n*-heptyl alcohol and acetyl chloride in ether solution in the presence of magnesium dust	By reduction of enanthic aldehyde, which is a distillation product of castor oil
Natural occurrence	Heptyl alcohol has been reported found in the essential oil of *Litsea zeylanica*; the corresponding ester has not yet been identified among the constituents of essential oils	Reported found in a few essential oils: hyacinth, violet leaves, *Litsea zeylanica*
Reported uses[3]	Non-alcoholic beverages 4.1 ppm Ice cream, ices, etc. 3.3 ppm Candy 4.9 ppm Baked goods 4.8 ppm	Non-alcoholic beverages 0.90 ppm Ice cream, ices, etc. 1.0–5.0 ppm Candy 3.0 ppm Baked goods 3.0 ppm
Regulatory status	FDA 121.1164; FEMA No. 2547	FDA 121.1164; FEMA No. 2548

REFERENCES

For References 1–5, see end of Part III.

	Heptyl butyrate	*n*-Heptyl cinnamate
Other names	*n*-Heptyl-*n*-butanoate *n*-Heptyl-*n*-butyrate	Heptyl-β-phenylacrylate Heptyl-3-phenyl propenoate
Empirical formula	$C_{11}H_{22}O_2$	$C_{16}H_{22}O_2$
Structure	$CH_3-(CH_2)_2-\overset{\overset{\displaystyle O}{\parallel}}{C}-O-(CH_2)_6-CH_3$	$CH_3-(CH_2)_5-CH_2-$ (ring with O)
Physical/chemical characteristics[5]		
Appearance	Colorless liquid	Colorless to pale straw-yellow liquid; tends to polymerize
Molecular weight	186.30	246.36
Melting point	−57/−58°C	
Boiling point	105°C (226°C) at 10–12 mm Hg	
Specific gravity	0.8555 at 30°C	
Refractive index	1.4231 at 20°C	
Solubility	Almost insoluble in water; soluble in alcohol; soluble in propylene glycol	Insoluble in water; soluble in alcohol; almost insoluble in propylene glycol
Organoleptic characteristics	Fruity, camomile-like odor; sweet, green, tea-like taste suggestive of plum	Green, leafy odor with a secondary hyacinth note
Synthesis	From *n*-heptyl alcohol and butyric acid in the presence of HCl[6]	By esterification of *n*-heptanol with cinnamic acid
Natural occurrence	Not reported found in nature	Not reported found in nature
Reported uses[3]	Non-alcoholic beverages 0.66 ppm Ice cream, ices, etc. 0.74 ppm Candy 2.7 ppm Baked goods 2.4 ppm	Non-alcoholic beverages 3.3 ppm Ice cream, ices, etc. 1.0–2.0 ppm Candy 1.0–6.0 ppm Baked goods 1.0 ppm Chewing gum 270 ppm
Regulatory status	FDA 121.1164; FEMA No. 2549	FDA 121.1164; FEMA No. 2551

REFERENCES

For References 1–5, see end of Part III.

6. **Bilterys and Gisseleire,** *Bull. Soc. Chim. Belg.*, 44, 569, 1935.

	Heptyl formate	2-Heptylfuran
Other names		1-(2-Furyl)-heptane
Empirical formula	$C_8H_{16}O_2$	$C_{11}H_{18}O$
Structure	$H-C(=O)-O-CH_2-(CH_2)_5-CH_3$	(2-furyl ring with heptyl chain ending in CH_3)
Physical/chemical characteristics[5]		
Appearance	Colorless liquid	
Molecular weight	144.21	166.26
Boiling point	76–77 C at 25 mm Hg (176.7–178 C)	
Specific gravity	0.88277 at 15 C; 0.894 at 0 C	
Refractive index	1.4140 at 20 C	
Solubility	1:5 in 70% alcohol; insoluble in water; soluble in ether and most organic solvents	
Organoleptic characteristics	Fruity-floral odor with an orris-rose undertone; sweet taste reminiscent of plum	
Synthesis	From *n*-heptyl bromide and formamide at the boil[6]	
Natural occurrence	Not reported found in nature	Reported found in heated beef[7]
Reported uses[3]	Non-alcoholic beverages 1.7 ppm Ice cream, ices, etc. 0.87 ppm Candy 3.6 ppm Baked goods 3.3 ppm	Non-alcoholic beverages 0.02 ppm Ice cream, ices, etc. 0.02 ppm Baked goods 0.06 ppm Gelatins and puddings 0.05 ppm Chewing gum 0.06 ppm Milk, dairy products 0.02 ppm
Regulatory status	FDA 121.1164; FEMA No. 2552	FEMA No. 3401

REFERENCES

For References 1–5, see end of Part III.

6. E. I. du Pont de Nemours & Co., U.S. Patent 2,375,301, 1942.
7. Volatile Compounds in Food, T.N.O. Report No. 4030, 3rd ed., Centraal Instituut vorr Voedingsonderzoek, Zeist, Netherlands.

	n-Heptyl octanoate	Heptyl isobutyrate
Other names	Heptyl caprylate Heptyl octylate	*n*-Heptyl dimethylacetate *n*-Heptyl isobutanoate *n*-Heptyl-2-methylpropanoate
Empirical formula	$C_{15}H_{30}O_2$	$C_{11}H_{22}O_2$
Structure	$CH_3-(CH_2)_6-C(=O)-O-(CH_2)_6-CH_3$	$(H_3C)_2CH-C(=O)-O-CH_2-(CH_2)_5-CH_3$
Physical/chemical characteristics[5]		
Appearance	Colorless, oily liquid	Colorless liquid
Molecular weight	242.41	186.30
Melting point	−10°C	
Boiling point	160°C at 14 mm Hg (291°C)	98°C at 10 mm Hg
Specific gravity	0.8520 at 30°C	0.8625 at 15°/15°C
Refractive index	1.4340 at 20°C	1.4190 at 20°C
Solubility	1:3 in 90% alcohol; insoluble in water	Insoluble in water; soluble in most organic solvents
Organoleptic characteristics	Oily, green odor and flavor	Characteristic woody odor with distinctly herbaceous, sweet undernotes; sweet, herbaceous, fruity, and slightly warm, floral taste
Synthesis	By direct esterification of *n*-heptanol with *n*-octanoic acid	By esterification of *n*-heptanol with isobutyric acid
Natural occurrence	Not reported found in nature	Reported found in the essential oil of hops[6]
Reported uses[3]	Non-alcoholic beverages 1.0 ppm	Non-alcoholic beverages 1.2 ppm Ice cream, ices, etc. 0.82 ppm Candy 2.6 ppm Baked goods 3.0 ppm
Regulatory status	FDA 121.1164; FEMA No. 2553	FDA 121.1164; FEMA No. 2550

REFERENCES

For References 1–5, see end of Part III.

6. Buttery et al., *Nature*, 4933(202), 701, 1964.

	1-Hexadecanol	ω-6-Hexadecenlactone
Other names	Alcohol C-16 Cetyl alcohol	Ambrettolide Cyclohexadecen-7-olide 6-Hexadecenolide 16-Hydroxy-6-hexadecenoic acid, ω-lactone 16-Hydroxy-Δ7-hexadecenoic acid, lactone
Empirical formula	$C_{16}H_{34}O$	$C_{16}H_{28}O_2$
Structure	$CH_3-(CH_2)_{14}-CH_2OH$	HC=HC ring: HC–$(CH_2)_7-CH_2$–O–C(=O)–$(CH_2)_5$–CH
Physical/chemical characteristics[5]		
Appearance	White solid	Colorless liquid
Assay		98% min
Molecular weight	242.45	252.39
Melting point	47°C (49°C)	
Boiling point	344°C	185–190°C at 16 mm Hg
Congealing point	46°C	
Specific gravity	0.8152 at 55°C	0.955–0.957
Refractive index	1.4283 at 78.9°C	1.480–1.481
Solubility	Insoluble in water; soluble in ether; 1 gm in 3 ml of 90% alcohol	1:9 in 80% alcohol; 1:1 in 90% alcohol
Organoleptic characteristics	Odorless	Very strong, musk-like odor
Synthesis	See Reference 7.	Synthetically prepared by Baudart;[6] from bromohexadecenoic acid, dihydroxypalmitic acid, aleuritic acid, or juniperic acid
Natural occurrence	Reported as a major constituent of spermaceti oil, where it is present chiefly as cetyl palmitate	Reported found in ambrette seed oil (*Hibiscus abelmoschus* L.); identified by Kerschbaum (1927) and studied by Stoll and Gardner (1934)
Reported uses[3]	Ice cream, ices, etc. 2.0 ppm Candy 2.0 ppm	Non-alcoholic beverages 0.32 ppm Ice cream, ices, etc. 0.18 ppm Candy 0.16 ppm Baked goods 0.19 ppm Gelatins and puddings 0.007 ppm Chewing gum 0.70 ppm
Regulatory status	FDA 121.1164; FEMA No. 2554	FDA 121.1164; FEMA No. 2555

REFERENCES

For References 1–5, see end of Part III.

6. **Baudart,** *C. R. Séances Acad. Sci. Paris,* 221, 205, 1945.
7. For a more complete list of the physical properties, see Maison G. de Navarre, *The Chemistry and Manufacture of Cosmetics,* 2nd ed., Vol. 2, D. Van Nostrand Co., 1962, 97.

	trans,trans-2,4-Hexadienol	γ-Hexalactone
Other names	Sorbic aldehyde	γ-Caprolactone Ethyl butyrolactone γ-Ethyl-*n*-butyrolactone Hexanolide-1,4 4-Hydroxy hexanoic acid γ-lactone Tonkalide
Empirical formula	C_6H_8O	$C_6H_{10}O_2$
Structure	$CH_3-CH=CH-CH=CH-CHO$	$CH_3-CH_2-CH-CH_2-CH_2-C=O$ (ring closed from CH to C through O)
Physical/chemical characteristics		
Appearance		Colorless liquid
Molecular weight	96.12	114.15
Melting point		−18°C
Boiling point	173–174°C at 754 mm Hg 76°C at 30 mm Hg 64–66°C at 11 mm Hg	220°C
Congealing point	−17.5 to −16.5°C	
Specific gravity	0.9087 at 22°C	
Refractive index	1.5372 at 22°C	
Solubility		Very slightly soluble in water; soluble in alcohol and propylene glycol
Organoleptic characteristics		Warm, powerful, herbaceous, sweet odor; sweet, coumarin-caramel taste
Synthesis	Can be obtained by slowly heating crotonaldehyde with acetaldehyde in the presence of pyridine at 85–90°C	By reduction of sorbic acid using Zn, Sn, or $SnCl_2$ and concentrated HCl in acetic acid solution at 85°C;[19] from ethylene oxide and sodio-malonic ester; also from propyl alcohol and methylacrylate in the presence of di-*tert*-butyl peroxide
Natural occurrence	In olive,[7] roasted peanuts,[8-11] tomato,[12-14] caviar and fish,[15] autooxidized salmon oil,[16] tea[17,18]	Identified among the volatile components of pineapple[21] and peach var. Red Globe[21,22]
Reported uses[3]	Non-alcoholic beverages 4.0 ppm Alcoholic beverages 1.0 ppm Ice cream, ices, etc. 4.0 ppm Candy 6.0 ppm Gelatins and puddings 4.0 ppm Condiments 4.0 ppm Jellies, preserves, spreads 2.0 ppm	Non-alcoholic beverages 7.0 ppm Ice cream, ices, etc. 0.07–84 ppm Candy 21 ppm Baked goods 21 ppm
Regulatory status	FEMA No. 3429	FDA 121.1164; FEMA No. 2556

REFERENCES

For References 1–5, see end of Part III.

6. **Kuhn et al.,** *Chem. Ber.*, 63, 2168, 1930; 69, 102, 1936.
7. **Flath et al.,** *J. Agric. Food Chem.*, 21, 948, 1973.
8. **Brown et al.,** *J. Agric. Food Chem.*, 20, 703, 1972.
9. **Cobb et al.,** *Peanut – Culture and Uses*, Ch. 6.
10. **Brown et al.,** *J. Agric. Food Chem.*, 21, 463, 1973.
11. **Brown et al.,** *J. Am. Oil Chem. Soc.*, 50, 16, 1973.
12. **Viani et al.,** *Helv. Chim. Acta*, 52, 887, 1969.
13. **Buttery et al.,** *J. Agric. Food Chem.*, 19, 524, 1971.
14. **Johnson et al.,** *Chem. Ind. (London)*, p. 1212, 1971.
15. **Golovnya et al.,** *J. Chromatogr.* 61, 65, 1971.
16. **Wyatt et al.,** *J. Food Sci.*, 28, 305, 1963.
17. **Bricout et al.,** *Helv. Chim. Acta*, 50, 1517, 1967.
18. **Bondarovich et al.,** *J. Agric. Food Chem.*, 15, 36, 1967.
19. **Ingold,** *J. Chem. Soc.*, p. 2036, 1929.
20. **Silverstein et al.,** *J. Food Sci.*, 30, 668, 1965.
21. **Jennings and Sevenants,** *J. Food Sci.*, 29, 796, 1964.
22. **Jennings and Sevenants,** *J. Food Sci.*, 31, 81, 1966.

	Δ-Hexalactone	Hexanal
Other names	5-Hydroxyhexanoic acid, δ-lactone δ-Caprolactone Tetrahydro-6-methyl-2H-pyran-2-one	Aldehyde C-6 *n*-Caproaldehyde Caproic aldehyde Hexoic aldehyde
Empirical formula	$C_6H_{10}O_2$	$C_6H_{12}O$
Structure	(ring structure: H_3C–substituted six-membered lactone, O, =O)	$CH_3{-}CH_2{-}CH_2{-}CH_2{-}CH_2{-}CH{=}O$
Physical/chemical characteristics[5]		
Appearance	Colorless oily liquid	Liquid
Assay		96% min
Molecular weight	114.14	100.16
Melting point	31°C	
Boiling point	110–112°C at 15 mm Hg 59–60°C at 0.04 mm Hg	128–128.2°C
Flash point		25°C
Specific gravity	1.0162 at 20°C 1.037 at 21°C	0.8180–0.8210 at 15°C; 0.81392 at 20°C
Refractive index	1.4425 at 20°C 1.4564 at 25°C	1.4035–1.4050 at 20°C
Optical rotation	–47.2° at 24°C (as is) –39.1° (c = 4.1, in anhydrous ethanol)	
Acid value		7 max
Solubility	Soluble in ether	Very slightly soluble in water; soluble in alcohol and ether; 1:1.0–1.5 in 70% alcohol
Organoleptic characteristics		Characteristic fruity odor and taste (on dilution)
Synthesis	By oxidation of 1-substituted cycloalkanes[6]	Prepared from the calcium salt of caproic acid and formic acid[10]
Natural occurrence	Reported found as a volatile flavor component in coconut oil[7] and in heated milk fat;[8] reportedly isolated from high-quality Australian butter oil by degassation under high vacuum[9]	Reported found in some natural aromas of: apple,[11] strawberry,[12] camphor oil,[13] tea extracts,[14] tobacco leaves, *Eucalyptus globulus*, dwarf pine, and bitter orange;[15] also in coffee
Reported uses[3]	Non-alcoholic beverages 10.0 ppm Ice cream, ices, etc. 10.0 ppm Candy 10.0 ppm Baked goods 10.0 ppm Gelatins and puddings 10.0 ppm Margarine 10.0 ppm Salad oil 10.0 ppm Shortening 10.0 ppm	Non-alcoholic beverages 1.3 ppm Ice cream, ices, etc. 2.8 ppm Candy 3.6 ppm Baked goods 4.2 ppm Gelatins and puddings 2.0–2.5 ppm Chewing gum 3.0 ppm
Regulatory status	FEMA No. 3167	FDA 121.1164; FEMA No. 2557

REFERENCES

For References 1–5, see end of Part III.

6. **Lakodey, Weiss and Mathias,** German Patent 1,952,848, 1970.
7. **Allen,** *Chem. Ind.* (London), 36, 1560, 1965.
8. **Parliment, Nawar and Fagerson,** *J. Dairy Sci.*, 48(5), 615, 1965.
9. **Forss, Stark and Urbach,** *J. Dairy Sci.*, 34(2), 131, 1967.
10. **Lieben and Janeček,** *Justus Liebigs Ann. Chem.*, 187, 126, 1877.
11. **White,** *Food Res.*, 15, 68, 1950.
12. **Winter et al.,** *Perfum. Essent. Oil Rec.*, 49, 250, 1958.
13. **Ono and Imoto,** *J. Chem. Soc. Jap.*, 59, 1014, 1938.
14. **Yamamoto and Ito,** *Parfums Fr.*, 16, 200, 1938.
15. **Mehlitz and Minas,** *Ind. Obst Gemueseverwert.*, 50, 861, 1965.

	2,3-Hexanedione	3,4-Hexanedione
Other names	Acetyl butyryl Acetyl-*n*-butyryl Methyl propyl diketone	Dipropionyl 3,4-Dioxohexane Diethyl-α,β-diketone
Empirical formula	$C_6H_{10}O_2$	$C_6H_{10}O_2$
Structure	$CH_3-C(=O)-C(=O)-CH_2-CH_2-CH_3$	$CH_3-CH_2-C(=O)-C(=O)-CH_2-CH_3$
Physical/chemical characteristics[5]		
Appearance	Yellow, oily liquid	Yellow oil
Molecular weight	114.15	114.04
Melting point		-10°C
Boiling point	128°C	130°C 49°C at 13 mm Hg
Specific gravity	0.93	0.941 at 21°C
Refractive index		1.4130 at 21°C
Solubility	Slightly soluble in water; soluble in propylene glycol, alcohol, and oils	Almost insoluble in water; soluble in alcohol and oils; very soluble in propylene glycol
Organoleptic characteristics	Powerful, creamy, sweet, and buttery odor (less than diacetyl); buttery, cheese taste	Pungent, unpleasant odor reminiscent of butter; taste similar to that of diacetyl
Synthesis	From propionyl aldehyde condensed over ethyl acetylacetate. The resulting product is then oxidized (H_2O_2 and sodium tungstate), hydrolyzed, and finally decarboxylated to 2,3-dexanedione;[7] from methyl butyl ketone or ethyl propyl ketone by way of the monoxime; also from acetoxy mesityl oxide	By condensation of ethyl propionate in the presence of sodium metal, followed by oxidation of the resulting propionin with copper acetate or ferric chloride[6]
Natural occurrence	Reported found in fermented soybean together with diacetyl and acetyl propionyl[8]	Not reported found in nature
Reported uses[3]	Non-alcoholic beverages 6.6 ppm Ice cream, ices, etc. 4.8 ppm Candy 7.3 ppm Baked goods 6.6 ppm	Non-alcoholic beverages 10.0 ppm Ice cream, ices, etc. 10.0 ppm Candy 10.0 ppm Gelatins and puddings 10.0 ppm
Regulatory status	FDA 121.1164; FEMA No. 2558	FEMA No. 3168

REFERENCES

For References 1–5, see end of Part III.

6. **Wegmann and Dahn,** *Helv. Chim. Acta,* 29, 101, 1946.
7. **Igarashi and Midorikawa,** *J. Org. Chem.,* 29, 3113, 1964.
8. **Yasuo et al.,** *Chem. Abstr.,* 60, 11290d, 1964.

	Hexanoic acid	3-Hexanol
Other names	*n*-Caproic acid Hexoic acid *n*-Hexylic acid Pentane-1-carboxylic acid	Ethyl propyl carbinol 3-Hydroxyhexane
Empirical formula	$C_6H_{12}O_2$	$C_6H_{14}O$
Structure	$CH_3{-}CH_2{-}CH_2{-}CH_2{-}CH_2{-}COOH$	$CH_3{-}CH_2{-}CH_2{-}CH(OH){-}CH_2{-}CH_3$
Physical/chemical characteristics[5]		
Appearance	Oily, colorless liquid	
Molecular weight	116.16	102.18
Melting point	−3°C	
Boiling point	205.8°C	*l*-, 131–133°C *dl*- and *l*-, 135°C
Congealing point	−3.24°C	
Specific gravity	0.93136 at 15°C	*dl*- and *l*-, 0.8213 at 20°C *dl*-, 0.8182 at 20°C
Refractive index	1.41877 at 15°C; 1.4150 at 20°C	*d*-, 1.4150 at 20°C *l*-, 1.4140 at 22°C *dl*-, 1.4167 at 20°C
Optical rotation		*dl*-, +6.81 at 20° C (in chloroform) *l*-, −7.17 at 18°C
Solubility	Insoluble in water; slightly soluble in alcohol and ether	Slightly soluble in water; soluble in alcohol and acetone; soluble in ether in all proportions
Organoleptic characteristics	Unpleasant odor reminiscent of copra oil; exhibits an acrid taste	
Synthesis	By fractionation of the volatile fatty acids of coconut oil	By hydroboration of *cis*-3-hexene;[6,7] from 3-hexyne;[8] from 1-(and 2-) hexenes[9]
Natural occurrence	A secondary product of butyric fermentation. Reported found in the essential oils of: lavender, camphor, palmarosa, lemongrass, and *Juniperus phoenicea*; in a few fruital aromas: apple, currant, and strawberry.[1 3] Also identified among the constituents of petitgrain lime oil[1 4]	Reported found in lavender oil,[10] as a volatile flavor component in pineapple,[11] and in the odorous portion of *Vaccinium macrocarpon*[14]
Reported uses[3]	Non-alcoholic beverages 1.8 ppm Ice cream, ices, etc. 4.3 ppm Candy 28 ppm Baked goods 22 ppm Chewing gum 1.5 ppm Condiments 450 ppm	Non-alcoholic beverages 1.0 ppm Ice cream, ices, etc. 1.0 ppm Candy 2.0 ppm Baked goods 2.0 ppm Gelatins and puddings 1.0 ppm Preserves and spreads 1.0 ppm
Regulatory status	FDA 121.1164; FEMA No. 2559	FEMA No. 3351

REFERENCES

For References 1–5, see end of Part III.

6. **Brown, Ayyangar and Zweifer,** *J. Am. Chem. Soc.*, 86(3), 397, 1964.
7. **Brown,** U.S. Patent 3,078,313, 1963.
8. **Brown and Zweifer,** *J. Am. Chem. Soc.*, 83, 3834, 1962.
9. **Brown and Subba Rao,** *J. Org. Chem.*, 22, 1137, 1957.
10. **Staikov, Chingova and Kalaidzhiev,** *Soap Perfum. Cosmet.*, 42(12), 883–887, 1969.
11. **Näf-Müller and Willhalm,** *Helv. Chim. Acta*, 54, 1880, 1971.
12. **Anjou and von Sydow,** *Acta Chem. Scand.*, 21(8) 2076, 1967.
13. **Peyron,** *Riv. Ital. EPPOS*, 8, 413, 1965.
14. **Willhalm et al.,** *Helv. Chim. Acta*, 49, 66, 1966.

	3-Hexanone			
Other names	Ethyl propyl ketone 3-Oxohexane			
Empirical formula	$C_6H_{12}O$			
Structure	$CH_3—CH_2—CH_2—\overset{O}{\overset{\|\|}{C}}—CH_3—CH_3$			
Physical/chemical characteristics[5]				
Molecular weight	100.16			
Boiling point	118°C at 745 mm Hg 123–124°C			
Specific gravity	0.8118 at 20°C			
Refractive index	1.4004 at 20°C			
Solubility	Slightly soluble in water; soluble in acetone; soluble in alcohol and ether in all proportions			
Synthesis	From organo-mercury compounds;[6] by dehydrogenation of 3-hexanol in the presence of Ni-zinc oxide-phosphate catalyst;[7] by hydrogenolysis of 2-ethylfuran;[8] by passing β-furylpropionic acid (or 2-furanpropionic acid) over 5% Pt-C catalyst to 250 to 300°C;[9] by two patented processes[10,11]			
Natural occurrence	Reported found in coffee,[12] as a component in the volatile by-product developed during the catalytic hydrogenation of soybean oil using Ni catalysts;[13] in the scent of green-vegetable bug *Nezara viridula*;[14] a volatile flavor component in pineapple[15]			
Reported uses[3]	Non-alcoholic beverages	20.0 ppm	Meat, meat sauces, soups	20.0 ppm
	Candy	20.0 ppm	Milk, dairy products	20.0 ppm
	Baked goods	20.0 ppm	Cereals	20.0 ppm
	Gelatins and puddings	20.0 ppm		
Regulatory status	FEMA No. 3290			

REFERENCES

For References 1–5, see end of Part III.

6. **Kurts, Beletskaya, Savchenko and Rentov,** *J. Organomet. Chem.*, 17(2), P21, 1969.
7. **Johnson,** British Patent, 1,112,242, 1968.
8. *Chem. Abstr.*, 52, 5369a, 1958.
9. **Bel'skii, Shuinin, Shostakovskii, and An,** *Dokl. Akad. Nauk U.S.S.R.*, 156(4), 861, 1964; *Chem. Abstr.*, 61, 6915d, 1964.
10. **Rheinpreussen A.-G.,** German Patent 1,130,800, 1962.
11. Cincinnati Milling Machine Co., Netherlands Patent Appl. 6,616,520, 1967.
12. **Stoll et al.,** *Helv. Chim. Acta*, 50(2), 628, 1967.
13. **Kawada, Mookherjee and Chang,** *J. Am. Oil Chem. Soc.*, 43(4), 237, 1966.
14. **Gilby and Waterhouse,** *Proc. R. Soc. Ser. B*, 162(986), 105, 1965.
15. **Näf-Müller and Willhalm,** *Helv. Chim. Acta*, 54(7), 1880, 1971.

	2-Hexenal	4-Hexene-3-one
Other names	α,β-Hexylenaldehyde β-Propyl acrolein	
Empirical formula	$C_6H_{10}O$	$C_6H_{10}O$
Structure	$CH_3-CH_2-CH_2-CH=CH-C(=O)-H$	$CH_3-CH_2-C(=O)-CH=CH-CH_3$
Physical/chemical characteristics[5]		
Appearance	Oily liquid	
Molecular weight	98.15	98.14
Boiling point	43°C at 12 mm Hg; 47–48°C at 17 mm Hg	136–137°C 76.2°C at 8.5 mm Hg
Specific gravity	0.8491 at 20°C; 0.8740 at 17.9°C	0.8537 at 20°C
Refractive index	1.4462 at 18°C; 1.44602 at 17.9°C	1.4385 at 20°C
Solubility	Soluble in most organic solvents	
Organoleptic characteristics	Characteristic green, leafy odor	
Synthesis	For the synthesis see reference 8.	By hydration of 4-hexen-2-yne with aqueous H_2SO_4-$HgSO_4$;[6] by hydrogenation of 2-hexen-5-yn-4-one over Pd-$CaCO_3$ catalyst in methanol[7]
Natural occurrence	Reported found in the distillation waters of several plants, such as *Carpinus betulus*;[9] also identified among the constituents of tea (leaves) oil[10] and in citronella[11]	
Reported uses[3]	Non-alcoholic beverages 3.1 ppm Ice cream, ices, etc. 0.70 ppm Candy 15 ppm Baked goods 16 ppm	Non-alcoholic beverages 1.0 ppm Ice cream, ices, etc. 1.0 ppm Candy 1.0 ppm Baked goods 1.0 ppm Gelatins and puddings 1.0 ppm Chewing gum 1.0 ppm Condiments, pickles 1.0 ppm Meat, meat sauces, soups 1.0 ppm Milk, dairy products 1.0 ppm Cereals 1.0 ppm
Regulatory status	FDA 121.1164; FEMA No. 2560	FEMA No. 3352

REFERENCES

For References 1–5, see end of Part III.

6. **Favorskaya, Auvinen and Artsybasheva,** *Zh. Obshch. Khim.,* 28, 1785, 1958; **Favorskaya,** *Chem. Abstr.,* 55, 350g, 1961.
7. **Bondarev and Petrov,** *Zh. Org. Khim.,* 2(5), 782, 1966.
8. **Currius and Franzen,** *Annalen,* 390, 89, 1912.
9. **Janistyn, Hatanaka and Ono,** *Bull. Agric. Chem. Soc. Jap.,* 24, 532, 1960.
10. **Takei et al.,** *Chem. Zentralbl.,* 1, 2770, 1934.
11. **Bohnsack,** *Ber. Dtsch. Chem. Ges. B,* 76, 564, 1943.

	trans-2-Hexenoic acid	3-Hexenoic acid
Other names	Acrylic, β-propyl acid Hexen-2-oic acid α,β-Hexylenic acid α,β-Hexenoic acid	3-Hexenic acid Hydrosorbic acid β-Amylene-α-carboxylic acid 2-Pentene-1-carboxylic acid Propylidenepropionic acid
Empirical formula	$C_6H_{10}O_2$	$C_6H_{10}O_2$
Structure	$CH_3-(CH_2)_2-CH=CH-COOH$	$CH_3-CH_2-CH=CH-CH_2-COOH$
Physical/chemical characteristics[5]		
Appearance	Colorless needle-like crystals	
Molecular weight	114.15	114.15
Melting point	37°C	12°C
Boiling point	217°C 118°C at 19 mm Hg	208°C 118–119°C at 22 mm Hg 103°C at 9–10 mm Hg 81–82°C at 2 mm Hg
Specific gravity	0.965 at 20°C	0.964 at 23°C
Refractive index	1.4467 at 40°C	1.4935 at 20°C
Solubility	Slightly soluble in water; soluble in alcohol, propylene glycol, and oils; very soluble in ether	
Organoleptic characteristics	Pleasant fatty character, long-lasting	
Synthesis	By condensation of butryaldehyde with malonic acid[6]	By condensation of butyraldehyde and malonic acid;[8] also formed during the distillation of ethyl paraconic acid
Natural occurrence	Reported found in Japanese peppermint oil[7]	The *cis*-form has been reported found in yellow passion fruit[9]
Reported uses[3]	Non-alcoholic beverages 5.0 ppm Ice cream, ices, etc. 5.0 ppm Candy 5.0 ppm Baked goods 5.0 ppm Gelatins and puddings 5.0 ppm Chewing gum 1.0 ppm Salad dressings 50.0 ppm	Non-alcoholic beverages 10.0 ppm Ice cream, ices, etc. 10.0 ppm Candy 10.0 ppm Baked goods 10.0 ppm Gelatins and puddings 10.0 ppm Chewing gum 10.0 ppm Milk, dairy products 50.0 ppm
Regulatory status	FEMA No. 3169	FEMA No. 3170

REFERENCES

For References 1–5, see end of Part III.

6. **Boxer and Linstead,** *J. Chem. Soc.,* p.740, 1931.
7. **Walbaum and Rosenthal,** *J. Fr. Chim.,* 124, 63, 1930.
8. **Howton and Davis,** *J. Org. Chem.,* 16, 1405, 1951.
9. **Winter and Klöti,** *Helv. Chim. Acta,* 55(6), 1916, 1972.

2-Hexen-1-ol

Other names	2-Hexenol *trans*-2-Hexenol α,β-Hexenol Leaf alcohol γ-Propyl allyl alcohol
Empirical formula	$C_6H_{12}O$
Structure	$CH_3-CH_2-CH_2-C(H)=C(H)-CH_2OH$
Physical/chemical characteristics[5]	
Appearance	Colorless liquid
Molecular weight	100.16
Boiling point	61°C (155°C) at 12 mm Hg
Specific gravity	0.85
Refractive index	1.4410 at 20°C
Solubility	Very slightly soluble in water; soluble in alcohol and propylene glycol
Organoleptic characteristics	Powerful, fruity, green, wine-like, leafy odor; sweet, fruity flavor important to strawberry and orange juice
Synthesis	The *cis*-form can be prepared by hydrogenation of *cis*-2-hexenol, using an aqueous suspension of colloidal palladium;[6] from *cis*-4-chloro-2-butenol and magnesium ethyl bromide.[7,8] The commercial product is a mixture of the *cis*- and *trans*-isomers and can be prepared from propyl vinyl carbinol by heating with aluminum oxide.
Natural occurrence	Reported found as a constituent of fresh raspberry aroma;[9] also identified in Valencia orange juice[10] and apple aroma,[11] probably occurring as an ester.
Reported uses[3]	Non-alcoholic beverages 1.0 ppm Ice cream, ices, etc. 0.63 ppm Candy 3.8 ppm Baked goods 4.1 ppm
Regulatory status	FDA 121.1164; FEMA No. 2562

REFERENCES

For References 1–5, see end of Part III.

6. **Smets,** *Chem. Abstr.,* 56, 8315, 1950.
7. **Colonge and Poilane,** *Compt. Rend.,* 238, 1821, 1954.
8. **Colonge and Poilane,** *Bull. Soc. Chim. Fr.,* p. 954, 1955.
9. **Winter and Willhalm,** *Helv. Chim. Acta,* 47, 1215, 1964.
10. **Schultz et al.,** *J. Food Sci.,* 29, 790, 1964.
11. **Koch,** *Chem. Abstr.,* 62, 1509e, 1965.

	3-Hexen-1-ol
Other names	Blätter alcohol *cis*-3-Hexenol β,γ-Hexenol *m*-β-Hexenol Leaf alcohol (name also used for *trans*-2-Hexenol)
Empirical formula	$C_6H_{12}O$
Structure	$CH_3-CH_2-CH=CH-CH_2-CH_2OH$ (H, H on the double-bonded carbons)
Physical/chemical characteristics[5]	
Appearance	Colorless, oily liquid
Molecular weight	100.16
Boiling point	156–157°C (55–56°C at 9 mm Hg)
Specific gravity	0.8508 at 15°C
Refractive index	1.4803 at 20°C
Solubility	Soluble in most organic solvents; slightly soluble in water
Organoleptic characteristics	Intense, green odor, not as strong as the corresponding aldehyde; characteristic herbaceous, leafy odor on dilution
Synthesis	Extracted from various essential oils and purified by reacting it to the corresponding phthalate or allophanate.[6] It was synthesized by Ruzička and Schinz,[7] who also clarified its chemical structure.[8] Stoll and Rouve reported on the most significant differences between the natural and the synthetic products.[9]
Natural occurrence	Main constituent of the oil distilled from the infusion of fermented tea leaves;[10] reported found as the corresponding ester of phenylacetic acid in the oil of Japanese mint (*Mentha arvensis*).[11] The volatile oil of *Thea chinensis* contains approximately 26–35% 3-hexen-1-ol, whereas larger amounts are reported in the oils of *Morus bombycis, Robinia pseudacacia,* and *Raphanus sativus.* Probably occurring also in several green leaves and herbs. Reported found in the fruit juices of raspberry,[12] grapefruit,[13] and others.
Reported uses[3]	Non-alcoholic beverages 1.0 ppm Ice cream, ices, etc. 3.7 ppm Candy 5.0 ppm Baked goods 5.0 ppm
Regulatory status	FDA 121.1164; FEMA No. 2563

REFERENCES

For References 1–5, see end of Part III.

6. **Ruzička,** *Helv. Chim. Acta,* 27, 1561, 1944.
7. **Ruzička and Schinz,** *Helv. Chim. Acta,* 17, 1602, 1934.
8. **Ruzička and Schinz,** *Helv. Chim. Acta,* 27, 156, 1944.
9. **Stoll and Rouve,** *Helv. Chim. Acta,* 21, 1542, 1938; *Ber. Dtsch. Chem. Ges.,* 73, 1358, 1940.
10. **Von Romburgh,** *Versl. Plantentuin Buitenzorg,* p. 119, 1895.
11. **Wahlbaum,** *J. Prakt. Chem.,* 2(96), 245, 1917.
12. **Bohnsack,** *Ber. Dtsch. Chem. Ges.,* p. 75, 1942.
13. **Kirchner and Miller,** *Chem. Abstr.,* 47, 8924, 1953.

	4-Hexen-1-ol	2-Hexen-1-yl acetate
Other names	2-Hexen-ol-6 4-Hexenyl alcohol	
Empirical formula	$C_6H_{12}O_2$	$C_8H_{14}O_2$
Structure	$CH_3-CH=CH-CH_2-CH_2-CH_2-OH$	$CH_3-(CH_2)_2-CH=CH-CH_2-O-\overset{\overset{O}{\parallel}}{C}-CH_3$
Physical/chemical characteristics		
Appearance		Liquid
Molecular weight	100.16	142.19
Melting point	69.5°C at 17 mm Hg	
Boiling point	*cis*-, 158–159°C *trans*-, 159–159.4°C	67–68°C at 16 mm Hg
Specific gravity	0.8513 at 20°C	0.898 at 20°C
Refractive index	*cis*-, 1.4420 at 20°C *trans*-, 1.4402 at 20°C	1.42765 at 20°C
Organoleptic characteristics	Pungent oily odor	Pleasant, fruity odor and corresponding taste
Synthesis	By treating *cis*- or *trans*-3-chloro-2-methyl-tetrahydropyrone with sodium in ether[6,7]	By heating at the boil 1-bromohexen-2-ol with sodium acetate and acetic acid
Natural occurrence	The *cis*- and *trans*-forms occur in banana,[8-10] the *cis*-form in passion fruit;[11] in beans (isomer not specified)[12]	Reported found in *Fragaria vesca* and other fruits; only the *trans*-form is known
Reported uses[3]	Non-alcoholic beverages 2.0 ppm Ice cream, ices, etc. 2.0 ppm Candy 4.0 ppm Baked goods 4.0 ppm Gelatins and puddings 2.0 ppm Vegetables 2.0 ppm	Non-alcoholic beverages 0.28 ppm Ice cream, ices, etc. 0.40 ppm Candy 1.7 ppm Baked goods 1.7 ppm
Regulatory status	FEMA No. 3430	FDA 121.1164; FEMA No. 2564

REFERENCES

For References 1–5, see end of Part III.

6. **Brandon et al.,** *J. Am. Chem. Soc.*, 72, 2120, 1950.
7. **Crombie et al.,** *J. Chem. Soc.*, p. 1707, 1957.
8. **Murray et al.,** *J. Food Sci.*, 33, 632, 1968.
9. **Wick et al.,** *J. Agric. Food Chem.*, 17, 751, 1969.
10. **Maroni et al.,** *Riv. Ital. EPPOS*, 55, 168, 1973.
11. **Murray et al.,** *Aust. J. Chem.*, 25, 1921, 1972.
12. **Stevens et al.,** *J. Am. Soc. Hort. Sci.*, 91, 833, 1967.

	cis-3-Hexen-1-yl acetate	*cis*-3-Hexenyl butyrate
Other names	Verdural®	β,γ-Hexenyl-*n*-butyrate *cis*-3-Hexenyl butanoate Leaf butyrate
Empirical formula	$C_8H_{14}O_2$	$C_{10}H_{18}O_2$
Structure	$CH_3-CH_2-CH=CH-CH_2-CH_2-O-C(=O)-CH_3$	$CH_3-CH_2-CH_2-C(=O)-O-CH_2-CH_2-CH=CH-CH_2-CH_3$
Physical/chemical characteristics[5]		
Appearance		Colorless liquid
Molecular weight	142.19	170.25
Boiling point	66°C at 16 mm Hg	192°C at 760 mm Hg 103°C at 25 mm Hg
Specific gravity		0.899 at 25°C
Refractive index		1.4318 at 20°C
Solubility		Almost insoluble in water; soluble in alcohol and propylene glycol; miscible with oils
Organoleptic characteristics	Powerful green, floral note reminiscent of banana	Green, fruity, somewhat buttery aroma
Synthesis	By acetylation of the corresponding alcohol[6]	By esterification of *cis*-3-hexenol with *n*-butyric acid under azeotropic conditions[8]; also [structure: OH (hexenol) + HO–C(=O)–butyl →]
Natural occurrence	Reported found in apple, bilberry, guava, strawberry, and in black and green tea; a flavor component of yellow passion fruit (*Passiflora edulis* f. *flavicarpa*)[7]	Reported found in passion fruit[9]
Reported uses[3]	Non-alcoholic beverages 0.40 ppm Ice cream, ices, etc. 0.40 ppm Candy 0.50 ppm Baked goods 0.50 ppm	Non-alcoholic beverages 0.5 ppm Ice cream, ices, etc. 5.0 ppm Candy 10.0 ppm Gelatins and puddings 5.0 ppm
Regulatory status	FEMA No. 3171	FEMA No. 3402

REFERENCES

For References 1–5, see end of Part III.

6. **Bohnsack,** *Chem. Ber.,*, 75, 74, 1942.
7. **Winter and Klöti,** *Helv. Chim. Acta,* 55(6), 1916, 1972.
8. **Arctander,** *Perfume and Flavor Chemicals,* Montclair, N.J., 1969, No. 1613.
9. **Eckert et al.,** *J. Agric. Food Chem.*, 20, 104, 1972.

	3-Hexenyl formate		*cis*-3-Hexenyl formate[6]	
Other names	3-Hexenyl methanoate			
Empirical formula	$C_7H_{12}O_2$		$C_7H_{12}O_2$	
Structure	$H_3C-CH_2-CH=CH-CH_2-CH_2-O-CHO$		$H-C(=O)-O-CH_2-CH_2-CH=CH-CH_2-CH_3$	
Physical/chemical characteristics[5]				
Appearance	Colorless liquid			
Molecular weight	128.17		128.17	
Boiling point	155°C 85°C at 75 mm Hg			
Congealing point	53–55°C at 12 mm Hg			
Specific gravity	0.908 at 25°C/25°C			
Refractive index	1.4270 at 20°C			
Solubility	Almost insoluble in water; soluble in alcohol, propylene glycol, and oils			
Organoleptic characteristics	Green, vegetable-like odor		Light topnate; fruity fresh odor	
Synthesis	From *cis*-3-hexenol and formic acid[6]			
Reported uses[3]	Non-alcoholic beverages	1.0 ppm	Non-alcoholic beverages	0.15 ppm
	Ice cream, ices, etc.	1.0 ppm	Ice cream, ices, etc.	0.2 ppm
	Candy	1.0 ppm	Candy	0.5 ppm
	Baked goods	2.0 ppm	Baked goods	0.5 ppm
	Gelatins and puddings	2.0 ppm	Gelatins and puddings	0.2 ppm
	Chewing gum	1.0 ppm	Milk, dairy products	0.2 ppm
	Preserves and spreads	1.0 ppm		
Regulatory status	FEMA No. 3353		FEMA No. 3431	

REFERENCES

For References 1–5, see end of Part III.

6. **Bohnsack,** *Chem. Ber.*, 75, 74, 1942.

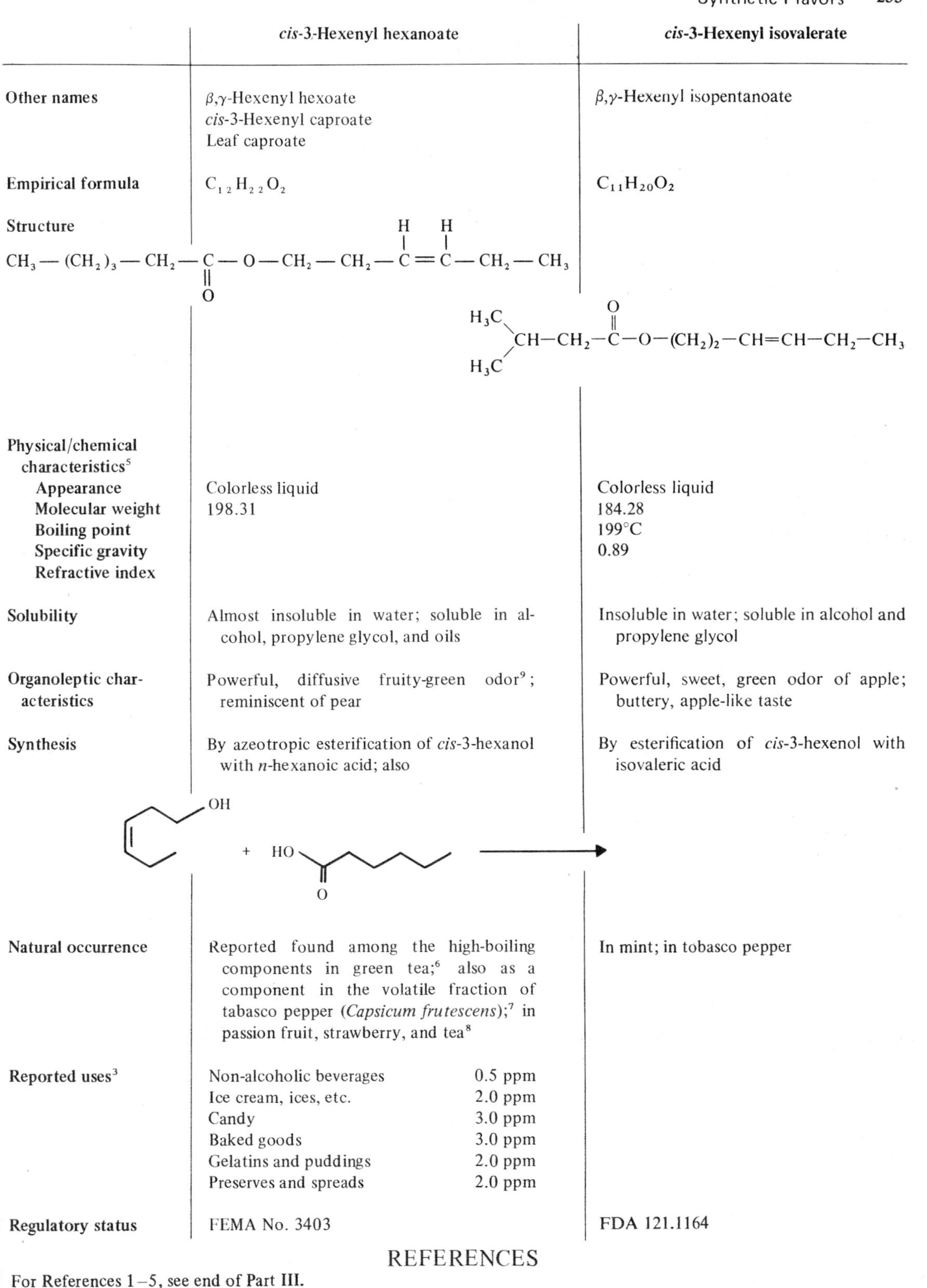

	cis-3-Hexenyl hexanoate	*cis*-3-Hexenyl isovalerate
Other names	β,γ-Hexenyl hexoate *cis*-3-Hexenyl caproate Leaf caproate	β,γ-Hexenyl isopentanoate
Empirical formula	$C_{12}H_{22}O_2$	$C_{11}H_{20}O_2$
Structure	$CH_3-(CH_2)_3-CH_2-C(=O)-O-CH_2-CH_2-CH{=}CH-CH_2-CH_3$	$(H_3C)_2CH-CH_2-C(=O)-O-(CH_2)_2-CH{=}CH-CH_2-CH_3$
Physical/chemical characteristics[5]		
Appearance	Colorless liquid	Colorless liquid
Molecular weight	198.31	184.28
Boiling point		199°C
Specific gravity		0.89
Refractive index		
Solubility	Almost insoluble in water; soluble in alcohol, propylene glycol, and oils	Insoluble in water; soluble in alcohol and propylene glycol
Organoleptic characteristics	Powerful, diffusive fruity-green odor[9]; reminiscent of pear	Powerful, sweet, green odor of apple; buttery, apple-like taste
Synthesis	By azeotropic esterification of *cis*-3-hexanol with *n*-hexanoic acid; also OH + HO … O →	By esterification of *cis*-3-hexenol with isovaleric acid
Natural occurrence	Reported found among the high-boiling components in green tea;[6] also as a component in the volatile fraction of tabasco pepper (*Capsicum frutescens*);[7] in passion fruit, strawberry, and tea[8]	In mint; in tobasco pepper
Reported uses[3]	Non-alcoholic beverages 0.5 ppm Ice cream, ices, etc. 2.0 ppm Candy 3.0 ppm Baked goods 3.0 ppm Gelatins and puddings 2.0 ppm Preserves and spreads 2.0 ppm	
Regulatory status	FEMA No. 3403	FDA 121.1164

REFERENCES

For References 1–5, see end of Part III.

6. **Nose,** *Agric. Biol. Chem.*, 35, 261, 1971.
7. **Wendell et al.,** *J. Agric. Food Chem.*, 19, 1131, 1971.
8. Volatile Compounds in Food, T.N.O. Report No. 4030, 3rd ed., Central Instituut voor Voedingsonderzoek, Zeist, Netherlands.
9. **Arctander,** *Perfume and Flavor Chemicals*, Montclair, N.J., 1969, No. 1616.

	cis-3-Hexenyl-2-methylbutyrate	Hexyl acetate
Other names	*cis*-3-Hexenyl-α-methylbutyrate	
Empirical formula	$C_{11}H_{20}O_2$	$C_8H_{16}O_2$
Structure	$CH_3-CH_2-\underset{}{\overset{H}{C}}=\overset{H}{C}-(CH_2)_2-O-\overset{O}{\overset{\parallel}{C}}-\overset{CH_3}{\overset{\mid}{CH}}-CH_2-CH_3$	$CH_3-(CH_2)_4-CH_2-O-\overset{O}{\overset{\parallel}{C}}-CH_3$
Physical/chemical characteristics[5]		
Appearance	Colorless liquid	Oily, colorless liquid
Molecular weight	184.28	144.21
Melting point		−81°C
Boiling point		171–172°C; 61–62°C at 12 mm Hg
Specific gravity		0.8718 at 20°C
Refractive index		1.4092 (1.4120) at 20°C
Solubility	Insoluble in water; soluble in alcohol	Insoluble in water; soluble in alcohol and ether
Organoleptic characteristics	Strong, warm, fruity odor of apple and pineapple; sweet, apple-like taste	Pleasant fruity odor and bittersweet taste suggestive of pear
Synthesis	From *cis*-3-hexenol via esterification with α-methylbutyric acid	From *n*-hexyl alcohol and excess acetic anhydride at the boil or with an excess of acetic acid in the presence of concentrated sulfuric acid
Natural occurrence	In mint	Reported found in fruital aromas (e.g., *Fragaria vesca*) and essential oils
Reported uses[3]		Non-alcoholic beverages 4.6 ppm Ice cream, ices, etc. 4.6 ppm Candy 26 ppm Baked goods 26 ppm Chewing gum 3.0 ppm
Regulatory status	FDA 121.1164	FDA 121.1164; FEMA No. 2565

REFERENCES

For References 1–5, see end of Part III.

	2-Hexyl-4-acetoxytetrahydrofuran*	Hexyl alcohol
Other names	2-Hexyl-tetrahydrofuran-4-yl acetate	Alcohol C-6 1-Hexanol
Empirical formula	$C_{12}H_{22}O_3$	$C_6H_{14}O$
Structure	$H_3C-C(=O)-O-$ (tetrahydrofuran ring, H substituents) $CH_2-(CH_2)_4-CH_3$	$CH_3-(CH_2)_4-CH_2OH$
Physical/chemical characteristics[5]		
Appearance	Colorless liquid	Colorless liquid
Molecular weight	214.31	102.17
Melting point		−51.6°C
Boiling point		157.2–157.8°C
Specific gravity		0.8186 at 20°C
Refractive index		1.4179 at 20°C
Solubility	Slightly soluble in water; soluble in alcohol	Rather soluble in water; soluble in most organic solvents
Organoleptic characteristics	Sweet, floral-fruity odor; sweet, powerful, fruity (peach-apricot) taste	Fruity odor and aromatic flavor
Synthesis		By reduction of *n*-caproic acid. The *n*-hexyl alcohol represents one of the 14 possible isomers of this alcohol.
Natural occurrence	Not reported found in nature	Reported found among the constituents of several essential oils and aromas: apple, strawberry, tea, violet (leaves and flowers), Java citronella, Bourbon geranium, lavender, lavandin, spike,[7] *Litsea zeylanica;*[8] also identified in bitter orange[9]
Reported uses[3]	Non-alcoholic beverages 1.0 ppm Ice cream, ices, etc. 3.0 ppm Candy 3.0 ppm Baked goods 3.0 ppm	Non-alcoholic beverages 6.6 ppm Ice cream, ices, etc. 26 ppm Candy 21 ppm Baked goods 18 ppm Gelatins and puddings 0.22–0.28 ppm
Regulatory status	FDA 121.1164; FEMA No. 2566	FDA 121.1164; FEMA No. 2567

* 2-Hexyl-4-acetoxytetrahydrofuran has been found not to have the identity indicated by this name, but instead to contain several non-isomeric components. In view of this and shifting considerations of technological value, it is being dropped from the GRAS list.[6]

REFERENCES

For References 1–5, see end of Part III.

6. **Hall and Oser,** *Food Technol.,* 24, 533, 1970.
7. **Girard,** *Chem. Abstr.,* 44, 4635, 1950.
8. **Sharma et al.,** *Chem. Abstr.,* 47, 11667, 1953.
9. **Mehlitz and Minas,** *Ind. Obst Gemueseverwert.,* 50, 861, 1965.

	n-Hexyl-2-butenoate	Hexyl butyrate
Other names	Hexyl crotonate	
Empirical formula	$C_{10}H_{20}O_2$	$C_{10}H_{20}O_2$
Structure	$CH_3-CH=CH-C(=O)-O-(CH_2)_5-CH_3$	$CH_3-(CH_2)_2-C(=O)-O-(CH_2)_5-CH_3$
Physical/chemical characteristics[5]		
Appearance		Liquid
Molecular weight	172.27	172.27
Melting point		−78°C
Boiling point		208°C
Specific gravity		0.8567 at 30°C
Organoleptic characteristics		Characteristic fruity (apricot) odor and sweet taste suggestive of pineapple
Synthesis		From butyric acid and *n*-hexyl alcohol in the presence of HCl[6]
Natural occurrence	Reported as occurring naturally in banana, passion fruit, wine, cocoa, coffee and soybean	Reported found in the essential oils of lavender and lavandin[7] and in the oil from fruits of *Heracleum giganteum*[8]
Reported uses[3]	Non-alcoholic beverages 8.0 ppm Ice cream, ices, etc. 8.0 ppm Candy 10.0 ppm Baked goods 10.0 ppm Gelatins and puddings 8.0 ppm Preserves 8.0 ppm	Non-alcoholic beverages 2.6 ppm Ice cream, ices, etc. 2.1 ppm Candy 7.8 ppm Baked goods 8.6 ppm
Regulatory status	FEMA No. 3354	FDA 121.1164; FEMA No. 2568

REFERENCES

For References 1–5, see end of Part III.

6. Bilterys and Gisseleire, *Bull. Soc. Chim. Belg.*, 44, 569, 1935.
7. Peyron and Benezet, *C. R. Séances Acad. Sci. Paris*, 262, 1105, 1966.
8. Franchimont and Zincke, *Ann. Chem.*, 163, 198, 1872.

	α-Hexyl cinnamaldehyde	Hexyl formate
Other names	2-Benzylidene-octanal α-*n*-Hexyl cinnamic aldehyde α-*n*-Hexyl-β-phenyl acrolein	
Empirical formula	$C_{15}H_{20}O$	$C_7H_{14}O_2$
Structure	$C_6H_5-CH=C(-CHO)-(CH_2)_5-CH_3$	$H-C(=O)-O-CH_2-(CH_2)_4-CH_3$
Physical/chemical characteristics[5]		
Appearance	Pale-yellow liquid	Colorless liquid
Assay	95% min[2]	
Molecular weight	216.33	130.18
Melting point	4°C	
Boiling point	174–176°C at 15 mm Hg	176.7°C
Specific gravity	0.953–0.959 at 25°/25°C;[2] (0.95 at 24°C)	0.8789 at 20°C; (0.898 at 0°C)
Refractive index	1.5480–1.5520 at 20°C;[2] (1.5268 at 25°C)	1.4071 at 20°C
Acid value	Not more than 5.0[2]	
Solubility	1:1 and more in 90% alcohol[2]	Very slightly soluble in water; miscible with alcohol and ether
Organoleptic characteristics	Jasmine-like odor, particularly on dilution	Fruity, apple-like or unripe-plum odor; corresponding sweet taste
Synthesis	By condensation of octylaldehyde with benzaldehyde	By prolonged boiling of *n*-hexyl alcohol and formic acid,[6] or by azeotropic distillation of the alcohol and isopropyl formate[7]
Natural occurrence	Not reported found in nature	In peach, grape, and apple
Reported uses[3]	Non-alcoholic beverages 0.80 ppm Ice cream, ices, etc. 2.6 ppm Candy 6.5 ppm Baked goods 2.4 ppm Gelatins and puddings 0.05 ppm	Non-alcoholic beverages 12 ppm Ice cream, ices, etc. 45 ppm Candy 39 ppm Baked goods 52 ppm
Regulatory status	FDA 121.1164; FEMA No. 2569	FDA 121.1164; FEMA No. 2570

REFERENCES

For References 1–5, see end of Part III.

6. **Vogel,** *J. Chem. Soc.,* p. 628, 1948.
7. **Cousort,** Electrochemical Industries, German Patent 721,300, 1939.

	Hexyl hexanoate	Hexyl isobutyrate
Other names	Hexyl caproate Hexyl capronate Hexyl hexylate	
Empirical formula	$C_{12}H_{24}O_2$	$C_{10}H_{20}O_2$
Structure	$CH_3-(CH_2)_4-C(=O)-O-(CH_2)_5-CH_3$	$(H_3C)_2CH-C(=O)-O-CH_2-(CH_2)_4-CH_3$
Physical/chemical characteristics[5]		
Appearance	Oily liquid	Colorless liquid
Assay	97% min	
Molecular weight	200.32	172.27
Melting point	−55°C	
Boiling point	245°C; 84–86°C at 3 mm Hg	199°C
Flash point	68°C	
Specific gravity	0.8541 at 30°C; 0.8750–0.8780 at 15°C	0.87
Refractive index	1.4070–1.4090 at 20°C; 1.4140 at 20°C	
Acid value	0.2 max	
Solubility	1:2–2.5 in 70% alcohol	Almost insoluble in water; soluble in alcohol and propylene glycol; miscible with oils
Organoleptic characteristics	Herbaceous odor	Powerful, somewhat harsh fruity aroma
Synthesis	By passing *n*-hexyl alcohol over CuO + UO_3 catalyst at 220–310°C,[6] or by treating *n*-hexyl alcohol with $Ca(BrO_3)_2$ and diluted aqueous HBr at 30°C[7,8]	By direct esterification of *n*-hexanol with isobutyric acid
Natural occurrence	Reported found in several natural products	Reported found in lavender oil,[9] in hops oil[10–12] and in sherry[13]
Reported uses[3]	Non-alcoholic beverages 2.5–3.0 ppm Ice cream, ices, etc. 2.5 ppm Candy 3.6–10 ppm Baked goods 10 ppm	Non-alcoholic beverages 20.0 ppm Ice cream, ices, etc. 20.0 ppm Candy 20.0 ppm Gelatins and puddings 20.0 ppm Chewing gum 20.0 ppm
Regulatory status	FDA 121.1164; FEMA No. 2572	FEMA No. 3172

REFERENCES

For References 1–5, see end of Part III.

6. Iwannikow, *Chem. Abstr.*, 34, 7874, 1940.
7. Farkas and Schachter, U.S. Patent 2,444,924, 1945.
8. Farkas and Schachter, *J. Am. Chem. Soc.*, 71, 2828, 1949.
9. Steltenkamp and Casazza, *J. Agric. Food Chem.*, 15(6), 1063, 1967.
10. Buttery, Black and Kealy, *J. Chromatogr.*, 18(2), 399, 1965.
11. Buttery, McFadden and Black, *Brew. Chem. Proc.*, p. 137, 1964.
12. Buttery, Black, Kealy and McFadden, *Nature*, 202(4933), 701, 1964.
13. Webb, Kepner and Galetto, *Am. J. Enol. Vitic.*, 15(1), 1, 1964.

	Hexyl isovalerate	Hexyl 2-methylbutyrate
Other names	*n*-Hexyl isopentanoate	2-Methylbutanoic acid, *n*-hexylester
Empirical formula	$C_{11}H_{22}O_2$	$C_{11}H_{22}O_2$
Structure	$(H_3C)_2CH-CH_2-C(=O)-O-(CH_2)_5-CH_3$	$CH_3-CH_2-CH(CH_3)-C(=O)-O-C_6H_{13}$
Physical/chemical characteristics[5]		
Appearance	Colorless liquid	Colorless liquid
Molecular weight	186.30	186.30
Boiling point	215°C	
Specific gravity	0.87	
Solubility	Insoluble in water; soluble in alcohol	Almost insoluble in water; soluble in alcohol
Organoleptic characteristics	Odor suggestive of unripe fruit; mainly employed in perfumery (earthy and tobacco notes)	Strong, green, fruity odor; sweet, fruity taste reminiscent of unripe strawberry
Synthesis	By esterification of *n*-hexanol with isovaleric acid	By esterification of *n*-hexanol with 2-methylbutanoic acid
Regulatory status	FDA 121.1164	FDA 121.1164

REFERENCES

For References 1–5, see end of Part III.

	Hexyl octanoate	Hexyl propionate
Other names	Hexyl caprylate *n*-Hexyl-*n*-octanoate *n*-Hexyl-*n*-octoate *n*-Hexyl octylate	*n*-Hexyl propanoate
Empirical formula	$C_{14}H_{28}O_2$	$C_9H_{18}O_2$
Structure	$CH_3-(CH_2)_6-C(=O)-O-(CH_2)_5-CH_3$	$CH_3-CH_2-C(=O)-O-(CH_2)_5-CH_3$
Physical/chemical characteristics[5]		
Appearance	Liquid	Liquid
Molecular weight	228.37	158.24
Boiling point		180°C
Melting point	−31°C	
Specific gravity	277°C (260°C)	0.880 at 15.5°C
Refractive index	0.8527 (0.87) at 30°C	1.4105 at 20°C
Solubility	Insoluble in water; soluble in alcohol	Almost insoluble in water; soluble in alcohol and propylene glycol
Organoleptic characteristics	Fresh vegetable and slightly fruity odor; sweet, green, fruity taste	Earthy, acrid odor suggestive of rotting fruits; sweet, metallic-fruity taste
Synthesis	By esterification of *n*-hexanol with caproic acid	By esterification of *n*-hexanol with propionic acid
Natural occurrence	Reported found in several natural products (apple, strawberry, and banana)	Reported found in hops and apple
Reported uses[3]	Non-alcoholic beverages 1.0 ppm Gelatins and puddings 0.70 ppm	Non-alcoholic beverages 5.7 ppm Ice cream, ices, etc. 23 ppm Candy 21 ppm Baked goods 22 ppm
Regulatory status	FDA 121.1164; FEMA No. 2575	FDA 121.1164; FEMA No. 2576

REFERENCES

For References 1–5, see end of Part III.

	1-Hydroxy-2-butanone		Hydroxycitronellal	
Other names	2-Oxo-1-butanol Propionyl carbinol Ethyl hydroxymethyl ketone 1-Butanol-2-one		3,7-Dimethyl-7-hydroxy octanal 7-Hydroxy-3,7-dimethyl octan-1-al Laurine	
Empirical formula	$C_4H_8O_2$		$C_{10}H_{20}O_2$	
Structure	$CH_3-CH_2-C(=O)-CH_2OH$		(structure: CH_3, CHO, H_3C, CH_3)	
Physical/chemical characteristics[5]			The physical constants for commercial products vary, depending on the source and the method of preparation.	
Appearance	Colorless liquid		Colorless, viscous liquid	
Assay			95% min[2] (97%) (94–96%)	
Molecular weight	88.02		172.27	
Boiling point	152–154°C		94–96°C at 1 mm Hg	
Flash point	1.0186 at 20°C		132°C	
Specific gravity	1.4200 at 20°C		0.917–0.921 at 25°/25°C;[2] 0.920; 0.925–0.930 at 15°C	
Refractive index			1.4470–1.4500 at 20°C;[2] 1.449; 1.448–1.450	
Acid value			Not more than 5.0[2]	
Optical rotation			+9° to +10.5° (Java citronella type) +0.5 to −0.5 (*Eucalyptus citriodora* type)	
Solubility	Miscible with water, alcohol, and ether		1:1 and more in 50% alcohol;[2] slightly soluble in water and glycerol; soluble in essential oils	
Organoleptic characteristics			Intense, sweet, floral, lily-type odor	
Synthesis	From 1-chlorobutan-2-one by hydrolysis or by heating the chloro compound with potassium formate in methanol;[6] the ethyl ester may be prepared by bacterial oxidation of the corresponding glycol with *Aspergillus niger*[7]		By hydration of natural citronellal obtained from Java citronella or from *Eucalyptus citriodora*. β-Pinene is converted to myrcene, which on hydration may yield either linalool or a mixture of geraniol and nerol. The latter mixture can be hydrogenated to citronellol and subsequently converted to citronellal and hydroxycitronellal.[6] Also by hydrogenation of 3,7-dimethyl-7-hydroxy-2-octen-2-al over palladium carbon in ethyl acetate solution[7]	
Natural occurrence			Not reported found in nature	
Reported uses[3]	Non-alcoholic beverages	80.0 ppm	Non-alcoholic beverages	3.5 ppm
	Ice cream, ices, etc.	80.0 ppm	Ice cream, ices, etc.	13 ppm
	Candy	80.0 ppm	Candy	9.4 ppm
	Gelatins and puddings	80.0 ppm	Baked goods	10 ppm
			Gelatins and puddings	0.30 ppm
			Chewing gum	16 ppm
Regulatory status	FEMA No. 3173		FDA 121.1164; FEMA No. 2583	

REFERENCES

For References 1–5, see end of Part III.

6. **Damilev et al.,** *J. Gen. Chem. U.S.S.R.*, 18, 1956, 1948.
7. **Walti,** *J. Am. Chem. Soc.*, 56, 2723, 1934.
8. **Erni** (ABRAC), *Oru. Perfum. Cosmet.*, 78, 37, 1963.
9. Hoffmann-LaRoche & Co., Belgian Patent 636,402, 1964; *Chem. Abstr.*, 61, 13354c, 1964.

For further information on hydroxycitronellal see the following:

Webb (Glidden Co.), U.S. Patent 3,028,431, 1962; *Chem. Abstr.*, 61, 14727b, 1964.

	Hydroxycitronellal diethyl acetal	Hydroxycitronellal dimethyl acetal
Other names	8,8-Diethoxy-2,6-dimethyl-octanol-2	8,8-Dimethoxy-2,6-dimethyl-octanol-2
Empirical formula	$C_{14}H_{30}O_3$	$C_{12}H_{26}O_3$
Structure	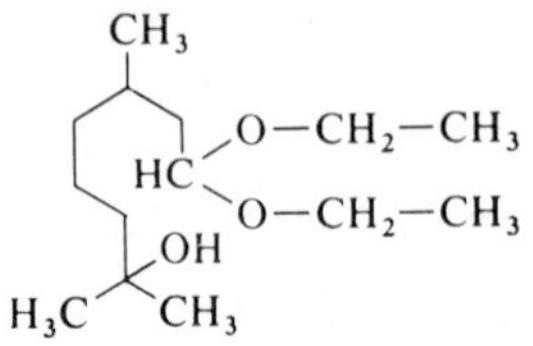 CH_3; HC(O—CH_2—CH_3)(O—CH_2—CH_3); OH; H_3C CH_3	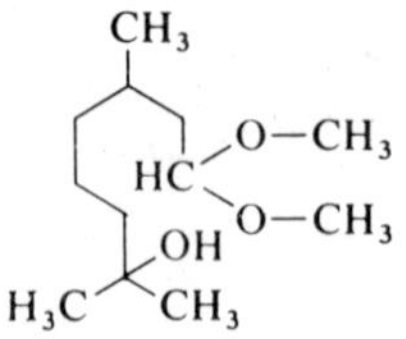 CH_3; HC(O—CH_3)(O—CH_3); OH; H_3C CH_3
Physical/chemical characteristics[5]		
Appearance	Colorless, slightly oily liquid	Colorless liquid
Assay		95% min[2]
Molecular weight	246.39	218.34
Boiling point	Approx 260°C	
Flash point		>100°C
Specific gravity		0.925–0.930 at 25°/25°C[2]
Refractive index		1.4410–1.4440 at 20°C[2]
Acid value		Not more than 1[2]
Free aldehyde		Not more than 3% (as hydroxycitronellal)[2]
Solubility	Almost insoluble in water; soluble in alcohol	1:2 in 50% alcohol[2]
Organoleptic characteristics	Delicate, green-floral taste with almost no odor	Light, green, flowery odor similar to hydroxycitronellal
Synthesis	By condensation of hydroxycitronellal with ethanol using dry HCl	Usually prepared from hydroxycitronellal and methyl alcohol in the presence of a catalyst
Natural occurrence	Not reported found in nature	Not reported found in nature
Reported uses[3]	Non-alcoholic beverages 2.7 ppm Ice cream, ices, etc. 0.50–1.0 ppm Candy 7.3 ppm Baked goods 2.2 ppm	Non-alcoholic beverages 10 ppm Ice cream, ices, etc. 0.50 ppm Candy 24 ppm Baked goods 0.50–20 ppm
Regulatory status	FDA 121.1164; FEMA No. 2584	FDA 121.1164; FEMA No. 2585

REFERENCES

For References 1–5, see end of Part III.

	Hydroxycitronellol	4-Hydroxy-2,5-dimethyl-3(2H)-furanone
Other names	3,7-Dimethyloctane-1,7-diol 3,7-Dimethyl-1,7-octanediol	2,5-Dimethyl-4-hydroxy-2,3-dihydro-furan-3-one
Empirical formula	$C_{10}H_{22}O_2$	$C_6H_8O_3$
Structure	CH_3; CH_2OH; OH; H_3C; CH_3	O; OH; H_3C; O; CH_3
Physical/chemical characteristics[5]		
Appearance	Colorless, viscous oil	Colorless or white solid
Molecular weight	174.29	128
Boiling point	156°C at 15 mm Hg; 124–125°C at 1 mm Hg	
Melting point		70°C (75–78°C)
Flash point	>100°C	
Specific gravity	0.922–0.930 at 25°/25°C; 0.935 at 20°C	
Refractive index	1.4550–1.4600 (1.4610) at 20°C	
Optical rotation	+1°9′ at 20°C	
Aldehyde content	Not more than 10% (as hydroxycitronellal)	
Solubility	Slightly soluble in toluene and benzene; 1:1 in 60% alcohol	Slightly soluble in water; soluble in alcohol and oils
Organoleptic characteristics	Odor reminiscent of rose and grape hyacinth	Fruity caramel or "burnt pineapple" aroma
Synthesis	By hydrogenation (under pressure in presence of Raney Ni) of 3,7-dimethyl-7-hydroxy-octan-l-al;[6] also in good yields by catalytic hydrogenation (using Raney Ni) of 1,2-epoxy-3,7-dimethyl-octan-7-ol[7]	From dimethyl-3,4-dihydroxyfuran-2,5-dicarboxylate[8]
Natural occurrence	Not reported found in nature	Reported as a constituent of pineapple aroma
Reported uses[3]	Non-alcoholic beverages 2.0 ppm Ice cream, ices, etc. 1.6 ppm Candy 3.6 ppm Baked goods 3.5 ppm Gelatins and puddings 0.30 ppm Chewing gum 0.30 ppm	Ice cream, ices, etc. 5.0 ppm Candy 10.0 ppm Baked goods 10.0 ppm Gelatins and puddings 5.0 ppm
Regulatory status	FDA 121.1164; FEMA No. 2586	FEMA No. 3174

REFERENCES

For References 1–5, see end of Part III.
6. **Palfray, Sabetay, and Rangel,** *Compt. Rend.*, 212, 912, 1941.
7. Webb (Glidden Co.) U.S. Patent 3,028,431, 1962.
8. **Henry and Silverstein,** *J. Org. Chem.*, 31, 2391, 1966.

	6-Hydroxy-3,7-dimethyl octanoic acid, lactone	N-(4-Hydroxy-3-methoxybenzyl)-8-methyl-6-nonenamide
Other names	6-Hydroxy-3,7-dimethyl caprylic acid, lactone 4-Methyl-7-isopropyl-2-oxoepanone	Capsaicin Isodecenoic acid, vanillylamide *trans*-8-Methyl-N-vanillyl-6-nonenamide 7-Methyloct-5-ene-1-carboxylic acid, vanillylamide
Empirical formula	$C_{10}H_{18}O_2$	$C_{18}H_{27}NO_3$
Structure	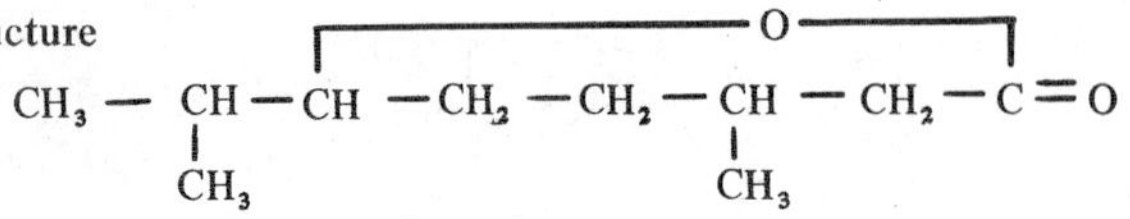 $CH_3-CH(CH_3)-CH-CH_2-CH_2-CH(CH_3)-CH_2-C{=}O$ (ring closed through O)	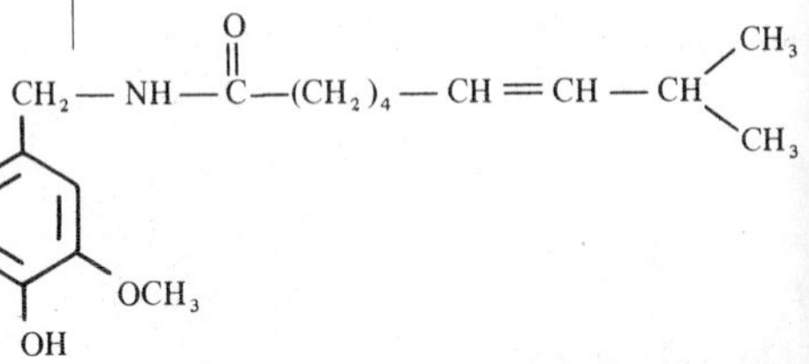 $CH_2-NH-C(=O)-(CH_2)_4-CH{=}CH-CH(CH_3)_2$ on 4-hydroxy-3-methoxyphenyl ring (OH, OCH_3)
Physical/chemical characteristics[5]		
Appearance		Monoclinic rectangular scales or platelets (from petroleum ether)
Molecular weight	170.24	305.42
Melting point		65°C
Boiling point		210–220°C at 0.01 mm Hg
Solubility		Insoluble in water at room temperature; slightly soluble in warm water and carbon disulfide; soluble in alcohol, ether, benzene, and chloroform
Organoleptic characteristics		Mild, warm-herbaceous odor; burning pungent taste (10 ppm);[12] used in compounded flavors for sauces where the pungent note is desired
Synthesis		From 3-chloro-2-isopropyltetrahydropyran;[6] biosynthesis from *Capsicum frutescens*;[8,9] separation from *cis*-capsaicin, pelargonic acid vanilamide, and dihydrocapsaicin,[10] reaction of capsaicin;[11] 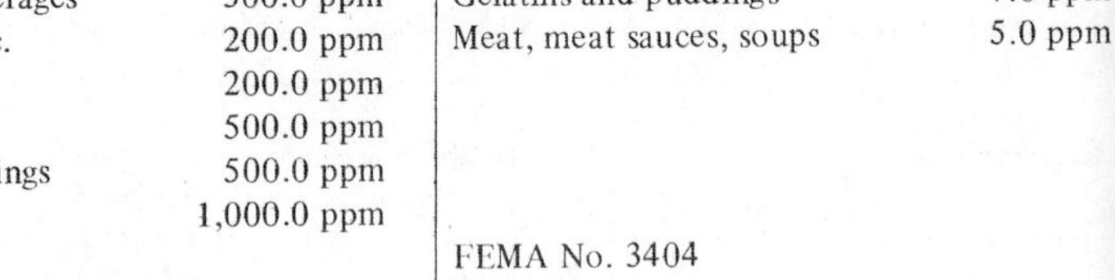
		(vanillylamine: CH_2-NH_2, OH, OCH_3 ring) + (Cl–C(=O)–chain–CH=CH–CH(CH_3)$_2$) $\xrightarrow{B^\ominus}$
Natural occurrence		The pungent principle in the fruits of various *Capsicum* species (Solanaceae)[7]
Reported uses[3]	Non-alcoholic beverages 300.0 ppm Ice cream, ices, etc. 200.0 ppm Candy 200.0 ppm Baked goods 500.0 ppm Gelatins and puddings 500.0 ppm Syrups 1,000.0 ppm	Gelatins and puddings 7.0 ppm Meat, meat sauces, soups 5.0 ppm
Regulatory status	FEMA No. 3355	FEMA No. 3404

REFERENCES

For References 1–5, see end of Part III.

6. Crombie et al., *J. Chem. Soc.*, p. 1025, 1955.
7. Mathew and Nambudiri, *Flavour Ind.*, 2, 691, 1971.
8. Bennett and Kirby, *J. Chem. Soc.*, p. 442, 1968.
9. Leete and Louden, *J. Am. Chem. Soc.*, 90, 6837, 1968.
10. Rancouonwala, *J. Chromatogr.*, 41, 265, 1969.
11. Walker, *Manuf. Chem.*, 39, 35, 1968.
12. Arctander, *Perfume and Flavor Chemicals*, Montclair, N.J., 1969, No. 567.

	n-(4-Hydroxy-3-methoxybenzyl)-nonanamide	2-Hydroxymethyl-6,6-dimethyl bicyclo[3,1,1]hept-2-enyl formate
Other names	*n*-Nonanoyl-4-hydroxy-3-methoxybenzyl-amide Nonanoyl vanillylamide Pelargonyl vanillylamide	Myrtenyl formate
Empirical formula	$C_{17}H_{27}NO_3$	$C_{11}H_{16}O_2$
Structure	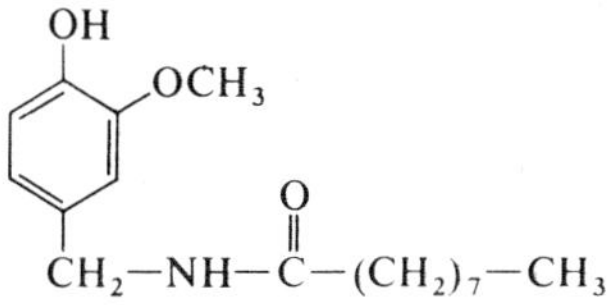 OH, OCH_3 (benzene ring) $CH_2-NH-C(=O)-(CH_2)_7-CH_3$	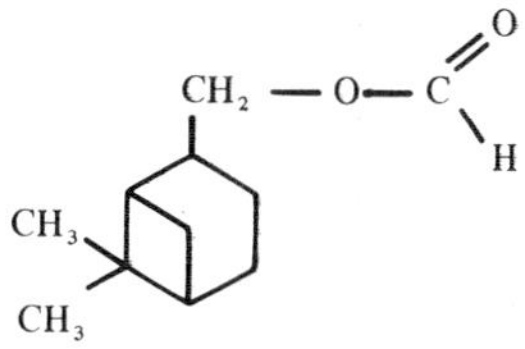 $CH_2-O-C(=O)H$; CH_3, CH_3 (bicyclic ring)
Physical/chemical characteristics		
Appearance	White powder (colorless crystal)	
Molecular weight	293.41	180.24
Solubility	Slightly soluble in water; soluble in alcohol	
Organoleptic characteristics	Odorless; pungent, burning taste	
Synthesis	From nonanyl chloride and vanillylamine	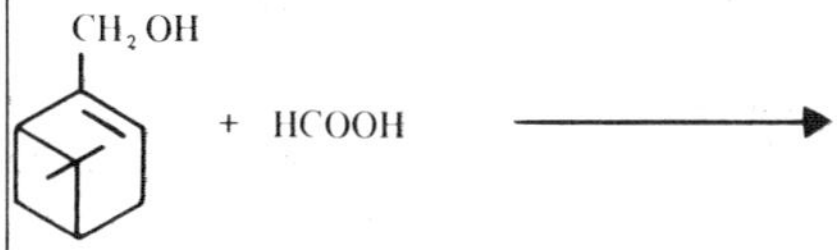 CH_2OH (bicyclic ring) + HCOOH ⟶
Natural occurrence	Not reported found in nature	
Reported uses[3]		Non-alcoholic beverages 3.0 ppm Ice cream, ices, etc. 3.0 ppm Candy 3.0 ppm Baked goods 3.0 ppm Gelatins and puddings 3.0 ppm Condiments, pickles 3.0 ppm Preserves and spreads 3.0 ppm
Regulatory status	FDA 121.1164	FEMA No. 3405

REFERENCES

For References 1–5, see end of Part III.

	3-(Hydroxymethyl)-2-octanone		Hydroxynonanoic acid, Δ-lactone	
Other names			δ-Nonalactone α,*n*-Butyl-δ-valerolactone δ-Hydroxypelargonic acid, lactone 6-Butyltetrahydro-2H-pyran-2-one	
Empirical formula	$C_9H_{18}O_2$		$C_{19}H_{16}O_2$	
Structure	$CH_3—(CH_2)_4—CH(CH_2OH)—CO—CH_3$			
Physical/chemical characteristics[5]				
Appearance	Colorless oily liquid		Colorless, very viscous liquid	
Molecular weight	158.24		156.23	
Boiling point			137.0–137.5°C at 11 mm Hg	
Congealing point			73°C at 0.095 mm Hg	
Specific gravity			0.9871 at 20°C	
Refractive index			1.4562 at 20°C	
Solubility	Almost insoluble in water; soluble in alcohol and oils		Almost insoluble in water; soluble in alcohol and oils	
Organoleptic characteristics	Musty-herbaceous, sweet, slightly earthy odor		Mild, nut-like odor; fatty, milk-creamy taste	
Synthesis	By condensation of methyl hexyl ketone with formaldehyde, followed by hydrogenation		By microbiological reduction of the corresponding keto acids;[6,7] by a patented process[8]	
Natural occurrence			Reported as responsible for the off-flavor in milk-fat[9]	
Reported uses[3]	Non-alcoholic beverages	0.3 ppm	Non-alcoholic beverages	2.0 ppm
	Ice cream, ices, etc.	0.3 ppm	Ice cream, ices, etc.	2.0 ppm
	Candy	5.0 ppm	Candy	4.0 ppm
	Baked goods	10.0 ppm	Baked goods	4.0 ppm
	Condiments, pickles	10.0 ppm	Gelatins and puddings	2.0 ppm
			Margarine	2.0 ppm
			Preserves	2.0 ppm
Regulatory status	FEMA No. 3292		FEMA No. 3356	

REFERENCES

For References 1–5, see end of Part III.

6. **Tuynenburg-Muys, van der Van and de Jonge,** *Appl. Microbiol.*, 11(5), 389, 1963.
7. Lever Brothers Co., U.S. Patent 3,076,750, 1963.
8. **Rosenmund and Bach,** U.S. Patent 3,048,599, 1962.
9. **Keeney,** *J. Am. Oil Chem. Soc.*, 34, 356, 1957.

	5-Hydroxy-4-octanone	4-(*p*-Hydroxyphenyl)-2-butanone
Other names	Butyroin 5-Octanol-4-one	*p*-Hydroxybenzyl acetone
Empirical formula	$C_8H_{16}O_2$	$C_{10}H_{12}O_2$
Structure	$CH_3-(CH_2)_2-CH(OH)-C(=O)-(CH_2)_2-CH_3$	$HO-C_6H_4-CH_2-CH_2-C(=O)-CH_3$ (*para*)
Physical/chemical characteristics[5]		
Appearance	Yellowish liquid	Crystalline solid
Molecular weight	144.21	164.21
Melting point		82–83 °C (84–85 °C)
Boiling point	80–82 °C (182 °C) at 10 mm Hg	
Specific gravity	0.9231 (0.91) at 20 °C	
Refractive index	1.4290 at 20 °C	
Solubility	Almost insoluble in water; soluble in alcohol	Soluble in alcohol and ether; insoluble in ligroin at room temperature
Organoleptic characteristics	Sweet, slightly pungent, buttery, nut-like odor; sweet, buttery, oily taste	
Synthesis	By reacting sodium metal with ethyl butyrate in boiling ether[6,7] or in the same fashion starting from methyl butyrate[8]	By catalytic hydrogenation of *p*-hydroxybenzylidene acetone in the presence of platinum black in ether or methanol[9] or in the presence of palladium absorbed on charcoal[10]
Natural occurrence	Not reported found in nature	Reported found in raspberry aroma[11]
Reported uses[3]	Non-alcoholic beverages 0.50–5.0 ppm Ice cream, ices, etc. 1.0–20 ppm Candy 10 ppm Baked goods 7.8 ppm	Non-alcoholic beverages 16 ppm Ice cream, ices, etc. 34 ppm Candy 44 ppm Baked goods 54 ppm Gelatins and puddings 5.0–50 ppm Chewing gum 40–320 ppm
Regulatory status	FDA 121.1164; FEMA No. 2587	FDA 121.1164; FEMA No. 2588

REFERENCES

For References 1–5, see end of Part III.

6. **Corson, Benson and Goodwin,** *J. Am. Chem. Soc.*, 52, 3991, 1930.
7. **Bernhauer and Hoffmann,** *J. Prakt, Chem.*, 2(149), 322, 1937.
8. **Hansley,** *J. Am. Chem. Soc.*, 57, 2305, 1935.
9. **Nomura and Nozawa,** *Chem. Zentralbl.*, 1, 1017, 1921.
10. **Maninch and Merz,** *Arch. Pharm.* (Weinheim), p. 22, 1927.
11. **Ruys,** *Fr. Ses Parfums*, 9, 34, 1966.

	Indole
Other names	1-Benzazole 2,3-Benzopyrrole
Empirical formula	C_8H_7N
Structure	
Physical/chemical characteristics[5]	
Appearance	White, low-melting, crystalline solid
Molecular weight	117.14
Melting point	52°C
Boiling point	253–254°C; 123°C at 5 mm Hg
Congealing point	>51°C (after being dried over sulfuric acid)[2]
Solubility	Readily soluble in most organic solvents; 1 gm in 2 ml of 70% alcohol[2]
Organoleptic characteristics	Almost floral odor when highly purified; otherwise exhibits characteristic odor of feces. Not very stable on exposure to light (turns red)
Synthesis	Obtained from the 220–260°C boiling fraction of coal tar[6] or by heating sodium phenylglycine-*o*-carboxylate with NaOH, saturating the aqueous solution of the melt with CO_2, and finally reducing with sodium amalgam;[7,8] can be prepared also by the reduction of indoxyl, indoxyl carboxylic acid, or indigo[2]
Natural occurrence	Reported occurring in several natural products as a complex compound that decomposes during enfleurage or steam distillation yielding free indole.[9,10] Reported found in the essential oil from flower of *Jasminum grandiflorum*,[11] in neroli oil,[12] and in the oil extracted from flowers of bitter orange;[13] also reported in the flowers of several plants: lemon and coffee,[14] *Hevea brasiliensis*, and *Randia formosa*,[15] in the oil extracted from flowers of *Jasminum odoratissinium* L.,[16] and in the oil of *Narcissus jonquilla*[17]
Reported uses[3]	Non-alcoholic beverages 0.26 ppm Ice cream, ices, etc. 0.28 ppm Candy 0.50 ppm Baked goods 0.58 ppm Gelatins and puddings 0.02–0.40 ppm
Regulatory status	FDA 121.1164; FEMA No. 2593

REFERENCES

For References 1–5, see end of Part III.

6. **Kruber,** German Patent 454,696, 1927.
7. BASF, German Patent 152,682, 1902.
8. BASF, German Patent 260,327, 1912.
9. **Hesse,** *Ber. Dtsch. Chem. Ges.*, 32, 2612, 1899; 33, 1587, 1900; 34, 2929, 1901; 37, 1457, 1904.
10. **Soden,** *J. Prakt. Chem.*, 2(69), 268, 1904.
11. **Cerighelli,** *Compt. Rend.*, 179, 1193, 1924.
12. **Giuffrè,** *Riv Ital. EPPOS*, 9, 285, 1925.
13. **Hesse and Zeitschel,** *J. Prakt. Chem.*, 2(66), 504, 1902.
14. **Sack,** *Chem. Zentralbl.*, 1, 1367, 1911.
15. **Sack,** *Chem. Zentralbl.*, 2, 695, 1911.
16. **Tsuchihashi and Tasaki,** *Schimmel Ber.*, 29, April–October 1929.
17. **Soden,** *J. Prakt. Chem.*, 2(110), 278, 1925.

	α-Ionone	β-Ionone
Other names	α-Cyclocitrylideneacetone 4-(2,6,6-Trimethyl-2-cyclohexen-l-yl)-3-buten-2-one	β-Cyclocitrylideneacetone 4-(2,6,6-Trimethyl-1-cyclohexen-1-yl)-3-buten-2-one
Empirical formula	$C_{13}H_{20}O$	$C_{13}H_{20}O$
Structure	H_3C, CH_3, $CH=CH-C(=O)-CH_3$, CH_3 (cyclohexene ring, double bond in ring)	H_3C, CH_3, $CH=CH-C(=O)-CH_3$, CH_3 (cyclohexene ring, double bond at ring junction)
Physical/chemical characteristics[5]	The composition of commercial ionones varies widely in the proportion of alpha and beta isomers. The specifications refer to the best grade of alpha and beta ionones and to a commercial grade that contains both alpha and beta ionones.[2]	
Appearance	Colorless to pale-yellow liquid	Colorless to pale-yellow liquid
Assay	>85%[2] (as alpha isomer); >60%, comm.[2]	>95%[2] (as beta isomer)
Molecular weight	192.30	192.30
Boiling point	121–122°C at 10 mm Hg	128–129°C at 10 mm Hg
Flash point	115°C	
Specific gravity	0.927–0.933 at 25°/25°C;[2] 0.927–0.936, comm.[2]	0.941–0.947 at 25°/25°C;[2] 0.927–0.936, comm.[2]
Refractive index	1.4970–1.5020 at 20°C;[2] 1.4970–1.5060, comm.[2]	1.5190–1.5215 at 20°C;[2] 1.4970–1.5060, comm.[2]
Ketone content	>98%;[2] >90%, comm.[2]	>98%;[2] >90%, comm.[2]
Solubility	1:3 in 70% alcohol[2]	1:3 in 70% alcohol[2]
Organoleptic characteristics	Characteristic violet-like odor	Characteristic violet-like odor, more fruity and woody than α-ionone
Synthesis	By condensing citral with acetone to form pseudoionone, which is then cyclized by acid-type reagents[2]	By condensing citral with acetone to form pseudoionone, which is then cyclized by acid-type reagents[2]
Natural occurrence	Not too common in nature; probably occurring in the absolute essence of *Acaciá farnesiana*. Reported in the essential oil of *Sphaeranthus indicus* L. The *d*-, *l*-, and *dl*-isomers are known.[7]	Reported found in raspberry, in the distillate from flowers of *Boronia megatisma* Nees., and in few other essences
Reported uses[3]	Non-alcoholic beverages 2.5 ppm Ice cream, ices, etc. 3.6 ppm Candy 12 ppm Baked goods 6.7 ppm Gelatins and puddings 3.6 ppm Chewing gum 39 ppm Icings 50 ppm	Non-alcoholic beverages 1.6 ppm Ice cream, ices, etc. 3.4 ppm Candy 7.6 ppm Baked goods 5.2 ppm Gelatins and puddings 5.8 ppm Chewing gum 89 ppm Maraschino cherries 10 ppm
Regulatory status	FDA 121.1164; FEMA No. 2594	FDA 121.1164; FEMA No. 2595

REFERENCES

For References 1–5, see end of Part III.

6. **Baslas,** *Miltizer Ber.*, p. 111, 1960.
7. **Sobotka et al.,** *J. Am. Chem. Soc.*, 65, 2061, 1943.

For further information see the following:

For α-ionone: Theimer and Lemberg, *Proc. Sci. Sect. Toilet Goods Assoc.*, 35, 19, 1961.

For α- and β-ionone: Naves and Bachmann, *Helv. Chim. Acta*, 26, 2151, 1943; 27, 645, 1944.

For β-ionone: Naves, *Perfum. Essent. Oil Rec.*, 10, 658, 1964.

For β-ionone: Mousseron-Canet, Mousseron and Legendre, *Bull. Soc. Chim. Fr.*, 1, 50, 1964; *Compt. Rend.*, 257(25), 3782, 1963; *Bull. Soc. Chim. Fr.*, 2, 379, 1963.

	γ-Ionone	α-Irone
Other names	4-(2,2-Dimethyl-6-methylenecyclohexyl)-3-buten-2-one	*cis*-(2,6)-*cis*-(2[1],2[2])-α-Irone 6-Methyl-α-ionone 4-(2,5,6,6-Tetramethyl-2-cyclohexen-l-yl)-3-buten-2-one
Empirical formula	$C_{13}H_{20}O$	$C_{14}H_{22}O$
Structure	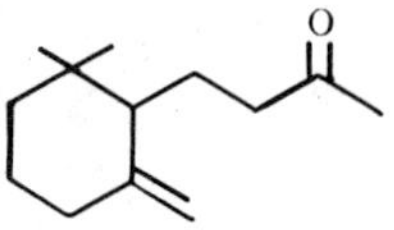	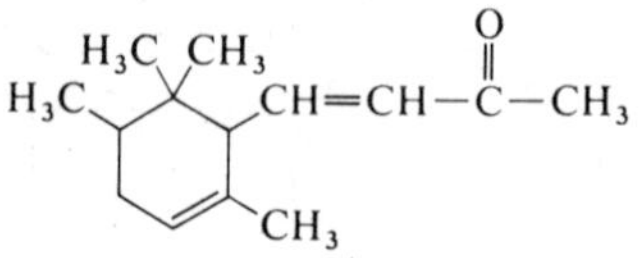 H_3C CH_3 H_3C $CH{=}CH{-}\overset{O}{\overset{\Vert}{C}}{-}CH_3$ CH_3
Physical/chemical characteristics[5]		The physical constants refer to those usually encountered for a commercial product.
Appearance	Liquid	Yellowish liquid
Assay		98% min
Molecular weight	192.30	206.33
Boiling point		110–112°C at 3.2 mm Hg (synthetic)
Specific gravity	0.932 at 20°C	0.932–0.936 at 25°C; 0.9349 at 20°C from orris root; 0.9355 at 20°C synthetic
Refractive index	1.4999 at 25°C	1.5020–1.5030 at 20°C; 1.5003 at 30°C from orris root; 1.4970 at 20°C synthetic
Optical rotation		+226°0′ at 20°C (from orris root)
Solubility		1:4 in 70% alcohol
Organoleptic characteristics		Characteristic orris and violet-like odor
Synthesis	By condensation of cyclogeranic aldehyde with acetone, followed by separation from the β-ionone-rich reaction mixture[6]	See references 8, 9, and 10. The commercial product corresponds generally to the α-isomer
Natural occurrence	Not reported found in nature; dihydro-γ-ionone has been reported as a constituent in ambergris oil[7]	For structure and isomers see References 11, 12, and 13. Occurs in orris root, raspberry, and some flowers of *Pittosporum* sp.
Reported uses[3]	Non-alcoholic beverages 10.0 ppm Candy 10.0 ppm Gelatins and puddings 10.0 ppm	Non-alcoholic beverages 1.2 ppm Ice cream, ices, etc. 2.3 ppm Candy 4.1 ppm Baked goods 5.4 ppm Chewing gum 1.4 ppm
Regulatory status:	FEMA No. 3175	FDA 121.1164; FEMA No. 2597

REFERENCES

For References 1–5, see end of Part III.

6. Willhalm, Steiner and Schinz, *Helv. Chim. Acta,* 41, 1359, 1958.
7. Ružička, Seidel and Pfeiffer, *Helv. Chim. Acta,* 31, 827, 1948.
8. Schinz et al., *Helv. Chim. Acta,* 30, 1810, 1947.
9. Barton and Mousseron-Canet, *J. Chem. Soc.* (London), p. 271, 1960.
10. Eschinazi, *J. Am. Chem. Soc.,* 81, 2905, 1959.
11. Ruzicka et al., *Helv. Chim. Acta,* 30, 2168, 1947.
12. Naves and Bachmann, *Helv. Chim. Acta,* 30, 2222, 1957.
13. Gildemeister and Hoffmann, *Die Aetherischen Oele,* Vol. 3c, Akademie Verlag, 1963, 305.

	Isoamyl acetate	Isoamyl acetoacetate
Other names	Common amyl acetate β-Methylbutyl acetate (In commercial practice *amyl* invariably means *isoamyl*, unless it is prefaced by the *n*- for normal.)	Isoamyl β-ketobutyrate Isoamyl 3-oxobutanoate
Empirical formula	$C_7H_{14}O_2$	$C_9H_{16}O_3$
Structure	$CH_3-C(=O)-O-CH_2-CH_2-CH(CH_3)_2$	$(H_3C)_2CH-(CH_2)_2-O-C(=O)-CH_2-C(=O)-CH_3$
Physical/chemical characteristics[5]		
Appearance	Colorless liquid	Colorless liquid
Assay	95% min[2]	
Molecular weight	130.18	172.23
Melting point	−78.5°C	
Boiling point	142.5°C; 30°C at 2 mm Hg	222–224°C
Specific gravity	0.868–0.878[2] at 25°/25°C	0.954 at 10°C
Refractive index	1.4000–1.4040 at 20°C;[2] 1.4017 at 18°C	
Acid value	Not more than 1[2]	
Solubility	1:3 and more in 60% alcohol; almost insoluble in water; soluble in ether and most common organic solvents	Insoluble in water; soluble in alcohol
Organoleptic characteristics	Powerful, fruity odor with a bittersweet taste reminiscent of pear; if impure, the odor is strong, penetrating, and almost shocking	Ethereal, sweet, winey odor; green-apple flavor
Synthesis	Usually prepared by esterification of commercial isoamyl alcohol with acetic acid	By transesterification of ethyl acetylacetate with isoamyl alcohol in the presence of sodium
Natural occurrence	Reported found in the volatile portion of banana fruit and cocoa bean	
Reported uses[3]	Non-alcoholic beverages 28 ppm Ice cream, ices, etc. 56 ppm Candy 190 ppm Baked goods 120 ppm Gelatins and puddings 100 ppm Chewing gum 2700 ppm	
Regulatory status	FDA 121.1164; FEMA No. 2055	FDA 121.1164

REFERENCES

For References 1–5, see end of Part III.

	Isoamyl alcohol	Isoamyl benzoate
Other names	Isobutyl carbinol Isopentyl alcohol 3-Methyl-l-butanol	Isopentyl benzoate
Empirical formula	$C_5H_{12}O$	$C_{12}H_{16}O_2$
Structure	$(H_3C)_2CH-CH_2-CH_2OH$	$C_6H_5-O-C(=O)-CH_2-CH(CH_3)_2$
Physical/chemical characteristics[5]		
Appearance	Oily, clear liquid	Colorless liquid
Assay		99% min
Molecular weight	88.15	192.26
Melting point	−117.2°C	
Boiling point	132 C	261–262°C
Flash point		>100°C
Specific gravity	0.812 at 20°C	0.986–0.990 at 25°/25°C; 0.9910 at 17.4 C
Refractive index	1.4084 at 17.8°C	1.492–1.495 at 20°C; 1.4950 at 17.4°C
Acid value		1 max
Solubility	Soluble in water; soluble in most organic solvents	Insoluble in water; soluble 1:3 in 80% alcohol
Organoleptic characteristics	Characteristic pungent odor and repulsive taste	Fruity, slightly pungent odor
Synthesis	Industrially prepared by rectification of fusel oil	By transesterification of methyl benzoate and isoamyl alcohol in the presence of potassium isoamylate;[9] also by heating benzoyl chloride and isoamyl acetate[10]
Natural occurrence	Constitutes the major portion of fusel oil; also known as fermentation amyl alcohol. It has been identified as an ester among the constituents of Roman camomile oil,[6] and in several essential oils: French peppermint, Java citronella, Reunion geranium, tea, *Teucrium chamaedrys*,[7] *Eucalyptus amigdalina*, *Achillea ageratum*, *Artemisia camphorata*, and others; also in the aromas of strawberry and raspberry; identified in rum.[8]	In vinegar; in cocoa
Reported uses[3]	Non-alcoholic beverages 17 ppm Alcoholic beverages 100 ppm Ice cream, ices, etc. 7.6 ppm Candy 52 ppm Baked goods 24 ppm Gelatins and puddings 46 ppm Chewing gum 300 ppm	Non-alcoholic beverages 3.0 ppm Ice cream, ices, etc. 2.5 ppm Candy 3.5 ppm Baked goods 7.4 ppm Gelatins and puddings 4.6 ppm Chewing gum 200 ppm
Regulatory status	FDA 121.1164; FEMA No. 2057	FDA 121.1164; FEMA No. 2058

REFERENCES

For References 1–5, see end of Part III.

6. **Köbig,** *Justus Liebigs Ann. Chem.*, 195, 92, 1879.
7. **Rovesti,** *Ind. Parfum.*, 12, 334, 1957.
8. **Fenaroli et al.,** *Riv Ital. EPPOS*, 9, 484, 1965; 2, 75, 1966.
9. **Reimer and Downes,** *J. Am. Chem. Soc.*, 43, 949, 1921.
10. **Zetsche,** *Helv. Chim. Acta*, 9, 181, 1926.

	Isoamyl butyrate	Isoamyl cinnamate
Other names	Isopentyl butyrate	Isoamyl β-phenylacrylate Isoamyl 3-phenylpropenoate Isopentyl cinnamate
Empirical formula	$C_9H_{18}O_2$	$C_{14}H_{18}O_2$
Structure	$CH_3-(CH_2)_2-C(=O)-O-(CH_2)_2-CH(CH_3)_2$	$C_6H_5-CH=CH-C(=O)-O-(CH_2)_2-CH(CH_3)_2$
Physical/chemical characteristics[5]		
Appearance	Colorless liquid	Colorless to pale-yellow liquid
Assay	98% min[2]	98% min[2]
Molecular weight	158.24	218.29
Boiling point	177–178°C	
Flash point	65°C	
Specific gravity	0.860–0.864 at 25°/25°C;[2] 0.8642 at 20°C	0.992–0.997 at 25°/25°C;[2] 0.9995–1.0015 at 15°C
Refractive index	1.4090–1.4140 at 20°C;[2] 1.4110 at 20°C	1.5350–1.5390[2] (1.5360–1.5375) at 20°C
Acid value	Not more than 1[2]	Not more than 1.0[2] (0.2)
Solubility	1:4 and more in 70% alcohol[2]	1:7 in 80% alcohol[2]
Organoleptic characteristics	Strong, characteristic, fruity (pear-like) odor; sweet, corresponding taste	Balsamic odor reminiscent of cocoa
Synthesis	Usually prepared by esterification of commercial isoamyl alcohols with butyric acid by heating in the presence of Twitchell reagent;[6] or by fermentation from butyric acid and isoamyl alcohol[7,8]	By esterification of cinnamic acid with commercial isoamyl alcohols, which vary in isomer distribution according to source
Natural occurrence	Reported found in the oil of *Eucalyptus macarthuri* and in coconut oil; also identified in cocoa bean and in cider	Reported found in several natural products
Reported uses[3]	Non-alcoholic beverages 13 ppm Ice cream, ices, etc. 34 ppm Candy 79 ppm Baked goods 51 ppm Gelatins and puddings 60 ppm Chewing gum 570 ppm	Non-alcoholic beverages 3.1 ppm Ice cream, ices, etc. 4.2 ppm Candy 13 ppm Baked goods 13 ppm
Regulatory status	FDA 121.1164; FEMA No. 2060	FDA 121.1164; FEMA No. 2063

REFERENCES

For References 1–5, see end of Part III.

6. **Zaganiaris and Varvoglis,** *Ber. Dtsch. Chem. Ges.*, 69, 2280, 1936.
7. **Rona and Ammon,** *Werner Biochem,* 221, 385, 1930; **Rona and Mühlbock,** *Werner Biochem.*, 223, 139, 1930.
8. **Mardaschew,** *Arh. Biol. Nauka,* 37, 399, 1935.

	Isoamyl formate	Isoamyl-2-furanbutyrate
Other names	Isopentyl formate	2-Furanbutyric acid, isoamylester Isoamyl furfurylpropionate α-Isoamyl furfurylpropionate Isopentyl-2-furanbutyrate
Empirical formula	$C_6H_{12}O_2$	$C_{13}H_{20}O_3$
Structure	$H-C(=O)-O-(CH_2)_2-CH(CH_3)_2$	$C_4H_3O-(CH_2)_3-C(=O)-O-(CH_2)_2-CH(CH_3)_2$
Physical/chemical characteristics[5]		
Appearance	Colorless liquid	Pale-yellow liquid
Assay	92% min (97% min)	
Molecular weight	116.16	224.30
Boiling point	123–124°C	
Flash point	53°C	
Specific gravity	0.878–0.885 at 25°C; 0.8870–0.8900 at 15°C	
Refractive index	1.3960–1.4000 (1.3970–1.3990) at 20°C	
Acid value	2 max (0.9)	
Solubility	Soluble in alcohol; slightly soluble in water; miscible with ether	Insoluble in water; soluble in alcohol
Organoleptic characteristics	Fruity, characteristic odor suggestive of black currant and plum with a corresponding sweet taste	Sweet, buttery, fruity odor; caramel-like flavor
Synthesis	By reacting concentrated H_2SO_4 over a mixture of isoamyl alcohols and sodium formate[6]	From furfural
Natural occurrence	Reported found in several natural products	Not reported found in nature
Reported uses[3]	Non-alcoholic beverages 8.4 ppm Ice cream, ices, etc. 14 ppm Candy 22 ppm Baked goods 16 ppm Gelatins and puddings 2.0–28 ppm Chewing gum 250 ppm	Non-alcoholic beverages 0.03–5.0 ppm Ice cream, ices, etc. 2.8 ppm Candy 6.0 ppm Baked goods 0.50–8.0 ppm Gelatins and puddings 5.0 ppm
Regulatory status	FDA 121.1164; FEMA No. 2069	FDA 121.1164; FEMA No. 2070

REFERENCES

For References 1–5, see end of Part III.
6. **Wagner,** German Patent 529,135, 1929.

	Isoamyl-2-furanpropionate	Isoamyl hexanoate
Other names	2-Furanpropionic acid, isoamylester Isoamyl furfurhydracrylate α-Isoamyl furfurylacetate Isopentyl-2-furanpropionate	Isoamyl caproate Isoamyl capronate Isoamyl hexylate Isopentyl hexanoate
Empirical formula	$C_{12}H_{18}O_3$	$C_{11}H_{22}O_2$
Structure	(2-furyl)$-(CH_2)_2-C(=O)-O-(CH_2)_2-CH(CH_3)_2$	$CH_3-(CH_2)_4-C(=O)-O-(CH_2)_2-CH(CH_3)_2$
Physical/chemical characteristics[5]		
Appearance	Pale-yellow liquid (colorless when freshly made)	Colorless liquid
Assay		98% min[2]
Molecular weight	210.27	186.30
Boiling point		94–96°C at 10 mm Hg
Flash point		88°C
Specific gravity		0.858–0.863 at 25°/25°C;[2] 0.8650–0.8670 at 15°C
Refractive index		1.4180–1.4220[2] (1.4195–1.4205) at 20°C
Acid value		Not more than 1.0[2] (0.2)
Solubility	Insoluble in water; soluble in alcohol	1:3 in 80% alcohol;[2] insoluble in water
Organoleptic characteristics	Sweet, green, slightly floral odor with a distinct fruital note; similar flavor	Fruity odor
Synthesis	From furfurhydracrylic acid with isoamyl alcohol	By esterification of caproic acid with the isomeric amyl alcohols obtained from fusel oil and other sources
Natural occurrence	Not reported found in nature	Reported found in fermentation products (wine) and the rind of California orange
Reported uses[3]	Non-alcoholic beverages 0.02–0.33 ppm Ice cream, ices, etc. 0.33–0.65 ppm Candy 1.6–3.6 ppm Baked goods 1.6–3.6 ppm	Non-alcoholic beverages 7.8 ppm Ice cream, ices, etc. 14 ppm Candy 17 ppm Baked goods 15 ppm Gelatins and puddings 3.7 ppm
Regulatory status	FDA 121.1164; FEMA No. 2071	FDA 121.1164; FEMA No. 2075

REFERENCES

For References 1–5, see end of Part III.

	Isoamyl isobutyrate	Isoamyl isovalerate
Other names		Isopentyl isovalerate
Empirical formula	$C_9H_{18}O_2$	$C_{10}H_{20}O_2$
Structure	$(H_3C)_2CH-C(=O)-O-(CH_2)_2-CH(CH_3)_2$	$(H_3C)_2CH-CH_2-C(=O)-O-(CH_2)_2-CH(CH_3)_2$
Physical/chemical characteristics[5]		
Appearance	Liquid	Liquid
Assay	98% min	97.5% min
Molecular weight	158.24	172.27
Boiling point	170°C	193°C
Flash point		72°C
Specific gravity	0.8620–0.8690 at 15°C; 0.8627 at 20°C	0.8600–0.8625 at 15°C; 0.8583 at 18.7°C
Refractive index	1.4070–1.4110 at 20°C	1.4125–1.4135 at 20°C; 1.4100 at 25°C
Acid value	0.09 max	0.09 max
Solubility		1:6.5–8 in 70% alcohol; insoluble in water; soluble in most organic solvents
Organoleptic characteristics	Fruity odor with an undertone reminiscent of apricot and pineapple	Characteristic fruity odor and sweet apple-like flavor
Synthesis	By passing vapors of isobutyl alcohol and isoamyl alcohol over an Ag-activated Cu-MnO catalyst[6]	By passing vapors of isoamyl alcohol and isovaleric aldehyde over a copper uranium based catalyst at 240–280°C in the presence of hydrogen[7]
Natural occurrence	Not reported found in nature	Reported found in banana fruits,[8] in the ether oil from leaves of *Eucalyptus microcorys,*[9] in wine, oil of hop, and olive oil
Reported uses[3]		Non-alcoholic beverages 8.5 ppm Ice cream, ices, etc. 14 ppm Candy 33 ppm Baked goods 41 ppm Gelatins and puddings 1.0–61 ppm Chewing gum 390 ppm Jellies 10 ppm
Regulatory status	FDA 121.1164	FDA 121.1164; FEMA No. 2085

REFERENCES

For References 1–5, see end of Part III.

6. **Wasskewitch and Bulanova,** *Zh. Obshch. Khim.,* 8, 1096, 1938.
7. **Abramow and Dolgow,** *Chem. Abstr.,* 34, 7847, 1940; 42, 1561, 1948.
8. **Kothenbach and Eberlein,** *Chem. Zentralbl.,* 1, 1105, 1905.
9. **Jones and Lahey,** *Proc. R. Soc. Queensl.,* 50, 44, 1939.

	Isoamyl laurate	Isoamyl nonanoate
Other names	Isoamyl dodecanoate Isoamyl dodecylate Isopentyl laurate	Isoamyl nonylate Isoamyl pelargonate Isopentyl nonanoate Nonate
Empirical formula	$C_{17}H_{34}O_2$	$C_{14}H_{28}O_2$
Structure	$CH_3-(CH_2)_{10}-\overset{O}{\overset{\|}{C}}-O-(CH_2)_2-CH(CH_3)_2$	$CH_3-(CH_2)_7-\overset{O}{\overset{\|}{C}}-O-(CH_2)_2-CH(CH_3)_2$
Physical/chemical characteristics[5]		
Appearance	Colorless, oily liquid	Colorless, oily liquid
Molecular weight	270.46	228.37
Boiling point		260–265°C
Specific gravity		0.86
Solubility	Insoluble in water; soluble in alcohol	Insoluble in water; soluble in alcohol
Organoleptic characteristics	Very faint, oily, fatty odor; fatty flavor	Nutty, oily, apricot-like odor; fruity, winey, cognac-rum flavor
Synthesis		From isoamyl alcohol and pelargonic acid
Natural occurrence	Reported found at the 3% level in fusel oil from raisins[6,7]	In banana and strawberry
Reported uses[3]	Non-alcoholic beverages 0.04–3.0 ppm Ice cream, ices, etc. 0.16–6.0 ppm Candy 0.50–6.0 ppm Baked goods 0.50–6.0 ppm Also used as a solvent for various flavors and fragrances	Non-alcoholic beverages 1.5 ppm Ice cream, ices, etc. 3.3 ppm Candy 3.0 ppm Baked goods 4.0 ppm
Regulatory status	FDA 121.1164; FEMA No. 2077	FDA 121.1164; FEMA No. 2078

REFERENCES

For References 1–5, see end of Part III.

6. **Kepner and Webb,** *Am. J. Enol. Vitic.*, 12, 159, 1961.
7. **Kepner and Webb,** *Chem. Abstr.*, 56, 7798b, 1962.

	Isoamyl octanoate	Isoamyl phenylacetate
Other names	Isoamyl caprylate Isoamyl octylate Isopentyl octanoate	Isoamyl α-toluate Isopentyl phenylacetate
Empirical formula	$C_{13}H_{26}O_2$	$C_{13}H_{18}O_2$
Structure	$CH_3-(CH_2)_6-C(=O)-O-(CH_2)_2-CH(CH_3)_2$	$C_6H_5-CH_2-C(=O)-O-CH_2-CH(CH_3)-CH_2-CH_3$
Physical/chemical characteristics[5]		
Appearance	Colorless liquid	Liquid
Assay	98% min[2]	98% min
Molecular weight	214.35	206.28
Boiling point		265–266°C at 723 mm Hg
Flash point	104°C	>100°C
Specific gravity	0.855–0.861 at 25°/25°C;[2] 0.8620–0.8660 at 15°C	0.975–0.977 at 25°/25°C; 0.982 at 20°C
Refractive index	1.4250–1.4290 at 20°C[2]	1.4850–1.4870 at 20°C; 1.4872 at 21°C
Acid value	Not more than 1.0[2] (0.2)	1 max
Solubility	1:7–10 in 80% alcohol[2]	1:3 in 80% alcohol
Organoleptic characteristics	Fruity odor	Sweet, pleasant odor reminiscent of cocoa with a slight birch-tar undertone
Synthesis	By esterification of octanoic acid with commercial isoamyl alcohols, which vary in isomer distribution according to source	By esterification of phenylacetic acid with isoamyl alcohol in the presence of concentrated sulfuric acid;[6] by heating benzyl nitrile and isoamyl alcohol in the presence of excess concentrated H_2SO_4
Natural occurrence	Reported found in wine, banana, strawberry, and apple	Not reported found in nature
Reported uses[3]	Non-alcoholic beverages 6.6 ppm Ice cream, ices, etc. 5.1 ppm Candy 7.4 ppm Baked goods 3.5 ppm Gelatins and puddings 2.1 ppm	Non-alcoholic beverages 5.0 ppm Ice cream, ices, etc. 16 ppm Candy 12 ppm Baked goods 14 ppm Gelatins and puddings 0.15–3.4 ppm Toppings 0.25–0.80 ppm
Regulatory status	FDA 121.1164; FEMA No. 2080	FDA 121.1164; FEMA No. 2081

REFERENCES

For References 1–5, see end of Part III.
6. **Guye and Chavanne,** *Bull. Soc. Chim. Fr.*, 3(15), 292, 1896.

	Isoamyl propionate	Isoamyl pyruvate
Other names	Isopentyl propionate	Isoamyl α-ketopropionate Isoamyl 2-oxopropanoate Isopentyl pyruvate
Empirical formula	$C_8H_{16}O_2$	$C_8H_{14}O_3$
Structure	$CH_3-CH_2-\overset{\overset{O}{\parallel}}{C}-O-(CH_2)_2-CH(CH_3)_2$	$CH_3-CO-\overset{\overset{O}{\parallel}}{C}-O-CH_2-CH_2-CH(CH_3)_2$
Physical/chemical characteristics[5]		
Appearance	Colorless liquid	
Assay	98% min[2]	
Molecular weight	144.21	158.20
Boiling point	160°C	185°C; 86°C at 14 mm Hg
Flash point	41°C	
Specific gravity	0.866–0.871 at 25°/25°C;[2] 0.8720–0.8760 at 15°C	0.978 at 17°C
Refractive index	1.4050–1.4090 at 20°C[2]	
Acid value	Not more than 1.0[2] (0.2)	
Solubility	1:3 in 70% alcohol;[2] insoluble in water; soluble in most organic solvents	
Organoleptic characteristics	Apricot-pineapple odor. The taste is somewhat harsh when freshly distilled but subsequently yields a pleasant apricot-plum bittersweet flavor.	
Synthesis	By esterification of propionic acid with the isomeric amyl alcohols obtained from fusel oil and other sources	By air oxidation of boiling isoamyl lactate; also by reacting methyl magnesium iodide over isoamyl α-hydroxy isobutyrate
Natural occurrence	Reported found in cocoa bean and Bulgarian peppermint	Not reported found in nature
Reported uses[3]	Non-alcoholic beverages 3.8 ppm Ice cream, ices, etc. 13 ppm Candy 38 ppm Baked goods 6.1 ppm Gelatins and puddings 0.80–3.7 ppm Chewing gum 750 ppm	Non-alcoholic beverages 4.7 ppm Ice cream, ices, etc. 8.1 ppm Candy 9.2 ppm Baked goods 12 ppm
Regulatory status	FDA 121.1164; FEMA No. 2082	FDA 121.1164; FEMA No. 2083

REFERENCES

For References 1–5, see end of Part III.

	Isoamyl salicylate	Isoborneol
Other names	Isoamyl 2-hydroxybenzoate Isoamyl *o*-hydroxybenzoate Isopentyl salicylate (Known commercially as amyl salicylate)	Isobornyl alcohol Isocamphol
Empirical formula	$C_{12}H_{16}O_3$	$C_{10}H_{18}O$
Structure	OH, O, C—O—CH₂—CH₂—CH(CH₃)(CH₃)	CH₃, H, OH, H₃C, CH₃
Physical/chemical characteristics[5]		
Appearance	Colorless liquid	White, crystalline solid
Assay	98% min[2] (99% min)	92% min[2]
Molecular weight	208.26	154.24
Melting point		212–213°C (*d*, 214–217°C; *l*, 212–214°C)
Boiling point	273°C; 151–152°C at 15 mm Hg	
Flash point	133°C	
Specific gravity	1.047–1.053 at 25°/25°C;[2] 1.065 at 15°C	
Refractive index	1.5050–1.5085 at 20°C;[2] 1.505 at 20°C	
Acid value	Not more than 1[2]	
Optical rotation		−30.6° at 19°C (*l*-form from alcohol); +33.89°C (*d*-form from alcohol); −19.6° at 15°C (*l*-form from toluene)
Solubility	1:2 to 1:3 in 90% alcohol;[2] very slightly soluble in water	1 gm in 2 ml of 70% alcohol[2]
Organoleptic characteristics	Characteristic aromatic, strong, herbaceous, persistent odor; bittersweet taste reminiscent of strawberry	Piney, camphoraceous odor
Synthesis	By esterification of salicylic acid with the isomeric amyl alcohols obtained from fusel oil and other sources	By the hydrolysis of isobornyl acetate,[2] or by catalytic reduction of camphor (both *d*- and *l*-isomers). The optically inactive compound can be prepared by treating camphene with a 1:1 mixture of sulfuric acid and glacial acetic acid[6] and then hydrolyzing the isobornyl acetate.
Natural occurrence	Not reported found in nature	Rarely found in nature (in the oil of *Abies sibirica* and a few other essences); probably present in the essential oil from roots of *Chamaecyparis formosensis*[7]
Reported uses[3]	Non-alcoholic beverages 1.4 ppm Ice cream, ices, etc. 2.9 ppm Candy 3.0 ppm Baked goods 3.0 ppm	Non-alcoholic beverages 6.2 ppm Ice cream, ices, etc. 23 ppm Candy 11 ppm Baked goods 8.3 ppm Chewing gum 0.80 ppm
Regulatory status	FDA 121.1164; FEMA No. 2084	FDA 121.1164; FEMA No. 2158

REFERENCES

For References 1–5, see end of Part III.

	Isobornyl acetate	Isobornyl formate
Other names	2-Camphanyl acetate	
Empirical formula	$C_{12}H_{20}O_2$	$C_{11}H_{18}O_2$
Structure	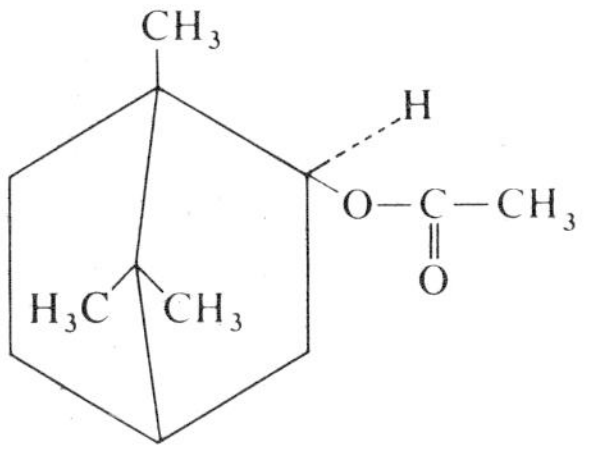	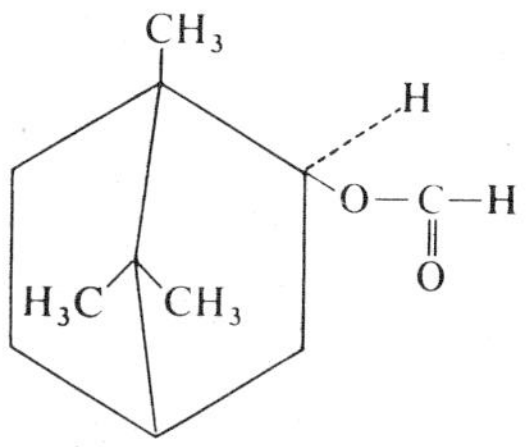
Physical/chemical characteristics[5]		
Appearance	Clear, colorless liquid; tends to yellow slightly on aging	Liquid
Assay	97% min[2]	
Molecular weight	196.29	182.26
Boiling point	102–103°C at 12–13 mm Hg	110°C at 20 mm Hg
Specific gravity	0.980–0.984 at 25°/25°C;[2] 0.9905 at 15°C	1.000 at 18°C
Refractive index	1.4620–1.4650 at 20°C;[2] 1.46589 at 14.4°C	1.4717 at 18°C
Optical rotation	+1° to −1°[2]	
Acid value	Not more than 1[2]	
Solubility	1:3 in 70% alcohol;[2] 1:1 in 90% alcohol; readily soluble in most organic solvents	Insoluble in water; soluble in most organic solvents
Organoleptic characteristics	Pleasant, camphor-like odor reminiscent of some varieties of pine needles and hemlock; fresh, burning taste	Characteristic aromatic, pine-needles odor
Synthesis	By treatment of camphene with acetic acid, usually in the presence of a catalyst; also by acetylation of isoborneol. Depending on the starting material (*d*-camphene or *l*-camphene), the resulting acetate may exhibit a slight optical activity. The commercial product is considered to be optically inactive	By reaction of formic acid with camphene in the presence of a catalyst;[6,7] also in the absence of a catalyst. The resulting products do not exhibit well defined optical characteristics[8] The corresponding optically active esters have been prepared from *d*- or *l*-camphene and formic acid in the presence of phthalic anhydride.
Natural occurrence	Not reported found in nature	Not reported found in nature
Reported uses[3]	Non-alcoholic beverages 9.6 ppm Ice cream, ices, etc. 12 ppm Candy 3.9 ppm Baked goods 9.5 ppm Gelatins and puddings 70 ppm	Non-alcoholic beverages 0.06–1.0 ppm Ice cream, ices, etc. 0.03–1.0 ppm Candy 0.74 ppm Baked goods 0.80 ppm
Regulatory status	FDA 121.1164; FEMA No. 2160	FDA 121.1164; FEMA No. 2162

REFERENCES

For References 1–5, see end of Part III.

6. **Brus and Vebra,** *Compt. Rend.,* 191, 267, 1930.
7. **Schering and Kahlbaum,** German Patent 573,979, 1925.
8. **Pariselle,** *Compt. Rend.,* 180, 1833, 1925.

	Isobornyl isovalerate	Isobornyl propionate
Other names		
Empirical formula	$C_{15}H_{26}O_2$	$C_{13}H_{22}O_2$
Structure	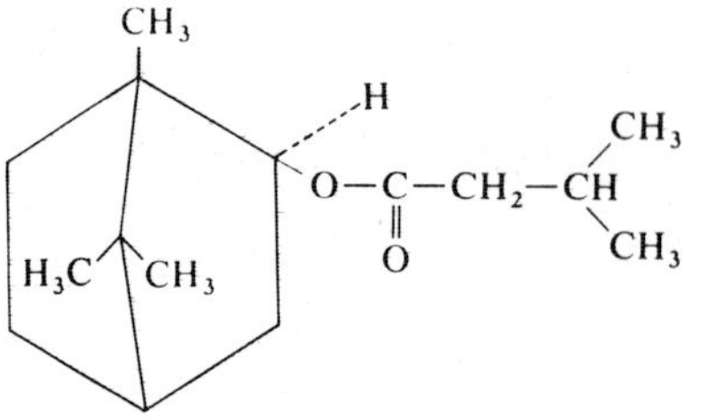	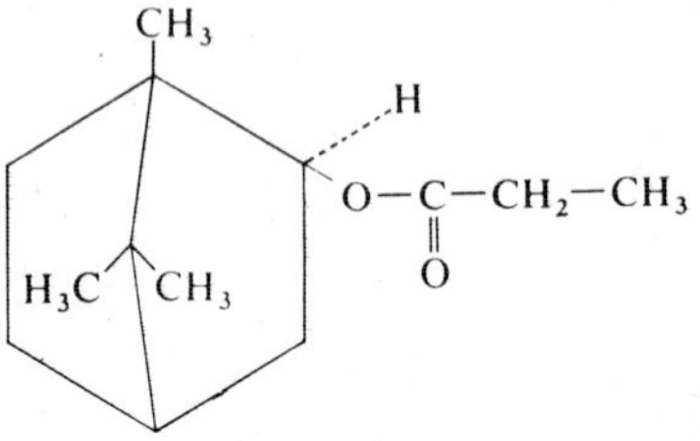
Physical/chemical characteristics[5]		
Appearance	Colorless liquid	Colorless, oily liquid
Molecular weight	238.37	210.31
Boiling point		245°C
Specific gravity		0.978 at 15.5°C
Refractive index		1.4640
Solubility	Insoluble in water; soluble in alcohol; very slightly soluble in glycerin and propylene glycol	Almost insoluble in water; poorly soluble in propylene glycol and glycerin; soluble in alcohol
Organoleptic characteristics	Warm, herbaceous, camphoraceous odor; slightly green and woody	Characteristic, definite turpentine odor but less pungent than the corresponding acetate
Synthesis		From isoborneol and propionic acid
Natural occurrence	Not reported found in nature	Not reported found in nature
Reported uses[3]	Non-alcoholic beverages 0.60–1.0 ppm Ice cream, ices, etc. 0.30–1.0 ppm Candy 0.90 ppm Baked goods 0.80–2.0 ppm	Non-alcoholic beverages 0.01–1.0 ppm Ice cream, ices, etc. 0.80–1.0 ppm Candy 1.2 ppm Baked goods 1.8 ppm
Regulatory status	FDA 121.1164; FEMA No. 2166	FDA 121.1164; FEMA No. 2163

REFERENCES

For References 1–5, see end of Part III.

	Isobutyl acetate	Isobutyl acetoacetate
Other names		Isobutyl-β-ketobutyrate Isobutyl-3-oxobutanoate
Empirical formula	$C_6H_{12}O_2$	$C_8H_{14}O_3$
Structure	$(H_3C)_2CH-CH_2-O-C(=O)-CH_3$	$(H_3C)_2CH-CH_2-O-C(=O)-CH_2-CO-CH_3$
Physical/chemical characteristics[5]		
Appearance	Colorless liquid	Colorless liquid
Assay	95% min	
Molecular weight	116.16	158.20
Boiling point	116.5°C	84.5°C at 11 mm Hg
Flash point	30°C (19°C)	
Specific gravity	0.8710–0.8785 (0.8921) at 15°C	0.9697 at 25°C
Refractive index	1.3890–1.3910 at 20°C	1.4219 at 25°C
Acid value	0.2 max	
Solubility	Soluble in most organic solvents	Insoluble in water; soluble in alcohol
Organoleptic characteristics	Fruity (currant-pear), floral (hyacinth-rose) odor; characteristic ether-like, slightly bitter flavor	Sweet, winey, brandy-like odor; sweet and slightly fruity flavor
Synthesis	By direct esterification of isobutyl alcohol with acetic acid	From isobutyl acetate by heating in the presence of sodium isobutylate
Natural occurrence	Occurring in a few natural products, such as *Endoconidiophora coerulescens,*[6] raspberry, pear, and pineapple	
Reported uses[3]	Non-alcoholic beverages 11 ppm Ice cream, ices, etc. 16 ppm Candy 36 ppm Baked goods 35 ppm Gelatins and puddings 170 ppm Chewing gum 860 ppm Icings 5.5 ppm	Non-alcoholic beverages 4.0 ppm Ice cream, ices, etc. 7.0 ppm Candy 25 ppm Baked goods 25 ppm
Regulatory status	FDA 121.1164; FEMA No. 2175	FDA 121.1164; FEMA No. 2177

REFERENCES

For References 1–5, see end of Part III.

6. Birkinshaw and Morgan, *Biochem. J.*, 47, 55, 1950.

	Isobutyl alcohol	Isobutyl angelate
Other names	Isobutanol 2-Methylpropanol	Isobutyl *cis*-α,β-dimethylacrylate Isobutyl *cis*-2-methyl-2-butenoate Isobutyl *cis*-α-methylcrotonate
Empirical formula	$C_4H_{10}O$	$C_9H_{16}O_2$
Structure	$(H_3C)_2CH-CH_2OH$	$(H_3C)_2CH-CH_2-O-C(=O)-C(CH_3)=CH-CH_3$
Physical/chemical characteristics[5]		
Appearance	Liquid	Colorless liquid
Molecular weight	74.42	156.23
Melting point	−108°C	
Boiling point	108.4°C	177°C
Specific gravity	0.803 at 20°C	
Refractive index	1.396 at 20°C	
Solubility	Soluble in water (forming azeotropes); miscible with most common organic solvents	Almost insoluble in water; soluble in alcohol
Organoleptic characteristics	Disagreeable odor	Faint, pleasant, camomile-like odor and flavor
Synthesis	From isobutylene; by reduction of isobutyraldehyde with sodium amalgam or in the presence of a catalyst; by fermentation of isobutyraldehyde; isolated during fermentation of carbohydrates	By esterification of isobutyl alcohol with angelic acid
Natural occurrence	Reported found (free or esterified) in the essential oils of: Java citronella, tea, *Eucalyptus amygdalina*, and a few others; also present in apple and currant aromas; identified among the volatile constituents in rum;[6] present in fusel oil from beets and potatoes	Reported found in the essential oil of Roman camomile[7]
Reported uses[3]	Non-alcoholic beverages 12 ppm Alcoholic beverages 1.0 ppm Ice cream, ices, etc. 7.0 ppm Candy 34 ppm Baked goods 32 ppm Cream 4.0 ppm	Non-alcoholic beverages 1.5 ppm Ice cream, ices, etc. 1.5 ppm Candy 5.0 ppm Icings 3.0–100 ppm
Regulatory status	FDA 121.1164; FEMA No. 2179	FDA 121.1164; FEMA No. 2180

REFERENCES

For References 1–5, see end of Part III.

6. **Fenaroli et al.,** *Riv. Ital. EPPOS*, 9, 484, 1956; 2, 75, 1966.
7. **Fittig et al.,** *Justus Liebigs Ann. Chem.*, 195, 79, 1879.

	Isobutyl anthranilate	Isobutyl benzoate
Other names	Isobutyl 2-aminobenzoate Isobutyl *o*-aminobenzoate	
Empirical formula	$C_{11}H_{15}O_2N$	$C_{11}H_{14}O_2$
Structure	NH_2, $C(=O)-O-CH_2-CH(CH_3)_2$ on benzene ring	$C(=O)-O-CH_2-CH(CH_3)_2$ on benzene ring
Physical/chemical characteristics[5]		
Appearance	Colorless liquid	Colorless liquid
Assay		98% min
Molecular weight	193.25	178.23
Boiling point	169–170°C at 13.5 mm Hg	241–242°C
Flash point		96°C
Specific gravity		0.994–0.997 at 25°/25°C; 0.9896 at 30°C
Refractive index		1.493–1.496 at 20°C
Acid value		1 max
Solubility		1:2 in 80% alcohol; insoluble in water; miscible with alcohol and ether
Organoleptic characteristics	Faint, orange-flowers odor	Floral-leafy odor reminiscent of rose and geranium
Synthesis	By heating the anhydride of *N*-carboxyl anthranilic acid with sodium isobutoxide and subsequently hydrolyzing the reaction product	From benzoic acid and isobutyl alcohol in the presence of HCl catalyst;[6] or by ester exchange between methylbenzoate and isobutyl alcohol in the presence of potassium isobutylate[7]
Natural occurrence	Not reported found in nature	In cocoa; in banana
Reported uses[3]	Non-alcoholic beverages 2.0 ppm Ice cream, ices, etc. 4.0 ppm Candy 12 ppm Baked goods 12 ppm Chewing gum 5.0–1,700 ppm	Non-alcoholic beverages 2.0–9.0 ppm Ice cream, ices, etc. 7.9 ppm Candy 12 ppm Baked goods 10–23 ppm
Regulatory status	FDA 121.1164; FEMA No. 2182	FDA 121.1164; FEMA No. 2185

REFERENCES

For References 1–5, see end of Part III.

6. Stohmann, Rodatz and Herzberg, *J. Prakt. Chem.*, 2(36), 4, 1887.
7. Reimer and Downes, *J. Am. Chem. Soc.*, 43, 949, 1921.

	Isobutyl-2-butenoate	Isobutyl butyrate
Other names	Isobutyl crotonoate	2-Methyl propanyl butyrate
Empirical formula	$C_8H_{14}O_2$	$C_8H_{16}O_2$
Structure	$CH_3-CH=CH-C(=O)-O-CH_2-CH(CH_3)-CH_3$	$(H_3C)_2CH-CH_2-O-C(=O)-(CH_2)_2-CH_3$
Physical/chemical characteristics[5]		
Appearance		Colorless liquid
Assay		98% min[2]
Molecular weight	142.19	144.21
Boiling point	171°C	157–158°C; 51–52°C at 12 mm Hg
Specific gravity	0.8910 at 20°C	0.858–0.863 at 25°/25°C;[2] 0.8619 at 20°C
Refractive index	1.4291 at 20°C	1.4020–1.4050 at 20°C[2]
Acid value		Not more than 1.0[2]
Solubility		1:8 in 60% alcohol;[2] slightly soluble in water
Organoleptic characteristics		Fruity odor reminiscent of apple or pineapple; sweet flavor reminiscent of rum
Synthesis	By heating the corresponding acid and alcohol with sulfuric acid[6]	By esterification of butyric acid with isobutyl alcohol in the presence of concentrated H_2SO_4;[8] or by enzymatic action of hog pancreas on a mixture of isobutyl alcohol and butyric acid in the absence of water[9]
Natural occurrence	In grapes[7]	Reported found in several natural products
Reported uses[3]	Non-alcoholic beverages 3.0 ppm Ice cream, ices, etc. 3.0 ppm Candy 5.0 ppm Baked goods 5.0 ppm Gelatins and puddings 3.0 ppm Chewing gum 25.0 ppm Jellies, preserves, spreads 3.0 ppm Milk, dairy products 3.0 ppm	Non-alcoholic beverages 8.3 ppm Alcoholic beverages 2.0 ppm Ice cream, ices, etc. 16 ppm Candy 25 ppm Baked goods 24 ppm Gelatins and puddings 14 ppm Chewing gum 2,000 ppm
Regulatory status	FEMA No. 3432	FDA 121.1164; FEMA No. 2187

REFERENCES

For References 1–5, see end of Part III.

6. Schgåmberg, *Z. Physik Chem.*, 174, 466, 1935.
7. Stern et al., *J. Agric. Food Chem.*, 15, 1100, 1967.
8. Grünzweig, *Chem. Ann.*, 162, 207, 1872.
9. Rona and Ammon, *Biochem. Z.*, 221, 386, 1930.

	Isobutyl cinnamate	Isobutyl formate
Other names	Isobutyl-3-phenylpropenoate Isobutyl-β-phenylacrylate	Tetryl formate
Empirical formula	$C_{13}H_{16}O_2$	$C_5H_{10}O_2$
Structure	$C_6H_5-CH=CH-C(=O)-O-CH_2-CH(CH_3)_2$	$(H_3C)_2CH-CH_2-O-CH(=O)$
Physical/chemical characteristics[5]		
Appearance	Colorless liquid	Colorless liquid
Assay	98% min	
Molecular weight	204.27	102.13
Melting point		−95.3°C
Boiling point	145°C at 13 mm Hg	97–98.2°C
Flash point	>100°C	
Specific gravity	1.001–1.004 at 25°C	0.8798 (0.875) at 20°C
Refractive index	1.539–1.541 at 20°C	1.3855 at 20°C
Acid value	1 max	
Solubility	1:3 in 80% alcohol	Slightly soluble in water; completely miscible with alcohol and ether; soluble in most organic solvents
Organoleptic characteristics	Sweet, fruity, balsamic odor; sweet taste reminiscent of currant	Fruity, ether-like odor; sweet taste reminiscent of rum
Synthesis	By heating cinnamyl chloride and isobutyl alcohol[6]	From isobutyl alcohol and carbon monoxide in the presence of sodium isobutylate at 110°C and 400 atm[7]
Natural occurrence	Not reported found in nature	In pineapple and apple; in vinegar
Reported uses[3]	Non-alcoholic beverages 1.3 ppm Alcoholic beverages 2.0 ppm Ice cream, ices, etc. 3.4 ppm Candy 5.4 ppm Baked goods 5.4 ppm	Non-alcoholic beverages 2.2 ppm Ice cream, ices, etc. 7.1 ppm Candy 19 ppm Baked goods 8.2 ppm Gelatins and puddings 5.0 ppm
Regulatory status	FDA 121.1164; FEMA No. 2193	FDA 121.1164; FEMA No. 2197

REFERENCES

For References 1–5, see end of Part III.
6. **Sudborough and Thompson,** *J. Chem. Soc.*, 83, 676, 1903.
7. E. I. du Pont de Nemours & Co., U.S. Patent 2,117,600, 1935.

	Isobutyl-2-furanpropionate	Isobutyl heptanoate
Other names	Isobutyl furfurylacetate Isobutyl-3-(2-furyl)-propanoate Isobutyl furylpropionate	Isobutyl heptoate Isobutyl heptylate
Empirical formula	$C_{11}H_{16}O_3$	$C_{11}H_{22}O_2$
Structure	(2-furyl)$-(CH_2)_2-C(=O)-O-CH_2-CH(CH_3)_2$	$(H_3C)_2CH-CH_2-O-C(=O)-(CH_2)_5-CH_3$
Physical/chemical characteristics[5]		
Appearance	Colorless to pale, straw-yellow liquid	Colorless liquid
Molecular weight	196.25	186.30
Boiling point		209°C (95–97°C at 12 mm Hg)
Specific gravity		0.8593 at 20°C
Solubility	Almost insoluble in water; soluble in alcohol	Soluble in most organic solvents
Organoleptic characteristics	Fruity, winey, brandy-like odor; pungent taste above 50 ppm; sweet, dry, brandy flavor at low levels	Characteristic green odor and corresponding taste when freshly prepared
Synthesis	By esterification of 2-furanpropionic acid with isobutanol	By direct esterification of heptanoic acid with isobutyl alcohol or by treating heptanoic acid and isobutyl alcohol in the presence of defatted *Chelidonium majus* seeds at 20°C for 3 days[6]
Natural occurrence	Not reported found in nature	Not reported found in nature
Reported uses[3]	Non-alcoholic beverages 8.1 ppm Ice cream, ices, etc. 14 ppm Candy 17 ppm Baked goods 21 ppm Gelatins and puddings 4.0–30 ppm Chewing gum 12 ppm Icings 20 ppm	Non-alcoholic beverages 0.50–1.5 ppm Ice cream, ices, etc. 2.4–10 ppm Candy 7.0–25 ppm Baked goods 7.0–25 ppm
Regulatory status	FDA 121.1164; FEMA No. 2198	FDA 121.1164; FEMA No. 2200

REFERENCES

For References 1–5, see end of Part III.

6. **Bournot,** *Biochem. Z.*, 65, 156, 1914.

	Isobutyl hexanoate	Isobutyl isobutyrate
Other names	Isobutyl caproate Isobutyl capronate Isobutyl hexylate	
Empirical formula	$C_{10}H_{20}O_2$	$C_8H_{16}O_2$
Structure	$(H_3C)_2CH-CH_2-O-C(=O)-(CH_2)_4-CH_3$	$(H_3C)_2CH-C(=O)-O-CH_2-CH(CH_3)_2$
Physical/chemical characteristics[5]		
Appearance	Colorless liquid	Colorless liquid
Assay	98% min	98% min
Molecular weight	172.27	144.21
Melting point		−80 to −81°C
Boiling point		144–151°C; 148–149°C
Flash point	76°C	49°C
Specific gravity	0.854–0.858 at 25°C	0.875 at 0°C
Refractive index	1.412–1.416 at 20°C	1.3999 at 20°C
Acid value	1 max	
Solubility	1:2 in 80% alcohol; 1:1 in 90% alcohol	Insoluble in water; soluble in most organic solvents
Organoleptic characteristics	Fruity odor with cocoa-like undertone	Odor and taste reminiscent of pineapple
Synthesis	By direct esterification	By catalytic reaction passing vapors of isobutyl alcohol over $CuO + Al_2O_3$, $ZnO + Al_2O_3$, or $CuO + ZnO + Al_2O_3$ at 350–400°C under pressure[6]
Natural occurrence	In banana, grapes, and wine	Reported found in the essential oil of hops[7] and in *Vitis vinifera*
Reported uses[3]	Non-alcoholic beverages 5.4 ppm Ice cream, ices, etc. 3.9 ppm Candy 8.1 ppm Baked goods 8.3 ppm Chewing gum 2.0 ppm	Non-alcoholic beverages 7.5 ppm Alcoholic beverages 2.0 ppm Ice cream, ices, etc. 7.4 ppm Candy 16 ppm Baked goods 17 ppm Gelatins and puddings 3.3–10 ppm
Regulatory status	FDA 121.1164; FEMA No. 2202	FDA 121.1164; FEMA No. 2189

REFERENCES

For References 1–5, see end of Part III.

6. **Ipatieff and Haensel,** *J. Org. Chem.*, 7, 194, 1942.
7. **Buttery et al.,** *Nature,* 200, 435, 1936.

	2-Isobutyl-3-methoxypyrazine	2-Isobutyl-3-methylpyrazine
Other names	2-Methoxy-3-(2-methylpropyl)pyrazine	2-Methyl-3-isobutyl pyrazine 2-Methyl-3-(2-methylpropyl)-pyrazine
Empirical formula	$C_9H_{14}ON_2$	$C_9H_{14}N_2$
Structure	[pyrazine ring with $CH_2-CH(CH_3)_2$ and OCH_3]	[pyrazine ring with CH_3 and $CH_2-CH(CH_3)_2$]
Physical/chemical characteristics		
Assay	C: 65.03%; H: 8.49%; O: 9.63%; N: 16.85%	C: 71.96%; H: 9.39%; N: 18.65%
Molecular weight	166.22	150.22
Boiling point		74°C at 10 mm Hg
Mass spectra	40(3), 41(2), 53(3), 66(1), 68(3), 80(2), 81(6), 94(15), 95(9), 106(2), 109(3), 124(100), 125(8), 151(22), 152(2), 165(4), 166(9)[11]	41(17), 42(13), 53(5), 60(1), 67(12), 82(1), 93(4), 108(100), 135(10), 150(5)[14]
IR spectra[7]	S (3.39, 3.43, 3.50, 6.49, 6.85, 6.9, 7.2, 7.7, 8.6, 9.9) M (3.28, 3.46, 7.4, 8.4, 9.4, 11.8) W (3.33, 6.34, 7.9, 8.1, 9.2, 10.5, 10.8, 11.2, 13.3)	
Organoleptic characteristics	The extremely powerful odor has a reported threshold value of 2 parts per trillion in water; diluted water solutions exhibit a characteristic odor of green bell peppers[7]	
Synthesis	By condensation of leucine amide with glyoxal, followed by methylation with CH_2N_2[6]	By condensation of ethylenediamine with 5-methyl-2,3-hexanedione[12]
Natural occurrences	Reported present as a flavor constituent in green bell pepper, coffee, galbanum oil, potato products, peas[8-10]	Reported present in coffee and potato products[13,15]
Reported uses[3]	Non-alcoholic beverages 0.05 ppm Ice cream, ices, etc. 0.05 ppm Candy 0.05 ppm Baked goods 0.05 ppm Gelatins and puddings 0.05 ppm Chewing gum 0.05 ppm Condiments 0.05 ppm Meat, meat sauces, soups 0.05 ppm Milk, dairy products 0.05 ppm Cereals 0.05 ppm	Non-alcoholic beverages 5.0 ppm Ice cream, ices, etc. 5.0 ppm Candy 5.0 ppm Baked goods 5.0 ppm Gelatins and puddings 5.0 ppm
Regulatory status	FEMA No. 3132	FEMA No. 3133

REFERENCES

For References 1–5, see end of Part III.

6. **Buttery, Seifert, Lundin, Guadagni and Ling.,** *Chem. Ind.* (London), 15, 490, 1969.
7. **Buttery et al.,** *J. Agric. Food Chem.*, 17(6), 1322, 1969.
8. **Gianturco et al.,** *J. Agric. Food Chem.*, 19(3), 530, 1971.
9. **Murray et al.,** *Chem. Ind.* (London), 27, 897, 1970.
10. **Maga and Sizer,** *Crit. Rev. Food Technol.*, 4(1), 39, 1973.
11. **Friedel et al.,** *J. Agric. Food Chem.*, 19, 530, 1971.
12. **Flament and Stoll,** *Helv. Chim. Acta,* 50, 1754, 1967.
13. **Stoll et al.,** *Helv. Chim. Acta,* 50, 1754, 1967.
14. **Buttery et al.,** *J. Agric. Food Chem.*, 19, 1045, 1971.
15. **Maga and Sizer,** *Crit. Rev. Food Technol.*, 4(1), 39, 1973.

	α-Isobutylphenethyl alcohol	Isobutyl phenylacetate
Other names	Benzyl isoamyl alcohol Isobutyl benzylcarbinol α-Isobutyl phenylethyl alcohol 4-Methyl-1-phenyl-2-pentanol	Isobutyl α-toluate
Empirical formula	$C_{12}H_{18}O$	$C_{12}H_{16}O_2$
Structure	C_6H_5–CH_2–CH(OH)–CH_2–CH(CH_3)$_2$	C_6H_5–CH_2–C(=O)–O–CH_2–CH(CH_3)$_2$
Physical/chemical characteristics[5]		
Appearance	Colorless, oily liquid	Colorless liquid
Assay		98% min[2] (99%)
Molecular weight	178.27	192.26
Boiling point	250°C	253°C
Flash point		116°C
Specific gravity	0.96	0.984–0.988 at 25°/25°C[2]
Refractive index		1.4860–1.4880 at 20°C[2]
Acid value		Not more than 1[2] (0.02%)
Solubility	Insoluble in water; soluble in alcohol	1:2 and more in 80% alcohol[2]
Organoleptic characteristics	Green, floral, somewhat herbaceous odor; buttery, oily, caramellic flavor	Sweet, musk-like fragrance; sweet, honey-like flavor
Synthesis	By reacting isobutyl aldehyde with benzyl magnesium chloride[6]	By esterification of phenylacetic acid with isobutyl alcohol
Natural occurrence		Not reported found in nature
Reported uses[3]	Non-alcoholic beverages 1.0–10 ppm Alcoholic beverages 50 ppm Ice cream, ices, etc. 38 ppm Candy 54 ppm Baked goods 15–50 ppm	Non-alcoholic beverages 2.8 ppm Ice cream, ices, etc. 2.8 ppm Candy 5.5 ppm Baked goods 5.0 ppm Gelatins and puddings 5.0 ppm Maraschino cherries 3.0 ppm
Regulatory status	FDA 121.1164; FEMA No. 2208	FDA 121.1164; FEMA No. 2210

REFERENCES

For References 1–5, see end of Part III.

6. **Petyunin,** *Chem. Abstr.,* 38, 950, 1944.

	Isobutyl propionate	Isobutyl salicylate
Other names		Isobutyl *o*-hydroxybenzoate
Empirical formula	$C_7H_{14}O_2$	$C_{11}H_{14}O_3$
Structure	$(H_3C)_2CH-CH_2-O-C(=O)-CH_2-CH_3$	o-$HO-C_6H_4-COO-CH_2-CH(CH_3)_2$
Physical/chemical characteristics[5]		
Appearance	Colorless liquid	Colorless liquid
Assay		98% min[2] (98.5%)
Molecular weight	130.18	194.23
Melting point		−5.9°C
Boiling point	137–138°C; 66.5°C at 60 mm Hg	259°C
Flash point		121°C
Specific gravity	0.8600–0.8635 at 20°/20°C	1.062–1.066 at 25°/25°C;[2] 1.0700–1.0720 at 15°C
Refractive index	1.3975 at 20°C	1.5070–1.5100 at 20°C[2]
Acid value		Not more than 1[2] (0.2)
Solubility	Insoluble in water; soluble in most organic solvents	1:9 in 80% alcohol;[2] 1:1–2 in 90% alcohol
Organoleptic characteristics	Odor similar to the corresponding ethyl ester but more refined and fresher; bitter flavor	Orchid, wintergreen-like odor similar to amyl salicylate; bitter taste
Synthesis	By direct esterification or by reacting propionyl chloride with isobutyl alcohol in the presence of Mg dust in ether solution[6]	By esterification of isobutyl alcohol and salicylic acid
Natural occurrence	Probably occurring in grape (var. white Sauvignon)[7], apple, pear, and rum	Not reported found in nature
Reported uses[3]	Non-alcoholic beverages 5.4 ppm Ice cream, ices, etc. 4.2 ppm Candy 25 ppm Baked goods 35 ppm	Non-alcoholic beverages 3.5 ppm Ice cream, ices, etc. 1.8 ppm Candy 2.6 ppm Baked goods 5.0 ppm
Regulatory status	FDA 121.1164; FEMA No. 2212	FDA 121.1164; FEMA No. 2213

REFERENCES

For References 1–5, see end of Part III.

6. **Spassow,** *Ber. Dtsch. Chem. Ges.,* 70, 1928, 1937.
7. **Chaundhary et al.,** *Am. J. Enol. Vitic.,* 15, 190, 1964.

	2-Isobutyl thiazole	Isobutyraldehyde
Other names		Isobutyl aldehyde Isobutyric aldehyde 2-Methyl propanal
Empirical formula	$C_7H_{11}NS$	C_4H_8O
Structure	(thiazole ring)–CH_2 — $CH(CH_3)_2$	$(H_3C)_2CH-CH=O$
Physical/chemical characteristics		
Appearance		Colorless liquid
Molecular weight	141.23	72.10
Boiling point	172–180°C	63.5°C at 754 mm Hg
Specific gravity	0.9953 at 25°C	0.7904 at 20°C
Refractive index	1.4939 at 25°C	1.3738 at 20°C
Solubility		Forms azeotropic mixtures with water
Organoleptic characteristics		Characteristic odor
Synthesis	Via cyclization of aliphatic amides with α,β-dibromo ether[6]	By oxidation of isobutyl alcohol with potassium dichromate and concentrated sulfuric acid[6]
Natural occurrence	Reportedly identified as the odorous component in tomatoes[7,8] and in processed tomatoes;[9] a flavor constituent of tomatoes	Reported found in apple and currant aromas and in the essential oils from tobacco leaves[7] and tea leaves;[8] also in the essential oils of: *Pinus jeffreyi* Murr. leaves,[9] *Citrus aurantium* leaves,[10] and *Datura stramonium*[11]
Reported uses[3]	Non-alcoholic beverages 1.0 ppm Ice cream, ices, etc. 1.0 ppm Candy 1.0 ppm Baked goods 1.0 ppm Gelatins and puddings 1.0 ppm Meat, meat sauces, soups 1.0 ppm Sauces 1.0 ppm Vegetables 1.0 ppm	Non-alcoholic beverages 0.30 ppm Alcoholic beverages 5.0 ppm Ice cream, ices, etc. 0.25–0.50 ppm Candy 0.67 ppm Baked goods 0.50–1.0 ppm
Regulatory status	FEMA No. 3134	FDA 121.1164; FEMA No. 2220

REFERENCES

For References 1–5, see end of Part III.

6. **Cottet, Gallo and Metzger,** *Bull. Soc. Chim. Fr.*, 12, 4499, 1967.
7. **Viani et al.,** *Helv. Chim. Acta*, 52(4), 887, 1969.
8. **Stevens,** *J. Am. Soc. Hort. Sci.*, 95(1), 9, 1970; *Chem. Abstr.*, 72, 131227p, 1970.
9. **Neipawer et al.,** *Flavour Ind.*, 2, 465, 1971.
10. **Pfeiffer,** *Ber. Dtsch. Chem. Ges.*, 5, 699, 1872.
11. **Onishi and Yamasaki,** *Bull. Agric. Chem. Soc. Jap.*, 19, 137, 1955.
12. **Yamamoto and Kato,** *J. Agric. Chem. Soc. Jap.*, 10, 661, 1934.
13. **Sondern,** *J. Am. Pharm. Assoc.*, 22, 114, 1933.
14. **Peyron,** *Soap Perfum. Cosmet.*, 33, 373, 1960.
15. **Ssivolobow,** *Chem. Zentralbl.*, 2, 227, 1916.

	Isobutyric acid	Isoeugenol
Other names	Isopropylformic acid 2-Methylpropanoic acid	1-Hydroxy-2-methoxy-4-propenylbenzene 2-Methoxy-4-propenylphenol 4-Propenyl guaiacol
Empirical formula	$C_4H_8O_2$	$C_{10}H_{12}O_2$
Structure	H_3C CH—COOH H_3C	OH OCH_3 $CH{=}CH{-}CH_3$
Physical/chemical characteristics[5]		
Appearance	Colorless liquid	Pale-yellow, rather viscous liquid
Assay		99.5% min[2]
Molecular weight	88.10	164.21
Melting point	−46.1°C	14–18°C
Boiling point	154.7°C	>12°C[2]
Congealing point		
Flash point	76°C	>100°C
Specific gravity	0.95296 at 15°C; 0.948 at 15.5°C	1.079–1.085 at 25°/25°C[2]
Refractive index	1.39525 at 15°C	1.5720–1.5770 at 20°C[2]
Solubility	1:5 in water; soluble in alcohol and ether	1:5 in 50% alcohol;[2] 1:1 in 70% alcohol
Organoleptic characteristics	Odor and flavor similar to *n*-butyric acid	Floral odor reminiscent of carnation
Synthesis	By oxidation of isobutyl alcohol	By the alkaline isomerization of eugenol obtained from essential oils high in eugenol[6]
Natural occurrence	Reported found in several essential oils: *Arnica montana*, Roman camomile, *Laurus nobilis*, imperatoria, and in carob fruits (*Siliqua dulcis*); also identified in the essence of *Seseli tortuosum,*[11] *Artemisia transiliensis,*[12] and in strawberry aroma[13]	The commercial product is a mixture of *cis*- and *trans*-isomers. Reported found in the essential oils of ylang-ylang[7] and nutmeg;[8] also in the oil from flowers of *Michelia champaca*[9] and in the oil from seeds of *Nectandra puchury*[10]
Reported uses[3]	Non-alcoholic beverages 4.1 ppm Ice cream, ices, etc. 12 ppm Candy 41 ppm Baked goods 38 ppm Chewing gum 470 ppm Margarine 30 ppm	Non-alcoholic beverages 3.7 ppm Ice cream, ices, etc. 3.8 ppm Candy 5.0 ppm Baked goods 11 ppm Chewing gum 0.30–1,000 ppm Condiments 1.0 ppm
Regulatory status	FDA 121.1164; FEMA No. 2222	FDA 121.1164; FEMA No. 2468

REFERENCES

For References 1–5, see end of Part III.

6. **Tiemann,** *Ber. Dtsch. Chem. Ges.,* 24, 2870, 1891.
7. Schimmel and Co., *Chem. Zentralbl.,* 2, 1907, 1901.
8. **Power and Salvay,** *J. Chem. Soc.,* 91, 2041, 1907.
9. **Brooks,** *J. Am. Chem. Soc.,* 33, 1764, 1911.
10. *Chem. Zentralbl.,* 1, 360, 1922.
11. **Salgues,** *C. R. Séances Acad. Sci. Paris,* 241, 677, 1956.
12. **Goryaev and Pugachev,** *Chem. Abstr.,* 50, 1684, 1956.
13. **Willhalm et al.,** *Helv. Chim. Acta,* 49, 66, 1966.

	Isoeugenyl acetate	Isoeugenyl benzyl ether
Other names	Acetyl isoeugenol 2-Methoxy-4-propenylphenyl acetate	Benzyl isoeugenol 2-Methoxy-4-propenylphenyl ether
Empirical formula	$C_{12}H_{14}O_3$	$C_{17}H_{18}O_2$
Structure	O—CO—CH$_3$; OCH$_3$; CH=CH—CH$_3$	O—CH$_3$; —CH$_2$—O— ; —CH=CH—CH$_3$
Physical/chemical characteristics[5]		
Appearance	White crystals	White to ivory-colored crystalline powder
Assay	98% min[2]	
Molecular weight	206.24	254.33
Melting point	Approx. 80°C	58–59°C; 48°C
Boiling point		282°C
Congealing point	>78°C[2]	>57.0°C[2]
Isoeugenyl content		Not more than 1.0%[2]
Acid value	1.0 max[2]	
Solubility	Soluble in most organic solvents	Approx 1:50 in 90% alcohol[2]
Organoleptic characteristics	Weak, rose-carnation, somewhat spicy odor; initially burning, then sweet taste	Faint, floral odor of rose-carnation
Synthesis	From isoeugenol and acetic acid by esterification	By benzylation of isoeugenol; also by alkaline isomerization of benzyleugenol[6]
Natural occurrence	Not reported found in nature	Not reported found in nature
Reported uses[3]	Non-alcoholic beverages 0.44 ppm Ice cream, ices, etc. 2.1 ppm Candy 17 ppm Baked goods 17 ppm Chewing gum 100 ppm	
Regulatory status	FDA 121.1164; FEMA No. 2470	FDA 121.1164

REFERENCES

For References 1–5, see end of Part III.

6. **Pond and Beers,** *J. Am. Chem. Soc.*, 19, 828, 1897.

	Isoeugenyl ethyl ether	Isoeugenyl formate
Other names	1-Ethoxy-2-methoxy-4-propenylbenzene 2-Ethoxy-5-propenylanisole Ethyl isoeugenol	2-Methoxy-4-propenyl phenyl formate 4-Propenyl-2-methoxyphenylformate
Empirical formula	$C_{12}H_{16}O_2$	$C_{11}H_{12}O_3$
Structure	O—C₂H₅, O—CH₃, CH=CH—CH₃ (structural formula)	O=CH—O, O—CH₃, CH=CH—CH₃ (structural formula)
Physical/chemical characteristics[5]		
Appearance	Crystalline solid	Colorless to pale straw-yellow, viscous liquid
Molecular weight	192.26	192.22
Melting point	63–64°C (with sublimation)	
Boiling point	245°C	282°C
Congealing point		1.038 (1.12) at 15.5°C
Isoeugenol content		1.5660 at 20°C
Solubility	Highly soluble in ether and benzene; soluble in alcohol and acetic acid	Almost insoluble in water; soluble in alcohol; slightly soluble in propylene glycol
Organoleptic characteristics	Odor similar to isoeugenol	Faint, orris-like, green, sweet, woody odor with clove-like undertones; warm, spicy flavor
Synthesis	From isoeugenol by boiling with ethyl *p*-toluenesulfonate[6]	By formylation of isoeugenol
Natural occurrence	Not reported found in nature	Not reported found in nature
Reported uses[3]	Non-alcoholic beverages 7.8 ppm Ice cream, ices, etc. 0.50 ppm Candy 17 ppm Baked goods 1.0–3.5 ppm	Condiments 0.20 ppm
Regulatory status	FDA 121.1164; FEMA No. 2472	FDA 121.1164; FEMA No. 2474

REFERENCES

For References 1–5, see end of Part III.

6. **van Duin,** *Recl. Trav. Chim. Pays-Bas,* 45, 351, 1926.

	Isoeugenyl methyl ether	Isoeugenol phenylacetate
Other names	1,2-Methoxy-4-propenylbenzene Methyl isoeugenol 4-Propenyl veratrole	Isoeugenol α-toluate 2-Methoxy-4-propenylphenyl phenylacetate 4-Propenylguaiacyl phenylacetate
Empirical formula	$C_{11}H_{14}O_2$	$C_{18}H_{18}O_3$
Structure	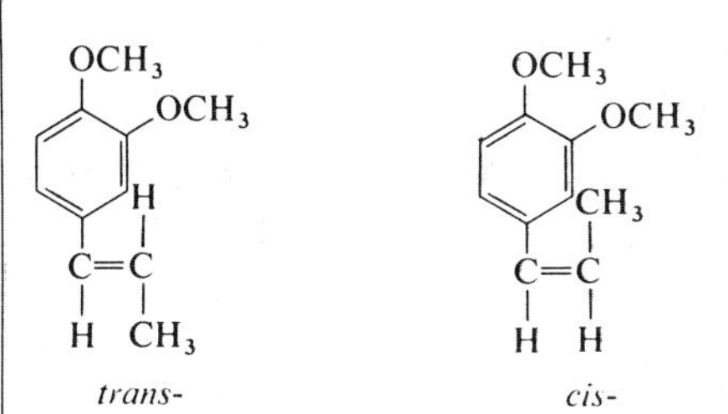	O O—C—CH2— OCH3 CH=CH—CH3
Physical/chemical characteristics[5]		
Appearance	Colorless to pale-yellow liquid	Yellowish, viscous liquid 95% min
Molecular weight	178.23	282.34
Melting point	*trans*-, 16–17°C	
Boiling point	262–264°C; *cis*-, 138–140°C at 12 mm Hg; *trans*-, 143–144°C at 11 mm Hg	$>$100°C
Specific gravity	1.047–1.053 at 25°/25°C[2]	1.113–1.117
Refractive index	1.5660–1.5690 at 20°C[2]	1.575–1.577
Isoeugenol content		1 max
Acid value		
Solubility	1:2 in 70% alcohol;[2] 1:6.5–7.5 in 60% alcohol. Forms azeotropes with glycerin (75%), acetamide (31%), and phenylacetic acid (40%)	1:2 in 95% alcohol; insoluble in water
Organoleptic characteristics	Delicate, clove-carnation odor; burning, bitter taste	Intensely sweet odor reminiscent of carnation; vanilla, clove-like, and honey odor; sweet, spicy, slightly fruity flavor
Synthesis	By methylation of isoeugenol with methyl sulfate in alkaline solution[9]	From phenylacetyl chloride plus sodium isoeugenol
Natural occurrence	The natural and the commercial products are a mixture of *cis*- and *trans*-isomers. Originally isolated in the oil from roots of *Asarum arifolium* Michx.;[6] in the oil of *Cymbopogon javanensis*,[7] *Orthodon methylisoeugenoliferum* F. (approx 53%),[8] and others	Not reported found in nature
Reported uses[3]	Non-alcoholic beverages 4.0 ppm Ice cream, ices, etc. 7.7 ppm Candy 13 ppm Baked goods 18 ppm Gelatins and puddings 0.10 ppm Chewing gum 110 ppm	Non-alcoholic beverages 0.05 ppm Ice cream, ices, etc. 0.20 ppm Candy 3.0 ppm Baked goods 2.0–3.0 ppm
Regulatory status	FDA 121.1164; FEMA No. 2476	FDA 121.1164; FEMA No. 2477

REFERENCES

For References 1–5, see end of Part III.

6. **Miller,** *Arch. Pharm.*, 240, 371, 1902.
7. **Hoffman,** *Chem. Zentralbl.*, 3, 886, 1919.
8. **Fujita,** *Chem. Abstr.*, 41, 3585, 1947.
9. **Francesconi and Puxeddu,** *Gazz. Chim. Ital.*, 1, 207, 1909.

	Isojasmone	DL-Isoleucine
Other names	2-Hexyl-cyclopenten-2-one-1	2-Amino-3-methylpentanoic acid α-Amino-β-methylvaleric acid
Empirical formula	$C_{11}H_{18}O$	$C_6H_{13}O_2N$
Structure	(cyclopentenone ring with hexyl chain, CH_3, O)	$CH_3-CH_2-CH(CH_3)-CH(NH_2)-COOH$
Physical/chemical characteristics[5]		
Appearance	Yellow to yellow-brown liquid	Rhombic or monoclinic platelets (from diluted alcohol)
Assay	80% min	
Molecular weight	166.26	131.11
Melting point	144°C at 10 mm Hg	275°C (sealed tube) 292°C (decomposes)
Flash point	>100°C	
Specific gravity	0.908–0.920 at 25°C; 0.911 at 22°C	
Refractive index	1.4750–1.4800 at 20°C	
Acid value	2 max	
Solubility	1:10 in 70% alcohol; 1:1 in 80% alcohol	Slightly soluble in water
Organoleptic characteristics	Green odor reminiscent of jasmine	Flavorless
Synthesis		From the corresponding aldehyde by Bucherer's reaction[6]
Natural occurrence	Not reported found in nature	
Reported uses[3]		Non-alcoholic beverages 50.0 ppm Baked goods 50.0 ppm Condiments, pickles 50.0 ppm Meat, meat sauces, soups 50.0 ppm Milk, dairy products 50.0 ppm Cereals 50.0 ppm
Regulatory status	FDA 121.1164	FEMA No. 3295

REFERENCES

For References 1–5, see end of Part III.

6. *Ullmanns Enzyklopädie der Technischen Chemie,* 3rd ed., Vol. 3, Urban and Schwarzenberg, 1952, 514.

	Isopropenylpyrazine		Isopropyl acetate	
Other names	2-Isopropenyl-1,4-diazine 2-(1-Methylvinyl)pyrazine 2-Isopropenylpyrazine			
Empirical formula	$C_7H_8N_2$		$C_5H_{10}O_2$	
Structure	pyrazine ring with $-C(CH_3)=CH_2$ substituent		$(H_3C)_2CH-O-C(=O)-CH_3$	
Physical/chemical characteristics				
Appearance			Colorless liquid	
Assay	C: 69.97%; H: 6.71%; N: 23.31%			
Molecular weight	120.16		102.13	
Boiling point			89.5°C	
Specific gravity			0.8718 at 20°C	
Refractive index			1.3773 at 20°C	
Mass spectra[7]	39 (35), 41 (18), 52 (23), 67 (24), 94 (11), 119 (100), 120 (84)			
Organoleptic characteristics			Intense, fruity odor; on dilution a sweet apple-like flavor	
Synthesis	By alkylation of the Mannich base obtained by reacting formaldehyde and dimethylamine with ethylpyrazine, followed by the Hoffmann degradation[8]		By direct acetylation of isopropyl alcohol in the presence of various catalysts: concentrated H_2SO_4, diethyl sulfate, chlorosulfonic acid, and boron trifluoride.[6,7,8]	
Natural occurrence	Reported found as a steam-volatile component in roasted Spanish peanuts[6]		In pineapple, pear, and cocoa	
Reported uses[3]	Non-alcoholic beverages	10.0 ppm	Non-alcoholic beverages	16 ppm
	Ice cream, ices, etc.	10.0 ppm	Ice cream, ices, etc.	17 ppm
	Candy	10.0 ppm	Candy	58 ppm
	Gelatins and puddings	10.0 ppm	Baked goods	75 ppm
	Meat, meat sauces, soups	10.0 ppm		
	Milk, dairy products	10.0 ppm		
	Cereals	10.0 ppm		
Regulatory status	FEMA No. 3296		FDA 121.1164; FEMA No. 2926	

REFERENCES

For References 1–5, see end of Part III.

6. **Walradt et al.,** *J. Agric. Food Chem.,* 19, 972, 1971.
7. **Walradt et al.,** *J. Agric. Food Chem.,* 18, 926, 1971.
8. **Kamal et al.,** *J. Org. Chem.,* 27, 1355, 1962.
9. Dow Chemical Co., U.S. Patent 2,021,851, 1932.
10. **Dorris, Sowa and Nieuwland,** *J. Am. Chem. Soc.,* 56, 2689, 1934.
11. **Vogel,** *J. Chem. Soc.,* p. 629, 1948.

	α-Isomethylionone	**Isopentylamine**
Other names	α-Cyclocitrylidene butanone Methyl-γ-ionone (so called) 4-(2,6,6-Trimethyl-2-cyclohexen-1-yl)-3-methyl-3-buten-2-one	Isoamylamine 3-Methylbutylamine Isoamino pentane Isobutyl carbylamine
Empirical formula	$C_{14}H_{22}O$	$C_5H_{13}N$
Structure	(structure: H_3C, CH_3, O, CH_3, CH_3, CH_3)	CH_3—CH(CH_3)—CH_2—CH_2—NH_2
Physical/chemical characteristics[5]		
Appearance	Yellowish liquid	Colorless to pale straw-colored mobile liquid
Assay	85–95% (98% min)	
Molecular weight	206.30	87.17
Boiling point	93°C at 3.1 mm Hg	96.5–97.5°C
Specific gravity	0.9304 at 22°C; 0.925–0.929 at 25°/25°C	0.7491 at 20°C
Refractive index	1.5000–1.5020 at 20°C; 1.4990 at 25°C	1.4083 at 20°C
Solubility		Miscible with water, alcohol, propylene glycol, glycerine, and oils
Organoleptic characteristics	Fine violet and orris-like odor	Unpleasant ammoniacal odor
Synthesis	By condensation of citral with methyl ethyl ketone controlling the ratio between the *n*- and the *iso*-forms. The methyl *pseudo*-ionone obtained is then subjected to ring closure, using strong alkali and high temperature (thus favoring the formation of the normal α- and β-methylionones) or with quaternary ammonium base (thus favoring the formation of *iso*-forms). See also monographs of α- and β-ionone and methyl α-, β-, and γ-ionones.	From isoamyl chloride and sodamide in liquid ammonia;[6,7] by reduction of isoamyl nitrile[8,9]
Natural occurrence	Not reported found in nature	Reported found in rye ergot and in many botanical sources (Rosaceae, Saxifragaceae, Umbelliferae, Caprifoliaceae);[10] also commonly found among the degradation products of proteins
Reported uses[3]	Non-alcoholic beverages 0.97 ppm Ice cream, ices, etc. 0.98 ppm Candy 4.9 ppm Baked goods 4.3 ppm Gelatins and puddings 0.05 ppm Chewing gum 0.80 ppm	Non-alcoholic beverages 0.10 ppm Ice cream, ices, etc. 0.10 ppm Candy 0.10 ppm Baked goods 0.10 ppm Gelatins and puddings 0.10 ppm
Regulatory status	FDA 121.1164; FEMA No. 2714	FEMA No. 3219

REFERENCES

For References 1–5, see end of Part III.

6. **Shreve and Rothenberger,** *Ind. Eng. Chem.*, 29, 1361, 1937.
7. **Shreve and Burtsfield,** *Ind. Eng. Chem.*, 33, 220, 1941.
8. **Seigle and Hass,** *Ind. Eng. Chem.*, 31, 649, 1939.
9. **Miduno and Takayama,** *Bull. Chem. Soc. Jap.*, 17, 137, 1942.
10. **Steiner and Loffler,** *Z. Wiss. Bot.*, 71, 523, 1929.

	p-Isopropylacetophenone	Isopropyl alcohol
Other names	Acetocumene *p*-Acetyl cumol 1,4-Acetyl-isopropyl benzol 1-Isopropyl-4-acetylbenzene *p*-Isopropyl acetylbenzol Methyl *p*-isopropylphenyl ketone	Dimethylcarbinol Isopropanol 2-Propanol
Empirical formula	$C_{11}H_{14}O$	C_3H_8O
Structure	$(H_3C)_2CH-C_6H_4-C(=O)-CH_3$	$CH_3-CH(OH)-CH_3$
Physical/chemical characteristics[5]		
Appearance	Colorless liquid	Colorless liquid
Molecular weight	162.23	60.09
Melting point		−81 to −83°C
Boiling point	252–254°C (250°C)	82.3–82.4°C
Flash point		15°C
Specific gravity	0.9755 at 15°C; 0.975 at 15.5°C	0.7854 at 20°C
Refractive index		1.37757 at 20°C
Solubility	Insoluble in water; soluble in alcohol	Miscible with water, alcohol, ether, and chloroform
Organoleptic characteristics	Odor reminiscent of orris and basil; warm, spicy flavor	Alcoholic, somewhat unpleasant odor; burning taste
Synthesis	By condensing cymene with acetyl chloride in the presence of $AlCl_3$;[6] also from benzol, acetyl chloride, and isopropyl chloride via Friedel-Crafts	Synthetically prepared from acetylene or propylene
Natural occurrence	In oil of lavandin	Reported found in apple and cognac aromas (esterified)
Reported uses[3]	Non-alcoholic beverages 0.08 ppm Ice cream, ices, etc. 0.10 ppm Candy 0.50 ppm Baked goods 1.0 ppm Pickles 5.0 ppm	Non-alcoholic beverages 25 ppm Candy 10–75 ppm Baked goods 75 ppm
Regulatory status	FDA 121.1164; FEMA No. 2927	FDA 121.1164; FEMA No. 2929

REFERENCES

For References 1–5, see end of Part III.

6. **Widman,** *Ber. Dtsch. Chem. Ges.*, 21, 225, 1888.

	Isopropyl benzoate	***p*-Isopropylbenzyl alcohol**
Other names		Cumic alcohol Cuminic alcohol Cuminol Cuminyl alcohol *p*-Cymen-l-ol
Empirical formula	$C_{10}H_{12}O_2$	$C_{10}H_{14}O$
Structure	$C_6H_5-C(=O)-O-CH(CH_3)_2$	CH_2OH / $CH(CH_3)_2$ (para-substituted benzene)
Physical/chemical characteristics[5]		
Appearance	Liquid	From colorless to yellow liquid
Molecular weight	164.21	150.22
Boiling point	218–219°C	248°C
Specific gravity	1.0263 at 4°C	0.978 at 20°C
Refractive index		1.522 at 24°C
Solubility	Insoluble in water; readily soluble in alcohol and ether	Slightly soluble in water; miscible with alcohol and ether
Organoleptic characteristics		Intense, persistent, caraway-like odor; aromatic, burning taste
Synthesis	By heating benzoyl chloride and isopropyl alcohol[6]	From cuminaldehyde with H_2 or NH_3 under pressure in methanol solution and in the presence of Raney Ni catalyst; or by catalytic reduction of methyl *p*-isopropylbenzoate
Natural occurrence	In apple, pear, cocoa and honey	Reported found in the oil from fruits of *Cuminum cyminum* and *Carum carvi* L.; also in the oils of *Eucalyptus bakeris* (esterified) and French lavender
Reported uses[3]	Non-alcoholic beverages 0.50 ppm Ice cream, ices, etc. 1.0 ppm Candy 1.0 ppm Baked goods 1.0 ppm	Non-alcoholic beverages 11 ppm Ice cream, ices, etc. 0.47 ppm Candy 33 ppm Baked goods 35 ppm
Regulatory status	FDA 121.1164; FEMA No. 2932	FDA 121.1164; FEMA No. 2933

REFERENCES

For References 1–5, see end of Part III.

6. **Perkin,** *J. Chem. Soc.*, 69, 1174, 1896.

	Isopropyl butyrate	Isopropyl cinnamate
Other names		Isopropyl β-phenylacrylate Isopropyl 3-phenylpropenoate
Empirical formula	$C_7H_{14}O_2$	$C_{12}H_{14}O_2$
Structure	$(H_3C)_2CH-O-C(=O)-CH_2-CH_2-CH_3$	$C_6H_5-CH=CH-C(=O)-O-CH(CH_3)_2$
Physical/chemical characteristics[5]		
Appearance	Colorless liquid	Colorless, viscous liquid; solidifies when cold; tends to polymerize on becoming odorless
Molecular weight	130.18	190.24
Boiling point	129–131°C	268°–270°C
Specific gravity	0.8588 at 20°C	1.017 (1.03) at 18°C
Refractive index	1.3936 at 20°C	
Solubility		Insoluble in water; soluble in alcohol
Organoleptic characteristics	Pleasant odor reminiscent of butyric acid	Balsamic, sweet, dry amber-like odor; fresh, fruity flavor
Synthesis	From butyric acid and propylene in presence of concentrated H_2SO_4 at 125°C in sealed tube;[6,7] also from butyric acid and isopropyl alcohol in the presence of HCl[8] or *p*-toluenesulfonic acid[9]	By esterification of isopropanol with cinnamic acid
Natural occurrence	Reported found in strawberry	Not reported found in nature
Reported uses[3]	Non-alcoholic beverages 9.7 ppm Ice cream, ices, etc. 21 ppm Candy 39 ppm Baked goods 39 ppm	Non-alcoholic beverages 0.52 ppm Ice cream, ices, etc. 0.75 ppm Candy 1.3 ppm Baked goods 2.3 ppm
Regulatory status	FDA 121.1164; FEMA No. 2935	FDA 121.1164; FEMA No. 2939

REFERENCES

For References 1–5, see end of Part III.
6. Dow Chemical Co., U.S. Patent 2,021,851, 1932.
7. **Fichter and Bürgin,** *Helv. Chim. Acta,* 14, 95, 1931.
8. **Kohlrausch and Sabathy,** *Monatsh. Chem.*, 72, 310, 1939.
9. **Salmi and Leimu,** *Suom. Kemistil. B,* 20, 46, 1947.

	Isopropyl formate	Isopropyl hexanoate
Other names		Isopropyl caproate Isopropyl capronate Isopropyl hexylate
Empirical formula	$C_4H_8O_2$	$C_9H_{18}O_2$
Structure	$(H_3C)_2CH{-}O{-}\overset{O}{\overset{\|\|}{C}}H$	$(H_3C)_2CH{-}O{-}\overset{O}{\overset{\|\|}{C}}{-}(CH_2)_4{-}CH_3$
Physical/chemical characteristics[5]		
Appearance	Colorless liquid	Colorless liquid
Molecular weight	88.10	158.24
Boiling point	67–68°C	74°C (176°C) at 20 mm Hg
Specific gravity	0.8774 at 20°C; 0.883 at 0°C	0.8570 at 20°C
Refractive index	1.3678 at 20°C	
Solubility	Slightly soluble in water; completely miscible with alcohol, ether, and most organic solvents	Almost insoluble in water; soluble in alcohol
Organoleptic characteristics	Characteristic fruity, ether-like odor; sweet taste reminiscent of plum	Sweet, delicate, fruity odor of pineapple; fresh, sweet, berry-like taste
Synthesis	By direct esterification	By esterification of hexanoic acid with isopropyl alcohol in benzene solution in the presence of trace amounts of concentrated H_2SO_4 and subsequent azeotropic distillation[6]
Natural occurrence	Reported found in several natural products	In strawberry; in bleu cheese
Reported uses[3]	Non-alcoholic beverages 18–25 ppm Ice cream, ices, etc. 18–25 ppm Candy 55–100 ppm Baked goods 60–100 ppm	Non-alcoholic beverages 0.50 ppm Ice cream, ices, etc. 5.5–10 ppm Candy 20–40 ppm Baked goods 20–40 ppm
Regulatory status	FDA 121.1164; FEMA No. 2944	FDA 121.1164; FEMA No. 2950

REFERENCES

For References 1–5, see end of Part III.

6. Commercial Solvents Corp., U.S. Patent 2,076,111, 1934.

	Isopropyl isobutyrate	Isopropyl isovalerate
Other names		
Empirical formula	$C_7H_{14}O_2$	$C_8H_{16}O_2$
Structure	$(H_3C)_2CH-C(=O)-O-CH(CH_3)_2$	$(H_3C)_2CH-CH_2-C(=O)-O-CH(CH_3)_2$
Physical/chemical characteristics[5]		
Appearance	Liquid	Liquid
Molecular weight	130.18	144.21
Boiling point	121°C	68–70°C at 55 mm Hg
Specific gravity	0.8687 at 0°C	
Refractive index		1.3960 at 20.5°C
Solubility	Insoluble in water; soluble in most organic solvents	Insoluble in water; soluble in most organic solvents
Organoleptic characteristics	Intense, fruity, ether-like odor	Ether-like odor similar to the *n*-propyl ester; sweet, apple-like taste
Synthesis	By boiling isobutyryl chloride and isopropyl alcohol[6]	From isobutene and isopropyl alcohol in the presence of CO and HF under pressure[7]
Natural occurrence	Reported found in several natural products	Not reported found in nature
Reported uses[3]	Non-alcoholic beverages 12–25 ppm Ice cream, ices, etc. 18–25 ppm Candy 58–100 ppm Baked goods 60–100 ppm	Non-alcoholic beverages 3.4 ppm Ice cream, ices, etc. 3.4 ppm Candy 11 ppm Baked goods 11 ppm
Regulatory status	FDA 121.1164; FEMA No. 2937	FDA 121.1164; FEMA No. 2961

REFERENCES

For References 1–5, see end of Part III.
6. **Kohlrausch and Skrabal,** *Monatsh. Chem.,* 70, 393, 1937.
7. **Friedmann and Cotton,** U.S. Patent 3,052,698, 1962.

	2-Isopropyl-5-methyl-2-hexenal	*p*-Isopropylphenylacetaldehyde
Other names	*iso*-Dihydrovandulyl aldehyde	Cuminic acetaldehyde *p*-Cymen-7-carboxaldehyde Homo-cuminic aldehyde 4-Isopropyl phenylacetaldehyde
Empirical formula	$C_{10}H_{18}O$	$C_{11}H_{14}O$
Structure	$CH_3-CH(CH_3)-CH_2-CH=C(CH(CH_3)_2)-CHO$	$(H_3C)_2CH-C_6H_4-CH_2-CHO$
Physical/chemical characteristics[5]		
Appearance		Colorless liquid
Molecular weight	154.24	162.23
Boiling point		230°C (243°C)
Specific gravity		0.955 at 15.5°C
Refractive index		1.5200 at 20°C
Solubility		Almost insoluble in water; soluble in alcohol
Organoleptic characteristics		Characteristic bark odor; citrus, bitter-sweet, fruity flavor
Synthesis	$(H_3C)_2CH-CH_2-CHO + (H_3C)_2C=CH-CHO \xrightarrow{B^{\ominus}}$	From cumyl magnesium chloride and ethyl formate or triethyl *o*-formate followed by acid hydrolysis of the acetal
Natural occurrence	Reported found in cocoa[6]	Not reported found in nature
Reported uses[3]	Non-alcoholic beverages 1.0 ppm Ice cream, ices, etc. 1.0 ppm Candy 2.0 ppm Baked goods 2.0 ppm Gelatins and puddings 2.0 ppm Milk, dairy products 0.5 ppm	Non-alcoholic beverages 0.10 ppm Ice cream, ices, etc. 0.50 ppm Candy 0.50 ppm
Regulatory status	FEMA No. 3406	FDA 121.1164; FEMA No. 2954

REFERENCES

For References 1–5, see Introduction to Part II.

6. **van Praag et al.,** *J. Agric. Food Chem.*, 16, 1005, 1968.

	Isopropyl phenylacetate	3-(*p*-Isopropylphenyl)-propionaldehyde
Other names	Isopropyl α-toluate	Cuminyl acetaldehyde *p*-Cymyl propanal *p*-Isopropylhydrocinnamaldehyde 3-(*p*-Isopropylphenyl)-propionic aldehyde
Empirical formula	$C_{11}H_{14}O_2$	$C_{12}H_{16}O$
Structure	$C_6H_5-CH_2-C(=O)-O-CH(CH_3)_2$	$(H_3C)_2CH-C_6H_4-CH_2-CH_2-CHO$
Physical/chemical characteristics[5]		
Appearance	Liquid	Colorless, viscous liquid
Molecular weight	178.23	176.26
Boiling point	253 C	
Specific gravity	1.0096 at 20°C	
Solubility		Insoluble in water; soluble in alcohol
Organoleptic characteristics	Fragrant, rose-like scent; sweet, honey-like flavor with a wine undertone	Powerful, sweet, green, floral odor; peculiar, sweet, green, fruity flavor
Synthesis		By condensation of cumin aldehyde with acetaldehyde, followed by hydrogenation to the saturated aldehyde; also from cumyl magnesium chloride and propyl formate, followed by acid hydrolysis of the acetal
Natural occurrence	Reported found in several natural products	Not reported found in nature
Reported uses[3]	Non-alcoholic beverages 0.20–0.50 ppm Ice cream, ices, etc. 1.8 ppm Candy 0.50–8.0 ppm Baked goods 3.0–8.0 ppm	Non-alcoholic beverages 0.80 ppm Ice cream, ices, etc. 0.80 ppm Candy 1.3 ppm Baked goods 3.0 ppm Chewing gum 5.0 ppm
Regulatory status	FDA 121.1164; FEMA No. 2956	FDA 121.1164; FEMA No. 2957

REFERENCES

For References 1–5, see end of Part III.

	Isopropyl propionate	Isopropyl tiglate
Other names		Isopropyl-α-methyl crotonate Isopropyl-3-methyl-2-butenoate
Empirical formula	$C_6H_{12}O_2$	$C_8H_{15}O_2$
Structure	$(H_3C)_2CH-O-C(=O)-CH_2-CH_3$	$CH_3-CH=C(CH_3)-C(=O)-O-CH(CH_3)_2$
Physical/chemical characteristics[5]		
Appearance	Liquid	
Molecular weight	116.16	143.20
Boiling point	108–110°C	
Specific gravity	0.8660 at 20°C	
Refractive index	1.3872 at 20°C	
Solubility	Completely miscible with alcohol and with diluted alcohol	
Organoleptic characteristics	Bittersweet taste reminiscent of plum	
Synthesis	By direct esterification in benzene solution and in the presence of *p*-toluenesulfonic acid[8]	By photolysis of 2,2,5,5-tetramethyldihydro-3-furanone in methanol;[6] also by esterification of 3-methyl-2-butenoic acid with isopropyl alcohol in the presence of concentrated sulfuric acid[7]
Natural occurrence	In raspberry and currant	
Reported uses[3]	Non-alcoholic beverages 9.7 ppm Ice cream, ices, etc. 5.0–50 ppm Candy 40–50 ppm Baked goods 30–50 ppm	Baked goods 5.0 ppm Gelatins and puddings 1.0 ppm Soups 1.0 ppm Salad oil 10.0 ppm
Regulatory status	FDA 121.1164; FEMA No. 2959	FEMA No. 3229

REFERENCES

For References 1–5, see end of Part III.

6. **Hagens, Wasacz, Joullie and Yates,** *J. Org. Chem.*, 35, 3682, 1970.
7. **Hagens et al.,** *J. Org. Chem.*, 35, 3682, 1970.
8. **Salmi and Leimu,** *Chem. Abstr.*, 42, 4031, 1948

	Isopulegol	Isopulegone
Other names	*p*-Menth-8-en-3-ol 1-Methyl-4-isopropenylcyclohexan-3-ol	*p*-Menth-8-en-3-one 1-Methyl-4-isopropenylcyclohexan-3-one 1-Methyl-4-isopropenyl-3-cyclohexanone
Empirical formula	$C_{10}H_{18}O$	$C_{10}H_{18}O$
Structure	CH_3 / —OH / H_3C C CH_2	CH_3 / =O / H_3C C CH_2
Physical/chemical characteristics[5]		
Appearance	Colorless liquid	Colorless liquid
Molecular weight	154.24	152.23
Boiling point	*l*-iso, 94°C at 14 mm Hg; *d*-neoiso, 95°C at 17 mm Hg; *d*-neoiso, 96°C at 15 mm Hg	101–102°C at 17 mm Hg
Specific gravity	*l*-iso, 0.9110 at 20°C; *d*-neoiso, 0.9107 at 25°C	0.92177 at 20.4°C
Refractive index	*l*-iso, 1.4723 at 20°C; *d*-neoiso, 1.4686 at 25°C; *d*-neoiso, 1.4775 at 10°C	1.46787 at 20.4°C
Optical rotation	*l*-iso, −25.9°; *d*-neoiso, +39.3°	
Synthesis	Several stereoisomers are possible. Only *l*-isopulegol and *d*-α-isopulegol have been isolated from mixtures of alcohols obtained by cyclization of *d*-citronellal[6]	The structure of isopulegone was defined by Tiemann and Schmidt[7] as well as by Hugh, Kon, and Linstead;[8] usually prepared from isopulegol.[8]
Natural occurrence	*l*-Isopulegol has been reported found in the essences of lemongrass. East African geranium, and *Eucalyptus citriodora*. *d*-Isopulegol is present in the oils of *Backhousia* and *Baeckea citriodorae*. *d*-Neoisopulegol is found in *Mentha rotundifolia*.	Natural isopulegone is dextrorotatory. Reported found in the essential oils from leaves of *Bystropogon mollis*, *Mentha timija*, and others
Reported uses[3]	Non-alcoholic beverages 7.4 ppm Ice cream, ices, etc. 29 ppm Candy 23 ppm Baked goods 23 ppm	Non-alcoholic beverages 4.0 ppm Ice cream, ices, etc. 12 ppm Candy 16 ppm Baked goods 16 ppm
Regulatory status	FDA 121.1164; FEMA No. 2962	FDA 121.1164; FEMA No. 2964

REFERENCES

For References 1–5, see end of Part III.
6. **Pickard et al.,** *J. Chem. Soc.* (London), 117, 1248, 1920.
7. **Tiemann and Schmidt,** *Ber. Dtsch. Chem. Ges.*, 29, 914, 1896.
8. **Hugh, Kon and Linstead,** *J. Chem. Soc.*, (London), p. 2585, 1927.

	Isopulegyl acetate	Isoquinoline
Other names		2-Benzazine Benzo(*c*)pyridine
Empirical formula	$C_{12}H_{20}O_2$	C_9H_7N
Structure	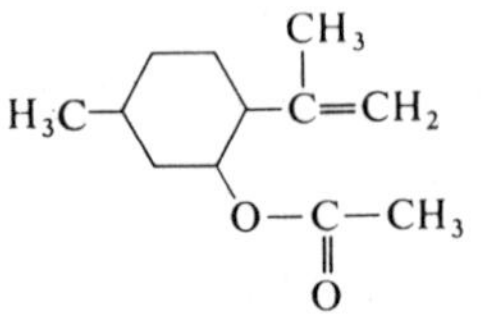	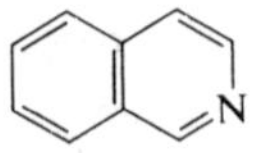
Physical/chemical characteristics[5]		
Appearance	Colorless liquid	Low-melting, crystalline solid
Assay	95% min	
Molecular weight	196.29	129.15
Melting point		26.5°C
Boiling point	104–105°C at 10 mm Hg	243°C
Flash point	87°C	
Specific gravity	0.932–0.936 (0.9394) at 20°C	1.0980 at 20°C
Refractive index	1.4572 at 20°C	1.61478 at 20°C
Optical rotation	−1.1°	
Acid value	1 max	
Solubility	1:4 in 70% alcohol; 1:2 in 80% alcohol	Slightly soluble in water; soluble in most organic solvents
Organoleptic characteristics	Sweet, mint-like odor	Odor reminiscent of benzaldehyde and anise
Synthesis	By prolonged heating of citronellal with acetic anhydride (with or without sodium acetate)	Obtained from coal tar (238–250°C boiling fraction). It is isolated as the sulfate or as is by repeated freezing.[6]
Natural occurrence	Not reported found in nature	Not reported found in nature; occurs in coal tar, where it is formed during the dry distillation of coal
Reported uses[3]	Non-alcoholic beverages 5.8 ppm Ice cream, ices, etc. 22 ppm Candy 19 ppm Baked goods 19 ppm	Non-alcoholic beverages 0.25 ppm Ice cream, ices, etc. 0.25 ppm Candy 1.0 ppm Baked goods 0.004–1.0 ppm
Regulatory status	FDA 121.1164; FEMA No. 2965	FDA 121.1169; FEMA No. 2978

REFERENCES

For References 1–5, see end of Part III.
6. Kjellman, U.S. Patent 2,483,420, 1949.

	Isosafrole	Isovaleric acid
Other names	3,4-Methylenedioxy-l-propenylbenzene	Delphinic acid Isobutyl formic acid Isopropylacetic acid β-Methyl butyric acid Valerianic acid
Empirical formula	$C_{10}H_{10}O_2$	$C_5H_{10}O_2$
Structure	O—CH₂ / O (methylenedioxy ring on benzene), CH=CH—CH₃	$(H_3C)_2CH-CH_2-COOH$
Physical/chemical characteristics[5]		
Appearance	Colorless liquid	Dense, colorless liquid
Molecular weight	162.18	102.13
Melting point	6.7–6.8°C	−30°C
Boiling point	252°C (127–128°C at 15 mm Hg)	176.5°C
Flash point		70°C
Specific gravity	1.122 at 20°C	0.93080 at 15°C
Refractive index	1.5777 at 20°C	1.4064 at 15°C
Solubility	Soluble in most organic solvents	Soluble in water (forms the hydrate $C_5H_{10}O_2 \cdot H_2O$ with 30 parts of water), alcohol, ether, and chloroform
Organoleptic characteristics	Anise odor	Characteristic disagreeable odor, extremely penetrating and persistent; sour taste
Synthesis	By alkaline isomerization of safrole using KOH at the boil[6] or an alcoholic NaOH solution at room temperature under pressure[7]	By oxidation of isoamyl alcohol or isovaleric aldehyde
Natural occurrence	Of the two isomers (*cis-* and *trans-*), the *trans*-form is the more stable and has been isolated in the pure state, probably occurring in the essential oil of ylang-ylang. It has been identified in the oils of *Illicium religiosum*[8] and *Ligusticum acutilobum* Sieb. and Zucc.[9]	Of the three possible isomers of *n*-valeric acid, only isovaleric acid finds extensive application in flavoring. Originally reported in seal and dolphin fat; subsequently isolated from valerian. Also reported found in the essential oils of: cypress, citronella, laurel leaves, cajeput, *Cymbopogon javanensis*, hops, *Persea pubescens*, geranium, American peppermint, niaouli, spearmint, rosemary, lemongrass, *Eucalyptus goniocalyx* and other sp., tobacco, *Monarda fistulosa*, *Thymus mastichina*, *Artemisia frigida*, and probably in lavender. Finally, reported among the constituents of petitgrain lemon[10]
Reported uses[3]		Non-alcoholic beverages 1.2 ppm Ice cream, ices, etc. 14 ppm Candy 12 ppm Baked goods 5.5 ppm Cheese 2.4 ppm
Regulatory status	Use in foods is not permitted in U.S.A.	FDA 121.1164; FEMA No. 3102

REFERENCES

For References 1–5, see end of Part III.

6. **Wagner,** *Riechst. Aromen,* p. 52, 1926.
7. **Nagai,** *Soc. Chem. Ind.* (Japan), 29, 364, 1926.
8. **Chen,** *Chem. Zentralbl.,* 1, 1163, 1930.
9. **Noguchi et al.,** *Chem. Zentralbl.,* 2, 4050, 1937.
10. **Peuron,** *Riv. Ital. EPPOS,* 8, 413, 1965.

	Jasmone	2-Keto-4-butanethiol
Other names	3-Methyl-2-(2-pentenyl)-2-cyclopenten-1-one	4-Mercapto-2-butanone
Empirical formula	$C_{11}H_{16}O$	C_4H_8OS
Structure	(cyclopentenone ring with CH_3, =O and CH_2–CH=CH–CH_2–CH_3 side chain)	$CH_3-C(=O)-CH_2-CH_2SH$
Physical/chemical characteristics[5]		
Appearance	Yellow oil	
Molecular weight	164.25	104.17
Boiling point	134–135°C at 12 mm Hg 109–110°C at 5 mm Hg 257–258°C at 755 mm Hg	59–60°C at 15 mm Hg
Specific gravity	0.9437 at 22°C 0.9466 at 15°C	
Refractive index	1.4979 at 23°C 1.50058 at 15°C	
Solubility	Slightly soluble in water; soluble in alcohol, ether, and carbon tetrachloride	
Organoleptic characteristics	Jasmine odor	
Synthesis	A review and classification on the synthesis of jasmone is available[10]	By reduction of 4-bromo-2-butanone with NaHS;[6] by reacting methyl vinyl ketone with hydrogen sulfide;[7] from methyl vinyl ketone and thioacetamide;[8] from methyl vinyl ketone and hydrogen sulfide at 12 to 15°C by a patented process[9]
Natural occurrence	The *cis*-form is reportedly occurring naturally; it is found in the essential oils of jasmine (3%), jonquil, and *Pittosporum glabratum,* in neroli oil, in peppermint (*Mentha piperita*), and in bergamot	
Reported uses[3]	Non-alcoholic beverages 10.0 ppm Ice cream, ices, etc. 10.0 ppm Candy 10.0 ppm Gelatins and puddings 10.0 ppm	Baked goods 0.20 ppm Meat, meat sauces, soups 0.20 ppm
Regulatory status	FEMA No. 3196	FEMA No. 3357

REFERENCES

For References 1–5, see end of Part III.

6. **Asinger, Thiel and Höringklee,** *Annalen,* 610, 1, 1957.
7. **Murata and Arai,** *J. Chem. Soc. Jap.,* 59, 129, 1956.
8. **Barrett et al.,** *J. Chem. Soc.,* p. 788, 1964.
9. **Kostyukoskii and Slavachevskaya,** U.S.S.R. Patent 319,591.
10. **Anonymous,** *Fr. Ses Parfums,* 1(3), 39, 1958.

	Lactic acid	Lauric aldehyde
Other names	2-Hydroxypropanoic acid	Aldehyde C-12 1-Dodecanal *n*-Dodecyl aldehyde Lauraldehyde
Empirical formula	$C_3H_6O_3$	$C_{12}H_{24}O$
Structure	$CH_3-CH(OH)-COOH$	$CH_3-(CH_2)_{10}-CHO$
Physical/chemical characteristics[5]		
Appearance	Syrupy, rather hygroscopic liquid	Colorless to yellow liquid, solidifying at low temperature
Assay		92% min[2] (96.5%)
Molecular weight	90.08	184.32
Melting point	18°C; *d,l*: 52–53°C	11°C (8°C min)
Boiling point	122°C at 15 mm Hg	237°C; 99.5–100°C at 3.5 mm Hg
Flash point		82°C (119°C)
Specific gravity	1.249 at 15°C; *d*, 1.2485 at 20°C	0.826–0.836 at 25°/25°C;[2] 0.8352 at 15°C
Refractive index		1.4330–1.4390 at 20°C;[2] 1.4328 at 24.7°C
Optical rotation	*l*, −2.6°; *d*, +2.6° (from water) at 21–22°C	
Acid value		10 max[2] (3.8)
Solubility	Completely miscible with water, alcohol, and glycerin; readily soluble in ether	Insoluble in water; soluble in alcohol and most organic solvents; 1:2 in 80% alcohol; 1:0.5 in 90% alcohol
Organoleptic characteristics	Slightly sour odor and taste	Characteristic fatty odor reminiscent of violet on dilution; fatty, woody taste
Synthesis	By fermentation from sugar or starch. The commercial product is the racemic form.	By oxidation of the corresponding alcohol or reduction of the acid
Natural occurrence	Reported found in *Papaver somniferum* L., apples and other fruits, tomato juice, and several plants. It is a constituent of wine and sour milk. The two optically active isomers are found in muscular tissues and are formed by the action of lactic-acid bacteria in several fermentation processes.	Reported found in the essential oils of: lemon, sweet orange, bitter orange, mandarin, Mexican lime, and petitgrain bigarade and in the turpentine of some varieties of pine
Reported uses[3]	Non-alcoholic beverages 34 ppm Ice cream, ices, etc. 66 ppm Candy 130 ppm Baked goods 89 ppm Gelatins and puddings 14–25 ppm Chewing gum 610 ppm Toppings 300 ppm Pickles and olives 1,200–24,000 ppm	Non-alcoholic beverages 0.93 ppm Ice cream, ices, etc. 1.5 ppm Candy 2.4 ppm Baked goods 2.8 ppm Gelatins and puddings 0.10 ppm Chewing gum 0.20–110 ppm
Regulatory status	FDA GRAS (general); FEMA No. 2611	FDA 121.1164; FEMA No. 2615

REFERENCES

For References 1–5, see end of Part III.

	Lauryl acetate	Lauryl alcohol
Other names	Acetate C-12 Dodecanyl acetate Dodecyl acetate	Alcohol C-12 Dodecyl *n*-Dodecyl alcohol 1-Dodecanol
Empirical formula	$C_{14}H_{28}O_2$	$C_{12}H_{26}O$
Structure	$CH_3-(CH_2)_{10}-CH_2-O-\overset{\overset{\displaystyle O}{\Vert}}{C}-CH_3$	$CH_3-(CH_2)_{10}-CH_2OH$
Physical/chemical characteristics[5]		
Appearance	Colorless liquid	Colorless liquid at room temperature; crystalline or flakes below 20°C
Assay	98% min	97% min[2] (98.5%)
Molecular weight	228.37	186.34
Melting point		26°C (21–24°C)
Boiling point		150°C at 20 mm Hg
Congealing point		>21°C[2]
Flash point		>82°C
Specific gravity	0.860–0.865 at 25°C	0.830–0.836 at 25°/25°C[2]
Refractive index	1.4320–1.4360 at 20°C	1.4400–1.4440 at 20°C[2]
Acid value		Not more than 1[2]
Solubility	1:4 in 80% alcohol; soluble in most organic solvents	1:3 and more in 70% alcohol;[2] 1:2 in 80% alcohol; insoluble in water but soluble in most organic solvents
Organoleptic characteristics	Characteristic citrus-rose odor; the corresponding flavor develops only on dilution	Characteristic fatty odor; unpleasant at high concentrations but delicate and floral on dilution; fatty, waxy flavor
Synthesis	By acetylation of lauryl alcohol	Commercially prepared by hydrogenation of lauric acid; normally employed as a replacement for the corresponding aldehyde
Natural occurrence	Reported found in several natural products	Reported found in the oil of Mexican lime and in the oil from flowers of *Furcraea gigantea*
Reported uses[3]	Non-alcoholic beverages 2.3 ppm Ice cream, ices, etc. 1.7 ppm Candy 4.6 ppm Baked goods 5.6 ppm	Non-alcoholic beverages 2.0 ppm Ice cream, ices, etc. 1.0 ppm Candy 2.8 ppm Baked goods 1.7 ppm Chewing gum 16–27 ppm Syrups 7.0 ppm
Regulatory status	FDA 121.1164; FEMA No. 2616	FDA 121.1164; FEMA No. 2617

REFERENCES

For References 1–5, see end of Part III.

	Lepidine	L-Leucine
Other names	Cincholepidine 6-Methylquinoline* γ-Methylquinoline	α-Aminoisocaproic acid 2-Amino-4-methylpentanoic acid 2-Aminoisobutylacetic acid 2-Amino-4-methylvaleric acid
Empirical formula	$C_{10}H_9N$	$C_6H_{13}O_2N$
Structure	(4-methylquinoline ring structure with CH_3 and N)	$(CH_3)_2CH-CH_2-CH(NH_2)-COOH$
Physical/chemical characteristics[5]		
Appearance	Colorless, oily liquid	Hexagonal platelets (from diluted alcohol)
Molecular weight	143.18	131.18
Melting point		293–295°C (decomposes)
Boiling point	261–263°C	145–148°C (with sublimation)
Congealing point	0°C	
Specific gravity	1.086 at 20°C	1.293 at 18°C
Optical rotation		–10.42° at 20°C in water
Solubility	Slightly soluble in water; miscible with alcohol, benzene, and ether	Slightly soluble in water; insoluble in alcohol and ether
Organoleptic characteristics	Quinoline odor; tends to darken on exposure to light	Slightly bitter taste
Synthesis	From coal tar, by synthesis from *p*-aminobenzaldehyde and acetone	By bromination, followed by amination of isocaproic acid;[6] via the acetamidomalonic ester;[7] by isolation from gluten, casein, keratin;[8] from hydantoin[9]
Natural occurrence		Reported found as a constituent in proteins; also present in the free state in the human body
Reported uses[3]	In whisky 0.22 ppm Ice cream, ices, etc. 1.4 ppm Candy 1.8 ppm Baked goods 1.8 ppm	Non-alcoholic beverages 50.0 ppm Baked goods 50.0 ppm Condiments, pickles 50.0 ppm Meat, meat sauces, soups 50.0 ppm Milk, dairy products 50.0 ppm Cereals 50.0 ppm
Regulatory status	FDA 121.1164; FEMA No. 2744	FEMA No. 3295

*The term *p*-methylquinoline, although sometimes used, is a misnomer.

REFERENCES

For References 1–5, see end of Part III.

6. **Horning,** *Organic Syntheses,* Coll. Vol. 3, Wiley & Sons, New York, 1955, 523.
7. **Albertson and Archer,** *J. Am. Chem. Soc.,* 67, 308, 1945.
8. **Barnett,** *J. Biol. Chem.,* 100, 543, 1933; U.S. Patent 2,009,868.
9. **White,** U.S. Patent 2,557,920, assigned to Dow Chemical Co., 1951.

	Levulinic acid
Other names	β-Acetylpropionic acid γ-Ketovaleric acid 4-Oxopentanoic acid
Empirical formula	$C_5H_8O_3$
Structure	$CH_3{-}\overset{\displaystyle O}{\overset{\|}{C}}{-}CH_2{-}CH_2{-}COOH$
Physical/chemical characteristics[5]	
Appearance	White, crystalline solid
Molecular weight	116.12
Melting point	37°C
Boiling point	139–140°C at 8 mm Hg
Specific gravity	1.147 at 18°C (over cooled liquid)
Refractive index	1.4342 at 40°C
Solubility	Readily soluble in water, alcohol, and ether
Synthesis	By action of more or less concentrated HCl on sucrose, glucose, or fructose; hence, its reported presence in caramels
Natural occurrence	Reported found in several natural products
Reported uses[3]	Non-alcoholic beverages 14 ppm Ice cream, ices, etc. 14 ppm Candy 53 ppm Baked goods 53 ppm Gelatins and puddings 4.0 ppm
Regulatory status	FDA 121.1164; FEMA No. 2627

REFERENCES

For References 1–5, see end of Part III.

	Limonene (*d*-, *l*-, and *dl*-)
Other names	Cajeputene Carvene Dipentene Kautschin 1,8(9)-*p*-Menthadiene *p*-Mentha-1,8-diene 1-Methyl-4-isopropenyl-1-cyclohexene
Empirical formula	$C_{10}H_{16}$
Structure	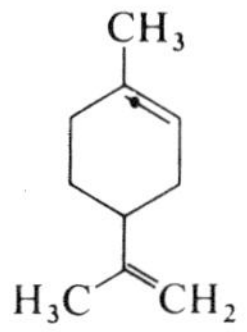 CH_3 ; H_3C CH_2
Physical/chemical characteristics[5]	
Appearance	Colorless liquid
Molecular weight	136.23
Boiling point	*dl*-, 176–178°C
Specific gravity	*d*-, 0.838–0.843 at 25°/25°C;[2] *d*- and *l*-, 0.8403 at 20°C; *dl*-, 0.8481 at 15°C
Refractive index	*d*-, 1.4710–1.4740 at 20°C;[2] *d*- and *l*-, 1.4701–1.4706 at 25°C
Optical rotation	*d*-, +96° to +104°;[2] *d*-, +126.6°; *l*-, −126.3°
Peroxide value	*d*-, not more than 2.0
Solubility	Insoluble in water; miscible with alcohol
Organoleptic characteristics	Pleasant, lemon-like odor free from camphoraceous and turpentine-like notes[2]
Synthesis	*d*-Limonene may be obtained by steam distillation of citrus peels and pulp resulting from the production of juice and cold-pressed oils, or from deterpenation of citrus oils. It is sometimes redistilled.
Natural occurrence	The most important and widespread terpene. It is known in the *d*- and *l*- optically active forms and in the optically inactive *dl*-form (known as dipentene). It has been reported found in more than 300 essential oils in amounts ranging from 90–95% (lemon, orange, mandarin) to as low as 1% (palmarosa).[6] The most widespread form is the *d*-limonene, followed by the racemic form, and then *l*-limonene.
Reported uses[3]	Non-alcoholic beverages 31 ppm Ice cream, ices, etc. 68 ppm Candy 49 ppm Baked goods 120 ppm Gelatins and puddings 48–400 ppm Chewing gum 2,300 ppm
Regulatory status	FDA GRAS; FEMA No. 2633 (*d*-form)

REFERENCES

For References 1–5, see end of Part III.
6. Gildemeister and Hoffmann, *Die Aetherischen Oele,* Vol. IIIa, Akademie Verlag, 1960, pp. 45–46.

Linalool

Other names	Coriandrol (*d*-linalool from coriander oil) 3,7-Dimethyl-1,6-octadien-3-ol Licareol (*l*-linalool from bois de rose oil) *dl*-Linalool (synthetic)
Empirical formula	$C_{10}H_{18}O$
Structure	H_3C OH, CH_2, H_3C CH_3
Physical/chemical characteristics[5]	The following constants refer to the synthetic product. These may vary within an ample range for a product of natural origin, depending on the source.
Appearance	Colorless liquid
Assay	>95%[2]
Molecular weight	154.24
Boiling point	198°C
Flash point	78°C
Specific gravity	0.858–0.862 at 25°C[2]
Refractive index	1.4600–1.4630 at 20°C[2]
Saponification number	Not more than 1.5[2]
Solubility	1:4 in 60% alcohol;[2] insoluble in water; miscible with alcohol and ether
Organoleptic characteristics	Typical floral odor free from camphoraceous and terpenic notes.[2] Synthetic linalool exhibits a cleaner and fresher note than the natural product.
Synthesis	It can be prepared synthetically starting from myrcene or from dehydrolinalool. It can be obtained by fractional distillation and subsequent rectification from the oils of: Cajenne rosewood (*Licasia guaianensis Ocotea caudata*), Brazil rosewood (*Ocotea parviflora*), Mexican linaloe, shiu (*Cinnamomum camphora* Sieb. var. *linalooifera*), and coriander seeds (*Coriandrum sativum* L.).
Natural occurrence	The optically active forms (*d*- and *l*-) and the optically inactive form occur naturally in more than 200 oils from herbs, leaves, flowers, and wood. The *l*-form is present in the largest amounts (80–85%) in the distillates from leaves of *Cinnamomum camphora* var. *orientalis* and *Cinnamomum camphora* var. *occidentalis* and in the distillate from Cajenne rosewood. It also has been reported in: champaca, ylang-ylang, neroli, Mexican linaloe, bergamot, lavandin, and others. A mixture of *d*- and *l*-linalool has been reported in Brazil rosewood (85%). The *d*-form has been found in palmarosa, mace, sweet orange-flower distillate, petitgrain, coriander (60–70%), marjoram, *Orthodon linalooliferum* (80%), and others. The inactive form has been reported in clary sage, jasmine, and *Nectandra elaiophora*.
Reported uses[3]	Non-alcoholic beverages 2.0 ppm Ice cream, ices, etc. 3.6 ppm Candy 8.4 ppm Baked goods 9.6 ppm Gelatins and puddings 2.3 ppm Chewing gum 0.80–90 ppm Condiments 40 ppm
Regulatory status	FDA GRAS; FEMA No. 2635

REFERENCES

For References 1–5, see end of Part III.

	Linalyl acetate	Linalyl anthranilate
Other names	Bergamol 3,7-Dimethyl-1,6-octadien-3-yl acetate	3,7-Dimethyl-1,6-octadien-3-yl anthranilate Linalyl-2-aminobenzoate Linalyl-*o*-aminobenzoate
Empirical formula	$C_{12}H_{20}O_2$	$C_{17}H_{23}NO_2$
Structure	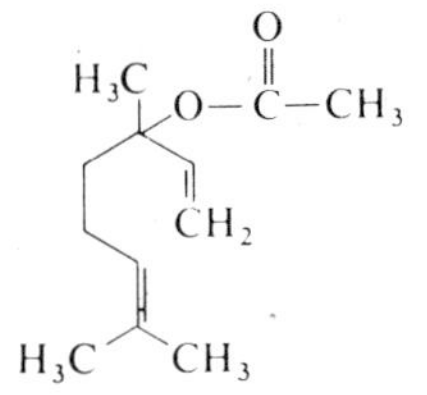	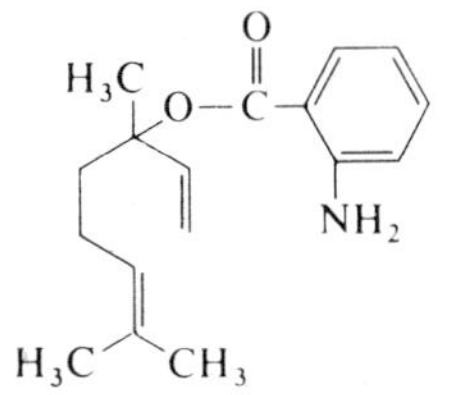
Physical/chemical characteristics[5]	The physical constants differ within an ample range, depending on the natural source or on the synthetic route.	
Appearance	Colorless liquid	Liquid, almost colorless or very pale straw-yellow
Assay	95% (syn)	
Molecular weight	196.29	273.38
Boiling point	115–116°C at 25 mm Hg (nat)	350°C
Flash point	85°C (nat)	
Specific gravity	0.9060 at 15°C (nat); 0.895–0.905 at 25°C (syn)	1.055 at 15.5°C
Refractive index	1.4500 at 20°C (nat); 1.4495–1.4512 at 20°C (syn)	1.4970 at 20°C
Optical rotation	−7 42′ to +8 18′ (nat); ±0° (syn)	
Acid value	0.5 max (syn)	
Solubility	1:3–5 in 70% alcohol; soluble in essential oils and most common organic solvents	Insoluble in water; almost insoluble in propylene glycol; soluble in alcohol
Organoleptic characteristics	Characteristic bergamot-lavender odor and persistent sweet, acrid taste	Sweet orange-like flavor; fragrance varies, depending on the source of linalool
Synthesis	Normally prepared by direct acetylation of linalool;[6] another method starts from myrcene hydrochloride, anhydrous sodium acetate, and acetate anhydride in the presence of a catalyst.[7] All synthetic methods tend to avoid the simultaneous formation (because of isomerization) of terpenyl and geranyl acetate.	From linalool and isatoic anhydride with catalytic amounts of sodium hydroxide; from linalyl formate plus methylanthranilate with sodium linalool
Natural occurrence	Reported found in the essential oils of bergamot, lavender, clary sage, and lavandin; also identified among the constituents of the essential oils of: *Salvia officinalis*, petitgrain, sassafras, neroli, lemon, Italian lime, jasmine, *Mentha citrata*, *Mentha aquatica*, *Thymus mastichina*, etc. Also reported in abundant quantities in the essential oil from flowers, leaves, and stems of *Tagetes patula*,[8] in the distillate from leaves of *Citrus aurantifolia*[9] from India, and in the essential oil of *Mentha arvensis*.[10]	Not reported found in nature
Reported uses[3]	Non-alcoholic beverages, 1.9 ppm; ice cream, ices, etc., 3.8 ppm; candy, 11 ppm; baked goods, 8.9 ppm; gelatins and puddings, 3.8 ppm; chewing gum, 13 ppm	Non-alcoholic beverages, 1.8 ppm; ice cream, ices, etc., 0.72 ppm; candy, 4.7 ppm; baked goods, 0.20–8.0 ppm
Regulatory status	FDA GRAS; FEMA No. 2636	FDA 121.1164; FEMA No. 2637

REFERENCES

For References 1–5, see end of Part III.

6. **Tiemann,** *Chem. Ber.*, 31, 839, 1898.
7. U.S. Patent 2,794,826, 1957.
8. **Dhingra and Dhingra,** *Perfum. Essent. Oil. Rec.*, 47, 391, 1956.
9. **Banerji and Kumar,** *Perfum. Essent. Oil Rec.*, 48, 343, 1957.
10. **Chagawets and Borisyuk,** *Miltizer Ber.*, p. 61, 1958.

	Linalyl benzoate	Linalyl butyrate
Other names		Linalyl-*n*-butyrate
Empirical formula	$C_{17}H_{22}O_2$	$C_{14}H_{24}O_2$
Structure	H_3C–C(–O–C(=O)–C$_6$H$_5$)(CH=CH$_2$)–CH$_2$–CH$_2$–CH=C(CH$_3$)$_2$	H_3C–C(–O–C(=O)–CH$_2$–CH$_2$–CH$_3$)(CH=CH$_2$)–CH$_2$–CH$_2$–CH=C(CH$_3$)$_2$
Physical/chemical characteristics[5]		
Appearance	Yellow or yellow-brownish liquid; high-quality grades almost colorless	Colorless or yellowish liquid
Assay	75% min	95% min
Molecular weight	258.37	224.34
Boiling point	263°C	232°C
Flash point	98°C	>100°C
Specific gravity	0.980–0.999 at 25°C	0.893–0.904 at 25°/25°C; 0.8890–0.8903 at 20°C
Refractive index	1.505–1.520 at 20°C	1.451–1.456 at 20°C; 1.4483–1.4492 at 25°C
Optical rotation		±1° (optically inactive if obtained from synthetic linalool)
Acid value	5 max	1 max
Solubility	1:1 in 90% alcohol	1:3 in 80% alcohol
Organoleptic characteristics	Heavy odor reminiscent of genet, bergamot, and lily	Fruity note with a bergamot-like undertone flavor reminiscent of certain types of honey on dilution; moderately stable
Synthesis	From linalool and benzoyl chloride in the presence of pyridine	Analogous to that of geranyl butyrate
Natural occurrence	Reported found in the essential oils of ylang-ylang and tuberose[6]	Reported found in the oil of *Artemisia porrecta* var. *coerulea* and in lavender oil[7,8]
Reported uses[3]	Non-alcoholic beverages 0.31 ppm Ice cream, ices, etc. 0.42 ppm Candy 1.2 ppm Baked goods 1.6 ppm Gelatins and puddings 0.28 ppm	Non-alcoholic beverages 1.2 ppm Ice cream, ices, etc. 4.3 ppm Candy 2.2 ppm Baked goods 13 ppm Gelatins and puddings 0.09 ppm
Regulatory status	FDA 121.1164; FEMA No. 2638	FDA 121.1164; FEMA No. 2639

REFERENCES

For References 1–5, see end of Part III.

6. **Labaune,** *Traites Chim. Org.*, 22, 1047, 1953.
7. **Bürger,** *Riechstoffindustrie*, p. 132, 1926.
8. **Bürger,** *Riv. Ital. EPPOS*, 7, 450, 1966.

	Linalyl cinnamate	Linalyl formate
Other names	Linalyl β-phenylacrylate Linalyl 3-phenylpropenoate	3,7-Dimethyl-1,6-octadien-3-yl formate
Empirical formula	$C_{19}H_{24}O_2$	$C_{11}H_{18}O_2$
Structure	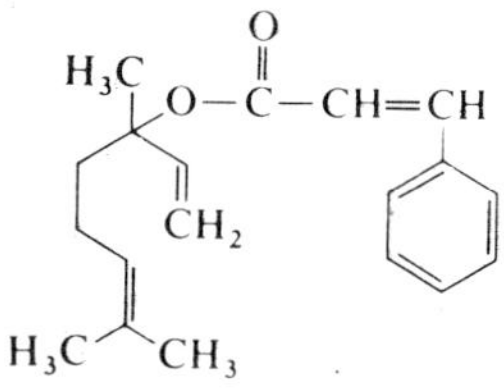 H_3C, O–C(=O)–CH=CH–(phenyl); CH_2; H_3C CH_3	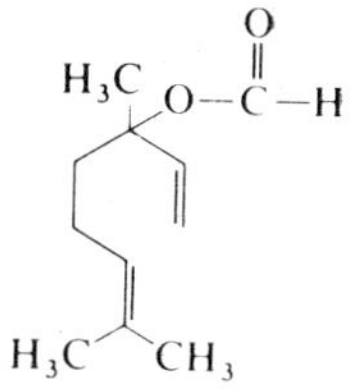 H_3C, O–C(=O)–H; H_3C CH_3
Physical/chemical characteristics[5]		
Appearance	Colorless liquid	Colorless to yellowish liquid
Assay		90% min[2]
Molecular weight	284.40	182.26
Boiling point	353°C	102°C (202°C) at 10 mm Hg
Flash point		87.5°C
Specific gravity	0.980 at 15.5°C	0.910–0.918[2] (0.915–0.924) at 25/25°C
Refractive index	1.5330 at 20°C	1.453–1.458[2] (1.4550–1.4580) at 20°C
Acid value		Not more than 3.0[2] (1)
Solubility	Insoluble in water; soluble in alcohol	1:6 in 70% alcohol;[2] 1:1 in 80% alcohol; insoluble in water; soluble in most organic solvents
Organoleptic characteristics	Sweet, persistent odor with soft floral notes; clearly balsamic; slightly fruity flavor	Fruity, floral (rose) odor reminiscent of bergamot and vervain; bittersweet flavor with pineapple-like undernote
Synthesis	From linalyl formate plus methylcinnamate and sodium linalool; by "Roche" method from dehydrolinalool and cinnamic acid via the dehydro ester, which is hydrolyzed to the subject compound	By the esterification of linalool with formic acid
Natural occurrence	Not reported found in nature	Reported found in nature: *Lavandula, Prunus armeniaca, Prunus persica,* and *Salvia sclarea,* peach, oil of lime, petitgrain
Reported uses[3]	Non-alcoholic beverages 0.57 ppm Ice cream, ices, etc. 0.59 ppm Candy 2.0 ppm Baked goods 2.1 ppm	Non-alcoholic beverages 1.3 ppm Ice cream, ices, etc. 12 ppm Candy 3.8 ppm Baked goods 13 ppm Chewing gum 1.2 ppm
Regulatory status	FDA 121.1164; FEMA No. 2641	FDA 121.1164; FEMA No. 2642

REFERENCES

For References 1–5, see end of Part III.

	Linalyl hexanoate	Linalyl isobutyrate
Other names	Linalyl caproate Linalyl capronate Linalyl hexylate	3,7-Dimethyl-1,6-octadien-3-yl isobutyrate
Empirical formula	$C_{16}H_{28}O_2$	$C_{14}H_{24}O_2$
Structure	H_3C, $O-C(=O)-(CH_2)_4-CH_3$, $=CH_2$, H_3C CH_3	H_3C, $O-C(=O)-CH(CH_3)_2$, $=CH_2$, H_3C CH_3
Physical/chemical characteristics[5]		
Appearance	Colorless, oily liquid	Colorless to slightly yellow liquid
Assay		95% min[2]
Molecular weight	252.40	224.34
Boiling point	252°C	
Flash point		>100°C
Specific gravity	0.90	0.882–0.888 at 25°/25°C;[2] 0.8926 at 15°C
Refractive index		1.446–1.451[2] (1.4490) at 20°C
Acid value		Not more than 1.0[2]
Optical rotation		−10.54° at 20°C (undiluted)
Solubility	Insoluble in water; soluble in alcohol	1:3 in 80% alcohol[2]
Organoleptic characteristics	Green, warm, fruity odor with an animal-like undertone; pineapple-pear flavor	Light, fruity odor with a lavender note; sweet flavor reminiscent of black currant
Synthesis	By esterification of dehydrolinalool with *n*-hexanoic acid followed by hydrogenation at the dehydro ester	By esterification of linalool with isobutyric anhydride
Natural occurrence		Reported found in the essential oil of Ceylon cinnamon and in lavender oil.[6,7] A *dextro*-rotatory form has been reported in the oil from leaves of *Agathosoma gnidioides* (Schlechter)[9]
Reported uses[3]	Non-alcoholic beverages 3.2 ppm Ice cream, ices, etc. 6.0 ppm Candy 11 ppm Baked goods 15 ppm	Non-alcoholic beverages 1.7 ppm Ice cream, ices, etc. 2.8 ppm Candy 4.9 ppm Baked goods 13 ppm
Regulatory status	FDA 121.1164; FEMA No. 2643	FDA 121.1164; FEMA No. 2640

REFERENCES

For References 1–5, see end of Part III.

6. **Bürger,** *Riechstoffindustrie,* p. 132, 1926.
7. **Kauffmann and Kjelsberg,** *Riechstoffindustrie,* 2, 171.
8. **Smith and Roux,** *Trans. R. Soc. S. Afr.,* 31, 340, 1947; *Chem. Abstr.,* 41, 7673, 1947.

	Linalyl isovalerate	Linalyl octanoate
Other names	Linalyl isopentanoate	Linalyl caprylate Linalyl octoate Linalyl octylate
Empirical formula	$C_{15}H_{26}O_2$	$C_{18}H_{32}O_2$
Structure	H_3C, O—C(=O)—CH_2—CH(CH_3)$_2$; =CH_2; H_3C CH_3	H_3C, O—C(=O)—$(CH_2)_6$—CH_3; =CH_2; H_3C CH_3
Physical/chemical characteristics[5]		
Appearance	Liquid	Colorless, oily liquid
Assay	96% min	
Molecular weight	238.37	280.46
Boiling point		264°C
Specific gravity	0.8940–0.9000 at 15°C	0.88
Refractive index	1.4510–1.4540 at 20°C	
Acid value	0.09 max	
Solubility	1 : 5.5–7 in 80% alcohol	Insoluble in water; soluble in alcohol
Organoleptic characteristics	Stable, suave, fruity odor; sweet, apple-like taste somewhat reminiscent of plum and peach	Dry, fruity, herbaceous odor; sweet apricot and plum-like taste with a perfumey note
Synthesis	From dehydrolinalool and isovaleric acid followed by hydrogenation; also from linalool and isovaleric acid by azeotropic esterification	By esterification of linalool with *n*-octanoic acid under azeotropic conditions; also by hydrogenation of dehydrolinalyl octanoate
Natural occurrence	Reported found in nature (*Salvia officinalis* and *Salvia spinosa*)	Reported found in lemongrass
Reported uses[3]	Non-alcoholic beverages 0.96 ppm Ice cream, ices, etc. 0.91 ppm Candy 5.7 ppm Baked goods 5.6 ppm Gelatins and puddings 0.10 ppm	Non-alcoholic beverages 1.3 ppm Ice cream, ices, etc. 0.50–3.0 ppm Candy 0.50 ppm Baked goods 0.60–15 ppm
Regulatory status	FDA 121.1164; FEMA No. 2646	FDA 121.1164; FEMA No. 2644

REFERENCES

For References 1–5, see end of Part III.

	Linalyl propionate	Malic acid
Other names	3,7-Dimethyl-1,6-octadien-3-yl propanoate	"Apple acid" (*l*-form found in apple) 2-Hydroxy-1,4-butanedioic acid Hydroxysuccinic acid
Empirical formula	$C_{13}H_{22}O_2$	$C_4H_6O_5$
Structure	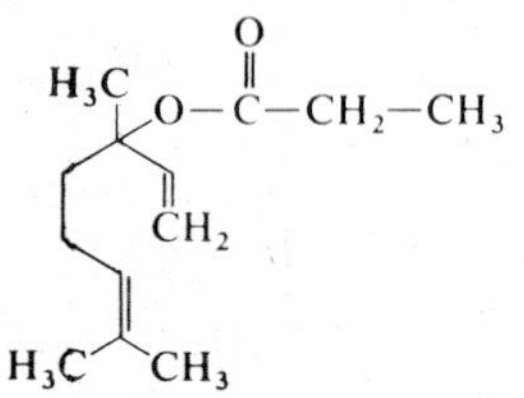 H_3C, O–C(=O)–CH_2–CH_3, =CH_2, H_3C CH_3	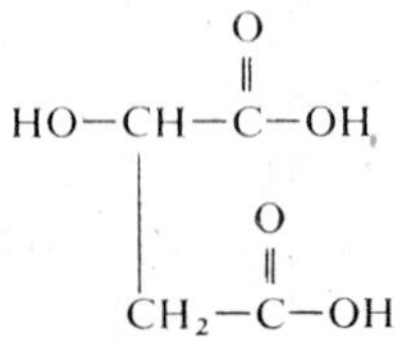 HO–CH–C(=O)–OH, CH_2–C(=O)–OH
Physical/chemical characteristics[5]		
Appearance	Almost colorless liquid	Crystalline solid
Assay	92% min[2] (95%)	
Molecular weight	210.31	134.09
Melting point		*l*-, 99–100°C; *d*-, 99–101°C
Boiling point	115°C (226°C) at 10 mm Hg	*l*-, approx 140°C (with decomposition)
Specific gravity	0.895–0.902[2] (0.8940–0.8954) at 25°/25°C	*l*-, 1.595 at 20°C
Refractive index	1.4500–1.4550[2] (1.4495–1.4510) at 20°C	
Optical rotation	±1°[2]	At 18°C (in acetone): *l*-, −5.7°; *d*-, +5.2°
Acid value	1 max[2] (0.5)	
Solubility	1:7 in 70% alcohol; very slightly soluble in water; slightly soluble in propylene glycol; insoluble in glycerin	Readily soluble in alcohol and water; slightly soluble in ether
Organoleptic characteristics	Sweet, floral odor reminiscent of bergamot oil; sweet taste reminiscent of black currant (also pear and pineapple)	Nearly odorless (sometimes a faint, acrid odor); tart, acidic taste; non-pungent
Synthesis	By esterification of linalool with propionic acid or propionic anhydride	By hydration of maleic acid; by fermentation from sugars
Natural occurrence	The *l*-form has been reported found in lavender and sage	Occurs in maple sap, apple, and many other fruits
Reported uses[3]	Non-alcoholic beverages 4.9 ppm Ice cream, ices, etc. 3.6 ppm Candy 5.3 ppm Baked goods 12 ppm Gelatins and puddings 4.4 ppm	Non-alcoholic beverages 380 ppm Ice cream, ices, etc. 390 ppm Candy 420 ppm Baked goods 0.60–1.5 ppm
Regulatory status	FDA 121.1164; FEMA No. 2645	FDA GRAS; FEMA No. 2655

REFERENCES

For References 1–5, see end of Part III.

	Maltol	*p*-Mentha-1,8-dien-7-ol
Other names	3-Hydroxy-2-methyl-(1,4-pyran) 3-Hydroxy-2-methyl-4H-pyran-4-one 3-Hydroxy-2-methyl-4-pyrone Larixinic acid 2-Methyl pyromeconic acid	1-Hydroxymethyl-4-isopropenyl-1-cyclohexene Perillyl alcohol
Empirical formula	$C_6H_6O_3$	$C_{10}H_{16}O$
Structure	O, OH, CH_3	CH_2OH, H_3C, C, CH_2
Physical/chemical characteristics[5]		The following values are for the *d*- and *l*-forms present in ginger grass oil; constants vary according to the source.[6,7,8]
Appearance	Monoclinic prisms from chloroform; orthorhombic bipyramidal crystals from alcohol; white crystals usually	Dense, oily liquid
Molecular weight	126.11	152.23
Melting point	159–162°C (164°C); volatile with steam; sublimes at 93°C	
Boiling point		*l*-, 92–93°C at 5 mm Hg; *d*-, 94–96°C at 4–5 mm Hg
Specific gravity		At 15°C: *l*-, 0.9510; *d*-, 0.9536
Refractive index		At 20°C: *l*-, 1.49629; *d*-, 1.49761
Optical rotation		*l*-, −13.9°; *d*-, +12.7°
Solubility	Soluble in alcohol and chloroform; 1.2% in water; sparingly soluble in benzene, ether, and petroleum ether. Soluble in NaOH solutions yielding CO_2; 1.2% in glycerin; 2.8% in propylene glycol	
Organoleptic characteristics	Warm, sweet, fruity odor; jam-like odor in solution	Characteristic odor similar to linalool and terpineol
Synthesis	By alkaline hydrolysis of streptomycin salts; also from piperidine to pyromeconic acid and subsequent methylation at the 2 position	The *l*-form is obtained starting from β-pinene or by reducing perillic aldehyde with zinc dust and acetic acid, followed by saponification of the acetate. For structural studies see reference 9.
Natural occurrence	Reported found in the bark of young larch trees (*Pinus larix*), pine needles (*Abies alba*), chicory, wood tars and oils, and roasted malt	The *d*- and *l*-forms occur naturally in ginger grass essential oil; the *l*-form is found in lavandin and bergamot, while the *d*-form is found in caraway; also reported found in either form in the essential oils of *Juniperus sabina* L., East Indian geranium, and others
Reported uses[3]	Non-alcoholic beverages 4.1 ppm Ice cream, ices, etc. 8.7 ppm Candy 31 ppm Baked goods 30 ppm Gelatins and puddings 7.5 ppm Chewing gum 90 ppm Jellies 15 ppm	Non-alcoholic beverages 0.50–1.0 ppm Ice cream, ices, etc. 0.50–1.0 ppm Candy 20 ppm Baked goods 10–50 ppm
Regulatory status	FDA 121.1164; FEMA No. 2656	FDA 121.1164; FEMA No. 2664

REFERENCES

For References 1–5, see end of Part III.

6. Semmler and Zaar, *Ber. Dtsch. Chem. Ges.*, 44, 460, 1911.
7. Semmler and Zaar, *Ber. Dtsch. Chem. Ges.*, 44, 54, 1911.
8. Elze, *Chem. Ztg.*, 34, 767, 1910.
9. Semmler and Zaar, *Ber. Dtsch. Chem. Ges.*, 44, 52, 460, 1911.

	p-Mentha-8-thiol-3-one		*p*-Menth-1-ene-9-al	
Other names	8-Mercapto-*p*-menthane-3-one		Carvomenthenal	
Empirical formula	$C_{10}H_{18}OS$		$C_{11}H_{18}O$	
Structure	CH_3; C—SH; H_3C CH_3; O		CH_3; CH; CH_2 CH_3; CHO	
Physical/chemical characteristics				
Molecular weight	185.31			
Boiling point	62°C at 0.1 mm Hg (mixture of stereoisomers)			
Refractive index	1.4948–1.4951 (mixture of stereoisomers)			
Optical rotation	(*trans*-form), −32° at 20°C (c = 0.117 g/100 ml in CH_3OH) (*cis*-form), +40° at 20°C (c = 0.127 g/100 ml in CH_3OH)			
Organoleptic characteristics	Black-currant flavor			
Synthesis	By reaction of pulegone or isopulegone with hydrogen sulfide excess and ethanolic potassium hydroxide[6]			
Natural occurrence	In buchu leaf oil[7]			
Reported uses[3]	Non-alcoholic beverages	1.0 ppm	Non-alcoholic beverages	2.0 ppm
	Ice cream, ices, etc.	1.0 ppm	Ice cream, ices, etc.	2.0 ppm
	Candy	1.0 ppm	Candy	2.0 ppm
	Baked goods	1.0 ppm	Gelatins and puddings	2.0 ppm
	Gelatins and puddings	1.0 ppm		
	Chewing gum	1.0 ppm		
Regulatory status	FEMA No. 3177		FEMA No. 3178	

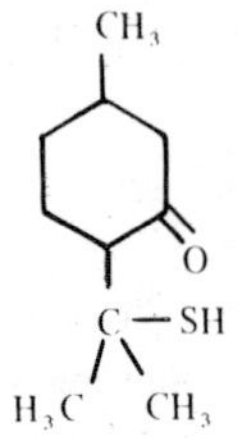

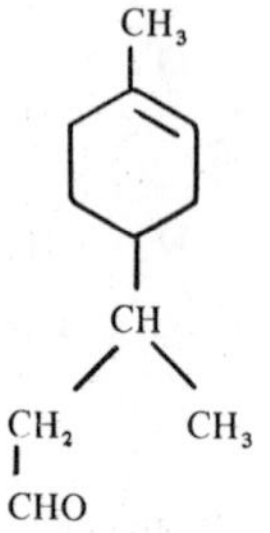

REFERENCES

For References 1–5, see end of Part III.

6. German Patent, 2,043,341.
7. **Lamparsky and Schudel**, *Tetrahedron Lett.*, 3, 3323, 1971.

	p-Menth-3-en-l-ol	Menthofuran
Other names	Δ_3-*para*-Menthen-1-ol 1-Methyl-4-isopropyl-3-cyclohexen-1-ol 1-Terpinenol	4,5,6,7-Tetrahydro-3,6-dimethylbenzofuran
Empirical formula	$C_{10}H_{18}O$	$C_{10}H_{14}O$
Structure	H_3C OH H_3C CH_3	H_3C O CH_3
Physical/chemical characteristics[5]		
Appearance	Colorless, oily liquid	Bluish liquid
Assay	*Note*: This ingredient is considered an undesirable component of commercial terpineol.	
Molecular weight	154.24	150.22
Boiling point	92–97°C (210°C) at 14 mm Hg	80°C at 18 mm Hg 90–92°C at 3 mm Hg
Specific gravity	0.9210 at 24°C	0.965 at 15°C
Refractive index	1.4778 at 24°C	1.4807 at 20°C
Optical rotation		+92.5 at 25°C
Solubility	Slightly soluble in water; soluble in alcohol, propylene glycol, and various oils	
Organoleptic characteristics	Dry, woody, somewhat musty odor	Odor very similar to that of menthol
Synthesis	Prepared by Wallach[6]	By thermal decomposition of the sulfone of 3-hydroxy-3,8(9)-methadiene-9-sulfonic acid; the latter is obtained by the sulfona-pulegone in acetic anhydride[6]
Natural occurrence	Reported found in camphor essential oil	Reported found in peppermint oil[7]
Reported uses[3]		Non-alcoholic beverages 10.0 ppm Ice cream, ices, etc. 10.0 ppm Candy 10.0 ppm
Regulatory status	FDA 121.1164	FEMA No. 3235

REFERENCES

For References 1–5, see end of Part III.

6. **Wallach,** *Annalen,* 356, 218, 1907; 362, 269, 1908.
7. **Eastman,** *J. Am. Chem. Soc.,* 72, 5313, 1950.
8. **Wienhaus,** *A. Angew. Chem.,* 47, 415, 1934.

	Menthol
Other names	Hexahydrothymol 3-*p*-Menthanol *p*-Menthan-3-ol *l*-Menthol 1-Methyl-4-isopropylcyclohexan-3-ol
Empirical formula	$C_{10}H_{20}O$
Structure	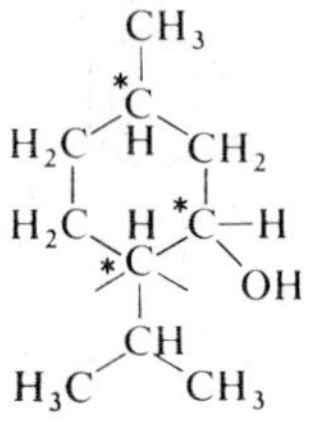 CH_3 / *C H / H_2C CH_2 / H_2C H *C—H / *C OH / CH / H_3C CH_3
Physical/chemical characteristics[5]	
Appearance	Crystals or granules
Assay	See Reference 6
Molecular weight	156.27
Melting point	42–43°C for *l*-form
Boiling point	216.5°C for *l*-form; 111°C at 20 mm Hg for *l*-form
Specific gravity	0.904 at 15°/15°C; 0.9007 at 20°C; 0.8911 at 30°C
Refractive index	1.461 at 20°C; 1.460 at 22°C; 1.4580 at 25°C; 1.4461 at 60°C
Optical rotation	–50° 0′ (in alcohol) at 20°C; –49° 0′ (c-2.0 in alcohol) at 16°C
Solubility	Slightly soluble in water; very soluble in alcohol, chloroform, ether, pet ether, glacial acetic acid, and liquid petrolatum
Organoleptic characteristics	Mint-like odor; fresh, cooling taste
Synthesis	By hydrogenation of thymol (see References 7 and 8). For a complete review of the various menthols, see Reference 9.
Natural occurrence	In peppermint and other mint oils (e.g., *M. arvensis*)
Reported uses[3]	Non-alcoholic beverages 35 ppm Ice cream, ices, etc. 68 ppm Candy 400 ppm Baked goods 130 ppm Chewing gum 1,100 ppm
Regulatory status	FDA 121.1164; FEMA No. 2665

REFERENCES

For Reference 1–5, see end of Part III.

6. *Food Chemical Codes*, 1st ed., National Academy of Sciences, 1963, 423.
7. *Beilstein's Encyclopedia of Organic Chemistry*, B6², Springer Verlag, p. 49.
8. *Beilstein's Encyclopedia of Organic Chemistry*, B6², Springer Verlag, p. 39.
9. Guenther, *The Essential Oils*, Vol. 2, D. Van Nostrand Co., 1949, 216.

	Menthone	
Other names	2-Isopropyl-5-methyl-cyclohexanone *p*-Menthan-3-one	
Empirical formula	$C_{10}H_{18}O$	
Structure	CH_3 / cyclohexanone ring =O / H_3C–C–CH_3	
Physical/chemical characteristics	*Racemic form*	*Optically active form*
Appearance	Colorless liquid	Colorless liquid
Assay	98%	95%
Molecular weight	154.24	154.24
Boiling point	Approx 210°C	Approx 210°C
Specific gravity	0.890–0.894 at 25°C	0.888–0.895 at 25°C
Refractive index	1.4480–1.4520 at 20°C	1.4480–1.4530 at 20°C
Optical rotation	±2°	−15° to +15°
Solubility	Insoluble in water; soluble 1:3 in 70% alcohol; soluble in alcohol and most common organic solvents	
Organoleptic characteristics	Characteristic odor similar to menthol	
Synthesis[7]	By oxidation of menthol;[6] see also Reference 7 and 8	
Natural occurrence	Several stereoisomers are reported found in nature. *l*-Menthone is a constituent of the essential oils of Russian and American peppermint, geranium, *Andropogon fragrans, Mentha timija, Mentha arvensis,* and others. *d*-Menthone is present in the essential oils of *Barosma pulchellum, Nepeta japonica* Maxim., and others. *d*-Isomenthone has been reported isolated from *Micromeria abissinica* Benth., *Pelargonium tomentosum* Jacquin., etc. *l*-Isomenthone has been identified in Reunion geranium, *Pelargonium capitatum,* and others.	
Reported uses[3]	Non-alcoholic beverages 7.7 ppm Ice cream, ices, etc. 33 ppm Candy 71 ppm Baked goods 52 ppm Chewing gum 8.7 ppm	
Regulatory status	FDA 121.1164; FEMA No. 2667	

REFERENCES

For References 1–5, see end of Part III.

6. **Moriya,** *J. Chem. Soc.* (London), 39, 77, 1881.
7. **Anders and Andreef,** *Ber. Dtsch. Chem. Ges.*, 25, 617, 1892.
8. **Beckmann and Pleissner,** *Annalen,* 21, 262, 1891.

	l-Menthyl acetate	Menthyl isovalerate
Other names	*l*-*p*-Menth-3-yl acetate	*p*-Menth-3-yl isovalerate Validol
Empirical formula	$C_{12}H_{22}O_2$	$C_{15}H_{28}O_2$
Structure	(menthyl ring with CH_3, isopropyl H_3C CH_3) $-O-\overset{O}{\overset{\|\|}{C}}-CH_3$	(menthyl ring with CH_3, isopropyl H_3C CH_3) $-O-\overset{O}{\overset{\|\|}{C}}-CH_2-CH(CH_3)_2$
Physical/chemical characteristics[5]		
Appearance	Colorless liquid	Colorless, oily liquid
Assay	95% min	96% min
Molecular weight	198.31	240.39
Boiling point	227–228°C	140–146°C (241°C) at 22 mm Hg
Flash point	92°C	120°C
Specific gravity	0.9220–0.9310 at 25°C; 0.9264 at 20°C	0.9040–0.9080 (0.907) at 15°C
Refractive index	1.4440–1.4500 (1.4472) at 20°C	1.4470–1.4480 at 20°C
Optical rotation	±2° (commercial product); *l*-, −72°47′ to −73°18′	Approx −36° (commercial product) (−64°12′)
Acid value		0.09 max
Solubility	Soluble in essential oils and most common solvents; 1:15 in 65% alcohol; 1:6 in 70% alcohol; 1:1 in 80% alcohol	Insoluble in water; 1:8 to 1:10 in 80% alcohol
Organoleptic characteristics	Fresh odor similar to mint and rose (on dilution); characteristic, fresh, pungent flavor, different from menthol (being much milder); cool mouthfeel with only a trace of mint flavor	Odor similar to both menthol and isovaleric acid—sweet and herbaceous, somewhat balsamic or root-like; bitter, wood-like taste
Synthesis	By reacting acetic anhydride with menthol in the presence of anhydrous sodium acetate	Prepared by heating a mixture of *l*-menthol and isovaleric acid at 100–110°C in the presence of trace concentrated H_2SO_4[6] or at 100°C in the presence of HCl[7]
Natural occurrence	Reported as a normal constituent of peppermint oil in varying amounts, depending on the source. The synthetic product is optically *laevo*-rotatory; the commercial product is optically inactive.	Reported found in several peppermint essential oils: American,[8] French,[9,10] English,[11] and Russian[12] peppermint oil
Reported uses[3]	Non-alcoholic beverages 5.5 ppm Ice cream, ices, etc. 4.0 ppm Candy 26 ppm Baked goods 24 ppm Chewing gum 5.2 ppm	Non-alcoholic beverages 5.0 ppm Ice cream, ices, etc. 2.9 ppm Candy 7.4 ppm Baked goods 18 ppm
Regulatory status	FDA 121.1164; FEMA No. 2668	FDA 121.1164; FEMA No. 2669

REFERENCES

For References 1–5, see end of Part III.

6. Senderens and Aboulenc, *Ann. Chim. Paris*, 9(18), 184, 1922.
7. Bobranski, *J. Pharm. Chim.*, 8(20), 437, 1934.
8. Power and Kleber, *Arch. Pharm.*, 232, 652, 1894.
9. Charabot, *Bull Soc. Chim. Fr.*, 3(19), 117, 1898.
10. Roure-Bertrand fils, *Giornale*, p. 17, March 1900.
11. Umney, *Chem. Zentralbl.*, 1, 710, 1896.
12. Schindelmeiser, *Chem. Zentralbl.*, 2, 1764, 1906.

	3-Mercapto-2-butanone	2-Mercaptomethylpyrazine
Other names		Pyrazine methanethiol Pyrazinyl methylmercaptan
Empirical formula	C_4H_8OS	$C_5H_6N_2S$
Structure	$CH_3-C(=O)-CH(SH)-CH_3$	(pyrazine ring)$-CH_2-SH$
Physical/chemical characteristics		
Appearance	Becomes cloudy after short storage; forms water droplets within 5 hrs at room temperature	
Assay		C: 47.59%; H: 4.79%; N: 22.20%; S: 25.41%
Molecular weight	104.17	126.18 (calculated = 126.025171)
Melting point		
Boiling point	46–70°C at 15 mm Hg 50°C at 18 mm Hg	44–45°C at 0.07 mm Hg
Specific gravity	1.0156 at 20°C	
Refractive index	1.4694 at 20°C	
Solubility	With difficulty soluble in water; easily soluble in alcohol, ether, pyridine, and aqueous alkalis	
Organoleptic characteristics		Roasted, meat-like flavor
Synthesis	From 2,4,5-trimethyl-2-ethyl-3-thiazoline with butyl and hydrolysis.[6] From oxo-compound with sulfur or polysulfides and ammonia at room temperature[7]	Chloromethylpyrazine is added to sodium hydrogen sulfide[8]
Reported uses[3]	Non-alcoholic beverages 0.2 ppm Candy 0.2 ppm Baked goods 0.2 ppm Meat, meat sauces, soups 0.2 ppm Milk, dairy products 0.2 ppm Cereals 0.2 ppm	Baked goods 10.0 ppm Condiments 10.0 ppm Meat, meat sauces, soups 10.0 ppm
Regulatory status	FEMA No. 3298	FEMA No. 3299

REFERENCES

For References 1–5, see end of Part III.

6. **Thiel et al.,** *Ann. Chem.,* 613, 128, 1958.
7. **Asinger, Thiel and Pallas,** East German Patent 13,954, 1957; *Chem. Abstr.,* 53, 11235e, 1959.
8. **Winter et al.,** U.S. Patent 3,702,253, 1972.

	3-Mercapto-2-pentanone		Mesityl oxide	
Other names			4-Methyl-3-penten-2-one Isopropylidene acetone Methyl isobutenyl ketone	
Empirical formula	$C_5H_{10}OS$		$C_6H_{10}O$	
Structure	$CH_3-CH_2-CH(SH)-CO-CH_3$		$(H_3C)_2C=CH-C(=O)-CH_3$	
Physical/chemical characteristics				
Appearance			Colorless oily liquid	
Molecular weight	118.19		98.15	
Melting point			-52.85°C	
Boiling point			129.76°C 72.1°C at 100 mm Hg 41°C at 23 mm Hg -8.7°C at 1 mm Hg	
Flash point			+30.6°C	
Specific gravity			0.8653 at 20°C	
Refractive index			1.4440 at 20°C	
Solubility			Slightly soluble in water, acetone, and propylene glycol; soluble in alcohol and ether in all proportions	
Organoleptic characteristics			The commercial grade exhibits an unpleasant odor; the pure material has a pleasant honey-like odor	
Synthesis			By reacting acetone or diacetone alcohol with iodine or other dehydrating agent;[6,7] by condensation of acetone over sulfonated polystyrene-divinylbenzene resin used as an ion exchange catalyst[8]	
Natural occurrence			Reported found as a constituent in coffee aroma[9]	
Reported uses[3]	Non-alcoholic beverages	0.2 ppm	Ice cream, ices, etc.	0.75 ppm
	Candy	0.2 ppm	Candy	1.12 ppm
	Baked goods	0.2 ppm	Baked goods	2.25 ppm
	Meat, meat sauces, soups	0.2 ppm	Gelatins and puddings	0.50 ppm
	Cereals	0.2 ppm	Milk, dairy products	0.40 ppm
	Milk, dairy products	0.2 ppm		
Regulatory status	FEMA No. 3300		FEMA No. 3368	

REFERENCES

For References 1–5, see end of Part III.

6. **Conant and Tuttle,** *Organic Syntheses,* Coll. Vol. 1, Gilman, H. and Blatt, A. H., Eds., Wiley & Sons, New York, 1941, 345.
7. *Ullmanns, Enzyklopädie der Technischen Chemie,* 3rd ed., Vol. 9, Urban and Schwarzenberg, 1958, 555.
8. **Klein and Banchero,** *Ind. Eng. Chem.,* 48, 1278, 1956.
9. **Stoll et al.,** *Helv. Chim. Acta,* 50, 628, 1967.

	D,L-Methionine		*o*-Methoxybenzaldehyde
Other names	2-Amino-4-(methylthio)butanoic acid 2-Amino-γ-(methylthio)butyric acid		*o*-Anisaldehyde Methyl salicylaldehyde Salicylaldehyde methyl ether
Empirical formula	$C_5H_{11}O_2NS$		$C_8H_8O_2$
Structure	SCH_3 and NH_2 substituents: $CH_2—CH_2—CH—COOH$ (SCH₃ on first CH₂, NH₂ on CH)		CHO on benzene ring, ortho $—O—CH_3$
Physical/chemical characteristics[5]			
Appearance	Platelets (from alcohol)		Crystals, colorless or cream-colored
Molecular weight	149.21		136.15
Melting point	281°C (decomposes)		35–37°C (38°C)
Boiling point			243–244°C; 122°C at 20 mm Hg
Specific gravity	1.340		1.1326 at 20°C
Refractive index			1.560 at 20°C
Solubility	Very soluble in water; insoluble in ether, benzene, and petroleum ether; soluble in diluted acids and alkalis; slightly soluble in alcohol		Insoluble in water; soluble in alcohol and benzene; very soluble in ether and chloroform
Organoleptic characteristics			Faint, sweet, floral odor; blends well with cassia; spice-like flavor, quite bitter above 30–40 ppm
Synthesis	By addition of methanethiol to acrolein;[6] by chemical conversion of methylthiopropionic aldehyde[7-9]		Prepared from salicylaldehyde and dimethyl sulfate in weak alkaline solution
Natural occurrence			Reported found in the oil of *Cinnamomum cassia* Nees ex Bl.[10]
Reported uses[3]	Non-alcoholic beverages	50.0 ppm	
	Baked goods	50.0 ppm	
	Condiments, pickles	50.0 ppm	
	Meat, meat sauces, soups	50.0 ppm	
	Milk, dairy products	50.0 ppm	
	Cereals	50.0 ppm	
Regulatory status	FEMA No. 3301		FDA 121.1164

REFERENCES

For References 1–5, see end of Part III.

6. British Patent 605,311, 1948; *Chem. Abstr.*, 43, 677a, 1949.
7. **Pierson et al.,** *J. Am. Chem. Soc.*, 70, 1450, 1948.
8. **Pierson and Tishler,** U.S. Patent, 2,584, 496, assigned to Merck & Co., 1952.
9. **Weiss,** U.S. Patent 2,732,400, assigned to American Cyanamid Co., 1956.
10. **Dodge,** *Chem. Zentralbl.*, 1, 747, 1919.

	p-Methoxybenzaldehyde	*o*-Methoxycinnamaldehyde
Other names	*p*-Anisaldehyde Anisic aldehyde Aubepine	*o*-Cumaric aldehyde methyl ether β-(*o*-Methoxyphenyl)-acrolein 3-(*o*-Methoxyphenyl)-propenal 3-(*o*-Methoxyphenyl)propen-2-al-1
Empirical formula	$C_8H_8O_2$	$C_{10}H_{10}O_2$
Structure	CHO (benzene ring) O—CH₃	CH=CH—CHO (benzene ring) O—CH₃
Physical/chemical characteristics[5]		
Appearance	Colorless to slightly yellow liquid	Pale-yellow crystal flakes (from alcohol)
Assay	97.5% min[2] (97–99%)	
Molecular weight	136.15	162.19
Melting point	2.5°C	45–46°C
Boiling point	247°C	295°C with decomposition; 160–161°C at 12 mm Hg
Congealing point	−1°C min	
Flash point	121°C	
Specific gravity	1.119–1.123 at 25°/25°C;[2] 1.123 at 20°C	
Refractive index	1.5710–1.5740 at 20°C[2]	
Acid value	6 max[2]	
Solubility	1:7 in 50% alcohol; soluble in most organic solvents; at least 99% soluble in sodium bisulfite solution[2]	Soluble in alcohol, ether, and chloroform; very slightly soluble in water
Organoleptic characteristics	Characteristic hawthorn odor; pungent, anise-like flavor; bitter flavor above 30–40 ppm	Sweet, warm, spicy-floral odor with fruity undertones; sweet, spicy, warm flavor; somewhat pungent above 200–300 ppm
Synthesis	By methylation and oxidation of *p*-cresol and also by oxidation of anethol	It is formed after a long contact period between salicylaldehyde and acetaldehyde in diluted alkaline solution.[7] It can be formed by condensation of salicylaldehyde methylether with acetaldehyde under alkaline conditions. From the corresponding oxime[9]
Natural occurrence	Reported found in essential oils and extracts: vanilla, *Acacia farnesiana* Willd., *Magnolia salicifolia* Maxim., *Erica arborea*, *Pirus communis*, *Boswellia serrata*, and others;[6] also in anise, fennel, star anise (especially when aged due to the oxidation of anethol)	Reported found in cinnamon essential oil[8] (*Cinnamomum cassia* Nees ex Bl.), from which it is separated as stearoptene; in cinnamon[10]
Reported uses[3]	Non-alcoholic beverages 6.3 ppm Ice cream, ices, etc. 5.6 ppm Candy 14 ppm Baked goods 16 ppm Gelatins and puddings 0.50–30 ppm Chewing gum 18–76 ppm	Candy 80.0 ppm Baked goods 40.0 ppm Chewing gum 450.0 ppm
Regulatory status	FDA 121.1164; FEMA No. 2670	FDA 121.1164; FEMA No. 3181

REFERENCES

For References 1–5, see end of Part III.

6. **Honda,** *Miltizer Ber.*, p. 36, 1958.
7. **Darlington and Christensen,** *J. Am. Pharm. Assoc.*, 33, 298, 1944.
8. **Bertrams and Kürstens,** *J. Prakt. Chem.*, 2(51), 316, 1895.
9. **Brady et al.,** *J. Chem. Soc.*, 125, 1418, 1924.
10. **Heide,** *J. Agric. Food Chem.*, 20, 747, 1972.

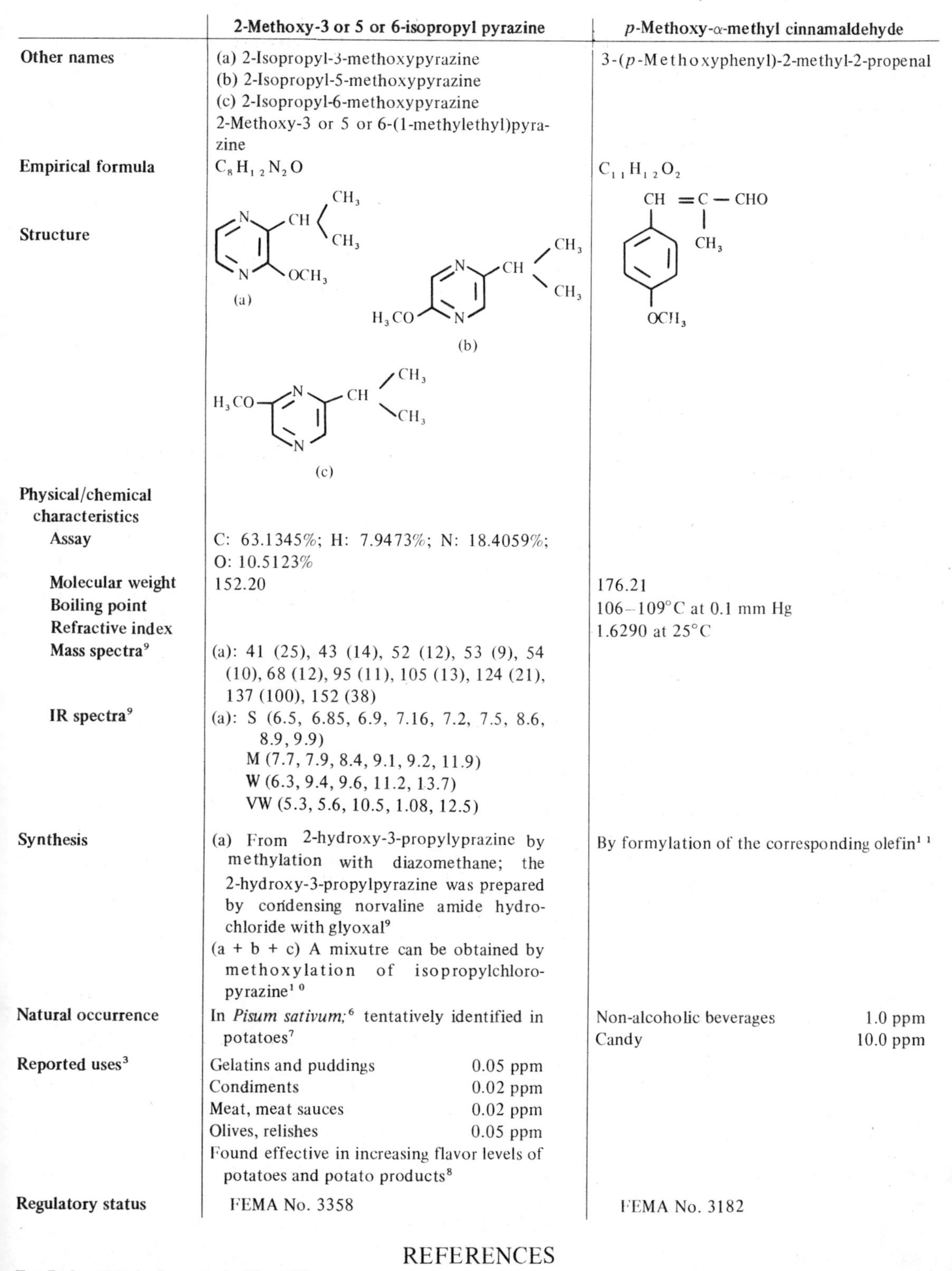

	2-Methoxy-3 or 5 or 6-isopropyl pyrazine	*p*-Methoxy-α-methyl cinnamaldehyde
Other names	(a) 2-Isopropyl-3-methoxypyrazine (b) 2-Isopropyl-5-methoxypyrazine (c) 2-Isopropyl-6-methoxypyrazine 2-Methoxy-3 or 5 or 6-(1-methylethyl)pyrazine	3-(*p*-Methoxyphenyl)-2-methyl-2-propenal
Empirical formula	$C_8H_{12}N_2O$	$C_{11}H_{12}O_2$
Structure	(a) (b) (c)	
Physical/chemical characteristics		
Assay	C: 63.1345%; H: 7.9473%; N: 18.4059%; O: 10.5123%	
Molecular weight	152.20	176.21
Boiling point		106–109°C at 0.1 mm Hg
Refractive index		1.6290 at 25°C
Mass spectra[9]	(a): 41 (25), 43 (14), 52 (12), 53 (9), 54 (10), 68 (12), 95 (11), 105 (13), 124 (21), 137 (100), 152 (38)	
IR spectra[9]	(a): S (6.5, 6.85, 6.9, 7.16, 7.2, 7.5, 8.6, 8.9, 9.9) M (7.7, 7.9, 8.4, 9.1, 9.2, 11.9) W (6.3, 9.4, 9.6, 11.2, 13.7) VW (5.3, 5.6, 10.5, 1.08, 12.5)	
Synthesis	(a) From 2-hydroxy-3-propylprazine by methylation with diazomethane; the 2-hydroxy-3-propylpyrazine was prepared by condensing norvaline amide hydrochloride with glyoxal[9] (a + b + c) A mixutre can be obtained by methoxylation of isopropylchloropyrazine[10]	By formylation of the corresponding olefin[11]
Natural occurrence	In *Pisum sativum*;[6] tentatively identified in potatoes[7]	Non-alcoholic beverages 1.0 ppm Candy 10.0 ppm
Reported uses[3]	Gelatins and puddings 0.05 ppm Condiments 0.02 ppm Meat, meat sauces 0.02 ppm Olives, relishes 0.05 ppm Found effective in increasing flavor levels of potatoes and potato products[8]	
Regulatory status	FEMA No. 3358	FEMA No. 3182

REFERENCES

For References 1–5, see end of Part III.

6. **Murray et al.,** *Chem. Ind.* (London), 27, 897, 1970.
7. **Buttery et al.,** *J. Agric. Food Chem.*, 18, 538, 1970.
8. **Guadagni et al.,** *J. Food Chem.*, 36, 363, 1971.
9. **Seifert et al.,** *J. Agric. Food Chem.*, 18(2), 246, 1970.
10. **Winter,** U.S. Patent 3,622,346, 1967.
11. **Schimdle and Barnett,** *J. Am. Chem. Soc.*, 78, 3209, 1956.

	2-Methoxy-4-methylphenol	2-Methoxy-3-(1-methylpropyl)pyrazine
Other names	Creosol 4-Hydroxy-3-methyl-1-methyl benzene 2-Methoxy-*p*-cresol 4-Methylguaiacol	2-Methoxy-3-*sec*-butylpyrazine
Empirical formula	$C_8H_{10}O_2$	$C_9H_{14}N_2O$
Structure		
Physical/chemical characteristics[5]		
Appearance	Colorless to yellowish liquid; solidifies when cold	
Assay		C: 65.03%; H: 8.49%; O: 9.63%; N: 16.85%
Molecular weight	138.16	166.22 (calculated = 166.110612)
Melting point	5.5°C	
Boiling point	220–221°C (222°C)	
Specific gravity	1.098 (1.10) at 20°C	
Refractive index	1.5353 at 25°C	
Mass spectra[12]		105(14), 123(12), 124(100), 137(50), 138(90), 151(46), 152(12)
IR spectra[12]		S (3.4, 6.9, 7.2, 7.4, 8.6) M (3.5, 6.5, 7.7, 8.4, 8.9, 9.9, 11.9) W (3.3, 6.4, 7.9, 9.0, 11.5)
Solubility	Slightly soluble in water; soluble in alcohol, ether, benzene, chloroform, and acetic acid	
Organoleptic characteristics	Characteristic aromatic odor (sweet, spicy, slightly vanilla-like); somewhat bitter taste (vanilla-like)	
Synthesis	By methylation of homopyrocatechin using dimethyl sulfate and alkali;[6,7] also by hydrogenation of vanillin	By condensation of isolencine amide with glyoxal, followed by methylation with CH_2N_2[11]
Natural occurrence	Occurs in beechwood tar; also identified in the essential oils of ylang-ylang,[8] jasmine,[9] and anise seeds. It is one of the active constituents of creosote.[10]	
Reported uses[3]	Non-alcoholic beverages 10–21 ppm Alcoholic beverages 0.02 ppm Ice cream, ices, etc. 0.05 ppm Candy 0.77 ppm Baked goods 1.0 ppm	Non-alcoholic beverages 0.05 ppm Alcoholic beverages 20.0 ppm Ice cream, ices, etc. 0.05 ppm Candy 0.05 ppm Baked goods 0.05 ppm Gelatins and puddings 0.05 ppm Vegetables 0.05 ppm
Regulatory status	FDA 121.1164; FEMA No. 2671	FEMA No. 3433

REFERENCES

For References 1–5, see end of Part III.

6. Fachberg, List and Co., German Patent 258,185.
7. Fachberg, List and Co., *Chem. Zentralbl.*, 1, 1481, 1913.
8. Schimmel and Co., *Chem. Zentralbl.*, 1, 1087, 1913.
9. **Naves and Grampoloff,** *Helv. Chim. Acta,* 25, 1500, 1942.
10. **Monod et al.,** *Chem. Abstr.*, 45, 3124, 1951.
11. **Buttery et al.,** *Chem. Ind.*, 15, 490, 1969.
12. **Kolor et al.,** personal communication, 1973.

	2 or 5 or 6-Methoxy-3-methylpyrazine (mixture of isomers)
Other names	(a) 2-Methoxy-3-methylpyrazine (b) 2-Methoxy-5-methylpyrazine (c) 2-Methoxy-6-methylpyrazine
Empirical formula	$C_6H_8N_2O$
Structure	(a) (b) (c)
Physical/chemical characteristics	
Assay	C: 58.05%; H: 6.50%; N: 22.57%; O: 12.89%
Molecular weight	124.14
Boiling point	48–50°C at 15 mm Hg (mixture)
Refractive index	n_D^{23} 1.5055 d_4^{23} 1.082 (mixture)
Mass spectra[8]	(a): 40 (3), 52 (4), 53 (3), 67 (3), 68 (12), 80 (8), 81 (2), 94 (15), 95 (23), 106 (33), 108 (36), 123 (28), 124 (100)
NMR spectra[9]	(a): δ 2.42 (3, S, –CH_3), 3.91 (3, S, –O– CH_3), 7.87 (1, d, ring –H), 7.95 (1, d, ring –H) (b): δ 2.40 (3 S, –CH_3); 3.86 (3, S, –O– CH_3), 7.86 (1, d, ring –H), 8.04 (1, d, ring –H) (c): δ 2.37 (3, S, –CH_3), 3.86 (3, S, –O–CH_3), 7.88 (1, S, ring –H), 7.94 (1, S, ring –H)
Organoleptic characteristics	Aroma reminiscent of hazelnut, almond, and peanut
Synthesis	From 2-methylpyrazine by various routes[6]
Reported uses[3]	Ice cream, ices, etc. 2.0 ppm Candy 4.0 ppm Baked goods 4.0 ppm Gelatins and puddings 2.0 ppm The methyl methoxypyrazines are used as nut-like flavoring agents for foods and beverages;[6] the 2-methoxy-3-methyl derivative is used as a coffee aroma enhancing agent[7]
Regulatory status	FEMA No. 3183

REFERENCES

For References 1–5, see end of Part III.
6. **Winter,** U.S. Patent 3,622,346, 1967.
7. **Parliment et al.,** French patent 2,062,168.
8. **Friedel et al.,** *J. Agric. Food Chem.,* 19, 530, 1971.
9. **Nakel and Haynes,** *J. Agric. Food Chem.,* 20(3), 683, 1972.

	4-(*p*-Methoxyphenyl)-2-butanone	1-(4-Methoxyphenyl)-4-methyl-1-penten-3-one
Other names	Anisyl acetone "Frambinone methylether" *p*-Methoxy phenylbutanone (common commercial name) "Raspberry ketone methylether"	α,α-Dimethylanisylacetone "Homo e·hone" *p*-Methoxystyryl isopropyl ketone
Empirical formula	$C_{11}H_{14}O_2$	$C_{13}H_{16}O_2$
Structure	$CH_2-CH_2-C(=O)-CH_3$ on benzene ring, para $O-CH_3$	$CH=CH-C(=O)-CH(CH_3)_2$ on benzene ring, para $O-CH_3$
Physical/chemical characteristics[5]		
Appearance	Colorless liquid; solidifies below 10°C	Colorless, oily liquid
Molecular weight	178.23	204.27
Melting point	9–10°C	28°C
Boiling point	277°C (160°C at 22 mm Hg)	217–219°C at 40 mm Hg
Specific gravity	1.0504 at 22°C	
Solubility	Very slightly soluble in water; soluble in alcohol; poorly soluble in propylene glycol	Very slightly soluble in water; soluble in alcohol; poorly soluble in propylene glycol
Organoleptic characteristics	Intensely sweet, floral, fruity odor; cherry-raspberry flavor at low concentrations	Caramel, fruity odor with buttery top-note; sweet taste with fruity note
Synthesis	By condensing acetone with anisaldehyde to yield anisylidene acetone and subsequent hydrogenation in the presence of Pd catalyst[6]	By condensing methyl isopropyl ketone with anisaldehyde in the presence of diluted alkali solution[8]
Natural occurrence	Contained (approximately 53% level) in the odorous principle obtained by extraction and hydrolysis from aloe wood (*Aquilaria agallocha* Roxb.)[7]	Not reported found in nature
Reported uses[3]	Non-alcoholic beverages 12 ppm Ice cream, ices, etc. 10–12 ppm Candy 28 ppm Baked goods 26 ppm Gelatins and puddings 25 ppm	
Regulatory status	FDA 121.1164; FEMA No. 2672	FDA 121.1164

REFERENCES

For References 1–5, see end of Part III.

6. **Zemplén, Bognár and Boskovitz,** *Ber. Dtsch. Chem. Ges. B.*, 77, 784, 1944.
7. **Itikawa and Heiken Yo,** *J. Chem. Soc. Jap.*, 60, 1247, 1939.
8. **Vorländer and Knötzsch,** *Ann. Chem.*, 294, 334, 1897.

	1-(*p*-Methoxyphenyl)-1-penten-3-one	1-(*p*-Methoxyphenyl)-2-propanone
Other names	"Ethone" *p*-Methoxystyryl ethyl ketone α-Methyl anisylacetone α-Methylanisylidene acetone	Anisic ketone Anisketone Anisyl methyl ketone *p*-Methoxyphenylacetone
Empirical formula	$C_{12}H_{14}O_2$	$C_{10}H_{12}O_2$
Structure	$CH{=}CH{-}C({=}O){-}CH_2{-}CH_3$ on benzene ring, para $O{-}CH_3$	$CH_2{-}C({=}O){-}CH_3$ on benzene ring, para $O{-}CH_3$
Physical/chemical characteristics[5]		
Appearance	White-yellowish crystals	Colorless, oily liquid
Assay	99% min	
Molecular weight	190.24	164.21
Melting point	60°C (min)	
Boiling point	278°C at 12 mm Hg	264–265°C; 136–138°C at 12 mm Hg
Specific gravity		1.0814 at 0°C; 1.067 at 18°C
Refractive index		1.5253 at 20°C
Solubility	1 gm in 5 ml of 95% alcohol; insoluble in water; soluble in alcohol; poorly soluble in propylene glycol	Almost insoluble in water; soluble in alcohol and ether
Organoleptic characteristics	Odor similar to butter, sweet and lasting; sweet, fruity taste at low levels; slightly burning taste at higher levels	Odor and taste similar to anise
Synthesis	By saturation at room temperature of a mixture of anisaldehyde and methyl ethyl ketone with HCl,[6] or by treating α-methyl-α-(4-hydroxy)benzylidene acetone with methyl iodide in alkaline solution[7]	Obtained by boiling α- or β-anetholglycol with a 20% solution of sulfuric acid,[8] or from 1-(*p*-methoxyphenyl)-propan-1,2-ol by treatment with diluted sulfuric acid
Natural occurrence	Not reported found in nature	Reported found in Chinese star anise[9] and in Russian star anise;[10] also reported found in sweet and common fennel
Reported uses[3]	Non-alcoholic beverages 2.3 ppm Ice cream, ices, etc. 2.3 ppm Candy 28 ppm Baked goods 12 ppm	Non-alcoholic beverages 0.60–2.8 ppm Ice cream, ices, etc. 1.2–2.8 ppm Candy 4.4–6.0 ppm Baked goods 4.4–6.0 ppm
Regulatory status	FDA 121.1164; FEMA No. 2673	FDA 121.1164; FEMA No. 2674

REFERENCES

For References 1–5, see end of Part III.

6. **Iwamoto,** *Chem. Zentralbl.*, 1, 2730, 1927; *Bull. Chem. Soc. Jap.*, 2, 54, 1927.
7. **Iwamoto,** *Bull. Chem. Soc. Jap.*, 2, 55, 1927.
8. **Toenies,** *Ber. Dtsch. Chem. Ges.*, 20, 2982, 1887.
9. **Tardy,** *Bull. Soc. Chim. Fr.*, 3(27), 990, 1902.
10. **Bouchard and Tardy,** *C. R. Séances Acad. Sci. Paris*, 122, 198, 1896.

	Methoxypyrazine	2-Methoxy-4-vinylphenol
Other names	2-Methoxy-1,4-diazine 2-Methoxypyrazine	4-Hydroxy-3-methoxystyrene *p*-Vinylguaiacol
Empirical formula	$C_5H_6N_2O$	$C_9H_{10}O_2$
Structure	[pyrazine ring with OCH_3]	[benzene ring with OH, OCH_3, $CH{=}CH_2$]
Physical/chemical characteristics		
Assay	C: 54.54%; H: 5.49%; N: 25.44%; O: 14.53%	
Molecular weight	110.12	150.18
Boiling point	60–61°C at 29 mm Hg	100°C at 5 mm Hg
Organoleptic characteristics	Odor threshold: 700 ppb in water[7]	
Synthesis	From 2-hydroxypyrazine by various routes[6]	From vanillin by reacting with acetic anhydride and sodium acetate yielding 3-methoxy-4-hydroxycinnamic acid on subsequent hydrolysis; the latter, heated with quinoline in the presence of hydroquinone, is decarboxylated to 4-vinylguaiacol[8]
Natural occurrence		Reported found among the steam-distillable phenols from the alcoholic fermentation of cereal grains, especially corn; probably formed from ferulic acid[9]
Reported uses[3]	Non-alcoholic beverages 10.0 ppm Ice cream, ices, etc. 10.0 ppm Candy 10.0 ppm Baked goods 10.0 ppm Gelatins and puddings 10.0 ppm Meat, meat sauces, soups 10.0 ppm Milk, dairy products 10.0 ppm Cereals 10.0 ppm	Non-alcoholic beverages 0.25–3.0 ppm Ice cream, ices, etc. 0.25–11 ppm Candy 1.0–8.0 ppm Baked goods 1.0–8.0 ppm
Regulatory status	FEMA No. 3302	FDA 121.1164; FEMA No. 2675

REFERENCES

For References 1–5, see end of Part III.

6. **Phillips and Phillips,** *J. Chem. Soc.*, p. 1294, 1956.
7. **Seifert et al.,** *J. Agric. Food Chem.*, 18, 246, 1970.
8. **Toshiyuki et al.,** *Nippon Kagaku Zasshi,* 80, 1337, 1959.
9. **Steinke and Paulson,** *J. Agric. Food Chem.*, 4(12), 381, 1964.

	Methyl acetate	4′-Methylacetophenone
Other names		*p*-Acetotoluene *p*-Methylacetophenone 1-Methyl-4-acetyl benzene Methyl *p*-tolyl ketone
Empirical formula	$C_3H_6O_2$	$C_9H_{10}O$
Structure	$H_3C-C(=O)-O-CH_3$	$H_3C-C_6H_4-C(=O)-CH_3$
Physical/chemical characteristics[5]		
Appearance	Colorless liquid	Colorless or nearly colorless liquid; tends to solidify at room temperature
Assay	96% min	98% min[2]
Molecular weight	74.08	134.17
Melting point	−98°C	−23°C
Boiling point	56°C	222–226°C at 756 mm Hg
Flash point	Approx 15°C (−16°C)	
Specific gravity	0.9415–0.9455 at 15°C; 0.933 at 20°C	1.001–1.004 at 25°/25°C;[2] 1.0016 at 20°C
Refractive index	1.3610–1.3630 (1.3619) at 20°C	1.5320–1.5345[2] (1.5335) at 20°C
Acid value	0.9 max	
Solubility	Soluble in water; completely miscible with alcohol and ether	1:10 in 50% alcohol;[2] 1:3 in 60% alcohol; insoluble in water; sparingly soluble in most organic solvents
Organoleptic characteristics	Pleasant, fruity odor and slightly bitter flavor	Fruity, floral odor resembling acetophenone; sweet, strawberry-like flavor
Synthesis	By boiling acetic acid and methanol in the presence of acid catalysts; or by heating methanol with an excess carbon monoxide under pressure in the presence of a catalyst (phosphoric acid, cobalt salts)	By slow addition of acetyl chloride to a mixture of toluene and $AlCl_3$ in an ice bath and under vacuum, maintaining the temperature at +5°C and then letting it increase to +20°C
Natural occurrence	Reported found in several fruits, such as grape, banana, and *Citrus maxima*, and in coffee; also identified in rum ether[6]	Reported found in the essential oil distilled from the wood of *Myrocarpus fastigiatus*,[7,8] *Myrocarpus frondosus*,[9] bois de rose, and probably mimosa extracts[10,11]
Reported uses[3]	Non-alcoholic beverages 28 ppm Alcoholic beverages 0.20 ppm Ice cream, ices, etc. 29 ppm Candy 11 ppm Baked goods 14 ppm Gelatins and puddings 0.10 ppm	Non-alcoholic beverages 1.1 ppm Ice cream, ices, etc. 1.6 ppm Candy 5.2 ppm Baked goods 4.9 ppm Chewing gum 870 ppm Condiments 5.8 ppm Maraschino cherries 8.0 ppm
Regulatory status	FDA 121.1164; FEMA No. 2676	FDA 121.1164; FEMA No. 2677

REFERENCES

For References 1–5, see end of Part III.

6. **Fenaroli et al.,** *Riv. Ital. EPPOS,* 9, 484, 1965; 2, 75, 1966.
7. **Verley,** *Bull. Soc. Chim. Fr.,* 3(17), 909, 1897.
8. **Verley,** *Bull. Soc. Chim. Fr.,* 3(19), 138, 1898.
9. **Naves,** *Helv. Chim. Acta,* 31, 44, 1948.
10. *Parfum. Mod.,* 5, 99, 1926.
11. *Chem. Drug.,* 105, 201, 1926.

	1-Methyl-2-acetyl pyrrole	2-Methylallyl butyrate
Other names	Methyl 1-methylpyrrol-2-yl ketone	Isopropenyl carbinyl-*n*-butyrate Methallyl butyrate β-Methylallyl-*n*-butyrate 2-Methyl-2-propen-1-yl butyrate
Empirical formula	C_7H_9NO	$C_8H_{14}O_2$
Structure	(pyrrole ring with N–CH_3 and 2-CO–CH_3)	$CH_3-(CH_2)_2-C(=O)-O-CH_2-C(CH_3)=CH_2$
Physical/chemical characteristics[5]		
Appearance		Colorless liquid
Molecular weight	123.15	142.19
Boiling point	88–93°C at 22 mm Hg	168°C
Solubility		Insoluble in water; soluble in alcohol
Organoleptic characteristics		Powerful, penetrating, fruity, ethereal odor; sharp, acrid taste above 20 ppm, becoming sweet, pineapple-, apple-, and plum-like
Synthesis	From N-methyl pyrrole and Grignard reagent;[6] by acetylation of 1-methylpyrrole[7]	By direct esterification of β-methylallyl alcohol with butyric acid
Natural occurrence	Reported found as a component in coffee aroma[8]	Not reported found in nature
Reported uses[3]	Non-alcoholic beverages 10.0 ppm Ice cream, ices, etc. 10.0 ppm Candy 10.0 ppm Gelatins and puddings 10.0 ppm	Non-alcoholic beverages 0.20 ppm Baked goods 0.20 ppm
Regulatory status	FEMA No. 3184	FDA 121.1164; FEMA No. 2678

REFERENCES

For References 1–5, see end of Part III.

6. **Hertz,** *J. Org. Chem.*, 22, 1260, 1957.
7. **Anderson,** *Can. J. Chem.*, 35, 21, 1957.
8. **Gianturco, Gianmarino, Friedel and Flanagan,** *Tetrahedron Lett.*, 20, 2951, 1964.

	Methyl anisate	*o*-Methylanisole
Other names	Methyl *p*-methoxybenzoate	*o*-Cresyl methyl ether 2-Methoxy toluene *o*-Methoxy toluene Methyl *o*-tolyl ether
Empirical formula	$C_9H_{10}O_3$	$C_8H_{10}O$
Structure	$H_3C{-}O{-}C{=}O$ on benzene ring, *para* $O{-}CH_3$	Benzene ring with $O{-}CH_3$ and adjacent CH_3
Physical/chemical characteristics[5]		
Appearance	Low-melting solid	Colorless liquid
Molecular weight	166.18	122.17
Melting point	48–49°C	
Boiling point	254–255°C	166–167°C; 63–64°C at 14 mm Hg (172°C)
Specific gravity		0.9853 at 15.5°C
Refractive index		1.5199 at 15.3°C
Solubility	Insoluble in water; soluble in alcohol	Insoluble in water; soluble in alcohol
Organoleptic characteristics	Herbaceous, anise-like odor; sweet taste similar to melon	Pungent, warm, floral odor with earthy, walnut undertones; sweet, fruity, nut-like flavor at low levels
Synthesis	By esterification of anisic acid with methanol or from sodium anisate, dimethyl sulfate, and small amounts of methanol	By methylation of *o*-cresol using dimethyl-sulfate in caustic soda at 40°C[7]
Natural occurrence	Reported found in the mushroom variety *Trametes graveolens*[6]	Reported found in several natural products
Reported uses[3]	Non-alcoholic beverages 2.7 ppm Ice cream, ices, etc. 3.0 ppm Candy 8.0 ppm Baked goods 6.2 ppm	Non-alcoholic beverages 1.7 ppm Ice cream, ices, etc. 1.0 ppm Candy 2.3–4.0 ppm Baked goods 4.0 ppm
Regulatory status	FDA 121.1164; FEMA No. 2679	FDA 121.1164; FEMA No. 2680

REFERENCES

For References 1–5, see end of Part III.

6. **Aye,** *Arch. Pharm.*, p. 246, 1931.
7. **Boyd and Hardy,** *J. Chem. Soc.*, p. 637, 1928.

	p-Methylanisole	Methyl anthranilate
Other names	*p*-Cresyl methyl ether *p*-Methoxy toluene Methyl *p*-cresol Methyl *p*-tolyl ether	Methyl 2-aminobenzoate Methyl *o*-aminobenzoate
Empirical formula	$C_8H_{10}O$	$C_8H_9NO_2$
Structure	O—CH₃ on benzene ring, para CH₃	NH₂ and C(=O)—O—CH₃ on benzene ring (ortho)
Physical/chemical characteristics[5]		
Appearance	Colorless liquid	Colorless to pale-yellow liquid with bluish fluorescence
Assay		98% min[2]
Molecular weight	122.17	151.17
Melting point		24–25°C
Boiling point		132°C at 14 mm Hg
Congealing point		23.8°C (24°C)
Specific gravity	0.966–0.970 at 25°/25°C[2]	1.161–1.169 at 25°/25°C;[2] 1.1640 at 25°C
Refractive index	1.5100–1.5130 at 20°C[2]	1.5820–1.5840 at 20°C;[2] 1.5802 at 25°C
Cresol content	Not more than 0.5%[2]	
Solubility	1:3 in 80% alcohol;[2] soluble in most organic solvents; 1:7 in 70% alcohol	1:5 in 60% alcohol;[2] soluble in water and most organic solvents
Organoleptic characteristics	Pungent odor suggestive of ylang-ylang	Characteristic orange-flower odor, and slightly bitter, pungent taste
Synthesis	By methylation of *p*-cresol	By heating anthranilic acid and methyl alcohol in the presence of sulfuric acid and subsequent distillation
Natural occurrence	Reported found in the oils of ylang-ylang, cananga, and others	Reported found in several essential oils: neroli, orange, bergamot, lemon, mandarin, jasmine, tuberose, gardenia, champaca, ylang-ylang, and others; also in the juice and oil of *Vitis labrusca*[6,7]
Reported uses[3]	Non-alcoholic beverages 2.7 ppm Ice cream, ices, etc. 2.7 ppm Candy 4.8 ppm Baked goods 7.6 ppm Gelatins and puddings 0.50–4.0 ppm Condiments 2.0 ppm Syrups 8.0 ppm	Non-alcoholic beverages 16 ppm Alcoholic beverages 0.20 ppm Ice cream, ices, etc. 21 ppm Candy 56 ppm Baked goods 20 ppm Gelatins and puddings 23 ppm Chewing gum 2,200 ppm
Regulatory status	FDA 121.1164; FEMA No. 2681	FDA GRAS; FEMA No. 2682

REFERENCES

For References 1–5, see end of Part III.

6. **Mattick et al.,** *J. Agric. Food Chem.*, 4, 334, 1963.
7. **Roger,** *Food Technol.*, 6, 309, 1961.

	Methyl benzoate	Methylbenzyl acetate (mixed *o*-,*m*-,*p*-)
Other names	Niobe oil Oil niobe	Tolyl acetate
Empirical formula	$C_8H_8O_2$	$C_{10}H_{12}O_2$
Structure	$C_6H_5-C(=O)-O-CH_3$	$CH_3-C_6H_4-CH_2-O-C(=O)-CH_3$
Physical/chemical characteristics[5]		
Appearance	Colorless liquid	Colorless liquid
Assay	98% min[2]	98% min[2]
Molecular weight	136.15	164.20
Melting point	−12 to −13°C	
Boiling point	199–200°C	
Flash point	82°C	
Specific gravity	1.082–1.088 at 25°/25°C;[2] 1.0785 at 30°C	1.030–1.035 at 25°/25°C[2]
Refractive index	1.5140–1.5180 at 20°C;[2] 1.5181 at 16°C	1.5015–1.5040 at 20°C[2]
Acid value	Not more than 1[2]	Not more than 1[2]
Solubility	1:4 and more in 60% alcohol;[2] 1:1.5 in 70% alcohol; insoluble in water; miscible with alcohol and ether	1:2 in 70% alcohol;[2] soluble in benzyl benzoate, mineral oil, and most fixed oils; partly soluble in propylene glycol; insoluble in glycerin
Organoleptic characteristics	Fruity odor, also similar to cananga	
Synthesis	By heating benzoic acid and dimethyl sulfate to high temperature;[6] or by exchange between ethyl benzoate and methanol in KOH solution[7]	By acetylation of the alcohol
Natural occurrence	Reported in the oils of tuberose (flowers),[8] ylang-ylang,[9] clove,[10] *Polianthes tuberosa* L. (flowers),[11] and *Narcissus jonquilla* L. (flowers)[12]	
Reported uses[3]	Non-alcoholic beverages 2.2 ppm Ice cream, ices, etc. 4.5 ppm Candy 8.4 ppm Baked goods 9.9 ppm Chewing gum 61 ppm	
Regulatory status	FDA 121.1164; FEMA No. 2683	FDA 121.1164

REFERENCES

For References 1–5, see end of Part III.

6. **Simon,** *Compt. Rend.*, 176, 585, 1923.
7. **Reimer and Downes,** *J. Am. Chem. Soc.*, 43, 949, 1921.
8. *Schimmel Ber.*, April 1903.
9. *Schimmel Ber.*, October 1901.
10. *Schimmel Ber.*, April 1903.
11. **Elze,** *Chem. Zentralbl.*, 2, 2198, 1928.
12. **Soden,** *J. Prakt. Chem.*, 110, 277, 1925.

	α-Methylbenzyl acetate	α-Methylbenzyl alcohol
Other names	Gardenol Methyl phenylcarbinyl acetate Styralyl acetate Styrolene acetate	Methylphenylcarbinol 1-Phenylethanol α-Phenylethyl alcohol Phenyl methyl carbinol Styralyl alcohol Styrolyl alcohol
Empirical formula	$C_{10}H_{12}O_2$	$C_8H_{10}O$
Structure	C_6H_5–CH(CH_3)–C(=O)–O–CH_3	C_6H_5–CH(CH_3)–OH
Physical/chemical characteristics[5]		
Appearance	Colorless liquid	Colorless liquid congealing below room temperature
Assay	97% min[2]	97% min[2]
Molecular weight	164.21	122.17
Melting point		20°C
Boiling point	213°C	205°C
Congealing point		19°C min[2]
Flash point	81°C	93°C
Specific gravity	1.023–1.026 at 25°/25°C;[2] 1.024 at 25°C	1.009–1.014 at 25°/25°C;[2] 1.013 at 20°C
Refractive index	1.4935–1.4970 (1.4948) at 20°C[2]	1.5250–1.5290 at 20°C[2]
Acid value	2.0 max[2]	
Ketone content		1% max[2] (as acetophenone)
Solubility	1:7 in 60% alcohol;[2] soluble in most organic solvents	1:3 in 50% alcohol;[2] 1:16 in 30% alcohol; insoluble in water; soluble in most organic solvents
Organoleptic characteristics	Intensive green odor suggestive of gardenia; bitter, acrid taste, interesting on dilution	Mild hyacinth-gardenia odor
Synthesis	By acetylation of methyl phenyl carbinol; from benzaldehyde by reacting with magnesium methyl bromide and subsequent acetylation; from 1-bromoethylbenzene and silver acetate in acetic acid	By oxidation of ethylbenzene or by reduction of acetophenone
Natural occurrence	Reported found in gardenia flower oil	Not reported found in natural oils. Two optically active isomers exist. The commercial product is the racemic form
Reported uses[3]	Non-alcoholic beverages 3.9 ppm Ice cream, ices, etc. 5.4 ppm Candy 12 ppm Baked goods 17 ppm Chewing gum 0.80 ppm Topping 30 ppm	Non-alcoholic beverages 4.6 ppm Ice cream, ices, etc. 3.8 ppm Candy 6.8 ppm Baked goods 9.0 ppm Gelatins and puddings 4.0 ppm Chewing gum 0.30 ppm
Regulatory status	FDA 121.1164; FEMA No. 2684	FDA 121.1164; FEMA No. 2685

REFERENCES

For References 1–5, see end of Part III.

	α-Methylbenzyl butyrate	α-Methylbenzyl formate
Other names	Methyl phenylcarbinyl-*n*-butyrate Styralyl butyrate	Methyl phenylcarbinyl formate "Styralyl formate"
Empirical formula	$C_{12}H_{16}O_2$	$C_9H_{10}O_2$
Structure	$C_6H_5-CH(CH_3)-O-C(=O)-CH_2-CH_2-CH_3$	$C_6H_5-CH(CH_3)-O-C(=O)-H$
Physical/chemical characteristics[5]		
Appearance	Colorless, oily liquid	Colorless liquid
Molecular weight	192.26	150.18
Solubility	Almost insoluble in water; soluble in alcohol	Insoluble in water; soluble in alcohol
Organoleptic characteristics	Fruital-floral, jasmine-like odor; apricot/apple-like flavor notes	Woody odor of mimosa and gardenia; dry notes and fruity taste
Synthesis	From methylphenylcarbinol and *n*-butyric acid by esterification	By esterification of methylphenylcarbinol with formic acid in the presence of acetic anhydride
Natural occurrence		Not reported found in nature
Reported uses[3]	Non-alcoholic beverages 4.0–5.0 ppm Ice cream, ices, etc. 4.0–10 ppm Candy 10–20 ppm Baked goods 10–20 ppm	Non-alcoholic beverages 2.0–5.0 ppm Ice cream, ices, etc. 3.0–5.0 ppm Candy 10–20 ppm Baked goods 10–20 ppm
Regulatory status	FDA 121.1164; FEMA No. 2686	FDA 121.1164; FEMA No. 2688

REFERENCES

For References 1–5, see end of Part III.

	α-Methylbenzyl isobutyrate	α-Methylbenzyl propionate
Other names	Methyl phenylcarbinyl isobutyrate Styralyl isobutyrate	Methyl phenylcarbinyl propionate Styralyl propionate
Empirical formula	$C_{12}H_{16}O_2$	$C_{11}H_{14}O_2$
Structure	$C_6H_5-CH(CH_3)-O-C(=O)-CH(CH_3)_2$	$C_6H_5-CH(CH_3)-O-C(=O)-CH_2-CH_3$
Physical/chemical characteristics[5]		
Appearance	Colorless liquid	Colorless liquid
Molecular weight	192.26	178.23
Solubility	Very slightly soluble in water; soluble in alcohol	Insoluble in water; soluble in alcohol
Organoleptic characteristics	Jasmine-like, floral odor	Floral, sweet, green odor typical of gardenia and jasmine
Synthesis	By esterification (under azeotropic conditions) of methylphenylcarbinol with isobutyric acid	By direct esterification of methylphenylcarbinol with propionic acid under azeotropic conditions
Natural occurrence	Not reported found in nature	Not reported found in nature
Reported uses[3]	Non-alcoholic beverages 2.0 ppm Ice cream, ices, etc. 10 ppm Candy 10 ppm Baked goods 10 ppm	Non-alcoholic beverages 4.0–5.0 ppm Ice cream, ices, etc. 4.0–5.0 ppm Candy 10–15 ppm Baked goods 10–15 ppm
Regulatory status	FDA 121.1164; FEMA 2687	FDA 121.1164; FEMA No. 2689

REFERENCES

For References 1–5, see end of Part III.

	2-Methyl-1-butanethiol	3-Methyl-2-butanethiol
Other names	Amyl mercaptan 2-Methylbutyl mercaptan Thioamyl alcohol	Isopentyl mercaptan Isoamyl mercaptan Isopentanethiol Isoamyl thioalcohol Isoamyl sulfhydrate
Empirical formula	$C_5H_{12}S$	$C_5H_{12}S$
Structure	$CH_3-CH_2-CH(CH_3)-CH_2SH$	$CH_3-CH(CH_3)-CH_2-CH_2-SH$
Physical/chemical characteristics[5]		
Appearance		Colorless liquid
Molecular weight	104.22	104.22
Boiling point	118.2°C	118°C
Specific gravity	0.8420 at 20°C	0.8350 at 20°C
Refractive index	1.4440 at 20°C	1.4418 at 20°C
Optical rotation	+3.21° at 23°C	
Solubility		Insoluble in water; soluble in alcohol and ether in all proportions
Organoleptic characteristics		Repulsive, characteristic mercaptan-like odor
Synthesis	From 2-methyl-1-butyl isothiourea picrate by conversion to bis [(s)-2-methylbutyl] disulfide, followed by reduction to the corresponding thiol, using sodium metal in liquid ammonia[6]	From isoamyl chloride and potassium sulfhydrate[8] also, the corresponding sodium salt can be prepared from diisopentyl-disulfide and sodium metal in liquid ammonia[9]
Natural occurrence	Reported as occurring in the petroleum fraction boiling between 111°C and 150°C[7]	Not reported found in nature
Reported uses[3]	Non-alcoholic beverages 0.8 ppm Candy 0.8 ppm Baked goods 0.8 ppm Meat, meat sauces, soups 0.8 ppm Milk, dairy products 0.8 ppm Cereals 0.8 ppm	Non-alcoholic beverages 1.2 ppm Candy 1.2 ppm Baked goods 1.2 ppm Meat 1.2 ppm Milk, dairy products 1.2 ppm Cereals 1.2 ppm
Regulatory status	FEMA No. 3303	FEMA No. 3304

REFERENCES

For References 1–5, see end of Part III.

6. **Polak, Bregant and Balenovic,** *Bull. Sci. Cons. Acad. RSF Yougosl. Sect. A,* 13(1–2), 1968; *Chem. Abstr.,* 69, 43345h, 1968.
7. **Colenian, Thompson, Hopkins and Rall,** *J. Chem. Eng. Data,* 10(1), 80, 1965.
8. *Merck Index,* 8th ed., Merck & Co., Rahway, N.J., 1968, 579.
9. **Williams and Gebauer-Fuelnegg,** *J. Am. Chem. Soc.,* 53, 355, 1931.

	2-Methyl-2-butenal	3-Methylbutyl aldehyde
Other names	2-Methyl crotonaldehyde Tiglic aldehyde 2,3-Dimethylacrolein	Isovaleraldehyde 3-Methylbutanal 3-Methylbutyraldehyde
Empirical formula	C_5H_8O	$C_5H_{10}O$
Structure	$CH_3=CH-C(CH_3)-CHO$	$CH_3-CH(CH_3)-CH_2-CHO$
Physical/chemical characteristics[5]		
Appearance	Colorless liquid; readily oxidizes in the presence of air	Colorless liquid
Molecular weight	84.11	86.14
Boiling point	118°C 116.5–117.5°C at 738 mm Hg 63–65°C at 119 mm Hg	92.5°C
Specific gravity	0.8710 at 20°C	0.7977 at 20°C
Refractive index	1.4475 at 20°C	1.3902 at 20°C
Solubility	Slightly soluble in water; miscible with alcohol and ether; soluble in oils	Very slightly soluble in water; soluble in alcohol, propylene glycol, and oils
Organoleptic characteristics	Penetrating, powerful green-ethereal odor	Powerful, penetrating, acrid odor; at very low levels the flavor is warm, herbaceous, slightly fruity, and nut-like
Synthesis	By condensation of acetaldehyde and propionaldehyde[6,7] $H_3O\text{–}CH_2CH_2\text{–}CHO + H_3C\text{–}CHO \xrightarrow{B^-}$	
Natural occurrence	Reported found in geranium oil;[8] also found in onion, garlic, mint, cooked chicken, coffee, roasted filberts, and roasted peanuts[9]	
Reported uses[3]	Non-alcoholic beverages 5.0 ppm Ice cream, ices, etc. 5.0 ppm Candy 5.0 ppm Baked goods 5.0 ppm Gelatins and puddings 5.0 ppm Chewing gum 5.0 ppm Meat, meat sauces, soups 2.5 ppm	
Regulatory status	FEMA No. 3407	FDA 121.1164

REFERENCES

For References 1–5, see end of Part III.

6. **Morgan et al.,** *Chem. Ind.*, p. 519, 1933.
7. **Bernhauer et al.,** *J. Prakt. Chem.*, 155, 315, 1940.
8. **Timmer et al.,** *J. Agric. Food Chem.*, 19(6), 1066, 1971.
9. Volatile Compounds in Food, T.N.O. Report No. 4030, 3rd ed., Centraal Instituut voor Voedingsonderzoek, Zeist, Netherlands.

	2-Methylbutyl isovalerate	2-Methylbutyl-2-methyl butyrate
Other names	*d-sec*-Butylcarbinyl isopentanoate	α,β-Methylbutyl-*dl*-2-methyl butanoate
Empirical formula	$C_{10}H_{20}O_2$	$C_{10}H_{20}O_2$
Structure	$CH_3-CH_2-CH(CH_3)-CH_2-O-C(=O)-CH_2-CH(CH_3)_2$	$CH_3-CH_2-CH(CH_3)-C(=O)-O-CH_2-CH(CH_3)-CH_2-CH_3$
Physical/chemical characteristics[5]		
Appearance	Colorless liquid	
Molecular weight	172.27	172.27
Boiling point		184–187°C
Solubility	Insoluble in water; soluble in alcohol	
Organoleptic characteristics	Herbaceous, fruity, somewhat earthy fragrance; sweet, herbaceous, fruity flavor	
Synthesis	By direct esterification of 2-methylbutanol with isovaleric acid	By esterification of 2-methylbutanoic acid with 2-methyl-1-butanol;[6] by condensation of isobutyraldehyde at 120°C[7]
Natural occurrence		Reported found as a component in the essential oil of hops (*Humulus lupulus*);[8-10] also as the odorous component in *Vaccinium vitis idaea*[11] and in cocoa beans[12]
Reported uses[3]	The compound has been reported used in fabricating artificial peppermint essential oil, contributing the "non-minty" portion of the flavor.	Non-alcoholic beverages 5.0 ppm Ice cream, ices, etc. 5.0 ppm Candy 8.0 ppm Baked goods 8.0 ppm Gelatins and puddings 5.0 ppm Preserves 5.0 ppm
Regulatory status	FDA 121.1164	FEMA No. 3359

REFERENCES

For References 1–5, see end of Part III.

6. Bailey and Hass, *J. Am. Chem. Soc.*, 91, 63, 1969.
7. Hagemeyer, Jr. and Wright, Jr., U.S. Patent 3,081,344, 1963.
8. Naya and Kotake, *Nippon Kagaku Zasshi*, 88(12), 1302, 1967.
9. Buttery, Black, Guadagni and Kealy, *Am. Soc. Brew. Chem. Proc.*, 103, 1965.
10. Buttery, Black, Kealy and McFadden, *Nature*, 202(4933), 701, 1964.
11. Anjou and von Sydow, *Acta Chem. Scand.*, 21(4), 945, 1967.
12. Dietrich, Lederer, Winter and Stoll, *Helv. Chim. Acta*, 47(6), 1581, 1964.

	Methyl *p-tert*-butylphenylacetate	2-Methyl butyraldehyde
Other names	*p-tert*-Butylphenylacetic acid, methyl ester	2-Methylbutanal-l α-Methyl butyraldehyde Methyl ethyl acetaldehyde
Empirical formula	$C_{13}H_{18}O_2$	$C_5H_{10}O$
Structure	$H_3C-C(CH_3)_2-C_6H_4-CH_2-C(=O)-O-CH_3$	$CH_3-CH_2-CH(CH_3)-CHO$
Physical/chemical characteristics[5]		
Appearance	Colorless, oily liquid	Colorless liquid
Molecular weight	206.28	86.13
Boiling point		93°C
Specific gravity		0.80
Solubility	Almost insoluble in water; soluble in alcohol	Slightly soluble in water; soluble in alcohol and propylene glycol
Organoleptic characteristics	Sweet, woody, and camphoraceous odor; roasted, chocolate-like flavor	Powerful, choking odor; peculiar cocoa and coffee-like flavor when diluted; sweet slightly fruity, chocolate-like taste
Synthesis	By esterification of *p-tert*-butylphenylacetic acid with methanol	By oxidation of *sec*-butylcarbinol isolated from fermented fusel oil; the *dl*-form from *sec*-butyl magnesium bromide and formaldehyde; by reduction of methylethylacetic acid
Natural occurrence		
Reported uses[3]	Non-alcoholic beverages 0.50 ppm Ice cream, ices, etc. 0.35–1.0 ppm Candy 2.0 ppm Baked goods 2.0 ppm	Non-alcoholic beverages 1.5–2.0 ppm Ice cream, ices, etc. 2.0–8.0 ppm Candy 6.6 ppm Baked goods 5.7 ppm
Regulatory status	FDA 121.1164; FEMA No. 2690	FDA 121.1164; FEMA No. 2691

REFERENCES

For References 1–5, see end of Part III.

	Methyl butyrate	2-Methyl butyric acid
Other names		Butane-2-carboxylic acid 2-Methylbutanoic acid α-Methyl butyric acid Methylethyl acetic acid Optically active isovaleric acid
Empirical formula	$C_5H_{10}O_2$	$C_5H_{10}O_2$
Structure	$CH_3{-}O{-}C(=O){-}(CH_2)_2{-}CH_3$	$CH_3{-}CH_2{-}CH(CH_3){-}COOH$
Physical/chemical characteristics[5]		
Appearance	Colorless, mobile liquid	Colorless liquid
Assay		Exists in the *d*- (natural), *l*-, and *dl*-forms
Molecular weight	102.13	102.13
Melting point		*d*-, 176°C; *l*-, 176–177°C; *dl*-, 173–174°C
Boiling point	102°C (103°C)	
Flash point	14°C	
Specific gravity	0.8981 at 20°C; 0.905 at 15.5°C	At 20°C: *d*-, 0.934; *l*-, 0.934; *dl*-, 0.9332
Refractive index	1.3873 (1.3900) at 20°C	At 21.2°C: *d*-, 1.4044
		d-, +16° to +21°; *l*-, −6° to −18°
Solubility	1:60 in water; miscible in alcohol and ether	Slightly soluble in water; soluble in alcohol and propylene glycol; slightly soluble in glycerin
Organoleptic characteristics	Apple-like odor and corresponding sweet taste; not very powerful; below 100 ppm may have banana-pineapple flavor	Pungent, acrid odor similar to roquefort cheese; acrid taste; at low dilutions pleasant, fruity taste
Synthesis	From methyl alcohol and butyric acid in the presence of concentrated H_2SO_4[6]	By decarboxylation of methyl ethyl malonic acid (with heat); also by oxidation of fermentation amyl alcohol (fusel oil)
Natural occurrence	Reported found in wood oil[7]	Occurring as the *d*-, *l*-, and *dl*-isomers. The racemic form has been reported found in angelica root oil and coffee. The *d*-isomer in the ester form has been identified in lavender oil.[8]
Reported uses[3]	Non-alcoholic beverages 17 ppm Ice cream, ices, etc. 31 ppm Candy 86 ppm Baked goods 48–200 ppm	Non-alcoholic beverages 0.50 ppm Ice cream, ices, etc. 3.0 ppm Candy 5.0 ppm
Regulatory status	FDA 121.1164; FEMA No. 2693	FDA 121.1164; FEMA No. 2695

REFERENCES

For References 1–5, see end of Part III.

6. **Vogel,** *J. Chem. Soc.*, p. 629, 1948.
7. **Fraps,** *Am. Chem. J.*, 25, 37, 1901.
8. **Seidel et al.,** *Helv. Chim. Acta*, 27, 663, 1944.

	α-Methylcinnamaldehyde	Methyl cinnamate
Other names	α-Methylcinnamal α-Methyl cinnamic aldehyde 2-Methyl-3-phenyl-2-propen-l-al 3-Phenyl-2-methyl acrolein	Methyl-3-phenyl propenoate
Empirical formula	$C_{10}H_{10}O$	$C_{10}H_{10}O_2$
Structure	C_6H_5–CH=C(CH_3)–CHO	C_6H_5–CH=CH–C(=O)–O–CH_3
Physical/chemical characteristics[5]		
Appearance	Yellow, oily liquid	White to slightly yellow, crystalline solid; the *cis-* form (which is not a commercial material) is a liquid
Assay	97% min[2]	98% min[2]
Molecular weight	146.18	162.18
Melting point		*trans-*, 36°C; *cis-*, −3°C
Boiling point	137–138°C at 22 mm Hg	*trans-*, 254–255°C; *cis-*, 49°C
Congealing point		>33.8°C[2]
Specific gravity	1.035–1.039 at 25°/25°C;[2] 1.0538 at 20°C	At 50°C: *trans-*, 1.0573
Refractive index	1.6025–1.6070[2] (1.6081) at 20°C	At 20°C: *trans-*, 1.5670; *cis-*, 1.5582
Acid value	Not more than 5[2]	Not more than 2[2]
Solubility	1:2 in 70% alcohol;[2] insoluble in water and glycerin	1:4 and more in 80% alcohol;[2] insoluble in water; soluble in propylene glycol, mineral oil, and other oils
Organoleptic characteristics	Characteristic cinnamon-type odor; soft, spicy flavor	Fruity, balsamic odor similar to strawberry; corresponding sweet taste
Synthesis	By condensing benzaldehyde with propionic aldehyde in the presence of a 1% caustic soda solution; also by the controlled hydrogenation of α-methylcinnamic aldehyde	By esterification of cinnamic acid with methanol using HCl as catalyst;[6] or by adding HCl to a boiling solution of cinnamyl nitrile in methanol[7]
Natural occurrence		Reported found in the oil from rhizomes of *Alpinia malaccensis*;[8] in the oil from leaves of *Ocimum canum* Sims.;[9,10] in the oil of *Narcissus jonquilla* L.;[11] in the oil from rhizomes of *Gastrochilus panduratum* Ridl.[12] Two isomers (*cis-* and *trans-*) are known.
Reported uses[3]	Non-alcoholic beverages 0.50–11 ppm Ice cream, ices, etc. 1.0–15 ppm Candy 26 ppm Baked goods 27 ppm Chewing gum 430 ppm	Non-alcoholic beverages 1.9 ppm Ice cream, ices, etc. 3.8 ppm Candy 8.7 ppm Baked goods 13 ppm Gelatins and puddings 1.7–14 ppm Chewing gum 2.7–40 ppm Condiments 0.40 ppm
Regulatory status	FDA 121.1164; FEMA No. 2697	FDA 121.1164; FEMA No. 2698

REFERENCES

For References 1–5, see end of Part III.

6. **Weger,** *Chem. Ann.*, 27, 74, 1885.
7. **Pfeiffer, Engelhardt and Alfuss,** *Chem. Ann.*, 467, 178, 1928.
8. **Van Romburgh,** *K. Ned. Akad. Wet. Versl. Gewone Vergad. Afd. Natuurkd.*, p. 445, 1900.
9. **Roure-Bertrand fils,** *Chem. Zentralbl.*, 1, 1654, 1914.
10. Schimmel and Co., *Chem. Ber.*, 1, 1654, 1914.
11. **Soden,** *J. Prakt. Chem.*, 2(110), 277, 1925.
12. **Ultee,** *Versl. Akad. Amsterdam*, 36, 1262, 1927; *Chem. Zentralbl.*, 99, 2018, 1928.

	p-Methylcinnamic aldehyde	Methyl citronellate
Other names	3-(*p*-Methylphenyl)-propenal	Methyl-3,7-dimethyl-6-octenoate
Empirical formula	$C_{10}H_{10}O$	$C_{11}H_{20}O_2$
Structure	CH=CH−CHO; CH_3	CH_3; $COOCH_3$; H_3C; CH_3
Physical/chemical characteristics[5]		
Appearance	Crystals from diluted alcohol	Colorless mobile liquid
Molecular weight	146.18	184.28
Melting point	41.5°C	216°C
Boiling point	154°C at 25 mm Hg	78°C at 3 mm Hg
Specific gravity		0.8973 at 20°C
Refractive index		1.4415 at 20°C
Optical rotation		+5.45°
Solubility		Very slightly soluble in water; soluble in alcohol and oils
Organoleptic characteristics		Fruity (apple), brandy-like odor
Synthesis		From citronellic acid, methanol, and sulfuric acid;[6] from citronellic acid treated with diazomethane in ether solution[7]
Natural occurrence		Reported as naturally occurring in the oil from the leaves of *Calytrix tetragona*[8]
Reported uses[3]		Non-alcoholic beverages 1.5 ppm Baked goods 3.0 ppm Condiments, pickles 1.5 ppm
Regulatory status	FDA 121.1164	FEMA No. 3361

REFERENCES

For References 1–5, see end of Part III.

6. Tatsuzo and Okazawa *J. Chem. Soc. Jap.*, 64, 501, 1943.
7. Lukeš, Zobáčová and Plešek, *Croat. Chem. Acta*, 29, 201, 1957; *Chem. Abstr.*, 53, 17898, 1959.
8. Panfold, Ramage and Simonsen, *J. Proc. R. Soc. N. S. W.*, 68, 80, 1935.

	6-Methyl coumarin	3-Methylcrotonic acid
Other names	Cocodescol 6-Methylbenzopyrone Pralina Toncair Toncarine (Tonkarin)	Senecioic acid 3,3-Dimethylacrylic acid β,β-Dimethylacrylic acid
Empirical formula	$C_{10}H_8O_2$	$C_5H_8O_2$
Structure	H_3C– (6-methyl coumarin ring structure)	$(H_3C)_2C{=}CH{-}COOH$
Physical/chemical characteristics[5]		
Appearance	White, crystalline solid	Prisms (from water)
Molecular weight	160.17	100.12
Melting point	74.6–75 C	70°C (sublimes)
Boiling point	303°C (305 C) at 725 mm Hg	199°C 114°C at 40 mm Hg
Congealing point	> 73.5°C	
Solubility	Very slightly soluble in hot water; insoluble in petroleum ether; soluble in oils	Soluble in most common organic solvents
Organoleptic characteristics	Somewhat dry, herbaceous (tonka-like) odor; also characterized as a delicate fig or date sweetness; almost bitter taste above 50 ppm, turning sweet and vanilla-like at lower levels	
Synthesis	By heating 6-methyl coumarin-3-carboxylic acid to 300–340 C; by condensation of *p*-cresoldisulfonic acid with fumaric acid in the presence of H_2SO_4; by condensation of *p*-homosalicylic aldehyde with malonic acid in the presence of aniline, followed by heating to form the lactone; from salicylaldehyde with propionic anhydride and sodium propionate	By the action of alkalis or bases on the ester of α-bromoisovaleric acid; by condensation of acetone with malonic acid or bromoacetic ester;[6] by oxidation of mesityl oxide with sodium hypochlorite
Natural occurrence	Not reported found in nature	Reported found in the rhizomes of *Senecio kaempferi* Sieb.,[7] the corresponding ester occurs naturally in wasteswort
Reported uses[3]	Non-alcoholic beverages 5.2 ppm Ice cream, ices, etc. 4.8 ppm Candy 21 ppm Baked goods 24 ppm Gelatins and puddings 39 ppm Chewing gum 0.80–15 ppm	Non-alcoholic beverages 5.0 ppm Ice cream, ices, etc. 5.0 ppm Candy 5.0 ppm Gelatins and puddings 5.0 ppm
Regulatory status	Status not fully defined by FDA; FEMA No. 2699	FEMA No. 3187

REFERENCES

For References 1–5, see end of Part III.

6. Kon and Linstead, *J. Chem. Soc.*, 127, 624, 1925.
7. Asahina, *Arch. Pharm.*, 251, 355, 1913.

	2-Methyl-1, 3-cyclohexadiene	1-Methyl-2,3-cyclohexadione
Other names	Dihydrotoluene (1,3) Dihydrotoluene (Δ1,3)	3-Methyl-1,2-cyclohexanedione 2-Methyl-3,4-cyclohexanedione
Empirical formula	C_7H_{10}	$C_7H_{10}O_2$
Structure	(structure: cyclohexadiene ring with CH_3)	(structure: cyclohexanedione ring with CH_3, O, O)
Physical/chemical characteristics[5]		
Appearance	Liquid	
Molecular weight	94.16	126.15
Melting point	107–108°C (110°C)	61–63°C
Boiling point		69–72°C at 1 mm Hg
Specific gravity	0.8354 (0.835) at 20°C	
Refractive index	1.4662 at 18°C	
Solubility	Insoluble in water; very soluble in alcohol; soluble in ether	
Synthesis		By the method described by Wallach;[9] from the ethyl ester of γ-propionylbutyric acid and sodium ethylate[10]
Natural occurrence	By dehalogenation of 1,2-dibromo-1-methyl cyclohexane with sodium ethoxide;[6,7,8] purified by distillation over metallic sodium.	Reported found as a volatile constituent in coffee[11]
Reported uses[3]		Non-alcoholic beverages 0.6 ppm Candy 9.0 ppm Baked goods 6.0 ppm Gelatins and puddings 0.2 ppm
Regulatory status	FDA 121.1164	FEMA No. 3305

REFERENCES

For References 1–5, see end of Part III.

6. **Mousseron and Winternitz,** *Bull. Soc. Chim.*, 12, 70, 1945.
7. **Mousseron and Winternitz,** *Compt. Rend.*, 219, 210, 1944.
8. *Beilstein's Encyclopedia of Organic Chemistry*, Vol. 5, Springer Verlag, p. 115.
9. **Wallach,** *Liebigs Ann. Chem.*, 414, 314, 1916–1917; 437, 180, 1924.
10. **Lapin and Horeau,** *Chimia*, 15, 551, 1961.
11. **Gianturco, Gianmarino and Friedel,** *Nature*, 210(5043), 1358, 1966.

	3-Methyl-2-cyclohexen-1-one	3-Methyl-1-cyclopentadecanone
Other names	3-Methyl-Δ^2-cyclohexenone	*d, l*-Muscone Methyloxaltone
Empirical formula	$C_7H_{10}O$	$C_{16}H_{30}O$
Structure		
Physical/chemical characteristics[5]		
Molecular weight	110.05	238.42
Melting point	Approximately -21°C	
Boiling point	94.5–95.5°C at 22 mm Hg 78–79°C at 12 mm Hg 53–53.5°C at 2 mm Hg	130°C at 0.5 mm Hg
Specific gravity	0.9693 at 20°C	0.9221 at 17°C
Refractive index	1.4947 at 20°C	1.4802 at 17°C
Optical rotation		-13°, O' at 17°C
Solubility		Very slightly soluble in water; miscible with alcohol
Organoleptic characteristics	Miscible with water in all proportions	Soft, sweet, tenacious musky odor
Synthesis	By acid hydrolysis and decarboxylation of the corresponding 4-carbetoxy derivative;[6] by oxidation of 1-methyl-cyclohex-1-ene with chromium trioxide in acetic acid;[7] by cyclization of 3-carbetoxy-6-chlorohept-5-en-2-one with sulfuric acid[8]	From the condensation of dodecamethylene-α,ω-dimethylketone hexadecane);[10,11] for preparation of *d*- and *l*-forms, see References 12 and 13
Natural occurrence	Reported found in the oil of *Mentha pulegium*[9]	In natural musk
Reported uses[3]	Non-alcoholic beverages 0.02 ppm Ice cream ices, etc. 0.05 ppm Candy 0.08 ppm Baked goods 0.15 ppm Gelatins and puddings 0.05 ppm Milk, dairy products 0.02 ppm	Non-alcoholic beverages 0.03 ppm Alcoholic beverages 0.02 ppm Ice cream, ices, etc. 0.03 ppm Candy 0.05 ppm Baked goods 0.05 ppm Gelatins and puddings 0.02 ppm
Regulatory status	FEMA No. 3360	FEMA No. 3434

REFERENCES

For References 1–5, see end of Part III.

6. **Smith and Ronalt,** *J. Am. Chem. Soc.*, 65, 631, 1943.
7. **Whitmore and Pedlow,** *J. Am. Chem. Soc.*, 63, 758, 1941.
8. **Wichterle et al.,** *Chem. Abstr.*, 42, 8162, 1948.
9. **Naves,** *Helv. Chim. Acta*, 26, 162, 1943.
10. **Stoll et al.,** *Helv. Chim. Acta*, 30, 2019, 1947.
11. **Blonquist et al.,** *J. Am. Chem. Soc.*, 70, 34, 1948.
12. **Ställberg-Stenhagen,** *Ark. Kemi*, 3, 517, 1951.
13. **Felix et al.,** *Helv. Chim. Acta*, 54, 2896, 1971.

	Methylcyclopentenolone	1-Methyl-1-cyclopenten-3-one
Other names	Cyclotene 2-Hydroxy-3-methyl-2-cyclopenten-1-one Kentonarome 3-Methylcyclopentan-1,2-dione	3-Methyl-2-cyclopenten-1-one
Empirical formula	$C_6H_8O_2$	C_6H_8O
Structure	=O, =O, CH_3 or =O, OH, CH_3	O, CH_3
Physical/chemical characteristics[5]		
Appearance	Crystalline solid	
Molecular weight	112.13	96.12
Melting point	105–107°C (anhydrous)	
Boiling point		157–158°C
Specific gravity		0.9712 at 26°C
Refractive index		1.4714 at 26°C
Solubility	Readily soluble in boiling water, alcohol, acetone, and chloroform; slightly less soluble in ether	Very soluble in water
Organoleptic characteristics	Sweet flavor somewhat similar to licorice	
Synthesis	See references 6 and 7. Sublimes and distills at atmospheric pressure with decomposition	By dehydrohalogenation of 2-chloro-1-methyl-cyclopentan-3-one[10]
Natural occurrence	Reported formed during the dry distillation of wood; found also in the corresponding tar oil;[8] identified in fenugreek[9]	
Reported uses[3]	Non-alcoholic beverages 11 ppm Ice cream, ices, etc. 5.6 ppm Candy 18 ppm Baked goods 13 ppm Gelatins and puddings 14 ppm Chewing gum 8.0–15 ppm Syrups 10–30 ppm	Non-alcoholic beverages 2.0 ppm Baked goods 3.0 ppm Condiments 2.0 ppm Meat, sauces, soups 2.0 ppm Cereals 2.0 ppm
Regulatory status	FDA 121.1164; FEMA No. 2700	FEMA No. 3435

REFERENCES

For References 1–5, see end of Part III.

6. **Nazarov and Akbrem,** *Izv. Akad. Nauk SSSR Otd. Khim. Nauk*, p. 1383, 1956.
7. **Krimen and Norman,** U.S. Patent 2,856,962, 1958.
8. **Meyerfeld,** *Chem. Ztg.*, 36, 549, 1912.
9. **Ruys,** *Fr. Ses Parfums*, 47(9), 34, 1966.
10. **Godchot et al.,** *Compt. Rend.*, 156, 1780, 1913.

5H-5-Methyl-6,7-dihydrocyclopentapyrazine

Other names	
Empirical formula	$C_8H_{10}N_2$
Structure	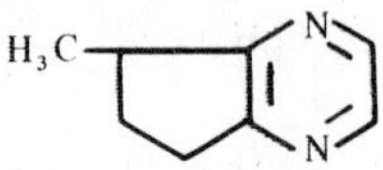
Physical/chemical characteristics	
Appearance	Colorless crystals
Assay	C: 68.82%; H: 8.25%; N: 22.93%
Molecular weight	134.18
Melting point	121–123°C at 1 atm
Boiling point	47°C at 0.7 mm Hg 78–80°C at 10 mm Hg
Mass spectra[7]	119 (100), 134 (48), 133 (25), 39 (21), 27 (18), 52 (16), 78 (12), 41 (12)
IR spectra[7]	3050, 2960, 2930, 2870, 1455, 1430, 1385, 1330, 1155, 1125, 1090, 1075, 1015, 870, 845
UV spectra[7]	λ_{max} (MeOH): 277, 282, ($E = 9.7 \times 10^{-3}$), 308 nm
Synthesis	From condensation of 2-hydroxy-3-methyl-2-cyclopentene-1-one and ethylene diamine.[7] By condensing a cyclopentenolone with alkylenediamines, or alternatively, an aliphatic α-diketone with 1,2-diamino cyclopentane; the resulting product is dehydrogenated in the presence of palladium over activated charcoal or copper chromite[8]
Natural occurrence	Reported to be present in peanuts and filberts[6]
Reported uses[3]	Non-alcoholic beverages 0.045 ppm Ice cream, ices, etc. 0.25 ppm Candy 1.0 ppm Chewing gum 0.15 ppm Protein foods 0.5 ppm
Regulatory status	FEMA No. 3306

REFERENCES

For References 1–5, see end of Part III.

6. Maga and Sizer, *Crit. Rev. Food Technol.*, 4(1), 39, 1973.
7. Pittet et al., *J. Agric. Food Chem.*, 22(2), 275, 1974.
8. Flament et al., *Helv. Chim. Acta*, 56(2), 610, 1973.

	Methyl dihydrojasmonate
Other names	2-Amylcyclopentanone acetic acid, methyl ester Methyl hydrojasmonate Methyl-92-amyl-3-oxocyclopentyl) Methyl-2(-pentyl-3-oxo-1-cyclopentyl) acetate
Empirical formula	$C_{13}H_{22}O_3$
Structure	
Physical/chemical characteristics[5]	
Appearance	Pale-straw-colored to yellowish oily liquid
Molecular weight	226.32
Boiling point	> 300°C 109–112°C at 0.2 mm Hg[6] 100°C at 0.03 mm Hg[7]
Specific gravity	0.9968 at 21°C
Refractive index	1.4583 at 20°C
Solubility	Very slightly soluble in water; soluble in alcohol and oils
Organoleptic characteristics	Powerful sweet-floral, jasmine-like, somewhat fruity odor[8]
Synthesis	By condensation of 2-pentyl-2-cylcopenten-1-one with ethyl malonate, followed by hydrolysis, decarboxylation, and methylation;[6] also[9]

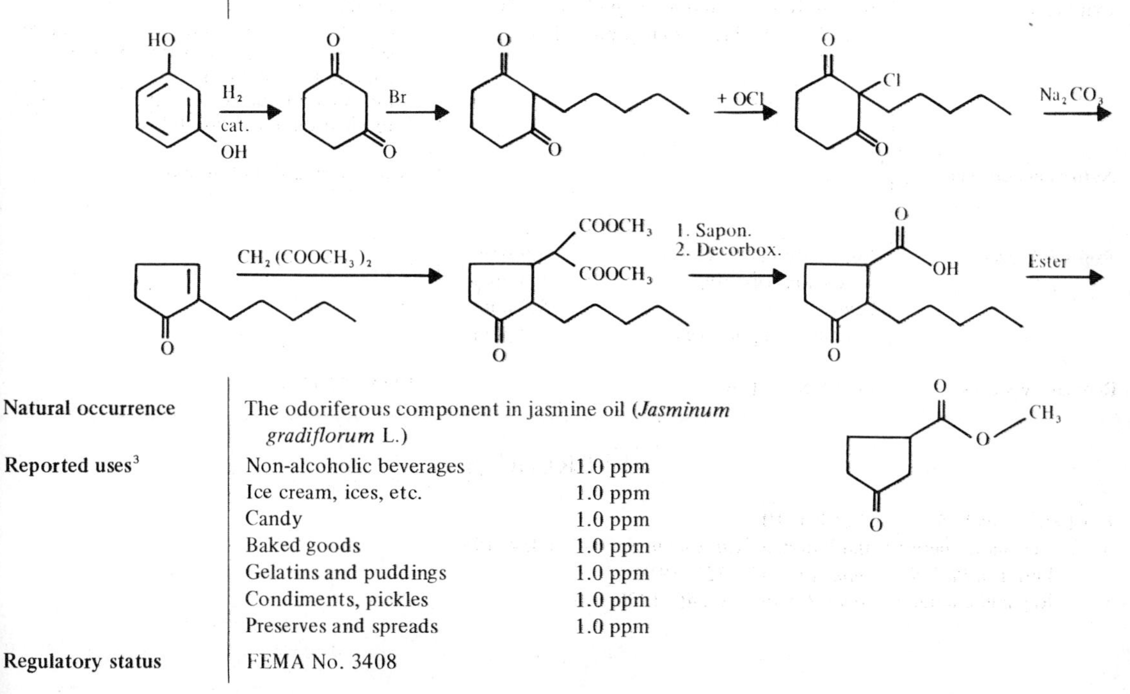

Natural occurrence	The odoriferous component in jasmine oil (*Jasminum gradiflorum* L.)	
Reported uses[3]	Non-alcoholic beverages	1.0 ppm
	Ice cream, ices, etc.	1.0 ppm
	Candy	1.0 ppm
	Baked goods	1.0 ppm
	Gelatins and puddings	1.0 ppm
	Condiments, pickles	1.0 ppm
	Preserves and spreads	1.0 ppm
Regulatory status	FEMA No. 3408	

REFERENCES

For References 1–5, see end of Part III.

6. **Demole et al.,** *Helv. Chim. Acta,* 45, 685, 1962.
7. **Hikino et al.,** *Yakugaku Zasshi,* 83, 219, 1963; 85, 179, 1965; *Chem. Abstr.,* 62, 12969, 1965.
8. **Arctander,** *Perfume and Flavor Chemicals,* Montclair, N.J., 1969, No. 2076.
9. German Patent 2,160,646.

	p-Methyl diphenyl	Methyl disulfide
Other names	4-Methyl biphenyl Phenyl-*p*-tolyl *p*-Phenyltoluene	Dimethyl disulfide *Note:* Do not confuse with dimethyl sulfide
Empirical formula	$C_{13}H_{12}$	$C_2H_6S_2$
Structure	(biphenyl structure with CH_3 in para position)	$H_3C{-}S{-}S{-}CH_3$
Physical/chemical characteristics[5]		
Appearance	Platelets (from methanol or ligroin)	Pale-yellow liquid
Molecular weight	168.24	94.19
Melting point	49–50°C	
Boiling point	267–268°C 134–136°C at 15 mm Hg	116–118°C (112°C)
Specific gravity	1.015 at 27°C	1.0647 at 20°C
Refractive index		1.5260 at 20°C
Solubility	Insoluble in water: soluble in alcohol and ether; very soluble in organic solvents	Very slightly soluble in water; soluble in alcohol and oils
Organoleptic characteristics		Diffuse intense onion odor; nonlachrymatory
Synthesis	By catalytic reduction of *p*-phenyl benzyl alcohol obtained from *p*-phenyl benzoic acid[6]	From magnesium methyl iodide and S_2Cl_2,[7] or from Na_2S_2 and sodium methyl sulfate,[8] also from methyl bromide and sodium thiosulfate, after which the resulting sodium methylthiosulfate is heated to yield dimethyl disulfide
Natural occurrence		Not reported found in nature
Reported uses[3]	Non-alcoholic beverages 5.0 ppm Ice cream, ices, etc. 5.0 ppm Candy 5.0 ppm Gelatins and puddings 5.0 ppm	
Regulatory status	FEMA No. 3186	FDA 121.1164

REFERENCES

For References 1–5, see end of Part III.

6. Newman, Eberwein and Wood, *J. Am. Chem. Soc.*, 81, 6454, 1959.
7. Ferrario, *Bull. Soc. Chim. Fr.*, 4(7), 524, 1910.
8. Ray and Gupta, *Z. Anorg. Chem.*, 187, 40, 1930.

	2-Methyl-3-furanthiol	5-Methyl furfural
Other names		5-Methyl-2-furaldehyde α-Methylfurfural
Empirical formula	C_5H_6OS	$C_6H_6O_2$
Structure	[structure: furan ring with SH at 3-position, CH_3 at 2-position]	[structure: H_3C–furan–$\overset{O}{\overset{\|\|}{C}}$–H]
Physical/chemical characteristics		
Appearance		Colorless liquid
Molecular weight	114.16	110.11
Boiling point		187°C; 79–81°C at 12 mm Hg
Specific gravity		1.1072 at 18°C
Refractive index		1.5263 at 20°C
Solubility		Soluble in water (3.3%); very soluble in alcohol; soluble in propylene glycol and oils
Organoleptic characteristics	Odor and taste of roasted meat	Sweet, spicy, warm odor; sweet caramel-like flavor
Synthesis	By a patented process[6]	See *Beilstein* B 17², 315; from sucrose with HCl, followed by treating with stannous chloride; from various methylpentoses by distillation with acid
Reported uses[3]	Baked goods 0.25 ppm Condiments 0.25 ppm Meat, meat sauces, soups 0.25 ppm	Non-alcoholic beverages 0.13 ppm Ice cream, ices, etc. 0.13 ppm Candy 0.03–0.13 ppm Baked goods 0.03 ppm
Regulatory status	FEMA No. 3188	Status not fully defined by FDA; FEMA No. 2702

REFERENCES

For References 1–5, see end of Part III.

6. Evers, International Flavors and Fragrances, Inc., German Patent 2,003,525, 1970.

	Methyl furfuryl disulfide		2-Methyl-3 or 5 or 6-furfuryl thiopyrazine	
Other names			1. 2-Furfurylthio-3-methylpyrazine 2. 2-Furfurylthio-5-methylpyrazine 3. 2-Furfurlythio-6-methylpyrazine	
Empirical formula	$C_6H_8OS_2$		$C_{10}H_{10}N_2OS$	
Structure	$CH_3-S-S-CH_2-$(2-furyl)		1. (pyrazine with S—R at 2, CH_3 at 3) 2. (pyrazine with S—R at 2, H_3C at 5) 3. (pyrazine with S—R at 2, H_3C at 6) R = $-CH_2-$(2-furyl)	
Physical/chemical characteristics				
Assay			C: 58.23%; H: 4.89%; N: 13.58%; 0: 7.76%; S: 15.54%	
Molecular weight	128.19		206.27 (calculated = 206.051385)	
Boiling point			153–156°C at 10 mm Hg (mixture)[6]	
Refractive index			n_D^{20} = 1.5970 n_4^{20} = 1.2164	
Organoleptic characteristics			Roasted, coffee-like flavor	
Synthesis			A mixture of 2-methyl 3-, 5-, and 6-chloropyrazine is prepared by chlorination of 2-methylpyrazine;[7] sodium furfuryl mercaptide is added to this mixture to give the three isomers[6]	
Natural occurrence	In bread (unpublished communication)			
Reported uses[3]	Non-alcoholic beverages	1.0 ppm	Non-alcoholic beverages	1.0 ppm
	Ice cream, ices, etc.	1.0 ppm	Ice cream, ices, etc.	1.0 ppm
	Candy	2.0 ppm	Baked goods	1.0 ppm
	Baked goods	2.0 ppm	Condiments	1.0 ppm
	Gelatins and puddings	1.0 ppm	Sauces	1.0 ppm
	Meat, meat sauces, soups	1.0 ppm		
Regulatory status	FEMA No. 3362		FEMA No. 3189	

REFERENCES

For References 1–5, see end of Part III.

6. **Winter et al.**, U.S. Patent 3,702,253, 1972.
7. **Hirschberg and Spoerri**, *J. Org. Chem.*, 26, 2356, 1961.

	Methyl furoate		3-(5-Methyl-2-furyl) butanal	
Other names	Furan-α-carboxylic acid, methyl ester Methyl-2-furoate Methyl pyromucate		3-(5-Methyl-2-furyl) butyraldehyde	
Empirical formula	$C_6H_6O_3$		$C_9H_{12}O_2$	
Structure	(furan ring)–C(=O)–O–CH_3		H_3C–(furan ring)–CH(CH_3)–CH_2–CHO	
Physical/chemical characteristics[5]				
Appearance	Colorless liquid; tends to yellow on exposure to light			
Molecular weight	126.11		152.19	
Boiling point	181–182°C; 81–82°C at 20 mm Hg			
Specific gravity	1.1786 at 21.4°C			
Refractive index	1.4860 at 20°C			
Solubility	Insoluble in water; soluble in alcohol and most organic solvents			
Organoleptic characteristics	Pleasant, fruity odor similar to mushroom, fungus or tobacco; sweet, tart, fruity taste, quite heavy			
Synthesis	From 2-furoic acid in methanol solution in the presence of HCl as catalyst[6] (oxidation or Carmizzaro reaction)			
Natural occurrence	Not reported found in nature		Reported not found in nature	
Reported uses[3]	Non-alcoholic beverages	0.61 ppm	Non-alcoholic beverages	0.1 ppm
	Ice cream, ices, etc.	0.06–1.3 ppm	Ice cream, ices, etc.	0.4 ppm
	Candy	0.66 ppm	Candy	1.2 ppm
	Baked goods	1.0–1.3 ppm	Chewing gum	2.5 ppm
	Condiments	0.02 ppm	Protein foods	0.1 ppm
Regulatory status	Status not fully defined by FDA; FEMA No. 2703		FEMA No. 3307	

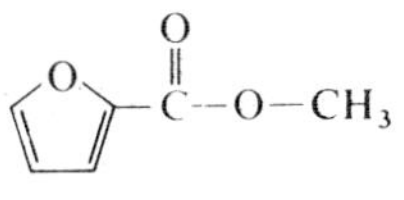

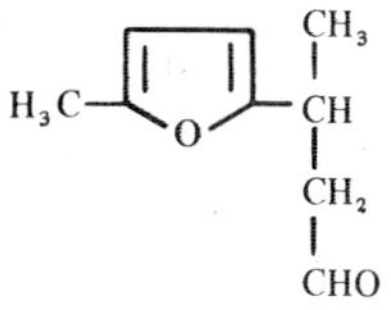

REFERENCES

For References 1–5, see end of Part III.
6. Gennari, *Gazz. Chim. Ital.*, 24(1), 249, 1894.

	6-Methyl-3,5-heptadien-2-one	Methyl heptanoate
Other names	2-Methyl-hepta-2,4-dien-6-one Methyl heptadienone 1-Acetyl-4-methyl-1,3-pentadiene	Methyl heptoate Methyl heptylate
Empirical formula	$C_8H_{12}O$	$C_8H_{16}O_2$
Structure	$CH_3-C(=O)-CH=CH-CH=C(CH_3)-CH_3$	$CH_3-(CH_2)_5-C(=O)-O-CH_3$
Physical/chemical characteristics[5]		
Appearance	Colorless to pale yellow oily liquid	Colorless liquid
Molecular weight	124.18	144.21
Melting point		−55.8°C
Boiling point	190°C 83.5°C at 9 mm Hg	173.8°C
Specific gravity	0.8980 at 20°C	0.87115 at 20°C
Refractive index	1.5306 at 20°C	1.41334 at 20°C
Solubility	Almost insoluble in water; soluble in alcohol and oils	Soluble in most common organic solvents
Organoleptic characteristics	Cinnamon-like odor with a coconut undertone	Strong, almost fruity, orris-like odor; currant-like flavor
Synthesis	By reacting 2-methylbuten-2-al with acetone in the presence of sodium ethylate or sodium hydroxide;[6] by pyrolysis of tertiary acetylenic carbinyl acetoacetates in the presence of an acid catalyst[7]	By treating heptanoic acid with methyl alcohol in the presence of HCl or H_2SO_4[9-11]
Natural occurrence	Reported as occurring in the oil of lavandin[8]	Not reported found in nature
Reported uses[3]	Non-alcoholic beverages 0.05 ppm Ice cream, ices, etc. 0.10 ppm Candy 0.15 ppm Baked goods 0.30 ppm Gelatins and puddings 0.05 ppm Milk, dairy products 0.05 ppm	Non-alcoholic beverages 0.80 ppm Ice cream, ices, etc. 0.83 ppm Candy 0.33 ppm Baked goods 0.50–0.60 ppm
Regulatory status	FEMA No. 3363	FDA 121.1164; FEMA No. 2705

REFERENCES

For References 1–5, see Introduction to Part II.

6. **Fischer and Löwenberg,** *Annalen.*, 494, 279, 1932.
7. **Lacey,** British Patent 741,047; *Chem. Abstr.*, 50, 16839, 1956.
8. **Stadler.** *Helv. Chim. Acta*, 43, 1601, 1960.
9. **Gartenmeister,** *Justus Liebigs Ann. Chem.*, 233, 281, 1886.
10. **Bilterys and Gisseleire,** *Bull. Soc. Chim. Belg.*, 44, 572, 1935.
11. **Vogel,** *J. Chem. Soc.* (London), p. 638, 1948.

	2-Methylheptanoic acid	6-Methyl-5-hepten-2-one
Other names	Hexane-2-carboxylic acid Isocaprylic acid Isooctanoic acid Methylamylacetic acid 2-Methyloenanthic acid	2-Methyl heptenone 2-Methyl-2-hepten-6-one Methyl hexenyl ketone
Empirical formula	$C_8H_{16}O_2$	$C_8H_{14}O$
Structure	$CH_3-(CH_2)_4-CH(CH_3)-C(=O)-OH$	$(CH_3)_2-C=CH-CH_2-CH_2-C(=O)-CH_3$ $CH_2=C(CH_3)-(CH_2)_3-C(=O)-CH_3$*
Physical/chemical characteristics[5]		
Appearance	Colorless, oily liquid; solidifies in the cold to colorless or white scaly crystals	Colorless to slightly yellow liquid
Assay		97% min[2]
Molecular weight	144.21	126.19
Melting point	15°C	−67.1°C
Boiling point	*dl*-, 121–122°C at 13 mm Hg; *d*-, 94–96°C at 1 mm Hg; 220°C	173.1°C (174°C)
Specific gravity	*l*-, 0.902 at 25°C	0.846–0.854 at 25°/25°C;[2] 0.8602 at 20°C
Refractive index	At 25°C: *l*-, 1.4233; *d*-, 1.4235	1.4380–1.4420[2] (1.4445) at 20°C
Optical rotation	*d*-, +8°9′; *l*-, −15°6′	
Solubility	Very slightly soluble in water; soluble in alcohol and oils	1:2 in 70% alcohol;[2] insoluble in water; soluble in most common organic solvents. Yields azeotropes with pentachloroethane (93%), *o*-cresol (15%), cineol (52%), and others
Organoleptic characteristics	Fatty, sour odor (rancid-like); sour, fruity, and nut-like flavor	Strong, fatty, green, citrus-like odor, and bittersweet taste reminiscent of pear
Synthesis	By decarboxylation of methylamyl malonic acid. Two optically active isomers and an optically inactive form can be prepared	From oil of lemongrass or from citral by refluxing for 12 hours in aqueous solution containing K_2CO_3, and subsequent distillation and vacuum fractionation; from acetoacetic ester and methylbuten-3-ol-2 with aluminum alcoholate in Carroll's reaction followed by pyrolysis of the ester.
Natural occurrence	Not reported found in nature	Originally identified in lemongrass.[6,7] It has been reported found in the essential oils of palmarosa, lemon, citronella, vervain, geranium, *Ocimum canum*, *Artemisia scoparia*, *Urtica dioica*, and others.
Reported uses[3]	Non-alcoholic beverages 1.0 ppm Ice cream, ices, etc. 10 ppm Candy 10 ppm Baked goods 10 ppm	Non-alcoholic beverages 1.1 ppm Ice cream, ices, etc. 1.1 ppm Candy 1.1 ppm Baked goods 1.3 ppm Gelatins and puddings 1.3 ppm
Regulatory status	FDA 121.1164; FEMA No. 2706	FDA 121.1164; FEMA No. 2707

* Some products are 80% 6-methyl-5-hepten-2-one with 20% 2-methyl-1-hepten-6-one.

REFERENCES

For References 1–5, see end of Part III.
6. **Barbier and Bouveault,** *C. R. Séances Acad. Sci. Paris,* 118, 983, 1894.
7. **Tiemann, et al.,** *Ber. Dtsch. Chem. Ges.,* 28, 2126, 1895.

	5-Methyl-2,3-hexanedione	Methyl hexanoate
Other names	2-Methyl-4,5-hexanedione Acetyl isovaleryl Isobutyl methyl diketone Isobutyl methyl glyoxal Acetyl isopentanoyl	Methyl caproate Methyl capronate Methyl hexylate
Empirical formula	$C_7H_{12}O_2$	$C_4H_{14}O_2$
Structure	$CH_3-C(=O)-C(=O)-CH_2-CH(CH_3)-CH_3$	$CH_3-(CH_2)_4-C(=O)-O-CH_3$
Physical/chemical characteristics[5]		
Appearance	Yellowish liquid	Liquid
Molecular weight	128.17	130.18
Melting point		−71°C
Boiling point	138°C	151°C; 56°C at 20 mm Hg
Specific gravity	0.908 at 22°C	0.8850 at 20°C
Refractive index	1.4119 at 20°C	1.4049 at 20°C
Solubility	Slightly soluble in water; miscible with alcohol, propylene glycol, and glycerine	Insoluble in water; soluble in alcohol
Organoleptic characteristics	Powerful oily-buttery odor; sweet taste in aqueous solution	Ether-like odor reminiscent of pineapple
Synthesis	By bromination of mesityl oxide, followed by treatment of the resulting dibromide with sulfuric acid[6]	By reacting methyl alcohol with hexanoic acid at 130–140°C in the presence of concentrated H_2SO_4 and distilling the ester from the reaction mixture[7,8]
Natural occurrence		Reported found in pineapple and *Fragaria vesca*
Reported uses[3]	Non-alcoholic beverages 20.0 ppm Candy 20.0 ppm	Non-alcoholic beverages 4.1 ppm Ice cream, ices, etc. 8.5 ppm Candy 5.3 ppm Baked goods 20 ppm
Regulatory status	FEMA No. 3190	FDA 121.1164; FEMA No. 2708

REFERENCES

For References 1–5, see end of Part III.

6. **Dufraisse and Moureu,** *Compt. Rend.,* 184, 99, 1927; *Bull. Soc. Chim. Fr.,* 41(4), 855, 1375, 1927.
7. Commercial Solvents Corp., U.S. Patent 2,029,694, 1934.
8. **Vogel,** *J. Chem. Soc.,* p. 630, 1948.

	2-Methylhexanoic acid	Methyl 2-hexenoate
Other names	2-Methylcaproic acid 2-Butylpropionic acid Butyl methylacetic acid Hexane-2-carboxylic acid	Methyl-α,β-hexenoate Methyl-β-propylacrylate
Empirical formula	$C_7H_{14}O_2$	$C_7H_{12}O_2$
Structure	$CH_3-CH_2-CH_2-CH_2-CH(CH_3)-COOH$	$CH_3-(CH_2)_2-CH=CH-C(=O)-O-CH_3$
Physical/chemical characteristics[5]		
Appearance	Colorless oil	Colorless, mobile liquid
Molecular weight	130.19	128.17
Boiling point	105°C at 5 mm Hg	56–58°C at 13 mm Hg; 53–56°C at 7 mm Hg; 169°C
Specific gravity	0.909 at 25°C	
Refractive index	1.4189 at 25°C	
Optical rotation	*d*-, +19.6 at 22°C in ether *l*-, −15.25 at 25°C in ether	
Solubility	Either optically active form is miscible with water, ether, alcohol, and chloroform, as well as with other solvents	Very slightly soluble in water; soluble in alcohol and oils
Organoleptic characteristics		Green, musty, earthy, sweet, fruity odor; very powerful, sharp, green-fruity flavor
Synthesis	By catalytic hydrogenation of hexadien-2-carboxylic acid[6]	By boiling the corresponding acid with methanol in the presence of concentrated H_2SO_4 in chloroform and subsequent elimination of water from the vapors using $MgSO_4$;[7] also from 3-chloropropionic acid methyl ester and anhydrous sodium acetate in acetic acid[8]
Natural occurrence		Not reported found in nature
Reported uses[3]	Candy 8.0 ppm Baked goods 2.0 ppm Gelatins and puddings 2.0 ppm	Non-alcoholic beverages 0.03–0.12 ppm Candy 0.03 ppm
Regulatory status	FEMA No. 3191	FDA 121.1164; FEMA No. 2709

REFERENCES

For References 1–5, see end of Part III.

6. **Reichstein and Trivelli,** *Helv. Chim. Acta,* 15, 528, 1932.
7. **Baker et al.,** *J. Org. Chem.,* 12, 144, 1947.
8. **Guest,** *J. Am. Chem. Soc.,* 69, 301, 1947.

	Methyl-3-hexenoate		5-Methyl-3-hexen-2-one	
Other names	Methyl hydrosorbate		Isobutylidene acetone	
Empirical formula	$C_7H_{12}O_2$		$C_7H_{12}O$	
Structure	$CH_3-CH_2-CH=CH-CH_2-C(=O)-O-CH_3$		$CH_3-C(=O)-CH=CH-CH(CH_3)-CH_3$	
Physical/chemical characteristics[5]				
Molecular weight	128.17		112.17	
Boiling point			77.5°C at 50 mm Hg	
Specific gravity			0.8407 at 22°C	
Refractive index			1.4395 at 22°C	
Synthesis	By selective hydrogenation of methyl sorbate;[6] from 3-hexenoic acid and methanol in the presence of BF_3[7]		Acetone + isobutyraldehyde (CHO) $\xrightarrow{B^-}$	
Natural occurrence	Reported found as a volatile component in smooth Cayenne pineapple[8]		Reported found in roasted filberts[6]	
Reported uses[3]	Non-alcoholic beverages	2.0 ppm	Non-alcoholic beverages	2.0 ppm
	Ice cream, ices, etc.	2.0 ppm	Ice cream, ices, etc.	2.0 ppm
	Candy	4.0 ppm	Candy	4.0 ppm
	Baked goods	4.0 ppm	Baked goods	4.0 ppm
	Gelatins and puddings	2.0 ppm	Gelatins and puddings	4.0 ppm
	Preserves and spreads	2.0 ppm	Milk, dairy products	1.0 ppm
			Cereals	2.0 ppm
Regulatory status	FEMA No. 3364		FEMA No. 3409	

REFERENCES

For References 1–5, see end of Part III.

6. **Frankel, Selke and Glass,** *J. Am. Chem. Soc.*, 90(9), 2446, 1968.
7. **Fell, Bendel, Lauscher and Huebner,** *J. Chromatogr.*, 24(1), 161, 1966.
8. **Flath and Forrey,** *J. Agric. Food Chem.*, 18(2), 306, 1970.
9. Volatile Compounds in Food, T.N.O. Report No. 4030, 3rd ed., Central Instituut voor Voedingsonderzoek, Zeist, Netherlands.

	5-Methyl-5-hexen-2-one	Methyl-α-ionone
Other names	Methallyl acetone 4-Acetyl-2-methyl-1-butene	α-Cetone α-Cyclocitrylidene butanone α-Cyclocitrylidene methyl ethyl ketone α-*n*-Methylionone Raldeine 5-(2,6,6-Trimethyl-2-cyclohexen-1-yl)-4-penten-3-one
Empirical formula	$C_7H_{12}O$	$C_{14}H_{22}O$
Structure	$CH_3-C(=O)-CH_2-CH_2-C(CH_3)=CH_2$	
Physical/chemical characteristics[5]		
Appearance		Yellowish, oily liquid
Molecular weight	112.17	206.30
Boiling point	148–149°C	97°C at 2.6 mm Hg; 238°C
Specific gravity	0.8475 at 25°C	0.9210 at 25°C
Refractive index	1.4279 at 25°C	1.4938 at 25°C
Organoleptic characteristics		Odor similar to α-ionone but lighter and more rounded; pleasant, fruity, and sweet nuance
Synthesis	From methallyl alcohol, the corresponding methallyl acetoacetone is prepared, dissolved in diphenyl ether, and finally heated to 200–215°C[6]	By condensation of citral with methyl ethyl ketone (see methyl γ-ionone or α-isomethylionone)
Natural occurrence		Not reported found in nature
Reported uses[3]	Non-alcoholic beverages 0.12 ppm Ice cream, ices, etc. 0.25 ppm Candy 0.40 ppm Baked goods 0.75 ppm Gelatins and puddings 0.25 ppm Milk, dairy products 0.12 ppm	Non-alcoholic beverages 1.7 ppm Ice cream, ices, etc. 2.4 ppm Candy 6.6 ppm Baked goods 6.5 ppm Chewing gum 0.60 ppm Jellies 0.21 ppm
Regulatory status	FEMA No. 3365	FDA 121.1164; FEMA No. 2711

REFERENCES

For References 1–5, see end of Part III.

6. Kimel and Cope, *J. Am. Chem. Soc.*, 65, 1922, 1943.

	Methyl-β-ionone	*iso*-Methyl-β-ionone
Other names	β-Cetone β-Cyclocitrylidene butanone β-*n*-Methylionone Raldeine 5-(2,6,6-Trimethyl-1-cyclohexen-1-yl)-4-penten-3-one	Isomethyl-β-ione β-Isomethylionone "*delta*"-Methylionone 4-(2,6,6-Trimethyl-3-cyclohexen-1-yl)-3-methyl-3-buten-2-one
Empirical formula	$C_{14}H_{22}O$	$C_{14}H_{22}O$
Structure	H_3C, CH_3, O, CH_3, CH_3	H_3C, CH_3, O, CH_3, CH_3, CH_3
Physical/chemical characteristics[5]		
Appearance	Yellowish liquid	Almost colorless to slightly yellow oil
Molecular weight	206.30	206.33
Boiling point	102°C at 2.6 mm Hg; 242°C	232°C; 76°C at 0.3 mm Hg
Specific gravity	0.9338 at 25°C	
Refractive index	1.5140 at 25°C	1.4969 at 25°C
Solubility		Insoluble in water; soluble in alcohol
Organoleptic characteristics	Odor similar to β-ionone with a Bulgarian leather note	Light, warm, floral odor; berry-like flavor
Synthesis	By condensation of citral with methyl ethyl ketone (see α-isomethylionone). It is usually present in low levels (6–7%) in commercial products.	See Reference 6
Natural occurrence	Not reported in nature	Not reported found in nature
Reported uses[3]	Non-alcoholic beverages 2.0 ppm Ice cream, ices, etc. 2.2 ppm Candy 7.5 ppm Baked goods 5.9 ppm	Non-alcoholic beverages 0.61 ppm Ice cream, ices, etc. 0.89 ppm Candy 3.2 ppm Baked goods 2.8 ppm
Regulatory status	FDA 121.1164; FEMA No. 2712	FDA 121.1164; FEMA No. 2713

REFERENCES

For References 1–5, see end of Part III.

6. **Bedoukian,** *Perfumery and Flavoring Synthetics,* American Elsevier, 1967.

	Methyl isobutyrate	2-Methyl-3-(*p*-isopropylphenyl)-propionaldehyde
Other names	Methyl dimethylacetate Methyl-2-methylpropanoate	Cyclamen aldehyde *p*-Isopropyl-α-methylhydrocinnamaldehyde *p*-Isopropyl-α-methylphenylpropylaldehyde α-Methyl-*p*-isopropylhydrocinnamaldehyde
Empirical formula	$C_5H_{10}O_2$	$C_{13}H_{18}O$
Structure	$CH_3-O-C(=O)-CH(CH_3)_2$	$CH_2-CH(CH_3)-CHO$ attached to benzene ring, para $CH(CH_3)-CH_3$
Physical/chemical characteristics[5]		
Appearance	Colorless, mobile liquid	Colorless to pale-yellow liquid
Assay		90% min[2] (50–96%)
Molecular weight	102.13	190.28
Melting point	−85°C	
Boiling point	92–93°C	270°C
Specific gravity	0.891 at 20°C	0.946–0.952 at 25°/25°C;[2] 0.950 at 20°C
Refractive index	1.3840 at 20°C	1.5030–1.5080[2] (1.500) at 20°C
Acid value		Not more than 5[2]
Solubility	Slightly soluble in water; miscible with most common organic solvents	1:1 in 80% alcohol;[2] soluble in most common organic solvents
Organoleptic characteristics	Fruity (apple-pineapple) odor; sweet flavor reminiscent of apricot	Strong, flowery odor
Synthesis	By direct esterification of methanol with isobutyric acid	By condensation of cuminic aldehyde and propionaldehyde followed by hydrogenation in the presence of a catalyst[8]
Natural occurrence	Reported found in Russian champagnes[6] and among the volatile components of strawberry juice[7]	Not reported found in nature
Reported uses[3]	Non-alcoholic beverages 22 ppm Ice cream, ices, etc. 38 ppm Candy 48–200 ppm Baked goods 48–200 ppm	Non-alcoholic beverages 0.30 ppm Ice cream, ices, etc. 0.45 ppm Candy 0.99 ppm Baked goods 1.2 ppm
Regulatory status	FDA 121.1164; FEMA No. 2694	FDA 121.1164; FEMA No. 2743

REFERENCES

For References 1–5, see end of Part III.

6. **Rodopulo and Egorov,** *Chem. Abstr.*, 61, 11291g, 1964.
7. **Roy et al.,** *J. Food Sci.*, 28, 478, 1963.
8. **Mousseron et al.,** *Fr. Ses Parfums,* 3, 28, 1958.

	Methyl isovalerate
Other names	
Empirical formula	$C_6H_{12}O_2$
Structure	$(H_3C)_2CH-CH_2-C(=O)-O-CH_3$
Physical/chemical characteristics[5]	
Appearance	Colorless liquid
Molecular weight	116.16
Melting point	
Boiling point	116–117°C
Specific gravity	0.8807 at 20°C
Refractive index	1.3927 at 20°C
Solubility	Insoluble in water; soluble in most organic solvents
Organoleptic characteristics	Strong, pungent, apple-like odor; bitter flavor
Synthesis	By esterification of isovaleric acid with methyl alcohol at the boil in the presence of concentrated H_2SO_4[6]
Natural occurrence	Reported found in pineapple fruits[7,8] and in the juice of a few varieties of Florida oranges[9]
Reported uses[3]	Non-alcoholic beverages 9.3 ppm Ice cream, ices, etc. 26 ppm Candy 26 ppm Baked goods 30 ppm Chewing gum 35 ppm
Regulatory status	FDA 121.1164; FEMA No. 2753

REFERENCES

For References 1–5, see end of Part III.
6. **Vogel,** *J. Chem. Soc.,* p. 630, 1948.
7. **Haagen-Smit et al.,** *J. Am. Chem. Soc.,* 67, 1647, 1945.
8. **Connel,** *Aust. J. Chem.,* 1(17), 130, 1964.
9. **Wolford et al.,** *J. Food Sci.,* 28, 320, 1963.

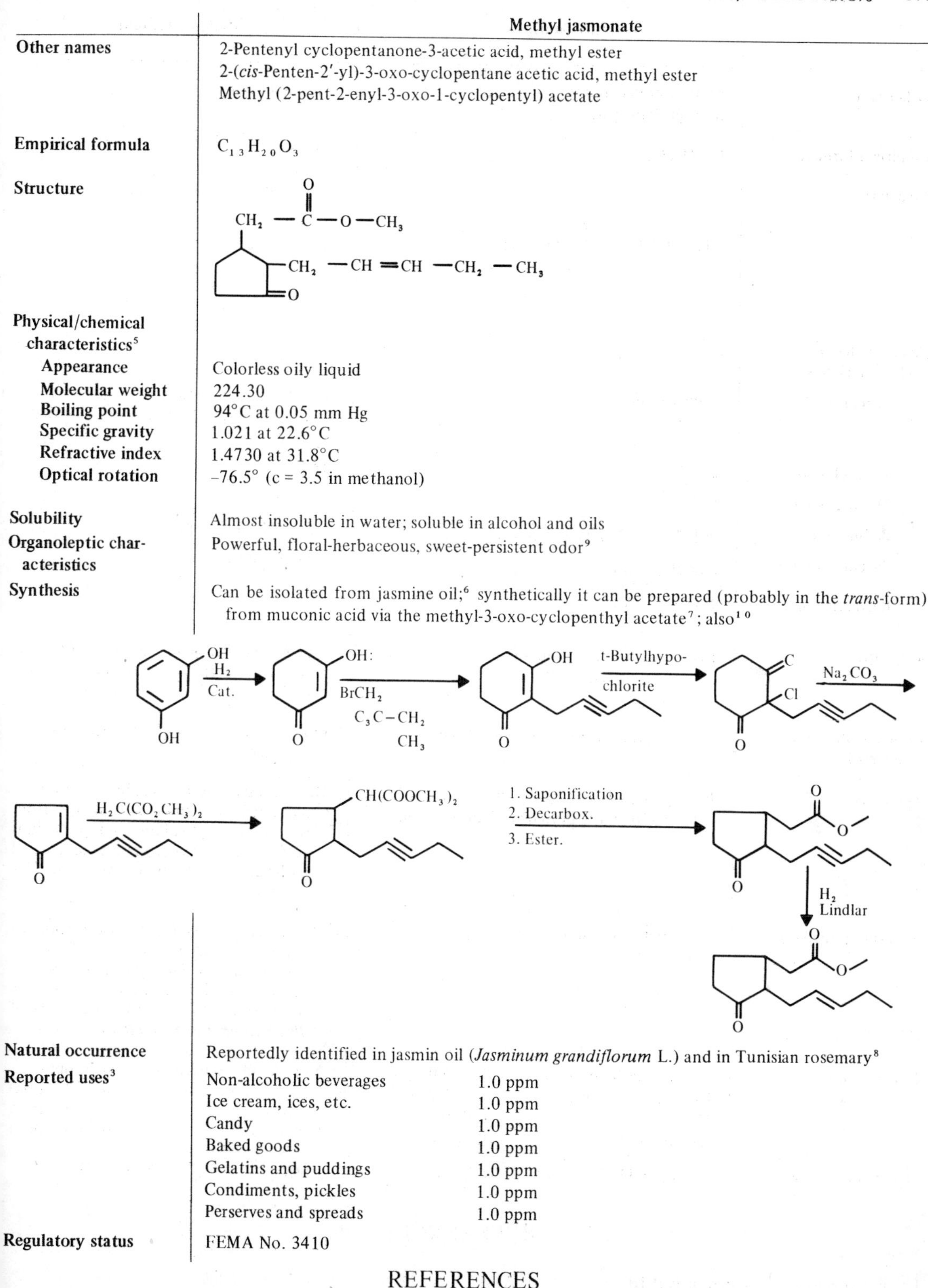

	Methyl jasmonate
Other names	2-Pentenyl cyclopentanone-3-acetic acid, methyl ester 2-(*cis*-Penten-2′-yl)-3-oxo-cyclopentane acetic acid, methyl ester Methyl (2-pent-2-enyl-3-oxo-1-cyclopentyl) acetate
Empirical formula	$C_{13}H_{20}O_3$
Structure	
Physical/chemical characteristics[5]	
Appearance	Colorless oily liquid
Molecular weight	224.30
Boiling point	94°C at 0.05 mm Hg
Specific gravity	1.021 at 22.6°C
Refractive index	1.4730 at 31.8°C
Optical rotation	–76.5° (c = 3.5 in methanol)
Solubility	Almost insoluble in water; soluble in alcohol and oils
Organoleptic characteristics	Powerful, floral-herbaceous, sweet-persistent odor[9]
Synthesis	Can be isolated from jasmine oil;[6] synthetically it can be prepared (probably in the *trans*-form) from muconic acid via the methyl-3-oxo-cyclopenthyl acetate[7]; also[10]
Natural occurrence	Reportedly identified in jasmin oil (*Jasminum grandiflorum* L.) and in Tunisian rosemary[8]
Reported uses[3]	Non-alcoholic beverages 1.0 ppm Ice cream, ices, etc. 1.0 ppm Candy 1.0 ppm Baked goods 1.0 ppm Gelatins and puddings 1.0 ppm Condiments, pickles 1.0 ppm Perserves and spreads 1.0 ppm
Regulatory status	FEMA No. 3410

REFERENCES

For References 1–5, see end of Part III.

6. **Demole et al.,** *Helv. Chim. Acta,* 45, 675, 1962.
7. **Demole et al.,** *Helv. Chim. Acta,* 45, 692, 1962.
8. **Crabalone,** *Compt. Rend.* Paris, C264, 2074; 1967; *Chem. Abstr.,* 67, 84752r, 1967.
9. **Arctander,** *Perfume and Flavor Chemicals,* Montclair, N.J., 1969, No. 2093.
0. German Patent 2,160,646.

	Methyl laurate	Methyl mercaptan
Other names	Methyl dodecanoate Methyl dodecylate	Methanethiol
Empirical formula	$C_{13}H_{26}O_2$	CH_4S
Structure	$CH_3-(CH_2)_{10}-\overset{\overset{\displaystyle O}{\|}}{C}-O-CH_3$	H_3C-SH
Physical/chemical characteristics[5]		
Appearance	Colorless liquid	Colorless gas; liquid under compression. The hydrate is a white crystalline material.
Molecular weight	214.35	48.11
Melting point	5°C	−121°C
Boiling point	261–262°C; 127°C at 8 mm Hg	6.1–6.2°C at 727 mm Hg
Specific gravity	0.8695 at 20°C	0.894 (liquid)
Refractive index	1.4320 at 20°C	
Solubility	Insoluble in water; soluble in most common organic solvents	Gas is slightly soluble in water (decomposes slowly); soluble in alcohol, ether, and most common organic solvents
Organoleptic characteristics	Fatty, floral odor reminiscent of wine	Objectionable odor of decomposing cabbage
Synthesis	By prolonged boiling of lauric acid with methanol in the presence of sulfuric acid[10]	By heating an aqueous solution of potassium methyl sulfate and KHS; from sodium methyl sulfate and potassium sulfhydrate; also from methanol and hydrogen sulfide in the presence of a catalyst.
Natural occurrence	Reported found in orris absolute[11]	Originally identified in the fresh roots of *Rafanus sativus* L.;[6] reported found in the red algae of *Polysiphonia fastigiata*.[7] Also reported in the leaves of *Coprosma foetidissima* Forst.[8] and the seeds of *Brassica napus* var. *oleifera*[9]
Reported uses[3]	Non-alcoholic beverages 0.50–5.0 ppm Ice cream, ices, etc. 0.50–5.0 ppm Candy 0.02–0.50 ppm Baked goods 1.0 ppm	Non-alcoholic beverages 0.56 ppm Ice cream, ices, etc. 0.13–1.0 ppm Candy 0.13–1.0 ppm Baked goods 0.15–1.0 ppm
Regulatory status	FDA 121.1164; FEMA No. 2715	FDA 121.1164; FEMA No. 2716

REFERENCES

For References 1–5, see end of Part III.

6. **Nakamura,** *Biochem. Z.*, 164, 31, 1925.
7. **Haas,** *Biochem. J.*, 29, 1297, 1935.
8. **Sutherland,** *Br. Abstr. A3*, p. 645, 1949.
9. **Andre,** *Chem. Abstr.*, 44, 2262, 1950.
10. **Vogel,** *J. Chem. Soc.*, p. 631, 1948.
11. **Naves,** *Helv. Chem. Acta*, 32, 2306, 1949.

	Methyl *o*-methoxybenzoate	1-Methyl-3-methoxy-4-isopropylbenzene
Other names	Dimethyl salicylate Methyl *o*-anisate	3-Methyl-*p*-cymene Thymol methylether
Empirical formula	$C_9H_{10}O_3$	$C_{11}H_{16}O$
Structure	O=C(–O–CH3) on benzene ring, para O–CH3	CH3; OCH3; H3C CH3
Physical/chemical characteristics[5]		
Appearance	Low-melting solid	
Molecular weight	166.18	164.24
Boiling point	48–49°C	215°C
	254–255°C	94–96° C at 15 mm Hg
Specific gravity		0.9388 at 14°C
Refractive index		1.5062 at 20°C
Solubility	Insoluble in water; soluble in alcohol	
Organoleptic characteristics	Herbaceous, anise-like odor; sweet taste reminiscent of melon	
Synthesis	By esterification of anisic acid with methanol; or from sodium anisate and dimethylsulfate in the presence of methanol	From thymol and dimethylsulfate in alkaline solution;[6] from sodium thymolate with dimethylsulfate in water[7]
Natural occurrence	Reported found in a mushroom variety, *Trametes graveolens*[15]	In the oils of French *Christinium maritinium*,[8] *Orthodon madai*,[9] *O. hirtum*,[10] *O. tenuicaule*,[11] *O. japonicum albiflorum*,[12] *O. pseudohirtum*,[12] *O. goshizanese*,[12] *Eupatorium fortunei*,[13] and *Monarda punctata*[14]
Reported uses[3]	Non-alcoholic beverages 2.7 ppm Ice cream, ices, etc. 3.0 ppm Candy 8.0 ppm Baked goods 6.2 ppm	Non-alcoholic beverages 2.0 ppm Candy 3.0 ppm Baked goods 3.0 ppm Condiments 2.0 ppm Meat, meat sauces, soups 2.0 ppm
Regulatory status	FDA 121.1164; FEMA No. 2679	FEMA No. 3436

REFERENCES

For References 1–5, see end of Part III.

6. **Kuroda et al.,** *Sci. Pap. Inst. Phys. Chem. Res.*, 18, 61, 1932.
7. **Caujolle et al.,** *Compt. Rend.*, 218, 571, 1944.
8. **Delepine,** *Compt. Rend.*, 150, 1061, 1910; *Bull. Soc. Chim. Fr.*, [4] 7, 468, 1910.
9. **Fujita,** *J. Chem. Soc. Jap.*, 65, 166, 1944; *Chem. Abstr.*, 41, 3585, 1947.
10. **Fujita,** *J. Chem. Soc. Jap.*, 59, 500, 1938; *Chem. Abstr.*, 61, 137, 1940.
11. **Fujita,** *J. Chem. Soc. Jap.*, 61, 782, 1940.
12. **Fujita,** *Report of the Osaka Industrial Research Institute* No. 306, 1955, 8.
13. **Liang et al.,** *Chem. Abstr.*, 54, 5019, 1960.
14. **Schimmel and Co.,** *Ber. Schimmel*, No. 46, 1922.
15. **Aye,** *Arch. Pharm.*, p. 246, 1931.

	2-Methyl-5-methoxy thiazole	Methyl-*n*-methylanthranilate
Other names		Dimethyl anthranilate Methyl 2-methylaminobenzoate Methyl *o*-methylaminobenzoate
Empirical formula	C_5H_7NOS	$C_9H_{11}O_2N$
Structure		
Physical/chemical characteristics[5]		
Appearance		Pale-yellow liquid with slight bluish fluorescence
Assay		98% min[2] (99%)
Molecular weight	141.18	165.19
Melting point		18.5–19.5°C
Boiling point		256°C; 130–131°C at 13 mm Hg
Congealing point		>12.0°C[2] (12.5–15.5°C)
Flash point		91°C
Specific gravity		1.126–1.132 at 25°/25°C;[2] 1.3545–1.1375 at 15°C
Refractive index		1.5785–1.5810[2] (1.5800–1.5810) at 20°C
Acid value		0.2 max
Solubility		1:3 in 80% alcohol;[2] 1:8–11 in 70% alcohol; insoluble in water; soluble in most common organic solvents
Organoleptic characteristics		Orange and mandarin-peel-like odor; musty, grape-like flavor; somewhat more berry-like than grape
Synthesis	Can be prepared from chloro- (or bromo-) methoxyacetaldehyde and thioacetamide[1,2]	By methylation of methyl anthranilate or esterification of *N*-methylanthranilic acid
Natural occurrence		Reported found in mandarin essential oil[6] and mandarin-leaves essential oil (50–76.5%);[7,8] also reported in the oil from bulbs of *Kaempferia ethelae* L.,[9] in orange petitgrain,[10] and in the oil from hyacinth flowers[11]
Reported uses[3]	Non-alcoholic beverages 2.0 ppm Baked goods 4.0 ppm Meat, soups 2.0 ppm	Non-alcoholic beverages 5.1 ppm Ice cream, ices, etc. 5.0 ppm Candy 18 ppm Baked goods 17 ppm Jellies 4.0 ppm
Regulatory status	FEMA No. 3192	FDA 121.1164; FEMA No. 2718

REFERENCES

For References 1–5, see end of Part III.

6. **Walbaum,** *Chem. Zentralbl.*, 2, 677, 1900.
7. **Charabot,** *Chem. Zentralbl.*, 2, 879, 1903.
8. **Schimmel and Co.,** *Chem. Zentralbl.*, 2, 1785, 1930.
9. **Goulding and Roberts,** *J. Chem. Soc.* (London), 107, 314, 1915.
10. **Glichitch and Naves,** *Chem. Zentralbl.*, 1, 604, 1930.
11. **Hoejenbons and Coppens,** *Chem. Zentralbl.*, 2, 2746, 1932.
12. **Pitten et al.,** German Patent 2,152,557; *Chem. Abstr.*, 55, 88488, 1972.

	Methyl-2-methylbutyrate	Methyl-4-(methylthio) butyrate
Other names	Methyl-2-methylbutanoate Methyl methylethylacetate	Mixture of methyl 9,12-octadecadienoate and methyl 9,12,15- octadecatrienoate Methyl γ-methyl mercapto butyrate
Empirical formula	$C_6H_{12}O_2$	$C_6H_{12}O_2S$
Structure	$CH_3-CH_2-CH(CH_3)-C(=O)-O-CH_3$	$CH_3-S-CH_2-CH_2-CH_2-C(=O)-O-CH_3$
Physical/chemical characteristics		
Appearance	Colorless liquid	
Molecular weight	116.16	148.21
Boiling point	115–116°C	
Specific gravity	0.8851 at 20°C	
Refractive index	1.3942 at 20°C	
Solubility	Very slightly soluble in water; soluble in alcohol	
Organoleptic characteristics	Pungent, fruity odor; sweet, fruity, apple-like taste at low levels	
Synthesis	A mixture of esters including methyl 2-methylbutyrate is obtained from isobutene, CO, and methanol under pressure, or from butene, CO, and an acid catalyst under pressure[6]	Acrylic acid (OH, O) → 1. SH; 2. CH_3OH
Natural occurrence	Reported found in fresh strawberry juice.[7] The laevorotatory and the racemic forms are known.	
Reported uses[3]	Non-alcoholic beverages 5.0 ppm Ice cream, ices, etc. 10 ppm Candy 10 ppm Baked goods 10 ppm	Baked goods 10.0 ppm Meat, meat sauces 10.0 ppm Sauces 10.0 ppm
Regulatory status	FDA 121.1164; FEMA No. 2719	FEMA No. 3412

REFERENCES

For References 1–5, see end of Part III.

6. **Eidus and Kaal,** *Zh. Obshch. Khim.,* 33, 3283, 1963; 35, 120, 1965.
7. **Willhalm, Pauly and Winter,** *Helv. Chim. Acta,* 49, 65, 1966.

	Methyl-3-methylthiopropionate
Other names	Methylmercaptomethylpropionate Methyl-β-methylmercaptopropionate Methyl-β-methylthiopropionate β-Methylthiopropionic acid, methyl ester
Empirical formula	$C_4H_{10}O_2S$
Structure	$CH_2-CH_2-C(=O)-O-CH_3$, with $S-CH_3$ on the first CH_2
Physical/chemical characteristics[5]	
Appearance	Liquid
Molecular weight	134.20
Boiling point	69°C (181°C) at 11 mm Hg
Refractive index	1.4600 at 32°C
Solubility	Very slightly soluble in water; soluble in alcohol
Organoleptic characteristics	Extremely powerful, penetrating, sweet odor; onion-like at high concentrations; sweet, pineapple flavor at high dilutions
Synthesis	Obtained in 84% yields by reacting $CH_2=CH-C(=O)-OCH_3$ with CH_3SH.[6] The corresponding acid also has been synthesized.[7]
Natural occurrence	Reported found in pineapple fruits[8]
Reported uses[3]	Non-alcoholic beverages 0.35 ppm Ice cream, ices, etc. 0.37 ppm Candy 0.74 ppm Baked goods 1.0 ppm Syrups 0.05 ppm
Regulatory status	FDA 121.1164; FEMA No. 2720

REFERENCES

For References 1–5, see end of Part III.

6. **Haagen-Smit et al.,** *J. Am. Chem. Soc.*, 67, 1646, 1945.
7. **Lester et al.,** *J. Am. Chem. Soc.*, 70, 3667, 1948.
8. **Barger and Coyne,** *Biochem. J.*, 22, 1417, 1928.

2-Methyl-3 or 5 or 6-methylthiopyrazine

Other names

(a) 2-Methyl-3-methylthiopyrazine
(b) 2-Methyl-5-methylthiopyrazine
(c) 2-Methyl-6-methylthiopyrazine
(Methylthio) methylpyrazine

Empirical formula

$C_8H_8N_2S$

Structure

N CH$_3$ N S—CH$_3$

(a)

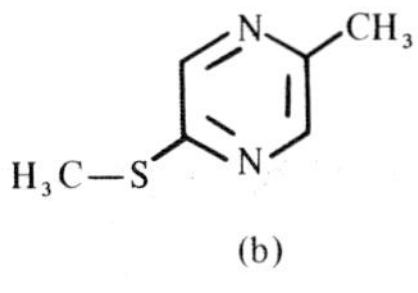

(b)

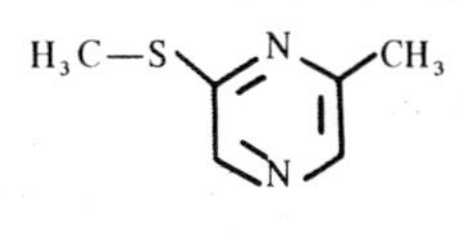

(c)

Physical/chemical characteristics

Appearance Colorless liquid

Assay C: 51.40%; H: 5.75%; N: 19.98%; S: 22.87%

Molecular weight 140.21

Boiling point 105–106°C at 12 mm Hg

Specific gravity 1.142–1.145

Refractive index 1.579–1.583

Mass spectra (a): 84 (14), 93 (14), 94 (10), 106 (18), 107 (31), 125 (27), 140 (100)

IR spectra[7] (a): 9.15, 7.37, 7.26, 9.07, 8.53, 9.40, 11.88 μm

Solubility Soluble in 1 part 70% alcohol or 5 parts 50% alcohol

Organoleptic characteristics Roasted almonds, hazelnuts

Synthesis From 2-methylchloropyrazine; a mixture of 2-methyl-3-methylthiopyrazine, 2-methyl-5-methylthiopyrazine, and 2-methyl-6-methylthiopyrazine is obtained[6]

Reported uses[3]

Ice cream, ices, etc.	2.0 ppm
Candy	2.0 ppm
Baked goods	4.0 ppm
Gelatins and puddings	2.0 ppm
Coffee aroma enhancing agent[8]	

Regulatory status FEMA No. 3208

REFERENCES

For References 1–5, see end of Part III.
6. Firmenich & Co., French Patent 1,391,212, 1965.
7. Friedel et al., *J. Agric. Food Chem.*, 19, 530, 1971.
8. Parliment et al., French Patent 2,062,168.

	Methyl-4-methylvalerate	Methyl myristate
Other names	Methyl isobutrylacetate Methyl isocaproate Methyl-4-methyl pentanoate	Methyl *n*-tetradecanoate
Empirical formula	$C_7H_{14}O_2$	$C_{15}H_{30}O_2$
Structure	$(H_3C)_2CH-CH_2-CH_2-C(=O)-O-CH_3$	$CH_3-(CH_2)_{12}-C(=O)-O-CH_3$
Physical/chemical characteristics[5]		
Appearance	Colorless liquid	Colorless, oily liquid or white, waxy solid
Molecular weight	130.18	242.41
Melting point		18°C (22°C)
Boiling point	142–143°C	300°C (295°C)
Specific gravity		0.870 at 15.5°C
Refractive index		1.4875 at 20°C
Solubility	Insoluble in water; soluble in alcohol and oils	Insoluble in water; soluble in alcohol and oils
Organoleptic characteristics	Sweet, pineapple-like flavor	Honey and orris-like odor; corresponding flavor at trace levels
Synthesis	From isocapric acid and methanol at the boil in the presence of H_2SO_4[6]	By direct esterification of methanol with myristic acid in the presence of gaseous HCl
Natural occurrence	Reported found in pineapple fruits[7]	Reported found in violet roots[8]
Reported uses[3]	Non-alcoholic beverages 11 ppm Ice cream, ices, etc. 44 ppm Candy 33 ppm Baked goods 33 ppm	Non-alcoholic beverages 0.25–0.50 ppm Ice cream, ices, etc. 0.25–0.50 ppm Candy 2.4 ppm Baked goods 0.30–2.0 ppm Gelatins and puddings 0.24 ppm
Regulatory status	FDA 121.1164; FEMA No. 2721	FDA 121.1164; FEMA No. 2722

REFERENCES

For References 1–5, see end of Part III.

6. **Kohlrausch et al.,** *Z. Phys. Chem.*, 22, 366, 1933.
7. **Haagen-Smit et al.,** *J. Am. Chem. Soc.*, 67, 1647, 1945.
8. **Tiemann and Krüger,** *Ber. Dtsch. Chem. Ges.*, 26, 2677, 1893.

	1-Methylnaphthalene	Methyl β-naphthyl ketone
Other names	α-Methylnaphthalene	2′-Acetonaphthone 2-Acetylnaphthalene Methyl naphthyl ketone β-Naphthyl methyl ketone
Empirical formula	$C_{11}H_{10}$	$C_{12}H_{10}O$
Structure	CH_3	O ‖ C−CH_3
Physical/chemical characteristics[5]		
Appearance	Oily liquid	White or nearly white crystalline solid
Assay		99% min[2]
Molecular weight	142.20	170.21
Melting point		56°C (52–54°C)
Boiling point	244°C 110°C at 12 mm Hg 107.4°C at 10 mm Hg	300–301°C >53.0°C[2]
Congealing point	−22°C	
Flash point		129°C
Specific gravity	1.0202 at 20°C	
Refractive index	1.6170 at 20°C	
Solubility	Insoluble in water; soluble in benzene; very soluble in alcohol and ether	1 gm in 5 ml of 95% alcohol;[2] insoluble in water; soluble in most common organic solvents
Organoleptic characteristics		Odor suggestive of orange blossom; strawberry-like flavor
Synthesis	Generally obtained from coal tar and petroleum oils[6]	Prepared by Friedel-Crafts reaction of naphthalene, acetyl chloride and $AlCl_3$[7]
Natural occurrence		Not reported found in nature
Reported uses[3]	Non-alcoholic beverages 1.0 ppm Ice cream, ices, etc. 1.0 ppm Candy 1.0 ppm Gelatins and puddings 1.0 ppm	Non-alcoholic beverages 0.50 ppm Ice cream, ices, etc. 0.75 ppm Candy 5.3 ppm Baked goods 2.0 ppm Gelatins and puddings 2.2–3.0 ppm Chewing gum 480–700 ppm
Regulatory status	FEMA No. 3193	FDA 121.1164; FEMA No. 2723

REFERENCES

For References 1–5, see end of Part III.

6. *Ullmanns Enzyklopädie der Technischen Chemie,* Vol. 2, Urban and Schwarzenberg, 1960, 588.
7. **Leroy,** *Bull. Soc. Chim. Fr. Ser. 3,* 7, 649, 1892.

	Methyl nonanoate	Methyl-2-nonenoate
Other names	Methyl pelargonate	"Neofolione"
Empirical formula	$C_{10}H_{20}O_2$	$C_{10}H_{18}O_2$
Structure	$CH_3{-}(CH_2)_7{-}\overset{\overset{\displaystyle O}{\|}}{C}{-}O{-}CH_3$	$CH_3{-}(CH_2)_5{-}CH{=}CH{-}\overset{\overset{\displaystyle O}{\|}}{C}{-}O{-}CH_3$
Physical/chemical characteristics[5]		
Appearance	Colorless, oily liquid	Colorless or light-yellow liquid
Assay	>96%	95%
Molecular weight	172.27	170.25
Melting point	−35°C	
Boiling point	213.5–214°C	
Specific gravity	0.8331 at 0°C; 0.877 at 18°C	0.893–0.898 at 25°C
Refractive index	1.4302 at 25°C	1.4400–1.4440 at 20°C
Solubility	Insoluble in water; soluble in alcohol and ether	1:4 in 70% alcohol; insoluble in water
Organoleptic characteristics	Wine and coconut-like odor; below 10 ppm sweet, coconut-like flavor	Characteristic violet-like odor
Synthesis	By heating pelargonic acid with methyl alcohol in the presence of concentrated sulfuric acid and subsequent rectification; or by hydrogenation of 1,5-octadien-carboxylic acid (?) methyl ester using palladium chloride in methanol solution	Obtained in quantitative yields by treating methyl β-chlorocaproate with sodium acetate, or by dehydrogenating the corresponding saturated ester[6]
Natural occurrence	Reported found in orris derivatives	Not reported found in nature
Reported uses[3]	Non-alcoholic beverages 3.9 ppm Ice cream, ices, etc. 3.6 ppm Candy 6.2 ppm Baked goods 7.1 ppm	Non-alcoholic beverages 3.2 ppm Ice cream, ices, etc. 12 ppm Candy 9.9 ppm Baked goods 13 ppm
Regulatory status	FDA 121.1164; FEMA No. 2724	FDA 121.1164; FEMA No. 2725

REFERENCES

For References 1–5, see end of Part III.

6. **Melanetton, Brown and Henry,** *J. Org. Chem.*, 28, 3261, 1963.

	Methyl 2-nonynoate	2-Methyloctanal
Other names	Methyl octine carbonate Methyl octyne carbonate	Methyl hexyl acetaldehyde
Empirical formula	$C_{10}H_{16}O_2$	$C_9H_{18}O$
Structure	$CH_3-(CH_2)_5-C\equiv C-\overset{\overset{\displaystyle O}{\|}}{C}-O-CH_3$	$CH_3-(CH_2)_5-\overset{\overset{\displaystyle CH_3}{\|}}{CH}-CHO$
Physical/chemical characteristics[5]		
Appearance	Colorless, oily liquid	Colorless, oily liquid
Molecular weight	168.24	142.24
Boiling point	232°C; 122°C at 20 mm Hg; 220°C	82–83°C at 20 mm Hg
Specific gravity	0.916–0.918 at 20°C; 0.932 at 15.5°C	0.8411 at 0°C
Refractive index	1.448–1.450 at 20°C (1.4515)	1.42335 at 14°C
Solubility	Insoluble in water; soluble in alcohol	Insoluble in water; soluble in alcohol and oils
Organoleptic characteristics	Characteristic violet-like odor; sweet taste reminiscent of fresh peach, unripe banana, and cucumber peel	Delicate, floral odor (rose or lily-like)
Synthesis	From 2-octene by way of 1-octyne and octyne carboxylic acid	By heating methoxymethyl hexyl carbinol with oxalic acid[6,7] or with anhydrous formic acid.[8,9] Also from hexanal and propionic aldehyde. Darzen's synthesis of reacting methyl hexyl ketone and ethyl chloroacetate with sodium ethylate; the resulting ester is then hydrolyzed to the acid and this heated under vacuum.
Natural occurrence	Not reported found in nature	Not reported found in nature
Reported uses[3]	Non-alcoholic beverages 0.69 ppm Ice cream, ices, etc. 0.28 ppm Candy 0.61 ppm Baked goods 2.2 ppm Gelatins and puddings 0.02–0.12 ppm Condiments 10 ppm	Non-alcoholic beverages 1.0 ppm Ice cream, ices, etc. 1.0 ppm Candy 2.0 ppm Baked goods 2.0 ppm
Regulatory status	FDA 121.1164; FEMA No. 2726	FDA 121.1164; FEMA No. 2727

REFERENCES

For References 1–5, see end of Part III.

6. **Behal and Sommelet,** *Compt. Rend.*, 138, 92, 1904.
7. **Behal and Sommelet,** *Chem. Zentralbl.*, 2, 1791, 1906.
8. **Sommelet,** *Ann. Chem. Paris,* 8(9), 557, 1906.
9. **Sommelet,** *Bull. Soc. Chem. Fr.*, 4(1), 407, 1907.

	Methyl octanoate	Methyl-*cis*-4-octenoate
Other names	Methyl caprylate Methyl octoate Methyl octylate	
Empirical formula	$C_9H_{18}O_2$	$C_9H_{16}O_2$
Structure	$CH_3-(CH_2)_6-C(=O)-O-CH_3$	$CH_3-CH_2-CH_2-CH=CH-CH_2-CH_2-C(=O)-O-CH$
Physical/chemical characteristics[5]		
Appearance	Colorless, oily liquid	
Molecular weight	158.24	156.22
Boiling point	200°C (194–195°C)	90°C at 11 mm Hg
Specific gravity	0.887 at 15.5°C; 0.8784 at 20°C	
Refractive index	1.4180 (1.4170) at 20°C	
Solubility	Insoluble in water; soluble in alcohol	Soluble in ether
Organoleptic characteristics	Powerful, winey, fruity, and orange-like odor; oily, somewhat orange taste	
Synthesis	From coconut fatty acids by alcoholysis in the presence of gaseous HC1[6]	From *cis*-4-octenoic acid ethyl ester and methanol in the presence of sodium catalyst[9]
Natural occurrence	Reported found in orris essential oil[7] in the fruits of *Ananas sativus* Lindl.,[8] in strawberry, pineapple, and pear	Reported found in pineapple[10]
Reported uses[3]	Non-alcoholic beverages 0.02–1.0 ppm Ice cream, ices, etc. 1.0–10 ppm Candy 13 ppm Baked goods 1.0–40 ppm	Non-alcoholic beverages 2.0 ppm Ice cream, ices, etc. 2.0 ppm Candy 4.0 ppm Baked goods 4.0 ppm Gelatins and puddings 2.0 ppm Preserves 2.0 ppm
Regulatory status	FDA 121.1164; FEMA No. 2728	FEMA No. 3367

REFERENCES

For References 1–5, see end of Part III.

6. **Haller and Youssoufian,** *Compt. Rend.*, 143, 805, 1906.
7. **Naves,** *Helv. Chim. Acta,* 32, 2306, 1949.
8. **Haagen-Smit et al.,** *J. Am. Chem. Soc.*, 67, 1647, 1945.
9. **Näf-Müller and Willhalm,** *Helv. Chim. Acta,* 54(7), 1880, 1971.
10. **Creveling, Silverstein and Jennings,** *J. Food Sci.*, 33(3), 284, 1968.

	Methyl-2-octynoate	2-Methyl pentanal
Other names	Methyl heptine carbonate Methyl heptyne carbonate Methyl-*n*-hept-1-yne-1-carboxylate	2-Methyl valeraldehyde
Empirical formula	$C_9H_{14}O_2$	$C_6H_{12}O$
Structure	$CH_3(CH_2)_4{-}C{\equiv}C{-}O{-}\overset{O}{\overset{\Vert}{C}}{-}CH_3$	$CH_3{-}CH_2{-}CH_2{-}\underset{CH_3}{\underset{\vert}{CH}}{-}C{\genfrac{}{}{0pt}{}{{/\!\!/}O}{\diagdown H}}$
Physical/chemical characteristics[5]		
Appearance	Colorless to slightly yellow liquid	
Assay	96% min[2] (98%)	
Molecular weight	154.21	100.15
Melting point	215°C (217°C)	
Boiling point		116°C at 760 mm Hg
Flash point	Approx 89°C (95°C)	
Specific gravity	0.919–0.924 at 25°/25°C;[2] 0.9275–0.9375 at 15°/15°C	
Refractive index	1.4460–1.4490[2] (1.4460–1.4475) at 20°C	1.4120 at 20°C
Acid value	Not more than 1[2] (0.3)	
Solubility	1:5 and more in 70% alcohol;[2] 1:2 in 80% alcohol; insoluble in water	
Organoleptic characteristics	Odor similar to violets on dilution; if concentrated, powerful and unpleasant odor; wine-berry flavor (muscatel)	
Synthesis	From heptaldehyde via heptyne and heptyne carboxylic acid; the acid is subsequently esterified	From 2-methylpentanol by catalytic oxidation[6]

$$2\ H_3C\diagup\!\!\diagdown CHO \xrightarrow{B^{\ominus}} H_3C\diagup\!\!\diagdown\!\!=\!\!\underset{}{\overset{CH_3}{\vert}}\diagdown CHO \xrightarrow{H_2}$$

	Methyl-2-octynoate		2-Methyl pentanal	
Natural occurrence	Not reported found in nature		A volatile component in onion,[7] garlic, milk, meat, and roasted peanuts[8]	
Reported uses[3]	Non-alcoholic beverages	0.15 ppm	Non-alcoholic beverages	2.0 ppm
	Ice cream, ices, etc.	0.30 ppm	Ice cream, ices, etc.	2.0 ppm
	Candy	1.4 ppm	Candy	2.0 ppm
	Baked goods	1.4 ppm	Baked goods	2.0 ppm
	Gelatins and puddings	1.7 ppm	Gelatins and puddings	2.0 ppm
	Chewing gum	13–20 ppm	Chewing gum	2.0 ppm
	Jellies	0.23 ppm		
Regulatory status	FDA 121.1164; FEMA No. 2729		FEMA No. 3413	

REFERENCES

For References 1–5, see end of Part III.

6. **Komarewsky et al.,** *J. Am. Chem. Soc.*, 66, 1117, 1944.
7. **Boelens et al.,** *J. Agric. Food Chem.*, 19, 984, 1971.
8. Volatile Compounds in Food, T.N.O. Report No. 4030, 3rd ed., Centraal Instituut voor Voedingsonderzoek, Zeist, Netherlands.

	4-Methyl-2,3-pentanedione	3-Methylpentanoic acid
Other names	Acetyl isobutyryl	3-Methylvaleric acid
Empirical formula	$C_6H_{10}O_2$	$C_6H_{12}O_2$
Structure	$CH_3-C(=O)-C(=O)-CH(CH_3)_2$	$CH_3-CH_2-CH(CH_3)-CH_2-COOH$ (printed as $CH_2-CH_2-CH(CH_3)-CH_2-COOH$)
Physical/chemical characteristics[5]		
Appearance	Yellow oil	
Molecular weight	114.15	116.16
Melting point	−2.4°C	
Boiling point	116°C	*d*-, 110°C at 30 mm Hg *d*-, 92–93°C at 10 mm Hg *l*-, 196–197°C *l*-, 105°C at 30 mm Hg *d,l*-, 197.2–197.8°C *d, l*-, 91–92°C at 10 mm Hg *d, l*-, −41°C
Specific gravity	0.9215 at 11°C	*d*-, 0.9276 at 20.5°C *l*-, 0.9230 at 25°C *d, l*-, 0.9262 at 20°C
Refractive index		*d*-, 1.4158 at 20.5°C *l*-, 1.4152 at 25°C *d, l*-, 1.4159 at 20°C
Optical rotation		*d*-, +8.5° at 20°C in ethanol *l*-, −8.9° at 20°C in ethanol
Organoleptic characteristics	Characteristic pungent odor	Sour, herbaceous, slightly green odor
Synthesis	Prepared from isonitroso methyl isobutyl ketone by treatment with concentrated H_2SO_4.[6] Other methods employ mesityl oxide as the starting material.[7,8]	From the diethylester of *sec*-butyl-malonic acid[9]
Natural occurrence	In coffee and beer	In the oil from tobacco leaves (*d*-form);[10-12] in cheese[13]
Reported uses[3]	Non-alcoholic beverages 7.6 ppm Ice cream, ices, etc. 5.6 ppm Candy 6.2 ppm Baked goods 8.3 ppm Gelatins and puddings 1.2–18 ppm	Non-alcoholic beverages 2.0 ppm Ice cream, ices, etc. 2.0 ppm Candy 5.0 ppm Baked goods 5.0 ppm Gelatins and puddings 2.0 ppm
Regulatory status	FDA 121.1164; FEMA No. 2730	FEMA No. 3437

REFERENCES

For References 1–5, see end of Part III.

6. **Otto and Peckmann,** *Ber. Dtsch. Chem. Ges.,* 22, 2121, 1889.
7. **Pauly,** *Ber. Dtsch. Chem. Ges.,* 33, 503, 1900.
8. **Dufraisse and Moureau,** *C. R. Séances Acad. Sci. Paris,* 184, 99, 1927.
9. **Vliet et al.,** *Org. Synth.,* 11, 76, 1931.
10. **Sabetay et al.,** *C. R. Hebd. Séances Acad. Sci.,* 225, 887, 1947.
11. **Onishi et al.,** *Bull. Agric. Chem. Soc. Jap.,* 20, 68, 1956; 21, 82, 1957.
12. **Onishi et al.,** *Koryo,* 55, 9, 1960; *Chem. Abstr.,* 55, 12778, 1961.
13. **Panouse et al.,** *Ind. Aliment. Agric.* p. 133, 1972.

	4-Methyl-2-pentanone	2-Methyl-2-pentenal
Other names	Methyl isobutyl ketone	3-Ethyl-2-methylacraldehyde 2-Propylidene propionaldehyde Methyl ethyl acrolein 2,4-Dimethyl crotonaldehyde Homotiglic aldehyde Isohexenal
Empirical formula	$C_6H_{12}O$	$C_6H_{10}O$
Structure	$(H_3C)_2CH-CH_2-C(=O)-CH_3$	$CH_3-CH_2-CH=C(CH_3)CHO$
Physical/chemical characteristics[5]		
Appearance	Colorless liquid	Colorless mobile liquid
Molecular weight	100.16	98.15
Melting point	−83.5°C; −80.4°C	
Boiling point	117°C; 115.8°C	134–136°C at 745 mm Hg 77–78.5°C at 100 mm Hg 38–39°C at 18 mm Hg
Specific gravity	0.8062 at 20°C	0.8581 at 20°C
Refractive index	1.3959 at 20°C	1.4488 at 20°C
Solubility	Insoluble in water; slightly miscible with alcohol, ether, and benzene	Insoluble in water; soluble in alcohol, ether, benzene, and methanol
Organoleptic characteristics	Pleasant odor	Powerful grassy-green, slightly fruity odor
Synthesis	By hydrogenation of mesityl oxide over Ni at 160–190°C;[7] also by oxidation of methyl isobutyl carbinol	By aldolic condensation of propionaldehyde[6]
Natural occurrence	In orange and grape; in vinegar	Reported found as a volatile component in onion
Reported uses[3]	Non-alcoholic beverages 6.3 ppm Ice cream, ices, etc. 6.3 ppm Candy 6.3 ppm Baked goods 6.3 ppm	Non-alcoholic beverages 80.0 ppm Ice cream, ices, etc. 80.0 ppm Candy 80.0 ppm Baked goods 80.0 ppm Gelatins and puddings 80.0 ppm Chewing gum 80.0 ppm
Regulatory status	FDA 121.1164; FEMA No. 2731	FEMA No. 3194

REFERENCES

For References 1–5, see end of Part III.

6. **Rodd,** *Chemistry of Carbon Compounds,* 2nd ed., Vol. 1, American Elsevier, New York, 1965, 59.
7. **Darzens,** *C. R. Séances Acad. Sci. Paris,* 140, 152, 1905.

	2-Methyl-2-pentenoic acid	β-Methyl phenethyl alcohol
Other names	3-Ethyl-2-methylacrylic acid β-Amylene-β-carboxylic acid 2-Pentene-2-carboxylic acid 2-Propylidenepropionic acid	Hydratropyl alcohol Hydrotropic alcohol 2-Phenylpropan-1-ol β-Phenylpropyl alcohol
Empirical formula	$C_6H_{10}O_2$	$C_9H_{12}O$
Structure	$CH_3-CH_2-CH=C(CH_3)-COOH$	$C_6H_5-CH(CH_3)-CH_2OH$
Physical/chemical characteristics[5]		
Appearance	Prismatic crystals (*trans*-form)	Viscous liquid
Molecular weight	114.04	136.19
Melting point	24.4°C	
Boiling point	*trans*-, 112°C at 12 mm Hg *trans*-, 213°C at 750 mm Hg *trans*-, 106°C at 10 mm Hg *trans*-, 123–125°C at 29.5 mm Hg *cis*-, 94°C at 10 mm Hg *cis*-, –42°C	113–114°C at 14 mm Hg
Specific gravity	0.9812 at 25°C	1.015 at 15.5°C
Refractive index	1.4578 at 25°C (1.4488 at 25°C)	1.5270 at 20°C
		d-, +15.25°; *l*-, −15.16°
Solubility	Soluble in ether, chloroform, and carbon disulfide; slightly soluble in water.	
Organoleptic characteristics		Faint, aromatic odor reminiscent of hyacinth
Synthesis	By distillation of 2-hydroxy-2-methylvaleric acid (*cis*-form)[6]	By reduction of the corresponding aldehyde with zinc and acetic acid and subsequent saponification of the sodium salt. Two optically active isomers (*d*- and *l*-) are known.
Natural occurrence		
Reported uses[3]	Candy 1.0 ppm Baked goods 1.0 ppm Condiments 1.0 ppm Meat, meat sauces, soups 1.0 ppm	Non-alcoholic beverages 1.1 ppm Ice cream, ices, etc. 0.42 ppm Candy 1.2 ppm Baked goods 0.92 ppm
Regulatory status	FEMA No. 3195	FDA 121.1164; FEMA No. 2732

REFERENCES

For References 1–5, see end of Part III.

6. **Lucas and Prater,** *J. Am. Chem. Soc.*, 59, 1684, 1937.

	α-Methylphenethyl butyrate		Methylphenethyl ether	
Other names	1-Phenyl-2-propyl butyrate		Phenylethyl methyl ether Methyl phenylethyl ether	
Empirical formula	$C_{13}H_{18}O_2$		$C_9H_{12}O$	
Structure	$CH_3-CH_2-CH_2-C(=O)-O-CH(CH_3)-CH-CH_2-C_6H_5$		$C_6H_5-CH_2-CH_2-OCH_3$	
Physical/chemical characteristics[5]				
Appearance			Colorless mobile oil	
Molecular weight	206.27		136.20	
Boiling point			185–187°C 100–104°C at 25 mm Hg	
Solubility			Almost insoluble in water; soluble in alcohol and oils	
Organoleptic characteristics			Powerful diffusive and penetrating odor with a warm, floral note; "grassy" and pungent on dilution[9]	
Synthesis			By catalytic reduction of phenylacetaldehyde dimethyl acetal;[6] from phenylethyl alcohol and methanol[7]	
Natural occurrence	Reported not found in nature		Reported as occurring in *Pandanus odoratissimus*[7] and in the oil of *Tagetes signata*[8]	
Reported uses[3]	Non-alcoholic beverages	0.60 ppm	Non-alcoholic beverages	20.0 ppm
	Ice cream, ices, etc.	3.0 ppm		
	Candy	6.0 ppm		
	Gelatins and puddings	3.0 ppm		
Regulatory status	FEMA No. 3197		FEMA No. 3198	

REFERENCES

For References 1–5, see end of Part III.

6. **Arctander,** *Perfume and Flavor Materials of Natural Origin,* 1961, 504.
7. **Deshapande,** *J. Indian Chem. Soc.,* 15, 509, 1938.
8. **Chopra, Nigam, Kapoor and Manda,** *Soap Perfum. Cosmet.,* 36, 686, 1963.
9. **Chopra et al.,** *Perfum. Rec.,* 54, 238, 1963.

	Methyl phenylacetate	3-Methyl-4-phenyl-3-butene-2-one
Other names	Methyl α-toluate	3-Benzylidene-butan-2-one Benzylidene methyl acetone
Empirical formula	$C_9H_{10}O_2$	$C_{11}H_{12}O$
Structure	$C_6H_5-CH_2-C(=O)-O-CH_3$	$C_6H_5-CH=C(CH_3)-C(=O)-CH_3$
Physical/chemical characteristics[5]		
Appearance	Colorless or nearly colorless liquid	Crystalline solid
Assay	98% min[2]	
Molecular weight	150.18	160.22
		38–40°C
Boiling point	215°C	124–125°C at 10 mm Hg
Flash point	89°C	
Specific gravity	1.061–1.067 at 25°/25°C;[2] 1.0633 at 16°C	1.0072 at 40.2°C
Refractive index	1.5050–1.5090 at 20°C;[2] 1.5091 at 16°C	1.5720 at 40.2°C
Acid value	1 max[2]	
Solubility	1:6 in 60% alcohol;[2] 1:2 in 70% alcohol; insoluble in water; soluble in most organic solvents	
Organoleptic characteristics	Intense odor suggestive of honey and jasmine; sweet, honey-like flavor	Camphor-like odor
Synthesis	By methanolic esterification of the corresponding acid or nitrile; by reacting HCl over a solution of benzyl nitrile in methanol[6]	By condensation of benzaldehyde with butanone in the presence of dry, gaseous HCl at low temperature[7,8]
Natural occurrence	In cocoa, coffee, and strawberry	Not reported found in nature
Reported uses[3]	Non-alcoholic beverages 3.9 ppm Ice cream, ices, etc. 2.5 ppm Candy 13 ppm Baked goods 12 ppm Gelatins and puddings 0.10 ppm Chewing gum 11 ppm Syrups 37 ppm	Non-alcoholic beverages 0.59 ppm Ice cream, ices, etc. 2.0 ppm Candy 2.8 ppm Baked goods 2.0 ppm
Regulatory status	FDA 121.1164; FEMA No. 2733	FDA 121.1164; FEMA No. 2734

REFERENCES

For References 1–5, see end of Part III.

6. **Rising and Sieglitz,** *J. Am. Chem. Soc.,* 40, 725, 1918.
7. **Rateb and Soliman,** *J. Chem. Soc.,* p. 1426, 1960.
8. **Stiles et al.,** *J. Am. Chem. Soc.,* 81, 628, 1959.

	2-Methyl-4-phenyl-2-butyl acetate	2-Methyl-4-phenyl-2-butyl isobutyrate
Other names	Dimethyl phenylethyl carbinyl acetate 1,1-Dimethyl-3-phenylpropan-1-yl acetate	Dimethyl phenylethyl carbinyl isobutyrate Phenylethyl dimethylcarbinyl isobutyrate
Empirical formula	$C_{13}H_{18}O_2$	$C_{15}H_{22}O_2$
Structure	$CH_3-C(CH_3)(O-C(=O)-CH_3)-CH_2-CH_2-C_6H_5$	$C_6H_5-CH_2-CH_2-C(CH_3)_2-O-C(=O)-CH(CH_3)_2$
Physical/chemical characteristics[5]		
Appearance	Liquid	Colorless liquid
Assay	97.5%	
Molecular weight	206.28	234.34
Melting point		
Boiling point	275°C	250°C
Flash point	101°C	
Specific gravity	0.9865–0.9900 at 15°C; 0.990 at 15.5°C	
Refractive index	1.4880–1.4900 at 20°C	
Acid value	0.2 max	
Solubility	1:5–6 in 70% alcohol; insoluble in water	Insoluble in water; soluble in alcohol
Organoleptic characteristics	Jasmine and hyacinth-like odor with a slightly rose undertone	Herbaceous, tea-like, sweet odor; fruit-juice flavor
Synthesis		By esterification of dimethyl phenylethyl carbinol with isobutyric acid or anhydride
Natural occurrence	Not reported found in nature	
Reported uses[3]	Non-alcoholic beverages 1.8 ppm Ice cream, ices, etc. 0.5 ppm Candy 0.50–10 ppm Baked goods 0.50–10 ppm	Non-alcoholic beverages 0.50–11 ppm Ice cream, ices, etc. 1.0–44 ppm Candy 11 ppm Baked goods 2.0–30 ppm
Regulatory status	FDA 121.1164; FEMA No. 2735	FDA 121.1164; FEMA No. 2736

REFERENCES

For References 1–5, see end of Part III.

	3-Methyl-2-phenylbutyraldehyde	Methyl-4-phenylbutyrate
Other names	α-Isopropyl phenylacetaldehyde α-Phenyl isovaleraldehyde	γ-Phenylbutyric acid, methyl ester
Empirical formula	$C_{11}H_{14}O$	$C_{11}H_{14}O_2$
Structure	$(H_3C)(CH_3)CH-CH(C_6H_5)-CHO$	$C_6H_5-CH_2-CH_2-CH_2-C(=O)-O-CH_3$
Physical/chemical characteristics[5]		
Appearance	Colorless, oily liquid	Colorless liquid
Molecular weight	162.23	178.23
Boiling point	238°C	
Solubility	Insoluble in water; soluble in alcohol	Very slightly soluble in water; soluble in alcohol
Organoleptic characteristics	Green, fruity odor when dilute. Below 5 ppm flavor is fruity; above 20 flavor is bitter.	Powerful, sweet, fruity, floral odor; extremely sweet, fruity flavor below 10 ppm
Synthesis	From phenyl isopropylglycidic acid; by rearrangement of 2-phenylpentane-1,2-diol	By direct esterification of γ–butyric acid
Natural occurrence	Not reported found in nature	Not reported found in nature
Reported uses[3]	Non-alcoholic beverages 0.10 ppm Ice cream, ices, etc. 0.50 ppm Candy 0.32–0.50 ppm	Non-alcoholic beverages 0.56 ppm Ice cream, ices, etc. 0.52 ppm Candy 1.6 ppm Baked goods 1.4 ppm
Regulatory status	FDA 121.1164; FEMA No. 2738	FDA 121.1164; FEMA No. 2739

REFERENCES

For References 1–5, see end of Part III.

	3-Methyl-3-phenyl glycidic acid ethyl ester	5-Methyl-2-phenyl-2-hexenal
Other names	Aldehyde C-16 Ethyl α,β-epoxy-β-methyl-hydrocinnamate Ethyl α,β-epoxy-β-methylphenyl propionate Ethyl 2,3-epoxy-3-phenyl-butanoate Ethyl methyl-phenyl-glycidate "Strawberry aldehyde"	
Empirical formula	$C_{12}H_{14}O_3$	$C_{14}H_{18}O$
Structure	$H_3C-C(C_6H_5)(-O-)CH-C(=O)-O-CH_2-CH_3$	$CH_3-CH_2-CH(CH_3)-CH_2-CH=C(C_6H_5)-CHO$
Physical/chemical characteristics[5]		
Appearance	Colorless to pale-yellow liquid	
Assay	98% min[2] (99%)	
Molecular weight	206.24	202.28
Boiling point	150–153°C at 8 mm Hg	96–100°C at 0.7 mm Hg
Flash point	134°C	
Specific gravity	1.088–1.112 at 25°/25°C;[2] 1.0965–1.0985 at 15°/15°C	
Refractive index	1.5040–1.5130[2] (1.5050–1.5075) at 20°C	
Acid value	Not more than 2[2] (0.9)	
Ketone content	1% max (as C_8H_8O)	
Solubility	1:3 and more in 70% alcohol;[2] 1:2 in 80% alcohol; insoluble in water; soluble in most organic solvents	
Organoleptic characteristics	On dilution, strong, fruity odor suggestive of strawberry; characteristic, slightly acid taste reminiscent of strawberry	Cocoa aroma; often used in chocolate and cocoa-type flavors[8]
Synthesis	By reaction of acetophenone and the ethyl ester of monochloroacetic acid in the presence of an alkaline condensing agent[2]	From phenyl acetaldehyde and isopropyl acetaldehyde by aldolic condensation[6]
Natural occurrence	Not reported found in nature	Reported found in the neutral volatile fraction of roasted peanuts[7]
Reported uses[3]	Non-alcoholic beverages 5.6 ppm Ice cream, ices, etc. 6.7 ppm Candy 21 ppm Baked goods 18 ppm Gelatins and puddings 13 ppm Chewing gum 470 ppm	Non-alcoholic beverages 25.0 ppm Ice cream, ices, etc. 25.0 ppm Candy 25.0 ppm Baked goods 25.0 ppm Gelatins and puddings 25.0 ppm Chewing gum 25.0 ppm
Regulatory status	FDA GRAS; FEMA No. 2444	FEMA No. 3199

REFERENCES

For References 1–5, see end of Part III.

6. **Van Praag and Stein,** German Patent 1,921,560, assigned to International Flavors and Fragrances Inc., 1969.
7. **Van Praag and Stein,** German Patent 1,921,560, 1969.
8. **Van Praag and Stein,** German Patent 1,921,560, 1969.

	4-Methyl-1-phenyl-2-pentanone	4-Methyl-2-phenyl-2-pentenal
Other names	Benzyl isobutyl ketone Isobutyl benzyl ketone	
Empirical formula	$C_{12}H_{16}O$	$C_{12}H_{14}O$
Structure	$C_6H_5-CH_2-C(=O)-CH_2-CH(CH_3)_2$	$CH_3-CH(CH_3)-CH=C(C_6H_5)-CHO$
Physical/chemical characteristics[5]		
Appearance	Colorless, oily liquid	
Molecular weight	176.26	174.23
Boiling point	250–251°C	82–87°C at 0.7 mm Hg
Specific gravity	0.969 at 0°C	
Solubility	Very slightly soluble in water; poorly soluble in propylene glycol; soluble in alcohol and oils	
Organoleptic characteristics	Sweet, woody, spicy odor; at low levels sweet and tart, fruity flavor	Cocoa aroma; often used in chocolate and cocoa-type flavors[9]
Synthesis	By passing phenylacetic acid and isovaleric acid over ThO_2 catalyst at 450–470°C[6,7,8]	By aldolic condensation of the corresponding aldehydes in the presence of a basic catalyst[10]
Natural occurrence	Not reported found in nature	
Reported uses[3]	Non-alcoholic beverages 1.0 ppm Ice cream, ices, etc. 5.0 ppm Candy 0.06–5.0 ppm Baked goods 5.0 ppm	Non-alcoholic beverages 1.5 ppm Ice cream, ices, etc. 1.5 ppm Candy 1.5 ppm Baked goods 1.5 ppm Gelatins and puddings 1.5 ppm Chewing gum 1.5 ppm
Regulatory status	FDA 121.1164; FEMA No. 2740	FEMA No. 3200

REFERENCES

For References 1–5, see end of Part III.

6. **Senderens,** *Compt. Rend.*, 150, 1338, 1910.
7. **Senderens,** *Bull. Soc. Chim. Fr.*, 4(7), 648, 1910.
8. **Senderens,** *Ann. Chim. Paris,* 8(28), 322, 1913.
9. **Van Praag and Stein,** German Patent 1,921,560 assigned to International Flavors and Fragrances Inc., 1969.
10. **Van Praag and Stein,** German Patent 1,921,560, 1969.

	Methyl-3-phenylpropionate	Methyl propionate
Other names	Methyl dihydrocinnamate Methyl hydrocinnamate Methyl phenyl propionate	Methyl propanoate
Empirical formula	$C_{10}H_{12}O_2$	$C_4H_8O_2$
Structure	$C_6H_5-CH_2-CH_2-C(=O)-O-CH_3$	$CH_3-O-C(=O)-CH_2-CH_3$
Physical/chemical characteristics[5]		
Appearance	Colorless liquid	Colorless liquid
Assay	98% min	
Molecular weight	164.21	88.11
Boiling point	238–239°C	79°C (80°C)
Flash point	>100°C	−2°C
Specific gravity	1.037–1.042 at 25°C; 1.0473 at 0°C	0.915 at 15.5°C; 0.9148 at 20°C
Refractive index	1.4990–1.5030 at 20°C	1.3750 at 20°C; 1.37767 at 18.5°C
Acid value	1 max	
Solubility	1:3 in 70% alcohol; almost insoluble in water	Slightly soluble in water (6.5%); completely miscible with alcohol, ether, and propylene glycol
Organoleptic characteristics	Strong floral-fruity odor	Fruity odor reminiscent of rum; sweet flavor suggestive of black currant
Synthesis	By reduction of methyl cinnamate in methanol solution[8] with hydrogen and Ni under pressure	By direct esterification of the acid with methanol in the presence of concentrated H_2SO_4[6]
Natural occurrence	Not reported found in nature	Probably occurring in California orange (var. Valencia) juice[7] and honey
Reported uses[3]	Non-alcoholic beverages 0.46 ppm Ice cream, ices, etc. 0.56 ppm Candy 1.7 ppm Baked goods 0.70–4.0 ppm	Non-alcoholic beverages 20 ppm Ice cream, ices, etc. 29 ppm Candy 96 ppm Baked goods 130 ppm
Regulatory status	FDA 121.1164; FEMA No. 2741	FDA 121.1164; FEMA No. 2742

REFERENCES

For References 1–5, see end of Part III.

6. **Vogel,** *J. Chem. Soc.*, p. 629, 1948.
7. **Schultz et al.,** *J. Food Sci.*, 29, 790, 1964.
8. **Brochet and Bauer,** *Compt. Rend.*, 159, 191, 1914.

	3-Methyl-5-propyl-2-cyclohexen-1-one	Methyl propyl disulfide
Other names	"Celery ketone" 3-Methyl-5-propyl-2-cyclohexenone 1-Methyl-5-*n*-propyl-1-cyclohexen-3-one	
Empirical formula	$C_{10}H_{16}O$	$C_4H_{10}S_2$
Structure	[structure: cyclohexenone ring with =O, H_3C–CH₂–CH₂ substituent and CH_3]	$CH_3-S-S-CH_2-CH_2-CH_3$
Physical/chemical characteristics		
Appearance	Pale-yellow to colorless liquid	Pale-yellow, mobile liquid
Molecular weight	152.23	122.26
Boiling point	242–244°C; 122–123°C at 12 mm Hg	40–41°C at 9 mm Hg
Specific gravity	0.9267 at 22.3°C	
Refractive index		n_D^{20} 1.5075
Solubility	Insoluble in water; soluble in ether, alcohol, and oils	Almost insoluble in water; soluble in alcohol and oils
Organoleptic characteristics	Warm, spicy, woody odor; similar flavor at low levels.	Odor and taste reminiscent of cooked onion
Synthesis	By condensation of butyric aldehyde with ethyl acetoacetate in the presence of diethylamine and subsequent saponification with a 10% KOH solution; or by condensation in the presence of piperidine in ethanol solution[1,4]	From Bunte salt and the corresponding thiol[6]
Natural occurrence	Not reported found in nature	Reported found in onion oil[7,8] and in *Allium* species[9-13]
Reported uses[3]		Baked goods 1.0 ppm Condiments, pickles 1.0 ppm Meat, meat sauces, soups 1.0 ppm
Regulatory status	FDA 121.1164	FEMA No. 3201

REFERENCES

For References 1–5, see end of Part III.

6. **Milligan and Swan,** *J. Chem. Soc.*, p. 6008, 1963.
7. **Jacobs,** *Am. Perfum.*, 70(5), 53, 1957.
8. **Carson and Wong,** *Am. Chem. Soc. Div. Pet. Chem. Prepr.*, 2(4), D115, 1957.
9. **Bernhard,** *Qual. Plant. Mater. Veg.*, 18(1–3), 72, 1969.
10. **Wahlroos and Virtanen,** *Acta Chem. Scand.*, 19(6), 1327, 1965.
11. **Saghir, Mann and Yamaguchi,** *Plant Physiol.*, 40(4), 681, 1965.
12. **Saghir, Mann, Bernhard and Jacobson,** *Proc. Am. Soc. Hortic. Sci.*, 84, 386, 1964.
13. **Jacobsen et al.,** *Arch. Biochem. Biophys.*, 104(3), 473, 1964.
14. **Whitmore and Roberts,** *J. Org. Chem.*, 13, 31, 11948.

	2-Methylpropyl-3-methyl butyrate	2-(2-Methylpropyl)-pyridine
Other names	Isobutyl isovalerate	2-Isobutyl pyridine
Empirical formula	$C_9H_{18}O_2$	$C_9H_{13}N$
Structure	$(H_3C)_2CH-CH_2-C(=O)-O-CH_2-CH_2-CH_2-CH_3$	2-pyridyl $-CH_2-CH(CH_3)_2$
Physical/chemical characteristics[5]		
Molecular weight	158.24	135.20
Boiling point	171.4°C 60–62°C at 12 mm Hg	181°C 110–111°C at 55 mm Hg
Specific gravity	0.8736 at 20°C	0.8973 at 20°C
Refractive index	1.4057 at 20°C	1.4831 at 20°C
Solubility	Insoluble in water; very soluble in acetone; soluble in alcohol and ether in all proportions	
Organoleptic characteristics	Aroma reminiscent of apple/raspberry	
Synthesis	By oxidation of the corresponding ketone[6]	From 2-pycolyllithium and isopropyl bromide;[6] from 2-methyl pyridine and isopropyl bromide in liquid ammonia in the presence of potassium metal and ferric nitrate[7]
Natural occurrence	Reported found in the oil from fruits of *Xanthoxylum piperitum*[7] and in the oil of *Elsholtzia ciliata*[8]	
Reported uses[3]	Non-alcoholic beverages 5.0 ppm Ice cream, ices, etc. 5.0 ppm Candy 8.0 ppm Baked goods 8.0 ppm Gelatins and puddings 5.0 ppm Preserves and spreads 5.0 ppm	Condiments, pickles 0.5 ppm Meat, meat sauces, soups 0.2 ppm
Regulatory status	FEMA No. 3369	FEMA No. 3370

REFERENCES

For References 1–5, see end of Part III.

6. **Emmons and Lucas,** *J. Am. Chem. Soc.,* 77, 2287, 1955.
7. **Sakai, Yoshihara and Hirose,** *Bull. Chem. Soc. Jap.,* 41(8), 1945, 1968.
8. **Fujita, Tanaka and Iwamura,** *Nippon Kagaku Zasshi,* 88(7), 763, 1967.
9. **Osuch and Levine,** *J. Am. Chem. Soc.,* 78, 1723, 1956.
10. **Pines and Notari,** *J. Am. Chem. Soc.,* 82, 2209, 1960.

	3-(2-Methylpropyl)-pyridine		2-(1-Methylpropyl)thiazole	
Other names	3-Isobutyl pyridine		2-*sec*-Butyl thiazole	
Empirical formula	$C_9H_{13}N$		$C_7H_{11}NS$	
Structure	(pyridine ring)–CH_2–CH(CH_3)(CH_3)		(thiazole ring, S)–CH(CH_3)–CH_2–CH_3	
Physical/chemical characteristics[5]				
Molecular weight	135.19		141.23	
Boiling point	68–68.5°C at 8 mm Hg			
Refractive index	1.4912 at 20°C			
Synthesis	From β-picoline with $NaNH_2$ and alkyl chloride or bromide[6]			
Reported uses[3]	Non-alcoholic beverages	0.1 ppm	Ice cream, ices, etc.	0.1 ppm
	Ice cream, ices, etc.	0.2 ppm	Candy	0.5 ppm
	Candy	0.5 ppm	Baked goods	0.5 ppm
	Baked goods	1.0 ppm	Gelatins and puddings	0.1 ppm
	Gelatins and puddings	0.3 ppm	Condiments, pickles	0.5 ppm
	Cereals	1.0 ppm	Meat, meat sauces, soups	0.2 ppm
Regulatory status	FEMA No. 3371		FEMA No. 3372	

REFERENCES

For References 1–5, see end of Part III.
6. **Wibaut and Hoogzand,** *Chem. Weekbl.*, 52, 357, 1956.

	Methyl propyl trisulfide	
Other names	Propyl methyl trisulfide	
Empirical formula	$C_4H_{10}S_3$	
Structure	$CH_3-S-S-S-CH_2-CH_2-CH_3$	
Physical/chemical characteristics		
Appearance	Pale-yellow, mobile liquid	
Molecular weight	154.32	
Boiling point	52°C at 1.2 mm Hg	
Refractive index	N_d^{21} 1.5578	
Organoleptic characteristics	Powerful, penetrating warm-herbaceous odor somewhat reminiscent of onion	
Synthesis	By an adaptation of Westlake's procedure[6]	
Natural occurrence	Reported found in cocoa;[7] also in the volatile portion of fresh onion oil	
Reported uses[3]	Baked goods	2.0 ppm
	Condiments, pickles	2.0 ppm
	Meat, meat sauces, soups	2.0 ppm
Regulatory status	FEMA No. 3308	

REFERENCES

For References 1–5, see end of Part III.

6. **Westlake et al.,** *J. Am. Chem. Soc.,* 72, 436, 1950; *Chem. Abstr.,* 44, 4739a, 1950.
7. **Marion, et al.,** *Helv. Chim. Acta,* 50(6), 1509, 1967.

2-Methylpyrazine

Other names	2-Methyl-1,4-diazine Methylpyrazine
Empirical formula	$C_5H_6N_2$
Structure	(pyrazine ring with CH_3 substituent)
Physical/chemical characteristics	
Assay	C: 63.81%; H: 6.43%; N: 29.76%
Molecular weight	94.12
Boiling point	136–137°C
Specific gravity	1.0290 at 20°C
Refractive index	1.5067 at 18.7°C
Mass spectra[10]	39 (21), 40 (18), 52 (7), 53 (17), 66 (3), 67 (55), 79 (1), 94 (100), 95 (5)
IR spectra[10]	S (6.8, 6.9, 7.1, 8.6, 9.5, 9.8, 12.0) M (3.3, 6.3, 6.5, 7.7, 8.0, 8.5, 10.2) W (5.2, 5.7)
UV spectra[10]	(2.66.0, 272.0 nm)
Solubility	Soluble in water, alcohol, and ether in all proportions; soluble in acetone
Organoleptic characteristics	Odor threshold: 105,000 ppb in water,[9] 60,000 ppb in water[11]
Synthesis	From the corresponding carboxylic acid;[6] by catalytic dehydrogenation of 2-methylpiperazine;[7] by condensation of methylglyoxal with ethylenediamine[12]
Natural occurrence	Reported to be present in the following foods: bakery products, roasted barley, cocoa products, coffee, dairy products, meat, peanuts, filberts, pecans, popcorn, potato products, rum and whisky, soy products;[8] a volatile flavor component in roasted filberts;[13] also in toasted off-flavors[14]

Reported uses[3]

Non-alcoholic beverages	10.0 ppm	Gelatins and puddings	10.0 ppm
Ice cream, ices, etc.	10.0 ppm	Meat, meat sauces, soups	10.0 ppm
Candy	10.0 ppm	Milk, dairy products	10.0 ppm
Baked goods	10.0 ppm	Cereals	10.0 ppm

Regulatory status FEMA No. 3309

REFERENCES

For References 1–5, see end of Part III.

6. **Stoehr,** *J. Fr. Chim.,* 51(2), 449, 1895.
7. **Kitchen and Hanson,** *J. Am. Chem. Soc.,* 73, 1838, 1951.
8. **Maga and Sizer,** *Crit. Rev. Food Technol.,* 4(1), 39, 1973.
9. **Koehler et al.,** *J. Food Sci.,* 36, 816, 1971.
10. **Bondarovich et al.,** *J. Agric. Food Chem.,* 15, 1093, 1967.
11. **Guadagni et al.,** *J. Sci. Food Agric.,* 23, 1435, 1972.
12. **Ishiguro and Matsumura,** *Yakugaku Zasshi,* 78, 229, 1958.
13. **Sheldon et al.,** *J. Food Sci.,* 37(2), 313, 1972.
14. **Sapers et al.,** *J. Food Sci.,* 36(1), 93, 1971.

	Methyl-2-pyrrolyl ketone		5-Methylquinoxaline	
Other names	2-Acetyl pyrrole 2-Acetopyrrole			
Empirical formula	C_6H_7NO		$C_9H_8N_2$	
Structure	(pyrrole ring, N–H)–CO — CH_3		(quinoxaline ring with CH_3 at position 5)	
Physical/chemical characteristics[5]				
Appearance	Crystals (from water)			
Molecular weight	109.13		144.19	
Melting point	90°C		20–21°C	
Boiling point	220°C		120°C at 15 mm Hg	
Solubility	Soluble in water, alcohol and ether			
Synthesis	From pyrryl magnesium iodide and acetyl chloride;[6] a volatile flavor component in roasted filberts[8]			
Natural occurrences	Reported found in the essential oils of tobacco leaves and tea[7]		Reported found in coffee[6]	
Reported uses[3]	Non-alcoholic beverages	50.0 ppm	Non-alcoholic beverages	10.0 ppm
	Ice cream, ices, etc.	50.0 ppm	Ice cream, ices, etc.	10.0 ppm
	Candy	50.0 ppm	Candy	10.0 ppm
	Gelatins and puddings	50.0 ppm	Gelatins and puddings	10.0 ppm
Regulatory status	FEMA No. 3202		FEMA No. 3203	

REFERENCES

For References 1–5, see end of Part III.

6. **Oddo,** *Berichte,* 43, 1014, 1910.
7. **Onishi and Yamamoto,** *Bull. Agric. Chem. Soc.,* 21, 90, 1957.
8. **Sheldon et al.,** *J. Food Sci.,* 37(2), 313, 1972.
9. **Stoll et al.,** *Helv. Chim. Acta,* 50(2), 628, 1967.

	Methyl salicylate	Methyl sulfide
Other names	Methyl *o*-hydroxybenzoate	Dimethyl sulfide
Empirical formula	$C_8H_8O_3$	C_2H_6S
Structure	(benzene ring with OH and C(=O)–O–CH_3 in ortho positions)	$H_3C-S-CH_3$
Physical/chemical characteristics[5]		
Appearance	Colorless liquid	Colorless liquid
Molecular weight	152.14	62.04
Melting point	−8.3°C (−8.6°C)	−83.2°C
Boiling point	222.2°C (223.3°C)	37.3–37.5°C
Flash point	96°C	
Specific gravity	1.180–1.185 at 25°C	0.8449 at 20°C
Refractive index	1.5350–1.5380 at 20°C	
Solubility	1:7 in 70% alcohol; almost insoluble in water; soluble in most common organic solvents and glacial acetic acid	Insoluble in water; soluble in alcohol and ether
Organoleptic characteristics	Characteristic wintergreen-like odor	Unpleasant odor of wild radish, cabbage-like; can be tolerated as a green-vegetable note only at very low levels (0.1–3.0 ppm)
Synthesis	By extraction from natural sources; or by esterification of salicylic acid with methanol	By reaction of potassium sulfide with methyl chloride in methanol solution;[7] from potassium methyl sulfate and potassium sulfide
Natural occurrence	Reported found in the oil of wintergreen (*Gaultheria procumbens*);[6] subsequently found also in birch, tuberose, *Dianthus caryophyllus* L., *Acacia cavenia* Hook. and Arn., ylang-ylang, and several fruit juices—cherry, apple, raspberry, and others	Reported found in American peppermint oil,[8] the oil of Algerian geranium,[9] and butter[10]
Reported uses[3]	Non-alcoholic beverages 59 ppm Ice cream, ices, etc. 27 ppm Candy 840 ppm Baked goods 54 ppm Chewing gum 8,400 ppm Syrups 200 ppm	Non-alcoholic beverages 1.1 ppm Ice cream, ices, etc. 0.30 ppm Candy 1.4 ppm Baked goods 1.6 ppm Gelatins and puddings 0.13 ppm Syrups 0.50 ppm
Regulatory status	Status not fully defined by FDA; FEMA No. 2745	FDA 121.1164; FEMA No. 2746

REFERENCES

For References 1–5, see end of Part III.

6. **Cahours,** *Annalen,* 48, 60, 1843.
7. **Regnault,** *Annalen,* 34, 24, 1840.
8. *Schimmel Ber.*, p. 61, October 1896.
9. *Schimmel Ber.*, p. 50, April 1909.
10. **Ruys,** *Fr. Ses Parfums,* 47(9), 34, 1966.

	2-Methyltetrahydrofuran-3-one	4-Methyl-5-thiazoleethanol
Other names		4-Methyl-5-(β-hydroxyethyl)-thiazole 5-Hydroxyethyl-4-methylthiazole 5-(2-Hydroxyethyl)-4-methylthiazole
Empirical formula	$C_5H_8O_2$	C_6H_9NOS
Structure	(ring structure: =O, CH$_3$, O)	(ring structure: N, S, CH$_3$, CH$_2$—CH$_2$OH)
Physical/chemical characteristics[5]		
Appearance		Viscous oily liquid
Molecular weight	100.11	143.21
Boiling point	139°C	135°C at 7 mm Hg 103°C at 1 mm Hg
Specific gravity		1.196 at 24°C
Refractive index	1.4291 at 20°C	
Solubility		Very soluble in water; soluble in alcohol, ether, benzene, and chloroform
Organoleptic characteristics		Characteristic disagreeable odor of thiazole compounds; somewhat pleasant, nut-like on extreme dilution
Synthesis	By hydrolytic decarboxylation of 2-methyl-4-carbomethoxytetrahydrofuran-3-one;[6] by acid-catalyzed ring closure of β-alkoxy-diazoketones[7]	By reduction of ethyl-4-methylthiazole-5-acetate using $LiAlH_4$;[10] by condensation of thioformamide with bromoacetopropanol[11] or with γ,γ-dichloro-γ,γ-diacetodipropyl ether[12,13]
Natural occurrence	Reported found as a constituent in coffee aroma;[8] also as a volatile flavor component in roasted filbert nuts[9]	
Reported uses[3]	Non-alcoholic beverages 10.0 ppm Ice cream, ices, etc. 10.0 ppm Candy 10.0 ppm Baked goods 10.0 ppm Gelatins and puddings 10.0 ppm Chewing gum 10.0 ppm Condiments, pickles 10.0 ppm Meat, meat sauces, soups 10.0 ppm Milk, dairy products 10.0 ppm Cereals 10.0 ppm	Non-alcoholic beverages 55.0 ppm Ice cream, ices, etc. 55.0 ppm Candy 55.0 ppm Baked goods 55.0 ppm Gelatins and puddings 55.0 ppm Chewing gum 55.0 ppm Condiments 55.0 ppm Meat, meat sauces, soups 55.0 ppm Milk, dairy products 55.0 ppm Cereals 55.0 ppm
Regulatory status	FEMA No. 3373	FEMA No. 3204

REFERENCES

For References 1–5, see end of Part III.

6. **Gianturco, Giammarino and Friedel,** *Tetrahedron Lett.,* 20, 1763, 1964.
7. **Wynberg,** *Angew. Chem.,* p. 453, 1963.
8. **Stoll et al.,** *Helv. Chim. Acta,* 50(2), 628, 1967.
9. **Sheldon et al.,** *J. Food Sci.,* 37(2), 313, 1972.
10. **Eusebi, Brown and Cerecedo,** *J. Am. Chem. Soc.,* 71, 2931, 1949.
11. **Buchman,** *J. Am. Chem. Soc.,* 58, 1803, 1936.
12. **Stein and Stevens,** French Patent 945,198, British Patent 641,426, assigned to Merck & Co. Inc.
13. **Londergau and Schmitz,** U.S. Patent 2,654,760, assigned to E. I. du Pont de Nemours & Co., 1953.

	4-Methyl-5-thiazoleethanol acetate		2-Methyl thioacetaldehyde	
Other names	4-Methyl-5-(2-acetoxyethyl)-thiazole		C_3H_6OS	
Empirical formula	$C_8H_{11}O_2NS$			
Structure	(thiazole ring with CH_3) $-CH_2-CH_2-O-C(=O)-CH_3$		CH_3-S-CH_2-CHO	
Physical/chemical characteristics				
Molecular weight	185.24		90.14	
Boiling point	105–108°C at 4 mm Hg			
Synthesis	From 2-methylthio-4-methyl-5-(2-acetoxyethyl)-thiazole and AlHg;[6] by dehydrogenation with sulfur at 130°C of the β-acetyl derivative of 4-methyl-5-(β-hydroxyethyl)-3-thiazoline[7]		By hydrolysis of the diethyl acetal of 2-methyl thioacetaldehyde with diluted HCl; the diethyl acetal is prepared from diethylbromoacetal treated with sodium methyl mercaptide[8]	
Reported uses[3]	Non-alcoholic beverages	55.0 ppm	Non-alcoholic beverages	0.50 ppm
	Ice cream, ices, etc.	55.0 ppm	Ice cream, ices, etc.	0.50 ppm
	Candy	55.0 ppm	Candy	0.50 ppm
	Baked goods	55.0 ppm	Baked goods	0.50 ppm
	Gelatins and puddings	55.0 ppm	Gelatins and puddings	0.50 ppm
	Chewing gum	55.0 ppm	Chewing gum	0.50 ppm
	Condiments	55.0 ppm	Condiments	0.50 ppm
	Meat, meat sauces, soups	55.0 ppm	Meat, meat sauces, soups	0.50 ppm
	Milk, dairy products	55.0 ppm	Milk, dairy products	0.50 ppm
	Cereals	55.0 ppm	Cereals	0.50 ppm
Regulatory status	FEMA No. 3205		FEMA No. 3206	

REFERENCES

For References 1–5, see end of Part III.

6. *Chem. Abstr.*, 50, 5634c, 1956.
7. Thiel, Asinger and Stengler, *Annalen.*, p. 161, 1958.
8. Cope, Kovacic and Burg, *J. Am. Chem. Soc.*, 71, 3658, 1949.

	4-(Methylthio)butanal		1-(Methylthio)-2-butanone	
Other names	4-(Methylmercapto)butanal			
Empirical formula	$C_5H_{10}S$		$C_5H_{10}SO$	
Structure	$CH_3-S-CH_2-CH_2-CH_2-CHO$		$CH_3-S-CH_2-CO-CH_2-CH_3$	
Physical/chemical characteristics				
Appearance			Colorless liquid	
Molecular weight	106.18		118.20	
Boiling point			52–53°C at 8 mm Hg	
Specific gravity			0.9970 at 22°C	
Refractive index			1.4700 at 20°C	
Organoleptic characteristics			Odor reminiscent of mushroom, with a characteristic garlic undertone	
Synthesis			From chloro-1-butan-2-one[6]	
Natural occurrence			Reported found as a constituent in coffee aroma[6]	
Reported uses[3]	Non-alcoholic beverages	0.5 ppm	Condiments	1.0 ppm
	Ice cream, ices, etc.	0.5 ppm	Soups	1.0 ppm
	Candy	1.0 ppm	Sauces	1.0 ppm
	Baked goods	1.0 ppm		
	Meat, meat sauces, soups	5.0 ppm		
	Vegetables	5.0 ppm		
Regulatory status	FEMA No. 3414		FEMA No. 3207	

REFERENCES

For References 1–5, see end of Part III.

6. Stoll et al., *Helv. Chim. Acta,* 50, 628, 1967.

	4-Methylthio-2-butanone		3-Methylthiobutyraldehyde	
Other names	3-Methylmercapto-2-butanone		3-Methylthiobutanal	
Empirical formula	$C_5H_{10}OS$		$C_5H_{10}OS$	
Structure	$CH_3-S-CH_2-CH_2-C(=O)-CH_3$		$CH_3-S-CH(CH_3)-CH_2-CHO$	
Physical/chemical characteristics				
Molecular weight	118.19		118.19	
Boiling point	106°C at 55 mm Hg 77–78°C at 20 mm Hg		63°C at 10 mm Hg 80–92°C at 14 mm Hg	
Specific gravity			0.997 at 20°C	
Refractive index	1.4711 at 25°C		1.5078 at 20°C	
Organoleptic characteristics	Flavor like arrowhead			
Synthesis	From methylvinyl ketone and methylmercaptan;[6] from methylvinyl ketone and methyl-bis (methylthio)aluminum[7]		By addition of methanethiol to the corresponding unsaturated aldehyde (previously cooled to −20°C), using piperidine or copper acetate as catalyst;[8,10] by a patented oxo process[11]	
Reported uses[3]	Non-alcoholic beverages	1.0 ppm	Condiments, pickles	2.0 ppm
	Ice cream, ices, etc.	1.0 ppm	Vegetables	2.0 ppm
	Candy	2.0 ppm		
	Baked goods	2.0 ppm		
	Gelatins and puddings	1.0 ppm		
	Meat, meat sauces, soups	1.0 ppm		
Regulatory status	FEMA No. 3375		FEMA No. 3374	

REFERENCES

For References 1–5, see end of Part III.

6. **Reimer,** *J. Am. Chem. Soc.*, 78, 2132, 1956.
7. **Lalancette and Cole,** *Can. J. Chem.*, 49, 2983, 1971.
8. **Bonstany,** *J. Chem. U.A.R.*, 9(3), 317, 1966.
9. **Bateman, Cunneen and Ford,** *J. Chem. Soc.*, p. 3056, 1956.
10. **Reisner,** *J. Am. Chem. Soc.*, 78, 2132, 1956.
11. Badische Aniline und Soda Fabrik, British Patent 683,267.

	Methyl thiobutyrate		Methyl thiofuroate	
Other names	Thiobutyric acid, methyl ester Methanethiol *n*-butyrate		Thiofuroic acid, methylester	
Empirical formula	$C_5H_{10}OS$		$C_6H_6O_2S$	
Structure	$CH_3-CH_2-CH_2-C(=O)-S-CH_3$		2-furyl–$COSCH_3$ or 3-furyl–$COSCH_3$	
Physical/chemical characteristics				
Molecular weight	118.10		142.12	
Boiling point			92–93°C at 11 mm Hg	
Reported uses[3]	Non-alcoholic beverages	0.0005 ppm	Non-alcoholic beverages	3.0 ppm
	Ice cream, ices, etc.	0.05 ppm	Ice cream, ices, etc.	3.0 ppm
	Candy	0.5 ppm	Candy	3.0 ppm
	Baked goods	7.0 ppm	Baked goods	5.0 ppm
	Gelatins and puddings	0.06 ppm	Condiments	3.0 ppm
	Protein foods	5.0 ppm	Meat, soups	3.0 ppm
Regulatory status	FEMA No. 3310		FEMA No. 3311	

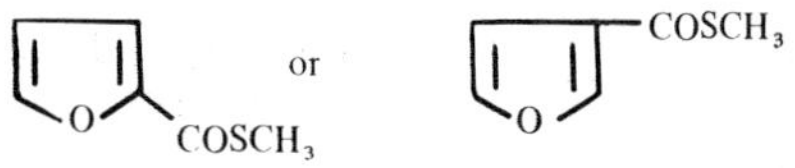

REFERENCES

For References 1–5, see end of Part III.

	3-Methylthio-1-hexanol		2-Methyl-5-thiomethylfuran	
Other names	3-Methylmercapto-1-hexanol		2-Methyl-4-(methylthio)furan	
Empirical formula	$C_7H_{16}OS$		C_6H_8OS	
Structure	$CH_3-CH_2-CH_2-CH(S-CH_3)-CH_2-CH_2-OH$		H_3C (furan ring, O) SCH_3	
Physical/chemical characteristics				
Appearance			Pale-yellow liquid	
Molecular weight	148.26		128.19	
Boiling point			80°C at 45–50 mm Hg 66–67°C at 23 mm Hg	
Specific gravity			1.0574 at 20°C	
Refractive index			1.5170 at 20°C	
Organoleptic characteristics			Very strong sulfurated odor	
Synthesis			From 2-methylfuran with CH_3SCl;[6] from 2-methylfuran with an organolithium compound[7]	
Natural occurrence			Constituent of coffee[8]	
Reported uses[3]	Non-alcoholic beverages	3.0 ppm	Baked goods	1.5 ppm
	Ice cream, ices, etc.	5.0 ppm	Meat, meat sauces, soups	1.5 ppm
	Baked goods	6.0 ppm		
	Gelatins and puddings	4.0 ppm		
	Condiments	4.0 ppm		
	Confections	5.0 ppm		
Regulatory status	FEMA No. 3438		FEMA No. 3366	

REFERENCES

For References 1–5, see end of Part III.

6. **Stoll et al.,** *Helv. Chim. Acta,* 50, 687, 1967.
7. **Goldfarb et al.,** *Dokl. Akad. Nauk SSSR,* 151(2), 2231, 1963; *Chem. Abstr.,* 59, 8681d, 1963.
8. **Stoll et al.,** *Helv. Chim. Acta,* 50, 687, 1967.

	4-Methylthio-4-methyl-2-pentanone	5-Methyl-2-thiophenecarboxaldehyde
Other names		5-Methyl-2-thenaldehyde
Empirical formula	$C_7H_{14}OS$	C_6H_6OS
Structure	$CH_3-C(CH_3)(SCH_3)-CH_2-C(=O)-CH_3$	H_3C–(thiophene ring, S)–CHO
Physical/chemical characteristics		
Molecular weight	146.25	126.17
Boiling point	84°C at 12 mm Hg 78°C at 15 mm Hg	52.5°C at 0.7 mm Hg 113–114°C at 25 mm Hg
Refractive index	1.4750 at 13°C	1.5742 at 20°C
Synthesis	From mesityl oxide by addition of methyl sulfide;[6] from acetone and methylbis(methylthio) aluminum[7]	From thiophene (or its derivatives) and formamide in the presence of $POCl_3$;[8] from N-(2-thenyl)formaldimines[9]
Natural occurrence		Reported as occurring in roasted peanuts[10]
Reported uses[3]	Non-alcoholic beverages 1.0 ppm Ice cream, ices, etc. 1.0 ppm Candy 2.0 ppm Baked goods 2.0 ppm Gelatins and puddings 1.0 ppm Meat, meat sauces, soups 1.0 ppm	Non-alcoholic beverages 0.50 ppm Ice cream, ices, etc. 0.50 ppm Candy 0.50 ppm Baked goods 0.50 ppm Gelatins and puddings 0.50 ppm Chewing gum 0.50 ppm Condiments 0.50 ppm Meat, meat sauces, soups 0.50 ppm Milk, dairy products 0.50 ppm Cereals 0.50 ppm
Regulatory status	FEMA No. 3376	FEMA No. 3209

REFERENCES

For References 1–5, see end of Part III.

6. **Bravo et al.,** *Gazz. Chim. Ital.*, 98, 1046, 1968.
7. **Lalancette et al.,** *Can. J. Chem.*, 49, 2983, 1971.
8. Abbott Laboratories, German Patent 953,082, 1956.
9. **Harbough and Dickert,** *J. Am. Chem. Soc.*, 71, 3922, 1949.
10. **Walradt et al.,** *J. Agric. Food Chem.*, 19(5), 972, 1971.

	2-Methylthiophenol		o-(Methylthio)phenol	
Other names	Erroneously called *o*-toluenethiol		2-Hydroxy-2-methylmercaptobenzene 2-Methylmercapto phenol Methyl-(2-hydroxyphenol)sulfide 1-Thioguaiacol	
Empirical formula	C_7H_8OS		C_7H_8OS	
Structure	(benzene ring with SCH_3 and OH)		(benzene ring with $S-CH_3$ and OH)	
Physical/chemical characteristics				
Appearance			Liquid	
Molecular weight	140.19		140.20	
Boiling point	105°C at 22 mm Hg		218–219°C 105°C at 22 mm Hg	
Specific gravity	1.168 at 25°C		1.168 at 25°C	
Synthesis	By diazotization of methyl-(2-amino-phenyl)-sulfide and decomposition by boiling[6]		From the diazonium salt of methyl-(2-aminofuryl)- sulfide[9]	
Natural occurrence	In coffee[7,8]		Reported found as a constituent in coffee aroma[10]	
Reported uses[3]	Baked goods	0.20 ppm	Non-alcoholic beverages	0.20 ppm
	Soups, sauces	0.20 ppm	Ice cream, ices, etc.	0.20 ppm
			Candy	0.20 ppm
			Gelatins and puddings	0.20 ppm
Regulatory status	FEMA No. 3240		FEMA No. 3210	

REFERENCES

For References 1–5, see end of Part III.

6. **Holt et al.,** *J. Am. Chem. Soc.,* 46, 2335, 1924.
7. **Goldmann,** *Helv. Chim. Acta,* 50, 628, 1967.
8. **Wolter et al.,** *Z. Ernährungswiss.,* 9, 123, 1969.
9. **Holt and Reid,** *J. Am. Chem. Soc.,* 46, 2335, 1924.
10. **Stoll et al.,** *Helv. Chim. Acta,* 50, 628, 1967.

	3-(Methylthio)propanol	3-Methylthiopropionaldehyde
Other names	Methionol 3-Methylthiol propyl alcohol γ-Hydroxypropyl methyl sulfide γ-Methyl mercaptopropyl alcohol Methyl-3-hydroxypropyl sulfide	Methional β-Methiopropionaldehyde (true name) Methyl-β-mercaptopropionaldehyde 3-Methylthiopropanol
Empirical formula	$C_4H_{10}OS$	C_4H_8OS
Structure	$CH_3-S-CH_2-CH_2-CH_2OH$	$CH_3-S-CH_2-CH_2-C(=O)H$
Physical/chemical characteristics		
Appearance	Pale-yellowish, mobile liquid	Pale-yellow, mobile liquid
Molecular weight	106.19	104.10
Boiling point	195°C 93–94°C at 17 mm Hg	165°C
Specific gravity	1.03	1.04
Refractive index	1.4832 at 20°C	
Solubility	Slightly soluble in water; soluble in alcohol, propylene glycol, and oils	Almost insoluble in water; soluble in alcohol and propylene glycol
Organoleptic characteristics	Powerful sweet soup or meat-like odor and flavor in high dilution	Powerful onion, meat-like odor; pleasant, warm, meat and soup-like flavor at low levels
Synthesis	From propylene chlorhydrin and sodium hydrosulfide[6] H_3C-SH + $CH_2=CH-CH_2OH$ → $CH_3-S-CH_2-CH_2-CH_2OH$	By transamination and decarboxylation of various amino acids; by oxidation of the alcohol
Natural occurrence	Identified in the volatile portion of soy sauce,[7-9] and in tomatoes and wine[10,11]	In potato and potato chips
Reported uses[3]	Non-alcoholic beverages 0.05 ppm Candy 0.1 ppm	Non-alcoholic beverages 0.35 ppm Ice cream, ices, etc. 0.01–1.0 ppm Candy 0.01–1.0 ppm Baked goods 0.66 ppm Condiments 0.62 ppm Meats 1.9 ppm
Regulatory status	FEMA No. 3415	FDA GRAS; FEMA No. 2747

REFERENCES

For References 1–5, see end of Part III.

6. **Kisner,** *J. Am. Chem. Soc.*, 50, 2352, 1928.
7. **Akabori et al.,** *Proc. Imp. Acad. Tokyo*, 12, 136, 1936.
8. **Buttery et al.,** *J. Food Chem.*, 19, 524, 1971.
9. **Johnson et al.,** *Chem. Ind.*, (London), p. 1212, 1971.
10. **Weurman** Report, 3rd ed.
11. **Müller et al.,** *Am. J. Enol. Vitic.*, 22, 156, 1971.

	3-Methylthiopropyl isothiocyanate	2-Methyl-3-tolyl propionaldehyde (mixed *o*-, *m*-, *p*-)
Other names	3-Methylmercaptopropyl isothiocyanate Ibervirin®	
Empirical formula	$C_5H_9NS_2$	$C_{11}H_{14}O$
Structure	$H_3C-S-CH_2-CH_2-CH_2-NCS$	CH_3 on $CH_2-CH-CHO$ … (as printed: CH_2-CH_2-CHO with CH_3 above, attached to benzene ring bearing CH_3)
Physical/chemical characteristics[5]		
Appearance	Liquid	Colorless, oily liquid
Assay		Commercial product mainly *para*-isomer with some *ortho*- and only traces of *meta*-isomer
Molecular weight	147.26	162.23
Boiling point	116°C at 9 mm Hg	
Solubility		Very slightly soluble in water; soluble in alcohol
Organoleptic characteristics	Characteristic radish odor	Extremely sweet, fruity odor; somewhat balsamic with bitter-almond notes
Synthesis	From 3-methylthiopropylamine and $CSCl_2$[6]	Commercially available as a mixture of the three isomers; from tolyl aldehyde (*o*-, *m*-, *p*- mixture) by condensation with propionaldehyde followed by hydrogenation
Natural occurrence	Reported found in the hydrolysates of crucifer seeds and *Lesquerella* seed meal,[8] and in *Iberis sempervirens* seeds[9]	Not reported found in nature
Reported uses[3]	Gelatins and puddings 4.0 ppm Frozen pies 4.0 ppm	Non-alcoholic beverages 0.05 ppm Ice cream, ices, etc. 1.0 ppm Candy 1.0 ppm Baked goods 1.0 ppm
Regulatory status	FEMA No. 3312	FDA 121.1164; FEMA No. 2748

REFERENCES

For References 1–5, see end of Part III.

6. **Kjaer, Gruelin and Larsen,** *Acta Chem. Scand.*, 9, 1143, 1955; *Chem. Abstr.*, 49, 12782f, 1955.
7. **Daxenbichler, VanEtten, Brown, and Jones,** *J. Agric. Food Chem.*, 12(2), 127, 1964.
8. **Daxenbichler, VanEtten, Zobel, and Wolff,** *J. Am. Oil Chem. Soc.*, 39, 244, 1962.
9. **Kjaer, Gruelin, and Larsen,** *Acta Chem. Scand.*, 9, 1143, 1955; *Chem. Abstr.*, 49, 12782f, 1955.

	2-Methyl undecanal	Methyl 9-undecenoate
Other names	"Aldehyde C-12, M.N.A." 2-Methylhendecanal Methyl nonyl acetaldehyde	Methyl 9-hendecenoate Methyl 9-undecylenate
Empirical formula	$C_{12}H_{24}O$	$C_{12}H_{22}O_2$
Structure	$CH_3-(CH_2)_8-CH(CH_3)-CHO$	$CH_3-CH=CH-(CH_2)_7-C(=O)-O-CH_3$
Physical/chemical characteristics[5]		
Appearance	Colorless to slightly yellow liquid	Colorless to pale straw-yellow, oily liquid
Assay	94% min[2]	
Molecular weight	184.32	198.31
Melting point		−27.5°C
Boiling point	114°C (232°C) at 10 mm Hg	247–248°C; 142°C at 26 mm Hg
Flash point	69°C	
Specific gravity	0.822–0.830 at 25°/25°C[2]	0.8861 at 20°C
Refractive index	1.4310–1.4360 at 20°C[2]	1.4388 at 20°C
Acid value	10 max[2]	
Solubility	Soluble in alcohol and most organic solvents; insoluble in water; 1:4 in 80% alcohol	Insoluble in water; soluble in alcohol
Organoleptic characteristics	Characteristic fatty odor assuming a floral note on dilution; acrid, fatty unpleasant flavor at high levels; at low levels the flavor is honey- and nut-like	Sweet, persistent, light, fatty, green odor reminiscent of wine and iron; oily, brandy-like flavor at low concentrations; unpleasant at high levels
Synthesis	By cracking the corresponding glycidic acid;[2] from α-nonyl acrolein; from methyl nonyl ketone and ethyl monochloroacetate in the presence of sodium ethylate	By direct esterification of the acid with methanol and concentrated H_2SO_4[6] or HCl;[7] by thermal decomposition of methyl ricinoleate[8]
Natural occurrence	Not reported found in nature	Not reported found in nature
Reported uses[3]	Non-alcoholic beverages 0.31 ppm Ice cream, ices, etc. 0.11 ppm Candy 0.94 ppm Baked goods 1.3 ppm Gelatins and puddings 0.50–2.5 ppm Chewing gum 0.20 ppm Jellies 0.33 ppm	Non-alcoholic beverages 3.7 ppm Ice cream, ices, etc. 6.7 ppm Candy 22 ppm Baked goods 22 ppm
Regulatory status	FDA 121.1164; FEMA No. 2749	FDA 121.1164; FEMA No. 2750

REFERENCES

For References 1–5, see end of Part III.

6. **Komppa,** *Ber. Dtsch. Chem. Ges.*, 34, 897, 1901.
7. **Bornwater,** *Recl. Trav. Chim. Pays-Bas*, 2g, 410, 1907.
8. **Grün and Wirth,** *Ber. Dtsch. Chem. Ges.*, 55, 2208, 1922; *Chem. Zentralbl.*, 2, 747, 1928.

	Methyl undecyl ketone	Methyl 2-undecynoate
Other names	Hendecyl methyl ketone 2-Tridecanone	Methyl decine carbonate Methyl decyne carbonate
Empirical formula	$C_{13}H_{26}O$	$C_{12}H_{20}O_2$
Structure	$CH_3-C(=O)-(CH_2)_{10}-CH_3$	$CH_3-(CH_2)_7-C\equiv C-C(=O)-O-CH_3$
Physical/chemical characteristics[5]		
Appearance	White wax or colorless crystalline platelets	Colorless, oily liquid
Molecular weight	198.34	196.29
Melting point	28–29°C	
Boiling point	263°C 195°C at 110 mm Hg	121°C at 5 mm Hg; 230°C at 760 mm Hg
Specific gravity	0.82168 at 30°C	0.9177 at 20°C; 0.940 at 15.5°C
Refractive index	1.43175 at 30°C	1.4465 (1.4530) at 20°C
Solubility	Insoluble in water; very soluble in alcohol, ether, acetone, and benzene	Insoluble in water; soluble in alcohol
Organoleptic characteristics	Warm-oily, herbaceous odor reminiscent of nut	Powerful, waxy, green, floral odor; sweet flavor reminiscent of violet
Synthesis	By heating a mixture of lauric acid and acetic acid over thorium oxide at 450°C[6]	By methylation of the corresponding acid at the boil in the presence of HCl[7] or at room temperature in the presence of H_2SO_4.[8] The acid can be prepared from methyl nonyl ketone by reaction with phosphorous pentachloride, which yields the acetylenic compound. Treatment of this with potassium carbonate yields undecyne, which is then converted to the acid by treatment with Na and methyl chlorocarbonate.
Natural occurrences	Reported found in coconut and palm oils; also in the oil of *Schizandra nigra* Max. (Matsubusa)	Not reported found in nature
Reported uses[3]	Ice cream, ices, etc. 3.0 ppm Candy 3.0 ppm Baked goods 5.0 ppm Gelatins and puddings 3.0 ppm Margarine 3.0 ppm	Non-alcoholic beverages 0.10–5.0 ppm Ice cream, ices, etc. 20 ppm Candy 15 ppm Baked goods 15 ppm
Regulatory status	FEMA No. 3388	FDA 121.1164; FEMA No. 2751

REFERENCES

For References 1–5, see end of Part III.

6. **Dreger et al.,** *Ind. Eng. Chem.*, 36, 610, 1944.
7. **Osskerko,** *Chem. Ber.*, 70, 56, 1937.
8. **Jeffery and Vogel,** *J. Chem. Soc.*, p. 678, 1948.

	Methyl valerate	2-Methylvaleric acid
Other names	Methyl pentanoate Methyl-*n*-valerate Methyl valerianate	2-Methylpentanoic-1-acid Methyl propyl acetic acid α-Methyl valeric acid Pentane-2-carboxylic acid
Empirical formula	$C_6H_{12}O_2$	$C_6H_{12}O_2$
Structure	$CH_3{-}O{-}\overset{\overset{\displaystyle O}{\Vert}}{C}{-}(CH_2)_3{-}CH_3$	$CH_3{-}CH_2{-}CH_2{-}\underset{\underset{\displaystyle CH_3}{\vert}}{CH}{-}\overset{\overset{\displaystyle O}{\Vert}}{C}{-}OH$
Physical/chemical characteristics[5]		
Appearance	Colorless, mobile liquid	Solidifies at room temperature in a mass similar to ice; otherwise a colorless liquid
Molecular weight	116.16	116.16
Boiling point	126–127°C	195–196°C (194°C)
Specific gravity	0.8895 at 20°C	0.9230 at 20°C
Refractive index	1.3969 at 20°C	1.4136 at 20°C
Optical rotation		*d*-, +18.5° at 16°C, undiluted; *l*-, −18.4° at 25°C, undiluted
Solubility	Very slightly soluble in water; soluble in propylene glycol, alcohol, and oils	Soluble in water; miscible with alcohol and some essential oils
Organoleptic characteristics	Pungent, green-fruity odor, apple and pineapple-like; corresponding flavor at 10–50 ppm	Powerful, pungent, acrid odor; below 10 ppm agreeable, sour, oily flavor; at higher concentrations, too acidic
Synthesis	By direct esterification of valeric acid with methanol in the presence of concentrated H_2SO_4.[6] *Note*: this ester hydrolyzes readily.	By catalytic oxidation of 2-methyl pentanealdehyde; from 2-chloropentane with sodium and CO_2 under pressure; by decarboxylation of methyl propyl malonic acid. Two optically active isomers are known
Natural occurrence	Reported found in small amounts in the essential oil from the fruits of *Ananas sativus* Schult [7]	In coffee, wine, and cheese
Reported uses[3]	Non-alcoholic beverages 9.1 ppm Ice cream, ices, etc. 25 ppm Candy 28 ppm Baked goods 39 ppm	Candy 0.80 ppm
Regulatory status	FDA 121.1164; FEMA No. 2752	FDA 121.1164; FEMA No. 2754

REFERENCES

For References 1–5, see end of Part III.

6. **Vogel,** *J. Chem. Soc.*, p. 630, 1948.
7. **Haagen-Smit et al.,** *J. Am. Chem. Soc.*, 67, 1647, 1945.

	2-Methyl-5-vinylpyrazine	4-Methyl-5-vinylthiazole
Empirical formula	$C_7H_8N_2$	C_6H_7NS
Structure	$H_2C{=}HC$–(pyrazine ring)–CH_3	$H_2C{=}HC$–(thiazole ring, S, N)–H_3C
Physical/chemical characteristics		
Assay	C: 69.97%; H: 6.71%; N: 23.31%	
Molecular weight	120.16	125.19
Boiling point	65–66°C at 12 mm Hg	78–82°C at 18–20 mm Hg
Mass spectra[8]	39 (16), 40 (7), 52 (60), 54 (29), 66 (3), 67 (2), 79 (10), 80 (1), 93 (12), 94 (9), 119 (17), 120 (100)	
IR spectra[8]	S (6.7, 8.6, 9.7, 10.1, 10.7) M (3.3, 7.1, 7.2, 7.4, 7.6, 10.8, 11.1) W (5.3, 6.1, 6.5, 7.8, 7.9, 8.2, 9.2, 11.4)	
UV spectra[8]	(233.3, 290.5, 297.0 nm)	
Solubility		Soluble in alcohol
Synthesis	By methylation of the Mannich base obtained by reacting formaldehyde and dimethylamine with dimethyl-2,5-pyrazine, followed by the Hoffman degradation[6]	By a patented process[6]
Natural occurrence	Reported to be present in the following foods: coffee, dairy products, peanuts, filberts, potato products[7]	Compound of cocoa aroma;[7] compound of yellow passion fruit aroma[8]
Reported uses[3]	Non-alcoholic beverages 10.0 ppm Ice cream, ices, etc. 10.0 ppm Candy 10.0 ppm Gelatins and puddings 10.0 ppm	Ice cream, ices, etc. 2.0 ppm Candy 2.0 ppm Baked goods 8.0 ppm Gelatins and puddings 8.0 ppm Meat, meat sauces, soups 6.0 ppm
Regulatory status	FEMA No. 3211	FEMA No. 3313

REFERENCES

For References 1–5, see end of Part III.

6. **Kamal et al.,** *J. Org. Chem.,* 27, 1355, 1962.
7. **Maga and Sizer,** *Crit. Rev. Food Technol.,* 4(1), 39, 1973.
8. **Bondarovich et al.,** *J. Agric. Food Chem.,* 15, 1093, 1967.
9. U.S.S.R. Patent 164, 878, 1964.
10. **Stoll, Dietrich, Sunolt and Winter,** *Helv. Chim. Acta,* 50(7), 2065, 1967; *Chem. Abstr.,* 61, 13809b, 1964; *Chem. Abstr.,* 66, 9499f, 1967.
11. **Winter and Klöti,** *Helv. Chim. Acta,* 55(6), 1916, 1972.

	Myrcene	Myristaldehyde
Other names	7-Methyl-3-methylene-1,6-octadiene	Aldehyde C-14 (true) Myristic aldehyde Tetradecanal *n*-Tetradecyl aldehyde Tetradecyl aldehyde
Empirical formula	$C_{10}H_{16}$	$C_{14}H_{28}O$
Structure		$CH_3-(CH_2)_{12}-CHO$
Physical/chemical characteristics[5]		
Appearance	Oily liquid	Colorless to slightly yellow liquid; polymerizes readily to a white mass, melting at 65°C
Assay		85–92%
Molecular weight	136.23	212.36
Melting point		20.5°C (23°C)
Boiling point	65–66°C (166–168°C) at 20 mm Hg	155°C (260°C) at 10 mm Hg
Specific gravity	0.8013 at 15°C	0.825–0.835 at 25°C
Refractive index	1.4650 at 20°C	1.4380–1.4450 at 20°C
Solubility	Insoluble in water; soluble in alcohol, chloroform, and ether	1:1 in 80% alcohol; insoluble in water and most common organic solvents
Organoleptic characteristics	Pleasant odor	Strong, fatty, orris-like odor; sweet, fatty, "citrus-peel" flavor (diluted)
Synthesis	From linalool.[6] For structure identification see references 7 and 8.	Industrially prepared from the corresponding myristic acid
Natural occurrence	Reported found in *Mircia acris* D.C.; in the distillates from leaves of *Rhus cotinus* and *Barosma venustum* (52% and 43%, respectively); in lemon grass, cypress, artemisia, orange, and lemon; in the fruits of *Phellodendron amurense* (92%) and *Phellodendron japonicum*; in the oils of *Picea balsamea*, *Tsuga canadensis*, *Abies balsamea*, clary sage, and others.	Reported found in the essential oils of *Ocotea usambarensis* Engl., *Pinus sabiniana* Dougl., and probably others
Reported uses[3]	Non-alcoholic beverages 4.4 ppm Ice cream, ices, etc. 6.4 ppm Candy 0.50–13 ppm Baked goods 4.9 ppm	Non-alcoholic beverages 2.7 ppm Ice cream, ices, etc. 0.06–8.0 ppm Candy 1.9 ppm Baked goods 0.08–24 ppm Gelatins and puddings 0.15 ppm
Regulatory status	FDA 121.1164; FEMA No. 2762	FDA 121.1164; FEMA No. 2763

REFERENCES

For References 1–5, see end of Part III.

6. **Arbusow and Abramow,** *Ber. Dtsch. Chem. Ges.*, 67, 1942, 1934.
7. **Ružička and Stoll,** *Helv. Chim. Acta*, 7, 272, 1924.
8. **Power and Kleber,** *Pharm. Rundsch.*, 13, 60, 1895.

	Myrtenol	β-Naphthyl ethyl ether
Other names	6,6-Dimethyl-2-oxymethlybicyclo[1.1.3]-hept-2-ene 10-Hydroxy-2-pinene 2-Pinen-10-ol	"Bromelia" 2-Ethoxynaphthalene Ethyl-2-naphthyl ether Ethyl-β-naphthyl ether β-Naphthol ethyl ether Nerolin II Nerolin bromelia
Empirical formula	$C_{10}H_{16}O$	$C_{12}H_{12}O$
Structure	CH_2OH, H_3C, H_3C — *l*-form; CH_2OH, H_3C, H_3C — *d*-form	$-O-C_2H_5$
Physical/chemical characteristics[5]		
Appearance		White, crystalline solid
Molecular weight	152.24	172.23
Melting point		37°C
Boiling point	218°C at 771 mm Hg	282°C
Congealing point	103–104°C at 11 mm Hg	>35.0°C[2]
Specific gravity	0.9763 at 20°C	1.064 at 36°C
Refractive index	1.4967 at 20°C	1.597 at 36°C
Optical rotation	+45, 45°	
Solubility		1:5 in 95% alcohol;[2] insoluble in water; poorly soluble in propylene glycol
Organoleptic characteristics		Odor suggestive of orange blossom with faint, fruity undertone; corresponding sweet taste suggestive of strawberry (on extreme dilution only)
Synthesis	Can be obtained in *d,l*-form from α-pinene with SeO_2 in ethanol[6,7]	By esterification of β-naphthol with ethyl alcohol and sulfuric acid;[13] from β-naphtholsodium and diethylsulfate in weak aqueous base; from β-naphthol-sodium and ethyl bromide
Natural occurrence	In cranberry and bilberry,[8] mint,[9] and hop oil[10-12]	Not reported found in nature
Reported uses[3]	Non-alcoholic beverages 10 ppm Ice cream, ices, etc. 10 ppm Candy 10 ppm Gelatins and puddings 10 ppm Milk, dairy products 10 ppm	Non-alcoholic beverages 0.65 ppm Ice cream, ices, etc. 0.74 ppm Candy 2.8 ppm Baked goods 3.6 ppm Gelatins and puddings 0.12 ppm
Regulatory status	FEMA No. 3439	Status not fully defined by FDA; FEMA No. 2768

REFERENCES

For References 1–5, see end of Part III.

6. **DuPont et al.,** *Bull. Soc. Chim. Fr.*, 50, 536, 1935.
7. **Schmidt,** *Chem. Abstr.*, 37, 4715, 1943.
8. **von Sydow et al.,** *Lebensm. Wiss. Technol.*, 2, 78, 1969.
9. **Canova,** 5th International Congress on Essential Oils, Sao Paulo, 1971.
10. **Lammens et al.,** *J. Inst. Brew.*, 74, 341, 1968.
11. **Drawert et al.,** *Brauwinenschaft,* 22, 169, 1969.
12. **Naya et al.,** *Bull. Chem. Soc. Jap.*, 45, 2887, 1972.
13. *Ullmanns Enzyklopädie der Technischen Chemie,* 3rd ed., Vol. 12, Urban and Schwarzenberg, 1960, 605.

	2-Naphthyl mercaptan	β-Naphthyl methyl ether
Other names	2-Naphthalenethiol 2-Mercaptonaphthalene 2-Thionaphthol β-Thionaphthol	2-Methoxynaphthalene Methyl-2-naphthyl ether Methyl-β-naphthyl ether β-Naphthol methyl ether Nerolin I Nerolin yara-yara "Yara-yara"
Empirical formula	$C_{10}H_8S$	$C_{11}H_{10}O$
Structure	(naphthalene ring)—SH	(naphthalene ring)—O—CH_3
Physical/chemical characteristics		
Appearance	Crystals (from ethanol)	White, crystalline solid
Molecular weight	160.24	158.20
Melting point	81°C	72°C (71.5–73.5°C)
Boiling point	285°C (decomposes) 210.5°C at 100 mm Hg 189°C at 50°C 162°C at 20°C	274°C (sublimes when heated)
Congealing point		71.5–73.0°C[2]
Specific gravity	1.1607 at 20°C	
Refractive index	1.6802 at 20°C	
Solubility	Slightly soluble in water and diluted alkalis; very soluble in alcohol and ether	1 gm in 25 ml of 95% alcohol;[2] soluble in ether, chloroform, and benzene
Organoleptic characteristics	Disagreeable mercaptan-like odor	Intensely sweet, floral odor suggestive of orange blossoms; free from naphthol by-odor; sweet, strawberry taste
Synthesis	By catalytic hydrogenation of a sulfonic acid derivative of naphthalene;[6] by reduction of naphthalenesulfonyl chloride with zinc[7]	From potassium β-naphthol and methyl chloride at 300°C; by methylation of β-naphthol with dimethyl sulfate or by direct esterification with methyl alcohol[8]
Natural occurrence		Not reported found in nature
Reported uses[3]	Baked goods 0.5 ppm Meat, meat sauces, soups 0.5 ppm Cereals 0.5 ppm	
Regulatory status	FEMA No. 3314	Status not fully defined by FDA

REFERENCES

For References 1–5, see end of Part III.

6. **Lazier and Signaigo,** U.S. Patent 2,402,641; *Chem. Abstr.*, 40, 5768, 1946.
7. **Holt,** U.S. Patent 2,216,840, assigned to E. I. duPont de Nemours & Co., 1940.
8. **Gattermann,** *Chem. Ann.*, 244, 72, 1888.

	d-Neomenthol
Other names	*d*-β-Pulegomenthol
Empirical formula	$C_{10}H_{20}O$
Structure	Stereoisomer of menthol (see Menthol)
Physical/chemical characteristics[5]	
Appearance	Liquid
Molecular weight	156.27
Melting point	−22°C
Boiling point	211.5–212.5°C at 755–760 mm Hg; 107–108°C at 20 mm Hg; 98°C at 16 mm Hg; 87°C at 8 mm Hg
Specific gravity	0.903 at 15°/15°C; 0.8995 at 20°C; 0.8970 at 22°C
Refractive index	1.4617 at 17°C; 1.4594–1.4603 at 20–22°C
Optical rotation	+17°48′ at 16°C; +19°42′ at 18°C; +20° 0′ at 20°C
Solubility	Insoluble in water; soluble in alcohol and acetone
Organoleptic characteristics	Menthol-like odor; used in mixtures of menthols as a substitute for *l*-menthol
Synthesis	See References 6 and 7
Natural occurrence	Small quantities found in *Mentha arvensis*
Reported uses[3]	Non-alcoholic beverages 10 ppm Ice cream, ices, etc. 31 ppm Candy 50 ppm Baked goods 48 ppm
Regulatory status	FDA 121.1164; FEMA No. 2666

REFERENCES

For References 1–5, see end of Part III.

6. *Beilstein's Encyclopedia of Organic Chemistry*, B6[3], Springer Verlag, p. 139.
7. **Guenther,** *The Essential Oils,* Vol. 2, D. Van Nostrand Co., 1949, 220.

Nerol

Other names	2-*cis*-3,7-Dimethyl-2,6-octadien-1-ol *Note*: see Geraniol for *trans*-form.
Empirical formula	$C_{10}H_{18}O$
Structure	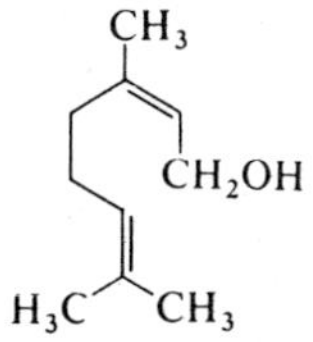 CH_3 / CH_2OH / H_3C CH_3
Physical/chemical characteristics[5]	
Appearance	Colorless liquid
Assay	95% min[10]
Molecular weight	154.24
Boiling point	225–226°C[9]
Specific gravity	0.8771–0.8788 at 15°C[9] 0.877–0.879 at 20/4°[10]
Refractive index	1.462 at 20°C;[9] 1.475–1.478 at 20°C[10]
Solubility	1:5 in 50% alcohol; insoluble in water; soluble in most common organic solvents
Organoleptic characteristics	Fresh, sweet, rose-like odor; bitter flavor
Synthesis	From pinene.[6] For selective synthesis of stereoisomers, see reference 7; for gas chromatographic analysis of geraniol and nerol, see reference 8.
Natural occurrence	Reported found in neroli oil (with geraniol) and in the essential oils of: lemon grass, Ceylon citronella, ylang-ylang, champaca, cajenne bois-de-rose, and bergamot; also in lemon, sweet orange, and petitgrain bergamot; in clary sage, lavandin, lavender, Mexican linaloe, myrrh, jasmine, Paraguay petitgrain. Also reported among the volatile constituents of currant aroma. *Helicrysum angustifolium* contains up to 30–50% nerol.
Reported uses[3]	Non-alcoholic beverages 1.4 ppm Ice cream, ices, etc. 3.9 ppm Candy 16 ppm Baked goods 19 ppm Gelatins and puddings 1.0–1.3 ppm Chewing gum 0.80 ppm
Regulatory status	FDA 121.1164; FEMA No. 2770

REFERENCES

For References 1–5, see end of Part III.

6. **Bedoukian,** *Perfumery and Flavoring Synthetics,* 2nd ed., American Elsevier, 1967.
7. **Yukawa, Hanafusa and Fujita,** *Bull. Chem. Soc. Jap.,* 2(37), 158, 1964.
8. **Naves and Odermatt,** *Bull. Soc. Chim. Fr.,* p. 377, 1958.
9. **Blumann and Zeitschell,** *Ber. Dtsch. Chem. Ges.,* 44, 2590, 1911.
10. The Givaudan Index, *Givaudan-Delawanna,* Inc., 1949.

	Nerolidol	Neryl acetate
Other names	3,7,11-Trimethyl-1,6,10-dodecatrien-3-ol	2-*cis*-3,7-Dimethyl-2,6-octadien-1-yl acetate
Empirical formula	$C_{15}H_{26}O$	$C_{12}H_{20}O_2$
Structure	*cis-* *trans-*	CH_3 ... $CH_2-O-C(=O)-CH_3$ ... H_3C CH_3
Physical/chemical characteristics[5]		
Appearance	Colorless to straw-yellow liquid	Colorless to slightly yellow, oily liquid
Assay	90% min	95% min
Molecular weight	222.36	196.29
Boiling point	89–100°C at 0.3 mm Hg; 145°C at 12 mm Hg	134°C (231°C) at 25 mm Hg
Specific gravity	0.870–0.880 at 25°/25°C;[2] 0.8788 at 16°C	0.903–0.906 at 15°C; 0.907–0.913 at 25°C
Refractive index	1.4789–1.4830 at 20°C;[2] 1.4801 at 16°C	1.4510–1.4540 (1.4600–1.4650) at 20°C
Optical rotation	Natural: +11° to +14°; synthetic: optically inactive	
Ester content	Not more than 0.5% (as nerolidyl acetate)	
Solubility	1:4 in 70% alcohol[2]	1:5–6 in 70% alcohol; soluble in the essential oils and most common organic solvents
Organoleptic characteristics	Faint, floral odor similar to rose and apple; unusually sweet, fresh, tenacious odor	Very sweet, floral, orange-blossom and rose-like odor; initially fresh and pungent, and then a honey-like flavor with a raspberry undertone
Synthesis	The natural product can be *dextro-* or *laevo*-rotatory, whereas the synthetic product is optically inactive. The double bond at positions 6–7 accounts for the *cis-* and *trans-* forms. For synthesis see references 6, 7, and 8.	By esterification of nerol with acetic acid; also prepared from myrcene by hydrobromination and esterification[11]
Natural occurrence	Reported found in over 25 natural sources, including the essential oils of neroli, ylang-ylang, Peru balsam; in currant aroma; also reported in *Dalbergia sissoo* (60%, *dl*-),[9] vervain, the distillation waters of petitgrain bigarade, *Helicrysum italicum*,[10] *Myrocarpus frondosus* and *Myrocarpus fastigiatus* (80%, *d*-), Tolu balsam, *Acacia farnesiana*, orange-flower water, Paraguay petitgrain, jasmine, *Melaleuca smithii*, and *Melaleuca viridiflora*	Reported found in the essential oils of lemon, neroli, and petitgrain bigarade
Reported uses[3]	Non-alcoholic beverages 0.91 ppm Ice cream, ices, etc. 0.92 ppm Candy 3.5 ppm Baked goods 2.0–8.0 ppm	Non-alcoholic beverages 1.3 ppm Ice cream, ices, etc. 1.6 ppm Candy 5.1 ppm Baked goods 15 ppm
Regulatory status	FDA 121.1164; FEMA No. 2772	FDA 121.1164; FEMA No. 2773

REFERENCES

For References 1–5, see end of Part III.

6. **Ružička,** *Helv. Chim. Acta,* 6, 488, 1923.
7. **Ružička and Capato,** *Helv. Chim. Acta,* 8, 259, 1925.
8. **Ofrier et al.,** *Helv. Chim. Acta,* 42, 2577, 1959.
9. **Kathpalia and Dutt,** *Indian Soap J.,* 18, 213, 1953.
10. **Rovesti,** *Riv. Ital. EPPOS,* 17, 19, 1935.
11. *Schimmel Ber.,* p. 200, 1955.

For further information on nerolidol see the following:

A. **Vlad and Sanček,** *Collect. Czech. Chem. Commun.,* 27(7), 1726, 1962.
B. **Naves,** *Bull. Soc. Chim. Fr.,* 18, 505, 1951; *Helv. Chim. Acta,* 30, 279, 1947.
C. **Naves and Odermatt,** *Bull. Soc. Chim. Fr.,* p. 377, 1958.
D. **Lebedewa et al.,** *J. Gen. Chem. USSR,* 21, 1183, 1951.
E. **Gildemeister and Hoffman,** *Die Aetherischen Oele,* Vol. 3b, Akademie Verlag, 1962, 239.

	Neryl butyrate	Neryl formate
Other names	2-*cis*-3,7-Dimethyl-2,6-octadien-1-yl butyrate Neryl-*n*-butyrate	2-*cis*-3,7-Dimethyl-2,6-octadien-1-yl formate
Empirical formula	$C_{14}H_{24}O_2$	$C_{11}H_{18}O_2$
Structure	$CH_2-O-\overset{O}{\overset{\Vert}{C}}-CH_2-CH_2-CH_3$ (neryl skeleton with CH_3, H_3C, CH_3)	$CH_2-O-\overset{O}{\overset{\Vert}{C}}-H$ (neryl skeleton with CH_3, H_3C, CH_3)
Physical/chemical characteristics[5]		
Appearance	Colorless liquid	Colorless liquid
Assay		91% min
Molecular weight	224.44	182.26
Boiling point	239°–240°C	215°C (225°C)
Specific gravity	0.8968–0.8986 at 15°/15°C; 0.895 at 15.5°C	0.9163–0.9169 at 15°C; 0.920 at 15.5°C
Refractive index	1.4539–1.4556 (1.4590) at 20°C	1.4558–1.4578 (1.4565) at 20°C
Solubility	1:6 in 80% alcohol; insoluble in water	1:10 in 70% alcohol; almost insoluble in water
Organoleptic characteristics	Very sweet, leafy, floral, neroli-like odor; sweet taste similar to cocoa	Sweet, herbaceous, green, penetrating odor suggestive of neroli and rose; sweet, tart, and fruity flavor
Synthesis	By azeotropic esterification of nerol with butyric acid or conventionally using butyric anhydride	By direct esterification of nerol. The pure ester tends to decompose readily.
Natural occurrence	Not reported found in nature	Not reported found in nature
Reported uses[3]	Non-alcoholic beverages 6.2 ppm Ice cream, ices, etc. 22 ppm Candy 16 ppm Baked goods 25 ppm	Non-alcoholic beverages 5.3 ppm Ice cream, ices, etc. 23 ppm Candy 17 ppm Baked goods 22 ppm
Regulatory status	FDA 121.1164; FEMA No. 2774	FDA 121.1164; FEMA No. 2776

REFERENCES

For References 1–5, see end of Part III.

	Neryl isobutyrate	Neryl isovalerate
Other names	2-*cis*-3,7-Dimethyl-2,6-octadien-1-yl isobutyrate	2-*cis*-3,7-Dimethyl-2,6-octadien-1-yl isovalerate Neryl-β-methylbutyrate
Empirical formula	$C_{14}H_{24}O_2$	$C_{15}H_{26}O_2$
Structure	CH_3; $CH_2-O-C(=O)-CH(CH_3)_2$; H_3C CH_3	CH_3; $CH_2-O-C(=O)-CH_2-CH(CH_3)_2$; H_3C CH_3
Physical/chemical characteristics[5]		
Appearance	Colorless liquid	Colorless liquid
Molecular weight	224.34	238.38
Boiling point	130°C (232°C) at 3 mm Hg	252°C
Specific gravity	0.8915–0.8936 at 15°C	0.8898 at 15°C
Refractive index	1.4508–1.4527 at 20°C	1.4531 at 20°C
Solubility	1:5 in 80% alcohol; almost insoluble in water	1:8 in 80% alcohol; insoluble in water
Organoleptic characteristics	Delicate, sweet, rose-like fragrance with a slightly fruity undertone; sweet, strawberry taste	Strong, rich, fruity odor reminiscent of bergamot, clary sage, and petitgrain; bitter taste; below 10 ppm flavor is berry-like, somewhat similar to black currant and apple.
Synthesis	By esterification of nerol using isobutyric acid or the anhydride	By esterification of nerol with isovaleric acid
Natural occurrence	In oil of hop	Not reported found in nature
Reported uses[3]	Non-alcoholic beverages 1.2 ppm Ice cream, ices, etc. 1.8 ppm Candy 3.3 ppm Baked goods 5.4 ppm	Non-alcoholic beverages 0.97 ppm Ice cream, ices, etc. 1.6 ppm Candy 4.0 ppm Baked goods 5.7 ppm
Regulatory status	FDA 121.1164; FEMA No. 2775	FDA 121.1164; FEMA No. 2778

REFERENCES

For References 1–5, see end of Part III.

	Neryl propionate	Nona-2-*trans*,-6-*cis*-dienal
Empirical formula	$C_{13}H_{22}O_2$	$C_9H_{14}O$
Structure	CH_3; $CH_2-O-C(=O)-CH_2-CH_3$; H_3C CH_3	$CH_3-CH_2-CH=CH-CH_2-CH_2-CH=CH-CHO$
Physical/chemical characteristics[5]		
Appearance	Colorless, oily liquid	Colorless to pale yellow oil
Molecular weight	210.31	138.21
Boiling point	233°C	187°C 85–87°C at 11 mm Hg 80°C at 4 mm Hg
Specific gravity	0.9044 at 15°/15°C	0.8678 at 20°C
Refractive index	1.4550 at 20°C	1.4460 at 20°C
Solubility	Very slightly soluble in water; soluble in alcohol; 1:15 in 70% alcohol	Almost insoluble in water; soluble in alcohol, propylene glycol, and oils
Organoleptic characteristics	Ether-like, sweet and intense fruity odor reminiscent of jasmine and rose; "jam-like" aroma of fruit preserves; sweet taste suggestive of plum	Odor reminiscent of green cucumber; the organoleptic character is attributed almost entirely to olfactory sensations
Synthesis	By esterification of nerol with propionic acid	From natural hexenol;[6-8] from steam distillation of fresh cucumber[9]
Natural occurrence	In oil of hop	Reported as naturally occuring in violet leaves and flowers, and in cucumber juice[10]
Reported uses[3]	Non-alcoholic beverages 6.3 ppm Ice cream, ices, etc. 23 ppm Candy 21 ppm Baked goods 21 ppm	Condiments 1.0 ppm Preserves and spreads 1.0 ppm
Regulatory status	FDA 121.1164; FEMA No. 2777	FEMA No. 3377

REFERENCES

For References 1–5, see end of Part III.

6. **Ružička, Schinz and Susz,** *Helv. Chim. Acta,* 27, 1561, 1944.
7. **Jutz,** *Chem. Ber.,* 92, 1983, 1959.
8. **Sondheimer,** *J. Am. Chem. Soc.,* 74, 4040, 1952.
9. **Takei et al.,** *J. Agric. Chem. Soc. Jap.,* 15, 193, 1939; *Chem Abstr.,* 33, 6524, 1939.
10. **Takei and Ono,** *J. Agric. Chem. Soc. Jap.,* 15, 193, 1939.

	2,4-Nonadienal	2,6-Nonadienal diethyl acetal
Empirical formula	$C_9H_{14}O$	$C_{13}H_{24}O_2$
Structure	$CH_3-(CH_2)_3-CH{=}CH-CH{=}CH-CHO$	$CH_3-CH_2-CH{=}CH-CH_2-CH_2-CH{=}CH-CH(OCH_2CH_3)_2$
Physical/chemical characteristics[5]		
Appearance		Colorless oily liquid
Molecular weight	138.20	212.34
Boiling point	72–74°C at 3 mm Hg 97–99°C at 10 mm Hg	67–70°C at 0.2 mm Hg
Specific gravity	0.862 at 25°C	
Refractive index	1.5184 at 27°C	1.4439 at 18°C
Solubility		Almost insoluble in water; soluble in alcohol and oils
Organoleptic characteristics		Odor reminiscent of green cucumber
Synthesis	By the method of Marshall and Whiting;[6] by reduction with $LiAlH_4$ of dienoic acid prepared by the Doebner synthesis, followed by oxidation of the resulting dienol with MnO_2 to the corresponding 2,4-dienol[7]	By direct hydrogenation of the corresponding acetylenic acetal, using a palladium/calcium carbonate catalyst (2-*cis*,6-*cis*-form)[6]
Natural occurrence	Reported found in the oxidized flavor of skim milk;[8] in peas by enzymatic formation from lipids;[9] in salmon oil;[10] in sunflower oil (*cis-*, *trans-*, and *trans, trans-*form);[11] in the autooxidation of lard;[12] in frozen peas;[13] in tomatoes;[14] and as a volatile component in fish products[15]	
Reported uses[3]	Baked goods 0.20 ppm Condiments 0.20 ppm Meat, meat sauces, soups 0.20 ppm Milk, dairy products 0.20 ppm Cereals 0.20 ppm	Ice cream, ices, etc. 0.1 ppm Candy 0.1 ppm Condiments, pickles 0.1 ppm Meat, meat sauces, soups 0.1 ppm
Regulatory status	FEMA No. 3212	FEMA No. 3378

REFERENCES

For References 1–5, see end of Part III
6. Pippen and Nonaka, *J. Org. Chem.*, 23, 1580, 1958.
7. Forss and Hancox, *Aust. J. Chem.*, 9, 420, 1956.
8. Forss, Pont and Stark, *J. Dairy Res.*, 22, 345, 1955.
9. Grosch, *Z. Lebensm. Unters. Forsch.*, 135(2), 75, 1967; *Chem. Abstr.*, 68, 58612w, 1968.
10. Wyatt and Day, *J. Food Sci.*, 28, 305, 1963.
11. Swoboda and Lee, *J. Sci. Food Agric.*, 16(11), 680, 1965.
12. Gaddis, Ellis and Curie, *J. Am. Oil Chem. Soc.*, 43(3), 147, 1966.
13. Whitfield and Shipton, *J. Food Sci.*, 31(3), 328, 1966.
14. Buttery et al., *J. Agric. Food Chem.*, 19(3), 524, 1971.
15. *J. Chromatogr.*, 61(1), 133, 1971.
16. Sandheimer, *J. Am. Chem. Soc.*, 74, 4040, 1952.

	2,6-Nonadien-l-ol	γ-Nonalactone
Other names	"Cucumber alcohol" Nonadienol *trans*-2-*cis*-6-Nonadienol (natural) *trans*-2-*trans*-6-Nonadienol (synthetic) "Violet leaf alcohol"	Aldehyde C-18 (so called) γ-Amyl butyrolactone Coconut aldehyde 4-Hydroxynonanoic acid, γ-lactone Nonanolide-1,4 Prunolide®*
Empirical formula	$C_9H_{16}O$	$C_9H_{16}O_6$
Structure	$CH_3-CH_2-CH{=}CH-(CH_2)_2-CH{=}CH-CH_2OH$	$CH_3-(CH_2)_4-CH$ (ring: O, C=O, H_2C-CH_2)
Physical/chemical characteristics[5]		
Appearance	Colorless, oily liquid	Colorless to very pale, straw-yellow liquid
Assay		97% min[2] (98%)
Molecular weight	140.23	156.23
Boiling point	96–98°C (196°C) at 12 mm Hg	243°C
Flash point		>100°C (132°C)
Specific gravity	0.87	0.958–0.966 at 25°/25°C;[2] 0.969–0.971 at 15°/15°C
Refractive index		1.4460–1.4500[2] (1.4465–1.4480) at 20°C
Acid value		Not more than 5[2]
Solubility	Very slightly soluble in water; soluble in alcohol, propylene glycol, and essential oils	1:5 in 50% alcohol;[2] 1:2 in 60% alcohol; insoluble in water; soluble in most common organic solvents
Organoleptic characteristics		Strong odor reminiscent of coconut; fatty, peculiar taste
Synthesis	Prepared from naturally occurring 3-hexen-1-ol (leaf alcohol)[6] to yield the natural *trans*-2-*cis* form, or from synthetic hexenol to yield the *trans*-2-*trans* form	By reacting methylacrylate and hexanol in the presence of ditertiarybutyl peroxide; by condensation of undecylenic acid and malonic acid; by lactonization of nonenoic acid
Natural occurrence	Originally isolated in the oils of violet leaf[7] and violet flowers[8] (hence, the name "violet leaf alcohol"). Its structure has been studied by L. Ružička.[9] It has been isolated also in cucumber.[10]	Reported found in nature—peaches and apricots,[11] roasted barley,[12] rum,[13] and tomato[14]
Reported uses[3]	Non-alcoholic beverages 0.01 ppm Alcoholic beverages 0.01 ppm Ice cream, ices, etc. 0.05 ppm Candy 0.05–0.50 ppm Baked goods 0.01 ppm	Non-alcoholic beverages 11 ppm Ice cream, ices, etc. 14 ppm Candy 33 ppm Baked goods 55 ppm Gelatins and puddings 28 ppm Chewing gum 15 ppm Icings 25 ppm
Regulatory status	FDA 121.1164; FEMA No. 2780	FDA 121.1164; FEMA No. 2781

* Givaudan-Delawanna, Inc., 330 W. 42nd St., New York, N.Y

REFERENCES

For References 1–5, see end of Part III.

6. **Takei et al.,** *Chem. Zentralbl.,* 2, 3396, 1938.
7. **Ružička and Schinz,** *Helv. Chim. Acta,* 18, 381, 1935.
8. **Ružička,** *Chem. Zentralbl.,* 2, 3170, 1938.
9. **Ružička et al.,** *Helv. Chim. Acta,* 27, 1561, 1944.
10. **Takei et al.,** *Chem. Zentralbl.,* 2, 3705, 1939.
11. **Ruys,** *Fr. Ses Parfums,* 9(47), 34, 1966.
12. **Wang et al.,** *J. Agric. Biol. Chem.,* 34(4), 561, 1970.
13. **Liebich et al.,** *J. Chromatogr.,* 8, 527, 1970.
14. **Buttery et al.,** *J. Agric. Food Chem.,* 19(13), 524, 1971.

	n-Nonanal	1,3-Nonanediol acetate, mixed esters
Other names	Aldehyde C-9 Nonanoic aldehyde *n*-Nonyl aldehyde α-Oxononane Pelargonic aldehyde	Hexylene glycol diacetate 3-Hexyl-1,3-propanediol acetate, mixed esters Nonanediol-1,3-acetate "Octylcrotonyl acetate" *Also*: diacetate; diasmol; diasmylacetate; drago-jasimia; jasmelia; jasmonyl; jersemal; nonane; tepylacetate; ysminia
Empirical formula	$C_9H_{18}O$	$C_{13}H_{24}O_4$
Structure	$CH_3-(CH_2)_7-CHO$	$CH_3-C(=O)-O-(CH_2)_2-CH(-O-C(=O)-CH_3)-(CH_2)_5-CH_3$
Physical/chemical characteristics[5]		
Appearance	Colorless to light-yellow liquid	Colorless to slightly yellow liquid
Assay	92% min[2]	
Molecular weight	142.24	244.34
Boiling point	91–92°C (191°C) at 22 mm Hg	
Flash point		>100°C
Specific gravity	0.820–0.830 at 25°/25°C;[2] 0.8280 at 16°C	0.964–0.970 (0.960–0.970) at 25°/25°C
Refractive index	1.4220–1.4290 at 20°C;[2] 1.4274 at 18°C	1.4410–1.4450 (1.4400–1.4500) at 20°C
Saponification value		280–320 (300–330)
Acid value	Not more than 10[2]	1 max
Solubility	Soluble in alcohol and most common organic solvents; 1:3 in 70% alcohol; insoluble in water	1:4 in 60% alcohol; 1:2 in 70% alcohol; slightly soluble in water
Organoleptic characteristics	Strong, fatty odor developing an orange and rose note on dilution; fatty, citrus-like flavor	Intense floral odor reminiscent of jasmine
Synthesis	By catalytic oxidation of the corresponding alcohol (*n*-nonanol) or reduction of the corresponding acid	The commercial product is a mixture of the two possible isomers; the starting material is usually 1-octene
Natural occurrence	Reported as a constituent in the oils of sweet and bitter orange, mandarin, lemon, lime, orris, Ceylon cinnamon, ginger, *Xanthoxylum rhetsa*, rose,[6] and clary sage;[7] in the turpentines from *Pinus jeffreyi* Grev. and Balf., *Pinus sabiniana* Dangl., and others	Not reported found in nature
Reported uses[3]	Non-alcoholic beverages 1.3 ppm Ice cream, ices, etc. 1.3 ppm Candy 4.1 ppm Baked goods 2.3 ppm Gelatins and puddings 6.0 ppm Chewing gum 0.20–38 ppm	Non-alcoholic beverages 0.30–1.0 ppm Ice cream, ices, etc. 0.50–1.0 ppm Candy 1.5–6.0 ppm Baked goods 1.5–4.0 ppm
Regulatory status	FDA 121.1164; FEMA No. 2782	FDA 121.1164; FEMA No. 2783

REFERENCES

For References 1–5, see end of Part III.

6. **Calvarano**, *Essenze Deriv. Agrum.*, 28, 157, 1958.
7. **Tesseire**, *Recherches*, 9, 10, 1959.

	Nonanoic acid	2-Nonanol
Other names	Nonoic acid *n*-Nonylic acid Octane-1-carboxylic acid Pelargonic acid	Methyl *n*-heptyl carbinol
Empirical formula	$C_9H_{18}O_2$	$C_9H_{20}O$
Structure	$CH_3-CH_2)_7-COOH$	$CH_3-(CH_2)_6-CH(OH)-CH_3$
Physical/chemical characteristics[5]		
Appearance	Colorless, oily liquid	Colorless liquid
Assay	>98%	
Molecular weight	158.24	144.26
Melting point	9°C	
Boiling point	255.6°C	*d*-, 105°C at 19 mm Hg *dl*-, 193–194°C *dl*-, 91°C at 12 mm Hg
Congealing point	>12.24°C	-35°C
Specific gravity	0.90552 at 20°C	*d*-, 0.8230 at 20°C *dl*-, 0.8471 at 20°C
Refractive index	1.4379 at 20°C	*d*-, 1.4299 at 20°C
Optical rotation		+8.98° at 19°C
Solubility	1:8 in 50% alcohol; 1:3 in 60% alcohol; insoluble in water; soluble in most organic solvents	Insoluble in water; soluble in alcohol and oils
Organoleptic characteristics	Fatty, characteristic odor; corresponding unpleasant taste	Powerful fruity-green odor
Synthesis	By oxidation of methylnonyl ketone; by oxidation of oleic acid; or from heptyl iodide via malonic ester synthesis	From nonene[10]
Natural occurrence	Reported found in several essential oils, either free or esterified: rose, geranium, orris, *Litsea cubeba, Artemisia arborescens* L., hops, *Chamaecyparis pisifera* Endl.,[6] *Eremocitrus glauca* L.,[7] French lavender,[8] and in oak musk[9]	Reported found in *Ruta pinnata* leaf oil[11] and in coconut oil[12]
Reported uses[3]	Non-alcoholic beverages 1.8 ppm Ice cream, ices, etc. 7.8 ppm Candy 6.6 ppm Baked goods 13 ppm Shortening 10 ppm	Baked goods 15.0 ppm Condiments, pickles 80.0 ppm Dips, spreads 12.0 ppm
Regulatory status	FDA 121.1164; FEMA No. 2784	FEMA No. 3315

REFERENCES

For References 1–5, see end of Part III.

6. **Uchida,** *Chem. Zentralbl.*, 2, 1577, 2198, 1928.
7. **Hitchcock and Jones,** *Chem. Zentralbl.*, 1, 2281, 1937.
8. **Seidel et al.,** *Helv. Chim. Acta*, 27, 663, 1944.
9. **Stoll and Scherrer,** *Chem. Zentralbl.*, 2, 2512, 1938.
10. **Kulesza et al.,** *Riechst. Aromen Koerperpflegem.*, 19(1), 8, 1969.
11. **Estevez-Reyes and Gonzalez-Gonzalez,** *An. Quim.*, 66(2), 167, 1970; 73:48483, 1970.
12. **Jasperson and Jones,** *J. Soc. Chem. Ind.*, 66, 15, 1967.

	2-Nonanone		3-Nonanone	
Other names	Methyl heptyl ketone		Ethyl hexyl ketone 3-Oxononanone	
Empirical formula	$C_9H_{18}O$		$C_9H_{18}O$	
Structure	$CH_3-\overset{O}{\overset{\parallel}{C}}-(CH_2)_6-CH_3$		$CH_3-CH_2-CH_2-CH_2-CH_2-CH_2-\overset{O}{\overset{\parallel}{C}}-CH_2-CH_3$	
Physical/chemical characteristics				
Appearance	Colorless, oily liquid; solidifies when cold			
Molecular weight	142.24		142	
Melting point	-15°C			
Boiling point	194–196°C		190°C 80°C at 20 mm Hg	
Congealing point			-8°C	
Specific gravity	0.8317 at 20°C		0.825 at 20°C	
Refractive index	1.4175 at 20°C		1.4208 at 20°C	
Solubility	Insoluble in water; soluble in alcohol			
Organoleptic characteristics	Characteristic rue odor; rose and tea-like flavor			
Synthesis	By dry distillation of barium caprylate and barium acetate at 0.5–2 mm Hg;[6] or by oxidation of methyl heptyl carbinol at room temperature using chromic acid.[7] It can be isolated also from natural products by fractional distillation		By prolonged boiling of *o*-propionic acid triethyl ester with excess hexylmagnesium bromide in ether under a nitrogen atmosphere[12]	
Natural occurrence	Reported found in the essential oil of rue (*Ruta angustifolia* Pers., *R. Bracteosa, R. montana* L., and *R. gravelolens*). Also identified in the oil of carnation,[8] in coconut oil,[9] and in the distillate from leaves of a variety of *Boronia ledifolia.*[10] Reported found in mixture with methyl nonyl ketone in the essential oil of *Ruta chalepensis.*[11]		In banana,[13] passion fruit,[14] and cooked beef[15]	
Reported uses[3]	Non-alcoholic beverages	0.55 ppm	Non-alcoholic beverages	20 ppm
	Ice cream, ices, etc.	0.10–1.0 ppm	Alcoholic beverages	20 ppm
	Candy	0.40–4.0 ppm	Candy	20 ppm
	Baked goods	0.40–4.0 ppm	Baked goods	20 ppm
			Gelatins and puddings	20 ppm
			Chewing gum	20 ppm
			Milk, dairy products	20 ppm
			Cereals	20 ppm
Regulatory status	FDA 121.1164; FEMA No. 2785		FEMA No. 3440	

REFERENCES

For References 1–5, see end of Part III.

6. **Morgan and Holmes,** *J. Chem. Ind.,* 44, 109T; *Chem. Zentralbl.,* 1, 2215, 1925.
7. **Van Gysegem,** *Chem. Zentralbl.,* 1, 530, 1907.
8. Schimmel and Co., *Chem. Zentralbl.,* 1, 1086, 1903.
9. **Haller et al.,** *Chem. Zentralbl.,* 2, 28, 1910.
10. **Penfold and Morrison,** *Chem. Abstr.,* 43, 3148, 1949.
11. **Costa and do Vale,** *Chem. Abstr.,* 48, 11732, 1954.
12. **Barré,** *Can. J. Res.,* 27, 65, 1949.
13. **Tressl et al.,** *Z. Naturforsch.,* Teil B, 24, 781, 1969.
14. **Winter et al.,** *Helv. Chim. Acta,* 55, 1916, 1972.
15. **Chang,** Proceedings, 3rd Nordic Aroma Symposium, Hämeenlinna, Finland, 1972, 203.

	3-Nonanon-l-yl acetate	2-Nonenal
Other names	"Compound 1051" 1-Hydroxy-3-nonanone acetate "Ketone alcohol ester" Methylol methyl hexyl ketone acetate 3-Oxononanyl acetate	3-Hexylacrolein β-Hexylacrolein Iris aldehyde® α-Nonenyl aldehyde Heptylidene aldehyde
Empirical formula	$C_{11}H_{20}O_3$	$C_9H_{16}O$
Structure	$CH_3(CH_2)_5-\overset{\overset{O}{\parallel}}{C}-CH_2-CH_2-O-\overset{\overset{O}{\parallel}}{C}-CH_3$	$CH_3-(CH_2)_5-CH=CH-CHO$
Physical/chemical characteristics[5]		
Appearance	Colorless, oily liquid	Almost colorless or pale straw-colored oily liquid; readily oxidizes to noneoic acid on exposure to air
Molecular weight	200.28	140.23
Boiling point		189°C 56–58°C at 0.1 mm Hg
Specific gravity		0.8418 at 25°C
Refractive index		1.4502 at 25°C
Solubility	Very slightly soluble in water; soluble in alcohol, propylene glycol, and oils	Insoluble in water; soluble in alcohol and oils
Organoleptic characteristics	Herbaceous, fruity odor	Very powerful, penetrating fatty odor in concentrated form; orris-like, waxy, and quite pleasant on dilution; odor reminiscent of dried orange peels
Synthesis	By condensation of methyl hexyl ketone with formaldehyde under mild alkaline conditions to yield β-ketol, which is subsequently acetylated	By oxidation of 9,10,12-trihydroxystearic acid (Criegee reaction)[6]
Natural occurrence	Not reported found in nature	Reported found in orris oil
Reported uses[3]	Non-alcoholic beverages 0.30 ppm Ice cream, ices, etc. 0.30 ppm Candy 0.80–5.0 ppm Baked goods 1.0 ppm Condiments 10 ppm	Baked goods 0.20 ppm Condiments 0.20 ppm Meat, meat sauces, soups 0.20 ppm Milk, dairy products 0.20 ppm Cereals 0.20 ppm
Regulatory status	FDA 121.1164; FEMA No. 2786 3-Nonanon-l-yl acetate has been found actually to consist of a mixture of three related compounds. In view of this, it is being dropped from the GRAS list pending possible future review by an expert panel.[7]	FEMA No. 3213

REFERENCES

For References 1–5, see end of Part III.

6. **Scaulan and Swern,** *J. Am. Chem. Soc.,* 62, 2309, 1940.
7. **Hall and Oser,** *Food Technol.,* 24, 533, 1970.

	trans-2-Nonen-1-ol	Nonyl acetate
Other names		Acetate C-9 Nonanol acetate *n*-Nonyl acetate Pelargonyl acetate
Empirical formula	$C_9H_{18}O$	$C_{11}H_{22}O_2$
Structure	$CH_3-(CH_2)_5-CH=CH-CH_2OH$	$CH_3(CH_2)_8-O-C(=O)-CH_3$
Physical/chemical characteristics[5]		
Appearance		Colorless liquid
Assay		97% min
Molecular weight	142.23	186.30
Boiling point	108–110°C at 17 mm Hg	212°C
Specific gravity	0.845 at 15°C	0.864–0.868 at 25°C; 0.8785 at 15°C
Refractive index	1.4476 at 15°C	1.4220–1.4260 (1.4328) at 20°C
Solubility		1:1 in 80% alcohol; 1:6 in 70% alcohol
Organoleptic characteristics		Floral, fruity (mushroom-gardenia) odor; corresponding flavor on dilution; bitter taste when concentrated
Synthesis	By hydrolysis of the corresponding acetate; the acetate is prepared from 1-bromo-2-nonene and acetic anhydride in acetic acid solution[6]	By direct esterification of *n*-nonyl alcohol with acetic acid
Natural occurrence	Reported found as a volatile component in Cucurbitaceae (cucumber, squash, pumpkin, etc.)[7,8]	Reported found in several natural products
Reported uses[3]	Non-alcoholic beverages 0.15 ppm Ice cream, ices, etc. 0.15 ppm Candy 0.2 ppm Baked goods 0.2 ppm Gelatins and puddings 0.15 ppm Meat, meat sauces, soups 0.15 ppm	Non-alcoholic beverages 0.81 ppm Ice cream, ices, etc. 0.81 ppm Candy 1.9 ppm Baked goods 3.1 ppm
Regulatory status	FEMA No. 3379	FDA 121.1164; FEMA No. 2788

REFERENCES

For References 1–5, see end of Part III.

6. **Delaby and Guillot-Allègre,** *Bull. Soc. Chim. Fr.,* [4] 53, 307, 1933.
7. **Kemp, Stoltz and Knavel,** *J. Agric. Food Chem.,* 20, 196, 1972.
8. **Kemp, Knavel and Stoltz,** *Phytochemistry,* 10(8), 1925, 1971.

	Nonyl alcohol	Nonyl isovalerate
Other names	Alcohol C-9 Nonanol-1 1-Nonanol Octyl carbinol Pelargonic alcohol	Nonanol isopentanoate
Empirical formula	$C_9H_{20}O$	$C_{14}H_{28}O_2$
Structure	$CH_3-(CH_2)_7-CH_2OH$	$CH_3(CH_2)_8-O-C(=O)-CH_2-CH(CH_3)_2$
Physical/chemical characteristics[5]		
Appearance	Colorless, mobile liquid	Colorless liquid
Assay	97% min	
Molecular weight	144.26	228.37
Melting point	−5°C	
Boiling point	213.5–215°C	264°C
Flash point	>82°C	
Specific gravity	0.824–0.830 at 25°C; 0.8274 at 20°C	
Refractive index	1.4310–1.4350 (1.43347) at 20°C	
Solubility	Very slightly soluble in water; soluble in most common organic solvents; 1:3 in 60% alcohol	Insoluble in water; soluble in alcohol
Organoleptic characteristics	Characteristic rose-orange odor; slightly fatty, bitter taste reminiscent of orange	Heavy fruit odor (apple-hazelnut with citrus notes); sweet, warm, floral taste
Synthesis	By reduction of ethyl pelargonate; from heptaldehyde via heptanol and heptylmagnesium bromide with ethylene oxide	By esterification of *n*-nonanol with isovaleric acid
Natural occurrence	Reported as occurring frequently in nature, free or esterified: in the essential oils of grapefruit, Guinea sweet orange, Italian and Israeli sweet orange, bitter orange;[6] in oak musk concrete and others. It has been identified also in cheese.	In rue (*Ruta graveolens* L.)
Reported uses[3]	Non-alcoholic beverages 0.70 ppm Ice cream, ices, etc. 0.61 ppm Candy 2.0 ppm Baked goods 1.9 ppm Chewing gum 18 ppm	Non-alcoholic beverages 0.50–1.0 ppm Ice cream, ices, etc. 0.50 ppm Candy 1.0–2.0 ppm Baked goods 1.4 ppm
Regulatory status	FDA 121.1164; FEMA No. 2789	FDA 121.1164; FEMA No. 2791

REFERENCES

For References 1–5, see end of Part III.

6. **Mehlitz and Minas,** *Ind. Obst Gemueseverwert.*, 50, 861, 1965.

	Nonyl octanoate	Nootkatone
Other names	Nonyl caprylate *n*-Nonyl octoate Nonyl octylate	4,4a,5,6,7,8-Hexahydro-6-isopropenyl-4,4a-dimethyl-2(3H)-naphthalenone 4a,5-Dimethyl-1,2,3,4,4a,5,6,7-keto-3-isopropenyl-naphthalene 5,6-Dimethyl-8-isopropenylbicyclo-(4,4,0)-dec-1-en-3-one
Empirical formula	$C_{17}H_{34}O_2$	$C_{15}H_{22}O$
Structure	$CH_3-(CH_2)_8-O-\overset{\overset{O}{\parallel}}{C}-(CH_2)_6-CH_3$	O, CH_3, CH_3, C, CH_2, H_3C, H
Physical/chemical characteristics[5]		
Appearance	Colorless, oily liquid	Crystals (from light petroleum fractions)
Molecular weight	270.46	218.34
Melting point		36–37°C
Boiling point	315°C	
Specific gravity	0.86	
Optical rotation		195.5° (c = 1.5 in chloroform)
Solubility	Insoluble in water; soluble in alcohol	Practically insoluble in water; soluble in alcohol and oils
Organoleptic characteristics	Sweet, rose odor with a mushroom note; very powerful flavor	Pleasant taste
Synthesis	By esterification of *n*-nonanol with *n*-octanoic acid	By oxidation of valencene (a sesquiterpene) with tertiary butyl chromate[6]
Natural occurrence	This ester is found in bitter orange oil	Reported found in grapefruit oil and juice; traces are also reported found in the oils of bergamot, lemon, lime, orange, and tangerine;[7] also reported formed in canned orange juice on storage[8]
Reported uses[3]	Non-alcoholic beverages 2.0 ppm Baked goods 0.06 ppm	Non-alcoholic beverages 10.0 ppm Candy 10.0 ppm
Regulatory status	FDA 121.1164; FEMA No. 2790	FEMA No. 3166

REFERENCES

For References 1–5, see end of Part III.

6. *J. Food Sci.*, 30, 876, 1965.
7. *J. Food Sci.*, 29, 565, 1964.
8. *Dfsh. Lebensm. Rundsch.*, 68(6), 173, 1972.

	9,12-Octadecadienoic acid (48%) plus 9,12,15-octadecatrienoic acid (52%)	**γ-Octalactone**
Other names	Methyl linoleate (48%) and methyl linolenate (52%) mixture	γ-*n*-Butyl-γ-butyrolactone 4-Hydroxyoctanoic acid, γ-lactone *n*-Octalactone Octanolide-1,4
Empirical formula	$C_{19}H_{34}O_2$ (48%) $C_{19}H_{32}O_2$ (52%)	$C_8H_{14}O_2$
Structure	$CH_3(CH_2)_4CH{=}CHCH_2CH{=}CH-(CH_2)_7-COOCH_3$ [48%] $CH_3CH_2CH{=}CHCH_2CH{=}CHCH_2CH{=}CH-(CH_2)_7-COOCH_3$ [52%]	$CH_3-CH_2-CH_2-CH_2-CH$ (ring: O, C=O, H_2C-CH_2)
Physical/chemical characteristics[5]		
Appearance	Colorless oil	Slightly yellow liquid
Assay		95% min
Molecular weight	294.46 (48%) 292.44 (52%)	142.20
Boiling point		132–133°C at 20 mm Hg; 83–86°C at 1 mm Hg (234°C)
Flash point		>100°C
Specific gravity		0.970–0.980 at 25°/25°C
Refractive index		1.443–1.447 at 20°C; 1.4610 at 15°C
Acid value		8 max
Solubility	Miscible in fats, oil, and most organic solvents	1:3 in 50% alcohol; 1:2 in 60% alcohol; 1:3 in 70% alcohol; slightly soluble in water; somewhat soluble in propylene glycol
Organoleptic characteristics		Strong, fruity odor reminiscent of coconut; very sweet taste
Synthesis	Prepared from safflower seed oil[6]	Prepared synthetically from epoxy-1,2-hexane and sodio-malonic ester. After saponification the oxyacid is extracted with ether and is lactonized.
Natural occurrence	In safflower seed[6]	Reported found in apricots and peaches[7]
Reported uses[3]	Non-alcoholic beverages 10 ppm Ice cream, ices, etc. 10 ppm Candy 10 ppm Baked goods 10 ppm Gelatins and puddings 10 ppm Condiments 10 ppm Jellies, preserves, spreads 10 ppm	Non-alcoholic beverages 4.8 ppm Ice cream, ices, etc. 16 ppm Candy 16 ppm Baked goods 17 ppm Gelatins and puddings 15 ppm Syrups 57 ppm
Regulatory status	FEMA No. 3411	FDA 121.1164; FEMA No. 2796

REFERENCES

For References 1–5, see end of Part III.

6. **Parker et al.,** *Biochem. Prep.*, 4, 88, (1955).
7. **Ruys,** *Fr. Ses Parfums,* 47(9), 34, 1966.

	Δ-Octalactone	
Other names	Tetrahydro-6-propyl-5-hydroxy-2H-pyran-2-one Octanoic acid, Δ-lactone	
Empirical formula	$C_8H_{14}O_2$	
Structure	$H_3C-CH_2-CH_2-CH-(CH_2)_3-C{=}O$ (CH and C=O joined through O, ring)	
Physical/chemical characteristics[5]		
Molecular weight	142.20	
Boiling point	115–117°C at 12 mm Hg 104°C at 3 mm Hg	
Specific gravity	0.972 at 25°C	
Refractive index	1.4480 at 25°C	
Synthesis	From 5-hydroxyoctanoic acid by reduction with $NaBH_4$ in water;[6] by microbiological reduction of the corresponding keto acid;[7,8] from 5-hydroxyoctanoic methyl amide by alkaline saponification, using $Ba(OH)_2$ [9]	
Natural occurrence	Reported found as a volatile flavor component in coconut oil,[10] in butterfat,[11] in milk fat,[12,13] in the essence of the Blenheim variety of apricot,[14] and in pineapple;[15] also reported found in yellow passion fruit[16]	
Reported uses[3]	Non-alcoholic beverages	20.0 ppm
	Ice cream, ices, etc.	20.0 ppm
	Candy	20.0 ppm
	Baked goods	20.0 ppm
	Gelatins and puddings	20.0 ppm
	Margarine	20.0 ppm
	Salad oil	20.0 ppm
	Shortening	20.0 ppm
Regulatory status	FEMA No. 3214	

REFERENCES

For References 1–5, see end of Part III.

6. **Stetter and Dierichs,** *Chem. Abstr.,* 46, 7520d, 1952.
7. U.S. Patent 3,076,750, assigned to Lever Brothers Co.; *Chem. Abstr.,* 58, P11928e, 1963.
8. **Muys, van der Ven and de Jouge,** *Appl. Microbiol.,* 11(5), 389, 1963.
9. **Lukěs and Cerný,** *Collect. Czech. Chem. Commun.,* 24, 2722, 1959.
10. **Allen,** *Chem. Ind.* (London), 36, 1560, 1965.
11. **van Beers and van der Zijden,** *Rev. Fr. Corps Gras,* 13(7), 463, 1966; *Chem. Abstr.,* 65, 6194a, 1967.
12. **Khatsi, Libbey and Day,** *J. Agric. Food Chem.,* 14(5), 465, 1966.
13. **Wyatt, Pereira and Day,** *Lipids,* 2(3), 208, 1967.
14. **Tang and Jennings,** *J. Agric. Food Chem.,* 16(2), 252, 1968.
15. **Näf-Müller and Willhalm,** *Helv. Chim. Acta,* 54(7), 1880, 1971.
16. **Winter and Klöty,** *Helv. Chim. Acta,* 55(6), 1916, 1972.

	n-Octanal	Octanal dimethyl acetal
Other names	Aldehyde C-8 Caprylaldehyde Caprylic aldehyde *n*-Octaldehyde *n*-Octylaldehyde	Aldehyde C-8 dimethyl acetal Caprylaldehyde dimethyl acetal 1,1-Dimethoxy octane Octaldehyde dimethyl acetal
Empirical formula	$C_8H_{16}O$	$C_{10}H_{22}O_2$
Structure	$CH_3-(CH_2)_6-CHO$	$CH_3-(CH_2)_6-CH(O-CH_3)(O-CH_3)$
Physical/chemical characteristics[5]		
Appearance	Colorless to light-yellow liquid	Colorless liquid
Assay	92% min[2]	
Molecular weight	128.22	174.29
Boiling point	171–173°C	185°C
Flash point	54°C	
Specific gravity	0.818–0.830 at 25°/25°C;[2] 0.8211 at 20°C	0.86
Refractive index	1.4170–1.4250[2] (1.42167) at 20°C	1.4153 at 22.5°C
Acid value	Not more than 10[2]	
Solubility	Soluble in alcohol and most common organic solvents; 1:2 in 70% alcohol; insoluble in water	Almost insoluble in water; soluble in alcohol
Organoleptic characteristics	Sharp, fatty, fruity odor; characteristic pungent, fatty flavor	Characteristic woody aroma reminiscent of cognac
Synthesis	By oxidation of the corresponding alcohol or reduction of the corresponding acid; also from coconut fatty acids via methyl-*n*-octoate	By condensation of octaldehyde with methanol using dry HCl as a dehydrating agent
Natural occurrence	Reported found in the essential oils of: sweet orange, bitter orange, mandarin, tangerine, grapefruit, Mexican lime, lemon, Taiwan citronella, rose, lemongrass, *Pinus sabiniana*, *Pinus jeffreyi*, *Xanthoxylum rhetsa*,[6] lime petitgrain,[7] clary sage,[8] lavandin, and others	Not reported found in nature
Reported uses[3]	Non-alcoholic beverages 1.4 ppm Ice cream, ices, etc. 1.6 ppm Candy 3.4 ppm Baked goods 4.4 ppm Gelatins and puddings 3.0–6.1 ppm Chewing gum 0.10 ppm	Non-alcoholic beverages 0.74 ppm Alcoholic beverages 3.0 ppm Ice cream, ices, etc. 0.78 ppm Candy 2.8 ppm Baked goods 2.8 ppm
Regulatory status	FDA 121.1164; FEMA No. 2797	FDA 121.1164; FEMA No. 2798

REFERENCES

For References 1–5, see end of Part III.

6. **Naves,** *Chem. Abstr.,* 45, 4889, 1951.
7. **Peyron,** *Fr. Ses Parfums,* 6(1), 17, 1958.
8. **Tesseire,** *Recherches,* 9, 10, 1959.

	Octanoic acid	1-Octanol
Other names	Caprylic acid C-8 acid *n*-Octoic acid *n*-Octylic acid	Alcohol C-8 *n*-Caprylic alcohol Heptyl carbinol Octyl alcohol
Empirical formula	$C_8H_{16}O_2$	$C_8H_{18}O$
Structure	$CH_3{-}(CH_2)_6{-}COOH$	$CH_3{-}(CH_2)_6{-}CH_2OH$
Physical/chemical characteristics[5]		
Appearance	Colorless liquid; solidifies to leafy crystals when cold	Colorless liquid
Assay		98% min[2]
Molecular weight	144.21	130.23
Melting point	16.3°C (17°C)	−15.4 to −16.3°C
Boiling point	239.7°C	194–195.2°C
Flash point		>84.5°C
Specific gravity	0.90884 at 20°C	0.822–0.830 at 25°/25°C;[2] 0.8246–0.8270 at 20°C
Refractive index	1.4335 (1.4327) at 20°C	1.4280–1.4310[2] (1.42920) at 20°C
Acid value		Not more than 1[2]
Solubility	Almost insoluble in cold water; 1:400 in hot water; soluble in alcohol, ether, chloroform, and others	1:5 in 50% alcohol;[2] 1:3 in 60% alcohol; 0.05% in water; soluble in propylene glycol
Organoleptic characteristics	Faint, fruity-acid odor; slightly sour taste	Fresh, orange-rose odor, quite sweet; oily, sweet, slightly herbaceous taste
Synthesis	By fermentation and fractional distillation of the volatile fatty acids present in coconut oil	By reduction of some caprylic esters such as methyl caprylate with sodium ethoxide
Natural occurrence	Reported as frequently occurring naturally in the essential oils of: *Cupressus torulosa*, *Cryptomeria japonica*, *Andropogon iwarancusa*, *Cymbopogon javanensis*, camphor, nutmeg, lemon grass, lime, tobacco (flowers), *Artemisia herba-alba*, camomile, hops, and others; also reported in apple aroma, coconut oil as glyceride, and wine as an ester. It has been identified (free and esterified) among the constituents of petitgrain lemon oil.[6]	Reported as frequently occurring in essential oils as an ester. The free alcohol has been reported found in the essential oils of: green tea, grapefruit, California orange, *Andropogon intermedius*, *Heracleum villosum*, violet leaves, *Anethum graveolens*, and bitter orange[6]
Reported uses[3]	Non-alcoholic beverages 2.9 ppm Ice cream, ices, etc. 2.0 ppm Candy 13 ppm Baked goods 18 ppm Condiments 12 ppm	Non-alcoholic beverages 2.9 ppm Ice cream, ices, etc. 0.91 ppm Candy 2.8 ppm Baked goods 3.0 ppm Gelatins and puddings 1.5 ppm Chewing gum 16–57 ppm
Regulatory status	FDA 121.1164; FEMA No. 2799	FDA 121.1164; FEMA No. 2800

REFERENCES

For References 1–5, see end of Part III.

6. **Peyron**, *Riv. Ital. EPPOS*, 8, 413, 1965.
7. **Mehlitz and Minas**, *Ind. Obst Gemueseverwert.*, 50, 861, 1965.

	2-Octanol	3-Octanol
Other names	*sec*-Capryl alcohol Methyl hexyl carbinol Hexyl methyl carbinol *sec-n*-Octyl alcohol	Amyl ethyl carbinol Ethyl *n*-amyl carbinol *d-n*-Octanol
Empirical formula	$C_8H_{18}O$	$C_8H_{18}O$
Structure	$CH_3-(CH_2)_5-CH(OH)-CH_3$	$CH_3-CH_2-CH(OH)-(CH_2)_4-CH_3$
Physical/chemical characteristics[5]		
Appearance	Colorless, oily liquid	Colorless liquid
Molecular weight	130.23	130.23
Melting point	−38.6°C	
Boiling point	*dl*-, 180°C; *l*-, 80°C at 11 mm Hg	178.5–179.5°C
Specific gravity	0.8193 at 20°C; *d*-, 0.8170 at 25°C	0.8247 at 20°C
Refractive index	1.42444 at 20°C; *l*-, 1.4256; *d*-, 1.4256	1.4252 at 20°C
Optical rotation	*l*-, −9.84°; *d*-, +9.79°	+11.3° at 20°C; 0.9791 gm in 20 cc alcohol +10.20° at 20°C; 1.0196 gm in 20 cc benzene
Solubility	Soluble in most common organic solvents	Very slightly soluble in water; soluble in alcohol
Organoleptic characteristics	Characteristic disagreeable but aromatic odor	Sweet, nutty, warm, herbaceous odor
Synthesis	By distilling sodium ricinoleate with an excess of sodium hydroxide[6-10]	Can be prepared in the racemic form by reduction of ethyl-*n*-amyl ketone with sodium in ether solution[12]
Natural occurrence	Two optically active isomers of the alcohol have been found. Reported in the oil of Reunion geranium;[11] also identified in a few species of mint and lavender	It has been identified (free and esterified) in a variety of mint;[13] also reported in lavender[14] and the essential oils of *Mentha arvensis* var. *piperascens*[15] and *Mentha spicata*[16]
Reported uses[3]	Ice cream, ices, etc 0.60 ppm Candy 3.0 ppm Baked goods 4.0 ppm Gelatins and puddings 2.0 ppm	
Regulatory status	FDA 121.1164; FEMA No. 2801	FDA 121.1164

REFERENCES

For References 1–5, see end of Part III.

6. **Bouis,** *Annalen,* 80, 303, 1851; 92, 395, 1854; 97, 34, 1855.
7. **Moschnin,** *Annalen,* 87, 111, 1853.
8. **Schorlemmer,** *Annalen,* 147, 222, 1868.
9. **Neison,** *J. Chem. Soc.* (London), 27, 507, 837, 1874.
10. **Freund and Schönfeld,** *Ber. Dtsch. Chem. Ges.,* 24, 3352, 1891.
11. **Bohnsack,** *Ber. Dtsch. Chem. Ges.,* 75, 502, 1942.
12. **Pickard and Kenyon,** *J. Chem. Soc. Fr.,* 103, 1923, 1913.
13. **Naves,** *Helv. Chim. Acta,* 26, 1034, 1943.
14. **Benezet,** *Chem. Abstr.,* 40, 6757, 1946.
15. **Carnero, Benezet and Igolen,** *Chem. Abstr.,* 43, 4427, 1949.
16. **Bryan, Burnett and Wearn,** *J. Am. Chem. Soc.,* 73, 1848, 1951.

	2-Octanone		3-Octanone	
Other names	*n*-Hexyl methyl ketone Methyl hexyl ketone		Amyl ethyl ketone E.A.K. Ethyl amyl ketone Ethyl-*n*-amyl ketone	
Empirical formula	$C_8H_{16}O$		$C_8H_{16}O$	
Structure	$CH_3-C(=O)-C_6H_{13}$		$CH_3-CH_2-C(=O)-(CH_2)_4-CH_3$	
Physical/chemical characteristics[5]				
Appearance	Colorless liquid		Colorless, mobile liquid	
Assay	98% min		98% min	
Molecular weight	128.21 −16°C		128.21	
Boiling point	175°C (172–173°C)		169–170°C (172°C)	
Flash point	61°C		Approx 50°C	
Specific gravity	0.8225–0.8255 at 15°C; 0.8185 at 20°C		0.817–0.821 at 23°C; 0.8254 at 15°C	
Refractive index	1.4150–1.4165 at 20°C; 1.41613 at 20°C		1.4130–1.4170 (1.41536) at 20°C	
Acid value	0.9 max			
Solubility	1:9–12 in 50% alcohol; 1:1–1.7 in 70% alcohol. Forms azeotropes with: *d*-limonene (57%), phenol (32%), phenethyl alcohol (5%), *o*-cresol (12%), ethylene glycol (76%), and cineol (55%)		1:7 in 60% alcohol; 1:3 in 70% alcohol; soluble in most common organic solvents; almost insoluble in water	
Organoleptic characteristics	Floral and bitter, green, fruity (unripe apple) odor; bitter, camphoraceous taste		Strong, penetrating, fruity odor reminiscent of lavender	
Synthesis	By oxidation of methyl hexyl carbinol with $K_2Cr_2O_7$ and sulfuric acid;[11] also by oxidation of 2-octanol over zinc oxide at 330–340°C[12]		It can be prepared by passing a mixture of vapors of caproic acid and acetic acid over ThO_2 at 400°C,[6] or by oxidation of *d*-ethyl *n*-amyl carbinol with chromates. Another synthetic route is reported.[7]	
Natural occurrence	Reported found in small quantities in the essential oil of *Ruta montana* L.[13] and in a few varieties of banana and citrus fruits		Reported identified in the low-boiling fraction of the essential oil of lavender;[8] also reported found in the essential oils of *Lavandula vera* (10%)[9] and French lavender[10]	
Reported uses[3]	Non-alcoholic beverages	0.10–1.0 ppm	Non-alcoholic beverages	3.3 ppm
	Ice cream, ices, etc.	0.20–1.0 ppm	Ice cream, ices, etc.	10 ppm
	Candy	0.40–4.0 ppm	Candy	11 ppm
	Baked goods	0.40–4.0 ppm	Baked goods	11 ppm
Regulatory status	FDA 121.1164; FEMA No. 2802		FDA 121.1164; FEMA No. 2803	

REFERENCES

For References 1–5, see end of Part III.

6. **Pickard et al.,** *J. Chem. Soc.* (London), p. 103, 1944.
7. **Behal,** *Bull. Soc. Chim. Fr. 2*, 50, 359, 1888.
8. Schimmel and Co., *Chem. Zentralbl.*, 1, 1086, 1903; 2, 1124, 1903.
9. **Williams et al.,** *Chem. Zentralbl.*, 2, 3852, 1936.
10. **Crabalona,** *Chem. Zentralbl.*, 1, 306, 1940.
11. **Schorlemmer,** *Annalen*, 147, 224, 1868.
12. **Ottensooser,** *Bull. Soc. Chim. Fr. 4*, 41, 324, 1927.
13. **St. Pfau,** *Helv. Chim. Acta*, 22, 382, 1939.

	3-Octanon-1-ol	2-Octenal
Other names	Caproyl ethanol Compound 1010 Hexanoyl ethanol γ-Ketoctanol "Ketone alcohol" Methylol methyl amyl ketone 3-Oxooctanol	2-Pentyl acrolein α-Amyl acrolein
Empirical formula	$C_8H_{16}O_2$	$C_8H_{14}O$
Structure	$CH_3-(CH_2)_4-C(=O)-CH_2-CH_2OH$	$CH_2=C(CHO)-(CH_2)_4-CH_3$
Physical/chemical characteristics[5]		
Appearance	Colorless liquid	Colorless liquid
Molecular point	144.21	126.20
Boiling point	94.6°C at 8–9 mm Hg	174°C 81°C at 48 mm Hg 72°C at 30 mm Hg
Specific gravity		0.8562 at 14°C
Refractive index		1.4373 at 20°C
Solubility	Very slightly soluble in water; soluble in alcohol	Insoluble in water; soluble in alcohol and oils
Organoleptic characteristics	Fruity, herbaceous, spicy odor; slightly bitter above 20 ppm; otherwise, warm and fruity	Peculiar green-leafy odor, less fatty than octanal; orange, honey-like, cognac-like aroma
Synthesis	By partial hydrogenation of the corresponding di-ketone at 100°C over Raney Ni in methanol solution; also by condensation of methyl amyl ketone with formaldehyde under mildly alkaline conditions	By prolonged heating of a solution of heptanal and formaldehyde in the presence of dimethylamino hydrochloride;[6] by boiling 4,5-diketo-3-pentyltetrahydrofuran under a nitrogen blanket[7]
Natural occurrence	Not reported found in nature	A volatile flavor component in potato chips[8]
Reported uses[3]	Non-alcoholic beverages 0.20 ppm Ice cream, ices, etc. 0.30 ppm Candy 0.80 ppm Baked goods 0.60–0.80 ppm Condiments 1.0 ppm	Candy 1.0 ppm Baked goods 1.0 ppm Condiments 1.0 ppm Meat, meat sauces, soups 1.0 ppm
Regulatory status	FDA 121.1164; FEMA No. 2804	FEMA No. 3215

REFERENCES

For References 1–5, see end of Part III.

6. **Marvel, Myers and Saunders,** *J. Am. Chem. Soc.*, 70, 1698, 1948.
7. **Hinder, Schinz and Seidel,** *Helv. Chim. Acta*, 30, 1500, 1947.
8. **Deck, Pokorny, and Chang,** *J. Food Sci.*, 38(2), 345, 1973.

	1-Octen-3-ol	3-Octen-2-one
Other names	Amyl vinyl carbinol Matsutake alcohol 3-Octenol *n*-Pentyl vinyl carbinol	
Empirical formula	$C_8H_{16}O$	$C_8H_{14}O$
Structure	$CH_3-(CH_2)_4-CH(OH)-CH=CH_2$	$CH_3-C(=O)-CH=CH-(CH_2)_3-CH_3$
Physical/chemical characteristics[5]		
Appearance	Colorless, oily liquid	
Molecular weight	128.21	126.19
Boiling point	175–175.2°C; 68–71°C at 12 mm Hg	
Specific gravity	0.84	
Optical rotation	−7°49′ (−13° to −18°) at 14°C	
Solubility	Insoluble in water; soluble in alcohol	
Organoleptic characteristics	Powerful, sweet, earthy odor with strong, herbaceous note reminiscent of lavender-lavandin, rose, and hay; sweet, herbaceous taste	
Synthesis	From magnesium amyl bromide and acrolein[6]	By bubbling air in a solution of an organoborane derivative of 1-butene and 1-butyn-2-one in tetrahydrofuran;[11] also O + CHO —B⁻→
Natural occurrence	Originally reported (Ito and Imai) found in the mushroom *Armillaria matsutake*, a parasite growing on the radical hairs of *Pinus densiflora* in the forests of Japan.[7] It has been isolated also in the essential oils of *Mentha pulegium* L.,[8] lavender,[9] and *Mentha timjia*.[10]	Reported found in roasted filberts[12]
Reported uses[3]	Non-alcoholic beverages 0.20 ppm Ice cream, ices, etc. 1.0 ppm Candy 2.0 ppm Baked goods 6.0 ppm Condiments 6.0 ppm Soups 6.0 ppm	Non-alcoholic beverages 1.0 ppm Ice cream, ices, etc. 1.0 ppm Candy 2.0 ppm Baked goods 2.0 ppm Gelatins and puddings 2.0 ppm Condiments, pickles 0.5 ppm
Regulatory status	FDA 121.1164; FEMA No. 2805	FEMA No. 3416

REFERENCES

For References 1–5, see end of Part III.

6. **Levene and Walti,** *J. Biol. Chem.*, 94, 593, 1931.
7. **Murahashi,** *Chem. Zentralbl.*, 1, 1958, 1937; 2, 1249, 1938.
8. **Naves,** *Helv. Chim. Acta*, 26, 1992, 1943.
9. **Crabalona,** *Bull. Soc. Chim. Fr.*, 11, 67, 1944.
10. **Sficas,** *Chem. Abstr.*, 48, 3934, 1954.
11. **Suzuki et al.,** *Hokkaido Daigaku Kagakubu Kenkyu Hokoku*, 45, 1971; *Chem. Abstr.*, 75, 129860h, 1971.
12. Volatile Compounds in Food, T.N.O. Report No. 4030, 3rd ed., Centraal Instituut voor Voedingsonderzoek, Zeist, Netherlands.

	1-Octen-3-yl acetate	Octyl acetate
Other names	3-Acetoxy octene "Amyl crotonyl acetate" Amyl vinyl carbinyl acetate Octenyl acetate β-Octenyl acetate *n*-Pentyl vinyl carbinyl acetate	Acetate C-8 Capryl acetate *n*-Octyl acetate *Note*: Do not confuse with 2-ethyl hexyl acetate
Empirical formula	$C_{10}H_{18}O_2$	$C_{10}H_{20}O_2$
Structure	$CH_3-(CH_2)_4-CH(-O-C(=O)-CH_3)-CH=CH_2$	$CH_3-(CH_2)_6-CH_2-O-C(=O)-CH_3$
Physical/chemical characteristics[5]		
Appearance	Colorless liquid	Colorless liquid
Assay		98% min
Molecular weight	170.25	172.27
Boiling point	80°C (190°C) at 15 mm Hg	203–208°C
Flash point		88.5°C
Specific gravity	0.8775 (0.89) at 20°C	0.865–0.868 at 25°C; 0.8730–0.8740 at 15°C
Refractive index	1.4253 at 20°C	1.4180–1.4210 (1.4190–1.4207) at 20°C
Solubility	1:4 in 70% alcohol	1:4 in 70% alcohol; 1:15–20 in 60% alcohol; soluble in most common organic solvents
Organoleptic characteristics	Characteristic odor reminiscent of lavender and lavandin	Fruity odor reminiscent of neroli and jasmine; fruity, slightly bitter taste suggestive of peach
Synthesis	By acetylation of amyl vinyl carbinol	By acetylation of the corresponding alcohol
Natural occurrence		Reported found in the essential oils of green tea and *Heracleum giganteum* L.; also identified in the essential oil of bitter orange[6]
Reported uses[3]		Non-alcoholic beverages 1.6 ppm Ice cream, ices, etc. 0.87 ppm Candy 4.7 ppm Baked goods 6.0 ppm
Regulatory status	FDA 121.1164	FDA 121.1164; FEMA No. 2806

REFERENCES

For References 1–5, see end of Part III.

6. **Mehlitz and Minas,** *Ind. Obst Gemueseverwert.*, 50, 861, 1965.

	3-Octyl acetate	Octyl butyrate
Other names		Octyl-*n*-butyrate
Empirical formula	$C_{10}H_{20}O_2$	$C_{12}H_{24}O_2$
Structure	$CH_3-C(=O)-O-CH(CH_2-CH_3)-(CH_2)_4-CH_3$	$CH_3-(CH_2)_7-O-C(=O)-(CH_2)_2-CH_3$
Physical/chemical characteristics[5]		
Appearance	Colorless liquid	Colorless liquid
Molecular weight	172.27	200.32
Melting point		−56°C
Boiling point	191–191.5°C; 74°C at 10 mm Hg	244°C (245°C)
Specific gravity	0.8641 at 20°C	0.8549 at 30°C
Refractive index	1.4152 at 20°C	1.4210 at 20°C
Solubility		Almost insoluble in water; soluble in alcohol
Organoleptic characteristics	Characteristic, complex aroma with a rose and jasmine note and an apple-lemon undertone; sweet, peach-like flavor	Green, herbaceous odor remotely reminiscent of orange but more so of galbanum and parsley; sweet, melon-like flavor
Synthesis		From *n*-octanol and butyric acid in the presence of HCl[6]
Natural occurrence	Reported found in the essential oil of *Mentha pulegium* L.[9]	Reported found in the essential oil of *Heracleum villosum* Fisch.[7] and in the oil from fruits of *Pastinaca sativa*[8]
Reported uses[3]		Non-alcoholic beverages 0.53 ppm Ice cream, ices, etc. 1.3 ppm Candy 2.9 ppm Baked goods 2.9 ppm
Regulatory status	FDA 121.1164	FDA 121.1164; FEMA No. 2807

REFERENCES

For References 1–5, see end of Part III.

6. **Biltery and Gisseleire,** *Bull. Soc. Chim. Belg.*, 44, 569, 1935.
7. **Rutowski and Vinogradowa,** *Chem. Zentralbl.*, 2, 1311, 1927.
8. **Renesse,** *Chem. Ann.*, 166, 80, 1873.
9. **Naves,** *Helv. Chim. Acta,* 26, 1036, 1999, 1943.

	n-Octyl formate		Octyl heptanoate	
Other names	Octyl formate Octyl methanoate		Octyl heptoate Octyl heptylate	
Empirical formula	$C_9H_{18}O_2$		$C_{15}H_{30}O_2$	
Structure	$CH_3-(CH_2)_7-O-\overset{\overset{\displaystyle O}{\|}}{C}-H$		$CH_3-(CH_2)_7-O-\overset{\overset{\displaystyle O}{\|}}{C}-(CH_2)_5-CH_3$	
Physical/chemical characteristics[5]				
Appearance	Colorless liquid		Colorless liquid	
Assay	96% min[2]			
Molecular weight	158.24		242.41	
Melting point			−21.5°C	
Boiling point	198–199°C; 73°C at 10 mm Hg		290.8°C	
Specific gravity	0.869–0.872 at 25°/25°C;[2] 0.8786 at 15°C		0.85200 at 20°C	
Refractive index	1.4180–1.4200 at 20°C;[2] 1.41813 at 15°C		1.43488 at 20°C	
Acid value	Not more than 1; may increase on storage[2]			
Solubility	1:5 and more in 70% alcohol;[2] insoluble in water; soluble in most common organic solvents		Soluble in most common organic solvents; insoluble in water	
Organoleptic characteristics	Fruity odor with a rose-orange note; bitter flavor, refreshingly fruity-green at low concentrations		Characteristic fruity, slightly fatty odor; corresponding flavor	
Synthesis	By esterification of *n*-octyl alcohol with formic acid; by boiling *n*-octyl chloride with formamide in the presence of small amounts of HCl[8]		By esterification of *n*-octyl alcohol with heptanoic acid in the presence of mineral acids[6,7]	
Natural occurrence	Not reported found in nature		Not reported found in nature	
Reported uses[3]	Non-alcoholic beverages	0.01–1.0 ppm	Non-alcoholic beverages	0.13–1.0 ppm
	Ice cream, ices, etc.	1.0 ppm	Ice cream, ices, etc.	0.13–1.0 ppm
	Candy	5.0 ppm	Candy	0.13–2.0 ppm
	Baked goods	7.0 ppm	Baked goods	0.20–2.0 ppm
Regulatory status	FDA 121.1164; FEMA No. 2809		FDA 121.1164; FEMA No. 2810	

REFERENCES

For References 1–5, see end of Part III.

6. **Gartenmeister,** *Justus Liebigs Ann. Chem.*, 233, 284, 1886.
7. **Bilterys and Gisseleire,** *Bull. Soc. Chim. Belg.*, 44, 578, 1935.
8. E. I. du Pont de Nemours & Co., U.S. Patent 2,375,401, 1942.

	Octyl isobutyrate	*n*-Octyl isovalerate
Other names		*n*-Octyl-3-methylbutyrate
Empirical formula	$C_{12}H_{24}O_2$	$C_{13}H_{26}O_2$
Structure	$CH_3-(CH_2)_7-O-C(=O)-CH(CH_3)_2$	$CH_3-(CH_2)_7-O-C(=O)-CH_2-CH(CH_3)_2$
Physical/chemical characteristics[5]		
Appearance	Colorless liquid	Colorless liquid
Molecular weight	200.32	214.35
Boiling point	237°C	245°C
Specific gravity	0.87	0.86
Solubility	Almost insoluble in water; soluble in alcohol	Insoluble in water; soluble in alcohol
Organoleptic characteristics	Fruity, fatty fragrance with a soft and humid undertone reminiscent of parsley and fern root; sweet flavor suggestive of grape	Strong odor reminiscent of fatty aldehyde with an apple-pineapple undertone
Synthesis	Esterification of *n*-octanol with isobutyric acid	By esterification of *n*-octanol with isovaleric acid
Natural occurrence	Reported found among the volatile components of hop[7]	Reported found among the volatile constituents of the aroma of fresh Florida orange juice[6]
Reported uses[3]	Non-alcoholic beverages 2.0 ppm Ice cream, ices, etc. 2.4 ppm Candy 3.5 ppm Baked goods 3.5 ppm Chewing gum 0.50 ppm	Non-alcoholic beverages 0.90 ppm Ice cream, ices, etc. 0.80–1.0 ppm Candy 1.0–4.0 ppm Baked goods 1.0–4.0 ppm
Regulatory status	FDA 121.1164; FEMA No. 2808	FDA 121.1164; FEMA No. 2814

REFERENCES

For References 1–5, see end of Part III.

6. **Wolford et al.,** *J. Food Sci.,* 28, 320, 1963.
7. **Buttery et al.,** *Nature,* 202, 701, 1964.

	Octyl octanoate	*n*-Octyl phenylacetate
Other names	Octyl caprylate *n*-Octyl octoate Octyl octylate	Octyl-α-toluate *n*-Octyl-α-toluate
Empirical formula	$C_{16}H_{32}O_2$	$C_{16}H_{24}O_2$
Structure	$CH_3-(CH_2)_6-\overset{O}{\overset{\Vert}{C}}-O-(CH_2)_7-CH_3$	$C_6H_5-CH_2-\overset{O}{\overset{\Vert}{C}}-O-CH_2-(CH_2)_6-CH_3$
Physical/chemical characteristics[5]		
Appearance	Colorless liquid	Colorless, viscous liquid
Molecular weight	256.44	248.37
Melting point	−18°C	315°C
Boiling point	306°C; 307°C; 192–193°C at 30 mm Hg	
Specific gravity	0.860 at 15.5°C; 0.8554 at 25°C	0.983 at 15.5°C
Refractive index	1.4925 (1.4352) at 20°C	
Solubility	Insoluble in water; soluble in alcohol	Insoluble in water; soluble in alcohol
Organoleptic characteristics	Faint, fatty odor reminiscent of green tea; oily, fruity, sweet, mildly green taste	Pleasant, slightly fatty, citrus-like odor
Synthesis	By esterification of octanoic acid with octyl alcohol in the presence of HCl catalyst;[6] or by passing vapors of octanoic acid and hydrogen over a copper chromium oxide catalyst at high temperature (320°C)[7]	By esterification of *n*-octanol with phenylacetic acid
Natural occurrence	Reported found in several natural products	Not reported found in nature
Reported uses[3]	Non-alcoholic beverages 0.50–1.0 ppm Ice cream, ices, etc. 0.50–1.0 ppm Candy 0.50–2.0 ppm Baked goods 0.50–2.0 ppm	Non-alcoholic beverages 1.3 ppm Ice cream, ices, etc. 1.0 ppm Candy 0.20–4.0 ppm Baked goods 4.0 ppm
Regulatory status	FDA 121.1164; FEMA No. 2811	FDA 121.1164; FEMA No. 2812

REFERENCES

For References 1–5, see end of Part III.
6. Ruhoff and Reid, *J. Am. Chem. Soc.*, 55, 3827, 1933.
7. Kessler Chemical Corp., U.S. Patent 1,997,172, 1934.

	Octyl propionate		3-Oxobutanal, dimethyl acetal	
Other names			3-Ketobutyraldehyde, dimethyl acetal Acetyl acetaldehyde, dimethyl acetal 1,1-Dimethoxy-3-butanone	
Empirical formula	$C_{11}H_{22}O_2$		$C_6H_{12}O_3$	
Structure	$CH_3-(CH_2)_7-O-C(=O)-CH_2-CH_3$		$CH_3-C(=O)-CH_2-CH(OCH_3)_2$	
Physical/chemical characteristics[5]				
Appearance	Colorless liquid			
Molecular weight	186.30		132.16	
Boiling point	228°C (225°C)			
Specific gravity	0.87044 at 15°C; 0.870 at 15°C			
Refractive index	1.4225 at 20°C			
Solubility	Insoluble in water; soluble in alcohol and propylene glycol			
Organoleptic characteristics	Complex, waxy odor reminiscent of myrtle berries with a pineapple undertone			
Synthesis	By esterification of *n*-octanol with propionic acid			
Natural occurrence	In oil of hop		Closely related materials (e.g., the diethylacetal) reported in roasted cocoa[6]	
Reported uses[3]	Non-alcoholic beverages	0.84 ppm	Ice cream, ices, etc.	80 ppm
	Ice cream, ices, etc.	0.57 ppm	Candy	80 ppm
	Candy	3.6 ppm	Baked goods	160 ppm
	Baked goods	2.0–4.0 ppm	Gelatins and puddings	80 ppm
			Condiments	500 ppm
			Meat, soups	2,000 ppm
Regulatory status	FDA 121.1164; FEMA No. 2813		FEMA No. 3381	

REFERENCES

For References 1–5, see end of Part III.

6. van der Wal et al., *J. Agric. Food Chem.*, 19(2), 276, 1971.

	ω-Pentadecalactone
Other names	Angelica lactone Exaltolide® 15-Hydroxypentadecanoic acid, ω-lactone Pentadecanolide "Thibetolide"
Empirical formula	$C_{15}H_{28}O_2$
Structure	CH_2—C=O ; $(CH_2)_{12}$; CH_2—O (ring)
Physical/chemical characteristics[5]	
Appearance	Plastic solid at room temperature; limpid and colorless when liquified
Assay	98% min
Molecular weight	240.39
Melting point	30–31°C to 36–37°C
Boiling point	150°C at 5 mm Hg (Muscolacton, Lactone M.C. 15) 176°C at 15 mm Hg
Congealing point	36°C min ("Thibetolide")
Specific gravity	0.9447 at 33°C
Refractive index	1.4669 at 33°C
Solubility	Insoluble in water; almost insoluble in glycerol; 1:6 in 80% alcohol; soluble in alcohol
Organoleptic characteristics	Extraordinarily persistent, musk-like odor
Synthesis	Commercially available under a variety of Trade names: Exaltolide®, "Thibetolide," "Muscolacton," "Lactone M.C. 15," and others. For synthesis see References 6 through 21.
Natural occurrence	Originally reported found in the essential oil from angelica roots[22-25]
Reported uses[3]	Non-alcoholic beverages, 0.27 ppm; alcoholic beverages, 0.50 ppm; ice cream, ices, etc., 0.68 ppm; candy, 1.4 ppm; baked goods, 1.5 ppm; gelatins and puddings, 0.10 ppm
Regulatory status	FDA 121.1164; FEMA No. 2840

REFERENCES

For References 1–5, see end of Part III.

6. **Ružička and Stoll,** *Helv. Chim. Acta,* 11, 1159, 1928.
7. **Stoll and Gardner,** *Helv. Chim. Acta,* 17, 1609, 1934.
8. **Stoll et al.,** *Helv. Chim. Acta,* 17, 1283, 1289, 1934.
9. **Stoll et al.,** *Helv. Chim. Acta,* 18, 1087, 1935.
10. **Stoll and Rouver,** *Parfum. Mod.,* 29, 207, 1935.
11. **Hunsdiecker and Erlbach,** *Ber. Dtsch. Chem. Ges.,* 80, 129, 1947.
12. **Hunsdiecker et al.,** Dutch Patent 53,032.
13. **Haarman and Reimer,** German Patent 449,217; *Chem. Zentralbl.,* 2, 2351, 1927.
14. **Stoll,** *Helv. Chim. Acta,* 30, 1393, 1947.
15. **Polak and Schwarz,** Dutch Patent 67,458.
16. **Carothers et al.,** *J. Am. Chem. Soc.,* 51, 51, 1929; 69, 129, 1947.
17. **Collaud,** *Helv. Chim. Acta,* 25, 965, 1942; 26, 849, 1155, 1943.
18. **Klouven and Koh,** *Parfum. Kosmet. II,* 43, 35, 1962.
19. **McCrae,** *Tetrahedron,* 20(7), 1173, 1964.
20. **Tesseire and Corbier,** *Recherches,* 11, 27, 1961.
21. **Roure-Bertrand and DuPont,** French Patent 1,005,766, 1947.
22. **Müller,** *Ber. Dtsch. Chem. Ges.,* 14, 2576, 1881.
23. **Beilstein and Wiegand,** *Ber. Dtsch. Chem. Ges.,* 15, 1741, 1882.
24. **Ciamician and Silber,** *Ber. Dtsch. Chem. Ges.,* 29, 1811, 1896,
25. **Kerschbaum,** *Ber. Dtsch. Chem. Ges.,* 60, 902, 1927.

	2,4-Pentadienal	2,3-Pentanedione
Other names		Acetyl propionyl
Empirical formula	C_5H_6O	$C_5H_8O_2$
Structure	$CH_2=CH-CH=CH-CHO$	$CH_3-C(=O)-C(=O)-CH_2-CH_3$
Physical/chemical characteristics[5]		
Appearance		Yellow liquid
Molecular weight	82.09	100.12
Melting point		−52°C
Boiling point	*cis*-, 29–30°C at 10 mm Hg *trans*-, 30°C at 10 mm Hg	110–112°C (180°C)
Specific gravity		0.955–0.959 at 15°C
Refractive index	*cis*-, 1.5115 at 19°C *trans*-, 1.5185–1.5190 at 19°C	1.4081 (1.40135) at 19°C
Solubility		Partially water soluble (1:15)
Organoleptic characteristics		Somewhat sweet odor similar to quinone
Synthesis	From 2-ethoxy-Δ^3-dihydropyrone and phosphoric acid;[6] the *cis*- and *trans*-isomers can be prepared from the corresponding penta-2,4-dien-1-ol (*cis*- or *trans*-form) in methylene chloride in the presence of manganese dioxide[7]	By oxidation of methyl propyl ketone with excess $NaNO_2$ and diluted HCl in the presence of hydroxylamine hydrochloride under nitrogen blanket
Natural occurrence		Identified in the essential oil of Finnish pine[8]
Reported uses[3]	Non-alcoholic beverages 1.0 ppm Ice cream, ices, etc. 1.0 ppm Candy 1.0 ppm Baked goods 1.0 ppm Gelatins and puddings 1.0 ppm Chewing gum 1.0 ppm Condiments 1.0 ppm Meat, meat sauces, soups 1.0 ppm Milk, dairy products 1.0 ppm Cereals 1.0 ppm	Non-alcoholic beverages, 0.60 ppm; ice cream, ices, etc., 3.3 ppm; candy, 5.9 ppm; baked goods, 9.6 ppm; gelatins and puddings, 0.28 ppm; toppings, 0.30 ppm
Regulatory status	FEMA No. 3217	FDA 121.1164; FEMA No. 2841

REFERENCES

For References 1–5, see end of Part III.

6. **Woods and Sanders,** *J. Am. Chem. Soc.*, 68, 2483, 1946.
7. **Boehm and Whiting,** *J. Chem. Soc.*, p. 2541, 1963.
8. **Aschau,** *Z. Angew. Chem.*, 20, 1815, 1907.

	2-Pentanol		2-Pentanone	
Other names	Methyl *n*-propyl carbinol *sec*-*n*-Amyl alcohol		Ethyl acetone Methyl propyl ketone	
Empirical formula	$C_5H_{12}O$		$C_5H_{10}O$	
Structure	$CH_3-CH_2-CH_2-CH(OH)-CH_3$		$CH_3-CH_2-CH_2-C(=O)-CH_3$	
Physical/chemical characteristics[5]				
Appearance	Colorless liquid		Liquid at room temperature	
Molecular weight	88.15		86.13	
Melting point			−77.75°C	
Boiling point	118–119°C 62°C at 60 mm Hg		102°C	
Specific gravity	*dl*-, 0.8053 at 25°C *d*-, 0.8101 at 20°C		0.810 at 15.5°C; 0.80435 at 25°C	
Refractive index	*dl*-, 1.4063 at 15°C *d*-, 1.4056 at 20°C		1.3985 (1.38946) at 20°C	
Optical rotation	*d*-, +13.9 at 20°C *l*-, −13.4 at 15°C			
Solubility	Very soluble in water; soluble in alcohol and ether		Slightly soluble in water; miscible with alcohol and ether	
Organoleptic characteristics			Wine, acetone-like, characteristic odor	
Synthesis	By catalytic reduction of methyl *n*-propyl ketone[10]		By dry distillation of a mixture consisting of calcium acetate and calcium butyrate;[6] also by oxidation of sodium or ammonium *n*-caproate with H_2O_2	
Natural occurrence			Reported found in wood spirit;[7,8] also reported present in *Ananas sativus*,[9] a few banana species, grape vines, and some citrus fruits	
Reported uses[3]	Non-alcoholic beverages	2.0 ppm	Non-alcoholic beverages	13 ppm
	Baked goods	2.0 ppm	Ice cream, ices, etc.	34 ppm
	Gelatins and puddings	2.0 ppm	Candy	32 ppm
	Condiments, pickles	2.0 ppm	Baked goods	32 ppm
	Meat, meat sauces, soups	2.0 ppm		
	Cereals	2.0 ppm		
Regulatory status	FEMA No. 3316		FDA 121.1164; FEMA No. 2842	

REFERENCES

For References 1–5, see end of Part III.

6. **Friedel,** *Annalen,* 108, 124, 1858.
7. **Vladesco,** *Ber. Dtsch. Chem. Ges.,* 23, 435, 1890.
8. **Buisine,** *Chem. Zentralbl.,* 1, 1020, 1899.
9. **Haagen-Smit et al.,** *J. Am. Chem. Soc.,* 67, 1646, 1945.
10. **Wilson,** *J. Am. Chem. Soc.,* 70, 1313, 1948.

	2-Pentenal	4-Pentenoic acid
Other names	2-Ethylacrylic aldehyde	Allyl acetate Allyl acetic acid
Empirical formula	C_5H_8O	$C_5H_8O_2$
Structure	$CH_3-CH_2-CH=CH-CHO$	$CH_2=CH-CH_2-CH_2-C(=O)-OH$
Physical/chemical characteristics[5]		
Appearance		Colorless liquid
Molecular weight	84.11	100.12
Melting point		−22 to −23°C
Boiling point	124°C 56°C at 65 mm Hg	189°C (93°C at 20 mm Hg)
Specific gravity	0.8532 at 21°C	0.9809 at 20°C
Refractive index	1.4439 at 21°C	1.4281 at 20°C
Solubility		Slightly soluble in water; readily soluble in alcohol and ether
Organoleptic characteristics		Sour, caramellic flavor with sweet aftertaste; acrid flavor and odor at high levels
Synthesis	By treating β-phenoxyacrolein with secondary amines, Grignard reagent, and active methylene compounds;[6] by heating 1-ethoxy-1,3-pentadiene with H_3PO_4 at 80°C;[7] from *cis-* and *trans*-but-2-ene-1,4-diol and $SOCl_2$, followed by treatment of the chloroalcohol with methyl magnesium bromide and final oxidation with MnO_2[8]	By decarboxylation by heating allylmalonic acid; or from the corresponding ethyl ester prepared by boiling ethyl 4-chloro-*n*-valerate in quinoline
Natural occurrence	Reported found in aqueous orange essence (*trans*-form);[9] as a flavor component in cognac oil;[10] in *Chrysocoris stolli;*[11] in tomato aroma;[12] in the autooxidation of cod liver oil;[13] in peas;[14] as a flavoring component in reverted soy bean oil;[15,16] in salmon oil;[17] in strawberry;[18] in potato chips during storage;[19] in raspberry;[20] in butterfat fishy flavor[21]	Not reported found in nature
Reported uses[3]	Non-alcoholic beverages 10.0 ppm Candy 10.0 ppm Gelatins and puddings 10.0 ppm	Non-alcoholic beverages 1.0 ppm Ice cream, ices, etc. 2.0 ppm Candy 5.0 ppm Baked goods 5.0 ppm Margarine 2.0 ppm
Regulatory status	FEMA No. 3218	FDA 121.1164; FEMA. No. 2843

REFERENCES

For References 1–5, see end of Part III.

6. **Gelin and Makula,** *Bull. Soc. Chim. Fr.*, (3), 1129, 1968.
7. **Nazarov et al.,** *Zh. Obshch. Khim.*, 29, 3692, 1959.
8. **Thomas and Warburton,** *J. Chem. Soc.*, p. 2988, 1965.
9. **Moshonas and Shaw,** *J. Food Sci.*, 38, 360, 1973.
10. **Schaefer and Timmer,** *J. Food Sci.*, 35(1), 10, 1970.
11. **Choudhuri and Das,** *Arch. Int. Physiol. Biochim.*, 77(4), 609, 1969.
12. **Schormueller and Kochmann,** *Z. Lebensm. Unters. Forsch.*, 141(1), 1, 1969.
13. **Fisher et al.,** *Lipids.* 3(1), 88, 1968.
14. **Grosch,** *Z. Lebensm. Unters. Forsch.*, 135(2), 75, 1967.
15. **Smouse and Chang,** *J. Am. Oil Chem. Soc.*, 44(8), 509, 1967.
16. **Iwata, Morita, and Ota,** *Yukagaku,* 14(5), 241, 1965.
17. **Wyatt and Day,** *J. Food Sci.*, 28, 305, 1963.
18. **Winter and Willhalm,** *Helv. Chim. Acta,* 47(5), 1215, 1964.
19. **Dornseifer and Powers,** *Food Technol.*, 17(10), 118, 1963.
20. **Winter and Sundt,** *Helv. Chim. Acta.* 45, 2195, 1962.
21. **Forss, Dunstone and Stark,** *J. Dairy Res.*, 27, 211, 1960.

	1-Penten-3-ol	1-Penten-3-one
Other names	Ethyl vinyl carbinol	Ethyl vinyl ketone Propionyl ethylene
Empirical formula	$C_5H_{10}O$	C_5H_8O
Structure	$CH_3-CH_2-CH(OH)-CH=CH_2$	$CH_3-CH_2-C(=O)-CH=CH_2$
Physical/chemical characteristics[5]		
Appearance	Liquid	Liquid; polymerizes readily with heat or in the presence of alkali
Molecular weight	86.13	84.04
Boiling point	114°C; 32°C at 15 mm Hg	68–70°C at 200 mm Hg 38°C at 60 mm Hg 31°C at 47 mm Hg
Specific gravity	0.8344 at 25°C	0.8468 at 20°C
Refractive index	1.4223 at 25°C	1.4192 at 20°C
Optical rotation	+10.5° (*d*, in ethanol); −7.1° (*l*, in ethanol)	
Solubility	Sparsely soluble in water; miscible with alcohol and ether	Insoluble in water; soluble in most organic solvents
Organoleptic characteristics		Powerful, penetrating odor
Synthesis	By prolonged contact of 1-chloro-2-pentene with NaOH solution.[6] For the synthesis of the optically active forms, see references 7, 8, and 9.	By reacting ethylene with a mixture of propionyl chloride, aluminum trichloride and carbon disulfide[10]
Natural occurrence	In orange, strawberry, and tomato	A flavor component of orange essence[11]
Reported uses[3]		Non-alcoholic beverages 5.0 ppm Ice cream, ices, etc. 5.0 ppm Candy 8.0 ppm Baked goods 8.0 ppm Gelatins and puddings 5.0 ppm Condiments, pickles 0.01 ppm Milk, dairy products 0.005 ppm Preserves 5.0 ppm
Regulatory status	FDA 121.1164	FEMA No. 3382

REFERENCES

For References 1–5, see end of Part III.

6. **Meisenheimer and Link,** *Chem. Ann.*, 479, 231, 1930.
7. **Kohler,** *J. Am. Chem. Soc.*, 38, 525, 1916.
8. **Hurd and McNamee,** *J. Am. Chem. Soc.*, 59, 104, 1937.
9. **Bouis,** *Ann. Chim. Paris,* 10(9), 497, 1928.
10. **MacMahon et al.,** *J. Am. Chem. Soc.*, 70, 2971, 1948.
11. **Moshonas and Shaw,** *J. Food Sci.*, 38(2), 360, 1973.

	3-Penten-2-one		2-Pentyl furan	
Other names	Ethylidene acetone Methyl propenyl ketone			
Empirical formula	C_5H_8O		$C_9H_{14}O$	
Structure	$CH_3-CH=CH-\overset{O}{\overset{\|\|}{C}}-CH_3$		O $-(CH_2)_4-CH_3$	
Physical/chemical characteristics[5]				
Appearance	Colorless liquid		Colorless liquid	
Molecular weight	84.31		138.08	
Boiling point	122°C		57–59°C at 10 mm Hg	
Specific gravity	0.861 at 15°C 0.8624 at 20°C		0.8837 at 23°C	
Refractive index	1.4350 at 20°C		1.4462	
Organoleptic characteristics	Fruity odor becoming pungent on storage; used in tobacco flavors		Fruity odor	
Synthesis	From 4-hydroxy-pentan-2-one refluxed with oxalic acid;[6] also O + CHO $\xrightarrow{B^-}$		From 2-furyl-1-pentene by catalytic reduction with Ni-Raney[9]	
Natural occurrence	Reported found among the volatile components of roasted peanuts;[7] also in cranberry, bilberry, rum, cocoa, coffee, tea, roasted filberts, roasted peanuts, and potato chips[8]		Reported found as a constituent in coffee aroma;[10] a volatile flavor component in roasted filberts[11]	
Reported uses[3]	Non-alcoholic beverages	2.0 ppm	Non-alcoholic beverages	3.0 ppm
	Ice cream, ices, etc.	2.0 ppm	Candy	3.0 ppm
	Candy	4.0 ppm	Baked goods	3.0 ppm
	Baked goods	4.0 ppm	Meat, meat sauces, soups	3.0 ppm
	Gelatins and puddings	4.0 ppm	Milk, dairy products	3.0 ppm
	Milk, dairy products	1.0 ppm	Cereals	3.0 ppm
	Cereals	2.0 ppm		
Regulatory status	FEMA No. 3417		FEMA No. 3317	

REFERENCES

For References 1–5, see end of Part III.

6. *Chem. Abstr.*, 70, 26505p, 1969.
7. **Walradt et al.,** *J. Agric. Food Chem.*, 19, 972, 1971.
8. Volatile Compounds in Food, T.N.O. Report No. 4030, 3rd ed., Centraal Instituut voor Voedingsonderzoek, Zeist, Netherlands.
9. **Stoll et al.,** *Helv. Chim. Acta,* 50, 628, 1967.
10. **Stoll et al.,** *Helv. Chim. Acta,* 50, 628, 1967.
11. **Sheldon et al.,** *J. Food Sci.,* 37(2), 313, 1972.

	Pentyl-2-furyl ketone		2-Pentyl pyridine	
Other names	2-Hexanoylfuran 1-(2-Furyl-1-hexanone)		2-Amyl pyridine	
Empirical formula	$C_{10}H_{14}O_2$		$C_{10}H_{15}N$	
Structure	(furan ring)–C(=O)–CH₂–CH₂–CH₂–CH₂–CH_3		(pyridine ring)–CH_2–CH_2–CH_2–CH_2–CH_3	
Physical/chemical characteristics				
Molecular weight	166		149.23	
Boiling point			82.5–84.5°C at 8–9 mm Hg 102–107°C	
Specific gravity			0.8881 at 20°C	
Refractive index			1.4834 at 20°C	
Synthesis			By alkylation of 2-pycollyl lithium;[6] by hydrogenation of 2-pentenyl pyridine[7]	
Natural occurrence			Reported as occurring in shallow-fried beef;[8] also as a steam-volatile component in roasted Spanish peanuts[9]	
Reported uses[3]	Non-alcoholic beverages	0.05 ppm	Ice cream, ices, etc.	0.1 ppm
	Ice cream, ices, etc.	0.05 ppm	Candy	0.5 ppm
	Baked goods	0.15 ppm	Baked goods	1.0 ppm
	Gelatins and puddings	0.10 ppm	Gelatins and puddings	0.1 ppm
	Chewing gum	0.15 ppm	Condiments, pickles	0.4 ppm
	Milk, dairy products	0.05 ppm	Meat, meat sauces, soups	0.5 ppm
Regulatory status	FEMA No. 3418		FEMA No. 3383	

REFERENCES

For References 1–5, see end of Part III.

6. **Osuch and Levine,** *J. Am. Chem. Soc.,* 78, 1723, 1956.
7. *Chem. Abstr.,* 71, 61162z, 1969.
8. **Watanabe et al.,** *J. Agric. Food Chem.,* 19, 1017, 1971.
9. **Walradt et al.,** *J. Agric. Food Chem.,* 19, 972, 1971.

	α-Phellandrene	Phenethyl acetate
Other names	1-Isopropyl-4-methyl-2,4-cyclohexadiene 4-Isopropyl-1-methyl-1,5-cyclohexadiene 5-Isopropyl-2-methyl-1,3-cyclohexadiene *p*-Mentha-1,5-diene 1-Methyl-4-isopropyl-1,5-cyclohexadiene 2-Methyl-5-isopropyl-1,3-cyclohexadiene	Benzyl carbinyl acetate 2-Phenylethyl acetate
Empirical formula	$C_{10}H_{16}$	$C_{10}H_{12}O_2$
Structure	CH_3 / H_3C CH_3	$CH_2-CH_2-O-\overset{O}{\overset{\parallel}{C}}-CH_3$
Physical/chemical characteristics[5]		
Appearance	Colorless, mobile oil	Colorless liquid
Assay		98% min[2]
Molecular weight	136.23	164.21
Boiling point	*dl*-, 175–176°C; *d*-, 175–176°C at 754 mm Hg; *l*-, 173–175°C at 754 mm Hg	232°C
Flash point		105°C
Specific gravity	*dl*-, 0.841 at 22°C; *d*-, 0.8565 at 15°C; *l*-, 0.848 at 15°C	1.030–1.034 at 25°/25°C[2]
Refractive index	*dl*-, 1.4760 at 22°C; *d*-, 1.4732 at 19°C; *l*-, 1.47694 at 20°C	1.4970–1.5010 at 20°C[2]
Acid value		Not more than 1.0[2]
Optical rotation	*d*-, +40°40′; *l*-, −84°10′	
Solubility	Insoluble in water; soluble in ether	1:2 and more in 70% alcohol;[2] soluble in most common organic solvents
Organoleptic characteristics	Pleasant, fresh, citrus, peppery odor with a discrete mint note	Floral odor reminiscent of rose with a honey-like undertone; sweet, fruit-like taste reminiscent of raspberry
Synthesis	The *l*-form by isolation from *Eucalyptus numerosa* and similar oils; also from the manufacture of synthetic menthol	By acetylation of phenyl ethyl alcohol
Natural occurrence	The *d*-form has been reported found in the essential oils of bitter fennel, Ceylon cinnamon leaves, star anise, *Curcuma longa*, and others. The *l*-form has been reported found in various eucalyptus oils, angelica, *Lantana camara*, and others.	Reported found in natural products
Reported uses[3]	Non-alcoholic beverages 10 ppm Ice cream, ices, etc. 28 ppm Candy 130 ppm Baked goods 41 ppm	Non-alcoholic beverages 1.4 ppm Ice cream, ices, etc. 2.2 ppm Candy 4.2 ppm Baked goods 5.6 ppm
Regulatory status	FDA 121.1164; FEMA No. 2856	FDA 121.1164; FEMA No. 2857

REFERENCES

For References 1–5, see end of Part III.

For further information on α-phellandrene see the following:

Berry et al., *J. Chem. Soc.* (London), p. 1418, 1939.
Semmler, *Ber. Dtsch. Chem. Ges.*, 36, 1749, 1903.
Smith et al., *J. Chem. Soc.* (London), 125, 930, 1924.

	Phenethyl alcohol	Phenethylamine
Other names	Benzyl carbinol 2-Phenylethanol Phenylethyl alcohol β-Phenylethyl alcohol	β-Phenylethylamine 1-Amino-2-phenylethane β-Aminoethyl benzene
Empirical formula	$C_8H_{10}O$	$C_8H_{11}N$
Structure	CH_2-CH_2-OH	$-CH_2-CH_2-NH_2$
Physical/chemical characteristics[5]		
Appearance	Colorless, viscous liquid	Liquid; absorbs CO_2 on exposure to air
Assay	99% min	
Molecular weight	122.17	121.18
Melting point	−25.8°C	
Boiling point	220–222°C at 740 mm Hg	194.5–195°C
Congealing point	101°C	
Flash point	1.017–1.020 at 25°/25°C;[2] 1.0242 at 15°C	
Specific gravity	1.5310–1.5340[2] (1.53212) at 20°C	0.9640
Refractive index		1.5290 at 25°C
Solubility	1:2 in 50% alcohol;[2] 1:1 in 60% alcohol; soluble (approx. 1:60) in water; soluble in most common organic solvents	Soluble in water; readily soluble in alcohol and ether
Organoleptic characteristics	Characteristic rose-like odor; initially a slightly bitter taste, then sweet and reminiscent of peach	Fishy odor
Synthesis	From toluene, benzene, or styrene	By reduction of benzyl cyanide with sodium metal in alcohol[6] or with Ni-Raney[7]
Natural occurrence	Reported found (as is or esterified) in several natural products: rose concrete, rose absolute (60% or more), and rose distillation waters; also found in the essential oils of neroli, ylang-ylang, narcissus, hyacinth, lily, tea leaves, *Michelia champaca*, *Pandamus odoratissimus*, Congo and Reunion geranium, tobacco, and others. It has been identified in wines.[8]	Reported found in the oil of bitter almonds
Reported uses[3]	Non-alcoholic beverages 1.5 ppm Ice cream, ices, etc. 8.3 ppm Candy 12 ppm Baked goods 16 ppm Gelatins and puddings 0.15 ppm Chewing gum 21–80 ppm	Non-alcoholic beverages 0.10 ppm Ice cream, ices, etc. 0.10 ppm Candy 0.10 ppm Baked goods 0.10 ppm Gelatins and puddings 0.10 ppm
Regulatory status	FDA 121.1164; FEMA No. 2858	FEMA No. 3220

REFERENCES

For References 1–5, see end of Part III.

6. **Johnson and Guest,** *Am. Chem. J.*, 42, 346, 1909.
7. **Robinson and Snyder,** in *Organic Syntheses*, Coll. Vol. 3, Horning, E. C., Ed., Wiley & Sons, New York, 1955, 720.
8. **Ayrapas,** *Natuer*, 5, 472, 1962.

For further information on phenethyl alcohol see the following:

Gurunadham, *J. Sci. Ind. Res. (India) Sect. B*, 20(8), 408, 1961.

Kishore and Lal, *Indian Oil Soap J.*, 6, 255, 1961.

	Phenethyl anthranilate	Phenethyl benzoate
Other names	Benzyl carbinyl anthranilate β-Phenylethyl-*o*-aminobenzoate	2-Phenylethyl benzoate
Empirical formula	$C_{15}H_{15}NO_2$	$C_{15}H_{14}O_2$
Structure	$C_6H_5-CH_2-CH_2-O-C(=O)-C_6H_4(NH_2)$	$C_6H_5-CH_2-CH_2-O-C(=O)-C_6H_5$
Physical/chemical characteristics[5]		
Appearance	Fused yellow-amber mass (poor quality); white, colorless, crystalline mass	Colorless to yellowish, oily liquid
Assay		98% min
Molecular weight	241.29	226.28
Melting point	42°C	
Boiling point	324°C	189°C (>300°C) at 21 mm Hg
Congealing point	>40°C	
Flash point		>100°C
Specific gravity	1.14 (liquid)	1.092–1.096 at 25°/25°C; 1.023–1.030 at 15°C
Refractive index		1.558–1.562 (1.5591–1.5605) at 20°C
Acid value	1 max	1 max
Solubility	Insoluble in water; 1 gm in 6 ml of 80% alcohol; 1 gm in 1 ml of 90% alcohol	1:1 in 90% alcohol; 1:18 in 80% alcohol; insoluble in water
Organoleptic characteristics	Pure material almost odorless; with other materials it has a characteristic odor reminiscent of neroli and grapefruit with a sweet, grape and orange undertone; not very stable	Sweet, rose and honey-like odor; used in trace quantities for fixation of honey, strawberry, and others
Synthesis	By direct esterification of anthranilic acid with phenylethyl alcohol; pure material from phenylethyl alcohol plus isatoic anhydride; also from phenylethyl alcohol and methylanthranilate	From phenethyl alcohol and benzoyl chloride in the presence of NaOH;[6] from phenylethyl alcohol and methylbenzoate; by esterification of phenylethyl alcohol with benzoic acid
Natural occurrence	Not reported found in nature	Reported found in the essential oil from flowers of rose and orange[7]
Reported uses[3]	Non-alcoholic beverages 1.4 ppm Ice cream, ices, etc. 1.9 ppm Candy 6.2 ppm Baked goods 5.8 ppm	Non-alcoholic beverages 1.0 ppm Ice cream ices, etc. 1.0 ppm Candy 2.0 ppm Baked goods 4.0 ppm Chewing gum 3.8 ppm
Regulatory status	FDA 121.1164; FEMA No. 2859	FDA 121.1164; FEMA No. 2860

REFERENCES

For References 1–5, see end of Part III.

6. **Klazes and Allendorf,** *Ber. Dtsch. Chem. Ges.*, 31, 1003, 1898.
7. **Grignard,** *Traite Chim. Org.*, 22, 1047, 1953.

	Phenethyl butyrate	Phenethyl cinnamate
Other names	β-Phenethyl-*n*-butanoate	Benzyl carbinyl cinnamate Phenylethyl cinnamate Phenylethyl-β-phenylacrylate β-Phenethyl-3-phenylpropenoate
Empirical formula	$C_{12}H_{16}O_2$	$C_{17}H_{16}O_2$
Structure	$C_6H_5-CH_2-CH_2-O-C(=O)-CH_2-CH_2-CH_3$	$C_6H_5-CH_2-CH_2-O-C(=O)-CH=CH-C_6H_5$
Physical/chemical characteristics[5]		
Appearance	Colorless liquid	White, crystalline solid
Assay	98% min	98–99% min
Molecular weight	192.26	252.32
Melting point		58°C
Boiling point	238°C	>300°C
Congealing point		>54–56°C
Flash point	>100°C	
Specific gravity	0.991–0.994 (1.00) at 25°/25°C	
Refractive index	1.488–1.492 at 20°C	
Acid value	1 max	
Solubility	1:2 in 80% alcohol; 1:1 in 90% alcohol; almost insoluble in water	1 gm in 10 ml of 95% alcohol with slight turbidity; soluble in hot alcohol
Organoleptic characteristics	Rose-like fragrance; sweet taste suggestive of honey; not as stable as the isobutyrate	Sweet, balsamic odor reminiscent of rose; at low levels sweet, plum-like taste
Synthesis	By esterification of phenylethyl alcohol with *n*-butyric acid	From phenylethyl alcohol and methylcinnamate
Natural occurrence	In passion fruit, grapes, and wine	Reported found in several natural products
Reported uses[3]	Non-alcoholic beverages 3.2 ppm Ice cream, ices, etc. 8.9 ppm Candy 13 ppm Baked goods 13 ppm	Non-alcoholic beverages 1.7 ppm Ice cream, ices, etc. 0.80 ppm Candy 3.2 ppm Baked goods 3.1 ppm Gelatins and puddings 0.10 ppm
Regulatory status	FDA 121.1164; FEMA No. 2861	FDA 121.1164; FEMA No. 2863

REFERENCES

For References 1–5, see end of Part III.

	Phenethyl formate	Phenethyl hexanoate
Other names	Benzyl carbinyl formate Phenethyl methanoate 2-Phenylethyl formate	Phenyl ethyl caproate β-Phenethyl hexoate Benzyl carbinyl hexylate
Empirical formula	$C_9H_{10}O_2$	$C_{14}H_{20}O_2$
Structure	$C_6H_5-CH_2-CH_2-O-C(=O)-H$	$C_6H_5-CH_2-CH_2-O-C(=O)-(CH_2)_4-CH_3$
Physical/chemical characteristics[5]		
Appearance	Colorless liquid	Colorless oily liquid
Assay	96% min (90%)	
Molecular weight	150.18	220.31
Boiling point	226°C	263°C
Flash point	91°C (approx 100°C)	
Specific gravity	1.0660–1.0700 at 15°C; 1.056–1.062 at 25°C	0.98
Refractive index	1.5065–1.5090 (1.5060–1.5100) at 20°C	
Acid value	0.1 max (1.0)	
Solubility	Partially soluble in water; soluble in most common organic solvents. For the 96% active product, 1:4 in 70% alcohol	Almost insoluble in water; soluble in alcohol and oils
Organoleptic characteristics	Rose-like odor reminiscent of hyacinth and chrysanthemum; bittersweet flavor suggestive of unripe plum; not very stable	Fresh, fruity-green odor reminiscent of pineapple; pleasant taste
Synthesis	By direct esterification of phenethyl alcohol with formic acid	See Reference 6
Natural occurrence	Reported found in several natural products	Reported found in yellow passion fruit[7]
Reported uses[3]	Non-alcoholic beverages 1.3 ppm Ice cream, ices, etc. 11 ppm Candy 13 ppm Baked goods 15 ppm	Non-alcoholic beverages 8.0 ppm
Regulatory status	FDA 121.1164; FEMA No. 2864	FEMA No. 3221

REFERENCES

For References 1–5, see end of Part III.

6. **Mentani,** *Koryo,* (61), 61, 1961; *Chem. Abstr.,* 58, 2400g, 1963.
7. **Winter and Klöty,** *Helv. Chim. Acta,* 55(6), 1916, 1972.

	Phenethyl isobutyrate	Phenethyl isovalerate
Other names	2-Phenylethyl isobutyrate	Benzyl carbinyl isovalerianate Phenethyl isopentanoate Phenethyl-3-methylbutyrate 2-Phenylethyl isovalerate
Empirical formula	$C_{12}H_{16}O_2$	$C_{13}H_{18}O_2$
Structure	$C_6H_5-CH_2-CH_2-O-C(=O)-CH(CH_3)_2$	$C_6H_5-CH_2-CH_2-O-C(=O)-CH_2-CH(CH_3)_2$
Physical/chemical characteristics[5]		
Appearance	Colorless to yellowish liquid	Colorless liquid
Assay	98% min	98% min
Molecular weight	192.26	206.28
Boiling point	230°C	263°C
Flash point	>100°C	>100°C
Specific gravity	0.987–0.990 (1.00) at 25°C	0.973–0.976 at 25°C; 0.9845 at 15°C
Refractive index	1.4860–1.4900 at 20°C	1.4840–1.4860 (1.4855) at 20°C
Acid value	1 max	1 max
Solubility	1:3 in 80% alcohol	1:12–14 in 80% alcohol; 1:11 in 70% alcohol; insoluble in water
Organoleptic characteristics	Fruity odor; bittersweet taste reminiscent of unripe plum, pineapple, and banana	Fruity (rose-peach-apricot) odor; bittersweet flavor reminiscent of peach; not very stable
Synthesis	By esterification of phenethyl alcohol with *n*-hexanoic acid	By esterification of phenethyl alcohol with isovaleric acid
Natural occurrence	Reported found in several natural products	Reported found in several natural products
Reported uses[3]	Non-alcoholic beverages 3.4 ppm Ice cream, ices, etc. 4.0 ppm Candy 13 ppm Baked goods 11 ppm	Non-alcoholic beverages 1.3 ppm Ice cream, ices, etc. 2.5 ppm Candy 5.9 ppm Baked goods 6.1 ppm Chewing gum 0.80–45 ppm
Regulatory status	FDA 121.1164; FEMA No. 2862	FDA 121.1164; FEMA No. 2871

REFERENCES

For References 1–5, see end of Part III.

	Phenethyl-2-methylbutyrate	Phenethyl octanoate
Other names	Benzyl carbinyl ethyl methyl acetate Phenethyl-α-methylbutanoate	Phenyl ethyl caprylate Phenethyl octoate Phenyl ethyl octanoate Benzyl carbinyl octylate
Empirical formula	$C_{13}H_{18}O_2$	$C_{16}H_{24}O_2$
Structure	$C_6H_5-CH_2-CH_2-O-\overset{\overset{O}{\|}}{C}-\underset{\underset{CH_3}{\vert}}{CH}-CH_2-CH_3$	$C_6H_5-CH_2-CH_2-O-\overset{\overset{O}{\|}}{C}-(CH_2)_6-CH_3$
Physical/chemical characteristics[5]		
Appearance	Colorless liquid	Colorless oily liquid
Molecular weight		248.37
Melting point	206.28	
Boiling point		295.5°C 143.5°C at 1.5 mm Hg
Specific gravity		0.9473 at 20°C
Refractive index		1.4801 at 20°C
Solubility	Insoluble in water; soluble in alcohol	Insoluble in water; soluble in alcohol and oils
Organoleptic characteristics	Sweet, floral, fruity odor with warm, herbaceous notes; sweet, slightly fruity taste	Mild fruity, wine-like odor
Synthesis	By esterification of phenethyl alcohol with α-methylbutyric acid	See Reference 6
Natural occurrence	In mint	
Reported uses[3]	The compound is used sometimes in imitation peppermint	Non-alcoholic beverages 5.0 ppm
Regulatory status	FDA 121.1164	FEMA No. 3222

REFERENCES

For References 1–5, see end of Part III.
6. **Mentani,** *Koryo,* (61), 61, 1961; *Chem. Abstr.,* 58, 2400g, 1963.

	Phenethyl phenylacetate	Phenethyl propionate
Other names	Benzyl carbinyl phenylacetate 2-Phenylethyl phenylacetate Phenethyl-α-toluate	Benzyl carbinyl propionate 2-Phenylethyl propionate
Empirical formula	$C_{16}H_{16}O_2$	$C_{11}H_{14}O_2$
Structure	$C_6H_5-CH_2-CH_2-O-C(=O)-CH_2-C_6H_5$	$C_6H_5-CH_2-CH_2-O-C(=O)-C_2H_5$
Physical/chemical characteristics[5]		
Appearance	Colorless liquid or white crystals at low temperature	Colorless, slightly oily liquid; sometimes yellowish
Assay	98% min	98% min
Molecular weight	240.31	178.23
Melting point	28°C	
Boiling point	325°C (177–178°C at 4–5 mm Hg)	245°C
Congealing point	>26°C	
Flash point	>100°C	>100°C
Specific gravity	1.079–1.082 (1.09) at 25°C	1.010–1.014 at 25°/25°C
Refractive index		1.493–1.496 at 20°C
Acid value	1 max	1 max
Solubility	1:4 in 90% alcohol; insoluble in water	1:4 in 70% alcohol; soluble in propylene glycol; insoluble in water
Organoleptic characteristics	Heavy, sweet, floral, and balsamic odor, somewhat rosy; sweet, honey-like flavor	Very sweet odor reminiscent of red rose with a fruity undertone; honey-like flavor; warm, sweet raspberry and strawberry-like flavor
Synthesis	By esterification of phenylacetic acid with phenethyl alcohol in the presence of H_2SO_4;[6] also by direct esterification in the presence of gaseous HCl	By esterification of phenethyl alcohol with propionic acid
Natural occurrence	Not reported found in nature	In peanut (*Arachis hypogaea*)
Reported uses[3]	Non-alcoholic beverages 2.3 ppm Ice cream, ices, etc. 4.2 ppm Candy 4.8 ppm Baked goods 5.3 ppm Maraschino cherries 10 ppm	Non-alcoholic beverages 3.6 ppm Ice cream, ices, etc. 11 ppm Candy 12 ppm Baked goods 16 ppm
Regulatory status	FDA 121.1164; FEMA No. 2866	FDA 121.1164; FEMA No. 2867

REFERENCES

For References 1–5, see end of Part III.

6. **Hibbert and Burt,** *J. Am. Chem. Soc.*, 47, 2242, 1925.

	Phenethyl salicylate	Phenethyl senecioate
Other names	Benzyl carbinyl salicylate Phenethyl-2-hydroxybenzoate Phenethyl-*o*-hydroxybenzoate 2-Phenylethyl salicylate	Phenethyl-3,3-dimethylacrylate Phenethyl-3-methyl-2-butenoate Phenethyl-3-methylcrotonate
Empirical formula	$C_{15}H_{14}O_3$	$C_{13}H_{16}O_2$
Structure	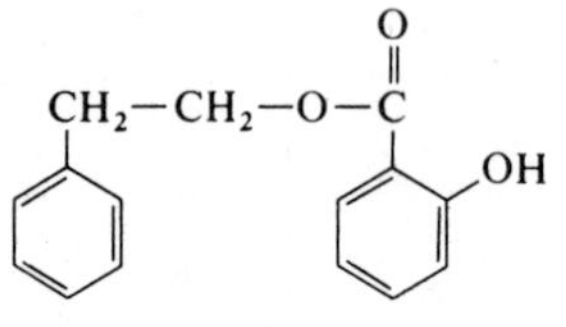 $C_6H_5-CH_2-CH_2-O-C(=O)-C_6H_4(OH)$	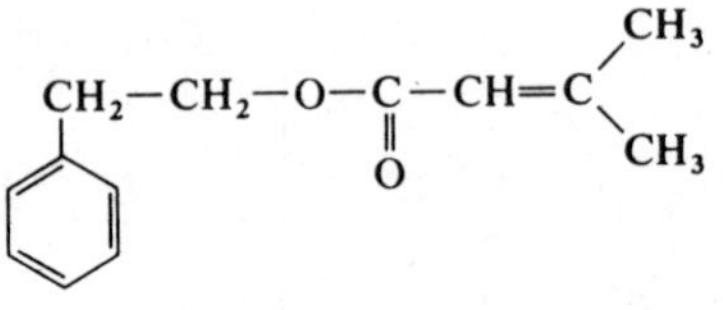 $C_6H_5-CH_2-CH_2-O-C(=O)-CH=C(CH_3)_2$
Physical/chemical characteristics[5]		
Appearance	White, crystalline solid	Colorless, oily liquid
Assay	98% min	
Molecular weight	242.28	204.27
Melting point	44°C	
Boiling point	370°C	
Congealing point	41°C	
Specific gravity	1.15 (liquid)	
Acid value	1 max	
Solubility	1:16 in 95% alcohol	Insoluble in water; soluble in alcohol
Organoleptic characteristics	Faint but persistent rose-like odor, somewhat hyacinth and carnation-like; sweet taste suggestive of peach, apricot, and honey	Faint, deep, sweet odor, somewhat herbaceous and winy; sweet, winy, and deep, fruity flavor
Synthesis	From phenethyl alcohol and methylsalicylate by interchange; also by esterification of phenethyl alcohol with salicylic acid	By esterification of phenethyl alcohol with 3-methylcrotonic acid
Natural occurrence	Not reported found in nature	Not reported found in nature
Reported uses[3]	Non-alcoholic beverages 0.75 ppm Ice cream, ices, etc. 0.67 ppm Candy 1.5 ppm Baked goods 2.0 ppm	Alcoholic beverages 5.0 ppm Ice cream, ices, etc. 5.0 ppm Candy 5.0 ppm
Regulatory status	FDA 121.1164; FEMA No. 2868	FDA 121.1164; FEMA No. 2869

REFERENCES

For References 1–5, see end of Part III.

	Phenethyl tiglate	Phenol
Other names	Benzyl carbinyl tiglate Phenethyl *trans*-2,3-dimethylacrylate Phenethyl *trans*-2-methylbutenoate Phenethyl *trans*-2-methylcrotonate Phenylethyl tiglate	Carbolic acid Benzenol Hydroxybenzene Phenic or phenylic acid Phenyl hydroxide Oxybenzene
Empirical formula	$C_{13}H_{16}O_2$	C_6H_5OH
Structure	$C_6H_5-CH_2-CH_2-O-C(=O)-C(CH_3)=CH-CH_3$	C_6H_5OH
Physical/chemical characteristics[5]		
Appearance	Colorless liquid	Colorless or white needles; turns pink on exposure to air
Assay		98% min.
Molecular weight	204.27	94.11
Melting point		43°C
Boiling point	259°C	181.75°C 70.86°C at 10 mm Hg
Congealing point		41°C
Flash point		79°C
Specific gravity	1.02	1.0722 at 20/20°C 1.0576 at 20/4°C
Refractive index		1.5509 at 21°C 1.5418 at 41°C
Solubility	Insoluble in water; soluble in alcohol	Soluble in water, alcohol, chloroform, and carbon disulfide; very soluble in ether; soluble in carbon tetrachloride, acetone, and benzene in all proportions
Organoleptic characteristics	Pleasant, warm, rose-like, sweet odor; sweet, winy taste	
Synthesis	From phenethyl alcohol and tiglyl chloride in the presence of pyridine	See Reference 6
Natural occurrence	Not reported found in nature	Formed in dry distillation of wood, peat, and coal; coal tar is one of the commercial sources of phenol and its homologs; reported found as a constituent in coffee aroma;[7] also reported found in cognac[8] and in yellow passion fruit[9]
Reported uses[3]	Non-alcoholic beverages 0.80–0.90 ppm Ice cream, ices, etc. 4.3 ppm Candy 10 ppm Baked goods 10 ppm	Non-alcoholic beverages 0.50 ppm Ice cream, ices, etc. 0.50 ppm Candy 0.50 ppm Baked goods 0.50 ppm Gelatins and puddings 0.50 ppm
Regulatory status	FDA 121.1164; FEMA No. 2870	FEMA No. 3223

REFERENCES

For References 1–5, see end of Part III.

6. **Dierichs and Rubička,** *Phenole und Basen, Vorkommen und Gewinnung,* Akademie-Verlag, Berlin, 1958.
7. **Stoll et al.,** *Helv. Chim. Acta,* 50, 628, 1967.
8. **Schaefer and Timmer,** *J. Food Sci.,* 35(1), 10, 1970.
9. **Winter and Klöty,** *Helv. Chim. Acta,* 55(6), 1916, 1972.

	Phenoxyacetic acid	2-Phenoxyethyl isobutyrate
Other names	Glycolic acid phenyl ether Phenoxyethanoic acid *o*-Phenylglycolic acid Phenylium	Ethyleneglycol monophenylether, isobutyrate 2-Phenoxyethyl isobutanoate Phenylcellosolve isobutyrate
Empirical formula	$C_8H_8O_3$	$C_{12}H_{16}O_3$
Structure	$C_6H_5-O-CH_2-COOH$	$C_6H_5-O-CH_2-CH_2-O-C(=O)-CH(CH_3)_2$
Physical/chemical characteristics[5]		
Appearance	Crystalline solid	Colorless liquid
Assay		97% min
Molecular weight	152.14	208.26
Melting point	98°C	
Boiling point	285°C (decomposes)	265°C
Flash point		>100°C
Specific gravity		1.044–1.048 at 25°/25°C
Refractive index		1.4920–1.4950 at 20°C
Acid value		1 max
Solubility	Slightly soluble in water; readily soluble in alcohol, ether, and other common organic solvents	1:2 in 70% alcohol
Organoleptic characteristics	Sour, sweet odor; honey-like taste	Faint, rose-like and honey odor; sweet, fruity (peach-like) taste
Synthesis	By reacting phenol and monochloroacetic acid	From the corresponding phenoxyethyl alcohol and isobutyric acid in the presence of boric acid in benzene solution[6]
Natural occurrence	Reported as naturally occurring in cocoa beans	Not reported found in nature
Reported uses[3]	Non-alcoholic beverages 0.37 ppm Ice cream, ices, etc. 1.0 ppm Candy 2.2 ppm Baked goods 2.2 ppm	Non-alcoholic beverages 0.90–5.0 ppm Ice cream, ices, etc. 5.0–30 ppm Candy 15–30 ppm Baked goods 15–30 ppm
Regulatory status	FDA 121.1164; FEMA No. 2872	FDA 121.1164; FEMA No. 2873

REFERENCES

For References 1–5, see end of Part III.

6. **Takaharu and Mentani,** *Chem. Abstr.*, 58, 2400g, 1963.

	Phenylacetaldehyde	Phenylacetaldehyde-2,3-butylene glycol acetal
Other names	Hyacinthin 1-Oxo-2-phenylethane α-Toluic aldehyde α-Tolualdehyde α-Tolyl aldehyde	4, 5-Dimethyl-2-benzyl-1,3-dioxolan
Empirical formula	C_8H_8O	$C_{12}H_{16}O_2$
Structure	CH_2-CHO (on benzene ring)	$C_6H_5-CH_2-CH$ (ring: O–CH(CH_3)–CH(CH_3)–O)
Physical/chemical characteristics[5]		
Appearance	Colorless to slightly yellow, oily liquid; becomes more viscous on aging	Colorless, viscous liquid
Assay	90% min[2] (95%)	
Molecular weight	120.14	192.26
Melting point	33–34°C	
Boiling point	193–194°C (206°C)	
Flash point	68°C	
Specific gravity	1.025–1.035[2] (1.023–1.040) at 25°/25°C	
Refractive index	1.5240–1.5320 at 20°C[2]	
Acid value	Not more than 5.0[2]	
Solubility	1:2 in 80% alcohol;[2] insoluble in water; soluble in most common organic solvents	Insoluble in water; soluble in alcohol
Organoleptic characteristics	Harsh, green odor reminiscent of hyacinth on dilution; unpleasant, pungent, bitter flavor, turning sweet and fruit-like at low levels	Sweet, floral, somewhat earthy fragrance; fruity flavor
Synthesis	By Darzen glycidic ester synthesis from benzaldehyde;[2] readily oxidizable to phenylacetic acid	By condensation of phenylacetaldehyde with 2,3-butylene glycol
Natural occurrence	Identified among the constituents of several essential oils: neroli, *Citrus sinensis* leaves, other citrus species (flowers and leaves), narcissus, magnolia, lily, rose, tea, and others	Not reported found in nature
Reported uses[3]	Non-alcoholic beverages 0.68 ppm Ice cream, ices, etc. 0.75 ppm Candy 1.6 ppm Baked goods 2.0 ppm Chewing gum 1.7–87 ppm	Candy 4.0 ppm
Regulatory status	FDA 121.1164; FEMA No. 2874	FDA 121.1164; FEMA No. 2875

REFERENCES

For References 1–5, see end of Part III.

	Phenylacetaldehyde diisobutyl acetal	Phenylacetaldehyde dimethyl acetal
Other names	1,1-Diisobutoxy-2-phenylethane	1,1-Dimethoxy-2-phenylethane α-Toluic aldehyde dimethylacetal
Empirical formula	$C_{16}H_{26}O_2$	$C_{10}H_{14}O_2$
Structure	$C_6H_5-CH_2-CH(O-CH_2-CH_2-CH_2-CH_3)(O-CH_2-CH_2-CH_2-CH_3)$	$C_6H_5-CH_2-CH(O-CH_3)(O-CH_3)$
Physical/chemical characteristics[5]		
Appearance	Colorless liquid	Colorless liquid
Assay		95% min[2]
Molecular weight	250.38	166.22
Boiling point	240°C	221°C
Flash point		88°C
Specific gravity	0.93	1.000–1.006 at 25°/25°C[2]
Refractive index		1.4930–1.4960 at 20°C[2]
Acid value		Not more than 1[2]
Free aldehyde content		3.0% max[2]
Solubility	Almost insoluble in water; soluble in alcohol and oils	1:2–10 in 70% ethanol; insoluble in water
Organoleptic characteristics	Sweet, floral, delicate-green odor	Strong, green odor with hyacinth-like note; at low levels sweet, green, spicy flavor, turning bitter at high levels
Synthesis	From benzyl magnesium chloride and butyl *o*-formate; also from phenyl acetaldehyde and butyl alcohol	By the cold reaction of the corresponding aldehyde with methanol or with orthoformic ester in the presence of acid[2]
Natural occurrence		In Bulgarian tobacco
Reported uses[3]	Non-alcoholic beverages 10.0 ppm Alcoholic beverages 25.0 ppm Ice cream, ices, etc. 30.0 ppm Candy 75.0 ppm Baked goods 50.0 ppm Gelatins and puddings 50.0 ppm	Non-alcoholic beverages 0.40 ppm Ice cream, ices, etc. 0.78 ppm Candy 1.4 ppm Baked goods 8.8 ppm Chewing gum 1.0 ppm
Regulatory status	FEMA No. 3384	FDA 121.1164; FEMA No. 2876

REFERENCES

For References 1–5, see end of Part III.

	Phenylacetaldehyde glyceryl acetal	Phenylacetic acid
Other names	Mixture of 60% 5-hydroxymethyl-2-benzyl-1,3-dioxolan and 40% 5-hydroxy-2-benzyl-1,3-dioxan	α-Toluic acid
Empirical formula	$C_{11}H_{14}O_3$	$C_8H_8O_2$
Structure	$C_6H_5-CH_2-CH(-O-CH_2-)(-O-CH(CH_2OH)-)$ (60%); $C_6H_5-CH_2-CH(-O-CH_2-CH(OH)-CH_2-O-)$ (40%)	$C_6H_5-CH_2-COOH$
Physical/chemical characteristics[5]		
Appearance	Colorless, viscous liquid	White, crystalline solid (leaf)
Assay		99–100%[2]
Molecular weight	194.23	136.15
Melting point		76–78°C[2]
Boiling point		265.5°C (266°C)
Flash point	95°C	
Specific gravity	1.1650–1.1680 at 15°C	1.080 at 15.5°C
Refractive index	1.5315–1.5345 at 20°C	
Aldehyde content	0.5% max	
Solubility	1:4–7 in 30% alcohol; insoluble in water	Slightly soluble (2%) in cold water; soluble in hot water, alcohol, ether, and propylene glycol
Organoleptic characteristics	Faint, sweet, rosy odor; sweet, green flavor	Geranium-leaf and rose-like odor in dilute solution; persistent and disagreeable in concentrated solution; sweet, honey-like flavor at high levels; at low levels, a sweetener
Synthesis	From phenylacetaldehyde and glycerol	By the treatment of benzyl cyanide with dilute sulfuric acid[6] and other procedures
Natural occurrence	Not reported found in nature	Reported found among the constituents of a few essential oils: tobacco, *Rosa centifolia*,[7] Bulgarian rose,[8] orange flowers absolute,[9] neroli, and *Mentha arvensis* of Japanese origin; also reported present among the volatile constituents of cocoa[10]
Reported uses[3]	Non-alcoholic beverages 5.0 ppm Ice cream, ices, etc. 20 ppm Candy 0.06–20 ppm	Non-alcoholic beverages 1.8 ppm Alcoholic beverages 0.10 ppm Ice cream, ices, etc. 5.3 ppm Candy 5.9 ppm Baked goods 12 ppm Gelatins and puddings 27 ppm Chewing gum 5.4–11 ppm Syrups 0.10 ppm
Regulatory status	FDA 121.1164; FEMA No. 2877	FDA 121.1164; FEMA No. 2878

REFERENCES

For References 1–5, see end of Part III.

6. **Blatt, A. H. (Ed.),** *Organic Syntheses,* Coll. Vol. 1, John Wiley & Sons, 1941, 436.
7. **Igolen,** *Chem. Zentralbl.,* 2, 2718, 1939.
8. **Guenther and Garnier,** *Chem. Zentralbl.,* 1, 2128, 1931.
9. **Elze,** *Chem. Zentralbl.,* 2, 660, 1926.
10. **Quesnel,** *Choc. Confiserie Fr.,* 213, 30, 1965.

	4-Phenyl-2-butanol	2-Phenyl-2-butenal
Other names	Methyl phenethyl carbinol Phenylethyl methyl carbinol	2-Phenyl crotonaldehyde
Empirical formula	$C_{10}H_{14}O$	$C_{10}H_{10}O$
Structure	$C_6H_5-CH_2-CH_2-CH(OH)-CH_3$	$CH_3-CH=C(C_6H_5)-CHO$
Physical/chemical characteristics[5]		
Appearance	Oily liquid	
Molecular weight	150.22	146.18
Boiling point	123–124°C at 15 mm Hg	177°C at 15 mm Hg
Specific gravity	0.976 at 19°C	1.045 at 18/0°C
Refractive index	1.513 at 17°C	1.5605 at 18°C
Organoleptic characteristics	Characteristic, aromatic, floral odor	
Synthesis	The optically inactive product can be prepared by hydrogenation of benzylidene acetone in alcohol solution; under pressure in the presence of platinum oxide, palladium oxide, or ferrous sulfate;[6] by reduction with magnesium in methanol[7]	From acetaldehyde and phenyl acetaldehyde[8]
Natural occurrence	Not reported found in nature	Reported as the odorous component in black tea[9] and in *Phallus impudicus*[10]
Reported uses[3]	Non-alcoholic beverages 0.12–0.90 ppm Ice cream, ices, etc. 0.60–6.0 ppm Candy 1.5–15 ppm Baked goods 1.5–15 ppm	Non-alcoholic beverages 2.0 ppm Ice cream, ices, etc. 2.0 ppm Candy 2.0 ppm Baked goods 2.0 ppm Gelatins and puddings 2.0 ppm Chewing gum 2.0 ppm
Regulatory status	FDA 121.1164; FEMA No. 2879	FEMA No. 3224

REFERENCES

For References 1–5, see end of Part III.

6. **Kern, Shriner and Adams,** *J. Am. Chem. Soc.*, 47, 1157, 1925.
7. **Zechmeister and Rom,** *Ann. Chem.*, 468, 126, 1929.
8. **Kuhn and Michel.,** *Ber. Dtsch. Chem. Ges.*, 71, 1119, 1938.
9. **Bricout, Viani, Muggler-Chavan, Marion, Reymond and Egli,** *Helv. Chim. Acta,* 50(6), 1517, 1967.
10. **List and Freund,** *Naturwissenschaften,* 53(22), 585, 1966.

	4-Phenyl-3-buten-2-ol	4-Phenyl-3-buten-2-one
Other names	Methyl styryl carbinol	Benzylidene acetone Cinnamyl methyl ketone Methyl styryl ketone
Empirical formula	$C_{10}H_{12}O$	$C_{10}H_{10}O$
Structure	$C_6H_5-CH=CH-CH(OH)-CH_3$	$C_6H_5-CH=CH-C(=O)-CH_3$
Physical/chemical characteristics[5]		
Appearance	Dense, oily liquid	Colorless to slightly yellow, crystalline solid
Assay		97% min
Molecular weight	148.20	146.18
Melting point		41–42°C
Boiling point	131°C at 12 mm Hg	260–262°C; 147–149°C at 20 mm Hg
Congealing point		39°C
Specific gravity	1.0134 at 22°C	1.0076 at 47.3°C
Refractive index		1.5824 at 47.3°C
Solubility	Insoluble in water; soluble in most organic solvents	Readily soluble in alcohol, ether, benzene, and chloroform; 1:3 in 60% alcohol
Organoleptic characteristics	Characteristic floral odor	Odor reminiscent of coumarin and rhubarb; sharp, pungent flavor; not very stable
Synthesis	From cinnamic aldehyde and magnesium methyl bromide in ether solution and subsequent hydrolysis of the ester[8]	By condensing benzaldehyde and acetone in the presence of an alkaline condensing agent.[6] For the synthesis of *cis*- and *trans*-benzylidene acetone, see reference 7.
Natural occurrence	Not reported found in nature	In hydrolyzed soy protein
Reported uses[3]	Non-alcoholic beverages 2.0 ppm Ice cream, ices, etc. 20 ppm Candy 0.03–20 ppm Baked goods 20 ppm	Non-alcoholic beverages 0.82 ppm Ice cream, ices, etc. 0.84 ppm Candy 3.7 ppm Baked goods 4.5 ppm Gelatins and puddings 2.1 ppm Shortening 0.20 ppm
Regulatory status	FDA 121.1164; FEMA No. 2880	FDA 121.1164; FEMA No. 2881

REFERENCES

For References 1–5, see end of Part III.

6. **Lawson et al.,** *J. Chem. Soc.,* 125, 636, 1924.
7. **Schinz,** *Riechst. Aromen,* 5, 67, 1955.
8. **Muskat and Ludeman,** *Chem. Ber.,* 62, 2286, 1929.

	4-Phenyl-2-butyl acetate	1-Phenyl-3-methyl-3-pentanol
Other names	Phenylethyl methyl carbinyl acetate	Methyl ethyl phenylethyl carbinol 3-Methyl-1-phenyl-3-pentanol Phenylethyl methyl ethyl carbinol
Empirical formula	$C_{12}H_{16}O_2$	$C_{12}H_{18}O$
Structure	$C_6H_5-CH_2-CH_2-CH(O-C(=O)-CH_3)-CH_3$	$C_6H_5-CH_2-CH_2-C(CH_2CH_3)(CH_3)-OH$
Physical/chemical characteristics[5]		
Appearance	Colorless liquid	Colorless, slightly viscous liquid
Molecular weight	192.26	178.27
Boiling point	123–124°C at 13 mm Hg; *d*, 130°C at 15 mm Hg	254°C
Specific gravity	0.991 at 16°C; *d*, 0.9874 at 17°C	0.97
Refractive index	1.4895 at 16°C; *d*, 1.5012 at 20°C	
Optical rotation	*d*, +0.83°	
Solubility	Insoluble in water; soluble in alcohol	Almost insoluble in water; soluble in alcohol
Organoleptic characteristics	Mild, green, fruity odor; sweet, fruity taste	Warm, rose-like odor; fruity, green taste
Synthesis	By acetylation of the corresponding alcohol. The racemic and the *dextro*-rotatory forms are known.	From phenethyl magnesium bromide and methyl ethyl ketone
Natural occurrence	Not reported found in nature	Not reported found in nature
Reported uses[3]	Non-alcoholic beverages 0.10–3.0 ppm Ice cream, ices, etc. 3.0 ppm Candy 3.0 ppm Baked goods 0.50–3.0 ppm	Non-alcoholic beverages 0.16 ppm Candy 0.16 ppm Gelatins and puddings 0.60 ppm
Regulatory status	FDA 121.1164; FEMA No. 2882	FDA 121.1164; FEMA No. 2883

REFERENCES

For References 1–5, see end of Part III.

	3-Phenyl-4-pentenal		1-Phenyl-1,2-propanedione	
Other names	β-Vinylhydrocinnamaldehyde		Acetyl benzoyl Methyl phenyl diketone Methyl phenyl glyoxal	
Empirical formula	$C_{11}H_{12}O$		$C_9H_8O_2$	
Structure	$CH_2=CH-CH(C_6H_5)-CH_2-CHO$		$C_6H_5-C(=O)-C(=O)-CH_3$	
Physical/chemical characteristics[5]				
Appearance			Yellow oily liquid	
Molecular weight	160.21		148.17	
Boiling point			222°C 101°C at 12 mm Hg	
Specific gravity			1.0065 at 20°C	
Refractive index			1.537 at 10°C	
Solubility			Soluble in water, alcohol, and ether	
Organoleptic characteristics			Pungent odor	
Synthesis			By oxidation of propiophenone or benzyl methyl ketone with selenium dioxide;[6,7] by the acid hydrolysis of oximinopropiophenone[8] or other synthetic routes[9]	
Natural occurrence	Not reported found in nature		Reported found as a constituent in coffee aroma[10]	
Reported uses[3]	Non-alcoholic beverages	1.0 ppm	Non-alcoholic beverages	10.0 ppm
	Candy	1.0 ppm	Ice cream, ices, etc.	10.0 ppm
	Baked goods	1.0 ppm	Candy	10.0 ppm
	Meat, meat sauces, soups	1.0 ppm	Gelatins and puddings	10.0 ppm
	Milk, dairy products	1.0 ppm		
	Cereals	1.0 ppm		
Regulatory status	FEMA No. 3318		FEMA No. 3226	

REFERENCES

For References 1–5, see end of Part III.

6. **Riley et al.,** *J. Chem. Soc.,* p. 1875, 1932.
7. **Wegmann and Dalm,** *Helv. Chim. Acta,* 29, 1247, 1946.
8. **Hartmann and Roll,** in *Organic Syntheses,* Coll. Vol. 2, Blatt, A. H., Ed., Wiley & Sons, New York, 1943.
9. **Rodd,** *Chemistry of Carbon Compounds,* Vol. 3, Part B, American Elsevier, New York, 1956, 898.
10. **Stoll et al.,** *Helv. Chim. Acta,* 50, 628, 1967.

	1-Phenyl-1-propanol	3-Phenyl-1-propanol
Other names	Dihydro isocinnamic alcohol Dihydro-α-phenyl allyl alcohol Ethyl phenyl carbinol Phenylethyl carbinol *sec*-Phenyl propyl alcohol	Benzyl ethyl alcohol Hydrocinnamyl alcohol Phenyl propyl alcohol
Empirical formula	$C_9H_{12}O$	$C_9H_{12}O$
Structure	C_6H_5–CH(OH)–CH_2–CH_3	C_6H_5–CH_2–CH_2–CH_2OH
Physical/chemical characteristics[5]		
Appearance	Colorless, oily liquid	Colorless, slightly viscous liquid
Assay		98% min[2]
Molecular weight	136.19	136.19
Boiling point	220°C	233–235°C at 740 mm Hg
Specific gravity	0.99	0.998–1.002 at 25°/25°C;[2] 1.006 at 24°C
Refractive index		1.5240–1.5280 at 20°C;[2] 1.5351 at 23°C
Aldehyde content		0.5% max (as phenylpropyl aldehyde)
Solubility	Insoluble in water; soluble in alcohol	1:1 in 70% alcohol;[2] 1:4 in 45% alcohol; 1:3 in 50% alcohol; soluble in water (1:300) and most common organic solvents
Organoleptic characteristics	Balsamic, floral fragrance; sweet, honey-like taste	Characteristic sweet, hyacinth-mignonette odor; sweet and pungent taste suggestive of apricot
Synthesis	By hydrogenation of phenethyl ketone	By hydrogenation of either cinnamic aldehyde or cinnamic alcohol[2]
Natural occurrence	Not reported found in nature	Frequently occurring in nature—storax, Sumatra benzoin, tea, Peru balsam, and other products (passion fruit, strawberry)
Reported uses[3]	Non-alcoholic beverages 0.50 ppm Ice cream, ices, etc. 0.50 ppm Candy 1.5 ppm Baked goods 1.5 ppm	Non-alcoholic beverages 0.73 ppm Alcoholic beverages 5.0 ppm Ice cream, ices, etc. 1.4 ppm Candy 2.8 ppm Baked goods 3.3 ppm Chewing gum 4.3 ppm
Regulatory status	FDA 121.1164; FEMA No. 2884	FDA 121.1164; FEMA No. 2885

REFERENCES

For References 1–5, see end of Part III.

	2-Phenylpropionaldehyde	3-Phenylpropionaldehyde
Other names	Hydratropaldehyde α-Methyl phenylacetaldehyde α-Methyl tolualdehyde 2-Phenylpropanal α-Phenyl propionaldehyde	Benzyl acetaldehyde Dihydrocinnamic aldehyde Hydrocinnamaldehyde Hydrocinnamic aldehyde β-Phenyl propionaldehyde Phenylpropyl aldehyde
Empirical formula	$C_9H_{10}O$	$C_9H_{10}O$
Structure	$CH_3-CH-CHO$ (with phenyl ring on CH)	CH_2-CH_2-CHO (with phenyl ring on first CH_2)
Physical/chemical characteristics[5]		
Appearance	Colorless to pale-yellow liquid	Colorless to pale-yellow liquid
Assay	95% min[2] (98% min)	96% min
Molecular weight	134.17	134.17
Boiling point	204°C	104–105°C (222°C) at 13 mm Hg
Flash point	80°C	97°C (95°C)
Specific gravity	0.998–1.006 at 25°/25°C;[2] 1.0050–1.0080 at 15°C	1.0160–1.0200 at 15°C; 1.010–1.020 at 25°C
Refractive index	1.5140–1.5190[2] (1.5155–1.5175) at 20°C	1.5205–1.5240 (1.5200–1.5220) at 20°C
Aldehyde content		3.7 max
Acid value	Not more than 5[2] (0.9)	
Solubility	1:3–10 in 80% alcohol;[2] 1:2–4 in 70% alcohol	Insoluble in water; soluble in most common organic solvents; 1:2 in 70% alcohol; 1:7 in 60% alcohol; 1:19 in 50% alcohol
Organoleptic characteristics	Intense, green, floral odor reminiscent of hyacinth	Strong, floral odor reminiscent of hyacinth; balsamic, green, warm (almost burning) flavor
Synthesis	By alkaline condensation of acetophenone and ethyl chloroacetate; also by distillation of methylphenyl ethylene glycol in 91% yields (under atmospheric pressure). See reference 6.	From phenyl propionitrile; also from cinnamic aldehyde diethylacetal
Natural occurrence	Not reported found in nature	Reported found in the essential oil of Ceylon cinnamon and in strawberry
Reported uses[3]	Non-alcoholic beverages 0.61 ppm Ice cream, ices, etc. 0.30 ppm Candy 0.85 ppm Baked goods 0.85 ppm	Non-alcoholic beverages 1.0 ppm Ice cream, ices, etc. 1.7 ppm Candy 5.0 ppm Baked goods 5.5 ppm Gelatins and puddings 4.3 ppm
Regulatory status	FDA 121.1164; FEMA No. 2886	FDA 121.1164; FEMA No. 2887

REFERENCES

For References 1–5, see end of Part III.

6. **Wagner,** *Die Riechstoffe und ihre Derivate – Die Aldehyde,* Hartleben's ed., 1929, 617.

	2-Phenylpropionaldehyde dimethyl acetal	3-Phenylpropionic acid
Other names	1,1-Dimethoxy-2-phenylpropane Hydratropic aldehyde dimethyl acetal	Benzylacetic acid Hydrocinnamic acid γ-Phenylpropionic acid
Empirical formula	$C_{11}H_{16}O_2$	$C_9H_{10}O_2$
Structure	$C_6H_5-CH(CH_3)-CH(O-CH_3)_2$	$C_6H_5-CH_2-CH_2-COOH$
Physical/chemical characteristics[5]		
Appearance	Oily liquid	White, crystalline solid
Molecular weight	180.25	150.18
Melting point		47–48°C (49°C)
Boiling point	240–241°C; 111–112°C at 14 mm Hg	280°C; 169–170°C at 28 mm Hg
Specific gravity	0.990 at 15.5°C; 0.9883 at 19°C	1.07 (liquid)
Refractive index	1.4990 at 20°C	
Solubility	Insoluble in water; soluble in alcohol	Slightly soluble in water (0.6%); soluble in alcohol, ether, propylene glycol, and glycerin
Organoleptic characteristics	Strong, warm, spicy odor reminiscent of walnut; warm, mushroom, nut-like taste	Faint, sweet odor, somewhat balsamic and coumarin-like; mildly sweet-sour taste, vanilla-like
Synthesis	By catalytic hydrogenation of cinnamyl aldehyde dimethyl acetal in the presence of colloidal palladium;[6] or from 2-phenylpropionaldehyde and a methanolic solution of HC1[7]	By reduction of cinnamic acid using sodium amalgam
Natural occurrence	Not reported found in nature	Reported found in raspberry and labdanum
Reported uses[3]	Non-alcoholic beverages 0.26 ppm Ice cream, ices, etc. 0.51 ppm Candy 1.5 ppm Baked goods 3.1 ppm Chewing gum 5.0 ppm Condiments 5.0 ppm	Non-alcoholic beverages 0.02–1.0 ppm Ice cream, ices, etc. 0.48–1.0 ppm Candy 0.80–4.0 ppm Baked goods 17 ppm Gelatins and puddings 1.2 ppm Dairy products 2.0 ppm Toppings 1.0 ppm
Regulatory status	FDA 121.1164; FEMA No. 2888	FDA 121.1164; FEMA No. 2889

REFERENCES

For References 1–5, see end of Part III.

6. **Strauss and Berkow,** *Ann. Chem.,* 401, 158, 1913.
7. **Fischer and Hoffa,** *Ber. Dtsch. Chem. Ges.,* 31, 1992, 1898.

	3-Phenylpropyl acetate	2-Phenylpropyl butyrate
Other names	Hydrocinnamyl acetate	Hydratropyl butyrate β-Methyl phenethyl butyrate α-Phenylpropyl alcohol, butyric ester 2-Phenylpropyl-*n*-butyrate
Empirical formula	$C_{11}H_{14}O_2$	$C_{13}H_{18}O_2$
Structure	$C_6H_5-CH_2-CH_2-CH_2-O-C(=O)-CH_3$	$H_3C-CH(C_6H_5)-CH_2-O-C(=O)-(CH_2)_2-CH_3$
Physical/chemical characteristics[5]		
Appearance	Colorless liquid	Colorless liquid
Assay	98% min	
Molecular weight	178.23	206.28
Boiling point	244–245°C	
Specific gravity	1.012–1.015 at 25°C; 1.0285 at 15°C	
Refractive index	1.4940–1.4970 (1.4975) at 20°C	
Solubility	1:2–3 in 70% alcohol; soluble in most common organic solvents	Insoluble in water; soluble in alcohol
Organoleptic characteristics	Characteristic floral, spicy odor reminiscent of phenylpropyl alcohol and of geranyl acetate; bittersweet, burning flavor suggestive of currant	Fruity, woody, very sweet odor; sweet flavor reminiscent of plum
Synthesis	By acetylation of the corresponding alcohol	By esterification of the alcohol with *n*-butyric acid
Natural occurrence	Reported found among the constituents of the essential oils of narcissus and *Heracleum candicans* and probably in cinnamon	Not reported found in nature
Reported uses[3]	Non-alcoholic beverages 3.2 ppm Ice cream, ices, etc. 4.8 ppm Candy 4.6 ppm Baked goods 6.3 ppm Chewing gum 10 ppm Condiments 0.10 ppm	Non-alcoholic beverages 1.0 ppm Ice cream, ices, etc. 1.0 ppm Candy 2.0 ppm Baked goods 2.0 ppm
Regulatory status	FDA 121.1164; FEMA No. 2890	FDA 121.1164; FEMA No. 2891

REFERENCES

For References 1–5, see end of Part III.

	3-Phenylpropyl cinnamate	3-Phenylpropyl formate
Other names	Hydrocinnamyl cinnamate 3-Phenylpropyl-β-phenylacrylate 3-Phenylpropyl-3-phenyl-2-propenoate	Hydrocinnamyl formate 3-Phenyl-1-propyl methanoate
Empirical formula	$C_{18}H_{18}O_2$	$C_{10}H_{12}O_2$
Structure	$C_6H_5-CH_2-CH_2-CH_2-O-C(=O)-CH=CH-C_6H_5$	$C_6H_5-CH_2-(CH_2)_2-O-C(=O)-H$
Physical/chemical characteristics[5]		
Appearance	Colorless to slightly yellow liquid	Colorless liquid
Assay	98% min	90% min
Molecular weight	266.33	164.21
Boiling point	> 300°C (decomposes)	238°C (approx 280°C)
Flash point	> 100°C	> 100°C
Specific gravity	1.074–1.078 at 25°C	1.035–1.038 at 25°C
Refractive index	1.585–1.588 at 20°C	1.5030–1.5060 at 20°C
Acid value	1 max	2 max
Solubility	Slightly soluble in 95% alcohol; soluble in benzyl benzoate and in diethylphthalate	Insoluble in water; soluble in most common organic solvents; 1:2 in 80% alcohol
Organoleptic characteristics	Sweet, heavy, floral-fruity odor with a balsamic note; sweet flavor suggestive of cocoa	Sweet, floral odor reminiscent of honey and hyacinth; powerful, sweet, fruity, herbaceous taste
Synthesis	From phenylpropyl alcohol and methylcinnamate	Cold formylation of phenylpropyl alcohol
Natural occurrence	Reported found in Oriental storax,[6-8] American storax,[9-10] Peru balsam from Honduras,[11-13] and Sumatra benzoin[14]	Not reported found in nature
Reported uses[3]	Non-alcoholic beverages 3.4 ppm Ice cream, ices, etc. 4.1 ppm Candy 4.3 ppm Baked goods 5.3 ppm	Non-alcoholic beverages 1.3 ppm Ice cream, ices, etc. 0.90–1.5 ppm Candy 3.0–5.0 ppm Baked goods 2.7 ppm
Regulatory status	FDA 121.1164; FEMA No. 2894	FDA 121.1164; FEMA No. 2895

REFERENCES

For References 1–5, see End of Part III.

6. **Miller,** *Chem. Ber.*, 9, 275, 1876.
7. **Miller,** *Chem. Ann.*, 188, 201, 1877; 189, 353, 1877.
8. **Tschirch and van Itallie,** *Arch. Pharm.*, 239, 513, 1901.
9. **Miller,** *Arch. Pharm.*, 220, 648, 1882.
10. **Tschirch and van Itallie,** *Arch. Pharm.*, 239, 536, 1901.
11. **Thoms and Blitz,** *Chem. Zentralbl.*, 2, 1047, 1904.
12. **Hellström,** *Arch. Pharm.*, 218, 243, 1881.
13. **Tschirch and Werdmüller,** *Arch. Pharm.*, 248, 421, 1910.
14. **Tschirch and Lüdy,** *Arch. Pharm.*, 231, 54, 1893.

	3-Phenylpropyl hexanoate	2-Phenylpropyl isobutyrate
Other names	Hydrocinnamyl hexanoate 3-Phenylpropyl caproate Phenylpropyl capronate Phenylpropyl hexylate	Hydratropyl isobutyrate β-Methyl phenethyl isobutyrate α-Phenylpropyl alcohol, isobutyric ester
Empirical formula	$C_{15}H_{22}O_2$	$C_{13}H_{18}O_2$
Structure	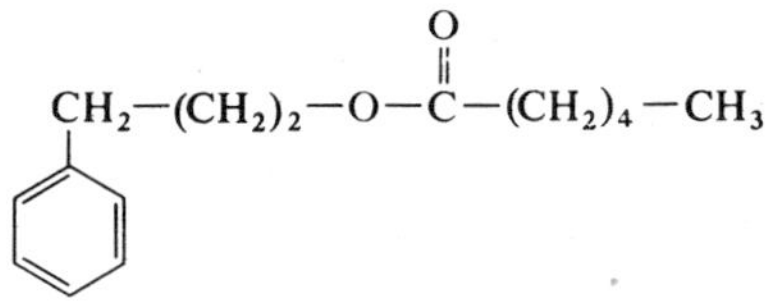 $C_6H_5-CH_2-(CH_2)_2-O-C(=O)-(CH_2)_4-CH_3$	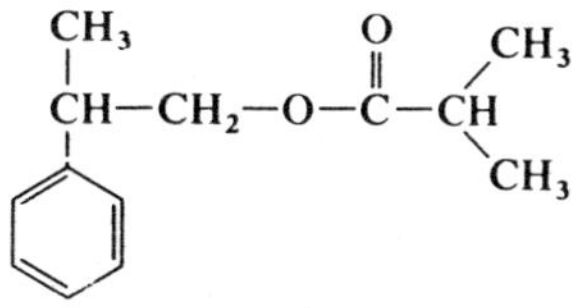 $C_6H_5-CH(CH_3)-CH_2-O-C(=O)-CH(CH_3)_2$
Physical/chemical characteristics[5]		
Appearance	Colorless liquid	Colorless liquid
Molecular weight	234.35	206.28
Boiling point	292°C	
Specific gravity	1.01	
Solubility	Insoluble in water; soluble in alcohol	Insoluble in water; soluble in alcohol
Organoleptic characteristics	Sweet, fruity, green odor; powerful, warm, fruity taste of peach and pineapple	Fruity, woody, highly sweet odor; fruity, sweet taste
Synthesis	By esterification of phenylpropyl alcohol with *n*-hexanoic acid	
Natural occurrence	Not reported found in nature	Not reported found in nature
Reported uses[3]	Non-alcoholic beverages 0.6 ppm Ice cream, ices, etc. 1.3 ppm Candy 3.3 ppm Baked goods 3.7 ppm	Non-alcoholic beverages 5.0 ppm Ice cream, ices, etc. 20 ppm Candy 20 ppm
Regulatory status	FDA 121.1164; FEMA No. 2896	FDA 121.1164; FEMA No. 2892

REFERENCES

For References 1–5, see end of Part III.

	3-Phenylpropyl isobutyrate	3-Phenylpropyl isovalerate
Other names	Hydrocinnamyl isobutyrate	Hydrocinnamyl isovalerate 3-Phenylpropyl-β-methylbutyrate
Empirical formula	$C_{13}H_{18}O_2$	$C_{14}H_{20}O_2$
Structure	$C_6H_5-CH_2-CH_2-CH_2-O-C(=O)-CH(CH_3)_2$	$C_6H_5-(CH_2)_3-O-C(=O)-CH_2-CH(CH_3)_2$
Physical/chemical characteristics[5]		
Appearance	Colorless liquid	Colorless liquid
Molecular weight	206.28	220.31
Boiling point	258°C	280°C
Specific gravity	0.99	1.00
Solubility	Insoluble in water; soluble in propylene glycol and alcohol	Almost insoluble in water; soluble in alcohol
Organoleptic characteristics	Sweet, balsamic odor, finer and fresher than the corresponding *n*-butyrate; bittersweet flavor reminiscent of peach	Complex, fruity (strawberry-raspberry) odor with a plum-like undertone; sweet, jam-like, fruity taste with nut-like undernotes
Synthesis	By esterification of hydrocinnamic alcohol with *n*-butyric acid	By esterification of phenylpropyl alcohol with isovaleric acid
Natural occurrence	Not reported found in nature	Not reported found in nature
Reported uses[3]	Non-alcoholic beverages 1.3 ppm Ice cream, ices, etc. 3.0 ppm Candy 5.0 ppm Baked goods 5.8 ppm	Non-alcoholic beverages 0.90 ppm Ice cream, ices, etc. 0.90 ppm Candy 1.8 ppm Baked goods 1.7 ppm
Regulatory status	FDA 121.1164; FEMA No. 2893	FDA 121.1164; FEMA No. 2899

REFERENCES

For References 1–5, see end of Part III.

	3-Phenylpropyl propionate	2-(3-Phenylpropyl)-tetrahydrofuran
Other names	Hydrocinnamyl propionate 3-Phenylpropyl propanoate β-Phenylpropyl propionate	2-Hydrocinnamyl tetrahydrofuran α-(3-Phenylpropyl)-tetrahydrofuran
Empirical formula	$C_{12}H_{16}O_2$	$C_{13}H_{18}O$
Structure	$C_6H_5-CH_2-CH_2-CH_2-O-C(=O)-CH_2-CH_3$	(tetrahydrofuran-2-yl)$-CH_2-CH_2-CH_2-C_6H_5$
Physical/chemical characteristics[5]		
Appearance	Colorless liquid	Colorless to pale, straw yellow liquid
Molecular weight	192.26	190.28
Boiling point	265°C (247°C)	105–107°C at 1 mm Hg
Specific gravity	1.010 at 15.5°C	
Refractive index	1.5125 at 20°C	1.5142 at 20°C
Solubility	Almost insoluble in water; soluble in alcohol	Very slightly soluble in water; soluble in alcohol
Organoleptic characteristics	Heavy floral, balsamic odor with hyacinth and mimosa undertone; sweet, fruity, slightly green taste at low concentrations	Sweet, fruity odor; honey-like, sweet flavor
Synthesis	By esterification of phenylpropyl alcohol with propionic acid	From β-[2-(5-phenyl furyl)]-propionic acid butyl ester by catalytic hydrogenation with copper chromite under pressure and subsequent ring closure by heating the intermediate 7-phenylheptan-1,4-diol in the presence of Al_2O_3[6]
Natural occurrence	Not reported found in nature	Not reported found in nature
Reported uses[3]	Non-alcoholic beverages 0.49 ppm Ice cream, ices, etc. 0.52 ppm Candy 2.0 ppm Baked goods 2.4 ppm Chewing gum 0.80–50 ppm	Non-alcoholic beverages 0.50 ppm Ice cream, ices, etc. 2.0 ppm Candy 0.03–2.0 ppm Gelatins and puddings 2.0 ppm Chewing gum 2.3 ppm
Regulatory status	FDA 121.1164; FEMA No. 2897	FDA 121.1164; FEMA No. 2898

REFERENCES

For References 1–5, see end of Part III.

6. **Thewalt and Rudolph,** *Ber. Dtsch. Chem. Ges.*, 96, 136, 1963.

	α-Pinene	β-Pinene
Other names	2-Pinene 2,6,6-Trimethylbicyclo-(3,1,1)-2-heptene	6,6-Dimethyl-2-methylene norpinane Nopinene 2(10)-Pinene
Empirical formula	$C_{10}H_{16}$	$C_{10}H_{16}$
Structure	CH_3; CH_3; C; H_3C	CH_2; CH_3; C; H_3C
Physical/chemical characteristics[5]		
Appearance	Mobile, colorless liquid	Colorless liquid
Assay		85–90%; 96% (1% α-pinene content)
Molecular weight	136.23	136.23
Melting point	−62.5°C	
Boiling point	156°C at 760 mm Hg	*d*-, 164–166°C; *l*-, 162–163°C
Specific gravity	0.8625 at 15°C	*d*-, 0.8660 at 20°C; *l*-, 0.8666 at 25°C
Refractive index	1.4680 at 15°C	*d*-, 1.4746 at 20°C; *l*-, 1.4762 at 25°C
Optical rotation	*d*-, +51°8′; *l*-, −51°17′, −58°	*d*-, +28.6°; *l*-, −23.4°
Solubility	Insoluble in water; soluble in alcohol, chloroform, and ether; almost insoluble in propylene glycol and glycerin	Insoluble in water; soluble in alcohol; almost insoluble in propylene glycol
Organoleptic characteristics	Characteristic odor of pine; turpentine-like; oxidized material rosin-like odor.	Characteristic turpentine odor; dry, woody or resinous aroma
Synthesis	From turpentine, by distillation. For total synthesis see Reference 6.	Isolated from American turpentine; also by conversion from α-pinene; as an intermediate extremely important for the manufacture of citral, citronellol, hydroxycitronellal, geraniol, citronellal, linalool, ionones, methylionones, and menthol
Natural occurrence	The structure would account for the presence of four optically active and two optically inactive isomers. Only *d*-, *l*-, and *dl*-α-pinene are known, however. Its presence in nature has been reported in more than 400 essential oils. In the largest amounts it has been reported found in: *Achillea millefolium* (*d*-), *Artemisia tridentata* (*d*-), Italian rosemary (*l*-), wild thyme (*l*-), French lavender (*l*-), coriander (*d*-, *dl*-), cumin (*d*-, *dl*-), labdanum (*l*-), neroli (*l*-), lemon, *Litsea cubeba* (*d*-), and ylang-ylang (*d*-).	Usually occurring together with α-pinene but in smaller amounts. The *d*- and *l*-forms are reported found in the essential oils of various *Artemisiae* and several *Cupressaceae*, in coriander and cumin. The *l*-form is a constituent of several citrus oils
Reported uses[3]	Non-alcoholic beverages 16–54 ppm Ice cream, ices, etc. 64 ppm Candy 48 ppm Baked goods 160 ppm Condiments 2.6–150 ppm	Non-alcoholic beverages 0.05–16 ppm Ice cream, ices, etc. 64 ppm Candy 48–600 ppm Baked goods 48–600 ppm
Regulatory status	FDA 121.1164; FEMA No. 2902	FDA 121.1164; FEMA No. 2903

REFERENCES

For References 1–5, see end of Part III.

6. **Komppa,** *Ann. Acad. Sci. Fenn. Ser. A*, 59, 3, 1943.

	Piperidine	Piperine
Other names	Hexahydropyridine Hexazane Pentamethylenimine	1-Piperoylpiperidine
Empirical formula	$C_5H_{11}N$	$C_{17}H_{19}NO_3$
Structure	(piperidine ring, N–H)	H_2C–O–(benzene ring)–O–CH=CH–CH=CH–C(=O)–N(piperidine ring)
Physical/chemical characteristics[5]		
Appearance	Colorless liquid	Colorless prisms (from alcohol and other solvents)
Assay	95–98%	
Molecular weight	85.15	285.32
Melting point	−7° to −9°C	128–129.5°C
Boiling point	106°C	
Congealing point	−13 to −17°C	
Flash point	16°C	
Specific gravity	0.8628 at 18.7°C (0.8615 at 20°C)	1.193 (liquid)
Refractive index	1.4535 at 18.7°C (1.4530 at 20°C)	
Optical rotation		
Solubility	Soluble in water, alcohol, ether, and chloroform	Almost insoluble in water; soluble in alcohol (10%), chloroform, and ether
Organoleptic characteristics	Heavy, sweet, floral, animal odor; burning peppery taste	Odorless; tasteless at first, but developing a burning aftertaste of pepper
Synthesis	Usually prepared by electrolytic reduction of pyridine	From piperoyl chloride and piperidine[12]
Natural occurrence	Reported found in black pepper,[6,7] *Piper officinarum*,[8] *Psilocaulon absimile* N.E. Br.,[9] tobacco,[10] and *Petrosimonia monandra*[11]	Reported found in black pepper; also in *Piper longum*, *Piper officinarum*, *Piper lowong* Bl.,[13] *Piper famechoni*,[14] *Piper chaba*,[15] and the leaves of *Rhododendron fauriae* var. *rupescens*[16]
Reported uses[3]	Non-alcoholic beverages 3.0 ppm Candy 5.0 ppm Baked goods 0.05–5.0 ppm Condiments 0.05 ppm Meats 0.05 ppm Soups 0.05 ppm	
Regulatory status	FDA 121.1164; FEMA No. 2908	FDA 121.1164; FEMA No. 2909

REFERENCES

For References 1–5, see end of Part III.

6. **Johnstone,** *Ber. Dtsch. Chem. Ges.*, 22, 61, 1889.
7. **Englaender,** *Ber. Dtsch. Chem. Ges.*, 68, 2218, 1935.
8. **Wangerin,** *Chem. Zentralbl.*, 2, 214, 1903.
9. **Rimington,** *Chem. Zentralbl.*, 1, 80, 1936.
10. **Späth et al.,** *Ber. Dtsch. Chem. Ges.*, 69, 2448, 1936.
11. **Juraschewski et al.,** *Chem. Zentralbl.*, 1, 1841, 1940.
12. **Rugheimer,** *Ber. Dtsch. Chem. Ges.*, 15, 1390, 1882.
13. **Peinemann,** *Arch. Pharm.*, 234, 204, 245, 1896.
14. **Barilli,** *Chem. Zentralbl.*, 2, 384, 1901.
15. **Bose,** *Chem. Zentralbl.*, 1, 4315, 1936.
16. **Kawaguchi et al.,** *Chem. Abstr.*, 44, 9634, 1950.

	Piperitol	*d*-Piperitone
Other names	1-Methyl-4-isopropyl-1-cyclohexen-3-ol *p*-Menth-1-en-3-ol Neopiperitol (*trans*-form)	*p*-Menth-l-en-3-one 4-Isopropyl-1-methyl-1-cyclohexen-3-one
Empirical formula	$C_{10}H_{18}O$	$C_{10}H_{16}O$
Structure	CH_3; OH; CH; H_3C; CH_3	CH_3; O; H_3C; CH_3
Physical/chemical characteristics[5]		
Appearance	Crystals; *cis*- and *trans*-form optically active	Colorless liquid
Molecular weight	154.24	152.23
Melting point	*cis*-, *l*-, 36°C *cis*-, *dl*-, 28°C	*l*-, −29°C
Boiling point	*trans*-, *dl*-, 57°C at 0.15 mm Hg *trans*-, *dl*-, 66–69°C at 0.2 mm Hg	*d*-, 116–118.5°C at 20 mm Hg; *dl*-, 235–237°C; *l*-, 232.5–234.7°C
Specific gravity	*trans*-, *d*-, 0.9203 at 25°C *trans*-, *dl*-, 0.9217 at 25°C	*d*-, 0.9344 at 20°C; *dl*-, 0.9375 at 19°C; *l*-, 0.9330 at 20°C
Refractive index	*trans*-, *d*-, 1.4762 at 25°C	*d*-, 1.4848 at 20°C; *dl*-, 1.4845 at 20°C; *l*-, 1.4815 at 20°C
Optical rotation	*cis*-, *l*-, −246° at 17°C (c = 2 in benzene) *trans*-, *d*-, +28° at 17°C (c = 2 in benzene)	*d*-, +49°8′ at 20°C; *l*-, −67.6° or −51.53°
Solubility		Insoluble in water; soluble in alcohol
Organoleptic characteristics	Fresh pungent odor and taste	Camphor-like odor; sharp, minty flavor
Synthesis	By reduction of racemic piperitone with lithium aluminum hydride or aluminum isopropoxide;[6] the *d*-form is isolated from natural sources[7]	Isolated from Japanese mint oil; *l*-form from *Eucalyptus dives* oil; by hydrogenation of diosphenol; by reduction of 5-methyl-2-isopropylanisole with sodium in liquid NH_4
Natural occurrence	Reported found in several naturally occurring oils[7,8] originally isolated from the oil of *Eucalyptus radiata*[7]	The structure has been defined by the work of several authors.[9-13] The *d*-, *l*-, and *dl*-forms are known. The *d*-form is found in the essential oils of *Mentha silvestris*, *Cymbopogon sennaarensis*, and Japanese peppermint oil. The *dl*-form is found in the essential oils of: *Andropogon iwarancusa*, *Mentha pulegium* var. *hirsuta*,[12-14] and, in mixture with the *l*-form, *Eucalyptus dives*. The *l*-form, the most abundant, is found in about 30 varieties of eucalyptus and in other plants.
Reported uses[3]	Non-alcoholic beverages 20.0 ppm Ice cream, ices, etc. 20.0 ppm Candy 20.0 ppm	Non-alcoholic beverages 1.0–11 ppm Ice cream, ices, etc. 18 ppm Candy 18 ppm Baked goods 18 ppm
Regulatory status	FEMA No. 3179	FDA 121.1164; FEMA No. 2910

REFERENCES

For References 1–5, see end of Part III.

6. **Macbeth, Milligan and Shannon,** *J. Chem. Soc.*, p. 901, 1953.
7. **Baker and Smith,** *Research on the Eucalyptus,* 2nd ed., Government of the State of New South Wales, Sydney, 1920, 373.
8. **Simonsen,** *Indian For. Rec.*, 10, 161, 1924.
9. **Penfold,** *J. Proc. R. Soc. N. S. W.*, 54, 40, 1920.
10. **Wallach and Meister,** *Annalen,* 362, 272, 1908.
11. **Simonsen,** *J. Chem. Soc.* (London), 119, 1646, 1921.
12. **Morani,** *Ann. Chim. Appl.*, 14, 293, 1924.
13. **Morani,** *Riv. Ital. EPPOS,* 7, 65, 1925.
14. **Romeo and Giuffre,** *Ann. Chim. Appl.*, 15, 368, 1925.

	Piperonal	Piperonyl acetate
Other names	Heliotropine 3,4-Methylenedioxybenzaldehyde Piperonylaldehyde Protocatechualdehyde methylene ether	Heliotropyl acetate 3,4-Methylenedioxybenzyl acetate
Empirical formula	$C_8H_6O_3$	$C_{10}H_{10}O_4$
Structure	[structure: benzene ring with CHO; O—CH₂—O methylenedioxy ring]	[structure: benzene ring with O—CH₂—O methylenedioxy ring; CH₂—O—C(=O)—CH₃]
Physical/chemical characteristics[5]		
Appearance	White, lustrous crystals; tends to yellow on exposure to air or light	Oily liquid (see references 12 and 13.)
Assay	99.0% min[2]	
Molecular weight	150.14	194.19
Melting point	37°C	
Boiling point	263°C (140°C at 15 mm Hg)	153–154°C at 14 mm Hg
Congealing point	35°C min[2]	
Specific gravity		1.240 at 18°C
Refractive index		1.528 at 18°C
Solubility	1 gm in 4 ml of 70% alcohol;[2] 0.2% in cold water; somewhat soluble in propylene glycol	Almost insoluble in water; soluble in alcohol
Organoleptic characteristics	Sweet, flowery odor reminiscent of heliotrope; bittersweet taste	Very sweet, floral odor with cherry-like undernotes; sweet, fruity flavor at low levels, bitter at high levels
Synthesis	By the oxidation of isosafrole with potassium dichromate and sulfuric acid and subsequent steam distillation of piperonal[6]	By acetylation of the corresponding alcohol;[10] or by boiling dimethylpiperonyl amine with acetic anhydride[11]
Natural occurrence	Reported found in the essential oils of *Robinia pseudo-acacia*[7] and *Eryngium poterium*;[8] in the oils of *Spirea ulmaria*[9] and of leaves of *Doryphora sassafras.* Also reported found in Tahitian vanilla, camphor wood oil, and violet flowers concrete and absolute	Not reported found in nature
Reported uses[3]	Non-alcoholic beverages 6.0 ppm Ice cream, ices, etc. 7.0 ppm Candy 7.4 ppm Baked goods 18 ppm Gelatins and puddings 5.8 ppm Chewing gum 36 ppm	Non-alcoholic beverages 27–50 ppm Ice cream, ices, etc. 80–110 ppm Candy 70–80 ppm Baked goods 55–80 ppm
Regulatory status	FDA GRAS; FEMA No. 2911	FDA 121.1164; FEMA No. 2912

REFERENCES

For References 1–5, see end of Part III.

6. **Ciamician and Silber,** *Gazz. Chim. Ital.,* 20, 571, 1890.
7. **Elze,** *Chem. Ztg.,* 34, 814, 1910.
8. **Salgue,** *C. R. Séances Acad. Sci. Paris,* 241, 987, 1955.
9. **Ikeda,** *J. Chem. Soc. Jap.,* 61, 583, 1940.
10. **Tiffenau,** *Bull. Soc. Chim. Fr.,* 4(9), 932, 1911.
11. **Tiffenau and Fuhrer,** *Bull. Soc. Chim. Fr.,* 4(15), 172, 1914.
12. **Vavon,** *Compt. Rend.,* 154, 361, 1912.
13. **Vavon,** *Ann. Chim. Paris,* 9(1), 166, 1914.

	Piperonyl isobutyrate	L-Proline
Other names	Heliotropyl isobutyrate 3,4-Methylenedioxybenzyl isobutyrate	2-Pyrrolidine carboxylic acid
Empirical formula	$C_{12}H_{14}O_4$	$C_5H_9NO_2$
Structure	(3,4-methylenedioxyphenyl)–CH_2–O–C(=O)–CH(CH_3)$_2$	
Physical/chemical characteristics[5]		
Appearance	Colorless, oily liquid	White needles or prisms
Molecular weight	222.24	115.13
Melting point		220–222°C (decomposes)
Optical rotation		–80.9° at 20°C (c = 1 in water) –52.6° at 20°C (c = 0.58 in 0.5*N* HCl) –93° at 20°C (c = 0.24 in 0.6*N* KOH)
Solubility	Almost insoluble in water; soluble in alcohol	Very soluble in water; slightly soluble in alcohol, acetone, and benzene; insoluble in ether
Organoleptic characteristics	Very sweet but mild, fruity, and berry-like fragrance with some jam-like qualities; intense, sweet, plum-like flavor	
Synthesis	By esterification of the corresponding alcohol with isobutyric acid	From α-piperidone,[6] from cyclopentanone,[7] or from *l*-glutamic acid[8]
Natural occurrence	Not reported found in nature	Reported found as a component in many proteins; also widely occurring as the free acid in natural products
Reported uses[3]	Non-alcoholic beverages 0.05–1.0 ppm Ice cream, ices, etc. 0.05 ppm Candy 0.05–3.5 ppm Baked goods 0.10–3.5 ppm	Non-alcoholic beverages 50.0 ppm Baked goods 50.0 ppm Condiments, pickles 50.0 ppm Meat, meat sauces, soups 50.0 ppm Milk, dairy products 50.0 ppm Cereals 50.0 ppm
Regulatory status	FDA 121.1164; FEMA No. 2913	FEMA No. 3319

REFERENCES

For References 1–5, see end of Part III.

6. **Heymons,** *Berichte.,* 66, 846, 1933.
7. **Schmidt,** *Chemistry of Amino Acids and Proteins,* 2nd ed., C C Thomas, Springfield, 1944, 89.
8. **Buyle,** *Chem. Ind.* (London), p. 380, 1966.

	Propenylguaethol	Propenyl propyl disulfide
Other names	6-Ethoxy-*m*-anol 1-Ethoxy-2-hydroxy-4-propenylbenzene 2-Ethoxy-5-propenylphenol Hydroxymethyl anethole 2-Propenyl-6-ethoxyphenol	
Empirical formula	$C_{11}H_{14}O$	$C_6H_{12}S_2$
Structure	HO; O—CH₂—CH₃; CH=CH—CH₃ (benzene ring structure)	$CH_3-CH_2-CH_2-S-S-CH=CH-CH_3$
Physical/chemical characteristics		
Appearance	Colorless, crystalline solid	
Molecular weight	162.23	148.29
Melting point	85–86°C (sublimes under vacuum)	
Boiling point		78–80°C at 13 mm Hg
Refractive index		n_D^{20} 1.5219
Solubility	10% in alcohol; 4% in propylene glycol; 4% in polysorbate 80; 20% in benzyl alcohol; slightly soluble in water	
Organoleptic characteristics	Vanilla-like odor and flavor	Odor and taste like that of cooked onions
Synthesis	By alkaline hydrolysis of 1-ethoxy-2-methoxy-4-propenyl benzene in ethanol or methanol solution at 150–190°C under pressure,[9] or in ethylene glycol solution at the same temperature under atmospheric pressure; by saponification and contemporary transposition of the double bond starting from 1-ethoxy-2-methoxy-4-allyl benzene	Treatment of 1-ethylthio-2-propenylthioethene with $NaNH_2$ in liquid ammonia results in formation of Na-1-propenethiolate, which reacts with S-propyl propanethiosulfonate to give propenyl propyl disulfide in good yield[6]
Natural occurrence	Not reported found in nature	Flavor component of onion oil (*trans*-form);[7] volatile component from chives (*Allium schoenoprosum*)[8]
Reported uses[3]	Non-alcoholic beverages 5.9 ppm Ice cream, ices, etc. 6.3 ppm Candy 20 ppm Baked goods 20 ppm Gelatins and puddings 2.5 ppm	Baked goods 2.0 ppm Condiments, pickles 2.0 ppm Meat, meat sauces, soups 2.0 ppm
Regulatory status	FDA 121.1164; FEMA No. 2922	FEMA No. 3227

REFERENCES

For References 1–5, see end of Part III.

6. **Wijers, Boelens and Van der Cjeu,** *Recl. Trav. Chim. Pays-Bas,* 88(5), 519, 1969.
7. **Brodnitz, Pollack and Vallon,** *J. Agric. Food Chem.,* 17(4), 760, 1969.
8. **Wahlroos and Virtanen,** *Acta Chem. Scand.,* 19(6), 1327, 1965.
9. **Junge,** *Riechstoffind. Kosmet.,* 8, 81, 1933.

	Propionaldehyde	Propyl acetate
Other names	Methylacetaldehyde Propanal Propyl aldehyde	*n*-Propyl acetate
Empirical formula	C_3H_6O	$C_5H_{10}O_2$
Structure	CH_3-CH_2-CHO	$CH_3-CH_2-CH_2-O-C(=O)-CH_3$
Physical/chemical characteristics[5]		
Appearance	Colorless liquid	Highly mobile, colorless liquid
Molecular weight	58.08	102.13
Melting point	−81°C	
Boiling point	47.5–49°C	101.6°C
Specific gravity	0.8074 at 21°C	0.8905–0.8992 at 15°C
Refractive index	1.3695 at 16.6°C	1.3840–1.3884 at 20°C
Solubility	Soluble in water (1:5 at 20°C); soluble in alcohol, ether, and chloroform	Fairly soluble in water (1:60); 1:17 in 30% alcohol; soluble in most common organic solvents
Organoleptic characteristics	Characteristic odor similar to acetaldehyde	Fruity (pear-raspberry) odor; pleasant, bittersweet flavor reminiscent of pear on dilution
Synthesis	By oxidation of propyl alcohol, or by dry distillation of barium propionate with calcium formate[6]	By direct acetylation of propyl alcohol
Natural occurrence	Reported found in apple aroma[7] and in the essential oils of: camphor,[8] *Rosa centifolia*,[9] clary sage,[10] *Pinus excelsa*, *Pinus silvestris*,[11] and others	In raspberry, banana, and tomato
Reported uses[3]	Non-alcoholic beverages 3.9 ppm Ice cream, ices, etc. 12 ppm Candy 11 ppm Baked goods 13 ppm	Non-alcoholic beverages 4.0 ppm Ice cream, ices, etc. 16 ppm Candy 12 ppm Baked goods 14 ppm
Regulatory status	FDA 121.1164; FEMA No. 2923	FDA 121.1164; FEMA No. 2935

REFERENCES

For References 1–5, see end of Part III.

6. **Limpricht,** *Justus Liebigs Ann. Chem.*, 97, 361, 1856.
7. **Huelin,** *J. Sci. Res. Aust.*, 5, 328, 1952.
8. **Ono and Imoto,** *J. Chem. Soc. Jap.*, 59, 1014, 1938.
9. **Calvarano,** *Essenze Deriv. Agrum.*, 28, 157, 1958.
10. **Panayotov,** *Perfum. Essent. Oil Rec.*, 49, 233, 1958.
11. **Aschau,** *Chem. Zentralbl.*, 1, 147, 1914.

	Propyl alcohol	*p*-Propyl anisole
Other names	Albacol Optal 1-Propanol Propylic alcohol *n*-Propyl alcohol	Dihydroanethole 1-Methoxy-4-*n*-propylbenzene Methyl *p*-propylphenyl ether Propylmethoxybenzene
Empirical formula	C_3H_8O	$C_{10}H_{14}O$
Structure	$CH_3{-}CH_2{-}CH_2OH$	OCH_3 (para-substituted benzene ring) $CH_2{-}CH_2{-}CH_3$
Physical/chemical characteristics[5]		
Appearance	Colorless liquid	Colorless to pale-yellow, oily liquid
Molecular weight	60.09	150.22
Melting point	−127°C	
Boiling point	97.2°C	212–213°C (104–107°C at 20 mm Hg)
Flash point	22°C	
Specific gravity	0.8044 at 20°C	0.940–0.943 at 25°/25°C;[2] 0.94718 at 20°C
Refractive index	1.38543 at 20°C	1.5025–1.5055[2] (1.5045) at 20°C
Solubility	Miscible with water, alcohol, ether, and propylene glycol	1:5 in 80% alcohol[2]
Organoleptic characteristics	Alcoholic odor; characteristic ripe, fruity flavor	Characteristic anise-type odor with a sassafras undertone
Synthesis	Can be isolated from the head fractions during the distillation of wine, or from the tail fractions during rectification of spirits	By partial hydrogenation of anethole in the presence of nickel at 60–95°C under pressure,[7–9] or under atmospheric pressure[10,11]
Natural occurrence	Reported found in the natural aromas of apple, cognac, and rum;[6] also forms during alcoholic fermentation	Not reported found in nature
Reported uses[3]	Non-alcoholic beverages 0.50–5.0 ppm Ice cream, ices, etc. 0.50 ppm Candy 0.50 ppm Baked goods 0.65 ppm	Non-alcoholic beverages 4.3 ppm Ice cream, ices, etc. 9.9 ppm Candy 64 ppm Baked goods 67 ppm
Regulatory status	FDA 121.1164; FEMA No. 2928	FDA 121.1164; FEMA No. 2930

REFERENCES

For References 1–5, see end of Part III.

6. **Fenaroli et al.,** *Riv. Ital. EPPOS,* 9, 484, 1965; 2, 75, 1966.
7. **Ipatjew,** *Chem. Ber.,* 46, 3590, 1913.
8. **Brochet and Bauer,** *Compt. Rend.,* 154, 192, 1914.
9. **Brochet and Bauer,** *Bull. Soc. Chim. Fr.,* 4(17), 53, 1915.
10. **Brochet and Cavaret,** *Comp. Rend.,* 159, 328, 1914.
11. **Brochet and Cavaret,** *Bull. Soc. Chim. Fr.,* 4(17), 58, 1915.

	Propyl benzoate	Propyl butyrate
Other names	*n*-Propyl benzoate	*n*-Propyl-*n*-butanoate *n*-Propyl butyrate
Empirical formula	$C_{10}H_{12}O_2$	$C_7H_{14}O_2$
Structure	$C_6H_5-C(=O)-O-CH_2-CH_2-CH_3$	$CH_3-(CH_2)_2-C(=O)-O-(CH_2)_2-CH_3$
Physical/chemical characteristics[5]		
Appearance	Colorless, oily liquid	Colorless liquid
Molecular weight	164.21	130.18
Melting point	−51 to −52°C	
Boiling point	230°C (230–231°C)	142–143°C
Specific gravity	1.039 at 15.5°C; 1.0274 at 15°C	0.8722 at 20°C
Refractive index	1.5100 at 20°C; 1.50139 at 15°C	1.3995 at 20°C
Solubility	Insoluble in water; soluble in alcohol	Almost insoluble in water; miscible with alcohol and ether
Organoleptic characteristics	Balsamic odor reminiscent of nut; sweet, fruity, nut-like taste	Pineapple and apricot-like odor; sweet, fruity flavor of banana and pineapple
Synthesis	By exchange between methyl benzoate and propyl alcohol in the presence of potassium propylate;[6] or by heating benzamide sulfate with propyl alcohol at 90–95°C[7]	From propyl alcohol and butyric acid in the presence of *p*-toluenesulfonic acid in benzene solution at the boil[8]
Natural occurrence	Reported found in several natural products	Reported found in several natural products
Reported uses[3]	Non-alcoholic beverages 11 ppm Ice cream, ices, etc. 44 ppm Candy 33 ppm Baked goods 33 ppm	Non-alcoholic beverages 6.8 ppm Ice cream, ices, etc. 4.6 ppm Candy 24 ppm Baked goods 16 ppm
Regulatory status	FDA 121.1164; FEMA No. 2931	FDA 121.1164; FEMA No. 2934

REFERENCES

For References 1–5, see end of Part III.

6. **Reimer and Downes,** *J. Am. Chem. Soc.,* 43, 949, 1921.
7. **Roessler** (Hasslacher Chemical Co.). German Patent 463,721, 1928.
8. **Salmi and Leimu,** *Suom. Kemistil. B,* 20, 46, 1947.

	Propyl cinnamate
Other names	*n*-Propyl cinnamate Propyl-β-phenyl acrylate Propyl-3-phenylpropenoate
Empirical formula	$C_{12}H_{14}O_2$
Structure	$C_6H_5-CH{=}CH-C({=}O)-O-CH_2-CH_2-CH_3$
Physical/chemical characteristics[5]	
Appearance	Colorless, viscous liquid; solidifies into crystalline mass when cold
Molecular weight	190.24
Melting point	13°C
Boiling point	283–284°C; 102–104°C at 1 mm Hg
Specific gravity	1.0435 at 0°/0°C
Refractive index	1.5510 at 20°C; 1.5530 at 14°C
Solubility	Insoluble in water; soluble in alcohol
Organoleptic characteristics	Characteristic peach-apricot flavor; wine-like odor
Synthesis	By esterification of cinnamic acid with propyl alcohol in the presence of HCl as a catalyst;[6] or by direct esterification in the presence of benzene
Natural occurrence	Not reported found in nature
Reported uses[3]	Non-alcoholic beverages 2.6 ppm Ice cream, ices, etc. 2.9 ppm Candy 4.9 ppm Baked goods 4.3 ppm Gelatins and puddings 0.07 ppm
Regulatory status	FDA 121.1164; FEMA No. 2938

REFERENCES

For References 1–5, see end of Part III.

6. **Weger,** *Chem. Ann.,* 221, 76, 1883.
7. **Challenger and Rawling,** *J. Chem. Soc.,* p. 872, 1937.

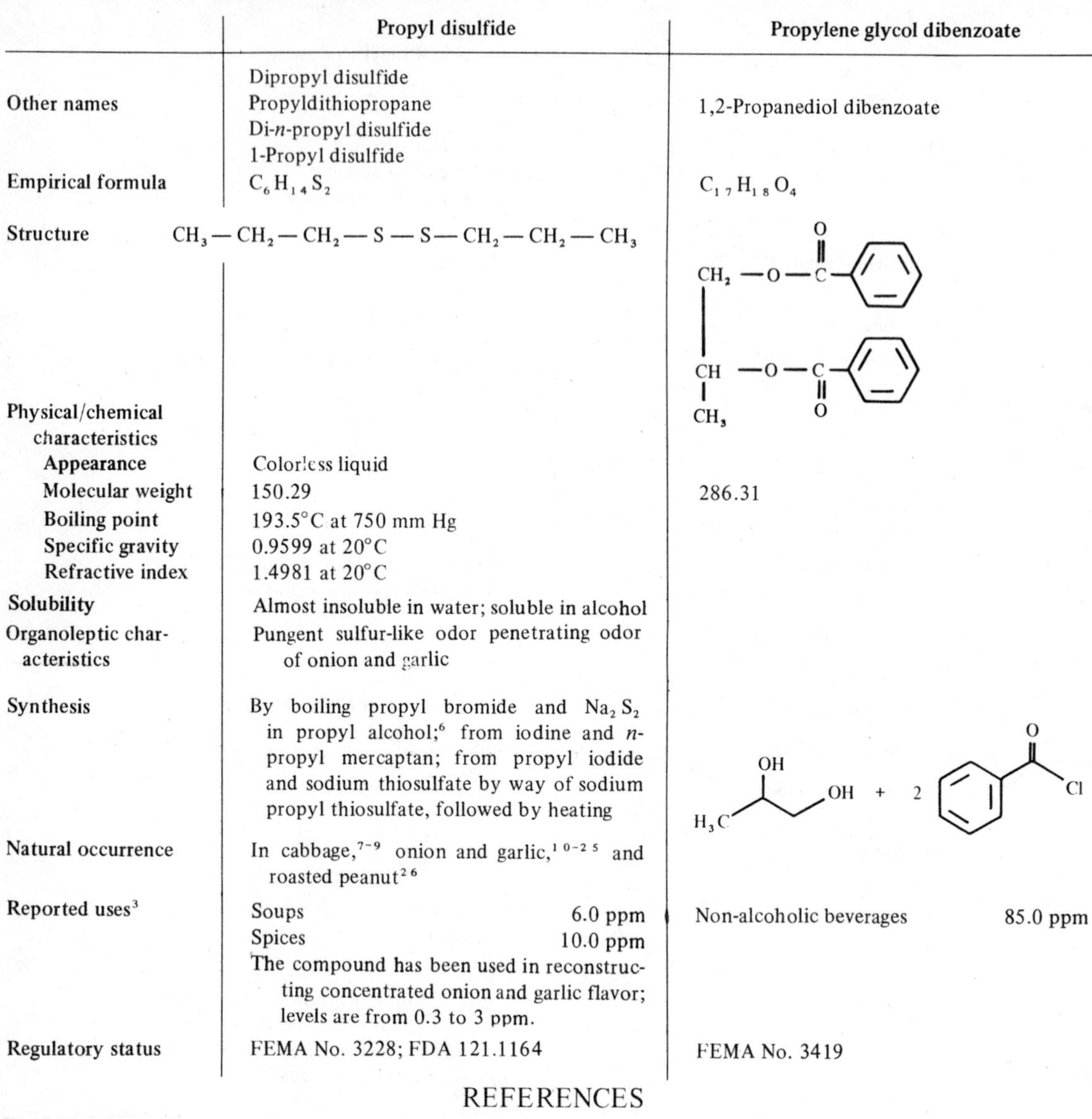

	Propyl disulfide	Propylene glycol dibenzoate
Other names	Dipropyl disulfide Propyldithiopropane Di-*n*-propyl disulfide 1-Propyl disulfide	1,2-Propanediol dibenzoate
Empirical formula	$C_6H_{14}S_2$	$C_{17}H_{18}O_4$
Structure	$CH_3-CH_2-CH_2-S-S-CH_2-CH_2-CH_3$	
Physical/chemical characteristics		
Appearance	Colorless liquid	
Molecular weight	150.29	286.31
Boiling point	193.5°C at 750 mm Hg	
Specific gravity	0.9599 at 20°C	
Refractive index	1.4981 at 20°C	
Solubility	Almost insoluble in water; soluble in alcohol	
Organoleptic characteristics	Pungent sulfur-like odor penetrating odor of onion and garlic	
Synthesis	By boiling propyl bromide and Na_2S_2 in propyl alcohol;[6] from iodine and *n*-propyl mercaptan; from propyl iodide and sodium thiosulfate by way of sodium propyl thiosulfate, followed by heating	
Natural occurrence	In cabbage,[7-9] onion and garlic,[10-25] and roasted peanut[26]	
Reported uses[3]	Soups 6.0 ppm Spices 10.0 ppm The compound has been used in reconstructing concentrated onion and garlic flavor; levels are from 0.3 to 3 ppm.	Non-alcoholic beverages 85.0 ppm
Regulatory status	FEMA No. 3228; FDA 121.1164	FEMA No. 3419

REFERENCES

For References 1–5, see end of Part III.

6. **Challenger et al.,** *J. Chem. Soc.,* p. 872, 1937.
7. **Bailey et al.,** *J. Food Sci.,* 26, 163, 1960.
8. **Hewit,** *J. Agric. Food Chem.,* 11, 14, 1963.
9. **Johnson et al.,** *Chem. Ind.* (London), p. 556, 1971.
10. **Niegisch et al.,** *Food Res.,* 21, 657, 1956.
11. **Carson et al.,** *Am. Chem. Soc. Div. Pet. Chem. Prepr.,* 2(4), Abstr., D 115, 1957.
12. **Carson et al.,** *J. Agric. Food Chem.,* 9, 140, 1961.
13. **Jacobsen et al.,** *Arch. Biochem. Biophys.,* 104, 473, 1964.
14. **Saghir et al.,** *Proc. Am. Soc. Hort. Sci.,* 84, 386, 1964.
15. **Wilkens,** *Memo No. 385, Cornell University, Agricultural Experimental Station,* Ithaca, N.Y., 1964.
16. **Wahlroos et al.,** *Acta Chem. Scand.,* 19, 1327, 1965.
17. **Schwob et al.,** *Ann. Nutr. Aliment.,* 19, 1327, 1965.
18. **Saghir et al.,** *Plant Physiol.,* 40, 681, 1965.
19. **Bernhard,** *Adv. Chem. Ser.,* 56, 131, 1966.
20. **Bernhard,** *J. Food Sci.,* 33, 298, 1968.
21. **Brodnitz et al.,** *J. Agric. Food Chem.,* 17, 760, 1969.
22. **Bernhard,** *Qual. Plant. Mater. Veg.,* 18, 72, 1969.
23. **Brodnitz et al.,** *Food Technol.,* 24, 78, 1970.
24. **Boelens et al.,** *J. Agric. Food Chem.,* 19, 984, 1971.
25. **Dubois et al.,** *Ind. Aliment. Agric.,* p. 127, 1972.
26. **Walradt et al.,** *J. Agric. Food Chem.,* 19, 972, 1971.

	Propyl formate	Propyl 2-furanacrylate
Other names	*n*-Propyl formate *n*-Propyl methanoate	Propyl β-furylacrylate Propyl-3-(2-furyl)-acrylate Propyl-3-furylpropenoate
Empirical formula	$C_4H_8O_2$	$C_{10}H_{12}O_3$
Structure	$CH_3-CH_2-CH_2-O-\overset{\overset{\displaystyle O}{\|}}{C}-H$	(2-furyl)$-CH=CH-\overset{\overset{\displaystyle O}{\|}}{C}-O-(CH_2)_2-CH_3$
Physical/chemical characteristics[5]		
Appearance	Colorless liquid	Colorless liquid (sometimes very pale-yellow)
Molecular weight	88.10	180.21
Melting point	−92.9°C	
Boiling point	81–81.3°C	119°C (236°C) at 7 mm Hg
Flash point	27°C	
Specific gravity	0.9039 at 20°C	1.0744 at 20°C
Refractive index	1.3769 at 20°C	1.5229 at 20°C
Solubility	Scarcely soluble in water; miscible with alcohol, ether, and most common organic solvents	Insoluble in water; soluble in alcohol
Organoleptic characteristics	Characteristic fruity (rum-plum) odor; corresponding bittersweet flavor	Light strawberry-, apple-, pear-like odor; fruity, caramellic flavor at low levels, turning bitter at higher levels
Synthesis	By action of sulfuric acid on a mixture of propyl alcohol, formic acid, and sodium formate;[6] also by distilling propyl alcohol with anhydrous formic acid in the presence of sodium formate[7]	By esterification of *n*-propanol with furanacrylic acid
Natural occurrence	Reported found in several natural products	Not reported found in nature
Reported uses[3]	Non-alcoholic beverages 20 ppm Ice cream, ices, etc. 57 ppm Candy 65 ppm Baked goods 85 ppm	Non-alcoholic beverages 3.0 ppm Candy 0.03 ppm
Regulatory status	FDA 121.1164; FEMA No. 2943	FDA 121.1164; FEMA No. 2945

REFERENCES

For References 1–5, see end of Part III.

6. **Wagner**, German Patent 546,806, 1929.
7. **Adickes**, *J. Prakt. Chem.*, 2(145), 236, 1936.

	Propyl heptanoate	*n*-Propyl hexanoate
Other names	*n*-Propyl heptoate *n*-Propyl heptylate	*n*-Propyl caproate *n*-Propyl-*n*-hexoate *n*-Propyl hexylate
Empirical formula	$C_{10}H_{20}O_2$	$C_9H_{18}O_2$
Structure	$CH_3{-}CH_2{-}CH_2{-}O{-}\overset{\overset{\displaystyle O}{\Vert}}{C}{-}(CH_2)_5{-}CH_3$	$CH_3{-}(CH_2)_2{-}O{-}\overset{\overset{\displaystyle O}{\Vert}}{C}{-}(CH_2)_4{-}CH_3$
Physical/chemical characteristics[5]		
Appearance	Colorless liquid	Colorless liquid
Molecular weight	172.27	158.24
Melting point	−64.8°C (from pentane)	−69°C
Boiling point	207.9°C (206°C)	187°C; 84–85°C at 20 mm Hg
Specific gravity	0.870 at 15.5°C; 0.85705 at 20°C	0.8632 at 25°C
Refractive index	1.4160 (1.41894) at 20°C	1.4109 at 25°C
Solubility	Soluble in most common organic solvents; insoluble in water	Insoluble in water; soluble in alcohol
Organoleptic characteristics	Characteristic, strong ether-like odor; corresponding flavor reminiscent of grape	Ether-like odor with a pineapple-blackberry undertone; sweet, fruity flavor of blackberry and plum
Synthesis	By treating heptanoic acid with propyl alcohol in the presence of mineral acids,[6,7] or in benzene solution in the presence of *p*-toluenesulfonic acid[8]	By esterification of hexanoic acid with propyl alcohol in benzene solution in the presence of *p*-toluenesulfonic acid[8]
Natural occurrence	Not reported found in nature	Reported found in several natural products
Reported uses[3]	Non-alcoholic beverages 3.8 ppm Alcoholic beverages 3.0 ppm Ice cream, ices, etc. 5.1 ppm Candy 5.9 ppm Baked goods 18 ppm	Non-alcoholic beverages 2.2 ppm Ice cream, ices, etc. 3.0 ppm Candy 8.0 ppm
Regulatory status	FDA 121.1164; FEMA No. 2948	FDA 121.1164; FEMA No. 2949

REFERENCES

For References 1–5, see end of Part III.

6. **Gartenmeister,** *Justus Liebigs Ann. Chem.*, 233, 282, 1886.
7. **Bilterys and Gisseleire,** *Bull. Soc. Chim. Belg.*, 44, 569, 572, 1935.
8. **Hoback, Parsons and Bartlett,** *J. Am. Chem. Soc.*, 65, 1606, 1943.

	3-Propylidene phthalide	Propyl isobutyrate
Other names		*n*-Propyl isobutyrate *n*-Propyl-2-methylpropanoate
Empirical formula	$C_{11}H_{10}O_2$	$C_7H_{14}O_2$
Structure	(phthalide ring)=CH—CH$_2$—CH$_3$	$CH_3-(CH_2)_2-O-C(=O)-CH(CH_3)_2$
Physical/chemical characteristics[5]		
Appearance	Colorless to very pale, yellow liquid	Colorless liquid
Molecular weight	174.20	130.18
Boiling point	169°C (273°C) at 12 mm Hg	134°C (135°C) at 752 mm Hg
Specific gravity	1.08	0.8643 at 20°C
Refractive index		1.3955 at 20°C
Solubility	Almost insoluble in water; soluble in alcohol	Insoluble in water; soluble in most common organic solvents, including alcohol
Organoleptic characteristics	Very powerful and warm, spicy odor; warm, spicy, herbaceous flavor	Pineapple-like odor; corresponding sweet flavor
Synthesis	By condensing at 175°C phthalic anhydride and butyric anhydride in the presence of sodium butyrate[6]	From propyl alcohol and isobutyric acid in benzene solution in the presence of concentrated H_2SO_4[7]
Natural occurrence	Not reported found in nature	Reported found in several natural products
Reported uses[3]	Ice cream, ices, etc. 5.0 ppm Candy 5.0 ppm Baked goods 5.0 ppm	Non-alcoholic beverages 6.8 ppm Ice cream, ices, etc. 4.8 ppm Candy 24 ppm Baked goods 20 ppm
Regulatory status	FDA 121.1164; FEMA No. 2952	FDA 121.1164; FEMA No. 2936

REFERENCES

For References 1–5, see end of Part III.

6. **Bromberg,** *Ber. Dtsch. Chem. Ges.*, 29, 1436, 1896.
7. **Vogel,** *J. Chem. Soc.*, p. 637, 1948.

	Propyl isovalerate	Propyl mercaptan
Other names	*n*-Propyl isovalerate *n*-Propyl-β-methylbutyrate	*n*-Propane thiol Propanthiol-1 *n*-Propyl mercaptan *n*-Thiopropyl alcohol
Empirical formula	$C_8H_{16}O_2$	C_3H_8S
Structure	$CH_3-(CH_2)_2-O-C(=O)-CH_2-CH(CH_3)_2$	$CH_3-CH_2-CH_2-SH$
Physical/chemical characteristics[5]		
Appearance	Colorless, mobile liquid	Colorless, mobile liquid
Molecular weight	144.21	76.165
Melting point		−111.5°C
Boiling point	156–157°C; 49°C at 13 mm Hg	67°–68°C
Flash point		−20°C
Specific gravity	0.8617 at 20°C	0.8408 at 20°C
Refractive index	1.4031 at 20°C	1.4380 at 20°C
Solubility	Insoluble in water; soluble in most common organic solvents, including alcohol	Forms binary azeotropic mixtures with propyl alcohol (91.35%) and a ternary azeotrope with water and propyl alcohol; slightly soluble in water; soluble in alcohol and propylene glycol
Organoleptic characteristics	Fruity odor; bittersweet flavor similar to apple	Characteristic odor of cabbage; below 2–3 ppm sweet onion and cabbage-like flavor
Synthesis	By prolonged boiling of propyl alcohol with isovaleric acid in benzene in the presence of concentrated H_2SO_4[6]	By reacting propyl alcohol and H_2S under particular conditions;[7] from propyl alcohol or propyl disulfide and naphthalene;[8,9] from *n*-propyl chloride and potassium hydrosulfide; also from *n*-propanol plus bromine plus red phosphorus in the presence of sodium sulfate
Natural occurrence	Reported found in several natural products	Reported found in *Allium cepa*[10]
Reported uses[3]	Non-alcoholic beverages 5.0 ppm Ice cream, ices, etc. 16 ppm Candy 17 ppm Baked goods 20 ppm	
Regulatory status	FDA 121.1164; FEMA No. 2960	FDA 121.1164

REFERENCES

For References 1–5, see end of Part III.

6. **Vogel,** *J. Chem. Soc.*, p. 640, 1948.
7. **Sabatier et al.,** C. R. *Séances Acad. Sci. Paris,* 150, 1219, 1910.
8. **Mereshkoski,** *Chem. Zentralbl.*, 86, 982, 1915.
9. **Faragher et al.,** *Chem. Zentralbl.*, 99, 119, 1928.
10. **Challenger and Greenwood,** *Biochem. J.*, 44, 87, 1949.

	Propyl phenylacetate	α-Propylphenylethyl alcohol
Other names	*n*-Propyl-α-toluate	Benzylbutyl alcohol Benzyl-*n*-propyl carbinol 1-Phenyl-2-pentanol 1-Phenylpentan-2-ol *n*-Propyl benzyl carbinol
Empirical formula	$C_{11}H_{14}O_2$	$C_{11}H_{16}O$
Structure	$C_6H_5-CH_2-C(=O)-O-CH_2-CH_2-CH_3$	$C_6H_5-CH_2-CH(OH)-CH_2-CH_2-CH_3$
Physical/chemical characteristics[5]		
Appearance	Colorless liquid	Colorless and somewhat oily liquid
Molecular weight	178.23	164.25
Boiling point	253°C	247°C
Specific gravity	0.990 at 15.5°C	0.98
Refractive index	1.4955 at 20°C	
Solubility	Almost insoluble in water; soluble in alcohol	Almost insoluble in water; soluble in alcohol
Organoleptic characteristics	Characteristic honey-like, apricot-rose odor; sweet, honey-like taste	Mild, green, sweet odor with an earthy undernote; tart, fruit-like, green, sweet taste
Synthesis	By esterification of *n*-propanol with phenylacetic acid	From phenyl acetaldehyde and propyl magnesium bromide; also by hydrogenation of propyl benzyl ketone
Natural occurrence	Reported found in several natural products	Not reported found in nature
Reported uses[3]	Non-alcoholic beverages 0.30–1.0 ppm Ice cream, ices, etc. 0.30–1.5 ppm Candy 2.7 ppm Baked goods 1.0–5.0 ppm	Non-alcoholic beverages 1.0 ppm Ice cream, ices, etc. 5.0 ppm Candy 5.0 ppm Gelatins and puddings 5.0 ppm
Regulatory status	FDA 121.1164; FEMA No. 2955	FDA 121.1164; FEMA No. 2953

REFERENCES

For References 1–5, see end of Part III.

	Propyl propionate		Propyl thioacetate	
Other names	*n*-Propyl propionate			
Empirical formula	$C_6H_{12}O_2$		$C_5H_{10}OS$	
Structure	$CH_3-CH_2-CH_2-O-C(=O)-CH_2-CH_3$		$CH_3-C(=O)-S-CH_2-CH_2-CH_3$ or $CH_3-C(=S)-O-CH_2-CH_2-CH_3$	
Physical/chemical characteristics[5]				
Appearance	Colorless liquid			
Molecular weight	116.16		118.96	
Boiling point	122–125°C		137–139°C	
Flash point	79°C			
Specific gravity	0.883 at 15.5°C			
Refractive index	1.3910 (1.3935) at 20°C			
Solubility	0.5% soluble in water; miscible with alcohol, propylene glycol, and ether			
Organoleptic characteristics	Complex, fruity odor reminiscent of apple, banana, and pineapple; somewhat pleasant, bitter flavor			
Synthesis	By esterification of propyl alcohol with the corresponding acid in the presence of concentrated H_2SO_4,[6,7] or in the presence of BF_3[8]		From acetyl chloride and phenyldipropylarsin sulfide at 80–90°C[10]	
Natural occurrence	Reported found among the volatile constituents of grape var. white Sauvignon[9]			
Reported uses[3]	Non-alcoholic beverages	6.0 ppm	Non-alcoholic beverages	1.0 ppm
	Ice cream, ices, etc.	12 ppm	Candy	1.0 ppm
	Candy	25 ppm	Condiments, pickles	1.0 ppm
	Baked goods	25 ppm	Meat, meat sauces, soups	1.0 ppm
			Cereals	1.0 ppm
Regulatory status	FDA 121.1164; FEMA No. 2958		FEMA No. 3385	

REFERENCES

For References 1–5, see end of Part III.

6. **Vogel,** *J. Chem. Soc.*, p. 629, 1948.
7. **Salmi and Leimu,** *Chem. Abstr.*, 42, 4031, 1948.
8. **Hinton and Nieuwland,** *J. Am. Chem. Soc.*, 54, 2018, 1932.
9. **Chandhery et al.,** *Am. J. Enol. Vitic.*, 15, 190, 1964.
10. *Chem. Abstr.* 69, 52239d, 1968.

	Pulegone	Pyrazine ethanethiol
Other names	*p*-Menth-4(8)-en-3-one 1-Methyl-4-isopropylidenecyclohexan-3-one	2-(Pyrazinyl)ethanethiol 2-Pyrazinyl ethylmercaptan
Empirical formula	$C_{10}H_{16}O$	$C_6H_8N_2S$
Structure	CH_3 … O … H_3C CH_3	N … CH_2—CH_2—SH … N
Physical/chemical characteristics		
Appearance	Colorless, oily liquid	
Assay		C: 51.40%; H: 5.75%; N: 19.98%; S: 22.87%
Molecular weight	152.23	140.04
Boiling point	*dl*, 220–222°C; *d*, 224°C at 750 mm Hg	57–69°C at 0.05 mm Hg
Specific gravity	*d*, 0.9405 at 15°C	
Refractive index	*d*, 1.48796 at 20°C; *dl*, 1.4846 at 18°C	
Optical rotation	*d*, +20°48′	
Solubility	Insoluble in water; miscible with alcohol, ether, and chloroform	
Organoleptic characteristics	Pleasant odor, somewhat similar to peppermint and camphor	
Synthesis	Isolated from pennyroyal oil (Moroccan or Spanish); synthesis from 3-methyl cyclohexanone	From 2-vinylpyrazine by reaction with thiolactic acid, followed by conversion of the thiolester to 2-(pyrazinyl) ethanethiol[6]
Natural occurrence	The structure has been defined by the work of several authors.[7-10] The *d*-, *l*-, and *dl*-forms are known. The *dl*-form is prepared synthetically and is not found in nature. The *l*-form[11] is found in the essential oils of *Agastache formosana*, Israeli orange, *Barosma betulina*, and *Barosma crenulata*. The *d*-form is the most abundant and is found in pennyroyal oils.	
Reported uses[3]	Non-alcoholic beverages 5.0–8.0 ppm Ice cream, ices, etc. 5.0–32 ppm Candy 17 ppm Baked goods 24–25 ppm	Non-alcoholic beverages 10.0 ppm Ice cream, ices, etc. 10.0 ppm Baked goods 10.0 ppm
Regulatory status	FDA 121.1164; FEMA No. 2963	FEMA No. 3230

REFERENCES

For References 1–5, see end of Part III.

6. **Bardwell and Hewett,** *J. Org. Chem.*, 22, 980, 1957.
7. **Beckmann and Pleissner,** *Annalen,* 262, 1, 1891.
8. **Semmler,** *Ber. Dtsch. Chem. Ges.*, 25, 3519, 1892.
9. **Wallach,** *Annalen,* 289, 337, 1896.
10. **Fredga and Leskinen,** *Ark. Kem. B,* 19, 1, 1944.
11. **Fujita,** *Bot. Mag.* (Tokyo), 64, 165, 1951.

	Pyrazinylmethylsulfide	Pyridine
Other names	Methylthiopyrazine	
Empirical formula	$C_5H_6N_2S$	C_5H_5N
Structure	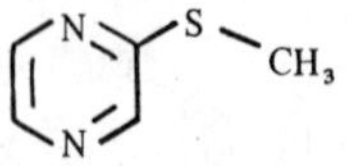	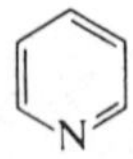
Physical/chemical characteristics		
Appearance		Flammable, colorless liquid
Assay	C: 47.59%; H: 4.79%; N: 22.20%; S: 25.41%	
Molecular weight	126.18	79.10
Melting point		−42°C
Boiling point		115.2°C
Flash point		21°C
Specific gravity		0.9808 at 21°C
Refractive index		1.5092 at 21°C
Solubility		Miscible with water, alcohol, ether, propylene glycol, and most oils
Organoleptic characteristics		Characteristic, penetrating odor; sharp taste
Synthesis	Sodium methyl mercaptide is added to 2-chloropyrazine, which is prepared from 2-hydroxypyrazine[6]	From coal-tar distillation
Natural occurrence		Reported found in wood oil, the leaves and roots of *Atropa belladonna,*[7,8] and in other plants (coffee, tobacco)
Reported uses[3]	Non-alcoholic beverages 1.0 ppm Ice cream, ices, etc. 1.0 ppm Baked goods 1.0 ppm	Non-alcoholic beverages 1.0 ppm Ice cream, ices, etc. 0.02–0.12 ppm Candy 0.40 ppm Baked goods 0.40 ppm
Regulatory status	FEMA No. 3231	FDA 121.1164; FEMA No. 2966

REFERENCES

For References 1–5, see end of Part III.

6. **Karmas and Spoerri,** *J. Am. Chem. Soc.,* 74, 1580, 1952.
7. **Goris and Larsonneau,** *Chem. Zentralbl.,* 1, 757, 1922.
8. **King and Ware,** *J. Chem. Soc.* (London), 331, 1941.

	2-Pyridine methanethiol		Pyroligneous acid, extract	
Other names	2-Pyridyl methanethiol 2-Pyridyl methyl mercaptan 4-(Methylthio)pyridine		"Liquid smoke" Pyroligneous vinegar Wood vinegar	
Empirical formula	C_6H_7SN		A mixture of material from wood distillation; chief component is water containing nearly 10% acid (formic, acetic, and propionic). The extract is free of water, acid, and tar.	
Structure	(pyridine ring)—CH_2—SH			
Physical/chemical characteristics[5]				
Appearance			Yellow to red liquid	
Molecular weight	125.19			
Boiling point	57–58°C at 0.6 mm Hg			
Specific gravity	1.1533 at 25°C		1.018–1.030	
Refractive index	1.5758 at 25°C			
Solubility			Miscible with water, alcohol, and propylene glycol but not oils. A tar-like phase can separate on standing; it is insoluble in alcohol and water.	
Organoleptic characteristics			Smoke odor	
Synthesis	By hydrolysis of 2-pyridylmethyl-*tert*-butyl sulfide with sulfuric acid in aqueous solution;[6] from 2-pyridylmethyl chloride[7]		Obtained by destructive distillation of wood, preferably birch. The crude product contains methanol, acetic acid, acetone, furfural, and various tar and related products. The extract is rendered free of water, acid, and tar by (1) alkali washing, followed by (2) re-acidification and (3) solvent extraction.	
Reported uses[3]	Non-alcoholic beverages Ice cream, ices, etc. Candy Gelatins and puddings	2.0 ppm 2.0 ppm 2.0 ppm 2.0 ppm	Alcoholic beverages Baked goods Meats	20 ppm 50–200 ppm 100–300 ppm
Regulatory status	FEMA No. 3232		FDA 121.1164; FEMA No. 2968	

REFERENCES

For References 1–5, see end of Part III.

6. **Loev and Olin,** U.S. Patent 3,069,472.
7. **Maneszewska-Wieczorkowska and Michalski,** *Rocz. Chem.,* 31, 543, 1957; *Chem. Abstr.,* 50, 12046h, 1956; *Chem. Abstr.,* 51, 11347a, 1957.

	Pyrrole	Pyruvaldehyde
Other names	Azole Imidole Divynyleneimine	Acetyl formaldehyde 1,2-Ketopropionic aldehyde α-Ketopropionic aldehyde Methyl glyoxal 2-Oxopropanal Pyruvic aldehyde
Empirical formula	C_4H_5N	$C_3H_4O_2$
Structure	[pyrrole ring, N-H]	$CH_3-C(=O)-CHO$
Physical/chemical characteristics[5]		
Appearance	Colorless liquid; tends to darken on aging	Mobile, yellow liquid; hygroscopic; in air readily polymerizes to a viscous, non-pourable glass
Molecular weight	67.09	72.06
Melting point	-24°C	
Boiling point	130°C at 761 mm Hg (decomposes)	72°C
Specific gravity	0.968–0.969 at 20°C	1.0455 at 24°C
Refractive index	1.5091–1.5095 at 20°C	1.4002 at 17.5°C
Solubility	Insoluble in water and in alkalis; soluble in alcohol, ether, and organic solvents; dissolves slowly in acids, with polymerization, gradual separation of "pyrrole red," and liberation of ammonia	Soluble in alcohol, ether, and benzene; yields colorless solution with alcohol and yellow solutions with the latter two solvents
Organoleptic characteristics	Sweet, warm-ethereal odor reminiscent of chloroform	Characteristic, pungent, stinging odor with a pungent, caramellic, sweet flavor
Synthesis	By fractional distillation of bone oil and subsequent purification via the corresponding potassium salt; by thermal decomposition of ammonium mucate in glycerol or mineral oil[6,7]	By distilling a dilute solution of dihydroxyacetone from calcium carbonate; by oxidation of acetone with selenium dioxide; by heating dihydroxy acetone with phosphorus pentoxide; by warming isonitroso acetone with diluted H_2SO_4
Natural occurrence	Reported found in coal tar and in bone oil (together with several methyl homologs) also reported found in lemon and orange trees and in vervain leaves	Reported found in the dry distillate of *Manilla copal*
Reported uses[3]	Ice cream, ices, etc. 3.0 ppm Candy 3.0 ppm Baked goods 3.0 ppm Gelatins and puddings 3.0 ppm Meat, meat sauces, soups 3.0 ppm Cereals 3.0 ppm	Non-alcoholic beverages 1.0 ppm Ice cream, ices, etc. 1.0 ppm Candy 0.03–5.0 ppm Baked goods 0.03–5.0 ppm
Regulatory status	FEMA No. 3386	FDA 121.1164; FEMA No. 2969

REFERENCES

For References 1–5, see end of Part III.

6. **Blicke and Powers,** *Ind. Eng. Chem.,* 19, 1334, 1927.
7. **McElrain and Bollinger,** in *Organic Syntheses,* Coll. Vol. 1, Gilman, H. and Blatt, A. H., Eds., Wiley & Sons, New York, 1941, 437.

	Pyruvic acid	Quinine
Other names	Acetylformic acid α-Ketopropionic acid 2-Oxopropanoic acid Pyroracemic acid	
Empirical formula	$C_3H_4O_3$	$C_{20}H_{24}N_2O \cdot 3\ H_2O$
Structure	$CH_3-C(=O)-C(=O)-OH$	HO–CH attached to 6-methoxyquinoline (H_3C-O) and quinuclidine ring (N, CH_2, CH_2) bearing $CH{=}CH_2$
Physical/chemical characteristics[5]		
Appearance	Colorless liquid; tends to darken and decompose unless kept free of minor contaminants and in tightly sealed containers	Triboluminescent, orthorhombic needles; colorless or white, turning brown on exposure to air. Solutions (dilute) show blue fluorescence
Molecular weight	88.06	324.43 (anhydrous); 378.47 (hydrous, 3 moles H_2O)
Melting point	12–14°C (11.8°C)	177°C (anhydrous); 57°C (hydrous)
Boiling point	165°C with decomposition; 65°C at 12 mm Hg	Sublimes in high vacuum at 120–180°C
Specific gravity	1.267 at 15°C	
Refractive index	1.4138 at 20°C	At 15°C: 169° (c = 2 in 97% alcohol) At 17°C: 117° (c = 1.5 in chloroform) At 25°C: 285° (c = 0.4 molar on 0.1 N H_2SO_4)
Solubility	Completely miscible with water, alcohol, and ether	0.05% in water at room temperature, 0.14% in boiling water; soluble in alcohol; slightly soluble in hydrocarbons
Organoleptic characteristics	Sour, acetic odor (similar to acetic acid); pleasant, sour taste with a burning, somewhat sweet note	Odorless; intense, bitter taste
Synthesis	By distillation of tartaric acid in the presence of potassium acid sulfate as a dehydrating agent; from acetyl chloride and potassium cyanide to yield the nitrile, which is subsequently acid hydrolyzed to the acid. Pyruvic acid must be rectified under vacuum.	By extraction from cinchona bark (*Cinchona officinalis*), where it is present at approximately 8%
Natural occurrence	Isolated from cane sugar fermentation broth and from a few plants; also reported found in peppermint[6]	In *Cinchona officinalis*
Reported uses[3]	Non-alcoholic beverages 0.25 ppm Ice cream, ices, etc. 0.25–20 ppm Candy 27 ppm Baked goods 30 ppm Chewing gum 110 ppm	The alkaloid as such is not used in food flavors; *see* bisulfate, sulfate, and hydrochloride salts. In the United States the salts are permitted in carbonated beverages at 83 ppm calculated as quinine alkaloid base
Regulatory status	FDA 121.1164; FEMA No. 2970	*See* bisulfate, sulfate, and hydrochloride salts.

REFERENCES

For References 1–5, see end of Part III.

6. Towers et al., *J. Am. Chem. Soc.*, 76, 2392, 1954.

	Quinine bisulfate	Quinine hydrochloride
Other names		Quinine chloride Quinine monohydrochloride
Empirical formula	$C_{20}H_{24}N_2O_2 \cdot H_2SO_4 + 7\ H_2O$	$C_{20}H_{24}N_2O_2 \cdot HCl + 2\ H_2O$
Structure	*See* Quinine	*See* Quinine
Physical/chemical characteristics[5]		
Appearance	White, crystalline powder; turns brown on exposure to air	Brilliant, silky needles; colorless to white
Molecular weight	548.62	
Solubility	10% in water; 4% in alcohol; 6% in glycerin	6% in water; 50% in alcohol; 12% in glycerin
Organoleptic characteristics	Odorless; bitter taste	Odorless; intensely bitter flavor
Uses	Non-alcoholic beverages 95–100 ppm[3]	Non-alcoholic beverages 110 ppm
Regulatory status	FDA 121.1081; FEMA No. 2975	FDA 121.1081; FEMA No. 2976

REFERENCES

For References 1–5, see end of Part III.

	Rhodinol	Quinine sulfate
Other names	*l*-Citronellol (3,7-Dimethyl-6-octen-1-ol) 3,7-Dimethyl-7-octen-1-ol	
Empirical formula	$C_{10}H_{20}O$	$(C_{20}H_{24}N_2O_2)_2 \cdot H_2SO_4 + 7\ H_2O$
Structure	The constitution of commercial rhodinol is still controversial. We define rhodinol as the isolate of Reunion geranium oil containing in larger portion *l*-citronellol.	*See* Quinine
Physical/chemical characteristics[5]		
Appearance	Colorless liquid	Dull-white, needle-like crystals
Assay	82% min[2]	
Molecular weight	156.26	873.05
Specific gravity	0.860–0.880 at 25°/25°C[2]	
Refractive index	1.4630–1.4730 at 20°C[2]	
Optical rotation	−4° to −9°;[2] −5° to −9°	
Ester content	Not more than 1% (as rhodinyl acetate)[2]	
Solubility	1:1.2 in 70% alcohol;[2] 1:1.5 in 60% alcohol; insoluble in water; soluble in most common organic solvents	0.12% in water; 0.9% in alcohol
Organoleptic characteristics	Very pleasant, rose-like odor	Odorless; intensely bitter flavor
Synthesis	By fractional distillation of Reunion geranium oil	
Natural occurrence	Geranium oil	
Reported uses[3]	Non-alcoholic beverages 2.0 ppm Ice cream, ices, etc. 2.1 ppm Candy 7.6 ppm Baked goods 8.1 ppm Gelatins and puddings 2.9 ppm Chewing gum 31 ppm Jellies 0.92 ppm	Non-alcoholic beverages 100 ppm
Regulatory status	FDA 121.1164; FEMA No. 2980	FDA 121.1081; FEMA No. 2977

REFERENCES

For References 1–5, see end of Part III.

	Rhodinyl acetate	Rhodinyl butyrate
Other names		Citronellyl butyrate
Empirical formula		$C_{14}H_{26}O_2$
Structure	Rhodinyl acetate is chiefly a mixture of acetates of various terpene alcohols derived from geranium oil, in which *l*-citronellyl acetate predominates[2]	CH_3 / $CH_2O-\overset{O}{\overset{\parallel}{C}}-CH_2-CH_2-CH_3$ / H_3C CH_3
Physical/chemical characteristics[5]		
Appearance	Colorless to slightly yellow liquid	Colorless to yellowish or greenish liquid
Assay	87% min[2]	90%–93% min
Molecular weight		226.36
Boiling point	Approx 237°C	137°C at 13 mm Hg
Flash point		>100°C
Specific gravity	0.895–0.908 at 25°/25°C;[2] 0.8777–0.8810 at 15°C	0.886–0.890 at 25°/25°C; 0.8924 at 15°C
Refractive index	1.4530–1.4580[2] (1.4668–1.4690) at 20°C	1.451–1.455 at 20°C; 1.4474 at 20°C[6]
Acid value	Not more than 1[2]	1 max
Optical rotation	−2° to −6°;[2] −6.5° to −8.2°	
Solubility	1:2 in 80% alcohol;[2] 1:3.5–5.5 in 60% alcohol; insoluble in water	1:8 in 80% alcohol; 1:1 in 90% alcohol
Organoleptic characteristics	Characteristic fresh, rose-like odor; analogous taste on dilution, bitter when concentrated	Fruity, sweet odor similar to rose; sweet taste reminiscent of blackberry
Synthesis	By acetylation of rhodinol from geranium oil	By esterification of rhodinol, as for geraniol and linalool
Natural occurrence		Reported found in geranium essential oil
	Non-alcoholic beverages 2.8 ppm Ice cream, ices, etc. 1.4 ppm Candy 9.4 ppm Baked goods 18 ppm	Non-alcoholic beverages 0.94 ppm Ice cream, ices, etc. 1.1 ppm Candy 3.0 ppm Baked goods 9.7 ppm Chewing gum 1.1 ppm
Regulatory status	FDA 121.1164; FEMA No. 2981	FDA 121.1164; FEMA No. 2982

REFERENCES

For References 1–5, see end of Part III.

6. **Bedoukian,** *Perfumery and Flavoring Synthetics,* 2nd ed., American Elsevier, 1967.

	Rhodinyl formate	Rhodinyl isobutyrate
Other names	Citronellyl formate	Citronellyl isobutyrate
Empirical formula	$C_{11}H_{20}O_2$	
Structure	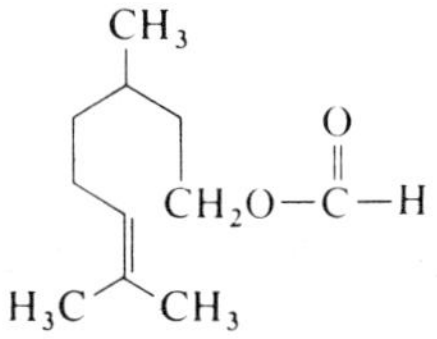 CH_3 … $CH_2O-\overset{O}{\overset{\parallel}{C}}-H$ … H_3C CH_3	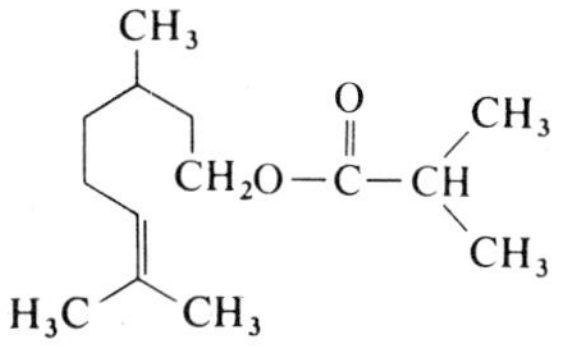 CH_3 … $CH_2O-\overset{O}{\overset{\parallel}{C}}-CH(CH_3)_2$ … H_3C CH_3
Physical/chemical characteristics[5]		
Appearance	Colorless to yellowish liquid	Colorless, oily liquid
Assay	85% min	
Molecular weight	184.26	226.26
Boiling point		Approx 260°C
Flash point	>100°C	
Specific gravity	0.901–0.906 at 25°C	0.89
Refractive index	1.453–1.458 at 20°C	
Acid value	2 max	
Solubility	1:2 in 80% alcohol	Insoluble in water; soluble in alcohol
Organoleptic characteristics	Rose-like, leafy odor; bittersweet taste suggestive of cherry	Fruity, sweet, floral odor; fruity, floral, pineapple-like flavor, intensely sweet
Synthesis	By treating rhodinol with formic acid; the rhodinol is extracted from geranium essential oil. Its composition is dependent on the geraniol-citronellol ratio existing in the rhodinol used for esterification.	From rhodinol and isobutyric anhydride
Natural occurrence	Geranyl and citronellyl formate have been found in geranium essential oil. Rhodinyl formate as such has not been reported found in nature.	Not reported found in nature
Reported uses[3]	Non-alcoholic beverages 1.3 ppm Ice cream, ices, etc. 1.8 ppm Candy 4.3 ppm Baked goods 4.9 ppm Gelatins and puddings 0.08 ppm	Non-alcoholic beverages 1.1 ppm Ice cream, ices, etc. 1.8 ppm Candy 3.3 ppm Baked goods 4.5 ppm Gelatins and puddings 0.01 ppm
Regulatory status	FDA 121.1164; FEMA No. 2984	FDA 121.1164; FEMA No. 2983

REFERENCES

For References 1–5, see end of Part III.

	Rhodinyl isovalerate	Rhodinyl phenylacetate
Other names	Citronellyl isovalerate Rhodinyl-β-methylbutyrate	Citronellyl phenylacetate Rhodinyl α-toluate
Empirical formula		$C_{18}H_{26}O_2$
Structure	CH_3; $CH_2O-C(=O)-CH_2-CH(CH_3)_2$; H_3C CH_3	CH_3; $CH_2O-C(=O)-CH_2-C_6H_5$; H_3C CH_3
Physical/chemical characteristics[5]		
Appearance	Colorless, oily liquid	Colorless to very pale yellow-greenish liquid
Assay		92%–95% min
Molecular weight	240.26	274.26
Boiling point	147°C (approx 270°C) at 14 mm Hg	Approx 340°C
Flash point		>100°C
Specific gravity	0.8829 at 15°C	0.965–0.972 at 25°/25°C
Refractive index	1.4465 at 20°C	1.5000–1.5040 at 20°C
Acid value		1 max
Solubility	Almost insoluble in water; soluble in alcohol	1:3 in 90% alcohol; insoluble in alcohol
Organoleptic characteristics	Lasting, deep, sensual, red rose-like fragrance; bittersweet flavor reminiscent of cherry	Odor reminiscent of rose absolute with a honey-like undertone; very sweet, rose-honey taste
Synthesis		By esterification of rhodinol with phenylacetic acid
Natural occurrence	Not reported found in nature	Not reported found in nature
Reported uses[3]	Non-alcoholic beverages 2.0 ppm Ice cream, ices, etc. 2.3 ppm Candy 7.2 ppm Baked goods 7.2 ppm	Non-alcoholic beverages 1.2 ppm Ice cream, ices, etc. 1.2 ppm Candy 3.8 ppm Baked goods 4.4 ppm
Regulatory status	FDA 121.1164; FEMA No. 2987	FDA 121.1164; FEMA No. 2985

REFERENCES

For References 1–5, see end of Part III.

	Rhodinyl propionate	Rose oxide
Other names	Citronellyl propionate	Tetrahydro-4-methyl-2-(2-methylpropen-1-yl)-pyran
Empirical formula		$C_{10}H_{18}O$
Structure	CH_3; $CH_2O-\overset{O}{\overset{\Vert}{C}}-CH_2-CH_3$; H_3C CH_3	*cis-* / *trans* (H$_3$C, H, O, CH=C, H$_3$C, CH$_3$)
Physical/chemical characteristics[5]		
Appearance	Colorless, oily liquid	Colorless mobile liquid
Molecular weight	212.26	154.24
Boiling point	122°C (255°C) at 14 mm Hg	70°C at 12 mm Hg (pure material)
Flash point	100°C	
Specific gravity	0.910 at 15.5°C; 0.8959 at 15°C[13]	0.8751 (pure material)
Refractive index	1.4570 at 20°C; 1.4452 at 20°C[13]	1.4566 (pure material)
Solubility	Almost insoluble in water; soluble in alcohol	Very slightly soluble in water; soluble in alcohol and oils
Organoleptic characteristics	Sweet odor reminiscent of rose and geranium with a verbena-like undertone; rich, sweet flavor	Powerful, distinctive geranium top-note
Synthesis	From rhodinol and propionic acid (or anhydride)	Several patented syntheses for the preparation of rose oxide are available[6-9]
Natural occurrence	Not reported found in nature	Reported as occurring naturally in the oils of rose (Bulgarian) and geranium (Reunion);[11] both the *cis-* and the *trans*-form have been reportedly isolated from geranium oil;[12] rose oxide exhibits different optical rotations, depending on the source and method of preparation; the highest value (−41.5°C) has been measured on material obtained from Bulgarian attar of roses
Reported uses[3]	Non-alcoholic beverages 1.8 ppm Ice cream, ices, etc. 2.4 ppm Candy 4.9 ppm Baked goods 5.8 ppm	Non-alcoholic beverages 0.40 ppm Candy 2.0 ppm Baked goods 2.0 ppm
Regulatory status	FDA 121.1164; FEMA No. 2986	FEMA No. 3236

REFERENCES

For References 1–5, see end of Part III.

6. **Eschenmoser and Seidel,** U.S. Patent 3,161,657, 1964.
7. **Eschinasi and Cotter,** U.S. Patent 3,163,658, 1964.
8. **Markus,** U.S. Patent 3,166,576, 1965.
9. **Schenk, Ohloff and Klein,** German Patent 1,137,730, 1962.
10. **Seidel and Stoll,** *Helv. Chim. Acta,* 42, 1830, 1959.
11. **Seidel et al.,** *Helv. Chim. Acta,* 44, 598, 1961.
12. **Naves et al.,** *Bull. Soc. Chim. Fr.,* p. 645, 1961.
13. **Bedoukian,** *Perfumery and Flavoring Synthetics,* 2nd ed., American Elsevier, 1967.

	Rum ether
Other names	Ethyl oxyhydrate
Empirical formula	Defined as the mixture resulting from the oxidation and the hydration of ethyl alcohol
Structure	
Physical/chemical characteristics[5]	
Appearance	Rum ether consists of at least 99% water, ethyl alcohol, ethyl acetate, methanol, ethyl formate, acetone, acetaldehyde, and formaldehyde. It all distills at a temperature not exceeding 100°C at atmospheric pressure and leaves no residue on evaporation. The methanol and formaldehyde content, combined, should not exceed 5%.
Boiling point	Rectified material, 65–87°C
Organoleptic characteristics	Rum-like odor and flavor
Synthesis	Place 95% ethanol and manganese dioxide in the reaction flask equipped with distillation apparatus. Add pyroligneous acid and finally sulfuric acid (66°Be) slowly with agitation, maintaining the final temperature *below* 40–50°C. The mixture is then distilled under atmospheric pressure; the fraction distilling between 60° and 100°C is collected. Finally, the collected fraction undergoes rectification. The rectified fraction exhibits a boiling range between 65 and 87°C. Average yield is 98% based on the amount of ethanol used. Sometimes the distillate is further divided into two additional fractions—one more, and one less rich in acetaldehyde. The amounts of starting material employed vary between the following limits: Ethanol (95%) kg 12–25 Pyroligneous acid (12% acetic acid) kg 10–16 Sulfuric acid (66°Be) kg 3–8 Manganese dioxide kg 2–5 A gas chromatographic analysis of rum ether has led to the identification of 22 components in varying concentrations (as low as 50 ppm).[6] Caramel is added to the rum ether for coloring purposes. In general, the addition of rum ether to the finished liquor does not exceed 2–3% for flavoring purposes.
Reported uses[3]	Non-alcoholic beverages 67 ppm Alcoholic beverages 80–1,600 ppm Ice cream, ices, etc. 110 ppm Candy 320 ppm Baked goods 230 ppm Gelatins and puddings 1.7 ppm Chewing gum 380 ppm
Regulatory status	FDA 121.1164; FEMA No. 2996

REFERENCES

For References 1–5, see end of Part III.

6. **Fenaroli, Poy and Maroni,** *Riv. Ital. EPPOS*, 9, 484, 495, 1965; 11, 593, 1965; 2, 75, 1966.

	Safranal	Safrole
Other names	2,6,6-Trimethylcyclohexa-1,3-dienyl methanal Dehydrocyclocitral	4-Allyl-1,2-methylenedioxybenzene *m*-Allylpyrocatechin methylene ether 3,4-Methylenedioxyallylbenzene
Empirical formula	$C_{10}H_{14}O$	$C_{10}H_{10}O_2$
Structure	H_3C, CH_3, CHO, CH_3	$CH_2{-}CH{=}CH_2$, O, O—CH_2
Physical/chemical characteristics[5]		
Appearance	Yellow oil	Colorless to slightly yellow, oily liquid
Molecular weight	150.09	162.18
Melting point		11°C
Boiling point	172°C 85–78°C at 12 mm Hg 70°C at 1 mm Hg	233–236°C
Congealing point		10.0–11.2°C[2]
Flash point		97°C
Specific gravity	0.9730 at 19°C	1.092–1.101 at 25°/25°C;[2] (1.100–1.107 at 15°C)
Refractive index	1.5281 at 19°C	1.5360–1.5385 at 20°C[2]
Optical rotation		−1° to +1°;[2] −0.30° to +0.30°
Solubility		1:3 in 90% alcohol;[2] insoluble in water; soluble in alcohol, ether, chloroform, and most common organic solvents
Organoleptic characteristics	Characteristic saffron-like odor and taste	Characteristic odor of sassafras
Synthesis	By acid hydrolysis of picrocrocin, the bitter principle in saffron;[6] by catalytic dehydrogenation of β-cyclocitral or α-cyclogeranic acid with selenium dioxide[8]	By distillation from the essential oils rich in safrole. An uneconomical synthesis starts with catechol methylene ether reacted with allyl chloride; also 4-allyl alcohol with methylene iodide
Natural occurrence	Occurring naturally as a glucoside (picrocrocin) in saffron[6]	Originally isolated in the oil from roots of *Sassafras officinale*; constituent of several essential oils, such as camphor, nutmeg, and cinnamon leaves. The essential oil from the rind of *Nemuaron humboldtii* contains up to 99% safrole;[6] Brazilian sassafras oil, up to 93%;[7] and American sassafras oil, up to 80%.[8]
Reported uses[3]	Candy 5.0 ppm Baked goods 10.0 ppm Condiments, pickles 5.0 ppm Preserves and spreads 10.0 ppm	
Regulatory status	FEMA No. 3389	Its use in foods is not permitted in USA.

REFERENCES

For References 1–5, see end of Part III.

6. **Kuhn and Winterstein,** *Chem. Ber.*, 67, 354, 1934.
7. **Kuhn and Wendt,** *Chem. Ber.*, 69, 1546, 1936.
8. **Bachli and Karrer,** *Helv. Chim. Acta*, 38, 1863, 1955.
9. **Chabeau,** *Chem. Zentralbl.*, 1, 3747, 1939.
10. **De Souza,** *Chem. Zentralbl.*, p. 2765, 1952.
11. **Hickey,** *J. Org. Chem.*, 13, 443, 1948.

	Salicylaldehyde	Santalol, α and β
Other names	2-Hydroxybenzaldehyde *o*-Hydroxybenzaldehyde Salicylic aldehyde	"Argeol" "Arheol" *d*-α-Santalol *l*-β-Santalol
Empirical formula	$C_7H_6O_2$	$C_{15}H_{24}O$
Structure	OH, CHO (structure)	α: CH_3, CH_3, OH, CH_3 (structure) β: CH_3, CH_2, CH_2OH, CH_3 (structure)
Physical/chemical characteristics[5]		
Appearance	Colorless, oily liquid; solidifies when cold	Colorless to slightly yellow, viscous liquid
Assay		95% min[2] (98% min)
Molecular weight	122.12	220.34
Melting point	2°C	
Boiling point	93°C (196°C) at 25 mm Hg	α, 129°C at 1 mm Hg; 302°C β, 133°C at 1 mm Hg; 309°C
Congealing point	1.6°C	
Specific gravity	1.1690 (1.17) at 20°C	0.965–0.975 at 25°/25°C;[2] 0.9769 at 20°C; 0.9750 at 20°C; 0.970–0.974
Refractive index	1.5740 at 19.7°C	1.5060–1.5090 at 20°C;[2] 1.5050–1.5080; α, 1.5023; β, 1.5115
Optical rotation		−11° to −19°;[2] −14° to −19°; α, +17.0° at 20 °C; β, −90.5° at 20°C
Solubility	Soluble in most organic solvents; slightly soluble in water	1:4 in 70% alcohol;[2] insoluble in water; soluble in most common organic solvents
Organoleptic characteristics	Pungent, irritating odor similar to benzaldehyde, acetophenone, and nitrobenzene, but with phenolic notes; nut-like, coumarin flavor at low levels	Sweet, sandalwood odor
Synthesis	From phenol, chloroform, and alkali according to Reimer-Tiemann method (1876); starting material for the manufacture of coumarin	The commercial product is generally a mixture of α- and β-santalol obtained by fractional distillation of sandalwood oil. Non-commercial syntheses are known.
Natural occurrence	Occurs rather frequently in nature; in the flowers of *Spirea ulmaria* and other *Spireae*, in the roots of *Crepis foetida* L., in the fruits of *Pinus avium*, in the rind of *Rauwolfia caffra*, in the leaves of *Ceanothus velutinus*, and in the essential oils of *Cinnamomum cassia* and of tobacco leaves[6]	The stereochemistry of santalols has been discussed by Brieger.[6] Santalols are found among the constituents of various sandalwood species (*Santalum album* L., *S. spicatum*, *S. autrocaledonicum*). The α- and β-isomers are present in varying ratios, with the α-isomer usually more abundant.
Reported uses[3]	Non-alcoholic beverages 0.55 ppm Alcoholic beverages 5.0 ppm Ice cream, ices, etc. 1.1 ppm Candy 1.8 ppm Baked goods 6.3 ppm Chewing gum 11–18 ppm Condiments 2.0 ppm	Non-alcoholic beverages 0.06–2.0 ppm Ice cream, ices, etc. 0.35–2.0 ppm Candy 1.0–1.0 ppm Baked goods 1.0–8.0 ppm Chewing gum 0.20 ppm
Regulatory status	FDA 121.1164; FEMA No. 3004	FDA 121.1164; FEMA No. 3006

REFERENCES

For References 1–5, see end of Part III.

6. **Onishi and Yamamoto,** *Chem. Abstr.*, 50, 15028, 1956.

	Santalyl acetate, α- and β-	**Santalyl phenylacetate, α and β**
Other names	α-Santalol, acetate β-Santalol, acetate	α-Santalyl α-toluate β-Santalyl α-toluate
Empirical formula	$C_{17}H_{26}O_2$	$C_{23}H_{30}O_2$
Structure	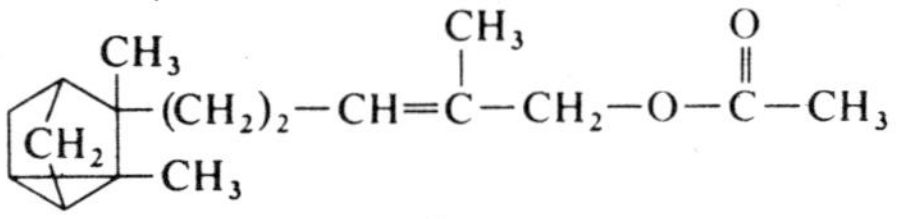 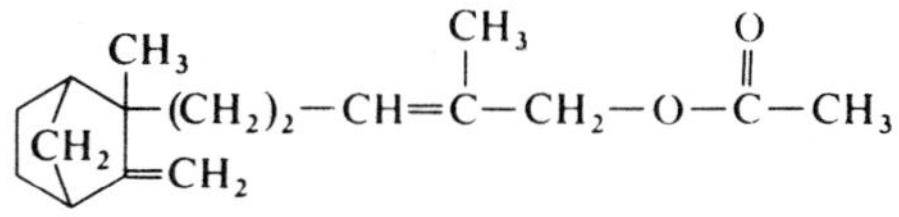 CH_3, CH_2, $-(CH_2)_2-CH=C(CH_3)-CH_2-O-C(=O)-CH_3$, $-CH_3$ α CH_3, CH_2, $-(CH_2)_2-CH=C(CH_3)-CH_2-O-C(=O)-CH_3$, $=CH_2$ β	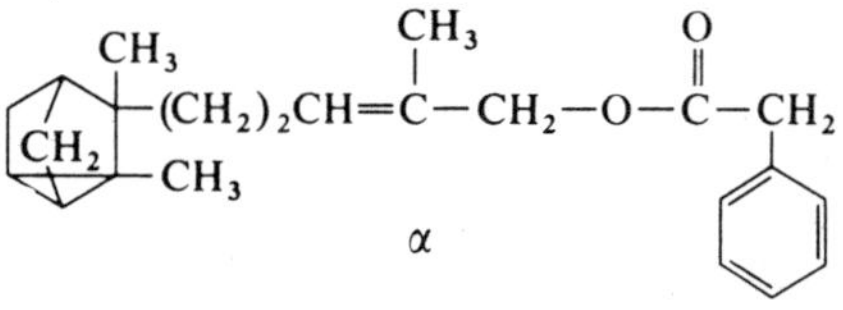 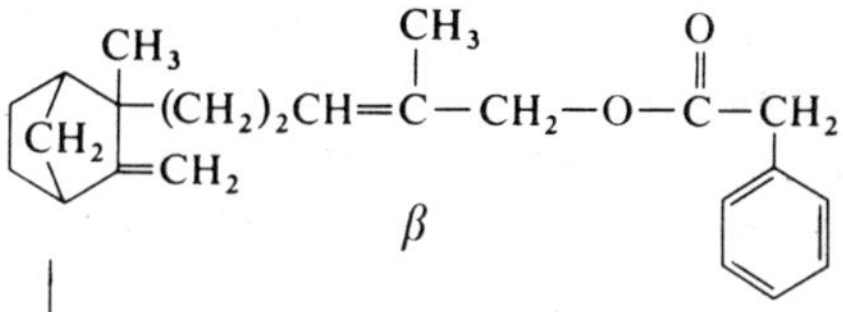 CH_3, CH_2, $-(CH_2)_2CH=C(CH_3)-CH_2-O-C(=O)-CH_2$ (phenyl), $-CH_3$ α CH_3, CH_2, $-(CH_2)_2CH=C(CH_3)-CH_2-O-C(=O)-CH_2$ (phenyl), $=CH_2$ β
Physical/chemical characteristics[5]		
Appearance	Colorless to yellowish liquid	Colorless liquid
Assay	95% min	
Molecular weight	262.40	338.49
Boiling point	Approx 20.8°C at 3 mm Hg	
Specific gravity	0.9884–0.9900 at 15°C; 0.981–0.985 at 25°C	1.024–1.026 at 25°/25°C
Refractive index	1.4894–1.4901 (1.4880–1.4910) at 20°C	
Optical rotation	Approx −16°	
Acid value		1 max
Solubility	1:1 in 90% alcohol; 1:9–11.5 in 80% alcohol; soluble in most common organic solvents	Slightly soluble in alcohol; soluble in benzyl benzoate and in diethyl phthalate
Organoleptic characteristics	Characteristic, sandalwood-like odor; pleasant bittersweet taste with an apricot-like undertone	Sandalwood-like odor; characteristic flavor with a fruity and honey-like undertone
Synthesis	By acetylation of santalol obtained from sandalwood oil that contains the two stereoisomer forms (α and β) of the alcohol. Therefore, the resulting ester is a mixture of the α- and β-isomers.	The commerical product consists of a mixture of the α- and β-isomers
Natural occurrence	Not reported found in nature	Not reported found in nature
Reported uses[3]	Non-alcoholic beverages 0.53 ppm Ice cream, ices, etc. 0.78 ppm Candy 2.0 ppm Baked goods 2.0 ppm Chewing gum 2.3 ppm	Non-alcoholic beverages 1.0 ppm Ice cream, ices, etc. 0.95 ppm Candy 2.0 ppm Baked goods 2.0 ppm
Regulatory status	FDA 121.1164; FEMA No. 3007	FDA 121.1164; FEMA No. 3008

REFERENCES

For References 1–5, see end of Part III.

	Skatole
Other names	3-Methylindole β-Methylindole
Empirical formula	C_9H_9N
Structure	(indole ring with CH_3 at position 3; N–H)
Physical/chemical characteristics[5]	
Appearance	White scales or powder (from pet ether); tends to darken on exposure to air
Molecular weight	121.18
Melting point	96°C (93–96°C)
Boiling point	265–266°C
Solubility	Soluble in boiling water; soluble in alcohol and other common organic solvents
Organoleptic characteristics	Characteristic fecal odor at high levels, becoming pleasant, sweet, and warm at very low levels; warm, over-ripe, fruity flavor below 0.1 ppm
Synthesis	Synthesized by Fischer and German[6] by heating glycerol and aniline in the presence of $ZnCl_2$; also by treating the phenylhydrazone of propionic aldehyde with a calculated amount of $ZnCl_2$ in an oil bath at 180°C;[7] or by fusing egg albumin with KOH
Natural occurrence	Reported as a constituent of feces,[8] nectandra wood, and civet
Reported uses[3]	Non-alcoholic beverages 0.75 ppm Ice cream, ices, etc. 1.0 ppm Candy 0.78 ppm Baked goods 0.80 ppm Gelatins and puddings 0.01 ppm Chewing gum 0.10 ppm
Regulatory status	FDA 121.1164; FEMA No. 3019

REFERENCES

For References 1–5, see end of Part III.

6. **Fischer and German,** *Ber. Dtsch. Chem. Ges.*, 16, 710, 1883.
7. **Fischer,** *Ber. Dtsch. Chem. Ges.*, 12, 1566, 1879.
8. **Brieger,** *Ber. Dtsch. Chem. Ges.*, 12, 1985, 1879.

Styrene

Other names	Styrol Phenylethylene Vinylbenzene
Empirical formula	C_8H_8
Structure	C_6H_5–$CH=CH_2$
Physical/chemical characteristics	
Appearance	Colorless to yellowish oily liquid
Molecular weight	104.14
Melting point	–33°C
Boiling point	145–146°C 48°C at 20 mm Hg
Flash point	38°C
Specific gravity	0.9090 at 20°C 0.9045 at 25°C
Refractive index	1.5463 at 20°C
Solubility	Soluble in alcohol, ether, acetone, and carbon disulfide; sparingly soluble in water
Organoleptic characteristics	Characteristic sweet, balsamic, almost floral odor, extremely penetrating
Synthesis	From ethylbenzene or from phenylethanol[6]
Natural occurrence	In cranberry and bilberry,[7] currants,[8-10] grape,[11,12] vinegar,[13] parsley,[14] milk and dairy products,[15] whisky,[16,17] cocoa,[18,19] coffee,[20-22] tea,[20] roasted filberts,[23] and roasted peanuts[24]
Reported uses[3]	Ice cream, ices, etc. 0.20 ppm Candy 0.20 ppm Baked goods 0.20 ppm
Regulatory status	FDA 121.1164; FEMA No. 3234

REFERENCES

For References 1–5, see end of Part III.

6. **Wilen et al.,** *J. Chem. Educ.,* 38, 304, 1961.
7. **Sydow et al.,** *Lebensm. Wiss. Technol.,* 2, 78, 1969; 3, 11, 1970.
8. **Andersson et al.,** *Acta Chem. Scand.,* 20, 522, 1966.
9. **Nursten et al.,** *Chem. Ind.* (London), p. 487, 1967; *J. Sci. Food Agric.,* 20, 613, 1969.
10. **Sydow et al.,** *Lebensm. Wiss. Technol.,* 4, 54, 1971.
11. **Stevens et al.,** *J. Agric. Food Chem.,* 14, 249, 1966.
12. **Stern et al.,** *J. Agric. Food Chem.,* 15, 1100, 1967.
13. **Kahn et al.,** *J. Agric. Food Chem.,* 20, 214, 1972.
14. **Kasting et al.,** *Phytochemistry,* 11, 2277, 1972.
15. **Siek et al.,** *J. Dairy Sci.,* 51, 1887, 1968.
16. **Kahn et al.,** *J. Food Sci.,* 33, 395, 1968; *J. Assoc. Of. Agric. Chem.,* 52, 1166, 1969; *J. Food Sci.,* 34, 587, 1969.
17. **Williams et al.,** *J. Sci. Food Agric.,* 23, 1, 1972.
18. **Van de Wal et al.,** *Recl. Trav. Chim. Pays-Bas,* 87, 238, 1968; *J. Agric. Food Chem.,* 19, 276, 1971.
19. **Rohan,** *Food Process. Mark.,* 38, 12, 1969.
20. **Heins et al.,** *J. Gas Chromatogr.,* 4, 395, 1966.
21. **Stoll et al.,** *Helv. Chim. Acta,* 50, 628, 1967.
22. **Goldmann et al.,** *Helv. Chim. Acta,* 50, 694, 1967.
23. **Walter et al.,** *Z. Ernährungswiss.,* 9, 123, 1969.
24. **Kinlin et al.,** *J. Agric. Food Chem.,* 20, 1021, 1972.

	Sucrose octaacetate	Tartaric acid
Other names	Octoacetyl sucrose	*l*-2,3-Dihydroxy butanedioic acid *d*-α-β-Dihydroxysuccinic acid *l*-Tartaric acid
Empirical formula	$C_{28}H_{38}O_{19}$	$C_4H_6O_6$
Structure	See below.	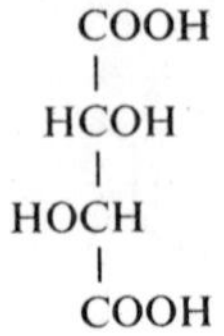 COOH–HCOH–HOCH–COOH

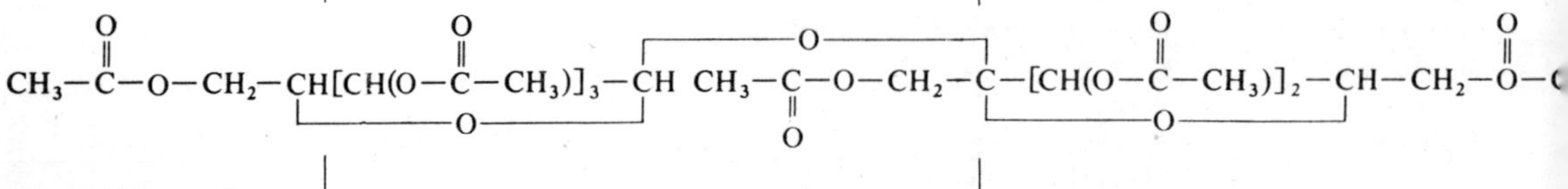

$$CH_3-\overset{O}{\overset{\|}{C}}-O-CH_2-CH[CH(O-\overset{O}{\overset{\|}{C}}-CH_3)]_3-CH\ \ CH_3-\underset{O}{\underset{\|}{C}}-O-CH_2-C-[CH(O-\overset{O}{\overset{\|}{C}}-CH_3)]_2-CH-CH_2-O-\overset{O}{\overset{\|}{C}}$$

	Sucrose octaacetate	Tartaric acid
Physical/chemical characteristics[5]		
Appearance	Needles from alcohol	Colorless crystals
Molecular weight	678.61	150.09
Melting point	84°C	*d*, 168–170°C; *dl*, 206°C; *meso*-, 165–166°C
Specific gravity		*d*, 1.7598 at 20°C; *dl*, 1.697 at 20°C
Optical rotation	+56.6° at 20°C (in chloroform)	*d*, +11.98°C
Solubility	0.1% in water; 9% in alcohol	Soluble in water and alcohol; slightly soluble in chloroform and ether
Organoleptic characteristics	Odorless; intensely bitter flavor influenced by food acids	Odorless; characteristic acid taste
Synthesis	By heating sucrose, acetamide, and sodium acetate;[7] or by acetylation of sucrose using acetic anhydride, zinc chloride, and acetic acid	*dl*-Tartaric acid can be produced by oxidation of fumaric and maleic acids; also from glyoxal and hydrocyanic acid, followed by acid hydrolysis
Reported uses	Non-alcoholic beverages, such as ginger ale, "quinine bitters," and others: 0.35–20 ppm	*d*-Tartaric acid occurs in many fruits or other parts of plants, free or combined with potassium, calcium, or magnesium. Produced from argols or wine-lees, which are formed in the manufacture of wine by extracting the potassium acid tartrate, transforming this into the calcium salt, and then acidifying with dilute sulfuric acid. Also by oxidation of *d*-glucose with nitric acid. The *dl*-tartaric acid is obtained by boiling the *d*-tartaric acid with an aqueous solution of NaOH or by oxidation of fumaric acid. The *l* and the *meso*-tartaric acid are also known but are less important. Non-alcoholic beverages 960 ppm Ice cream, ices, etc. 570 ppm Candy 5,400 ppm Baked goods 1,300 ppm Gelatins and puddings 60 ppm Chewing gum 3,700 ppm Condiments 10,000 ppm
Regulatory status	FDA 121.1164; FEMA No. 3038	FDA GRAS (general); FEMA No. 3044

REFERENCES

For References 1–5, see end of Part III.

6. **Herzfeld,** *Chem. Zentralbl.*, p. 749, 1887.

	α-Terpinene	α-Terpineol
Other names	*p*-Menthadiene-1,3 1,3-*p*-Menthadiene 1-Methyl-4-isopropyl-1,3-cyclohexadiene 1-Methyl-4-isopropylcyclohexadiene-1,3	1,4-*p*-Menthadiene 1-Methyl-4-isopropyl-1,4-cyclohexadiene
Empirical formula	$C_{10}H_{16}$	$C_{10}H_{16}$
Structure	CH_3 H_3C CH_3	CH_3 H_3C CH_3
Physical/chemical characteristics[5]		
Appearance	Colorless, oily liquid	Colorless, oily liquid
Molecular weight	136.23	136.23
Boiling point	173.5–174.8°C (176°C)	183°C
Specific gravity	0.8375 at 19.6°C	0.849 at 20°C
Refractive index	1.477 at 19.7°C	1.4765 at 14.5°C
Solubility	Miscible with alcohol and ether; insoluble in water	Insoluble in water; soluble in alcohol
Organoleptic characteristics	Characteristic lemon odor, lemony flavor, becoming bitter at high levels	Characteristic lemon odor (not as lemony as the α-isomer, but warmer); slightly bitter, herbaceous, citrus-like flavor
Synthesis	By isolation from the terpene fraction of sweet orange oil or orange terpenes (8–10% of the total monoterpenes); by isolation from fractions of American turpentine oil; from 1-methyl-4-isopropylcyclohexadien-1,3-one-2; also from terpinene dihydrochloride with aniline	Isolated from the essential oil of *Lantana camara;* also from *p*-cymene by the action of sodium and alcohol in liquid ammonia
Natural occurrence	Originally reported found in cardamom essential oil; also in orange, coriander, wormseed, *Eucalyptus australiana*, and in approximately 20 additional essential oils.	Reported found in several essential oils: *Mosla japonica, Eucalyptus dives. Orthodon perforatum* and other *Orthodon* species, lemon, pine, and others, often in mixture with the α-isomer; primarily in *Lantana camara*
Regulatory status	FDA 121.1164	FDA 121.1164

REFERENCES

For References 1–5, see end of Part III.

For further information on γ-terpinene see the following.

Marot, *Bull. Inst. Pin,* 38, 61, 1933.

Richter and Wolff, *Ber. Dtsch. Chem. Ges.*, 60, 477, 1927; 63, 1714, 1930.

α-Terpineol

Other names	*p*-Menth-1-en-8-ol 1-Methyl-4-isopropyl-1-cyclohexen-8-ol
Empirical formula	$C_{10}H_{18}O$
Structure	CH_3 / OH / H_3C / CH_3
Physical/chemical characteristics[5]	
Appearance	Colorless liquid, viscous at room temperature; the pure α-isomer is a white, crystalline powder
Assay	Commercial terpineol is a mixture of α- and β-isomers with possibly some γ- and other alcohols.[2] A good grade of α-terpineol contains about 3–5% β-isomer
Molecular weight	154.24
Melting point	*d*, 40–41°C; *l*, 37–38°C; *dl*, 35°C
Boiling point	214–224°C;[2] *d*, 115–116°C at 25 mm Hg; *l*, 219–221°C; *dl*, 218–219°C
Congealing point	+2°C[2]
Flash point	91°C
Specific gravity	0.930–0.936 at 25°/25°C;[2] *d*, 0.940 at 15°C; *dl*, 0.935–0.940 at 15°C
Refractive index	1.4825–1.4850 at 20°C;[2] *d*, 1.48322 at 18°C; *dl*, 1.48131 at 20°C; *dl*, 1.48084 at 20°C
Optical rotation	−0°10′ to +0°10′;[2] *d*, +95°9′; *l*, −106°
Solubility	1:8 or more in 50% alcohol;[2] soluble in propylene glycol; very slightly soluble in water
Organoleptic characteristics	Characteristic lilac odor; sweet taste reminiscent of peach on dilution
Synthesis	Obtained from terpin hydrate by splitting off water;[2] from pentane tricarboxylic acid by cyclization, followed by esterification to the hydroxy ester, then the unsaturated ester and Grignard to terpineol; also from isoprene and methyl vinyl ketone, using methyl magnesium iodide
Natural occurrence	Reported found in more than 150 derivatives from leaves, herbs, and flowers. The *d*-, *l*-, and *dl*-isomers are known. The *d*-form is found in the essential oils from *Cupressaceae* in general; also in the oils of *Elettaria cardamomum*, star anise, marjoram, clary sage, neroli, and others. The *l*-form is found in *Satureia montana*, lavandin, cajeput, lime, lemon, cinnamon leaves, and the distillates from *Pinaceae* (with the exception of *Pinus silvestris*, which contains *d*-terpineol together with the racemic form); likewise, *Nectandra elaiophora* (wood) and petitgrain bigarade. The racemic form is found in cajenne linalool, *Thymus caespititius*, cajeput, *Eucalyptus globulus*; mixed with the *l*-form it is found in petitgrain.[6] A non-defined form of terpineol has been reported in bitter orange.[7]
Reported uses[3]	Non-alcoholic beverages 5.4 ppm Ice cream, ices, etc. 16 ppm Candy 14 ppm Baked goods 19 ppm Gelatins and puddings 12–16 ppm Chewing gum 40 ppm Condiments 38 ppm
Regulatory status	FDA 121.1164; FEMA No. 3045

REFERENCES

For References 1–5, see end of Part III.
6. Peyron, *Fr. Ses Parfums,* 6(1), 17, 1958.
7. Mehlitz and Minas, *Ind. Obst Gemueseverwert.*, 50, 861, 1965.

	β-Terpineol	Terpinolene
Other names	*p*-Menth-8-en-1-ol Δ-8,9-*p*-Menthenol-1 1-Methyl-4-isopropenylcyclohexan-1-ol *Note*: the *trans*-β-form is the most common.	*p*-Menth-1,4(8)-diene 1-Methyl-4-isopropylidene-1-cyclohexene 1,4(8)-Terpadiene
Empirical formula	$C_{10}H_{18}O$	$C_{10}H_{16}$
Structure	H_3C, OH; H_3C, CH_2	CH_3; H_3C, CH_3
Physical/chemical characteristics[5]		
Appearance	Colorless liquid	Colorless to very pale, straw-yellow, oily liquid; polymerizes readily
Molecular weight	154.24	136.23
Melting point	*cis*-, 36°C; *trans*-, 32–33°C	
Boiling point	*cis*-, 78°C at 1.5 mm Hg; *trans*-, 90°C at 10 mm Hg; 210°C	183–185°C
Specific gravity	*cis*-, 0.9258 at 20°C; *trans*-, 0.923 at 15°C	0.8620 at 20°C
Refractive index	*cis*-, 1.4793 at 20°C; *trans*-, 1.4747 at 20°C	1.4900 at 20°C
Solubility	Very slightly soluble in water; soluble in alcohol	Insoluble in water; soluble in alcohol
Organoleptic characteristics	Pungent or earthy, woody odor	Sweet, pine odor; somewhat sweet, citrus flavor
Synthesis	From steam-distilled fractions of pine oil; from foreruns of the rectification of terpineol obtained from α-pinene and terpin hydrate; by reduction of limonene epoxide	The pure product was isolated and structurally identified by Baeyer;[6] by alcoholic H_2SO_4 treatment of pinene
Natural occurrence	In strawberry, raspberry, and blackberry	Reported found as a constituent of a few essential oils in small amounts: *Manilla elemi*, a few pine and fir varieties, *Nectandra elaiophora*, *Dacrydium colensoi*, and a few others
Reported uses[3]		Non-alcoholic beverages 16 ppm Ice cream, ices, etc. 64 ppm Candy 0.12–48 ppm Baked goods 49 ppm
Regulatory status	FDA 121.1164	FDA 121.1164; FEMA No. 3046

REFERENCES

For References 1–5, see end of Part III.
6. **Baeyer,** *Ber. Dtsch. Chem. Ges.*, 27, 443, 1894.

	Terpinyl acetate	Terpinyl anthranilate
Other names	*p*-Menth-1-en-8-yl acetate	*p*-Menth-1-en-8-yl anthranilate Terpinyl-2-aminobenzoate Terpinyl-*o*-aminobenzoate
Empirical formula	$C_{12}H_{20}O_2$	$C_{17}H_{23}NO_2$
Structure	α, β, γ	
Physical/chemical characteristics[5]		
Appearance	Clear, colorless liquid	
Assay	97% min[2]	
Molecular weight	196.29	273.38
Boiling point	α, 105°C at 11 mm Hg; α, *dl*-, 220°C	365°C
Flash point	α, *dl*-, 85°C; α, 100°C	
Specific gravity	0.953–0.962 at 25°/25°C;[2] α, *dl*-, 0.9704 at 15°C	1.055 at 15.5°C
Refractive index	1.4640–1.4670 at 20°C;[2] α, *dl*-, 1.4689 at 21°C	1.4830 at 20°C
Optical rotation	+9° to −9°[2]	
Solubility	1:5 in 70% alcohol;[2] soluble in essential oils and most common organic solvents	
Organoleptic characteristics	Lavender- and bergamot-like character; characteristic flavor—first slightly pungent, then slightly sweet—similar to currant	Complex fruity odor and bitter taste
Synthesis	Usually prepared by acetylation of α-terpineol or mixed isomeric terpineols, using acetic anhydride and anhydrous sodium acetate. The first synthesis dates back to 1888.[6] Another synthetic route starts with 1,8-cineol.[7] α-Terpinyl acetate is also the starting material for the synthesis of carvone.[8]	By esterification of anthranilic acid with terpineol[10]
Natural occurrence	Reported in the essential oils of cypress, Malabar cardamom, cajeput, niaouli, Siberian pine needles, pine, *Melaleuca trichostachya*, *Melaleuca pauciflora*, and others; also identified in the essential oil of bitter orange.[9]	Not reported found in nature
Reported uses[3]	Non-alcoholic beverages 3.5 ppm Ice cream, ices, etc. 3.2 ppm Candy 9.9 ppm Baked goods 1.5 ppm Chewing gum 14–260 ppm Condiments 15 ppm Meats 1.7–40 ppm	Non-alcoholic beverages 1.1 ppm Ice cream, ices, etc. 1.5–2.6 ppm Candy 6.3 ppm Baked goods 6.0 ppm
Regulatory status	FDA 121.1164; FEMA No. 3047	FDA 121.1164; FEMA No. 3048

REFERENCES

For References 1–5, see end of Part III.

6. **Lafont,** *Ann. Chim. Paris,* 15, 153, 1888.
7. **Matsuura et al.,** *Militizer Ber.,* p. 34, 1956.
8. **Suga,** *Bull. Chem. Soc. Jap.,* 7, 569, 1958.
9. **Mehlitz and Minas,** *Ind. Obst Gemueseverwert.,* 50, 861, 1965.
10. **Wagner,** *Chem. Ztg.,* 51, 855, 1927.

	Terpinyl butyrate	Terpinyl cinnamate
Other names	*p*-Menth-1-en-8-yl butyrate	*p*-Menth-1-en-8-yl cinnamate Terpinyl β-phenacrylate Terpinyl-3-phenyl propenoate
Empirical formula	$C_{14}H_{24}O_2$	$C_{19}H_{24}O_2$
Structure	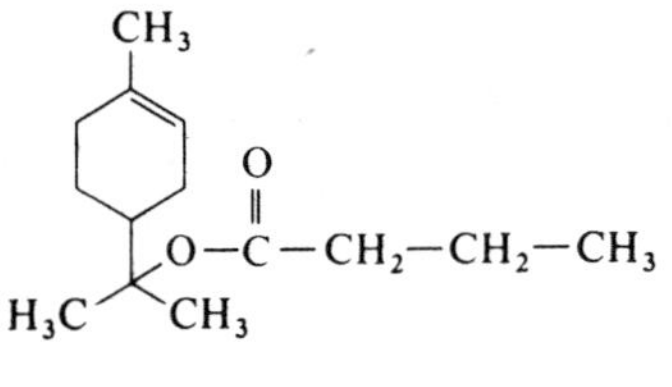 CH_3; O; $O-C-CH_2-CH_2-CH_3$; H_3C CH_3	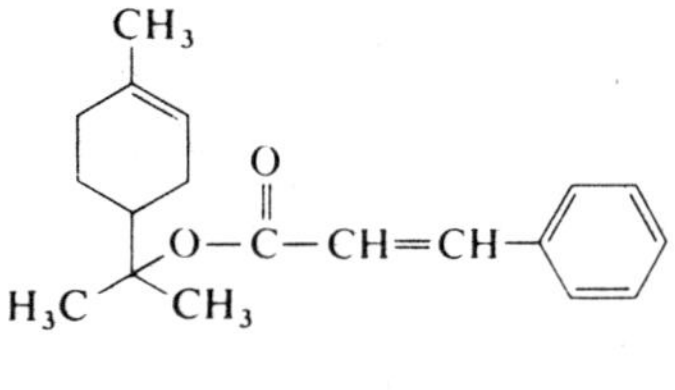 CH_3; O; $O-C-CH=CH-$; H_3C CH_3
Physical/chemical characteristics[5]		
Appearance		Colorless to straw-yellow, viscous liquid
Molecular weight	224.34	284.40
Boiling point	244°C	360°C
Specific gravity	0.965 at 15.5°C	1.030 at 15.5°C
Refractive index	1.4656 at 20°C	1.5500 at 20°C
Solubility		Insoluble in water; soluble in alcohol
Organoleptic characteristics	Characteristic, rosemary-like, acrid, fruity odor with a balsamic undertone; bitter-sweet taste reminiscent of plum	Persistent, sweet, balsamic odor reminiscent of wine moscato; heavy, fruity, somewhat spicy taste
Synthesis		By esterification from terpineol and cinnamic acid; also from methyl cinnamate, sodium terpineol, and terpinyl formate
Natural occurrence	Reported found in the essential oils of niaouli, cypress, and *Eucalyptus diversicolor*	Not reported found in nature
Reported uses[3]	Non-alcoholic beverages 6.4 ppm Ice cream, ices, etc. 9.2 ppm Candy 11 ppm Baked goods 9.5 ppm Chewing gum 210 ppm	Non-alcoholic beverages 0.50 ppm Ice cream, ices, etc. 2.6 ppm Candy 6.0 ppm Baked goods 6.0 ppm
Regulatory status	FEMA No. 3049	FDA 121.1164; FEMA No. 3051

REFERENCES

For References 1–5, see end of Part III.

	Terpinyl formate	Terpinyl isobutyrate
Other names	*p*-Menth-1-en-8-yl formate	*p*-Menth-1-en-8-yl isobutyrate
Empirical formula	$C_{11}H_{18}O_2$	$C_{14}H_{24}O_2$
Structure	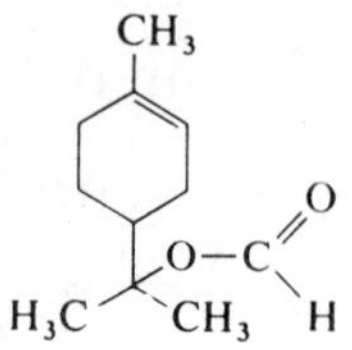 CH_3; $O-C(=O)H$; H_3C CH_3	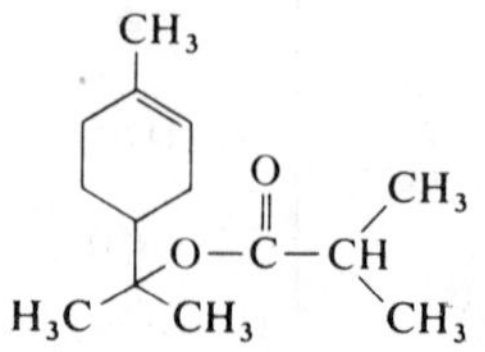 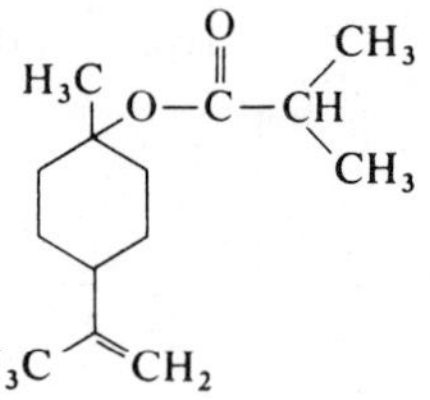 CH_3; $O-C(=O)-CH(CH_3)_2$; H_3C CH_3; α H_3C $O-C(=O)-CH(CH_3)_2$; H_3C CH_2; β
Physical/chemical characteristics[5]		
Appearance	Colorless liquid	Colorless, oily liquid
Molecular weight	182.26	224.34
Boiling point	*d*-α, 133–136°C; *l*-α, 135–138°C;[8] 225°C (commercial product)	Approx 242°C
Specific gravity	*d*-α, 0.971 at 36°C; *l*-α, 0.9986;[8] 0.990 at 15.5°C (commercial product)	0.93
Refractive index	*d*-α, 1.4712 at 20°C; 1.4700 at 20°C (commercial product)	
Optical rotation	*d*-α, undiluted, +88.32°; *l*-α, −69.25°[8]	
Solubility	1:6 in 60% alcohol; very slightly soluble in water	Almost insoluble in water; soluble in alcohol
Organoleptic characteristics	Pleasant, floral, citrus odor; bitter taste at high levels, becoming dry and fruity at lower levels	Fruity, floral, herbaceous odor; heavy, fruity-type flavor
Synthesis	From *d*-α-terpineol and formic acetic anhydride;[6] the racemic α-terpineol is also obtained by an analogous synthesis.[7]	By esterification of terpineol with isobutyric acid
Natural occurrence	Probably occurring in Ceylon cardamom essential oil	Not reported found in nature
Reported uses[3]	Non-alcoholic beverages 0.50–3.0 ppm Alcoholic beverages 1.0 ppm Ice cream, ices, etc. 2.6–5.0 ppm Candy 6.0–10 ppm Baked goods 6.0–10 ppm	Non-alcoholic beverages 0.90–2.4 ppm Ice cream, ices, etc. 5.0 ppm Candy 4.0–15 ppm Baked goods 5.0–15 ppm
Regulatory status	FDA 121.1164; FEMA No. 3052	FDA 121.1164; FEMA No. 3050

REFERENCES

For References 1–5, see end of Part III.

6. **Fuller and Kenyon,** *J. Chem. Soc.* (London), 125, 2306, 1924.
7. **Robert,** *Chem. Zentralbl.,* 2, 2295, 1927.
8. **Lafont,** *Bull. Soc. Chim. Fr.,* 2(49), 325, 1888; *Ann. Chim. Paris,* 6(15), 185, 1888.

	Terpinyl isovalerate	Terpinyl propionate
Other names	*p*-Menth-1-en-8-yl isovalerate *p*-Menth-1-en-8-yl-β-methylbutyrate	*p*-Menth-1-en-8-yl propionate
Empirical formula	$C_{15}H_{26}O_2$	$C_{13}H_{22}O_2$
Structure	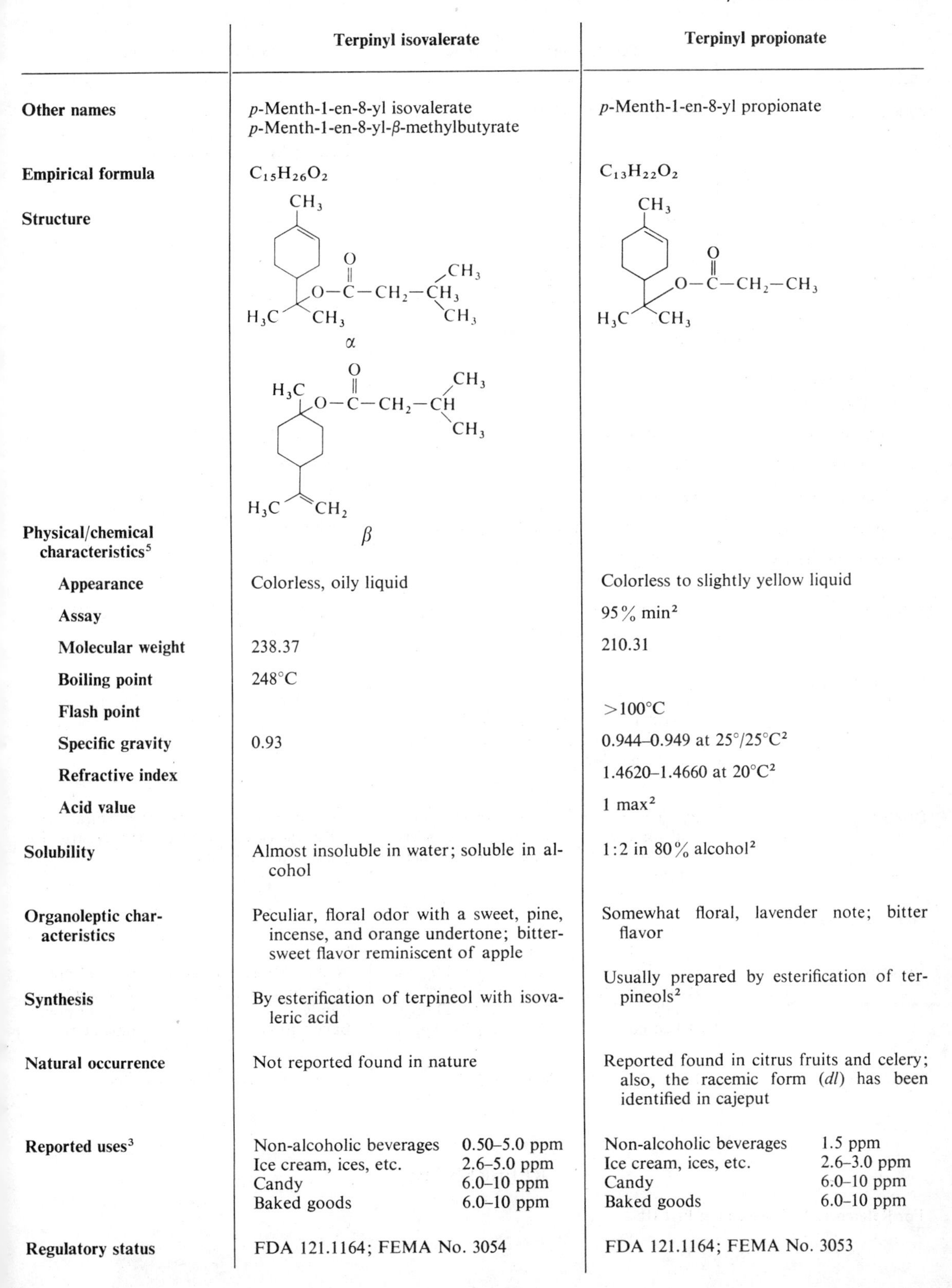	
Physical/chemical characteristics[5]		
Appearance	Colorless, oily liquid	Colorless to slightly yellow liquid
Assay		95% min[2]
Molecular weight	238.37	210.31
Boiling point	248°C	
Flash point		>100°C
Specific gravity	0.93	0.944–0.949 at 25°/25°C[2]
Refractive index		1.4620–1.4660 at 20°C[2]
Acid value		1 max[2]
Solubility	Almost insoluble in water; soluble in alcohol	1:2 in 80% alcohol[2]
Organoleptic characteristics	Peculiar, floral odor with a sweet, pine, incense, and orange undertone; bittersweet flavor reminiscent of apple	Somewhat floral, lavender note; bitter flavor
Synthesis	By esterification of terpineol with isovaleric acid	Usually prepared by esterification of terpineols[2]
Natural occurrence	Not reported found in nature	Reported found in citrus fruits and celery; also, the racemic form (*dl*) has been identified in cajeput
Reported uses[3]	Non-alcoholic beverages 0.50–5.0 ppm Ice cream, ices, etc. 2.6–5.0 ppm Candy 6.0–10 ppm Baked goods 6.0–10 ppm	Non-alcoholic beverages 1.5 ppm Ice cream, ices, etc. 2.6–3.0 ppm Candy 6.0–10 ppm Baked goods 6.0–10 ppm
Regulatory status	FDA 121.1164; FEMA No. 3054	FDA 121.1164; FEMA No. 3053

REFERENCES

For References 1–5, see end of Part III.

	Tetrahydrofurfuryl acetate	Tetrahydrofurfuryl alcohol
Other names	Tetrahydro-2-furyl methylacetate	Tetrahydro-2-furancarbinol Tetrahydro-2-furanmethanol Tetrahydro-2-furylmethanol THFA
Empirical formula	$C_7H_{12}O_3$	$C_5H_{10}O_2$
Structure	(tetrahydrofuran ring)$-CH_2-O-\overset{O}{\overset{\|\|}{C}}-CH_3$	(tetrahydrofuran ring)$-CH_2OH$
Physical/chemical characteristics[5]		
Appearance	Colorless liquid	Colorless liquid
Molecular weight	144.17	102.13
Boiling point	194–196°C	178–179°C
Specific gravity	1.061 at 20°/20°C (1.005 at 15.5°C)	1.0495 at 20°C
Refractive index	1.4475 at 20°C	1.4505 at 20°C
Solubility	Completely miscible with water, alcohol, ether, and chloroform	Completely miscible with water, alcohol, and ether
Organoleptic characteristics	Faint, fruity, ethereal odor, somewhat similar to acetic acid (reminiscent of methyl formate); sweet, ethereal, deep, fruit-like flavor	Faint, warm, oily, caramellic odor; coffee and nut-like flavor at very low levels (0.03–1 ppm)
Synthesis	By acetylation of the corresponding alcohol with acetic anhydride, acetyl chloride, or acetic acid and mineral acids	By catalytic reduction of furfural with Raney-Nickel; also by the destructive hydrogenation of lignin
Natural occurrence		
Reported uses[3]	Non-alcoholic beverages 1.3–2.0 ppm Ice cream, ices, etc. 8.0 ppm Candy 1.0–20 ppm Baked goods 1.0–20 ppm	Non-alcoholic beverages 0.03–14 ppm Ice cream, ices, etc. 0.03 ppm Candy 0.03–18 ppm Baked goods 0.04 ppm
Regulatory status	FDA 121.1164; FEMA No. 3055	FDA 121.1164; FEMA No. 3056

REFERENCES

For References 1–5, see end of Part III.

	Tetrahydrofurfuryl butyrate	Tetrahydrofurfuryl cinnamate
Other names	Tetrahydrofurfuryl-*n*-butyrate Tetrahydro-2-furylmethyl-*n*-butanoate	Cinnamic acid, tetrahydrofurfuryl ester
Empirical formula	$C_9H_{16}O_3$	$C_{14}H_{16}O_3$
Structure	(tetrahydrofuryl)$-CH_2-O-C(=O)-CH_2-CH_2-CH_3$	$C_6H_5-CH=CH-C(=O)-O-CH_2-$(tetrahydrofuryl)
Physical/chemical characteristics[5]		
Appearance	Colorless liquid	Colorless, slightly viscous liquid
Molecular weight	172.23	232.28
Boiling point	227°C	>300°C
Specific gravity	1.01	1.11
Solubility	Insoluble in water; soluble in alcohol	Insoluble in water; soluble in alcohol and oils
Organoleptic characteristics	Heavy, sweet odor reminiscent of apricot and pineapple; sweet, fruity flavor at low levels, becoming pungent at higher levels	Sweet, persistent, balmy-vinous odor
Synthesis	By esterification of the alcohol with *n*-butyric acid	By azeotropic esterification of the corresponding acid
Natural occurrence	Not reported found in nature	
Reported uses[3]	Non-alcoholic beverages 0.90 ppm Ice cream, ices, etc. 6.0 ppm Candy 15 ppm Baked goods 15 ppm	Non-alcoholic beverages 20.0 ppm Ice cream, ices, etc. 20.0 ppm Baked goods 20.0 ppm
Regulatory status	FDA 121.1164; FEMA No. 3057	FEMA No. 3320

REFERENCES

For References 1–5, see end of Part III.

	Tetrahydrofurfuryl propionate	Tetrahydrolinalool
Other names	Tetrahydro-2-furylmethylpropionate	3,7-Dimethyloctan-3-ol 3,7-Dimethyloctanol-3
Empirical formula	$C_8H_{14}O_3$	$C_{10}H_{22}O$
Structure	(tetrahydrofuran ring)$-CH_2-O-C(=O)-CH_2-CH_3$	H_3C, OH; H_3C, CH_3
Physical/chemical characteristics[5]		
Appearance	Colorless liquid	Colorless, oily liquid
Assay		98% min
Molecular weight	158.20	158.29
Boiling point	64–68°C (207°) at 3 mm Hg	87–88°C at 10 mm Hg; *l*, 89°C at 15 mm Hg
Flash point		Approx. 84°C
Specific gravity	1.04	0.825–0.830 at 25°/25°C; 0.8280; *l*, 0.8327 at 16°C
Refractive index	1.4370 at 23°C	1.4320–1.4360 (1.4335) at 20°C
Optical rotation		±1; *l*, −0.78°
Solubility	Slightly soluble in water; soluble in alcohol	Very slightly soluble in water; soluble in alcohols and other organic solvents; 1:2 in 70% ethanol
Organoleptic characteristics	Sweet, caramellic, fruity odor; sweet flavor, apricot and chocolate-like Synthetically prepared by esterification of the corresponding alcohol with acetic anhydride and melted sodium acetate	Sweet, oily, floral odor (more so than linalool); citrus, floral taste
Synthesis		By hydrogenation of *dl*-linalool in the presence of platinum black according to Barbier and Lacquin. Also from magnesium ethyl bromide and isoamylketone, or by hydrogenation of 2,6-dimethyl-2-octen-6-ol in the presence of nickel at 100°C. Optically active and racemic forms are expected because of the structure of this product
Natural occurrence	Not reported found in nature	Not reported found among the natural constituents of essential oils
Reported uses[3]	Non-alcoholic beverages 1.3–2.0 ppm Ice cream, ices, etc. 8.0 ppm Candy 1.0–20 ppm Baked goods 1.0–20 ppm	Non-alcoholic beverages 1.3 ppm Ice cream, ices, etc. 2.7 ppm Candy 5.6 ppm Baked goods 5.6 ppm
Regulatory status	FDA 121.1164; FEMA No. 3058	FDA 121.1164; FEMA No. 3060

REFERENCES

For References 1–5, see end of Part III.

	Tetrahydro-pseudoionone	5,6,7,8-Tetrahydroquinoxaline
Other names	Dihydrogeranyl acetone 6,10-Dimethyl-9-undecen-2-one	Cyclohexapyrazine Tetrahydroquinoxaline
Empirical formula	$C_{13}H_{24}O$	$C_8H_{10}N_2$
Structure		
Physical/chemical characteristics		
Appearance	Pale straw-yellow or colorless, oily liquid	
Assay		C: 71.61%; H: 7.51%; N: 20.88%
Molecular weight	196.33	134.18
Melting point		29–30°C
Boiling point	126–127°C (234°C) at 14 mm Hg	108–111°C
Mass spectra[8]		134(100), 133(46), 106(18), 52(17), 39(14), 119(14), 41(11)
IR spectra[8]		3040, 3000, 1475, 1445, 1405, 1165, 1120, 1070, 850, 755 cm^{-1}
UV spectra[8]		λ_{max} (MeOH): 273 mm, 278 nm (ϵ = 9.2 × 10^{-3}), 302 nm
NMR spectra[6]		($CDCl_3$) δ 7.6 (5,2, pyrazinyl–H) 2.4 (t,4, $CH_2-CH_2-CH_2-CH_2$) 1.4 (4,$-CH_2-CH_2-CH_2-CH_2$)
Specific gravity	0.87	1.1334 at 48°C
Refractive index		1.6231 at 48°C
Solubility	Almost insoluble in water; soluble in alcohol	
Organoleptic characteristics	Sweet, floral, balsamic odor with a rose to woody quality; sweet, fruity flavor, somewhat apple- and peach-like	
Synthesis	By hydrogenation of pseudoionone with colloidal palladium; this reaction leads to a product with considerable mixture. Also from linalool plus acetic anhydride and sodium ethoxide, followed by hydrogenation. Another method starts with geraniol via the acid chloride.	Condensation of an alicyclic α,β-diketone with an α,β-diamine;[8] condensation of an alicyclic α,β-diamine with an α,β-dicarbonyl[8]
Natural occurrence		Reported present in filberts and peanuts[7]
Reported uses[3]	Non-alcoholic beverages 0.50 ppm Ice cream, ices, etc. 0.60–2.4 ppm Candy 14 ppm Baked goods 14 ppm	Non-alcoholic beverages 5.0 ppm Candy 5.0 ppm Gelatins and puddings 5.0 ppm Baked goods 2.0 ppm Condiments 2.0 ppm Meat, meat sauces, soups 2.0 ppm Milk, dairy products 1.0 ppm Cereals 2.0 ppm
Regulatory status	FDA 121.1164; FEMA No. 3059	FEMA No. 3321

REFERENCES

For References 1–5, see end of Part III.

6. **Walradt et al.,** *J. Agric. Food Chem.*, 19, 972, 1971.
7. **Maga and Sizer,** *Crit. Rev. Food Technol.*, 4(1), 39, 1973.
8. **Pittet et al.,** *J. Agric. Food Chem.*, 22(2), 273, 1974.

	Tetramethyl ethylcyclohexenone	**2,3,5,6-Tetramethylpyrazine**
Other names	Mixture of 5-ethyl-2,3,4,5-tetramethyl-2-cyclohexen-1-one and 5-ethyl-3,4,5-6-tetramethyl-2-cyclohexen-1-one	Tetramethylpyrazine
Empirical formula	$C_{12}H_{20}O$	$C_8H_{12}N_2$
Structure	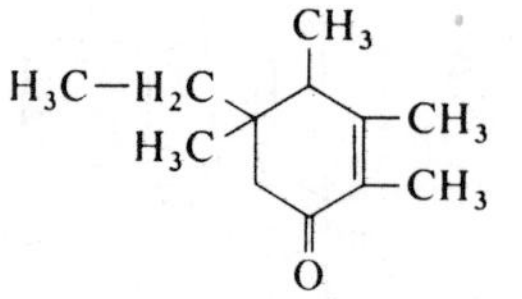 5-Ethyl-2,3,4,5-tetramethyl-2-cyclohexen-1-one; 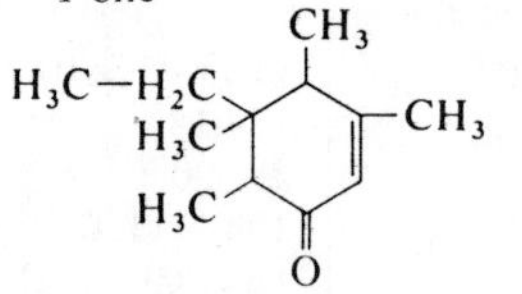 5-Ethyl-3,4,5,6-tetramethyl-2-cyclohexen-1-one	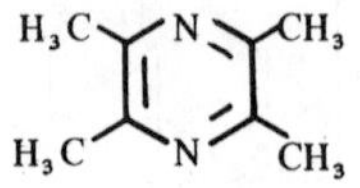
Physical/chemical characteristics		
Appearance	Almost colorless, oily liquid	
Assay	Approx. 30% 5-ethyl-2,3,4,5 and 70% 5-ethyl-3,4,5,6	C: 70.55%; H: 8.88%; N: 20.57%
Molecular weight	180.29	136.20
Mass spectra[6]		39(20), 42(37), 53(23), 54(100), 66(2), 68(2), 80(3), 94(3), 95(6), 121(3), 135(7), 136(85)
IR spectra[6]		7.06, 6.91, 10.08, 8.17, 3.42, 8.46, 8.32 μ
Solubility	Very slightly soluble in water; soluble in alcohol	
Organoleptic characteristics	Warm, caramellic, fruity, spicy odor; pleasant, creamy, caramellic, fruity flavor	Odor threshold: 10,000 ppb in H_2O[8]
Synthesis		From 2,5-dimethylpyrazine by ring alkylation with MeLi;[9] also by condensation of 2,3-butanedione with 2,3-butanediamine
Natural occurrence		Reported present in cocoa products, coffee, dairy products, galbanum oil, meats, peanuts, filberts, rum, whisky, and soy products[7]
Reported uses[3]	Non-alcoholic beverages 5.0 ppm Ice cream, ices, etc. 30 ppm Candy 30 ppm Baked goods 30 ppm	Non-alcoholic beverages 5.0 ppm Candy 5.0 ppm Baked goods 5.0 ppm Condiments 10.0 ppm Meat, meat sauces, soups 10.0 ppm Milk, dairy products 5.0 ppm
Regulatory status	FDA 121.1164; FEMA No. 3061	FEMA No. 3237

REFERENCES

For References 1–5, see end of Part III.

6. **Friedel et al.,** *J. Agric. Food Chem.*, 19, 530, 1971.
7. **Maga and Sizer,** *Crit. Rev. Food Technol.*, 4(1), 39, 1973.
8. **Koehler et al.,** *J. Food Sci.*, 30, 816, 1971.
9. **Rizzi,** *J. Org. Chem.*, 33(4), 1333, 1968.

	3,3′-Tetrathio-bis(2-methylfuran)		2-Thienyl disulfide	
Other names	Bis(2-methyl-3-furyl) tetrasulfide 2-Methyl-3-furyl tetrasulfide		2,2′-Dithiodithiophene	
Empirical formula			$C_8H_6S_4$	
Structure	[structure: two 2-methylfuran rings joined at C-3 by S–S–S–S; labels O, O, CH_3, HC_3]		[structure: two thiophene rings joined at C-2 by S–S; labels S, S]	
Physical/chemical characteristics				
Molecular weight	290.44		230.39	
Melting point			55–60°C	
Boiling point			115–45°C at 0.02 mm Hg	
Synthesis			By reduction of the corresponding sulfonyl chloride by red phosphorus in the presence of iodine as a catalyst;[6] by reduction of 2-thiocyano thiophene with $NaBH_4$ in methanol[7,8]	
Reported uses[3]	Baked goods	0.1 ppm	Baked goods	0.15 ppm
	Condiments, pickles	0.1 ppm	Condiments, pickles	0.15 ppm
	Meat, meat sauces, soups	0.1 ppm	Meat, meat sauces, soups	0.15 ppm
	Milk, dairy products	0.1 ppm	Cereals	0.15 ppm
	Cereals	0.1 ppm		
Regulatory status	FEMA No. 3260		FEMA No. 3323	

REFERENCES

For References 1–5, see end of Part III.
6. **Kiyoski and Kawahara,** *Yakugaki Zasshi,* 77, 963, 1957.
7. **Fedorov and Stoyanovich,** *Zh. Obshch. Khim.,* 32, 1518, 1962; *Chem. Abstr.,* 58, 4495h, 1963.
8. **Stoyanovich et al.,** *Izv. Akad. Nauk, SSSR Ser. Khim.,*(2), 387, 1969.

	2-Thienyl mercaptan	2,2′-(Thiodimethylene)-difuran
Other names	2-Mercapto-thiophene 2-Thienylthiol	2-Furfuryl monosulfide Difurfurylsulfide
Empirical formula	$C_4H_4S_2$	$C_{10}H_{10}O_2S$
Structure	(thiophene ring)–SH	(furan ring)–CH_2 — S — CH_2–(furan ring)
Physical/chemical characteristics		
Appearance	Oily, yellowish, or colorless liquid; tends to oxidize in air	
Molecular weight	116.21	194.25
Melting point		31–32°C
Boiling point	166°C	135–143°C at 14 mm Hg
Solubility	Very slightly soluble in water; soluble in alcohol	
Organoleptic characteristics	Very unpleasant, burnt caramellic and sulfuraceous odor; similar flavor	
Synthesis	By heating sodium sulfosuccinate with phosphorous trichloride;[6] also by reduction of thiophene-2-sulfonyl chloride	By reflexing a mixture of 2-furylmethanethiol in ethanolic KOH and furfuryl bromide in ether[7]
Natural occurrence	Not reported found in nature	In coffee[7-9]
Reported uses[3]	Candy 0.10 ppm Baked goods 0.10 ppm	Non-alcoholic beverages 1.0 ppm Ice cream, ices, etc. 1.0 ppm Candy 1.0 ppm Baked goods 1.0 ppm Gelatins and puddings 1.0 ppm Chewing gum 1.0 ppm Condiments 1.0 ppm Meat, meat sauces, soups 1.0 ppm
Regulatory status	FDA 121.1164; FEMA No. 3062	FEMA No. 3238

REFERENCES

For References 1–5, see end of Part III.

6. **Meyer and Neure,** *Ber. Dtsch. Chem. Ges.,* 20, 1756, 1887.
7. **Stoll et al.,** *Helv. Chim. Acta,* 50, 688, 1967.
8. **Gautschi et al.,** *J. Agric. Food Chem.,* 15, 15, 1967.
9. **Walter et al.,** *Z. Ernährungswiss.,* 9, 123, 1969.

	Thiolactic acid	Thujan-4-ol
Other names	2-Mercaptopropionic acid α-Thiolpropionic acid	Sabina hydrate Sabinene hydrate
Empirical formula	$C_3H_6O_2S$	$C_{10}H_{18}O$
Structure	$CH_3-CH(SH)-COOH$	(bicyclic structure: CH_3, OH; CH(CH_3)$_2$)
Physical/chemical characteristics[5]		
Appearance	Oily liquid	White crystals
Molecular weight	106	154.24
Melting point	10°C	*trans*-, 60–61°C *cis*-, 36.5–37.2°C
Boiling point	*d*-, 95–100°C at 16 mm Hg *dl*-, 98.5–99.0°C at 14 mm Hg	*trans*-, 193–198°C *cis*-, 195–200°C
Specific gravity	*l*-, 1.193 at 19.2°C	
Refractive index	*dl*-, 1.4823	*trans*-, 1.4430 at 60°C *cis*-, 1.4489 at 60°C *cis*-, 1.4632 at 25°C
Optical rotation	+38.32°C at 20°C –45.47	*trans*-, +47.2° at 25°C (natural source)
Solubility	Soluble in water, alcohol, and ether	
Organoleptic characteristics		Odor reminiscent of terpineol
Synthesis	By electrolysis of the corresponding sulfide, $S(SCHMeCO_2H)_2$[6]	The Grignard reaction with sabina ketone yields two isomers, of which one is identical to an alcohol isolated from American peppermint oil;[7] synthetic routes for the preparation of both the *cis*- and *trans*-form are also available[8]
Natural occurrence		Reported found in peppermint oil (*trans*-form)
Reported uses[3]	Meat, meat sauces, soups 50.0 ppm	Non-alcoholic beverages 10.0 ppm Ice cream, ices, etc. 10.0 ppm Candy 10.0 ppm Baked goods 10.0 ppm
Regulatory status	FEMA No. 3180	FEMA No. 3239

REFERENCES

For References 1–5, see end of Part III.
6. **Parsonn,** *Chem. Abstr.*, 22, 4470, 1928.
7. **Daly, Green and Eastman,** *J. Am. Chem. Soc.*, 80, 6330, 1958.
8. **Fanta and Eastman,** *J. Org. Chem.*, 33, 1656, 1968.

	Thymol	Tolualdehyde glyceryl acetal, mixed *o*-, *m*-, *p*-
Other names	3-*p*-Cymenol 3-Hydroxy-*p*-cymene *p*-Isopropyl-*m*-cresol 1-Methyl-3-hydroxy-4-isopropylbenzene 3-Methyl-6-isopropylphenol "Thyme camphor" *m*-Thymol	2-(*o*-,*m*-,*p*-Cresyl)-5-hydroxydioxan 2-(*o*-,*m*-,*p*-Cresyl)-5-hydroxymethyldioxolan 2-(Methylphenyl)-1,3-dioxan-5-ol, mixed *o*-, *m*-,*p*- "Tolyl glycerin"
Empirical formula	$C_{10}H_{14}O$	$C_{11}H_{14}O_3$
Structure	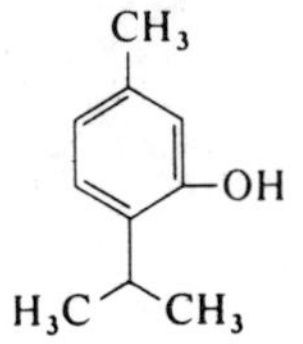 CH_3; OH; H_3C CH_3	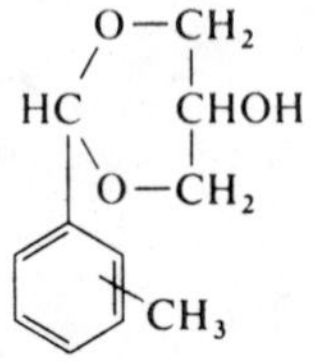 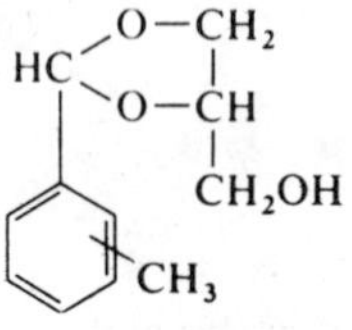 $O-CH_2$; HC; CHOH; $O-CH_2$; CH_3 — Approx. 40% HC; $O-CH_2$; $O-CH$; CH_2OH; CH_3 — Approx. 60%
Physical/chemical characteristics[5]		
Appearance	White crystals	Colorless, slightly viscous liquid
Assay		Commercial products are about 60% of the dioxolans and 40% dioxans.
Molecular weight	150.22	194.23
Melting point	48–51°C	
Boiling point	233°C	292°C
Specific gravity	0.972–0.979	1.14
Refractive index	1.523 at 20°C	
Solubility	Very slightly soluble in water (0.1%) and glycerol; soluble in alcohol, various organic solvents, and glacial acetic acid	Slightly soluble in water (decomposes slowly, especially under acidic conditions); soluble in alcohol
Organoleptic characteristics	Characteristic phenol-like, aromatic odor; sweet, medicinal, spicy flavor	Fresh material is almost odorless; mild, bitter-almond flavor
Synthesis	By action of potassium nitrite on thymyl amine in acetic acid solution and subsequent heating of the diazonium salt to boiling;[6] also by boiling 1 mole of 2,4-dibromomenthone with 6 moles of quinoline (Beckmann); from *m*-cresol and isopropyl chloride at −10°C according to Friedel-Crafts condensation.	From mixed tolylaldehydes and glycerol with phosphoric acid, finishing with an azeotropic distillation to remove water
Natural occurrence	Reported in the essential oils of *Monarda punctata*,[7] *Satureia thymera*,[8] *Origanum floribundum*, *Ocimum viride*, *Ocimum gratissimum*, and particularly in thyme (*Thymus vulgaris* L., *T. capitatus*, *T. serpillum* L.), where it is contained up to 50%	
Reported uses[3]	Non-alcoholic beverages 2.5–11 ppm Ice cream, ices, etc. 44 ppm Candy 9.4 ppm Baked goods 5.0–6.5 ppm Chewing gum 100 ppm	Non-alcoholic beverages 0.08–6.0 ppm Ice cream, ices, etc. 6.0–8.0 ppm Candy 12–15 ppm Baked goods 12–15 ppm
Regulatory status	FDA 121.1164; FEMA No. 3066	FDA 121.1164; FEMA No. 3067

REFERENCES

For References 1–5, see end of Part III.

6. **Widmann,** *Ber. Dtsch. Chem. Ges.*, 15, 170, 1882.
7. *Schimmel Ber.*, p. 20, September 1885.
8. *Schimmel Ber.*, pp. 55, October 1889.

	Tolualdehydes, mixed *o*-,*m*-,*p*-	*p*-Tolylacetaldehyde
Other names	Methyl benzaldehydes, mixed *o*-,*m*-,*p*- Tolyl aldehydes, mixed *o*-,*m*-,*p*-	*p*-Methyl phenylacetaldehyde Syringa aldehyde
Empirical formula	C_8H_8O	$C_9H_{10}O$
Structure	CHO, CH$_3$ (benzene ring)	CH$_3$ (benzene ring) CH$_2$—CHO
Physical/chemical characteristics[5]	The commercial product is a mixture of the three isomers	
Appearance	Colorless liquid	Colorless, oily liquid
Assay		95% min
Molecular weight	120.14	134.17
Boiling point	*m*-, 199°C; *o*-, 200°C; *p*-, 204°C	210°C
Flash point		70°C
Specific gravity	*m*-, 1.019 at 20°C; *o*-, 1.0386 at 19°C; *p*-, 1.015 at 15.5°C	1.010–1.016 at 25°C 1.5300–1.5350 at 20°C
Refractive index	*m*-, 1.5407 at 21.4°C; *o*-, 1.549 at 19°C; *p*-, 1.5469 at 16.6°C	
Solubility	Slightly soluble in water; soluble in alcohol and ether	1:1 in 80% alcohol; soluble in most common organic solvents; insoluble in alcohol
Organoleptic characteristics	Odor reminiscent of bitter almond	Characteristic odor reminiscent of bitter almond and corresponding flavor
Synthesis	By oxidation of *o*-, *m*-, or *p*-xylene (chemical or electrolytical oxidation)	From *p*-methylbenzaldehyde; also from *p*-methylbenzyl chloride and hexamethylene tetramine
Natural occurrence	Reported found in roasted nuts[6]	Not reported found in nature
Reported uses[3]	Non-alcoholic beverages 11 ppm Ice cream, ices, etc. 16 ppm Candy 25 ppm Baked goods 28 ppm Gelatins and puddings 8.3 ppm Chewing gum 430 ppm Maraschino cherries 100 ppm	Ice cream, ices, etc. 2.0 ppm Candy 0.03–2.0 ppm Baked goods 2.0 ppm
Regulatory status	FDA 121.1164; FEMA No. 3068	FDA 121.1164; FEMA No. 3071

REFERENCES

For References 1–5, see end of Part III.

6. *J. Agric. Food Sci.,* 20, 5, 1972.

	o-Tolyl acetate	*p*-Tolyl acetate
Other names	Acetyl *o*-cresol *o*-Cresyl acetate α-Cresylic acetate	Acetyl-*p*-cresol *p*-Cresyl acetate (true name) *Note*: this compound title is *not* true *p*-tolyl acetate.
Empirical formula	$C_9H_{10}O_2$	$C_9H_{10}O_2$
Structure	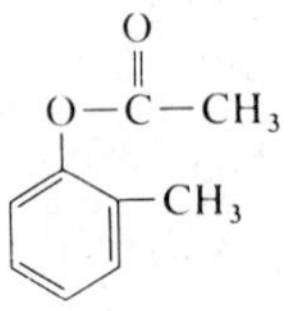	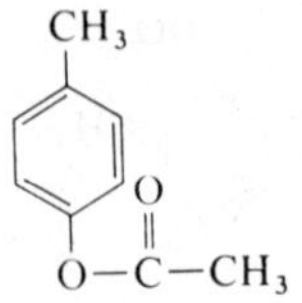
Physical/chemical characteristics[5]		
Appearance	Colorless liquid	Clear, colorless liquid
Assay		98% min[2]
Molecular weight	150.18	150.18
Boiling point	Approx 208°C (approx 87°C at 12 mm Hg)	208–209°C
Flash point		95°C
Specific gravity		1.044–1.050 at 25°/25°C;[2] 1.056 at 15.5°C; 1.0512 at 4°C
Refractive index		1.4990–1.5020[2] (1.5035) at 20°C
Acid value		1.0 max[2]
Free cresol		1% max[2]
Solubility	Almost insoluble in cold water; slightly soluble in hot water, essential oils, and most common organic solvents	1:2 in 70% alcohol;[2] 1:9–12 in 60% alcohol; soluble in most common organic solvents
Organoleptic characteristics	Powerful, fruity, medicinal odor that becomes sweet and floral on dilution; sweet and fruity flavor at low levels	Strong, floral odor (narcissus); characteristic honey-like flavor
Synthesis	From sodium *o*-cresol and acetic anhydride	Synthetically prepared by suitable acetylation of *p*-cresol
Natural occurrence	Not reported found in nature	Reported as a constituent of the essential oils of wallflower, cananga, and ylang-ylang
Reported uses[3]	Non-alcoholic beverages 2.8 ppm Ice cream, ices, etc. 2.6 ppm Candy 11 ppm Baked goods 9.0–10 ppm Gelatins and puddings 1.0 ppm Chewing gum 0.30–220 ppm	Non-alcoholic beverages 0.50–1.0 ppm Ice cream, ices, etc. 1.3 ppm Candy 4.3 ppm Baked goods 4.4 ppm Chewing gum 0.30–220 ppm Condiments 10 ppm
Regulatory status	FDA 121.1164; FEMA No. 3072	FDA 121.1164; FEMA No. 3073

REFERENCES

For References 1–5, see end of Part III.

	4-(*p*-Tolyl)-2-butanone	*p*-Tolyl isobutyrate
Other names	*p*-Methylbenzylacetone	*p*-Cresyl isobutyrate (true name)
Empirical formula	$C_{11}H_{14}O$	$C_{11}H_{14}O_2$
Structure	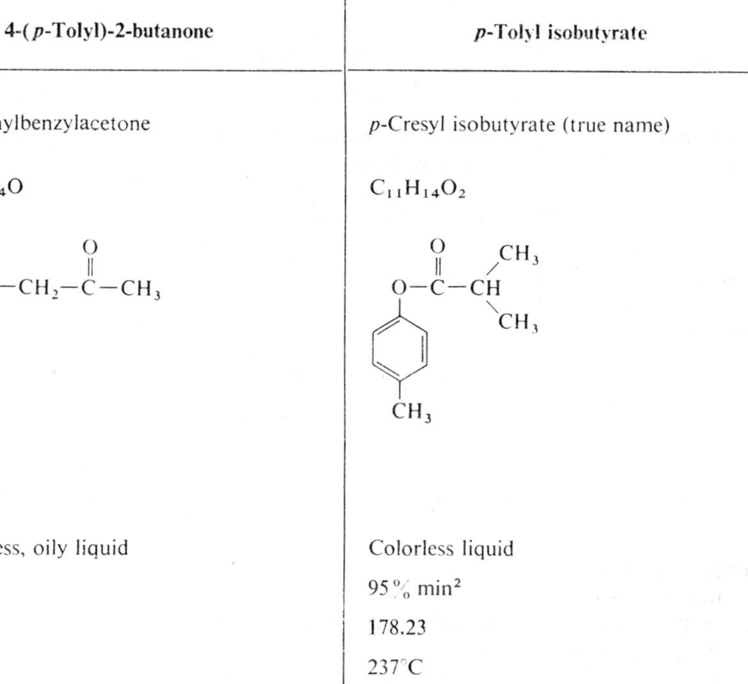	
Physical/chemical characteristics[5]		
Appearance	Colorless, oily liquid	Colorless liquid
Assay		95% min[2]
Molecular weight	162.23	178.23
Boiling point		237°C
Flash point		> 100°C
Specific gravity		0.991–0.996 at 25°/25°C;[2] 1.005 at 25°C
Refractive index		1.4860–1.4890[2] (1.4840) at 20°C
Acid value		1.0 max[2]
Free cresol		1.0% max[2]
Solubility	Very slightly soluble in water; soluble in alcohol	1:2 in 80% alcohol;[2] 1:7 in 70% alcohol; insoluble in water
Organoleptic characteristics	Very sweet, fruity, floral odor, with raspberry, plum, and gardenia notes; intense, sweet, deep fruit flavor	Very strong, lily-narcissus odor
Synthesis	By condensation of *p*-tolylaldehyde with acetone followed by hydrogenation	By esterification of *p*-cresol with isobutyric acid
Natural occurrence	Not reported found in nature. Its use as an insect attractant has been studied[6]	Not reported found in nature
Reported uses[3]	Non-alcoholic beverages 1.0 ppm Ice cream, ices, etc. 1.5 ppm Candy 6.0 ppm Baked goods 6.0 ppm	Non-alcoholic beverages 0.10–4.0 ppm Ice cream, ices, etc. 0.05 ppm Candy 0.12–6.0 ppm Baked goods 0.12–7.0 ppm
Regulatory status	FDA 121.1164; FEMA No. 3074	FDA 121.1164; FEMA No. 3075

REFERENCES

For References 1–5, see end of Part III.

6. *J. Agric. Food Chem.*, 10, 270, 1962.

	p-Tolyl laurate	*p*-Tolyl-3-methyl butyrate
Other names	*p*-Cresyl dodecanoate (true name) *p*-Cresyl laurate *p*-Tolyl dodecanoate *p*-Tolyl dodecylate	*p*-Cresyl isovalerate
Empirical formula	$C_{19}H_{30}O_2$	$C_{12}H_{16}O_2$
Structure	$CH_3-C_6H_4-O-C(=O)-(CH_2)_{10}-CH_3$	$(H_3C)_2CH-CH_2-C(=O)-O-C_6H_4-CH_3$
Physical/chemical characteristics[5]		
Appearance	Colorless, oily liquid	
Molecular weight	290.45	192.25
Solubility	Insoluble in water; soluble in alcohol	
Organoleptic characteristics	Mild, nondescript, floral, sweet odor and flavor	
Synthesis	From sodium *p*-cresol and lauryl chloride	
Natural occurrence	Not reported found in nature	Naturally occurring in passion fruit, raspberry, rum, whisky, coffee, tea, and soybean.
Reported uses[3]	Non-alcoholic beverages 1.0 ppm Ice cream, ices, etc. 1.0 ppm Candy 2.0 ppm Baked goods 2.0 ppm	Non-alcoholic beverages 1.0 ppm Ice cream, ices, etc. 1.0 ppm Candy 1.0 ppm Baked goods 2.0 ppm Gelatins and puddings 1.0 ppm
Regulatory status	FDA 121.1164; FEMA No. 3076	FEMA No. 3387

REFERENCES

For References 1–5, see end of Part III.

	p-Tolyl phenylacetate	2-(*p*-Tolyl)-propionaldehyde
Other names	*p*-Cresyl phenylacetate (true name) *p*-Tolyl α-toluate	*p*-Methylhydratropic aldehyde 2-(*p*-Methylphenyl)-propionaldehyde
Empirical formula	$C_{15}H_{14}O_2$	$C_{10}H_{12}O$
Structure	$C_6H_5-CH_2-C(=O)-O-C_6H_4-CH_3$	$H_3C-CH(C_6H_4CH_3)-CHO$
Physical/chemical characteristics[5]		
Appearance	White crystals	Colorless to slightly yellow liquid
Assay	98% min[2]	95% min
Molecular weight	226.28	148.20
Melting point	74–75°C	
Boiling point	310°C	222–224°C
Flash point		89°C
Specific gravity		0.979–0.985 at 25°C; 0.9941 at 13°C
Refractive index		1.5140–1.5160 at 20°C
Congealing point	> 73.5°C[2] (73.7°C)	
Acid value	1.0 max[2]	5 max
Solubility	1 gm in 45 ml of 95% alcohol;[2] insoluble in water; poorly soluble in propylene glycol	Insoluble in water; readily soluble in alcohol and ether
Organoleptic characteristics	Faint but tenacious lily, hyacinth, narcissus odor; sweet, honey-like flavor	Intense, sweet, refreshing odor similar to peppermint
Synthesis	From *p*-cresol and phenylacetic acid by esterification, or by heating phenylacetyl chloride with freshly distilled *p*-cresol[6]	By reaction of α-terpinene and chromyl chloride in carbon disulfide and subsequent hydrolysis of the resulting cymene-chromyl chloride addition product; also by heating *p*-tolyl glycidic acid and water
Natural occurrence	Not reported found in nature	
Reported uses[3]	Non-alcoholic beverages 1.6 ppm Ice cream, ices, etc. 0.87 ppm Candy 4.8 ppm Baked goods 5.4 ppm	Non-alcoholic beverages 0.13 ppm Alcoholic beverages 0.005 ppm Ice cream, ices, etc. 0.13 ppm Candy 0.13 ppm Baked goods 0.20 ppm
Regulatory status	FDA 121.1164; FEMA No. 3077	FDA 121.1164; FEMA No. 3078

REFERENCES

For References 1–5, see end of Part III.

6. **Raiford and Hildebrand,** *Am. J. Pharm.*, 101, 471, 1929.

	Triacetin	Tributyl acetylcitrate
Other names	"Acetin" "Enzactin" Glyceryl triacetate "Vanay"	Acetyl tributylcitrate Citroflex A-4 (Pfizer)
Empirical formula	$C_9H_{14}O_6$	$C_{20}H_{34}O_8$
Structure	$CH_2-O-C(=O)-CH_3$ $CH-O-C(=O)-CH_3$ $CH_2-O-C(=O)-CH_3$	$CH_3-C(=O)-O-C(-COO-C_4H_9)(CH_2-COO-C_4H_9)(CH_2-COO-C_4H_9)$
Physical/chemical characteristics[5]		
Appearance	Colorless liquid; solidifies at low temperature	Colorless, slightly viscous liquid
Molecular weight	218.20	402.46
Melting point	3.2°C	
Boiling point	130.5°C at 7 mm Hg (260°C)	172–174°C (>300°C) at 1 mm Hg
Flash point	138°C	
Specific gravity	1.1596 at 20°C	1.14
Refractive index	1.4306 at 20°C	
Solubility	7% soluble in water; soluble in alcohol and most common organic solvents	Insoluble in water; soluble in alcohol
Organoleptic characteristics	Very faint, fruity odor; mild, sweet taste, bitter above 0.05%	Very faint, sweet, herbaceous odor; at high levels (e.g., 1000 ppm emulsion in water) mild, fruity, non-descript flavor
Synthesis	By direct reaction of glycerol with acetic acid in the presence of Twitchell reagent;[6] or in benzene solution of glycerol and boiling acetic acid in the presence of a cationic resin (Zeo-Karb H) pretreated with diluted H_2SO_4[7]	From citric acid via the tributyl ester followed by acetylation
Natural occurrence	Reported found in several natural products	Non-alcoholic beverages 0.40 ppm (probably 20–40 ppm and higher)
Reported uses[3]	Non-alcoholic beverages 190 ppm Ice cream, ices, etc. 60–2,000 ppm Candy 560 ppm Baked goods 1,000 ppm Chewing gum 4,100 ppm	
Regulatory status	FDA GRAS (general); FEMA No. 2007	FDA 121.1164; FEMA No. 3080

REFERENCES

For References 1–5, see end of Part III.

6. **Zaganiaris and Varvoglis**, *Ber. Dtsch. Chem. Ges.*, 69, 2279, 1936.
7. **Sussman**, *Ind. Eng. Chem.*, 38, 1229, 1946.

	2-Tridecenal	Trimethylamine
Other names	3-Decylacrolein Tridecen-2-al-1	
Empirical formula	$C_{13}H_{24}O$	C_3H_9N
Structure	$CH_3-(CH_2)_9-CH=CH-C(=O)H$	$N(CH_3)_3$
Physical/chemical characteristics[5]		
Appearance	Colorless, oily liquid	
Molecular weight	196.33	59.1
Boiling point	106–108°C at 0.28 mm Hg; 115–118°C at 10 mm Hg; (232°C)	3.8°C
Melting point		-124°C
Specific gravity	0.8476 at 19°C	0.6709 at 0°C
Refractive index	1.4582 at 19°C	
Solubility	Almost insoluble in water; soluble in alcohol	Miscible with water and alcohol; soluble in ether, benzene, and chloroform
Organoleptic characteristics	Very strong, citrus-peel, waxy odor; bitter-sweet taste	Pungent fishy odor
Synthesis	From undecyl aldehyde by condensation with acetic aldehyde using sodamide in ether	From paraformaldehyde and ammonium chloride;[6] by the action of formaldehyde and formic acid on ammonia[7]
Natural occurrence	Identified in sunguli oil from Turkmenistan;[22] studied by Stoll[23]	In cheese,[8,9] caviar,[10] fish (also as the N-oxide),[11-15] beer,[16] whisky,[17] cocoa,[18-20] and coffee[21]
Reported uses[3]	Non-alcoholic beverages 0.10–0.30 ppm Ice cream, ices, etc. 1.6–6.0 ppm Candy 4.0–6.0 ppm Baked goods 4.0–6.0 ppm Chewing gum 0.10 ppm	Soups 10.0 ppm
Regulatory status	FDA 121.1164; FEMA No. 3082	FEMA No. 3241

REFERENCES

For References 1–5, see end of Part III.

6. **Adams et al.,** *Organic Syntheses,* Coll. Vol. 1, Gilman, H. and Blatt, A. H., Eds., Wiley & Sons, New York, 1941, 58.
7. **Sommelet et al.,** *Bull. Soc. Chim.,* [4]35, 446, 1924.
8. **Golovnja et al.,** *J. Anal. Chem. U.S.S.R.,* 23, 766, 1968.
9. **Panouse et al.,** *Ind. Aliment. Agric.,* p. 133, 1972.
10. **Golovnja et al.,** *J. Anal. Chem. U.S.S.R.,* 22, 612, 956, 1967.
11. **Herrit et al.,** 142nd National Meeting of the American Chemical Society, 1962.
12. **Jones,** 4th Symposium on Food, Oregon State University, 1965, Oregon State University Press, Corvallis, 1967, 267
13. **Wong et al.,** *J. Assoc. Of. Agric. Chem.,* 50, 8, 1967.
14. **Hard et al.,** *Naeringsmiddelindustrien,* 1/2, 46, 1970.
15. **Weurman et al.,** *J. Food Sci.,* 26, 239, 1961.
16. **Slaughter et al.,** *J. Inst. Brew.,* 77, 446, 1971.
17. **Kahn,** *J. Assoc. Of. Agric. Chem.,* 52, 1166, 1969.
18. **Weurman et al.,** *J. Food Sci.,* 26, 239, 1961.
19. **Marion et al.,** *Helv. Chim. Acta,* **50, 1509, 1967.**
20. **Rohan,** *Food Process. Mark.,* 38, 12, 1969.
21. **Walter et al.,** *Z. Ernährungswiss.,* 9, 123, 1969.
22. **Isskenderow,** *Chem. Zentralbl.,* 1, 1892, 1938.
23. **Stoll,** *Helv. Chim. Acta,* **30**:995, 1947.

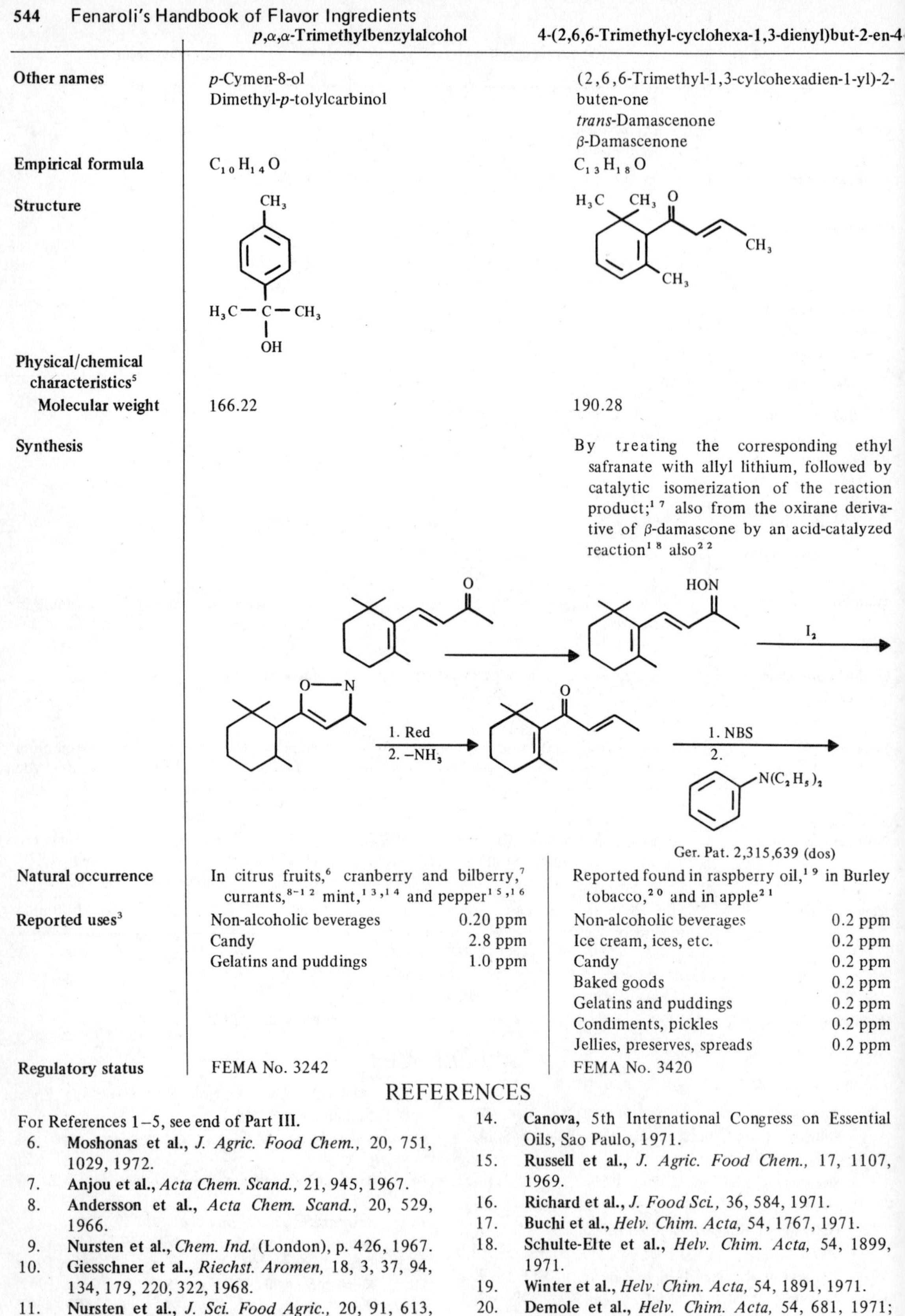

	p,α,α-Trimethylbenzylalcohol	4-(2,6,6-Trimethyl-cyclohexa-1,3-dienyl)but-2-en-4
Other names	*p*-Cymen-8-ol Dimethyl-*p*-tolylcarbinol	(2,6,6-Trimethyl-1,3-cylcohexadien-1-yl)-2-buten-one *trans*-Damascenone β-Damascenone
Empirical formula	$C_{10}H_{14}O$	$C_{13}H_{18}O$
Structure		
Physical/chemical characteristics[5]		
Molecular weight	166.22	190.28
Synthesis		By treating the corresponding ethyl safranate with allyl lithium, followed by catalytic isomerization of the reaction product;[17] also from the oxirane derivative of β-damascone by an acid-catalyzed reaction[18] also[22] Ger. Pat. 2,315,639 (dos)
Natural occurrence	In citrus fruits,[6] cranberry and bilberry,[7] currants,[8-12] mint,[13,14] and pepper[15,16]	Reported found in raspberry oil,[19] in Burley tobacco,[20] and in apple[21]
Reported uses[3]	Non-alcoholic beverages 0.20 ppm Candy 2.8 ppm Gelatins and puddings 1.0 ppm	Non-alcoholic beverages 0.2 ppm Ice cream, ices, etc. 0.2 ppm Candy 0.2 ppm Baked goods 0.2 ppm Gelatins and puddings 0.2 ppm Condiments, pickles 0.2 ppm Jellies, preserves, spreads 0.2 ppm
Regulatory status	FEMA No. 3242	FEMA No. 3420

REFERENCES

For References 1–5, see end of Part III.

6. **Moshonas et al.,** *J. Agric. Food Chem.*, 20, 751, 1029, 1972.
7. **Anjou et al.,** *Acta Chem. Scand.*, 21, 945, 1967.
8. **Andersson et al.,** *Acta Chem. Scand.*, 20, 529, 1966.
9. **Nursten et al.,** *Chem. Ind.* (London), p. 426, 1967.
10. **Giesschner et al.,** *Riechst. Aromen*, 18, 3, 37, 94, 134, 179, 220, 322, 1968.
11. **Nursten et al.,** *J. Sci. Food Agric.*, 20, 91, 613, 1969.
12. **Sydow et al.,** *Lebensm. Wiss. Technol.*, 4, 54, 1971.
13. **Schnelle et al.,** *Planta Med.*, 16, 48, 1968.
14. **Canova,** 5th International Congress on Essential Oils, Sao Paulo, 1971.
15. **Russell et al.,** *J. Agric. Food Chem.*, 17, 1107, 1969.
16. **Richard et al.,** *J. Food Sci.*, 36, 584, 1971.
17. **Buchi et al.,** *Helv. Chim. Acta*, 54, 1767, 1971.
18. **Schulte-Elte et al.,** *Helv. Chim. Acta*, 54, 1899, 1971.
19. **Winter et al.,** *Helv. Chim. Acta*, 54, 1891, 1971.
20. **Demole et al.,** *Helv. Chim. Acta*, 54, 681, 1971; 55, 1866, 1972.
21. **Narsten and Woolfe,** *J. Sci. Food Agric.*, 23, 803, 1972.
22. German Patent 2,315,639.

	2,6,6-Trimethylcyclohexa-1,3-dienyl methanal		2,6,6-Trimethylcyclohex-2-ene-1,4-dione	
Other names	2,6,6-Trimethyl-1-carboxy-1,3-cyclohexadiene		3,5,5-Trimethyl-2-cyclohexene-1,4-dione	
Empirical formula	$C_9H_{14}O$		$C_9H_{12}O_2$	
Structure			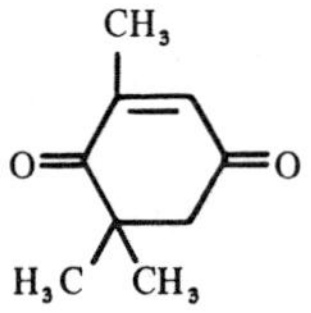	
Physical/chemical characteristics[5]				
Molecular weight	138.20		152.19	
Natural occurrence			Reported found among the volatile components in saffron[6] and in cigarette smoke[7]	
Reported uses[3]	Candy	5.0 ppm	Non-alcoholic beverages	1.0 ppm
	Baked goods	10.0 ppm	Candy	5.0 ppm
	Condiments, pickles	5.0 ppm		
	Preserves and spreads	10.0 ppm		
Regulatory status	FEMA No. 3389		FEMA No. 3421	

REFERENCES

For References 1–5, see end of Part III.
6. **Zarghami et al.,** *Photochemistry,* p. 2755, 1971.
7. **Neurath et al.,** *Beitr. Tabakforsch.,* (6)12, 1971.

	4-[(2,6,6)-Trimethyl-cyclohex-1-enyl]-but-2-en-4-one		3,5,5-Trimethyl-1-hexanol	
Other names	1-[(2,6,6)-Trimethyl-cyclohex-1-enyl]-but-2-en-1-one β-Damascone Dorinone® (Firmenich)		Isononyl alcohol *tert*-Butyl isopentanol Trimethyl hexyl alcohol	
Empirical formula	$C_{13}H_{20}O$		$C_9H_{20}O$	
Structure	$CO-C=C-CH_3$ on ring with H_3C, H_3C, CH_3		$(H_3C)_3C-CH_2-CH(CH_3)-CH_2-CH_2OH$	
Physical/chemical characteristics				
Appearance			Colorless oily liquid	
Molecular weight	192.30		144.26	
Boiling point	*trans*-, 55°C at 0.001 mm Hg[6] *cis*-, 52°C at 0.001 mm Hg[6]		142°C at 150 mm Hg 62°C at 14 mm Hg	
Specific gravity	*trans*-, 0.934 at 20°C[6] *cis*-, 0.930 at 20°C[6]		0.8350 at 20°C	
Refractive index	*trans*-, 1.4980 at 20°C[6] *cis*-, 1.4957 at 20°C[6]		1.4352 at 20°C 1.4300 at 25°C	
Solubility			Almost insoluble in water; soluble in alcohol, acetone, and ether	
Organoleptic characteristics			Strong, oily-herbaceous odor; becomes sweet on dilution	
Synthesis	Various methods;[6] recently from ionone izoxazoles;[7] also from 7,8-dehydro-β-ionole;[9] see various patents.[8]		By oxo-reaction of diisobutylene to the corresponding aldehyde, from which the alcohol is prepared by hydrogenation[12]	
Natural occurrence	In Burley tobacco oil;[10] in volatile fractions from leaves of *Carphephorus corymbosus* and *C. paniculatus*[11]			
Reported uses[3]	Non-alcoholic beverages	10.0 ppm	Baked goods	30.0 ppm
	Ice cream, ices, etc.	10.0 ppm	Condiments, pickles	30.0 ppm
	Candy	10.0 ppm		
	Baked goods	10.0 ppm		
	Gelatins and puddings	10.0 ppm		
	Chewing gum	10.0 ppm		
Regulatory status	FEMA No. 3243		FEMA No. 3324	

REFERENCES

For References 1–5, see end of Part III.

6. **Demole et al.,** *Helv. Chim. Acta,* 53, 541, 1970.
7. **Schulte-Elte et al.,** *Helv. Chim. Acta,* 56, 310, 1973.
8. German Patent 2,242,751; *Chem. Abstr.,* 79, 5053, 1972. German Patent 2,065,324; *Chem. Abstr.,* 79, 18207p, 1972. German Patent, 2,065,322; *Chem. Abstr.,* 18208g, 1972. German Patent 2,065,323; *Chem. Abstr.,* 79, 31582f, 1972. Swiss Patent 53,752; *Chem. Abstr.,* 79, 104808s, 1972. German Patent 2,305,140; *Chem. Abstr.,* 79, 115743t, 1972.
9. **Ohloff,** *Helv. Chim. Acta,* 56, 1503, 1973.
10. **Demole et al.,** *Helv. Chim. Acta,* 54, 681, 1971.
11. **Karlsson et al.,** *Acta Chem. Scand.,* 26, 3839, 1972.
12. **Keulmans, Kwantes and Van Bavel,** *Recl. Trav. Chim.,* 67, 298, 1948.

	2,2,4,-Trimethyl-1,3-oxacyclopentane	2,6,10-Trimethyl-2,6,10-pentadecatrien-14-one
Other names	2,2,4-Trimethyl-1,3-dioxolane Propylene glycol acetone ketal	6,10,14-Trimethyl-5,9,13-pentadecatrien-2-one Farnesyl acetone
Empirical formula	$C_6H_{12}O_2$	$C_{18}H_{30}O$
Structure	(five-membered ring: O, O; two H_3C groups on C2; CH_3 on C4)	$CH_3-C(CH_3)=CH-CH_2-CH_2-C(CH_3)=CH-CH_2-CH_2-C(CH_3)=CH-CH_2-CH_2-C(=O)-CH_3$
Physical/chemical characteristics		
Molecular weight	116.16	262.44
Boiling point	98–99°C	147–148°C at 0.5 mm Hg
Specific gravity	0.909 at 18.5°C	0.8900 at 20°C
Refractive index	1.4019 at 18.5°C	1.4812 at 20°C
Synthesis	From acetone and propylene glycol[6]	By treating farnesyl bromide with sodium acetate, followed by treating with methanolic KOH[7,8]
Natural occurrence		In tomato[9,10]
Reported uses[3]	Non-alcoholic beverages 4.0 ppm Alcoholic beverages 1.0 ppm Ice cream, ices, etc. 4.0 ppm Candy 6.0 ppm Gelatins and puddings 4.0 ppm Condiments 4.0 ppm Jellies, preserves, spreads 4.0 ppm	Non-alcoholic beverages 2.0 ppm Ice cream, ices, etc. 2.0 ppm Candy 3.0 ppm Gelatins and puddings 2.0 ppm Condiments 1.0 ppm Jellies, preserves, spreads 1.0 ppm
Regulatory status	FEMA No. 3441	FEMA No. 3442

REFERENCES

For References 1–5, see end of Part III.

6. **Böseken et al.,** *Recl. Trav. Chim. Pays-Bas,* 42, 1105, 1923.
7. **Ruzička et al.,** *Helv. Chim. Acta,* 22, 394, 1949.
8. **Karrer et al.,** *Helv. Chim. Acta,* 27, 1299, 1944.
9. **Buttery et al.,** *Chem. Ind.* (London), p. 238, 1969; *J. Agric. Food Chem.,* 19, 524, 1971.
10. **Johnson et al.,** *Chem. Ind.* (London), p. 1212, 1971.

	2,3,5-Trimethylpyrazine	2,4,5-Trimethyl thiazole
Other names	Trimethylpyrazine	
Empirical formula	$C_7H_{10}N_2$	C_6H_9SN
Structure	[structure: pyrazine ring with N, CH$_3$, H$_3$C, N, CH$_3$]	[structure: thiazole ring with H$_3$C, N, H$_3$C, S, CH$_3$]
Physical/chemical characteristics		
Assay	C: 68.82%; H: 8.25%; N: 22.93%	
Molecular weight	122.17	127.21
Boiling point		167°C 65–67°C at 20 mm Hg
Mass spectra[7]	39 (27), 42 (100), 53 (8), 54 (12), 66 (1), 67 (2), 80 (5), 81 (22), 108 (2), 122 (81), 123 (8)	
IR spectra[7]	S (3.4, 6.8, 7.2, 7.3, 8.6, 10.6) M (7.8, 7.9, 9.8, 10.0, 11.2) W (5.6, 6.5, 9.2, 10.2, 12.0)	
UV spectra[7]	(277.5 nm)	
Organoleptic characteristics	Baked-potato or roasted-peanut aroma; odor threshold: 9,000 ppb in water[8]	
Synthesis	From 2,5-dimethylpyrazine by ring alkylation with MeLi;[9] by condensing propylenediamine with 2,3-butanedione[10]	By decarboxylation of 2,4-dimethylthiazole-5-acetic acid;[11] by reacting acetamide and phosphorous pentasulfide with methyl-α-bromoethyl ketone[12]
Natural occurrences	Reported to be present in the following foods: bakery products, roasted barley, cocoa products, coffee, dairy products, meat, peanuts, filberts, pecans, popcorn, potato products, rum and whisky, soy products[6]	
Reported uses[3]	Non-alcoholic beverages 5.0 to 10.0 ppm Candy 5.0 to 10.0 ppm Baked goods 5.0 to 10.0 ppm Condiments 2.0 ppm Meat, meat sauces, soups 2.0 ppm Milk, dairy products 1.0 ppm Cereals 2.0 ppm	Ice cream, ices, etc. 2.0 ppm Candy 2.0 ppm Condiments, pickles 6.0 ppm Meat, meat sauces, soups 6.0 ppm
Regulatory status	FEMA No. 3244	FEMA No. 3325

REFERENCES

For References 1–5, see end of Part III.

6. Maga and Sizer, *Crit. Rev. Food Technol.*, 4(1), 39, 1973.
7. Bondarovich et al., *J. Agric. Food Chem.*, 15, 1093, 1967.
8. Koehler et al., *J. Food Sci.*, 36, 816, 1971.
9. Rizzi, *J. Org. Chem.*, 33(4), 1333, 1968.
10. Ishiguro and Matsumura, *Yakugaku Zasshi*, 78, 229, 1958.
11. Gregory and Wiggins, *J. Chem. Soc.*, p. 1400, 1947.
12. Kurkjy and Brown, *J. Am. Chem. Soc.*, 74, 5778, 1952.

	2,4-Undecadienal	2,3-Undecadione
Other names		Acetyl nonanoyl Acetyl nonyryl Acetyl pelargonyl
Empirical formula	$C_{11}H_{18}O$	$C_{11}H_{20}O_2$
Structure	$O{=}C(H)-CH{=}CH-CH{=}CH-(CH_2)_5-CH_3$	$CH_3-C({=}O)-C({=}O)-(CH_2)_7-CH_3$
Physical/chemical characteristics[5]		
Appearance	Colorless mobile oil	Oily, yellow liquid
Molecular weight	166.27	184.28
Boiling point		109–111 C at 10 mm Hg
Solubility	Insoluble in water, soluble in alcohol and oils[6]	Very slightly soluble in water; soluble in alcohol
Organoleptic characteristics	Powerful green odor	Strong, sweet-cream, warm odor with a coconut-oil note; sweet-cream and coconut-milk taste
Synthesis	H_3C–CH=CH–CHO + CHO–(CH₂)₅–CHO $\xrightarrow{B^-}$	By reacting 3-bromo-2-undecanone with sodium azide in diluted methanol and then boiling the resulting solution with concentrated HCl;[8] also from methyl nonyl ketone via the monoxime of the title material
Natural occcurrence	Reported found in heated beef and in roasted peanuts[7]	Not reported found in nature
Reported uses[3]	Baked goods 1.0 ppm Meat, meat sauces, soups 1.0 ppm Cereals 1.0 ppm	Non-alcoholic beverages 1.5 ppm Ice cream, ices, etc. 3.0 ppm Candy 3.0 ppm Baked goods 3.0 ppm
Regulatory status	FEMA No. 3422	FDA 121.1164; FEMA No. 3090

REFERENCES

For References 1–5, see end of Part III.

6. **Arctander,** *Perfume and Flavor Chemicals,* Montclair, N.J., 1969, No. 3022.
7. Volatile Compounds in Food, T.N.O. Report No. 4030, 3rd ed., Centraal Instituut vor Voedingsonderzoek, Zeist, Netherlands.
8. **I. G. Garbenindustrie,** German Patent 573,722, 1931.

	γ-Undecalactone	Δ-Undecalactone
Other names	"Aldehyde C-14" γ-Heptyl butyrolactone 4-Hydroxyundecanoic acid, γ-lactone Peach aldehyde Undecanolide-1,4	5-Hydroxyundecanoic acid, lactone α-*n*-Hexyl-δ-valerolactone Undecanolide-1,5
Empirical formula	$C_{11}H_{20}O_2$	$C_{11}H_{20}O_2$
Structure	O=C(–O–)CH–CH$_2$–(CH$_2$)$_5$–CH$_3$; ring: H$_2$C——CH$_2$	CH$_3$–(CH$_2$)$_4$–CH$_2$–CH; ring: C(=O), O, CH$_2$, CH$_2$, CH$_2$
Physical/chemical characteristics[5]		
Appearance	Colorless to very pale yellow, viscous liquid	Colorless, very viscous liquid
Assay	98% min[2]	
Molecular weight	184.28	184.28
Boiling point	297°C	152–155°C at 10.5 mm Hg
Flash point	137°C	
Specific gravity	0.942–0.945 at 25°/25°C[2]	0.96
Refractive index	1.4500–1.4540 at 20°C[2]	
Acid value	Not more than 5[2]	
Solubility	1:5 and more in 60% alcohol;[2] 1:2 in 70% alcohol; almost insoluble in water; soluble in most common organic solvents	Almost insoluble in water; soluble in alcohol and oils
Organoleptic characteristics	Strong, fruity odor suggestive of peach (particularly on dilution); pungent and sweet flavor also similar to peach	Creamy, peachy, somewhat fatty aroma reminiscent of coconut
Synthesis	By the action of sulfuric acid on undecylenic acid;[2] also prepared starting from castor oil;[6] from octanol-1 plus methylacrylate with di-*tert*-butylperoxide; from heptylethylene oxide and sodiomalonic ester	By intramolecular Cannizzaro type rearrangement of 2-hexylglutaraldehyde[8]
Natural occurrence	Reported found in hydrolyzed soy protein,[7] butter, peach, apricot, and passion fruit	Reportedly occurring in coconut flavor[9]
Reported uses[3]	Non-alcoholic beverages 4.4 ppm Ice cream, ices, etc. 3.0 ppm Candy 11 ppm Baked goods 7.1 ppm Gelatins and puddings 90 ppm	Non-alcoholic beverages 2.0 ppm Ice cream, ices, etc. 5.0 ppm Candy 5.0 ppm Baked goods 25.0 ppm Meat, meat sauces, soups 1.0 ppm Milk, dairy products 10.0 ppm Cereals 5.0 ppm
Regulatory status	FDA 121.1164; FEMA No. 3091	FEMA No. 3294

REFERENCES

For References 1–5, see end of Part III.
6. **Müller,** *Riechstoffindustrie,* p. 39, 1927.
7. **Manley et al.,** *J. Food Sci.,* 35(3), 286, 1970.
8. **Meerwein,** *Berichte,* 53, 1829, 1920.
9. **Lin and Wilkens,** *J. Food Sci.,* 35(5), 538, 1970.

Undecanal

Other names	Aldehyde C_{11} (saturated) Hendecanal α-Oxo-undecane *n*-Undecylic aldehyde
Empirical formula	$C_{11}H_{22}O$
Structure	$CH_3-(CH_2)_9-CHO$
Physical/chemical characteristics[5]	
Appearance	Colorless to slightly yellow liquid
Assay	92% min[2] (97%)
Molecular weight	170.30
Melting point	−4°C
Boiling point	118–120°C at 20 mm Hg
Flash point	96°C
Specific gravity	0.825–0.832 at 25°/25°C;[2] 0.8251 at 23°C; 0.8320–0.8350 at 15°C
Refractive index	1.430–1.435 at 20°C;[2] 1.4324 at 23°C; 1.4315–1.4330
Acid value	Not more than 10[2] (3.7)
Solubility	Insoluble in water; soluble in most common organic solvents; 1:1 in 80% alcohol; 1:5 in 70% alcohol
Organoleptic characteristics	Sweetish, fatty odor with an orange and rose undertone; characteristic flavor; tends to polymerize unless tightly sealed
Synthesis	Usually prepared by oxidation of the corresponding alcohol or reduction of the corresponding acid
Natural occurrence	Until a few years ago believed not to be found in nature. It has subsequently been identified in the essential oils of mandarin and lemon.[6]
Reported uses[3]	Non-alcoholic beverages 0.95 ppm Ice cream, ices, etc. 3.1 ppm Candy 2.0 ppm Baked goods 2.4 ppm Chewing gum 56 ppm
Regulatory status	FDA 121.1164; FEMA No. 3092

REFERENCES

For References 1–5, see end of Part III.

6. **Riganenis and Calvarano,** *Essenze Deriv. Agrum.*, 26, 167, 1956.

	Undecanoic acid	2-Undecanol
Other names	*n*-Undecylic acid Decane-α-carboxylic acid Hendecanoic acid	Methyl nonyl carbinol 2-Hendecanol *sec*-Undecyl alcohol
Empirical formula	$C_{11}H_{22}O_2$	$C_{11}H_{24}O$
Structure	$CH_3-(CH_2)_9-COOH$	$CH_3-(CH_2)_7-CH_2-CH(OH)-CH_3$
Physical/chemical characteristics[5]		
Appearance	Colorless crystals	Colorless liquid
Assay		
Molecular weight	186.30	172.31
Melting point	28–29°C 284°C	*dl*-, 0°C *d*-, 12°C
Boiling point	164.5°C at 15 mm Hg	*dl*-, 115°C at 10 mm Hg *d*-, 128°C at 20 mm Hg
Congealing point		*l*-, 231–233°C
Specific gravity	0.9948 at 20°C	*dl*-, 0.8263 at 19°C *d*-, 0.8270 at 20°C
Refractive index	1.4355 at 30°C 1.4294 at 45°C 1.4203 at 70°C	*d*-, 1.4369 at 20°C
Optical rotation		+10.29° at 20°C (c = 5.15 in benzene) +8.11° at 20°C (c = 5.11 in alcohol) –5.40° (from *Litsea odorifera;* 1 = 10 cm) –1.18° (from rue; 1 = 2.5 cm)
Solubility	Insoluble in water; very soluble in alcohol, chloroform, and acetone; soluble in benzene in all proportions	Insoluble in water; soluble in alcohol and ether
Organoleptic characteristics	Faint fatty, aldehydic odor	Fatty odor with a fruity note; a fruity taste is also detectable at low concentration
Synthesis	By reduction of undecylenic acid obtained from castor oil[6]	By reduction of methyl nonyl ketone with sodium metal in alcohol;[10] the *d*-form is isolated from the optically inactive material via the corresponding phthalate and its salt with strychnine and brucine[11]
Natural occurrence	Occurs naturally in butter[7] and in the oil of *Artemisia frigida* Willd; also reported found in the oils of *Chamaecyparis pisifera* Engl., wild thyme,[8] *Thymus marschollianus,* and coconut[9]	The *l*-form is reportedly found in rue oil[12] and in the essential oil of *Litsea odorifera;*[3] the *d*-form occurs in coconut[14]
Reported uses[3]	Baked goods 2.0 ppm	Baked goods 20.0 ppm
Regulatory status	FEMA No. 3245	FEMA No. 3246

REFERENCES

For References 1–5, see end of Part III.

6. **Rodd,** *Chemistry of Carbon Compounds,* 2nd ed., Vol. 1, American Elsevier, New York, 1965, 137.
7. **Martinenghi,** *Olii, Grassi e Derivati,* 3rd ed., Hoepli.
8. **Spiridonowa,** *Chem. Zentralbl.,* 1, 1812, 1937.
9. **Nobori,** *Chem. Abstr.,* 42, 6140, 1948.
10. **Mannich,** *Ber. Dtsch. Chem. Ges.,* 35, 2144, 1902.
11. **Pickard and Kempon,** *J. Chem. Soc.* (London), 55, 70, 99, 1911; 103, 1957, 1913; 105, 850, 1914.
12. **Power and Lees,** *J. Chem. Soc.* (London), 81, 1585, 1902.
13. **van Romburg,** *Chem. Zentralbl.,* 2, 1863, 1911.
14. **Haller and Lassieur,** *Chem. Zentralbl.,* 2, 1913, 1910.

	2-Undecanone
Other names	Methyl nonyl ketone
Empirical formula	$C_{11}H_{22}O$
Structure	$CH_3-C(=O)-(CH_2)_9-CH_3$
Physical/chemical characteristics[5]	
Appearance	Colorless to slightly yellowish liquid
Assay	95% min
Molecular weight	170.30
Melting point	12–13°C (13–15°C)
Boiling point	228–230°C; 105–107°C at 10 mm Hg
Specific gravity	0.8262 at 20°C; 0.822–0.826 at 25°C
Refractive index	1.4175 at 22°C; 1.4280–1.4330 at 20°C
Solubility	Insoluble in water; soluble in most common organic solvents; 1:3 in 70% alcohol
Organoleptic characteristics	Characteristic rue odor; sweet flavor reminiscent of peach (on dilution)
Synthesis	Can be isolated from natural oils by fractional distillation;[6] also by dry distillation of calcium acetate and calcium caprylate, or by boiling octylacetoacetic acid ethyl ester with an alcoholic KOH solution[7]
Natural occurrence	Originally reported found in the essential oil of *Ruta graveolens;*[8] subsequently was identified in the essential oils of *Citrus limetta* Risso., *Fagara xanthoxyloides* Lamm., and *Litsea odorifera* Val. (leaves). A method for the determination of methyl nonyl ketone in various rue species (*Ruta montana, Ruta bracteosa*) was devised.[9] It is also present in the essential oils of *Jaborandi* (leaves), *Hottuynia cordata, Phellodendron anaurense, Schizandar nigra* Maxim., and in coconut and palm oils. Also identified as the main constituent of the essential oil of the physiological variety of *Boronia ledifolia* Gai.[10] A 92% content level was reported in the essential oil of *Ruta chalepensis.*[11]
Reported uses[3]	Non-alcoholic beverages 2.8 ppm Ice cream, ices, etc. 0.54 ppm Candy 2.6 ppm Baked goods 3.1 ppm Gelatins and puddings 5.0 ppm
Regulatory status	FDA 121.1164; FEMA No. 3093

REFERENCES

For References 1–5, see end of Part III.

6. **Gorup et al.,** *Annalen,* 157, 275, 1871.
7. **Guthzeit,** *Annalen,* 204, 275, 1880.
8. **Williams,** *Annalen,* 107, 374, 1858.
9. **Roure-Bertrand fils,** *Chem. Zentralbl.,* 2, 139, 1911.
10. **Penfold and Morrison,** *Chem. Abstr.,* 43, 3148, 1949.
11. **Costa and do Vale,** *Chem. Abstr.,* 48, 11732, 1954.

	2-Undecenal	9-Undecenal
Other names	Undecen-2-al 3-Octylacrolein The commercial product is a mixture of several isomers of undecylenic aldehyde	Aldehyde C-11 undecylenic Hendecen-9-al 9-Undecen-1-al (correct name) Undecenoic aldehyde Undecylenic aldehyde
Empirical formula	$C_{11}H_{20}O$	$C_{11}H_{20}O$
Structure	OHC—CH=CH—$(CH_2)_7$—CH_3	CH_3—CH=CH—$(CH_2)_7$—CHO
Physical/chemical characteristics[5]		
Appearance	Colorless or very pale straw-colored liquid[11]	Colorless, pale yellow, oily liquid
Molecular weight	168.27	168.28
Boiling point	94°C at 2 mm Hg 229°C	
Solubility		Insoluble in water; soluble in alcohol
Organoleptic characteristics	Powerful fresh aldehydic odor	Orange peel-like, sweet odor; sweet, citrus flavor
Synthesis	By oxidation of 2-undecanol;[6] also H_3C(CH₂)₆CHO + CH₂=CH—O—CH_3 —BF_3→	
Natural occurrence	Reported found among the volatile components in fish products;[8] also formed in the oxidative decomposition of heated lard;[9] its presence in nature has been reported in cranberry and bilberry, milk, cooked chicken, heated beef and pork, roasted filberts, roasted peanuts, pecans, and potato chips;[10] exhibits fungistatic activity[7]	
Reported uses[3]	Baked goods 0.05 ppm Condiments, pickles 0.02 ppm Meat, meat sauces, soups 0.05 ppm Milk, dairy products 0.05 ppm	Non-alcoholic beverages 4.8 ppm Ice cream, ices, etc. 4.2 ppm Candy 4.5 ppm Baked goods 4.6 ppm
Regulatory status	FEMA No. 3423	FDA 121.1164; FEMA No. 3094

REFERENCES

For References 1–5, see end of Part III.

6. **Swift et al.,** *J. Am. Chem. Soc.,* 71, 1512, 1949.
7. **Wyss et al.,** *Arch. Biochem.,* 7, 422, 1945.
8. **Golovnya et al.,** *J. Chromatogr.,* 61, 65, 1971.
9. **Watanabe et al.,** *Nippon Chikusan Gakkai-Ho,* 42, 393, 1971; *Chem. Abstr.,* 76, 57867m, 1972.
10. Volatile Compounds in Food, T.N.O. Report No. 4030, 3rd ed., Centraal Instituut voor Voedingsonderzoek, Zeist, Netherlands.
11. **Arctander,** *Perfume and Flavor Chemicals,* Montclair, N.J., 1969, No. 3033.

	10-Undecenal	10-Undecenoic acid
Other names	Aldehyde C-11 undecylenic Hendecenal Undecylenic aldehyde	10-Hendecenoic acid Undecylenic acid
Empirical formula	$C_{11}H_{20}O$	$C_{11}H_{20}O_2$
Structure	$CH_2{=}CH{-}(CH_2)_8{-}CHO$	$CH_2{=}CH{-}(CH_2)_8{-}COOH$
Physical/chemical characteristics[5]		
Appearance	Colorless to light-yellow liquid	Colorless leaflets
Assay	90% min[2] (97%)	
Molecular weight	168.28	184.28
Melting point		24.5°C
Boiling point	103°C at 3 mm Hg; 101–103°C at 10 mm Hg	275°C 131°C at 1 mm Hg 165°C at 15 mm Hg
Flash point	92°C (100°C)	
Specific gravity	0.840–0.850 at 25/25°C;[2] 0.8496 at 21°C; 0.8470–0.8500 at 25°C	0.9072 at 24°C 0.910 at 15.5°C
Refractive index	1.4410–1.4470 at 20°C;[2] 1.4464 at 21°C; 1.4425–1.4440	1.4486 at 24°C
Acid value	6 max[2] (2.7)	
Iodine value		136–138
Solubility	1:1–2 in 80% alcohol;[2] insoluble in water; soluble in most common organic solvents	Insoluble in water; soluble in alcohol and ether
Organoleptic characteristics	Characteristic fatty, rose odor on dilution; unpleasant sweet, fatty taste; tends to polymerize	Characteristic pungent odor
Synthesis	By oxidation of the corresponding alcohol or reduction of the corresponding acid. By passing vapors of formic acid and undecylenic acid over TiO_2 at 250°C under pressure;[6] by passing vapors of undecylenic acid and methanol over a manganese dioxide based catalyst;[7] or by reduction of ethyl undecylenate with lithium hydride and sodium in tetrahydrofuran solution at very low temperature (−50°C)[8]	From malonic acid;[9] by pyrolysis of ricinoleic acid or castor oil[10,11]
Natural occurrence	Not reported found in nature	Reported found as a metabolite in *Rhodotorula glutinis* var. *lusitanica*;[11] naturally occurring in the essential oils of *Juniperus chinensis*[12] and *Thujopsis dolabrata*[13,14]
Reported uses[3]	Non-alcoholic beverages 0.05–1.0 ppm Ice cream, ices, etc. 0.20 ppm Candy 0.20 ppm	Non-alcoholic beverages 0.50 ppm Ice cream, ices, etc. 0.50 ppm Candy 0.50 ppm Baked goods 0.50 ppm
Regulatory status	FDA 121.1164; FEMA No. 3095	FEMA No. 3247

REFERENCES

For References 1–5, see end of Part III.

6. **Davies and Hodgson,** *J. Chem. Soc.*, p. 86, 1943.
7. **Kagon, Poljakova and Belov,** *Oel Fettind.*, 19, 25, 1954.
8. **Zakharkin et al.,** *Chem. Abstr.*, 61, 5505d, 1964.
9. **Linstead et al.,** *J. Chem. Soc.*, p. 1971, 1937.
10. **Vernon and Ross,** *J. Am. Chem. Soc.*, 58, 2431, 1936.
11. **Seher,** *Fette and Seifen*, 58, 1077, 1956.
12. **Prista,** *Chem. Abstr.*, 49, 11075, 1952.
13. **Uchida,** *Chem. Zentralbl.*, 2, 2198, 1928.
14. **Uchida,** *Chem. Zentralbl.*, 1, 948, 1929.

	Undecen-1-ol	10-Undecen-1-yl acetate
Other names	Alcohol C-11 undecylenic 10-Undecen-1-ol Undecylenic alcohol	Acetate C-11 10-Hendecenyl acetate Undecenyl acetate Undecelynic acetate
Empirical formula	$C_{11}H_{22}O$	$C_{13}H_{24}O_2$
Structure	$CH_2{=}CH{-}(CH_2)_8{-}CH_2OH$	$CH_3{-}C({=}O){-}O{-}(CH_2)_9{-}CH{=}CH_2$
Physical/chemical characteristics[5]		
Appearance	Clear, very mobile liquid	Colorless liquid
Assay	97–98.5% min	95% min
Molecular weight	170.30	212.33
Melting point	−3°C min	
Boiling point		272°C; 140–142°C at 15 mm Hg
Flash point	>82.5°C	
Specific gravity	0.8420–0.8460 at 25°C	0.876–0.880 at 25°C; 0.8808 at 20°C
Refractive index	1.4490–1.4540 at 20°C	1.4380–1.4420 (1.4389) at 20°C
Solubility	1:2 in 70% alcohol; insoluble in water; soluble in most common organic solvents	1:2 in 80% alcohol; soluble in most common organic solvents
Organoleptic characteristics	Fatty odor reminiscent of lemon; fatty, burning taste	Light, rose-like odor; fatty, somewhat agreeable flavor
Synthesis	Prepared from coconut fatty acids as starting material	By acetylation of the corresponding alcohol
Natural occurrence		Not reported found in nature
Reported uses[3]	Reported found in distilled essential oil from leaves of *Litsea odorifera* Valeton.	Non-alcoholic beverages 3.7 ppm Ice cream, ices, etc. 15 ppm Candy 12 ppm Baked goods 12 ppm
Regulatory status	FDA 121.1164	FDA 121.1164; FEMA No. 3096

REFERENCES

For References 1–5, see end of Part III.

	Undecyl alcohol		Valencene	
Other names	Alcohol C-11 Hendecanol 1-Undecanol		1,2,3,5,6,7,8,8a-Octahydro-1,8a-dimethyl-7-(1-methylethenyl)- naphthalene	
Empirical formula	$C_{11}H_{24}O$		$C_{15}H_{24}$	
Structure	$CH_3-(CH_2)_9-CH_2OH$		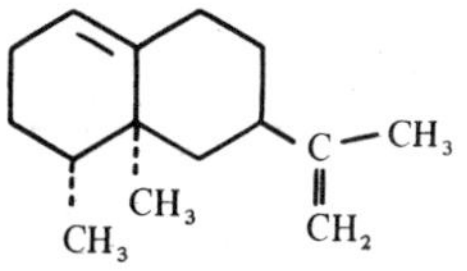	
Physical/chemical characteristics				
Appearance	Colorless liquid at room temperature			
Assay	97–98.5% min			
Molecular weight	172.31		204.36	
Melting point	15–19°C			
Boiling point			123°C at 11 mm Hg	
Flash point	>82°C			
Specific gravity	0.828–0.834 at 25°C			
Refractive index	1.4370–1.4430 at 20°C		1.5075 at 20°C	
Solubility	1:4 in 60% alcohol; insoluble in water; soluble in most common organic solvents			
Organoleptic characteristics	Floral, citrus-like odor; fatty flavor			
Synthesis	By reduction of the corresponding aldehyde		By a Wolf-Kishner reduction of nootkatone[6,7]	
Natural occurrence	Reported found in nature (citrus fruits)		In citrus fruits[8-10] and in cocoa[11,12]	
Reported uses[3]	Non-alcoholic beverages	2.9 ppm	Non-alcoholic beverages	0.9 ppm
	Ice cream, ices, etc.	15 ppm	Alcoholic beverages	0.9 ppm
	Candy	12 ppm	Ice cream, ices, etc.	0.9 ppm
	Baked goods	12 ppm	Candy	0.9 ppm
			Baked goods	0.9 ppm
			Gelatins and puddings	0.9 ppm
			Chewing gum	0.9 ppm
			Milk, dairy products	0.9 ppm
			Cereals	0.9 ppm
Regulatory status	FDA 121.1164; FEMA No. 3097		FEMA No. 3443	

REFERENCES

For References 1–5, see end of Part III.
6. **Hunter et al.**, *J. Food Sci.*, 30, 1, 1965.
7. **Mac Leod**, *Tetrahedron Lett.*, p. 4779, 1965.
8. **Hunter et al.**, *J. Food Sci.*, 30, 383, 1965.
9. **Attaway**, *Phytochemistry*, 5, 1273, 1966.
10. **Maroni**, *Riv. Ital. EPPOS*, 55, 168, 1973.
11. **Van de Wal**, *Recl. Trav. Chim. Pays-Bas*, 87, 238, 1968; *J. Agric. Food Chem.*, 19, 276, 1971.
12. **Rohan**, *Food Process Mark.*, 38, 12, 1969.

	n-Valeraldehyde	Valeric acid
Other names	Amylaldehyde *n*-Pentanal Valeral *n*-Valeric aldehyde	Pentanoic acid Propylacetic acid
Empirical formula	$C_5H_{10}O$	$C_5H_{10}O_2$
Structure	$CH_3-CH_2-CH_2-CH_2-CH=O$	$CH_3-(CH_2)_3-C(=O)-OH$
Physical/chemical characteristics[5]		
Appearance	Colorless liquid	Colorless liquid
Molecular weight	86.13	102.13
Melting point	−91.5°C	34°C
Boiling point	103.4°C	184–186°C (185.5–186.5°C)
Specific gravity	0.819 at 15°C	0.9459 at 20°C
Refractive index	1.3882 (1.3944) at 20°C	1.4100 at 20°C
Solubility	Very slightly soluble in water; soluble in alcohol, propylene glycol, and oils	Slightly soluble in water; soluble in alcohol and ether
Organoleptic characteristics	Powerful, acrid, pungent odor; at low levels taste is warm, slightly fruity, and nut-like	Unpleasant odor and flavor, similar to butyric acid
Synthesis	By distillation of calcium valerate and calcium formate;[6] by reduction of *n*-valeric acid	By oxidation of *n*-amyl alcohol or, together with other isomers, by distillation of valerian roots. Also by reacting butyl bromide and sodium cyanide with subsequent saponification of the formed butyl nitrile
Natural occurrence	Reported found among the constituents of several essential oils: Brazilian sassafras,[7] Bulgarian rose,[8] Bulgarian clary sage,[9] and others; also in the distillates from leaves of various *Eucalyptus* species: *E. cinerea*,[10] *E. globulus*,[11] *E. dives*,[12] *E. maideni*,[13] and *E. hemilampra*[14]	The acid is not too common in nature. Reported (as the corresponding ester) found in the essential oil of *Boronia anemonifolia*, in pineapple fruits, and in other plants; also identified as acid or the corresponding ester in the essential oil of lemon petitgrain[15]
Reported uses[3]	Non-alcoholic beverages 1.3 ppm Ice cream, ices, etc. 5.0 ppm Candy 4.2 ppm Baked goods 5.4 ppm	Non-alcoholic beverages 1.2 ppm Ice cream, ices, etc. 1.8 ppm Candy 2.5 ppm Baked goods 8.0 ppm
Regulatory status	FDA 121.1164; FEMA No. 3098	FDA 121.1164; FEMA No. 3101

REFERENCES

For References 1–5, see end of Part III.

6. **Lieben and Rossi,** *Justus Liebigs Ann. Chem.*, 159, 70, 1871.
7. **Paul and Jachan,** *Perfum. Essent. Oil Rec.*, 42, 416, 1951.
8. **Panayotov and Ivanov,** *Perfum. Essent. Oil Rec.*, 49, 231, 1958.
9. **Panayotov,** *Perfum. Essent. Oil Rec.*, 49, 233, 1958.
10. **Baker and Smith,** *Res. Eucalyptus Essent. Oils* (Sidney), p. 119, 1920.
11. *Schimmel Ber.*, 4, 47, 1904.
12. *Schimmel Ber.*, p. 46, 1928; p. 41, 1930.
13. **Rutowski et al.,** *Riechstoffindustrie,* 1, 159, 1926.
14. **Baker and Smith,** *Res. Eucalyptus Essent. Oils* (Sidney), p. 57, 1910.
15. **Peyron,** *Riv. Ital. EPPOS,* 8, 413, 1965.

	d,l-Valine	Vanillin acetate
Other names	α-Aminoisovaleric acid 2-Amino-3-methylbutanoic acid 2-Amino-3-methylbutyric acid 2-Isopropyl glycine	Acetyl vanillin 3-Methoxy-4-acetoxy benzaldehyde Vanillyl acetate
Empirical formula	$C_5H_{11}NO_2$	$C_{10}H_{12}O_4$
Structure	$CH_3-CH(CH_3)-CH(NH_2)-C(=O)-OH$	$CH_2-O-C(=O)-CH_3$ on benzene ring with $-O-CH_3$ and OH
Physical/chemical characteristics		
Appearance		Crystalline solid
Molecular weight	117.15	194.19
Melting point	156–157.5°C (decomposes at 293°C)	78–79°C
Optical rotation	–6,1° at 25°C (water) –23,6° at 25°C (1*N* HCl) –29,4° at 25°C (6*N* HCl) –39,1° at 25°C (acetone)	
Solubility	184 parts in water at 24°C	Slightly soluble in water; soluble in alcohol and ether
Organoleptic characteristics		Mild balsamic, floral odor; non-vanillin flavor
Synthesis	By the action of ammonia on α-bromoisovaleric acid;[6-8] also through a hydantoin intermediate[9,10]	By acetylation of vanillin or by oxidation of isoeugenyl acetate with chromic acid in the presence of sulfanilic acid;[11] also by electrolytic reduction of vanillin, followed by acetylation of vanillyl alcohol
Natural occurrence	In many fruit, plant, and animal tissues; in milk and dairy products	
Reported uses[3]	Non-alcoholic beverages 15 ppm Ice cream, ices, etc. 40 ppm Candy 80 ppm Baked goods 60 ppm Gelatins and puddings 80 ppm Chewing gum 200 ppm Milk, dairy products 40 ppm	Non-alcoholic beverages, 11 ppm; ice cream, ices, etc., 11 ppm; candy, 28 ppm; baked goods, 28 ppm
Regulatory status	FEMA No. 3444	FDA 121.1164; FEMA No. 3108

REFERENCES

For References 1–5, see end of Part III.

6. **Clark et al.,** *Annalen,* 139, 202, 1866.
7. **Schmidt et al.,** *Annalen,* 193, 101, 1878.
8. **Marvel,** *Org. Synth.,* 20, 106, 1940.
9. **Goldsmith et al.,** U.S. Patent 2,480,644, assigned to Merck & Co.
10. **White,** U.S. Patents 2,557,920 and 2,700,054, assigned to Dow Chemical Co.
11. **Fritzsch & Co.,** German Patent 207,702, August 29, 1905 (March 10, 1909); *Chem. Zentralbl.,* 80, 1207, 1909.

	Vanillin
Other names	4-Hydroxy-3-methoxybenzaldehyde Methyl protocatechuic aldehyde Protocatechualdehyde-3-methylether
Empirical formula	$C_8H_8O_3$
Structure	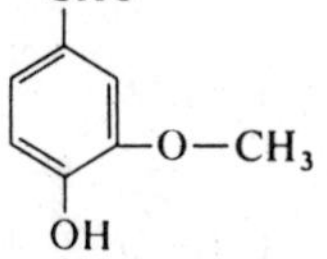 Benzene ring with CHO (position 1), –O–CH_3 (position 3), OH (position 4)
Physical/chemical characteristics[5]	
Appearance	White to yellowish crystalline powder
Molecular weight	152.14
Melting point	81°C (83°C); sublimes when heated
Boiling point	284–285°C (with decomposition)
Specific gravity	1.06 (liquid)
Relative humidity	0.5% max (weight loss %)
Solubility	1:3 in 70% alcohol; 1:2 in 95% alcohol; soluble in ether, chloroform, and hot ligroin; rather soluble in water and glycerol
Organoleptic characteristics	Characteristic strong vanilla-like odor; very sweet taste
Synthesis	From eugenol by direct oxidation or by previous transformation of eugenol in isoeugenol; the oxidation is carried out in alkaline solution after acetylation of the phenol group. From safrole obtained from camphor or sassafras oil by converting to isosafrole; by subsequent methylation and oxidation of isosafrole, a mixture of vanillin and isovanillin is obtained; the separation of the two isomers is a somewhat tedious operation. By introduction of the aldehyde group to the guaiacol molecule, a mixture of vanillin and isovanillin is also obtained.
Natural occurrence	Vanillin widely occurs in nature. It has been reported in the essential oil of Java citronella (*Cymbopogon nardus* Rendl.), in benzoin, Peru balsam, clove bud oil, and chiefly vanilla pods (*Vanilla planifolia, V. tahitensis, V. pompona*). More than 40 vanilla varieties are cultivated. Vanillin is also present in the plants as glucose and vanillin. Another source for vanilla is from the waste (liquor) of the wood-pulp industry. Vanillin is extracted with benzene after saturation of the sulfite waste liquor with CO_2.
Reported uses[3]	Non-alcholic beverages, 63 ppm; ice cream, ices, etc., 95 ppm; candy, 200 ppm; baked goods, 220 ppm; gelatins and puddings, 120 ppm; chewing gum, 270 ppm; syrups, 330–20,000 ppm; chocolate, 970 ppm; toppings, 150 ppm; margarine, 0.20 ppm
Regulatory status	FDA GRAS; FEMA No. 3107

REFERENCES

For References 1–5, see end of Part III.

	Veratraldehyde	*o*-Vinylanisole
Other names	3,4-Dimethoxybenzaldehyde 3,4-Dimethoxybenzenecarbonal Methyl vanillin Protocatechualdehyde dimethylether Vanillin methyl ether	1-Methoxy-2-vinylbenzene
Empirical formula	$C_9H_{10}O_3$	$C_9H_{10}O$
Structure	O=C(H)– ring with –OCH₃ (3) and –OCH₃ (4)	ring with OCH₃ and CH=CH₂
Physical/chemical characteristics[5]		
Appearance	Needles (from ether); oxidizes in air to odorless veratric acid	
Molecular weight	166.17	134.17
Melting point	43–45°C	
Solubility	Almost insoluble in cold water; soluble in hot water, alcohol, and oils	
Organoleptic characteristics	Very sweet, woody, vanilla-like odor; warm, sweet, vanilla-like taste	
Synthesis	Prepared from vanillin[6] by methylation of vanillin with dimethylsulfate under mildly alkaline conditions	
Natural occurrence	Reported among the constituents of the essential oils of *Cymbopogon javanensis*,[7] and *Eryngium poterium*[8]	In *Origanum vulgare*[9,10]
Reported uses[3]	Non-alcoholic beverages 9.0 ppm Ice cream, ices, etc. 9.2 ppm Candy 32 ppm Baked goods 30 ppm Gelatins and puddings 15 ppm	Non-alcoholic beverages 10.0 ppm Candy 10.0 ppm Baked goods 10.0 ppm
Regulatory status	FDA 121.1164; FEMA No. 3109	FEMA No. 3248

REFERENCES

For References 1–5, see end of Part III.

6. **Tiemann and Haarmann,** *Ber. Dtsch. Chem. Ges.,* 8, 1115, 1875.
7. **Hoffman,** *Chem. Zentralbl.,* 3, 886, 1919.
8. **Salgues,** *C. R. Séances Acad. Sci. Paris,* 241, 987, 1955.
9. **Maarse,** Ph.D. dissertation, University of Groningen, Germany, 1971.
10. **Maarse et al.,** *Flavour Ind.,* 4, 477, 1973.

	Vitamin B_1	2,6-Xylenol
Other names	3-(4′-Amino-2′-methyl-5′-pyrimidyl methyl)-5-(2-hydroxyethyl)-4-methyl thiazolium chloride Thiamine hydrochloride Thiamine chloride Aneurin and many other trade names	2,6-Dimethylphenol
Empirical formula	$C_{12}H_{17}ClN_4OS \cdot HCl$	$C_8H_{10}O$
Structure	H_3C, N, N, NH_2HCl, CH_2, S, N (+), CH_2-CH_2OH, CH_3, $Cl^{(-)}$	OH, H_3C, CH_3
Physical/chemical characteristics[5]		
Appearance	Plates in rosette-like clusters; mildly hygroscopic	
Molecular weight	337.28	122.2
Melting point	248°C (decomposes)	49°C
		212°C
Specific gravity		1.169 at 15°C
Solubility	Very soluble in water; slightly soluble in alcohol; insoluble in ether, benzene, hexane, and chloroform; soluble in propylene glycol	
Organoleptic characteristics	Odor slightly reminiscent of thiazole; bitter taste	
Synthesis	See References 6–9	From coal tar oil or coal hydrogenation[10]
Natural occurrence	Rice husks are reportedly the principal source of vitamin B_1; in variable amounts it is a constituent of yeast, milk, green leaves, roots, and tubers; it is also present in high concentration in seeds, and in lesser amounts in different animal organs and muscles	In whisky[11] and coffee[12,14]
Reported uses[3]	Non-alcoholic beverages 50.0 ppm Baked goods 50.0 ppm Condiments, pickles 50.0 ppm Meat, meat sauces, soups 50.0 ppm Milk, dairy products 50.0 ppm Cereals 50.0 ppm	Baked goods 1.0 ppm Condiments 1.0 ppm
Regulatory status	FEMA No. 3322	FEMA No. 3249

REFERENCES

For References 1–5, see end of Part III.

6. **Williams,** *J. Am. Chem. Soc.,* 57, 229, 1935; 58, 1063, 1936.
7. **Williams and Cline,** *J. Am. Chem. Soc.,* 58, 1504, 1936.
8. **Williams, Ruehle and Finkelstein,** *J. Am. Chem. Soc.,* 59, 526, 1937.
9. **Gravin,** *J. Appl. Chem. U.S.S.R.,* 16, 105, 1943.
10. **Böcher,** in *Ullmanns Encyklopädie der Technischen Chemie,* Vol. 13, Urban and Schwarzenberg, 1962, 440.
11. **Nishimura et al.,** *J. Food Sci.,* 36, 819, 1971.
12. **Goldmann et al.,** *Helv. Chim. Acta,* 50, 628, 1967.
13. **Stoffelsma et al.,** *J. Agric. Food Chem.,* 16, 1000, 1968.
14. **Walter et al.,** *Z. Ernährungswiss.,* 9, 123, 1969.

	Zingerone
Other names	4-Hydroxy-3-methoxy benzylacetone (4-Hydroxy-3-methoxyphenylethyl)methyl ketone 3-Methoxy-4-hydroxy benzylacetone 4-(3-Methoxy-4-hydroxyphenyl)-2-butanone Vanillyl acetone
Empirical formula	$C_{11}H_{14}O_3$
Structure	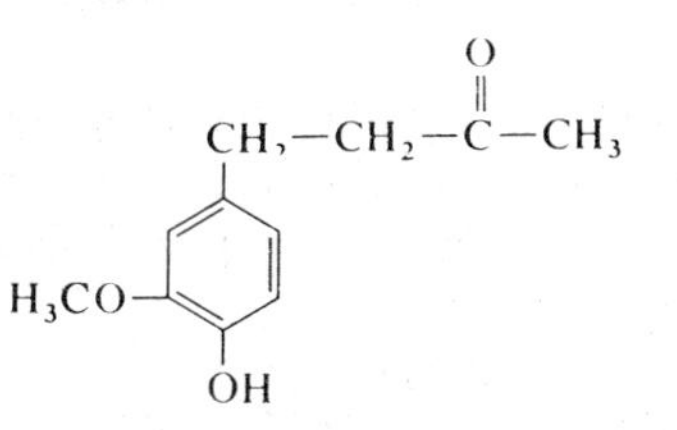
Physical/chemical characteristics[5]	
Appearance	Crystals (from ether and petroleum ether). The commercial product is a yellowish liquid.
Assay	93% min
Molecular weight	194.23
Melting point	40°C (approx. 41°C)
Boiling point	185.5°C at 13 mm Hg Approximately 102°C
Specific gravity	1.138–1.139 at 25°C
Refractive index	1.5440–1.5450 at 20°C
Solubility	1:1 in 50% alcohol
Organoleptic characteristics	Strong, pungent odor reminiscent of ginger; sharp taste similar to ginger
Synthesis	By condensation of vanillin with acetone followed by hydrogenation
Natural occurrence	Reported found in the essential oil of *Zingiber officinale.*[6] It was isolated by Lapwort and collaborators.[7]
Reported uses[3]	Non-alcoholic beverages 6.9 ppm Ice cream, ices, etc. 7.8 ppm Candy 11 ppm Baked goods 11 ppm Chewing gum 15 ppm
Regulatory status	FDA 121.1164; FEMA No. 3124

REFERENCES

For References 1–5, see end of Part III.

6. **Nomura,** *J. Chem. Soc.* (London), 111, 716, 769, 1917.
7. **Lapwort et al.,** *J. Chem. Soc.* (London), 111, 785, 1917.

REFERENCES FOR PART III

The following references have been used in Part III with respect to physical/chemical constants, methods, and use:

1. **Fenaroli, G.,** *Sostanze Aromatiche Naturali,* Vol. 1, Hoepli, Milan, Italy, 1963.
2. *EOA Specification and Standards,* Essential Oil Association of U.S.A., Inc., 60 East 42nd Street, New York, 10017.
3. FEMA (Flavor and Extract Manufacturers' Association) list of GRAS substances as published in various issues of *Food Technology,* starting with Vol. 19, 1965.
4. *Official Methods of Analysis of the Association of Official Agricultural Chemists,* Section 0.091, 10th edition, 1965.
5. **Fenaroli, G.,** *Sostanze Aromatiche Isolate e Sintetiche,* Vol. 2, Parts 1 and 2, Hoepli, Milan, Italy, 1968.

BIBLIOGRAPHY FOR PART III

Information on natural and synthetic flavoring ingredients is quite plentiful, but scattered throughout several texts, often in limited availability. Nonetheless, good central technical libraries should have these available as reference material. The following texts should be consulted for more detailed information:

Arctander, S., *Perfume and Flavor Chemicals,* Steffan Arctander, Montclair, N.J., 1969.
Arctander, S., *Perfume and Flavor Materials of Natural Origin,* Steffan Arctander, Elizabeth, N.J., 1960.
Bedoukian, P. Z., *Perfumery and Flavoring Synthetics,* 2nd ed., American Elsevier Publishing Co., 1967.
Chemicals Used in Food Processing, Publication 1274, National Academy of Sciences/National Research Council, 1965.
Clair, C., *Of Herbs and Spices,* Abelard-Schuman Ltd., 1961.
Gildemeister, E. and Hoffmann, F. R., *Die Aetherischen Oele,* Vols. 3a–3c, Akademie Verlag, 1960–1963.
Givaudan Index, 2nd ed., Givaudan-Delawanna, Inc., 1961.
Guenther, E., *The Essential Oils,* Vols. 1–6, D. Van Nostrand Co., 1948.
Merory, J. M., *Food Flavorings,* 2nd ed., Avi Publishing Co., 1968.

PART IV

The Use of Flavor Ingredients in Food

RELATIONSHIP OF CERTAIN FLAVOR INGREDIENTS TO TASTE AND FLAVOR

Before entering a discussion of complex flavor formulations representing specific food products, a classification of some important natural and synthetic flavor ingredients according to taste and corresponding flavor type may be helpful as a guideline. As indicated in Table 1, while each ingredient can be classified readily into a primary taste, the flavor type best represented by the ingredient can vary. In many instances the same ingredient can deliver several flavor impressions. For this reason and because food substrates are known to alter or modify flavor (*e.g.*, high-protein foods), the single ingredients presented in Table 1 are seldom, if ever, used in rendering a finished flavor.

TABLE 1

Classification of Flavor Ingredients by Primary Taste and Flavor Type

Flavor ingredient	Taste			Flavor type
	Sweet	Bittersweet	Bitter	
Acetophenone			x	—
Allyl anthranilate		x		Green leaves
Allyl benzoate		x		Cherry
Allyl butyrate		x		Apple, apricot
Allyl caproate		x		Pineapple
Allyl cyclohexylacetate		x		Pineapple
Allyl cyclohexylbutyrate		x		Pineapple
Allyl cyclohexylcaproate		x		Peach, apricot
Allyl cyclohexylpropionate		x		Pineapple
Allyl cyclohexylvalerate		x		Peach, apricot, apple
Allyl formate				Mustard
Allyl isovalerate		x		Apple, cherry
Allyl 2-nonylenate				Pineapple
Allyl pelargonate		x		Cognac, pineapple
Allyl phenoxyacetate		x		Pineapple, honey
Allyl phenylacetate			x	Honey
Allyl propionate		x		Apple, apricot
Allyl salicylate		x		Wintergreen, grape
Allyl undecylate		x		Coconut, peach
n-Amyl butyrate		x		Cherry, apple
Amyl phenylacetate	x			Apricot, peach
Anethol	x			Anise
Anisyl alcohol	x			Peach
Anisyl butyrate		x		Cherry, peach
Anisyl formate	x			Strawberry
Anisyl propionate		x		Cherry
Benzyl acetate			x	—
Benzyl butyrate	x			Pear
Benzyl cinnamate	x			Honey
Benzyl formate	x			Apricot, pineapple
Benzyl isobutyrate	x			Strawberry
Benzyl isovalerate	x			Apple
Benzyl propionate	x			Apricot, peach

TABLE 1 (Continued)

Classification of Flavor Ingredients by Primary Taste and Flavor Type

Flavor ingredient	Taste			Flavor type
	Sweet	Bittersweet	Bitter	
Benzyl salicylate	x			Raspberry
Bornyl acetate	x			Pineapple
n-Butyl acetate	x			Pineapple
Butyl formate				Plum
Butyl isobutyrate	x			Pineapple
Butyl isovalerate	x			Apple
Butyl propionate	x			Apricot
Butyl valerate				Apple
Carvacryl acetate		x		Honey
Cinnamaldehyde		x		Cinnamon, melon
Cinnamic acid	x			Apricot
Cinnamyl acetate	x			Pineapple
Cinnamyl alcohol			x	—
Cinnamyl anthranilate		x		Grape
Cinnamyl butyrate	x			Honey
Cinnamyl formate		x		Apple
Cinnamyl isobutyrate	x			Apple
Citral				Lemon
l-Citronellol	x			Peach
Citronellyl acetate	x			Apricot
Citronellyl butyrate	x			Plum
Citronellyl formate	x			Plum
Citronellyl isovalerate	x			Apple
Citronellyl propionate		x		Plum
Coumarin			x	—
p-Cresyl acetate				Honey
p-Cresyl ethyl ether	x			Honey
m-Cresyl phenylacetate	x			Honey
p-Cresyl phenylacetate	x			Honey
Cuminic alcohol		x		Strawberry
Cyclohexyl acetate		x		Apple, banana
Cyclohexyl butyrate		x		Banana, apple, currant
Cyclohexyl caproate		x		Peach, cognac
Cyclohexyl cinnamate				Peach, cherry
Cyclohexyl formate		x		Cherry
Cyclohexyl isovalerate		x		White apple
Cyclohexyl phenylacetate		x		Honey
Cyclohexyl propionate		x		Apple, banana
γ-Decalactone		x		Plum, apricot, peach
Decanal dimethyl acetal			x	Citrus
Decyl acetate	x			Pineapple
Decyl formate	x			Grape

TABLE 1 (Continued)

Classification of Flavor Ingredients by Primary Taste and Flavor Type

Flavor ingredient	Taste			Flavor type
	Sweet	Bittersweet	Bitter	
Diacetyl	x			Butter
Dimethylbenzylcarbinol			x	—
Dimethyl hydroquinone			x	—
Dimethyl phenethyl carbinyl acetate			x	—
Dimethyl phenethyl carbinyl propionate	x			Rose-like
Diphenyl ether	x			Black currant
γ-Dodecalactone		x		Apricot, peach
Ethyl acetate		x		Wine
2-Ethylbutyl acetate		x		Pear
Ethyl butyrate	x			Pineapple
Ethyl cinnamate	x			Apricot, peach
Ethyl formate			x	Rum
2-Ethyl-3-furylacrolein		x		Cola
Ethyl heptylate	x			Wine, pear
Ethyl hexadienoate		x		Pineapple, melon
Ethyl isovalerate	x			Apple
Ethyl methylphenylglycidate	x			Strawberry
Ethyl-2-octynoate	x			—
Ethyl phenoxyacetate		x		Pineapple, honey
Ethyl phenylacetate		x		Honey
Ethyl phenylglycidate	x			Strawberry
Ethyl undecylate		x		Coconut
Ethyl undecynoate			x	—
Ethyl valerate		x		Apple, banana
Ethyl vanillin		x		Vanilla
Eugenol			x	Clove buds
Geraniol			x	Rose-like
Geraniol "palmarosa"	x			Peach, apricot
Geranyl acetate			x	—
Geranyl anthranilate			x	—
Geranyl butyrate	x			Apricot
Geranyl formate			x	—
Geranyl isobutyrate	x			Apricot
Geranyl isovalerate	x			Apricot
Geranyl propionate			x	—
Guaiol acetate	x			Black currant, grape
Guaiol butyrate	x			Plum
Guaiol phenylacetate	x			Honey
Heptyl acetate	x			Apricot
Heptyl formate				Plum
Heptyl propionate		x		Apricot
Hexyl acetate		x		Pear
Hexyl butyrate	x			Pineapple

TABLE 1 (Continued)

Classification of Flavor Ingredients by Primary Taste and Flavor Type

Flavor ingredient	Taste			Flavor type
	Sweet	Bittersweet	Bitter	
Hexyl formate	x			Plum
Hexyl furan carboxylate		x		Pear, mushroom
α-Ionone	x			Raspberry
Isoamyl acetate		x		Pear
Isoamyl formate	x			Plum
Isoamyl isobutyrate				Pineapple
Isoamyl propionate		x		—
Isoamyl salicylate		x		Strawberry
Isobutyl acetate			x	—
Isobutyl anthranilate		x		Strawberry, grape
Isobutyl butyrate	x			Rum
Isobutyl cinnamate	x			Raspberry
Isobutyl formate	x			Rum
Isobutyl phenylacetate	x			Honey
Isobutyl propionate			x	—
Isobutyl salicylate			x	—
Isopropyl acetate	x			Apple
Isopropyl benzyl carbinol		x		Peach
Isopropyl formate	x			Plum
Isopropyl isovalerate	x			Apple
Isopropyl propionate		x		Plum
Isopropyl valerate		x		Apple
Isovalerophenone		x		Grape
Linalool	x			Plum
Linalyl acetate	x			Black currant
Linalyl anthranilate	x			Orange
Linalyl butyrate	x			Honey
Linalyl formate		x		Pineapple
Linalyl isobutyrate	x			Black currant
Linalyl isovalerate			x	—
Linalyl propionate	x			Black currant
Methyl acetate			x	—
Methylacetophenone	x			Strawberry
2-Methylallyl butyrate		x		Apple, plum
2-Methylallyl caproate		x		Pineapple
Methyl amyl ketone			x	Pear
Methyl anisate	x			Melon
Methyl anthranilate			x	—
Methylbenzyl propionate		x		Cherry
Methyl butyrate	x			Apple
Methyl cinnamate	x			Strawberry
Methyl eugenol			x	Clove
Methylheptenone			x	Pear

TABLE 1 (Continued)

Classification of Flavor Ingredients by Primary Taste and Flavor Type

Flavor ingredient	Taste			Flavor type
	Sweet	Bittersweet	Bitter	
Methyl ionone	x			Raspberry, black currant
Methyl isobutyrate	x			Apricot
Methyl isoeugenol			x	Clove
Methyl isovalerate			x	—
Methyl methylanthranilate		x		Peach
Methyl-β-methylpropionate		x		Pineapple
Methyl naphthyl ketone		x		Strawberry
Methyl nonyl ketone	x			Peach
Methyl octine carbonate	x			Peach
Methyl phenylacetate	x			Honey
Methyl phenyl carbinyl acetate				—
Methyl propionate	x			Black currant
Methyl undecylate				Pineapple
Methyl undecyl ketone		x		Coconut
Musk ambrette	x			Peach
Nerol			x	Rose-like
Nerolin	x			Strawberry
Neryl acetate	x			Raspberry
Neryl butyrate	x			Cocoa
Neryl formate			x	—
Neryl isobutyrate	x			Strawberry
Neryl isovalerate			x	—
Neryl propionate	x			Plum
γ-Nonalactone		x		Coconut
Nonyl acetate			x	—
Nonyl alcohol			x	—
γ-Octalactone		x		Peach, coconut, walnut
Octyl acetate	x			Peach
Octyl butyrate	x			Melon
Octyl formate		x		—
Octyl isobutyrate	x			Grape
2-Octynoate	x			—
Phenethyl acetate		x		Honey
Phenethyl alcohol			x	Peach, rose
Phenethyl butyrate	x			Honey
Phenethyl cinnamate			x	—
Phenethyl dimethyl carbinol	x			Apricot
Phenethyl dimethyl carbinyl isovalerate		x		Rose
Phenethyl formate		x		Green plum
Phenethyl isobutyrate		x		Green plum
Phenethyl isovalerate		x		Peach
Phenethyl phenylacetate	x			Honey
Phenethyl propionate				Honey
Phenethyl salicylate	x			Peach

TABLE 1 (Continued)

Classification of Flavor Ingredients by Primary Taste and Flavor Type

Flavor ingredient	Taste			Flavor type
	Sweet	Bittersweet	Bitter	
Phenylacetaldehyde dimethyl acetal		x		—
Phenylacetic acid	x			Honey
Phenylallyl alcohol		x		Plum, peach
Phenylglycidate	x			Strawberry
Phenylpropyl acetate		x		Grape
Phenylpropyl alcohol	x			Apricot
Phenylpropyl butyrate	x			Plum
Phenylpropyl cinnamate	x			Cocoa
Phenylpropyl ether		x		Grape
Phenylpropyl isobutyrate		x		Peach
Propenyl guaethol	x			Vanilla
Propyl acetate		x		Pear
Propyl cinnamate		x		Peach, apricot
Propyl formate		x		Plum
Propyl isobutyrate	x			Pineapple
Propyl phenylacetate	x			Honey
Propyl propionate			x	—
Rhodinol			x	Rose
Rhodinyl acetate			x	—
Rhodinyl butyrate	x			Whortleberry
Rhodinyl formate		x		Cherry
Rhodinyl isobutyrate	x			Peach
Rhodinyl isovalerate		x		Cherry
Santalol	x			Woody, raspberry
Santalyl acetate		x		Apricot
Santalyl phenylacetate	x			Honey
Styralyl acetate			x	Grapefruit
Terpenyl acetate	x			Raspberry
Terpenyl anthranilate			x	—
Terpenyl butyrate		x		Plum
Terpenyl cinnamate			x	—
Terpenyl formate				—
Terpenyl isovalerate	x			Apple
Terpenyl propionate		x		—
Terpineol		x		Peach
Tetrahydrofurfuryl propionate		x		Apricot, chocolate
Tetrahydrogeraniol			x	—
Tolualdehyde (*o,m,p*)		x		Cherry, almond
γ-Undecalactone	x			Apricot, peach
Undecynoate			x	—
Vanillin			x	Vanilla
Vanillylidene acetone		x		Vanilla
Yara yara	x			Strawberry

FLORAL AND CITRUS FLAVORS

In addition to specific information concerning the formulation of flavors for various applications, a separate discussion of the characteristics, functionality, and potential use of floral (rose, jasmine, violet, orris) and citrus (orange, lemon, mandarin, cedrat) flavors may be helpful. These products are not only important technically, but they also are important from a commercial viewpoint, since they represent a very sizable portion of the world market value for flavor ingredients.[1,2]

Floral Flavors

Rose

Among all of the floral derivatives, rose is undoubtedly the most important, mainly because it is used in a wide range of flavor formulations. In addition rose is used by its producing countries and in the Middle East as the basic flavoring component in a large variety of products. Rose (*Rosa damascena* and *R. centifolia*) derivatives include the concrete, absolute, essential oil (rose otto), and the cohobation waters. The concrete is not employed as such but used only for the preparation of the absolute. Cohobation waters have only a very limited use for flavoring finished products and are never used in formulations of high yields. Sometimes the waters are used to dilute soluble, essential-oil-based flavors and to prepare syrups used in pharmaceuticals. The absolute and the essential oil exhibit different flavoring strengths. The note of the absolute is related strictly to the large phenethyl alcohol content and bears close resemblance to the rose flower note. The absolute is preferred therefore in flavor work, while the distilled essential oil is more suitable for use in perfumery.

Rose, with its intensely scented flowers, has attracted the imagination of chefs in times past; the use of rose petals is called for frequently in old recipes for sauces and other delicacies. This is also true for some classic liqueur and sweet cordial formulations, a few of which have been retained until the present time (*e.g.*, rosolio, alkermes). Also rose-flavored sherbets have found a certain degree of acceptance.

On examination of rose-based flavors for various culinary and liqueur formulations, it is apparent that the most interesting results have been obtained from the following combinations:

1. Rose with citrus notes: cedrat, lemon, orange, bergamot
2. Rose with floral notes: orange flowers, violet, jasmine, orris
3. Rose with spicy notes: nutmeg, clove, cinnamon
4. Rose with other notes: angelica, rosemary, bitter almond

Rose is employed successfully in the formulation of *compounded flavors*, such as sour cherry, mango, raspberry, sweet cherry, strawberry, tutti-frutti, pistachio, and millefiori (thousand flowers). The mellow-flavored rose is a very important element, conferring roundness to flavors that often exhibit a harshness from the absence of certain unidentified constituents. In many instances economic considerations limit the use of rose absolute in flavors, and its replacement as part of the formulation is required. Replacement with other flavor ingredients should follow certain criteria; even an excellent rose composition formulated for use in fragrances cannot be adapted for flavoring food products. Many constituents that are highly functional in perfumes are unpalatable or exhibit at least a taste quite foreign to the aroma. Therefore, flavor substitutes for true rose must contain only those constituents of rose absolute with a pleasant flavor. A rose-like composition fulfilling the above requirements can be formulated according to the following general guidelines:

Imitation Rose Flavor

Ingredients	Parts by weight
Geraniol	4.0
Phenethyl alcohol	3.0
Linalool	1.25
Rhodinol	1.25
Citronellol	0.5
Aldehyde C_9	0.05
Citral	0.04

The resulting product can be used as is for flavoring baked goods, while the addition of 5–10% true rose absolute is advisable for candy; this should be increased further when flavoring liqueurs. Whenever a rose note is desired for products soluble in very low-strength alcohol, the use of a 3:1 mixture of phenethyl alcohol and citronellol may suffice if properly dosed.

Jasmine

Jasmine absolute and its substitutes (which must be formulated with the same precautions employed for rose absolute) have very limited use. The absolute should be employed only in minimal doses and limited mainly to liqueurs, candy, special tannings, and fruital aromas. Jasmine absolute also may be replaced entirely or partially by other less expensive flavor compositions. The following guideline formulation, especially suitable for baked goods, may be usefully employed:

Jasmine Formulation

Ingredients	Parts by weight
Benzyl acetate	10.0
Linalyl acetate	4.0
Benzyl butyrate	1.5
Phenethyl butyrate	1.5
Phenethyl alcohol	1.0
α-Amyl cinnamaldehyde	1.0
Hydroxycitronellal	0.3

Violet and Orris

In addition to the violet-flowers note (ionone and methylionone), the characteristic aroma exhibited by the derivatives of orris rhizomes (the butter, absolute, and resinoid) also should be considered for flavors. Violet and orris aromas are extremely important and find application in wines, liqueurs, and fruital flavors used for candy and pastry. While violet and orris aromas are caused chiefly by irone, the fragrance of violet leaf absolute consists of entirely different aromatic principles and does not find application in flavors; however, the aroma of orris derivatives is considered fundamental in flavors. Fresh violet flowers once were candied with a certain degree of acceptance; this practice has nearly disappeared with the exception of a few compounded aromas. Dried violet flowers are used currently to prepare tinctures for liqueurs and, when blended with other dried herbs, for compounded flavors used in carbonated beverages.

The aroma of fresh violet flowers can be approximated using a guideline formulation similar to the following:

Ingredients	Parts by weight
α-Ionone (colorless)	55–58%
Phenethyl alcohol	28–32%
Bergamot	5%
Vanillin	3%
Orris butter	2–3%
Ylang-ylang	1–2%
Guaiac wood	1%

The above formulation is rather simple, has a satisfactory taste, and can be employed either as a base flavor or as a modifier. It is also heat stable and has a harmonious bouquet, especially on dilution.

Citrus Flavors

Citrus essential oils require modification in order to be formulated into flavors that are readily absorbed by the food substrate without yielding objectionable separations or agglomerates during the shelf-life of the product. While food manufacturers have at their disposal more stable and functional essential oils with good solubility, this creates additional problems for the flavorist; the full flavor of the natural essential oil is deprived of a number of complementary notes that are no longer reproducible even when the soluble oil is used for flavoring in combination with large amounts of natural juices. There is no easy solution to this problem at the present time, and the possibility of reproducing the aroma of freshly expressed fruit juice using various aroma formulations does not seem readily attainable. Therefore, the following alternatives (equally applicable to the use of citrus aromas in flavors) remain open in the field of beverage concentrates (syrups, slushes, etc.) and finished ready-to-market carbonated beverages.

Terpeneless and Sesquiterpeneless Citrus Oils

Should citrus oils be employed as is, or should terpeneless and sesquiterpeneless oils be used instead? Paste flavors (for carbonated beverages) usually employ raw or rectified essential oils, while the soluble oils for syrups and liqueurs consist mainly of hydroalcoholic solutions of citrus oils largely deprived of terpene and sesquiterpene fractions by a manufacturing process similar to cold-extracted terpeneless oils. These two product types yield entirely different flavor values. Although flavor judgments are subjective, undoubtedly beverages flavored with paste flavors (all other flavor components remaining the same) duplicate the aroma of freshly expressed juices better than soluble oils. However, paste flavor formulations cannot be prepared with all citrus oils; flavorwise, the most successful paste flavor is orange. Because of the poor stability of lemon (and cedrat) essential oils, the use of the more stable soluble oil is preferred to obtain longer lasting and more consistent flavor effects.

The beverage industry, with a world consumption of several thousand tons of citrus oils, does not represent the only outlet for citrus-based flavors. Statistical figures on the exact ratio of citrus-oil consumption between the beverage and confection industries generally are not available. However, this ratio is estimated between 3:1 and 4:1 for the European market (*e.g.*, Italy) but perhaps 16:1 in the United States. The Italian pastry industry, for example, has attained world-wide acceptance with its important product panettone, a traditional Italian fruitcake. This has created a need for the characteristic flavor of panettone derived from citrus oils; the flavor duplicates the fruitcake aroma obtained by adding candied citron, raisins, etc. to the dough.

The panettone aroma is not the only type for which an industrial need exists; several other products that are employed for pastry flavoring can be considered variations of the panettone theme. All types may contain an additional butter-like note. Independent of the presence of the butter-like note and of the varying ratios of the constituents in the flavor, the characteristic notes of panettone and other similar flavors include the citrus and vanilla notes. The citrus note is subdivided further into citron-like and "fantasia" types. Citron-like essential oils are duplicated with a blend of lemon, orange, and bergamot essential oils. The formulation of such blends varies widely according to the composition of natural citron essential oil, but only limonene, dipentene, citral, and citroptene are important to the aroma. Therefore, a good lemon essential oil with a citral content that has been increased from 3–4% to 5–6% should suffice for good duplication.

In practice organoleptic requirements have led to more complex formulations. The organoleptic requirement, *i.e.*, the desired aromatic result, is usually satisfied by the use of flavor formulations with the following minimum and maximum values, together with all possible intermediate values:

Oil	Formulation types				
	A	B	C	D	E
Lemon	12	2	1	1	80
Orange	1	2	1	2	20
Bergamot	0.5	1	—	2	0.25

In the above formulations having the citrus note, type D can be eliminated from further contention as a flavor, since it is used only in perfumery; with slight modifications it is a good starting material for colognes. Type A is a good starting material for the preparation of a soluble oil useful for flavoring carbonated beverages. Type B is a magnificent flavor base for preparing compounded oils for liqueurs and syrups, especially if the lemon concentration is decreased slightly by 10% and the orange concentration increased by 30% (*i.e.*, final formulation 62, 22, 16). Finally, formulations E and C are extremely suitable for use in pastry. Mandarin oil also may be added to the above bases to sweeten the flavor note; when added to formulation A, B, C, or E, it should never exceed 0.5% based on the level of orange oil. Referring once again to panettone, this flavor type may be formulated with or without bergamot. In either case a rounding of the citrus-citron note is required by the addition of a good mandarin oil—one part to formulation E or one-half part to formulation C (with or without the use of bergamot). The finishing touches to the formulation are rendered by dosing with 1 part of vanillin (25% maximum).

The remarkable evolution in taste over the past 60 years for lemon-flavored carbonated beverages should be noted. At the beginning of this century, a typical lemon-flavored beverage consisted of a soluble lemon oil dissolved in ordinary syrup and diluted with carbonated water. Today cola beverages have attained widespread acceptance; they are bittersweet, fantasia-type, lemon-based beverages whose formulation originated in the United States at the end of the nineteenth century and was introduced into Europe during World War I. The flavor composition is complex and based chiefly on the following components: cola-nut extract, lemon oil or juice, lime, sweet orange, neroli, traces of spicy oils, caffeine, caramel, and phosphoric acid. The lime, lemon, and orange essential oils, modified to a slightly bitter flavor, represent the citrus note in the syrup. Other flavor ingredients only complement the characteristic note. The introduction of cola beverages has radically modified the consumers' taste and has created in a matter of a few years a much wider acceptance for bitter-flavored beverages. On the continent, for example, grapefruit and bitter-orange flavored beverages (scarcely accepted until a few years ago) and other bitter tonics based on a citrus undernote (vermouth-rhubarb) have captured large portions of the consumer market. The degree of carbonation has undergone a parallel change, and highly carbonated beverages are currently preferred.

Bitter and Sweet Orange

Sweet orange essential oil is employed as is for flavoring both non-alcoholic beverages and liqueurs. Bitter orange oil, however, is never used alone or in concentrations exceeding 50% of the aroma formulation (maximum levels are used in liqueurs). A blend of bitter and sweet orange oils containing approximately 50% each of the two oils (also ratios varying between 90:10 and higher) is used to prepare bitter-flavored beverages and aperitifs. In fact the flavor of certain aperitifs consists of more than 95% sweet and bitter orange together with the corresponding peel extracts, traces of vanillin, and complementary bitter notes. The sweet orange note is widely employed in compounded oils for liqueurs and fruital aromas (apricot, pineapple, benedictine, banana, cherry, chartreuse, strawberry, peach and benevento).

Lemon

The widespread acceptance of the new lemon-flavored beverages together with citron-like combinations has already been mentioned. Lemon essential oil is used alone, especially in beverages and candy. Alone or in combination with lime (distilled lime oil has an aroma entirely different from the cold-expressed product), lemon oil finds use in the flavor composition of several bittersweet beverages. In the formulation of compounded oils and fruital aromas, lemon oil is used less extensively than sweet orange oil, but it can be utilized (found) in the following liqueurs and fruital aromas: apricot, pineapple, benedictine, banana, curaçao, strawberry, apple, currant, and benevento.

Mandarin

The mandarin note is used alone to flavor candies and liqueurs. Usually it finds very limited use in the formulation of compounded aromas with the exception of the very important flavors for baked goods and pastry (panettone). The rum-mandarin note is a very interesting aroma combination for mandarin punch.

FRUITAL FLAVORS

Introduction

One of the most interesting areas of flavor formulation is the imitation of fruital aromas. It is well known that man retains a strongly conservative predilection for natural flavors. The attraction for fruital flavors probably occurs during youth, at which time most persistent organoleptic preferences are developed. This psychological need, associated with the lack of available natural fruit aromas with the strength to flavor food products or at least to partially duplicate the pleasant fruit flavor sensation, has compelled the flavorist to concentrate his efforts in the imitation of fruit aromas.

Originally the flavor ingredients of choice were very limited in number and consisted chiefly of essential oils. The resulting flavor was scarcely attractive. A renewed impulse to create fruital formulations was brought about by the introduction of the so-called "fruit ethers." These are the highly volatile esters readily prepared by esterification of aliphatic acids with ethyl, methyl, butyl, propyl, or amyl alcohols. For several years thereafter fruit flavor formulations were based almost exclusively on essential oil and fruit ether flavor blends.

The systematic identification of constituents in natural flavors is a more recent development. Until recently all attempts to reformulate a natural aroma were limited to the analytical organoleptic perceptiveness of the flavorist. Science was not interested in such marginal problems, and the enthusiasm of a few pioneers was not sufficient to make flavor chemistry progress at the same fast pace as other industry-oriented branches of chemistry. As investigations gradually expanded, new flavor ingredients became available to the flavorist. However, the accidental discovery of a flavor component with a definite aroma was kept highly secret by those few better-equipped researchers. Such products were made commercially available as complex mixtures under a variety of trade names or arbitrarily qualified as "aldehydes" (*e.g.*, aldehydes C_{15}, C_{17}, C_{19}, C_{21}, C_{25}, C_{27}, C_{29}, C_{30}, and C_{31}). Other suppliers traded the pure product under imaginative names or used the misnomer of aldehyde (*e.g.*, aldehyde C_{14} for γ-undecalactone, aldehyde C_{18} for γ-nonalactone, and aldehyde C_{16} for ethyl methylphenylglycidate). More advanced flavors could be formulated with the use of these new products, especially strawberry, peach, coconut, plum, and apricot. In recent years studies of the constitution of flavors have progressed markedly by the introduction of instrumental analysis; most of the chemically defined components of the more important fruital flavors and their relative ratios have been identified.

As a result of the acquired knowledge, increasing attention has been paid to a number of synthetic chemicals with regard to their potential use in small percentages in modern fruital flavors. A list of the most recurrent chemicals, grouped by functional classification, is shown below. When properly chosen and employed in appropriate amounts, these products play a very important role in the formulation of fruital flavors, such as apple, banana, bilberry, grape, grapefruit, guava, orange, peach, pear, pineapple, raspberry, strawberry, tomato, and tea. Often close duplication of the natural flavor notes is achieved. Those chemicals marked with an asterisk are either specific aroma components identified in one of the above-mentioned natural flavors or common to a number of the above flavors. In all cases, they are the most widely employed in the formulation of well-balanced fruital flavors.

Alcohols and Esters

1-Penten-3-ol*
cis-2-Pentenal*
cis-3-Pentenal*
Pentenyl caproate (hexanoate)*
1-Hexen-3-ol*
cis-2-Hexenol*
cis-2-Hexenyl acetate
cis-2-Hexenyl caproate (hexanoate)
trans-2-Hexenol*
trans-2-Hexenyl acetate*
trans-2-Hexenyl propionate
trans-2-Hexenyl butyrate*
trans-2-Hexenyl caproate (hexanoate)*

Alcohols and Esters *(continued)*

trans-2-Hexenyl caprylate (octanoate)*
cis-3-Hexenol*
cis-3-Hexenyl acetate*
cis-3-Hexenyl butyrate
cis-3-Hexenyl propionate*
cis-3-Hexenyl formate
cis-3-Hexenyl valerate
cis-3-Hexenyl isovalerate
cis-3-Hexenyl 2-methylbutyrate
cis-3-Hexenyl tiglate
cis-3-Hexenyl caproate (hexanoate)*
cis-3-Hexenyl hepatonoate
cis-3-Hexenyl caprylate (octanoate)*

Alcohols and Esters ***(continued)***

cis-3-Hexenyl caprate (decanoate)
cis-3-Hexenyl lactate
cis-3-Hexenyl pyruvate
cis-3-Hexenyl benzoate*
cis-3-Hexenyl phenylacetate
cis-3-Hexenyl cinnamate
cis-3-Hexenyl salicylate
cis-3-Hexenyl anthranilate
trans-3-Hexenol*
trans-3-Hexenyl acetate*
trans-3-Hexenyl caproate (hexanoate)
trans-3-Hexenyl caprylate (octanoate)*
β, α,Hexenol (mixture of *cis*- and *trans*-3-hexenol)
β, α-Hexenyl acetate
β, α-Hexenyl formate
β, α-Hexenyl isobutyrate
trans, trans-2,4-Hexadienol
trans, trans-2,4-Hexadienyl acetate
cis-3-Heptenol*
cis-3-Heptenyl acetate*
cis-4-Heptenol
1-Octen-3-ol*
1-Octen-3-yl acetate
1-Octen-3-yl butyrate
1-Octen-3-yl propionate
trans-2-Octenol*
trans-2, *cis*-6-Octadienol
trans-2, *cis*-6-Octadienyl acetate
2-Nonanol*
1-Nonen-3-ol*
1-Nonen-3-yl acetate
trans-2, *cis*-6-Nonadienol
trans-2, *cis*-6-Nonadienyl acetate
2-Decanol*

Acetals

Acetaldehyde ethyl-(*cis*-3-hexen)-yl acetal*
Acetaldehyde ethyl-(*trans*-3-hexen)-yl acetal*
Acetaldehyde (*cis*-3-hexen)-yl acetal*
trans-2-Hexenal diethyl acetal
cis-4-Heptenal diethyl acetal
trans, trans-2,4-Hexadienal diethyl acetal
trans-2, *cis*-6-Nonadienal diethyl acetal

α-Ethylenic Aldehydes

trans-2-Pentenal*
trans-2-Hexenal*
trans-2-Heptenal*
trans-2-Octenal*
trans-4-Octenal*
trans-2-Nonenal*
trans-2-Decenal*

α-Ethylenic Aldehydes ***(continued)***

trans-2-Undecenal*
trans-2-Dodecenal*

Monoethylenic Aldehydes

cis-4-Heptenal

Diethylenic Aldehydes

trans, trans-2,4-Hexadienal*
trans, trans-2,4-Heptadienal*
trans-2, *cis*-6-Octadienal
trans-2, *cis*-6-Nonadienal

Ketones

Nootkatone*

Acids and Esters

2-Methylbutyric acid*
Methyl 2-methylbutyrate*
Ethyl 2-methylbutyrate*
Propyl 2-methylbutyrate*
Butyl 2-methylbutyrate*
Isobutyl 2-methylbutyrate*
Amyl 2-methylbutyrate*
Isoamyl 2-methylbutyrate
Hexyl 2-methylbutyrate*
Octyl 2-methylbutyrate
Ethyl-3-hydroxy hexanoate*
trans-2-Hexenoic acid*
Ethyl *trans*-2-hexenoate
Methyl *trans*-2-hexenoate*
cis-3-Hexenoic acid*
Ethyl *cis*-3-hexenoate
trans-3-Hexenoic acid
Ethyl *trans*-3-hexenoate
Methyl *trans*-3-hexenoate*
Methyl *cis*-4-octenoate*
trans-2-Octenoic acid*
Ethyl *trans*-2-octenoate*
Methyl *trans*-2-octenoate*
trans-3-Octenoic acid
Ethyl *trans*-3-octenoate
Methyl *trans*-3-octenoate
trans-2-Decenoic acid
Ethyl *trans*-2-decenoate*
Methyl *trans*-2-decenoate*
trans-3-Decenoic acid
cis-4-Decenoic acid
Ethyl *cis*-4-decenoate*
Methyl *cis*-4-decenoate*
Ethyl *trans*-2, *cis*-4-decadienoate*
Ethyl *trans*-2, *trans*-4-decadienoate*
Methyl methylthiopropionate

Finally, the flavor effects obtainable by the addition to fruital formulation of flavor components such as C_6 to C_{12} gamma and delta lactones should not be overlooked. These components, occurring naturally in many pit fruits, contribute to create flavor fullness by arousing sensations reminiscent of fruit pulp. Therefore, they are considered essential in the formulation not only of peach and apricot flavors, but also of pear, raspberry, and many other fruital flavors.

Will it be possible with this detailed knowledge to duplicate exactly the various fruital flavors? A positive answer should be expected based on analogy with the excellent results obtained in duplicating essential oil flavors. In a few cases not only the flavor but exact physical-chemical constants have been elucidated. However, even perfectly rebuilt essential oils exhibit somewhat apparent organoleptic deficiencies. Similar results in the sector of rebuilt fruital aromas are probably quite a long distance away. For example, gas chromatographic analysis using capillary and semi-capillary columns, together with other refined instrumental techniques, yields flavor compositions that appear easily duplicable. In other cases the analysis reveals an exceedingly large number of components. Simply mixing components based on gas chromatographic data has led to disappointing results; on the other hand, results of a more general and positive nature have been obtained using these techniques. Such work has led to the formulation of flavor bases and in this respect contributed to products that duplicate the natural types much more closely than possible with prior flavor art.

The causes determining the organoleptic deficiencies in a rebuilt aroma, even when all of the constituents are identified and available, are manifold and complex. Formation of "artifacts" during isolation of the flavor complex or during the analytical investigation cannot be overlooked. Degradation products formed during the time elapsed between harvesting and working-up the fruit to isolate the aroma should also be considered, since the maturation process may have been interrupted, giving way to a degenerative process. Present knowledge does not allow definite conclusions. Considering, however, that natural flavor components are produced by constructive and destructive mechanisms functioning concomitantly, it has been impossible (at least to date) to rebuild flavors under conditions of biogenesis. Based on present knowledge, the only choice is to duplicate natural flavors as closely as possible with good imitations.

In the following pages the more important fruit flavors will be discussed together with specific formulation guidelines and aroma compositions, if available, as identified by more recent analytical studies. Formulations will be based on readily available raw materials; specific components whose use is subject to government regulations that vary from country to country will be omitted. The formulations list all probable components with potentially useful alternatives indicated in parentheses ( ) or as separate entries. *Only those marked with an asterisk (*) comprise the guideline formulation.* The formulation guideline should *not* be considered in any way finished; it must be further refined by the flavorist to meet specific end-use requirements.

Apple

The complexity and the continuing changes in the innumerable types of apple aromas have caused quite different approaches to the formulation of the aroma imitations. In addition to amyl acetate, amyl butyrate, amyl valerate, and ethyl butyrate, which comprise up to 50% of the total aroma formulation, a fairly large number of aliphatic esters (especially the C_5 to C_{10} acid esters) are often employed in the formulation. Although natural apple aroma contains many of the above esters, the proper blending of the esters in the formulation is extremely difficult, especially when aroma imitations of specific cultivars rather than generic and polyvalent aroma formulations must be prepared.

Products most frequently added to modify or characterize the various apple aroma formulations (in addition to the above-mentioned flavor ingredients) follow: ethyl acetate, ethyl valerate, ethyl isovalerate, ethyl pelargonate, vanillin, lemon essential oil, citral, citronellal, rose absolute, geraniol, orange essential oil, geranium essential oil, aldehyde C_{10}, ethyl heptanoate, acetaldehyde, so-called aldehydes C_{14} and C_{16}, styralyl acetate, dimethyl benzyl carbinyl acetate, benzyl formate, phenyl ethyl isobutyrate, cinnamyl isovalerate, anise essential oil, and the methyl esters of colophony and benzaldehyde.

The use of butyl isovalerate, isopropyl acetate, and methyl butyrate (naturally occurring products) has also been found of interest in apple aroma imitations. Encouraging results have been obtained by using the following products not yet identified in natural apple aroma but most likely occurring in other natural products: terpenyl isovalerate, isopropyl isovalerate, citronellyl isovalerate, geranyl isovalerate, benzyl isovalerate, cinnamyl formate, isopropyl valerate, butyl valerate, and methyl allyl butyrate.

Synthetic products not yet found in nature whose use in apple aroma formulations is governed (as the above compounds) by specific regulations in different countries include: cyclohexyl acetate, allyl butyrate, allyl cyclohexylvalerate, allyl isovalerate, and cyclohexyl butyrate. The aroma of these products blends very well with natural apple aroma.

Several analytical investigations have broadened the present knowledge of apple aroma. The first of these studies dates back to 1920. To date, about 40 researchers have contributed to the investigation. From the combined results 90 components have been identified in various natural apple aromas; these have been sub-divided in the following table according to their chemical functionality.

Components of Natural Apple Aroma

Hydrocarbons
- 1-Butoxy-1-ethoxyethane
- 1,-1-Diethoxyethane
- 1-Ethoxy-1-hexoxyethane
- 1-Ethoxy-1-methoxyethane
- 1-Ethoxy-1-(2-methylbutoxy)-ethane
- 1-Ethoxy-1-propoxyethane
- 1-Methyl-naphthalene
- 2-Methyl-naphthalene
- 2,4,5-Trimethyl-1,3-dioxolane

Free and Esterified Alcohols
- Butanol
- Ethanol
- Geraniol
- Hexanol
- *n*-Hexenol
- *trans*-2-Hexen-1-ol
- 3-Hexen-1-ol
- Methanol
- 2-Methylbutan-1-ol
- 3-Methylbutan-1-ol
- 2-Methylpropan-1-ol
- Pentanol
- Propanol
- 2-Propanol

Esters
- Benzyl acetate
- Butyl acetate
- Butyl butyrate
- Butyl hexanoate
- Butyl propionate

Esters *(continued)*
- Ethyl acetate
- Ethyl butyrate
- Ethyl formate
- Ethyl hexanoate
- Ethyl 2-methylbutyrate
- Ethyl 2-methylpropionate
- Ethyl octanoate
- Ethyl pentanoate
- Ethyl 2-phenylacetate
- Ethyl propionate
- *trans*-2-Hexen-1-yl-acetate
- Hexyl acetate
- Hexyl butyrate
- Hexyl propionate
- Methyl acetate
- 2-Methylbutyl acetate
- 3-Methylbutyl acetate
- 3-Methylbutyl 3-methylbutyrate
- Methyl butyrate
- Methyl formate
- Methyl hexanoate
- Methyl 2-methylbutyrate
- Methyl 3-methylbutyrate
- 2-Methylpropyl acetate
- 2-Methylpropyl propionate
- Pentyl acetate
- Pentyl butyrate
- Pentyl 2-methylbutyrate
- 2-Phenylethyl acetate
- Propyl acetate
- 2-Propyl acetate
- Propyl butyrate
- Propyl pentanoate
- Propyl propionate

Free and Esterified Acids
- Acetic
- Benzoic
- Butyric
- Formic
- Hexanoic
- *n*-Hexenoic
- 3-Methylbutyric
- 4-Methylpentanoic
- 2-Methylpropionic
- Octanoic
- Pentanoic
- Propionic

Aldehydes and Ketones
- Acetaldehyde
- Acetone
- Acetophenone
- Butanal
- 2-Butanone
- Diacetyl
- Formaldehyde
- Furfural
- Hexanal
- 2-Hexanone
- 2-Hexenal
- 2-Methylbutanal
- 3-Methylbutanal
- 2-Methylpropanal
- Nonanal
- Pentanal
- 2-Pentanone
- 3-Pentanone
- Propanal

Apple Flavor

Ingredient	Parts by weight	Ingredient	Parts by weight
Acetaldehyde	13.0	Ethyl isovalerate	
Acids (formic, acetic, butyric*)	4.0	Ethyl methylphenylglycidate*	12.0
Aldehyde C_{10}*	5.0	Ethyl pelargonate	
Allyl butyrate		Ethyl phenylacetate	
Allyl cyclohexylvalerate		Ethyl valerate	
Allyl isovalerate		Geraniol*	21.0
Amyl acetate*	240.0	Geranium essential oil*	14.0
Amyl butyrate*	170.0	Geranyl acetate*	100.0
Amyl valerate*	570.0	Geranyl isovalerate	
Anise essential oil (anethol*)	3.0	Isopropyl isovalerate (isopropyl valerate)	
Benzaldehyde		Lemon essential oil*	24.0
Benzyl formate*	1.1	Maltol*	7.0
Benzyl isovalerate		Methyl acetate	
Butyl valerate		Methyl allyl butyrate	
Cinnamyl formate		Orange essential oil*	11.2
Cinnamyl isovalerate*	35.7	Phenethyl isobutyrate*	28.4
Citral	15.0	Rhodinol*	5.0
Citronellal*	13.0	Rose, absolute*	0.6
Citronellyl isovalerate		Styralyl acetate*	10.0
Cyclohexyl acetate		Terpenyl isovalerate	
Cyclohexyl butyrate		γ-Undecalactone*	14.0
Dimethyl benzyl carbinyl acetate*	34.0	Vanillin* (ethyl vanillin)	50.0
Ethyl acetate*	80.0	Solvent	455.0
Ethyl butyrate			
Ethyl formate		**Total**	2,000.0
Ethyl heptanoate	64.0		

Notes:
1. An asterisk denotes the guideline formulation.
2. Parentheses indicate potentially useful alternatives.

Apricot

Apricot flavor, although varying somewhat from cultivar to cultivar, exhibits a characteristic note, *i.e.*, a very pronounced floral note with a citrus-like undernote that becomes more intense with incipient ripening. The various imitation apricot flavors have originated from an association with violet, rose, and jasmine, together with other suitable ingredients. Historically the formulation of imitation apricot flavor has evolved through the following stages:

1. Essential oils of bitter almond, orange, geranium, and cognac, vanilla, ethyl acetate, ethyl valerate, and ethyl butyrate were combined.
2. Amyl acetate, amyl valerate, and amyl butyrate were then added to the base ingredients in No. 1.
3. Coumarin was the next major ingredient to be added.
4. New formulations based on the ingredients in Nos. 1, 2, and 3 contained α-ionone, orris butter, jasmine absolute, ethyl caproate, and ethyl caprylate.
5. Finally ethyl methylphenylglycidate, γ-nonalactone, γ-undecalactone, traces of eugenol, angelica essential oil, and petitgrain bigarade were employed (all or only a few of these products in varying combinations).

Although the above combinations attained a certain degree of success, in most instances they are too complicated. The constituents of a rather simple but effective formulation include the following: amyl acetate, amyl butyrate, amyl formate, amyl valerate, benzaldehyde, cinnamaldehyde, ethyl caproate, ethyl heptylate, ethyl valerate, geraniol, α-ionone, neroli, γ-undecalactone, and vanillin.

An examination of current formulation capabilities not only confirms the usefulness of almost all the constituents employed in the evolution of the apricot aroma formulation, but it also points out the usefulness of a few products previously identified in nature and found to be extremely functional in the formulation. These include cinnamic acid, phenyl propyl alcohol, geraniol, ethyl cinnamate, citronellyl acetate, geranyl isobutyrate, geranyl butyrate, propyl butyrate, methyl isobutyrate, and butyl propionate. A few additional products, such as benzyl propionate, heptyl acetate, methyl isobutyrate, citronellyl isobutyrate, propyl cinnamate, santalyl acetate, and isoamyl phenylacetate, may also find practical use in the formulation; their natural occurrence, although very probable, has not yet been reported. New synthetic molecules have been prepared whose notes blend well with apricot aroma; among these the following are noteworthy: allyl cyclohexylvalerate, allyl cyclohexylcapronate, tetrahydrofurfuryl propionate, allyl butyrate, and allyl propionate. The use of such synthetic products is subject to government regulations by which the formulator must abide.

Studies of the composition of natural apricot aroma have not solved the problem of reconstituting the exact aroma. While the analytical results would indicate the sure presence of certain components in the aroma, these results are partially conditioned by the potential formation of artifacts. Gas chromatography, aided by other instrumental techniques, has identified about 15 of the approximately 40 components of a natural apricot aroma absorbed on activated carbon. The identified products include myrcene, limonene, *p*-cymene, *trans*-2-hexanol, acidic acid, linalool, terpinolene, geranial, 2-methylbutyric acid, α-terpineol, γ-octalactone, γ-decalactone, and γ-nonalactone. This preliminary identification may explain to some extent the floral note and citrus undernotes originally recognized in apricot aroma.

Apricot Flavor

Ingredient	Parts by weight	Ingredient	Parts by weight
Allyl butyrate		Isoamyl phenylacetate*	0.1
Allyl cyclohexylcaproate*	0.2	Jasmine, absolute (benzyl acetate*)	9.5
Allyl cyclohexylvalerate		Lemon essential oil* (citral)	5.0
Allyl propionate		Methyl cinnamate	
Almond, bitter* (benzaldehyde)	11.5	Methyl isobutyrate	
Amyl acetate*	7.5	Neroli* (petitgrain bigarade sur fleurs)	18.5
Amyl butyrate*	7.5		
Amyl formate*	10.0	γ-Nonalactone	
Amyl valerate*	15.0	γ-Octalactone	
Benzyl propionate		Orange essential oil*	10.5
Ceylon cinnamon (cinnamaldehyde)	0.5	Orris butter (α-irone)	
Citronellyl isobutyrate		Phenethyl butyrate	
γ-Decalactone		Propyl butyrate	
γ-Dodecalactone		Propyl cinnamate*	0.2
Ethyl acetate*	14.5	Rose, absolute*	3.0
Ethyl butyrate*	4.5	Tetrahydrofurfuryl propionate	
Ethyl hexanoate	10.0	γ-Undecalactone	200.0
Ethyl octanoate		Vanillin* (ethyl vanillin, propenyl guaethol, vanilla absolute)	85.0
Ethyl valerate*	50.0		
Geranium (rhodinol, geraniol, *l*-citronellol, linalool,* α-terpineol)	0.5	Solvent	527.0
Heptyl acetate			
α-Ionone*	9.5	**Total**	1,000.0

Notes:
1. An asterisk denotes the guideline formulation.
2. Parentheses indicate potentially useful alternatives.

Banana

Imitations of banana flavors have attained success for some time. Amyl and isoamyl acetate together with amyl butyrate (the presence of these components in the natural aroma was subsequently ascertained) were found to be extremely suitable in banana formulations and closely duplicated the natural aroma. The specific note of these esters is usually rounded off by the addition of a green, woody note together with a more volatile fruity note (amyl butyrate alone also exhibits a fruity note). Flavorists have attempted to further improve the aroma formulation by means of various esters, a few essential oils (mainly citrus oils), and vanillin.

A cursory examination of formulations perfected around 1920 for candy flavoring with a yield of approximately 3,000 shows that about 50% of the aroma formulations consisted of amyl acetate and amyl butyrate usually in a 2:3 ratio, while an additional 20% consisted of amyl valerate. Other esters, such as ethyl caprylate and ethyl heptanoate, were employed at lower levels; the addition of citrus oils and vanillin did not exceed a total of 10% in the formulation. The above ratios were largely increased in the absence of essential oils to the following upper limits: 44% amyl butyrate, 43% amyl acetate, 8% ethyl butyrate, 1% geranyl propionate, and 4% vanillin. Some flavorists recommended the use of approximately 10% acetaldehyde in formulations containing about 50% amyl acetate and, in addition, strong floral, rose-violet notes, coumarin, vanilla, and heliotropin (10% total).

Based on the above description, it is apparent that imitations of banana aroma were rather simple, containing only a limited number of components. In practice the basic schemes have not been abandoned altogether, although significant qualitative improvements have been introduced in the formulation subsequent to the commercial availability of esters having much higher purity.

A more recent approach to rounding off the banana aroma containing mainly amyl acetate, isoamyl acetate, and isoamyl butyrate consists of using benzyl butyrate, 2-ethylbutyl acetate, methyl heptanone, isopropyl benzyl carbinol, methyl amyl ketone, ethyl heptanoate, ethyl valerate, hexyl acetate, and cyclohexyl esters (butyrate, isovalerate, propionate). Not all of the above products have been identified in the constituents of the natural aroma; therefore, their use is conditioned by certain limitations. Butyl acetate and butyl butyrate might also be used.

Several studies of the constituents of banana aroma have been published, especially since 1950. Because the aroma of bananas from various cultivars and at varying stages of the ripening process were studied, analytical results were significantly different. Although the common constituents were always found to be present, their relative ratios in the aroma varied widely from type to type. Two banana cultivars (Gros Michel and Valery) have been the subject of extensive investigation. Approximately four–five constituents of the banana aroma remain unidentified; two are found among the low-volatile components. The following components have been identified in Gros Michel banana aroma:

Components of Gros Michel Banana Aroma

Amyl acetate	*n*-Hexyl acetate
1-Butanol	Isoamyl acetate
2-Butanol	Isoamyl alcohol
n-Butyl acetate	Isoamyl butyrate
Ethanol	Isobutyl acetate
Ethyl acetate	2-Pentanone
1-Hexanol	2-Pentyl acetate
trans-2-Hexenal	2-Pentyl butyrate
1-Hexyl acetate	1-Propanol

It is interesting to note the presence of alcohols in this aroma. Prior to their identification alcohols were never taken into consideration, although they produced interesting aromatic effects by imparting a woody, somewhat mold-like note often associated with the green note typical of several natural banana aromas.

Banana Flavor

Ingredient	Parts by weight	Ingredient	Parts by weight
Acetaldehyde*	0.75	Heliotropin*	7.20
Alcohol C_6		2-Hexenal*	0.5
Amyl acetate*	300.50	Hexyl acetate	
Amyl butyrate*	206.0	Isoamyl acetate*	70.0
Amyl valerate*	18.0	Isoamyl alcohol*	0.50
Benzyl butyrate		Isoamyl butyrate*	40.0
Benzyl propionate*	6.6	Isobutyl acetate	
Butyl acetate*	1.0	Isopropyl benzyl carbinol	
Butyl butyrate		Lemon essential oil*	1.30
Cyclohexyl butyrate		Methyl amyl ketone	
Cyclohexyl isovalerate		Methyl heptenone*	1.0
Cyclohexyl propionate*	2.0	Orange essential oil*	0.30
Ethyl acetate		Rose (linalool,* rhodinol, geraniol,	
2-Ethyl butyl acetate		citronellol)	12.0
Ethyl butyrate*	30.0	Vanillin*	9.0
Ethyl capronate*	7.20	Violet flowers	
Ethyl caprylate		(or α-ionone,* α-irone,	0.25
Ethyl heptanoate		△-methyl ionone*)	2.0
Ethyl valerate		Solvent	277.9
Geranyl propionate*	6.0		
		Total	1,000.00

Notes:
1. An asterisk denotes the guideline formulation.
2. Parentheses indicate potentially useful alternatives.

Currant

The two most characteristic notes of currant aroma, namely the red and black varieties, have been the subject of several attempts at imitation. The main effort was directed toward the imitation of black currant aroma, the type in the greatest demand throughout central European countries because of the large number of flavor applications. Very few chemically defined flavor ingredients exhibit organoleptic characteristics that blend well with currant flavor. In the case of black currant aroma, the naturally occurring methyl propionate can be employed together with linalyl propionate, whose presence in nature, although most likely, has not yet been confirmed. More generic flavor ingredients include linalyl acetate, guaiol acetate, and γ-ethylionone (the last product is synthetic). The use of diphenyl ether also may be used in the formulation.

Old formulations for non-specific imitation currant aromas (but mainly attempting to duplicate red currant aroma) were based chiefly on the use of citrus essential oils. Of the 100 parts of the aroma formulation, approximately 45 parts consisted of lemon essential oil complemented by 5–6 parts of sweet orange essential oil. Amyl acetate, amyl butyrate, amyl formate, and ethyl acetate rounded off by orris, rose, and vanilla notes were also employed; recent aroma imitations still retain most of these products. New formulations also contain the constituents of rose—namely citronellol, ethyl benzoate, ethyl butyrate, ethyl heptanoate, α-ionone, sometimes menthol of mint essential oil, petitgrain bigarade or neroli, rum ether, and strawberry complex aromas.

Grape

The preparation of a reconstructed grape flavor is an extremely difficult task because of the great variety of grape species, each having its own characteristic note. The flavor also can vary significantly within each species, depending on whether the whole fruit (including the peel) or only the pulp is utilized. The flavor of grape juice tends to change quickly (apart from fermentation) and, if not properly inhibited, produce off-flavors. With these factors in mind, it is apparent that there are only limited formulation possibilities, in light of present knowledge, for the reconstruction of specific grape flavors. Nonetheless, imitation grape flavors or grape juice fortified with approximately 50–100 ppm of complementary flavor ingredients is extremely important to the beverage industry.

Currant Flavors

Ingredient	Parts by weight	
	Red currant	Black currant
Amyl acetate*	430.0	184.0
Amyl butyrate	4.0	96.0
Amyl formate	2.0	
Benzyl acetate*		2.8
Ceylon cinnamon essential oil* (cinnamaldehyde)		1.0
Chinese cinnamon essential oil* (cassia)		40.0
Cognac essential oil* (ethyl heptanoate)	44.0	
Ethyl acetate*	64.0	64.0
Ethyl benzoate*		40.0
Ethyl butyrate*	146.0	120.0
Guaiol acetate		
α-Ionone*	17.0	
β-Ionone*		2.4
Isobutyl acetate		80.0

Ingredient	Parts by weight	
	Red currant	Black currant
Lemon essential oil*	40.0	44.6
Linalyl propionate		
γ-Methylionone		
Methylphenylglycidate*		3.2
Methyl propionate*		125.0
Methyl salicylate*	6.0	
Mint essential oil* (menthol)		20.0
Neroli essential oil (petitgrain bigarade sur fleurs)		22.0
Orange essential oil		64.0
Orris resinoid*	16.0	
Rose, absolute (citronellol*)	1.0	
Rum ether*	690.0	
Vanillin*	162.0	40.0
Solvent (alcohol)	378.0	1,051.0
Total	2,000.0	2,000.0

Notes:
1. An asterisk denotes the guideline formulation.
2. Parentheses indicate potentially useful alternatives.

Among the various types of grape aromas, the muscatel is considered the most characteristic and functional for various flavor applications. Muscatel aroma is in fact recurrent also in muscat wines. The characteristic flavor is imitated in sparkling wine-based beverage formulations (*spume*) typical of the Italian market by means of infusions of coriander seeds, violet flowers, clary sage flowering tops, and elder flowers. Imitations of muscat grape aroma can be obtained also by the use of essential oils.

Volatile Constituents of *Vitis vinifera* Aroma

Hydrocarbons
- Cyclohexane
- *p*-Ethyl toluene
- Hexane
- *n*-Propyl benzene
- Toluene
- *o*-Xylene
- *p*-Xylene

Alcohols
- 2-Heptanol
- 1-Hexanol
- *trans*-2-Hexen-1-ol

Alcohols *(continued)*
- *trans*-3-Hexen-1-ol or *cis*-3-hexen-1-ol
- 2-Methyl-1-butanol
- 3-Methyl-1-butanol

Esters
- *n*-Amyl acetate
- Butyl phthalate
- Ethyl acetate
- Ethyl butyrate
- Ethyl propionate

Aldehydes and Acetates
- 1,1-Diethoxyethane
- Hexanal
- *cis*-2-Hexanal or *trans*-2-hexanal
- 2-Methylbutanal
- 2,4,5-Trimethyl-1,3-dioxolane

Ketones
- Acetone
- 2-Heptanone
- 3-Heptanone
- 2-Pentanone

In addition to the essential oils of coriander and clary sage, which form the main body of the formulation, the essential oils of rose, cognac, celery, and orange, together with vanilla absolute are employed to introduce complementary notes. Whenever an ethereal note is desired, the following naturally occurring ingredients are commonly used: amyl acetate, ethyl butyrate, and ethyl isobutyrate. Additional products used for the same purpose but not yet identified in nature include cinnamyl anthranilate, phenyl *p*-tolyl ether, decyl formate, allyl salicylate, and isovalerophenone.

A few grape flavor formulations are based on the use of methyl anthranilate, ethyl capronate, benzyl propionate, benzylideneacetone or dimethyl anthranilate, ethyl acetate, ethyl anthranilate, methyl naphthyl ketone, rum ether, cinnamyl alcohol, so-called aldehydes C_{16} and C_{14}, ethyl heptanoate, α-ionone, and terpeneless lemon or lime essential oil. Isobutyl anthranilate blends well with the American Concord grape aroma (*Vitis labrusca* L.).

A typical grape flavor formulation carrying the Concord variety note (methyl anthranilate) and suitable for hard candies, confections, baked goods, and soft beverages includes the following ingredients:

Grape Flavor Formulation

Ingredient	Parts	Ingredient	Parts
Amyl isovalerate	8	Ethyl oenanthate	8
Cinnamyl alcohol	5	Ethyl pelargonate	3
Cinnamyl isovalerate	3	Hydroxycitronellal	1
Cinnamyl propionate	3	Methyl anthranilate	132
Citral	1	Methyl salicylate	12
Ethyl acetate	62	Paraguay petitgrain	1
Ethyl benzoate	3	Terpinyl acetate	10
Ethyl butyrate	15	Tolylaldehyde	3
Ethyl caproate	3	Solvent	718
Ethyl methylphenylglycidate	9		
		Total	1,000

Peach

Peach aromas vary widely depending on their cultivar. The various aromas, however, can be reduced to a single basic type; various formulations of imitation aromas are discussed below.

Originally the formulation called for use of the following constituents in suitable ratios: the essential oils of bay, bitter almond, cinnamon, clove, coriander, neroli, and sweet and bitter orange, with ethyl and methyl cinnamate when intended primarily for flavoring liqueurs rather than candy or pastry. The peach aroma imitation commonly offered for a long period for flavoring candy and pastry was a simple mixture consisting of aldehyde C_{14} (γ-undecalactone) and small amounts of amyl acetate and ethyl butyrate. An aroma formulation midway between the above compositions consisted of the following ingredients: aldehyde C_{14}, amyl acetate, amyl butyrate, amyl valerate, ethyl acetate, ethyl caprylate, ethyl cinnamate, ethyl heptanoate, ethyl nitrate, ethyl phenylacetate, ethyl vanillin, and the essential oils of Ceylon cinnamon, clary sage, lemon, pimenta, and sweet orange.

A step forward in peach aroma imitations was represented by the following formulation: aldehyde C_{14}, amyl acetate, amyl butyrate, amyl formate, ethyl capronate, ethyl valerate, vanillin, the essential oils of bitter almond, cinnamon, geranium, neroli, and valerian, and subsequently a small percentage of aldehyde C_{18} (more correctly defined as γ-nonalactone).

Considering that the flavorist had at his disposal few raw materials together with his organoleptic capability for evaluation (only a few studies of the constitution of natural peach aroma had been carried out), it is understandable that early flavorists could not attain a peach aroma beyond a certain quality. Formulations hinged chiefly on flavorists' ability of flowering, spicing, fruitening, and sweetening the fundamental benzaldehyde note, almost the only constituent positively identified in the natural peach aroma complex. This was achieved by using products not entirely compatible with the aroma formulation, such as ethyl and methyl cinnamate (exhibiting a honey or fruital raspberry-like note) or the newly available aldehyde C_{14}, which modified a slight ester note. Independent of the closeness of the imitation to the natural aroma, the following observations can be made by examining a list of constituents selected for various specific formulations: (1) citrus notes were largely used in formulations destined for flavoring liqueurs, and (2) the amount of volatile esters employed decreased for those formulations used in baked goods. The use of a lactone, together with a trace amount of esters, yielded functional formulations exhibiting the required chemical-physical compatibility between the various constituents of the aroma as well as between the aroma and food substrate. Among the flavor ingredients mentioned thus far, only γ-undecalactone, in addition to the fundamental benzaldehyde note, blends well with peach aroma. The other ingredients impart only secondary notes, especially when harmoniously modified by the addition of essential oils or even vanillin.

Products having notes similar to peach and considered suitable for blending in formulations include those identified as constitutents of natural peach aromas: methyl methylanthranilate, terpineol, octyl acetate, phenethyl alcohol, butyl propionate, anisyl butyrate, and *l*-citronellol. Others, most likely occurring in the natural products but not as yet identified, include propyl cinnamate, phenylethyl cinnamate, amyl phenylacetate, benzyl propionate, and phenyl salicylate. Several synthetics of value include allyl cyclohexylcapronate, allyl cyclohexylvalerate, methyl octyl carbonate, and phenyl allyl alcohol. All of the above products may be useful in the formulation of peach aroma imitations for liqueurs or pastry flavoring.

Recent gas chromatographic studies (conducted with the help of other instrumental techniques) of the aroma of Red Globe peaches indicate that while various peach aromas exhibit the same components, the relative ratios vary widely with different types. These studies give information on those constituents listed below in order of increasing retention time: acetaldehyde, methyl acetate, ethyl

Peach Flavor

Ingredient	Parts by weight	Ingredient	Parts by weight
Acetaldehyde		Ethyl phenylacetate	
Allyl cyclohexylcaproate		Ethyl valerate*	226.0
Allyl cyclohexylvalerate		Geranium essential oil (linalool*)	0.5
Amyl acetate*	140.0	γ-Heptalactone	
Amyl butyrate*	40.0	γ-Hexalactone	
Amyl formate*	82.0	Hexyl formate	
Amyl phenylacetate		α-Ionone*	10.0
Amyl valerate*	75.0	Lemon essential oil* (citral)	5.0
Anise essential oil (anethol)		Mandarin essential oil*	300.0
Anisyl butyrate		Methyl cinnamate	
Bay essential oil		Methyl methylanthranilate	
Benzaldehyde*	30.0	Methyl octine carbonate	
Benzyl acetate*	9.5	Neroli essential oil (petitgrain bigarade sur fleurs*)	136.0
Benzyl propionate		γ-Nonalactone	
Bitter almond essential oil*	30.0	γ-Octalactone	
Cinnamon essential oil (cinnamaldehyde*)	2.0	Octyl acetate	
Cognac essential oil (ethyl heptanoate)		Orange, bitter, essential oil*	
Coriander essential oil		Orange, sweet, essential oil*	310.0
γ-Decalactone		Phenethyl cinnamate	
Ethyl acetate*	16.0	Rose, absolute* (*l*-citronellol, phenethyl alcohol)	13.0
Ethyl butyrate*	4.5	Rum ether*	40.0
Ethyl capronate*	86.0	Terpineol (geraniol)	1.5
Ethyl caprylate		γ-Undecalactone*	900.0
Ethyl cinnamate		Vanillin* (ethyl vanillin)	440.0
Ethyl methylphenylglycidate		Solvent	1,103.0
Ethyl nitrate		**Total**	4,000.0

Notes:
1. An asterisk denotes the guideline formulation.
2. Parentheses indicate potentially useful alternatives.

acetate, ethyl alcohol, hexyl formate, hexyl acetate, *trans*-2-hexen-1-ol, benzaldehyde, isovaleric acid, ethyl benzoate, γ-hexalactone, benzyl acetate, γ-heptalactone, caproic acid, benzyl alcohol, γ-octalactone, γ-nonalactone, hexyl benzoate, γ-decalactone, α-pirone, and Δ-decalactone. The following comments can be made on examination of the gas chromatogram:

1. Nature has combined benzaldehyde with a low volatility nucleus consisting of γ-lactones (hexa- to deca- lactones).
2. Contemporarily, the presence of highly volatile components (*e.g.*, esters having an aroma outside the peach note) increases the total volatility of the peach aroma.
3. The presence of acids modifies the flavor of peach aroma.
4. The presence of a few aromatic compounds, such as benzoates, benzyl esters, and benzyl alcohol, is responsible for the floral fragrance of the natural aroma.

Pear

With a good quality amyl acetate, the formulation of an imitation pear aroma would appear to be a simple task. Amyl acetate is commercially available under the trade names "pear oil" and "banana oil." However, to obtain flavor results of real interest, more complex formulations are necessary. The problem is further complicated by the large number of pear aromas that vary widely, depending on the cultivar; for example, the characteristic aroma of Williams pears is used to prepare an excellent liqueur by direct fermentation and distillation. The large water content of the fruit yields an aromatic sensation so specific that even the transfer of a well formulated pear aroma from one food substrate to another causes a much greater negative shift in organoleptic sensations as compared to other imitation aromas.

In early formulations the amyl acetate note was rounded off by the addition of essential oils. Specifically, lemon and bergamot essential oils were blended into an imitation pear flavor consisting of more than 90% by weight of amyl acetate; vanillin, rose absolute, and sometimes a trace of raspberry aroma were also added to better balance the resulting formulation. A few esters—geranyl propionate, amyl valerate, ethyl acetate, and ethyl butyrate—were added for special effects. Some formulations also call for the use of aldehyde C_{14} (γ-undecalactone).

Today pear aroma imitations are still based on traditional schemes, together with the following useful, naturally occurring products: hexyl acetate, propyl acetate, methylheptanone, ethyl heptylate, and hexanoic, octanoic, and decanoic acid ethyl esters. Benzyl butyrate and hexyl phenyl carbonate (probably present but not yet identified in natural pear) may be of interest in the aroma formulation, together with 2-ethylbutyl acetate, a synthetic product not found in nature.

Components of Natural Pear Aroma

Amyl acetate	Hexanol
n-Butyl acetate	*cis*-Hexenyl acetate
Butyl alcohol	Hexyl acetate
Butyl *trans*-2-*cis*-6-dodecadienoate	Methyl acetate
trans-2-*trans*-4-Decadienoate	Methyl *cis*-2-*trans*-4-decadienoate
Ethyl acetate	Methyl *trans*-2-*cis*-4-decadienoate
Ethyl alcohol	Methyl decanoate
Ethyl *trans*-2-*trans*-4-decadienoate	Methyl *cis*-4-decanoate
Ethyl decanoate	Methyl 3-hydroxyoctanoate
Ethyl *cis*-4-decanoate	Methyl *trans*-2-octenoate
Ethyl *trans*-2-decenoate	Methyl *cis*-8-tetradecenoate
Ethyl dodecanoate	*n*-Octanol
Ethyl *trans*-2-dodecenoate	Octyl acetate
Ethyl *cis*-6-dodecenoate	Pentanol
Ethyl 4-hydroxy-*trans*-butenoate	Propyl acetate
Ethyl 3-hydroxyoctanoate	Propyl alcohol
Ethyl octanoate	Propyl *trans*-2-*trans*-4-decadienoate
Ethyl tetradecenoate	Terpene, unsaturated
Ethyl *cis*-8-tetradecenoate	
Heptyl acetate	

Studies of identification of the constituents of natural pear aroma are very recent. In the past ten years these studies have been concerned almost exclusively with Bartlett, Williams, and Winter pear cultivars. Components that have been identified are shown in the preceding table (p. 662).

Pear Flavor

Ingredient	Parts by weight	Ingredient	Parts by weight
Amyl acetate	1,340.0	α-Ionone (1% solution in alcohol*)	1.0
Amyl valerate	130.0	Isoamyl acetate	
Benzyl acetate (10% solution in alcohol)	25.0	Isoamyl valerate	
Benzyl butyrate		Lemon essential oil* (citral)	3.0
Bergamot essential oil		Methyl acetate	
n-Butyl acetate		Methyl heptenone	
Ethyl acetate*	80.0	Orris butter (orris resinoid, 1% solution in alcohol*)	5.0
Ethyl butyrate*	13.0	Propyl acetate	
Ethyl decanoate*	2.0	Rose, absolute (citronellol, 1% solution in alcohol*)	1.0
Ethyl heptylate*	2.0	γ-Undecalactone	
Ethyl hexylate*	1.0	Vanillin	5.0
Ethyl octanoate*	2.0	Solvent	288.0
Geranyl propionate*	100.0		
Hexyl acetate*	2.0		
		Total	2,000.0

Notes:
1. An asterisk denotes the guideline formulation.
2. Parentheses indicate potentially useful alternatives.

Pineapple

The exotic sour pineapple aroma (whose main components have been identified) found some acceptance in the past by flavorists in Europe, although consumer acceptance remained somewhat limited for quite a long time. The original aroma imitations were all based on the use of amyl and ethyl butyrate (more than 50% of the total formulation) together with traces of terpeneless lemon essential oil (sometimes citral) and approximately 2% by weight of vanillin. Subsequently, although amyl and ethyl butyrate were retained as basic raw materials, the aroma formulations became more complex. Butyric acid, amyl acetate, ethyl acetate, acetaldehyde, allyl capronate, allyl caprylate, ethyl caprylate, and orange and angelica essential oils found use in the aroma formulations.

Following the introduction of allyl cyclohexylpropionate, allyl phenoxyacetate, and other synthetic products, formulators further developed the aroma formulations. The following represented a typical aroma composition:

Ingredient	Per cent
Ethyl butyrate	18%
Ethyl isovalerate	18%
Allyl heptylate	14%
Butyl acetate	10%
Allyl capronate	10%
Ethyl propionate	8%
Allyl cyclohexylpropionate	6%
Ethyl heptanoate	6%
Amyl acetate	2%
Allyl phenoxyacetate	1%
Orange essential oil	
Lemon essential oil	
Vanillin traces	

The above formulation together with the addition of naturally occurring products, such as ethyl

Components of Pineapple Aroma

Acetaldehyde	Formaldehyde
Acetic acid	Furfural
Acetone	5-Hydroxy-2-methyl furfural
p-Allyl phenol	Isobutanol
Amyl *n*-capronate	Methanol
γ-Caprolactone	Methyl acetate
Diacetyl	Methyl *n*-butyrate
2,5-Dimethyl 4-hydroxy-3-(2H)-furanone	Methyl *n*-capronate
Ethanol	Methyl *n*-caprylate
Ethyl acetate	Methyl isocapronate
Ethyl acrylate	Methyl isovalerate
Ethyl *n*-butyrate	Methyl β-methylthiopropionate
Ethyl *n*-capronate	Methyl *n*-propyl ketone
Ethyl *n*-caprylate	Methyl *n*-valerate
Ethyl isovalerate	*n*-Pentanol
Ethyl lactate	*n*-Propanol
Ethyl β-methylthiopropionate	

Pineapple Flavor

Ingredient	Parts by weight	Ingredients	Parts by weight
Acetaldehyde		Ethyl butyrate*	45.0
Acetic acid		Ethyl capronate	
Allyl capronate*	21.4	Ethyl *n*-caprylate	
Allyl caprylate		Ethyl heptanoate	12.7
Allyl cyclohexylacetate		Ethyl hexandienoate	
Allyl cyclohexylbutyrate		Ethyl isovalerate*	36.0
Allyl cyclohexylpropionate*	16.4	Ethyl lactate	
Allyl heptylate		Ethyl propionate	16.0
Allyl-2-nonylenate		Formaldehyde	
p-Allyl phenol		Hexyl butyrate	
Allyl phenoxyacetate*	2.0	Isoamyl isobutyrate	
Allyl undecylenate		Lemon essential oil (citral*)	0.6
Amyl acetate*	28.6	Methyl acetate	
Amyl butyrate*	112.0	Methyl allyl caproate*	2.5
Amyl *n*-caprylate		Methyl butyrate	
Angelica essential oil		Methyl *n*-caproate	
Bornyl acetate*	0.1	Methyl *n*-caprylate*	1.0
Butyl acetate*	105.0	Methyl isocapronate	
Butyl isobutyrate		Methyl isovalerate	
Butyric acid*	8.0	Methyl-β-methylthiopropionate*	2.0
Cinnamyl acetate*	0.3	Methyl propyl ketone	
Citronellyl butyrate*	4.0	Methyl undecylenate	
Cognac essential oil* (ethyl heptanoate)	0.3	Methyl valerate	
		Orange essential oil*	22.5
Decyl acetate		Propyl isobutyrate	
Diethyl sebacate*	20.0	Vanillin*	5.6
Ethyl acetate*	376.0	Solvent	162.0
Ethyl acrylate			
		Total	1,000.0

Notes:
1. An asterisk denotes the guideline formulation.
2. Parentheses indicate potentially useful alternatives.

isovalerate and ethyl butyrate (main constituents of natural pineapple aroma), is of definite current interest (with the exception of allyl cyclohexylpropionate and ethyl phenoxyacetate).

The examination of a number of products identified in nature or prepared synthetically provides an additional list of flavor ingredients whose aroma is of interest in the formulation of pineapple flavors:

A. **Naturals**—Decyl acetate, butyl isobutyrate, methyl allyl capronate, cinnamyl acetate, *n*-butyl acetate, hexyl butyrate, propyl isobutyrate, methyl β-methylthiopropionate, and bornyl acetate.

B. **Synthetics**—Allyl cyclohexylacetate, allyl cyclohexylbutyrate, 2-allyl nonylenate, ethyl hexandienoate, methyl undecylenate, allyl undecylenate, and isoamyl isobutyrate.

Consultation with official regulatory lists on the use of the above products is highly recommended.

Since the original study of the composition of pineapple aroma by Haagen-Smit and collaborators in 1945, extensive knowledge of the components of the aroma has been acquired. An updated list of components positively identified in pineapple aroma appears on page 664.

Plum

Old formulations for imitation plum aroma duplicated the complex fragrance of the various plum types by a simple and unique association of 4 parts of amyl acetate, 1 part benzaldehyde, 1.33 parts ethyl heptanoate, 0.25 part vanillin, and a light trace note of orris butter. The resulting aroma imitation had a harsh flavor that would no longer find acceptance by the consumer who today is educated to extremely well formulated bouquets closely reminiscent of the various types of natural plum aromas, such as Mirabelle, Reine Claude (yellow plum), and blue plum. All of the original raw materials find application in today's formulations; only their relative ratios in the formulation have changed. Many

Plum Flavors

Ingredient	Parts by weight		
	Yellow plum	Blue plum	Common plum
Amyl acetate	52.0	55.0	—
Amyl butyrate*	38.0	—	20.0
Bitter almond essential oil* (benzaldehyde)	62.0	290.0	175.0
Butyl formate*	5.0	5.0	10.0
Butyric acid*	10.0	—	—
Citronellol	1.0	1.0	1.0
Citronellyl butyrate			
Citronellyl formate*	15.0	5.0	
Ethyl butyrate (ethyl acetate)	200.0	2.0	50.0
Ethyl heptanoate* (cognac essential oil)	3.5		
Ethyl pelargonate	22.0		
Ethyl valerate*	29.0		15.0
Heptyl butyrate			
Isoamyl formate*	6.0		
Isoamyl propionate*			10.0
Lemon essential oil* (citral)		15.0	
Linalool			

Ingredient	Parts by weight		
	Yellow plum	Blue plum	Common plum
Mandarin essential oil*	40.0	10.0	
Neroli essential oil* (petitgrain bigarade sur fleurs)		1.0	
γ-Nonalactone	2.0	32.0	5.0
Orange essential oil	20.5		15.0
Orris butter (orris resinoid)			
Phenethyl alcohol*		8.0	
Phenethyl butyrate			
Phenethyl formate			
2-Phenyl-1-propyl butyrate*	20.0	200.0	100.0
Propyl formate*	8.0		
Rose, absolute		1.0	
Rum ether*	36.0	28.0	—
Terpenyl butyrate*		5.0	
γ-Undecalactone	1.0	4.0	3.0
Vanillin*	44.0	2.0	10.0
Solvent	385.0	336.0	586.0
Total	1,000.0	1,000.0	1,000.0

Notes:
1. An asterisk denotes the guideline formulation.
2. Parentheses indicate potentially useful alternatives.

additional products also are used to impart characteristic notes. Mirabelle plum aroma, for instance, is formulated by adding rose, orange, and peach notes to the basic ingredients. The Reine Claude plum aroma has a characteristic wine-like note enhanced by adding ethyl butyrate, ethyl acetate, and some other esters. Coumarin derivatives, mandarin oil, and a few other ingredients play an important role in imparting the sweet, raisin-like background note characteristic of the aroma of dried plums.

In addition to the above products, several other naturally occurring constituents may find use in the formulation of various plum aroma formulations. These include isoamyl formate, isoamyl propionate, butyl formate, propyl formate, linalool, citronellyl butyrate, citronellyl formate, and phenethyl *n*-butyrate. Also used are constituents probably occurring in natural products but not yet identified, such as terpenyl butyrate, isopropyl propionate, phenethyl butyrate, neryl propionate, phenylpropyl butyrate, heptyl butyrate, and methyl allyl butyrate. The last product exhibits an aromatic note that blends very well with plum aroma.

Raspberry

The significant aroma difference existing between cultivated and wild raspberry has spurred the flavorists' efforts. At the onset, however, very few useful products were at the disposal of formulators to achieve satisfactory results. Among the available constituents orris butter represented an ideal base especially associated with rose and jasmine notes. However, the use of these, together with large amounts of amyl acetate and amyl butyrate in the presence of a little vanillin, was insufficient to produce aroma imitations similar to the valuable natural aroma. The use of ethyl methylphenylglycidate and α-ionone brought a decisive contribution to raspberry aroma formulations. These guidelines remain valid today.

By careful dosing of the ester content in the formulation and by working around the basic ionone note, it is possible to obtain satisfactory aroma results using the following ingredients in the proper ratios: orris butter, orris resinoid, α- and β-ionone, jasmine absolute, vanillin, aldehyde C_{14}, anethol, celery essential oil, aldehyde C_{16}, citral, methyl sulfide, maltol, dimethyl anthranilate, diethyl succinate, ethyl valerate, ethyl acetate, benzyl salicylate, isobutyl cinnamate, terpenyl acetate, and methyl β-naphthyl ketone.

The following table of naturally occurring components has been prepared by examining several studies of the constitution of natural raspberry aroma. It is interesting to note that current studies are now always in agreement with the presence of certain *esters*.

Components of Natural Raspberry Aroma

Acids	Carbonyl Compounds	Alcohols *(continued)*
Acetic	*(continued)*	1-Pentanol
Butyric	2-Hexenal	1-Penten-3-ol
Caproic	*cis*-3-Hexenal	
Caprylic	4-(*p*-Hydroxyphenyl)-2-butanone	Esters
Formic	(raspberry ketone)	Butyl acetate
2-Hexenoic	α-Ionone	Ethyl acetate
3-Hexenoic	β-Ionone	Ethyl butyrate
Isobutyric	2-Pentanone	Ethyl crotonate
Isovaleric	2-Pentenal	Ethyl propionate
Propionic	Propanal	2-Hexenyl acetate
Valeric		2-Hexenyl butyrate
	Alcohols	Hexyl acetate
Carbonyl Compounds	1-Butanol	Hexyl butyrate
Acetaldehyde	*trans*-2-Buten-1-ol	Isoamyl acetate
Acetone	Ethanol	Isopropyl butyrate
Acetyl methyl carbinol	Geraniol	Methyl butyrate
Acrolein	1-Hexanol	Methyl caproate
Diacetyl	*cis*-3-Hexen-1-ol	Methyl caprylate
β,β-Dimethylacrolein	Methanol	Propyl acetate
Hexanal	3-Methyl-3-buten-1-ol	

In the preceeding table note the following:

1. 4-(*p*-Hydroxyphenyl)-2-butanone exhibits the characteristic raspberry taste and odor.
2. The ratio of 2-hexenyl acetate, acetone, acetaldehyde, and the ethyl acetate esters are considered most important toward the aroma formulation.

Raspberry Flavor

Ingredient	Parts by weight	Ingredient	Parts by weight
Acetic aldehyde		Hexyl butyrate	
Acetone		4-(*p*-Hydroxyphenyl-2-butanone)	16.0
Acids (formic, acetic, butyric, valeric)		α-Ionone* (β-ionone)	210.0
Amyl acetate*	500.0	Isobutyl cinnamate	
Amyl butyrate*	15.0	Jasmine, absolute (10% solution in ethanol*)	7.0
Anethol*	0.5	Lemon essential oil (citral*)	13.0
Benzyl salicylate		Maltol*	3.0
Diacetyl (acetyl methyl carbinol)		Methyl butyrate	
Dimethyl anthranilate*	1.0	Methyl capronate	
Ethyl acetate*	60.0	Methyl disulfide*	1.0
Ethyl butyrate*	74.0	Methyl *p*-naphthyl ketone	
Ethyl methylphenylglycidate*	20.0	Orris butter (orris resinoid,* α-irone)	60.5
Ethyl succinate		Rose, absolute* (rhodinol)	10.0
Ethyl valerate		Terpenyl acetate	
Geraniol*	13.0	γ-Undecalactone*	10.0
Hexanal		Vanilla* (ethyl vanillin)	86.0
cis-3-Hexen-1-ol		Solvent	900.0
2-Hexenyl acetate			
2-Hexenyl butyrate			
		Total	**2,000.0**

Notes:
1. An asterisk denotes the guideline formulation.
2. Parentheses indicate potentially useful alternatives.

Sour Cherry

Morello sour-cherry aroma differs widely from marasca sour cherry. The aroma of morello cherries is largely imitated for use in pastry and beverages; concentrated morello cherry juice often is used in syrups, beverages, and other food products. The aroma of marasca cherries is not imitated, since the fruit distillate is preferred for flavoring liqueurs because of the particular fragrance.

Sour-cherry aroma imitations are based primarily on the use of benzaldehyde and bitter-almond essential oil, together with a few complementary flavor ingredients responsible for "fruitening" and sweetening the note of the aroma formulations. The choice and the per cent ratio of these complementary flavor ingredients are conditioned by the end-use of the aroma. Candy aromas may contain a high level and number of esters; the use of a few essential oils is recommended in the formulation of aromas for non-alcoholic beverages. For example, with benzaldehyde representing from 20–35% by weight of the aroma formulation, ethyl acetate is frequently employed in an approximately identical weight ratio. Additional esters commonly used include amyl acetate, amyl valerate, anisyl acetate, ethyl benzoate, and ethyl butyrate. Vanillin also is added in fairly large amounts (2–5%) together with a trace of heliotropin. Cinnamaldehyde, eugenol, rhodinol, citronellol, ethyl heptanoate, and α- or β-ionone also exhibit interesting notes and may be replaced (if necessary) by cinnamon, clove, and cognac essential oils, rose absolute, and orris butter. Some aroma formulations employ the essential oils of mace, coriander, and lemon, while the presence of γ-undecalactone has been reported in other cases. A few synthetic products not found in nature also exhibit cherry-like aromas highly suitable

Sour Cherry Flavor

Ingredient	Parts by weight
Allyl isovalerate*	
Amyl acetate*	103.0
Amyl butyrate*	1.0
Amyl valerate	
Anisyl acetate	
Bitter almond* (benzaldehyde)	380.0
Ceylon cinnamon essential oil* (cinnamaldehyde)	1.6
Clove bud essential oil* (eugenol)	0.8
Cognac oil*	
Coriander essential oil	0.5
Cyclohexyl cinnamate	
Cyclohexyl formate	
Ethyl acetate*	38.5
Ethyl benzoate	
Ethyl butyrate*	6.8
Ethyl heptanoate	7.4
Geranium essential oil (rhodinol, citronellol*)	2.6
Heliotropin	
α-Ionone*	4.0
β-Ionone	
Jasmine, absolute* (benzyl acetate)	1.0
Lemon essential oil* (citral)	10.0
Mace essential oil	
Neroli essential oil* (petitgrain bigarade sur fleurs)	1.8
Orange essential oil*	6.0
Orris butter (α-irone)	2.4
Orris resinoid*	16.4
Rose*	1.4
Rum ether*	90.0
o,m,p-Toluic aldehyde	10.0
γ-Undecalactone	
Vanillin*	55.0
Solvent	259.8
Total	1,000.0

Notes:
1. An asterisk denotes the guideline formulation.
2. Parentheses indicate potentially useful alternatives.

for blending with sour-cherry imitations; these are cyclohexyl cinnamate, cyclohexyl formate, *o*-, *m*-, and *p*-toluic aldehydes, and the less important allyl isovalerate. No significant studies of the identification of the natural constituents of sour-cherry aroma (morello or marasca) have been reported.

Strawberry

Millions of tons of strawberry aromas have been consumed by the candy and beverage industries. This large consumption has prompted the flavor industry to find the best solution to the vital problem of imparting a strawberry flavor to food products where it is impossible to use the fruit derivatives directly.

Originally, classic imitations of strawberry aroma consisted of the following cor pounds:

Ingredient	Parts
Amyl acetate	12.5
Amyl valerate	3.5
Bulgarian rose (or imitation)	2
Ethyl acetate	15
Ethyl butyrate	2
Ethyl nitrate	4
Methyl benzoate	1
Methyl cinnamate	0.5
Phenethyl alcohol	5
Yara yara (20% solution in benzyl benzoate)	3.5

Following the discovery of ethyl methylphenylglycidate (so-called aldehyde C_{16}, a product not found in nature), the flavor industry made commercially available literally hundreds of strawberry aromas. In addition to the above products, the following flavoring ingredients in varying ratios are employed: amyl butyrate, butyric acid, ethyl cinnamate, ethyl heptanoate, ethyl propionate, heliotropin, α-ionone, isobutyl butyrate, lemon and mint essential oils, methyl salicylate, neroli, orris butter, orris resinoid, raspberry base, rum ether, and vanillin.

When maltol (a product identified in 1964 in natural strawberry aroma) was synthesized and became commercially available, the aroma formulations acquired a different physiognomy. The large portions of aliphatic esters previously used were replaced by more complex products resulting in formulations with much higher yields. Formulations that became available consisted of the following:

A. Acetic acid, anethol, benzyl acetate, diacetyl, ethyl methylphenylglycidate, β-ionone, maltol, methyl anthranilate, methyl cinnamate, methyl heptine carbonate, methyl salicylate, γ-undecalactone, and vanillin.

B. Cinnamyl isobutyrate, cinnamyl isovalerate, diacetyl, dipropyl ketone, ethyl heptanoate, ethyl lactate, ethyl methylphenylglycidate, ethyl valerate, ethyl vanillin, maltol, methyl amyl ketone, and γ-undecalactone.

These aroma formulations also contain products not found in nature. Other flavoring ingredients might eventually find use in the above compositions, such as cumic alcohol (cuminol) and anisyl formate (among the naturals); isobutyl anthranilate, benzyl isobutyrate, neryl isobutyrate (probably occurring in nature but not as yet identified); and methyl naphthyl ketone, 4-methyl acetophenone, and ethyl 3-phenylglycidate (among the synthetics).

The economic importance of aroma has compelled the research chemist to know more about the flavor components present in the natural product. The following table of naturally occurring components in strawberry aroma is based on the results of such investigations. (Although the table has been prepared from carefully checked bibliographic references, it lends itself to some doubts because of disagreements among the various investigators concerning the presence of certain components. However, this may be explained by taking into account the different strawberry cultivars studied and the use of various analytical techniques for separating the natural aroma components.)

Naturally Occurring Components of Strawberry Aroma

Alcohols	Carbonyl Compounds	Esters *(continued)*	Free Acids *(continued)*
Benzyl alcohol	*(continued)*	Ethyl isobutyrate	Cinnamic
1-Borneol*	Acrolein	Ethyl isovalerate	Formic
Butanol*	*n*-Butanal	Ethyl α-methylbutyrate	Isobutyric
2-Butanol	Crotonal	Ethyl propionate	Isovaleric
Ethanol*	Diacetyl	Ethyl salicylate	α-Methylbutyric
2-Heptanol*	2-Heptanone	Ethyl valerate	Propionic
Hexanol*	*cis*-3-Hexal	*trans*-2-Hexenyl	Salicylic
trans-2-Hexenol*	Hexanal	*trans*-2-Hexenyl acetate	Succinic*
p-Hydroxyphenyl-2-ethanol	2-Hexenal	Hexyl acetate	*n*-Valeric
Isoamyl alcohol	Methyl-3-butanone	Hexyl butyrate	
Isobutanol	2-Pentanone	Isoamyl acetate	
Isofenchyl alcohol	2-Pentenal	Isopropyl butyrate	Others
Methanol*	Propanal	Methyl acetate	Acetals
1-Pentanol		Methyl butyrate	Acetoin
Penten-1-3-ol*	Esters	Methyl capronate	γ-Decalactone
Phenyl-2-ethanol	Butyl acetate	Methyl isobutyrate	1,1-Diethoxyethane
n-Propanol	Ethyl acetate	Methyl α-methylbutyrate	1,1-Dimethoxyethane
d-*l*-α-Terpineol	Ethyl acetoacetate	Propyl acetate	Dimethoxymethane
cis-Terpineol hydrate	Ethyl benzoate		Dimethyl sulfide
	Ethyl butyrate	Free Acids	1-Ethoxy-1-propoxyethane
Carbonyl Compounds	Ethyl capronate	Acetic	Hydrogen sulfide
Acetaldehyde	Ethyl cinnamate	Benzoic	Maltol
Acetophenone	Ethyl crotonate	Butyric	1-Methoxy-1-ethoxyethane
Acetone	Ethyl formate	Caproic	Methyl sulfide

* Asterisked alcohols are definitely present.

The following comments apply to the preceding table:

1. Based on more recent discussions, the number of alcohols found in the references should, in practice, be smaller. Those alcohols marked with an asterisk (*) are definitely present.
2. Several esters evidenced by gas chromatographic analysis have not as yet been identified.
3. The characteristic strawberry aroma is probably caused by carbonyl compounds, esters, and a few other products rather than the alcohols.

Studies of a complete identification of the strawberry aroma constituents continue, together with the search for an improved reconstruction of the aroma based on the increasing analytical information.

Strawberry Flavor

Ingredient	Parts by weight	Ingredient	Parts by weight
Amyl acetate*	34.0	α-Ionone (β-ionone*)	6.5
Amyl butyrate*	15.0	Isobutyl anthranilate	
Amyl valerate*	15.0	Isobutyl butyrate	
Anethol*	1.5	Lemon essential oil* (citral)	1.0
Anisyl formate		Maltol*	70.0
Benzyl acetate* (jasmine absolute)	85.0	4-Methylacetophenone	
Benzyl isobutyrate		Methyl anthranilate*	6.5
Butyric acid (acetic acid*)	15.0	Methyl benzoate	
Cinnamyl isobutyrate*	7.0	Methyl cinnamate*	35.5
Cinnamyl valerate*	9.5	Methyl heptine carbonate*	0.5
Cognac essential oil* (oenanthic ether)	1.5	Methyl naphthyl ketone	
Diacetyl*	10.0	Methyl salicylate*	6.5
Dipropyl ketone		Mint essential oil (menthol)	
Ethyl acetate*	50.0	Neroli essential oil* (petitgrain bigarade sur fleurs)	0.5
Ethyl amyl ketone	15.0	Nerolin (yara yara)	
Ethyl butyrate*	30.0	Neryl isobutyrate	
Ethyl cinnamate*	52.0	Orris butter* (orris resinoid)	1.5
Ethyl heptanoate		Phenethyl alcohol	
Ethyl heptylate*	2.5	Rose (rhodinol-citronellol)	
Ethyl lactate		Rum ether	
Ethyl methylphenylglycidate* (ethyl 3-phenylglycidate)	260.0	γ-Undecalactone*	58.5
Ethyl nitrate		Vanillin* (ethyl vanillin)	70.0
Ethyl propionate*	15.0	Solvent	1,060.0
Ethyl valerate*	60.0		
Heliotropin		**Total**	2,000.0
Hydroxyphenyl-2-butanone (10% solution in alcohol)	5.0		

Notes:
1. An asterisk denotes the guideline formulation.
2. Parentheses indicate potentially useful alternatives.

Sweet Cherry

Imitations of sweet-cherry aroma are frequently and erroneously prepared in an analogous manner to sour-cherry aroma by simply adding an excess of bitter-almond essential oil. It is true that the benzaldehyde note must be predominant in this flavor, but this must be modified by a fairly large amount of *p*-toluic aldehyde or by a mixture of *o*-, *m*-, and *p*-toluic aldehydes used in a 4:1–5:1 ratio and at maximum not to exceed 40% total in the formulation. Both *p*-toluic aldehyde and the mixture of *o*-, *m*-, and *p*-toluic aldehydes are not found in nature.

A formulation that has attained a certain degree of acceptance contains, in addition to the mixture of the above aldehydes, amyl butyrate, anisic aldehyde, anisyl acetate, benzyl acetate, cinnamaldehyde, ethyl butyrate, ethyl heptanoate, ethyl methylphenylglycidate (a product not found in nature), eugenol, and vanillin. Outside the above guidelines, satisfactory aromatic results become questionable; it is necessary to follow traditional formulation schemes consisting of bitter almond, the complex of the natural constituents of raspberry, a rose-like note, and vanillin. Other than the above products, there are no other natural products of interest for the imitation of sweet-cherry aroma.

Allyl benzoate is a potentially interesting product, but its toxicity is not yet sufficiently known. Anisyl propionate, anisyl butyrate, rhodinyl formate, and rhodinyl isovalerate, whose aromas blend well with sweet-cherry aroma, have not yet been identified in nature, although their presence is most likely. While other functional products, such as allyl isovalerate, cyclohexyl cinnamate, and cyclohexyl formate, have not yet been reported in the natural product, they may nonetheless prove useful.

Sweet Cherry Flavor

Ingredient	Parts by weight	Ingredient	Parts by weight
Allyl isovalerate		Ethyl methylglycidate*	100.0
Amyl butyrate*	200.0	Geranium (rhodinol-citronellol*)	60.0
Anisic aldehyde*	37.0	α-Ionone* (β-ionone)	4.0
Anisyl acetate*	25.0	Jasmine, absolute*	13.0
Anisyl butyrate		Lemon (citral)	1.0
Anisyl propionate		Maltol (5% solution in ethanol*)	1.0
Benzyl acetate*	50.0	Orris butter*	30.0
Bitter almond essential oil*		Orris resinoid*	160.0
(benzaldehyde)	4,640.0	Rhodinyl formate*	1.0
Ceylon cinnamon (cinnamaldehyde*)	18.0	Rhodinyl isovalerate	
Clove buds (eugenol*)	7.0	Rose, absolute*	1.0
Cyclohexyl cinnamate		*o*-, *m*-, and *p*-Toluic aldehydes*	500.0
Cyclohexyl formate		Vanillin*	400.0
Ethyl acetate*	680.0	Solvent	2,920.0
Ethyl butyrate*	152.0		
		Total	10,000.0

Notes:
1. An asterisk denotes the guideline formulation.
2. Parentheses indicate potentially useful alternatives.

"Fantasia" Formulations Based on Fruit Flavors

By properly compounding the various fruit aroma imitations, additional flavor products having a characteristic but normal taste can be formulated. An example of the most characteristic and intriguing fantasia aroma is the tutti-frutti. This aroma may be formulated by complex recipes, such as pineapple, banana, orange, lemon, raspberry, vanilla, or by using much less complex compositions. The following formulation of fruit blends is suitable for flavoring pastry and baked goods. Using natural fruit juices for beverage flavors, the following blend can be prepared: pear, apple, peach, apricot, raspberry,

Tutti-Frutti Flavor

Ingredient	Parts by weight	Ingredient	Parts by weight
Amyl acetate	300.0	Rose, imitation (10% solution in alcohol)	28.0
Amyl butyrate	48.0	Rum ether	100.0
Ethyl butyrate	36.0	γ-Undecalactone	18.0
α-Ionone	120.0	Vanillin	11.0
Jasmine (10% solution in alcohol)	1.0	Solvent	257.0
Lemon essential oil	1.0		
Orris resinoid	80.0	**Total**	1,000.0

strawberry, Mirabelle plum, and cherry. There is no limitation in preparing identical formulations using the corresponding fruit aroma imitations as long as a citrus note is added to the final formulation.

Another fantasia fruit aroma somewhat resembles pomegranate. This aroma is formulated for either pastry flavors or syrups using the following basic notes: vanilla, neroli, raspberry, and orange. Needless to say, imitation fruit aromas, essential oils, terpeneless oils, or natural juices may be employed for such formulations depending on the end-use; care should be taken to soften the more stable notes and to enhance the least stable ones in baked goods.

Pomegranate Flavor

Ingredient	Parts by weight
Amyl acetate	150.0
Amyl butyrate	2.4
Ethyl butyrate	20.0
Ethyl pelargonate	1.5
α-Ionone	6.0
Jasmine, imitation (10% solution in alcohol)	0.4
Orris resinoid	4.3
Petitgrain bigarade sur fleurs	2.5
Rose, imitation	1.4
Terpeneless sweet orange essential oil	100.0
Vanillin	15.5
Solvent	196.0
Total	500.0

BITTER FLAVORS

The creation of bitter aromas represents one of the most difficult and fascinating problems in formulating flavors. The flavorists' creativity as well as ingenuity and tradition become intriguingly entwined with the most advanced technology. The fields of application are vast; only a few can be covered by a general text on flavors. However, an attempt will be made to group and classify for ready reference not only various problem areas but also to suggest suitable solutions.

A word of caution with respect to bitter flavors. Probably no other flavor group is more bound by tradition than the bitter aroma. The term *bitters* is quite generic and hardly apt in describing the broad range of bitter flavors known the world over. To western European consumers the bitter aroma is often represented by a high degree of astringency over a strong wine base. The development of vermouths and aperitifs in France and Italy is a typical example. For many years such products were not very popular in the United Kingdom, and for the most part they were rejected in the United States. On the other hand, the bitter principle delivered by quinine has evolved through years of use from an apparently disagreeable medicinal to a universal mix for gin and vodka. This type of bitter is accepted throughout the English-speaking world. Additional examples become readily apparent when one considers more specific national preferences in bitter beverages.

To the flavorist the term *bitter* is intended as the *base* flavor or taste and really represents only one component in the formulation of bitters. In fact, flavorists speak of tonics as *aperitifs* or *digestives.* These concepts were discussed at the *Colloques des Plantes Aromatiques* held at Milly-LaForet in 1966.[3]

Table 2 illustrates the best known herbs and derivatives commonly employed in the formulation of bitters. (The individual plants, botanical sources, derivatives, and physical-chemical constants are detailed in Part II of this book.) The 72 botanical sources listed in Table 2 provide the utmost flexibility

TABLE 2

Herbs and Derivatives Used to Formulate Bitters

Common name	Botanical name	Parts of plant used
Aloe	*Aloe* species	Concentrated leaf juice
Ambrette	*Hibiscus abelmoschus* L.	Seeds
Angelica	*Angelica archangelica* L.	Roots
Angostura	*Galipea cusparia* DC.	Bark
Anise	*Pimpinella anisum* L.	Fruits
Artichoke	*Cynara scolymus* L.	Leaves
Balm (lemon balm)	*Melissa officinalis* L.	Leaves and flowering tops
Blessed thistle	*Cnicus benedictus* L.	Leaves and flowers
Calamus	*Acorus calamus* L.	Rhizomes
Calumba	*Jatrorrhiza palmata* (Lam.) Miers	Roots
Camomile, Hungarian or German	*Matricaria chamomilla* L.	Flowers
Camomile, Roman or English	*Anthemis nobilis* L.	Flowers
Cardamom	*Elettaria cardamomum* Maton	Fruits
Cascarilla	*Croton eluteria* Benn.	Bark
Catmint	*Nepeta cataria* L.	Flowering tops
Centaury	*Erythraea centaurium* Pers.	Whole plant
Chicory	*Cichorium intybus* L.	Roots
Chinotti	*Citrus myrtifolia* Risso	Peels or the whole fruit
Chirata	*Swertia chirata* (Roxb.) Buch.-Ham.	Whole plant
Cinchona	*Cinchona* species	Bark
Cinnamon, Ceylon	*Cinnamomum zeylanicum* Nees	Bark
Clary sage	*Salvia sclarea* L.	Flowering tops

TABLE 2 (continued)

Herbs and Derivatives Used to Formulate Bitters

Common name	Botanical name	Parts of plant used
Clove	*Eugenia caryophyllata* Thunb.	Buds
Condurango	*Marsdenia condurango* Reichenb. f.	Bark
Coriander	*Coriandrum sativum* L.	Fruits
Dandelion	*Taraxacum officinale* Weber	Leaves and roots
Dittany of Crete	*Origanum dictamnus* L.	Leaves and flowering tops
Elder	*Sambucus nigra* L.	Flowers
Elecampane	*Inula helenium* L.	Rhizomes
Galanga	*Alpinia officinarum* Hance	Rhizomes
Genepi	*Artemisia glacialis* L.	Whole plant
Gentian	*Gentiana lutea* L.	Rhizomes and roots
Gentian, stemless	*Gentiana acaulis* L.	Whole plant
Germander	*Teucrium chamaedrys* L.	Flowering tops
Ginger	*Zingiber officinale* Rosc.	Rhizomes
Grains of paradise	*Aframomum melegueta* Rosc.	Seeds
Hyssop	*Hyssopus officinalis* L.	Leaves and flowers
Imperatoria	*Peucedanum osthruthium* (L.) Koch.	Rhizomes
Juniper	*Juniperus communis* L.	Berries
Larch agaric	*Polyporus laricis* Jacq.	Inner portion of the thallus
Lemon	*Citrus limonum* (L.) Risso	Peels
Liatris (wild vanilla)	*Trilisa odoratissima* (Walt.) Cass.	Leaves
Licorice	*Glycyrrhiza glabra* L.	Roots
Mace	*Myristica fragrans* Houtt.	Arillodes
Marjoram, sweet	*Marjorana hortensis* Moench.	Flowering tops
Melilotus	*Melilotus officinalis* (L.) Lam.	Flowers
Mugwort	*Artemisia pontica* L.	Leaves and flowering tops
Myrrh	*Commiphora* species	Gum resin
Nutmeg	*Myristica fragrans* Houtt.	Fruits
Orange, bitter	*Citrus aurantium* L. subspecies *amara* L.	Peels
Orange, sweet	*Citrus sinensis* L. Osbeck	Peels
Orris	*Iris pallida* L. and *I. germanica* L.	Roots
Peppermint	*Mentha piperita* L.	Flowering tops
Quassia	*Picrasma excelsa* (Sw.) Planch.	Wood
Rhubarb	*Rheum* species	Rhizomes
Rue	*Ruta graveolens* L.	Leaves
Saffron	*Crocus sativus* L.	Stems
St. Johnswort	*Hypericum perforatum* L.	Flowering tops
Savory, summer	*Satureja hortensis* L.	Flowering tops
Southernwood	*Artemisia abrotanum* L.	Leaves and flowering tops
Star anise	*Illicium verum* Hook. f.	Fruits
Thyme	*Thymus vulgaris* L.	Whole flowering plant
Tonka bean	*Coumarona odorata* Aubl.	Seeds
Valerian	*Valeriana officinalis* L.	Rhizomes and roots
Vanilla	*Vanilla* species	Pods
Walnut	*Juglans regia* L.	Leaves and green nuts
Woodruff, sweet	*Asperula odorata* L.	Whole plant
Wormwood	*Artemisia absinthium* L.	Leaves and flowering tops
Wormwood, mountain	*Artemisia valesiaca* L.	Leaves and flowering tops

TABLE 2 (continued)

Herbs and Derivatives Used to Formulate Bitters

Common name	Botanical name	Parts of plant used
Yarrow	*Achillea millefolium* L.	Whole flowering plant excluding the root
Yarrow, musk	*Achillea moschata* Jacq.	Leaves and flowering tops
Zedoary	*Curcuma zedoaria* Rosc.	Bark

TABLE 3

Classification of Bitter Flavors

Bitter flavors		Complementary flavors		
Aromatic-bitter	**Bitter**	**Aromatic**	**Pungent**	**Sweet**
Angelica	Aloe	Ambrette	Cardamom	Anise
Balm	Angostura	Clary sage	Cinnamon, Ceylon	Licorice
Calamus	Artichoke	Coriander	Clove buds	Star anise
Camomile, Hungarian	Blessed thistle	Imperatoria	Ginger	Vanilla
Camomile, Roman	Calumba	Lemon	Grains of paradise	
Cascarilla	Centaury	Liatris (wild vanilla)	Juniper	
Catmint	Chicory	Melilotus	Mace	
Chinotti	Chirata	Myrrh	Nutmeg	
Condurango	Cinchona	Orange, sweet	Peppermint	
Dittany of Crete	Dandelion	Orris	Thyme	
Elder	Gentian	Saffron		
Elecampane	Gentian, stemless	St. Johnswort		
Galanga	Larch agaric	Savory, summer		
Genepi	Quassia	Tonka bean		
Germander	Rhubarb	Valerian		
Hyssop	Southernwood	Woodruff, sweet		
Marjoram, sweet	Walnut			
Mugwort				
Orange, bitter				
Rue				
Wormwood				
Wormwood, mountain				
Yarrow				
Yarrow, musk				
Zedoary				

in the formulation of flavors for bitters. In addition to the extracts and the tinctures, in some cases the soluble (terpeneless) essential oils are frequently employed (*e.g.*, bitter orange, sweet orange, and mint essential oils). The following essential oils are used occasionally as modifiers: angelica root, anise, calamus, cinnamon, clary sage, clove bud, coriander, ginger, lemon, and nutmeg.

The derivatives of the above listed products generally exhibit a definite bitter or aromatic-bitter flavor. Exceptions are vanilla, licorice, star anise, and anise, all of which have a characteristic *sweet* flavor and are used as either basic flavor ingredients or as sweeteners to yield contrasting effects in slightly *pungent* or *sharp* flavors, such as thyme, peppermint, Ceylon cinnamon, nutmeg, grains of paradise, clove buds, cardamom, juniper, mace, and ginger. The following 16 plants yield derivatives with a definite *aromatic* flavor: woodruff, liatris, melilotus, clary sage, coriander, summer savory, orris, ambrette, tonka bean, sweet orange, imperatoria, lemon, valerian, saffron, St. Johnswort, and myrrh.

The remaining 42 products all exhibit varying degrees of bitter flavor, of which 17 yield exclusively bitter flavors: southernwood, chirata, gentian, blessed thistle, cinchona, rhubarb, aloe, angostura, larch agaric, artichoke, chicory, calumba, walnut, quassia, dandelion, gentian (stemless), and centaury. These are sometimes defined as "characteristic," while the derivatives of the additional 25 plants exhibit an *aromatic-bitter* flavor.

Finally attention is called to some of the secondary flavor notes encountered in bitter principles, such as the *sweet notes* associated with the aromatic flavor of tonka bean, sweet orange, lemon, liatris, and melilotus. *Tonic action* is associated with the aromatic-bitter flavors of mugwort, bitter orange, marjoram, angelica root, cascarilla, catmint, condurango, and genepi or with the *bitter* flavor of centaury, angostura, blessed thistle, and cinchona. This classification is shown in Table 3.

ALCOHOLIC BEVERAGES

Vermouth

In 1786 Antonio and Benedetto Carpano prepared the first typical vermouth formulation, using herbs harvested in the mountains and the plains of Piedmont, Asti muscat wine, sugar, and alcohol. The name *vermouth*, today generic, is derived from *Wermuth*,* the original German name for wormwood (*Artemisia absinthium*), one of the aromatic herbs employed in the formulation.

According to definition and standards set by Italian law, the name *vermouth* is reserved exclusively for the finished product prepared using genuine wine of national production in combination with aromatic and bittering ingredients permitted under the regulations issued by the Italian Department of Health. The finished product must have an alcohol and sugar content within the following limits:

Red and white vermouths. Not less than 15.5% alcohol by volume and not less than 13% by weight total invert sugar (calculated following the inversion of sucrose).

Dry vermouth. Not less than 18% alcohol by volume and not more than 4% by weight total invert sugar (calculated following the inversion of sucrose).

In the preparation of Italian vermouths, the addition of sucrose, pure and rectified ethyl alcohol, and coloring matter (sucrose-caramel or burnt sugar) is permitted. The ingredients added (flavors, bittering substances, alcohol, sucrose, and caramel) must not exceed 30% by volume; the finished product also must be perfectly clear.

Constituents of Vermouth

Sucrose, free of active decolorizing agents, and good-tasting ethyl alcohol are obviously the products of choice for vermouth formulations. This leaves only the wine, the flavoring-bittering ingredients, and the caramel. The choice of the most suitable caramel type for .coloring is not limited simply to selecting the best caramel type prepared from sucrose for use in alcoholic beverages; it is usually complemented by a number of coloring, aging, and flavor panel tests to determine the following:

1. The total absence of secondary reactions among the various ingredients in the vermouth formulation that might cause flavor degradation, formation of deposits, or turbidity.
2. The color stability.
3. The palatability of the resulting flavor, remembering that caramel not only adds color but also flavor to the finished product. The latter is strongly dependent on the other ingredients present because of varying synergistic effects.

The choice of the wine deserves particular attention. Originally pure Asti muscat wine of a certain type was the only product prescribed for use in vermouths to obtain the desired flavor effects. Because of the increasing industrialization of vermouth production and the growing demand for the specific wine type, wine specialists selected certain cuts of wines from both Asti and other neighboring areas to formulate an alternative but nonetheless suitable product. This resulted in the *bianco carta* (paper white) that is used almost exclusively in today's qualified production of Italian vermouths. This compounded wine has a higher alcohol content than the original Asti variety.

The bitter flavor ingredients consist mainly of derivatives (tinctures, extracts, etc.) of the following herbs or dried spices:

---

* *Vermut-vein* is the medieval German name for wormwood-flavored wine.

Flavor Ingredients of Vermouth

Aromatic	Pungent	Sweet
Ambrette	Cardamom	Anise
Clary sage	Ceylon cinnamon	Licorice
Coriander	Clove	Vanilla
Ginger	Grains of paradise	
Imperatoria	Juniper	
Lemon	Mace	
Liatris (wild vanilla)	Nutmeg	
Melilotus	Peppermint	
Myrrh	Saffron	
Orange, sweet	Thyme	
Orris		
St. Johnswort		
Savory, summer		
Tonka bean		
Valerian		
Woodruff		

The choice and the dosage of the flavor ingredients together with the extraction procedures from the botanical source represent not only a typical challenging problem for the flavorist, but they are also responsible for the following classification of vermouths:

White dry vermouths
White, highly aromatic vermouths
White, lightly aromatic vermouths
Red vermouths
Red bitter vermouths
Cinchona-flavored red vermouths

Formulation of Vermouth Flavors

Before discussing criteria for vermouth formulations, some attention must be given to fundamental concepts of the extraction of the herbs, since the extractive technique employed directly affects the resulting flavor of the finished product.

Traditional methods for the preparation of vermouths prescribed the use of a blend of aromatic herbs in suitable predetermined ratios. The herbs were comminuted and subsequently macerated for a predetermined length of time in the wine of choice. The wine was fortified with added amounts of ethyl alcohol necessary to prepare a finished product having the required alcohol strength. Care was taken to account for evaporation losses and slight dilution effects by the residual moisture in the herbs. The sugar content was adjusted optionally before or after maceration. Following maceration and filtration of the herb residue, the infusion was aged and finally bottled.

Other methods of preparation consist of adding typical compounded flavor extracts to the fortified wine mixture. These compounded flavors are obtained (1) by hydro-alcoholic maceration of properly formulated blends of herbs or (2) by hydro-alcoholic extracts of the individual herbs that are subsequently blended together.

It is apparent from the above descriptions that the "flavor yield" of the individual aromatic herb depends on the different preparatory techniques. Since the basic operation consists of an extraction, not only the extractive power of the solvent but also the solubility of the flavor constituents of the herbs must be considered. In the original method a large number of synergistic effects may occur in the blend of herbs that are extracted with a low-strength solvent. In the second method the extraction is still carried out on a blend of herbs; the only difference is that a solvent with a higher strength is employed to prepare a flavor concentrate of the required yield. In the third method the extraction is carried out on the individual herbs using a solvent of strength similar to that used in the second method. Experience shows that if the same amounts and types of herb(s) are extracted in each of the three methods, three different flavors are obtained. Therefore, if one wishes to duplicate the flavor resulting from the first extraction method by using the second or third extraction procedure, respectively, it may be necessary to consider the higher solubility of the aromatic principles as well as the lower solubility of the water-soluble components.

Regardless of the flavorists' personal preference for the direct use of the aromatic herbs, the use of tinctures prepared from fluid extracts and distillates should not be neglected. Using tinctures not only insures a controlled reproducibility of the active flavor principles in the formulation, but it may be absolutely necessary in those formulations containing flavor ingredients of quite different solubility characteristics. Similarly distillates (derivatives usually distilled from alcoholic infusions of the aromatic herbs) are highly functional to impart specific flavor notes. They are used also in the preparation of white vermouths and special bitters. Approximately 1,000 kilograms (2,205 pounds) of finished vermouth or bitter may be prepared using a blend of aromatic herbs in amounts varying between 3–6 kilograms (7–13 pounds); *i.e.*, the aromatic herbs have a flavor yield of approximately 1 × 200 to 1 × 400.

Of the 72 bittering plants used by the flavor industry that were listed in Table 2 (see p. 673), only a few find use as flavor ingredients in vermouth formulations. The flavor ingredients employed in vermouth formulations and their preferential use in various types of vermouth are listed in Table 4.

The products listed in Table 4 should be interpreted with a certain degree of flexibility. The fact that specific herbs have not been reported used in certain types of vermouth does not imply their categoric exclusion from other formulations; a relative degree of freedom may be permitted to obtain specific flavor effects. For the sake of thoroughness (and with reference to the choice and blending of herbs for the various vermouth formulations shown in Table 4), it must be pointed out that the following herbs are included in the formulation of almost all vermouth types and are therefore the most important:

Angelica	Germander
Calamus	Marjoram
Cinnamon	Mugwort
Clove	Summer savory
Coriander	Wormwood
Dittany of Crete	Wormwood, mountain

Apart from the characteristic aromatic notes rendered by calamus, dittany of Crete, coriander, summer savory, cinnamon, marjoram, angelica, and clove, the reader should note the use of germander (an absolutely irreplaceable product of great potential), mountain wormwood, wormwood, and mugwort as basic flavor ingredients in vermouth formulations. For example, by keeping constant the desired concentration levels of germander and mountain wormwood, the concentration ratio of wormwood to mugwort can be increased or decreased to impart a more or less bitter-aromatic note to the resulting vermouth formulation.

TABLE 4

Flavor Ingredients in Vermouth Formulations

Flavor ingredients	Vermouths					
	White, dry	White, highly aromatic	White, lightly aromatic	Red	Red, bitter	Red, cinchona-flavored
Ambrette		x		x		
Angelica	x	x	x	x	x	x
Balm (lemon balm)				x		
Blessed thistle				x		
Calamus	x	x	x	x	x	x
Camomile		x		x		
Cascarilla*		x			x	
Centaury			x	x	x	x
Chirata*		x			x	x
Cinchona						x
Cinnamon	x	x	x	x	x	x
Clary sage	x		x	x		x
Clove		x	x	x	x	x
Coriander	x	x	x	x		x
Dittany of Crete	x	x	x	x	x	
Elder		x			x	
Elecampane		x	x		x	
Galanga			x		x	
Gentian			x		x	x
Gentian, stemless			x		x	
Germander		x	x	x	x	
Grains of paradise*		x	x		x	
Hyssop	x				x	
Imperatoria*					x	x
Licorice*				x		
Marjoram		x	x	x	x	x
Melilotus				x		
Mugwort	x	x	x	x	x	x
Nutmeg	x	x		x	x	
Orange, bitter	x	x	x			
Orange, sweet		x				
Orris		x				
Peppermint	x					
Savory, summer	x	x		x	x	x
Southernwood	x				x	
Tonka bean		x	x			
Thyme	x		x	x	x	x
Vanilla		x				
Wormwood	x	x	x	x	x	x
Wormwood, mountain	x	x	x	x	x	x
Yarrow			x	x	x	x
Yarrow, musk				x	x	x

* Rarely used.

A typical vermouth formulation is shown below. This formulation, however, is intended only as a guideline for the preparation of a white vermouth by direct infusion of the herbs in wine fortified with alcohol.

White Vermouth Formulation

Ingredient	Parts	Ingredient	Parts
Clary sage	2	Orange, bitter	1
Wormwood	2	Southernwood	1
Calamus	1.5	Thyme	1
Germander	1.5	Mugwort	0.5
Coriander	1	Summer savory	0.5
Dittany of Crete	1	Ceylon cinnamon	0.25
Hyssop	1	Nutmeg	0.25
Mountain wormwood	1	Vanilla	0.1

The comminuted herbs are macerated in the fortified wine for at least 15 days; the resulting infusion is pressed, sugar is added, and the infusion is finally filtered.

Bitters

This category of products is extensive and includes several characteristic examples. The following classification may be useful in defining this product category.

1. Alcoholic bitters (typical)
2. Alcoholic aperitifs with special flavors in mixture with citrus notes
3. Alcoholic aperitifs with a characteristic flavor note (*e.g.*, rhubarb, gentian, artichoke, etc.)
4. Tonics and digestives with a characteristic flavor note (anise, cinchona, genepi, etc.)
5. Tonics and digestives with a mixed flavor note (*e.g.*, centerbe, fernet, mint fernet, felsina bitter, etc.).

Following criteria similar to those used for vermouth formulations, a list of dried herbs useful for the formulation of bitters is given in Table 5. In addition to the herbs listed, soluble essential oils extracted from the same herbs are also employed in the formulation of special and digestive-tonic bitters. A "complex" typical bitter flavor can be added to round off the flavor note in tonics and digestives. It is also possible to employ flavor distillates in all bitter formulations to create special effects. Note that raspberry distillate and violet tincture are not included in Table 5 because of their very limited, specific use.

A guideline to a typical basic bitters formulation is shown below. A 40% tincture having an approximately 1 × 100 yield is prepared by maceration of the herbs for 15–20 days in approximately 60% alcohol.

Basic Bitters Formulation

Ingredient	Parts
Aloe	0.2
Cascarilla	3
Chinotto	2
Cinnamon	1
Gentian	16
Germander	11
Orange, bitter	11
Orange, sweet	13
Rhubarb	6
Wormwood	6

TABLE 5

Herbs Used in Bitters Formulations

Flavor ingredient	Typical bitters	Special bitters	Tonics and digestives	Flavor ingredient	Typical bitters	Special bitters	Tonics and digestives
Aloe	x	x	x	Imperatoria	x		x
Ambrette		x		Juniper	x	x	x
Angelica	x	x	x	Larch agaric		x	
Angostura		x		Lemon		x	
Anise			x				
				Liatris (wild vanilla)	x	x	
Artichoke		x		Licorice		x	
Balm (lemon balm)	x			Mace		x	
Calamus	x	x	x	Melilotus	x	x	
Calumba		x	x	Myrrh			x
Camomile, Hungarian		x	x				
				Nutmeg		x	
Camomile, Roman		x	x	Orange, bitter	x	x	x
Cardamom		x		Orange, sweet	x	x	
Cascarilla	x			Orris		x	
Catmint		x		Peppermint	x		x
Centaury	x	x	x				
				Quassia		x	
Chicory		x		Rhubarb	x	x	x
Chinotto	x	x	x	Rue		x	x
Chirata	x		x	Saffron			x
Cinchona	x	x	x	St. Johnswort			x
Cinnamon	x		x				
				Southernwood	x		
Clove	x	x	x	Tonka bean		x	
Condurango		x		Valerian		x	
Coriander			x	Vanilla		x	x
Dandelion		x		Walnut		x	x
Galanga	x						
				Woodruff		x	
Genepi		x	x	Wormwood	x		x
Gentian	x		x	Wormwood, mountain			
Gentian, stemless	x	x	x		x		
Germander	x			Yarrow	x		
Grains of paradise	x			Yarrow, musk	x	x	
Hyssop			x	Zedoary		x	x

In the aromatic blending of very bitter flavors, the following combinations are preferred:

Angelica with: balm, cardamom, coriander, hyssop, marjoram, mint, thyme, vanilla.
Calamus with: cardamom, cinnamon, mace or nutmeg, zedoary; **or:** calumba, camomile, cascarilla, cinchona, larch agaric, and rhubarb.
Camomile with: artichoke, bitter orange, cinchona, genepi, gentian, gentian (stemless), mint, summer savory.
Cascarilla with: bitter and sweet orange, calamus, chinotti, cinnamon, grains of paradise, lemon, nutmeg, thyme.
Centaury with: calamus, cinchona, condurango, gentian, gentian (stemless); **or:** bitter and sweet orange, cardamom, clove, lemon, licorice, mace.
Condurango with: bitter orange, cardamom, chicory, cinchona, dandelion, lemon, rhubarb.

The operative technique for the preparation of flavors for bitters consists of (1) production of the individual tinctures and their subsequent mixing to prepare the so-called "compounded alcoholic essence," having a yield varying between 3 × 100 and 1 × 100, and (2) preparation of the compound tincture by direct extraction of properly ratioed blends of the various herbs. All alcoholic extracts are added to water, alcohol, and sugar to prepare the finished products. The alcohol content of tinctures usually varies between 21–30%. In aromatic-bitter flavor formulations the total alcohol content should also be considered, as this may enhance certain notes. The sugar content in various formulations plays an important role; in all cases the sugar level is balanced against the bitter bouquet.

The aromatic note of so-called American bitters results from mixing two–three parts of good quality red vermouth with one part of bitter. The conclusions that may be drawn from this for formulative purposes are quite obvious.

The flavors obtained from bitter formulations vary widely as the result of carbonation. While flavor ingredients used are generally the same (*i.e.*, individual or compounded flavor extracts), the relative doses should be modified accordingly to counteract the two-fold effects of carbon dioxide. In addition to causing a partial desensitization of the taste buds following an initial temporary stimulative effect, carbon dioxide also reacts chemically with the constituents of the flavor complex. Therefore, the contribution by the reaction products to the resulting flavor must be taken into account, especially after prolonged storage. As in the case of non-carbonated bitters, the alcohol content plays an important role.

Regarding the formulation of aperitifs, tonics, and digestives, the topic differs somewhat from the basic considerations presented so far. As mentioned above, not only the various herbs and corresponding derivatives (extracts and tinctures) but also essential oils and sometimes distillates play an important role in the flavor of the final formulation. Therefore, formulations in most cases are hinged on the combination either of bitter flavors with citrus notes or of aromatic notes with bitter flavors. Specific examples of the above formulating techniques are represented by products such as aperol, select, and fernet. In the latter formulation the flavor contrast lies between mint and saffron and related variations, such as anise and saffron. In aperol, select, and other similar products, the basic flavor ingredient consists of a blend of soluble essential oils of sweet and bitter orange together with added amounts of vanilla or vanillin; the bitter principle consists of a complex bitter flavor formulated by using some of the herbs listed in Table 5.

In fernet, which is flavored with the characteristic aromatic note based on saffron and mint combinations, neither citrus nor other essential oils are used (except for mint essential oil). The bitter flavor usually is obtained with a few herbs used in suitable ratios, such as angelica (roots), calamus, calumba, camomile, centaury, cinchona, gentian, imperatoria, larch agaric, rhubarb, St. Johnswort, and zedoary to which aloe and myrrh resinoid are added. A large mint-to-bitter-complex ratio yields mint-flavored fernet types, whereas lower ratios yield bitters exhibiting flavor characteristics less reminiscent of typical fernets. The flavor of products with a definite basic note (anise, artichoke, cinchona, gentian, rhubarb, etc.) is rounded and upgraded using notes strictly dependent on the background note. This leaves very little room for variations and, therefore, permits only the addition of notes to refine and characterize the finished product.

Once the various flavors for alcoholic and non-alcoholic bitters and special aperitifs have been properly formulated, the suggested use levels for most typical products shown in Table 6 can be followed. The suggested formulations are suitable for the preparation of approximately 100 liters of finished product.

TABLE 6

Formulations of Various Bitters

Finished product	95% alcohol, l	Sugar, kg	Distilled water, l	Flavor complex	Color
Alpine herbs	33	15	59.2	q.s.*	q.s.*
Amarone	30	1	70.7	q.s.	q.s.
American, red and white	17	25	67.6	q.s.	q.s.
Artichoke	16.2	25	67.6	q.s.	q.s.
Centerbe	47	20	43.1	q.s.	q.s.
China-china	30	10	65	q.s.	q.s.
Cinchona bitter	30	10	65.5	q.s.	q.s.
Dutch bitter	32	25	54	q.s.	q.s.
Felsina bitter	31	14	61	q.s.	q.s.
Fernet	42	—	60	q.s.	q.s.
Fernet, mint	41.7	31	41	q.s.	q.s.
Ferro-china	21.5	10	73	q.s.	q.s.
Genepi	36	30	47	q.s.	q.s.
Milan bitter	26	24	60	q.s.	q.s.
Orange bitter	35	10	60.5	q.s.	q.s.
Tourin bitter	26	30	56	q.s.	q.s.
Venetia bitter	30	25	55.7	q.s.	q.s.
White bitter	36	20	53.2	q.s.	q.s.

* A sufficient quantity (*quantum sufficiat*).

Sweet Liqueurs

Parallel to bitters, the liquor industry manufactures a series of sweet liqueurs largely using traditional recipes from a great variety of ethnic sources. Sweet liqueurs consist of water, sugar, alcohol, and blends of flavor principles obtained by various extractive techniques from aromatic plants. Sweet liqueurs, therefore, differ from distilled liquors, which build up the aromatic principles and alcohol content during the fermentation of sugars present in the juices of fruit and other plants.

Prior to examining liqueurs from the standpoint of specific flavoring, a classification of the various categories of products is necessary. Only the most important products will be included in the following classification; wherever possible, synonyms will be used rather than trade names to designate commercially available products.

1. **Sweet liqueurs with definite fruity aromas:** apricot, banana, bitter and sweet orange, cedrat, cherry, mandarin, marasca cherry, peach, plum.
2. **Sweet liqueurs with definite floral, seedy, fruity, or herbaceous aromas:** anise, cocoa, coffee, kümmel, mint, rose, vanilla, vervain.
3. **Sweet liqueurs with a mixed, or fantasia, aroma:** alkermes, amaretto, arquebus, benedictine, benevento, certosa, cordial, goccia d'oro, goldwasser, millefiori, perfetto amore, ratafia.
4. **Elixirs (with a bittersweet note):** china (cinchona), rhubarb, walnut (nocino).

Basic Ingredients

Specific flavor ingredients (regardless of the preparation method) are the basic building blocks of liqueur formulations; consequently, the resulting flavor is strictly dependent on their use in the formulation. The flavorist also should be cautious in his choice of other basic ingredients, such as water, sugar, alcohol, and colorants. The functional importance of the above ingredients should in no way be underestimated.

Water. Although the choice of water is not as critical in the formulation of liqueurs as in carbonated beverages, the quality of water should be properly selected. Spring or tap water is usually good but not recommended; distilled or deionized water is preferred.

Sugar. Fine granulated sugar from reputable sources (cane or beet sugar) is preferred. Good manufacturing practice involves dissolving the sugar in distilled water, then adding colorants to the resulting syrup, and finally filtering.

Alcohol. The alcohol used must be free of off-flavors. Alcohol from the fermentation of fruit advantageously replaces alcohol from vinous fermentation in several liqueur types. The aromatic principles are dissolved or mixed in the required amount of alcohol; the resulting solution is added to the syrup. The resulting mixture is left standing for an extended period of time to allow an intimate blending of the flavor ingredients. Filtration of the finished liqueur prior to bottling is necessary to remove any residual haziness that is absolutely unacceptable in good commercial products.

Sweet Liqueurs with Definite Fruity Aromas

Apricot

In addition to alcoholic extracts (sometimes distilled) of fresh and dried fruits or concentrated juices, compound flavors prepared by mixing aldehyde C_{14} with various aromatic principles are frequently employed. These impart the following:

1. flowery notes—benzyl acetate, α-ionone, jasmine absolute, neroli essential oil, orris butter, and petitgrain bigarade essential oil;
2. walnut flavor—bitter almond, aldehyde C_{18}; and
3. slight ether notes—ethyl acetate.

Such flavor bases have very high yields ranging from 1 × 10,000 to 1 × 30,000, based on the weight of the finished product. Therefore, the use of tinctures prepared from fruit distillates is highly recommended to round off the flavor of the alcohol used for dilution.

Two basic types of liqueurs with an apricot flavor are prepared: (1) a sweet type, and (2) a somewhat less sweet type generally known as apricot brandy. Typical formulations for the preparation of 100 liters of finished product include the ingredients shown in the table below; the 96% alcohol can be replaced with a wine distillate (brandy, cognac) having a high (60–65%) alcohol content, with the water adjusted accordingly to maintain the desired alcoholic level in the finished product.

Apricot Liqueurs

Ingredient	Quantity	
	Sweet apricot	Apricot brandy
96% alcohol	30 l	33 l
Sugar	40 kg	36 kg
Distilled water	46.3 l	41.5 l
Compound flavor 1 × 100*	1 kg	1 kg
Concentrated apricot juice	3 kg	3 kg
Colorant	q.s.†	q.s.†

* The concentration of flavor ingredient is proportional to the yield.
† Sufficient quantity.

Banana

Banana liqueur is not currently in demand and is generally of limited consumption. The compounded essential oil used exclusively in the preparation of banana liqueur results from the blending of amyl and benzyl acetate and amyl and ethyl butyrate; the mixture is rounded off by the addition of essential oils (orange, lemon) and vanillin. The so-called "cream of banana" formulation consists of the following (recipe for 100 liters of finished product):

Cream of Banana

Ingredient	Quantity
96% alcohol	30 l
Sugar	40 l
Distilled water	46.4 l
Compound flavor 1 × 100*	1 kg
Colorant	q.s.

* The amount of flavor is proportional to the yield.

Bitter and Sweet Orange

Bitter and sweet orange are not commercially used names; curaçao and triple sec are those most frequently employed. These products can be prepared by directly using the tincture from the peels of bitter orange and mandarin, to which small amounts of chinotti tincture and spices, such as cinnamon and clove, can be added.

When the compounded essential oil (prepared from the individual essential oils and having a yield of approximately 1 × 10,000) is used for flavoring, a small amount of orange peel tincture can be added to improve the flavor. The compounded essential oil consists of a blend of sweet and bitter orange essential oils, rounded off by small additions of lemon, neroli, and petitgrain bigarade essential oils, and spiced with a flavor complex consisting of clove, cinnamon, coriander, and mace (or nutmeg) essential oils. A guideline to the formulation of compounded oil for flavoring curaçao follows:

Ingredient	Parts
Bitter orange essential oil	190
Cinnamon essential oil	4
Clove essential oil	3
Coriander essential oil	1
Lemon essential oil	22
Neroli essential oil	5
Nutmeg essential oil	2
Rum ether	5
Sweet orange essential oil	350
Alcohol	418
Total	1,000

A guideline to the formulation of orange-flavored liqueurs sufficient for the preparation of 100 liters of finished product is shown below:

Ingredient	Curaçao	Curaçao and triple sec
96% alcohol	36 l	36 l
Sugar	35 kg	38 kg
Distilled water	44.3 l	42.5 l
Compound flavor 1 × 200	0.5 kg	0.5 kg
Colorant	q.s.	q.s.

Sweet orange is also employed alone in the formulation of liqueurs; it is known under the generic name *cream of orange*. The soluble essence (used for flavoring syrups), more or less modified by the addition of tinctures, is usually employed for flavoring this type of liqueur. The following guideline formulation is suitable to prepare 100 liters of finished product:

Cream of Orange

Ingredient	Quantity
96% alcohol	30 l
Sugar	40 kg
Distilled water	46.4 l
Compound flavor 1 × 100	1 kg
Colorant	q.s.

A cream of mandarin is obtained by dilution similar to cream of orange. Mandarin soluble essence or the tincture is prepared from fresh mandarin peels, to which a small amount of vanilla is added.

Cedrat

The tincture from fresh peels is used in the preparation of this liqueur. Direct use of the essential oil is excluded because the resultant product is somewhat flavorless and because the essential oil is not produced industrially. The use of citrus-flavored blends prepared from the soluble essential oils of orange, lemon, and bergamot reminiscent of the cedrat flavor is of some interest from a technical viewpoint but is limited by the use of products of undefined origin.

The liqueur having a cedrat note is commercially available in two basic types: sweet citron water and strong citron water. These types usually are formulated by the following guideline formulations:

Cedrat-flavored Liqueurs

Ingredient	Quantity*	
	Sweet citron water	Strong citron water
96% alcohol	26.4 l	32 l
Distilled water	59.5 l	57.5 l
Compound flavor 1 × 200†	0.5 kg	0.5 kg
Colorant	q.s.	q.s.

* Preparation sufficient for 100 liters of finished product.
† The amount of compound flavor is proportional to the yield.

Cherry

Two techniques may be employed for the preparation of this liqueur, commercially available under the names *cherry* or *cherry brandy*. One method utilizes cherry juice, cherry distillate, alcohol, water, and sugar exclusively; the second method consists of diluting compounded essential oils in water, alcohol, and sugar, and adding cherry juice and cherry distillate. The amount of 96% alcohol in this formulation can be advantageously replaced, partially or totally, with a wine distillate, such as brandy or cognac, having a high (60–65%) alcohol content. The amount of water to be added should be adjusted based on the alcohol content in the finished product. The compounded oil contains bitter almond essential oil rounded off and finished with raspberry and rose notes (see discussion of sweet cherry flavor). The above-mentioned products may be prepared by the following guideline formulations:

Cherry-flavored Liqueurs

Ingredient	Quantity*	
	Cherry	Cherry brandy
96% alcohol	30 l	30 l
Distilled water	40 l	40 l
Compound flavor 1 × 100†	0.5 kg	0.5 kg
1:5 concentrated cherry or amarelle juice	—	ca. 3 kg
Colorant	q.s.	q.s.

* Preparation sufficient for 100 liters of finished product.
† The amount of compound flavor is proportional to the yield.

Mandarin

The use of mandarin essential oil is not always recommended for the preparation of the liqueur. The product resulting from the simple dilution of the essential oil with alcohol is rather flavorless; the use of alcoholic extract or tincture (usually 10% in 60% or 70% alcohol) prepared from fresh mandarin peels is preferred. The mandarin note can be rounded off by small additions of natural vanilla or vanillin solutions and neroli and petitgrain bigarade essential oils. The liqueur known by the generic name *cream of mandarin* can be prepared as follows:

Cream of Mandarin

Ingredient	Quantity*
96% alcohol	30 l
Sugar	40 kg
Distilled water	46.4 l
Compound flavor 1 × 100†	1 kg
Colorant	q.s.

* Quantities are sufficient for 100 liters of finished product.
† The amount of compound flavor is proportional to the yield.

Marasca Cherry

A typical formulation for this type of liqueur employs the distillate from Dalmatian marasca cherry, which is usually obtained by recovering the fractions distilling around 70°C from a tincture prepared as follows: (1) with water (20%), alcohol (60%), and marasca cherries (approx. 20%), and (2) directly from marasca cherries (20%) and 95% alcohol (80%). In the latter case the liqueur is prepared by diluting the distillate to the required alcohol content and adding sufficient amounts of sugar and flavor modifiers. The rose note plays a very important role in the aromatic maraschino complex.

When the use of compounded essential oils is desired (to decrease the use of marasca distillates), the corresponding product exhibits a basic bitter-almond note, as in cherry- and peach-flavored liqueurs; it can be formulated by using deacidified bitter almond or cherry laurel essential oils rounded off with neroli or petitgrain bigarade, citrus notes (orange, lemon), coumarin notes (tonka bean, melilotus, liatris, or coumarin), a slight cognac note, and rose absolute. The rose derivative is used generally at the 0.5% level based on the benzaldehyde content of the other ingredients (see discussion of sour cherry flavor, p. 668). When essential oils are used, the following guideline formulation is useful for preparing 100 liters of finished product:

Marasca Cherry Liqueur

Ingredient	Quantity
96% alcohol	33 l
Sugar	40 kg
Distilled water	44.1 l
Compound flavor 1 × 200	0.5 kg

Peach

The use of alcoholic peach extracts or tinctures yields only limited flavor results in peach liqueur. It is, therefore, necessary to use a compounded essential oil duplicating the flavor characteristics by modifying the bitter almond note fundamental in peach-flavored liqueurs. This compounded essential oil results from the blending of esters, such as methyl and ethyl cinnamate, with essential oils (clove, neroli, cinnamon, and coriander) imparting a floral, spicy note; it is sweetened by the addition of citrus essential oils, such as sweet and bitter orange, and sometimes small amounts of anise or anethol. The basic flavor resulting from a suitable blend of the above constituents has a yield of approximately 1 × 8,000. The use of aldehyde C_{14} yields excellent results in the formulation of flavors for pastry, even when used in massive doses; however, it may yield entirely different results in the formulation of liqueurs (see discussion of peach flavor, p. 660).

A typical product having a peach note (Persico Reale) may be formulated as follows:

Ingredient	Quantity*
96% alcohol	30 l
Sugar	40 kg
Distilled water	46.9 l
Compound flavor 1 × 200†	0.5 kg
Colorant	Only if required

* Quantities are sufficient for 100 liters of finished product.
† The amount of flavor is proportional to the yield.

Peach brandy can be formulated as follows:

Peach Brandy

Ingredient	Quantity*
96% alcohol	33 l
Sugar	36 kg
Distilled water	44.8 l
Compound flavor 1 × 100†	1 kg
Concentrated peach juice	3 kg
Colorant	q.s.

* Quantities are sufficient for the preparation of 100 liters of finished product.
† The amount of flavor is proportional to the yield.

Plum

Plum-flavored liqueurs are usually formulated by using compounded essential oils prepared by rounding off the bitter-almond note. The rounding off of this note employs several essential oils (*e.g.*, capsicum, cinnamon, clove, cognac), esters (amyl and ethyl acetate, ethyl butyrate, and ethyl and methyl cinnamate), ionones, and vanillin (see discussion of plum flavor, p. 665). The distillate and plum juice are also employed in the formulation of the best quality types, *i.e.*, sweet and dry plum liqueurs. Liqueurs formulated using essential oils exhibit varying compositions depending on the type. General guidelines are given below:

Plum-flavored Liqueurs

Ingredient	Quantity*			
	Plum, sweet	Plum, dry	Plum, regular (prunella)	Plum brandy
96% alcohol	30 l	50 l	30 l	33 l
Sugar	35 kg	1 kg	40 kg	36 kg
Distilled water	49.5 l	51.5 l	45.0 l	44.5 l
Compound flavor 1 × 100†	1 kg	0.5 kg	3 kg	3 kg
Concentrated plum juice	3 kg	—	—	3 kg
Colorant	q.s.	q.s.	q.s.	q.s.

* Quantities are sufficient for the preparation of 100 liters of finished product.
† The amount of flavor is proportional to the yield.

Sweet Liqueurs with Definite Floral, Seedy, Fruity, or Herbaceous Aromas

Anise

The liqueur industry uses anise flavor in a wide variety of finished products. The widespread acceptance of anise-flavored liqueurs is illustrated by the popularity of the following throughout the Mediterranean: Turkish raki, mastic of chio, Greek onyo, Egyptian arak-el balah, Lebanese arak, Moroccan mahia, and Tunisian nukha. Another classic series of anise-flavored liqueurs is *anisette*, originally prepared by distillation of green anise seeds macerated in a water-alcohol mixture. This liqueur also contains trace amounts of bitter principles or a floral note in order to impart a characteristic bouquet to the finished product.

In contrast to the traditional manufacturing techniques, there is regular use of compounded essential oils consisting mainly of anise essential oil (congealing point, 15°C), green anise essential oil, or anethol. Based on the percentage of anethol present in the formulation, the optional addition of trace quantities of characterizing flavors, the degree of the milky appearance of the liqueur on dilution with water, the amount of sugar, and the alcohol content, the various anise-flavored products may be classified into the following basic types: anisette, anice forte, sambuca, sambuca coffee, anesone, mastic of chio, anici lattanti, and sassolino. Typical formulations characteristic of the different type products are shown in the following table:

Anise-flavored Liqueurs

Ingredient	Anice forte	Anici lattanti	Anisette	Mastic of chio	Sambuca	Sassolino	Sambuca coffee	Anesone
96% alcohol	41.6 l	46.9 l	33 l	50 l	43.5 l	36 l	43.2 l	36 l
Sugar	1 kg	1 kg	40 kg	1 kg	38 kg	32 kg	38 kg	20 kg
Distilled water	60.3 l	55.1 l	44 l	51.6 l	35.3 l	45.7 l	35 l	53.7 l
Compound flavor 1 × 1000	0.5 kg	0.5 kg	0.5 kg	0.5 kg	0.5 kg	1 kg	1 kg	1 kg
Colorant	—	—	—	q.s.	—	—	q.s.	—

Anici lattanti flavored with wormwood was extremely popular during the nineteenth and early twentieth centuries for use in liqueurs in France. After the barring of wormwood, similar formulations were introduced in which wormwood was replaced by alternate bitter principles to satisfy the demand for this type of product. The use of cinnamon, camomile, lemon balm, lemon, orange, and coriander in such liqueurs as complementary flavor ingredients is maintained in such formulations. The effective yield of anethol, based on the finished liqueur "as is," is approximately 1 × 2,000.

Cocoa

This product is known commercially by the generic name *creme de cacao*; it is produced as both a colorless and a cocoa-like brown product. It is prepared directly from cocoa distillates or tinctures obtained either from roasted and ground cocoa beans or from bitter, defatted cocoa. Cocoa distillates are prepared by distilling an alcohol-water mixture of macerated cocoa beans (ca. 30%) and recovering the distilled fraction having a 60–65% alcohol content. The yield of the distillate, based on the finished liqueur, ranges between 1 × 25 and 1 × 33; distillates having a larger yield may be prepared by adding reinforcing flavors. The tincture consists of the above mixture after maceration and has a yield of 1 × 20 maximum.

The rounding off of the cocoa note using small amounts of vanilla (approximately 1 × 10,000 vanillin) in the finished liqueur is recommended. Compounded essential oils based on isolates or synthetic products are not commonly used in the formulation of liqueurs. A guideline formulation for creme de cacao is shown below; the formulation may also be used for the preparation of white or brown cream of chocolate liqueurs.

Creme de Cacao

Ingredient	Quantity*
96% alcohol	24 l
Sugar	40 kg
Distilled water	51.8 l
Compound flavor 1 × 100	1 kg
Colorant	If required

* This formulation yields 100 liters of finished liqueur.

Coffee

The discussion of coffee-flavored liqueurs parallels cocoa as to both the name *cream of coffee* and the use of tinctures and distillates. The amount of coffee beans used in the preparation of the flavor intermediates is analogous to cocoa beans, but greater emphasis is placed on the degree of roasting and the choice of coffee blends; highly aromatic and not too bitter coffee qualities are preferred. Modifying flavors are not usually added, except for a slight ether note (*e.g.*, traces of rum or vanilla flavor) in a few cases. The resulting formulation will be intensely colored by the addition of caramel. Similar considerations to those made for cocoa-flavored liqueurs are valid for the use of compounded essential oils or compound flavors with a definite coffee-like note. A guideline formulation for cream of coffee sufficient for 100 liters of finished product is given below:

Cream of Coffee

Ingredient	Quantity
96% alcohol	30 l
Sugar	40 kg
Distilled water	46.4 l
Compound flavor 1 × 100	1 kg
Colorant	q.s.

Crystallized Liqueurs

In a liqueur containing 42–43% alcohol, a 40% sugar content already yields crystallization. Whenever the alcohol content is decreased, the sugar content must be increased to insure crystallization. For example, in order to deposit sugar crystals on the branchlets of heath (as in a typical millefiori liqueur) without depositing it on the glass walls of the container, the amount of sugar must be very accurately dosed, based on the corresponding alcohol content so as to exceed only slightly the minimum amount yielding incipient crystallization. The following operative technique is suggested:

1. Dissolve the compound oil in the alcohol required for the preparation of a predetermined amount of finished liqueur. Aliquots of the alcohol solution are then divided into the various containers.
2. Prepare the sugar syrup (usually containing 550–650 g sugar per liter of finished liqueur) by heating; then add the colorant if required, and filter the solution immediately. Fill the bottles with the hot syrup and seal them temporarily.
3. Allow the bottles to cool without disturbing. Adjust the final volume by adding a small amount of hot syrup to replace volume losses caused by contraction of the mass on cooling.
4. Again cool the bottles, seal, and store them in a cool place for 15 days minimum without disturbance.

Kümmel

Kümmel is also known by other names, such as double kümmel or crystallized kümmel, based on the amount of sugar present in the finished liqueur. The basic ingredient for flavoring kümmel is caraway essential oil (rectified), imparting the characteristic anise-like note to the finished product. The preparation of the compounded oil calls for about 90% rectified caraway essential oil; the remaining 10% consists of a blend of spicy notes, such as cinnamon, nutmeg, clove, and coriander, together with small amounts of orange and lemon (sometimes in combination with a small amount of anethol).

Kümmel Flavor

Ingredient	Parts
Rectified caraway essential oil	350
Rum ether	20
Anethol	3
Rectified fennel essential oil	2
Lemon essential oil	2
Coriander essential oil	1
Solvent (alcohol)	122
Total	500

It is also possible to employ the alcohol distillate of caraway seeds as the basic ingredient. A guideline formulation for the preparation of 100 liters of liqueur is shown below:

Kümmel

Ingredient	Quantity
96% alcohol	38 l
Sugar	60 kg
Distilled water	28 l
Compound flavor 1 × 200	0.5 kg

Lemon

The production of lemon liqueur today is of limited domestic use only. The commercial product is often called by the corresponding French name *verveine*. In the latter case, however, the flavor is somewhat different and may be defined as a blend of flavors used in certain types of liqueurs, such as certosa and benedictine, together with a lemon-like flavor (lemon verbena). Homemade liqueurs are prepared by direct infusion of the green lemon verbena leaves in 40% alcohol without further addition of complementary flavors; the resulting infusion is sweetened with sugar (q.s.) to personal taste.

The liqueur flavored with essential oils can be prepared by using the guideline formulation shown below:

Lemon Verbena

Ingredient	Quantity*	
	Strong	**Sweet**
96% alcohol	32 l	32 l
Sugar	36 kg	40 kg
Distilled water	48.5 l	45 l
Compound flavor 1 × 200	0.5 kg	0.5 kg

* Formulation is sufficient for 100 liters of finished liqueur.

Mint

Mint-flavored liqueurs are usually prepared by diluting a good quality tri-rectified peppermint essential oil (approximate yield, 1 × 4,000) or the soluble essence (approximate yield, 1 × 2,000) with alcohol and syrup. Small amounts of flavor modifiers, such as citrus essential oils, may be added as optional ingredients; however, these should not exceed 3.5‰, based on the amount of peppermint essential oil employed, in order to obtain good flavor results.

Green and white mint liqueurs may be prepared by using the following guideline formulation, sufficient for 100 liters of finished product:

Mint Liqueurs

Ingredient	Quantity	
	Green mint	**White mint**
96% alcohol	32 l	32 l
Sugar	36 kg	40 kg
Distilled water	48.5 l	45 l
Compound flavor 1 × 200	0.5 kg	0.5 kg
Colorant	q.s.	—

Rose

The importance of this flavor ingredient is described in the discussion of Floral Flavors (see p. 573). The following description, therefore, will be limited to the formulation of creme de rose, or rosolio. The basic note in rose liqueurs is obviously the rose flavor that, when formulated as a compounded essential oil free of additional diluents, has an approximate yield of 1 × 8,000 based on the finished product. The choice of a rose absolute rich in phenethyl alcohol is recommended for the formulation of the compound flavors. The note of rose absolute alone is too sweet, however, and citrus notes (sweet orange) must be added in small amounts; this is rounded off by the addition of vanillin and spiced with a blend of cinnamon and nutmeg. If necessary to decrease formulating costs, only one part of true rose absolute can be employed together with massive doses of rose derivatives; it is advisable to add a small amount of rose distillation water to dilute the alcohol content. The following guideline formulation may be employed. Note that rosolio has approximately the same formulation as alkermes (see p. 622).

Rosolio

Ingredient	Quantity*
96% alcohol	30 l
Sugar	40 kg
Distilled water	46.4 l
Compound flavor 1 × 100	1 kg
Colorant	q.s.

* Formulation is sufficient for 100 liters of finished product.

Vanilla

White creme de vanilla is generally used as a flavor modifier when added in varying amounts to creme de mocha, rhubarb elixir, and a few other products, depending on traditional customs. The brown quality is used mainly as a liqueur. Creme de vanilla is prepared with vanilla tincture obtained by macerating vanilla pods in approximately 50% alcohol. The hydro-alcoholic solution of vanillin or vanilla absolute may be used for less expensive products. Both white and brown creme de vanilla can be prepared by the following formulation:

Creme de Vanilla

Ingredient	Quantity*	
	White	Brown
96% alcohol	24 l	24 l
Sugar	14 kg	40 kg
Distilled water	51.8 l	51 l
Compound flavor 1 × 100	1 kg	1 kg
Colorant	—	q.s.

* Formulation is sufficient for 100 liters of finished liqueur.

Sweet Liqueurs with Mixed, or Fantasia, Flavors

The formulation of these liqueurs varies widely and includes a certain number of basic flavor notes. The formulations discussed in the following pages represent examples of the individual categories of products apparently developed as variations of a single flavor theme. These liqueur varieties found wide acceptance at the beginning of the century and are still rather popular today.

Alkermes

The formulation of alkermes is rather typical. The characteristic bouquet results from a nearly equal-weight mixture of a strong citrus flavor (orange and lemon) together with a blend of spices (clove, cinnamon, and mace). The resulting flavor is strongly influenced by the spicy note, which can be reinforced for specific uses, as in *bagne** for pastry. A rose note that can be quite pronounced gives a flavor effect similar to rosolio. Other products also are used to round off the flavor note, such as coriander, calamus, and vanillin; the latter ingredient is usually present in a 1:15 ratio based on the citrus essential oils.

A guideline formulation for the preparation of a suitable compounded oil for alkermes is given below:

Ingredient	Parts
Clove essential oil	780
Lemon essential oil	400
Orange essential oil	300
Cinnamon essential oil	250
Mace essential oil	180
Vanillin	150
Neroli essential oil	10
Citronellol	2
Rose absolute	1
Solvent (alcohol)	927
Total	3,000

The liqueur is usually prepared according to the following formulation:

Alkermes

Ingredient	Quantity*
96% alcohol	30 l
Sugar	40 kg
Distilled water	46.8 l
Compound flavor 1 × 200	0.5 kg
Colorant	q.s.

* Formulation sufficient for 100 liters of finished liqueur.

Amaretto

Amaretto is a sweet, floral variation of bitter almond liqueurs. Its flavor results from using citrus flavors (*e.g.*, lemon, bitter orange, etc.) and then adding vanillin, cocoa distillate, and a trace of neroli. A guideline formulation for the preparation of amaretto follows:

Amaretto

Ingredient	Quantity*
96% alcohol	30 l
Sugar	40 kg
Distilled water	46.5 l
Compound flavor 1 × 100	1 kg
Colorant	q.s.

* Formulation for 100 liters of finished liqueur.

* *Bagne* are liqueur sauces used for soaking pound cakes, such as baba rum.

Arquebus

The characteristic flavor note of this product results from blending five basic notes: thyme, clary sage, rosemary, lavender, and anise. These might ultimately be blended in equal amounts. Other minor ingredients, such as fennel, orange or lemon, angelica, and oregano, enhance the thoroughness of the blend by bringing out certain characteristic notes of any one of the five basic flavors. The compounded essential oil is formulated by using only the individual essential oils; it has an approximate yield of 1 × 2,000. A guideline formulation is shown below:

Ingredient	Parts
Lavender essential oil	105
Clary sage essential oil	100
Rosemary essential oil	100
Thyme essential oil	90
Fennel essential oil	40
Mint essential oil	25.9
Angelica essential oil	5
Anise essential oil	5
Lemon essential oil	4
Wormwood essential oil	2
Cinnamon essential oil (10% in alcohol solution)	0.1
Solvent (alcohol)	523
Total	1,000.0

In place of the essential oils, the same liqueur can also be prepared by distillation of the hydro-alcoholic infusion made by macerating the corresponding dried herbs. The approximately 60% alcohol distillate has a yield of about 1 × 50 after the addition of small amounts of angelica and lavender essential oils in alcohol solution. A liqueur acceptable to traditional market requirements may be prepared by the following guideline formulation:

Arquebus

Ingredient	Quantity*
96% alcohol	46.9 l
Sugar	1 kg
Distilled water	55.1 l
Compound flavor 1 × 200	0.5 kg

* Formulation for 100 liters of finished liqueur.

Benedictine

Benedictine is prepared in countless flavor variations, especially in France, although it consists basically of the blending of three flavors: the citrus note (orange), angelica, and juniper together with coriander. The large number of flavor ingredients that may be added to the basic formulation greatly influences the flavor of the finished product. From an old formulation prepared by distillation of infusions of dried herbs comes the following list of flavor ingredients: coriander, cardamom, angelica seeds, angelica roots, lemon verbena, yarrow, arnica, hyssop, lemon balm, marjoram, cinnamon, mace, oregano, clove, juniper berries, pine buds, and peppermint. To obtain certain flavor results, the use of distillates and tinctures rather than the essential oils of the above herbs are preferred for benedictine formulations. Essential oils may be employed only in solution and in trace amounts to round off the flavor note. Benedictine is formulated according to the guideline values given for the preparation of green certosa liqueurs (see p. 625).

Benevento

The characteristic flavor of benevento results from cardamom, coriander, angelica, and some calamus in specific cases. To these basic ingredients, mint, cinnamon, ginger, a pronounced citrus note, and a touch of anise flavor must be added. Although the basic ingredients for the product have been listed, the potential variations of the basic formulation are numerous. A typical variation consists of angelica seeds, angelica roots, or mixtures of the two in varying ratios. The amount of calamus used depends on flavor variations.

Citrus flavors consisting of compound oils formulated exclusively with the individual essential oils must be properly selected. In specific cases the terpeneless oils are preferred because of their better solubility. Such compound flavors have an approximate yield of 1 × 8,000. A guideline for the preparation of the compound flavor is given below:

Ingredient	Parts
Cardamon essential oil	100
Coriander essential oil	100
Ginger essential oil	30
Angelica seeds	25
Angelica roots essential oil	20
Anethol	10
Lemon essential oil	8
Menthol	7
Ceylon cinnamon essential oil	5
Solvent (alcohol)	695
Total	1,000

The distillate from the hydro-alcoholic infusion prepared by macerating the corresponding dried herbs can also be used in the liqueur preparation. The distillate has a yield of approximately 1 × 50; the flavor is usually reinforced by small amounts of essential oils in alcoholic solution to enhance the sharp, citrus-anise note. The possible variations with respect to alcohol and sugar content in the formulation are somewhat limited and usually do not vary much from the following values:

Benevento

Ingredient	Quantity*
96% alcohol	36 l
Sugar	30 kg
Distilled water	47.5 l
Compound flavor 1 × 200	0.5 kg
Colorant	q.s.

* Suitable for the preparation of 100 liters of finished product.

Certosa

Certosa exhibits a flavor very similar to benedictine; in certosa the citrus flavors also are associated with angelica and juniper. The ingredient ratios, however, are entirely different from benedictine; two additional flavor notes are present—anise and wormwood. The latter can be replaced by a vermouth compound flavor.

Certosa liqueurs are produced in two types—green certosa and yellow certosa. The anise note is most pronounced in the yellow certosa, whereas it is somewhat subdued in green certosa. Rum and cognac notes may be used in green certosa formulations, while only trace amounts of these flavors should be used in the formulation of yellow certosa. Following are essential oils used exclusively for the preparation of a compound certosa flavor having a yield of approximately 1 × 5,000:

Compound Certosa Flavor

Ingredients	Parts
Lemon essential oil	60
Anethol	20
Juniper essential oil (rectified)	15
Clary sage essential oil	7.1
Hyssop essential oil	4
Bitter orange essential oil	3
Mint essential oil (tri-rectified)	3
Petitgrain bigarade	3
Coriander essential oil	2.5
Angelica essential oil	2.1
Mace essential oil	2.1
Wormwood essential oil	2.1
Sweet orange essential oil	2
Caraway essential oil	1
Ceylon cinnamon essential oil	1
Lavender essential oil	Trace

The above values should be considered as approximate, with anethol, lemon, and juniper comprising the bulk of the formulation.

Certosa may also be formulated by using the corresponding dried herbs in place of the essential oils. However, the resulting flavor effects will not be entirely satisfactory unless additional quantities of soluble essential oils are added to the compound flavor. In general, an infusion followed by distillation of a blend of cinnamon, mace, bitter orange, clary sage, juniper, angelica, coriander, wormwood, and hyssop could be prepared. The distillate then could be added with a suitable blend of essential oils, such as lemon, sweet orange, mint, petitgrain bigarade, and anise.

Yellow certosa and green certosa differ slightly as to flavor, color, and alcohol content. Both liqueurs may be prepared according to the following guideline formulations:

Yellow and Green Certosa

Ingredient	Quantity*	
	Yellow certosa	Green certosa
96% alcohol	36 l	38 l
Sugar	30 kg	30 kg
Distilled water	47.5 l	45.6 l
Compound flavor 1 × 200	0.5 kg	0.5 kg
Colorant	q.s.	q.s.

* Formulation for 100 liters of finished product.

Cordial

This liqueur exhibits chiefly floral and fruital flavors. In formulations considered nearly typical, the flavor results from a blend of violet and cherry notes. To obtain satisfactory flavor results, it is necessary to blend the characteristic irone note of orris butter present in raspberry distillate with that of kirsch and maraschino present in sweet and marasca cherry distillates, respectively. The addition of a small amount of cognac and sweet orange essential oil imparts a slightly tart, spicy, citrus-like flavor to the compounded essential oil that has an approximate yield of 1 × 4,000. A guideline formulation for the preparation of the compounded oil follows:

Ingredient	Parts
Rum ether	100
Vanillin	100
Cognac essential oil	62
Ethyl acetate	60
Sweet orange essential oil	35
Amyl acetate	20
Bitter almond essential oil	13
Cherry laurel essential oil	4
Orris butter	1.8
Lemon essential oil	1.4
α-Ionone	0.5
Rose absolute	0.3
Solvent (alcohol)	602
Total	1,000.0

A guideline formulation for the preparation of cordial follows:

Cordial

Ingredient	Quantity*
96% alcohol	36 l
Sugar	30 kg
Distilled water	47 l
Compound flavor 1 × 100	1 kg
Colorant	q.s.

* Formulation for 100 liters of finished liqueur.

Goldwasser

This type of liqueur is also known as gold-drop, goldwater, and danzig water. The name *goldwasser* derives from the tiny gold leaflets suspended in the finished liqueur. The characteristic flavor results from a mixed citrus note (orange and lemon, 1:2) blended with a spicy flavor complex consisting mainly of calamus, ginger, cinnamon, and nutmeg. Cardamom is sometimes added; lavender, neroli, and sometimes spearmint are frequently added to impart particular bouquets to the finished product. A guideline to the formulation of the compound flavor is given on the following page.

Goldwasser Flavor

Ingredient	Parts
Lemon essential oil	200
Orange essential oil	100
Cinnamon essential oil	100
Nutmeg essential oil	50
Calamus essential oil	25
Cardamon essential oil	12
Neroli essential oil	3
Rosemary essential oil	2
Solvent (alcohol)	508
Total	1,000

A guideline formulation for the preparation of 100 liters of finished liqueur follows:

Goldwasser

Ingredient	Quantity
96% alcohol	14 l
Sugar	25 kg
Distilled water	46.8 l
Compound flavor 1 × 200	0.5 kg
Colorant (to create gold leaflet effect)	q.s.

Millefiori

The name *millefiori* is intended to represent an association with countless floral notes. The basic flavor notes of this liqueur consist of bitter orange flowers (neroli), rose, orris, vanillin, and a trace amount of spices. Two liqueur types are formulated: one contains large amounts of sugar, the other has a small amount of sugar and is called "frosted." To create special effects, a small amount of ambrette seed tincture may be added to subdue the spicy cinnamon and cardamom notes (sometimes employed also in blends with coriander). A guideline formulation for the preparation of a suitable compounded oil is given below:

Ingredient	Parts
Neroli essential oil	115
Rose flavor imitation	75
Vanillin	60
Bitter orange essential oil	10
Clove essential oil	6
Ceylon cinnamon essential oil	5
Nutmeg essential oil	5
Orris butter	4
Solvent (alcohol)	720
Total	1,000

Following the guidelines discussed for the preparation of crystallized liqueurs (see p. 689), millefiori frosted or crystallized liqueurs may be formulated as follows:

Millefiori

Ingredient	Quantity*	
	Frosted	Crystallized
96% alcohol	35.5 l	30 l
Sugar	60 kg	40 kg
Distilled water	28.7 l	46.6 l
Compound flavor 1 × 100	1 kg	1 kg
Colorant	q.s.	q.s.

* Formulation for 100 liters of finished product.

Perfetto Amore

Liqueurs of this type have been formulated under the same name but with entirely different flavors in several European countries. Among the most characteristic formulations, one is specifically reminiscent of benevento liqueurs; it consists of lemon, coriander, mace, clove, and a trace of cinnamon. A few other products exhibit an anise flavor. The liqueur may be prepared according to the following formulation:

Perfetto Amore

Ingredient	Quantity*
96% alcohol	30 l
Sugar	40 kg
Distilled water	46.9 l
Compound flavor 1 × 200	0.5 kg
Colorant	q.s.

* Formulation for 100 liters of finished product.

Ratafia

The traditional formulation of this liqueur can be considered a spice-flavored variation of bitter almond liqueurs. Cherry laurel once was the basic ingredient in the compounded essential oil together with clove, cinnamon, mace, and coriander. The resulting flavor was complemented by a good red Barbera wine.

Although these product types have almost completely disappeared from the market, a flavor formulation free of cherry laurel (an oil that must be rendered completely free of prussic acid) can be prepared by replacing cherry laurel essential oil with that of bitter almond, to which small amounts of benzyl alcohol and citronellol are added. A guideline formulation follows:

Ratafia

Ingredient	Quantity*
96% alcohol	24 l
Wine	20 l
Sugar	40 kg
Distilled water	31.7 l
Compound flavor 1 × 100	1 kg
Colorant	q.s.

* Formulation for 100 liters of finished liqueur.

Elixirs

This category includes liqueurs prepared by dissolving the aromatic principles of specific herbs in water, alcohol, and sugar. The resulting flavor consists of a sugar flavor blended with the usually bitter note of the herb. The oldest and most typical liqueur formulations are representative of elixirs.

Rhubarb Elixir

The basic raw material for this preparation is Chinese rhubarb (flat or round). The comminuted rhizome is extracted with a mixture of water and alcohol (according to some preparation procedures: 1 part rhubarb, 1 part water, and 1 part alcohol). After a few days the tincture is drained off and filtered; small amounts of essential oils (calamus, coriander, wormwood, etc.) are added to impart the characteristic bouquet. Alternatively, a compound tincture of rhubarb with other herbs can be prepared using traditional preparatory methods. Rhubarb-flavored elixir is prepared according to the following guideline formulation:

Rhubarb Elixer

Ingredient	Quantity*
96% alcohol	16.5 l
Sugar	27 kg
Distilled water	65 l
Compound flavor 1 × 100	1 kg
Colorant	q.s.

* Formulation for 100 liters of finished product.

China (Cinchona-flavored Elixer)

Cinchona-flavored elixir is prepared usually from cinchona tinctures. Individual or compounded tinctures may be prepared by extracting the flavor principles from the individual herbs or from a blend. Cinchona bark is the most important raw material; characteristic flavor effects are obtained by adding clove, cinnamon, coriander, and sweet orange. Sometimes the bitter flavor is reinforced by the addition of gentian and chirata (the latter may be employed up to approximately equal weight of cinchona), whose flavors blend well with cinchona. Cinchona-flavored elixir may be prepared according to the following guideline formulation:

China

Ingredient	Quantity*
96% alcohol	32.2 l
Sugar	34 kg
Distilled water	47.4 l
Compound flavor 1 × 100	1 kg
Colorant	q.s.

* Formulation for 100 liters of finished product.

Nocino (Walnut-flavored Elixer)

Nocino, often found in old recipe books for liqueurs, today is produced primarily in Italy. The characteristic bitter flavor is obtained by the extraction of green walnuts harvested at the end of June according to Italian tradition. A recipe for a homemade liqueur for approximately 2.5 liters of finished product prescribes the extraction of about 30 walnuts in 70% alcohol, to which sufficient sugar is added. The addition of a few grams of cinnamon, clove, and lemon peel is optional.

By using specifically prepared compounded oils, 100 liters of liqueur may be formulated according to the following guideline:

Nocino

Ingredient	Quantity
96% alcohol	24 l
Sugar	40 kg
Distilled water	52.3 l
Compound flavor 1 × 200	0.5 kg
Colorant	q.s.

Distilled Liquor Imitations and Their Composition

The large market demand for distilled liquors has prompted the study and formulation of the corresponding imitations to satisfy widely divergent consumer needs. An additional factor that has stimulated the search for suitable imitations is the relevant difference in production costs between natural and imitation distillates; these high costs are attributed to the lengthy aging required by natural distillates together with the relatively large capital investment for cellar space and equipment. It should be stressed that the potential use of imitation distillates for flavoring pastry is high; well formulated imitations, in fact, have been proven to yield highly satisfactory flavor results. From the standpoint of regulations governing the production of natural distillates, a discussion of the comparative value of commercial distillate imitations is beyond the scope of the book.

Rum

Rums and tafias are obtained by distillation of raw or slightly cooked and concentrated fermented juices expressed from sugar canes (*Saccharum officinarum* L.) or by distillation of fermented molasses (black strap) residue from the refining of cane sugar. The production site of these distillates, together with the name *rum* or *tafia*, has always been employed to differentiate the various commercial qualities. Specific reference will be made, for example, in the text to the following types: Jamaica rum, Martinique rum, white or Antilles rum, and Demerara or British Guiana rum. Countries that produce rum distillates include Brazil, British Guiana, Cuba, Haiti, Jamaica, Puerto Rico, Santo Domingo, Trinidad, Madagascar, the United States (Kentucky, Massachusetts, and Hawaii), and the Virgin Islands. The Hawaiian production started only recently and follows production methods quite "new" from certain points of view, although concentrated sugar-cane juices are still used as the primary source of raw material.

The flavor complex of natural rum has been the subject of several analytical studies. The reader can refer to a comprehensive bibliography on the subject[7] and to a more recent gas-chromatographic study using capillary columns.[5,6] The characteristic rum flavor is dependent on the total impurities coefficient expressed as total impurities content in milligrams based on 100 cc anhydrous alcohol; the commercial classification based on the above coefficient is in practice the resulting sum of acids, ethers, alcohols, aldehydes, and furfural content. The most important free acids are formic, acetic, butyric, and caproic. For the weight composition of volatile components in rum flavor, the reader again is referred to the above study.[5,6]

The need for an imitation rum distillate was created in Europe by the unavailability of either cane sugar or molasses. From the end of the nineteenth century to the beginning of the twentieth century, sugar beet molasses, barley and other meals, and various fruit concoctions were employed as raw materials for the distillation together with flavor formulations containing old leather scrapings, black minced truffles, and other substances. Aging was carried out artificially by various expedients, such as singeing the inner walls of oak barrels with tarred straw. The evolution of the chemical studies led to preparation methods of a certain interest, such as the recipe below suitable for the preparation of 100 liters of 70% rum using only 8 liters of genuine Jamaica rum. The following ingredients are mixed in a still:

Imitation Rum Distillate

Ingredient	Quantity
90% alcohol	34 l
Water	20.5 l
Choice wine vinegar	1.5 l
Bleached sugar syrup	1 kg
Tannin	375 g
Sulfuric acid	190 g
Powdered carbon black	250 g

Preparation:

1. The mixture is distilled slowly until 35 liters of distillate is obtained; the following ingredients are added:

Peru balsam	130 g
Ethyl acetate	65 g
Candied brown sugar	2 kg
Genuine Jamaica rum	8 kg

2. The distillate is then diluted to 100 liters with water and sufficient alcohol to yield a 70% total alcoholic content.

Formulations similar to the above may be found by the hundreds in old formularies for liqueurs; all are different, however, because of the relentless search by flavorists for the characteristic flavor note. Some flavorists identified the note with Malaga wine flavor; a few others with that of green tea, vanilla, or cinnamon, but all failed to formulate a characteristic bouquet.

Rum Ether

A milestone in the formulation of rum imitations is the fabrication of rum ether, whose composition in volatile substances is qualitatively very similar to rum distillates.[8] To the flavorist this product represents the building block in the imitation of distillates. In the following sections various flavor ingredients suitable for addition to rum ether to formulate the most characteristic bouquets will be discussed.

Today's manufacture of rum ether is based on the use of the following ingredients: ethyl alcohol (95%), sulfuric acid (66°Be), manganese dioxide, and pirolígneous acid. The method of preparation consists of first adding the alcohol and manganese dioxide to the still, then the pirolígneous acid, and finally the sulfuric acid (slowly, with stirring, to prevent the temperature in the reaction mass from exceeding 40–50°C). The resulting mixture is distilled under atmospheric pressure, and the fraction distilling between 60–100°C is collected. The distillate is subsequently rectified.

A survey of the relative amounts of the various ingredients used in the preparation shows upper and lower limits (indicated as Type A and Type B, respectively, in the following table) as well as all possible intermediate values being used.

Ingredient	Type A		Type B
95% alcohol	12 kg	to	20 kg
Pirolígneous acid (12% sol in acetic acid)	10 kg	to	16 kg
Sulfuric acid	3 kg	to	8 kg
Manganese dioxide	2 kg	to	5 kg
Or, placing ethyl alcohol equal to 100			
95% ethyl alcohol	100 kg		100 kg
Pirolígneous acid (12% sol in acetic acid)	83 kg	to	80 kg
Sulfuric acid	25 kg	to	40 kg
Manganese dioxide	16.6 kg	to	25 kg

The average yield of rectified rum ether from various preparations is about 98% calculated from the amount of alcohol used. The rectified fraction has a distillation range between 65–87°C.

The distillate is sometimes divided into two main fractions suitable for specific applications in liqueur formulations based on the higher or lower acetaldehyde content. More often, two or more types of rum ether are distilled by varying the amount of the starting materials in order to prepare products more suitable in specific liqueur applications. A few manufacturers add ingredients such as starch, meals, birch tar, sugar beet or sugar cane molasses, sucrose, etc. to the basic mixture, but this is not commonly practiced. Some authors also suggest the addition of acetic acid, wood coal, and Peru balsam.[9]

Rum ethers prepared by different methods contain qualitatively the same volatile principles but in varying ratios. Therefore, the most suitable product can be selected for specific uses. The addition of other flavor ingredients will follow parallel guidelines by observing that the same ingredients are often used in more than one formulation. The ingredients most widely used in distillate imitations are as follows:

1. Tinctures: carob, raisin, saffron, vanilla (or vanillin)
2. Essential oils: cinnamon, cognac
3. Orris absolute

Among the above ingredients a few are recurrent in most formulations because of their functionality. Saffron tincture is the most significant example and is used to impart the flavor note characteristic of aged distillates. Cognac oil imparts the flavor experienced when swallowing the liquor, while cinnamon essential oil adds a characteristic spicy note.

Jamaica Rum

In addition to the above major ingredients, equal amounts of raisin and carob tincture and slightly less saffron tincture are used for flavoring Jamaica rum distillate imitation; for example, 1 part raisin tincture, 1 part carob tincture, and 0.04 part saffron tincture are added to 6 parts rum ether.

Martinique Rum

Natural Martinique rum is characterized by a low total impurities coefficient if distilled from fermented cane sugar juice and not from molasses. This is reproduced in the distillate imitation by avoiding the use of carob and raisin tinctures, while using only trace amounts of saffron tincture in proportioned blends with the essential oils and vanillin.

White Antilles Rum

The flavor composition of white Antilles rum is very similar to Martinique rum. The finished imitation should not be colored with caramel to avoid confusion with the less valuable Martinique rum from molasses. The ratio between rum ether and other flavor ingredients is usually 90:2 in this product as compared to the somewhat lower 90:3–3.5 in Martinique rum.

Demerara Rum

This distillate could be classified in flavor midway between Jamaica and Martinique rum. However, it exhibits a characteristic sweet, less spicy note; it also has a somewhat higher volatile content. Therefore, Demerara imitation distillates contain more than the usual amounts of carob tincture and rum ether, fewer essential oils, and more saffron tincture to improve the flavor of the blend, which is less aromatic than other rum types.

Following this rather cursory examination of the aroma notes characteristic of different rum distillates, it should be stressed that rum ether in blends with flavor ingredients (in suitable doses) may be diluted 200 to 400 fold with alcohol of the selected strength. Products with yields varying between approximately 1 × 100 and 1 × 200 on dilution with 70% alcohol are suitable for distillate imitations for pastry use.

Cognac and Brandy

Cognac is a distillate of French wines native to the Charentes region. From the white wine of another French district, namely the hilly area separating the Garonne from the Adour basin, Armagnac distillates are produced. The aging process of wine distillates in oak barrels is as important as the use of specific grapes and wines in order to obtain good distillates. Countries other than France also have started the production of wine distillates. Since the name *cognac* has been reserved exclusively for distillates of French production, several other names are used to designate wine distillates from other sources. The generic name *brandy* has been adopted to designate those distillates of other nations.

Aside from personal preferences for natural cognac or brandy types, there is no significant difference in the number and nature of the basic components in the volatile aroma of the various products. However, definite quantitative differences as well as characteristic undertones are detectable; these determine the personal preference and allow the classification of wine distillates according to quality. A quantitative comparison of a cognac and brandy with a 40% alcohol strength shows the following:

1. a large ester (methyl formate plus ethyl acetate) content in cognac
2. an approximately equal aldehyde content, which is sometimes slightly higher in brandy
3. equal acetal content (acetaldehyde plus propionaldehyde), and
4. a larger acetone, *n*-propyl alcohol, isobutyl alcohol, and isoamyl alcohol content in brandy.

These observations have been confirmed by gas-chromatographic analysis using capillary columns that may yield additional differences such as to pinpoint with sufficient precision why each individual cognac exhibits a characteristic bouquet.

Cognac and brandy have long been imitated by using formulations based on essential oils and spirits suitable for imparting flavors closely duplicating those typical to diluted alcohol solutions. The demand for cognac and brandy imitations is attributable mainly to economic considerations. Originally, the raw material widely employed for such imitations was cognac oil; subsequently cognac oil was replaced with oenanthic ether, commercially available as a mixture of esterified coconut fatty acids. Today both products are employed in liquor distillate imitations. While cognac oil consists mainly of ethyl caprylate and ethyl palmitate, oenanthic ether contains a larger number of ethyl esters, such as ethyl caproate, ethyl caprylate, ethyl caprate, ethyl laurate, ethyl myristate, and ethyl palmitate. This implies significant flavor differences, although the presence of trace free alcohols in the intimate blend of notes tends to level off such flavor differences.

Economic and qualitative considerations govern the use of the individual flavor ingredients or blends thereof. Usually the above flavor ingredients alone are not sufficient to rebuild the classic bouquet of the distillates; it is, therefore, necessary to add other flavor ingredients. The choice and the use level of complementary flavor ingredients is subordinate to the cognac or brandy being imitated. The selection of ingredients extends to several categories, such as essential oils, tinctures, distillates, isolates, and synthetic products, as follows:

Essential Oils. Cognac, bitter almond, bay, clove, cinnamon, orris, birch, and neroli.

Absolutes. Vanilla, tonka bean, and oakmoss.

Isolates and Synthetic Products. Vanillin, irone, oenanthic ether, ethyl pelargonate, benzaldehyde, coumarin, ethyl acetate, ethyl formate, ethyl butyrate, acetaldehyde, and rum ether.

Tinctures. Saffron, oak bark, raisin, carob, and tea.

The basic formulation guidelines are often modified by the individual creativity of the flavorist as follows:

1. To formulate a flavor having an approximate yield of 1 × 500 to 1 × 600, 1 kg cognac oil (or cognac oil and oenanthic ether mixture) is blended with approximately 200–250 g essential oils, 60–80 g vanillin plus coumarin (or equivalent products), 7–8 kg rum ether, 1 kg esters (or more if rum ether is not included in the formulation), and 12–15 kg tincture (saffron tincture is used only in trace amounts based on the finished product).
2. To formulate fantasia rum, cognac, or brandy imitations using essential oils, tinctures, and distillates, the formulations shown in the following table may be a useful starting point:

Ingredient	96% alcohol, l	Sugar, kg	Distilled water, l	Colorant and compound flavor
Various types fantasia rum (with/without the addition of natural rum distillate)	50	0.5–1	52	q.s.
Whisky (imitation)	41.6	1	60.3	q.s.
Various brandy types	45	1	56.6	q.s.
Vodka (imitation)	50	0.5	52	

A few products such as calvados, whisky, vodka, and arrak are not usually imitated. Calvados is a cider distillate aged in oak barrels.

Gin and Kirschwasser

Gin and kirschwasser are two liquors generally obtained by direct distillation of juniper berries and cherries, respectively; these liquors are successfully imitated by means of compound flavors and alcohol (both liquors contain only a very small amount of sugar). Juniper essential oil is obtained as a by-product in the distillation of gin. A soluble essential oil can be prepared by a very careful rectification of the essential oil, which can be dissolved in alcohol of the required strength to yield imitation gin. The formulation of the kirsch note is somewhat more complex; also in this case the flavor note is obtained by using cherry kernels distillate. A guideline formulation follows:

Ingredient	Quality*
96% alcohol	50 l
Sugar	1 kg
Distilled water	52.1 l
Compound flavor 1 × 200	0.5 kg

* Formulation for 100 liters of finished product.

Fruit Distillates

The imitation of fruit distillates and their flavoring are also important; grappa, for example, is obtained by distillation of the expressed grape residue. The most widely used flavors for the above applications are muscat wine, pear, gentian (bitter), mint, plum, and bitter almond.

Muscat Flavor. Muscat flavor is prepared by distillation of a water-alcohol infusion of coriander, clary sage, and elder. The 50-proof distillate flavor has a yield of 1 × 100 or more.

Almond-plum Flavor. Almond-plum flavor is prepared by blending a plum tincture with bitter-almond essential oil (free from prussic acid) or benzaldehyde (free from hydrochloric acid) rounded off by small additions of cherry distillate or cognac oil.

Pear Flavor. Pear flavor is a product of very limited use basically prepared with esters (amyl acetate and a few others having similar notes) imparting to the compound flavor a note reminiscent of Williams pear.

Bitter Flavor. The bitter flavor of rue is obtained by direct infusion of the herb in the distillate. Gentian bitter flavor is quite frequently encountered and is prepared by the addition of small amounts of gentian tincture to the distillate.

Mint Flavor. Trirectified mint essential oil is used.

Punch

Liquors of this type are diluted with hot boiling water immediately prior to use; therefore, an alcohol content suitable for dilution is required. From a flavor standpoint, the various punches may be considered as variations of a basic rum flavor note.

Rum-flavored punches represent the classic beverage consumed piping hot in cold winter months; all other products consist of rum flavor combined with citrus notes (orange and mandarin punch), spicy notes (alkermes punch), or other flavors such as coffee, camomile, etc. For the preparation of the above products, orange or mandarin soluble essential oils, the compounded essential oil of alkermes, or the tinctures of coffee or camomile are added to rum distillate. Guideline formulations are shown below:

Punch Formulations*

Ingredient	Rum	Orange	Mandarin	Alkermes	Coffee	Camomile
96% alcohol, l	33	33	33	33	33	33
Sugar, kg	25	25	25	25	25	25
Distilled water, l	53.5	53.5	53.5	53.5	46	53.5
Compound flavor 1 × 200, kg	0.5	0.5	0.5	0.5	1†	0.5
Colorant	q.s.	q.s.	q.s.	q.s.	q.s.	q.s.

* The formulations are sufficient for 100 liters of finished product.
† (1 × 100).

NON-ALCOHOLIC BEVERAGES

Syrups

Syrups for non-alcoholic beverages consist of solutions containing sufficient sugar and flavor to achieve a 4–5 fold dilution with cold water or soda immediately prior to use. Syrups must be clear, yield clear solutions (with the exception of products such as orzata*), and exhibit good storage stability in containers repeatedly uncorked during use. To fulfill these physical requirements and to have a pleasant taste, syrups should contain 65% sugar and 35% fruit juice. Since the manufacturer seldom has fresh fruit at his disposal for expression of the juice in which sugar may be dissolved at room temperature, it is common practice to use 5–7 fold concentrated juice diluted with the required amount of syrup prior to use.[10] When fresh juices are employed, the addition of colorant and, sometimes, citric acid may not be required. The above represents optimum guidelines for the preparation of syrups for beverages.

While the presence of 65% sugar inhibits the growth of molds and bacteria, fermentation or degradation of the syrup may still occur, even when the above precautions are taken.[11] Syrup degradation may occur because of the sugar or water quality employed, lack of cleanliness, and a disregard for elementary hygienic practice during manufacturing. While preparation of syrups is a rather simple task, it requires good manufacturing practice and good quality raw materials. Independent of existing regulations concerning manufacturing practice and marketing of all types of syrups, it is always advisable to practice the following basic rules:

1. Wash and sterilize all containers prior to bottling the syrup.
2. Wash and sterilize all containers used in the preparation, filtering, and bottling of syrups.
3. Use refined sugar.
4. Use distilled, purified, or boiled water.
5. Use pasteurized juices.
6. Thoroughly filter the syrup.

As previously mentioned fresh fruit juices are seldom available for the preparation of syrups. In some cases concentrated juices are used after diluting as required; in other cases juices are used only in limited amounts or omitted from the formulation. In the latter case the basic steps for the preparation of syrup are as follows:

1. Preparation of the sugar syrup
2. Acidification of the sugar syrup
3. Coloring the sugar syrup
4. Flavoring the syrups with or without the addition of concentrated juice
5. Addition of preservatives where permitted.

65 kg of sugar are soluble in 35 kg of water, yielding 100 kg of syrup (or 75.8 l)

or

85.7 kg of sugar are soluble in 46.2 l of water, yielding 100 l of syrup (or 131.9 kg)

Preparation of Sugar Syrups at 65° Brix

1. Refined sugar is dissolved in pure (preferably distilled) water in a stainless steel container equipped with a stirrer and steam jacket.
2. When distilled water is used, dissolution of the sugar can be carried out at a low temperature.
3. Otherwise, the water is heated to boiling, and the sugar is added and stirred at 80°C until completely dissolved. The solution is then allowed to cool to room temperature.
4. When a crystal-clear syrup (instead of a yellowish product) is desired, the residual color may be removed by mixing approximately 0.1–0.2 kg activated charcoal (of the same quality used in wines) with the syrup. The syrup is then cooled overnight and finally filtered using asbestos or felt-lined filters for complete removal of the charcoal.[1]

---

* Orzata is a mixture of comminuted almonds, sugar, and water yielding milky solutions.

The syrup prepared as above must have the following specific gravity: (1) by refractive index values, 65° Brix, or (2) by hydrometer values, 34.5° Be = 1.319 kg/l.

Preparation of Sugar Syrups at 68° Brix

Operate as in the preceding method, using the following formulation:

sugar	68.00 kg
water	31.95 kg
citric acid	0.05 kg
syrup	100.00 kg

The above syrup must have the following specific gravity: (1) by refractive index values, 68° Brix, or (2) by hydrometer values, 36° Be = 1.3324 kg/l.

Acidification of Sugar Syrups

Citric acid or tartaric acid is used to acidify sugar syrups. The amount of added acid varies from 150 g per ton in citron-mint flavored syrup to as much as 1,400 g per ton in champagne-flavored syrup. Usually the acid is dissolved in the minimum amount of boiling water; the resulting solution is added to the syrup with brisk stirring to insure uniform dispersion. Guidelines for the acidification of syrups are shown in Table 7.

TABLE 7

Guidelines to the Acidification of Syrups

Grams of Acid per 100 kg of Finished Syrup*

Syrup	Acid		Quantity, grams
Apricot, 35% juice	Citric acid or tartaric acid		500
Apricot, 10–15% juice	Citric acid or tartaric acid		700
Apricot, juiceless	Citric acid or tartaric acid		800
Amarelle, 35% juice	Citric acid or tartaric acid		500
Amarelle, 10–15% juice	Citric acid or tartaric acid		700
Amarelle, juiceless	Citric acid or tartaric acid		800
Pineapple, 35% juice	Citric acid	—	500
Pineapple, 10–15% juice	Citric acid	—	700
Pineapple, juiceless	Citric acid	—	800
Orange, 35% juice	Citric acid	—	700
Orange, 10–15% juice	Citric acid	—	1,000
Orange, juiceless	Citric acid	—	500
Chinotti, juiceless	Citric acid	—	750
Cherry, 35% juice	—	Tartaric acid	500
Cherry, 10–15% juice	—	Tartaric acid	700
Cherry, juiceless	—	Tartaric acid	800
Strawberry, 35% juice	Citric acid	—	500
Strawberry, 10–15% juice	Citric acid	—	700
Strawberry, juiceless	Citric acid	—	700

* The use of tartaric, ascorbic, malic, and other acids is regulated by present regulations. Modifications of the above listed doses may be required.

TABLE 7 (continued)

Guidelines to the Acidification of Syrups

Syrup	Acid		Quantity, grams
Raspberry, 35% juice	Citric acid	—	500
Raspberry, 10–15% juice	Citric acid	—	700
Raspberry, juiceless	Citric acid	—	700
Lemon, 35% juice	Citric acid	—	—
Lemon, 10–15% juice	Citric acid	—	500
Lemon, juiceless	Citric acid	—	1,300
Mandarin, 35% juice	Citric acid	—	700
Mandarin, 10–15% juice	Citric acid	—	1,000
Mandarin, juiceless	Citric acid	—	1,000
Apple, 35% juice	—	Tartaric acid	500
Apple, 10–15% juice	—	Tartaric acid	550
Apple, juiceless	—	Tartaric acid	600
Mint, juiceless	Citric acid (q.s.) or tartaric acid		30
Pear, 35% juice	Citric acid or tartaric acid		500
Pear, 10–15% juice	—	Tartaric acid	600
Pear, juiceless	—	Tartaric acid	600
Peach, 35% juice	Citric acid	—	500
Peach, 10–15% juice	Citric acid	—	600
Peach, juiceless	Citric acid	—	600
Plum, 35% juice	Citric acid	—	500
Plum, 10–15% juice	Citric acid	—	600
Plum, juiceless	Citric acid	—	600
Currant, 35% juice	Citric acid	—	—
Currant, 10–15% juice	Citric acid	—	500
Currant, juiceless	Citric acid	—	800
Rose, juiceless	Citric acid	—	35

Coloring of Sugar Syrups

The addition of colorants to syrup is permitted as long as certified colorants are employed. For best results it is advisable to weigh the colorant, place it in a glass-lined or enameled container with approximately 5–10 times its weight of distilled or deionized water, and then mix thoroughly with a glass rod until the colorant is completely dissolved. The resulting solution is added to the syrup. When crystalline caramel color must be used, follow the above procedure, using boiling water to achieve solution. If fluid caramel is used, mix the colorant thoroughly with a little syrup.

Flavoring of Sugar Syrups

The colored or colorless syrup (acidified or not) is added with the required amount of compound flavor or concentrated juice plus compound flavor to yield the desired aroma effects. For best flavor results and independent of the concentrated juice levels employed, soluble compound essential oils for syrups are added as flavor reinforcers. Such oils (having yields varying between 1 × 100 and 1 × 200) are formulated by the basic guidelines for typical fruit-flavored formulations and tailored specifically to yield the required solubility for use at room temperature. Prior to adding the flavor, it is important to make sure that the syrup is cooled completely. Then add the compound flavoring with agitation for homogeneous distribution of the flavor in the mass; avoid drawing air into the syrup. The latter may be accomplished by slow agitation without turbulence. After a few hours standing, the syrup is filtered over felt liners or filter-press. Orzata and tamarind syrups are usually filtered prior to flavoring.

Addition of Preservatives

To prevent oxidation of syrups, Italian law permits the use of ascorbic acid "300" or sodium *l*-ascorbate "301" (150 mg/l or 115 mg/kg syrup). These chemicals can be used also in combination with citric or tartaric acid during the acidification step, or they can be dissolved separately in hot water prior to addition to the finished syrup. Fruit juices stabilized with sulfur dioxide are also permitted (in Italy) provided that the amount of sulfur dioxide does not exceed 100 mg/kg of finished syrup. In the United States ascorbate can also be used for its antioxidant effect. When stabilizers are used, it is advisable to consult with the supplier of colorants. Both ascorbic acid and sulfur dioxide react with dyes (especially red and green dyes), causing color fading and, sometimes, complete decoloration of the syrup in a rather short length of time.

Guideline formulations for the preparation of syrups sufficient for about 100 kg of finished product are shown in Table 8.

Soft Drinks and Carbonated Beverages

Waters

This category includes products with different characteristics, such as tonic water, sparkling water, seltzer water, soda, Vichy water, carbonated water, and others.

Sparkling Waters

The generic term *sparkling waters* applies to all carbonated beverages consisting of carbonated water with or without the addition of salts (sodium carbonate, sodium bicarbonate, sodium chloride, sodium sulfate). The salt may be added as a concentrated solution during bottling, as in the case of

TABLE 8

Formulations for Syrups

Ingredient	30–35% fruit juice	10–15% fruit juice	Soluble[1] flavoring oils
Sugar syrup at 65° Brix, kg	90.2	96.3	98.2[4]
	92.9[2]	96.8[3]	99.5[4]
Colorant	q.s.	q.s.	q.s.
Acid (if required)	q.s.	q.s.	q.s.
Concentrated juice, 1:5–1:8 fold	q.s.	q.s.	—

[1] Soluble flavoring oils only having a yield of 1 × 100 to 1 × 200.
[2] 95.5 kg for lemon-flavored syrup.
[3] 97.3 kg for lemon-flavored syrup.
[4] 96 kg for tamarind-flavored syrup (yield, about 1 × 25).

syrups. Sparkling waters are carbonated by using carbon dioxide at high pressures. Club soda, Vichy waters, and slightly bitter-tonic-flavored waters (tonic water) are included in this group.

Carbonated Waters

The term *carbonated waters* applies only to those products consisting exclusively of distilled, deionized, or treated carbonated water. While these products may also be defined as sparkling waters, the name *soda* (which implies the addition of sodium salts) cannot be used. In England carbonated waters are also called aerated waters; in Italy this definition encompasses all carbonated waters.

Soda

Soda is a generic term commonly used in the United States and England as a synonym for carbonated soft beverages. More specifically the term defines carbonated waters containing soda (sodium bicarbonate). British soda contains 5 grains minimum of sodium bicarbonate (approximately 0.33 g per pint). In the manufacture of soda water, the hardness of the water used should be taken into account.

Seltzer Water

Originally seltzer water was an imitation of natural mineral waters. Therefore, the number of salts added is larger than in the case of soda waters. Typical seltzer water also contains considerably more sodium chloride than soda water.

Fruit-flavored Beverages

With regard to composition, these beverages can be subdivided into two categories: (1) beverages flavored with fruit juices (with or without fruit pulp), and (2) beverages flavored with fruit flavor imitations or with natural or mixed flavor ingredients. The first group includes citrus (orange, lemon, grapefruit) and other fruit-based beverages—grape, apple, pineapple, etc. All of these contain sugar, acids (citric or tartaric), and sometimes colorants.

Carbonated Beverage Ingredients

The preparation of carbonated beverages in general calls for use of the following main ingredients: water, carbon dioxide, sugar, salts, acids, flavors (including fruit juices), and colorants.

Water

Spring, well, and tap waters may be used in the manufacture of carbonated beverages provided that the salt composition is compatible (or rendered such) with the flavor ingredients being used. The amount of calcium and magnesium salts in the water is responsible for the hardness. This is measured in French degrees as grams of salt as calcium carbonate present in 100 liters of water or in German degrees as grams of salt as calcium oxalate present in 100 liters of water. Water hardness is classified as total, permanent, or temporary. *Total hardness* is the hardness determined directly in a sample of the water. *Permanent hardness* defines the hardness retained by the water boiled for about 30 minutes, *i.e.*, calcium and magnesium salts left in solution after boiling. *Temporary hardness* is the difference between total and permanent hardness, *i.e.*, the hardness removed with boiling. Salt composition and water hardness can affect the addition of organic acids as well as the solubility of certain flavor ingredients.

Carbon Dioxide

Like many other gases, the solubility of carbon dioxide in water increases with decreasing temperature: one unit volume of carbon dioxide is soluble in 0.6 unit volumes of water at 0°C, while 13 unit volumes of water is required to dissolve the same amount at 25°C. As shown in the following table, the solubility of carbon dioxide in sugar solutions varies depending on the sugar content.

Solubility of Carbon Dioxide in Sugar Solutions
At 15.5°C and 760 mm Hg

Dissolved sugar, g	Dissolved carbon dioxide, ml	Dissolved sugar, g	Dissolved carbon dioxide, ml
1.0	0.995	8.0	0.943
2.0	0.989	9.0	0.936
3.0	0.982	10.0	0.928
4.0	0.975	11.0	0.918
5.0	0.967	12.0	0.907
6.0	0.959	13.0	0.902
7.0	0.951		

Commercial carbon dioxide contains not less than 99% (by volume) CO_2.

Preparation of Carbon Dioxide

In the past carbon dioxide for use in the carbonated-beverage industry was produced directly by the beverage manufacturer by contacting some carbonates, such as gypsum, sodium bicarbonate, etc., with sulfuric or hydrochloric acid. The largest and most consistent yields were obtained using calcium bicarbonate, whereas varying yields were obtained with gypsum. As both liquid and solid carbon dioxide became commercially available several decades ago, they are no longer produced directly by the beverage manufacturer.

Carbon dioxide is obtained usually as a by-product in the following processes: (1) the manufacture of calcium, (2) the combustion of coke, coal, or natural gases, and (3) the fermentation of beer. Carbon dioxide from the fermentation of beer, unless further purified, retains a characteristic odor and can be used only for the carbonation of beer.

Carbon Dioxide Content in Beverages

The volume of carbon dioxide in the finished product is an extremely important factor determining, together with the temperature, the length of sparkling in the beverage. The carbonated beverage and bottling industries have adopted a standard unit volume, defined as the amount of gas in milliliters absorbed by a given volume of water at 15.5°C under atmospheric pressure (760 mm Hg). The above unit is used to measure the volumes of carbon dioxide in carbonated beverages.

Measurement of Carbon Dioxide Volume

In order to measure the volume of carbon dioxide dissolved in a carbonated beverage, it is necessary to know the temperature and the pressure in the container. Special measuring instruments are available based on the following principle: a needle perforates the cork of the bottle, allowing the outflow of the head pressure until a bubble develops in the liquid. The valve is then closed, and the container is shaken until the instrument gauge no longer registers an increase in pressure. The temperature of the liquid is measured at this point.

Addition of Carbon Dioxide

In view of the better solubility of carbon dioxide in cold water, several methods are available to the bottler for cooling the water. Usually liquid carbon dioxide is employed in the manufacture of carbonated beverages; the liquid is contained in a stainless steel cylinder connected to an automatic gassing equipment of metal tubings. A large water surface comes in contact with carbon dioxide under pressure in the gassing apparatus, allowing a fast rate of absorption of the gas by the cooled water. Carbonation is made continuous by means of a valve that regulates the gas flow in the gassing equipment as required. The carbonated water flows into the filling machine.

Gassing machines must be made of corrosion-resistant metal and specifically designed to avoid potential contamination. Two types of machines are commercially available. In the first type the syrup is cooled and carbonated in the gassing equipment, mixed with cooled and carbonated water in a separate mixing tank, and transferred to a collecting tank and then to the filling machine. In the second type the syrup is measured directly into the individual bottles and diluted with carbonated water. The latter system is usually employed in automatic and semi-automatic bottling equipment.

Pressure–Temperature Relationship

An increase in temperature causes a corresponding pressure increase in carbonated-beverage containers. This in turn may cause leakage through the bottle seal or deformation of cans. This danger increases when carbonated beverages are pasteurized directly in the container.

Losses of Carbon Dioxide

The decrease of gas pressure during the filling operation or subsequent to unsealing the container represents a major problem for the carbonated-beverage industry. Adequate cooling of the water and syrup prior to carbonation is helpful in preventing gas losses during filling. Gas losses upon unsealing the container are attributable to various factors, such as presence of nuclei, excessive air content in the beverage, excessive agitation, or inadequate storage.

Nuclei tend to form from unclean bottles, mold-covered seals, and residues from bottle-washing operations. It has been reported that black nylon brushes do not tend to accumulate particles, are easier to clean, and have readily identifiable bristles. Thorough cleaning and maintenance of the bottling equipment is of primary importance to decrease the formation of nuclei.

Sugar and Syrup

A good quality sugar (sucrose—specific gravity, 1.588; specific volume, 0.630) must yield clear solutions in the water used for syrup preparation. The preparation of the syrup is the first operation in the manufacture of carbonated beverages; acids, colorants, and flavor ingredients (including fruit juices) are then added to the syrup. The sugar solution must be prepared with stirring (avoiding the drawing of air into the solution) and at the lowest feasible temperature to avoid the appearance of a brown color from caramelization.

The acids, predissolved in a small amount of water, are added to the cooled sugar solution to avoid inversion. Concentrated colorant solutions, juices, and flavor ingredients also must be dispersed in the cooled syrup solutions.

Density of Sugar Solutions

Density	Degrees, Bé	Water, kg	Sugar, kg
1.1249	16	71	29
1.1418	18	67.5	32.5
1.1796	22	60	40
1.1898	23	58	42
1.2003	24	56	44
1.2328	27	50	50
1.2637	30	44.5	55.5
1.2958	33	39	61
1.3078	34	37	63
1.3199	35	35	65
1.3322	36	33	67

The various non-volatile substances dissolved in the syrup and, in turn, the beverage determine the *dry residue* in the beverage, which is expressed either in weight per cent (grams of residue on evaporation of 100 cc of beverage) or in degrees Brix. The sugar content in the syrup can be measured by a refractometer or a polarimeter. The sugar content also determines variations in the density of sugar solutions (also expressed in degrees Baumé).

Flavor Ingredients for Carbonated Beverages

The flavoring of carbonated beverages represents one of the most important tasks in aroma formulations. In addition to technical and specific formulation problems (*e.g.*, the formulation of bitters), the basic requirement of carbonated-beverage flavors is complete solubility in syrup. The flavored syrup must in turn be completely soluble in the carbonated water. The finished product must not develop haze on storage (except when this is desirable), even in cold weather.

There are several products that are useful for flavoring carbonated beverages: (1) individual or compounded soluble essential oils, (2) individual soluble extracts or compounded flavors from soluble tinctures, alcoholic extracts, and distillates from herbs (with or without the addition of soluble essential oils), and (3) flavoring pastes and fruit juices.

The various soluble oils and soluble extracts have yields ranging between 1 × 50 and 1 × 400 (based on the syrup). The relative amounts of sugar, acids, and carbon dioxide vary, depending on the flavor and the chemical nature of the flavor ingredient. Variations in the dry residue of various types of beverages are expected, depending on the amount of sugar, juices, salts, and acids employed. In certain types of juice-based beverages, minimum dry residue specifications are set by law.

Carbonated Beverage Formulations

Guideline formulations commonly employed in the preparation of a few carbonated-beverage types follow:

Gazzosa

Syrup

Sugar	50 kg
Citric acid	2.7 kg
Water	47 l
Soluble lemon essential oil	q.s. (depending on the yield)

Beverage

The beverage is usually prepared in 200 cc glass containers (saturation pressure = 5 atm).

1. By diluting 26 cc of the above syrup to 200 cc total, a carbonated beverage with approximately 8% dry residue is obtained.
2. By diluting 33 cc of the above syrup to 200 cc total, the resulting beverage contains approximately 10% dry residue.

Orange Soda

Syrup

Sugar	48 kg
Citric acid	0.8 kg
Water	40 l
Orange juice, 15° Brix (7.5 fold concentrated)	7.2 kg
Colorant	q.s.
Soluble oil or flavoring paste	q.s. (depending on the yield)

Beverage

The beverage is usually prepared in 200 cc glass containers (saturation pressure = 4 atm). By diluting approximately 40 cc of the above syrup to 200 cc total, a carbonated orange soda is obtained with an approximate 12% dry residue.

Citron Soda

Syrup

Sugar	50 kg
Citric acid	1 kg
Colorant	q.s.
Water	48.5 l
Soluble oil	q.s. (depending on the yield)

Beverage

The beverage usually is prepared in 200 cc glass containers (saturation pressure = 5 atm).

1. By diluting 36 cc of the above syrup to 200 cc total, a carbonated citron soda is obtained with an approximately 12% dry residue.
2. By diluting 32 cc of the above syrup to 200 cc total, the resulting beverage contains approximately 10% dry residue.

Cola

Syrup

Sugar	50 kg
Citric acid	0.5–0.8 kg
Caramel	q.s.
Water	46.5 l
Soluble flavor or extract	q.s. (depending on the yield)

Beverage

Cola is usually prepared in 200 cc glass containers (saturation pressure = 6 atm). By diluting 40 cc of the syrup to 200 cc total, a carbonated cola beverage is obtained with an approximate 12.5% dry residue.

Non-alcoholic Aperitifs

The formulation ingredients of non-alcoholic aperitifs duplicate closely those used in alcoholic aperitifs. The lack of a functional alcoholic content in the finished product, however, is of basic importance regarding the flavor results as well as the formation and preservation of the flavor in the beverage. Flavor formation in particular undergoes very pronounced changes.

The aromatic herbs used for alcoholic aperitifs are also employed in non-alcoholic aperitifs as individual or compounded hydro-alcoholic extracts; however, the yields are rather different. Often it may be necessary to reinforce the aroma by the addition of soluble essential oils to attain the desired flavor effects. In a few cases the soluble essential oils are the basic ingredients in the formulation and are reinforced by tinctures.

While finished alcoholic aperitifs are ideally neutral products, non-alcoholic aperitifs are usually acid. Citric (or tartaric) acid is present from 4–7% by weight, depending on the amount of sugar in the syrup and the degree of carbonation. The presence of citric acid may cause flavor problems that, if not readily overcome, may be further complicated by the presence of carbon dioxide. Acids and carbon dioxide, individually or synergistically, react with the flavor ingredients and often yield off-flavors that break the flavor balance in a fairly short period of time.

It is not possible to present general theoretical formulation guidelines, since both the quality and quantity of flavor ingredients may undergo unpredictable changes directly related to the chemical nature as well as to the varying acid and carbon dioxide ratios in the beverage.

Successful formulations used world-wide are based mainly on the tailored blend between a complex citrus background note and characteristic bitter flavor ingredients together with slightly spicy, sharp notes. The following flavor ingredients are usually employed: sweet and bitter orange, chinotti, vanilla, cinnamon, calamus, wormwood, gentian, and rhubarb.

The following procedure is usually employed in the preparation of non-alcoholic aperitifs:

Syrup

Sugar	50 kg
Citric acid	0.2 kg
Water	49.150 l
Colorant	q.s.
Soluble extract	q.s. (depending on the yield)

Beverage

The beverage is usually prepared in 100 cc glass containers (saturation pressure = 5 atm). By diluting 26 cc of the above syrup to 200 cc total, a non-alcoholic aperitif is obtained with an approximate 16% dry residue.

Bitter Non-alcoholic Beverages

In bitter, refreshing, non-alcoholic beverages, such as gingers, chinotto, spume, and tonic waters, the problem of the presence of carbon dioxide and citric acid is identical to aperitifs. The flavor ingredients used in the formulations are quite different from those used in bitter, alcoholic beverages. Guideline formulations for flavoring the above beverage types are given below.

Chinotto (Bitter Orange Soda)

The introduction of chinotto represents undoubtedly the new flavor line for carbonated beverages in Italy. The bittersweet characteristic flavor note represents a break from traditional citrus-flavored beverages as well as from the less aromatic but very well known cola beverages. Although the customer's taste has been educated to the bitter note through other commercial bitter-flavored beverages, the original creativity expressed in the beverage chinotto should not be underestimated.

The basic flavor of the beverage may be defined as the vermouth note (a red, slightly aromatic, and somewhat bitter vermouth) associated with the chinotti citrus flavor rounded off by the addition of an orange note.

Syrup

Sugar	48 kg
Citric acid	0.6–1.2 kg
Caramel for acids	q.s. (intense color)
Water	46 l
Soluble extract	q.s. (depending on the yield)

Beverage

The beverage is usually prepared in 200 cc glass containers (saturation pressure = 4 atm). By diluting approximately 40 cc of the above syrup to 200 cc total, a chinotto beverage with approximately 12% dry residue is obtained.

Tonic Waters

This beverage type exhibits a rather simple note obtained by a combination of bitter flavor with a citrus note. The flavor variations are somewhat limited, since the finished beverage must be colorless or only slightly colored. Quinine chlorhydrate (or a concentrated bitter tincture) is the water-soluble, bitter flavor principle. Soluble lemon essential oil in suitable dose (sometimes in combination with soluble orange essential oil) adds the complementary note to the aroma of the most simple beverage types of widespread consumption. In addition tonic waters yield an excellent refreshing cocktail when consumed with gin or vodka.

Syrup

Sugar	50 kg
Citric acid	2.35 kg
Water	46.65 l
Soluble flavor	q.s. (depending on the yield)

Beverage

The beverage is usually prepared in 200 cc glass containers (saturation pressure = 6 atm). By diluting 32 cc of the above syrup to 200 cc total, a tonic water with an approximate 10% dry residue is obtained.

Gingers

The spicy ginger note is indispensable to the flavor of certain internationally known gingers. The ginger flavor is associated with the citrus note in this beverage type. A typical ginger ale formulation, for example, employs ginger and capsicum oleoresin together with ginger essential oil, lime distilled essential oil, and cold-expressed lemon and orange essential oils.

The Italian type of ginger differs somewhat from the above flavor note; the following flavor ingredients are chiefly employed: sweet and bitter orange, lemon, vanilla, calamus, cinnamon, ginger, clove, and traces of additional bitter flavors.

Syrup

Sugar	50 kg
Citric acid	0.140 kg
Water	49 l
Colorant	q.s.
Soluble flavor	q.s. (depending on the yield)

Beverage

Ginger is usually prepared in 200 cc glass containers (saturation pressure = 5 atm). By diluting 48 cc of the above syrup to 200 cc total, a ginger beverage with approximately 15% dry residue is obtained. A ginger with approximately 14% dry residue is also prepared by diluting 46 cc of the syrup to 200 cc total.

Spume

Spume are non-alcoholic, carbonated, champagne-type beverages. They comprise a large number of white or dark carbonated beverages with a basic aromatic wine flavor (muscat) associated with a bitter note (sometimes a citrus undertone). It is also characterized by a slightly tart flavor. The possible flavor variations are countless. The basic muscat flavor note is imparted by the use of tinctures or distillates from clary sage, elder, coriander, violet, and orris. Clary sage is the most important flavor ingredient for this product.

White Spuma

Syrup

Sugar	50 kg
Citric acid	1 kg
Water	48 l
Caramel for acids	q.s. (traces only)
Soluble flavor extract	q.s. (depending on the yield)

Beverage

The beverage is usually prepared in 1,000 cc glass containers (saturation pressure = 5 atm). A white spuma with approximately 11% dry residue is obtained by diluting 180 cc of the above syrup to 1,000 cc total.

Dark Spuma

Syrup

Sugar	50 kg
Citric acid	0.7 kg
Water	47 l
Caramel for acids	q.s. (dark color)
Soluble flavor extract	q.s. (depending on the yield)

Beverage

The beverage is usually prepared in 1,000 cc glass containers (saturation pressure = 5 atm). By diluting 165 cc of the above syrup to 1,000 cc total, a dark spuma with an approximate 10% dry residue is obtained. A dark spuma with a 12% dry residue is also prepared by diluting 195 cc of the above syrup to 1,000 cc total.

Preservation and Alteration of Carbonated Beverages

The manufacture of carbonated beverages is not particularly difficult from a technological standpoint. The use of good quality raw materials and suitable bottling equipment is for the most part a guarantee to successful operation. However, the observance of strict hygienic conditions throughout the manufacturing process is indispensable. Carbonated beverages are extremely delicate and susceptible to degradation and fermentation due to several and quite different causes. These are not always predictable or controllable in spite of the extreme care observed during manufacture. The observance of absolutely hygienic manufacturing practices and the use of preservatives (wherever permitted) and/or pasteurization are of paramount importance.

In addition to the above precautions, the following are also recommended:

1. Avoid sugar inversion during the preparation of the syrup.
2. Add sufficient acid to the beverage.
3. Store finished products in cool, aerated places.
4. Avoid discoloration in the beverage from prolonged exposure to sunlight during shipping or storage.

In view of the commercial as well as the alimentary importance of carbonated beverages, a more detailed discussion of the potential causes of alteration in carbonated beverages encompasses very important theoretical and practical considerations.

Carbonated beverage consumption has increased steadily over the years. What was considered at one time a seasonal (summer) production today is distributed quite uniformly throughout the year. For readily apparent commercial reasons carbonated beverages may remain in storage for extended periods of time prior to consumption. Prolonged storage considerably increases the possibility of alteration, although the manufacturing technology has made remarkable advances in recent years.

Sometimes finished beverage stocks ready for shipment are stored by the manufacturer in refrigerators. Significant temperature changes during storage and subsequent shipment may affect beverage stability. In more recent years the carbonated-beverage industry has increased the consumption of fruit juices for the preparation of beverages. Juices are, in fact, extremely susceptible to deterioration from the action of various microorganisms. The addition of fruit juice enriches the content of vitamins, mineral salts, and nitrogenous compounds in the beverage. All of these ingredients create a favorable ground for bacterial growth with consequent spoilage of the beverage by bacteria. The potential causes of alteration may be present not only in the finished products but also during manufacturing, in the equipment and containers used, or even in the starting raw materials; all of these potential alterations can shorten shelf-life.

Alterations occurring in carbonated beverages are phenomena of physical, chemical, biochemical, and microbiological nature.

Physical Alterations

Light is undoubtedly the main physical source of alteration in carbonated beverages. Sunlight may induce terpenization of citrus flavors (especially orange and lemon flavors), which in turn causes the development of a characteristic, unpleasant, oily off-flavor reminiscent of turpentine. Light may also unpleasantly shift the acid taste or induce off-odors and off-flavors especially in beverages containing natural fruit aromas. Light also may cause partial discoloration of synthetic dyes, leading, especially in the case of dark or brown colors, to unsightly chromatic effects. Some food, drug, and cosmetic dyes are particularly sensitive to sunlight in the presence of ascorbic acid. Hence, it is necessary to preserve carbonated beverages (orange- and lemon-flavored beverages in particular) in dark places if extensive storage is required.

High temperatures may also affect the quality of carbonated beverages. Beverages accidentally subjected to temperatures higher than 40°C may undergo losses of carbon dioxide from the well known inverse solubility of carbon dioxide in water at higher rather than lower temperatures. Extended storage at low temperatures also may have adverse effects. Certain flavors and colorants, much less soluble at low temperatures, can readily precipitate out and do not redissolve completely even when the temperature is increased to normal values. Even when the components redissolve, the products in solution are different from the starting materials because of irreversible changes taking place during cooling.

Precipitation and flocculation may occur when hard water is employed unless the water is pretreated by resin (ion) exchange. The same source may also induce alteration of the colorants in the finished beverage, causing partial but highly objectionable discoloration.

An additional physical change may be caused by the rapid loss of carbon dioxide through imperfect seals, although the latter alteration should be considered as accidental only and seldom involves a large portion of the production. With canned carbonated beverages, corrosion of the lining is an important aspect.

Chemical Alterations

Sometimes various components present in carbonated beverages may react chemically with (1) the atmospheric oxygen dissolved in the water phase or present in the container, and (2) among themselves because of chemical incompatibility. The presence of the dispersing aqueous phase creates favorable conditions for the potential occurrence of chemical reactions throughout the beverage. For instance, the terpene or rancid flavor sometimes perceptible in beverages of old production is caused by the action of oxygen dissolved in the water phase or present in the head space of the container between the seal and the liquid surface. In addition to terpenes, a few aldehydes (present especially in fruit flavors) may readily undergo oxidation, thus accelerating the degradation process in beverages.

The presence of air in the carbon dioxide used for carbonation, caused by imperfect functioning of the automatic filling equipment or the non-observance of proper manufacturing techniques, may also be a problem. A recent survey reported that out of 340 carbon dioxide samples tested, 105 samples had an air content of 30% or more. Therefore, alterations attributable to this cause are in general much larger and occur much more frequently than expected from a cursory examination of the problem. It must be added that all oxidation reactions induced by atmospheric oxygen are catalyzed by sunlight and temperature. It is apparent that, to avoid such possibilities, it is necessary to deaerate the water for carbonation prior to use.

A very important role in the chemical alteration of beverages is attributed to heavy metals, such as iron and copper. Water used in the manufacture of beverages must contain less than 3 ppm iron. If larger amounts are present, iron may precipitate as a salt or a complex, or it may form yellow to reddish-brown rings consisting of iron hydroxide on the surface of the liquid. Iron also may react with dyes, causing the precipitation of insoluble lacquers, or with flavor ingredients, causing the development of disagreeable metallic off-flavors in the beverage. Therefore, the use of water containing iron, usually as soluble iron bicarbonate, is not recommended. The ferrous ion is readily oxidized to trivalent iron in the presence of atmospheric oxygen, yielding highly insoluble iron hydroxide. Whenever it is necessary to use waters containing iron in excess of 3 ppm, the iron can be easily removed by various methods including direct ion-exchange or complexed with chelating agents such as EDTA. Copper also plays an important role in catalyzing corrosion reactions in canned beverages.

Chlorides and alkaline substances may be detrimental to an effective preservation of carbonated beverages. Spoilage is also caused by an excessively acid pH that can cause sugar inversion. Invert sugar has a flavor quite different from regular sucrose. This last type of alteration is considered rare due to the length of time required for the inversion of sugar; it has been included only as an illustrative example of the many factors affecting the storage stability of carbonated beverages.

Excessive water hardness may cause chemical alterations (in addition to the previously mentioned physical changes) because of the presence of calcium. Calcium salts in the presence of tartaric acid used to acidify beverages may yield precipitation of insoluble calcium tartrates.

Biochemical Alterations

The main form of biochemical alteration consists of enzymatic activity. Enzymes are soluble or colloidal substances produced by living organisms and capable of catalyzing specific reactions. Chemically, enzymes may be considered simple or conjugated proteins. As to specific activity, only two enzymes are of interest to the carbonated beverage industry—carbohydrase and lipase. The former catalyzes hydrolysis carbohydrates or sugars, while the latter catalyzes the scission of fats and esters. One should remember that enzymes produced by the metabolism of microorganisms may exhibit activity long after the death of the enzyme-producing organism; they are therefore resistant to common preservatives.

Enzymes may be found in the raw materials employed by the beverage industry, such as fruit juices, fruit concentrates, and natural gums. The presence of lipase may be particularly harmful because of selectively attacking and breaking ester molecules (so-called esterase), a very important category of flavor ingredients in the formulation of aromas. Ester breakage by lipase may reduce the flavor yield or produce off-odors and flavors incompatible with the original note.

Microbiological Alterations

Undoubtedly alterations caused by the activity of microorganisms represent the gravest and most serious single source of preoccupation by the manufacturer of carbonated beverages. Microorganisms responsible for spoilage of beverages include fungi, yeasts, bacteria, algae, and protozoa. Statistically, yeasts are responsible for spoilage in most cases, whereas degradation from molds and bacteria (especially lactic acid bacteria) is encountered in relatively few instances. In general, the degradation phenomenon may be compared to an alcoholic fermentation characterized by (1) flavor changes, (2) a pressure increase in the container, (3) discoloration, and (4) the formation of sediments and precipitate usually considered the first and most visible indication of the incipient alteration of the beverage.

The largest number of alterations is attributable to yeasts, which require very few nutrients to survive and grow as compared to bacteria (autotrophic bacteria, however, are capable of surviving with the little nutrient present in water). In addition yeasts, although sensitive to high carbon dioxide pressures, are fairly resistant to the acid conditions prevalent in most carbonated beverages; certain types continue to grow even at pH 2.5 as compared to bacteria, which are much less resistant under acid conditions.

Any beverage based on fruit juice containing vitamins, proteins, and mineral salts provides an extremely favorable substrate for the growth of microorganisms and much more so than a clear beverage. Therefore, utmost attention must be given to the preparation of beverages flavored with natural juices. The various categories of microorganisms causing spoilage and the conditions influencing their specific activity are reviewed below.

Fungi

The groups of interest include ficomycetes, ascomycetes, and basidiomycetes. Spoilage caused by these fungi closely resembles that by yeasts and molds.

Yeast

This generic term defines a large class of substances. *Saccaromyces cervisiae* (beer yeast) is the best illustrative example in the class. Yeasts, when examined under the microscope, appear as very tiny, egg-shaped cells with small lateral humps (buds). They reproduce at an extremely high rate and grow freely in sugar syrups, acid solutions (optimum growth occurs at pH 4–4.5), and flavored non-alcoholic solutions. Reproduction of yeast on the above substrates occurs at a much faster rate than other categories of microorganisms.

F. Hale points out that 90% of the microbiological alterations occurring in beverages is attributable to yeasts.[12] The presence of this organism should not be underestimated, and great care must be exercised to control the yeast population for better preservation of beverages.

The vehicles through which yeast may find their way into beverages are countless—damp sugar bags (especially those made of canvas), dyes, pre-contaminated juices, dirty equipment and pipes, dust or dirt collected during manufacturing, and non-sterile containers and seals. The last factor, according to a study presented at the Third International Conference on the Microbiology of Food Products (Avian, 1960), is the major source of alteration in food products. Therefore, a strict control of the operative efficiency of the washing machines and a thorough sterilization of bottles and caps (leaving the least possible number of microorganisms, excluding the species capable of proliferating under acid conditions) is highly recommended.

The need for continuous microbiological analyses is the limiting factor to the practical implementation of the controls, since a well equipped laboratory and specialized personnel are required to perform the analyses. It is often sufficient, however, to overcome contamination problems by strictly implementing the hygienic measures required by good manufacturing practice.

In addition to observable phenomena caused by the growth of yeasts in the beverage, spoiled beverages also exhibit a typical fermentation flavor and a bitter taste. Flavor changes are strictly dependent on microbial activity and are, therefore, more or less directly dependent on the substances influencing growth, such as sugar, acid, and carbon dioxide.

Experiments conducted by U.S. scientists have shown that low-strength (5–10%) sugar solutions, such as carbonated beverages, provide an excellent substrate for the growth of yeasts proliferating to 400 fold the original value in a very short period of time (5 days). Such growth occurs at a much slower rate in syrups whose activity increases with the increasing sugar concentration.

As to the influence of pH, an increase in the concentration of inorganic acids (citric, tartaric, malic) added to a 10% sugar solution decreases the potential growth of yeasts. This is not absolutely valid, however, since yeasts show great adaptability to different environmental conditions. Therefore, the concept of improving the beverage stability by increasing the acidity is of limited application.

Studies by N. F. Insalata have shown that carbon dioxide has an inhibiting effect on yeast growth.[13] Three beverages having a 12% dry residue and a pH of 4.5 were carbonated with 3, 4, and 5 volumes, respectively, of carbon dioxide and inoculated with 150 cells of yeast. Sterilization occurred after 20, 13, and 10 days, respectively, thus proving that an increase in the carbon dioxide content corresponds to an increase of the anti-yeast activity in the beverage. It should be pointed out, however, that this study applied to yeast only; the results are of limited value to the manufacturer of carbonated beverages, who should in no way be induced to relax his regular hygienic controls.

Molds

This category of microorganisms is seldom encountered in spoiled beverages. Molds generally develop in oxygen-rich beverages, as oxygen is indispensable for their metabolism; mold growth is strongly inhibited by the presence of carbon dioxide. Usually molds proliferate on caps, canvas bags, and wood and cardboard containers used for packaging the raw materials; from these they subsequently are transferred into beverages. In addition to causing flocculation and precipitation, molds impart an unpleasant odor and flavor.

Bacteria

An additional potential source of contamination, although very rare in practice, is bacteria. Several species of bacteria are known, but only a few are pathogenic. Some bacterial species thrive only under optimum ambient conditions; others thrive also in oxygen-poor and carbon-dioxide-rich environments in the presence of rather acid pH values (3–4). The most visible phenomena in a beverage spoiled by bacteria are discoloration and the formation of off-flavors. The possibility of pathogenic bacterial contamination, such as *Escherichia coli* or *Salmonella typhosa*, in beverages should be considered remote if proper manufacturing practices are observed.

Algae and Protozoa

Algae comprise a heterogeneous group of substances that may be present as contaminants in the water used for bottling. Similarly, protozoa (microscopic unicellular organisms) may be introduced into the beverage through the use of contaminated well water. Both algae and protozoa may cause off-flavors.

Dry Beverage Bases

The formulation of dry beverage bases as a means of imparting an acceptable flavor to water represents an old challenge to the flavorist. Originally, dry beverage bases were developed to partially fill an increasing consumer's demand for prepared carbonated water that was usually high-priced and in scarce supply. A solution was found by using sodium bicarbonate and tartaric acid (prepared in individual packets in approximately stoichiometric ratios) dissolved in water shortly before use. Usually the packets were sufficient to prepare 1 liter of carbonated water. Tablets containing sodium bicarbonate and tartaric acid in a 3:1 ratio found use in the preparation of seltzer water imitations. The well known Vichy water was imitated by adding small amounts of other salts together with the above mentioned basic components in order to improve the flavor characteristics. A typical Vichy-water formulation consisted of the following:

Imitation Vichy Water

Ingredient	Quantity, grams
Packet 1	
Sodium bicarbonate	4.0
Sodium chloride	0.10
Sodium sulfate	0.25
Magnesium sulfate	0.07
Ferrous sulfate	0.05
Packet 2	
Tartaric acid	1.0

The above ingredients have remained basically unchanged. The carbonation principle based on the chemical reaction between sodium bicarbonate and tartaric acid yielding carbon dioxide is still valid. Only the slightly tart or salty flavor resulting from a deficiency or an excess in the amount of sodium bicarbonate used has slowly changed in order to impart the desired note to different products. Therefore, increasing emphasis was placed on the addition of salts, such as lithium carbonate, to the bicarbonate base. While the use of ascorbic and malic acids, together with tartaric acid was considered, citric acid did not make an inroad in these type formulations because of technical shortcomings. With the increasing availability of anhydrous chemicals, improved mixing and packaging techniques, and airtight seals, the need for individual packets was overcome; today single-package dry beverage bases are marketed.

As an extension of the above evolution in the preparation of Vichy water imitations, new ideas were developed to make carbonated water imitations into flavorful and thirst-quenching beverages replacing the syrup-based beverages and even carbonated waters.

A few products that served to create a favorable market for the development of dry beverage bases into flavored products included the so-called "effervescent granular mixtures" or "effervescent magnesia." No longer popular, these products consisted chiefly of an intimate mixture of sugar (about 50% by weight), citric or tartaric acid (sometimes also a blend of these acids), sodium bicarbonate, and terpeneless or soluble citrus oils dissolved and thoroughly mixed with a little hot syrup (usually not exceeding 85°C). The mixture was then manually or mechanically worked into granules and finally dried. Such formulations, together with almost countless modifications, were developed to produce inexpensive, well accepted granular dry flavor bases yielding sparkling, usually fruit-flavored, soft beverages. After several technically unsuccessful attempts to extend the concept of a two-packet product to the introduction of a third packet containing flavors, colorants, and sweeteners, single-package, flavored, dry beverage bases were developed by utilizing manufacturing techniques used for granular products.

Flavored dry beverage bases are formulated in two different types:

1. Dry beverage bases yielding carbon dioxide on dissolution in water.
2. "Still" dry beverage bases.

Both formulations may contain sugar (q.s.), synthetic sweeteners, or mixtures of natural and synthetic sweeteners. In order to generate carbon dioxide, a mixture of sodium bicarbonate and sodium tartate is used together with additional salts, acids, and colorants to complement the organoleptic characteristics of the flavored dry base.

The flavor ingredients are powders consisting either of spray-dried oils or lyophilized juices usually reinforced by spray-dried aromas. "Still" formulations are generally prepared from a large variety of fruit types, including the citrus flavors. They do not contain sodium bicarbonate, but only acids, colorants, and various salts. Citrus-flavored still beverages are prepared in the same manner as those producing carbonation, whereas spray-dried compounded flavors are used in place of the corresponding lyophilized fruit juices when the latter are unavailable. The amount of lyophilized juice must be properly dosed in the dry beverage base in order to reproduce the characteristic flavor of beverages prepared with natural fruit juices. It is worthwhile to note that flavor effects in dry beverage bases prepared with and without sugar or synthetic sweeteners (diet beverages) and containing identical flavor ingredients differ widely.

Teas and Tisanes

Teas

A certain confusion exists in the botanical classification of tea by both botanists and growers. The two varieties *Thea sinensis* L. and *Camelia sinensis* (L.) Kun. are considered almost equivalent with a slight preference for the latter. *Camelia sinensis* belongs to the family *Theaceae*. The plant is native to China; its cultivation and commercial development spread from China to Japan, then to India, Ceylon, Java, Africa (Kenya, Uganda, Tanzania), Sumatra, the Azores, Formosa, and Trans-Caucasian Russia (Georgia, Black Sea).

For the preparation of commercial tea, leaves harvested from three-year-old (minimum) plants are used. The plants are usually produced over a period of ten years. Commercial tea is classified into two important types: green and black teas. The differences consist in methods of treating the leaves prior to commercialization. Green tea is prepared by the following method: the harvested leaves are steam treated, dried, lightly roasted, rolled, again roasted, sieved, and aerated. The preparation of black tea follows the same procedure, with the addition of fermentation. This step is carried out after drying the leaves; fermentation favors the development of a particular flavor. The various leaf qualities (partially or completely ripe), the method of rolling the leaves (manually or mechanically), the age of the plant, and other factors are the elements on which an extremely fine classification of commercial green and black teas is based.

The method of preparation as well as the different sources of tea leaves tends to impart a characteristic aroma, which can be modified by the addition of complementary flavors to the finished product. Green tea, usually much less aromatic than fermented black tea, is sometimes flavored during manufacturing with jasmine or orange flowers. The flavoring operation is often continued by the consumer, who may add other herbs or dried flowers to the tea infusion. The flavoring of fermented black tea is carried out largely by the manufacturer. Well known specialties are flowery pekoe and orange pekoe, which are flavored with jasmine and orange flowers, respectively. Various tea manufacturers have marketed other products flavored with various characteristic aromas, including rose, lavender, ripe peach, mandarin, muscat grape, clove, orris, and sometimes a smoky aroma. Properly diluted essential oils, sprayed on the product according to traditional methods or to other undisclosed proprietary procedures, also are employed for tea flavoring.

Tisanes

This name designates those concoctions prepared by treating aromatic plants with hot water. Camomile, linden, mint, and karkade are the aromatic plants chiefly used for tisanes. Camomile has picked up new momentum in the consumer market in recent years. It is interesting to mention that, in a few cases, the characteristic camomile flavor has been modified by the addition of aromatic herbs in an attempt to create new flavors.

Cocoa

It is extremely probable that sugar, amino acids, and flavonoids are the precursors in the flavor complex of roasted cocoa. Defatted cocoa forms the familiar chocolate base. Often cocoa flavor is confused with chocolate flavor, although chocolate is derived from cocoa by subsequent treatments and the addition of other flavors, such as vanillin.

Cocoa flavor is not related to the so-called cocoa red, which is isolated as an oil from roasted arriba beans and has an extremely high flavoring strength (12–13 ppm). The extremely difficult reconstruction of cocoa flavor, although not as complex as coffee flavor, was originally obtained by tentatively reinforcing the flavor of extracts, tinctures, and cocoa distillates by addition of trace vanillin or aldehyde, benzyl butyrate, and amyl phenylacetate. These products together with aldehyde C_{18}, neryl butyrate, phenylpropyl cinnamate (floral note), ethyl benzoate, dimethylresorcin, linalool, cyclotene, diacetyl, maltol, dihydrocoumarin, resinoids, and balsams (Tolu and Peru) are the products most frequently used to formulate cocoa flavor imitations with high flavor yields. Careful attention must be given to the addition of vanilla in the formulation, as it tends not only to sweeten but also to shift the cocoa note toward chocolate. The potential use of phenylacetic acid and its esters should also be considered.

Studies of the constitution of natural cocoa aroma dating back to 1921 have received great impetus in more recent years. These studies have supplied a very extensive list of components reported present in solvent extracts, steam distillates, or cocoa butter; they have prompted considering the use of quaternary, alicyclic, aromatic, and heterocyclic compounds for cocoa flavor imitations. Products identified thus far in natural cocoa aroma are listed in the following table:[14]

Compounds in Natural Cocoa Aroma

Compounds	References
Hydrocarbons	
Undecane	37, 43
Dodecane	37
Tridecane	37
Tetradecane	37
Octadecane	37
β-Myrcene	37, 43
Limonene	41, 42, 43
β-Elemene	41, 42, 43
β-Pinene	41, 42, 43
Valencene	41, 42, 43
Caryophyllene	41, 42, 43
Benzene	29, 31, 35, 42
Toluene	29, 31, 35, 42
1,2-Dimethylbenzene	41, 42, 43
1,3-Dimethylbenzene	41, 42, 43
1,4-Dimethylbenzene	41, 42, 43
1,2,4-Trimethylbenzene (pseudocumene)	41, 42, 43
1,3,5-Trimethylbenzene (mesitylene)	41, 42, 43
1,2,3,5-Tetramethylbenzene	41, 42, 43
Vinylbenzene (styrene)	41, 42, 43
1-Ethyl-2-methylbenzene	41, 42, 43
1-Ethyl-3-methylbenzene	41, 43
1-Ethyl-4-methylbenzene	41, 42, 43
2-Ethyl-1,4-dimethylbenzene	41, 42, 43
2,4-Dimethyl-1-vinylbenzene	41, 42, 43
Propylbenzene	41, 42, 43
Isopropylbenzene (cumene)	41, 42, 43
1-Isopropyl-4-methylbenzene	41, 42, 43
Methylbiphenyl	37
Naphthalene	37, 43
2-Methylnaphthalene	37, 43
Dimethylnaphthalene	37
Trimethylnaphthalene	37
Tetramethylnaphthalene	37
Alcohols	
Methanol	28, 29, 31, 42
Ethanol	28, 29, 31, 35, 42
1-Propanol	28, 31, 35, 42
2-Propanol	28, 31, 35, 42
2-Methylpropan-1-ol (isobutanol)	28, 35, 39, 42
1-Butanol	31, 39
2-Butanol	31
3-Methylbutan-1-ol (isoamyl alcohol)	28, 35, 39, 40, 43
1-Pentanol	31, 35, 37, 39
2-Pentanol	39, 40
1-Hexanol	35, 39, 42
2-Heptanol	39
1-Octanol	39, 41, 42, 43
1-Octen-3-ol	35, 39, 42
Geraniol	28, 31, 35, 42, 43
Linalool	25, 28, 31, 35, 37, 40, 42, 43
Benzyl alcohol	35, 39, 43
1-Phenylethanol	35, 37, 39, 40, 43
2-Phenylethanol	35, 37, 39, 40, 43
2-Phenylpropan-2-ol	35, 39
2-Methyl-3-phenylpropan-2-ol	41, 42
Menthol	41, 42, 43
α-Terpineol	41, 42, 43
Terpineol-1	41, 42, 43
Terpineol-4	41, 42, 43
Borneol	41, 42, 43
Furfuryl alcohol	28, 31, 35, 39, 40, 42, 43
2,3-Butanediol	31, 35, 39, 42, 43
Carbonyls	
Acetaldehyde	28, 29, 31, 35, 40, 42
Propanal	29, 31, 35, 42
2-Propenal (acrolein)	35, 42
2-Methylpropanal	29, 31, 35, 40, 42, 43
2-Methylprop-2-enal (methacrolein)	41, 42
Butanal	29, 31, 35, 42
2-Butenal (crotonaldehyde)	35, 39, 42, 43
2-Methylbutanal	37, 43
3-Methylbutanal	29, 31, 35, 37, 40, 42, 43
Hexanal	41, 42, 43
5-Methyl-2-isopropylhex-2-enal	40
Octanal	41, 42, 43
2,4-Octadienal	35, 42
Nonanal	41, 42, 43
Decanal	41, 42, 43
Citronellal	28, 31, 35, 42, 43
Benzaldehyde	31, 35, 39, 40, 42, 43
Phenylacetaldehyde	35, 39, 40, 42, 43
Phenylbutenal	39
2-Phenylbut-2-enal	40
4-Methyl-2-phenylpent-2-enal	40
5-Methyl-2-phenylhex-2-enal	40
2-Propanone	29, 31, 35, 42, 43
Acetol acetate	35, 39, 42
2-Butanone	40
3-Hydroxybutan-2-one	31, 35, 37, 39, 40, 42, 43
2,3-Butanedione (diacetyl)	26, 28, 29, 31, 35, 40, 42, 43
trans-3-Penten-2-one	40
2,3-Pentanedione	35, 39, 40, 42
5-Methylhexan-2-one	41, 42, 43
2-Heptanone	37, 43
6-Methyl-5-hepten-2-one (2-methyl-2-hepten-6-one)	28, 31, 35, 37, 42
2-Octanone	37, 43
5-Hydroxyoctan-4-one	41, 42, 43

Compounds in Natural Cocoa Aroma (continued)

Compounds	References
4,5-Octanedione	41, 42, 43
2-Nonanone	37
2-Dodecanone	37
3-Heptadecanone	31, 35, 42
Acetophenone	31, 35, 37, 39, 40, 42, 43
2-Hydroxyacetophenone	35, 37, 39, 42, 43
4-Methylacetophenone	41, 42, 43
1-Phenylpropan-2-one	41, 42, 43
3-Methylcyclopentan-1,2-dione (cyclotene)	35, 39, 42
2-Acetyl-4-isopropenylcyclopent-1-ene	41, 42
Menthone	41, 42, 43
Camphor	41, 42, 43
Furfural	26, 35, 39, 40, 42
5-Methylfurfural	40, 41, 42, 43
2-Acetylfuran	35, 37, 39, 40, 42, 43
2-Acetyl-5-methylfuran	41, 42, 43
2-Propionylfuran	41, 42, 43
2-Methyltetrahydrofuran-3-one	35, 39, 40, 41 42, 43
Tetrahydrofuran-2-one	40
Acids	
Formic acid	35, 42, 43
Acetic acid	25, 31, 35, 40, 42, 43
Propanoic acid	25, 35, 40, 42, 43
2-Hydroxypropanoic acid (lactic acid)	31, 43
2-Methylpropanoic acid	39, 40
Butanoic acid	25, 31, 35, 42
2-Butenoic acid (crotonic acid)	40
2-Methylbutanoic acid	31, 35, 37, 42, 43
3-Methylbutanoic acid	37, 40, 43
2-Hydroxy-3-methylbutanoic acid	41, 42
Pentanoic acid	31, 35, 40, 42, 43
4-Methylpentanoic acid	40
2-Hydroxy-3-methylpentanoic acid	41, 42, 43
2-Hydroxy-4-methylpentanoic acid	41, 42, 43
Hexanoic acid	25, 31, 35, 40, 42, 43
Heptanoic acid	40
Octanoic acid	25, 31, 35, 40
Nonanoic acid	25, 35, 40, 42
Decanoic acid	31, 40, 42
Dodecanoic acid (lauric acid)	40
Tetradecanoic acid (myristic acid)	40
Hexadecanoic acid (palmitic acid)	40
2-Hydroxy-3-methylglutaric acid	41, 42, 43
Benzoic acid	40, 43
2-Methoxybenzoic acid	41, 42, 43
4-Hydroxybenzoic acid	30, 31, 32, 35, 42
4-Methoxybenzoic acid (anisic acid)	30, 31, 35, 42, 43
4-Hydroxy-3-methoxybenzoic acid (vanillic acid)	30, 31, 32, 35, 42
3,5-Dimethoxy-4-hydroxybenzoic acid	32, 35, 42
3,4-Dihydroxybenzoic acid (protocatechuic acid)	33, 35, 42
Phenylacetic acid	30, 31, 32, 35, 40, 42
2-Hydroxyphenylacetic acid	30, 31, 32, 35, 42
4-Hydroxyphenylacetic acid	30, 31, 32, 35, 42
4-Hydroxyphenylpropanoic acid	32
4-Hydroxycinnamic acid	32, 35, 42
4-Hydroxy-3-methoxycinnamic acid (ferulic acid)	32, 35, 42
2-Methoxyphenylacetic acid	30
4-Methoxyphenylacetic acid	30
Esters	
Methyl acetate	29, 31, 35, 42
Ethyl acetate	28, 31, 35, 40, 42
Propyl acetate	26, 28, 31, 35, 42
Isopropyl acetate	26, 28, 31, 35, 42
Butyl acetate	26, 28, 31, 35, 42
Isobutyl acetate	26, 28, 31, 35, 40, 42
2-Methylbutyl acetate	41, 42, 43
Amyl acetate	25, 26, 28, 31, 35, 42
Isoamyl acetate	35, 37, 40, 42, 43
2-Pentyl acetate	35, 40, 42
Neryl acetate	41, 42, 43
Geranyl acetate	41, 42, 43
Linalyl acetate	25, 28, 31, 35, 42, 43
Benzyl acetate	35, 37
Phenyl acetate	31, 35, 42
Phenethyl acetate	31, 35, 37, 39, 40, 42, 43
3-Phenylpropyl acetate	41, 42, 43
Furfuryl acetate	35, 39, 42
Ethyl propanoate	28, 31, 35, 42
Amyl propanoate	25, 28, 31, 35, 42
Hexyl propanoate	25, 35, 42
Ethyl 2-oxopropanoate	41, 42, 43
Amyl butanoate	25, 28, 31, 35, 42
Hexyl butanoate	25, 35, 42
Ethyl 2-butenoate (crotonate)	41, 42, 43
Ethyl 3-methylbut-2-enoate	41, 42, 43
Ethyl 4-oxopentanoate	41, 42, 43
Ethyl 4-methylpent-2-enoate	41, 42, 43
Ethyl 4-methylpent-3-enoate	41, 42, 43
Ethyl hexanoate	31, 35, 37, 39, 42, 43
Ethyl 3-heptenoate	41, 42, 43
Ethyl octanoate	37, 43
Ethyl decanoate	37, 43
Ethyl dodecanoate	37
Ethyl tetradecanoate	37

Compounds in Natural Cocoa Aroma (continued)

Compounds	References
Isobutyl benzoate	42, 43
Ethyl benzoate	37, 39, 42, 43
Isoamyl benzoate	40
Methyl phenylacetate	31, 35, 42, 43
Ethyl phenylacetate	31, 35, 39, 42, 43
Ethyl cinnamate	37
4-Hydroxybutanoic acid lactone	35, 39, 42
4-Hydroxy-2-methylbutanoic acid lactone	35, 42
4-Hydroxypentanoic acid lactone	35, 39, 42
4-Hydroxyhexanoic acid lactone	35, 39, 42
4-Hydroxynonanoic acid lactone	37
Bases	
Methylamine	27, 35, 42
Dimethylamine	27, 35, 42
Trimethylamine	27, 35, 42
Ethylamine	27, 35, 42
Triethylamine	27, 35, 42
Isobutylamine	27, 35, 42
sec-Butylamine	27, 35, 42
Isoamylamine	23, 31, 35, 42
Phenethylamine	31, 35, 42
Pyrrole	35, 39, 42
Methylpyrazine	35, 38, 39, 40, 42, 43
2,3-Dimethylpyrazine	35, 37, 38, 39, 40, 42, 43
2,5-Dimethylpyrazine	35, 37, 38, 39, 40, 42, 43
2,6-Dimethylpyrazine	31, 35, 37, 38, 39, 40, 42, 43
Trimethylpyrazine	35, 37, 38, 39, 40, 42, 43
Tetramethylpyrazine	31, 35, 37, 38, 39, 40, 42, 43
Ethylpyrazine	40
2-Methyl-5-ethylpyrazine	38, 39, 40, 43
2-Methyl-6-ethylpyrazine	35, 42, 43
2-Methyl-6-isoamylpyrazine	41, 42, 43
2-Methyl-6-(2-methylbutyl)-pyrazine	41, 42, 43
2,3-Dimethyl-5-ethylpyrazine	40, 43
2,3-Dimethyl-5-isoamylpyrazine	37
2,3-Dimethyl-5-(2-methylbutyl)-pyrazine	37
2,5-Dimethyl-3-ethylpyrazine	35, 37, 38, 39, 40, 42, 43
2,5-Dimethyl-3-propylpyrazine	37
2,5-Dimethyl-3,6-diethylpyrazine	41, 42, 43
2,6-Dimethyl-3-ethylpyrazine	38, 43
2,6-Dimethyl-3-isoamylpyrazine	41, 42, 43
2,6-Dimethyl-3,5-diethylpyrazine	41, 42, 43
Dimethyl-isopropylpyrazine	37
Trimethyl-ethylpyrazine	34, 39, 40, 43
Trimethyl-isoamylpyrazine	37

Compounds	References
S-Compounds	
Methylthiomethane	29, 35, 39, 40, 42
Methyldithiomethane	29, 35, 37, 40, 42, 43
Methyldithioisopropane	41, 42, 43
Methyldithiobenzene	41, 42, 43
Methyltrithiomethane	40, 42
Methyltrithiopropane	35, 43
Propyltrithiopropane	42
Isobutyl thiocyanate	41, 42, 43
3-Methylthiopropanal	40
2-Methylthio-2-methylpropanal	41, 42, 43
5-Methylfurfurylthiomethane	35, 39, 42
4-Methyl-5-(β-hydroxyethyl)-thiazole	36
4-Methyl-5-vinylthiazole	36
Benzothiazole	37, 43
Miscellaneous	
3-Methylbutane nitrile	41, 42, 43
Benzonitrile	35, 39, 43
Phenol	31, 35, 37, 39, 40, 42
Methylphenol (cresol)	31, 35, 42
2,3 Dimethylphenol	40
4-Ethylphenol	31, 35, 42
2-Methoxyphenol (guaiacol)	31, 35, 37, 39, 40, 42, 43
2-Methoxy-4-methylphenol	31, 35, 37
4-Allyl-2-methoxyphenol (eugenol)	31, 35, 37, 42
1-Allyl-3,4-methylenedioxy benzene (safrole)	41, 42, 43
6,7-Dihydroxycoumarin	32, 35, 42
1,4-Cineole	41, 42, 43
1,8-Cineole	41, 42, 43
2-Formylpyrrole	35, 39, 40, 42
2-Formyl-1-methylpyrrole	35, 39, 42, 43
2-Formyl-5-methylpyrrole	35, 39, 42
1-Ethyl-2-formylpyrrole	41, 42, 43
2-Formyl-1-pentylpyrrole	41, 42, 43
2-Acetylpyrrole	31, 35, 39, 40, 42, 43
2-Acetyl-1-pentylpyrrole	41, 42, 43
2-Propionylpyrrole	41, 42, 43
2-Methoxycarbonylpyrrole	41, 42, 43
2-Methyltetrahydrofuran	40
cis-5-(2-Hydroxy-isopropyl)-2-methyl-2-vinyltetrahydrofuran = *cis*-2,6-dimethyl-3,6-epoxy-2-hydroxyoct-7-ene = linalool-oxide II	35, 39, 42, 43
Furan	29, 35, 42
2-Methylfuran	29, 35, 42
3-Phenylfuran	41, 42, 43
3-Hydroxy-2-methyl-4-oxo-1,4-pyran (maltol)	31, 35, 42
Pentanal	43

Compounds in Natural Cocoa Aroma (continued)

Compounds	References	Compounds	References
2-Pentanone	43	Ethyl 2-furancarboxylate	43
3-Hexanone	43	2,5-Dimethyl-3-isobutyl-pyrazine	43
1-Acetyl-4-isopropenylcyclo-pent-1-ene	43	2,5-Dimethyl-3-(2-methylbutyl) pyrazine	43
4-Phenylbutan-2-one	43	2,5-Dimethyl-3-isoamyl pyrazine	37, 43
α-Terpinyl acetate	43	Isopropylpyrazine	43
Ethyl 2-hydroxypropanoate (lactate)	43	Methylthioethane	43
Ethyl 2,2-diethoxypropanoate	43	1,1-Diethoxyethane	43
Ethyl 3-ethoxypropanoate	43	1,1-Diethoxy-2-methylpropane	43
Ethyl 3-methylbutanoate	43	1,1-Diethoxy-2-methylbutane	43
Ethyl 2-hydroxy-3-methyl-butanoate	43	1,1-Diethoxy-3-methylbutane	43
Ethyl 4-methylpentanoate	43	1-Ethoxy-1-isobutoxy-3-methyl-butane	43
Ethyl 2-hydroxy-4-methyl-pentanoate	43	1,1-Diethoxy-3-oxobutane	43
Ethyl heptanoate	43	1,1-Dimethoxyphenylethane (acetal)	43
Ethyl nonanoate	43	Diisoamyl ether	43
Methyl 4-methoxybenzoate	43	Benzyl-ethyl ether	43
Ethyl acid succinate	43	Benzylcyanide	41, 42, 43
Diethyl succinate	43	*trans*-Linalooloxide I	43
Methyl 2-furancarboxylate	43		

Coffee

Whenever the distillate, tincture, or soft or dried extract of roasted coffee fails to adequately flavor any given food substrate because of technical limitations, it may be necessary to use coffee flavor imitations. The imitation of coffee flavor represents indeed an extremely difficult and complex problem to the flavorist. The green coffee bean almost lacks any flavor whatsoever; yet once roasted, it turns into an endless source of flavor sensations.

Furfural, long known to be present in roasted coffee aroma through the caramelization of carbohydrates and the decomposition of pentosans, is nearly useless for the reconstruction of natural coffee aroma. Coffee aroma, in fact, is the resulting product not only of the caramelization of sugars, but also of the (1) simultaneous drastic hydrolysis of plant tissues, (2) partial elimination of volatile products, (3) carbonization of cellulose, (4) decomposition of glucosides and fats, and (5) hydrolysis of proteins. All of these reactions cause the development of the characteristic coffee aroma; they are dependent both qualitatively and quantitatively on the method of roasting and the blend of coffee beans.

The initial attempts to imitate coffee aroma were limited to reinforcing the aroma of complex coffee extracts by the addition of a small amount of distillate to hydro-alcoholic tincture. Attempts were also made to strengthen the aroma of concentrated aqueous coffee extracts containing small amounts of propylene glycol or propylene glycol ethyl ester to improve the solubility of the flavor ingredients. The most functional products for such purposes were ethyl formate, acetaldehyde, diacetyl, furfural, pyridine, thional (condensation product of tolyl aldehyde with *n*-hexylmercaptan), and cyclotene or equivalent products (to sweeten the note). The formulations suggested by H. Staudinger and T. Reichstein provided guidelines for the preparation of the so-called "concentrated bases of mocha coffee imitations."[16] The following formulation represents a valuable starting point in the study of coffee-flavored imitations.

Imitation Coffee Formulation

Ingredient	Parts	Ingredient	Parts
Ethyl methyl glyoxal	4	Phenol	1
2-Methylbutanal	4	Isoeugenol	1
Acetaldehyde	3	Methyl mercaptan	0.6
Pyridine	3	α-Methylcyclopenten-	
α-Methylfurfural	2	olone	0.5
Isovaleric acid	2	Thioguaiacol	0.4
Diacetyl	1	Furfuryl mercaptan	0.3
Furfural	1	*n*-Octyl alcohol	0.3

Knowledge of the components present in coffee aroma does not imply the ability to rebuild the natural aroma via imitations. However, an examination of the components identified in coffee aroma by M. Stoll *et al.* may yield very valuable information.[17] According to the authors, the list may be quantitatively inaccurate and may also contain potential artifacts. The flavorist should keep in mind that coffee aroma is probably the most complex aroma and, therefore, the most difficult to imitate satisfactorily.

Components of Coffee Aroma

Hydrocarbons

p-Cymene
Dimethylnaphthalene
Diphenyl
2-Ethylnaphthalene
Fluorene
Indene
p-Isopropenyltoluene
p-Menthadiene (1,8)
3-Methyl diphenyl
1-Methylnaphthalene
2-Methylnaphthalene
Myrcene
Naphthalene
Styrene
1,2,4,5-Tetramethylbenzene
Tetramethylnaphthalene
Toluene
1,2,4-Trimethylbenzene
Trimethylnaphthalene

Alcohols and Esters

3,4-Dimethoxystyrene
Ethanol
2-Heptanol
1-Hexanol
Linalool
Methanol
3-Methyl-1-butanol
3-Methyl-2-buten-1-ol
3-Octanol
1-Octen-3-ol
1-Pentanol
α-Terpineol

Aldehydes

Acetaldehyde
Benzaldehyde
Butyraldehyde
Hexanal
Isobutyraldehyde
Isopentanal
2-Methyl-2-butanol
2-Methylbutyraldehyde
Phenyl acetaldehyde
Propionaldehyde
Salicylaldehyde
o-Toluic aldehyde

Ketones

Acetone
Butanone
Cyclopentanone
2-Decanone
6,10-Dimethyl-2-undecanone
2-Heptanone
3-Hexanone
Mesityl oxide
6-Methyl-5-hepten-2-one
2-Nonanone
2-Octanone
3-Octanone
2-Pentanone
3-Pentanone
Propiophenone (ethyl phenyl ketone)
2-Tridecanone
6,10,14-Trimethyl-2-pentadecanone
2-Undecanone

Ketoalcohols

1-Butanol-2-one
2-Butanol-2-one
3-Pentanol-2-one
2-Pentanon-3-one
1-Propanol-2-one

Diketones

2,3-Butanedione
3,4-Heptanedione
2,3-Hexanedione
3,4-Hexanedione
5-Methyl-3,4-heptanedione
6-Methyl-3,4-heptanedione
5-Methyl-2,3-hexanedione
4-Methyl-2,3-pentanedione
2,3-Octanedione
2,3-Pentanedione
1-Phenyl-1,2-propanedione

Acids and Anhydrides

Acetic acid
2,3-Dimethyl maleic anhydride
2-Ethyl-3-methyl maleic anhydride
Formic acid
Hexanoic acid
Isovaleric acid
2-Methylbutyric acid
Propionic acid

Esters and Lactones

Benzyl formate
γ-Butyrolactone
2-Isopentenyl acetate

Components of Coffee Aroma (continued)

Esters and Lactones *(continued)*

Methyl acetate
Methyl benzoate
Methyl formate
Methyl palmitate
2-Phenethyl formate

Mercaptans and Sulfides

Carbon disulfide
Methyl ethyl sulfide
Methyl mercaptan
Methyl sulfide
Methylthiobenzene
1-Methyl-2-thio-2-butanone
2-Methylthiophenol

Phenols

o-Cresol
2,3-Dimethylphenol (*o*-Xylenol)
2,5-Dimethylphenol (*p*-Xylenol)
2,6-Dimethylphenol (*m*-Xylenol)
3,4-Dimethylphenol (*o*-Xylenol, unsy.)
4-Ethyl-2-methoxyphenol
2-Ethylphenol
o-Ethylphenol
2-Hydroxyacetophenone
o-Methoxyphenol
5-Methyl-2-hydroxyacetophenone
Methyl salicylate
Phenol
2,3,5-Trimethylphenol

Furan Derivatives

2-Acetyl furane
2-Acetyl-5-methyl furane
2-Amyl furfurane
Benzofuran
2-Butyryl furane
2-Cyano-5-methyl furane
Difurfuryl ether
2,3-Dihydro benzofurane
Di-(5-methylfurfuryl) ether

Furan Derivatives *(continued)*

Furan
Furfural
Furfuryl acetate
Furfuryl alcohol
Furfuryl alcohol, methyl ester
Furfuryl butyrate
Furfuryl crotonate
Furfuryl, β,β-dimethyl acrylate
Furfuryl formate
2-Furfuryl furfurane
Furfuryl isobutyrate
Furfuryl isovalerate
Furfuryl mercaptan
Furfuryl 2-methylbutyrate
Furfuryl methylsulfide
Furfuryl propionate
Furfuryl sulfide
Furfuryl thioacetate
2-Furyl-1,2-butanedione
2-Furyl-2-butanone
2-Furyl-3-butanone
2-Furyl furfurane
Furyl methyl thiocarbonate
2-Furyl-1,2-propanedione
2-Furyl-2-propanone
5-Methyl furfural
2-Methyl furfurane
2-(5-Methylfurfuryl)furfurane
2-(5-Methylfurfuryl)5-methyl furfurane
5-Methylfurfuryl methylsulfide
5-Methylfuryl-1,2-butanedione
5-Methylfuryl-2-butanone
5-Methylfuryl-3-butanone
5-Methylfuryl-1,2-propanedione
5-Methylfuryl-2-propanone
2-Methyl-5-propionyl furane
3-Phenyl furfurane
2-Propionyl furane

Thiophene Derivatives

2-Acetyl-3-methyl thiophene
2-Acetyl-4-methyl thiophene

Thiophene Derivatives *(continued)*

2-Acetyl-5-methyl thiophene
Benzo-b-thiophene
4-Ethyl-2-methyl thiophene
2-Formyl-5-methyl thiophene
2-Formyl thiophene
2-Propionyl thiophene
Thienyl acetate
Thienyl alcohol
Thienyl formate
2-Thienyl-methyl carbonate
2-Thienyl-1,2-propanedione
3-Thienyl-1,2-propanedione
3,2[b] or 2,3[b]-Benzothiophene
3-Vinyl thiophene

Pyrrole Derivatives

1-Acetyl furfuryl-2-pyrrole
1-Ethylformyl-2-pyrrole
1-Formylfurfuryl-2-pyrrole
1-Furfuryl pyrrole
1-Isoamylformyl-2-pyrrole
1-Isoamyl pyrrole
1-Methyl acetyl-2-pyrrole
1-(2-Methyl butyl)formyl-2-pyrrole
2-Methyl butyl-1-pyrrole
1-Methylformyl-2-pyrrole
5-Methyl furfuryl-1-pyrrole
1-(2-Pyrrolyl)-1,2-butanedione

Thiazoles

2-Acetyl-4-methyl thiazole
2-Propionyl-4-methyl thiazole

Other Products

Anhydrous linalool oxide
cis-Linalool oxide
Maltol
2-Methyl-3-hydrofuranone
t-Methyl-quinoxaline
2-Methyl-3-tetrahydrothiophenone
Tetrahydrothiophene

Research on the identification of those components responsible for the aroma of roasted coffee has continued after 1967, date of the publication of the above list. An updated list of products identified thus far in coffee aroma is reported in the following table.

Recently Identified Components of Coffee Aroma

Compounds	References
Hydrocarbons	
Methane	58, 66
Ethylene	66
Butene	58, 66
2-Methylbuta-1,3-diene	46, 52, 55, 58, 61, 65, 66
Pentane	58, 66
Pentene	58, 66
Pentadiene	45, 58, 61, 66
Hexene	58, 66
Hexadiene	58, 66
Octane	58, 66
Octene	58, 66
Octyne	58, 66
Nonane	58, 66
Tetradecane	62, 66
Pentadecane	62, 66
Heptacosane	61, 66
Myrcene	59, 60, 66, 67
1-Methylcyclohexa-2,4-diene	58, 66
Limonene	60, 66
Benzene	56, 58, 66
Toluene	55, 56, 58, 60, 61, 66
Dimethylbenzene	56, 58, 66
1,2,4-Trimethylbenzene (pseudocumene)	60, 66
1,2,4,5-Tetramethylbenzene	60, 66
Ethylbenzene	58, 66
Vinylbenzene (styrene)	56, 60, 66
1-Isopropyl-4-methylbenzene	60, 66
1-Isopropenyl-4-methylbenzene	60, 66
Silvestrene	61, 66
Indene	60, 66
Biphenyl	60, 66
3-Methylbiphenyl	60, 66
Naphthalene	60, 61, 66
1-Methylnaphthalene	60, 66
2-Methylnaphthalene	60, 66
Dimethylnaphthalene	60, 66
Trimethylnaphthalene	60, 66
Tetramethylnaphthalene	60, 66
2-Ethylnaphthalene	60, 66
Fluorene	60, 66
Phenanthrene	57
Anthracene	57
Fluoranthrene	57
Pyrene	57
Chrysene	57
1,2-Benzanthrene	57
1,2-Benzopyrene	57
3,4-Benzopyrene	57
1,12-Benzperylene	57
Alcohols	
Methanol	44, 45, 46, 48, 52, 53, 55, 58, 60, 61, 65, 66

Compounds	References
Ethanol	44, 45, 46, 48, 52, 55, 58, 60, 61, 65, 66
2-Propanol	64, 65, 66
2-Methylpropan-1-ol (isobutanol)	58, 66
2-Methylpropan-2-ol	58, 66
2-Methylbutan-1-ol (isoamyl alcohol)	60, 66
3-Methylbut-2-en-1-ol	60, 65, 66
2-Methylbutan-2-ol	58, 66
1-Pentanol	58, 60, 66
1-Hexanol	56, 58, 60, 66
2-Heptanol	60, 66
3-Octanol	60, 66
1-Octen-3-ol	60, 66
Linalool	59, 60, 65, 66, 67
α-Terpineol	59, 60, 66
Furfuryl alcohol	51, 54, 55, 60, 61, 65, 66
Carbonyls	
Formaldehyde	55, 61, 66
Acetaldehyde	44, 45, 46, 47, 48, 52, 53, 54, 55, 58, 60, 61, 65, 66
Propanal	44, 45, 46, 47, 48, 52, 53, 54, 55, 56, 58, 60, 61, 65, 66
2-Propenal (acrolein)	44, 48, 61, 66
2-Methylpropanal	45, 46, 47, 52, 53, 54, 55, 56, 58, 60, 61, 65, 66
Butanal	45, 46, 52, 56, 58, 60, 61, 65, 66
2-Butenal (crotonaldehyde)	56, 66
2-Methylbutanal	45, 47, 53, 54, 55, 58, 60, 61 65, 66
3-Methylbutanal	44, 45, 46, 47, 48, 52, 53, 54, 60, 61, 65, 66
2-Methylbut-2-enal	44, 48, 60, 65, 66
Pentanal	45, 58, 61, 66
2-Methylpent-2-enal	44, 48, 61, 66
Hexanal	58, 60, 66
Benzaldehyde	60, 66, 67
2-Methylbenzaldehyde	60, 66
3-Methylbenzaldehyde	64, 65, 66
2-Hydroxybenzaldehyde (salicylaldehyde)	60, 66
Phenylacetaldehyde	60, 66

Recently Identified Components of Coffee Aroma (continued)

Compounds	References
2-Propanone	44, 45, 46, 47, 48, 52, 53, 54, 55, 56, 58, 60, 61, 65, 66
1-Hydroxypropan-2-one (acetol)	53, 55, 60, 61, 65, 66
Acetol acetate	53, 54, 55, 56, 61, 65, 66
2-Butanone	44, 45, 46, 48, 52, 53, 54, 55, 56, 58, 60, 61, 65, 66
3-Buten-2-one (1-buten-3-one)	44, 48, 61, 66
1-Hydroxybutan-2-one	60, 66
3-Hydroxybutan-2-one	53, 54, 55, 60, 61, 65, 66
3-Methylbutan-2-one	58, 66
2,3-Butanedione (diacetyl)	44, 45, 46, 48, 52, 53, 54, 55, 58, 60, 61, 65, 66
2-Pentanone	58, 60, 66
3-Pentanone	44, 49, 58, 60, 61, 66
trans-3-Penten-2-one (2-penten-4-one)	64, 65, 66
3-Hydroxypentan-2-one	55, 60, 61, 66
2-Hydroxypentan-3-one	55, 60, 61, 66
4-Methyl-3-penten-2-one (2-methyl-2-penten-4-one)	60, 66
2,3-Pentanedione	46, 52, 53, 54, 55, 56, 60, 61, 65, 66
2,4-Pentanedione	44, 48, 61, 66
4-Methylpentane-2,3-dione	60, 66
3-Hexanone	56, 60, 65, 66
2,3-Hexanedione	53, 54, 60, 65, 66
2,5-Hexanedione	64, 65, 66
3,4-Hexanedione	60, 66
5-Methylhexane-2,3-dione	60, 66
2-Heptanone	60, 66
6-Methyl-5-hepten-2-one (2-methyl-2-hepten-6-one)	60, 66
2,5-Heptanedione	64, 65, 66
3,4-Heptanedione	60, 66
5-Methylheptane-3,4-dione	60, 66
6-Methylheptane-3,4-dione	60, 66
2-Octanone	60, 66
3-Octanone	60, 66
2,3-Octanedione	60, 66
2-Nonanone	60, 66
trans-2-Nonenal	71
2-Decanone	60, 66
2-Undecanone	60, 66
6,10-Dimethylundecan-2-one	59, 60, 66
2-Tridecanone	60, 66
6,10,14-Trimethylpentadecan-2-one	59, 60, 66
2-Hydroxyacetophenone	60, 65, 66
2-Hydroxy-5-methylacetophenone	60, 66
2,3-Dihydroxyacetophenone	55, 61, 66
4-Hydroxy-3-methoxyacetophenone	61, 66
1-Phenylpropan-1-one (propiophenone)	60, 66
1-Phenylpropane-1,2-dione	60, 66
Cyclopentanone	55, 60, 61, 65, 66
Cyclopentan-1,2-dione	66
3-Methylcyclopentan-1,2-dione (cyclotene)	49, 55, 61, 65, 66
3,4-Dimethylcyclopentan-1,2-dione	49, 55, 61, 65, 66
3,5-Dimethylcyclopentan-1,2-dione	49, 55, 61, 66
3-Ethylcyclopentan-1,2-dione	49, 55, 61, 65, 66
Cyclopent-2-en-1-one	62, 66
2-Methylcyclopent-2-en-1-one	62, 66
2-Ethylcyclopent-2-en-1-one	62, 66
2,3-Dimethylcyclopent-2-en-1-one	67
2,3,5-Trimethylcyclopent-2-en-1-one	62, 66
3-Methylcyclohexan-1,2-dione	49, 55, 61, 65, 66
Cyclohex-2-en-1-one	67
2-Methylcyclohex-2-en-1-one	62, 66
2-Acetyltetrahydrofurane	67
Furfural	45, 51, 53, 54, 55, 58, 60, 61, 65, 66
5-Methylfurfural	51, 53, 54, 55, 60, 61, 65, 66
5-Hydroxymethoxyfurfural	68
3-(Furyl-2)propanal	64, 65, 66
3-(Furyl-2)prop-2-enal	58, 64, 65, 66
2-Acetylfuran	51, 53, 54, 55, 56, 60, 61, 65, 66
2-Acetyl-5-methylfuran	51, 55, 60, 61, 65, 66
2-Propionylfuran	60, 65, 66
5-Methyl-2-propionylfuran	60, 65, 66
1-(Furyl-2)propan-1,2-dione	51, 55, 60, 61, 65, 66
1-(5-Methylfuryl-2)propan-1,2-dione	51, 55, 60, 61, 65, 66
2-Acetonylfuran	60, 65, 66
2-Acetonyl-5-methylfuran	60, 66
2-Furfuryloxyacetone	64, 65, 66
2-Butyrylfuran	60, 65, 66

Recently Identified Components of Coffee Aroma (continued)

Compounds	References
2-Isobutyrylfuran	65
1-(Furyl-2)butan-2-one	60, 66
1-(5-Methylfuryl-2)butan-2-one	60, 66
1-(Furyl-2)butan-3-one	60, 65, 66
1-(5-Methylfuryl-2)butan-3-one	60, 65, 66
1-(Furyl-2)but-1-en-3-one	64, 65, 66
1-(Furyl-2)butan-1,2-dione	51, 55, 60, 61, 65, 66
1-(5-Methylfuryl-2)butan-1,2-dione	60, 66
2-Methyltetrahydrofuran-3-one	50, 53, 54, 55, 56, 60, 61, 65, 66
2,5-Dimethyl-(2H)-furan-3-one	67
2,4,5-Trimethyl-(2H)-furan-3-one	62, 66
Acids	
Formic acid	45, 55, 60, 61, 63, 65, 66
Acetic acid	45, 53, 55, 60, 61, 63, 65, 66
Propanoic acid	45, 53, 55, 60, 61, 63, 65, 66
2-Methylpropanoic acid	55, 61, 65, 66
2-Methylpropenoic acid	64, 65, 66
Butanoic acid	63, 65, 66
cis- and *trans*-2-Butenoic acid	55, 61, 64, 65, 66
2-Methylbutanoic acid	55, 60, 61, 66
3-Methylbutanoic acid	53, 55, 60, 61, 65, 66
2-Methylbut-2-enoic acid (tiglic acid)	64, 65, 66
3-Methylbut-2-enoic acid (senecioic acid)	55, 61, 65, 66
Pentanoic acid	63, 66
Hexanoic acid	60, 62, 66
Heptanoic acid	63, 66
Octanoic acid	63, 66
Nonanoic acid	63, 66
Decanoic acid	63, 66
Hexadecanoic acid (palmitic acid)	61, 66
3-Ethyl-2-methylmaleic anhydride	55, 60, 61, 64, 65, 66
2,3-Dimethylmaleic anhydride	55, 60, 61, 65, 66
Esters	
Methyl formate	44, 45, 46, 48, 52, 53, 54, 55, 56, 58, 60, 61, 65, 66
Ethyl formate	45, 46, 52, 61, 65, 66
Isopropyl formate	64, 65, 66
Benzyl formate	60, 66
Phenethyl formate	60, 65, 66
Furfuryl formate	51, 55, 60, 61, 65, 66
Methyl acetate	44, 45, 46, 48, 52, 53, 54, 55, 56, 58, 60, 61, 65, 66
Ethyl acetate	55, 58, 61, 65,
Butyl acetate	56, 66
2-Oxobutyl-1 acetate	62, 64, 65, 66
3-Methylbut-2-enyl acetate	60, 66
2-Oxopentyl-1 acetate	64, 65, 66
3-Oxopentyl-2 acetate	64, 65, 66
Isoamyl acetate	64, 65, 66
Furfuryl acetate	51, 53, 54, 55, 56, 60, 61, 65, 66
Methyl propanoate	56, 58, 65, 66
Acetonyl propanoate	62, 66
Furfuryl propanoate	60, 65, 66
Furfuryl 2-methylpropanoate	60, 66
Furfuryl butanoate	60, 66
Furfuryl 2-butenoate	60, 66
Furfuryl 3-methylbut-2-enoate	60, 65, 66
Methyl-3-methylbutanoate	56, 66
Furfuryl 3-methylbutanoate	60, 66
Methyl hexadecanoate (palmitate)	60, 66
Methyl benzoate	60, 66
Methyl phenylacetate	64, 65, 66
Methyl salicylate	60, 65, 66
Methyl nicotinate	55, 61, 65, 66
4-Hydroxybutanoic acid lactone	53, 54, 55, 60, 61, 65, 66
4-Hydroxy-2-methylbutanoic acid lactone	64, 65, 66
4-Hydroxybut-2-enoic acid lactone (crotonic)	55, 61, 66
4-Hydroxy-2,3-dimethylbut-2-enoic acid lactone	64, 65, 66
4-Hydroxypentanoic acid lactone	64, 65, 66
4-Hydroxy-3-methylpent-2-enoic acid lactone	64, 65, 66
4-Hydroxy-2,3-dimethylpent-2-enoic acid lactone	64, 65, 66
Furfuryl 2-methylbutanoate	60, 66
Bases	
Ammonia	66
Methylamine	66
Trimethylamine	66
Pyrrole	44, 48, 51, 55, 61, 65, 66
Dimethylpyrrole	44, 48, 58, 66
1-Methylpyrrole	44, 48, 51, 53, 54, 55, 56, 58, 61, 65, 66
2-Ethylpyrrole	56, 66
1-(2-Methylbutyl)pyrrole	60, 66
1-Butylpyrrole	58, 66
1-Pentylpyrrole	62, 66

Recently Identified Components of Coffee Aroma (continued)

Compounds	References
1-(3-Methylbutyl-1)pyrrole	60, 66
Pyrazine	53, 54, 55, 56, 60, 61, 62, 65, 66, 70
Methylpyrazine	53, 54, 55, 56, 60, 61, 62, 65, 66, 70
2,3-Dimethylpyrazine	53, 54, 55, 61, 62, 65, 66, 70
2,5-Dimethylpyrazine	53, 54, 55, 60, 61, 62, 65, 66, 70
2,6-Dimethylpyrazine	55, 60, 61, 62, 65, 66, 70
Trimethylpyrazine	60, 62, 65, 66, 70
Tetramethylpyrazine	64, 85, 66, 67, 70
2-Methyl-3-ethylpyrazine	60, 62, 65, 66, 70
2-Methyl-5-ethylpyrzaine	60, 62, 65, 66, 70
2-Methyl-6-ethylpyrazine	60, 62, 65, 66, 70
2-Methyl-5-vinylpyrazine	60, 62, 66, 70
2-Methyl-6-vinylpyrazine	60, 62, 66, 70
2-Methyl-5-propylpyrazine	67, 70
2-Methyl-6-propylpyrazine	60, 66, 70
2-Methyl-5-isopropylpyrazine	60, 62, 66, 70
2-Methyl-6-isopropylpyrazine	70
2-Methyl-3-isobutylpyrazine	60, 66, 70
2,3-Dimethyl-5-ethylpyrazine	67
2,5-Dimethyl-3-ethylpyrazine	60, 62, 65, 66, 70
2,6-Dimethyl-3-ethylpyrazine	60, 62, 65, 66, 70
2,5-Dimethyl-3-isobutylpyrazine	67, 70
Trimethylethylpyrazine	64, 65, 66, 67, 70
Ethylpyrazine	60, 62, 66, 70
Vinylpyrazine	62, 66, 70
2-Ethyl-6-vinylpyrazine	67, 70
2-Ethyl-6-propylpyrazine	67, 70
2,3-Diethyl-5-methylpyrazine	62, 65, 66, 70
2,5-Diethylpyrazine	60, 66, 67, 70
2,5-Diethyl-3-methylpyrazine	67
2,6-Diethylpyrazine	60, 62, 66, 70
3,5-Diethyl-2-methylpyrazine	62, 66, 70
Propylpyrazine	60, 62, 66, 70
trans-1-Propenylpyrazine	62, 66, 70
2-Methyl-5-(*trans*-1-propenyl) pyrazine	62, 66, 70
2-Methyl-6-(*trans*-1-propenyl) pyrazine	62, 66, 70
2-Isobutyl-3-methoxypyrazine	67, 70
2-(Furyl-2)pyrazine	67, 70
Pyridine	53, 54, 55, 56, 58 60, 61, 65, 66

Compounds	References
2-Methylpyridine (α-picoline)	60, 66
3-Methylpyridine (β-picoline)	55, 60, 61, 65, 66
3-Ethylpyridine	60, 66
5-Methylquinoxaline	59, 60, 66
S-Compounds	
Hydrogensulfide	46, 55, 61, 65, 66
Carbondisulfide	44, 45, 46, 48, 52, 58, 60, 61, 65, 66
Methanethiol	45, 46, 47, 52, 55, 58, 60, 61, 65, 66
Ethanethiol	47, 64, 65, 66
1-Propanethiol	64, 65, 66
Methylthiomethane	44, 45, 46, 47, 48, 52, 55, 56, 60, 61, 65, 66
Methylthioethane	44, 60, 61, 65, 66
Methyldithiomethane	44, 48, 56, 58, 65, 66
Methyldithioethane	44, 48, 56, 58, 61, 66
Ethyldithioethane	56, 66
Methylthiobenzene	60, 66
1-Methylthiobutan-2-one	60, 66
Thiophene	45, 55, 56, 58, 59, 61, 65, 66
2-Methylthiophene	56, 66
3-Vinylthiophene	60, 66
2-Methyl-4-ethylthiophene	60, 66
Benzo(b)thiophene	60, 66
2-Furylmethanethiol	56, 60, 61, 66
Furfurylthiomethane	51, 55, 59, 60, 61, 65, 66
5-Methylfurfurylthiomethane	59, 60, 66
(5-Methylfuryl-2)thiomethane	59, 60, 66
Furoylthiomethane	59, 60, 66
Furfurylthiolacetate	59, 60, 66
Difurfurylsulfide	59, 60, 66
(2-Thienyl)methanol (thenyl alcohol)	56, 60, 65, 66
Thenyl formate	60, 66
Thenyl acetate	60, 65, 66
2-Formylthiophene	60, 65, 66
5-Methyl-2-formylthiophene	60, 66
Tetrahydrothiophen-2-one	64, 65, 66
Tetrahydrothiophen-3-one	60, 66
2-Methyl-tetrahydrothiophen-3-one	60, 66
2-Acetylthiophene	55, 60, 61, 65, 66
3-Acetylthiophene	64, 65, 66
2-Acetyl-3-methylthiophene	60, 65, 66
2-Acetyl-4-methylthiophene	60, 66

Recently Identified Components of Coffee Aroma (continued)

Compounds	References
2-Acetyl-5-methylthiophene	60, 65, 66
2-Propionylthiophene	60, 65, 66
1-(Thienyl-2)propane-1,2-dione	60, 66
1-(Thienyl-3)propane-1,2-dione	60, 66
Methyl 2-thiophenecarboxylate	60, 66
2-(Methylthio)phenol	60, 66
2-Thiophenothiophene	59, 60, 66
2-Acetyl-4-methylthiazole	60, 66
2-Propionyl-4-methylthiazole	60, 66
2-Methyl-3-oxa-8-thiabicyclo-(3.3.0) 1,4-Octadiene (kahweofuran)	69
Miscellaneous	
1,1-Dimethoxyethane	61, 66
Propenenitril	44, 48, 56, 61, 66
3-Butenenitril	44, 48, 61, 66
Phenol	53, 54, 55, 56, 60, 61, 65, 66
2-Methylphenol (*o*-cresol)	60, 65, 66
3-Methylphenol	55, 60, 61, 66
4-Methylphenol	60, 66
2,3-Dimethylphenol	60, 66
2,5-Dimethylphenol	60, 66
2,6-Dimethylphenol	60, 65, 66
3,4-Dimethylphenol	60, 66
2,3,5-Trimethylphenol	60, 66
2-Ethylphenol	60, 66
4-Ethylphenol	60, 66
2-Methoxyphenol (guaiacol)	53, 54, 55, 61, 65, 66
4-Ethyl-2-methoxyphenol	55, 60, 61, 65, 66
2-Methoxy-4-vinylphenol	55, 59, 60, 61, 65, 66
4-Allyl-2-methoxyphenol (eugenol)	61, 66
1,2-Dihydroxybenzene	61, 66
1,3-Dihydroxybenzene (resorcinol)	66
1,4-Dihydroxybenzene (hydrochinone)	66
4-Vinyl-1,2-dihydroxybenzene	66
3,4-Dimethoxystyrene	54, 60, 66
2-Formylpyrrole	51, 55, 56, 61, 65, 66
2-Formyl-1-methylpyrrole	51, 55, 60, 61, 65, 66
2-Formyl-5-methylpyrrole	51, 55, 61, 65, 66
1-Ethyl-2-formylpyrrole	60, 65, 66
1,5-Dimethyl-2-formylpyrrole	51, 55, 61, 65, 66
2-Formyl-1-(2-methylbutyl-1) pyrrole	60, 66
2-Formyl-1-(3-methylbutyl-1) pyrrole	60, 66
2-Formyl-1-furfurylpyrrole	60, 66
2-Acetylpyrrole	51, 53, 55, 61, 65, 66
2-Acetyl-1-methylpyrrole	51, 55, 60, 61, 65, 66
2-Acetyl-1-ethylpyrrole	64, 65, 66
2-Acetyl-1-furfurylpyrrole	60, 66
2-Propionylpyrrole	51, 55, 61, 66
1-(Pyrryl-2)butan-1,2-dione	60, 66
1-Furfurylpyrrole	50, 51, 55, 60, 61, 65, 66
1-(5-Methylfurfuryl) pyrrole	60, 65, 66
Tetrahydrofuran	56, 64, 65, 66
2-Methyltetrahydrofuran	64, 65, 66
5-Isopropenyl-2-methyl-2-vinyltetrahydrofuran	59, 60, 66
cis-(2-Hydroxy-isopropyl)-2-methyl-2-vinyltetrahydrofuran = *cis*-2,6-dimethyl-3,6-epoxy-2-hydroxyoct-7-ene = *cis*-linaloöloxyde[48]	59, 60, 65, 66, 67
trans-linaloöloxyde[48]	64, 65, 66
Furan	44, 45, 46, 48, 51, 52, 53, 54, 55, 56, 58, 59, 60, 61, 65, 66
2-Methylfuran	44, 45, 46, 48, 51, 52, 53, 54, 55, 56, 58, 59, 60, 61, 65, 66
3-Methylfuran	56, 66
2,5-Dimethylfuran	44, 48, 51, 55, 56, 58, 61, 65, 66
2-Ethylfuran	56, 66
2-Propylfuran	44, 48, 51, 56, 58, 61, 66
3-Propylfuran	56, 66
2-Butylfuran	44, 48, 61, 66
2-Pentylfuran	60, 66
3-Phenylfuran	59, 60, 65, 66
Bi-(furyl-2)	59, 60, 66
Bi-(furyl-2)methane	51, 55, 59, 60, 61, 65, 66
(Furyl-2)-(5-methylfuryl-2) methane	59, 60, 65, 66
Bi-(5-methylfuryl-2)methane	59, 60, 66
Benzofuran	60, 66
2,3-Dihydrobenzofuran	60, 66
2-Methylbenzofuran	60, 66
Methylfurfuryl ether	60, 65, 66
Difurfuryl ether	59, 60, 65, 66
5-Methylfurfuryl-furfuryl ether	59, 60, 66
2-Methyl-5-cyanofuran	60, 66
3-Hydroxy-2-methyl-4-oxo-1,4-pyran (maltol)	53, 55, 60, 61, 65, 66
5-Acetyl-2-methyloxazole	64, 65, 66
N, α-Dimethylbutanimide	64, 65, 66
Butyraldoxime	67

BAKED GOODS

Bread

The characteristic fragrance of freshly baked bread has intrigued investigators of natural aromas not only to identify the precursors but also to rebuild flavors capable of imparting this aroma to other food products and to bread itself (when the particular flavor is lacking for various reasons). By examining the nature of bread ingredients, it is assumed that the transformation process of the various flavor components is dependent on the manufacturing process that consists of two basic operations—fermentation and baking. Bread flavor after fermentation, which is quite different from that of freshly mixed dough, yields the characteristic freshly baked bread flavor on baking. Although bread loses a portion of its flavor on cooling, the flavor may be partially restored by reheating. This peculiar behavior is attributed to a "locking in" process (with subsequent liberation on reheating) of a certain portion of the volatile constituents by the linear fraction of wheat starch. The flavor of stale bread is believed to be caused by the oxidation of the aldehyde fraction present in fresh bread and the formation of large amounts of acids. During the fermentation of the dough, amino acids and sugars are formed together with alcohols, whose presence may also be detected organoleptically in the brew. By baking the leavened dough, Maillard's reaction between amino acids and sugars is triggered; this takes place preferentially in the bread crust.

Investigation of bread flavor has thus far identified the following components:

Components of Bread Flavor

Acids
- Acetic
- Benzilic
- Butyric
- Capric
- Caproic
- Formic
- Heptanoic
- Hydrocinnamic
- Isobutyric
- Isocaproic
- Itaconic
- Lactic
- Levulinic
- Nonanoic
- Octanoic
- Propionic
- Pyruvic
- Valeric

Alcohols
- *d*-Amyl alcohol
- Ethanol
- Isoamyl alcohol
- Isobutanol
- *n*-Propanol

Esters
- Ethyl acetate
- Ethyl hydrocinnamate
- Ethyl itaconate
- Ethyl lactate
- Ethyl levulinate
- Ethyl pyruvate
- Ethyl succinate

Aldehydes
- Acetaldehyde
- Benzaldehyde
- 2-Butanal
- Crotonaldehyde
- 2-Ethylhexanal
- Formaldehyde
- Furfural
- *n*-Hexanal
- Hydroxymethylfurfural
- Isobutanaldehyde
- Isovaleral
- 2-Methylbutanal
- Propionaldehyde
- Pyruvaldehyde
- *n*-Valeraldehyde

Ketones
- Acetoin
- Acetone
- Diacetyl
- Ethyl *n*-butyl
- Maltol
- Methyl ethyl
- Methyl ethyl *n*-butyl

Studies of the reconstruction of bread flavors have been partially oriented toward the formation of complex flavors by Maillard's reaction between certain amino acids and glucose. Several experimental data are reported by Colombo.[18] The addition of amino acids to the bread dough may enhance the bread flavor during baking. In our opinion, however, the esters and alcohols directly related to the fermentation process should also be considered in the reconstruction of the bread flavor.

Cookies

The gamut of food products belonging to this category is quite extensive and varied. These products have attained an importance almost equal to that of bread. Uniformity in size, manufacturing and packaging, minimum moisture content, extended shelf-life without loss of appetence (although dependent on the raw materials used, baking procedure, and adequate storage facilities) are the basic requirements that determine the widespread universal acceptance of cookies of various types and shapes.

The formulation, consistency of the dough (hard, soft, and liquid), and manufacturing process differ for various types of cookies. Although these differences are somewhat marginal, they are sufficient, together with the differences in shape, appearance, friability, and palatability, to differentiate the various products. The flavoring of almost all products is substantially indentical and based on the use of vanilla and butter. Amaretti (made with minced almonds and sugar) belongs to the category of tea pastry and includes croccanti biscuits (crunchy, made of almonds and sugar) whose flavoring is almost exclusively obtained by the use of vanilla and other specific ingredients such as bitter and sweet almond, custards, zabaglione, and chocolate. Savoiardi and Savoy biscuits (spongecake) belong, instead, to the category of pound cakes.

Wafers are cookies consisting of thin layers and filling. The wafer is prepared with water, flour, milk, sugar, and eggs; the filling consists of a pound custard or fat that requires flavoring. Although the types of flavor in the filling (and therefore in the wafers) are countless, three basic flavors are usually employed: vanilla, lemon, or chocolate. Sometimes chocolate is used for a partial or total frosting of the wafer. A salty type of wafer is also produced whose flavoring is more a gastronomic problem.

Frosted cakes, egg cookies, and sweets in general are not included in this work, since their flavoring is the specific domain of the baker rather than of the flavorist. The flavoring of vanilla and lemon custards is obtained using vanillin and lemon essential oil, respectively; chocolate custards are flavored by direct addition of cocoa or chocolate.

Pound Cakes

In this category are included all those bakery products obtained by mixing eggs and sugar until a complete, light, well aerated suspension is formed. Flour, leavening, molten fats, and flavors are added to the suspension by simple blending. The resulting mixed product is poured into molds and baked. The oven temperature is kept at approximately 160°C, and the "pounding" is carried out for about 25 minutes. Recently products have been introduced on the market useful for the industrial manufacture of pound cakes by cutting the pounding time to 4–5 minutes; the basic ingredients and the accessory ingredients are mixed in one step. The pound mix does not "break" readily and allows better flexibility for the mechanical pouring into molds. Well known products in this category include plum cakes, angel food, Savoy biscuits, and spongecake. For products containing fats, the fats are emulsified together with sugar, then the eggs are added, and finally the flour.

Pound cakes can be consumed as is and therefore require adequate flavoring during production. Vanillin and candied fruit (plum cakes) complemented by a vanilla-flavored sugar dusting may be employed for flavoring or may be used as a base for more elaborate flavoring. Finishing requires filling and decoration by the use of custards containing fats and of bagne sauces that are spread on the whole or the cut pound cake. Aside from the innumerable custard formulations containing milk, eggs, and butter, the most frequently used flavors are vanilla and lemon, obtained by the use of vanillin and soluble lemon essential oil, respectively. Whipped cream (vanilla-flavored), chocolate, and zabaglione may also be employed. *Bagne* (sauces) are alcoholic (about 16%) products of various flavors obtained by dilution of commercial liquors or bagne bases, which are high-proof liquors for pastry use. Pastry flavoring is quite varied, but the most frequently used flavors in the filling are rum, strega, curaçao, alkermes, and maraschino. Bagne-based formulations are typical of the corresponding liqueur imitations.

Leavened Doughs

Fruit cake, brioche, croissant, krapfen, and baba are products included in this category. The leavened doughs owe their name to the leavening effect of the fermenting flour dough itself or to that of barm and baking powders (dry yeasts) added to the dough. A perfect knowledge of the delicate natural leavens requiring particular attention during preparation and storage (the leaven must be freshened every 3–4 hours by addition of flour and tepid water to retain its strength) is extremely difficult. Dough preparation is strongly time-dependent and consequently dependent also on the temperature, moisture content, and other factors of both the atmosphere and the raw materials.

Freshly prepared yeast must be elastic and must be kept stored at 25–30°C under odorless and low-humidity conditions to prevent weakening and the development of a strong acid flavor. Yeast yields acetic and butyric acids instead of alcohols during fermentation. A good yeast has an alcohol-like odor. Sugar is added to weak yeasts (thus increasing the carbon dioxide content), whereas the addition of flour dilutes the activity of strong yeasts. A gluten-rich flour is required to prepare leavened doughs, since it must raise a mixture rich in eggs, fats, and sugar in addition to candied fruit. As salt stops the leavening process, its addition must be carefully controlled. Industrial yeasts, barm, and baking powders do not exhibit the above shortcomings but they yield results inferior to natural yeasts.

A typical example of leavened dough is fruit cake, such as panettone Milano. The ingredients are whole wheat flour, natural yeast, butter, sugar, whole eggs, egg yolk, raisins, citron, orange peel, candied fruit, water, salt, milk, and powdered dry malt. The most recent manufacturing process calls for the use of two doughs prepared separately. The first dough contains yeast, flour, sugar, and butter; it is allowed to rise and is then introduced into a mechanical dough mixer with the second dough, which consists of flour, sugar, butter, eggs, salt, raisins, candied fruit, malt, milk, and flavor ingredients. The flavor ingredients are added last and dosed according to the instructions for use of the compounded flavor. The mixture is allowed to rise further, and then the fruit cake is baked. Guidelines for the preparation of compound flavors suitable for panettone flavoring are shown below:

Panettone Flavor

Ingredient	Parts
Sweet orange essential oil	180
Lemon essential oil	160
Mandarin essential oil	160
Bergamot essential oil	30
Ethyl vanillin	80
Solvent	890
Total	1,500

Note: Sometimes vanilla is also added to the above compound flavor.

Similar manufacturing processes using somewhat different techniques and raw materials are employed for the preparation of *colomba* and *pandoro*. As in other baked goods of this type, compounded flavors are used. The classical note in all such compound flavors is citrus together with vanilla and butter notes. The ingredient ratios in the flavor vary from product to product yielding characteristic flavor effects. The citron note is predominant in panettone, and butter (cocoa butter) is predominant in pandoro; a slight lemon or orange note characterizes the colomba flavor. The following guidelines may be useful for the preparation of pandoro and colomba compound flavors:

Colomba Flavor

Ingredient	Parts
Sweet orange essential oil	180
Lemon essential oil	130
Mandarin essential oil	40
Ethyl vanillin	25
Solvent	125
Total	500

Pandoro Flavor

Ingredient	Parts
Colomba aroma	100
Cream flavor, 1 × 2,000	3
Cocoa distillate, 65%	22
Solvent	125
Total	250

Specific flavor ingredients added to the dough during the mixing are required for other products included in this category. The most interesting flavor combinations include lemon essential oil for *Bussola Bresciana*; vanilla for *Veneziana*; orange, clove, cinnamon, orris, and kirsch for *Focaccia Veneziana*; fennel for *Pan Dolce di Genova*; lemon, citron, or vanilla in flavored marsala wine for *Panettone* and *Focaccia Piemontese*; citron, cinnamon, clove, coriander, and nutmeg for *Pan Giallo alla Romana*; nutmeg, cardamom, vanilla, and lemon for *Dresden stollen*; and orange for *Provenzal Focaccia.*

The above flavor recipes are often prepared at home with the addition of ground spices and fresh citrus peels. Brioches, krapfens, and croissants usually are not flavored or only slightly flavored with panettone flavors. The baked babas are usually soaked in aromatic wines and frequently in rum and cognac.

Crackers

Crackers are similar in composition to cookies, except that they taste salty rather than sweet. In the last few decades crackers with their countless varieties have attained world-wide acceptance. Originally salted crackers and unsalted, unsweetened crackers were produced only as a replacement for bread. Subsequently, salted crackers, either naturally leavened or prepared with yeast, were established in the consumer market. Those leavened with yeast lend themselves to many flavor variations, among which are the well known salted cheese-flavored crackers (biscuits au Rochefort). The introduction of the more easily digested Anglo-Saxon salted crackers took over the consumer market; these products may be classified as follows:

1. *Soda crackers*: leavened, containing sodium bicarbonate and salt
2. *Salt crackers*: similar to soda crackers, but having a saltier flavor
3. *Cream crackers*: containing whipped or butter cream

Crackers are usually small in size, thin, and sometimes leavened with barm. English salt crackers differ from the U.S. soda-cracker counterpart in their larger salt content (almost twice as much). However, the basic raw materials are present in almost identical ratios. Hydrogenated fats and special shortenings are used to prepare the dough. Sometimes malt is added together with a thin fatty layer (caprate esters) spread on the cracker surface to impart an appealing shiny appearance.

In the last two decades there has been a growing market for new crackers to serve with aperitifs and cocktails. These products have surpassed the sales of soda crackers and are continuously presented to the consumer with new and original flavors, such as cheese, soybean, bacon, ham, tomato, onion, caraway, and roasted sesame seeds. Flavor ingredients plated on a salt carrier, spice oleoresins, and compounded fantasia flavors may find interesting applications in the preparation of the above products.

BREAD FLAVOR

Joseph A. Maga
Department of Food Science and Nutrition
Colorado State University
Fort Collins, Colorado

INTRODUCTION

Various forms of bread-like products have been consumed for thousands of centuries by the vast majority of the world popluation. Thus, the acceptability of bread products is well documented. One of the key factors in promoting this acceptability is the influential role of bread flavor and odor, especially of baking or freshly baked products. Through the centuries bread types have evolved mainly because of their unique flavor and odor and the successful production of most breads was considered to be an art. However, during this century bread production, by the necessity of the modern world, has become more of a science and technology related food production and distribution system. In the transition, scientists have attempted to unravel the complex nature of bread flavor and odor. Few scattered reports of these studies are available before 1955. However, during the following decade numerous reports appeared, all contributing a small piece to the puzzle. These studies were promoted by the introduction of controversial mechanical techniques into the baking industry in an effort to more rapidly and economically mass produce acceptable bread products.

Most all would agree that the pleasing aroma and flavor of freshly baked bread is quite shortlived; the flavor chemist has found himself attempting to duplicate these delicate and fleeting sensory properties, which has resulted in a better understanding of bread flavor. However, to date no imitation bread product has had the commercial success of other imitation foods, indicating that additional work in the bread flavor area is needed to more fully understand and appreciate total bread flavor.

One of the major problems in understanding bread flavor is that it may originate from a wide variety of factors in the bread making process, including such areas as ingredients, fermentation, and baking. Researchers have had to devote much time and effort to a portion of the entire bread flavor forming process, only to discover that their area of interest was responsible for but a fraction of total bread flavor. Such discoveries have discouraged some investigators from continuing research in the area.

The relative importance of bread flavor to the food scientist and technologist can best be appreciated by citing a portion of the vast number of reviews that have appeared dealing with various phases of bread flavor. These include the works of Coffman (1967a,b), Collyer (1964), DeFigueiredo (1964), Drapron (1970a,b; 1971), Jackel (1969), Jakubezyk and Haber (1968), Johnson (1967), Johnson et al. (1966), Pomeranz and Shellenberger (1971), Pence (1967), Otterbacher (1959) and Wiseblatt (1961a,b).

This review will attempt to cover the subject by acknowledging both American and European workers, and by covering all areas of bread flavor formation, information lacking in most previous reviews.

Specific areas to be covered include:

1. The flavor components reportedly isolated, and where reported; the quantities found and their possible flavor contribution based on available threshold data.
2. Methods of bread flavor isolation, including comments on the merits and limitations of the techniques employed; suggestions for improved techniques that could be applicable in bread flavor research.
3. Formation of bread flavor, including detailed discussion on the role of ingredients, fermentation, mechanical and enzymatic degradation, mixing procedures, and baking techniques.
4. Effects of compositional and processing factors on impaired bread flavor.
5. Methods of enhancing bread flavor, including the use of condensates, addition of flavoring materials and precursors, and the pretreatment of ingredients.

FLAVOR COMPONENTS

Introduction

The whole area of bread flavor is clouded by the fact that researchers have reported on the isolation of numerous volatile flavor compounds from various phases of bread systems, but no one has attempted to tie all of the pieces together. For example, numerous data exist on the presence of volatile flavor compounds found in brews of pre-ferments, bread dough, oven gases, and bread crust and crumb. A great majority of these compounds may only serve as other flavor precursors, or may not even be present in the final product. The volatiles emitted from a loaf of bread during baking may not have any direct bearing on the flavor properties of the same bread eaten three days later. The author does not mean to imply that studies of this nature are not valuable; they can shed light on volatile flavor compound formation resulting from the baking process, which may in turn assist the baker in producing a more flavorful loaf of bread. However, it would appear that at this stage in the study of bread flavor a research group would be able to follow bread flavor formation and destruction through the entire breadmaking process. Research efforts with this goal in mind are to be encouraged.

Another area of bread flavor investigation that requires interpretation with reservation is that few researchers have quantitated their findings. Thus, in reading the literature one has difficulty establishing the possible flavor or odor contribution of specific compounds. To date, over 100 volatile flavor compounds have been detected in bread and only one study (Mulders, 1973) has attempted interpreting the overall flavor role of certain compounds found in bread.

Two areas of bread flavor that have received little attention include the possible role of nonvolatile compounds and the flavor of toast. Perhaps separation problems associated with nonvolatile compounds have discouraged researchers in this area. In regard to toast, it would be a simple task to isolate and quantitate flavor compounds formed during toasting, including those lost to the atmosphere and those retained in the product.

Various classes of compounds have been detected and the occurrence of each individual class reported during the bread making process will be discussed.

Acids

As can be seen from Table 1, most of the acids reported in bread systems have been isolated in pre-ferments, while few investigators have detected acidic compounds from oven vapors. The value of investigating the composition of oven vapors may be questioned in that it becomes extremely difficult to establish the relative importance of compounds that are no longer in the product in question.

Due to the large number of reports involved, only several in each bread system area will be discussed. In the area of pre-ferments, one of the most detailed studies was that reported by Hunter et al. (1961). They employed programmed flash exchange gas-liquid chromatography and found at least 44 compounds. They concluded that these compounds included mono-, di-, and other types of carboxylic acids. Also, dicarboxylic acids appeared both as mono- and di-esters. Among the compounds identified were 17 acids ranging from methanolic (formic) to hexadecanoic (palmitic). The authors concluded that these acids could significantly contribute to bread flavor possibly in two ways. First, the higher-boiling fatty acids could hinder the volatility of lower-boiling compounds, thus increasing the time required for their evaporation from bread. Secondly, it was noted that the acid fraction found, when heated, gave an aroma similar to bread. However, no actual sensory or analytical data were provided to substantiate these claims. It might well be that the solvent used (dichloromethane) in the original organic acid extraction from the pre-ferment also extracted other volatile compounds that were detected in the separation and contributed to the characteristic flavor of the extract.

Another interesting study dealing with the keto acids associated with pre-ferments was reported by Cole et al. (1966). A total of eight keto acids were identified in pre-ferments. The major keto acids were pyruvic, α-ketoisovaleric, and α-ketoglutaric acid. The concentrations of these three acids as influenced by time of fermentation were measured and these data are summarized in Table 2.

Pyruvic acid was the most prevalent and was also found before fermentation. Its high level is probably due to a decarboxylase deficiency in the yeast (Suomalainen and Oura, 1959). In most cases maximum keto acid concentrations were reached in three to five hours.

TABLE 1

Acids Reported in Bread Systems

Acid	Pre-ferment	Oven vapor	White bread	Rye products
Methanoic	Hunter et al. (1961)	–	Mulders and Dhont (1972)	Markova (1970)
Ethanoic	Johnson (1925)	Baker et al. (1953)	Wiseblatt (1960b)	von Sydow and Anjou (1969)
	Johnson and Miller (1957)	Wiseblatt (1960a)	Wick et al. (1964)	Markova (1970)
	Johnson et al. (1958)	Wiseblatt (1960b)	Mulders and Dhont (1972)	
	Drews (1960)		Mulders et al. (1973)	
	Drews (1960)			
	Wiseblatt (1960b)			
	Hunter et al. (1961)			
Propanoic	Hunter et al. (1961)	–	Mulders and Dhont (1972)	–
2–Methylpropanoic	Hunter et al. (1961)	–	Mulders and Dhont (1972)	von Sydow and Anjou (1969)
2–Oxopropanoic	Cole et al. (1962)	–	Wiseblatt and Kohn (1960)	Markova (1970)
	Cole et al. (1966)			
Butanoic	Wiseblatt (1960b)	–	Wiseblatt (1960b)	von Sydow and Anjou (1969)
	Hunter et al. (1961)		Mulders and Dhont (1972)	
	Korolkova et al. (1968a,b)			
3–Methylbutanoic	Wiseblatt (1960b)	–	Wiseblatt (1960b)	von Sydow and Anjou (1969)
	Hunter et al. (1961)		Mulders and Dhont (1972)	
3–Oxobutanoic	–	Baker et al. (1953)	–	–
2–Butenoic	Hunter et al. (1961)	–	–	–
Pentanoic	Hunter et al. (1961)	–	Mulders and Dhont (1972)	von Sydow and Anjou (1969)
4–Methylpentanoic	Hunter et al. (1961)	–	–	–
4–Oxopentanoic	–	–	Wiseblatt and Kohn (1960)	–
Hexanoic	Wiseblatt (1960b)	–	Mulders and Dhont (1972)	von Sydow and Anjou (1969)
	Hunter et al. (1961)			
Heptanoic	Hunter et al. (1961)	–	–	–
Octanoic	Hunter et al. (1961)	–	Mulders and Dhont (1972)	von Sydow and Anjou (1969)
Nonanoic	Hunter et al. (1961)	–	–	–
Decanoic	Hunter et al. (1961)	–	Mulders and Dhont (1972)	von Sydow and Anjou (1969)
Dodecanoic	Hunter et al. (1961)	–	Mulders and Dhont (1972)	von Sydow and Anjou (1969)
Tetradecanoic	Hunter et al. (1961)	–	–	–
Hexadecanoic	Hunter et al. (1961)	–	–	–
Benzoic	–	–	Mulders and Dhont (1972)	–

TABLE 1 (Continued)

Acids Reported in Bread Systems

Acid	Pre-ferment	Oven vapor	White bread	Rye products
2–Hydroxypropanoic	Johnson and Miller (1957) Johnson et al. (1958) Drews (1960) Drews (1961) Korolkova et al. (1968a,b) Permilovskaya and Berzina (1973)	–	Mulders and Dhont (1972)	Drews (1960)
Tetrahydroxypentanoic	–	–	–	Markova (1970)
Hydroxyethanolic	–	–	–	Markova (1970)
2–Hydroxy–1,2,3–propanetricarboxylic	Drews (1960) Drews (1961) Kazanskaya and Bezruchenko (1963) Korolkova et al. (1968a) Permilovskaya and Berzina (1973)	–	–	Kazanskaya and Bezruchenko (1963) Markova (1970)
Ethanedioic	Korolkova et al. (1968b) Permilovskaya and Berzina (1973)	–	–	–
Propanedioic	Korolkova et al. (1968b)	–	–	–
Butanedioic	Kazanskaya and Bezruchenko (1963) Permilovskaya and Berzina (1973)	–	–	Kazanskaya and Bezruchenko (1963) Markova (1970)
Hydroxybutanedioic	Drews (1960) Drews (1961) Kazanskaya and Bezruchenko (1963) Korolkova et al. (1968a) Permilovskaya and Berzina (1973)	–	–	Kazanskaya and Bezruchenko (1963)

TABLE 1 (Continued)

Acids Reported in Bread Systems

Acid	Pre-ferment	Oven vapor	White bread	Rye products
Dihydroxybutanedioic	Kazanskaya and Bezruchenko (1963); Korolkova et al. (1968a); Korolkova et al. (1968b); Permilovskaya and Berzina (1973)	–	–	Kazanskaya and Bezruchenko (1963)
Cis–butenedioic	–	–	–	Markova (1970)
Trans–butenedioic	Permilovskaya and Berzina (1973)	–	–	Markova (1970)
Propanedioic	Korolkova et al. (1968b)	–	–	–
α–Ketoglutaric	Cole et al. (1962); Suomalainen and Ronkainen (1963); Cole et al. (1966)	–	–	–
α–Ketoisovaleric	Cole et al. (1966)	–	–	–
β–Hydroxypyruvic	Cole et al. (1966)	–	–	–
α–Keto–γ–methiol–butyric	Cole et al. (1966)	–	–	–
ρ–Hydroxyphenyl–pyruvic	Cole et al. (1966)	–	–	–
α–ketoisocaproic	Cole et al. (1966)	–	–	–
α–Keto–β–Methyl–valeric	Cole et al. (1966)	–	–	–

TABLE 2

Influence of Fermentation Time on Keto Acid Production in Pre-ferments

Time (hr)	Pyruvic	α–Ketoisovaleric	α–Ketoglutaric
0	0.11*	0	0
1	7.1	0.21	0.17
2	14.0	0.48	0.20
3	18.0	0.80	0.22
4	20.0	0.74	0.25
5	19.0	0.62	0.26
6	17.0	0.78	0.25

*Mmol/liter of pre-ferment.
From Cole et al. (1966).

TABLE 3

Amino Acids Formed by Hydrogenation of Keto Acids

Keto acid	Amino acid
Pyruvic	Alanine
α–Ketoglutaric	Glutamic acid
α–Ketoisovaleric	Valine
β–Hydroxypyruvic	Serine
α–Keto–γ–methiolbutyric	Methionine
ρ–Hydroxyphenylpyruvic	Tyrosine
α–Ketoisocaproic	Leucine
α–Keto–β–methylvaleric	Isoleucine

From Cole et al. (1966).

TABLE 4

Volatile Organic Acids in Dough and Bread

Acid	Dough	Bread
Acetic	198.0 mg/1,000 g flour	150.0 mg/1,000 g flour
Butyric	23.1	10.6
Isovaleric	13.1	6.1
Caproic	6.2	–

From Wiseblatt (1960b).

Another phase of their study dealt with the hydrogenation of the keto acid derivatives isolated from pre-ferments into their corresponding amino acids. This information is summarized in Table 3.

They reported that alanine, glutamic acid, and valine occurred in sizeable amounts. They postulated that although most of the keto acids were present in small quantities, they could contribute to the sensory properties of bread, especially in the case of sulfur-containing keto acid, α-keto-γ-methiolbutyric acid.

Various investigators have reported on the presence of 15 acids in baked bread. The most comprehensive study in this area was by Mulders and Dhont (1972). They identified 13 acids, 8 of which had not been previously reported. It should be noted that one of the acids they reported finding was benzoic acid, this is the only report of benzoic acid being a bread constituent. They postulated that it was probably derived from the bleaching agent benzoyl peroxide, which was added to the flour used in the study.

In an earlier study, Wiseblatt (1960b) used gas-liquid chromatography and found acetic, butyric, and isovaleric acids in bread. Attempts were made to quantitate these acids and these values are summarized in Table 4.

There was a reduction in concentration of organic acids during baking. Wiseblatt (1960b) postulated that these reductions were probably due to hydrolysis during baking, or possibly due to esterification. He also pointed out that the role of these acids in bread flavor varied with changes during storage of baked bread.

In the case of acids in rye products, two major reports have appeared. The findings of von Sydow and Anjou (1969) are summarized in Table 5.

Of the nine acids found, caproic acid was the most prevalent. The entire acid fraction represented approximately 13% of the total volatile flavor fraction recovered. They postulated that the aliphatic acids probably are important to the aroma of rye crispbread, but no supportive data were given.

Another interesting report associated with the acids in rye was authored by Markova (1970). A portion of these data is summarized in Table 6.

The greatest number of acids was found in the flour itself. The role of flour on bread flavor will be discussed in a later section. Three of the ten acids were found in the flour only. Generally, acid levels decreased during the baking of the dough.

TABLE 5

Aliphatic Acid Composition of Rye Crispbread

Compound	Percent*
Acetic acid	2.8
Isobutyric acid	2.1
Butyric acid	0.3
Isovaleric acid	–
Valeric acid	0.05
Caproic acid	3.4
Caprylic acid	1.4
Capric acid	1.0
Lauric acid	1.7

*Percent of flavor fraction isolated.
From von Sydow and Anjou (1969).

TABLE 6

Influence of Baking Procedures on Organic Acid Distribution in Rye, Flour, Dough, and Bread

Compound	Flour	Dough	Bread
Acetic	0.05*	0.36	0.37
Arabonic	0.04	0.03	0.09
Lactic	0.10	1.22	0.76
Glycolic	0.02	0.14	0.05
Formic	0.10	0.11	0.15
Succinic	0.02	–	0.16
Malic	0.08	–	–
Citric	0.05	–	–
Fumaric	0.04	–	–
Pyruvic	0.01	0.18	0.03
Total:	0.51	2.04	1.61

*Percent dry matter.
From Markova (1970).

TABLE 7

Amino Acid Precursors in Alcohol Production

Amino acid	Resulting alcohol
Tyrosine	Tyrosol
Tyrptophan	Tryptophol
Phenylalanine	β–Phenylethanol
Valine	2–Methyl–1–propanol
Leucine	1–Pentanol
Alanine	Ethanol

From Otterbacher (1959).

Alcohols

The fact that ethanol and carbon dioxide are the primary end products of bread fermentation is well established. It is also known that alcohols can result from amino acid fermentation. Typical examples of such transformations are summarized in Table 7.

As can be seen, an alcohol containing one less carbon results, along with ammonia which, in turn, is utilized by the yeast. Alcohols such as 2-methyl-1-propanol and 1-pentanol, as well as those possessing an aromatic ring structure, have characteristic odors and tastes which in turn can contribute to the aroma and flavor of bread. A total of 19 alcohol derivatives have been reported in various bread systems. These are summarized in Table 8.

Among the 27 compounds they detected, Smith and Coffman (1960) found seven alcohols in bread pre-ferments. The most predominant were ethanol and two isomeric forms of 2,3-butanediol. The two butanediols amounted to approximately 85% of the extractable neutral fraction. Another uncommon alcohol that they found was 2-phenylethanol.

Coffman (1966) detected four alcohols in oven vapors. These were 1-propanol, 2-methyl-1-propanol, 1-pentanol and β-phenylethanol in the approximate ratios of 1:9:30:2, respectively.

In the case of bread, the group headed by Mulders has reported on the occurrence of 15 alcohols (Mulders et al., 1972; Mulders, 1973; Mulders et al., 1973). To date they are the only group reporting on the occurrence of ten such compounds in bread. The report of von Sydow and Anjou (1969) has dominated concerning the alcohols found in rye products. They have reported on the presence of nine such compounds, including the compound eugenol (4-allylguaiacol). A portion of their data is shown in Table 9.

TABLE 8

Alcohols Reported in Bread Systems

Alcohol	Pre-ferments	Oven vapor	White bread	Rye products
Ethanol	Smith and Coffman (1960) Miller et al. (1961) Cole et al. (1962) Permilovskaya and Berzina (1973)	Baker et al. (1953) Wiseblatt (1960a)	Ocker (1961) Wiseblatt and Kohn (1960) Ocker (1964) Wick et al. (1964) Mulders et al. (1972) Rothe et al. (1972) Mulders (1973)	Rothe (1971) Rothe et al. (1972)
1–Propanol	Smith and Coffman (1960)	Coffman (1966)	Wick et al. (1964) Mulders et al. (1972) Mulders (1973) Mulders et al. (1973)	–
2–Propanol	–	–	Wiseblatt and Kohn (1960)	–
2–Methyl–1–propanol	Smith and Coffman (1960) Miller et al. (1961)	Coffman (1966)	Wick et al. (1964) Mulders et al. (1972) Mulders (1973) Mulders et al. (1973)	von Sydow and Anjou (1969)
1–Butanol	–	–	Mulders et al. (1972)	–
2–Methyl–1–butanol	–	–	Wick et al. (1964) Mulders et al. (1973)	
3–Methyl–1–butanol	Smith and Coffman (1960) Miller et al. (1961)	–	Wick et al. (1964) Mulders et al. (1972) Mulders (1973) Mulders et al. (1973)	von Sydow and Anjou (1969)
1–Pentanol	–	Baker et al. (1953) Coffman (1966)	Mulders et al. (1972) Mulders et al. (1973)	von Sydow and Anjou (1969)
2–Pentanol	–	–	Mulders et al. (1973)	–
3–Pentanol	–	–	Mulders et al. (1973)	–
1–Hexanol	–	–	Mulders et al. (1972) Mulders et al. (1973)	von Sydow and Anjou (1969)

TABLE 8 (Continued)

Alcohols Reported in Bread Systems

Alcohol	Pre-ferments	Oven vapor	White bread	Rye products
1–Octanol	–	–	–	von Sydow and Anjou (1969)
2,3–Butanediol	Smith and Coffman (1960)	–	Mulders et al. (1973)	–
1–Penten–3–ol	–	–	–	von Sydow and Anjou (1969)
Benzyl alcohol	–	–	Mulders et al. (1973)	von Sydow and Anjou (1969)
Furfuryl alcohol	–	–	Mulders et al. (1972) Mulders et al. (1973)	–
2–Phenylethanol	Smith and Coffman (1960)	Coffman (1966)	Mulders et al. (1973)	von Sydow and Anjou (1969)
1,3–Propanediol monoacetate	Smith and Coffman (1960)	–	–	–
4–Allylguaiacol	–	–	–	von Sydow and Anjou (1969)

TABLE 9

Alcohol Distribution in Rye Crispbread

Alcohol	Percent*
2–Methyl–1–propanol	0.9
3–Methyl–1–butanol	2.9
1–Pentanol	0.1
1–Hexanol	0.3
1–Octanol	0.02
1–Penten–3–ol	0.2

*Percent of total flavor fraction evaluated.
From von Sydow and Anjou (1969).

In relation to the total flavor fraction, the alcoholic fraction represented 4.4% of the total, with 3-methyl-1-butanol being the most prevalent alcohol found. It should be noted that the shorter chain alcohols (ethanol-butanol) present in white bread were not detected in the rye product, probably due to the volatility of these compounds.

Aldehydes

This class of compounds contains the largest number (33) of flavor compounds reported in various bread systems. The specific compounds are summarized in Table 10.

Major emphasis has been placed by the bread flavor chemist on aldehydes and ketones as being primarily responsible for bread flavor. This is probably due to several reasons. First, they have been consistently found in bread systems by numerous investigators. Second, the characteristic sensory properties are generally both potent and pleasing. Third, their presence, formation, and decomposition can be explained chemically as arising from a variety of reactions that take place during the bread making process. For example, they can arise from amino acids, as can be seen in Table 11.

Almost all amino acids, except perhaps proline, are quite reactive in the formation of aldehydes. Formation mechanisms will be discussed in more detail in a later section. At this stage it is sufficient to say that aldehydes are probably the best understood class of compounds associated with bread flavor.

A total of 11 aldehydes, with acetaldehyde being the most commonly reported, have been found in pre-ferments, thus suggesting the influence of fermentation on their formation. As would be expected, the vast majority (27) of the compounds outlined in Table 10 have been found in bread itself. The most commonly reported aldehydes in bread are in the formaldehyde to heptanal area and include saturated, as well as unsaturated, compounds.

Aldehydes have also been assumed to be one of the key flavor ingredients in rye products; a total of 27 aldehydes have been isolated in this area. In rye crispbread von Sydow and Anjou (1969) reported that *trans,trans*-2,4-decadienal was the main aldehyde (Table 12).

The authors also mentioned the influential role that fat can play in aldehyde formation since most of the aldehydes they found in rye crispbread can be considered characteristic of fat oxidation products. For example, they postulated that hexanal and 2,4-decadienal could have originated from linoleic acid and 2,4-heptadienal from linolenic acid.

Esters

One must assume that the primary pathway for ester formation in bread systems is through the interaction of alcohols and acids. Since ethanol is the predominant alcohol found in a bread system, one would also assume that most of the esters would be of the ethyl variety. Likewise, commonly occurring acids such as formic and acetic would be logical choices for ester formation. The esters reported in various bread systems are summarized in Table 13.

In pre-ferments ethyl acetate is the most common ester. A total of seven esters have been found in bread; two of these are furfuryl esters (Mulders et al., 1973), which are derived from furfuryl alcohol. For the most part only relatively simple short chain acid-based esters have been found in bread.

TABLE 10

Aldehydes Reported in Bread Systems

Aldehyde	Pre-ferments	Oven vapor	White bread	Rye products
Formaldehyde	Miller et al. (1961) Linko et al. (1962a)	–	Ng et al. (1960) Lindo et al. (1962b) Hrdlicka et al. (1963) Linko and Johnson (1963) Wick et al. (1964) Davidek et al. (1965) Rothe (1966) El-Dash and Johnson (1967) Salem et al. (1967) Lorenz and Maga (1972) Mulders and Dhont (1972)	Hampl et al. (1964) Hrdlicka and Honischova (1967)
Acetaldehyde	Smith and Coffman (1960) Kohn et al. (1961) Miller et al. (1961) Linko et al. (1962a) Borovikova and Roiter (1971a)	Baker et al. (1953) Wiseblatt (1960a)	Ng et al. (1960) Ocker (1961) Wiseblatt and Kohn (1960) Linko et al. (1962b) Hrdlicka et al. (1963) Linko and Johnson (1963) Rothe (1963a) Rothe (1963b) Wick et al. (1964) Ocker (1964) Rothe (1966) El-Dash and Johnson (1967) Salem et al. (1967) Korolkova and Rakhmankulova (1968) Johnson and El-Dash (1969) Markianova and Roiter (1969) El-Dash and Johnson (1970) Borovikova and Roiter (1971a,b,c) Borovikova et al. (1971) Markianova et al. (1971) Lorenz and Maga (1972) Mulders and Dhont (1972) Mulders et al. (1972) Roiter and Borovikova (1972) Rothe et al. (1972) Mulders (1973)	Rothe (1960) Hampl et al. (1964) Rothe (1966) Tokareva and Smirnova (1966) Hrdlicka and Honischova (1967) Rothe and Boehme (1967) Rothe (1971) Rothe et al. (1972)

TABLE 10 (Continued)

Aldehydes Reported in Bread Systems

Aldehyde	Pre-ferments	Oven vapor	White bread	Rye products
Propanal	Linko et al. (1962a) Borovikova and Roiter (1971a)	–	Linko et al. (1962b) Hrdlicka et al. (1963) Linko and Johnson (1963) Wick et al. (1964) El-Dash and Johnson (1967) Korolkova and Rakhmankulova (1968) Johnson and El-Dash (1969) Markianova and Roiter (1969) El-Dash and Johnson (1970) Borovikova and Roiter (1971a,b,c) Borovikova et al. (1971) Markianova et al. (1971) Lorenz and Maga (1972) Roiter and Borovikova (1972)	Hampl et al. (1964) Bratovanova (1965) Tokareva and Smirnova (1966) Hrdlicka and Honischova (1967)
2–Methylpropanal	Miller et al. (1961) Linko et al. (1962a) Borovikova and Roiter (1971a)	Ng et al. (1960)	Linko et al. (1962b) Hrdlicka et al. (1963) Linko and Johnson (1963) Rothe (1963a) Wick et al. (1964) El-Dash and Johnson (1967) Salem et al. (1967) Korolkova and Rakhmankulova (1968) Johnson and El-Dash (1969) Markianova and Roiter (1969) El-Dash and Johnson (1970) Borovikova and Roiter (1971a) Borovikova et al. (1971) Markianova et al. (1971) Mulders and Dhont (1972) Mulders et al. (1972) Roiter and Borovikova (1972) Rothe et al. (1972) Mulders (1973)	Rothe (1960) Hampl et al. (1964) Rothe (1966) Tokareva and Smirnova (1966) Hrdlicka and Honischova (1967) Rothe and Boehme (1967) von Sydow and Anjou (1969) Rothe (1971) Rothe et al. (1972)

TABLE 10 (Continued)

Aldehydes Reported in Bread Systems

Aldehyde	Pre-ferments	Oven vapor	White bread	Rye products
2–Oxopropanal	Kohn et al. (1961) Borovikova and Roiter (1971a) Kohn et al. (1961) Miller et al. (1961)	Baker et al. (1953)	Wiseblatt and Kohn (1960) Hrdlicka et al. (1963) Rothe (1963a) Borovikova and Roiter (1971a) Mulders and Dhont (1972) Roiter and Borovikova (1972) Rothe et al. (1972)	Hampl et al. (1964) Rothe (1966) Hrdlicka and Honischova (1967) Rothe and Boehme (1967) Rothe et al. (1972)
2–Propenal	–	–	Borovikova and Roiter (1971a) Borovikova et al. (1971) Markianova et al. (1971) Mulders and Dhont (1972)	Hampl et al. (1964) Hrdlicka and Honischova (1967)
2–Phenylpropenal	–	–	Korolkova and Rakhmankulova (1968)	–
Butanal	Kohn et al. (1961) Miller et al. (1961) Borovikova and Roiter (1971a)	–	Wick et al. (1964) El-Dash and Johnson (1967) Johnson and El-Dash (1969) El-Dash and Johnson (1970) Borovikova and Roiter (1971b) Borovikova et al. (1971) Markianova et al. (1971) Lorenz and Maga (1972) Mulders and Dhont (1972) Roiter and Borovikova (1972)	Hrdlicka and Honischova (1967)
2–Methylbutanal	Miller et al. (1961) Linko et al. (1962a)	–	Ng et al. (1960) Linko et al. (1962b) Hrdlicka et al. (1963) Linko and Johnson (1963) Rothe (1963a) Rothe (1963b) Rothe (1966) El-Dash and Johnson (1967) Korolkova and Rakhmankulova (1968) Johnson and El-Dash (1969)	Rothe (1960) Hampl et al. (1964) Rothe (1966) Tokareva and Smirnova (1966) Hrdlicka and Honischova (1967) Rothe and Boehme (1967) von Sydow and Anjou (1969) Rothe et al. (1972)

TABLE 10 (Continued)

Aldehydes Reported in Bread Systems

Aldehyde	Pre-ferments	Oven vapor	White bread	Rye products
2-Methylbutanal (continued)			El-Dash and Johnson (1970) Rothe et al. (1972) Mulders et al. (1973)	
3–Methylbutanal	Kohn et al. (1961) Miller et al. (1961) Linko et al. (1962a) Borovikova and Roiter (1971a,b,c)	–	Linko et al. (1962b) Hrdlicka et al. (1963) Linko and Johnson (1963) Wick et al. (1964) El-Dash and Johnson (1967) Salem et al. (1967) Johnson and El-Dash (1969) Markianova and Roiter (1969) El-Dash and Johnson (1970) Borovikova and Roiter (1971a,b,c) Borovikova et al. (1971) Markianova et al. (1971) Mulders and Dhont (1972) Roiter and Borovikova (1972) Rothe et al. (1972)	Rothe (1960) Hampl et al. (1964) Hrdlicka and Honischova (1967) von Sydow and Anjou (1969)
2–Butenal	–	–	Wiseblatt and Kohn (1960) Wick et al. (1964)	Hrdlicka and Honischova (1967)
Pentanal	Miller et al. (1961) Linko et al. (1962a) Borovikova and Roiter (1971a)	Ng et al. (1960)	Linko et al. (1962b) Linko and Johnson (1963) Wick et al. (1964) El-Dash and Johnson (1967) Johnson and El-Dash (1969) El-Dash and Johnson (1970) Borovikova and Roiter (1971a,b,c) Borovikova et al. (1971) Markianova et al. (1971) Mulders and Dhont (1972) Roiter and Borovikova (1972)	Hrdlicka and Honischova (1967) von Sydow and Anjou (1969)
2–Methylpentanal	–	–	El-Dash and Johnson (1967) Johnson and El-Dash (1969) El-Dash and Johnson (1970)	Hrdlicka and Honischova (1967)

TABLE 10 (Continued)

Aldehydes Reported in Bread Systems

Aldehyde	Pre-ferments	Oven vapor	White bread	Rye products
Hexanal	Kohn et al. (1961) Linko et al. (1962a) Borovikova and Roiter (1971a)	–	Ng et al. (1960) Linko et al. (1962b) Linko and Johnson (1963) El-Dash and Johnson (1967) Johnson and El-Dash (1969) El-Dash and Johnson (1970) Borovikova and Roiter (1971b) Borovikova et al. (1971) Markianova et al. (1971) Lorenz and Maga (1972) Mulders and Dhont (1972) Mulders et al. (1972) Rother and Borovikova (1972) Mulders et al. (1973)	Hrdlicka and Honischova (1967) von Sydow and Anjou (1969)
2–Ethylhexanal	–	–	Wiseblatt and Kohn (1960) Wick et al. (1964) El-Dash and Johnson (1967)	–
Heptanal	–	–	El-Dash and Johnson (1967) Korolkova and Rakhmankulova (1968) Lorenz and Maga (1972) Mulders and Dhont (1972)	von Sydow and Anjou (1969)
Octanal	–	–	Mulders and Dhont (1972)	Hrdlicka and Honischova (1967)
2–Octenal	–	–	–	Von Sydow and Anjou (1969)
Nonanal	–	–	Lorenz and Maga (1972) Mulders and Dhont (1972)	Hrdlicka and Honischova (1967)
2–Nonenal	–	–	Mulders and Dhont (1972)	von Sydow and Anjou (1969)
Dodecanal	–	–	–	Hrdlicka and Honischova (1967)
2,4–Heptadienal	–	–	–	von Sydow and Anjou (1969)
Trans, cis–2,4–decadienal	–	–	Mulders et al. (1973)	von Sydow and Anjou (1969)
Trans, trans–2,4–decadienal	–	–	–	von Sydow and Anjou (1969)

TABLE 10 (Continued)

Aldehydes Reported in Bread Systems

Aldehyde	Pre-ferments	Oven vapor	White bread	Rye products
Benzaldehyde	Kohn et al. (1961) Borovikova and Roiter (1971a)	–	Borovikova and Roiter (1971b) Borovikova et al. (1971) Markianova et al. (1971) Mulders et al. (1972) Mulders et al. (1973)	von Sydow and Anjou (1969)
4–Hydroxybenzaldehyde	–	–	Mulders and Dhont (1972)	–
Phenylacetaldehyde	Borovikova and Roiter (1971a)	–	Borovikova and Roiter (1971a) Borovikova et al. (1971) Markianova et al. (1971) Mulders and Dhont (1972)	Rothe (1960) Hampl et al. (1964) Bratovanova (1965) Hrdlicka and Honischova (1967) von Sydow and Anjou (1969)
2–Furaldehyde	Borovikova and Roiter (1971a)	Baker et al. (1953)	Ng et al. (1960) Wiseblatt and Kohn (1960) Hrdlicka et al. (1963) Linko and Johnson (1963) Rothe (1963a) Wick et al. (1964) Salem et al. (1967) Korolkova and Rakhmankulova (1968) Markianova and Roiter (1969) Borovikova and Roiter (1971c) Borovikova et al. (1971) Markianova et al. (1971) Mulders and Dhont (1972) Mulders et al. (1972) Rothe et al. (1972) Mulders et al. (1973)	Rothe (1960) Hampl et al. (1964) Bratovanova (1965) Rothe (1966) Tokareva and Smirnova (1966) Hrdlicka and Honischova (1967) Rothe and Boehme (1967) Rothe (1971) Rothe et al. (1972)
3–Furaldehyde	–	–	Mulders et al. (1973)	–
5–Methyl–2–furaldehyde	–	–	Mulders and Dhont (1972) Mulders et al. (1973)	–

TABLE 10 (Continued)

Aldehydes Reported in Bread Systems

Aldehyde	Pre-ferments	Oven vapor	White bread	Rye products
5–Hydroxymethyl–2–furaldehyde	–	–	Kretovich and Takareva (1950) Hrdlicka et al. (1963) Linko and Johnson (1963) Salem et al. (1967) Daniels (1971) Mulders and Dhont (1972) Rothe et al. (1972)	Kretovich and Takareva (1950) Rothe (1960) Hampl et al. (1964) Hrdlicka and Honischova (1967) Rothe (1971) Rothe et al. (1972)
2–Furan–dialdehyde	–	Coffman (1966)	–	–
4–Hydroxy–3–methoxy–benzaldehyde	–	–	–	von Sydow and Anjou (1969)

TABLE 11

Influence of Amino Acid Additions on the Formation of Various Aldehydes in Bread Crust

Amino acid added	Formaldehyde	Acetaldehyde	2-Methylpropanal	2-Methylbutanal
Control	0.49*	1.06	1.10	1.33
Glycine	1.88	1.73	1.56	1.41
Alanine	0.90	3.44	3.26	1.27
Valine	1.30	2.38	8.57	2.61
Leucine	0.51	1.18	1.22	2.78
Isoleucine	0.91	1.90	2.27	3.94
Glutamic acid	1.12	2.03	1.83	1.68
Histidine	1.43	3.59	2.72	2.62
Lysine	1.74	3.75	3.68	3.33
Phenylalanine	0.61	1.67	1.22	1.68
Proline	0.56	0.78	1.13	1.08
Arginine	0.36	1.31	0.97	1.96
Methionine	0.49	1.43	1.43	1.22

*Mg/100 g
From Salem et al. (1967).

TABLE 12

Aldehyde Distribution in Rye Crispbread

Aldehyde	Percent*
2-Methylpropanal	–
2-Methylbutanal	0.2
3-Methylbutanal	0.2
Pentanal	0.2
Hexanal	0.2
Heptanal	0.05
2-Octenal	0.05
2-Nonenal	0.09
2,4-Heptadienal	0.09
trans,cis-2,4-Decadienal	0.07
trans,trans-2,4-Decadienal	0.3

*Percent of total flavor fraction evaluated.
From von Sydow and Anjou (1969).

Von Sydow and Anjou (1969) found predominantly ethyl esters in rye crispbread. A notable exception was the occurrence of two β-phenylethanol alcohol-based esters. A total of ten esters have been reported for rye products. Von Sydow and Anjou (1969) also reported on the occurrence of a series of long chain acid-based esters (C:10-C:16), as can be seen in Table 14.

The main ester they found was ethyl palmitate; the total ester fraction amounted to approximately 3.5% of the total flavor fraction. This group felt that none of the esters they reported play an important role in rye crispbread flavor. However, in the case of white bread, one may speculate that since many simple esters have potent and pleasing aromas, they could influence bread odor and flavor.

Ether Derivatives

Only one ether derivative has been reported to be present in white bread. This compound is ethyl furfuryl ether and was found by Mulders et al. (1972, 1973). Difurfuryl ether has been reported by von Sydow and Anjou (1969) to be present in rye crispbread.

TABLE 13

Esters Reported in Bread Systems

Ester	Pre-ferments	White bread	Rye products
Ethyl formate	Smith and Coffman (1960)	Mulders et al. (1972) Mulders et al. (1973)	–
Ethyl acetate	Johnson and Miller (1957) Johnson et al. (1958) Miller et al. (1961) Wick et al. (1964)	Mulders et al. (1972) Mulders (1973)	Tokareva and Smirnova (1966)
Ethyl lactate	Smith and Coffman (1960)	–	–
Ethyl pyruvate	–	Ng et al. (1960) Wiseblatt and Kohn (1960)	– –
Ethyl levulinate	–	Wiseblatt and Kohn (1960)	–
1,3-Propanediol monoacetate	Smith and Coffman (1960)	–	–
Furfuryl formate	–	Mulders et al. (1973)	–
Furfuryl acetate	–	Mulders et al. (1973)	–
Acetonyl acetate	–	Mulders et al. (1973)	–
Ethyl caprinate	–	–	von Sydow and Anjou (1969)
Ethyl laurate	–	–	von Sydow and Anjou (1969)
Ethyl myristinate	–	–	von Sydow and Anjou (1969)
Ethylpentadecanoate	–	–	von Sydow and Anjou (1969)
Ethyl palmitate	–	–	von Sydow and Anjou (1969)
Glycol diacetate	–	–	von Sydow and Anjou (1969)
Ethyl phenylacetate	–	–	von Sydow and Anjou (1969)
β-Phenylethyl formate	–	–	von Sydow and Anjou (1969)
β-Phenylethyl acetate	–	–	von Sydow and Anjou (1969)
Amyl bensoate	–	–	von Sydow and Anjou (1969)

TABLE 14

Esters Associated with Rye Crispbread

Ester	Percent*
Ethyl caprinate	0.2
Ethyl laurate	0.1
Ethyl myristinate	0.09
Ethyl pentadecanoate	0.02
Ethyl palmitate	2.7
Glycol diacetate	0.2
Ethyl phenylacetate	0.03
β-Phenylethyl formate	0.09
β-Phenylethyl acetate	0.1
Amylbensoate	–

*Percent of total flavor fraction evaluated.
From von Sydow and Anjou (1969).

Furan Derivatives

It is generally agreed that most furan derivatives are probably thermally produced through carbohydrate breakdown. This can best be appreciated by viewing the data summarized in Table 15.

In the case of the crumb portion of bread, no 2-furaldehyde was found and the 5-hydroxymethyl-2-furaldehyde level remained essentially constant. However, in the crust 2-furaldehyde was detected, and the level of 5-hydroxymethyl-2-furaldehyde increased proportionally with increased sucrose content. These two compounds are also known to be reactive intermediates in the Maillard reaction (Hodge, 1953) and as such can undergo condensation with free amino acids. The influence of amino acids and also carbohydrate sources on 2-furaldehyde and 5-hydroxymethyl-2-furaldehyde (HMF) production was evaluated by Salem et al. (1967). Their findings are summarized in Table 16.

The ability of amino acids to react with 2-furaldehyde and HMF varied, but essentially all amino acids except methionine reduced the levels of these two compounds. The most reactive amino acids were leucine

TABLE 15

Influence of Sucrose Concentration on 2-Furaldehyde and 5-Hydroxymethyl-2-furaldehyde (HMF) Levels in Bread

	Crust		Crumb	
Percent sucrose	2-Furaldehyde	HMF	2-Furaldehyde	HMF
4	0.325*	4.05	–	0.675
5	0.338	6.65	–	0.723
6	0.350	8.31	–	0.614
7	0.386	8.37	–	0.677
8	0.435	9.09	–	0.638

*Mg/100 g
From Linko et al. (1962b).

TABLE 16

Influence of Added Amino Acids and Carbohydrate Sources on 2-Furaldehyde and 5-Hydroxymethyl-2-Furaldehyde (HMF) Levels in Bread Crust

	Glucose		Xylose	
Amino acid	2-Furaldehyde	HMF	2-Furaldehyde	HMF
Control	0.08	1.11	0.45	1.05
Glycine	0.03	0.83	0.12	0.95
Alanine	0.10	0.80	0.06	0.99
Valine	0.02	0.95	0.08	1.09
Leucine	0.02	0.47	0.06	0.60
Isoleucine	0.01	0.46	0.10	0.64
Glutamic Acid	0.04	1.05	0.09	0.92
Histidine	0.04	1.09	0.07	0.66
Lysine	0.02	1.04	0.08	0.76
Phenylalanine	0.06	1.03	0.14	0.76
Proline	0.04	0.91	0.12	0.66
Arginine	0.15	0.79	ND	ND
Methionine	0.11	1.65	ND	ND

*Mg/100 g
ND: not determined.
From Salem et al. (1967).

and isoleucine. Apparently carbohydrate source was not an important factor since little difference between glucose and xylose was noted.

A complete listing of furan derivatives reported in bread systems can be found in Table 17. To date few reports have identified such derivatives in any unheated bread system. Two reports concern the presence of furan derivatives in oven vapors; Baker et al. (1953) found 2-furaldehyde and Coffman (1966) reported the presence of 2,5-furan dialdehyde.

Of the 24 furan derivatives reported in bread systems, 17 have been found in bread itself, with the most commonly reported ones being 2-furaldehyde and 5-hydroxymethyl-2-furaldehyde. Numerous furan derivatives have been reported by Mulders and co-workers (Mulders and Dhont, 1972; Mulders et al., 1972; Mulders, 1973; Mulders et al., 1973) and include the parent compound furan. An interesting compound found by Mulders et al. (1973) was 2,5-dimethyl-2(2H)-furanone. This compound was earlier synthesized by Eugster et al. (1961), who claimed it possessed a strong odor of freshly baked bread. However, Mulders et al. (1973) did not find that this compound was typical of bread aroma. Perhaps concentration was a factor, or possibly an artifact was present in the original synthesis to explain this discrepancy. In addition, Mulders et al. (1973) reported that the above compound, and also 3-furaldehyde, were, to their knowledge, for the first time reported in the aroma of a food product.

With rye products von Sydow and Anjou (1969) were the first to report 14 furan derivatives in rye crispbread (see Table 18). The most predominant compound was furfuryl alcohol, which amounted to over 15% of the total volatile fraction.

Hydrocarbons

One report has appeared (Mulders et al. 1973) about the occurrence of the hydrocarbons toluene and limonene in white bread crust. Gas-liquid chromatography and mass spectral data were used to secure identification. However, no information about the relative or actual amounts detected, nor about the possible formation pathways for these compounds, was discussed. It should be noted that a seven-hour extraction using pentane and diethyl ether was used; there may be a possibility that the reported hydrocarbons were contaminants associated with the pentanes used. It is assumed that freshly redistilled solvents were employed, but this was not specifically indicated in the study.

Ketones

Maillard-type browning is the primary source of ketones, especially in bread crust. This is dramatized in the work reported by Salem et al. (1967), as shown in Table 19.

Essentially all amino acids except arginine were found to promote the production of 2-propanone (acetone) in bread crust. As can be seen, alanine and valine were highly reactive in the formation of acetone.

A total of 24 ketones, as outlined in Table 20, has been reported in various bread systems. Since a number of ketones have been found in pre-ferments, it is apparent that fermentation can also result in their formation. The most common ketones found in pre-ferments include acetone and methyl ethyl ketone (2-butanone). Baker et al. (1953) have also reported finding these two ketones along with 2,3-butanedione (diacetyl) in oven vapors.

Most of the ketones reported have been found in bread with almost every report dealing with ketones finding acetone to be present. Other commonly reported ketones in bread include methyl ethyl ketone, diacetyl and acetoin (3-hydroxy-2-butanone).

In rye products, acetone and diacetyl are the predominant ketones to be reported. In rye crispbread acetone was absent and acetoin represented the most abundant ketone (see Table 21).

Lactone Derivatives

Several groups have reported on the presence of lactone derivatives in bread products. In 1960 Smith and Coffman isolated butyrolactone from a bread pre-ferment system. This was followed by a report by Coffman (1966) who detected butyrolactone in oven gases. A report by von Sydow and Anjou (1969) identified five lactones associated with the aroma of rye crispbread. The compounds and relative amounts found are reported in Table 22.

TABLE 17

Furan Derivatives Reported in Bread Systems

Furan derivatives	Oven vapor	White bread	Rye products
2-Furaldehyde	Baker et al. (1953)	Ng et al. (1960) Wiseblatt and Kohn (1960) Linko et al. (1962b) Linko and Johnson (1963) Wick et al. (1964) Salem et al. (1967) El-Dash and Johnson (1970) Mulders and Dhont (1972) Mulders et al. (1972) Mulders et al. (1973)	Tokareva and Smirnova (1966) von Sydow and Anjou (1969)
3-Furaldehyde	–	Mulders et al. (1973)	–
5-Methyl-2-furaldehyde	–	Mulders and Dhont (1972) Mulders et al. (1973)	von Sydow and Anjou (1969)
5-Hydroxymethyl-2-furaldehyde	–	Kretovich and Tokareva (1950) Linko et al. (1962b) Linko and Johnson (1963) Salem et al. (1967) El-Dash and Johnson (1970) Daniels (1971) Mulders and Dhont (1972)	Kretovich and Tokareva (1950)
Furan	–	Mulders et al. (1972) Mulders (1973)	–
2-Methylfuran	–	Mulders et al. (1972) Mulders (1973)	–
2-Acetylfuran	–	Mulders et al. (1973)	von Sydow and Anjou (1969)
2-Acetyl-5-methylfuran	–	–	von Sydow and Anjou (1969)
2,5-Furan dialdehyde	Coffman (1966)	–	
2-Pentylfuran	–	Mulders et al. (1973)	von Sydow and Anjou (1969)
2-Propionylfuran	–	–	von Sydow and Anjou (1969)
1-(2-Furyl)-2-propanone	–	Mulders et al. (1973)	von Sydow and Anjou (1969)
1-(2-Furyl)-1,2-propanedione	–	Mulders et al. (1973)	–

TABLE 17 (Continued)

Furan Derivatives Reported in Bread Systems

Furan derivatives	Oven vapor	White bread	Rye products
1-(2-Furyl)-1,2-butanedione	–	–	von Sydow and Anjou (1969)
1-(5-Methyl-2-furyl)-1,2-propanedione	–	–	von Sydow and Anjou (1969)
Dihydro-2-methyl-3-(2H)-furanone	–	Mulders et al. (1973)	–
2,5-Dimethyl-3-(2H)-furanone	–	Mulders et al. (1973)	–
Ethyl furfuryl ether	–	Mulders et al. (1972) Mulders et al. (1973)	–
Difurfuryl ether	–	–	von Sydow and Anjou (1969)
Furfuryl alcohol	–	Mulders et al. (1972) Mulders et al. (1973)	von Sydow and Anjou (1969)
Furfuryl formate	–	Mulders et al. (1973)	von Sydow and Anjou (1969)
Furfuryl acetate	–	Mulders et al. (1973)	von Sydow and Anjou (1969)
1-Furfuryl-2-formylpyrrole	–	–	von Sydow and Anjou (1969)
1-Furfurylpyrrole	–	Mulders et al. (1973)	–

TABLE 18

Furan Derivatives Reported in Rye Crispbread

Furan derivative	Percent*
2-Pentylfuran	0.04
2-Furaldehyde	1.1
5-Methyl-2-furaldehyde	0.2
2-Acetylfuran	0.6
2-Acetyl-5- methylfuran	0.02
2-Propionylfuran	0.03
Furfuryl alcohol	15.2
Difurfuryl ether	–
Furfuryl formate	0.02
Furfuryl acetate	0.03
1-(2-Furyl)-1,2-propanedione	0.02
1-(2-Furyl)-1,2-butanedione	0.03
1-(5-Methyl-2-furyl)-1,2-propanedione	0.07
1-Furfuryl-2-formylpyrrole	0.07

*Percent of total flavor fraction evaluated.
From von Sydow and Anjou (1969).

TABLE 19

Influence of Amino Acid Additions on the Formation of 2-Propanone in Bread Crust

Amino acid added (0.02M)	Mg/100 g fresh crust weight 2-propanone found
Control	1.40
Glycine	3.62
Alanine	6.60
Valine	6.25
Leucine	2.10
Isoleucine	3.18
Glutamic acid	2.88
Histidine	5.40
Lysine	4.73
Phenylalanine	3.33
Proline	2.20
Arginine	1.44
Methionine	1.98

From Salem et al. (1967).

Even though the congregate amount of lactones found was approximately 1.5% of the total flavor fraction evaluated, the authors felt none of these compounds played a decisive role in bread flavor.

Two additional lactone derivatives were reported by Mulders et al. (1973) as present in white bread crust. The authors stated that they were not aware of any other report identifying 4-hydroxy-3-pentanoic acid lactone (β-angelica lactone) as being isolated from a food system. No amounts or possible flavor contribution of lactone derivatives were discussed in this report.

The lactones isolated in various bread systems are summarized in Table 23.

It should be noted that certain lactones have potent and characteristically unique flavor properties, and the occurrence of certain lactones in bread systems would be indicative that others may be present. In the future, the complete role of lactones in bread flavor may be elucidated.

Pyrazines

Pyrazines are heterocyclic nitrogen-containing compounds possessing unique flavor properties; they have been reported present in a wide variety of food products (Maga and Sizer, 1973a,b). To date, three investigative groups have reported on the presence of pyrazine derivatives in bread systems.

A report appeared in 1969 (von Sydow and Anjou) dealing with the aroma of rye crispbread. One of the aroma fractions investigated included the pyrazines, which were approximately 4.5% in volume of the total aroma mixture. The pyrazines found and their relative amounts in rye crispbread are summarized in Table 24.

Methylpyrazine was the most prevalent pyrazine derivative detected. In addition to the twelve compounds identified, the authors reported six related unidentified compounds, representing an additional 0.3%. Based on the characteristic odor of these compounds, the authors suggested that pyrazine derivatives were important to bread aroma. It was suggested that these products were the result of non-enzymatic sugar-amino acid reactions. However, it may be possible that certain pyrazines may actually be present in some of the original formulation materials, such as the flour or the dry skim milk.

Mulders et al. (1973) reported on the identification of 52 compounds isolated from the crust of white bread baked from flour, salt, yeast, and water only. Included in the compounds identified were methyl-, ethyl-, 2,3-dimethyl- and 2,5-dimethylpyrazine. Neither concentrations nor their possible flavor contribution was discussed. As mentioned previously, milk products may serve as a pyrazine source. Since in this study no milk product was used, the pyrazines were either naturally present or were formed during baking.

TABLE 20

Ketones Reported in Bread Systems

Ketone	Pre-ferments	Oven vapor	White bread	Rye products
2-Propanone	Smith and Coffman (1960) Kohn et al. (1961) Miller et al. (1961) Linko et al. (1962a) Linko et al. (1962b) Borovikova and Roiter (1971a)	Baker et al. (1953)	Ng et al. (1960) Ocker (1961) Wiseblatt and Kohn (1960) Hrdlicka et al. (1963) Linko and Johnson (1963) Rothe (1963a) Wick et al. (1964) Rothe (1966) El-Dash and Johnson (1967) Salem et al. (1967) Johnson and El-Dash (1969) Markianova and Roiter (1969) El-Dash and Johnson (1970) Borovikova and Roiter (1971a,b,c) Borovikova et al. (1971) Markianova et al. (1971) Lorenz and Maga (1972) Mulders and Dhont (1972) Roiter and Borovikova (1972) Rothe et al. (1972) Mulders (1973)	Hampl et al. (1964) Hrdlicka and Honischova (1967) Rothe et al. (1972)
2-Butanone	Miller et al. (1961) Linko et al. (1962a) Linko et al. (1962b) Borovikova and Roiter (1971a)	Baker et al. (1953)	Ng et al. (1960) Linko and Johnson (1963) Wick et al. (1964) El-Dash and Johnson (1967) Johnson and El-Dash (1969) El-Dash and Johnson (1970) Borovikova and Roiter (1971a,b,c) Borovikova et al. (1971) Markianova et al. (1971) Lorenz and Maga (1972) Mulders and Dhont (1972) Roiter and Borovikova (1972)	–

TABLE 20 (Continued)

Ketones Reported in Bread Systems

Ketone	Pre-ferments	Oven vapor	White bread	Rye products
3-Hydroxy-2-butanone	Visser't Hooft and deLeeuw (1935) Smith and Coffman (1960)	–	Visser't Hooft and deLeeuw (1935) Hrdlicka et al. (1963) Rothe (1963a) Rothe (1963b) Wick et al. (1964) Rothe et al. (1972) Mulders et al. (1973)	von Sydow and Anjou (1969) Rothe et al. (1972)
2,3-Butanedione	Smith and Coffman (1960) Borovikova and Roiter (1971a)	Baker et al. (1953)	Wiseblatt and Kohn (1960) Rothe (1963a) Wick et al. (1964) Johnson and El-Dash (1969) Borovikova and Roiter (1971a,b,c) Borovikova et al. (1971) Markianova et al. (1971) Mulders and Dhont (1972) Mulders et al. (1972) Roiter and Borovikova (1972) Rothe et al. (1972) Mulders (1973)	Hampl et al. (1964) Rothe (1966) Hrdlicka and Honischova (1967) Rothe et al. (1972)
2-Pentanone	–	–	Wick et al. (1964) Mulders et al. (1973)	–
3-Penten-2-one	Borovikova and Roiter (1971a)	–	Wiseblatt and Kohn (1960) Borovikova and Roiter (1971c) Mulders et al. (1972) Mulders et al. (1973)	–
4-Methyl-3-penten-2-one	–	–	–	von Sydow and Anjou (1969)
2-Cyclopenten-1-one	–	–	Mulders et al. (1973)	–
2,3-Pentanedione	–	–	Mulders et al. (1973)	–
2-Hexanone	Kohn et al. (1961)	–	Wick et al. (1964) El-Dash and Johnson (1967) Lorenz and Maga (1972) Mulders and Dhont (1972)	Hrdlicka and Honischova (1967)

TABLE 20 (Continued)

Ketones Reported in Bread Systems

Ketone	Pre-ferments	Oven vapor	White bread	Rye products
3-Hexanone	–	–	Borovikova and Roiter (1971b,c) Borovikova et al. (1971)	–
2-Heptanone	–	–	Lorenz and Maga (1972) Mulders and Dhont (1972) Mulders et al. (1973)	Hrdlicka and Honischova (1967)
3-Heptanone	–	–	Wiseblatt and Kohn (1960)	–
3,4-Heptanone	–	–	El-Dash and Johnson (1967)	–
2-Octanone	–	–	–	von Sydow and Anjou (1969)
6,10,14,-Trimethyl-2-pentadecanone	–	–	–	von Sydow and Anjou (1969)
2-Heptadecanone	–	–	–	von Sydow and Anjou (1969)
Dihydro-2-methyl-3-(2H)-furanone	–	–	Mulders et al. (1973)	–
2,5-Dimethyl-3-(2H)-furanone	–	–	Mulders et at. (1973)	–
2-Acetylfuran	–	–	Mulders et al. (1973)	–
1-(2-Furyl)-2-propanone	–	–	Mulders et al. (1973)	–
1-(2-Furyl)-1,2-propanedione	–	–	Mulders et al. (1973)	–
3-Hydroxy-2-methyl-4-pyrone	–	–	Mulders et al. (1973)	–
2-Methyl-3-oxolanone	–	–	–	von Sydow and Anjou (1969)

TABLE 21

Ketone Distribution in Rye Crispbread

Ketone	Percent*
3-Hydroxy-2-butanone	1.6
2-Octanone	0.04
2-Heptadecanone	0.08
6,10,14-Trimethyl-2-pentadecanone	0.2
4-Methyl-3-penten-2-one	0.08

*Percent of total flavor fraction evaluated.
From von Sydow and Anjou (1969).

TABLE 22

Lactone Derivatives Found in Rye Crispbread

Compound	Common name	Percent*
4-Hydroxybutanoic acid lactone	Butyrolactone	0.3
4-Hydroxypentanoic acid lactone	Valerolactone	0.1
4-Hydroxyhexanoic acid lactone	δ-Caprolactone	0.2
4-Hydroxynonanoic acid lactone	δ-Nonalactone	0.5
Dihydroactinidiolide	–	0.4
	Total:	1.5

*Percent of all flavor fractions evaluated.
From von Sydow and Anjou (1969).

TABLE 23

Lactone Derivatives Isolated from Bread Systems

Lactone derivative	Pre-ferments	Oven vapor	White Bread	Rye products
4-Hydroxybutanoic acid lactone	Smith and Coffman (1960)	Coffman (1966)	Mulders el al. (1973)	von Sydow and Anjou (1969)
4-Hydroxy-2-butenoic acid lactone	–	–	Mulders et al. (1973)	–
4-Hydroxypentanoic acid lactone	–	–	–	von Sydow and Anjou (1969)
4-Hydroxy-3-pentenoic acid lactone	–	–	Mulders et al. (1973)	–
4-Hydroxyhexanoic acid lactone	–	–	Mulders et al. (1973)	von Sydow and Anjou (1969)
4-Hydroxynonanoic acid lactone	–	–	–	von Sydow and Anjou (1969)
Dihydroactinidiolide	–	–	–	von Sydow and Anjou (1969)

TABLE 24

Pyrazine Derivatives Found in Rye Crispbread

Compound	Percent*
Pyrazine	0.01
Methylpyrazine	1.4
2,3-Dimethylpyrazine	0.5
2,5-Dimethylpyrazine	0.4
Ethylpyrazine	0.9
Trimethylpyrazine	0.6
2-Ethyl-3-methylpyrazine	0.2
2-Ethyl-6-methylpyrazine	0.2
3-Ethyl-2,5-dimethylpyrazine	0.3
Propylpyrazine	0.03
Vinylpyrazine	0.03
2-Vinyl-6-methylpyrazine	0.08
Total:	4.6

*Percent of all flavor fractions evaluated.
From von Sydow and Anjou (1969).

TABLE 25

Influence of Baking Time on Pyrazine Concentrations (ppm) in Bread Crust

Compound	Min. baking time at 218 °C				
	10	15	20	25	30
Methylpyrazine	23	7.6	6.0	3.1	2.3
Ethylpyrazine	0.20	0.46	0.21	0.35	0.45
2,3-Dimethylpyrazine	–	–	0.03	0.04	0.06
2-Ethyl-3-methylpyrazine	0.09	0.30	0.29	1.0	1.2
2-Methyl-6-propylpyrazine	0.30	0.33	0.26	0.24	0.23

From Sizer et al. (1973).

Sizer et al. (1973) reported on the influence of baking time on pyrazine concentration in the crust of white bread. A portion of their data is summarized in Table 25.

This group found methylpyrazine the main pyrazine derivative in white bread crust, just as von Sydow and Anjou (1969) found it most prevalent in rye crispbread. Baking time did influence pyrazine concentration. Compounds such as methylpyrazine dramatically decreased with increased baking time, while others, such as 2-ethyl-3-methylpyrazine, increased with baking time. The decrease of certain pyrazine compounds can be explained in several manners. Volatility may reduce the concentration of certain pyrazines with baking time; certain pyrazines may actually be destroyed, or they may serve as precursors for other pyrazine compounds. Based on the amounts found as compared to their odor thresholds reported in the literature, the authors concluded that pyrazines can significantly contribute to the sensory properties of bread.

Whether certain pyrazine compounds occur naturally in grains and flours is not known. With current pyrazine isolation and quantitation techniques, it would be a relatively easy task to answer this question.

The role of fermentation on pyrazine formation in bread has not been investigated. In other foods, such as cocoa beans and Natto, fermentation results in pyrazine compound formation.

A list summarizing the pyrazine compounds found to date in bread products can be found in Table 26.

TABLE 26

Pyrazine Compounds Isolated from Bread Products

Pyrazine	White bread	Rye Products
Unsubstituted pyrazine	–	von Sydow and Anjou (1969)
2-Methylpyrazine	Mulders et al. (1973) Sizer et al. (1973)	von Sydow and Anjou (1969)
2,3-Dimethylpyrazine	Mulders et al. (1973) Sizer et al. (1973)	von Sydow and Anjou (1969)
2,5-Dimethylpyrazine	Mulders et al. (1973)	von Sydow and Anjou (1969)
2-Ethylpyrazine	Mulders et al. (1973) Sizer et al. (1973)	von Sydow and Anjou (1969)
Trimethylpyrazine	–	von Sydow and Anjou (1969)
2-Ethyl-3-methylpyrazine	Sizer et al. (1973)	von Sydow and Anjou (1969)
2-Ethyl-6-methylpyrazine	–	von Sydow and Anjou (1969)
Propylpyrazine	–	von Sydow and Anjou (1969)
2-Methyl-6-propylpyrazine	Sizer et al. (1973)	–
3-Ethyl-2,5-dimethylpyrazine	–	von Sydow and Anjou (1969)
Vinylpyrazine	–	von Sydow and Anjou (1969)
2-Vinyl-6-methylpyrazine	–	von Sydow and Anjou (1969)

TABLE 27

Pyrrole Derivatives Found in Rye Crispbread

Compound	Percent*
Pyrrole	0.1
Acetylpyrrole	1.9
1-Methyl-2-formylpyrrole	0.05
1-Methyl-2-acetylpyrrole	0.07
5-Methyl-2-formylpyrrole	0.1
1-Furfuryl-2-formylpyrrole	0.07
Indole	0.2
Total:	2.5

*Percent of all flavor fractions evaluated.
From von Sydow and Anjou (1969).

Pyrrole Derivatives

A total of 7 pyrrole derivatives was reported by von Sydow and Anjou (1969) in rye crispbread. The compounds and the relative quantities found in the above study are summarized in Table 27.

Acetylpyrrole represented 80% of the total pyrrole fraction. It was suggested by the authors that due to their unique odors, pyrroles could be important to bread aroma. It was also postulated that they were derived from thermal reactions during baking involving non-enzymatic browning.

An additional three compounds were reported by Mulders et al. (1973) in white bread crust. This group also postulated that pyrroles are derived by non-enzymatic browning reactions or possibly by Maillard reactions. Mulders et al. (1973) also pointed out that other investigators have reported certain pyrrole derivatives possess cracker or bread-like odors. Such observations were reported by Kobayasi and Fujimaki (1965) for l-acetonylpyrroline; 2-acetyl-1-methylpyrrolidine and l-azabicyclo (3.3.0) octan-4-one by Hunter et al. (1966); and 2-acetyl-1,4,5,6-tetrahydropyridine by Hunter et al. (1969). However, none of these specific compounds were found in the current study (Mulders et al. 1973). Mulders and Dhont (1972) also reported the presence of the compound 3-hydroxy-2-methyl-4-pyrone in bread.

A summary of pyrrole derivatives found in bread products can be found in Table 28.

TABLE 28

Pyrrole Derivatives Isolated from Bread Products

Pyrrole derivative	White Bread	Rye Products
Pyrrole	Mulders et al. (1973)	von Sydow and Anjou (1969)
1-Methylpyrrole	Mulders et al. (1973)	–
2-Formylpyrrole	Mulders et al. (1973)	–
2-Acetylpyrrole	Mulders et al. (1973)	von Sydow and Anjou (1969)
1-Furfurylpyrrole	Mulders et al. (1973)	–
1-Methyl-2-formylpyrrole	–	von Sydow and Anjou (1969)
1-Methyl-2-acetylpyrrole	–	von Sydow and Anjou (1969)
5-Methyl-2-formylpyrrole	–	von Sydow and Anjou (1969)
1-Furfuryl-2-formylpyrrole	–	von Sydow and Anjou (1969)
Indole	–	von Sydow and Anjou (1969)
3-Hydroxy-2-methyl-4-pyrone	Mulders and Dhont (1972)	–

TABLE 29

Sulfur Compounds Isolated from Bread Products

Sulfur compound	White bread	Rye products
Dimethyl sulfide	Mulders et al. (1972) Mulders (1973)	–
Dimethyl disulfide	Mulders et al. (1972) Mulders (1973)	von Sydow and Anjou (1969)
3-Acetylthiophene	Mulders (1973)	–
2,3,4-Trithiapentane	–	von Sydow and Anjou (1969)
2-Formylthiophene	–	von Sydow and Anjou (1969)
Methane thiol	Ocker and Rotsch (1959)	Rotsch and Dorner (1957) Rotsch and Dorner (1958) Rothe and Thomas (1959)
Hydrogen sulfide	–	Rotsch and Dorner (1957) Rotsch and Dorner (1958)

Sulfur Compounds

In 1959 Ocker and Rotsch reported on the identification of methane thiol in white bread. The sulfur containing compounds dimethyl sulfide and dimethyl disulfide were detected in the vapors above white bread three hours after it was baked (Mulders et al., 1972). A nitrogen stripping technique was used to isolate the volatiles. In a later study, Mulders (1973) reported identifying the same two sulfur compounds using a different volatile vapor isolation technique with white bread. However, the quantities found in this later study were too small to even be calculated on the basis of relative amounts present. Another sulfur-based compound was identified by Mulders et al. (1973) as 3-acetylthiophene. It was found in a pentane-diethyl ether extract of bread crust. By viewing the published GLC scan, it is apparent that the relative proportion of 3-acetylthiophene detected was insignificant in quality. No postulations were made regarding the source of this latter sulfur-based compound.

With regard to rye bread, methane thiol has been reported present by Rotsch and Dorner (1957, 1958) and Rothe and Thomas (1959). Rotsch and Dorner (1957, 1958) also detected hydrogen sulfide in rye bread. Several sulfur-based compounds have been identified (von Sydow and Anjou, 1969) in rye crispbread that had been steam-distilled and ether-extracted. The sulfur compounds included dimethyl disulfide, 2,3,4-trithiapentane and 2-formylthiophene. The three compounds had a relative concentration to the other volatiles isolated of 0.16%, with 2-formylthiophene representing half this amount. Again, no formation pathways for these compounds were discussed, although one may assume that they originate from the thermal decomposition and rearrangement of sulfur-bearing amino acids. As can be seen in Table 29, sulfur compounds have only been reported in finished bread products.

TABLE 30

Miscellaneous Compounds Reported in Bread Products

Miscellaneous compounds	White bread	Rye products
Maltol	Bache (1910)	–
Isomaltol	Bache (1910)	–
1,1-Diethoxyethane	Mulders et al. (1972) Mulders (1973)	
Tetramethylbenzene	–	von Sydow and Anjou (1969)
Methylnaphthalene	–	von Sydow and Anjou (1969)
Dimethylnaphthalene	–	von Sydow and Anjou (1969)
Diphenylacetylene	–	von Sydow and Anjou (1969)
Phenol	–	von Sydow and Anjou (1969)
Cresol	–	von Sydow and Anjou (1969)
Ethylphenol	–	von Sydow and Anjou (1969)
2-Methoxyphenol	–	von Sydow and Anjou (1969)
4-Vinyl-2-methoxyphenol	–	von Sydow and Anjou (1969)
O-Hydroxydiphenyl	–	von Sydow and Anjou (1969)

Miscellaneous Compounds

Table 30 represents a compilation of reported volatile compounds that do not easily fit into any of the chemical classes just discussed.

The majority of these compounds are associated with rye crispbread. Von Sydow and Anjou (1969) cautioned that the first four compounds listed in Table 30 may possibly be artifacts, but they did not go into further detail. The compound 4-vinyl-2-methoxyphenol comprised 8.2% of the total volatile fraction in rye crispbread.

In conclusion, the complexity of bread flavor can be better appreciated by the fact that over thirteen chemical classes, comprised of over 100 compounds, have been reported present in various bread systems. It should also be noted that extensive and reliable quantitative data are painfully absent, which makes understanding bread flavor more difficult.

For the convenience of the reader all of the compounds that have been reported present in bread are summarized by compound class in Table 31.

FLAVOR ISOLATION TECHNIQUES

Introduction

If a major criticism has to be made of the isolation techniques used by bread flavor investigators, it is that few techniques were used by most investigative groups. Although some of the techniques had severe limitations, which were acknowledged by most groups, each succeeding investigator continued to use similar techniques. Admittedly, no one technique is adequate to extract the entire bread flavor spectrum, but the bread flavor chemists could have taken heed of the isolation techniques used by other food flavor chemists.

The results of numerous bread flavor investigations began to appear in the early 1960's; for the following decade similar procedures were used by all groups. However, recently innovative methods used by two groups in particular have resulted in the publication of more detailed and useful information about bread flavor than was produced by the previous reports combined.

This section will briefly outline the techniques used by choosing a representative report for each technique and pointing out some of the merits and limitations involved.

Volatile Reducing Substances

In 1949 Farber published a procedure to measure volatile organic substances. Baker et al. (1953) used this procedure for determining the volatile substances in bread crumb. Air at a flow rate of 1 liter per min for 10 min was passed through 2 g of bread crumb. The exit gases were passed through two test tubes in a

TABLE 31

Summary of Volatile Compounds Reported in Bread

Acids
Methanoic
Ethanoic
Propanoic
2–Methylpropanoic
2–Hydroxypropanoic
2–Oxopropanoic
Butanoic
3–Methylbutanoic
Pentanoic
4–Oxopentanoic
Hexanoic
Octanoic
Decanoic
Dodecanoic
Benzoic
*Tetrahydroxypentanoic
*Hydroxyethanolic
*2–Hydroxy–1,2,3–propantricarboxylic
*Butanedioic

Alcohols
Ethanol
1–Propanol
2–Propanol
2–Methyl–1–propanol
1–Butanol
2–Methyl–1–butanol
3–Methyl–1–butanol
1–Pentanol
2–Pentanol
3–Pentanol
1–Hexanol
*1–Octanol
2,3–Butanediol
*1–Penten–3–ol
Benzyl alcohol
Furfuryl alcohol
2–Phenylethanol
*4–Allylguaiacol

Aldehydes
Formaldehyde
Acetaldehyde
Propanal
2–Methylpropanal
2–Oxopropanal
2–Propenal
2–Phenylpropenal
Butanal
2–Methylbutanal
3–Methylbutanal
2–Butenal
Pentanal
2–Methylpentanal
Hexanal
2–Ethylhexanal
Heptanal
Octanal
*2–Octenal
Nonanal
2–Nonenal
*Dodecanal
*2,4–Heptadienal
Trans,cis–2,4–decadienal
**Trans,trans*–2,4–decadienal
Benzaldehyde
4–Hydroxybenzaldehyde
Phenylacetaldehyde
2–Furaldehyde
3–Furaldehyde
5–Methyl–2–furaldehyde
5–Hydroxymethyl–2–furaldehyde
2–Furan dialdehyde
4–Hydroxy–3–methoxybenzaldehyde

Esters
Ethyl formate
Ethyl acetate
Ethyl pyruvate
Ethyl levulinate
Furfuryl formate
Furfuryl acetate
Acetonyl acetate
*Ethyl caprinate
*Ethyl laurate
*Ethyl myristinate
*Ethyl pentadecanoate
*Ethyl palmitate
*Glycol diacetate
*Ethyl phenylacetate
*β–phenylethyl formate
*β–phenylethyl acetate
*Amyl bensoate

Ethers
Ethyl furfuryl ether
*Difurfuryl ether

Furan derivatives
2–Furaldehyde
3–Furaldehyde
5–Methyl–2–furaldehyde
5–Hydroxymethyl–2–furaldehyde
Furan
2–Methylfuran
2–Acetylfuran
*2–Acetyl–5–methylfuran
2–Pentylfuran
*2–Propionylfuran
1–(2–Furyl)–2–propanone
1–(2–Furyl)–1,2–propanedione
*1–(2–Furyl)–1,2–butanedione
*1–(5–Methyl–2–furyl)–1,2–propanedione
Dihydro–2–methyl–3–(2H)–furanone
2,5–Dimethyl–3–(2H)–furanone
Ethyl furfuryl ether
*Difurfuryl ether

TABLE 31 (Continued)

Summary of Volatile Compounds Reported in Bread

Furfuryl alcohol
Furfuryl formate
Furfuryl acetate
*1–Furfuryl–2–formylpyrrole
1–Furfurylpyrrole

Hydrocarbons
Limonene
Toluene

Ketones
2–Propanone
2–Butanone
3–Hydroxy–2–butanone
2,3–Butanedione
2–Pentanone
3–Penten–2–one
*4–Methyl–3–penten–2–one
2–Cyclopenten–1–one
2,3–Pentanedione
2–Hexanone
3–Hexanone
2–Heptanone
3–Heptanone
3,4–Heptanone
*2–Octanone
*6,10,14–Trimethyl–2–pentadecanone
*2–Heptadecanone
Dihydro–2–methyl–3–(2H)–furanone
2,5–Dimethyl–3–(2H)–furanone
2–Acetylfuran
1–(2–Furyl)–2–propanone
1–2(–Furtyl)–1,2–propanedione
3–Hydroxy–2–methyl–4–pyrone
*2–Methyl–3–oxolanone

Lactone derivatives
4–Hydroxybutanoic acid lactone
4–Hydroxy–2–butenoic acid lactone
*4–Hydroxypentanoic acid lactone
4–Hydroxy–3–pentenoic acid lactone
4–Hydroxyhexanoic acid lactone
*4–Hydroxynonanoic acid lactone
*Dihydroactinidiolide

Pyrazines
*Pyrazine
2–Methylpyrazine
2,3–Dimethylpyrazine
2,5–Dimethylpyrazine
2–Ethylpyrazine
*Trimethylpyrazine
2–Ethyl–3–Methylpyrazine
*2–Ethyl–6–methylpyrazine
*Propylpyrazine
2–Methyl–6–propylpyrazine
*3–Ethyl–2,5–dimethylpyrazine
*Vinylpyrazine
*2–Vinyl–6–methylpyrazine

Pyrrole derivatives
Pyrrole
1–Methylpyrrole
2–Formylpyrrole
2–Acetylpyrrole
1–Furfurylpyrrole
*1–Methyl–2–formylpyrrole
*1–Methyl–2–acetylpyrrole
*5–Methyl–2–formylpyrrole
*1–Furfuryl–2–formylpyrrole
*Indole
3–Hydroxy–2–methyl–4–pyrone

Sulfur compounds
*Hydrogen sulfide
Dimethyl sulfide
Dimethyl disulfide
Methane thiol
3–Acetylthiophene
2,3,4–Trithiapentane
2–Formylthiophene

Miscellaneous compounds
Maltol
Isomaltol
1,1–Diethoxyethane
*Tetramethylbenzene
*Methylnaphtalene
*Dimethylnaphtalene
*Diphenylacetylene
*Phenol
*Ethylphenol
*2–Methoxyphenol
*4–Vinyl–2–methoxyphenol
*O–Hydroxydiphenyl
*Cresol

*Only in rye products.

TABLE 32

Influence of Baking Conditions on Measured Volatile Reducing Substances in Bread

Type of bake	Volatile reducing substances
Normal bake	420*
Overbake	365
Underbake	460
Quick bake	440
Slow bake	390

*Micro-oxidation equivalents.
From Baker et al. (1953).

TABLE 33

Influence of Post-baking Conditions on Measured Volatile Reducing Substances in Bread

Treatment	Volatile reducing substances
15 min after baking	440*
60 min after baking	420
Vacuum cooled, 7 min	310
Exposed to air, 1 day	13
Wrapped, 1 day	411
Wrapped, 3 days	315
Wrapped, 9 days	103
Wrapped and frozen, 7 days	414

*Micro-oxidation equivalents.
From Baker et al. (1953).

series, each containing 10 ml of 0.1N potassium permanganate in 1.0N sodium hydroxide. The amount of reduced permanganate was then measured iodimetrically and converted to micro-oxidation equivalents. Typical data obtained by this method are shown in Table 32.

If one looks quickly at these data, they appear logical; one could reason that overbaking will drive off more volatiles and the overbaked product has a lower value. However, if this were the case, the data presented comparing quick and slow baking would be difficult to explain. One might assume that prolonged baking increases bread flavor and the baking method is not a true indicator of residual bread flavor. The authors did concede that the method primarily measured ethanol. Another limitation of the method is the size (2 g) of crumb sample used; such a small sample could be influenced by its location within the loaf.

Even with these severe limitations, additional data presented by Baker et al. (1953) using the method in relation to post-baking conditions do appear logical (Table 33) in that with more adverse storage conditions, the amount of volatile substances decreased.

The only limitation not explained with the second set of data is, that upon cooling and storage, there is a migration of volatile compounds inward. Thus, sample location would be critical.

A somewhat similar procedure was proposed by Tokareva and Kretovich (1961), which they recommended as an index of bread quality. It involved extracting the sample with 0.1% solution of sodium bicarbonate, filtering, and adding an excess of sodium bicarbonate titrated with 0.1N and 0.01N iodine solution. The bound bisulfite was released with a standard solution of sodium carbonate, then titrated with 0.1N iodine. The bisulfite binding ability was expressed as ml of 0.01N iodine per 100 g of bread.

Oven Vapor Condensation

As discussed elsewhere in this review, the major limitation of this technique is that compounds no longer an active part of the baked product are being measured. Admittedly, the aroma of baking bread is quite pleasing and can be considered an important part of its sensory properties. However, in today's world the aroma of baking bread mass-produced at central points has little effect on consumer acceptance.

A typical example as outlined by Baker et al. (1953) will be discussed. The oven vent was connected by a glass tube to a series of Dry Ice® traps attached in turn to an aspirator. As baking proceeded, the gases formed condensed in the traps, and the condensates combined to be evaluated later.

DNPH Derivatives

Various groups have isolated carbonyl compounds through the formation of their 2,4-dinitrophenylhydrazone (DNPH) derivatives. For example, when working with pre-ferments, Linko et al. (1962a) prepared a 1% solution of 2,4-dinitrophenylhydrazine in 5N sulfuric acid. Then the fermentation gases evolved from the pre-ferment system were passed through the DNPH reagent; in the process, hydrazones of the volatile carbonyl compounds were formed. The resulting hydrazones were then extracted with several aliquots of chloroform and saved for future analysis.

This method is relatively specific for carbonyl compounds, and it can serve as a means of separating a complex mixture of volatile flavor compounds. However, since the reaction takes place in an acid state, the possibility exists that other compounds are being degraded, which can result in the formation of additional carbonyls.

Steam Distillation

A typical example of this method was reported by von Sydow and Anjou (1969). A representative sample of the product was suspended in water and steam distilled for 3.5 hr at atmospheric pressure. The severity of this treatment may be questioned because if any reactive materials were present in the baked product, they were given ample opportunity to form additional, or increased amounts, of compounds which are to be identified later. The distillate was then saturated with ammonium sulfate and extracted with three portions of distilled diethyl ether. Water present in the extract was removed by drying over sodium sulfate, and the solvent was removed by evaporation through Vigreux columns until a volume of 1.5 ml remained. The sample was further concentrated, by an unspecified procedure, to a volume of 0.1 ml and stored at -20°C until evaluated.

Vacuum steam distillation, with appropriate precautions for trapping excessively volatile compounds, or vacuum solvent distillation are recommended instead of atmospheric steam distillation. These methods would minimize the possibility of artifact formation.

Direct Solvent Extraction

In a portion of their study on pre-ferments Miller et al. (1961) used a direct solvent extraction method. Essentially, this involved the addition of the extracting solvent to the product. They extracted 200 ml portions of pre-ferments with 40 ml portions of diethyl ether in a separatory funnel and repeated the process until 1,200 ml of pre-ferment had been extracted. The ether extracts were then dried over sodium sulfate and analyzed as such since concentrating the extract was not mentioned. Thus, any losses due to volatilization during concentration were eliminated. In most cases some form of concentration is advised to facilitate detecting compounds that may be present in small quantities. Actually, solvent extraction, when performed using a freshly distilled and purified solvent, produces few artifacts, especially when compared to steam distillation.

A direct dual-solvent extraction technique was reported by Mulders et al. (1972). The first extraction was with water, thus assuming all bread flavor compounds are water soluble. The supernatant liquid was distilled using a climbing film evaporator and a liquid-vapor separator. Retention time in this system was estimated at 8 seconds at a maximum temperature of 32°C and a pressure of 30 mm of mercury. This reduced the original 17 liter volume to 1.7 liter which were then freeze concentrated to 150 ml. This concentrated product was then extracted with a 2:1 mixture of pentane-diethyl ether. The resulting extract was dried over sodium sulfate and the solvent removed using a Vigreux distillation column. The process itself is relatively gentle, but there are approximately five time-consuming steps which make the reliability of any quantitative data attempted questionable.

Direct Vapor Analysis

Another relatively rapid and simple technique is direct vapor analysis; a typical example would be that reported by Mulders (1973). A 50 g slice of bread was cut into small cubes and placed in a 360 ml flask which was then equilibrated in a 36°C water bath for 45 min. At that time vapor samples were removed by a preheated glass syringe and analyzed. The possibility of artifact formation using this procedure is minimal since neither excessive temperature nor other chemicals or solvents are employed. Also, losses due to extractions, transfers and concentrations are eliminated. The primary disadvantage of this method is that only the most volatile compounds are detected, and the complete flavor profile of the product may be lacking. There is also no guarantee that the volatile compounds found by this method are those specifically associated with the characteristic flavor being evaluated.

Nitrogen Flush and Precolumn Trapping

Undoubtedly the mildest isolation procedure was that reported by Mulders et al. (1972). A sample of bread was cut into small cubes and placed in a glass container through which prepurified nitrogen was passed at a flow rate of 10 ml/min for 50 to 75 min. The exit gases were trapped in a Dry Ice cooled 30 cm by 3mm stainless steel column filed with uncoated glass beads 280 to 320 μm in diameter. The precolumn containing the stripped volatile was then removed and used for analysis. In this procedure no heat, solvents or other chemicals were used. One of its only limitations is that such a mild procedure probably does not remove all of the flavor compounds associated with bread flavor, and accordingly cannot be considered quantitative. Some investigators may also question cutting up the bread sample since compounds may escape in the process and increased exposure to air may have some consequences. A major advantage of the method is that it is relatively rapid to perform. Also, the efficiency of the system can be checked by sniffing the exit port of the precolumn to detect any volatiles not being trapped, and then smelling the stripped bread to determine if a characteristic odor still remains.

FLAVOR SEPARATION AND IDENTIFICATION TECHNIQUES

Introduction

For any flavor data to be meaningful, methods must be employed that simply and effectively separate the complex isolated mixtures into their individual components. Other prerequisites include the positive identification of the compounds and, ultimately, their quantitation. Some of today's advanced methods have assisted in separation and identification, but problems still face those investigators who are truly interested in determining concentrations. At this point some of the techniques employed will be discussed.

Traditional Methods

These procedures have been widely used for years by chemists; they supply information on the chemical characteristics of individual compounds which then can be used for identification purposes or quantitation.

An example would be the method used by Wiseblatt (1960a) to quantitate the amount of acetic acid present in the condensate obtained from oven vapors. Essentially the author used an acid-base type titration to arrive at an index of acidic compounds expressed as acetic acid. In reality other acids were probably also present, perhaps along with some compounds serving as buffering agents during the titration.

Another example would be the refractometer determination of ethanol from a pre-ferment mixture that probably contained other alcohols (Cole et al., 1962).

In the case of carbonyls several research groups have used a colorimetric procedure introduced by Lappin and Clark (1951). The problem with this procedure is that individual carbonyls give different colors and different color intensities; the more complex, the more unquantitative the procedure becomes.

Other traditional methods that have been used include specific gravity, melting point, and specific compound color reaction determinations.

Paper Chromatography

The use of paper chromatography has been reported in the separation of DNPH derivatives. However, the procedure is quite time-consuming and generally not all of the compounds present in a bread mixture are resolved through the use of only one solvent system. This is apparent from Figure 1. Thus, two solvent systems are required for more efficient separation.

Column Chromatography

This method has been used mainly to separate DNPH derivatives. For example, Ng et al. (1960) used a 2:1 silicic acid-celite column and various proportions of diethyl ether-Skellysolve-F mixtures. However, not all carbonyl derivatives present in the mixture were separated.

Thin-Layer Chromatography

Especially in the case of volatile compounds that have been derivatized, thin-layer chromatography has been used by numerous bread flavor investigators. If the mixtures are not too complex, it provides a relatively simple and inexpensive means of separating, identifying and roughly quantitating results. Attempts have been made to separate entire bread flavor mixtures, but, as can be seen in Figure 2, even with two-dimensional techniques complete separation was not achieved.

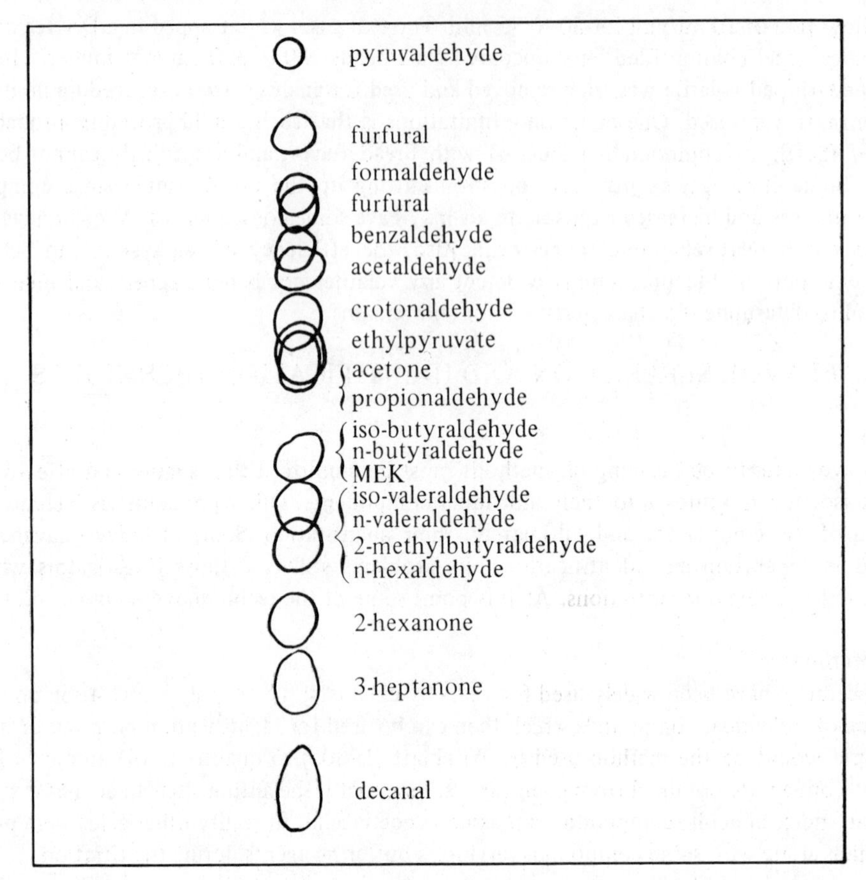

FIGURE 1. Typical paper chromatogram of pre-ferment carbonyl derivatives. (From Linko et al. (1962a)).

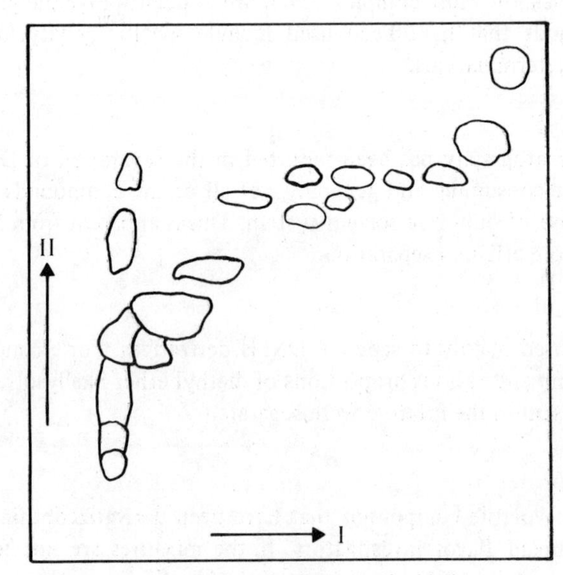

FIGURE 2. Typical two-dimensional thin-layer chromatogram. (From Daniels (1971)).

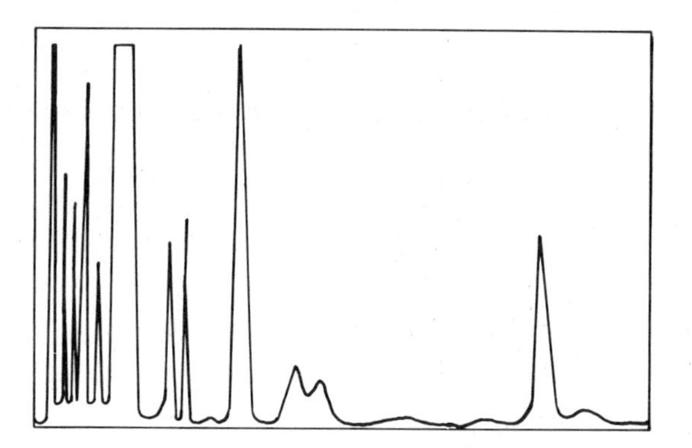

FIGURE 3. Typical gas-liquid direct vapor chromatogram. (From Mulders et al. (1972)).

Gas-Liquid Chromatography

The gas-liquid chromatograph has been a tremendous asset in the evaluation of bread flavor, especially since the introduction of temperature programming techniques. By selecting the appropriate separation conditions relative to temperature, flow rate, size of column and type of packing, a wide variety of classes of compounds can be effectively separated. Traditionally, compound identification using gas-liquid chromatography was based on matching retention times of authentic and unknown compounds separated under similar conditions on several types of columns. However, no definite identification can be made by this technique alone; time-consuming trapping techniques were introduced to obtain sufficient sample for verification identification by other analytical methods. The introduction of coupled gas-liquid chromatography – mass spectrometry analyses has been a tremendous asset, cost being the primary limiting factor of the technique. With the combined use of these two instruments, the structural identity of compounds incompletely separated by gas-liquid chromatography, or only appearing as minor peaks, can be resolved. Typical examples would be the works reported by von Sydow and Anjou (1969) and the Mulders group (1972). A typical scan using this technique is shown in Figure 3.

Most of the recent bread flavor works involving gas-liquid chromatography have attempted separations of total volatiles from bread systems. Perhaps more detailed separations could be obtained if the flavor concentrates obtained were first divided into at least acidic and basic fractions before gas-liquid chromatographic separation. This would simplify chromatograms which would be easier to interpret. Also, column packing materials specific for certain classes of compounds are currently available; work with these coatings would be informative.

A procedure introduced by Ralls (1960) and modified by Stephen and Teszler (1960) involving flash exchange gas-liquid chromatography was used by numerous groups in conjunction with the evaluation of DNPH derivatives of bread carbonyls. A typical example is that reported by Linko et al. (1962a). A mixture consisting of Celite and the derivative mixture was blended with alpha-ketoglutaric acid and placed in a capillary tube. The open end of the capillary tube was inserted in the septum of the gas chromatograph and the mixture was volatilized by the action of the alpha-ketoglutaric acid when an oil bath at 250°C was applied for 30 sec to the tube. Several limitations of the method include the possibility that complete regeneration is not achieved, especially of the longer chain carbonyls, and there may be product decomposition during the flashing process. Therefore, the procedure cannot be considered quantitative.

A summary of gas chromatographic conditions that have been applied to bread flavor systems is shown in Table 34.

TABLE 34

Gas-Liquid Chromatography Conditions Reported in Bread Flavor Investigations

Type of detector	Column size	Liquid phase	Solid phase	Reference
Thermal	10 ft × 1/4 in	Silicone-stearic acid	–	Wiseblatt (1960b)
–	5 ft × –	Carbowax® 400	Firebrick	Wiseblatt and Kohn (1960)
Thermal	6 ft × 1/4 in	20% Diisodecyl phthalate	Celite-545	Miller et al. (1961)
Thermal	6 ft × 1/4 in	30% Di(2–ethyl–hexyl) phthalate	Celite-545	Miller et al. (1961)
Thermal	6 ft × 1/4 in	25% Tetraethyleneglycol dimethyl ether	Firebrick	Miller et al. (1961)
–	10 ft × 1/4 in	Ucon® LB–1715	Chromosorb	Smith and Coffman (1961)
–	10 ft × 1/4 in	Ucon 50–HB–5100	Chromosorb	Smith and Coffman (1961)
–	10 ft × 1/4 in	Ucon 75–H–90,000	Chromosorb	Smith and Coffman (1961)
Thermal	6 ft × 1/4 in	Diisodecyl phthalate	Celite-545	Linko et al. (1962a)
Thermal	2m × 4mm	20% Quadrol®	Firebrick	Wick et al. (1964)
Thermal	2m × 8mm	20% Ucon–LB–1715	Chromosorb	Wick et al. (1964)
Thermal	2m × 8mm	20% DEGS	Chromosorb	Wick et al. (1964)
Flame	2m × 2mm	5% OPN	Gas Chrom A	Wick et al. (1964)
Flame	2m × 2mm	1% Ucon 50–HB–2,000	Gas Chrom A	Wick et al. (1964)
–	2m × 4mm	LAC–1–R–296®	Firebrick	Wick et al. (1964)
Flame	24 ft × 1/4 in	15% Ucon 75–H–90,000	Chromosorb-W	Coffman (1966)
–	46 ft × 1/4 in	15% Ucon 75–H–90,000	Chromosorb-W	Coffman (1967)
–	12 ft × 1/4 in	15% XE–60	Chromosorb-W	Coffman (1967)
Flame	10 ft × 1/8 in	20% Carbowax 20M	Chromosorb-P	Johnson and El-Dash (1969)
Flame	6 ft × 1/4 in	6% Apiezon	Anakrom-ABS	McWilliams and Mackey (1969)
Flame	6 ft × 1/4 in	6% DEGS	Anakrom-ABS	McWilliams and Mackey (1969)
Flame	6 ft × 1/4 in	3% Free fatty acid	Anakrom-ABS	McWilliams and Mackey (1969)
Flame	5.6m × 3/8 in	12% SF–96 (50)	Chromosorb-W	von Sydow and Anjou (1969)
Flame	3.6m × 3/8 in	3% Carbowax 20M	Chromosorb-G	von Sydow and Anjou (1969)
Flame	10 ft × 1/8 in	20% Carbowax 20M	Chromosorb-P	El-Dash and Johnson (1970)
Flame	8 ft × 1/8 in	20% Carbowax 20M	Chromosorb-P	Lorenz and Maga (1972)
Flame	1.3m × 4mm	45% D6–550+5% acid stearic	Celite-545	Mulders and Dhont (1972)
Flame	2m × 3mm	5% EGA	Gas Chrom-Q®	Mulders and Dhont (1972)
Flame	3.75m × 4mm	15% LAC–1–R–296	Chromosorb-W	Mulders et al. (1972)
Flame	3.75m × 4mm	10% SF–96 (50)	Chromosorb-W	Mulders et al. (1972)
–	50m × 0.25 mm	Ucon 50–LB–550–X	–	Mulders et al. (1972)
Flame	4m × 4mm	15% LAC–1–R–296	Chromosorb-W	Mulders (1973)
Flame	4m × 4mm	15% Carbowax–400	Chromosorb-W	Mulders (1973)
Thermal	4m × 4mm	21% LAC–1–R–296	Chromosorb-W	Mulders et al. (1973)
Thermal	4m × 4mm	15% Apiezon–M	Chromosorb-W	Mulders et al. (1973)

FACTORS AFFECTING THE SENSORY PROPERTIES OF BREAD

Introduction

Various researchers have traditionally stressed that pleasing bread flavor arises from these three primary factors associated with the bread making process: ingredients, fermentation, and baking. When studying bread flavor, one quickly learns to appreicate the chemistry involved in changing a mixture of individual ingredients, through the mixing fermentation, and baking, into a unique product possessing little physical or sensory resemblance to the original starting materials.

Due to the extremely close interaction among the various steps involved, one cannot simply state which phase is the key to the entire area of bread flavor. This portion of the review will cite literature which demonstrates the influential role each step can contribute to the sensory properties of bread and related products. The term sensory properties includes aroma, taste and color since the acceptance or rejection of most food items can be highly correlated to their visual properties.

TABLE 35

Volatile Compound Species Differences in Various Cereal Grains

Grain species	Butanal (%)	Hexanal (%)
Wheat	7.8	10.0
Triticale	9.2	5.7
Rye	16.6	3.4

From Hougen et al. (1971).

TABLE 36

Differences in Free Amino Acids in Wheat and Rye Flour

Amino acid(s)	Wheat flour	Rye flour
Aspartic acid	9.4*	19.7
Glutamic acid	10.5	20.7
Serine	1.2	3.4
Glycine	2.3	4.5
Alanine	4.9	19.2
Glutamine	2.0	1.2
γ–Aminobutyric + tyrosine	3.3	8.3
Valine	2.0	8.2
Leucine + isoleucine	3.6	8.8
Phenylalanine	2.2	2.8
Threonine	2.4	5.8
Methionine	0	0.3

*Mg % of dry weight.
From Ketovich and Ponomareva (1961).

TABLE 37

Comparative Sensory Properties of Various Grain Flours and Protein Supplements

	Sensory properties		
Product	Powder odor	Liquid odor	Liquid flavor
Whole wheat flour	5.5*	4.9	4.6
Corn flour	4.0	5.6	6.5
Rice flour	5.5	7.5	5.5
Non-fat dry milk	7.3	8.3	7.0
Sodium caseinate	6.9	6.8	5.0
Buttermilk solids	5.8	7.2	5.2
Defatted soy flour	4.3	4.2	1.7
Full-fat soy flour	3.1	4.9	2.5
Isolated soy proteinate	7.0	7.0	3.0
Fish protein concentrate	1.9	1.0	1.7
Peanut flour	2.8	3.2	2.9
Cottonseed flour	7.6	6.1	4.5

*Mean score 1-strong; 10-mild.
From Maga and Lorenz (1973c).

Flour and its Components

Since flour is the main ingredient in bread, it makes a definite flavor contribution. The flavor of flour milled from sound wheat is mild; the starch and enzymes in flour, when combined with salivary amylase, account for a portion of the characteristic sweetness in bread (Otterbacker, 1959).

From a compositional standpoint, wheat contains many classes of flavor reactive compounds, some of which may be present only in trace amounts. However, from a grain variety standpoint, the volatile composition of various grains are quite similar (Hougen et al., 1971), but the components differ in quantity. This can be appreciated by viewing the data summarized in Table 35. Triticale, a cross of wheat and rye, had butanal and hexanal levels between the two parents.

As will be discussed in more detail in a later section, free amino acids can play an influential role in the development of bread flavor. It can be concluded from the data in Table 36 that some of the characteristic flavor difference between wheat and rye products is due to rye generally having significantly higher levels of free amino acids.

The relative odor and flavor properties associated with whole wheat flour are compared to other grain flours and protein supplements in Table 37.

In the powdered form the odor of whole wheat flour was rated approximately halfway between strong and mild. When the product was suspended at a 3% level in water, its odor and flavor intensity increased slightly; milder odors and flavors were noted for corn and rice flours in the reconstituted form.

TABLE 38

Influence of Wheat Class on Bread Sensory Properties

	Odor		Taste	
Wheat class	Best (%)	Poorest (%)	Best (%)	Poorest (%)
Soft red winter	17	59	24	56
Hard red winter	38	24	27	27
Hard red spring	45	17	49	17

From King et al. (1937).

TABLE 39

Influence of Flour Extraction Rate on Bread Sensory Properties

	Odor		Taste	
Extraction rate (%)	Best (%)	Poorest (%)	Best (%)	Poorest (%)
100	20	44	20	41
85	41	26	44	26
70	39	30	37	33

From King et al. (1937).

TABLE 40

Influence of Flour Bleaching on Bread Sensory Properties

	Odor		Taste	
Bleach treatment	Best (%)	Poorest (%)	Best (%)	Poorest (%)
Bleach no. 1	26	30	27	34
Bleach no. 2	37	35	41	32
Unbleached	37	35	32	32

From King et al. (1937).

An early study (Table 38) indicated that wheat class influenced the odor and flavor of bread.

These data demonstrated that the panel preferred the odor and flavor of bread made from hard red spring wheat over breads made from hard red winter or soft red winter wheat. These data were just reported as preferences and no comments were included as to the specific superior odor and flavor properties associated with bread baked from hard red spring wheat.

In this same study King et al. (1937) reported on the sensory properties of bread baked from flours varying in extraction rates. These data are summarized in Table 39.

This study was done in the late 1930's when an extraction rate of 85% was probably preferred from both an odor and flavor standpoint, which would probably not be the case today.

The influence of bleaching on bread sensory properties was also investigated (Table 40).

In the case of odor no clear-cut preference for bleached or unbleached flour was observed. However, in the case of flavor there was a clear preference for one of the bleached flours. These data were probably psychological in nature in that the panel associated a better flavor with the whiter bleached product. The low flavor acceptance of the other bleached flour was perhaps due to a residual flavor from the bleaching agent.

TABLE 41

Typical Wheat Flour Fatty Acid Composition

Fatty acid	Chain Length	Percent
Myristic	C–14:0	0.1
Palmitic	C–16:0	24.5
Palmitoleic	C–16:1	0.8
Stearic	C–18:0	1.0
Oleic	C–18:1	11.5
Linoleic	C–18:2	56.3
Linolenic	C–18:3	3.7
Arachidic	C–20:0	0.8
Other	–	1.1

From Nelson et al. (1963).

TABLE 42

Influence of Storage on Wheat Flour Fatty Acid Class Distribution

Lipid class	Control (–20°C)	25°C for 4 months	15°C for 6 months
Triglycerides	162*	94	89
Diglycerides	97	119	126
Monoglycerides	20	37	50
Free fatty acids	76	144	159

*Mg/100 g of flour.
From Clayton and Morrison (1972).

TABLE 43

Reactive Carbohydrates Reported in Wheat Flour

Carbohydrate	Percent carbohydrate
Glucose	0.04
Fructose	Trace
Maltose	Trace
Sucrose	0.27
Raffinose	0.17

From MacKenzie (1958).

At this point the role of specific chemical classes on flour flavor will be expanded. Lipids will be discussed first. As can be seen from Table 41, unsaturated fatty acids comprise approximately 70% of the total wheat flour lipids. From a flavor standpoint this is quite interesting since, due to their unsaturation, these lipids are less stable when stored and processed. The most common decomposition products of unsaturated fatty acids are carbonyls which possess their own characteristic flavors or serve as precursors for other flavor related compounds.

Free fatty acids are another important class of lipids. In Table 42 free fatty acids are shown naturally present in wheat lipids, their level increasing from the decomposition of triglycerides upon storage. There was an approximate doubling in the amount of free fatty acids present as the result of room temperature storage for four months. Thus, free fatty acids can also contribute to the flavor of wheat flour.

The role of carbohydrates on wheat flavor will now be discussed. Most of the carbohydrate material in wheat is in the form of starch and contributes little to overall flavor. However, a small portion of the carbohydrate material does occur in more reactive forms. These are represented in Table 43.

TABLE 44

Changes in Wheat Carbohydrates as Affected by Storage Conditions

Carbohydrate	Air	Nitrogen	Carbon dioxide
Glucose	7*	24	23
Fructose	5	18	16
Galactose	3	9	9
Maltose	1	4	3
Sucrose	21	39	36

*Mg/10 g dry weight.
From Lynch et al. (1962).

TABLE 45

Influence of Milling and Baking on Amino Acid Composition

Amino acid	Wheat	Flour	Bread
Alanine	3.3	2.8	2.9
Arginine	4.7	3.8	3.6
Aspartic acid	5.1	4.1	4.6
Cystine	2.0	2.1	1.9
Glutamic acid	28.5	34.5	31.7
Glycine	3.9	3.2	3.2
Histidine	1.9	1.9	1.9
Isoleucine	3.9	4.3	4.3
Leucine	6.5	7.0	7.1
Lysine	2.7	2.1	2.5
Methionine	1.8	1.7	1.9
Phenylalanine	4.4	4.9	4.8
Proline	9.9	11.7	11.1
Serine	5.1	5.4	5.5
Threonine	3.0	2.8	3.0
Tryptophan	1.1	1.0	1.0
Tyrosine	3.1	3.3	3.3
Valine	4.5	4.5	4.7

*G/16 g nitrogen.
From Hepburn et al. (1957).

Reducing sugars such as glucose and fructose are quite reactive from a bread flavor precursor standpoint; their roles will be discussed in a later section. Although reducing sugars or their precursors are generally added during bread making, some are naturally present in wheat.

Storage of flour can influence the amount of available reactive sugars, especially with regard to non-enzymatic browning reactions (Table 44).

Storage under nitrogen and/or carbon dioxide protected the reactive carbohydrates, whereas under atmospheric conditions the levels of all reactive sugars decreased, probably due to their interaction in the formation of non-enzymatic browning compounds.

Various segments of protein composition influence wheat's flavor properties. Table 45 shows that the total amino acid composition undergoes various changes from wheat to bread.

Certain nitrogenous compounds, specifically peptides and free amino acids, are highly reactive, primarily as precursors in the formation of numerous volatile compounds during fermentation and baking. Few free amino acids can be found in the crust of bread due to their participation in non-enzymatic browning reactions. This can best be appreciated by viewing the free amino acid composition as influenced by various phases of bread production (Table 46).

Free amino acids in the original flour can serve as an important source of bread flavor precursor material.

Few data exist on the amount and types of short-chained peptides present in wheat flour. Since most flavor chemists agree that peptides generally have a disagreeable, astringent flavor, an interesting study would be to isolate and quantitate peptides in flour as influenced by storage and processing.

Another class of compounds that until recently received little attention is the simple phenolics associated with wheat flour. The simple phenolic acids reported to date in various wheat fractions are summarized in Table 47. Aside from free phenolic acids, bound forms are also present in wheat flour; when released, the total amount present approaches 150 ppm (Table 48).

The role of milling on free phenolic acid distribution has also been investigated and a portion of the data obtained with triticale is summarized in Table 49.

The bran fraction contained the highest level of free phenolics. However, approximately 75% of the original amount present in grain was found in the flour fraction. Similar results were obtained for wheat samples.

TABLE 46

Free Amino Acid Distribution During Various Phases of Bread Production

Amino acid(s)	Fraction				
	Flour	Unfermented dough	Fermented dough	Crumb	Crust
Alanine	14.34[a]	86.98	22.94	29.05	2.12
Arginine	2.11	6.66	15.30	11.43	Trace
Aspartic acid	14.17	60.53	14.91	10.92	0.83
Cystine	0.56	3.25	1.16	1.01	Trace
Glutamic acid	8.85	71.85	46.83	51.39	1.13
Glycine	4.08	11.74	10.43	12.43	0.84
Histidine	0.95	–	–	–	–
Isoleucine	4.13	6.60	1.58	3.07	0.52
Leucine	5.54	8.66	1.73	4.79	0.27
Lysine	2.39	32.31	25.53	12.76	Trace
Methionine	0.69	2.11	0.95	1.83	Trace
Phenylalanine	3.31	5.86	0.18	1.95	1.64
Proline	4.66	21.65	17.11	11.21	0.44
Serine and threonine	27.50	68.14	19.22	22.88	1.24
Tyrosine	3.12	5.99	0.49	1.09	–
Valine	6.83	16.15	4.60	6.41	0.62

a: μmol/100 g dry basis.
From El-Dash, 1971.

TABLE 47

Simple Free Phenolic Acids Reported in Wheat Fractions

Phenolic acid	Grain	Germ	Flour
ρ–Hydroxybenzoic acid	–	–	Maga and Lorenz (1973b)
Salicylic acid	–	–	Maga and Lorenz (1973b)
Vanillic acid	–	–	Maga and Lorenz (1973b)
ρ–Coumaric acid	El-Basyouni and Towers (1964)	–	Maga and Lorenz (1973b)
o–Coumaric acid	–	–	Maga and Lorenz (1973b)
Gentisic acid	–	–	Maga and Lorenz (1973b)
Iso-ferulic acid	–	–	Maga and Lorenz (1973b)
Ferulic acid	El-Basyouni and Towers (1964)	King (1962)	Maga and Lorenz (1973b)
Caffeic acid	–	–	Maga and Lorenz (1973b)
Protocatechuic acid	–	–	Maga and Lorenz (1973b)
Sinapic acid	–	–	Maga and Lorenz (1973b)
Syringic acid	–	King (1962)	Maga and Lorenz (1973b)
Iso-chlorogenic acid	–	–	Maga and Lorenz (1973b)
Chlorogenic acid	–	–	Maga and Lorenz (1973b)

TABLE 48

Free and Total Phenolic Acid Distribution in Wheat Flour

Phenolic acid	Amount present (ppm) Free	Total
ρ–Hydroxybenzoic acid	5	6
Salicylic acid	4	4
Vanillic acid	15	39
ρ–Coumaric acid	17	23
o–Coumaric acid	7	8
Gentisic acid	3	4
Iso-ferulic acid	1	4
Ferulic acid	18	32
Caffeic acid	–	3
Protocatechuic acid	3	6
Sinapic acid	–	15
Syringic acid	5	5
Iso-chlorogenic acid	–	1
Chlorogenic acid	–	5

From Maga and Lorenz (1973b).

TABLE 49

Influence of Milling on Free Phenolic Acid Distribution in Triticale

Phenolic acid	Amount present (ppm)			
	Grain	Bran	Shorts	Flour
ρ–Hydroxybenzoic acid	14	17	15	10
Salicylic acid	7	12	5	5
Vanillic acid	40	50	42	35
ρ–Coumaric acid	40	37	40	36
o–Coumaric acid	10	21	12	8
Gentisic acid	5	12	7	3
Iso-ferulic acid	5	7	7	4
Ferulic acid	34	43	35	30
Caffeic acid	5	10	7	3
Protocatechuic acid	4	8	7	2
Sinapic acid	14	16	15	12
Syringic acid	7	8	6	5
Iso-chlorogenic acid	2	4	2	1
Chlorogenic acid	7	10	7	5

From Maga and Lorenz (1973b).

TABLE 50

Possible Flavor Contribution of Free Phenolic Acids Found in Wheat Flour

Phenolic acid	Taste threshold (ppm)*	Possible flavor contribution**
ρ–Hydroxybenzoic acid	40	–
Salicylic acid	90	–
Vanillic acid	30	+
ρ–Coumaric acid	40	(+)
o–Coumaric acid	25	(+)
Gentisic acid	90	–
Ferulic acid	90	–
Caffeic acid	90	–
Protocatechuic acid	30	–
Syringic acid	240	–

*From Maga and Lorenz (1973a).
**–: Present in amounts significantly below reported threshold.
(+): Present in amounts near reported threshold.
+: Present in amounts at or above reported threshold.
From Maga and Lorenz (1973b).

Phenolic compounds possess strongly astringent organoleptic properties; in most food systems they would be considered objectionable. Therefore, the data presented in Table 50 demonstrates that, due to the amounts found, certain phenolic acids can contribute to the flavor properties of wheat.

It should also be noted that these data do not consider the possibility of synergistic effects among the phenolic compounds, which would further lower their taste threshold levels.

In the case of corn, heating increased the amount of simple phenolic acids (Steinke and Paulson, 1964). It would be interesting to measure phenolic composition as influenced by baking procedures to determine if the amount of simple phenolic acids actually increases. If this were so, perhaps phenolic compounds play an influential role in the somewhat drying and astringent sensation noted in bread.

Considering the role cereal grains play in the world food chain, relatively few studies have been published dealing with the volatile flavor compounds associated with these products. In the case of wheat flour, the data summarized in Table 51 are presented.

The majority of the volatile compounds reported in wheat flour have also been reported present in bread. Wheat itself is a direct source for many of the flavor compounds found in bread. Naturally the levels found in the two products may not be comparable; because of this, factors such as fermentation and baking have traditionally been identified as primary sources of bread flavor components.

Regarding the influence of storage on the flavor properties of wheat flour, Fellers and Bean (1970) found that storage for up to six months even at room temperature decreased bread acceptability. Their data are shown in Table 52.

Shortening

Shortening is responsible for crumb tenderization, improved texture, loaf volume, and flavor (Baker and Mize, 1939). A brief report by Baker and Mize (1939) mentioned that minor flavor differences were noted when breads were baked with no shortening, ordinary shortening, mineral oil, and ordinary shortening plus mineral oil. No specific data were presented.

Using levels of 0, 3, and 5% shortening, King et al. (1937) reported that a panel preferred the odor and taste of bread baked with the highest shortening level. In fact, bread odor and taste preferences increased with increasing shortening levels.

Shortening may also affect bread flavor by serving as a precursor for the formation of highly flavorful carbonyl compounds during thermal breakdown while baking. Kleinschmidt et al. (1963) have proposed the use of 0.5% peroxidized oil, based on flour weight, as a means of increasing bread flavor. The procedure involved the formation of peroxidized soybean oil by the action of soybean lipoxidase to obtain a peroxide

TABLE 51

Volatile Compounds Reported in Wheat Flour

Compound	Reference
*Methanol	Hougen et al. (1971)
*Ethanol	Hougen et al. (1971)
*3-Methyl-1-butanol	McWilliams and Mackey (1969)
*1-Pentanol	McWilliams and Mackey (1969)
*Acetaldehyde	McWilliams and Mackey (1969) Hougen et al. (1971)
*Propanal	Hougen et al. (1971)
*2-Methylpropanal	McWilliams and Mackey (1969) Hougen et al. (1971)
*Butanal	McWilliams and Mackey (1969) Hougen et al. (1971)
*2-Methylbutanal	McWilliams and Mackey (1969)
*Pentanal	McWilliams and Mackey (1969) Hougen et al. (1971)
*Hexanal	McWilliams and Mackey (1969) Hougen et al. (1971)
*Heptanal	McWilliams and Mackey (1969) Hougen et al. (1971)
*Octanal	McWilliams and Mackey (1969)
*2-Butenal	McWilliams and Mackey (1969)
*Phenylacetaldehyde	McWilliams and Mackey (1969)
*2-Propanone	Hougen et al. (1971)
*2-Butanone	McWilliams and Mackey (1969) Hougen et al. (1971)
3-Methyl-2-butanone	McWilliams and Mackey (1969)
2,2-Dimethyl-3-pentanone	McWilliams and Mackey (1969)
*2,3-Butanedione	McWilliams and Mackey (1969)
Cyclopentanone	McWilliams and Mackey (1969)
*Ethyl acetate	McWilliams and Mackey (1969)

*Also reported in bread

TABLE 52

Influence of Wheat Flour* Storage on Bread Acceptability

Storage temperature (°F)	Storage time (months)	Panel liking bread (%)
0	1	75
0	3	70
0	6	80
70	1	78
70	3	78
70	6	69
90	1	69
90	3	80
90	6	50
100	1	64
100	3	68
100	6	28

*70 parts flour, 30 parts wheat protein concentrate.
From Fellers and Bean (1970).

TABLE 53

Influences of Fat Content on Bread Sensory Preferences

	Odor		Taste	
Fat (%)	Best (%)	Poorest (%)	Best (%)	Poorest (%)
0	15	56	14	64
3	35	24	34	20
5	50	20	52	15

From King et al. (1937).

TABLE 54

Influence of the Addition of Peroxidized Oil* on Bread Flavor

Peroxide content (%)	Bread flavor
< 0.02	No flavor
0.02–0.03	Slight flavor
0.03–0.04	Definite flavor
0.05–0.06	Strong flavor

*0.5% Enzyme-peroxidized oil added.
From Kleinschmidt et al. (1963).

TABLE 55

Influence of Salt Level on Its Sensory Detection in Bread

Percent salt compared*	No. in panel	No. correct (triangle test)
0% vs. 1.25%	43	39**
0.25% vs. 1.25%	42	32**
0.50% vs. 1.25%	42	34**
0.75% vs. 1.25%	42	18
1.0% vs. 1.25%	43	14

*Percent on weight of finished bread.
**Significantly different at 5% level.
From Collyer (1966).

value of 0.03 to 0.04%. This material was then added to the formulation and the resulting bread flavor was evaluated.

Peroxide values of less than 0.03% did not impart enough flavor, while additives having peroxide values of more than 0.04% resulted in an objectionable flavor. Thus, the lipoxidase enzymatically breaks down the unsaturated fatty acids of the soybean oil substrate resulting in the formation of flavorful volatile compounds.

Salt

It has been reported by Dunn (1947) that salt serves three main functions in a bread formulation. These include stabilizing fermentation, strengthening the gluten, and developing flavor. He also stated that the proper function of salt in most foods is not to make the product taste salty, but to develop and bring out the natural flavors of the product. Thus, salt in bread system serves mainly in a flavor potentiation role.

Using sensory panel techniques Collyer (1966) determined panel sensitivity and salt level preferences in white bread. Some of these data are summarized in Table 55.

Although no data are presented to substantiate the fact, the author stated that there was a strong preference for the bread containing 1.25% salt based on the weight of the baked product. With this level serving as the control it was found that the untrained panel was able to significantly detect differences in salt levels of from 0 to 0.5% salt when compared to the control. However, no significant difference in flavor was detected when levels of 0.75 and 1.0% salt were compared to the 1.25% level. Therefore, it appears that when consumers are given a comparative choice, they prefer the higher salt level (1.25%). However, statistically no significant difference in flavor due to salt was detected when approximately half the control salt level was used.

An interesting project would be to analytically evaluate flavor extracts using gas-liquid chromatography of breads made from doughs having varying salt levels. It seems unlikely that salt could actively enter into a chemical reaction to form flavor compounds. However, salt may affect compound volatility by affecting the vapor of the product.

TABLE 56

Influence of Yeast on the Thermal Properties of Baking Bread

	Temperature (°F)	
Minutes of Baking	No yeast	Fermented
2	102	102
4	122	122
6	143	141
8	163	158
10	188	172
12	207	188
14	211	198
16	212	205
18	212	210
20	212	212

From Baker (1957).

Yeast

Yeast can influence bread flavor if excessively high levels are used; a characteristic yeast flavor will predominate in the final product. Numerous pure and mixed yeast cultures were obtained from commercial sources by Wardall (1910) and the resulting bread quality, including flavor, evaluated. No attempt was made to identify the yeasts present in the mixed cultures, but they generally produced bread of better flavor than pure strain yeasts.

Baker (1957) presented some interesting data which are summarized in Table 56.

He demonstrated that in an indirect manner yeast can influence bread flavor. The product containing no yeast culture achieved a higher temperature in a shorter period of time during baking as compared to bread containing yeast. For example, in the no-yeast product a temperature of 212°F was achieved in 16 min, whereas in the yeast containing bread, a temperature of 212°F was not obtained until after 20 min of baking. From a thermally induced flavor standpoint, yeast could tend to retard flavor formation. The difference in temperature between the two products was explained as follows: the gas formed from the yeast addition had an insulating effect, which caused the temperature to rise slower. Perhaps it would be interesting to study and evaluate the influence of various levels of yeast on thermally produced bread flavor in conjunction with the rate of temperature rise during baking. It is possible that from a total flavor standpoint, lower levels of yeast would result in greater amounts of thermally induced bread flavor components.

Markianova et al. (1971) prepared breads using a variety of hybrid yeasts, then compared the amount of volatile flavor compounds found against a compressed yeast control. Their data are summarized in Table 57.

All of the yeast hybrids they evaluated produced more total volatiles than the compressed yeast control. However, they concluded that all of the yeast hybrids produced bread of inferior sensory properties. The organoleptic properties of the bread made from the conventional yeast were described as pleasant, probably due to the presence of greater amounts of specific compounds such as valeraldehyde. Another possible explanation for this preference is that trace and measurable amounts of two unknown compounds were found in the hybrid yeast samples but not in the compressed yeast control. Also, perhaps the total volatile level was highly concentrated enough to make the aroma objectionable.

Bacterial Additions

The possible application of specific bacterial cultures in modifying or improving bread flavor was proposed by Fornet (1951) and investigated by Robinson et al. (1958a). They postulated that the proper amount of bacteria capable of producing short-chain organic acids would enhance bread flavor. Various pre-ferments were prepared and the total and individual bacterial populations determined. Total bacterial

TABLE 57

Influence of Yeast Hybrids on Bread Carbonyl Composition

Mg/100 g of product

Compound	Yeast hybrid 203	Yeast hybrid 196/6	Yeast hybrid 202	Yeast hybrid 14	Compressed yeast
Methylglyoxal	0.008	0.003	0.002	0.008	0.008
Glyoxal	0.021	0.025	0.007	0.012	0.031
Acetaldehyde	0.063	0.027	0.025	0.009	0.016
Propionaldehyde	0.001	0.001	0.001	0.002	0.004
Isobutyraldehyde	0.009	0.009	0.003	0.004	0.007
Acrolein	0.012	0.032	0.014	0.013	0.021
Acetone	0.017	0.036	0.024	0.018	0.027
Butyraldehyde	0.041	0.027	0.019	0.008	0.013
Isovaleraldehyde	0.005	0.006	0.003	0.003	0.002
Methyl ethyl ketone + diacetyl	0.015	0.021	0.015	0.003	0.012
Valeraldehyde	0.051	0.043	0.041	0.029	0.091
Caproaldehyde	0.043	0.036	0.030	0.023	0.047
Ethyl propyl ketone	0.004	0.011	0.009	0.007	0.011
Unknown	Trace	Trace	0.003	0.005	–
Unknown	Trace	Trace	0.003	0.001	–
Benzaldehyde	3.518	2.973	3.922	3.519	2.960
Oxybenzaldehyde	1.914	1.641	1.441	1.620	1.627
Phenylacetylaldehyde	0.148	0.206	0.209	0.124	0.166
Furfural	2.235	2.105	2.178	2.037	1.627
Total	8.10	7.20	7.95	7.45	6.67

From Markianova et al. (1971).

populations in pre-ferments were found to increase slightly for up to two hours of incubation; to decrease with 2 to 6 hr of incubation; and again to increase after 6 hr. This study also demonstrated that the majority of the 65 different aerobic microorganisms originated from the yeast, both in number and quantity, and that the microflora present in the flour was vastly different in the pre-ferment. The most common genera isolated from the pre-ferments included *Micrococcus, Aerobacter, Proteus, Bacillus* and *Lactobacillus.* Some of the typical odor descriptions of breads baked with specific microorganisms are summarized in Table 58.

They concluded that the best flavored product resulted from the use of *Lactobacillus bulgaricus 09* or *L. Bulgaricus 09,* plus a buttermilk culture. The use of certain microorganisms evaluated in this study may have application in the production of novelty bread items such as cheese-flavored bread.

To explain the bacterial population decrease noted by Robinson et al. (1958a) during incubation of pre-ferments, Robinson et al. (1958b) measured variables such as acid and alcohol production. During this study they isolated two inhibitory substances that they postulated were elaborated by the yeast during fermentation. No positive identifications of these inhibitory substances were made.

The possible role that bacteria play on the formation of carbonyl compounds in dough fermentation was investigated by Kohn et al. (1961); they used dough containing a normal bacterial load and a specially treated low-bacteria dough. Their data are summarized in Table 59.

Data indicated that bacterial count has no effect on carbonyl composition of dough. It would have been interesting to compare carbonyl levels as influenced by dough bacterial levels after baking. As Robinson et al. (1958a) pointed out, bacteria can act as a good source of organic acids which when baked interact to form carbonyls.

The relatively nonsignificant role of added bacteria on bread flavor was also demonstrated by Linko et al. (1962a). They measured the amounts of certain carbonyls found in both the resulting pre-ferments and fermentation gases (Table 60).

They concluded that the addition of *Pediococcus cerevisiae* resulted in the formation of vast amounts of acetone.

TABLE 58

Odor Descriptions of Bread Baked With Added Microorganisms

Microorganism added	Odor description of fresh bread
Standard loaf pre-ferments:	
Aerobacter aerogenes	Similar to standard
Bacillus coagulans	Stronger than standard
Bacillus laterosporus	Slightly stronger than standard
Lactobacillus plantarum	Strong, like fermentation
Micrococcus caseolyticus	Strong, like fermentation
Lactobacillus leichmannii	Mild, like fermentation
Lactobacillus casei	Mild, like fermentation
Aged sponge dough:	
Lactobacillus fermenti	Mild, like standard
Lactobacillus casei	Mild, like standard
Lactobacillus brevis	Cheese-like
Lactobacillus bulgaricus	Cheese-like
Lactobacillus plantarum	Cheese-like
Lactobacillus lactis	Strong, not acceptable

From Robinson et al. (1958a).

TABLE 59

Influence of Dough Bacterial Count on Carbonyl Concentration

Carbonyl (s)	Normal dough*	Low-bacteria dough**
2-hexanone and hexanal	246 ppm	263 ppm
Acetaldehyde	33	15
Butyraldehyde	22	25
Isovaleraldehyde	24	22
Acetone	70	49

*52 × 10^6 bacteria/g of dough.
**1,200 bacteria/g of dough.
From Kohn et al. (1961).

TABLE 60

Influence of Added Bacteria on Carbonyl Production in Bread Fermentation Gases

Species added	Formaldehyde	Acetaldehyde	Propanal+ 2-propanone	Butanol+ 2-methylpropanal	Pentanal + 3-methylbutanal+ 2-methylbutanal
Control	21*	460	41	63	39
P. *cerevisiae*	36	490	1010	94	84
L. *bulgaricus*	27	510	51	63	39
S. *thermophilus*	27	500	41	49	29
A. *aerogenes*	27	490	41	49	29
S. *diacetilactic*	36	490	41	57	32
B. *brevis*	Trace	430	39	54	49
L. *plantarum*	Trace	560	41	60	36

*Mg/100 ml of pre-ferment.
From Linko et al. (1962a).

TABLE 61

Influence of Added Protease Enzyme Levels on Total Carbonyl Compound Production in Bread Crust

Protease level (H.U.) /700 g flour	Mg carbonyl/100 g bread crust	
	Papain	Rhozyme J-25
0	17.92	17.99
75	22.25	19.81
150	35.10	20.52
300	32.37	23.96
500	35.15	23.94
700	18.49	23.38
1,000	14.35	23.49

From El-Dash and Johnson (1967).

Enzyme Additions

El-Dash and Johnson (1967) postulated that induced proteolysis of wheat proteins during sponge fermentation would provide additional free amino groups, which could react in turn with reducing sugars to form a more flavorful bread product. Levels of 0 to 1,000 hemoglobin units (H.U.)/700 g of flour of papain and Rhozyme® J-25 were used, and various flavor related parameters measured. For example, the levels of water-soluble nitrogen in fermented doughs were measured and it was found that papain was quite effective in forming water-soluble compounds. In fact, it was so effective that at the higher levels of incorporation, the dough turned to a liquid mass. Thus, they were able to demonstrate that protease enzymes can increase the amount of reactive amino groups in a fermentation system.

Another variable measured was the effect of enzyme addition on crust color. To a certain enzyme addition level, crust color did darken, indicating a more reactive flavor system. However, at higher levels, especially with papain, crust color actually lightened. This happened because at the higher papain levels, loaf volume decreased to the point that the crusts were below the top edge of the baking pans. The total carbonyl levels in the bread crust were measured and these data are summarized in Table 61.

Papain at a level up to 500 H.U./700 g of flour increased total crust carbonyl content to approximately twice that of the non-enzyme control, and then the carbonyl level dramatically decreased. Rhozyme J-25 increased total crust carbonyl levels to a lesser degree, but the carbonyl level was maintained even at the highest levels of addition.

Protease type was also found to influence the levels of specific carbonyls formed. The predominant carbonyl produced from papain addition was propanal, while in the case of Rhozyme J-25, heptanal was predominant. Neither enzyme influenced the amount of formaldehyde formed. Limited sensory data are presented reporting that neither enzyme affected the taste of bread, yet the aroma of the bread containing 150 H.U./700 g of flour was significantly different from the nonenzyme control. El-Dash and Johnson concluded that the carbonyl concentration affected the aroma of bread, but not its taste. Throughout the study no data were presented concerning possible crumb flavor differences.

Roiter and Borovikova (1972) evaluated the volatile compounds of bread crumb and crust prepared with the addition of 0.001% of the enzyme preparation Amylorisin Pox to the dough, they concluded that bread prepared with this enzyme additive had better aroma and flavor properties than a control bread prepared without the enzyme addition. In Table 62 no qualitative differences were apparent.

They found that enzyme application increased volatile carbonyl compound concentration 2.3-fold in the crumb and 3.2-fold in the crust. Aliphatic compounds associated with the enzyme treated product increased 9 and 15 times in the crumb and crust, respectively.

Sugars

Sugar amount and source can affect the sensory properties of bread in several ways. For example, as shown in Table 63, sugar amounts can influence the formation rate of specific flavor related compounds

TABLE 62

Influence of Enzyme Addition to Dough on Carbonyl Compounds in Bread

	Mg/100 g of specific volume			
	With enzyme		No enzyme	
Compound	Crumb	Crust	Crumb	Crust
Methylglyoxal	0.16	0.28	0.06	0.06
Glyoxal	0.16	0.14	0.01	0.01
Acetaldehyde	0.17	0.47	0.04	0.06
Propionaldehyde	0.01	0.07	Trace	Trace
Isobutyraldehyde	0.01	0.13	0.01	0.01
Acrolein	0.04	0.07	0.01	0.04
Acetone	0.09	0.23	0.01	0.09
Butyraldehyde	0.01	0.18	0.01	0.01
Isovaleraldehyde	0.01	0.08	Trace	Trace
Unknown	0.04	0.36	Trace	Trace
Methyl ethyl ketone	0.02	0.04	Trace	Trace
Diacetyl	0.04	0.05	Trace	Trace
Valeraldehyde	0.25	0.46	0.02	0.03
Caproaldehyde	0.22	0.24	0.01	0.01
Ethyl propyl ketone	0.07	0.20	Trace	Trace
Furfural	6.35	18.30	2.62	4.70
Benzaldehyde	0.65	0.94	0.29	0.50
Oxybenzaldehyde	5.20	3.50	2.95	2.75
Phenylacetylaldehyde	0.90	1.66	0.14	0.30
Total	14.40	27.40	6.15	8.45

From Roiter and Borovikova (1972).

TABLE 63

Influence of Sugar Level and Fermentation Time on Acetoin Formation in Bread Dough

	Mg acetoin/150 g dough Hours			
Sugar (%)	1	2	3	4
0	1.9	0.8	0.2	0.1
2	2.0	2.5	1.9	1.1
5	2.1	3.9	3.5	2.6

From Visser't Hooft and de Leeuw (1935).

during fermentation by providing an energy source for yeast activity, which is influential in flavor compound formation.

Sugar sources vary in their ability to form thermally induced flavor compounds as demonstrated in Table 64.

Certain disaccharides such as sucrose, if not completely hydrolyzed and utilized during fermentation, are capable of forming relatively high levels of flavor compounds.

Aside from flavor contributions, sugars can also influence crumb and crust color. Ponte et al. (1963) demonstrated that adding pentosans had a detrimental effect on bread color, even though they were substituted at levels equivalent to more commonly used sugars. An approximate doubling in crust thickness was also noted when pentosans such as arabinose and xylose were included (Table 65).

To better understand the role of sugars in fermentation and bread flavor (Table 66), the levels of sugars as influenced by fermentation have been studied by Lee et al. (1959).

TABLE 64

Influence of Sugar Types on Carbonyl Composition in Bread Crust

Carbonyl (s)	Fructose	Galactose	Glucose	Lactose	Maltose	Raffinose	Sucrose	Xylose
Formaldehyde	1.3*	1.4	1.1	1.1	1.i	1.2	1.2	2.1
Acetaldehyde	2.1	3.5	2.3	2.2	2.2	2.0	2.1	2.0
Acetone + Propanal	15.1	15.9	8.4	10.8	19.4	18.9	20.9	7.4
Isobutyraldehyde + Methyl ethylketone	0.3	0.7	0.4	0.4	0.3	0.2	0.2	0.2
Isovaleraldehyde + valeraldehyde + 2-methylbutanal + hexanal	2.8	3.3	2.7	2.7	2.4	2.6	2.8	3.2

*mg/100 g dry matter
From Linko and Johnson (1963).

TABLE 65

Influence of Added Pentosans on Bread Color and Crust Thickness

Carbohydrates added	Crumb color	Crust color	Crust thickness (mm)
5.5% Glucose	Creamy white	Golden brown	2
3.0% Glucose + 2.5% arabinose	Dull	Chestnut	5
3.0% Glucose + 2.5% xylose	Dull	Chestnut	5.5
8.0% Glucose	Greamy white	Chestnut	2.5
3.0% Glucose + 5% arabinose	Dull	Dark chestnut	5
3.0% Glucose + 5% xylose	Dull	Dark chestnut	5.5

From Ponte et al. (1963).

TABLE 66

Changes in Sugar Concentrations in Sponges, Doughs and Breads as Influenced by Fermentation Time*

System	Fermentation time (min.)	Glucose (%)	Fructose (%)	Maltose (%)
Sponge	4	0.30	0.46	1.11
	90	0.11	0.32	1.24
	180	0.00	0.00	0.24
	235	0.00	0.00	0.05
Dough	6	0.12	0.14	0.28
	30	0.09	0.10	0.25
	90	0.00	0.00	0.14
Bread	–	0.00	0.00	0.28

*No sugar added in formulation.
From Lee et al. (1959).

TABLE 67

Influence of Added Sugars on Sugar Concentrations as Influenced by Fermentation Time in Doughs and Breads

System	Sugar added (5%)	Fementation time	Percentage found		
			Glucose	Fructose	Maltose
Dough	Glucose	6	2.70	0.10	0.38
		30	2.60	0.14	0.36
		90	1.75	0.12	0.20
Bread		–	1.53	0.10	0.15
Dough	Fructose	6	0.19	2.74	0.45
		30	0.17	2.45	0.34
		90	0.11	2.20	0.16
Bread		–	0.10	1.66	0.10
Dough	Sucrose	6	1.55	1.52	0.47
		30	1.46	1.45	0.40
		90	1.18	1.37	0.19
Bread		–	0.98	1.21	0.18
Dough	Maltose	6	0.08	0.12	2.56
		30	0.08	0.13	2.22
		90	0.00	0.00	2.08
Bread		–	0.00	0.00	1.79

From Lee et al. (1959).

When no sugar was added to the formulation, they found that glucose and fructose were absent in bread. These monosaccharides were not present in dough after approximately 90 min of fermentation, probably because they were utilized by yeast. Maltose was still present in the dough after 90 min of fermentation and its level actually increased during baking. Based on these data, no glucose or fructose were available for thermal reaction.

When five percent of various sugars were added, they found that approximately half of the sugar remained after only 6 min of fermentation. Added residual sugars were also found in the bread (Table 67).

Since carbohydrates and amino acids react readily to form a wide variety of compounds, some highly pigmented, Rubenthaller et al. (1963) measured their reactivity based on their ability to darken bread crust (Table 68).

They added equimolar concentrations of either glycine, lysine or glutamic acid to breads, along with carbohydrates representative of pentosans, hexoses, disaccharides and trisaccharides. Of the three amino acids, glutamic acid was found to be the least active.

In a similar study Linko and Johnson (1963) added various amino acids to bread and measured the resulting amino acid composition of both crumb and crust (Table 69).

As expected, crust amino acid levels, independent of sugar incorporated, were quite low when compared to the crumb portions.

Protein Additions

The methods and merits of fortifying bread with protein are not the subject of this review; it should be sufficient to say that the majority of proposed additives have certain detrimental sensory effects on bread quality. Flóur substitution levels above 15% are not common. The data reported by Kim and de Ruiter (1968), as shown in Table 70, explain the situation (Table 70).

TABLE 68

Influence of Equimolar Concentrations* of Carbohydrates and Amino Acids on Bread Crust Color

	Amino Acid			
Carbohydrate	None	Glycine	Lysine	Glutamic Acid
None	41.0**	27.0	34.5	44.5
Pentoses				
Arabinose	15.5	13.0	11.5	14.0
Ribose	15.0	11.0	10.5	12.5
Xylose	15.5	11.0	10.0	13.0
Hexoses				
Galactose	17.5	10.5	11.5	13.0
Glucose	21.5	14.5	15.0	23.0
Levulose	20.0	14.0	14.0	21.0
Mannose	19.5	14.5	14.5	19.5
Sorbose	16.0	11.0	11.0	14.0
Rhaminose	23.0	15.0	14.0	18.5
Disaccharides				
Cellobiose	20.0	14.0	10.5	17.0
Lactose	17.0	11.0	10.5	15.5
Melibiose	15.5	11.0	10.0	15.0
Sucrose	21.5	15.5	14.0	17.0
Maltose	25.0	18.0	16.0	21.0
Trehalose	31.0	22.5	25.5	35.0
Trisaccharides				
Melezitose	26.0	20.5	18.0	21.0
Raffinose	21.5	10.0	9.0	10.5

*Glycine 0.2 g and hexoses 2 g.
**The higher the number, the lighter the color.
From Rubenthaller et al. (1963).

TABLE 69

Influence of Sugar Type on Selected Amino Acid Composition in Bread Crumb and Crust

	Aspartic Acid		Glutamic Acid		Alanine		Isoleucine + leucine	
Sugar	Crumb	Crust	Crumb	Crust	Crumb	Crust	Crumb	Crust
Arabinose	3.0*	0.7	2.8	0.8	1.8	0.9	0.7	0.3
Cellubiose	3.0	0.2	4.0	1.0	1.5	0.6	0.3	0.1
Fructose	3.4	0.2	4.4	0.9	1.2	0.4	0.2	0.1
Galactose	3.6	1.4	3.6	1.2	1.4	0.8	0.3	0.1
Glucose	3.6	0.8	5.2	1.6	1.4	0.6	0.4	0.1
Lactose	3.6	0.6	2.8	0.7	1.6	0.8	0.4	0.1
Maltose	4.6	0.8	2.8	0.5	1.6	0.6	0.4	0.1
Mannose	4.2	0.6	4.0	1.2	1.4	0.5	0.3	0.1
Melibiose	3.4	0.5	2.8	0.4	1.5	0.8	0.2	0.1
Raffinose	4.2	0.4	4.0	0.9	1.5	0.5	0.2	0.1
Rhamnose	4.4	1.0	3.0	0.6	2.0	1.2	0.4	0.1
Sorbose	5.8	0.6	2.8	0.4	1.7	0.5	0.4	0.1
Sucrose	3.8	0.4	4.4	0.9	1.5	0.5	0.2	0.1
Xylose	7.0	0.6	3.2	0.2	2.8	0.8	0.8	0.1

*Mg/100 g dry matter.
From Linko and Johnson (1963).

TABLE 70

Sensory Properties of Protein Supplemented Non-wheat-based Breads*

Protein additive	Crumb color	Taste description
Low-fat peanut	Rather white	Good, fairly neutral
Whole fish	Gray-green	Somewhat fishy
Torula	Ochre yellow	Spoiled fish
Cottonseed	Greenish yellow	Fairly good
Lipoprotein	Light gray	Fairly good
Peanut proteinate	Light gray	Fairly good
Defatted peanut	Grayish brown	Good, malt-like
Full-fat soy	Lemon yellow	Fairly good, neutral
Enzymatic soy	Grayish cream	Salty, fairly good
Defatted soy	Grayish cream	Fairly neutral, good
Soy proteinate	Rather white	Good, neutral
Flaked sesame	Rather white	Fairly good
Defatted coconut	Gray	Bad
Coconut concentrate	Light gray	Coconut Taste
Low-fat sesame	Light gray	Good
Low-fat safflower	Dirty gray	Spicy

*Yam flour base.
From Kim and de Ruiter (1968).

TABLE 71

Effect of Casein and Gliadin Additions on Bread Odor and Taste

Additive*	Sample location	Sensory property	Triangle test level of significance**	Panel preference
Casein	Crumb	Odor	0.05	6 Control, 3 casein
Gliadin	Crumb	Odor	0.01	9 Control, 1 gliadin
Gliadin	Crumb	Taste	0.01	7 Control, 2 gliadin
Gliadin	Crust	Taste	0.05	5 Control, 4 gliadin

*Level was 1.22g/100 g of dough.
**Compared to no-additive control.
From Wick et al. (1964).

These data are for non-wheat-based products, and the color imparted by most additives detracts from their use. The other major sensory limitation is the resulting flavor; some protein additives can be highly detrimental from a flavor standpoint.

Most common bread formulations have a protein additive incorporated. Usually this additive is dry milk solids; traditionally, the use of dry milk solids has been promoted because of their functional and sensory contributions. Specifically, the use of dry milk solids results in a finished product with better flavor and color.

Wick et al. (1964) reported on the odor and taste of bread to which 1.22 g of either casein (milk protein) or gliadin (corn protein) was added to 100 g of dough (Table 71).

In all cases the panel preferred the sensory properties of the controls baked without additives. Although protein additions were at a level closely approximating the protein level if dry milk solids were used, significant differences in preferences were noted. In the case of dry milk solids the lactose present can significantly contribute to flavor properties associated with bread.

Amino Acid Additions

Some nutritionists have suggested that the fortifying of bread with certain amino acids, especially lysine, would be a significant advance in improving the nutritional value of bread products. In 1959 Ehle et al. reported on the taste properties of white bread, cake, and donuts to which varying amounts of lysine monohydrochloride were added. They reported the taste threshold concentration of the lysine additive in white bread was 0.5 g/100 g of flour. This concentration was approximately double the level recommended from a nutritional improvement standpoint. In the case of donuts, which had ingredients such as vanilla, cinnamon, and nutmeg added, taste threshold levels approached 1.5 g/100 g of flour. At concentrations above taste threshold panel members described the product as tasting salty, bitter or having a peculiar aftertaste. The characteristic lysine taste was most pronounced in the crust. At the recommended supplementation level this group reported no significant difference in taste between lysine fortified and unfortified breads, either in the crust or crumb.

A major disadvantage of adding lysine was that it had a dramatic crust-browning effect which would certainly influence acceptability of the product. The authors suggested longer baking time at a lower temperature to overcome this problem. However, a slowdown in the baking process might be resisted because of economic considerations. A minor reservation about this study is that the amount of lysine remaining in the product was not measured. If baking losses are considered, perhaps higher levels of lysine would be required to satisfy nutritional requirements; the higher levels might more closely approximate taste threshold levels in the product. For example, Rosenberg and Rohdenburg (1951) reported that lysine losses in white bread ranged from 9.5 to 23.8%, the average loss as a direct result of baking being approximately 15%. They reported that toasting resulted in another 10 to 15% loss of lysine and staling brought about another 5 to 10% loss. Similar results on lysine destruction due to baking were obtained by McDermott and Pace (1957); they found minimal lysine losses in the crumb, whereas lysine loss in the crust approximated 15%. Pomeranz (1962) found minimal lysine losses in bread crumb ranging from 2 to 10% depending on the type of flour utilized. Lysine destruction in bread crust can be critical, especially in bread products that have a large crust area to crumb area, such as rolls. The problem is also complicated since if additional lysine is added to compensate for baking losses, objectionable flavors and colors may result.

Zentner (1961) reported on using lysine to improve the crust color of certain flour types normally producing bread with pale, greyish crust. In this study lysine was not added for nutritional considerations, but because it would react with reducing sugars to form desirable pigment products.

In addition to their nutritional value, free amino acids can also influence bread flavor and color. The possibility that certain amino acids may be useful as bread flavor enhancers was appreciated by various groups investigating bread flavor. For example, Wiseblatt and Zoumut (1963) reported on the aroma properties from the interaction of numerous amino acids with dihydroxyacetone (Table 72).

They concluded that the reaction of the amino acids proline, isoleucine and serine resulted in bread related aromas.

A similar approach to bread flavor was reported by Morimoto and Johnson (1966); they reacted various amino acids with glucose. Their observations are shown in Table 73.

They likewise observed a bread-like aroma from the reaction of proline. Other amino acids judged to possess bread-like aromas included aspartic acid, alanine, leucine, isoleucine and valine.

Earlier, Wick et al. (1964) had evaluated bread odor and taste properties after the direct addition of 100 mg of certain amino acids to 100 g of dough. In most cases, the panel preferred the sensory properties of the control breads containing no amino acid additions. One exception was proline; the panel statistically preferred the odor of the proline additive product. No difference in flavor preference was noted between the proline product and a control with no amino acid added (Table 74).

The use of amino acids as bread flavor precursors will be discussed in a later section.

Rooney et al. (1967) expanded earlier reports on the role of amino acids in bread flavor by measuring levels of carbonyl compounds formed in a model system through the interaction of amino acids and glucose (Table 75).

Proline reacted to form relatively low levels of volatile carbonyls in relation to some of the more reactive amino acids. Thus, degree of reactivity may not be critical, but proper individual, as well as total, concentrations could be the key to desirable bread flavor.

Their model system work was expanded (Table 76) to include a starch paste system.

The most significant difference between this table and the previous one was that quantities detected

TABLE 72

Aromas Resulting from the Action of Amino Acids and Dihydroxyacetone

Amino acid	Resulting aroma
Proline	Very strong, crackers, crust, toast
Lysine	Strong, dark corn syrup
Valine	Strong, yeasty, protein hydrolysate
Alanine	Weak, caramel
Glutamic acid	Moderate, chicken broth
Aspartic acid	Very weak
Arginine	Very weak
Cysteine	Mercaptan, hydrogen sulfide
Hydroxyproline	Weak, like proline
Phenylalanine	Very strong, hyacinth
Leucine	Strong, cheesy, baked potato
Isoleucine	Moderate crust
Serine	Weak, vaguely like bread
Threonine	Very weak
Methionine	Baked potato
Glycine	Baked potato
Histidine	Very weak

From Wiseblatt and Zoumut (1963).

TABLE 73

Aroma Properties Resulting from the Reaction of Glucose and Amino Acids

Amino acid reacted	Resulting aroma
Serine	Moderate
Glutamic acid	Moderate
Hydroxyproline	Weak
Aspartic acid	Strong, breadcrust
Tyrosine	Weak
Alanine	Moderate, breadcrust
Arginine	Weak
Isoleucine	Moderate, breadcrust
Valine	Moderate, breadcrust
Tryptophan	Strong
Histidine	Moderate
Proline	Strong, breadcrust, cracker
Threonine	Weak
Cystine	Strong, hydrogen sulfide
Methionine	Strong, baked potato
Glycine	Moderate
Phenylalanine	Strong, flower
Lysine	Weak
Leucine	Moderate, breadcrust

From Morimoto and Johnson (1966).

TABLE 74

Effect of Amino Acid Additions on Bread Odor and Taste

Additive – level	Location	Property	Level of Significance*	Preference
Alanine 100 mg/100 g dough	Crust	Odor	< 0.001	18 Control, 8 alanine
Alanine 100 mg/100 g dough	Crust	Taste	< 0.001	16 Control, 9 alanine
Arginine 100 mg/100 g dough	Crumb	Odor	0.01	11 Control, 10 arginine
Histidine 100 mg/100 g dough	Crumb	Taste	0.05	8 Control, 5 histidine
Proline 100 mg/100 g dough	Crumb	Odor	0.05	4 Control, 13 proline
Proline 100 mg/100 g dough	Crust	Odor	0.001	7 Control, 15 proline
Proline 100 mg/100 g dough	Crust	Taste	0.001	12 Control, 11 proline
Leucine 50 mg/100 g dough	Crumb	Odor	0.03	2 Control, 1 leucine
Valine 100 mg/100 g dough	Crust	Odor	0.01	16 Control, 3 valine
Valine 100 mg/100 g dough	Crust	Taste	0.03	15 Control, 2 valine

*Compared to no-additive control.
From Wick et al. (1964).

were dramatically reduced in the latter system. The variation in volatile compounds produced among all amino acids evaluated was quite small in the starch paste system.

Because free amino acids are flavor reactive, estimates of free amino acid levels as influenced by the baking process have been made. As seen in Table 77, and discussed previously, flour can serve as a free amino acid source.

Adding yeast and mixing resulted in a 300% increase in the amount of free amino acids. However, as fermentation progressed the level was approximately halved. Thermal reactions account for the low amount of free amino acids remaining in bread crust.

The same general trend was reported by El-Dash (1971) for specific free amino acids (Table 78).

TABLE 75

Carbonyl Formation from Glucose-Amino Acid Interactions in a Model System

	Mg compound formed				
Amino acid	Formaldehyde	Acetaldehyde	2-Propanone	2-Methylpropanal	3-Methyl butanal
None	0.40	0.16	0.08	–	–
Glycine	0.60	2.06	0.56	0.34	0.30
Alanine	0.48	13.72	0.95	1.26	0.29
Valine	0.82	3.16	4.93	40.1	4.08
Leucine	0.79	2.13	2.00	1.32	82.3
Isoleucine	0.73	2.64	1.75	2.64	79.0**
Glutamic acid	0.15	1.41	0.45	0.37	–
Histidine	0.18	0.59	0.47	0.29	–
Lysine	0.30	1.88	0.98	0.51	0.40
Arginine	0.30	0.88	1.24	0.36	–
Phenylalanine	0.56	2.19	0.85	–	6.28***
Tryptophan	0.39	2.11	6.08	1.32	0.30
Proline	0.09	0.68	0.34	0.23	–
Methionine	0.59	7.60*	2.09	0.63	0.59

*Includes methional.
**2-Methylbutanal.
***Includes Phenylacetaldehyde.
From Rooney et al. (1967).

TABLE 76

Carbonyl Formation from Glucose and Amino Acid Interactions in a Starch Paste System

Amino acid	Formaldehyde	Acetaldehyde	2-Propanone	2-Methylpropanal	3-Methylbutanal
None	0.03	0.01	0.05	0.09	–
Glycine	0.06	0.33	0.14	0.28	0.16
Alanine	0.06	0.87	0.25	0.34	0.23
Valine	0.08	0.45	0.26	3.15	0.32
Leucine	0.07	0.27	0.18	0.43	1.36
Isoleucine	0.04	0.32	0.14	0.26	5.29***
Glutamic acid	0.06	0.28	0.16	0.24	0.29
Histidine	0.06	0.29	0.12	0.26	0.06
Lysine	0.05	0.44	0.15	0.31	0.21
Phenylalanine	0.06	0.43*	0.18	0.14	0.15
Proline	0.03	0.33	0.14	0.25	0.16
Arginine	0.04	0.29	0.14	0.26	0.10
Methionine	0.09	0.06**	0.21	0.27	0.19

*Includes phenylacetaldehyde.
**Includes methional.
***2-Methylbutanal.
From Rooney et al. (1967).

Measurable amounts of the free amino acid lysine were lacking in bread crust.

As mentioned previously, free amino acids can also influence crust color. This is apparent from data in Table 79.

Adding any amino acid darkened crust color. Certain amino acids, such as methionine and arginine, were highly reactive in the presence of glucose. The range of darkening with xylose as the carbohydrate source was smaller than with glucose, but xylose alone initially produced a darker crusted product than did glucose used alone.

TABLE 77

Total Free Amino Acid Content in Various Bread Systems

System	Free amino acids (mg/100 g)
Flour	13
Unfermented dough	52
Fermented dough	23
Unfermented bread crumb	37
Fermented bread crumb	22
Unfermented bread crust	13
Fermented bread crust	2

From El-Dash and Johnson (1970).

TABLE 78

Influence of the Baking Process on the Free Amino Acid Content

Amino acid (s)	Flour	Unfermented dough	Fermented dough	Bread crumb	Bread crust
Alanine	14.3*	87.0	23.0	29.1	2.1
Arginine	2.1	6.7	15.3	11.4	Trace
Aspartic acid	14.2	60.5	14.9	10.9	0.8
Cystine	0.6	3.3	1.2	1.0	Trace
Glutamic acid	8.9	71.9	46.8	51.4	1.1
Glycine	4.1	11.7	10.4	12.4	0.8
Histidine	1.0	–	–	–	–
Isoleucine	4.1	6.6	1.6	3.1	0.5
Leucine	5.5	8.7	1.7	4.8	0.3
Lysine	2.4	32.3	25.5	12.8	Trace
Methionine	0.7	2.1	1.0	1.8	Trace
Phenylalanine	3.3	5.9	0.2	2.0	1.6
Proline	4.7	21.7	17.1	11.2	0.4
Serine and threonine	27.5	68.1	19.2	22.9	1.2
Tyrosine	3.1	6.0	0.5	1.1	–
Valine	6.8	16.2	4.6	6.4	0.6

*μmol/100 g dry basis.
From El-Dash (1971).

TABLE 79

Influence of Amino Acid and Sugar Additions on Crust Color (% Reflectance)

Amino acid	Glucose	Xylose
None	30.3*	21.8
Glycine	18.5	12.2
Alanine	19.5	14.5
Valine	22.2	15.0
Leucine	20.0	13.8
Isoleucine	21.8	18.0
Glutamic acid	19.0	17.5
Histidine	17.0	10.6
Lysine	17.6	12.2
Phenylalanine	18.3	12.0
Proline	23.5	18.5
Arginine	8.5	–
Methionine	7.0	–

*The smaller the number, the darker the color.
From Salem et al. (1967).

TABLE 80

Influence of Dough Absorption on Bread Preferences

Dough absorption	% Water	% Preference*	% Preference*	% Preference**
Stiff	56	39	72	59
Normal	62	31		
Slack	68	30	28	41

*Tested 24 hr after baking.
**Tested 72 hr after baking.
From Cathcart (1937).

TABLE 81

Effect of Dough Mixing Time on Crust Color and Flavor

Mixing time	Crust color	Flavor
Slightly mixed	Light, gray	Raw dough, flour
Mixed to optimum	Brown	–
Overmixed	Full, deep brown	Full, rich

From Baker et al. (1953).

Absorption

One could reason that the percent of water absorption in bread dough could affect final bread flavor by influencing the crumb and crust temperature of the baking bread. For example, a dough having a low absorption would heat more rapidly, attaining a higher crust temperature since there would be less water to evaporate from the system. Likewise, a dough with high absorption would heat more slowly, and the amount of thermally induced flavor compounds would be lower.

In Cathcart's (1937) report, a sensory panel studied the influence of absorption on baked bread (Table 80).

In one series of this study the percent of water added created stiff, normal and slack doughs that were evaluated for bread preference. Thirty-nine percent of the panel preferred the bread made from the stiff dough. When only stiff and slack doughs were compared, the majority of the panel (72%) preferred the stiff dough product. The influence of storage time was also evaluated with regard to degree of absorption; after 72 hr of storage, 59% of the panel still preferred the lower absorption product. The water absorption of dough had a definite effect on bread flavor; the panel selected the product with the least absorption which should have been the most flavorful.

Mixing

Baker et al. (1953) reported that mixing time can influence the crust color and flavor of bread (Table 81).

They observed that undermixing resulted in a light gray crust and raw dough flavor, whereas overmixing produced a full, deep, brown color and a full, rich flavor. With undermixing, the ingredients are not fully dispersed and conditions are not optimum for color and flavor formation. The opposite is true for an overmixed dough; more chemical breakdown has occurred, which enhances the flavor.

Mixing is a critical function in bread production and much research has been devoted to evaluating mixing techniques. For example, Fortmann (1959) evaluated various mixing techniques and their effects on the amount of carbonyls found in bread crumb. His data are summarized in Table 82.

This was the era when much controversy existed concerning the flavor quality of conventional vs continuous mix bread. Fortmann (1959) found that the continuous mix product had approximately half the crumb carbonyl level as conventional mix bread. No data were presented comparing the crust carbonyl

TABLE 82

Influence of Type of Mixing on Carbonyl Content in Bread Crumb

Mixing – additive	mg/g crumb
Conventional bread	0.096
Continuous mix	0.041
Continuous mix + oxidized lard	0.244
Laboratory bread – open mixer	0.044
Laboratory bread – oxygen mix	0.084
Laboratory bread – oxygen mix + lipoxidase	0.120

From Fortmann (1959).

TABLE 83

Influence of Fermentation Technique and Time on Acetoin Production in Bread Dough

	Mg acetoin/150 g flour				
	Hour				
Fermentation technique	1	2	3	4	5
Straight-dough	2.5	1.5	0.6	0.3	–
Sponge-dough	0.0	0.0	0.0	1.1	3.2

From Visser't Hooft and de Leeuw (1935).

TABLE 84

Influence of Fermentation Temperature and Time on Acetoin Formation in Bread Dough

	Mg acetoin/150 g dough			
	Hours			
Fermentation temperature (°F)	1	2	3	4
77	0.2	1.3	3.4	2.5
86	2.0	2.5	2.2	1.1
96	2.0	0.0	0.1	0.2

From Visser't Hooft and de Leeuw (1935).

levels in the two products. He demonstrated quite dramatically that crumb carbonyl level could be greatly changed by the type of mixing employed. The use of oxygen and lipoxidase during mixing resulted in the highest crumb carbonyl level.

Fermentation Procedures

Until recently fermentation was considered one of the key stages in the proper development of desirable bread flavor. However, there are indications that fermentation can be minimized, or even eliminated. Historically, the straight- and sponge-dough methods of fermentation have been used for the batch production of bread. An early study demonstrated that the levels of certain bread flavor related compounds could be influenced by the fermentation method employed (Table 83).

For example, if the straight-dough process were used, acetoin levels gradually decreased as fermentation time increased; in sponge-doughs acetoin did not appear until after four hours of fermentation. The same study also pointed out the importance of fermentation temperature and time on acetoin production (Table 84).

With a temperature of 77°F, acetoin production was slow, peaking after three hours of fermentation. With an intermediate fermentation temperature of 86°F, acetoin production reached its maximum after two hours. At a temperature of 96°F, peak production was achieved in one hour.

King et al. (1937) also evaluated the sensory properties of bread as influenced by fermentation temperature (Table 85).

They found that both the odor and taste of bread fermented at 86°F was preferred to bread fermented at 78°F. Presumably, the higher fermentation temperature resulted in more characteristic bread flavor.

The role of fermentation time on total organic acid and carbonyl levels was investigated by Cole et al. (1962) (Table 86).

No organic acids were detected at the beginning of fermentation, but their level dramatically increased after one to two hours. The level of organic acids continued to rise gradually through the remaining time. In

TABLE 85

Influence of Fermentation Temperature on Bread Sensory Properties

	Odor		Taste	
Fermentation temperature (°F)	Best (%)	Poorest (%)	Best (%)	Poorest (%)
86	58	42	57	43
78	42	58	43	57

From King et al. (1937).

TABLE 86

Influence of Fermentation Time on Total Acid and Carbonyl Concentration

Fermentation time (hr)	Total organic acids*	Total carbonyl compounds**
0	0.00	2.8
1	0.04	11.0
2	0.69	20.0
3	0.91	24.0
4	0.91	23.0
5	1.10	21.7
6	1.25	19.3

*Meq sodium hydroxide to neutralize acids in 50 ml pre-ferment.
**μmol acetophenone/l pre-ferment.
From Cole et al. (1962).

TABLE 87

Influence of Fermentation Method on the Odor and Taste of Bread

		Odor		Taste	
Test	Method of mixing	Best (%)	Poorest (%)	Best (%)	Poorest (%)
1	Straight dough	51	49	42	57
2	Straight dough	37	62	27	72
1	Sponge	49	51	57	42
2	Sponge	62	37	72	27

From King et al. (1937).

the case of carbonyls, there was a gradual rise in concentration through three hours of fermentation and then the level gradually dropped for the remaining time.

King et al. (1937) evaluated the odor and taste preferences of bread made from the straight- and sponge-dough procedures (Table 87).

They presented data from two separate tests. In the case of odor, no clear-cut preference was noted in one test, whereas in the other series, the odor of bread prepared by the sponge process was preferred. In both taste tests a definite preference for the sponge process product resulted.

Slightly different results (Table 88) were obtained in a much later study by Jackel and Ersoy (1961).

When straight- and sponge-dough processes were compared, the fermentation aromatics and acids resulting from the straight-dough procedure were judged more intense than those observed from the sponge-dough process.

TABLE 88

Influence of Fermentation Method on the Sensory Properties of Bread

	Sensory property		
Fermentation method	Fermentation aromatics	Fermentation acids	Yeasty taste
Straight dough	4*	4	4
Sponge dough	3	3	4

*Intensity scale from 0 to 4.
From Jackel and Ersoy (1961).

TABLE 89

Influence of Fermentation Time on Acid Production in Pre-ferments

Fermentation time (hr)	Lactic acid	Acetic acid
2	65*	15
4	70	35
8	80	40
16	100	50
24	130	50

*Mg/100 ml of pre-ferment.
From Johnson et al. (1958).

Johnson et al. (1958) also found that acid production (lactic and acetic) increased as fermentation time increased. Over the rather extended fermentation times studied, lactic acid concentration increased approximately twofold and that of acetic acid threefold (Table 89).

Linko et al. (1962b) compared the carbonyl composition in the crust and crumb of bread prepared by four different fermentation methods (Table 90).

In the crust portion, the total carbonyl level was highest in bread produced by the sponge method (29.8 mg/100 g) and lowest in the no-time fermentation method (21.7 mg/100 g). In the crumb portion the highest total carbonyl levels was found using the pre-ferment method, while the lowest level was associated with the straight-dough fermentation method.

The interaction of fermentation time on free amino acid distribution is summarized by the work of Morimoto (1966) in Table 91.

For most free amino acids there was a decrease, or actual disappearance, after six hours of fermentation. However, the free glycine level was found to increase.

The exact role of fermentation in relation to certain volatile carbonyls (Table 92) was measured by El-Dash and Johnson (1970).

Varied results were obtained, dependent on individual compounds. In most cases concentrations increased, but the concentrations of several major compounds decreased. These data were complicated because crust was evaluated and losses due to volatility may not have been considered (Table 93).

A similar study was performed by Roiter and Borovikova (1971), except fewer compounds were evaluated and only the exact influence of fermentation was measured. Again, the concentration of most compounds measured increased.

The importance of fermentation on final bread flavor has been seriously questioned by numerous researchers. Their main argument is that although the concentration of a compound has been found to develop or increase due to fermentation, there is no definite evidence that its presence in bread can be directly related to bread flavor. This philosophy is especially common among proponents of shortened or

TABLE 90

Influence of Fermentation Technique on Carbonyl Production

	Fermentation Method							
	Straight		Sponge		No-time		Pre-ferment	
Carbonyl(s)	Crust	Crumb	Crust	Crumb	Crust	Crumb	Crust	Crumb
Formaldehyde	0.99*	0.20	0.98	0.20	0.86	0.17	1.02	0.14
Acetaldehyde	2.20	0.32	2.17	0.35	1.65	0.29	1.82	0.35
Acetone + propionaldehyde	12.8	0.75	17.1	0.85	10.7	0.81	15.6	2.11
Iso-butyraldehyde + methyl ethyl ketone	0.82	0.14	0.97	0.15	1.47	–	0.70	0.23
Iso-valeraldehyde + valeraldehyde + 2-methylbutanal + hexaldehyde	2.02	0.51	1.60	0.76	1.18	0.86	3.23	0.62
Furfural	0.16	–	0.34	–	0.31	–	0.04	–
Hydroxymethylfurfural	3.19	0.65	6.65	0.72	5.43	0.59	3.29	0.56
Total carbonyl compounds	22.2	2.6	29.8	3.0	21.7	2.8	25.7	4.0

*Mg/100 g.
From Linko et al. (1962b).

TABLE 91

Influence of Fermentation Time and Baking on Free Amino Acid Content

	Dough		French-type bread	
Free amino acid(s)	0 Hour	6 Hour	Crumb	Crust
Aspartic acid	0.55*	0.05	0.07	0.03
Threonine and serine	1.08	0.13	0.18	0.07
Glutamic acid	0.58	0.11	0.28	0.10
Proline	0.08	0.10	0.13	0.08
Glycine	0.17	0.26	0.32	0.14
Alanine	1.08	0.07	0.18	0.08
Valine	0.21	–	0.05	0.02
Methionine	0.04	0.01	0.01	0.002
Isoleucine	0.10	–	0.03	0.01
Leucine	0.14	0.04	0.05	0.02
Tyrosine	0.08	–	0.02	0.005
Phenylalanine	0.09	–	0.02	0.008
Tryptophan	0.30	0.05	0.03	0.02
Lysine	0.13	0.11	0.18	0.08
Arginine	0.14	0.10	0.06	0.05

*μm/g (dry basis).
From Morimoto (1966).

no-time fermentation methods. The author feels that this attitude is somewhat unrealistic because if this assumption were expanded to include the flavor properties of other food systems, few data would be valid today.

Sensory data have shown that acceptable bread products can be made using abbreviated mixing and fermentation techniques. For example, Butterworth (1935) reported that unless cut drastically, fermentation time had little influence on the resulting odor and flavor of bread. It should be noted that his data represented the totaled scores of five judges.

TABLE 92

Influence of Fermentation on Carbonyl Content of Bread Crust

Compound(s)	Unfermented crust	Fermented crust
Ethanal	37.4*	27.0
Propanal	21.6	21.6
2-Propanone + 2-methylpropanal	26.9	30.0
Unknown	15.6	17.4
Butanal	Trace	0.5
2-Butanone	53.2	29.1
2-Methylbutanal + 3-methylbutanal	1.4	1.1
Unknown	1.1	1.3
Pentanal	0.7	0.8
2-Methylpentanal	8.4	9.9
Hexanal	4.6	4.8
Furfural + hydroxymethylfurfural	14.2	16.1

*mol/100 g.
From El-Dash and Johnson (1970).

TABLE 93

Influence of Fermentation on the Levels of Certain Carbonyls

	Mg/100 g of specific volume	
Compound	Before fermentation	After fermentation
Furfural	1.12	1.70
Benzaldehyde	2.28	4.80
Oxybenzaldehyde	0.80	1.07
Phenylacetylaldehyde	0.15	0.06

From Roiter and Borovikova (1971).

Cornford (1962), using duo-trio sensory methods, reported no statistical difference in flavor intensities between breads fermented for 3 or 18 hr using bulk fermentation, as these breads were compared to breads made by the Chorleywood process, which utilizes intense mechanical mixing in place of bulk fermentation. He also reported that bread made from conventionally mixed dough undergoing no fermentation had the same flavor intensity as normally fermented bread. However, flavor was judged only on the basis of intensity; possible differences in overall flavor impression or perception were not mentioned.

Collyer (1966) also reported on the sensory properties of breads made by using various fermentation techniques. For example, she evaluated crumb flavor differences in brew-bread and conventional bulk-fermented bread as influenced by storage. These data are summarized in Table 94.

She reported a significant difference in flavor between the two fermentation methods. She also concluded that freshness had no significant effect on preference influenced by fermentation method.

The same fermentation methods were used to evaluate the crumb and crust portions of bread stored for 16 hr. When crumb and crust portions were evaluated independently, no significant differences in flavor were noted. However, it was pointed out that a trend toward preference for the bulk-fermented bread was noted (Table 95).

The odor properties of freshly baked laboratory bread fermented by the bulk and Chorleywood processes were compared, and in four out of five triangle tests performed no significant flavor difference was noted.

A further flavor evaluation regarding storage influence was performed comparing three hour commercially bulk-fermented bread with Chorleywood process bread (Table 96).

TABLE 94

Influence of Fermentation Method and Time on Bread Preference

Bread age (hr)	Triangle test significance	Preference		
		Brew bread	Control	No preference
4	Significant 5% level	7	10	5
28	Significant 5% level	9	12	0

From Collyer (1966).

TABLE 95

Influence of Fermentation Method on Crumb and Crust Bread Preferences

Portion – fermentation methods compared	Triangle test significance	Preference		
		Brew bread	Bulk	No preference
Crumb-brew vs. bulk	Not significant at 5% level	3	10	2
Crust-brew vs. bulk	Not significant at 5% level	4	7	1

From Collyer (1966).

TABLE 96

Flavor Comparisons of Breads Prepared by Various Fermentation Methods

Fermentation method compared	Storage time (hr)	Triangle test significance	Preference		
			Bulk	Chorleywood	No preference
Bulk vs. Chorleywood	6	Not significant at 5%	4	0	5
Bulk vs. Chorleywood	30	Not significant at 5%	7	10	2

From Collyer (1966).

These data indicated that no significant difference in flavor was detected; it was concluded that the chemical products of yeast fermentation are not of prime importance in dictating bread flavor.

A Canadian group has also reported similar results, i.e. mechanically developed bread was not significantly distinguishable in aroma and taste from bread fermented by the traditional sponge-dough method (Table 97).

However, it should be noted that the formulations used incorporated high protein flours (12.8 to 16.0%). This factor would tend to minimize any flavor differences.

Proofing

The amount of time permitted for a product to proof could influence its flavor since proofing can be considered an extension of fermentation. An early study by Cathcart (1937) bears out this assumption (Table 98).

TABLE 97

Aroma and Flavor Comparisons of Bread as Influenced by Fermentation Method

Sensory property	Age of bread	Triangle test significance	Preference		
			Sponge-and dough	Mechanical development	No preference
Aroma	2	Not significant at 5%	11	4	1
Aroma	20	Not significant at 5%	5	5	6
Taste	20	Not significant at 5%	11	3	4

From Kilborn and Tipples (1968).

TABLE 98

Influence of Proof Time on Bread Preference

Proof time	% Preference
25 min underproofed	40
Normal proof time	60
25 min overproofed	38
Normal proof time	62

From Cathcart (1937).

TABLE 99

Influence of Proofing on Organic Acid Content in Dough and Bread

System	Meq. total acids
Dough, not proofed	10.1
Dough, proofed	11.8
Bread	13.1

From Cole et al. (1963).

TABLE 100

Influence of Proofing on Volatile Alcohol Content in Dough and Bread

System	Ml. volatile alcohols
Dough, not fermented	1.74
Dough, fermented	2.80
Bread	1.00

From Cole et al. (1963).

TABLE 101

Influence of Proofing on Total Carbonyl Content in Dough and Bread

System	Mmol. total carbonyls
Dough, not proofed	2.3
Dough, proofed	2.8
Bread	3.8

From Cole et al. (1963).

The panel expressed a strong preference for products that were proofed for a normal period of time over products that were either over- or underproofed. With overproofing, harsh flavors develop; with underproofing, a full bread flavor is lacking.

Cole et al. (1963) attempted to demonstrate the changes influenced by proofing in specific classes of bread flavor compounds. In addition, they measured these same compound classes after baking. In this way the relative importance of bread flavor formation up to the proofing step in bread production was shown in relation to the actual baking step.

In the case of organic acids (Table 99), they reported that proofing increased the level by approximately 17% over that of a no-proof control.

However, baking was found to increase the total organic acid level 30% over a no-proof control. Although proofing increased organic acid levels, the effect of the actual baking process was more influential.

When the volatile alcohol content was measured (Table 100), again proofing was found to increase the overall concentration, in this case by approximately 60% over that of the no-proof control. However, as would be expected, the baking process resulted in a significant decrease in the amount of alcohols, probably due to volatility.

Proofing resulted in an approximate increase of 20% in the total amount of carbonyls measured (Table 101).

TABLE 102

Influence of Proofing on the Levels of Certain Carbonyls

Compound	Mg carbonyl 100 g of product	
	Before proofing	After proofing
Furfural	1.70	1.76
Benzaldehyde	4.80	4.88
Oxybenzaldehyde	1.07	4.25
Phenylacetylaldehyde	0.06	0.13

From Roiter and Borovikova (1971).

However, carbonyl content increased 65% due to baking when compared to the no-proof control. It can be appreciated that proper proofing can be important to the overall flavor of bread.

In a later study, Roiter and Borovikova (1971) followed the change in concentration of four specific flavor compounds as influenced by proofing. This information is summarized in Table 102.

There was an increase in concentration for all compounds. In some cases the increase was four fold (oxybenzaldehyde). The exact reasoning following changes in composition of such compounds as furfural and benzaldehyde, predominantly formed under thermal conditions, was not discussed.

Loaf Size

It is well established that the temperature rise within a dough mass during baking is a direct function of oven temperature and the size of dough mass (Bailey and Munz, 1938). It is also known that the interior temperature of a dough mass will not exceed the boiling point of water, but within the crust, much higher temperatures are attained. This is especially true after most of the evaporation cooling effect has subsided. For thermal flavor formation the conditions are ideal, especially in the crust. The importance of dough size on its thermal properties is explained in Figure 4.

Dough from the same batch was made into three different size loaves (regular – 510 g; pup – 200 g; micro – 40 g) and proofed to height before baking under the same conditions. Just before baking, thermocouples were carefully inserted into the center and immediately below the surface of the loaves. The temperature rise was obtained for five different bakes with a recording potentiometer. These values were averaged to form Figure 4. The temperature rise was most rapid with the micro loaves and least rapid with the regular sized product. For example, it took 5 min for the micro loaf crumb to reach 180°F, while the pup loaf required 12 min and the regular loaf 18 min to reach the same temperature. The smaller dough masses underwent more extensive thermal treatments, which theoretically could result in increased flavor formation.

In this same study crust weights were averaged; for the regular, pup and micro loaves the weights were 17.4, 20.6 and 29.1%, respectively. From this standpoint the smaller loaf sizes could also be expected to contain more flavor compounds.

Rothe (1971) evaluated the influence of loafsize on the levels of certain compounds present in the crumb and crust of rye bread (Table 103).

Significantly higher levels of ethanol were found in the crumb than in the crust. This was expected due to its volatility. The opposite was found for furfural and hydroxymethylfurfural, which would verify their thermally induced formation. Regarding loaf size, conflicting data are presented. For example, although one would expect the smaller loaf to contain higher levels of the heat induced compounds, the opposite was reported. No explanation was offered for this apparent discrepancy.

Type of Oven

Another factor that could influence bread flavor is the type of oven used. Presumably this is due to the efficiency differences in the heating and heat retention rates of various ovens. If bread were baked in two different ovens for the same amount of time, sensory differences could be apparent. In a study appearing in

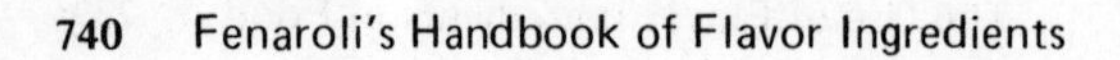

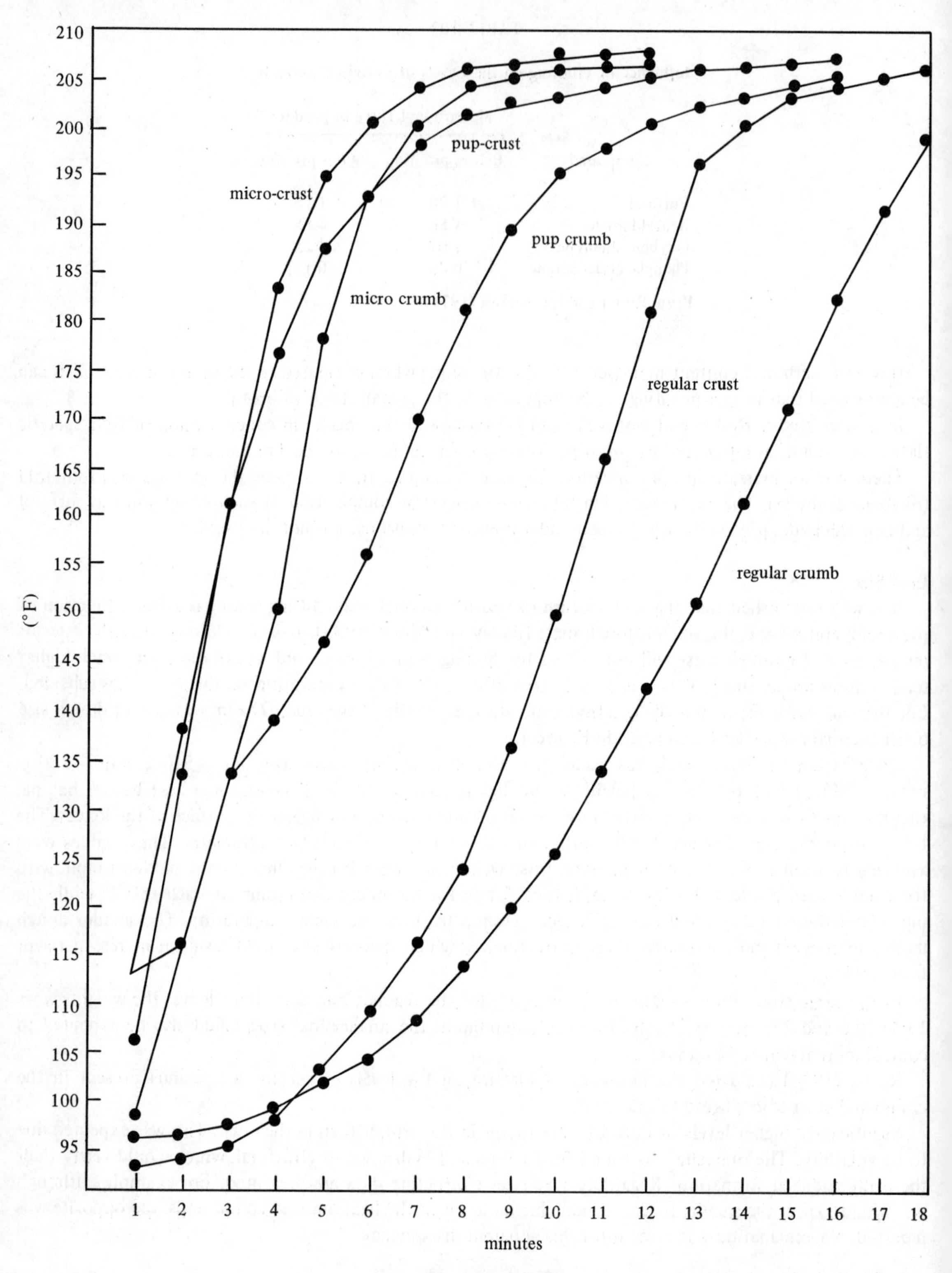

FIGURE 4. Influence of loaf size on the thermal properties of baking bread. (From Maga (unpublished data)).

TABLE 103

Influence of Loaf Size on the Flavor Compound Composition (ppm) in Rye Bread

	Crust		Crumb	
Compound	3.0 kg loaf	1.5 kg loaf	3.0 kg loaf	1.5 kg loaf
Furfural	25	12	0.7	0.3
Hydroxymethyl furfural	660	370	18	18
Isopentanal	27	20	1.6	1.1
Ethanol	890	630	3870	3350

From Rothe (1971).

TABLE 104

Effect of Oven type, Temperature, Baking and Time on Relative Carbonyl Composition

	Type of oven-time-temperature			
	Grate (150-200 °C)		Steam-tube drawplate (180 °C)	Traveling tray (230-270 °C)
Compound	60 min	80 min	56-60 min	48 min
Methylglyoxal	1*	1	1	1
Hydroxymethylfurfural	1	1	1	1
Furfural	2	3	3	2
Not identified	–	1	1	1
Formaldehyde	2	3	3	2
Diacetyl	1	1	1	1
Methional	3	2	3	2
Acetaldehyde	3	2	3	2
Phenylacetaldehyde	1	1	1	1
Propenal	1	1	1	1
Butenal	1	1	1	1
Acetone + propanal	3	2	3	2
Butanal	1	1	2	2
Methylpropanal	1	1	1	1
2-Methylbutanal	3	2	3	3
3-Methylbutanal	3	2	3	3
Pentanal	3	3	3	2
Methylpentanal	3	2	3	2
Hexanal	3	1	3	2
2-Hexanone	2	1	2	2
2-Heptanone	2	1	2	1
Octanal	2	1	2	1
Decanal	2	1	1	1
Dodecanal	2	1	1	1

*1: Small; 2: medium; 3: large.
From Hrdlicka and Honischova (1967).

the foreign literature (Hrdlicka and Honischova, 1967) blends of wheat and rye flour (60/40) were used for bread baked in three different types of ovens operated at various temperatures. The resulting breads were analyzed for volatile carbonyl compounds and for their sensory properties. The relative intensities of the compounds found are summarized in Table 104.

The same compounds are identified in all breads but ratios of individual compounds varied. The authors'

TABLE 105

Odor and Flavor Properties of Bread as Influenced by Oven Type and Time of Baking

Oven type	Baking time (min)	Oven temp. (°C)	Odor rating	Flavor rating	Odor of crumb
Grate	60	203	3*	3	Acid-aromatic
Grate	80	106	3	3	Acid-aromatic
Grate	60	215	4	3	Acid-aromatic
Grate	80	172	3	3	Acid-aromatic
Steam-tube drawplate	56	180	3	3	Slightly acid-agreeable
Traveling tray	48	230-270	2	3	Slightly acid

*0: Disagreeable; 5: excellent.
From Hrdlicka and Honischova (1967).

TABLE 106

Influence of Baking Temperature on Bread Sensory Properties

	Odor		Taste	
Baking temperature (°F)	**Best (%)**	**Poorest (%)**	**Best (%)**	**Poorest (%)**
450	31	69	35	65
425	69	31	65	35

From King et al. (1937).

interpretation of these data was that bread baked in the grate-type oven had a well-balanced composition of volatile compounds. As expected, baking in the grate-type oven for longer periods of time decreased the relative amounts of certain volatile compounds. The steam-tube draw-plate oven, which is similar in construction to the grate-type oven, produced similar results. In the case of the traveling-tray oven, an inferior product, one with an unbalanced ratio of volatile compounds, resulted. They concluded that ovens with lower heat favored more desirable bread flavor.

The odor and flavor properties of these breads were also judged on a six point scale ranging from zero, for strong and disagreeable, to five for excellent. Some of these data are summarized in Table 105.

These data indicate that no differences in the flavor rating were detected. The grate-type oven, in one instance, produced bread with a superior odor, while the traveling-tray oven's product had an inferior odor. Perhaps an expert panel would be able to detect minor sensory property differences due to oven type, which is purely academic. In a normal commercial operation the manufacturer has control over baking time and temperature, which are probably more important to final bread flavor than the oven used.

Baking Time and Temperature

Knowledge of heat's influence on bread flavor has established that baking time and temperature can definitely influence the sensory properties of bread. However, as shown by King et al. (1937) in Table 106, higher baking temperatures are not always preferable.

They found that 69% of the panel preferred the odor, and 65% preferred the taste, of the bread baked at the lower temperature. Actually, the lower temperature used in this study was close to a normal baking temperature; the panel may have been influenced by this factor. Perhaps higher baking temperatures than normal can result in excessive loss of pleasing volatile compounds and in the formation of bitter flavor complexes.

The influential role of baking time can be appreciated by viewing data obtained by Rothe (1966) with rye bread crust and summarized in Table 107.

TABLE 107

Influence of Baking Time on Carbonyl Composition (mg%) and Crust Properties in Rye Bread Crust

	Baking time (min)		
	30	50	70
Crust thickness (mm)	1	3	4.5
Sensory properties	mild, pleasing	slightly bitter	very bitter
Compound			
Acetaldehyde	0.95	1.56	1.89
2-Methylpropanal	0.02	0.19	0.70
2-Oxypropanal	0.60	1.00	1.30
2-Methylbutanal	0.50	2.03	3.60
2-Furaldehyde	0.33	0.74	1.67

From Rothe (1966).

TABLE 108

Influence of Baking Time on the Composition of Flavor Components in Rye Bread

Baking time (min)	Crust thickness (mm)	Crust furfural (ppm)	Crust acetaldehyde (ppm)	Crumb ethanol (ppm)
30	2.7	5.4	16.3	2900
40	3.2	9.8	18.5	2600
50	3.7	15.2	21.0	2350
60	4.1	18.5	22.7	2300
70	4.6	21.7	26.2	2300

From Rothe (1971).

The overall sensory properties became more objectionable as baking time increased from 30 to 70 min. Levels of the five compounds evaluated also increased with additional baking time. In the case of 2-methylpropanal, the level increased 35 times when baking time was extended from 30 to 70 min. A compound that can have pleasing sensory properties at low concentrations can become quite objectionable at higher concentrations. Another point brought out in this study was the dramatic increase in crust size because of increased baking time.

Similar data were presented by Rothe (1971); his study dealt with the influence of baking time on the composition of certain compounds in the crust and crumb of rye bread (Table 108).

Increasing baking time from 30 to 70 min resulted in approximately four times the furfural in the crust. As expected the amount of ethanol in the crumb decreased, due to volatility.

A detailed study dealing with the changes baking time caused in carbonyl composition was reported by Borovikova and Roiter (1971d). As can be seen from Table 109, no qualitative changes were observed.

Quantitative changes, however, were quite dramatic. For example, the total carbonyl concentration in the crumb dropped from a high of 13.4 mg/100 g of product to approximately half this amount after 55 min of baking. From this it would appear that the thermal production of carbonyls is not prevalent in the formation of bread crumb flavor. Data presented for individual crumb compounds indicated that most compounds were being volatilized, or destroyed, as baking time increased. For bread crust they reported a gradual increase in total carbonyl content as baking time increased, primarily due to the thermal formation of compounds. It is somewhat surprising that more dramatic differences were not found in the crust considering the wide range in baking times evaluated. Perhaps volatility was quite important; after a certain amount of time of the volatiles were lost to the atmosphere.

TABLE 109

Influence of Baking Time on Carbonyl Content of Crumbs and Crust

	Mg carbonyl 100 g of product							
	Crumb Baking time				Crust Baking time			
Compound	15	30	45	55	15	30	45	55
Methylglyoxal	0.05	0.02	0.09	0.05	0.08	0.12	0.11	0.04
Glyoxal	0.17	0.07	0.33	0.39	0.85	0.72	0.86	0.46
Acetaldehyde	5.25	1.93	1.98	0.65	1.43	0.93	0.07	0.07
Propionaldehyde	0.03	0.03	0.02	0.01	0.01	0.01	0.01	0.01
Isobutyraldehyde	0.11	0.04	0.04	0.02	0.12	0.04	0.04	0.02
Acrolein	0.07	0.04	0.13	0.05	0.23	0.16	0.18	0.09
Acetone	0.14	0.07	0.17	0.08	0.17	0.21	0.10	0.09
Butyraldehyde	0.08	0.06	0.15	0.06	0.16	0.15	0.09	0.05
Isovaleraldehyde	0.06	0.02	0.01	0.01	0.06	0.04	0.04	0.01
Unknown	–	0.01	0.06	0.01	0.01	0.03	0.01	0.06
Methyl ethyl ketone	0.02	0.01	0.03	0.02	0.04	0.01	0.01	0.05
Diacetyl	0.05	0.01	0.06	0.02	0.05	0.70	0.01	0.05
Valeraldehyde	0.29	0.08	0.06	0.02	0.87	0.77	0.68	0.39
Caproaldehyde	0.24	0.08	0.49	0.14	0.43	0.49	0.41	0.16
Ethyl propyl ketone	0.04	0.04	0.04	0.02	0.11	0.12	0.09	0.03
Furfural	1.80	0.55	0.43	0.82	1.42	1.90	2.16	5.30
Benzaldehyde	0.89	0.44	0.29	0.82	0.90	1.55	4.50	5.60
Oxybenzaldehyde	4.00	4.60	3.46	2.68	6.92	7.80	6.80	4.55
Phenylacetylaldehyde	0.10	0.08	0.05	0.14	0.20	0.80	1.26	1.35
Approx. Total	13.4	8.17	8.43	6.21	14.04	16.55	17.42	18.30

From Borovikova and Roiter (1971c).

TABLE 110

Influence of Freshness on Bread Preference

Age (hr)	% Preference
4	40
12	60
4	43
24	57
4	62
48	38
12	81
24	19
12	68
48	32

From Cathcart (1937).

Storage Conditions

The aroma and flavor of bread changes rapidly after baking. Both manufacturer and consumer continually search for bread products that have a lasting, desirable fresh-baked flavor.

Several early investigators demonstrated that storage has a detrimental effect on the sensory properties of bread. In the five sets of data reported by Cathcart (1937), three sets clearly demonstrated the harmful effect of storage on bread acceptance (Table 110).

TABLE 111

Influence of Bread Freshness on Bread Sensory Preferences

	Odor		Taste	
Age in hr	Best (%)	Poorest (%)	Best (%)	Poorest (%)
12	47	22	44	20
24	22	38	30	34
48	31	40	25	46

From King et al. (1937).

TABLE 112

Influence of Storage on Volatile Content in Breads

	V.R.S.* meq	
Product	Fresh	24 Hours
Rye bread	114	4.6
White bread A	179	6.3
White bread B	275	4.4

*Volatile reducing substances.
From Farber (1949).

TABLE 113

Influence of Storage on the Sensory Properties of Bread*

	Sensory properties		
Hours after baking	Fermentation aromatics	Fermentation acids	Yeasty taste
1	4.0**	4.0	4.0
12	2.0	3.5	4.0
24	0.5	3.1	4.0
48	0.25	3.0	4.0

*Made by the straight-dough procedure
**Intensity scale from 0 to 4.
From Jackel and Ersoy (1961).

In the case of 4-hour-old bread vs 12-and 14-hour bread, the older breads were preferred. Since the panel was asked to judge overall preference, the fresh bread may have been judged too moist and, as such, downgraded.

King et al. (1937) also evaluated the influence of storage time on the odor and taste properties of bread. However, their freshest product was already 12 hr old when evaluated; by then the majority of flavor changes had probably already occurred. From the standpoint of odor and taste, the freshest product was judged best (Table 111).

A later study by Farber (1949) demonstrated that there was a dramatic change within a 24-hr period in the amount of volatile reducing substances in various bread products (Table 112).

The same general trend was reported by Jackel and Ersoy (1961), who used a rather limited five point intensity scale to rate the fermentation aromatics, fermentation acids and yeasty taste detected with bread storage (Table 113).

The aromatic level was most dramatically affected by storage; after 12 hr only half of the original amount was considered present. The acid intensity dropped slightly, and after 48 hr of storage, 75% of the original level was still present. No change in yeasty taste was noted throughout the storage period.

Borovikova and Roiter (1971c) provided detailed volatile flavor compound analyses of the crumbs and crusts of breads that were stored for varying periods of time, with and without wrappers (Table 114).

Initial readings were taken after three hours, and the total volatile compound levels were 19.2 mg/100 g of specific volume in the crust, and 3.2 mg/100 g in the crumb. In the case of unwrapped bread, carbonyl content decreased with storage by 85% in the crust and 22% in the crumb. With wrapped bread the aliphatic carbonyl levels in both crust and crumb increased for the first 12 hr and then decreased; the aromatic carbonyls increased in the crumb and decreased in the crust. These results were attributed to the slower cooling rate of the wrapped product.

With the Flavor Profile Analysis method of sensory evaluation, Caul and Vaden (1972) attempted to follow the sensory changes associated with bread storage. Data they obtained in the analysis of crust aroma, crumb aroma, crust flavor and crumb flavor are summarized in Tables 115-118.

A possible criticism of this study is that the first analyses were performed when the product was already 24 hr old. Most of the dramatic changes in odor and flavor had probably already occurred. However, the initial sampling schedule closely approximated consumer consumption patterns. From these data the authors concluded that the degree of flavor blending and complexity decreased as bread aged. With storage they observed that some flavor characteristics disappeared and others appeared.

Lorenz and Maga (1972) also reported on the influence of storage on the sensory preferences of bread (Table 119).

The fresh bread received the highest acceptability rating; after four days of storage an average ranking of "dislike slightly" was recorded. In addition, the changes in relative concentrations of certain volatile carbonyls were followed (Table 120).

In the carbonyl mixture, aldehydes accounted for approximately 73% of the total in fresh bread, but after five days of room temperature storage, their composite level had decreased to 15%.

BREAD FLAVOR APPLICATIONS

Introduction

Gross chemical composition and ingredients of bread are well-defined by law. Consequently, manufacturers rely primarily on sensory factors as promotional selling points for their product. Thus the modification, improvement and retention of desirable bread flavor is vital to sales. In the area of flavor research most basic research has some degree of application and most of the information previously discussed has been used in modern bread production. Perhaps the true value of any knowledge about the chemistry of bread flavor can be measured by the patent literature generated in the area. However, one problem is that the relative importance of the flavor claims made in a majority of patents can be difficult to evaluate because of descriptive terms used. Examples of such terms include "an improvement in bread flavor" and "a bread-like flavor resulted."

Selected Bread Flavor Patents

Based on his early observations on the composition of oven vapor condensates, Wiseblatt (1960a) formulated a synthetic mixture which, when added to warm water (1 ml/150 ml), produced a pleasant aroma similar to the original oven gases isolated (Table 121).

From a practical standpoint this approach, when applied to bread flavor enhancement, is of little value. The compounds in question are quite volatile and, if added to a dough system, would be volatilized during baking.

Considerable effort has been devoted to bread flavoring materials suitable for addition to chemically leavened breads since these products normally lack characteristic bread aroma and flavor. For example, Moriarty and Newlin (1962) patented the addition of flavor-enhancing agents formed as condensation products resulting from the reaction among certain amino acids and sugars. They claimed that leucine and sugars formed typical flavors and aromas described as toasty and bread-like, associated with fresh

TABLE 114

Influence of Storage Conditions on the Carbonyl Composition of Bread Mg Carbonyl Compounds/100 g of Product

Compound	Unwrapped								Wrapped in polyethylene							
	Crumb Hours				Crust Hours				Crumb Hours				Crust Hours			
	3	12	24	48	3	12	24	48	3	12	24	48	3	12	24	48
Methylglyoxal	0.08	0.07	0.03	0.01	0.06	0.06	0.06	0.05	0.08	0.08	0.03	0.02	0.06	0.20	0.35	0.06
Glyoxal	0.10	0.19	0.11	0.01	0.28	0.10	0.20	0.19	0.10	0.07	0.19	0.12	0.28	0.73	0.72	0.20
Acetaldehyde	0.30	0.30	0.03	0.01	0.14	0.08	0.03	0.02	0.30	1.87	0.50	0.30	0.14	0.60	0.26	0.07
Propionaldehyde	0.01	0.01	0.01	0.01	0.01	0.04	0.01	0.01	0.01	Tr.	Tr.	Tr.	0.01	0.01	0.01	0.05
Isobutyraldehyde	0.02	0.02	0.01	0.01	0.02	0.02	0.01	0.01	0.02	0.08	0.03	0.02	0.02	0.10	0.08	0.03
Acrolein	0.06	0.04	0.02	0.01	0.05	0.04	0.06	0.02	0.06	0.07	0.04	0.04	0.05	0.65	0.21	0.06
Acetone	0.08	0.06	0.03	0.09	0.08	0.10	0.04	0.07	0.08	0.05	0.03	0.03	0.08	0.47	0.21	0.10
Butyraldehyde	0.08	0.04	0.04	0.01	0.04	0.06	0.06	0.04	0.08	0.05	0.04	0.03	0.04	0.24	0.19	0.05
Isovaleraldehyde	0.08	0.04	0.01	0.01	0.01	0.02	0.02	0.02	0.08	0.11	0.03	0.03	0.01	0.12	0.03	0.01
Unknown	0.04	0.01	0.01	0.01	0.01	Tr.	Tr.	0.01	0.04	0.01	Tr.	Tr.	0.01	0.19	Tr.	Tr.
Methyl ethyl ketone	0.04	0.01	0.01	0.01	0.01	0.01	0.01	0.01	0.04	0.02	0.01	0.01	0.01	0.04	0.02	0.02
Diacetyl	0.04	0.02	0.01	0.01	0.03	0.03	0.02	0.01	0.04	0.01	0.01	0.01	0.03	0.04	0.01	0.01
Valeraldehyde	0.36	0.24	0.14	0.04	0.28	0.22	0.29	0.21	0.36	0.17	0.15	0.05	0.28	1.01	0.97	0.30
Caproaldehyde	0.18	0.09	0.06	0.01	0.05	0.19	0.14	0.09	0.18	0.06	0.11	0.02	0.05	0.87	0.59	0.12
Ethyl propyl ketone	0.04	0.02	0.01	0.01	0.03	0.03	0.16	0.02	0.04	0.07	0.03	0.02	0.03	0.11	0.10	0.02
Furfural	0.43	0.34	0.23	0.25	8.40	2.59	1.62	0.76	0.43	0.42	0.48	0.97	8.40	2.16	1.63	1.46
Benzaldehyde	0.27	0.21	0.20	0.14	4.80	0.17	1.24	0.56	0.27	0.40	0.46	0.67	4.80	4.70	4.34	3.67
Oxybenzaldehyde	0.96	0.93	0.90	0.82	4.80	1.78	1.48	1.30	0.96	1.20	1.32	1.50	4.80	5.40	3.44	3.36
Phenylacetylaldehyde	0.05	0.03	0.03	0.02	0.08	0.10	0.09	0.11	0.05	0.21	0.55	1.06	0.08	0.11	0.06	0.06
Approx. total	3.22	2.67	1.89	1.49	19.20	5.64	5.54	3.51	3.22	4.69	4.00	4.88	19.20	17.75	13.22	9.65

From Borovikova and Roiter (1971b).

TABLE 115

Flavor Profile Analysis of Bread Crust Aroma as Influenced by Storage

24 Hr (Intensity 1½)		48 Hr (Intensity 1)		72 Hr (Intensity 1)		96 Hr (Intensity 1)	
Note	Level	Note	Level	Note	Level	Note	Level
Caramel	1	Caramel	1	Browned Flour	1	Browned Flour	1
Sweet	)(	Browned Flour	1	Caramel	)(	Caramel	)(
Browned Flour	1	Wheaty	)(				
Wheaty	)(						

)(: Threshold.
1: Slight.
From Caul and Vaden (1972).

TABLE 116

Flavor Profile Analysis of Bread Crumb Aroma as Influenced by Storage

24 Hr (Intensity 1)		48 Hr (Intensity 1)		72 Hr (Intensity 1)		96 Hr (Intensity 1)	
Note	Level	Note	Level	Note	Level	Note	Level
Sweet	)(	Sweet	)(	Yeasty (old)	1	Sour	½
Yeasty	1	Yeasty	1	Sour	)(	Yeasty	1
Ethanol	½	Wheaty	)(	Wheaty	)(	Wheaty-starchy	1
Wheaty	½	Doughy	)(	Starchy (old)	1		

)(: Threshold.
½: Threshold-slight.
1: Slight.
From Caul and Vaden (1972).

TABLE 117

Flavor Profile Analysis of Bread Crust Flavor as Influenced by Storage

24 Hr (Intensity 1)		48 Hr (Intensity 1)		72 Hr (Intensity 1)		96 Hr (Intensity 1)	
Note	Level	Note	Level	Note	Level	Note	Level
Browned flour	1	Browned flour	1	Browned flour	1	Papery-browned flour	1
Caramel	½	Caramel	)(	Papery	)(	Bitter	)(
Bitter	)(	Bitter	)(	Bitter	)(		

)(: Threshold.
½: Threshold-slight.
1: Moderate.
From Caul and Vaden (1972).

TABLE 118

Flavor Profile Analysis of Bread Crumb Flavor as Influenced by Storage

24 Hr (Intensity 1)		48 Hr (Intensity 1)		72 Hr (Intensity 1)		96 Hr (Intensity 1)	
Salty	)(	Sweet	)(	Starchy	1	Starchy	1
Doughy	½	Doughy	½	Sour	½	Salty	)(
Yeasty	1	Yeasty	1	Yeasty (old)	1	Yeasty (old)	1
Starchy	)(	Sour	)(			Sour	1

)(: Threshold.
½: Threshold-slight.
1: Slight.
From Caul and Vaden (1972).

TABLE 119

Influence of Bread Age on Panel Acceptability

Storage time	Average panel score*
2 hr	1.95
2 days	3.55
4 days	4.50

*Eight point scale – 1: like extremely; 8: dislike extremely
From Lorenz and Maga (1972).

TABLE 120

Influence of Storage on Relative Carbonyl Composition in Bread

Compound	Freshly baked	5-Day-old
Formaldehyde	2.5*	0.2
Acetaldehyde	4.0	0.9
Acetone	4.5	1.7
Propanal	13.8	1.3
Butanal	22.7	0.8
2-Butanone	11.0	14.7
2-Hexanone	8.2	67.4
Hexanal	8.9	2.4
2-Heptanone	3.4	1.1
Heptanal	1.3	–
Nonanal	14.2	9.5
Unknown	5.5	–

*Relative %
From Lorenz and Maga (1972).

TABLE 121

Synthetic Bread Aroma Formulations

Compound	Amount (mg)*	Amount (ml)**
Ethanol	4	100
Acetaldehyde	10	0.1
Acetic acid	40	–
Furfural	20	0.5
Acetylmethylcarbinol	6	–
Pyruvic acid	–	2.0
2-Ethylhexanal	–	0.2
Diacetyl	–	0.015
2-Butenal	–	0.2
Methylglyoxal	–	0.8

*From Wiseblatt (1960a).
**From Wiseblatt and Kohn (1960).

TABLE 122

Sensory Properties of Non-fermented Breads

Leavening system	Sensory properties: Fermentation aromatics	Fermentation acids	Yeasty taste
Yeast (control)	4*	4	4
Glucono-delta-lactone (GDL)	0	0.5	0
GDL + inactive dry yeast	0	0.5	4
GDL + yeast extracts	0	0.5	1-4
GDL + protein hydrolysates	0	0.5	0-2
GDL + carbohydrates	0	0.5	0
GDL + organic acids	0	2-4	0
GDL + amino acids	0	0.5	2-4

*Intenstiy scale from 0 to 4.
From Jackel and Ersoy (1961).

TABLE 123

Influence of the Addition of a Dry Fermentation Flavor Additive (DFF) on the Sensory Properties of Straight Dough Breads

Conditions	Sensory properties: Fermentation aromatics	Fermentation acids	Yeasty taste
3 Rises, no DFF	4*	4	4
3 Rises + 0.375% DFF	4+	4+	4+
2 Rises, no DFF	3	3.5	4
2 Rises + 0.375%	3.8	4	4+
1 Rise, no DFF	1.5	2.5	4
1 Rise + 0.375% DFF	2.9	3.7	4+

*Intensity scale from 0 to 4+.
From Jackel and Ersoy (1961).

yeast-raised bread. The amino acids arginine and histidine formed buttery notes. Their patent covered the formation of flavor compounds either prior to incorporation into the bread dough, or the addition of the precursors to the dry mix, resulting in flavor compound formation during the baking process. They acknowledged that the cost of certain pure amino acids would make the process prohibitive, and they proposed proteinaceous material as the amino acid source. Products such as cottonseed flour and soybean meal were either hydrolyzed with hydrochloric acid or with enzymes. They reported that the optimum amount of total free amino acids needed to form the flavor enhancing agents ranged from 0.2 to 0.4% dry weight.

The approach used by Jackel (1963) was to form a flavor concentrate suitable for addition to a chemically leavened product by preparing a fermentation brew. The brew was composed of 15 to 25% yeast solids, 4 to 8% of a fermentable carbohydrate source, preferably dextrose, and 4 to 6% of non-fat dry milk solids. Under anaerobic conditions at 86°F, the fermentation was carried out for 4 to 16 hr. Optimum development of the flavor compounds was monitored by measuring the titratable acidity. Jackel claimed that a titratable acidity within the range of 0.48 to 0.58 produced optimum flavor. If the acidity fell much lower, a cheese-like off-flavor was noted. The fermentation process was halted by pasteurization at 150°F for 15 min and the resulting product was dried and used as a bread flavoring agent.

The information required for the granting of the above patent was reported by Jackel and Ersoy in 1961. Their preliminary data with the use of glucono-delta-lactone in place of fermentation demonstrated that common bread flavor characteristics were lacking (Table 122).

The use of glucono-delta-lactone and other classes of common bread ingredients also resulted in inferior sensory properties. Next they added dry fermentation flavor extract. As seen in Table 123, they

TABLE 124

Influence of the Addition of a Dry Fermentation Flavor (DFF) Additive Levels on the Sensory Properties of Non-fermented Breads

	Sensory property		
Leavening – % additive	Fermentation aromatics	Fermentation acids	Yeasty taste
Yeast	4*	4	4
Glucono-delta-lactone (GDL)	0	0.5	0
GDL + 0.5% dry fermentation flavor (DFF)	0.5	0.9	1.0
GDL + 1.0% DFF	0.9	1.5	1.9
GDL + 1.5% DFF	1.5	3.1	3.4
GDL + 2.0% DFF	2.2	3.9	3.9
GDL + 2.5% DFF	2.8	4.2	4.3
GDL + 3.0% DFF	3.5	4.7	5
GDL + 3.5% DFF	3.9	5	5

*Intensity scale from 0 to 5.
From Jackel and Ersoy (1961).

TABLE 125

Influence of Storage and Leavening System on the Sensory Properties of Non-fermented Bread

		Sensory properties		
Leavening	Hours after baking	Fermentation aromatics	Fermentation acids	Yeasty taste
Yeast	1	4*	4	4
GDL + 2% DFF	1	2.3	4	4
Yeast	12	2	3.5	4
GDL + 2% DFF	12	1.2	3.5	4
Yeast	24	0.5	3	4
GDL + 2% DFF	24	0.5	3	4
Yeast	48	–	3	4
GDL + 2% DFF	48	0.4	3	4

GLD: Glucono-delta-lactone.
DFF: Dry fermentation flavor.
*Intensity scale from 0 to 4.
From Jackel and Ersoy (1961).

demonstrated that the addition of 0.375% of the dry extract significantly increased the sensory properties of bread made by the straight-dough procedure.

When they evaluated the effect of adding glucono-delta-lactone and the dry fermentation flavor material to nonfermented bread as shown in Table 124, flavor intensity was quite strong.

It should be noted that a level of at least 2.0% of the flavor extract was required. The influence of storage on nonfermented breads containing glucono-delta-lactone and 2% flavor extract is summarized in Table 125.

No significant difference in flavor intensity was noted with storage time; both samples lost flavor intensity quite readily.

A yeast-levened flavor resulted from a mixture of yeast autolyzate and maltol (Lendvay, 1970), which could be incorporated into chemically leavened products. Essentially a mixture of yeast, powdered sugar and maltol was fermented at room temperature until the volume increased 35%. Then additional sugar and maltol were added and the temperature raised to 55°C for 16 hr to permit autolysis. The final step involved heating the mixture to 90°C and holding for two hours before packaging. The resulting brown liquid was then added to dough and baked, or dried and incorporated into the product.

TABLE 126

Sensory Properties of 0.1% Maltol in Yeast Rolls

Panel of 36 flavor descriptions	Control	Maltol
Odor:		
Fresh bread	13	18
Yeasty, doughy, sour	16	11
Stale, musty, moldy	6	4
Sweet	0	5
Taste:		
Fresh bread	15	21
Sweet	5	5
Sour, fermented	7	2
Stale, musty	3	0
Flat, bland	3	4
Bitter	3	3
Preference (two with no preference)	12	22

From Hodge and Moser (1961).

TABLE 127

Sensory Properties of 0.1% Isomaltol in Yeast Rolls

Panel of 34 flavor descriptions	Control	Isomaltol
Odor:		
Fresh bread	12	21
Yeasty, doughy	15	6
Stale, musty, moldy	2	4
Sweet	0	3
Taste:		
Fresh bread	12	16
Sweet	3	5
Sour	3	2
Stale, musty	5	2
Salty	2	0
Cucumber	0	1
Preference (four with no preference)	13	17

From Hodge and Moser (1961).

The incorporation of maltol and related products as bread flavor enhancers was evaluated by Hodge and Moser (1961). They added 0.1% maltol or 0.1% isomaltol to yeast-roll formulations and, according to Tables 126 and 127, samples containing the additives were preferred in both odor and flavor over controls baked without additives.

The formulation of a synthetic bread flavor derived from the pretreatment of whey followed by yeast and bacterial fermentations was patented by Bundus and Luksas (1969 a,b,c,d). A mixture of bakers yeast and nontoxic bacteria of the coccus family was inoculated into a whey medium and held at 86°F for 14 to 18 hr with air bubbled through the system during the growth period. The resulting product was dried and used as a bread-flavoring additive at ranges from 0.5 to 15% of the flour weight. A level of 6% was found to be optimum. Their 1969b patent provided means of pretreating the whey. A high-heat treatment of the whey resulted in better flavor development since low-heat treatment resulted in slow yeast growth. Also, high-heat treatment of the whey probably induced the formation of additional flavor compounds. They also disclosed the use of enzymatically treated whey, created by lactase action and the direct addition of lactic acid, to improve the resulting flavoring material. The 1969c disclosure used a whey substrate low in lactose by simply diluting the whey with water. Their last patent (1969d) provided another method of pretreating whey to remove a portion of the lactose.

Earlier observations prompted Hunter and Walden (1969) to obtain a bread flavor patent based on the reaction products formed from heating a proline and glycerol to 100 to 150°C range in the ratio of 1 to 10 parts of glycerol per part of proline. While heating for 5 to 30 min the volatiles formed were flushed from the reaction with nitrogen and trapped in an acid trap, thus forming the nonvolatile salts. The flavor material was then added at levels of 0.1 to 1%, based on flavor weight. It was noted that other alkaline polyols were equally as effective as glycerol and that the reaction was promoted by alkaline conditions maintained by the addition of sodium hydroxide or carbonate. The flavor compounds formed supposedly provided a richer, longer-lasting fresh bread aroma in day-old bread, as compared to a day-old control containing no additive, and they did not alter bread color and textural properties.

The above patent resulted from preliminary data reported by Wiseblatt and Zoumut (1963) and Hunter et al. (1966, 1969). Wiseblatt and Zoumut (1963) noted a cracker-like aroma when proline and dihydroxyacetone were heated. Hunter et al. (1966) prepared bread flavor-like additives by heating a mixture of proline, glycerol and dihydroxyacetone. They then added the resulting product to dough at various levels. At levels near 0.2% of the flour weight, the additive reinforced the normal crust aroma of baked bread. Likewise, an improvement in crumb aroma and flavor was noted. However, when added at levels near 2 to 3%, an objectionable, peanut-like aroma and flavor resulted.

Later, Hunter et al (1969) described the preparation and identification of the compound they thought was responsible for the observed cracker odor. The compound was identified as 1,4,5,6-tetrahydro-2-acetopyridine. They stressed that similar compounds in the mixture also had cracker-like characteristics. It was reported that when one part of the synthetic compound was added to 100,000 parts of fresh bread extract, the crusty aroma was enhanced. They also sprayed a solution containing 6 ppm of the compound on week-old bread and, after air-drying, noted that fresh bread odor was restored to the crust.

Later Buchi and Wuest (1971) reported on the synthesis of the compound in question. They confirmed that the compound had sensory properties similar to bread. Another point mentioned was the compound was highly sensitive to air and precautions had to be taken in its preparation and storage.

The early bread flavor work of Wiseblatt also led to several patents. In 1966, Wiseblatt was granted a patent for forming bread flavor related compounds by the thermal interaction of aliphatic ketols with acids or certain amino acid decarboxylation products. The example given reported the reaction of proline with 1,3-dihydroxyacetone. Flavor application in chemically leavened instant bread mix resulted in acceptable flavor. It was also noted that the flavor improvement was even more pronounced when the bread was toasted. Several additional claims were granted to Wiseblatt in a 1967 patent.

A patent was granted to Cort (1970) for forming a bread flavor enhancer by heating a mixture of proline, glycine, valine, glutamine and glycerol. It was claimed that, by varying the ingredient ratios, either a cracker-like, biscuit-like or bread-like flavor and aroma resulted.

Several patents have been granted in the area of incorporating flavored shortening products to enhance bread flavor. For example, Nakel (1967) claimed that the use of a flavored fat or oil containing pyroline could impart an enhanced crusty flavor to bread. It was pointed out that both concentration and pH influenced the resulting flavor. In the case of bakery products, levels of 1 to 6 ppm based on flour weight, and a pH range of 5 to 6.5, was suggested.

A flavor-based shortening patent was also granted to Dirks and Nakel (1967). The resulting crusty flavor was formed by the reaction of piperidine, proline and dextrose by various methods. Final levels of the unspecified reaction product in baked bread ranged from 700 to 1,300 ppm based on flour weight.

The production of flavored shortening was also the basis of a patent granted to Dirks and Nakel (1968). The process involved fermenting yeast and sugar to develop typical fermentation flavor; then vegetable oil was added to the mixture and heated. During the heating process the oil absorbed the fermentation flavor. The mixture was then decanted, resulting in a fermentation-flavored oil which could then be used in a normal bakery operation.

TABLE 128

Influence of Added Organic Acids on the Flavor of Non-fermented Bread

Product	Moles organic acids /700 g flour	Flavor rank*
Fermented	0	+++
Non-fermented	0	–
Non-fermented	1.0×10^{-5}	+
Non-fermented	2.0×10^{-5}	++
Non-fermented	3.0×10^{-5}	+++
Non-fermented	4.0×10^{-5}	++++

*Exact definition not given.
From Johnson and Sanchez (1972).

TABLE 129

Influence of Added Organic Acids* and Free Amino Acids on the Flavor of Non-fermented Bread

Product	% Added amino acids**	Flavor rank***
Fermented	0	+++
Non-fermented	0	–
Non-fermented	0.1	++
Non-fermented	0.2	+++
Non-fermented	0.4	++++

*3.0×10^{-5} mol/700 g flour added to non-fermented products.
**Mixture of 20 amino acids in unspecified ratios.
***Exact definition not given.
From Johnson and Sanchez (1972).

The addition of an edible, enzyme-peroxidizable fatty material and enzyme source was the basis for a patent granted to Higashiuchi et al. (1963). The mixture resulted in the formation of peroxidized fatty material for which claims of product flavor improvement were made. Later Ferrari (1965) was granted a patent for the addition of enzyme-peroxidized oils which performed the same flavor function.

Recently, another bread flavor researcher has applied for a patent based on the data presented in Tables 128 and 129.

The proposed additive is a mixture of 20 amino acids obtained by the hydrolysis of wheat gluten and salts of the monocarboxylic acids found in normally fermented dough. As seen in Table 128, organic acids at a level of 3.0×10^{-5} moles/700 g of flour were judged to result in the same flavor intensity as normally fermented bread. In Table 129 an amino acid level of 0.2% based on flour weight was required to achieve the same flavor intensity as the control. This additive is being proposed for use with a no-time fermentation method.

CONCLUSION

The information summarized in this review represents only a portion of that published on the flavor properties of bread. Flavor chemistry is quite complex, in the case of bread flavor. The mechanisms involved in the formation and retention of bread flavor are far from completely understood.

Perhaps the work reported by Mulders (1973), a portion of which is summarized in Table 130, can be cited as an example of the basic information that is still needed to more completely understand the complexity of bread flavor.

TABLE 130

Composition, Odor Threshold Values and Odor Values of a Synthetic Bread Flavor

Compound	Concentration (ppm)	Odor threshold (ppm)	Odor value*
Ethanol	10^4	900	11.1
1-Propanol	9	40	0.2
2-Methyl-1-propanol	7	3.2	2.2
3-Methyl-1-butanol	7.5	0.77	9.7
Acetaldehyde	6.8	0.12	57
2-Methylpropanal	0.1	0.01	10
3-Methylbutanal	0.02	0.007	2.9
2-Propanone	1.4	300	< 1
2,3-Butanedione	0.005	0.0065	0.8
Ethylformate	0.6	17	< 1
Ethyl acetate	0.25	6.2	< 1
1,1-Diethoxyethane	0.15	0.042	3.6
Furan	0.016	4.5	< 1
2-Methylfuran	0.013	3.5	< 1
Dimethyl sulfide	0.005	0.001	5
Dimethyl disulfide	0.0025	0.00016	15.6
Total mixture	10,032.8615	80	125.4

*Concentration divided by odor thresholds.
From Mulders (1973).

Based on quantitative data obtained from bread volatiles, he formulated a synthetic mixture which resulted in a gas-liquid chromatogram identical to an authentic bread head-space sample. However, the odor of the synthetic mixture more closely resembled that of dough than that of bread; he concluded that neither the components detected in the original bread, nor their ratios, are completely characteristic of the odor of bread. However, he did not say that they are not important contributors to bread flavor. Based on their odor values, certain compounds could definitely contribute to a major portion of overall bread flavor.

Another conclusion reached by Mulders is that a particular component, due to its extremely low concentration but potent characteristic, was not detected by the analytical method used. Perhaps a sample larger than the 50 g used would have been advisable. The above conclusion was supported by the facts that there were areas in the scan where no apparent peak was observed, but odor sensations described as "golden brown crust," "bread" and "smell of bakery" were noted. Work is currently underway to identify these extremely potent and elusive compounds. Perhaps when identified, they will be the missing piece in providing the complete bread flavor profile.

ACKNOWLEDGMENT

The author wishes to express his thanks to Dr. James Long, History Department, Colorado State University, for translating the Russian articles contained in this review.

REFERENCES

Bache, A., A new compound in food products, *Comp. Rend.*, 150, 540, 1910.
Bailey, C. H. and Munz, E., The march of expansion and temperature in bread baking, *Cereal Chem.*, 15, 413, 1938.
Baker, J. C., The effects of yeast on bread flavor, *Baker's Dig.*, 31(5), 64, 1957.
Baker, J. C. and Mize, M. D., Some observations regarding the flavor of bread, *Cereal Chem.*, 16, 295, 1939.
Baker, J. C., Parker, H. K., and Fortmann, K. L., Flavor of bread, *Cereal Chem.*, 30, 22, 1953.

Borovikova, L., Sedova, L., and Roiter, I., Determination of the threshold sensitivity of carbonyl compounds and the aromatic number of foods, *Izv. Vyssh. Uchebn. Zaved. Pishch. Tekhnol.*, 6, 36, 1971.
Borovikova, L. and Roiter, I., Levels of aliphatic carbonyl compounds in bread during various stages of production, *Izv. Vyssh. Uchebn. Zaved. Pishch. Tekhnol.*, 4, 43, 1971a.
Borovikova, L. and Roiter, I., Changes of the content of carbonyl compounds in bread during storage, *Prikl. Biokhim. Mikrobiol.*, 7, 324, 1971b.
Borovikova, L. and Roiter, I., Effect of the length of baking on the level of carbonyl compounds in bread, *Khlebopek. Konditer. Prom.*, 15(9), 5, 1971c.
Bratovanova, B., Determination of carbonyl compounds in the crust of wheat bread of sofia type, *Khranitelna Prom.*, 14(6), 23, 1965.
Buchi, G. and Wuest, H., Synthesis of 2-acetyl-1,4,5,6-tetrahydropyridine, a constituent of bread aroma, *J. Org. Chem.*, 36, 609, 1971.
Bundus, R. H. and Luksas, A. J., Bread flavor, U.S. Patent 3,466,174, Sept. 9, 1969a.
Bundus, R. H. and Luksas, A. J., Bread flavor concentrate, U.S. Patent 3,466,177, Sept. 9, 1969b.
Bundus, R. H. and Luksas, A. J., Dough flavor concentrate, U.S. Patent 3,485,641, Dec. 23, 1969c.
Bundus, R. H. and Luksas, A. J., Manufacture of synthetic bread flavor, U.S. Patent 3,466,176, Sept. 9, 1969d.
Butterworth, S. W., Fermentation and bread flavor, *Nat. Ass. Rev.*, 957, 1935.
Cathcart, W. H., Experiments in determining bread flavor, *Cereal Chem.*, 14, 735, 1937.
Caul, J. F. and Vaden, A. G., Flavor of white bread as it ages, *Baker's Dig.*, 41(1), 39, 1972.
Clayton, T. A. and Morrison, W. R., Influence of storage on wheat flour fatty acid composition, *J. Sci. Food Agric.*, 23, 721, 1972.
Coffman, J. R., Presence of 2:5 furan-dialdehyde in bread oven gases, *Cereal Sci. Today*, 11, 471, 1966.
Coffman, J. R., Bread flavor and aroma – A review, *Baker's Dig.*, 46(1), 39, 1972.
Coffman, J. R., Bread flavor, in *Chemistry and Physiology of Flavors,* Westport, Conn., 1967b, 182.
Cole, E. W., Hale, W. S., and Pence, J. W., The effect of processing variations on the alcohol, carbonyl, and organic acid contents of pre-ferments for bread baking, *Cereal Chem.*, 39, 114, 1962.
Cole, E. W., Hale, W. S., and Pence, J. W., The effect of proofing and baking on concentrations of organic acids, carbonyl compounds, and alcohols in bread doughs prepared from pre-ferments, *Cereal Chem.*, 40, 260, 1963.
Cole, E. W., Halmke, V., and Pence, J. W., Alpha-keto acids in bread pre-ferments, *Cereal Chem.*, 43, 357, 1966.
Collyer, D. M., Bread flavor, *Baker's Dig.*, 38(1), 43, 1964.
Collyer, D. M., Fermentation products in bread flavour and aroma, *J. Sci. Food Agric.*, 17, 440, 1966.
Cornford, S. J., Role of fermentation on bread flavor, *Milling*, 138, 224, 1962.
Cort, W., Bread flavor and aroma enhancement composition, U.S. Patent 3,547,659, Dec. 15, 1970.
Daniels, D. G. H. Fractionation of aqueous ethanolic extracts from bread, *J. Sci. Food Agric.*, 22, 136, 1971.
Davidek, J., Hampl, J., and Pelankova, I., Carbonyl compounds of bread. I. Determination of carbonyl compounds, *Brot. Gebaeck*, 19(2), 35, 1965.
DeFigueiredo, M. P., Volatile components of bread, *Baker's Dig.*, 38(6), 45, 1964.
Dirks, B. M. and Nakel, G. M., Oleaginous composition and method for making same, U.S. Patent 3,336,140, Aug. 15, 1967.
Dirks, B. M. and Nakel, G. M., Flavored fatty material and process for preparing the same, U.S. Patent 3,394,013, July 23, 1968.
Drapron, R., Taste of bread, *Riv. Ital. Essenze Profumi Pianti Offic. Aromi Saponi Cosmet. Aerosol*, 52(10), 598, 1970a.
Drapron, R., Bread flavor, *Fr. Ses Parfums*, 13(69), 239, 1970b.
Drapron, R., Bread flavor, *Bull,. Anc. Eleves Ec. Fr. Meun.*, 244, 127, 1971.
Drews, E., Analytical investigations on rye bread production by varying fermentation methods, *Ber. Baeckerei-Tagung Detmold*, 6(8), 40, 1960.
Drews, E., The fate of organic acids characteristic of flour in the sourdough fermentation and the effects of citric, tartaric and malic acids on the lactic and acetic acid content of the sourdough bread, *Ber. Getridechemiker Tagung Detmold*, 183, 1961.
Dunn, J. A., The importance of salt in baking, *Baker's Dig.*, 21(4), 25, 1947.
Ehle, S. R., Hause, N. L., and Paul, D. E., Taste and texture evaluation of lysine-supplemented baked products, *Cereal Sci. Today*, 4, 74, 1959.
El-Basyouni, S. and Towers, G. H. N., The phenolic acids in wheat. I. Changes during growth and development, *Can. J. Biochem.*, 42, 203, 1964.
El-Dash, A. A., The precursors of bread flavor: effect of fermentation and proteolytic activity, *Baker's Dig.*, 45(6), 26, 1971.
El-Dash, A. A. and Johnson, J. A., Protease enzymes: effect of bread flavor, *Cereal Sci. Today*, 12, 282, 1967.
El-Dash, A. A. and Johnson, J. A., Influence of yeast fermentation and baking on the content of free amino acids and primary amino groups and their effect on bread aroma stimuli, *Cereal Chem.*, 47, 247, 1970.
Eugster, C. H., Allner, K., and Rosenkranz, R. E., Synthesis of 2,5-dimethyl-3 (2H)-furanone, *Chimia*, 15, 516, 1961.
Farber, L., Chemical evaluation of odor intensity, *Food Technol.*, 3, 300, 1949.
Fellers, D. A. and Bean, M. A., Flour blend A. II. Storage studies, *Baker's Dig.*, 44(6), 42, 1970.

Ferrari, C. G., Enzyme-peroxidized flavored oils, U.S. Patent 3,174,867, March 23, 1965.

Fornet, A. W., Can selected bacteria improve bread flavor, *Baker's Dig.*, 25(3), 31, 1951.

Fortmann, K. L., Technology of continuous dough processing, *Cereal Sci. Today*, 4, 290, 1959.

Hampl, J., Hrdlicka, J., and Ocenaskova, A., Effects of some technological factors on the formation of carbonyl compounds in bread flavor, *Sb. Vys. Sk. Chem.-Technol. Pr. Potravin Technol.*, 8, 131, 1964.

Hepburn, F. N., Lewis, E. W., and Elvehjem, C. A., The amino acid content of wheat, flour, and bread, *Cereal Chem.*, 34, 312, 1957.

Higashiuchi, K., Kleinschmidt, A. W., and Ferrari, C. G., Enzyme treated oil flavor additive, U.S. Patent 3,108,878, Oct. 29, 1963.

Hodge, J. E., Chemistry of browning reactions in model systems, *J. Agric. Food Chem.*, 1, 928, 1953.

Hodge, J. E. and Moser, H. A., Flavor of bread and pastry upon addition of maltol, isomaltol, and galactosylisomaltol, *Cereal Chem.*, 38, 221, 1961.

Hougen, F. W., Quilliam, M. A., and Curran, W. A., Headspace vapors from cereal grains, *J. Agric. Food Chem.*, 19, 182, 1971.

Hrdlicka, J., Hampl, J., and Tvrznik, K., Aromatic substances in bread. I. Production of carbonyl compounds in various kinds of bread during a week cycle, *Sb. Vys. Sk.-Technol. Pr. Potraviny. Technol.*, 7(2), 293, 1963.

Hrdlicka, J. and Honischova, E., Study of changes in thermal and hydro-thermal processes. X. Effect of various types of ovens in the origination of volatile carbonyl compounds in bread during the baking process, *Sb. Vys. Sk. Chem.-Technol. Pr. Potraviny*, 16, 63, 1967.

Hunter, I. R., Ng, H., and Pence, J. W., Volatile organic acids in pre-ferments for bread, *J. Food Sci.*, 26, 578, 1961.

Hunter, I. R. and Walden, M. K., Bread flavoring additive and use thereof, U.S. Patent 3,425,840, Feb. 4, 1969.

Hunter, I. R., Walden, M. K., McFadden, W. H., and Pence, J. W., Production of bread-like aromas from proline and glycerol, *Cereal Sci. Today*, 11, 493, 1966.

Hunter, I.R., Walden, M. K., Scherer, J. R., and Lundin, R. E., Preparation and properties of 1,4,5,6-tetrahydro-2-acetopyridine, a cracker-odor constituent of bread aroma, *Cereal Chem.*, 46, 189, 1969.

Jackel, S. S., Bread flavoring composition, U.S. Patent 3,102,033, Aug. 27, 1963.

Jackel, S. S., Fermentation flavors of white bread, *Baker's Dig.*, 43(5), 24, 1969.

Jackel, S. S. and Ersoy, E., Bread fermentation flavor: development and evaluation, *Baker's Dig.*, 35(4), 46, 1961.

Jakubezyk, T. and Haber, T., Bread flavor and conditions for its formation, *Przem. Spozyw.*, 22(1), 16, 1968.

Johnson, A. H., Identification and estimation of the organic acids produced during bread dough and cracker dough fermentation, *Cereal Chem.*, 2, 345, 1925.

Johnson, J. A., Bread aroma, *Brot. Gebaeck*, 21(4), 72, 1967.

Johnson, J. A. and El-Dash, A., Role of nonvolatile compounds in bread flavor, *J. Agric. Food Chem.*, 17, 740, 1969.

Johnson, J. A. and Miller, B. S., Pre-ferments: their role in bread making, *Baker's Dig.*, 29(3), 31, 1957.

Johnson, J. A., Miller, B. S., and Curnutte, B., Organic acids and esters produced in pre-ferments, *J. Agric. Food Chem.*, 6, 384, 1958.

Johnson, J. A., Rooney, L., and Salem, A., Chemistry of bread flavor, in *Flavor Chemistry*, Am. Chem. Soc., No. 56, 1966, 153.

Johnson, J. A. and Sanchez, C. R. S., New no-fermentation process controls bread flavor and costs, *Baker's Dig.*, 46(4), 30, 1972.

Kazanskaya, C. and Bezruchenko, L., Identification of non-volatile organic acids in bread baking products by chromatography, *Khlebopek. Konditer Prom.*, 7(7), 10, 1963.

Ketovich, V. L. and Ponomareva, A. N., Participation of amino acids in the reaction of melanoidin formation in bread making, *Biokhimiya*, 26(2), 237, 1961.

Kilborn, R. H. and Tipples, K. H., Sponge-and-dough-type bread from mechanically developed doughs, *Cereal Sci. Today*, 13, 25, 1968.

Kim, J. C. and de Ruiter, D., Bread from non-wheat flours, *Food Technol.*, 22, 867, 1968.

King, H. G. C., Phenolic compounds of commercial wheat germ, *J. Food Sci.*, 27, 446, 1962.

King, F. B., Coleman, D. A., and LeClerc, J. A., Report of the U.S. Department of Agriculture bread flavor committee, *Cereal Chem.*, 14, 49, 1937.

Kleinschmidt, A. W., Higashiuchi, K., Anderson, R., and Ferrari, C. G., Flavor improvement of continuously mixed bread with the lipoxidase of soy flour, *Baker's Dig.* 37(5), 44, 1963.

Kobayasi, N. and Fujimaki, M., Odor properties of 1-acetonylpyrroline, *Agric. Biol. Chem.*, 29, 1959, 1965.

Kohn, F. E., Wiseblatt, L., and Fosdick, L. S., Some volatile carbonyl compounds arising during panary fermentation, *Cereal Chem.*, 38, 165, 1961.

Korolkova, G. and Rakhmankulova, R., Carbonyl compound content in wheat bread prepared by various methods, *Khlebopek. Konditer. Prom.*, 12(9), 8, 1968.

Korolkova, G., Rakhmankulova, R., and Loshkareva, L., Effect of various factors on the non-volatile organic acids of wheat dough and bread, *Izv. Vyssh. Uchebn. Zaved. Pishch. Tekhnol.*, 5, 22, 1968a.

Korolkova, G., Rakhmankulova, R., Tipograf, D., and Shokina, T., Precursors of organic acids in wheat bread, *Khlebopek. Konditer. Prom.*, 12(12), 10, 1968b.

Kretovich, V. L. and Tokareva, R. R., Content of hydroxymethylfurfural in bread and hops, *Dokl. Akad. Nauk S.S.S.R.*, 74, 533, 1950.

Lappin, G. R. and Clark, L. C., Colorimetric method for determination of traces of carbonyl compounds, *Anal. Chem.*, 23, 541, 1951.

Lee, J. W., Cuendet, L. S., and Geddes, W. F., The fate of various sugars in fermenting sponges and doughs, *Cereal Chem.*, 36, 522, 1959.

Lendvay, A. T., Method of improving the flavor of baked goods and goods made by the method, U.S. Patent 3,499, 765, Mar. 10, 1970.

Linko, Y., Miller, B. S., and Johnson, J. A., Quantitative determination of certain carbonyl compounds in pre-ferments, *Cereal Chem.*, 39, 263, 1962a.

Linko, Y., Johnson, J. A., and Miller, B. S., The origin and fate of certain carbonyl compounds in white bread, *Cereal Chem.*, 39, 468, 1962b.

Linko, Y. and Johnson, J. A., Changes in amino acids and formation of carbonyl compounds during baking, *J. Agric. Food Chem.*, 11, 150, 1963.

Lorenz, K. and Maga, J. A., Staling of white bread: changes in carbonyl composition and GLC headspace profiles, *J. Agric. Food Chem.*, 20, 211, 1972.

Lynch, B. T., Glass, R. L., and Geddes, W. F., Role of storage on wheat carbohydrate changes, *Cereal Chem.*, 39, 256, 1962.

MacKenzie, R. M., Wheat carbohydrates, Soc. Chem. Ind. Monograph, No. 3, 1958.

McDermott, E. E. and Pace, J., The content of amino acids in white flour and bread, *Br. J. Nutr.*, 11, 446, 1957.

McWilliams, M. and Mackey, A. C., Wheat flavor components, *J. Food Sci.*, 34, 493, 1969.

Maga, J. A. and Lorenz, K., The role of phenolic acids on the flavor properties of certain flours, grains and oilseeds, *Cereal Sci. Today*, 18, 326, 1973a.

Maga, J. A. and Lorenz, K., Phenolic acid composition and distribution in wheat flours and various triticale milling fractions, submitted for publication in *Food Sci. Technol.*, 1973b.

Maga, J. A. and Lorenz, K., Flavor quality of various grain flours and oilseed milk, and marine protein supplements, *J. Milk Food Technol.*, 36, 232, 1973c.

Maga, J. A. and Sizer, C. E., Pyrazines in foods – A review, *J. Agric. Food Chem.*, 21, 22, 1973a.

Maga, J. A. and Sizer, C. E., Pyrazines in foods, *CRC Crit. Rev. Food Technol.*, 4(1), 39, 1973b.

Markianova, L., Borovikova, L., and Permilovskaya, Z., Action of yeast hybrids on the level of carbonyl compounds in bread, *Khlebopek. Konditer, Prom.*, 15(12), 18, 1971.

Markianova, L. and Roiter, I., Determination of volatile carbonyl compounds in bread by paper chromatography, *Izv. Vyssh. Uchebn. Zaved. Pishch. Tekhnol.*, 1, 156, 1969.

Markova, J., Qualitative and quantitative analysis of organic acids separated from rye flour, pre-ferment, dough and mix bread, *Sb. Vys. Sk. Chem.-Technol. Pr. Potraviny*, 29, 123, 1970.

Miller, B. S., Johnson, J. A., and Robinson, R. J., Identification of carbonyl compounds produced in pre-ferments, *Cereal Chem.*, 38, 507, 1961.

Moriarty, J. H. and Newlin, A. C., Chemically leavened bread process, U.S. Patent 3,060,031, Oct. 23, 1962.

Morimoto, T., Studies on free amino acids in sponges, doughs, and baked soda crackers and bread, *J. Food Sci.*, 31, 736, 1966.

Morimoto, T. and Johnson, J. A., Studies on the flavor fraction of bread crust absorbed by cation exchange resin, *Cereal Chem.*, 43, 627, 1966.

Mulders, E. J., The odour of white bread. IV. Quantitative determination of constituents in the vapour and their odour values, *Z. Lebensm.-Unters.-Forsch.*, 151, 310,1973.

Mulders, E. J. and Dhont, J. H., The odour of white bread. III. Identification of volatile carbonyl compounds and fatty acids, *Z. Lebensm.-Unters.-Forsch.*, 150, 228, 1972.

Mulders, E. J., Maarse, H., and Wueurman, C., The odour of white bread. I. Analysis of volatile constituents in the vapour and aqueous extracts, *Z. Lebensm.-Unters.-Forsch.*, 150, 68, 1972.

Mulders, E. J., Ten Noever de Brauw, M. C., and van Straten, S., The odour of white bread. II. Identification of components in pentane-ether extracts, *Z. Lebensm.-Unters.-Forsch.*, 150, 306, 1973.

Nakel, G. M., Oleanginous composition and method for making same, U.S. Patent 3,336,138, Aug. 15, 1967.

Nelson, J. H., Glass, R. L., and Geddes, W. F., Wheat flour fatty acid composition, *Cereal Chem.*, 40, 337, 1963.

Ng, H., Reed, D. J., and Pence, J. W., Identification of carbonyl compounds in an ethanol extract of fresh white bread, *Cereal Chem.*, 37, 638, 1960.

Ocker, H. D., Volatile flavor constituents of bread investigated by gas chromatography, *Ber. Getreidechemiker Tagung Detmold*, 172, 1961.

Ocker, H. D., Highly volatile baking aroma, *Qual. Plant. Mater. Veg.*, 11, 269, 1964.

Ocker, H. D. and Rotsch, A., Bread flavor studies, *Brot. Gebaeck*, 13, 165, 1959.

Otterbacher, T. J., A review of some technical aspects of bread flavor, *Baker's Dig.*, 33(3), 36, 1959.

Pence, J. W., Factors affecting bread flavor, *Baker's Dig.*, 41(2), 34, 1967.

Permilovskaya, Z. and Berzina, N., Effect of dough preparation process on the degree of loss of dry substances and on the formation of organic acids, *Karchova Prom.*, 1, 34, 1973.

Pomeranz, Y., The lysine content of bread supplemented with soya flour, wheat gluten, dry yeast and wheat germ, *J. Sci. Food Agric.*, 13, 78, 1962.

Pomeranz, Y. and Shellenberger, J. A., Flavor, in *Bread Science and Technology*, Avi Publishing, Westport, Conn., 1971, 180.

Ponte, J. G., Titcomb, S. T., and Cotton, R. H., Effect of L-arabinose and D-xylose on dough fermentation and crust browning, *Cereal Chem.*, 40, 78, 1963.

Ralls, J. W., Rapid method for semiqualitative determination of volatile aldehydes, ketones and acids. Flash exchange chromatography, *Anal. Chem.*, 32, 332, 1960.

Robinson, R. J., Lord, T. H., Johnson, J. A., and Miller, B. S., The aerobic microbiological population of pre-ferments and the use of selected bacteria for flavor production, *Cereal Chem.*, 35, 295, 1958a.

Robinson, R. J., Lord, T. H., Johnson, J. A., and Miller, B. S., Studies on the decrease of the bacterial population on pre-ferments, *Cereal Chem.*, 35, 306, 1958b.

Roiter, I. and Borovikova, L., Level of aromatic volatile carbonyl compounds at various stages in the production of bread, *Izv. Vyssh. Uchebn. Zaved. Pishch. Tekhnol.*, 5, 68, 1971.

Roiter, I. and Borovikova, L., Level of volatile carbonyl compounds in bread during the addition of enzyme preparations to the leavened dough, *Khlebopek. Konditer. Prom.*, 1, 14, 1972.

Rooney, L. W., Salem, A., and Johnson, J. A., Studies of the carbonyl compounds produced by sugar-amino acids reactions. I. Model systems, *Cereal Chem.*, 44, 539, 1967.

Rosenberg, H. R. and Rohdenburg, E. L., The fortification of bread with lysine, *J. Nutr.*, 45, 593, 1951.

Rothe, M., Volatile aromatic substances in rye bread, *Ernaehrungsforschung*, 5, 131, 1960.

Rothe, M., Paper chromatographic investigations of some bread flavor carbonyl compounds, *Wiss. Veroeffentl. Deut. Ges. Ernaehrung*, 10, 128, 1963a.

Rothe, M., Development of odor in ropy bread, *Ernaehrungsforschung*, 8(3), 432, 1963b.

Rothe, M., Aroma determination in bread, *Brot. Gebaeck*, 20(10), 189, 1966.

Rothe, M., Influence of technology on the taste of rye bread, *Ernaehrungsforschung*, 16, 287, 1971.

Rothe, M. and Boehme, E., Determination of individual carbonyl compounds in grain products, *Ernaehrungsforschung*, 12(2), 295, 1967.

Rothe, M. and Thomas, B., The formation, composition and determination of the aromatic substances of bread, *Nahrung*, 3, 1, 1959.

Rothe, M., Tunger, L., and Woelm, G., Shortened dough development with white bread and its influence upon flavor. I. Carbonyl compounds and ethanol, *Nahrung*, 16(5), 507, 1972.

Rotsch, A. and Dorner, H., Bread flavor studies, *Brot. Gebaeck*, 11, 173, 1957.

Rotsch, A. and Dorner, H., Investigations on aroma substances arising during baking, *Brot. Gebaeck*, 12, 138, 1958.

Rubenthaller, G., Pomeranz, Y., and Finney, K. F., Effects of sugars and free amino acids on bread characteristics, *Cereal Chem.*, 40, 658, 1963.

Salem, A., Rooney, L. W., and Johnson, J. A., Studies of the carbonyl compounds produced by sugar-amino acid reactions. II. In bread systems, *Cereal Chem.*, 44, 576, 1967.

Sizer, C. E., Maga, J. A., and Lorenz, K., Formation of pyrazines in bread crust, *J. Food Sci.*, in print, 1973.

Smith, D. E. and Coffman, J. R., Separation and identification of the neutral components from bread pre-ferment liquid, *Anal. Chem.*, 32, 1733, 1960.

Steinke, R. D. and Paulson, M. C., Production of steam volatile phenols during the cooking and alcoholic fermentation of grain, *J. Agric. Food Chem.*, 12, 381, 1964.

Stephen, R. L. and Teszler, A. P., Quantitative estimation of low boiling carbonyls by a modified alpha-ketoglutaric acid-2,4-dinitrophenylhydrazone exchange method, *Anal. Chem.*, 32, 1047, 1960.

Suomalainen, H. and Oura, E., Changes in the decarboxylase activity of baker's yeast during the growth phase, *Biochem. Biophys. Acta*, 31, 115, 1959.

Suomalainen, H. and Ronkainen, P., Keto acids in baker's yeast and in fermentation solution, *J. Inst. Brew.*, 69, 478, 1963.

Tokareva, R. R. and Kretovich, V., Method for determining aromatic substances in bread, *Khlebopek. Konditer. Prom.*, 5, 11, 1961.

Tokareva, R. R. and Smirnova, G. M., Formation of aromatic substances in the fermentation of rye dough, *Prikl. Biokhim. Mikrobiol.*, 2(4), 439, 1966.

Visser't Hooft, F. and de Leeuw, F. J. G., The occurrence of acetylmethylcarbinol in bread and its relation to bread flavor, *Cereal Chem.*, 12, 213, 1935.

Von Sydow, E. and Anjou, K., The aroma of rye crispbread, *Lebensm. Wiss. u. Technol.*, 2, 15, 1969.

Wardall, R. A., On the relation of yeast to flavor in bread, *J. Home Econ.*, 2, 75, 1910.

Wick, E. L., deFigueiredo, M., and Wallace, D. H., The volatile components of white bread prepared by a pre-ferment method, *Cereal Chem.*, 41, 300, 1964.

Wiseblatt, L., Some aromatic compounds present in oven gases, *Cereal Chem.*, 37, 728, 1960a.

Wiseblatt, L., The volatile organic acids found in dough, oven gases, and bread, *Cereal Chem.*, 37, 734, 1960b.

Wiseblatt, L., Flavor research in the bread-baking field, *Cereal Sci. Today*, 6, 298, 1961a.

Wiseblatt, L., Bread flavor research, *Baker's Dig.*, 35(5), 60, 1961b.

Wiseblatt, L., Processes which comprise reacting proline or pyrrolidine with an oxo group containing compound and products thereof, U.S. Patent 3,268,555, Aug. 23, 1966.

Wiseblatt, L., Breadstuff flavoring compositions and methods of making them, U.S. Patent 3,304,184, Feb. 14, 1967.

Wiseblatt, L. and Kohn, F. E., Some volatile aromatic compounds in fresh bread, *Cereal Chem.*, 37, 55, 1960.

Wiseblatt, L. and Zoumut, H., Isolation, origin and synthesis of a bread flavor constituent, *Cereal Chem.*, 40, 162, 1963.

Zentner, H., Influence of glycine and lysine on dough fermentation, *J. Sci. Food Agric.*, 12, 812, 1961.

CONFECTIONS

Hard Candies

Hard candies are solid solutions containing a minimum amount of water and forming hard, glassy masses. Alternatively they can be defined as solidified sugar syrup with or without the addition of carbohydrates having a solids content of 98–98.3% dissolved in 0.3–2% water with optionally added flavors and colorants. Hard candies are made in countless varieties. Their composition varies from pure sucrose to a mixture of carbohydrates with the addition of acid flavors and colorants. Depending on the method of preparation, they can be classified as transparent, granular, or satinate (silky look). Candy can be shaped by cylinders, cutters, or molds.

In the production of hard candy, care must be taken to avoid caramelization and the inversion of the sugar. Suitable raw materials and adequate manufacturing techniques must be accurately selected for each type of candy. Even the small amount of water employed can play an important role; alkaline water favors degradation of sucrose, whereas acid water favors sugar inversion. The commonly used acidulants (citric, malic, and tartaric) that are a complementary and indispensable addition to the flavor ingredients must be properly dosed as not to negatively interfere with the preparation of the mix. Too large amounts of reducing sugars cause candy to be sticky, while low levels hinder the formation of granules. A suitable addition of reducing sugars regulates moisture gain; excessive moisture gains or losses by candy are the most decisive factors causing deterioration of the finished product. The optimum percentage of reducing sugar based on the anhydrous product varies between 13–17% depending on climatic conditions.

The old method of cooking sugar under atmospheric pressure in the presence or absence of preheaters has been replaced by vacuum, high-vacuum, or semi-continuous vacuum techniques. The sucrose-glucose ratio governing the baking temperature can vary from 80:20, 70:30, 65:35, and 60:40 to more exceptional ratios, depending on particular requirements or specific formulations.

Once the candy mix is prepared, flavor ingredients and colorants are added. Flavoring also includes the addition of acidulants. The dispersion of the required ingredients must be made at the lowest possible temperature and effected by hand or mechanical mixing. When satinate candies are desired, "stretching" of the mix is effected by allowing air to be drawn into the mixture. The mix is finally pulled into strands that are passed through a molding machine.

From an examination of the various operations connected with the manufacture of hard candy, it is apparent that the commercial value of hard candy (as well as the products based on sugar alone, such as filled and gum candies, fondants, and tablets) is based primarily on adequate flavoring. Candy and similar products with a slow melting process flavor saliva and thus extend the pleasant taste sensation. The choice of flavors and the dosage of acids represent the most delicate stage of the manufacturing process; a suitable choice of the above ingredients justifies by itself the production of candy. Candy flavors should not be too aggressive as to cause violent flavor sensations; the flavor should be released slowly and uniformly to further the desire for a continuing flavor sensation. Obviously the preference for a variety of flavors and the relative flavor intensity may determine the manufacture of candy products having particular flavor characteristics.

Candy flavors are countless; only a few (such as orange, lemon, and mandarin) account for the largest segment of candy production. Mint-flavored candy is prepared using trirectified peppermint essential oil, since the use of menthol is not advisable. A small amount of menthol together with trirectified oil can be used to strengthen the flavor of certain candy types. Anethol or Chinese anise (m.p., 21–22°C) is preferred for use in anise-flavored candies; the essential oils and citric acid are used in citrus-flavored candy. The compounded essential oil and citric acid are used for citron flavors. Compounded essential oils are used for fruit or fantasia flavors, since concentrated juices interfere with candy manufacture because of their contribution of residual moisture. The optimum yield of flavors for hard candy is 1 kg per ton (1 × 1,000). Dilution of the more concentrated compounded flavors must be carried out beforehand using high-boiling and anhydrous solvents. Propylene glycol, ethylene glycol ethyl ester, and carbitols are preferred to 95% alcohol.

Filled Candies

Filled candy consists of two basic parts—the shell and the filling. This candy type can be considered derived from Neapolitan candies—hard candies poured into molds containing a piece of candied fruit or a coffee bean in the center. However, filled candies have a much thinner, more fragile shell and a more fluid filling.

Filled candy must provide a pleasant total flavor sensation when the shell is crushed in the mouth and the filling is exposed. Manufacturing of filled candies follows to a certain extent the basic techniques for the manufacturing of hard candies (preparation of the mix, baking of the shell, and final molding). The types vary widely because of filling formulations, formation of the strand, etc. The shell can be neutral or slightly flavored with a flavor compatible with that of the filling; it may or may not contain colorants and, if flavored, proportional amounts of acids. It can be transparent or satinate. The filling may vary in consistency; it can be gelatinous (fruit jellies or fondants), thick and syrupy (concentrated fruit juices or pastes containing suitable amounts of glucose syrup), or fluid (liqueurs or corresponding flavor imitations).

A description of the operative techniques for the preparation of the above products goes beyond the scope of the book; therefore, only the importance of the flavoring techniques is underlined. The flavoring of the shell follows the same guidelines set forth as in hard candy; the flavoring of the filling follows the criteria used for syrups and liqueurs in the choice of the most suitable flavors.

Taffy

This type of soft candy contains approximately 4–5% moisture and is prepared by a controlled Maillard reaction between sugar and milk proteins. Small scale production of taffy calls for boiling sugar, glucose, fresh milk or cream, butter, and only a complementary flavor together with a little vanillin. Flavoring is developed essentially by the caramelization of sugar in the mix. The industrial production of taffy, although based in general on the above principle, uses various basic mixes that are flavored with suitable compound flavors prior to molding. The formulation of the compounded flavors (usually having a 1 × 1,000 to 1 × 2,000 yield) falls in the category of cream and butter flavor imitations, which blend extremely well in the candy mix.

More complex and functional flavors are those that not only exhibit the butter note but also caramel and, sometimes, coffee notes. These flavors are often available in a powdered form having a 1 × 500 yield.

Fondants

Fondants represent a very important category among those products manufactured with cooked sugar (hard candy, taffy, etc.). The French name indicates the characteristic melting property that is primarily a function of the moisture content (10–12%) in the product. Moisture content alone, however, does not govern the candy consistency; this is determined by several other factors. Fondants are essentially semi-worked products very useful in the manufacture of confections, *e.g.*, as a filling for pralines, dragées, and other candies; as a frosting for pastries, cakes, and cookies; and as a special flavor ingredient in taffy. Classic fondants consist of molded or powdered boiled sugar covered with chocolate, adequately flavored, colored, and garnished.

The manufacturing technique is rather simple but follows strict basic rules. The consistency of the mass is dependent on the cooking temperature and the sucrose to glucose ratio, which should be kept at a 10:1 minimum. First the sugar is heated and complctely dissolved in water; then glucose is added to the syrup in the required amount. The mixture is boiled to about 110°C, yielding a supersaturated solution in which the excess sucrose crystallizes on rapid cooling. The cooling mass is stirred to cause crystallization of sucrose into very thin (less then 10 mm) crystals.

Whenever the fondant is to be used as candy rather than an ingredient in other products, it is flavored at this stage, colored with suitable dyes, thoroughly mixed, and poured into molds. Flavoring is extremely important to fondants; good results are achieved when the same flavor ingredients employed in hard candies are utilized. The flavor yields, however, are almost double in strength, using the same flavor ingredients, since there is no thermal or vacuum treatment involved. The flavor sensation also is perceived more rapidly because of the softness of fondants. The basic flavors are very similar to those employed for hard candies with the exception of anise, which is not used; the acid content is usually about 50% less than that used in hard candy. Good flavor results may also be obtained using fruit-juice-based flavors or fruit pastes by formulating the fondants to a suitable consistency. The flavoring of fondants to be used in other products should be chosen according to the end-use and should follow the guidelines indicated for typical fondants.

Gum Drops

Gum drops, gums, and gelatins are three classic candy types often designated by several other names causing a great deal of confusion. These products are prepared with natural gums (*e.g.*, gum arabic), mixtures of gum and gelatin, or gelatin alone, together with sucrose, glucose, glycerol, and water. Colorants, flavors, and acids complement the basic formulation. The sugar content or the degree of sweetness, the acid content, and the flavor intensity in such products vary somewhat from country to country. The products are characterized by their consistency: gum drops exhibit a rather hard consistency, gums are somewhat softer, and gelatins are quite plastic. All such products exhibit a transparent mass (foaming should be avoided during stirring to prevent opalescence) and are shaped by pouring the fluid mass into powdered (starch) molds. The product surface can be shiny, or it can be covered with crystalline sugar. The operative techniques and the mixtures of various ingredients are specific for each product; they are usually prepared not only for confections but also for pharmaceutical purposes (cough drops, laxatives, etc.).

A typical preparation consists of heating the sugar and the gum solutions separately, mixing the two solutions together with the desired colorants, acids, and flavors, pouring the mixture into powdered molds, cooling, drying, and finally polishing. The inversion of sugar must be controlled and dosed according to the desired characteristics of the final product; this is especially true in the preparation of gelatins. Because of their intrinsic composition, these products may be flavored with a wide variety of ingredients, such as fruit juice and pulp, essential oils, isolates, and compounded oils. Flavor components are limited only by a suitable choice of the gum mix. Also wine (Bordeaux, Marsala, Madeira, port, Malaga, etc.) and liqueurs (curaçao, kirsch) are used to flavor gelatins. Fruit juices most frequently employed include apple, strawberry, raspberry, orange, red and black currant, cherry, blueberry, grape, and licorice. The last juice, with added menthol, anethol, eucalyptol, or mixtures thereof, is extensively used to flavor gum drops and gums. Citrus essential oils are very suitable for flavoring gelatins; the corresponding imitation fruit flavors may also be used when fruit juices cannot be added because of technical and/or economic limitations. The flavors used for hard candies are considered generally suitable, except that the yield is almost doubled, *i.e.*, a hard candy flavor having a yield of 1 × 1,000 will exhibit an approximately 1 × 2,000 yield in gelatins.

Tablets

Flavored tablets and disks of various shapes and sizes for use as is and as pharmaceutical bases are obtained by compressing powdered mixes in the presence of agglutinating ingredients. To improve the compactness and the tablet consistency, the ingredients are prepared in granules with or without the addition of water and small amounts of gum (arabic, tragacanth) or starch. Flavoring is usually carried out with spray-dried flavor ingredients.

Confetti (Hard, Sugar-coated Candies)

Confetti consists of cores of varying shapes and sizes coated with sugar frosting. The candies are processed in special rotating drums to obtain a uniform, smooth sugar coating. The core usually consists of one of the following ingredients: almond, hazelnut, pistachio; spice fragments (clove, cinnamon); flavored and molded fondant; molded, minced almond paste; flavored gum-based gels and tablets; and candied fruit. An enormous variety of confections (and also pharmaceutical products) can be prepared by the above manufacturing technique, together with special operative techniques required by specific products. Usually the core is coated with natural gums to favor the adhesion of the sugar coating; the sugar is introduced into the rotating equipment partially as syrup and partially as finely granulated sugar. The sugar coating may be colored by adding dye dissolved in syrup toward the end of the coating operation. Vanilla or other suitable flavor ingredients can be added to the sugar syrup. The coating for certain products may consist also of chocolate, syrup, and sugar. Cores consisting of candied fruit, spices, and almond paste require no additional flavoring; tablet or fondant cores are flavored in the manufacturing process.

Cocoa and Chocolate Candies

The seeds of *Theobroma cacao*, a plant native to Central America, were carried to Europe about the sixteenth century. The industrial use of cocoa seeds is, therefore, over 400 years old. Although chocolate, as known today, was first produced toward the beginning of the nineteenth century, it has become one of the basic food ingredients and flavors. Cocoa derivatives include cocoa powder (sweet or bitter, with varying degrees of solubility), chocolate drink (obtained by dissolving cocoa powder in milk or water with or without the addition of sugar), chocolate bars (bitter and milk chocolate), and cocoa butter. All the above products exhibit different characteristics directly related to the fabrication methods, type, origin, and pretreatment of the seeds. The commercial qualities are, therefore, dependent on many variables.

Briefly, cocoa seeds are treated according to the following operations: fermentation (which influences the aromatic characteristics; the length and the method of fermentation vary depending on the growing site), drying, roasting, peeling, and crushing. The seeds are partially defatted (removal of cocoa butter) by hot pressing. Grinding and specific additional treatments follow to prepare the final cocoa powder. To prepare chocolate bars, the powder is mixed with other ingredients (sugar, milk, nougat, vanilla flavor); the mixture is refined and freed from lumps; it is then ripened at different temperatures and for various lengths of time depending on the quality, tempered, and finally molded.

Basic flavoring is introduced by mixing the flavor ingredients into the chocolate mixtures. Vanilla flavor is the most common and is obtained by using vanillin (also ethyl vanillin and propenylguaethol). Several other complementary flavors are employed; however, their addition is dependent on the use of chocolate in combination with other products. Chocolate bars and candy can be filled with flavored fondants, liqueur-soaked fruit, dry, roasted, or candied fruit, liqueurs, etc. The function of chocolate as a frosting and a flavoring for various filling ingredients follows the specific guidelines previously indicated for individual products.

Nougat

Nougat is a confection consisting mainly of honey, sugar, egg white, peeled sweet almonds (sometimes combined with pistachio, nougat, and roasted almond), candied fruit, glucose, and chocolate (for the so-called brown nougat). It is prepared by mixing the ingredients under heat using special techniques and is formed by pouring the mixture into wafer-coated molds. Typical hard nougat is usually vanilla-flavored, but other flavors are also used, such as lemon, orange, mint, cinnamon, clove, and coffee. The flavor ingredients are added toward the end of the mixing operation, immediately after the addition of the nougat, almond, pistachio, and similar products.

DAIRY PRODUCTS

Introduction

Countless pastry and food products require flavors similar to milk or its more important derivatives. These flavors must be sufficiently concentrated and stable when added to semi-finished or finished food products in place of the less stable natural dairy flavors, from which it is almost impossible to extract the active flavor principles in a form suitable for use.

Milk

Imitation milk flavors can be prepared using cream-flavor imitations in those food products containing powdered milk. Some flavor results are readily obtained in this manner, although the duplication is far from satisfactory. More recently the demand for specific milk flavors for use with liquid and powered products in which the above flavoring technique cannot be employed has increased steadily. Natural milk flavor is rather unstable and varies widely, depending not only on the animal's feeding habits but also on the animal itself (cow, sheep, goat). Generally, milk flavor is defined as the flavor of fresh cow's milk.

To prepare a milk flavor imitation, the following notes are usually blended: vanilla (vanillin, veratric aldehyde, ethyl vanillin), coumarin, melilotus, liatris, dihydrocoumarin (where use is permitted), diacetyl, and trace amounts of butyric acid and esters (ethyl capronate, cinnamyl butyrate, etc.). Methyl acetyl carbinol may also yield a positive contribution to the flavor as a total or partial replacement for diacetyl.

The resulting flavor imitations should be used with great care; when properly dosed, they may satisfy the odor requirements but are somewhat lacking in taste. The unavailability of the basic flavor principles present in natural milk (which also determine the characteristic flavor in a food product through primary or secondary reactions with the food substrate) makes a perfect imitation of the natural flavor somewhat problematic.

In order to better define the flavorist's search for milk flavor imitations, it is extremely helpful to know the exact composition of natural milk flavor. The studies of milk-flavor constituents conducted from 1938 to present allow the following observations:

1. The sweet taste of milk is attributable to lactose; the salty taste, to sodium chloride.
2. Fresh cow's milk flavor can be defined as *normal* and *abnormal*. A *normal* flavor is the flavor characteristic of freshly recovered milk formed through regular physiological processes. An *abnormal* milk flavor is caused by extraneous factors, such as the composition of the animal's feed and chemical degradations following milking.

Keeping in mind that feeding affects milk flavor, the milk flavor produced by cows fed with odorless and purified feed (cellulose, starches, sucrose, urea, inorganic ammonium salts, A and D vitamins, corn oil) has been taken as a reference standard. The following components present at low concentration have been identified in normal milk flavor: acetone, 2-butanone, 2-pentanone, 2-hexanone, 2-heptanone, formaldehyde, acetaldehyde, saturated straight chain C_3 to C_{12} aldehydes, saturated straight and branched chain C_{13} to C_{16} aldehydes, butyric and other low-molecular-weight fatty acids, and methylmercaptan.

A few components identified in other fresh dairy products may also be considered as probably occurring in milk flavor: diacetyl (in fresh butter), isovaleric aldehyde (in fresh cream), *cis*-4-heptenal (in the cream flavor of fresh butter), and δ-lactones (in butter, cream, and pasteurized milk).

The study of *abnormal* milk flavor can be defined according to the following scheme:

Abnormal milk flavor caused by extraneous factors	1. Feed components 2. Chemical degradations following milking	Souring Oxidation Photo-oxidation Heating Concentration Drying

Abnormal milk flavors may be classified into two types: flavors similar to, and flavors different from the flavor of the feed. This is a somewhat thin-lined classification, since strongly aromatic herbs (artemisia, hops, etc.) may be present occasionally in the feed; other products may also form in the feed during storage. This may determine the additional presence in normal milk flavor of alcohols, aldehydes, low-molecular-weight esters, and ketones. Operations subsequent to milking together with storage conditions may also affect the resulting milk flavor.

Souring (Rancid Flavor)

The souring of milk with the formation of the characteristic sour flavor is usually determined by an enzymatic (lipase) scission of the fatty acids present in milk. Lipolysis may occur spontaneously, or it may be induced by excessive agitation (pumping, homogenizing, etc.), separation and clarification, heating to 30°C with subsequent recooling, freezing with subsequent thawing, and adding small quantities of raw milk to pasteurized milk. All such conditions or manipulations can cause the formation of low-molecular-weight free fatty acids, chiefly butyric acid.

Oxidation

The so-called oxidized flavor of milk is believed to be caused by enzymatic and non-enzymatic conditions. The following components have been identified in oxidized milk flavor: saturated C_5 to C_{16} aldehydes, hexenal, 2-hepten-1-al, 2-octen-1-al, 2-nonen-1-al, 2-decen-1-al, 2-undecen-1-al, 2,4-nonadien-1-al, 2,4-decadien-1-al, 2,4-undecadien-1-al, and 2,4-dodecadien-1-al. A comparison of the analytical findings of an oxidized and a non-oxidized milk flavor indicates that the two flavors have a very similar qualitative composition (2-hexen-1-al, 2-hepten-1-al, 2,4-decadien-1-al, 2,4-undecadien-1-al, and 2,4-dodecadien-1-al are present only in the oxidized flavor) but a very different quantitative composition. In fact, a remarkable increase (almost double) of the flavor component ratios and a five-fold increase of the hexenal content (together with a few insignificant decreases of other components) is observed in the oxidized milk flavor. Trace metals such as copper and iron can accelerate non-enzymatic oxidation.

Photo-oxidation

The oxidative action of light on milk flavor is twofold: (1) it causes the oxidation of lipids with consequential formation of the oxidized flavor having a composition similar to the spontaneously formed (or copper-induced) oxidized flavor, and (2) it causes the formation of the so-called activated flavor caused primarily by proteolytic degradation of methionine (and probably of other sulfur-containing amino acids). This brings about the much disputed presence of methional (β-methylmercapto propionaldehyde), methyl mercaptan, dimethyl sulfide, and acrolein.

Heating

Heating normally applied during pasteurization may cause a more or less pronounced cooked flavor, depending on the length and the intensity of the heat cycle. This operation causes a decrease of the redox potential, which is reflected by the formation in the flavor of hydrogen sulfide and other volatile sulfides formed by the liberation of SH groups present in lactoglobulins. When milk undergoes unusually high heat cycles, it develops a pronounced caramelized flavor from the presence of fermentation products and the dehydration of sugars.

Concentration and Drying

Milk undergoes more or less extensive heat cycles during concentration and drying operations. Therefore, cooked and caramelized off-flavors may develop in a manner similar to those discussed above.

Analysis of concentrated and dried milk having off-flavors confirm the presence of the following:

1. **Sugar derivatives** (by Maillard reaction): furfural, hydroxymethylfurfural, diacetyl, and maltol.
2. **Decomposition products from amino acids** (Strecker's degradation): methylpropanal, and 3-methylbutanal.
3. **Derivatives formed by autoxidation**: formaldehyde, acetaldehyde, propionaldehyde, *n*-valeraldehyde, hexyl aldehyde, heptyl aldehyde, nonyl aldehyde, decyl aldehyde,

dodecyl aldehyde, and tetradecyl aldehyde (aldehydes are also present in untreated milk at different concentration levels, however).

4. **Derivatives formed by hydrolysis**: capronic, caprilic, capric, lauric, myristic, and palmitic acids.
5. **Derivatives formed by unknown processes**: acetone, 2-butanone, 2-pentanone, 2-hexanone, 2-heptanone, 2-nonanone, 2-undecanone, 2-tridecanone, 2-pentadecanone, δ-decalactone, δ-dodecalactone, δ-tetradecalactone, γ-dodecalactone. The presence of *p*-aminoacetophenone derived from indican present in milk is also probable.

Some authors have reported the presence of additional flavor components in milk flavor; the presence of benzaldehyde is traced to the type of feed, whereas vanillin (reported in pasteurized milk) may be traced to the heat treatment of the lignin eventually present in milk. In conclusion it is apparent that not one but several milk flavors occur naturally.

Cream

Most of the considerations concerning milk flavor are also applicable to fresh cream flavor. Cream is mechanically separated from milk and may contain from 18–48% butter fats.

Skimming, *i.e.*, the separation of fat from the aqueous layer of milk, determines flavor concentration together with the total or partial removal of the sour note. Cream flavor is sweeter and mellower than milk flavor. To prepare cream flavor imitations, the same ingredients listed for milk flavor imitations are usually applicable with the frequent addition of citrus notes (orange, lemon, citron, mandarin), floral notes (heliotropin, clove, rose), and undernotes based on sulfur-containing compounds (methyl sulfide, etc.). By the proper compounding of the ingredients and by using more modern techniques, it is possible to avoid the massive use of coumarin (or similar products), once a basic flavor component for cream flavor imitations. The use of coumarin (where permitted), in addition to yielding an excellent cream flavor, makes these products quite acceptable; the total coumarin content in the finished product usually does not exceed 20–22 ppm. Similar considerations are also applicable to those formulations strictly related to the use of cream flavor (taffy, Dutch cream, etc.). A guideline formulation for cream flavor imitations consists of the following:

Cream Flavor

Approximate Yield: 1 × 2,000

Ingredient	Parts
Glyceryl tributyrate	200
Melilotus tincture (15% in 65% alcohol)	170
Methyl acetyl carbinol	115
Ethyl vanillin	50
Woodruff tincture (15% in 65% alcohol)	35
Diacetyl	30
Heliotropin	30
Ethyl butyrate	20
Butyric acid (10% aqueous solution)	11
Ethyl capronate	9
Mandarin essential oil (10% solution in alcohol)	4
Lemon essential oil (10% solution in alcohol)	3
γ-Undecalactone (1.0% solution in alcohol)	3
Citronellol (10% solution in alcohol)	1
Neroli essential oil (10% solution in alcohol)	1
γ-Nonalactone (1.0% solution in alcohol)	1
Solvent	9,317
Total	10,000

Ice Cream and Sherbet

Typically, ice cream is a dairy product congealed and prepared from cream, milk, skim milk, condensed milk, butter, or a combination of these ingredients with the addition of sugar, various flavors, stabilizers, and colorants (optional). Air is drawn into the mixture during the congealing process. The product, originated in Europe by famous Sicilian and Neapolitan pastry masters, today is industrially produced throughout the world. The production of creamless or butterless ice cream dates back to the fifteenth century and was discovered probably by accident. The first production of a cream-based ice cream is mentioned in *The Experienced English Housekeeper*, published in 1769. In 1800 the homogenizer was invented in France. It was only between 1940 and 1945 that the production of ice cream packages suitable for maintenance at very low temperatures were developed and therefore suitable for delayed consumption of the product.

Classification

A standard classification for ice cream, sherbets, and frozen cream (cream gelats) is not yet possible because of the countless variations in the composition, ingredients, and production methods. The U.S. manufacturers have grouped products into classes, such as regular or Philadelphia ice cream, French or Neapolitan ice cream, sherbets, ices, and more recently ice milk. Keeping in mind the fundamental difference between true ice cream containing cream or its derivatives in addition to the regular ingredients of sherbets (sugar, fruit juices, flavors, and sometimes milk), all other specialities could be included in one of the above categories, depending on their basic formulation. Taking into account also the possible use of milk in sherbets, three fundamental categories should suffice to group all of the various ice-cream products:

1. Ice cream, containing cream or its derivatives
2. Milk sherbets, containing whole, powdered, or condensed milk
3. Ices, formulated without milk or milk derivatives.

Ingredients

The generic term *ice cream* is commonly used to define a congealed product consisting of a mixture of dairy products, sugar, flavors or liqueurs, colorants, and, if required, stabilizers. The finished product may or may not contain eggs, fresh, preserved, or dried fruit, fruit juices, etc.; it is thoroughly homogenized by pounding or mixing during the congealing process. Regular ice cream generally contains no more than 5% colorants (where permitted) and flavors calculated on the volume of the ice-cream mixture prior to congealing.

Ice sherbets are prepared with fruit juices, sugar, stabilizers, with or without fruit, acids, colorants, flavors, and water. Usually they contain approximately 28–30% sugar and no milk or dairy products.

Milk sherbets (ice milks) are prepared using fruit juices, sugar, stabilizers, and dairy products (whole, skim, condensed, or powdered milk) that totally or partially replace the water used in ice sherbets.

The use of stabilizers is governed by regulatory agencies in the various countries. Among the stabilizers employed are agar-agar, sodium alginate, calcium sulfate, gelatin, carob seed gum, guar gum, oatmeal gum, gum tragacanth, Irish moss and extracts, lecithin, carboxymethyl cellulose, and a few others. Emulsifiers consist usually of mono- and diglycerides obtained by glycerolysis of edible fats. The use of emulsifiers is also subject to specific regulations in various countries.

Ice-cream composition consists of the uncongealed mixture of all ingredients with the exception of air and flavors. The composition is usually expressed as a per cent of the various ingredients, *e.g.*, per cent milk fats, per cent non-fat milk solids, per cent sugar, per cent egg yolk solids, per cent stabilizers, and per cent total solids. The per cent of milk fats varies more than all other constituents, ranging from 8–24%. When the per cent of milk fats is increased, the per cent of non-fat milk solids must decrease accordingly. Usually a well balanced commercial ice cream formulation contains 36–42% total solids and 20–26% total milk solids.

Flavor is considered one of the most important features of ice cream; the taste includes the almost tactile sensations caused by the consistency of the product as well as the true flavor. Ice-cream flavor results from all of the ingredients; some ingredients do not exhibit a very pronounced and detectable flavor per se but contribute to the overall flavor sensation. Delicate and bland flavors may be used quite freely, since they do not put too much stress on the consumer's acceptance; strong, definite notes tend to wear out a consumer's acceptance faster even at rather low levels.

Vanilla is, without any doubt, the most accepted and popular ice-cream flavor. Statistics show that 75% of the total ice-cream production contains vanilla flavor. Ideal flavor ingredients would be the extracts of vanilla pods, but the demand for more concentrated products is increasing very rapidly. Vanillan together with vanilla in suitable mixture can be employed for ice-cream flavoring. Cocoa and chocolate flavors follow vanilla in popularity. For chocolate ice-cream flavoring approximately 2 kg cocoa or 3 kg chocolate per 50 kg of ice-cream mixture is employed with an equal weight of sugar. Cocoa or chocolate together with sugar must be added with the other dry ingredients prior to pasteurization.

Other types of flavor ingredients employed are almost exclusively geared toward fruit flavors—raspberry, strawberry, banana, lemon, orange, mandarin, pistachio, apricot, peach, etc. A few fantasia flavors, such as nougat and liqueur flavors, have also found remarkable acceptance. To obtain good flavor formulations, it is advisable, whenever feasible, to use the best available fruit juices. In a few cases (if permitted by local regulations on the manufacture of ice cream), it might prove useful to employ natural or synthetic flavors formulated according to the guidelines for the flavoring of syrups (see p. 707). Excessive amounts of volatile esters or pronounced coumarin-type notes should be avoided in ice-cream flavor formulations.

Production

Both in batch and continuous ice-cream production, the ingredients are added approximately in the same order. The liquid ingredients are placed together in the container, agitated, and heated. The dry ingredients, including powdered milk and egg, cocoa, sugar, and stabilizers, are added while stirring prior to heating at temperatures exceeding 49°C. Butter, frozen cream, and other frozen products are crushed in small pieces prior to pasteurization, allowing sufficient time to melt. Subsequently, pasteurization is carried out at high temperatures (109–117°C); this favors the kill of bacteria, imparts improved body and better flavor, and protects the product from oxidation. The mixture is then transferred to a homogenizer; homogenization consists of passing the mixture through a small orifice at high speed with pressure exerted by means of a piston pump forcing the liquid through tiny valves. A permanent and uniform suspension of fat is thus obtained by reducing the droplet size to less than 2 microns. The resulting ice cream is extremely uniform and smooth. The homogenizing temperature varies between 63–77°C. After homogenization, the mixture is immediately cooled to 0–4°C and allowed to "set" in suitable containers until ready for use. The *setting*, or ripening process, has been practiced since the very beginning of ice-cream production. The fat solidifies during setting, while the stabilizer swells (if gelatinous) and combines with water.

Proteins in the mixture may alter slightly, and the viscosity increases because of this. Undoubtedly the consistency of the products improves, as does its resistance to melting. Until not too long ago, ripening lasted a minimum of 24 hours; more recently 3–4 hours has been established as sufficient for ice-cream ripening. During setting the temperature must be kept around +2°C and sometimes as low as −2°C to −1°C.

At this stage, ice cream is congealed as quickly as possible, since the ice crystals that form during quick freezing are much finer than those obtained in a slow-freezing cycle. Therefore, congealing and withdrawal from the freezer must be effected very rapidly. In a continuous freezer the operation takes only a few seconds; in a batch freezer congealing takes approximately 6–10 minutes. Ice cream is finally transferred to freezers to harden; the crystals formed during this congealing stage are much larger.

The flavoring of so-called ice-cream flavor compositions follows the same guidelines mentioned above for ice cream and sherbets. Liquid flavors are not usually employed in the preparation of such products; good powders are usually available commercially. Individual and compounded spray-dried fruit flavors and lyophilized fruit juices are important flavor forms for ice-cream products.

Yogurt

There are numerous techniques for the preparation of yogurt, ranging from home to industrial production. To obtain products of consistent quality, cow's milk concentrated to about two-thirds of the original volume or regular cow's milk containing a little powdered milk is employed. The milk is coagulated directly in the distribution containers by inoculating with *Thermobacterium bulgaricum* cultivated in symbiosis with *Streptococcus thermophilus*. Coagulation is completed after 3–5 hours at 40–45°C. After this operation, yogurt is ready and must be cooled to low temperature to stop the fermentation of lactic acid, whose content in the finished product should be kept around 1%.

Yogurt prepared as described is white, does not separate serum, and has a pleasant, cheese-like appearance with a slightly sour flavor. Yogurt, consumed as is or with the addition of a little sugar, has won growing acceptance throughout the world and owes its widespread acceptance to complementary flavors. The acid yogurt flavor blends well with fruit flavors. Fruit-flavored yogurts are obtained by adding a suitable amount of flavored syrup (raspberry, strawberry, whortleberry, apricot, mandarin) to the yogurt immediately after coagulation. The best production employs syrups based on natural fruit juices or, simply, concentrated juices when sweetening is not required. Syrups flavored with strengthened fruit juices or with compounded flavors of the same type used in syrups may also be employed. Preliminary trials using lyophilized juices have given excellent results from a technical viewpoint.

Butter

Once established that the cream flavor is highly appreciated in milk and that cream is used to make butter, it is logical that the same cream flavor imitations with only slight modifications are the basis for most simple butter flavor formulations. The same ingredients are therefore used in the traditional formulations in slightly different proportions. These usually contain larger amounts of butyric, valeric, acetic, and lactic acids; esters (cinnamyl and ethyl butyrate, ethyl capronate, butyl butyryl lactate); trace amounts of additional products (cinnamaldehyde, benzylideneacetone, benzaldehyde, tributyrin, triacetin); various light notes of essential oils and extracts (nutmeg, civet, and fenugreek); and acetyl propionyl in products of the butterscotch type. All of the above ingredients are rather simple and intended primarily for pastry flavoring. Should more complex and more functional formulations be required (kept within the limitations imposed by government regulations and toxicological considerations), the study of the composition of natural butter flavor and the use of products with a long-established record of safety may very well supply interesting leads. It must be kept in mind that butter flavor has various notes and varying compositions depending on the following types: fresh butter, melted butter, roasted butter, or butter cooked in the presence of fats, sugar, flour, and other ingredients used in pastry and food products.

An inquiry into butter flavor formulations by the user must clearly specify the desired type; it is the flavorist's task to prepare formulations suitable for the various applications. Ultimately and ideally, a fresh butter flavor imitation should be capable of duplicating the natural product following the specific manufacturing treatment. The most important constituents of fresh butter are given below:

Components of Fresh Butter

Ingredient	Parts, ppm
Acetaldehyde	<0.1
Acetone	<0.1
Acetonin (acetyl methyl carbinol)	19
Bovolide (2,3-dimethyl-2,4-nonadien-4-olide)	0.45
Diacetyl	4.5
Formaldehyde	<0.1
2-Heptanone	<0.1
n-Hexyl aldehyde	<0.1
Isobutyraldehyde	<0.1
Isovaleral	1
2-Nonanone	<0.1
n-Nonyl aldehyde	<0.1
Phenylacetic aldehyde	<0.1

The aldehydes may be considered derivatives of the oxidative enzymatic decarboxylation of amino acids; the ketones, as probable derivatives from the decarboxylation of β-ketoacids; acetyl methyl carbinol, as a precursor rather than a derivative of diacetyl; bovolide, present in all butter types, could be dependent on the feed. Additional products, already encountered in heated milk, have also been identified in butter, *e.g.*, aldehydes, ketones, alcohols, aliphatic methyl and ethyl esters, lactones, and dimethyl sulfide.

Carbonyl compounds appear to play only a secondary role in the constitution of fresh butter flavor; δ-lactones appear to have been identified only in melted butter. They impart a well accepted flavor to melted or cooked butter, but their presence is totally objectionable in fresh butter flavor. A few formulations useful for the preparation of butter flavors are simply based on lactic and acetic acid, acetaldehyde, dimethyl sulfide, and diacetyl. A few authors have suggested the use of α- or γ-carboxy lactones (odorless C_6 to C_{12}) as precursors of the lactones present in melted butter flavor.[19]

Margarine

The discovery of margarine is commonly attributed to the French chemist Hippolyte Mege-Mouries, who patented the product in France and England in July, 1869. Mege-Mouries had observed that starving cows, although losing weight and giving less milk, still secreted fats in their milk. He came to the conclusion that cows consume their own body fat and that the *oleomargarine* in the cow's body is transformed into butter by enzymatic action of pepsin. A further deduction was that the odorous volatile coloring matter becoming rancid was not originally present in natural (white) fat, but it developed by action on organized tissues by fermentation, heat, and chemical agents. Mege-Mouries concluded that milk fats (so-called butter) were derived from modified fats in the body of the animal. By applying the above theories, the inventor first prepared a virgin tasteless and odorless fat and then a kind of true butter.

Margarine was originally produced in France and shortly thereafter in Holland, Denmark, Austria, Hungary, Germany, Great Britain, and the United States. Margarine production was conditioned by the availability of inexpensive fats suitable for the purpose. Therefore, the perfecting of the hydrogenation process of oils gave great impetus to the industrial production of margarine.

Fats Used in Margarine

Through the process of hydrogenation, the liquid glycerides (fats) are transformed into a solid fat at room temperature. The process allows the use of almost all vegetable fats for margarine production; this fact started the refining of coconut, linseed, soybean, safflower, palm, and whale oils. Today the margarine industry, taking advantage of both the hydrogenation and deodorizing processes as well as other oil-refining techniques, may use almost any edible oil or raw material. The largest portion of the raw materials consists of vegetable oils (and whale oil for the balance) in the English production of margarine. Vegetable oils are used as the starting material in 98.5% of U.S. margarine, the balance consisting of animal fats.

While in the United States the fats employed consist almost exclusively of linseed and soybean oils, at least four different fatty raw materials are used in English production. These include coconut oil, palm seed oil, peanut oil, palm oil, and small percentages of cottonseed, sesame, and safflower oils. Margarine production requires the use of fresh and fairly pure fatty raw materials. The fat must be physiologically and hygienically suitable and free of toxic constituents. For instance, castor oil is not suitable.

With the exception of cold-expressed olive oil and safflower seed oil, the extracted fats exhibit an unpleasant taste from the presence of non-glyceride impurities and the formation of acids by glyceride decomposition. Raw fats, therefore, require refining prior to use. The fats also must be non-nutritive. Normally natural fats are readily digested as long as their melting points are not far above the human body temperature. Their chemical composition must be suitable for the purpose for which they are intended. A few fats, such as hardened palm oil, confer remarkable plastic properties to the finished product, *i.e.*, they render the product readily spreadable. Others, such as coconut oil, enhance the friability of the product. The fat must be free of oxidizable products, and the physical properties must be correspondent to the end-use for which the product is intended.

Natural fats are mostly liquid and must be hydrogenated to reach the optimum melting point. These requirements are never met by raw fats, which usually contain impurities responsible for the dark color, turbidity, and disagreeable odors. Fats must undergo various refining processes prior to being transformed into suitable raw materials for margarine production. Only freshly extracted and refined fats should be employed in the production of margarine, since raw and paste fats (even the refined and hardened types) tend to deteriorate during storage and transport.

Production of Margarine

Margarine is an emulsion consisting of fat and water. Although the amount of dispersed water varies from country to country, it is usually about 16%. Margarine is a stable solid emulsion at room temperature with a consistency, plasticity, appearance, color, and flavor resembling butter. Colorants and flavors are added during the manufacturing process.

The production cycle can be outlined as follows:

1. Preparation of the aqueous phase
2. Preparation of the fatty phase
3. Emulsification of the two phases
4. Cooling
5. Plastification
6. Packaging

The aqueous phase may consist of pure sterilized water, aqueous lactose, lactic acid, or solutions of other ingredients. Most of the producing countries employ milk ripened by a bacteriological process similar to that employed for the production of butter. In this way margarine acquires a bland butter flavor. The synthetic flavor ingredients normally added to margarine do not impart the same agreeable taste as ripened milk.

Milk also contains emulsifying substances that help the formation of margarine emulsion. It is common practice, however, to add emulsion stabilizers when milk is used. Four different milk products can be used in the aqueous phase: cream (34–45% fats), whole milk (3–5% fats), skim milk (almost fat-free), and powdered skim milk reconstituted with water. The first two products are almost never used because of their high cost. Pasteurized and cooled milk is inoculated with a strain of bacteria to induce the ripening process, during which lactic acid is formed from lactose as well as flavor substances from fermented citric acid accompanying the lactic fermentation. These two fermentations are necessary for the formation of butter flavor. When ripening is complete, the acid content ranges between 0.75–0.80%.

With regard to the fatty phase, countless combinations of different fats may be used. Usually hydrogenated fats with a high melting point are mixed with soft fats. Deodorized fats are kept in aluminum containers just above their melting point. The ingredients are subsequently pumped into containers, placed over scales, and then mixed with suitable equipment.

From here the mixture is transferred into an emulsifying tank. The aluminum container, the scale-controlled mixer, and the emulsifying tank must be connected in such a way as to render the process almost continuous. At this stage the aqueous phase is at a temperature of 10°C, and the fatty phase, at 50°C maximum; higher temperatures might cause oxidation. First, the fatty phase is introduced, and ground ice is added to yield the required water ratio in the finished product and to allow a faster hardening of fats. At this point sodium chloride, sodium benzoate, EDTA, other stabilizers, and the eventual water-soluble emulsifiers are added. The dispersion of water into the fatty phase should not be too fine in order to avoid an excessive oily appearance in the finished product.

Cooling was at one time a separate operation. More recent techniques permit emulsification and cooling operations to be performed simultaneously in a single piece of equipment. In the last 20 years a method of cooling in pipes has been perfected. The machine emulsifies, cools, and mixes so that the product comes out of the pipes with a consistency suitable for direct molding and packaging.

The main additives, in addition to the above mentioned ones, are emulsifiers, anti-spattering agents, chelating agents, vitamins, flavor ingredients, colorants, antioxidants, and preservatives.

Flavor is an extremely important factor to the margarine consumer. The objective of margarine producers is to duplicate a butter flavor, which does not exist as a definite form since it varies enormously. The butter flavor in margarine is introduced mainly with the ripening process of milk added during the manufacturing process. It is well known that the main component influencing butter flavor is diacetyl, which is formed by bacteriological action on milk. Chemically, diacetyl is 2,3-butandione (CH_3—CO—CO—CH_3); it can be prepared by oxidation of acetyl methyl carbinol. Diacetyl is also produced synthetically by oxidation of methyl ethyl ketone with nitric acid and subsequent decomposition of the resulting isonitroso compound.

Whenever ripened milk is not added during the production of margarine for specific technical or economic reasons, the addition of butter flavors or compounded margarine flavors is definitely advisable. The formulation of such products may be realized in two versions, *i.e.*, hydrosoluble or liposoluble types. However, this is complicated by patents tending to limit the use of the potential flavor components. The formulation makes use of diacetyl and acetyl methyl carbinol as basic materials; substitutes for diacetyl could be the homologous ketones, *e.g.*, acetyl propionyl and dipropionyl. It has been definitely ascertained that margarine flavor is significantly improved by the addition of 0.2% monosodium glutamate, which is another flavor potentiator for margarine active at ppm levels.

Cheese

Each cheese type exhibits a peculiar and characteristic flavor dependent on the flavor of the milk used, the various curdling and manufacturing methods (salting, heat cycle, etc.), the type of fermentation, aging, and even the climatic conditions prevalent on the production site. In order to tackle a problem that presents an enormous number of variables, it is necessary first to classify the several hundred cheese products according to basic criteria (independent of the cheese flavor and palatability but based only on the industrial method of preparation); then the most representative examples from each category will be discussed. The following classification may be adopted:

1. Cheese made of whey plus curd (ricotta, mysost)
2. Cheese made of curd alone

These two classes may be further subdivided as follows:

1. Fermented cheese	Hard, grating cheese (Romano, Parmesan, etc.)
Ripened, aged	Hard (Cheddar, Swiss, Gruyere, Provolone, etc.)
	Semi-soft (Roquefort, Blue, Gorgonzola)
	Soft texture (Camembert, Brie, etc.)
2. Non-fermented cheese	High fat (cream, Neufchâtel)
Unripened	Low fat (cottage, baker's)
3. Melted cheese	

Cheese may also be classified as cooked, semi-cooked, and raw.

The imitation of cheese flavors is extremely problematic because there are so many varieties, especially when complete flavor (odor and taste) results are required. Flavor imitations in the majority of cases attempt to duplicate odor only; the following products find application for this purpose: butyl butyryl lactate, isovaleric and capronic acids, ethyl butyrate, and methyl *n*-amyl ketone. In certain cases (*e.g.*, Roquefort cheese), the use of ammonium isovalerate may be of interest, whereas trace amounts of the most important cream flavor components may be usefully employed in the formulation of melted cheese flavors. For more complex and functional flavor formulations, the study of the components of natural cheese flavors may prompt valuable information. Although there are no available analyses of the constitution of non-fermented cheese flavors, it appears likely that the latter must best be closely related to (1) milk flavor, (2) thermally degraded milk flavor, or (3) microbiologically degraded milk flavor.

A discussion of fermented cheese flavors becomes even more complex. While components of milk flavor "as is" may be considered as playing only a secondary role, products derived from the degradation of proteins, fats, and carbohydrates must be considered responsible for the characteristic flavor.

Proteins serve as a substrate for microorganisms. A few peptides are also responsible for the bitter taste. Amino acids formed by proteolysis are not responsible for the typical cheese flavor but provide a background note and probably function as flavor enhancers. A positive cheese-flavoring action is exerted by the decomposition products of amino acids.

Fats play a primary role as modifiers and solvents for aromatic compounds. However, the precursors of lactones, methyl ketones, esters, alcohols, and fatty acids formed by enzymatic (lipase) hydrolysis are directly implicated and have been identified among the components of the characteristic cheese flavor. Carbohydrates and lactose are a source of energy for lactic acid bacteria. The unused carbohydrate portions end up in whey, and only slight traces are left in the curd; traces may be very important flavor precursors, however.[20]

The following ingredients are present in cheese flavor:

Components of Cheese Flavor

Ingredient	Parts, ppm
Acetaldehyde	1.40
Acetone	1.60
1-Butanol	0.70
2-Butanol	0.30
Butanone	0.30
Diacetyl	0.80
Dimethyl sulfide	0.11
Ethanol	16.30
Ethyl butyrate	0.60
2-Heptanone	0.45
2-Methylbutyraldehyde	0.42
Methyl caproate	1.50
2-Pentanone	0.98
1-Propanol	2.90

The characteristic flavor of blue cheese has been the subject of intense research.[21-24] To formulate blue-cheese flavor imitations, the following products may be found useful:

1. Acids: acetic, benzoic, butyric, caproic, caprylic, hydroxybenzoic, *p*-hydroxyphenylacetic
2. Ketones: acetone, 2-heptanone, 2-nonanone, 2-pentanone, 2-undecanone
3. Alcohols: 2-heptanol, 2-nonanol, 2-pentanol, 2-phenylethanol
4. Esters and aldehydes: ethyl butyrate, methyl caproate, methyl caprylate, and a few aliphatic aldehydes

Acids must be employed in amounts equal to more than 80% of the total mixture.

THE ROLE OF ENZYMES IN FOOD FLAVORS

Author: **Basant K. Dwivedi**
Foster D. Snell, Inc.
Florham Park, New Jersey

INTRODUCTION

Proteins, carbohydrates, and lipids are the three main sources of food flavors. Some minor flavor precursors such as polyphenols, nucleotides, and the pigments of carotenoid series also contribute to a certain extent in flavor development.[1] Naturally occurring food enzymes and enzymes involved in the metabolic process of microorganisms used in ripened and fermented foods play a major role in the development of food flavors from these precursors.

Hewitt and co-workers[2,3] reported that the typical flavor of a number of food products, such as milk, celery, spinach, tomatoes, oranges, strawberries, etc., lost in food processing may be restored by the addition of the right enzyme preparation. In a typical experiment, Hewitt et al.[2] prepared a blanched and dehydrated product from fresh watercress. The dehydrated material was quite flavorless. Reconstitution of dried watercress did not improve the flavor and it tasted and smelled like hay. However, when a tasteless and odorless enzyme preparation from white mustard (white mustard and watercress both belong to the *Bressica* family) was added to the reconstituted product, the typical odor and taste of watercress was restored within a few minutes. The dramatic effect of mustard enzymes in restoring watercress flavor as proposed by Hewitt et al.[3] is reproduced in Figures 1 and 2.

The work of Hewitt et al.[2,3] amply demonstrates the importance of enzymes in the flavor development of food products. Unfortunately, comparatively little work has been done along this line to provide better insight of the various steps involved in flavor development by enzymatic action. Such information could be of great advantage in designing process details in the commercial processing of foods. Severe flavor losses during processing may be compensated by preserving the precursors and restoring the original flavor of processed food by enzymatic action. Additional research and developmental work are required on the structure of various flavor precursors, the nature of enzymes which act on these flavor precursors, and the reactions which take place. Some information on the nature of enzymes involved in the development of important flavor compounds is known. However, in order to study the enzymatic reactions and

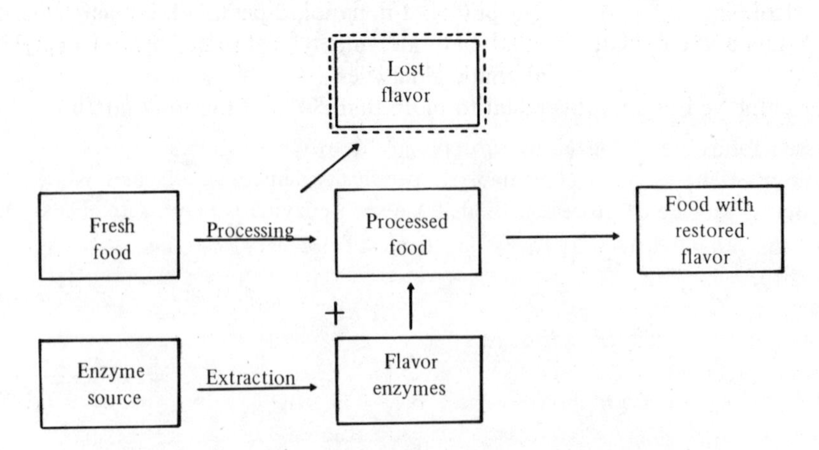

FIGURE 1. Restoration of natural flavor to processed foods. (From Hewitt, E. J. et al., in *Chemistry of Natural Food Flavors,* Mitchell, J. H., Jr., Leinen, N. L., Mrak, E. M., and Bailey, S. D., Eds., Chicago Quartermaster Food and Container Institute, Chicago, 1957, 88. With permission.)

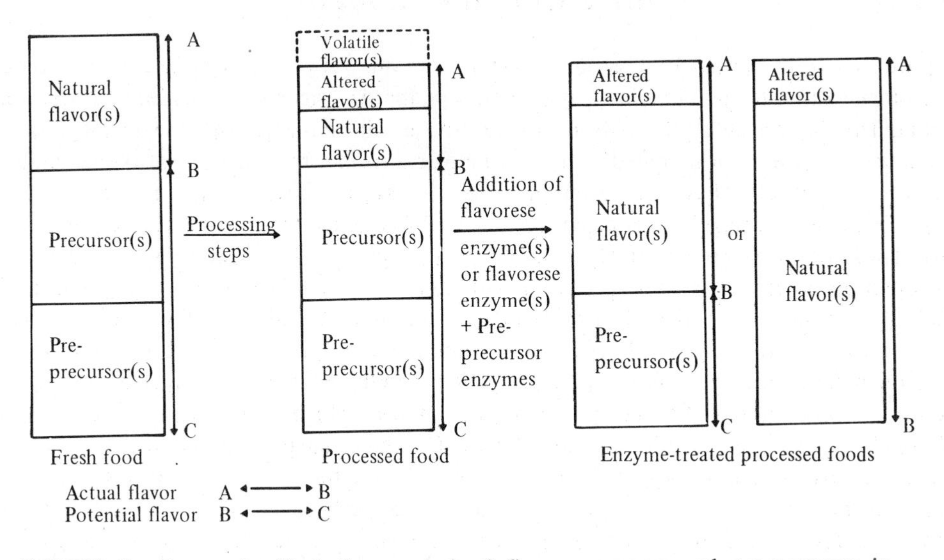

FIGURE 2. Barographs illustrating survival of flavor precursors and pre-precursors in conventional heat processing and action of enzymes on flavor precursors and pre-precursors in processed foods. (From Hewitt, E. J. et al., in *Chemistry of Natural Food Flavors,* Mitchell, J. H., Jr., Leinen, N. L., Mrak, E. M., and Bailey, S. D., Eds., Chicago Quartermaster Food and Container Institute, Chicago, 1957, 90. With permission.)

intermediate steps precisely and to adopt the information in preserving food flavors during processing, purified enzyme extracts would be required. The purified enzymes may be studied for their specificities and concentrations using model systems. The information obtained from these studies may then be applied to food systems and the influence of other enzymes and ingredients present in the food on the basic enzymatic reaction can be followed. In addition, as suggested by Hewitt et al.,[2,3] it would be desirable to study the flavor precursors and perhaps the flavor pre-precursors. These relatively nonvolatile compounds may be separated from the bulk of the food products by appropriate methods, such as column and ion-exchange chromatography, and identified by using a combination of infrared, nuclear magnetic resonance, pyrolytic gas chromatography, and mass spectrometry. Finally, the action of enzymes, believed to be involved in flavor development, on the isolated precursors could be studied and the liberated flavor examined by the sensory panel and instrumental techniques.

The climatic conditions, method and the time of harvesting, geographic location, and the species of plants are known to influence the flavor of foods, especially fresh foods. The flavor enzyme and precursor study would give us some information on the nature of enzymes and precursors in the plants that are influenced by these factors. The day can be foreseen when foodstuffs may be processed according to their flavor precursor content, rather than for their flavor.[4]

The processed foods are stabilized by substantially destroying or inactivating all the enzymes present. This inactivation, often done by blanching, is specifically aimed at enzymes which induce off-flavor development during storage of processed foods. One such enzyme is peroxidase. Thus, when blanching is carried out to the point where peroxidase activity is substantially destroyed, most of the other enzymes present in foodstuffs are also destroyed or inactivated. In other words, in inactivating the enzymes that are deleterious to food stability, the enzymes involved in flavor formation are also affected. The fresh flavor of the food is altered or partially destroyed by processing because many flavor compounds are volatile or heat-labile. Flavor precursors seem to survive processing, thereby representing a potential or latent source of fresh flavor.[2,3] The precursor and enzyme study may also provide information for selective destruction or control of enzymes and/or precursors responsible for undesirable flavors produced during storage of food products.

MILK AND MILK PRODUCTS

Milk produced from normal healthy cows contains a wide variety of enzymes. In their report on the Enzyme Nomenclature, to the Manufacturing Section of the American Dairy Science Association, Shahani et al.[5] compiled the available information on the bovine milk enzymes. All the major data concerning these enzymes were consolidated in a tabular form by these workers and can be conveniently referred to. They also attempted to discuss the physiological as well as technological significance of these enzymes. Milk enzymes are believed to originate in the secretory epithelial cells of the mammary gland and their presence may be regarded as the "spilling over" from these cells and serum during the milk secretion process.[6,7] It is surprising that there are few readily observed changes in milk directly attributed to enzymes other than lipase,[8] which causes lipolysis of milk fat, and possibly xanthine oxidase,[9] implicated with oxidative deterioration of milk. Zittle[10] believes that the pasteurization of milk as a common practice may be one of the reasons. Many milk enzymes are heat-sensitive and loose their catalytic activity when heated. Second, the enzymatic changes which do occur may not be obvious if the physical stability and flavor of milk are not altered. On the other hand, some heat-inactivated enzymes regain their activity during storage and may continue to act for a long time if a suitable substrate is available, as in the case of protease.[11] The reaction products in such cases continue to build up and may bring about spoilage or other detectable changes even though the concentration of residual or reactivated enzyme may be negligibly small. It is, however, difficult to distinguish between changes brought about by enzymes inherent in milk and the ones produced by microorganisms during prolonged storage.[10]

Some of the milk enzymes and their approximate concentrations in fresh cow milk that may have some role in the development of desirable or undesirable flavors are given in Table 1. It may be noted from this table that milk contains significant amounts of amylase, lipase, xanthine oxidase, and peroxidase.

TABLE 1

Partial List of Bovine Milk Enzymes[6,12]

Enzyme	Concentration in milk[a] Fresh	Pasteurized[b]	Sterilized[c]	Specificity, substrate	Optimum pH	Optimum temp. °C
Aldolase (4.1.2.13)				Ketose mono- and diphosphate, fructose 1,6-diphosphate	Assayed at pH 7.0	35
α-amylase (E.C.3.2.1.1)	11,800	950	620	α-1,4-D Glucan linkages in polysaccharides, starch glycogen, polysaccharides, and oligosaccharides	7.4	34
A-esterase (3.1.1.2)				Aryl esters	8.0	37
B-esterase (3.1.1.1)				n-Methyl esters of fatty acids, triglycerides	8.0	37
C-esterase (3.1.1.8)				Choline esters, tributyrin	8.0	37
Lipase (3.1.1.3)	109	–	–	Glycerol esters in emulsion	9.0 to 9.2	37
Peroxidase (1.11.1.7)	22,100	5,400	0	H_2O_2 and some organic peroxides	6.0 to 7.0	20
Protease (3.4.4.-)				Tryptic and/or chymotryptic specificity	8.0 6.5	37
Xanthine oxidase (1.2.3.2)	175	145	0	Aldehydes, oxypurines, pterin, DPN, xanthine	6 to 9 6 to 9	37

[a]All enzyme units are expressed in international units (IU)
One IU = moles of substrate transformed per minute per 1,000 ml milk
[b]15 sec at 74°C
[c]15 min at 115°C

Esterase

This enzyme catalyzes the hydrolysis of esters. Forster et al.[13] reported the presence of three esterases, A, B, and C, in milk. A-esterase hydrolyzes aromatic esters such as phenyl acetate. B-esterase hydrolyzes aliphatic and aromatic esters but not choline esters, whereas C-esterase splits choline esters more rapidly than aliphatic or aromatic esters.[14] Unlike milk lipases, very little work has been done on milk esterases for a possible role in the development of desirable or undesirable flavors in dairy products. Jensen[15] has reported that B-esterase behaves identically to the lipase in whole milk and contains glycerol hydrolase activity. Recently Kitchen[16] provided some insight into the nature and distribution of milk esterases. An important property in relation to the presence of milk esterases in Cheddar cheese is that the esterases are soluble at pH 4.6. It is unlikely, therefore, that any high proportion of milk esterases would be present in the curd, although, as some mechanical occlusion of whey proteins occurs during cheese making, some enzyme would be present in the final pressed cheese. Over the lengthy period of ripening, this small amount of enzyme might still be significant in relation to flavor development. The concentration of these enzymes in Cheddar cheese is so small that is is not possible to determine it experimentally and provide evidence for any contribution by these enzymes in flavor development. The contribution of these enzymes in the flavor of other dairy products is not known.

Lipase

Bovine milk contains lipases which hydrolyze milk fat and liberate fatty acids. Milk fat is especially rich in short chain fatty acids such as butyric acid. Liberation of these fatty acids from milk fat by the action of milk lipases imparts undesirable flavor, termed "rancid," to dairy products. Milk products such as cream and butter prepared from raw milk often develop rancidity on storage. An important use of milk lipases is in the manufacture of blue and Roquefort cheeses. "Peppery taste" of blue cheese has been attributed to short chain fatty acids and their hydrolyzable salts which are produced by the action of native milk lipases and lipases produced by microorganisms during cheese ripening.

It is believed that milk contains at least two different lipases, one irreversibly adsorbed on the fat globule membrane when milk is cooled, and the other associated with the casein of milk plasma.[17] Purified slime lipase shows an apparent specificity for short chain fatty acid-containing glycerides.[18] This enzyme shows no intramolecular specificity and hydrolyzes both short and long chain fatty acids at the same rate provided the fatty acids occupy the 1- and 3-positions in the same molecule.[19] However, it exhibits positional specificity, i.e., fatty acids in the 1- and 3-positions are hydrolyzed at a faster rate compared to fatty acids in the 2-position. The slime lipase that exists in association with casein is the major lipase in milk. Chandan and Shahani[20] isolated this enzyme in pure form and studied its properties (Table 2).[18,21,22]

Milk from some cows (up to 22%) in a herd is known to become rancid quickly.[23] The term "spontaneous" milk has been coined for such milks. The "spontaneous" milk does not require any treatment other than cooling when drawn or shortly after milking to 15 to 20°C to hasten lipolysis.[24] On the other hand, cooling and aging do not accelerate the lipolysis significantly in nonspontaneous or normal milk. Lipases from normal milk may be activated by homogenization of raw milk between 37.7 and 54.4°C for a short time. Other forms of agitation, such as shaking raw milk, churning raw milk or cream, and pumping, accelerate lipolysis. Increase in surface area of the substrate is the most important reason for the increase in lipolytic activity.[25] Thermal manipulations and certain chemicals, i.e., $CaCl_2$, $MgCl_2$, pseudoglobin, euglobin, lactalbumin, etc., also activate milk lipases. The spontaneous rancidity can be prevented or reduced by mixing with four to five times its volume of nonspontaneous milk.[27] Since only one out of five cows in a herd produces spontaneous milk, this defect is automatically eliminated in market milk. The mechanism involved in the inhibition of lipolysis in spontaneous milk by dilution with nonspontaneous milk is not known. In his review on the deterioration of milk lipids, Schwartz[26] speculated the presence in nonspontaneous milk of a labile inhibitor of spontaneous milk lipase.

TABLE 2

Properties of Purified Slime Lipase (Milk)[18,21,22]

		Reference
Optimum pH	9.0	18
Optimum temperature	37°C	18
Optimum substrate concentration	10%	18
Effect of		
Oxygen	Inhibitory	18
Light	Inhibitory	18
Heat	Inhibitory	18
Certain salts	Inhibitory	21
Glutathione	Stabilizing	21
p-Chloromercurobenzoate	Inhibitory	21
N-Ethylmaleimide	Inhibitory	21
Iodoacetate	Inhibitory	21
H_2O_2	Inhibitory	21
Substrate specificity		
Milk fat	Lipolyzed	22
Vegetable oils	Lipolyzed	22
Triglycerides	Lipolyzed	22
Simple esters	Not lipolyzed	18
Positional specificity	1,3 position	22

Peroxidase

Peroxidases are defined as enzymes catalyzing the following reaction:

$$ROOH + AH_2 = H_2O + ROH + A$$

The compound ROOH may be HOOH or some other organic peroxide. Milk peroxidase, often referred to as lactoperoxidase, has been isolated in pure form.[28,29] It appears to be associated with the albumin or whey protein components of milk. It constitutes about 1% of the total serum protein content and is fairly heat stable compared to other milk enzymes.[30] The number and nature of organic compounds that can serve as hydrogen acceptors for lactoperoxidase have not been fully investigated. Peroxidases, in general, use hydrogen peroxide, methyl hydrogen peroxide, ethyl hydrogen peroxide, and similar compounds as hydrogen acceptors. The hydrogen donors may be phenols, amines, or other organic compounds. The reaction mechanism seems to be based on the formation of enzyme-hydrogen donor complexes and two univalent oxidation steps, as shown below, where AH_2 stands for hydrogen donor and A for oxidized acceptor.

$$\text{Lactoperoxidase} + H_2O_2 = \text{Complex I};$$
$$\text{Complex I} + AH_2 = \text{Complex II} + AH;$$
$$\text{Complex II} + AH = \text{Lactoperoxidase} + A$$

The last step in this chain seems to be rate limiting. Milk peroxidase may be involved in oxidative deterioration of milk lipids. This enzyme should be thoroughly investigated for its physicochemical properties and its role, if any, in off-flavor development in dairy products.

Protease

Proteases hydrolyze the peptide linkages of proteins. Some of the peptides and amino acids released by the action of protease serve as precursors of flavoring compounds. Milk contains a very small amount of protease. Kaminogawa et al.[31] reported that milk protease has a sharp pH optimum at 8.0 with an accompanying shoulder at 6.5, shows tryptic and chymotryptic substrate specificity, and is most effectively inhibited by diisopropylfluorophosphate. They also found that when the protein was in more purified state, it could be destroyed by heating to 80°C for 10 min; however, when present in aseptically canned milk, considerable proteolysis occurred after storage for 30 days at 30°C. Among different casein fractions, β-casein has been found most susceptible to proteolysis followed by α- and k-casein.[31,32] Raw milk under normal conditions does not undergo spontaneous proteolysis, although protease exists in association with k-casein in milk. Murthy et al.[11] observed proteolysis in stored sterilized milk and attributed it to reactivation of protease. The presence of residual enzyme in pasteurized and evaporated milk has not been established. It is quite possible that upon long storages the residual or reactivated enzyme may initiate flavor and texture defects.[33]

Xanthine Oxidase

Xanthine oxidase is a nonspecific enzyme since it catalyzes the oxidation of purines, pyrimidines, and aldehydes. It oxidizes aldehydes to acids with the hydrated aldehyde as the intermediate step.[26,30,34] A similar hydration occurs at a carbon-nitrogen double bond in xanthine and on subsequent transfer of hydrogens, uric acid is formed. Thus, the action of xanthine oxidase on purines and aldehydes is more similar than one would anticipate from a quick look at the formulas. The hydrogens, when activated on the enzyme, can be transferred to other molecules such as redox dyes and to oxygen.[10] The term dehydrogenase is often used for this enzyme. The presence of xanthine oxidase in microsomal particles has been reported.[35] Xanthine oxidase is a metallo-protein, containing molybdenum and iron. It has a molecular weight of about 300,000 to 400,000 and pI 5.3 to 5.4. It is fairly stable to heat and, in contrast to other enzymes, its activity in milk increases with several heat treatments, homogenization, and by protease and lipase action.[5] The variation in xanthine oxidase content of milk produced in different localities apparently is related to the molybdenum content of the forage consumed by the cow.[37]

Xanthine oxidase, because of its broad range of activities, has been considered as perhaps responsible for some of the oxidative deterioration in the flavor of milk. The evidence, however, has been conflicting. Studies of Rajan et al.[38] and Smith and Dunkley[39] could not establish a correlation between xanthine oxidase content and oxidative flavor, based on chemical test or taste panel scores. Aurand and co-workers,[40-42] on the other hand, observed that the spontaneously oxidized flavor could be prevented either by the use of heat or an enzyme inhibitor. Subsequently they showed that the occurrence of spontaneously oxidized flavor could be correlated with a high level of xanthine oxidase activity and that specific inhibitors for xanthine oxidase were effective in preventing off-flavor development. They concluded that (1) high xanthin oxidase activity can be correlated with spontaneously oxidized flavor and (2) lability of the unsaturated fatty acids toward enzymatic oxidation was related to degree of unsaturation, i.e., linolenic acid was oxidized rapidly, linoleic acid at a very slow rate, and oleic acid unoxidized.

In contrast, other workers[39,43-45] have reported that ascorbic acid and copper were the essential reactants for the spontaneous flavor development and they were unable to obtain results consistent with xanthine oxidase theory.

Kiermeier and Grassmann[9] studied the relationship between xanthine oxidase and oxidized flavor in milk and reported that the milk containing active xanthine oxidase and buffered linoleic acid oxidized xanthine 1.5 to 2 times and adenine about 10% faster than normal milk. Various aldehydes tested could not act as substrate for xanthine oxidase.

Other milk enzymes that may have some role in flavor development of dairy products include aldolase and amylase. However, no information to this effect about these enzymes is available.

CULTURED DAIRY PRODUCTS

The contribution of microorganisms in the development of "typical" flavor of cultured dairy products is well known. However, detailed studies to provide deeper insight into the genesis of these flavors are still needed. Microorganisms naturally present or added during manufacture of cultured dairy products and cheeses metabolize carbohydrates, lipids, and proteins for energy and growth. The volatile flavor compounds are produced as by-products of these metabolic processes, catalyzed by microbial enzymes.

Several excellent reviews on the flavor of cultured dairy products have appeared in recent years.[46-49] A number of species of *Lactobacilli, Streptococci,* and *Leuconostoc* are used in starters for the manufacture of cultured dairy products. These aroma-producing lactic acid bacteria belong to the *Lactobacillaceae* family and may be homofermentative or heterofermentative. A thorough treatment of lactic acid bacteria may be found elsewhere.[50,51] These organisms are Gram-positive, nonsporulating cocci or rods. *Leuconostoc citrovorum, Leuconostoc dextranicum,* and *Streptococcus diacetilactis* are the main species used for aroma production in starter cultures. Diacetyl is the most important flavor compound in these products. Other compounds, such as acetaldehyde, acetylmethylcarbinol, 2,3-butanediol, acetic, and propionic acids, are also produced.

Homofermentative

Lactic acid bacteria belonging to this group produce as much as 1.8 mol of lactic acid per mol of glucose with minor amounts of acetic acid, ethanol, and carbon dioxide.

Embden-Meyerhof-Parnas (EMP) pathway is the main metabolic pathway for carbohydrate metabolism[52] (Figure 3). The factors affecting the nature of end products of fermentation of homofermentative lactic acid bacteria have not been studied thoroughly. In the case of *Streptococcus faecalis,* the end products depend very much on culture conditions, i.e., at pH 5.0 to 6.0 pyruvate is reduced by lactate dehydrogenase to D-, L-, or D,L-lactic acid. At neutral or slightly alkaline pH, however, pyruvate metabolism leads to the production of formic and acetic acid and ethanol in the ratio 2:1:1.[53]

Homofermentative lactic acid bacteria also use fructose, mannose, galactose, and disaccharides such as lactose, maltose, and sucrose as substrates. These sugars are converted into intermediates of EMP pathway by inducible enzymes.[50]

Lees and Jago,[54] quoted by Doelle,[50] have found a new pathway in certain *Streptococci,* i.e., *Streptococcus diacetilactis, Streptococcus lactis,* and *Streptococcus cremoris.* These bacteria utilize lactose and threonine from casein hydrolysate with the formation of glycine and acetaldehyde (Figure 4). The enzyme catalyzing the utilization of threonine is aldolase (L-threonine acetaldehydelyase, E.C. 4.1.2.5), which require pyridoxal phosphate as a cofactor. Acetaldehyde itself can be metabolized further via three different pathways, depending upon the availability of three enzymes as well as the utilization of lactose via the EMP pathway:[50] (1) Acetaldehyde and glyceraldehyde 3-phosphate are joined under the catalytic action of deoxyriboaldolase to form deoxyribose phosphate. (2) These bacteria also possess an enzyme which combines pyruvate with acetaldehyde with the formation of acetoin:

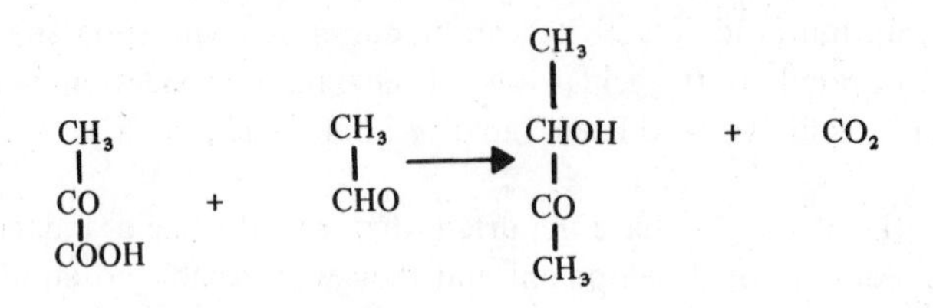

The action of this enzyme has been called a synthetase reaction by Lees and Jago, who gave the name "acetoin synthetase." (3) The third possible acetaldehyde metabolism is directed by an active alcohol dehydrogenase (alcohol:NAD oxidoreductase, E.C. 1.1.1.1) which converts acetaldehyde to ethanol:

$$CH_3CHO + NADH + H^+ \rightleftharpoons CH_3CH_2OH + NAD^+$$

There may be, of course, a number of *Streptococci* which do not follow this metabolism, as was shown with the glycine requiring *Streptococcus cremoris* Z8. The strain lacks the threonine aldolase.[50]

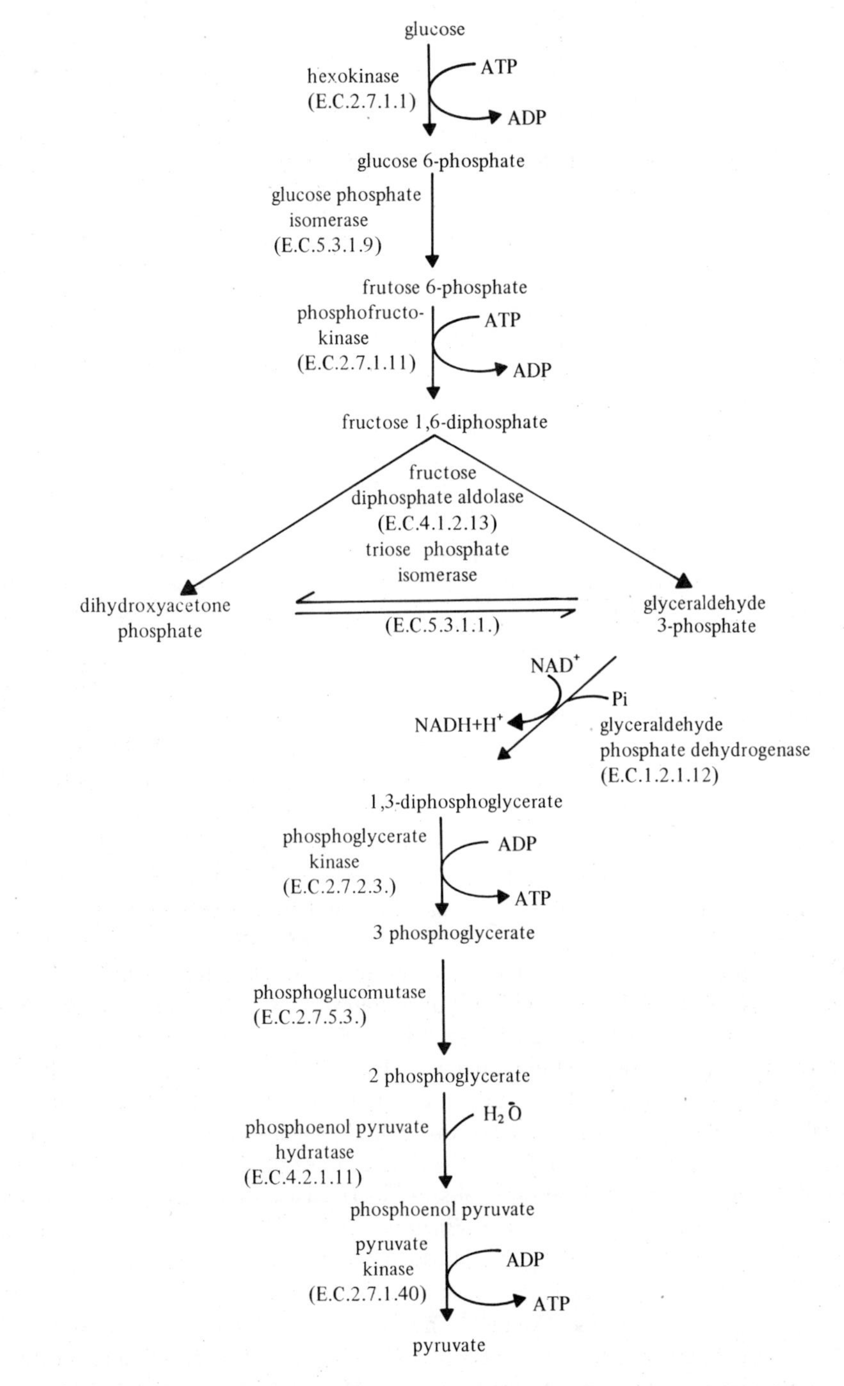

FIGURE 3. Embden-Meyerhof-Parnas (EMP) pathway of carbohydrate metabolism.

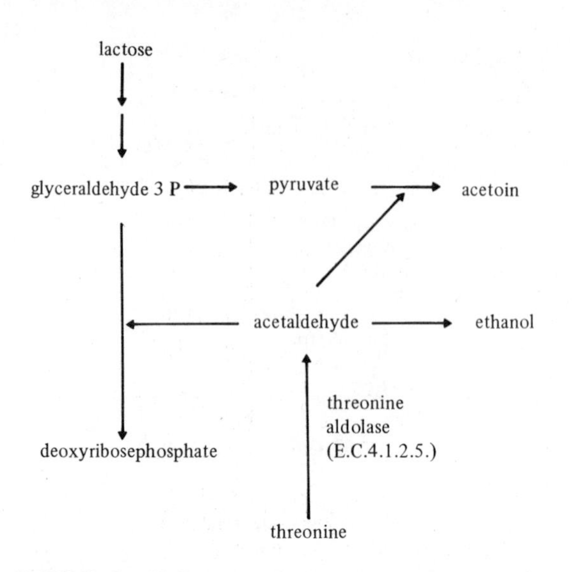

FIGURE 4. Utilization of threonine and lactose by certain *Streptococci.*[53]

Heterofermentative

Organisms that ferment hexoses with the formation of less than 1.8 mol of lactic acid per mol of glucose and in addition ethanol, acetate, glycerol, mannitol, and carbon dioxide are called "heterofermentative." The main group of organisms are strains of *Leuconostoc,* some strains of *Lactobacilli* and others.[52]

Leuconostoc-type heterofermentation results in the formation of 0.8 mol of lactic acid, 0.1 to 0.2 mol of acetate, 0.9 mol of CO_2, 0.8 mol of ethanol, and 0.2 to 0.4 mol of glycerol.

Leuconostoc species possess phosphoketolase pathway (Figure 5). All three other pathways, Embden-Meyerhof-Parnas pathway, hexose monophosphate pathway, and Entner-Doudoroff pathway are found to be absent.[50]

Investigations into the presence of enzymes in heterofermentative microorganisms have revealed the presence of hexokinase (ATP:D-hexose 6-phosphotransferase, E.C. 2.7.1.1), phosphoglycerate kinase (ATP:3-phospho-D-glycerate 1-phosphotransferase, E.C. 2.7.2.3), glycerophosphate dehydrogenase (L-glycerol-3-phosphate: (acceptor) oxidoreductase, E.C. 1.1.99.5), D-lactate dehydrogenase (D-lactate:NAD oxidoreductase, E.C. 1.1.1.28), ethanol dehydrogenase (alcohol:NAD oxidoreductase, E.C. 1.1.1.1),[55] acetaldehyde dehydrogenase, and acetoin dehydrogenase (acetoin:NAD oxidoreductase, E.C. 1.1.1.5). Fructose diphosphate aldolase (E.C. 4.1.2.13) and triose phosphate isomerase (D-glyceraldehyde 3-phosphate ketol isomerase, E.C. 5.3.1.1) are absent.[50] The lack of the latter two enzymes makes it impossible for the EMP pathway to function. Enzymes of the HMP pathway present are a glucose 6-phosphate dehydrogenase (E.C. 1.1.1.49) and 6-phosphogluconate dehydrogenase (E.C. 1.1.1.43).[56,57]

Two mechanisms have been proposed for the synthesis of diacetyl by lactic acid bacteria.[58-60] The mechanism proposed by Speckman and Collins[58] involves the condensation of active acetaldehyde and acetyl CoA (Figure 6). Using cell-free extracts of *Streptococcus diacetilactis* and *Leuconostoc,* these workers reported that only acetoin was formed from pyruvate in the absence of coenzyme A. Thiamine pyrophosphate (TPP) and Mg^{++} or Mn^{++} were required for acetoin synthesis. Addition of acetyl CoA to these solutions restored the ability to produced diacetyl. The second mechanism proposed by Seitz et al.[59] involves the oxidative decarboxylation of α-acetolactic acid (Figure 7). The latter mechanism of diacetyl synthesis has also been detected in yeast.[60] Citric acid fermentation and biosynthesis of diacetyl, acetoin, and 2,3-butanediol have been discussed in detail by Lindsay[46] and Collins.[48] In their study on the role of citrate and acetoin in the metabolism of *Streptococcus diacetilactis,* Harvey and Collins[61] noticed that this organism does not produce diacetyl and acetoin when grown in broth, on glucose, or lactose with no citrate or substitute of citrate. On the other hand, addition of citrate to milk enhances the production of acetoin and diacetyl. Based on these observations, these workers concluded that acetoin and diacetyl are produced only as a result of overproduction of pyruvate from citrate, which in turn is a consequence of the fact that

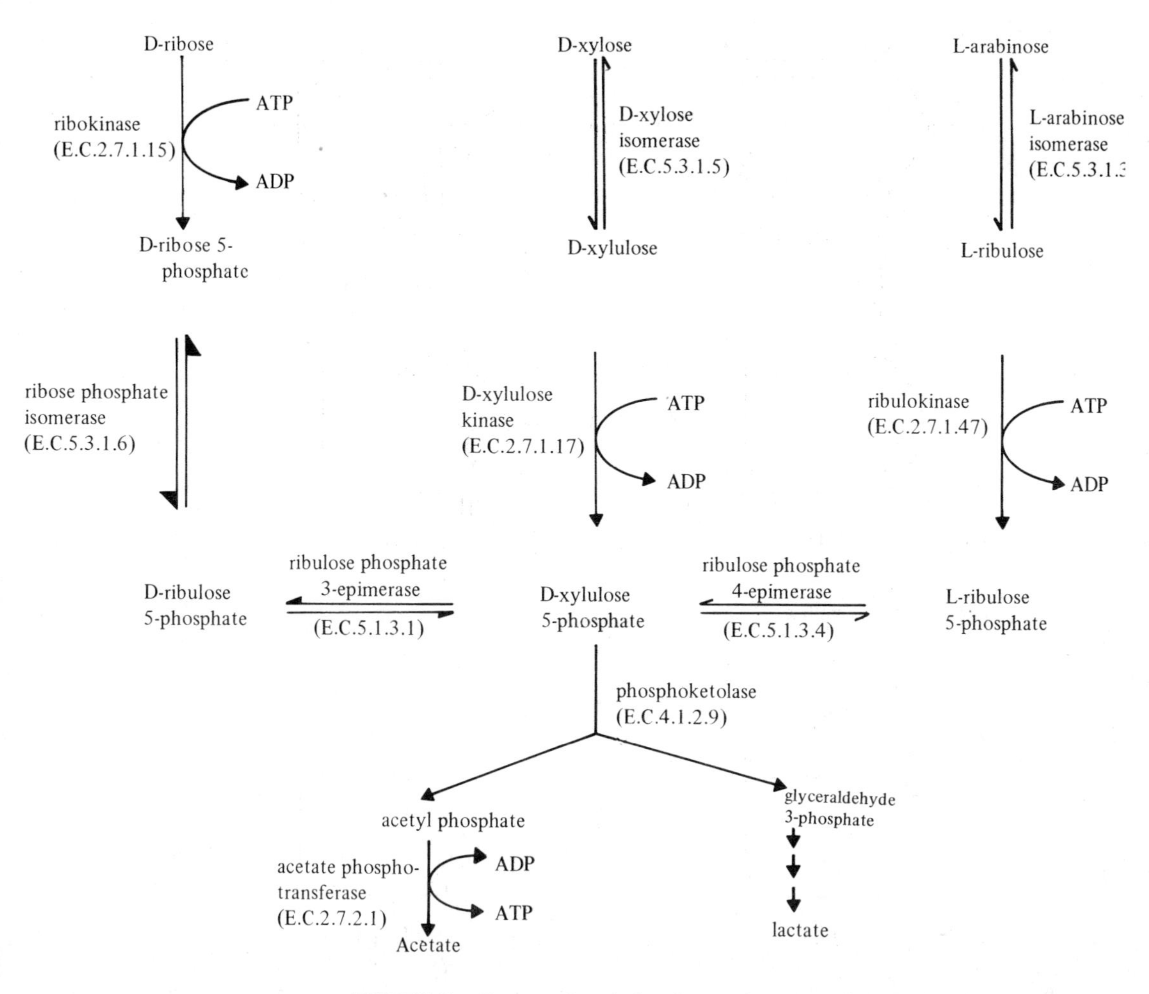

FIGURE 5. Pentose phosphoketolase pathway.

the entry of citrate into cells is uncontrolled. The absence of acetoin and diacetyl in citrate-devoid medium has been explained on the basis that in the normal carbohydrate metabolism 1 mol of glucose produces 2 mol of pyruvic acid and 2 mol of NAD (nicotinamide adenine dinucleotide) are reduced. This reduced NAD needs to be oxidized back to NAD for glucose metabolism to continue, and lactic acid bacteria use pyruvate as hydrogen acceptor to reoxidize NAD and convert it to lactic acid. However, these microorganisms do not need NAD for converting citrate to pyruvate:

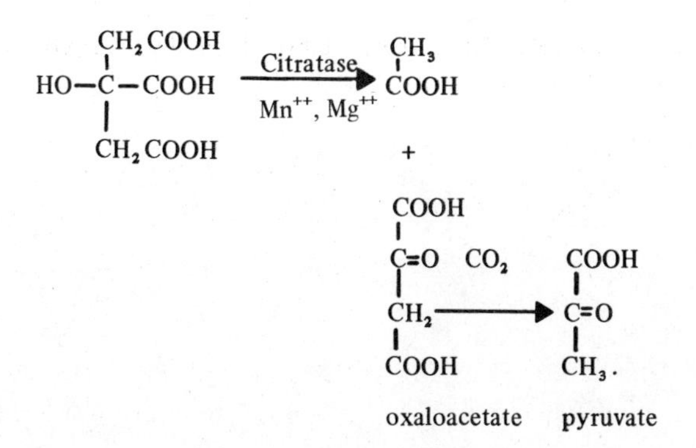

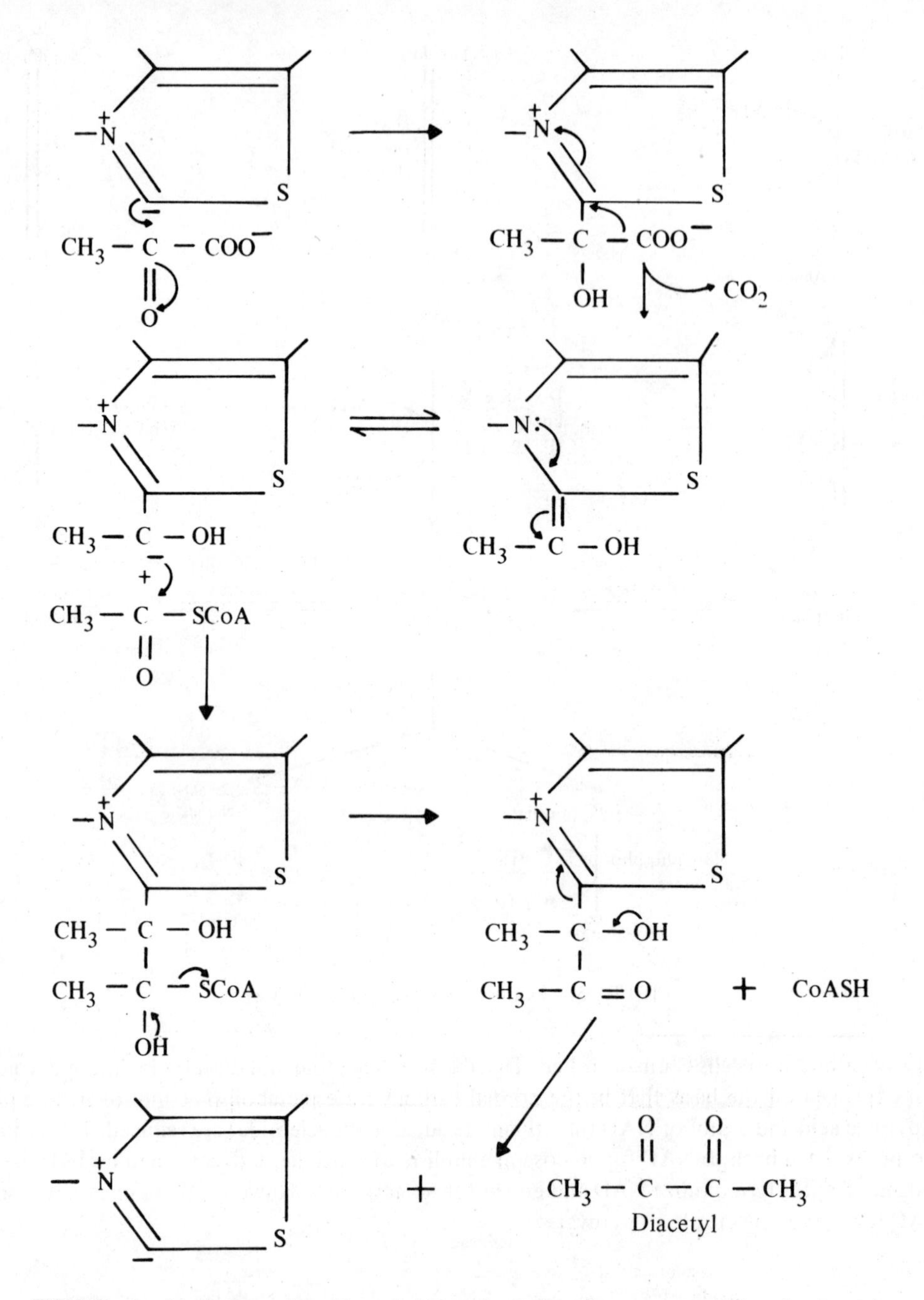

FIGURE 6. Mechanism for diacetyl biosynthesis by *Streptococcus diacetilactis* and *Leuconostoc citrovorum.*[57]

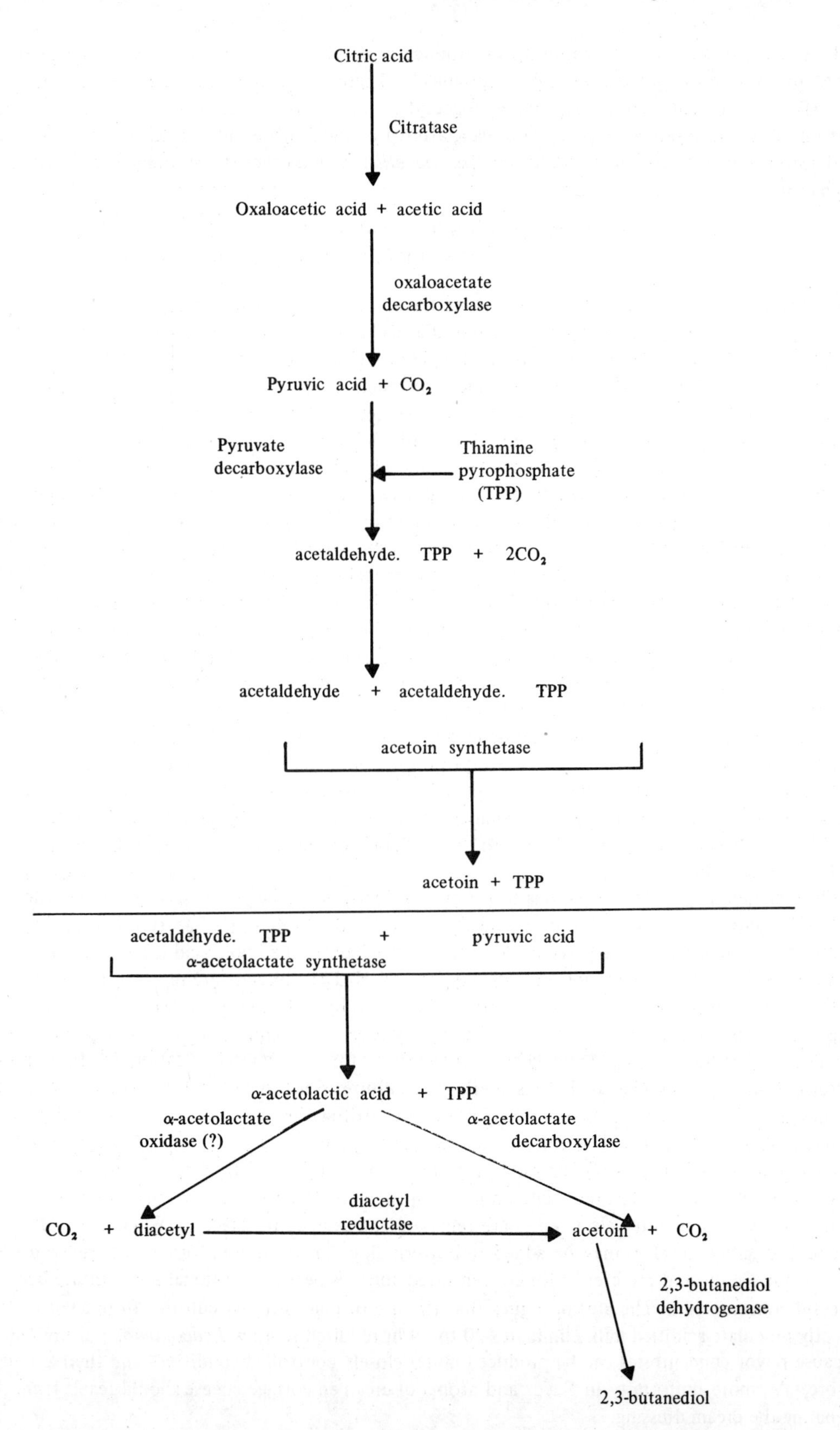

FIGURE 7. Pathways for conversion of citric acid by *Streptococcus diacetilactis.* (From Speckman, R. A. and Collins, E. B., *J. Bacteriol.*, 95, 174, 1968. With permission.)

It is believed that these microorganisms produce diacetyl and acetoin primarily as means of converting excess pyruvate into nontoxic neutral compounds.[61] These compounds are also involved in the reoxidation of NAD.[48] In the cultured dairy products diacetyl concentration increases for some time, as the bacterial cells multiply and produce diacetyl. However, after prolonged incubation diacetyl is converted to acetoin by diacetyl reductase present in *Streptococcus diacetilactis* and other starter organisms[59] by the following mechanism:

$$CH_3.CO.CO.CH_3 + NADH + H^+ \rightarrow$$
$$CH_3.CO.CHOH.CH_3 + NAD^+$$

Since the presence of diacetyl causes serious flavor defects in citrus juices, beer, and distilled liquors, advantage may be taken of the irreversibility of diacetyl reductase reaction to eliminate buttermilk flavor in these products.[47] The enzyme has an optimum pH of 7.0, which makes it unsuitable for use in citrus juices with pH level around 4.0. However, a method of protecting the enzyme and regenerating reduced NAD has been developed so that commercial use of enzymes, especially in beer, seems likely.[62]

Diacetyl production is pH dependent. Above pH 5.5 little diacetyl is produced. At low pH diacetyl reductase activity is slight[63] and at the same time citrate permease required for the active transport of citrate into the cell is much more active.[64] As discussed earlier, lactic cultures normally require a source of pyruvate in addition to carbohydrate for the production of considerable amounts of diacetyl, acetoin, and 2,3-butanediol. Consequently, sparing effect exerted by $NADH_2$ oxidase on the pyruvate produced from carbohydrate may not be of great importance. It is probably limited because the oxidases of *Streptococcus diacetilactis* and other organisms that have been studied, function optimally near neutrality.[64] However, $NADH_2$ oxidase undoubtedly has an important influence on the ratio of diacetyl to acetoin since diacetyl, acetoin, and 2,3-butanediol are structurally related and represent three levels of oxidation of same four-carbon skeleton. A limitation on the accumulation of diacetyl in cultures is the reduction of diacetyl to acetoin and acetoin to 2,3-butanediol. Since each of these reductions requires $NADH_2$ and a donor of electrons, oxidation of $NADH_2$ by oxidase results in larger accumulation of diacetyl.[64]

Aeration of cultures also increases diacetyl production. *Streptococcus diacetilactis* possesses an enzyme that oxidizes reduced NAD in the presence of oxygen.[64] Aeration stimulates the production of this enzyme and results in the decrease of pyruvate that must be converted to lactic acid; thus, more pyruvate becomes available for diacetyl production. Also with the oxidation of reduced NAD, there is less need for conversion of diacetyl to acetoin.[48]

Another importance of oxygen has to do more directly with the synthesis of diacetyl. The synthesis of diacetyl requires acetyl-CoA.[58] Lipoic acid is involved in the formation of acetyl-CoA from hydroxyethyl-thiamine pyrophosphate and coenzyme A.[65] In this mechanism the lipoic acid is reduced. NAD is necessary for reoxidation of lipoic acid, and in turn oxygen and $NADH_2$ oxidase are required for the reoxidation of NAD.[64] Thus oxygen influences the formation of diacetyl by influencing the formation of acetyl-CoA.

Based on the available information on diacetyl synthesis, Gilliland[66] developed a process for preparing concentrated cultures using *Leuconostoc citrovorum* for flavor intensification in dairy products like buttermilk and cottage cheese. In this method, a culture is grown in a broth medium. When the culture reaches maximum population, cells are harvested by centrifugation and resuspended in small volume of 10% nonfat milk solids (NFMS). The culture concentrate so prepared is dispensed in 1-g quantities and stored under liquid nitrogen (−196°C). This concentrate is inoculated into 10% NFMS to yield a population equal to an 18-hr milk culture. The inoculated milk is adjusted to pH 4.5 (optimum for diacetyl production) with citric acid. At various intervals samples are removed and assayed for diacetyl. When diacetyl concentration reaches a desirable level, it may be added to buttermilk or to cream used for cottage cheese dressing. Taste panel studies indicated preference for concentrated cultures over conventional milk culture because of high diacetyl concentration. The author argues that the use of concentrated cultures from a culture supplier to directly inoculate acidified milk eliminates 20 to 24 hr required to grow *Leuconostoc* culture in dairy plant. Because flavor concentrates can be produced under closely controlled conditions and their activity checked before sale, more uniformity in flavor and aroma of creamed cottage cheese should result from their use in preparing the cream dressing.

In his review on the cultured dairy products, Lindsay[46] has given a good account of flavor compounds which give a typical flavor character to cultured dairy products. Among carboxylic acids produced by lactic acid bacteria, acetic acid, being the major volatile acid in cultured products, is important in the flavor. The acidic taste of these products is largely due to lactic acid. Lactic acid being nonvolatile does not influence the aroma of these products. Very little attention, from a flavor standpoint, has been paid to other volatile acids such as formic, propionic, butyric, and valeric acids found in very small amounts in cultured dairy products. These acids may have some contribution in overall flavor balance of these products.

A small concentration of acetaldehyde is considered desirable in cultured dairy products.[46] The flavor balance of butter cultures was reported to be closely related to the diacetyl:acetaldehyde ratio. Desirable full-flavored cultures exhibit ratios from 3:1 to about 5:1, and a green flavor if the ratio drops below 3:1, and when the ratio exceeds 5:1, a harsh diacetyl flavor becomes apparent:

$$13{:}1 \longleftrightarrow 5{:}1 \longleftrightarrow 3{:}1 \longleftrightarrow 0.4{:}1$$

harsh good green

All *Streptococci* used in starter cultures produce acetaldehyde[67] and give a green flavor defect when overgrowth of *Streptococci* takes place. Bills and Day[68] showed that the concentration of ethanol varies from 0.2 to 36.8 ppm among starter strains and lower ethanol concentrations were accompanied by high acetaldehyde concentrations, consistent with the following reaction:

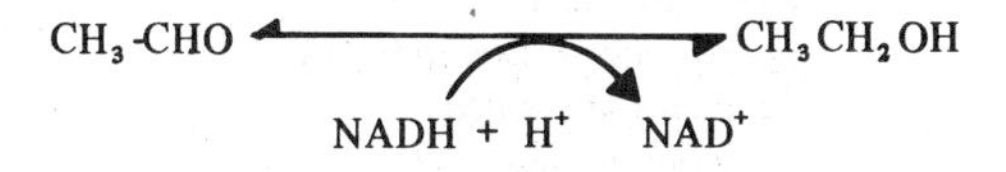

Following this, Keenan et al.[69] showed that *Leuconostoc* strains utilized acetaldehyde and that the green flavor defect could be overcome by incorporating 25 to 50% *Leuconostoc* in lactic starter cultures.

Culture-related flavor defects have been reviewed by Sandine et al.[70] Work done in this area indicates that the lack of diacetyl flavor in cottage cheese, cultured buttermilk, and sour cream is a major problem. Starter cultures and other contaminants contain diacetyl reductase which destroys diacetyl. This reaction proceeds more rapidly at higher temperature[71] and so proper refrigeration in retail stores becomes important. Use of hydrogen peroxide and catalase in buttermilk and sour cream improves the diacetyl production and its stabilization.[72] The possible role of oxygen in this connection has been discussed earlier. Morgan et al.[73] studied the malty defect of milk cultures. 3-Methylbutanol, 2-methylpropanol, 3-methylbutanal, and 2-methylpropanal were identified from the milk culture of *Streptococcus lactis* var. *maltigenes.* These compounds were absent in the cultures of nonmalty strains of *Streptococcus lactis*. Although the volatiles of milk cultures with malty defect contained more of the branched chain alcohols than the corresponding aldehydes, organoleptic comparison of the aroma of the untreated culture with that in which the aldehydes had been removed by treatment with hydroxylamine indicated that the typical malty aroma was principally due to aldehydes. Morgan[74] recently reported that the malty defect of milk cultures could be simulated by adding 0.34 ppm 3-methylbutanal to good quality homogenized pasteurized milk and used as a standard for judging this defect by an expert flavor panel. Similarly, the fruity defect caused by *Pseudomonas fragi* could be simulated in milk by the additon of a mixture of pure ethyl butyrate and ethyl hexanoate in 1,2-propanediol to give concentration of 0.35 and 0.50 ppm of the respective esters.[75]

CHEESE FLAVOR

During cheese ripening a number of chemical reactions proceed simultaneously, representing a dynamic system. Many of the reactions are mediated by microbial and native milk enzymes. During cheese ripening, protein, fat, and carbohydrate are degraded to varying degrees to yield a complex mixture of compounds, some of which contribute to the characteristic cheese flavor.[72] The microflora of cheese is complex and variable because different types of microorganisms predominate at different stages of cheese ripening. A list of microorganisms and enzymes used in cheese making is given in Table 3.[73]

TABLE 3

Microorganisms and Enzymes Used in Cheese Making[77]

Streptococcus lactis
Streptococcus cremoris
Streptococcus citrovorum
Streptococcus faecalis
Streptococcus zymogenes

Streptococcus durans
Streptococcus liquefaciens
Streptococcus thermophilus
Streptococcus diacetilactis
Leuconostoc cremoris

Lactobacillus caseii
Lactobacillus lactis
Lactobacillus bulgaricus
Bacterium lineus

Micrococcus caseolyticus
Pseudomonas fragi
Propionibacterium shermanii
Trametes sanguinae
Penicillium roqueforti

Penicillium camemberti
Penicillium frequentans
Calf lipase
Lactase
Enzymes from filbert nuts and
Saccharomyces cerevisiae
Extracted enzymes from selected
milk and cheese bacteria

Cheddar Cheese

A large number of compounds have been identified from Cheddar cheese. Amino acids and peptides, produced primarily by proteases of cheese microflora during cheese ripening, provide the brothy background to cheese flavor and as such play an important role in the taste of Cheddar cheese. Various amines such as tyramine, cadavarine, putrescine, and histamine have also been identified from normal Cheddar cheese.[76] Amino acids and amines produced from proteins also act as precursors of a number of aldehydes reported in Cheddar cheese and are produced probably by transamination and decarboxylation reactions. The Strecker degradation of various amino acids to corresponding aldehyde of one less carbon atom also contributes to cheese flavor. Although the reaction is nonenzymatic, the proteolytic activity of cheese would influence it.

$$R.CO.CO.R + R'.CHNH_2.COOH =$$

$$R'.CHO + CO_2 + R.CHNH_2.CO.R$$

Keeney and Day[78] studied individual amino acids for the character of odor yielded on Strecker degradation that would be of significance in "toasted" or "heated" cheese products. Glycine, tryptophane, arginine, histidine, lysine, aspartic acid, serine, threonine, and thyrosine yielded no odor. The following amino acids yielded the indicated odors:

alanine	–	malty
valine	–	apple
leucine	–	malty
isoleucine	–	malty-apple
proline	–	mushroom
phenylalanine	–	violets
cystine	–	H_2S
methionine	–	cheesy-brothy
glutamic acid	–	bacterial agar

The degradation products of lipids are essential for Cheddar flavor, as Cheddar cheese made from skim milk is practically devoid of Cheddar character.[79] Patton[80] reported that the 2:0, 4:0, 6:0, and 8:0 fatty acids are of primary importance in Cheddar cheese aroma. He suggested that acetic acid is of special importance in the unique aroma of cheese. In the absence of acetic acid, the volatile fatty acids found in cheese would promote hydrolytic rancid character. Bills and Day[81] determined the free-fatty acid concentrations of several cheese samples (Table 4) and concluded that in view of the variation in individual free-fatty acid concentrations and lack of direct relationship of free-fatty acids with the flavor of good quality Cheddar cheese, the balance between free-fatty acids and other flavor constituents is more important to Cheddar flavor than the concentration of free-fatty acids alone. Recently, Ohren and Tuckey[82] reported that the typical flavor of Cheddar cheese was related to the balance between free-fatty acids and acetate (Table 5). According to these workers, the microbial content of milk has a significant influence on the acetic and free-fatty acid content of cheese (Table 6).

Day et al.[83] identified several methyl ketones containing odd-numbered carbon atoms from Cheddar cheese and suggested that these may be the products of the normal curing process. The concentrations of these ketones are quite small and may also be formed by the degradation of β-keto esters normally found in milk fat.[84] Butanone, which is found in relatively large concentration in Cheddar cheese, may be produced from reduction of acetoin to 2,3-butylene glycol, which is in turn dehydrated to the ketone.[85] Citrate-fermenting bacteria produce a small amount of diacetyl in Cheddar cheese. The mechanisms of diacetyl production by these organisms have been discussed under cultured dairy products section.

Ethanol and 2-butanol are the two major alcohols in Cheddar cheese.[85] Day and Libbey[86] identified traces of secondary alcohols corresponding to the methyl ketone analogs. According to Day,[76] apparently some of the bacteria associated with ripening possess alcohol dehydrogenase capable of reducing the ketone and aldehyde analogs. Ethanol is probably produced as the end product of EMP or Entner-Doudoroff

TABLE 4

Summary of the Analysis of Free-fatty Acids in 14 Samples of Cheddar Cheese[81]

Fatty acid	Average concentration in mg/kg cheese	Concentration range in mg/kg cheese	Analytical precision (percent ±)
2:0	865	275–1,325	1.79
4:0	115	71–207	2.16
6:0	38	18–79	0.99
8:0	41	21–74	1.86
10:0	49	40–80	1.64
12:0	81	55–123	1.54
14:0	218	134–315	0.90
16:0	503	300–689	0.81
18:0	172	104–246	1.05
18:1	467	312–630	0.85
18:2	69	37–125	1.07
18:3	40	21–62	1.36

TABLE 5

Relationship Between Free-fatty Acids + Acetic Acid in Determining the Flavor of Cheddar Cheese

Flavor	Free-fatty acid + acetate (μmol)	Moles free-fatty acids/μmol acetate
Fine, Cheddar	12 to 28	0.55 to 1.0
Fermented, rancid, fruity, unclean	28	1.0
Flat	12	0.55

(From Sandine, W. E. and Elliker, P. R., *J. Agric. Food Chem.*, 18, 557, 1970. With permission.)

TABLE 6

Relationship of Standard Plate Count (SPC) of Milk to the Level of Free-fatty Acids + Acetic Acid Found in Cheddar Cheese Analyzed after 3 Months

SPC/ml		Free-fatty acids + acetate (μmol/g)	
Raw	H_2O_2	Raw	H_2O_2
1,000	300	18	11
7,000	150	21	17
20,000	9,000	20	18
20,000,000	50,000	45	24
21,000,000	16,000,000	59	53
32,000,000	10,000,000	80	26

(From Sandine, W. E. and Elliker, P. R., *J. Agric. Food Chem.*, 18, 557, 1970. With permission.)

pathway, depending upon the bacteria used.[57] Scarpellino and Kosikowski[85] postulated that 2-butanol was a terminal product in the breakdown of acetoin and proposed the following mechanism for its formation:

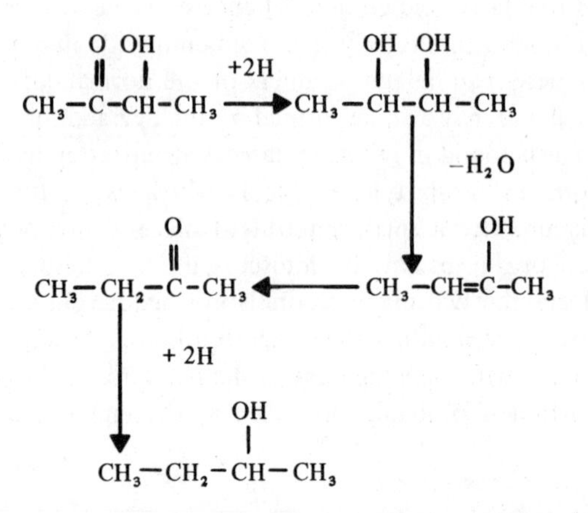

Bills et al.[87] reported ethyl butyrate and ethyl hexanoate as compounds primarily responsible for the fruity flavor defect of Cheddar cheese. Fruity samples contained two to ten times greater concentrations of ethyl butyrate and ethyl hexanoate compared to normal Cheddar cheese. Levels of ethanol were 6 to 16 times greater in fruity cheeses. It was suggested that the extensive production of ethanol may be responsible for the accentuated esterification of free fatty acids, thus resulting in higher levels of ethyl esters.

Mold-ripened Cheeses

The chemistry of flavor developments in mold-ripened cheeses is very complex. In addition to the processes leading to the formation of normally ripened cheeses, the metabolism of mold developing on or within such cheeses contributes to the characteristic flavor. In some of the mold-ripened cheeses, such as Roquefort, Gorgonzola, blue, and Stilton, the enzymes produced by *Penicillium roqueforti* play an important part in flavor development, in addition to the softening of the curd and the formation of various breakdown products. *Penicillium roqueforti* shows a very high lipase activity which leads to the rapid production and accumulation of a number of fatty acids, mainly caprylic acid but also capric and caproic acids.[49] These fatty acids are involved in the formation of methyl ketones which impart characteristic flavor to these cheese varieties.

In the ripening of Camembert cheese besides *Penicillium camemberti,* various film yeasts and *Geotrichum* species play an important role in flavor development by reducing the acidity of the cheese surface before the main mold becomes established. Information on the chemical compounds involved in Camembert maturation and the specific contribution of each group of organism to the development of the characteristic Camembert flavor is still very meager.[47]

The Embden-Meyerhof-Parnas (EMP) scheme (Figure 3) represents a pathway whereby many molds metabolize sugars. Several enzymes of the EMP pathway have been reported to occur in *Penicillium chrysogenum.*[88] The pentose phosphate pathway (Figure 4) is also an important pathway in the dissimilation of sugars. Heath and Koffler[89] estimated that two thirds of glucose metabolized by *Penicillium chrysogenum* was through the oxidative pathway. The Entner-Doudoroff pathway, which is so important in the metabolism of some bacteria, seems to have little place in fungal metabolism.[90]

Little information is available on the role of proteolysis and its effects on blue cheese flavor. *Penicillium roqueforti* has active proteolytic enzyme(s). Imamura and Tsugo[91] studied the changes of milk proteins by *Penicillium roqueforti* and found that although albumin was most readily decomposed of all the milk proteins, the changes in casein were of greater significance. The large quantities of amino acids released by proteolytic enzymes impart a "brothy" taste which constitutes an important background on which the typical flavor is superimposed.[76]

Several investigators have studied the lipase(s) of *Penicillium roqueforti.*[92-94] Eitenmiller et al.[92] partially purified the lipase of *Penicillium roqueforti* and reported that this enzyme had an optimum pH of 8.0 and an optimum temperature of 37°C. Maximum lipolytic activity occurred with 5% butter oil emulsion as the substrate. $MnCl_2$ and $MgCl_2$ stimulated the enzyme activity. The lipase was thermolabile, being inactivated completely within 10 min at 50°C. It hydrolyzed tributyrin, tricaprylin, tricaprin, tripropionin, and triolein in decreasing order. The fatty acids released by the action of *Penicillium roqueforti* and milk lipases have long been recognized as important components of blue cheese flavor. The short chain fatty acids, i.e., 4:0, 6:0, and 8:0, impart the characteristic "peppery" taste and act as methyl ketone precursors.

The methyl ketones, especially 2-heptanone and 2-nonanone, are generally considered the key flavor components of blue cheese.[76] The formation and metabolism of methyl ketones by fungi during the ripening process in mold-ripened cheeses appear to involve four main enzymic mechanisms:[95]

a. the liberation of free-fatty acids from the triglycerides of milk fat by lipases;
b. oxidation of the free-fatty acids to β-keto acids;
c. decarboxylation of β-keto acids to methyl ketones; and
d. reduction of methyl ketones to secondary alcohols.

The pathway of fatty-acid oxidation and methyl ketone formation is given in Figure 8.[95] *Penicillium roqueforti* spores[96] as well as mycelium[97,98] are capable of producing methyl ketones from fatty acids. Studies on the fatty acid metabolism[99] by the spores of *Penicillium roqueforti* indicate that both short chain as well as long chain fatty acids are metabolized into homologous series of methyl ketones and CO_2.

A large number of primary and secondary alcohols have been isolated from blue cheese.[100,101] Of the primary alcohols identified, methanol and ethanol appear to be important in flavor formation.[76] The secondary alcohol analogs of methyl ketones found in blue cheese result from reduction of corresponding ketone by mold mycelium by the following mechanism:[102]

$$RCOCH_3 \underset{DPN+}{\overset{DPNH + H^+}{\rightleftharpoons}} RCHOHCH_3$$

In recent years several attempts have been made to reduce the time needed for maturation of cheese by adding flavors produced by microorganisms and changing various parameters of cheese manufacture. Cort and Riggs[103] claim that cheese of the Roquefort or Camembert type may be produced by adding enzymes extracted from *Penicillium roqueforti* and *Penicillium camemberti* to the milk or to the curd, respectively, during manufacture. The method is claimed to be applicable to other cheese varieties, e.g., Cheddar, where the enzyme extract is prepared from a mature Cheddar cheese. Hedrick et al.[104] recently patented a process of blue cheese preparation by adding a blue cheese mold to the cheese and ripening the curd in a

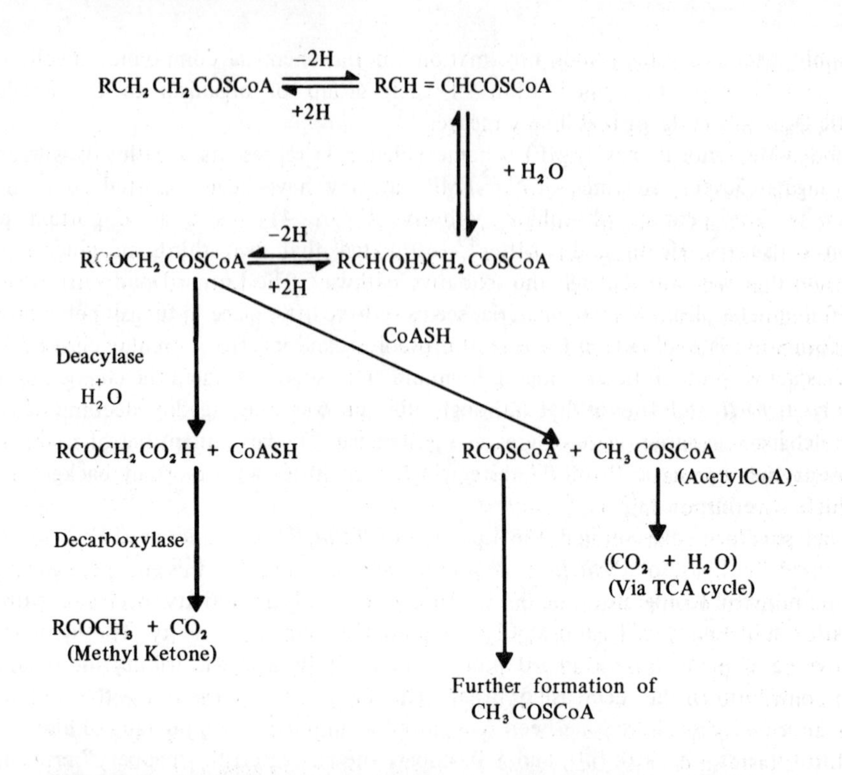

FIGURE 8. Mechanism of methyl ketone formation from fatty acids. (From Heath, E. C. and Koffler, H., *J. Bacteriol.*, 71, 174, 1956. With permission.)

divided condition. Ripening is said to be accelerated and slime formation minimized. A Dutch patent[105] on cheese flavor makes use of microorganisms that metabolize fat and produce volatile compounds closely simulating cheese flavor. A medium high in animal or plant fat (40 to 60%) and low in protein (not more than 3%) is used for flavor production. The medium is also fortified with carbohydrates, minerals, trace elements, vitamins, and other growth substances. The pH is preferably held at 4.5 and the incubation period is 48 hr at temperature 20 to 40°C. *Pencillium* (especially *P. roqueforti*), *Candida lipolytica, Oidium lactis, Cladosporium, Micrococcus,* and *Lipomyces* are recommended.

Processes have also been developed for preparing cheese flavorings. Knight[106] patented a process for producing blue cheese flavor by treating milk or cream with lipase to convert milk fat to fatty acids and fermenting the resulting milk-fatty acid mixture with *Penicillium requeforti* under submerged aerobic conditions. In this process free-fatty acids are converted to methyl ketones and the composition smells like blue cheese. Watts and Nelson[107] patented a similar process for blue cheese flavor manufacture. Recently, Dwivedi and Kinsella[108] developed a continuous process for blue cheese flavor manufacture by submerged fermentation of lipolyzed milk fat with the mycelium of *Penicillium roqueforti.*

MISCELLANEOUS ENZYMES

Enzymes such as lipases and proteases from sources other than milk and bacterial starter cultures are finding greater use in the modified dairy flavors. Nelson,[109] in his review on enzymatically produced flavors from fatty systems, has given an interesting account of the use of lipolyzed milk in chocolate. Rancidity, although considered a highly undesirable character in most dairy products, is finding use as a flavor enhancer and is used in developing desirable flavor notes. Anti[110] has produced a variety of tastes ranging from cream to butter from lipolyzed flavors formed by the action of lipase on butter fat emulsions. Recipes are given for their use in butterscotch, caramel, and butterscotch kisses. These confections are said to have improved flavor characteristics. Tanabe Seiyaku C. Ltd. of Japan has recently patented a process[111] of manufacturing highly flavorful yogurt by the addition of lipase at the start of the process. start of the process.

Major applications for enzyme modified milk fat include:[109]

a. chocolate products, including milk chocolate, compound coatings, and chocolate flavor syrups and beverages;

b. butter flavors, including margarine, butter creams, and butter sauces;

c. milk and cream flavors, including coffee whiteners, imitation sour cream, and imitation milks;

d. cheese flavor, particularly when the flavor profile of Italian-type cheese is desired. Neelakatan et al.[112] reported that the type and amount of lipase used in Italian cheeses play an important role in flavor development and provide some of the better means of flavor control. In a similar study Langler and Day[113] investigated major free-fatty acids in Swiss cheese and discussed their importance.

Flavor profiles from lipolyzed milk fat can exert a considerable range of effects on food flavor. At very low addition levels, a sensation of richness is imparted without any detectable free-fatty acid flavor character. As addition levels are increased, the flavors imparted resemble cream or butter. When addition levels are relatively high, the flavor imparted suggests cheese.[109] Lipases from different sources give different free-fatty acid profile in lipase-treated milk fat. This property or lipases may help in simulating flavors precisely.

Proteolytic activity during the ripening of cheeses in also extensive. Apart from the participation of starter cultures, it depends in part on the coagulating enzyme system which is used in the formation of curd. The extent of proteolytic breakdown in cheeses can be judged from the production of free-amino acids which contribute to the flavor of the cheese. The extent of hydrolysis varies with the nature of the coagulating enzyme. Vegetable rennets and pancreatin show a high rate of proteolysis and frequently produce a bitter taste in cheese. Deaminases, decarboxylases, aminases, transaminases, and other enzymes are active in cheese and lead to the production of a variety of amino acids, amines, keto acids, aldehydes, ammonia, and other derivatives. Although very little information is available, it is quite possible to influence the flavor of cheese by changing the concentration and the source of proteolytic enzyme used for clotting the milk.

Richardson and Nelson[114] developed a screening method for rapidly indicating the sensitivity of various cheese ripening agents. Freshly ground mild cheese curd was inoculated with high levels of enzyme preparations, incubated at 21°C for 3 days and evaluated organoleptically. This method could be used in evaluating the role of various microorganisms and their enzyme systems in cheese ripening. Similar methods could be developed for cultured dairy products.

REFERENCES

1. **Ohloff, G.,** Classification and genesis of food flavors, in *Proc. SOS/70. Third International Congress Food Science and Technology, Washington, D. C.,* Institute of Food and Technologists, Chicago, 1970, 368.
2. **Hewitt, E. J., MacKay, D. A. M., Konigsbacher, K. S., and Hasselstrom, T.,** The role of enzymes in food flavors, *Food Technol.,* 10, 487, 1956.
3. **Hewitt, E. J., MacKay, D. A. M., Konigsbacher, K. S.,** Flavor propagation through enzymatic action, in *Chemistry of Natural Food Flavors,* Mitchell, J. H., Jr., Leinen, N. L., Mrak, E. M., and Bailey, S. D., Eds., Chicago Quartermaster Food and Container Institute, Chicago, 1957, 86.
4. Little (Arthur D.), Inc., *Flavor Research and Food Acceptance,* Reinhold, New York, 1957, 285.
5. **Shahani, K. M., Harper, W. J., Jensen, R. G., Parry, R. M., Jr., and Zittle, C. A.,** Enzymes in bovine milk, *J. Dairy Sci.,* 56, 531, 1973.
6. **Dowben, R. M. and Brunner, J. R.,** The relation of enzyme composition and distribution in milk to the mammary cell, Proc. 146th Am. Chem. Soc. Meeting, 1964, 11A.
7. **Kitchen, B. J., Taylor, G. C. and White, I. C.,** Milk enzymes – their distribution and activity, *J. Dairy Res.,* 37(2), 279, 1970.
8. **Jensen, R. G.,** The dynamic state of milk lipolysis, *J. Dairy Sci.,* 46, 210, 1963.
9. **Kiermeier, F. and Grassmann, E.,** Relationship between xanthine oxidase and oxidized flavor in milk, *Z. Lebensm.-Unters. Forsch.,* 133, 310, 1967.

10. **Zittle, C. A.,** Dynamic state of milk: xanthine oxidase, acid phosphatase, ribonuclease and protease, *J. Dairy Sci.,* 47, 202, 1964.
11. **Murthy, L., Herreid, E. O., and Whitney, R. McL.,** Electrophoretic properties of casein from sterile milk stored at different temperatures, *J. Dairy Sci.,* 41, 1324, 1958.
12. **Wüthrich, S., Richterich, R., and Hostettler, H.,** Studies of milk enzymes, I. Enzymes in human and cows' milk, *Z. Lebensm.-Unters, Forsch.,* 124, 336, 1964.
13. **Forster, T. L., Bendixen, H. A., and Montgomery, M. W.,** Some esterases of cows' milk, *J. Dairy Sci.,* 42, 1903, 1959.
14. **Forster, T. L., Montgomery, M. W., and Montoure, J. E.,** Some factors influencing the activity of the A-, B-, and C-esterases in bovine milk, *J. Dairy Sci.,* 44, 1420, 1961.
15. **Jensen, R. G.,** Lipolysis, *J. Dairy Sci.,* 47, 210, 1964.
16. **Kitchen, B. J.,** Bovine milk esterases, *J. Dairy Res.,* 38, 171, 1971.
17. **Tarassuk, N. P. and Frankel, E. N.,** The specificity of milk lipase. IV. Partition of the lipase system in milk, *J. Dairy Sci.,* 40, 418, 1957.
18. **Jensen, R. G., Sampugna, J., Parry, R. M., Jr., Shahani, K. M., and Chandan, R. C.,** Lipolysis of synthetic triglycerides and milk fat by a lipase concentrate from milk, *J. Dairy Sci.,* 45, 1527, 1962.
19. **Jensen, R. G.,** *Progress in the Chemistry of Fats and Other Lipids,* Vol. 11, Pergamon Press, New York, 1971, 349.
20. **Chandan, R. C. and Shahani, K. M.,** Purification and characterization of milk lipase. I. Purification, *J. Dairy Sci.,* 46, 275, 1963.
21. **Chandan, R. C. and Shahani, K. M.,** Purification and characterization of milk lipase. II. Characterization of the purified enzyme, *J. Dairy Sci.,* 46, 503, 1963.
22. **Chandan, R. C. and Shahani, K. M.,** Role of sulfhydral groups in the activity of the milk lipase, *J. Dairy Sci.,* 48, 1413, 1965.
23. **Hileman, J. L. and Courtney, E.,** Seasonal variations in the lipase content of milk, *J. Dairy Sci.,* 18, 247, 1935.
24. **Tarassuk, N. P. and Richardson, G. A.,** The significance of lipolysis in the curd tension and rennet coagulation of milk. I. The role of fat globule adsorption "membrane." II. The effect of addition of certain fat acids to milk, *J. Dairy Sci.,* 24, 667, 1941.
25. **Schwartz, D. P.,** Unpublished work, 1973.
26. **Schwartz, D. P.,** Lipids of milk: Deterioration. Part I., in *Fundamentals of Dairy Chemistry,* Webb, B. H. and Johnson, A. H., Eds., Avi Publishing Co., Westport, Conn., 1965, 170.
27. **Tarassuk, N. P. and Henderson, J. L.,** Prevention of development of hydrolytic rancidity in milk, *J. Dairy Sci.,* 25, 801, 1942.
28. **Morrison, M. and Hultquist, D. E.,** Lactoperoxidase. II. Isolation, *J. Biol. Chem.,* 238, 2847, 1963.
29. **Polis, B. D. and Shmukler, H. W.,** Crystalline lactoperoxidase. I. Isolation by displacement chromatography. II. Physicochemical and enzymatic properties, *J. Biol. Chem.,* 201, 475, 1953.
30. **Jenness, R. and Patton, S.,** *Principles of Dairy Chemistry,* John Wiley & Sons, New York, 1959, 182.
31. **Kaminogawa, S., Yamuchi, K., and Tsugo, T.,** Properties of milk protease concentrated from acid-precipitated casein, *Jap. J. Zootech. Sci.,* 40, 559, 1969.
32. **Chen, J. H. and Ledford, R. A.,** Purification and characterization of milk protease, *J. Dairy Sci.,* 54, 763, 1971.
33. **Zittle, C. A.,** Purification of protease in cow's milk, *J. Dairy Sci.,* 48, 771, 1965.
34. **Sumner, J. B. and Somers, G. F.,** *Chemistry and Methods of Enzymes,* Academic Press, New York, 1953, 284.
35. **Morton, R. K.,** The lipoprotein particles in cow's milk, *Biochem. J.,* 57, 231, 1954.
36. **Shahani, K. M.,** Milk enzymes: their role and significance, *J. Dairy Sci.,* 49, 907, 1966.
37. **Kiermeier, F. and Capellari, K.,** Correlation between the molybdenum content and xanthine oxidase activity of cow's milk, *Biochem. Z.,* 330, 160, 1958.
38. **Rajan, T. S., Richardson, G. A., and Stein, R. W.,** Xanthine oxidase activity of milks in relation to stage of lactation, feed and spontaneous oxidation, *J. Dairy Sci.,* 45, 933, 1962.
39. **Smith, G. J. and Dunkley, W. L.,** Xanthine oxidase and incidence of spontaneous oxidized flavor in milk, *J. Dairy Sci.,* 43, 278, 1960.
40. **Aurand, L. W., Woods, A. E., and Roberts, W. M.,** Some factors involved in the development of oxidized flavor in milk, *J. Dairy Sci.,* 42, 961, 1959.
41. **Aurand, L. W. and Woods, A. E.,** Role of xanthine oxidase in the development of spontaneously oxidized flavor in milk, *J. Dairy Sci.,* 42, 1111, 1959.
42. **Aurand, L. W., Chu, T. M., Singleton, J. A., and Shen, R.,** Xanthine oxidase activity and development of spontaneously oxidized flavor in milk, *J. Dairy Sci.,* 50, 465, 1967.
43. **King, R. L. and Dunkley, W. L.,** Relation of natural copper in milk to incidence of spontaneous oxidized flavor, *J. Dairy Sci.,* 42, 420, 1959.
44. **Smith, G. J. and Dunkley, W. L.,** Prooxidants in spontaneous development of oxidized flavor in milk, *J. Dairy Sci.,* 45, 170, 1962.
45. **Smith, G. J. and Dunkley, W. L.,** Ascorbic acid oxidation and lipid peroxidation in milk, *J. Food Sci.,* 27, 127, 1962.
46. **Lindsay, R. C.,** Cultured dairy products, in *Chemistry and Physiology of Flavors,* Schultz, H. W., Day, E. A., and Libbey, L. M., Eds., Avi Publishing Co., 1967, 315.

47. **Sandine, W. E. and Elliker, P. R.**, Microbially induced flavors and fermented foods, *J. Agric. Food Chem.*, 18, 557, 1970.
48. **Collins, E. B.**, Biosynthesis of flavor compounds by microorganisms, *J. Dairy Sci.*, 55, 1022, 1972.
49. **Margalith, P. and Schwartz, Y.**, Flavor and microorganisms, *Adv. Appl. Microbiol.*, 12, 64, 1970.
50. **Doelle, H. W.**, Lactic acid bacteria, in *Bacterial Metabolism*, Academic Press, New York, 1969, 333.
51. **Keenan, T. W. and Bills, D. D.**, Metabolism of volatile compounds by lactic starter culture microorganisms. A review, *J. Dairy Sci.*, 51, 1561, 1968.
52. **DeLey, J.**, Comparative biochemistry and enzymology in bacterial classification, *Symp. Soc. Gen. Microbiol.*, 12, 164, 1962.
53. **Gunsallus, I. C., Horecker, B. L., and Wood, W. A.**, Pathways of carbohydrate metabolism in microorganisms, *Bacteriol. Rev.*, 19, 79, 1955.
54. **Lees, G. J. and Jago, G. R.**, Unpublished work reported by Doelle, 1967.
55. **DeMoss, R. D.**, A triphosphopyridine nucleotide dependent alcohol dehydrogenase from *Leuconostoc mesenteroides*, *J. Bacteriol.*, 68, 252, 1954.
56. **Hurnitz, J.**, Pentose phosphate clevage by *Leuconostoc mesenteroides*, *Biochim. Biophys. Acta*, 28, 599, 1958.
57. **Wood, W. A.**, Fermentation of carbohydrates and related compounds, in *The Bacteria*, Vol. 2, 2nd ed., Gunsallus, I. C. and Stanier, R. Y., Eds., Academic Press, New York, 1961, 59.
58. **Speckman, R. A. and Collins, E. B.**, Diacetyl biosynthesis in *Streptococcus diacetilactis* and *Leuconostoc citrovorum*, *J. Bacteriol.*, 95, 174, 1968.
59. **Seitz, E. W., Sandine, W. E., Elliker, P. R., and Day, E. A.**, Studies on diacetyl biosynthesis by *Streptococcus diacetilactis*, *Can. J. Microbiol.*, 9, 431, 1963.
60. **Swomadainen, H. and Ronkeinen, P.**, Mechanism of diacetyl formation in yeast fermentation, *Nature* (London), 220, 792, 1968.
61. **Harvey, R. J. and Collins, E. B.**, Role of citrate and acetoin in the metabolism of *Streptococcus diacetilactis*, *J. Bacteriol.*, 86, 1301, 1963.
62. **Whinery, J.**, M.S. Thesis, Oregon State University, Corvallis, 1969.
63. **Harvey, R. J. and Collins, E. B.**, Citrate transport system of *Streptococcus diacetilactis*, *J. Bacteriol.*, 83, 1005, 1962.
64. **Bruhn, J. C. and Collins, E. B.**, Reduced microtinamide adenine dinucleotide oxidase of *Streptococcus diacetilactis*, *J. Dairy Sci.*, 53, 857, 1970.
65. **Collins, E. B. and Bruhn, J. C.**, Role of acetate and pyruvate in the metabolism of *Streptococcus diacetilactics*, *J. Bacteriol.*, 103, 541, 1970.
66. **Gilliland, S. E.**, Flavor intensification with concentrated cultures, *J. Dairy Sci.*, 55, 1028, 1972.
67. **Keenan, T. W., Lindsay, R. C., Morgan, M. E., and Day, E. A.**, Acetaldehyde production by single-strain lactic *Streptococci*, *J. Dairy Sci.*, 49, 10, 1966.
68. **Bills, D. D. and Day, E. A.**, Dehydrogenase activity of lactic *Streptococci*, *J. Diary Sci.*, 49, 1473, 1966.
69. **Keenan, T. W., Lindsay, R. C., and Day, E. A.**, Acetaldehyde utilization by *Leuconostoc* species, *Appl. Microbiol.*, 14, 802, 1966.
70. **Sandine, W. E., Daly, C., Elliker, P. R., and Vedamuthu, E. R.**, Causes and control of culture related flavor defects in cultured dairy products, *J. Dairy Sci.*, 55, 1031, 1972.
71. **Pack, M. Y., Vedamuthu, E. R., Sandine, W. E., and Elliker, P. R.**, Effect of temperature on growth and diacetyl production by aroma bacteria in single and mixed strain lactic cultures, *J. Dairy Sci.*, 51, 339, 1968.
72. **Pack, M. Y., Vedamuthu, E. R., Sandine, W. E., and Elliker, P. R.**, Hydrogen peroxide – catalase milk treatment for enhancement and stabilization of diacetyl in lactic starter cultures, *J. Dairy Sci.*, 51, 511, 1968.
73. **Morgan, M. E., Lindsay, R. C., Libbey, L. M., and Pereiva, R. L.**, Identity of additional aroma constituents in milk cultures of *Streptococcus lactis* var. *maltigenes*, *J. Dairy Sci.*, 49, 15, 1966.
74. **Morgan, M. E.**, Microbial defects in dairy products and methods for their simulation. I. Malty flavor, *J. Dairy Sci.*, 53, 270, 1970.
75. **Morgan, M. E.**, Microbial defects in dairy products and methods for their simulation. II. Fruity flavor, *J. Dairy Sci.*, 53, 273, 1970.
76. **Day, E. A.**, Cheese flavor, in *Chemistry and Physiology of Flavors*, Schultz, H. W., Day, E. A., and Libbey, L. M., Eds., Avi Publishing Co., 1967, 331.
77. **Scott, R.**, Cheese making – enzymology or bacteriology? *Proc. Biochem.*, 7(11), 33, 1972.
78. **Keeney, M. and Day E. A.**, Probable role of the Strecker degradation of amino acids in development of cheese flavor, *J. Dairy Sci.*, 40, 874, 1957.
79. **Day, E. A.**, Role of milk lipids in flavors of dairy products, in *Flavor Chemistry*, Adv. in Chem., Ser., Gould, R. F., Ed., American Chemical Society, Washington, D. C., 1966, 94.
80. **Patton, S.**, Volatile acids and aroma of Cheddar cheese, *J. Dairy Sci.*, 46, 856, 1963.
81. **Bills, D. D. and Day, E. A.**, Determination of the major free-fatty acids of Cheddar cheese, *J. Dairy Sci.*, 47, 733, 1964.
82. **Ohren, J. A. and Tuckey, S. L.**, Relation of flavor development in Cheddar cheese to chemical changes in the fat of cheese, *J. Dairy Sci.*, 52, 598, 1969.
83. **Day, E. A., Bassette, R., and Keeney, M.**, Identification of volatile carbonyls from Cheddar cheese, *J. Dairy Sci.*, 43, 463, 1960.

84. **van der Ven, B., Begemann, P. H., and Schogt, J. C. M.**, Precursors of methyl ketones in butter, *J. Lipid Res.*, 4, 91, 1963.
85. **Scarpillino, R. and Kosikowski, F. V.**, Evolution of volatile compounds in ripened raw and pasteurized milk Cheddar cheese observed by gas chromatography, *J. Dairy Sci.*, 45, 343, 1962.
86. **Day, E. A. and Libbey, L. M.**, Cheddar cheese flavor: gas chromatographic and mass spectral analysis of the neutral components of the aroma fraction, *J. Food Sci.*, 29, 583, 1964.
87. **Bills, D. D., Morgan, M. E., Libbey, L. M., and Day, E. A.**, Identification of compounds responsible for fruity flavor defects of experimental Cheddar cheese, *J. Dairy Sci.*, 48, 1168, 1965.
88. **Sih, C. J. and Knight, S. G.**, Carbohydrate metabolism of *Penicillium chrysogenum, J. Bacteriol.*, 72, 694, 1956.
89. **Heath, E. C. and Koffler, H.**, Biochemistry of filamentous fungi. II. Quantitative significance of an "oxidative pathway" during the growth of *Penicillium chrysogenum, J. Bacteriol.*, 71, 174, 1956.
90. **Wang, C. H., Stern, I., Gilmour, C. M., Klungsoyr, S., Reed, D. J., Bialy, J. J., Christensen, B. E., and Cheldelin, V. H.**, Comparative study of glucose catabolism by radiorespirometric method, *J. Bacteriol.*, 76, 207, 1958.
91. **Imamura, T. and Tsugo, T.**, Studies on the chemical changes of milk components produced by *Penicillium roqueforti*, I. Changes in milk fat, *Int. Dairy Congr.*, 4, 622, 1953.
92. **Eitenmiller, R. R., Vakil, J. R., and Shahani, K. M.**, Production and properties of *Penicillium roqueforti* lipase, *J. Food Sci.*, 35, 130, 1970.
93. **Imamura, T. and Kataoka, K.**, Biochemical studies on the manufacture of Roquefort-type cheese. III. Isolation of lipases from mold-culture, *Jap. J. Dairy Sci.*, 15, A138, 1966.
94. **Imamura, T. and Kataoka, K.**, Biochemical studies on the manufacture of Roquefort-type cheese. II. Characteristics of lipases produced by *Penicillium roqueforti, Jap. J. Zootech. Sci.*, 34, 349, 1963.
95. **Hawke, J. C.**, The formation and metabolism of methyl ketones and related compounds, *J. Dairy Res.*, 33, 225, 1966.
96. **Gehrig, R. F. and Knight, S. G.**, Formation of methyl ketones from fatty acids by spores of *Penicillium roqueforti, Nature* (London), 182, 1237, 1958.
97. **Lawrence, R. C. and Hawke, J. C.**, The oxidation of fatty acids by mycelium of *Penicillium roqueforti, J. Gen. Microbiol.*, 51, 289, 1968.
98. **Dwivedi, B. K. and Kinsella, J. E.**, Carbonyl production from lipolyzed milk fat by the continuous mycelial culture of *Penicillium roqueforti, J. Food Sci.*, 39, 83, 1974.
99. **Darty, C. K. and Kinsella, J. E.**, Oxidation of sodium palmitate into carbonyl compounds by *P. roqueforti* spores, *J. Agr. Food Chem.*, in press.
100. **Anderson, D. F. and Day, E. A.**, Action of microorganisms in Blue cheese on 2-pentanone and pentanol, *J. Dairy Sci.*, 48, 784, 1965.
101. **Day, E. A. and Anderson, D. F.**, Gas chromatographic and mass spectral identification of neutral components of the aroma fraction of Blue cheese, *J. Agric. Food Chem.*, 13, 2, 1965.
102. **Franke, W., Platzeck, A., and Eichhorn, G.**, Fatty acid decomposition by molds, IV. Study of an additional reaction of methyl ketone, *Arch. Microbiol.*, 41, 154, 1962.
103. **Cort, W. M. and Riggs, L. K.**, Enzymatic process of making cheese and cheese products, U. S. Patent 3,295,991, 1967.
104. **Hedrick, T. I., Kondrup, E., and Williamson, W. T.**, Preparation of blue cheese by adding to the cheese a blue cheese mold and ripening the curd in a divided condition, U. S. Patent, 3,365,303, 1968.
105. **Unilever, N. V.**, Cheese aroma, Netherlands Patent, 6:515,288, 1966.
106. **Knight, S. G.**, Process for preparing flavoring compositions, U. S. Patent, 3,100,153, 1963.
107. **Watts, J. C. and Nelson, J. H.**, Cheese flavoring process, U. S. Patent, 3,072,488, 1963.
108. **Dwivedi, B. K. and Kinsella, J. E.**, Continuous production of blue cheese flavor by submerged fermentation of *Penicillium roqueforti, J. Food Sci.*, 39, 620, 1974.
109. **Nelson, J. H.**, Enzymatically produced flavors for fatty systems, *J. Am. Oil Chem. Soc.*, 49, 559, 1972.
110. **Anti, A. W.**, Lipolyzed acids used as flavor enhancers reduce formulation costs, *Candy Ind. Confect. J.*, 135(4), 25, 1970.
111. **Tanabe Seiyaku C. Ltd.**, Yogurt production, Japanese Patent 3107/71, 1971.
112. **Neelakantan, S., Shahani, K. M., and Arnold, R. G.**, Lipases and flavor development in some Italian cheese varieties, *Food Prod. Dev.*, 5(7), 52, 1971.
113. **Langler, J. E. and Day, E. A.**, Quantitative analyses of the major free fatty acids in Swiss cheese, *J. Dairy Sci.*, 49, 91, 1966.
114. **Richardson, G. H. and Nelson, J. H.**, Rapid evaluation of milk coagulating and flavor producing enzymes for cheese manufacture, *J. Dairy Sci.*, 51, 1502, 1968.

SAUCES, DRESSINGS, CONDIMENTS, AND POWDERED FLAVORS

Introduction

Sauces, dressings, condiments, and powdered flavors represent technological difficulties in classifying because of the countless flavor ingredients employed and the manner in which they are combined. These products are really new flavors developed through the centuries to impart characteristic flavor to traditional cuisines. Generally, this category of flavor ingredients includes products specifically designed for flavoring fried, boiled, broiled, or roasted meat and fish, cooked or raw vegetables, and pasta and cereal products.

Taste, tradition, availability of certain products, and dietary requirements have all been influential factors in the choice of flavor ingredients and their relative ratio in the flavoring mixtures. Sometimes these can be very simple, such as oil and vinegar dressings, or much more complex, such as vinegar sauces, tomato sauces (tomato ketchup, tomato chutney), Worcestershire sauce, the countless fondue sauces, and mixed spice powders (*e.g.*, curry).

The topic is too complex for an extensive discussion and goes beyond the scope of the book. The following is, therefore, a limited presentation of information concerning the flavoring of the most important products.

Sauces and Salad Dressings

Tables 9 and 10 list some traditional dressings, sauces, butters, etc. and indicate the *base* ingredients together with additional flavor ingredients used to augment the product. It is important to note that in almost all instances, the *base* contributes not only form, texture, mouth-feel, etc., but also a flavor background. In commercial products shelf-life requirements necessitate using emulsifiers, thickeners, antioxidants, and preservatives; these have not been included in the ingredients list, since they rarely contribute flavor to the finished product.

Mayonnaise

The Duc de Richelieu is credited with introducing mayonnaise into France in 1757. Although the original sauce, native to Minorca, was heavily dosed with garlic, legend tells that the Duc separated the garlic flavor before introducing the sauce into France.

Modern mayonnaise formulations are based on egg yolk, olive or light salad oils, lemon juice or wine vinegar, and salt; they are spiced with mustard, white pepper, cayenne pepper, and sometimes ginger and paprika. Other ingredients employed in mayonnaise include stabilizers, chelating agents, and preservatives to improve shelf-life; in the United States mayonnaise has standards of identity that specify the product. Essential oils may also be employed in flavoring mayonnaise. A characteristic formulation consists of the essential oils of mustard (4 parts), lemon (3 parts), celery (1 part), nutmeg (1 part), and black pepper (1 part).

TABLE 9

Salad Dressings

Name	Base	Flavor ingredients
Avocado	Mashed ripe avocado, salt, sugar, whipped heavy cream	Lemon juice, lemon peel
Celery seed	Salad oil, lemon juice, vinegar, salt, sugar	Onion (grated), honey, celery seed, paprika, dry mustard
Chili mayonnaise	Mayonnaise	Chili sauce
Citrus	Beaten egg, whipped heavy cream, sugar	Orange peel, lemon peel, lemon juice
Cooked dressing	Enriched flour, beaten egg yolk, milk, butter, vinegar, salt, sugar	Dry mustard, cayenne
Creamy mayonnaise	Mayonnaise, heavy cream	
French, clear	Salad oil, vinegar, lemon juice, salt, sugar	Dry mustard, paprika, cayenne

TABLE 9 (continued)

Salad Dressings

Name	Base	Flavor ingredients
French, creamy	Salad oil, egg, vinegar, salt, sugar	Paprika, cayenne
French, fruit	Salad oil, lemon juice, vinegar, sugar, salt	Orange juice, paprika, onion (grated)
French, garlic	Salad oil, vinegar, salt, sugar	Dry mustard, garlic (crushed)
French, snowy	Salad oil, white vinegar, blue cheese, salt, sugar	Onion juice, dry mustard, white pepper
Italian, classic	Olive oil, wine vinegar (also distilled), lemon juice	
Italian, spiced	Olive oil, white vinegar, salt	White pepper, celery salt, cayenne, dry mustard, garlic (minced), Tabasco sauce
Lemon-honey	Honey, salt	Lemon juice
Lime-honey	Honey, salt	Lime juice
Marshmallow	Mayonnaise, marshmallow creme	Orange juice, lemon juice
Pineapple	Unsweetened pineapple juice, corn starch, beaten eggs, cream cheese, salt, sugar	Orange juice, lemon juice
Pink fruit	Mayonnaise, salt	Cranberry juice, almond (toasted)
Russian	Cooked sugar and water, tomato ketchup, vinegar, salt, salad oil	Celery seed, paprika, lemon juice, Worcestershire sauce, onion
Shawano	Salad oil, tomato ketchup, sugar, vinegar, salt	Paprika, dry mustard, onion, garlic, steak sauce
Thousand island	Mayonnaise, chopped and sieved hard-boiled egg, salt	Chili sauce, green pepper, celery, onion, paprika
Tomato	Salad oil, condensed tomato soup, vinegar, sugar, salt	Worcestershire sauce, dry mustard, paprika, garlic (minced), onion
Vinaigrette	Clear French dressing, minced hard-boiled egg, chopped green olive	Chives, pimento
Yogurt	Mayonnaise, yogurt, sugar, salt	Lemon juice, lime juice

TABLE 10

Sauces

Name	Base	Flavor ingredients
Aioli, classic	Egg yolk, olive oil, salt	Garlic, lemon juice
Aioli (Greek Skordalia)	Egg yolk, olive oil, salt, bread crumbs	Garlic, lemon juice, blanched ground almonds, walnuts, or pine nuts
Ali-oli, Spanish	Olive oil, crumbled crustless white bread, salt	Garlic, lemon juice
Allemande	Hot veal, chicken, or fish broth, flour, butter, egg yolk, heavy cream, salt	Mushroom, white pepper, powdered nutmeg, lemon juice
Avgolemono	Hot lamb, chicken, or vegetable broth, salt, egg yolk	White pepper, lemon juice
Bagna Cauda	Olive oil, butter	Garlic, anchovy (minced)
Barbecue (Gaucho)	Tomato ketchup, water, sugar, salt, vinegar	Worcestershire sauce, Tabasco, celery seed
Barbecue (for meats)	Tomato ketchup, vinegar, honey, salt	Prepared mustard, Tabasco
Barbecue (for poultry)	Salad oil, Worcestershire sauce	Lemon juice, garlic
Bâtarde	Butter, flour, boiling water, egg yolk, heavy cream, salt	Black pepper, white pepper, lemon juice
Béarnaise	Shallots (minced), tarragon wine vinegar, egg yolk, butter, salt	Tarragon (chopped), chervil, cayenne pepper

TABLE 10 (continued)

Sauces

Name	Base	Flavor ingredients
Béchamel	Butter, flour, milk, salt	Onion, cayenne pepper
Bercy	Butter, hot fish broth, flour, heavy cream	Onion, dry white wine, parsley
Beurre Manie	Soft butter, flour, cream	
Beurre Meuniére	Clarified butter (browned)	Lemon juice
Beurre Montpellier	Yolks of hard-boiled egg, soft butter, mayonnaise, salt	Minced watercress, parsley, tarragon, chives, chervil, shallots, sour gherkin, garlic, anchovy, cayenne pepper
Beurre Nair	Clarified butter (browned)	Red wine vinegar, parsley, capers
Blueberry	Cooked blueberries, cornstarch	Lemon juice
Bordelaise, classic	Dry red wine, sauce Espagnole, meat glaze, sliced marrow (cooked), butter, salt	Shallots, pepper, parsley, lemon juice, bay, thyme
Bordelaise, sweet	Sherry, brown sauce, diced beef marrow (cooked), salt	Green onion, parsley
Brown, classic	Beef broth, butter, flour, salt	Tomato puree, pepper
Brown, spiced	Beef broth, butter, flour, salt	Tomato puree, pepper, browned onion, minced carrot, celery
Butters (*also see* Beurre)		
Anchovy	Butter (soft)	Anchovy paste (or fillets)
Bercy	Butter (soft), poached beef marrow, dry white wine, salt	Lemon juice, pepper, shallots, parsley
Chivry	Butter (soft)	Chives, parsley, chervil, tarragon, shallots
Garlic	Butter (soft)	Blanched, mashed garlic
Indian ghee	Melted butter separated from oil and white residue	
Lemon	Butter (soft), salt	Lemon juice, parsley, pepper
Lobster	Butter (soft), cooked lobster, roe, and strained lobster meat	
Maitre d'hôtel	Butter (soft), prepared mustard, salt	Pepper, parsley, tarragon, lemon juice
Most daintie	Butter (soft)	Cinnamon (or paprika), nutmeg (or mace)
Russian nutmeg	Butter (soft)	Cinnamon, nutmeg, chives, white pepper, lemon juice
Snail	Butter (soft), finely minced shallots (sauteed), salt	Garlic, parsley, pepper, nutmeg, thyme
Caper	Mayonnaise, sour pickles, capers	Chopped capers, prepared mustard, parsley
Chaud-froid	Chicken broth, flour, heavy cream, egg yolk, gelatin, aspic, salt, butter	Dry white port, onion, diced carrot, celery, pepper, tarragon leaves, bouquet garni (parsley sprigs, parsley root, bay leaves, dried thyme)
Chili	Tomatoes, brown sugar, cider vinegar, green peppers, salt	Celery, onion, cloves, dry mustard, cinnamon
Chinese mustard	Water (boiling), dry English mustard, salad oil, salt	
Chocolate, regal	Unsweetened chocolate, corn syrup, sugar, water, evaporated milk	Vanilla
Cocktail	Mayonnaise, chili sauce	Lemon juice, Worcestershire sauce, Tabasco, horseradish, paprika
Coney	Ground beef (browned), water, tomato sauce, salt, monosodium glutamate	Onion, garlic, chili powder
Cranberry	Cooked cranberries, sugar, water	
Cranberry, spiced	Jellied cranberry sauce, steak sauce, brown sugar, salad oil	Prepared mustard

TABLE 10 (continued)

Sauces

Name	Base	Flavor ingredients
Cumberland	Melted currant jelly, port wine	Orange juice, lemon juice, orange and lemon rind (simmered), shallots, prepared mustard, powdered ginger, cayenne pepper
Creole	Seasoned tomato sauce, broiled mushrooms, green pepper	
Custard sauce	Egg yolk, sugar, salt, scalded milk	Vanilla
Dal	Lentils, butter, light beef broth, salt	Onion, hot chili pepper (green), dried chili pepper (red), turmeric (or curry powder), mustard seed, coriander
Demi-glace	Sauce Espagnole, meat glaze, madeira or tawny port	
Diable	Brown sauce, Worcestershire sauce	Onion, sauterne, peppercorns (crushed)
Dill	Flour, butter, chicken broth, sugar, vinegar, egg yolk	Dill, dill weed
Duxelles	Sauce Espagnole, tomato puree, olive oil, butter, salt	Mushroom, shallots, onion, pepper, white wine, parsley
Espagnole	Butter, brown sauce, dry white wine, beef broth, salt, pepper	Tomato paste, onion, carrot, bouquet garni (bay, thyme, peppercorns, parsley)
Fossen Flodepeberrod (Danish frozen horseradish cream)	Whipped heavy cream, sugar, salt, white pepper	Lemon juice, horseradish
Hollandaise	Egg yolk, cold water, butter, salt	Lemon juice, cayenne pepper
Hummer Sao (Norwegian lobster sauce)	Allemande sauce of fish stock, lobster meat (diced), butter, salt	Pepper, cognac (or sherry)
Jalapena (Mexican hot sauce)	Plum tomato (drained), olive oil, beef broth, sugar	Jalapena chili peppers, thyme, garlic, onion
Maltaise	Egg yolk, butter, salt	Orange juice, cayenne pepper
Mint, classic	Mint leaves, salt, sugar	Vinegar
Mint, popular	Mint jelly, salt	Vinegar
Mornay	Béchamel sauce	Gruyere and Parmesan cheese
Mousseline	Hollandaise sauce, heavy cream	
Nantua (French lobster sauce)	Béchamel sauce made with fish stock, milk, heavy cream (scalded), lobster butter, diced lobster, salt	Pepper
Orange	Butter, sugar, orange juice	Orange peel and oils
Pesto	Basil leaves, olive oil, Parmesan cheese, salt	Garlic, pine nuts, pepper
Pomme d'amour	Tomato, olive oil, butter, salt	Bouquet garni, wine vinegar, garlic, basil, sugar, pepper
Raisin	Beef broth, brown sugar, salt, cornstarch	Red wine vinegar, currant jelly, madeira wine, seedless white or black raisins, bouquet garni
Ravigote	Vinaigrette dressing	Shallots, onions, capers, chives, chervil, parsley, tarragon
Rémoulade	Mayonnaise, minced capers, minced sour pickles, onions	Dijon mustard, parsley, tarragon, chives, chervil, anchovy paste
Rémoulade, hot	Olive oil, onion, brown sauce, dry white wine, lemon, salt	Garlic, tarragon, chervil, basil, parsley, cloves, capers, Dijon mustard
Robert	Butter, onion, sauce Espagnole, salt, pepper	White wine, sugar, Dijon mustard

TABLE 10 (continued)

Sauces

Name	Base	Flavor Ingredients
Rouille de Marseille	Pimento, egg yolk, olive oil, hot fish broth	Chili pepper sauce, garlic
Salsa verde (Italian green sauce)	Parsley, olive oil, salt	Garlic, black pepper, capers, lemon juice
Soubise	Onions, butter, béchamel sauce, salt, pepper, heavy cream	Nutmeg, sugar
Suprême	Velouté (chicken), strong chicken broth, butter, heavy cream, salt	Mushroom, pepper
Sweet and sour, Chinese	Peanut oil, soy sauce, cornstarch, chicken broth, sugar	Garlic, scallions, mushrooms, carrot, bamboo shoots, ginger, vinegar
Tartare	Mayonnaise, sweet gherkins, Dijon mustard	Scallions, chives, capers, olive
Velouté	Butter, hot veal, chicken or fish broth, flour	Onion
Verte (green French sauce)	White bread crumbs, dry white wine, sorrel or spinach, parsley, heavy cream, salt	Tarragon, sage

Mustard

Mustard is probably the most widespread among table sauces. Several countries boast the production of specialty products, among which French, English, and German types are the best known. The various mustard types differ mainly in sharp or aromatic flavor. French mustard (Dijon) has a sharp taste and contains vinegar, white mustard, flour, salt, turmeric, cayenne pepper, clove buds, and pimento. English mustard sauce contains, in addition to the above ingredients, sugar, tapioca, black mustard, and white pepper. German mustard sauce is made with approximately the same ingredients with the addition of wine.

In England Piccalilli type sauces are widely used. These consist of mustard sauces containing chopped vegetables (*e.g.*, cauliflower, cucumber, onion) and flavored with additional ingredients, such as cinnamon, cumin, fenugreek, nutmeg, and ginger.

Cremones mustard, frequently used in northern Italy but otherwise little known, is a condiment for boiled meat. The product originated and is manufactured in Cremona, a town well known for its gastronomic and pastry specialties. It consists of a fruit-based mustard prepared as follows: whole or chopped fruits are initially steeped in sucrose syrup; the syruped fruit then is transferred into mustard-flavored syrup in containers suitable for distribution. The flavoring consists of the addition of natural or synthetic mustard essential oil to sugar syrup. The mustard essential oil distributes uniformly in the fruit during a ripening period prior to sale. The final product contains approximately 35% syrup (by weight of the syruped fruit) and is flavored by approximately 0.01% mustard essential oil. The use of black mustard flour has been practically abandoned.

Ketchups and Chutneys

Ketchups and chutneys consist chiefly of fruit concentrates or pulp or flavored vegetables. The most important products in the category are tomato ketchup and tomato chutney. Aside from the type of fruit pulp or fruit concentrate used, specific flavor formulations based on essential oils that yield the characteristic spice note in the finished product have become quite popular. These formulations usually contain approximately 50% clove essential oil; the balance is made up of suitable blends of cinnamon, nutmeg, pimento, mace, mustard, and celery.

There is only a very slight difference in the flavoring of tomato ketchup and chutney. However, they differ quite substantially in consistency due to the use of thickening agents (*e.g.*, tragacanth gum) and to the use of more or less concentrated tomato bases. Although countless flavor variations are possible, they are all confined within specific formulation guidelines. Tomato ketchup is prepared with tomato concentrate, sugar, water, vinegar, onion, salt, tragacanth gum, and spices such as nutmeg, cardamom, cinnamon, clove buds, and coriander. Ingredients of tomato chutney include tomato concentrate and pulp, sugar, water, vinegar, onion, salt, spices (coriander, cayenne pepper, cinnamon, cardamom, pimento, celery), reinforced by the addition of essential oils (garlic, clove bud, etc.).

Mushroom ketchup is vinegar-free and contains hydrolyzed proteins, soy sauce, and water. Clove buds, pimento, black pepper, ginger, and cardamom are used for flavoring. Mushroom ketchup is used also as a flavor ingredient in thin-sauce formulations, such as Worcestershire.

Walnut ketchup is made basically with vinegar, soy sauce, onion, anchovies, and hydrolyzed proteins. Clove buds, ginger, mace, and black pepper are used as flavor ingredients. Walnut ketchup also is used as a flavor ingredient in Worcestershire sauce.

Apricot and onion chutneys (of Anglo-Saxon origin) contain similar flavor ingredients, such as ginger, cinnamon, paprika, cayenne pepper, pimento, and coriander. Apricot chutney calls for the use of a 6:1 ratio of onion to apricot; the onion-apricot mixture is pickled in vinegar, sweetened with sugar, and thickened with tragacanth gum and tapioca. Onion chutney is prepared with onion, tomato, grape, tamarind, and dates in a 7:2: 1:1 ratio, respectively; small levels of other flavor ingredients are added (sometimes a lemon note is also added).

Thick Sauces

These sauces exhibit a fairly thick consistency; they are based mainly on fruits (date, apple, mango, tamarind, lemon, and grape), tomato, molasses or sugar syrup, tapioca and similar flours, vinegar, salt, hydrolyzed proteins, anchovies, tragacanth gum, and a large variety of spices. Countless formulation blends are possible; mushroom ketchup is added to several formulations. A popular thick sauce available in the United States lists the following ingredients: tomato paste, distilled vinegar, corn syrup, raisins, salt, orange base, orange peel, caramel, dehydrated garlic, dehydrated onions, gum tragacanth, herbs, spices, and water.

Thin Sauces

Among the more fluid sauces Worcestershire is undoubtedly the best known and most widely imitated product. The formulation of Worcestershire sauce is complicated by the presence of finished products, such as mushroom and walnut ketchups and soy sauce. (For the preparation of mushroom and walnut ketchups, see the preceding discussion of Ketchups and Chutney.) Soy sauce, also used in the preparation of these ketchups, consists of a liquid brown sauce of Japanese (Soyu) or Chinese (Chiang Yu) origin prepared by prolonged fermentation of soy followed by salt-pickling. Worcestershire sauce also contains vinegar, sherry and brandy, cane sugar, salt, tamarind pulp, pork liver, and minced spices (mace, coriander, pimento, black or cayenne pepper). The cooking of pork liver (8–10 hours), the partial boiling of the various ingredients, and the mixing and final refining of the sauce represent only a few among several complicated steps. The complex preparatory technique makes the imitation of Worcestershire sauce extremely difficult. In some recipes for Worcestershire-type sauces, lemon pulp, mushroom powder, anchovies, hydrolyzed proteins, garlic, and caramel are used.

Anchovy Sauce and Paste

Regular anchovy paste is made without the addition of flavor ingredients. It is prepared by grinding cleaned and boned anchovy fillets and then adding a sufficient amount of salt to the fish paste. Anchovy sauce of Anglo-Saxon production has a much more complex formulation. Depending on the type, it contains vinegar (together with acetic acid), water, salt, tragacanth gum, cayenne pepper, mace, apple pulp, tamarind, tomato, ginger syrup, caramel, dates, raisins, garlic, lemon, citric acid, clove buds, bay, and scallions.

Tomato Sauces

In addition to the countless types of tomato-based sauces that do not require warming and are used for flavoring meats and vegetables, there are also tomato sauces that are served hot with macaroni, spaghetti, rice, etc. Tomato sauces served hot with pasta products and featuring typical recipes of the Italian cuisine have been available commercially for many years. The best known types follow:

1. Neapolitan-style tomato sauce containing, in addition to both precooked tomato paste and peeled tomatoes, fats, salt, a little sugar, celery, carrot, pepper, and onion.
2. Bolognese-style tomato sauce (ragout) that, in addition to containing the above ingredients, contains cooked ground meat flavored with the same ingredients used in the sauce.

Flavored Vinegars

The use of flavored vinegars is still quite prevalent in various countries; they are used either as a complementary condiment or as an alternate for salad vinegar or in other food products. In the past flavored vinegars were homemade by careful infusion of herbs and spices in pure wine vinegar. Formulations of such condiments were developed from gourmet preferences in various regions and are countless in number. The following recipe for truffle-flavored vinegar is mentioned as an example.

Truffle-flavored Vinegar

Ingredients:

White truffles	20 g
Garlic	5 g
Paprika	5 g
Geranium leaves	3 g
Wine vinegar	1 l

Preparation:

1. Crush the first four ingredients in a mortar containing a few drops of the wine vinegar.
2. Transfer the mixture into a bottle filled with the remainder of the vinegar.
3. Keep the mixture in a sealed bottle for at least one month, agitating periodically, until ready for use.

Truffle-flavored vinegar is also prepared by other recipes employing different doses, preparation methods, and ingredients (black truffles, black pepper).

Following are a few interesting blends of flavored vinegars:

1. Vinegar, capers, rosemary, sage, basil, garlic, and elder flowers.
2. Vinegar, lemon, garlic, onion, tarragon, and celery root.
3. Vinegar, tarragon, basil, elder flowers, mint, thyme, bay leaf, rosemary, garlic, paprika, clove, cinnamon, pepper, and capers.
4. Vinegar, tarragon, bay leaf, musk yarrow, clove, and nutmeg.

The last formulation above represents an imitation of true tarragon-flavored vinegar and has found widespread acceptance.

Those relatively few flavored vinegars that are manufactured are usually prepared according to the following methods:

1. Tarragon-flavored vinegar by an infusion of tarragon in vinegar.
2. Capsicum-flavored vinegar by an infusion of cayenne pepper in vinegar.
3. Spice-flavored vinegar by an infusion of cayenne pepper, clove buds, pimento, ginger, coriander, black mustard, and black pepper in vinegar.
4. Rosemary-flavored vinegar by infusion of rosemary, peppermint, and angelica root in wine vinegar.
5. French-type flavored vinegar by infusion of tarragon, basil, elder, thyme, rosemary, scallions, lemon peel, white pepper, pimento, bay leaf, and radish in wine vinegar with the addition of salt and sucrose.

In the industrial preparation of flavored vinegars, it is more convenient to use alcoholic tinctures, essential oils, oleoresins or fluid extracts of the various herbs or spices. The much higher flavor yields of these flavor ingredients, together with their degree of solubility in vinegar, must be taken into account. For insoluble products it may be convenient to use alcoholic extracts of the various flavor ingredients.

Compounded Table and Cooking Spices

Ground spices frequently are used for flavoring freshly cooked food immediately prior to consumption. The most frequently used spices include white and black peppers, paprika, and celery-flavored salt, all of which are fairly simple products. Compounded ground spices are also employed. Typical formulations call for the use of a blend of the following finely comminuted spices:

1. Paprika (6 parts), ginger (30 parts), pimento (30 parts), mustard (60 parts), black pepper (60 parts), fenugreek (90 parts), turmeric (130 parts), coumarin (60 parts), coriander (720 parts), salt (12 parts), vinegar (30 parts).
2. Black pepper (38 parts), ginger (7.5 parts), turmeric (7.5 parts), black mustard (4 parts), cumin (40 parts), nutmeg (3 parts), clove (4 parts), anise (1.8 parts), sugar (4 parts), dried and ground sweet almond (45 parts); the minced spices are mixed with tartaric acid (4 parts).
3. Coriander (1 part), paprika (1 part), cumin (2 parts), mustard (1 part), grape (3 parts), ginger (3 parts), white pepper (6 parts), turmeric (16 parts).
4. Mustard (40 parts), cinnamon (5 parts), cardamom (1 part), ginger (10 parts), paprika (3 parts), fenugreek (5 parts).
5. Marjoram (15 parts), parsley (21 parts), thyme (25 parts), lemon peel (12 parts), celery (4 parts), basil (12 parts), bay leaf (5 parts).

Curry is a compounded spice of Oriental origin. Its composition varies depending on its origin—India, Red Sea countries, Indonesia, Hong Kong, etc. A typical simplified curry formulation includes the following: cinnamon (1 part), clove buds (1 part), ginger (1 part), fenugreek (1 part), mustard (1 part), chili (1 part), coriander (1 part), turmeric (1 part), cumin (1 part), cardamom (1 part); approximately 30 ingredients are present in a true curry formulation.

SOUPS AND BROTHS

Bouillons for Broths and Condiments

The preparation of bouillons has had a decisive growth following World War II. At one time they were considered auxiliary culinary products or surrogate products of only limited interest. Today bouillons are considered indispensable ingredients in food preparation both at home and in restaurants because of their twofold function—the imparting of and the enhancing of food flavors. The reasons for the increased acceptance of these products are numerous, the most important of which is probably the substantial changes in manufacturing practice from empiric to scientific on an industrial scale. This has in turn offered guarantees of high quality required by products of large consumption.

Early bouillon preparations consisted mainly of animal or vegetable proteins hydrolyzed by hydrochloric acid and subsequently neutralized with sodium carbonate to form sodium chloride. Yeast residues from distilleries, dextrin, and vegetable and animal broth concentrates complemented the protein mixture. True meat extracts represented only a minor portion of such products with the exception of a few high-quality preparations. The transition to current products has occurred gradually. Parallel to the increasing demand for such products, formulations have been improved through continuing experience; today they are perfectly matched to specific end-use requirements. To arrive at these results, however, old formulations based on chemical reactions have been abandoned; the new bouillon formulations consist mainly of blends. Today's bouillon production is based on the following basic types:

1. Bouillons with strong, aromatic, vegetable flavor, also used in condiments.
2. Lean beef bouillons with meat and vegetable flavor.
3. Fat beef bouillons with meat and vegetable flavor.
4. Mixed flavor bouillons, spiced and seasoned for use with condiments in the presence and/or absence of tomatoes.
5. Chicken bouillons.

Basic raw materials in the preparation of bouillons include monosodium glutamate and other flavor enhancers, hydrogenated fats, concentrated meat and vegetable extracts, soybean and other flour types, powdered milk, lactose, yeast extract, sodium chloride, and flavors. The relative doses in the blends are important and directly related to the technological skill of the manufacturer. Flavoring must be compatible with the blend, but it is usually geared toward a few specific trends. In bouillons a typical flavor base is celery, onion, and carrot. Turmeric is often added, especially to fatty-type products. Flavors may be added as powdered, dried vegetables, concentrated extract, essential oil, oleoresin, or oleoresin plated over a salt carrier. Nutmeg also yields good flavoring results when added at suitable levels to vegetable-flavored bouillons (No. 1 above).

In bouillons for condiments, the inventiveness of the formulator finds extensive room for implementation with the limitation that pepper, ginger, capsicum, and similar products must be used only as minor ingredients and not as basic flavors. Rosemary and laurel flavors also may be used to characterize the product, but their presence limits the number of end-use applications.

Finished Soup Products

The demand for products suitable for the quick preparation of finished soups has increased remarkably in recent years, not only in quantity but also in variety. The quality of such products has been continuously perfected, justifying their widespread acceptance at all social levels as basic components of daily food. Products have been formulated to satisfy the taste of a very large majority of consumers, and formulations have been patterned after reknowned culinary recipes.

Technically these products (containing all the necessary ingredients) are formulated so as to reduce the cooking time to a minimum. It is sufficient to open the package, disperse the product in the required amount of water, then heat to boiling; the resulting product is ready for consumption. To arrive at such results, all ingredients normally requiring prolonged cooking must be precooked and then dehydrated to permit extended storage and packaging in a minimum amount of space. As in all soups, including homemade varieties, various flavor ingredients characteristic of each type of product are usually added.

Since most products consist of coarsely ground ingredients, powdered anhydrous flavors are usually employed. Dispersions of liquid flavors on suitable carriers are normally employed. When the direct use of dried, powdered herbs is not feasible or convenient, oleoresins plated on a salt carrier or similar flavor blends may be more functional. Monosodium glutamate and other flavor potentiators, yeast extracts, and similar products are almost exclusively utilized together with salt to complete and enhance the flavor. The precooked ingredients generally consist of farinas, cereal products, and vegetables. Spray-dried, dried, and dehydrated vegetables prepared by methods most suitable for the individual ingredients are used; to these are added hydrogenated fats, dry skim milk, various inorganic salts, and sometimes sugar. Flavor ingredients are used following the same criteria employed in the preparation of specific homemade products. It should be remembered that neutral blends require a larger number of flavor ingredients. It may be necessary at times to mask the specific flavor of certain ingredients or to reinforce the corresponding notes of certain dried vegetables. Since prolonged cooking of the finished, ready-to-eat soup is not required, precooked food flavors should be considered; for example, the flavor of roasted onion is entirely different from that of fresh onion.

DESSERTS

Puddings and Gelatins

In the past the preparation of dessert products such as puddings and gelatins were almost exclusively left to the extemporaneous expertise of home cooking. However, prepared gelatins and puddings have attained a remarkable importance in recent years by entering the category of industrial food products; their consumption has been increasing constantly. Reasons for their increased popularity include ease of preparation, mouth-feel, and storage stability; most important is the variety and ease of flavoring.

Generally, pudding and gelatin formulations should be prepared with flavor materials suitable for prolonged storage. The ingredients, depending on local government regulations, are chosen in general from among the following: rice or corn starches, roasted flours, milk powders, sugars, thickeners (alginate, agar, pectins), colorants, and flavors. Pudding powders can be added to milk or water, while gelatin-based powders are used only with water. Sometimes dry powdered milk is blended in the pudding formulation, in which case the addition of water is necessary. As to the flavor ingredients, pudding powders are most frequently flavored with vanilla, chocolate, Dutch custard, almond, pistachio, mandarin, and zabaglione; fruit flavors, such as strawberry, raspberry, apricot, banana, pineapple, pear, plum, apple, red currant, peach, together with citrus, rose, and pistachio flavors, are usually used in gelatin powders.

Pudding and gelatin flavors must be in powder form for better dispensibility with the other powders in the blend. Vanillin or natural vanilla extract may be used for vanilla flavors. Cocoa powder and a little vanilla may be used directly for chocolate flavors, whereas compound spray-dried or plated flavor ingredients may be used to impart other flavors. The use of lyophilized fruit juices may also give excellent results in the preparation of certain flavors for gelatin products.

Honey

Honey flavor, although it follows certain basic guidelines, often exhibits characteristic floral notes, depending on the flowers on which the bees feed. A few honey types are distinguished by their specific floral note, such as lavender, rose, genet, linden, etc. These various honey types are originally from areas where flowers of the corresponding type are cultivated or grow wild in the proximity of the hive. Honey flavor imitations do not take into account these particular flavor notes and attempt to duplicate only the basic honey note formulated with a more or less pronounced and generic floral note.

Considering that honey may find use not only as a sweetener but also as a flavor ingredient in pastry products and in candy, honey flavor imitations are prepared to exhibit characteristically high flavor yields. This permits the replacement of natural honey in formulations such as powdered products where natural honey cannot be employed because of technical reasons. In addition to pastry, natural and imitation honey flavors are popularly employed in tobacco and drugs.

The basic component on which honey flavor imitations are based is phenylacetic acid. A minute addition of formic acid is generally advisable. In addition to phenylacetic acid, the addition of esters to the formulation is important. Ethyl pelargonate, amyl acetate, ethyl acetate, and amyl valerate also may be added. However, flavor imitations based on these esters only (even with the addition of methyl acetophenone) exhibit a particular note of very limited use. The use of acetates, such as phenyl, ethyl, phenethyl, isobutyl, allyl, *p*-cresyl, cyclohexyl, and guayl, imparts more specific notes. Phenethyl acetate, benzyl cinnamate, phenethyl propionate, linalyl butyrate, and cinnamyl butyrate are also excellent auxiliary flavor ingredients when used with carvacryl acetate, *p*-cresyl acetate, *p*-cresyl ethyl ether, ethyl phenoxyacetate, and traces of amyl acetate.

Once the exact ratio between phenylacetic acid and the esters has been chosen, it is necessary to add vanillin, sometimes heliotropin, and a generic floral note that is normally introduced with essential oils. The most functional oils for this type of formulation include rose, clary sage, celery, jasmine, lavender, camomile, neroli, cardamom, together with genet and mimosa absolute. To better define the use of essential oils, it is important to point out that excellent flavor effects are already obtained by using only rose and clary sage essential oils, provided vanillin is also present in sufficient amounts. A guideline formulation for imitation honey can be represented by the following:

Imitation Honey Flavor

Ingredients	Parts
Phenylacetic acid	555
Amyl valerate	230
Amyl acetate	170
Ethyl acetate	100
Ethyl phenylacetate	90
Rum ether	25
Vanillin	25
Methyl phenylacetate	15
Ethyl formate	7
Lemon balm essential oil	3
Maltol (1% solution in alcohol)	3
Clary sage essential oil	2
Formic acid	2
Orange essential oil	2
Rose absolute	1
Solvent (ethylene glycol ethyl ester)	770
Total	2,000

For tobacco a rather simple honey formulation consists of phenylacetic acid, ethyl vanillin, methyl phenylacetate, methyl acetophenone, and geranium essential oil. It is not advisable in this instance to add heliotropin or hydroxycitronellal to strengthen the floral note.

Fruit Preserves

Marmalades (*confettura*), jellies, jams, and preserves are prepared by blending concentrated fruit pulp and/or juice with sugar and sometimes thickeners, such as pectin. The resulting product must be a gel sufficiently thick for easy spreading.

Preparation is usually performed by cooking and concentrating in either open or vacuum concentrators sometimes equipped with automatic devices for determining dry residue. In general, the dry residue of fruit preserves is about 60%. The initial method of preparation of the pulp or juice is normally used to classify various products. For example, marmalade is prepared from mashed pulp after removal of the peel, seeds, residual leaves, and stems by compressing the cooked fruit through a sieve. Preserves are prepared from peeled and de-seeded fruit cut into pieces. Jellies are prepared from fruit juice; jellies usually do not contain water insoluble substances.

Specific technological procedures are employed for the preparation of these products. However, the basic concept governing the preparation of gelatins and the addition and choice of sugars should be considered perfectly valid. Sucrose tends to invert because of heat and the presence of acid, while thickeners are more active at low pH values. In the preparation of marmalades, two-thirds of the required sugar is added prior to cooking, while the remainder is added toward the end; the thickener (pectin) is then added and finally acidulants as required. By this method an excessive inversion of sucrose to dextrose with subsequent potential crystallization of the dextrose is avoided, while a partial inversion of sucrose is effective to prevent the crystallization of sucrose. The optimum ratio in the resulting product should be two-thirds sucrose and one-third inverted sugar. The amount of sugar usually varies between 60–65%, based on the weight of the pulp and depending on the content of natural acids and pectin. In addition to beet (alkaline) and cane (acid) sugars, corn syrup, malt extracts, and glucose are also used in mixture with the sugar. The use of sorbitol in place of sugar is conditioned by the specific dietary requirements of consumers.

Preserves usually do not require additional flavoring; in several countries flavoring of preserves is not permitted. Marmalades, preserves, and jellies are used as flavor ingredients in confections and pastry. Very interesting flavor effects are obtained by blending these products (particularly fluid preserves) with yogurt or whipped cream.

MEATS

Meats

Fresh Meats

Of all aromas the generic aroma of cooked meat most stimulates the appetite of a large majority of mankind. The reconstruction of meat flavor for flavoring other food products represents an extremely important goal for the flavor industry. Until recently, however, thoroughly functional flavor imitations have not been formulated. In the continuing investigation of meat flavors, probably more emphasis has been placed on odor (resulting from the most important volatile components) rather than on flavor. Analytical studies of the natural flavor of precooked meat have uncovered similarities in the flavors of red meats from various animals—beef, lamb, pork, etc. The differences between these and poultry flavor are considerable.

Red Meats

In beef, lamb, and pork the flavor similarity is only in the "background note." The different flavor sensations experienced when tasting the various red meats are familiar to everyone. All red meats, independent of condiments and seasoning, exhibit a characteristic aroma and flavor that may be classified (tentatively) in relation to the animal source. Several other factors, such as age, type of feed, breeding, etc., strongly affect the resulting aroma of the various meats.

The flavor of cooked red meat (not mentioning the specific characteristics for each type of meat not yet investigated) develops from amino acids, reducing sugars, and fat precursors. Three fundamental reactions involving the above constituents may occur during cooking:

1. Strecker's degradation (deamination and decarboxylation) of β-amino acids, forming the following carbonyl compounds, acids, etc.:

Acetaldehyde	Lactic acid
Acetic acid	Methylamine
Ammonia	3-Methylbutanal
Dimethyl sulfide	Methyl mercaptan
Ethyl mercaptan	2-Methylpropanal
Formaldehyde	2-Methylpropionic acid
Formic acid	Propanal
Hydrogen sulfide	Propionic acid

2. Maillard's reaction, which leads to a darkening of the meat caused by the reaction between amino acids and reducing sugars, with the observed formation of the following alcohols and mono- and dicarbonyl compounds:

   Acetone
   2-Butanone
   Diacetyl
   Ethyl alcohol
   Methyl alcohol

3. Oxidation of free fatty acids and triglycerides. Fats (which act as a sink* for aromatic substances liberated by heat) yield the following carbonyl compounds:

Acetaldehyde	2-Hepten-1-al
Acetone	Hexanal
Aldehydes, di-unsaturated	2,4-Nonadien-1-al
Aldehydes, mono-unsaturated	Nonanal
Aldehydes, saturated (mainly hexanal)	Octanal
	Propanal
2,4-Decadien-1-al	Undecenal (?)

* It is probable that fats absorb the characteristic aromas exhibited by the above types of meat.

2-Decen-1-al	2-Undecen-1-al
2,4-Heptadien-1-al	

The presence of 5′-nucleotides in meat flavor has also been reported.[25]

Poultry

From the large number of studies of the identification of the constituents of chicken flavor (which consists of approximately 200 components, of which only a few have been identified), one might conclude that the formation mechanism of chicken flavor has not yet been clarified. While the oxidation of lipids is extremely probable, the degradation of amino acids (Strecker's reaction), although highly probable, has not been clearly demonstrated. The latter should occur concomitantly with more complex mechanisms, however.

Chicken flavor consists of a complex chicken-meaty flavor. Characteristic components of the meaty flavor probably consist of sulfur-containing compounds. Components responsible for the chicken flavor are probably the carbonyl compounds that follow:

Acetaldehyde	2-Hepten-1-al
Acetoin	Hexanal
Acetone	2-Hexen-1-al
Butanal	Nonanal
2-Butanone	Octanal
2,4-Decadien-1-al	Pentanal
2-Decen-1-al	2-Pentanal
Diacetyl	Propanal
2,4-Heptadien-1-al	2-Undecen-1-al

The presence of hydrogen sulfide and alkyl mercaptans up to C_6 has been confirmed, while the presence of symmetric and asymmetric alkyl sulfides and disulfides is extremely probable. Esters, hydrocarbons, benzene, and furan derivatives, resulting from not yet clarified mechanisms, have also been reported as probably present.

Preservation of Meats

The preservation of meat has been practiced for centuries as a means of stocking foodstuffs. With the improvement of preservation techniques, the practice has been slowly turned over to the food industry. In the past drying, salting, and smoking were the most common and simple methods used to preserve meat. Air drying was the most commonly used method by primitive people. This method is still practiced today for the preparation of tasajo and pemmican, dried and salted red meats preserved under corn flour or molten fat.

The preservation of meat is strictly related to the use of flavor ingredients that not only exhibit a flavor action of their own but also enhance the flavoring due to salting. Even in the most simple case of smoking dried or pre-salted meat, not only is the type of wood (birch, maple, olive, etc.) that is burned important to flavoring but also the leaves, berries, etc. of aromatic plants (bay leaf, juniper, origanum, etc.) that are burned. Flavor ingredients may also be added during drying; they are added sometimes to the pickling solution prior to drying. The preservation techniques that contribute most toward flavoring are salting and sterilization.

Salt-cured Meats

Salting (sometimes complementary to the drying and smoking operations) is carried out by pickling in a salt solution or by covering the meat with salt. It is used to preserve whole, chopped, or ground meat, either raw or precooked. From a technical flavoring viewpoint, salting is the most interesting operation in the preservation of meat. The addition of salt (1) decreases the free moisture content that is absorbed on the salt as water of hydration, (2) causes continuing dehydration of the meat during slow drying, and (3) improves the organoleptic characteristics during ripening (also because of the addition of about 1% potassium nitrate to salt).

Salt-cured meats may be classified in two broad categories—salted whole meats and cured meats. Salted meats include *prosciutto*, cooked ham, and bacon. Cured meats consist of raw or cooked ground meat, such as salami, sausages, etc. A few salted whole and cured meats, such as *prosciutto*, sausage, etc., also may be smoked.

Each product requires a specific flavoring usually imparted either by maceration in flavored wine, by direct addition of flavor ingredients to ground meat, or by both of these together with the addition of salt. Flavor ingredients are usually spices or ground herbs, oleoresins, and essential oils plated on salt or other similar carriers. Pepper, cinnamon, clove buds, nutmeg, cardamom, coriander, capsicum, and ginger are frequently employed together with garlic, sage, thyme, bay, caraway, and origanum.

Smoking may be replaced by a suitable flavoring that, wherever permitted, is effected with cade oil, to which birch, laurel, or pepper oil dispersed in the ground paste or in the maceration solution is sometimes added.

Canned Meats

Canned meats are usually sterilized and require the addition of flavor ingredients. Sterilization is conducted by thermal treatment (120°C) in autoclaves on shredded, precooked, or raw meat (red and also mixed). Precooked meat is best flavored using a mixture of celery, carrot, laurel, nutmeg, pepper, and garlic tailored to yield a neither too sharp nor too strong final flavor. Flavor ingredients are usually added during preliminary cooking.

MEAT FLAVOR

Author: **Basant K. Dwivedi**
Foster D. Snell, Inc.
Florham Park, New Jersey

INTRODUCTION

The flavor of meat is attributed to a complex mixture of compounds produced by heating a heterogeneous system containing nonodorous precursors, and it is comprised of:

1. Volatile compounds with odor properties.
2. Nonvolatile compounds with taste and tactile properties.
3. Potentiators and synergists.

The chemistry of the compounds responsible for the flavor of cooked meat, and how these flavor compounds are produced during the process of cooking, have been the subject of considerable research in the last two decades. Two obvious facts about cooked meat flavor are

1. The nature of the flavor depends upon the way in which meat is cooked.
2. Each meat has its own characteristic flavor.

Since all types of meat consist of broadly similar components – such as proteins and other amino compounds, lipids, carbohydrates, and mineral salts – it is apparent that there must be a significant difference either in the proportion in which these compounds are present or in the way in which they interact. The differences in flavor due to cooking are probably a direct function of temperature and degree of moisture in the meat.[1] Because it is necessary to heat meat to produce the desired flavor, it follows that meat must contain substances which interact or degrade during cooking to produce compounds responsible for the characteristic flavor of meat. These substances are termed flavor precursors. A general discussion on the importance of flavor precursors and enzymes in the development of various food flavors was given in an earlier review.[2] This review deals with the nature of volatile and nonvolatile compounds, flavor precursors, and enzymes that may, directly or indirectly, influence the flavor of cooked meat and ripened sausage.

NATURE OF MEAT FLAVOR

Volatile Compounds

The list of volatile compounds isolated from cooked meat is quite extensive and their contribution to the 'typical' flavor of meat uncertain. Tables 1 and 2 list most of the volatile compounds identified so far from beef and pork products. Most of these volatiles are produced during cooking from nonvolatile precursors. Unlike fruit, dairy, and fermented food flavors, enzymes have little significance in the development of cooked meat aroma from precursors.

The nature of volatile compounds responsible for the characteristic aroma of meat has not been resolved. However, significant advances have been made in the last 15 years. Hornstein et al.[6,7] in 1960 reported the isolation of two major fractions of volatile compounds from a lean beef slurry:

1. Highly volatile fraction containing ammonia, traces of methylamine, hydrogen sulfide, methyl mercaptan, formaldehyde, acetaldehyde, acetone, and other unidentified highly volatile compounds. A rather disagreeable aroma was noticed in this fraction.

TABLE 1

Volatile Compounds Isolated from Heat-treated Beef

Compound	Meat sources and treatment*	References
Acids		
Formic	9	15
Acetic	9,13	15,24
Propionic	9	15
Butanoic	2,6,13	10,16,24
Hexanoic	7	3
2-Methylpropionic	1,9	4,5,15
Aldehydes (aliphatic)		
Formaldehyde	3,6	6,7,10
Acetaldehyde	1,3,4,6,7	3,4,5,6,7,8,10,11
Propanal	1,3,6,7	3,4,5,6,7,10,11
Pentanal	7,8,12,13	3,12,13,14,24
Hexanal	3,7,8,12,13	3,6,7,12,13,14,24
Heptanal	7,8,12,13	3,12,13,14,24
Octanal	7,8,12,13	3,12,13,14,24
Nonanal	3,7,8,12,13	3,6,7,12,13,14,24
Decanal	12,13	24
Undecanal	8,12,13	12,13,14,24
Dodecanal	12,13	24
Tridecanal	13	24
Hexadecanal	7	3
Methional	8,14	12,13,14,25
2-Methylpropanal	1,6	4,5,10
2-Methylbutanal	12,13	24
3-Methylbutanal	1,4,7,8,12,13	3,4,5,12,13,14,24
2-Hexenal	13	24
2-Heptenal	12,13	24
2-Octenal	7,12,13	3,24
2-Nonenal	12,13	24
2-Decenal	12,13	24
2-Undecenal	12,13	24
2-Dodecenal	13	24
2-Tridecenal	13	24
6-Methylhepten-1-al	7	3
2,4-Octadienal	12,13	24
2,4-Nonadienal	13	24
2,4-Decadienal	7,12,13	11,24
2,4-Undecadienal	13	24
2,4-Dodecadienal	12,13	24
Esters		
Methyl formate	2	6
Ethyl acetate	7	3
Acetol acetate	14	25
Ethers		
Pentyl ether	7	3
Pyrroles		
Acetylpyrrole	13	24
1-Methyl-2-acetylpyrrole	14	25

TABLE 1 (continued)

Volatile Compounds Isolated from Heat-treated Beef

Compound	Meat sources and treatment*	References
Alcohols		
Methanol	1	4,5
Ethanol	1,7,8,9	3,4,5,12,13,14,15
Propanol	7,13	3,24
2-Propanol	8	12,13,14
2-Methylpropanol	7	3
Butanol	6,7,8,12,13	3,10,12,13,14,24
3-Methylbutanol	8	12,13,14
Pentanol	7,8,12,13	3,12,13,14,24
Hexanol	8,12	12,13,14,24
Heptanol	8,12,13	12,13,14,24
2-Heptanol	14	25
3-Heptanol	12,13	24
4-Heptanol	13	24
Octanol	7,8,12,13	3,12,13,14,24
3-Octanol	13,14	24,25
4-Octanol	13,14	24,25
Nonanol	14	25
Decanol	14	25
Undecanol	14	25
2-Butoxyethanol	12	24
2-Hexenol	7	3
1-Pentene-3-ol	7	3
1-Octene-3-ol	7	3
2-Octenol	12	24
2,3-Butanediol	12,13	24
Ketones		
Acetone	2,3,4	6,7,8,16
Propanone	1,3,9	4,5,6,7,15
2-Butanone	1,6,8	4,5,10,12,13,14
Acetoin	7,8	3,12,13,14
Diacetyl	5,7,9	3,9,15
2-Heptanone	12,13	24
3-Heptanone	13	24
2-Octanone	12,13	24
3-Octanone	12,13	24
4-Octanone	7	3
2-Nonanone	12,13	24
3-Nonanone	7	3
2-Decanone	12,13	24
2-Undecanone	12,13	24
2-Dodecanone	13	24
3-Dodecanone	7	3
2-Tridecanone	12,13	24
2-Pentadecanone	13	24
3-Hydroxy-2-butanone	12,13	24
2,3-Butanedione	12	24
2,3-Pentanedione	12	24
2,3-Octanedione	12	24
2-Furfuryl methyl ketone	14	25

TABLE 1 (continued)

Volatile Compounds Isolated from Heat-treated Beef

Compound	Meat sources and treatment*	References
Hydrocarbons		
Butane	5	9
Pentane	5	9
Hexane	5,7	3,9
Heptane	5	9
Octane	5	9
Nonane	12,13	24
Decane	12,13	24
Undecane	12,13	24
Dodecane	7,12,13	3,24
Tridecane	12,13	24
Tetradecane	12,13	24
Pentadecane	7,12,13	3,24
Hexadecane	7,12,13	3,24
Heptadecane	12,13	24
Octadecane	7,13	3,24
6(?)-Methyltetradecane	12	24
1-Pentadecene	7	3
1-Heptadecene	13	24
1-Undecene	7	3
Benzene compounds		
Benzene	7,8	3,12,13,14
Trimethylbenzene	12,13	24
Diethylbenzene	12	24
1,4-Dimethylbenzene	12,13	24
1,2-Dimethylbenzene	12	24
Propylbenzene	7	3
Butylbenzene	12	24
Benzaldehyde	4,7,8,12,13	3,12,13,14,24,29
3-Methylbenzaldehyde	7	3
Phenylacetaldehyde	8,13	12,13,14,24
Paradichlorobenzene	7	3
Toluene	7,12,13	3,24
Lactones		
δ-Valerolactone	13	24
γ-Valerolactone	7	3
γ-Butyrolactone	12,13	24
γ-Hexalactone	13	24
δ-Heptalactone	13	24
γ-Heptalactone	13	24
δ-Octalactone	13	24
γ-Octalactone	13	24
γ-Nonalactone	12,13	24
δ-Nonalactone	13	24
γ-Decalactone	13	24
Furans		
2-Acetylfuran	14	25
2-Pentylfuran	7,12,13	3,24
2-Hexylfuran	13	24
2-Heptylfuran	12,13	24
2-Octylfuran	13	24
2-Methyltetrahydrofuran-3-one	7	3
4-Hydroxy-2,5-dimethyl-3(2H)-furanone	4	28
4-Hydroxy-5-methyl-3(2H)-furanone	4	28

TABLE 1 (continued)

Volatile Compounds Isolated from Heat-treated Beef

Compound	Meat sources and treatment*	References
Sulfur compounds		
Hydrogen sulfide	1,2,3,7,9	4,5,6,7,11,15,16
Methyl mercaptan	1,2,4,7,10	4,5,8,11,15,16
Ethyl mercaptan	1,2,4,7	4,5,8,11,16
Propyl mercaptan	2	16
Butyl mercaptan	2,7	3,16
Isobutyl mercaptan	10	17
Dimethyl sulfide	1,4,7,9,10,12	3,4,5,8,15,16,17,24
Dimethyl disulfide	8,10	12,13,14,17
Ethyl methyl disulfide	10	17
Methyl vinyl disulfide	10	17
Diethyl disulfide	10	17
Methyl propyl sulfide	7	3
Methyl allyl sulfide	7	3
Diallyl sulfide	7	3
Thiophene	10	17
2-Methyl thiophene	10	17
2-Ethyl thiophene	10	17
2-*n*-Butyl thiophene	10	17
2-*n*-Amyl thiophene	10	17
n-Octyl thiophene	10	17
n-Tetradecyl thiophene	10	17
2-Thiophene carboxaldehyde	10	17
5-Methyl-2-thiophene carboxaldehyde	10	17
2,5-Dimethyl-3-thiophene carboxaldehyde	10	17
2-Thiophene acrolein	10	17
2-Acetyl thiophene	10	17
3-Acetyl thiophene	10	17
5-Methyl-2-acetylthiophene	10	17
1-(2-Thienyl)-1-propanone	10	17
1-(2-Methyl-5-thienyl)-1-propanone	10	17
2-Thiophene methanol	10	17
Tetrahydrothiophen-3-one	10	17
2-Methyl tetrahydrothiophen-3-one	10	17
Thiazole	10	17
2-Methylthiazole	10	17
4-Methylthiazole	10	17
2,4-Dimethylthiazole	10	17
5-Ethyl-4-methylthiazole	10	17
4-Ethyl-2-methylthiazole	10	17
2,4,5-Trimethylthiazole	10	17
2,4-Dimethyl-5-vinylthiazole	10	17
2-Acetylthiazole	10	17
Benzothiazole	10,14	17,25
3,5-Dimethyl-1,2,4-trithiolane	4,7,10	17,23,29
5,6-Dihydro-2,4,6-trimethyl-1,3,5-dithiazine	10	17
Trithioacetaldehyde	10	17
Trithioacetone	10	17
Methylthioacetate	10	17
1,1-bis-Methylthioethane	10	17
Dimethyl sulfone	12,13	24
2-Acetyl-2-thiazoline	4	27
1-Methylthio-ethanethiol	4	29
Thialdine	4	29

TABLE 1 (continued)

Volatile Compounds Isolated from Heat-treated Beef

Compound	Meat sources and treatment*	References
Nitrogen compounds		
Ammonia	3,9	6,7,15
Methylamine	3	6
2,4,5-Trimethyloxazoline	7	23
Pyrazine	11	18
Methylpyrazine	11,14	18,26
2,5-Dimethylpyrazine	11,14	18,26
2,6-Dimethylpyrazine	11,14	18,26
2,3-Dimethylpyrazine	11,14	18,26
Ethylpyrazine	11	18
2,5-Dimethyl-3-ethylpyrazine	14	26
2-Ethyl-6-methylpyrazine	11	18
2-Ethyl-5-methylpyrazine	11,14	18,26
Trimethylpyrazine	11,14	18,26
2,6-Diethyl-3-methylpyrazine	14	26
Tetramethylpyrazine	11,14	18,26
5-Ethyl-2,3-dimethylpyrazine	11	18
2-Ethyl-3,5-dimethylpyrazine	11	18
2-Ethyl-3,6-dimethylpyrazine	11	18
2,6-Diethylpyrazine	11	18
Methylpropylpyrazine	11	18
2-Methyl-3,5-diethylpyrazine	11	18
2-Methyl-3,6-diethylpyrazine	11	18
5-Methyl-2,3-diethylpyrazine	11	18
3,6-Diethyl-2,5-dimethylpyrazine	11	18
Triethylpyrazine	11	18
6,7-Dihydro-5(H)-cyclopentapyrazine	11	18
5-Methyl-6,7-dihydro-5(H)-cyclopenta-pyrazine	11	18
2(3),5-Dimethyl-6,7-dihydro-5(H)-cyclopenta-pyrazine	11	18
5,6,7,8-Tetrahydroquinoxaline	11	18
2-Methyl-5,6,7,8-tetrahydroquinoxaline	11	18
Vinylpyrazine	11	18
2-Methyl-6-vinylpyrazine	11	18
Isopropenylpyrazine	11	18
Acetylpyrazine	11	18
2-Methyl-5-acetylpyrazine	11	18
2-Ethyl-5-acetylpyrazine	11	18
1-Pyrazinyl-2-propanone	11	18

*Volatile compounds were isolated from the following sources:

1. Ox meat during boiling for meat extract.
2. Canned beef in broth.
3. Lyophilized lean muscle extract.
4. Boiled beef in broth.
5. Enzyme inactivated beef in cans.
6. Ground beef heated in water or fat.
7. Boiled beef.
8. Enzyme inactivated beef.
9. Boiled beef in slurry.
10. Pressure cooked beef (163° and 182°C) in slurry.
11. Pressure cooked beef (162.7°C) in slurry.
12. Roast beef heated to 163°C.
13. Roast beef drippings.
14. Shallow fried beef.

TABLE 2

Volatile Compounds from Processed Pork and Ham

Compound	Meat sources and treatment*	References
Acids		
Formic	2	20
Acetic	1,2	19,20
Propionic	2	20
Butyric	2	20
Isocapric	2	20
Aldehydes		
Formaldehyde	2	20,22
Acetaldehyde	1,2,3,4	6,19,20,21,22
Propanal	1,2,3,4	6,19,20,21,22
Butanal	1,2,3	19,21,22,23
Pentanal	1,2	19,20,22
Hexanal	1,2,4	6,19,22
Heptanal	1,2	19,22
Octanal	1,2,4	6,19,22
Nonanal	1,2,4	6,19,22
Decanal	2	22
Undecanal	2,4	6,22
Dodecanal	1,2	19,22
Tridecanal	1	19
Tetradecanal	1	19
Pentadecanal	1	19
Hexadecanal	1	19
Heptadecanal	1	19
Octadecanal	1	19
Isobutyraldehyde	1,2,3	19,20,21
Isovaleraldehyde	1,2,3	19,20,21
2-Methylbutanal	1,2,3	19,20,21
2-Butenal	1	19
2-Methyl-2-butenal	1	19
2-Hexenal	2	22
2-Heptenal	1,2,4	6,19,22
2-Octenal	1,2,4	6,19,22
2-Nonenal	2,4	6,22
2-Decenal	2,4	6,22
2-Undecenal	2,4	6,22
2-Dodecenal	2	22
2,4-Heptadienal	2,4	6,22
2,4-Nonadienal	2,4	6,22
2,4-Decadienal	2,4	6,22
2,4-Undecadienal	2	22
2,4-Dodecadienal	2	22
2-Phenyl-2-butenal	1	19
5-Methyl-2-phenyl-2-hexenal	1	19
Benzaldehyde	1	19
Phenylacetaldehyde	1	19
Ketones		
Acetone	1,2,3,4	6,19,20,21
2-Butanone	1,2	19,20
Diacetyl	2	20
2-Heptanone	1	19
2-Octanone	1	19
2-Decanone	1	19

TABLE 2 (continued)

Volatile Compounds from Processed Pork and Ham

Compound	Meat sources and treatment*	References
Ketones (cont.)		
2-Tridecanone	1	19
2-Heptadecanone	1	19
3-Pentanone	1	19
3-Octanone	1	19
4-Methyl-2-pentanone	1	19
3-Buten-2-one	1	19
3-Methyl-3-buten-2-one	1	19
3-Penten-2-one	1	19
2,3-Butanedione	1	19
2,3-Pentanedione	1	19
Cyclopentanone	1	19
2-Cyclopenten-1-one	1	19
1-Phenyl-2-propanone	1	19
1-Hydroxy-2-propanone	1	19
3-Hydroxy-2-butanone	1	19
3-Hydroxy-2-pentanone	1	19
2-Hydroxy-3-pentanone	1	19
Esters		
Methyl formate	1	19
Ethyl formate	1	19
Ethyl acetate	1	19
Pentyl acetate	1	19
Acetol acetate	1	19
Ethyl propionate	1	19
Ethyl isovalerate	1	19
Ethyl dodecanoate	1	19
Ethyl tetradecanoate	1	19
Methyl hexadecanoate	1	19
Alcohols		
Methanol	1	19
Ethanol	1	19
2-Propanol	1	19
Butanol	1	19
2-Methyl-1-propanol	1	19
2-Butanol	1	19
3-Methylbutanol	1	19
2-Methyl-2-butanol	1	19
2-Methyl-3-buten-2-ol	1	19
3-Octenol	1	19
Hydrocarbons		
2-Pentene	1	19
Heptane	1	19
Pentadecene	1	19
Pentadecane	1	19
Hexadecene	1	19
Hexadecane	1	19
Heptadecane	1	19
Benzene	1	19
Ethylbenzene	1	19
Hexylbenzene	1	19
Toluene	1	19
m-Xylene	1	19
Limonene	1	19

TABLE 2 (continued)

Volatile Compounds from Processed Pork and Ham

Compound	Meat sources and treatment*	References
Lactones		
δ-Nonalactone	1	19
δ-Decalactone	1	19
Furans		
2-Methylfuran	1	19
2-Ethylfuran	1	19
2-Butylfuran	1	19
2-Pentylfuran	1	19
2-Furaldehyde	1	19
5-Methyl-2-furaldehyde	1	19
Furfuryl alcohol	1	19
2-Acetylfuran	1	19
Propionylfuran	1	19
Furfuryl methyl ketone	1	19
2-Methyltetrahydrofuran-3-one	1	19
Furfuryl formate	1	19
Furfuryl acetate	1	19
Furfuryl propionate	1	19
Furfuryl butyrate	1	19
Furfuryl pentanoate	1	19
Furfuryl hexanoate	1	19
Ethyl furoate	1	19
Ethyl furfuryl ether	1	19
Furfuryl ether	1	19
2,2′-Methylenedifuran	1	19
5-Methyl-2,2′-methylenedifuran	1	19
2(or 3)-Phenylfuran	1	19
Thiazoles		
Thiazole	1	19
4-Methylthiazole	1	19
2-Acetylthiazole	1	19
Thiophenes		
2-Methylthiophene	1	19
2-Thiophenecarboxaldehyde	1	19
3-Thiophenecarboxaldehyde	1	19
3-Methyl-2-thiophenecarboxaldehyde	1	19
5-Methyl-2-thiophenecarboxaldehyde	1	19
2,5-Dimethyl-3-thiophenecarboxaldehyde	1	19
2-Acetylthiophene	1	19
3-Acetylthiophene	1	19
5-Methyl-2-acetylthiophene	1	19
Thiopheneacrolein	1	19
Sulfur compounds		
Hydrogen sulfide	2,3	21
Methanethiol	1,2,3	19,21
Methyl sulfide	1	19
Methyl disulfide	1	19
Methyl sulfone	1	19
Methyl sulfoxide	1	19
Methyl thioacetate	1	19
Methyl thiopropionate	1	19

TABLE 2 (continued)

Volatile Compounds from Processed Pork and Ham

Compound	Meat sources and treatment*	References
Sulfur compounds (cont.)		
Ethyl thiopropionate	1	19
Benzyl methyl sulfide	1	19
Furfuryl methyl sulfide	1	19
Furfuryl methyl disulfide	1	19
Pyrroles		
Pyrrole-2-carboxaldehyde	1	19
5-Methylpyrrole-2-carboxaldehyde	1	19
Acetylpyrrole	1	19
2-Acetylpyrrole	1	19
2-Propionylpyrrole	1	19
Pyrazines		
Pyrazine	1	19
Methylpyrazine	1	19
Ethylpyrazine	1	19
2,3-Dimethylpyrazine	1	19
2,5-Diethylpyrazine	1	19
2,6-Dimethylpyrazine	1	19
2-Ethyl-5-methylpyrazine	1	19
2-Ethyl-6-methylpyrazine	1	19
2,5-Diethylpyrazine	1	19
Trimethylpyrazine	1	19
2-Ethyl-3,5-dimethylpyrazine	1	19
2-Ethyl-3,6-dimethylpyrazine	1	19
5-Ethyl-2,3-dimethylpyrazine	1	19
2,3-Diethyl-5-methylpyrazine	1	19
2,5-Diethyl-3-methylpyrazine	1	19
3,5-Diethyl-2-methylpyrazine	1	19
Triethylpyrazine	1	19
Tetramethylpyrazine	1	19
2-Methyl-3-vinylpyrazine	1	19
2-Methyl-5-vinylpyrazine	1	19
Acetylpyrazine	1	19
2-Methyl-5(or 6)-acetylpyrazine	1	19
2-Ethyl-3-acetylpyrazine	1	19
2-Ethyl-5(or 6)-acetylpyrazine	1	19
6,7-Dihydro-5H-cyclopentapyrazine	1	19
2-Methyl-6,7-dihydro-5H-cyclopentapyrazine	1	19
5-Methyl-6,7-dihydro-5H-cyclopentapyrazine	1	19
2(or 3),5-Dimethyl-6,7-dihydro-5H-cyclopentapyrazine	1	19
Quinoxaline	1	19
2-Methylquinoxaline	1	19
6-Methylquinoxaline	1	19
5,6,7,8-Tetrahydroquinoxaline	1	19
2-Methyl-5,6,7,8-tetrahydroquinoxaline	1	19
(2-Furyl)pyrazine	1	19
2-(2-Furyl)-5(or 6)-methylpyrazine	1	19

TABLE 2 (continued)

Volatile Compounds from Processed Pork and Ham

Compound	Meat sources and treatment*	References
Miscellaneous compounds		
Ethyl vinyl ether	1	19
Diethyl ether	1	19
Acetaldehyde, diethyl acetal	1	19
Phenol	1	19
Trimethyloxazole	1	19
Vanillonitrile	1	19

*Volatile compounds were isolated from the following sources:

1. Pressure cooked (157°C) pork liver.
2. Dry cured ham.
3. Uncured ham.
4. Pork lipids heated at 100°C.

2. Relatively less volatile fraction having a pleasant meaty aroma. Lactic acid and its ammonium salt were the only compounds identified in this fraction.

Of the large number of volatile compounds identified in meat products, Herz and Chang[30] believe that lactones, furan ring compounds that do not contain sulfur, and aliphatic sulfur compounds make a direct contribution to the meat flavor profile. They, however, concede that none of the representatives of the above classes of compounds isolated from cooked meat by their group[3,23,30] has a characteristic meaty aroma. On the other hand, these researchers have reported the presence of four volatile compounds in cooked beef that have a meaty aroma.[3,23] These compounds include: 5-thiomethylfurfural, thiophencarboxy-2-aldehyde, 2,5-dimethyl-1,3,4-trithiolane, and 2,4,5-trimethyl-3-oxazoline.

Watanabe and Sato[26] reported that in shallow fried beef, the most likely compounds responsible for the typical roasted flavor are alkyl-substituted pyrazines and pyridines. This deduction was based on their observation that the removal of basic compounds from the total flavor extract in diethyl ether with hydrochloric acid solution resulted in a considerable loss of the roasted flavor. In a subsequent study with shallow fried beef, Watanabe and Sato[25] isolated a volatile fraction with a typical heated beef flavor and observed that no single compound isolated in this fraction had a typical heated beef flavor. It was suggested that the heated beef flavor might be a complex sensation resulting from a mixture of several different compounds: methional, 2-acetylfuran, 2-furfuryl methyl ketone, 1-methyl-2-acetylpyrrole, benzothiazole, and others.

It appears likely that the question of the nature of compounds responsible for the characteristic cooked meat flavor would remain unresolved for some time to come. It may, however, be pointed out that most of the odorous compounds present in cooked meat are likely to play a significant role in determining its flavor character. It is also common knowledge that odorous compounds, when presented in a mixture, often give an odor sensation somewhat different from the one predicted on the basis of odor characteristics of the individual components in the mixture. It may, therefore, be futile to look for the "characteristic" compounds responsible for the typical meat flavor. Most likely, the subtle quantitative balance of the various odorous components in cooked meat determines the aroma character of meat. For example, the aromas of cooked beef and cooked pork liver are easily distinguishable, although a cursory look at the volatile compounds identified in them (Tables 1 and 2) shows a striking similarity.

Unlike the cooked meat aroma produced from their precursors during heat treatment, the characteristic aroma of dry or fermented sausages is related, in part, to the hydrolytic and oxidative changes occurring in the lipid fraction during ripening.[31,32] The bacterial,[33-35] muscle, and adipose tissue[36] lipases liberate fatty acids from the sausage lipids. The unsaturated fatty acids thus liberated undergo nonenzymatic oxidation and produce lipid peroxides and carbonyl compounds.[31,37] The lipid peroxides are probably further metabolized by the sausage microflora to carbonyl compounds and fatty acids.[38,39] Demeyer et al.[40] have recently investigated the specificity of lipolysis during dry sausage ripening and report that unsaturated fatty acids are selectively liberated by the lipases from the sausage lipids. In this study, linoleic acid was liberated into the free fatty acid fraction at a much faster rate than all other acids. The rate of lipolysis decreased in the order: linoleic > oleic > stearic > palmitic acid. Since pork fat is known to have most of the linoleic and oleic acids (50 to 60%) at position 3 of triglyceride molecule,[41] it has been suggested by Demeyer and co-workers[40] that the lipases occurring in sausages have specificity for position 3 of triglycerides. The difference observed for linoleic and oleic acids may be related to specificity of the enzyme for the fatty acid structure, as both positional and structural specificity is known to occur in microbial lipases.[41] Since only unsaturated fatty acids undergo nonenzymatic oxidation and produce lipid peroxides and carbonyl compounds, the specificity of microbial lipases for position 3 of triglycerides, which in the case of pork lipids contains a greater percentage of unsaturated fatty acids than positions 1 and 2 of triglyceride molecule, appears to be an important factor in the development of ripened sausage flavor. Micrococci[32,33] and to a lesser degree lactobacilli[42,43] have been reported to be responsible for lipolysis in dry sausage.

During sausage ripening, the carbonyl concentration of the product increases severalfold. In a study carried out by Demeyer et al.,[40] carbonyl compounds increased in the sausage during the first week of ripening, decreased after smoking, and again increased to final values of about 300 μmol/100 g of dry matter. The initial increase was attributed to compounds formed in carbohydrate fermentation,[44] while the increase in the last stages might have been due to further metabolism of lipid peroxides.[39]

The volatile fatty acids produced by the microorganisms during sausage ripening are very important in the dry sausage aroma. Significant concentrations of acetic, propionic, and butyric acids have been reported in fermented sausages, along with large amounts of nonvolatile lactic acid which affects the taste and reduces the pH of sausage from an average of 5.8 to 4.8.[44] The homo- and hetero-fermentative metabolic pathways, which may be involved in the formation of volatile fatty acids and carbonyl compounds by the microorganisms during dry sausage ripening, were discussed in detail in the earlier review.[2] Among the volatile fatty acids mentioned above, acetic acid may be formed from aspartic acid, alanine, serine, glycine, cystine, threonine, and glutamic acid by the bacterial metabolism. Propionic acid may originate from alanine and serine. Butyric acid may be formed from threonine and glutamic acid. These fatty acids may also be synthesized from fats and carbohydrates as shown below:

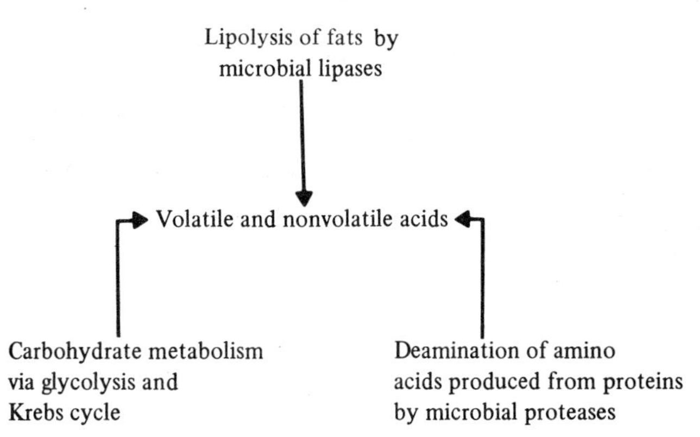

Nonvolatile Flavor Compounds and Flavor Potentiators

The overall composition of meat extracts and the possible taste effect of various compounds identified have been extensively investigated.[45-62] Inosine-5′-monophosphate (IMP) is a major component of meat extracts. It is considered a key flavor component responsible for the strong meaty taste.[58,63,64] Other nonvolatile components with taste properties include amino acids, peptides, and other nucleotides. The high content of IMP in meat extracts is due to the presence of adenosine-5′-triphosphate (ATP) as a major mononucleotide in the muscle of living animals. After slaughter there is a rapid transformation of this nucleotide to adenosine-5′-monophosphate (AMP), which is then deaminated to IMP.[64] A partial list of compounds isolated from meat extracts is presented in Table 3. Several of the compounds listed in this table act as flavor precursors. The meat flavor precursors are discussed in detail later in this article. The contribution of peptides and amino acids to the taste of foodstuffs has been investigated by Kirimura and co-workers.[65] The organoleptic characteristics of amino acids reported by these workers are reproduced in Table 4.

TABLE 3

Potential Flavor Precursors Isolated from Meat Extracts

Alanine	Fructose-6′-phosphate	Lysine
β-Alanine	Glucose	Methionine
quaternary Amines	Glucose-6′-phosphate	Methylhistidine
Ammonia	Glutamic acid	Nicotinamide-adeninedinucleotide
Anserine	Glutamine	Ornithine
Arginine	Glutathione	Peptides
Asparagine	Glycerophosphoethanolamine	Phenylalanine
Aspartic acid	Glycine	Phosphoethanolamine
Carnitine	Glycoproteins	Phosphoserine
Carnosine	Histidine	Proline
Citrulline	Hydroxyproline	Purine-nucleosides
Creatine	Hypoxanthine	Purine-nucleotides
Creatinine	Inosine-5′-monophosphate	Ribose
Cysteine	and other nucleotides	Ribose-5′-phosphate
Cystine	Isoleucine	Serine
Fructose	Leucine	Taurine
		Threonine
		Tyrosine
		Urea
		Valine

TABLE 4

Organoleptic Characteristics of Amino Acids

	Relative taste intensities[a]				
Amino acid	Sweet	Salty	Sour	Bitter	MSG-like
Sweet amino acids					
Hydroxyproline	***			**	
Lysine·HCl	**			**	
Alanine	***				
Glycine	***				
Serine	***		*		*
Glutamine	*				*
Threonine	***		*	*	
Proline	***			***	
Sour and MSG-like amino acids					
Aspartic acid			***		*
Glutamic acid			***		**
Histidine·HCl			***		
Asparagine			**	*	
Monosodium glutamate(MSG)	*	*			***
Na-aspartatate		**			**

TABLE 4 (continued)

Organoleptic Characteristics of Amino Acids

Amino acid	Relative taste intensities[a]				
	Sweet	Salty	Sour	Bitter	MSG-like
Bitter amino acids					
Histidine				**	
Arginine·HCl				***	
Methionine				***	*
Valine				***	
Arginine				***	
Isoleucine				***	
Phenylalanine				***	
Tryptophan				***	
Leucine				***	

[a]Asterisks denote relative taste intensities.

From Kirimura, J., Shimizu, A., Kimizuka, A., Ninomiya, T., and Katsuya, N., *J. Agric. Food Chem.*, 17, 689, 1969. With permission.

ORIGIN OF MEAT FLAVOR

Patents on Synthetic Meat Flavors

The importance of the flavor precursors, listed in Table 3, in the formation of meat flavor may be easily recognized by the fact that a large number of patents on meat flavoring compositions, which are based on the heat treatment of flavor precursors, have been awarded in the last few years. Some of the patents are based on the Maillard reactions between sugars and amino compounds. Other make use of the strong flavor potentiation of nucleotide derivatives, along with sugar-amine interactions. A sample of the important patents issued on synthetic meat flavors, in the last few years, is summarized below:

A patent issued to Unilever[66] in 1966 details a method of preparing flavoring compounds by heating a mixture of carbohydrates containing at least one amino acid in the presence of a fatty acid. The carbohydrate may be a pentose or a hextose or a polysaccharide, or a mixture of these. There may be a single amino acid or a mixture of two or more amino acids. For example, a mixture of 480 g wheat gluten hydrolyzate, 56 g cysteine-HCl, 70 g xylose, 50 g oleic acid, and 1,500 g water was heated for 2 hr with stirring in an oil bath at 120°C, then cooled to 50°C, and a slurry of 400 g cornmeal in 600 g water was added with vigorous stirring. A yellowish brown powder was obtained that had the flavor and aroma of roast mutton when dispersed in a small quantity of water.

A similar Unilever patent[67] deals with mixtures of amino acids including glutamic acid, nucleotides (preferably GMP and IMP), succinic acid, and hydroxycarboxylic acids such as lactic acid in a defined ratio to have more pronounced meaty flavor.

Another Unilever patent[68] is concerned with the meat flavorings containing at least one of the optical antipodes of pyrrolidonecarboxylic acid. In flavorings also containing other additives, the preferred weight ratios were: pyrrolidonecarboxylic acid/glutamine/succinic acid/5′-ribonucleotides = 1/(0–25)/(0–1)/(0–5).

In 1970, Van der Ouweland and Peer[69] were awarded a patent, assigned to Unilever, on a synthetic meat flavor utilizing the capability of certain heterocyclic ketones such as 4-hydroxy-5-methyl-2,3-dihydrofuran-3-one, to react with inorganic or organic sulfur compounds, and to form compounds with an aroma similar to that of cooked or roast meat.

A recent patent awarded to Brinkman and Van der Heyden,[70] assigned to Lever Brothers Co., claims that the addition of thiaalkanethiols, where two aliphatic groups are methyl- or ethyl-radical, or their salts, in an amount such that the food contains 0.1 to 1,000 mg compound/1,000 kg product results in a rich meat flavor. 2-Thiabutane- 3-thiol was cited as an example of thiaalkanethiol in the patent.

Kyowa Fermentation Industry was awarded a patent on meat flavoring composition[71] prepared by applying the Maillard reaction to ribose phosphate and an amino acid, such as lysine, hydrolyzed protein, or a casein hydrolyzate at 150 to 180°C.

Kitada et al.[72] were awarded a patent, assigned to Ajinomoto Co., on a Maillard reaction product having a meat-like flavor. Meat flavor was produced by heating a solution containing cystine and/or cysteine and a reducing sugar at 50 to 120°C at a pressure of 70 to 200 kg/cm^2.

Another Ajinomoto patent[73] discloses the synthetic compounds which enhance the meat flavor. Such compounds were obtained by mixing amino acids, organic acids, phosphates, 5′-nucleotides, and analogous compounds in water, adjusting pH of the solution to about 5 and then cooking it for a few minutes.

International Flavors & Fragrances patented a flavoring material[74] with a strong meat flavor by reacting thiamine or 3-acetyl-3- mercaptopropanol-1 in the presence of an organic acid.

Giacino patent,[75] assigned to International Flavors & Fragrances, discloses the preparation of meat-like flavors by heating a mixture comprised of taurine and thiamine and then adding it to a source of free amino acids.

Thomas patent,[76] assigned to Pfizer Inc., discloses synthetic ham- and bacon-flavored compositions prepared by heating an aqueous mixture of hexose or pentose, cystine or cysteine, and glycine. After cooling the mixture, monosodium glutamate, protein hydrolyzate, sucrose, edible fats, ribonucleotide, and hickory smoke flavor are incorporated to obtain the desired flavor.

Broderick and Marcus patent,[77] assigned to Kohnstamm and Co., discloses a procedure of preparing artificial flavoring substances with the aroma and taste of cooked meat by causing a ribose moiety-containing ribonucleotide to react at elevated temperature with cystine or cysteine in the presence of water.

The use of thiamine as one of the ingredients used in the preparation of synthetic meat flavors is not surprising. Thiamine, like cystine and cysteine, on heating produces a number of sulfur compounds that contribute significantly in meat-like synthetic flavor formulations. It should be noted here that meat flavor being a very characteristic and complex flavor has not yet lent itself to synthetic duplication. The patents on synthetic meat flavors discussed above and the others may be regarded as extenders of meat flavor, rather than the true meat flavor. However, the synthetic preparation of the characteristic flavors of different cuts of meat and species flavor, such as pork, lamb, and beef flavor, are within the realm of being feasible.

The Role of Enzymes in the Development of Flavor Precursors

The nature and concentration of meat flavor precursors depends, to a large extent, on the metabolic activities in the animal body and the post-mortem storage conditions. The biochemical reactions in a living system are catalyzed by enzymes. The nature and concentration of the enzymes found in different parts of animal body are directly responsible for the accumulation or depletion of metabolites, some of which act as flavor precursors, in muscle. The enzymatic reactions as related to the development of flavor precursors have not been studied to any significant extent. However, a considerable amount of indirect evidence is available. This evidence is presented under the headings: ante-mortem and post-mortem factors.

Ante-mortem Factors

The following observations suggest the importance of enzymes in the development of flavor precursors:

Age of the animal at the time of slaughter is related to cooked meat flavor. Barbella et al.[78] and more recently Lowe and Kastelic[79] reported that young calves up to about 11 months lack the typical flavor quality and intensity of beef. The change in metabolism with age is well known and it is an important factor in the development of a normal meat flavor. The change in the composition of muscle with age is another important factor. It is safe to assume that if flavor is related to the concentration of metabolites in muscle, an increase in intensity and a change in profile up to an age of a little metabolic change should be expected. A similar observation was made by Crocker[80] who reported that odor of cooked meat generally increases with age of the animal, as the distinctive taste of beef and insipid taste of veal testify. It is well known that myoglobin content of muscle increases with age. However, this increase is not likely to influence the flavor. It is more likely that this difference may be due to an increase in the concentration of certain amino acids; for example, Gruhm[81] has reported that the relative concentration of valine, methionine, leucine, and isoleucine increases with age, or to some constituents of intramuscular fat which alter with increasing maturity.

An interesting study made on 10 pairs of bull and steer twins by Bryce-Jones et al.[82] showed that steers were definitely more tender and also better than bulls in juiciness and flavor.

The effect of breed on flavor is less widely recognized, but scientific studies have revealed that this factor certainly has an influence. European-type cattle give a better flavored meat than do Brahmin cattle,[83] though it is not clear to what extent this difference may be affected by the undoubted toughness of the latter. Again, the introduction of Hampshire blood into a large white pig stock has been reported to favor an enhanced tendency to rancidity and thus to an off flavor development.

The sex of animal also influences the flavor of meat to some extent. According to Weir[84] mature males have a "staggy" odor. It is probably in regard to off flavor that the influence of the sex of an animal is most clearly noted. Thus, male pigs at maturity are frequently associated with an unpleasant "boar odor" on cooking. Patterson[85,86] has recently shown that the substance responsible is probably 5-androst-16-ene-3-one and, what is particularly interesting, this is closely related to a corresponding alcohol in the salivary gland of mature boar which acts as a stimulant to female at the time of mating.

Differences in meat flavor arising from the type of joint are widely recognized. Thus, the rump and topside yield roasts with markedly greater flavor though also greater toughness than those from the sirloin.[54]

Electric shock treatment during the 24 hr preceding slaughter at 20 min intervals invariably increases the pH of cooked steaks from Angus or Hereford steers, compared with control steaks from paired animals receiving no shock treatment.[87]

Feeding of antibiotics or tranquilizers could affect meat flavor. The effect on rumen or intestinal flora may be important in the case of antibiotics, because the digestion of ruminants relies on the action of bacteria, and the fatty acids produced by these microorganisms can modify the lipids of the animal. Harris et al.[88] reported that meat from conventionally raised chicken had a more characteristic flavor than meat from the chicken born and raised in a germfree environment.

A relatively high pH results from injecting adrenaline just ante-mortem.[89] Ante-mortem feeding of sucrose, on the other hand, lowers pH and increases lactate content of pork from pigs that had been transported 340 mi in 24 hr prior to slaughtering; the same effect was not noticed in animals transported only 3.1 mi.[90]

Proteolytic enzymes have been successfully injected ante-mortem into the bloodstream in order to tenderize the meat. Although no qualitative differences in lower boiling aroma volatiles of beef steak from the animal injected with papain were observed, some differences in quantities of the individual compounds were noticed.[91]

Biochemical state of the muscle is also important. Post-mortem all muscles convert their stores of glycogen into lactic acid causing the pH to fall to an ultimate value of about 5.5. If glycogen store is lower than 0.6% of the muscle weight, the extent of this post-mortem glycolysis results in an ultimate pH value between 5.5 and 7. A relationship between the ultimate pH of muscle and its flavor has been reported.[92]

Let us now examine how the flavor precursors are produced in an animal body. It is quite clear from the patent literature cited earlier and other reported findings that meat flavor precursors include amino acids, small peptides, carbohydrates, particularly hexoses and pentoses, and nucleotides (Table 3). As illustrated in Figure 1, amino acids found in blood and tissues may result from their absorption from intestines, breakdown of tissue proteins, and synthesis of amino acids, particularly in the liver. These amino acids are used by the animal body in a number of ways. The major pathways and products of amino acid metabolism are given in Table 5. Considering various metabolic functions of different animal tissues, it appears likely that most of the amino acids found in meat are produced by the breakdown of tissue proteins. In that case,

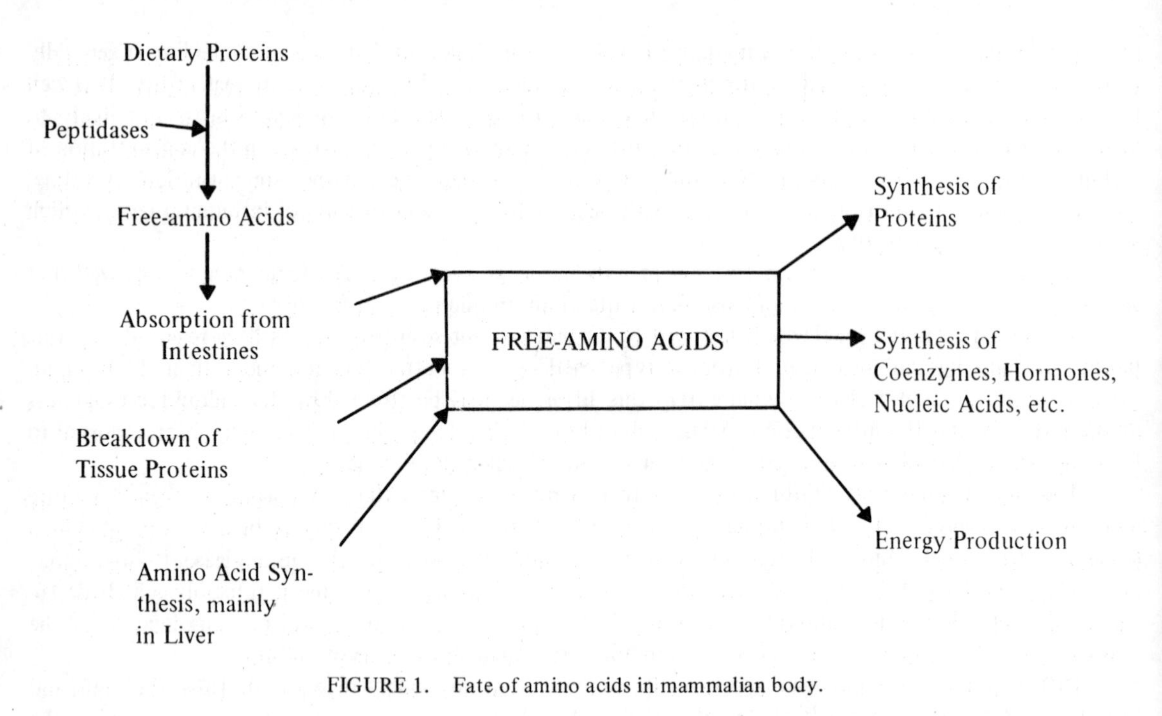

FIGURE 1. Fate of amino acids in mammalian body.

the nature and concentration of amino acids in the muscle would depend on the genetic make up and physiological state of the animal at the time of slaughter. Post-mortem, tissue proteins continue to degrade and release amino acids by the action of native muscle and microbial proteases.

Other metabolic pathways which may have a role in the development of meat flavor precursors are given in Figures 2 to 8.

Figure 2 refers to the mammalian fatty acid metabolism. The compounds isolated from meat extracts, which may originate from one or more of the metabolic schemes described, are acetoin, diacetyl, formic acid, acetic acid, hydroxybutyric acid, and succinic acid. It is widely recognized that the species flavor of meat develops from adipose tissues. The depot fat and water soluble constituents of adipose tissues probably interact to produce the species flavor. The nature of depot fat is governed by the genetic make up and diet of the animal. The various pathways involving fatty acid, glycerolipid, sphingolipid, and sterol metabolism, which are called upon by the living system and influence the make up of depot fat, are outlined in Figures 2 to 5. The role of depot lipids in the development of meat flavor is discussed later in this review.

The carbohydrates present in raw meat interact with amino compounds and produce volatile flavor compounds via Maillard reactions. These reactions are discussed in detail in a later section in this article. Sugars and sugar phosphates identified from meat extracts include: fructose, fructose-6′-phosphate, fructose-1,6′-diphosphate, glucose, glucose-6′-phosphate, ribose, and ribose-5′-phosphate (Table 3). The pathways of carbohydrate metabolism and interrelationship of sugars and sugar phosphates discussed above, are shown in Figure 6. These metabolic schemes also depict the synthesis of lactic acid from glycogen of the muscle. The amount of lactic acid present in meat is important in its characteristic taste.

Figure 7 gives the metabolic interrelationship of lipids, carbohydrates, and proteins. These metabolic schemes show how a metabolite such as acetate may originate from proteins, lipids, or carbohydrates.

Figure 8 details pathways of nucleoprotein catabolism. Nucleoproteins are composed of basic proteins (histones or protamines) associated with nucleic acids in a salt linkage. The reactions presented are for ribonucleic acid (RNA). A parallel series occurs with deoxyribonucleic acid (DNA). The breakdown products of nucleic acids, produced enzymatically post-mortem, are very important in meat flavor, particularly inosine 5′-monophosphate. The specific reactions involved in the formation of this important nucleotide are discussed in detail in a later section.

TABLE 5

Pathways of Amino Acid Metabolism

Amino acid	Product of oxidative deamination or transamination	Product of decarboxylation	Pathways and products of metabolism
1. L-Alanine	Pyruvic acid		
2. L-Arginine	α-Keto-δ-guanidovaleric acid	Agmatine	Arginine → ornithine + urea; arginine → citrulline + NH_3; arginine + glycine ↔ guanidoacetic acid + ornithine
3. L-Asparagine	α-Ketosuccinamic acid		Asparagine ⇌ aspartic acid + NH_3; α-ketosuccinamic acid → NH_3 + oxaloacetic acid
4. L-Aspartic acid	Oxaloacetic acid	α-Alanine	Aspartic acid + carbamylphosphate → P_i + carbamylaspartic acid →→→ pyrimidines; aspartic acid ↔ fumaric acid + NH_3; aspartic acid → aspartic semialdehyde → homoserine → (i) threonine, (ii) methionine, or (iii) lysine; aspartic acid → nitrogen of purine ring (*see* purine metabolism); aspartic acid + IMP → adenylosuccinate → AMP + fumarate, *See also* asparagine.
5. L-Citrulline	α-Keto-δ-carbamidovaleric acid		Citrulline + aspartic acid → argininosuccinic acid ↔ arginine + fumaric acid; citrulline + P_i ↔ ornithine + carbamylphosphate; carbamylphosphate + ADP ↔ CO_2 + NH_3 + ATP
6. L-Cysteine and L-cystine	β-Mercaptopyruvic acid		Cysteine → H_2S + NH_3 + pyruvic acid; β-mercaptopyruvic acid → pyruvic acid + S; cysteine → cysteine sulfinic acid* → (i) cysteic acid → CO_2 + taurine, (ii) CO_2 + hypotaurine, or (iii) via transamination → β-sulfinylpyruvate → pyruvate + SO_3^{2-}; 2 cysteine ↔ cystine. *See also* methionine for biosynthesis of cysteine.
7. L-Glutamic acid	α-Ketoglutaric acid	γ-Aminobutyric acid	Glutamic acid ↔ glutamic semialdehyde ↔ (i) ornithine or (ii) → Δ′-pyrroline-5-carboxylate ↔ proline; glutamic acid → *N*-acetylglutamic acid ↔ *N*-acetylglutamic semialdehyde ↔ *N*-acetylornithine → ornithine. *See also* glutamine, histidine.
8. L-Glutamine	α-Ketoglutaramic acid		α-Ketoglutaramic acid → NH_3 + α-ketoglutarate; glutamine ⇌ NH_3 + glutamic acid; glutamine + 5′-phospho-α-D-ribosylpyrophosphate → glutamate + 5′-phospho-β-D-ribosylamine →→ 5′-phosphoribosyl-*N*-formylglycineamide + glutamine → glutamate + 5′-phosphoribosyl-*N*-formylglycineamidine →→→ purines; glutamine + xanthylic acid → guanylic acid + glutamate; glutamine + fructose-6-phosphate → glucosamine-6-phosphate + glutamate. *See also* histidine.

TABLE 5 (continued)

Pathways of Amino Acid Metabolism

Amino acid	Product of oxidative deamination or transamination	Product of decarboxylation	Pathways and products of metabolism
9. Glycine	Glyoxylic acid	5,10-Methylene tetrahydrofolate + NH_3 + 2H	Glyoxylate → formate + CO_2; glycine + 5,10-methylene tetrahydrofolate ↔ serine + tetrahydrofolate; glycine + succinyl CoA → δ-aminolevulinate →→→ porphyrins; glycine + 5′-phospho-β-D-ribosylamine → 5-phosphoribosyl-glycineamide →→ purines; glycine → glycocholic acid; glycine → hippuric acid. *See also* serine, threonine, arginine.
10. L-Histidine	β-Imidazolepyruvic acid	Histamine	Histidine → NH_3 + urocanic acid → imidazolone propionate → *N*-formimino-glutamate → glutamate + 5-formimino tetrahydrofolate; histidine → (i) anserine, or (ii) carnosine; histamine → imidazoleacetic acid → (i) imidazole-acetic acid ribonucleoside or (ii) NH_3 + formylaspartic acid; histamine → methylhistamine; ATP + 5-phosphoribosyl-1-phosphate → phosphoribosyl-ATP → phosphoribosyl-AMP + glutamine → 5-amino-1-ribosyl-4-imidazolecarbox-amide-5′-phosphate + imidazoleglycerol phosphate → histidinol phosphate → histidinol → histidine.
11. Hydroxyproline	α-Keto-2-hydroxy-δ-aminovaleric acid		Hydroxyproline → Δ′-pyrroline-3-hydroxy-5-carboxylate → γ-hydroxyglutamate ↔ α-keto-2-hydroxyglutarate ↔ pyruvate + glyoxylate. *See also* proline.
12. L-Isoleucine	α-Keto-β-methylvaleric acid	2-Methylbutylamine	α-Keto-β-methylvaleric acid → CO_2 + α-methylbutyryl CoA ↔ tiglyl CoA ↔ α-methyl-β-hydroxybutyryl CoA ↔ α-methylacetoacetyl CoA ↔ acetyl CoA + propionyl CoA; pyruvate + α-ketobutyrate (*see* threonine) → CO_2 + α-aceto-α-hydroxybutyrate → α-β-dihydroxy-β-methylvalerate ↔ α-keto-β-methylvalerate ↔ isoleucine.
13. L-Leucine	α-Ketoisocaproic acid	3-Methylbutylamine	α-Ketoisocaproic acid → CO_2 + isovaleryl CoA ↔ β-methylcrotonyl CoA + CO_2 ↔ β-methylglutaconyl CoA ↔ β-hydroxy-β-methylglutaryl CoA → (i) acetoacetic acid + acetyl CoA, or (ii) →→→ cholesterol; α-ketoisovaleric acid (*see* valine) + acetyl CoA → β-carboxy-β-hydroxyisocaproic acid → α-hydroxy-β-carboxyiso-caproic acid → α-keto-β-carboxyisocaproic acid ↔ α-ketoisocaproic acid ↔ leucine.
14. L-Lysine	α-Keto-ε-aminocaproic acid	Cadaverine	α-Keto-ε-aminocaproic acid → Δ′-dehydropipecolate → L-pipecolate → Δ^6-dehydropipecolate → α-aminoadipate semialdehyde ↔ α-aminoadipic acid ↔ α-ketoadipic acid → glutaryl CoA → glutaconyl CoA → crotonyl CoA (fatty acid degradation); aspartic semialdehyde + pyruvate →→→ diaminopimelic acid → lysine; α-ketoglutarate + acetyl CoA →→→→ α-ketoadipic acid →

TABLE 5 (continued)

Pathways of Amino Acid Metabolism

Amino acid	Product of oxidative deamination or transamination	Product of decarboxylation	Pathways and products of metabolism
15. L-Methionine	α-Keto-γ-methiolbutyric acid		Methionine + ATP → PP_i + P_i + *S*-adenosylmethionine (active methyl donor); *S*-adenosylmethionine + guanidoacetate (*see* arginine) → creatine + *S*-adenosylhomocysteine; *S*-adenosylhomocysteine → adenosine + homocysteine; homocysteine + serine → cystathionine → cysteine + homoserine; homoserine → α-ketobutyrate + NH_3; α-ketobutyrate → CO_2 + propionyl CoA → methylmalonyl CoA → succinyl CoA; homoserine (*see* aspartic acid) + cysteine → cystathionine → serine + homocysteine; homocysteine + one C donor → methionine; homocysteine → (i) homocysteic acid or (ii) NH_3 + H_2S + α-ketobutyrate.
16. L-Ornithine	Glutamic semialdehyde, or α-keto-δ-aminovaleric acid	Putrescine	Ornithine ↔ proline; ornithine ↔ glutamic semialdehyde ↔ glutamate; ornithine + carbamylphosphate → citrulline. *See also* citrulline, glutamic acid.
17. L-Phenylalanine	Phenylpyruvic acid	Phenylethylamine	Phenylalanine → tyrosine; phenylpyruvic acid → (i) phenyllactic acid or (ii) CO_2 + phenylacetic acid; phosphoenolpyruvate + erythrose-4-phosphate → 3-deoxy-D-arabino-heptulosonic acid-7-phosphate →→→ 5-dehydroquinate → 5-dehydroshikimate → shikimate → shikimate-5-phosphate →→→ prephenate → (i) phenylpyruvate → phenylalanine or (ii) → *p*-hydroxyphenylpyruvate → tyrosine. *See also* tyrosine.
18. L-Proline	Glutamic semialdehyde or α-keto-δ-aminovaleric acid		Proline ↔ glutamate; proline ↔ ornithine; proline → hydroxyproline. *See also* glutamic acid, ornithine.
19. L-Serine	β-Hydroxypyruvic acid		Serine → pyruvate + NH_3; hydroxypyruvate ↔ D-glycerate ⇌ (i) 2-phosphoglycerate or (ii) 3-phosphoglycerate; 3-phosphoglycerate ↔ β-phosphohydroxypyruvate ↔ phosphoserine → serine; phosphatidyl ethanolamine + serine → ethanolamine + phosphatidyl serine; phosphatidyl serine → CO_2 + phosphatidyl ethanolamine; palmitaldehyde + serine → dihydrosphingosine. *See also* glycine, methionine, tryptophan.
20. L-Threonine	α-Keto-β-hydroxybutyric acid		Threonine → NH_3 + α-ketobutyrate; α-ketobutyrate → CO_2 + propionyl CoA → methylmalonyl CoA → succinyl CoA; threonine → (i) glycine + acetaldehyde or (ii) CO_2 + aminoacetone. *See also* aspartic acid.

TABLE 5 (continued)

Pathways of Amino Acid Metabolism

Amino acid	Product of oxidative deamination or transamination	Product of decarboxylation	Pathways and products of metabolism
21. L-Tryptophan	β-Indolepyruvic acid	Tryptamine	Tryptophan → *N*-formylkynurenine → formate + kynurenine → 3-hydroxykynurenine → alanine + 3-hydroxyanthranilate → α-amino-β-carboxymuconate semialdehyde → CO_2 + α-aminomuconate semialdehyde → α-aminomuconate → α-ketoadipate (*see* lysine for further degradation); tryptophan → 5-hydroxytryptophan → CO_2 + 5-hydroxytryptamine (serotonin) → 5-hydroxyindoleacetic acid; indole-3-glycerol phosphate + serine → tryptophan.
22. L-Tyrosine	*p*-Hydroxyphenylpyruvic acid	Tyramine	*p*-Hydroxyphenylpyruvic acid → CO_2 + homogentisic acid → maleylacetoacetic acid → fumarylacetoacetic acid → fumarate + acetoacetate; tyrosine → 3,4-dihydroxyphenylalanine → CO_2 + 3,4-dihydroxyphenylethylamine → noradrenalin → adrenalin. *See also* phenylalanine.
23. L-Valine	α-Ketoisovaleric acid	2-Methylpropylamine	α-Ketoisovaleric acid → CO_2 + isobutyryl CoA ↔ methacrylyl CoA ↔ β-hydroxybutyryl CoA → β-hydroxyisobutyrate ↔ methylmalonic semialdehyde → (i) β-aminoisobutyric acid or (ii) methylmalonyl CoA (?); 2 pyruvic acid → α-acetolactate → α,β-dihydroxyisovaleric acid → α-ketoisovaleric acid ↔ valine.

*Desulfination pathway for cysteine sulfinic acid → alanine + SO_2. Abbreviations: CoA = coenzyme A; ADP = adenosine diphosphate; AMP = adenosine phosphate; ATP = adenosine triphosphate; IMP = inosine phosphate; P_i = inorganic orthophosphate; PP_i = inorganic pyrophosphate.

From Altman, P. L. and Dittmer, D. S., Eds., in *Metabolism*, Federation of American Societies for Experimental Biology, Bethesda, Maryland, 1968, 426. With permission.

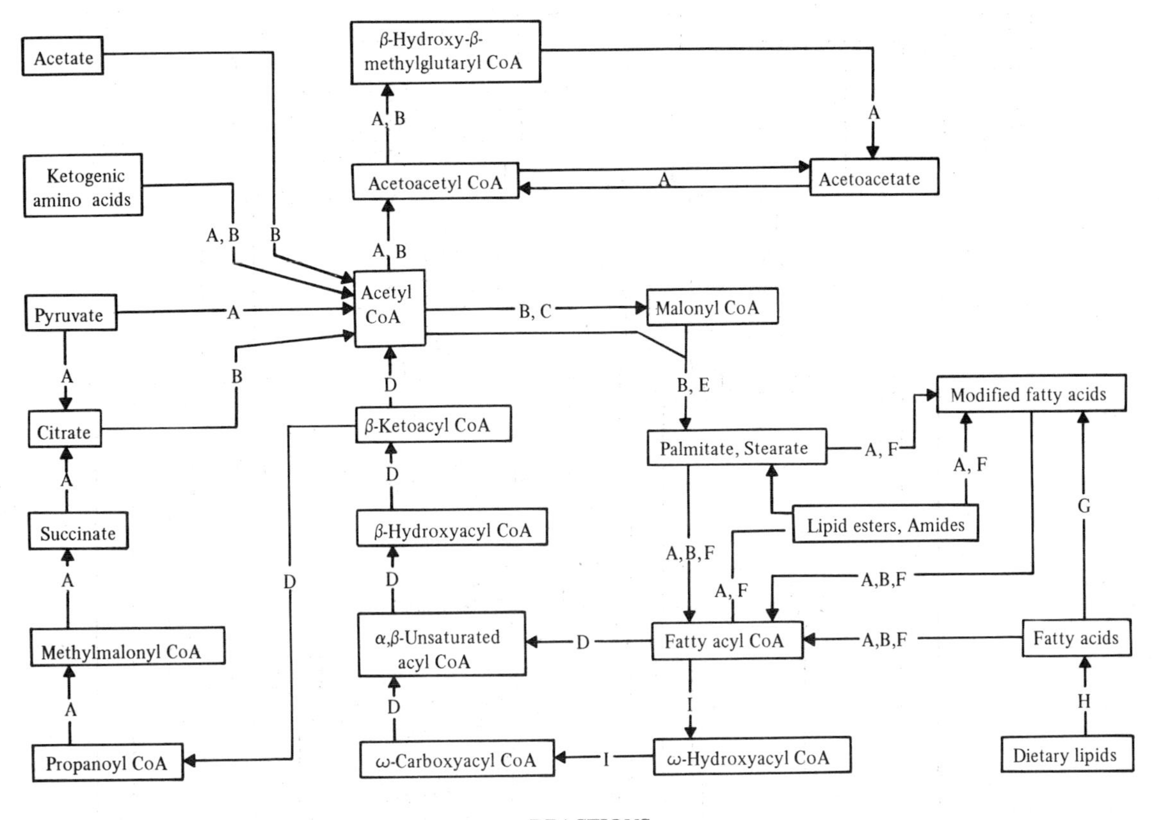

REACTIONS

A: Occurs in mitochondria.
B: Occurs in cytoplasm with soluble enzymes.
C: Requires biotin, ATP, Mg^{2+}, and HCO_3^-; citrate and isocitrate enhance carboxylation of acetyl CoA by complexing with and activating the carboxylase enzyme.
D: β-Oxidation in mitochondria; carnitine esters are involved in transporting fatty acids into mitochondria; requires FAD and NAD^+; ATP or GTP may participate in preliminary extramitochondrial formation of fatty acyl CoA.
E: Involves transacylases, ACP's β-ketoacyl-ACP synthetase, β-ketoacyl-ACP reductase, β-hydroxyacyl-ACP dehydrase, and enoyl-ACP reductase; requires NADPH (furnished mainly by pentose phosphate pathway and by oxidation of oxalacetate to pyruvate via malic acid; NADH required for the latter produced through glycolysis); yields mainly palmitic acid.
F: Occurs in microsomes.
G: Includes chain elongation in microsomes involving malonyl CoA and fatty acyl CoA (analogous to de novo synthesis), chain elongation in mitochondria involving acetyl CoA and fatty acyl CoA, shortening of carbon chain by 2 carbons in microsomes, and desaturation of fatty acyl CoA coupled with chain elongation to yield polyunsaturated acids (requires NADH or NADPH) in microsomes.
H: Involves mainly pancreatic and intestinal lipases and bile salts in intestinal lumen.
I: ω-Oxidation in intestinal lumen.

FIGURE 2. Mammalian fatty acid metabolism. (From Altman, P. L. and Dittmer, D. S., Eds., in *Biological Data Book*, Vol. III, Federation of American Societies for Experimental Biology, Bethesda, Maryland, 1973, 1538. With permission.)

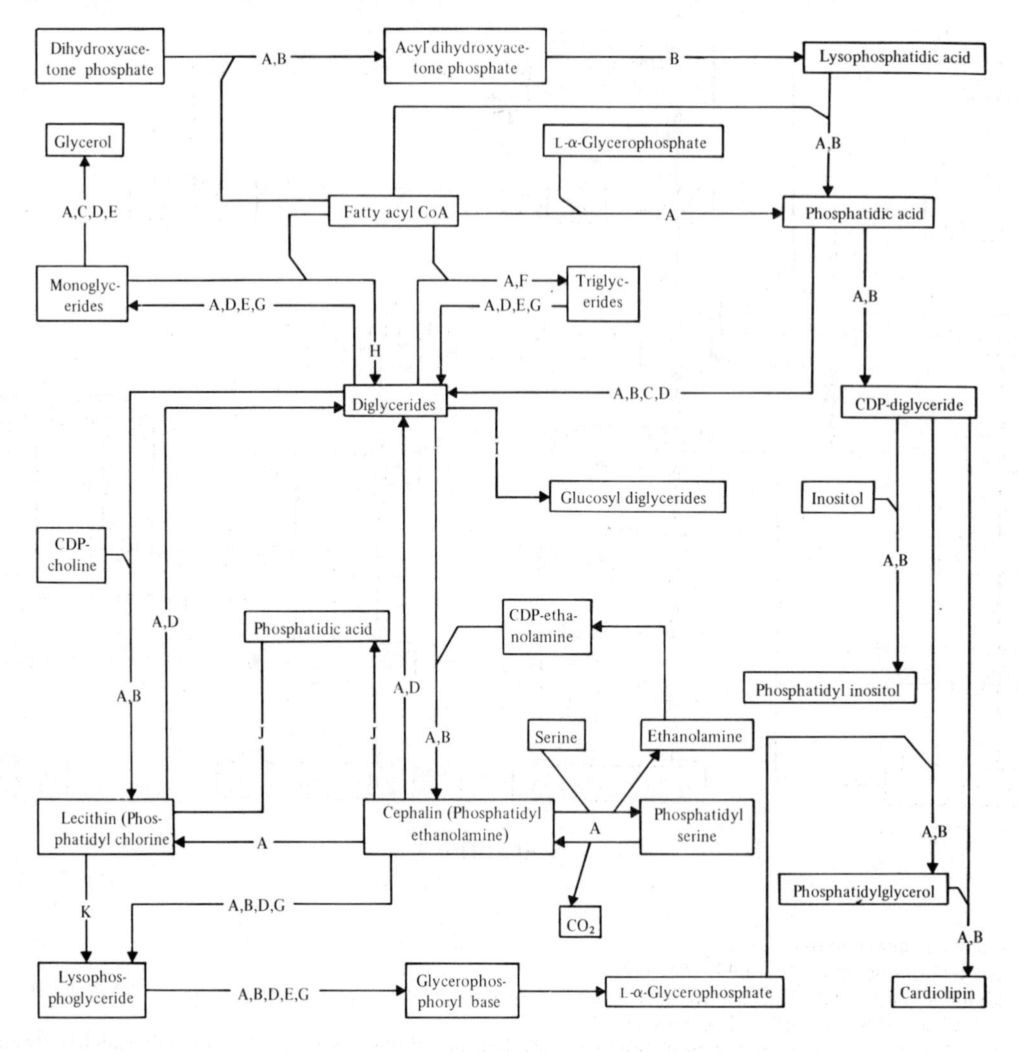

REACTIONS

A: Occurs in microsomes.
B: Occurs in mitochondria.
C: Occurs in cytoplasm with soluble enzymes.
D: Occurs in lysosomes.
E: Occurs in cell membrane.
F: 1,2-Diglycerides, but not 1,3-diglycerides.
G: Occurs in the intestinal lumen.
H: Acylation involves mainly a primary hydroxyl group; occurs in microsomes of the intestinal mucosa and mammary glands.
I: Occurs in microsomes of brain.
J: Occurs in plants.
K: Requires cholesterol as a fatty acid receiver; occurs in microsomes, mitochondria, lysosomes, and cell membrane of intestinal lumen and plants, and in blood plasma.

FIGURE 3. Mammalian glycerolipid metabolism. (From Altman, P. L. and Dittmer, D. S., Eds., in *Biological Data Book*, Vol. III, Federation of American Societies for Experimental Biology, Bethesda, Maryland, 1973, 1539. With permission.)

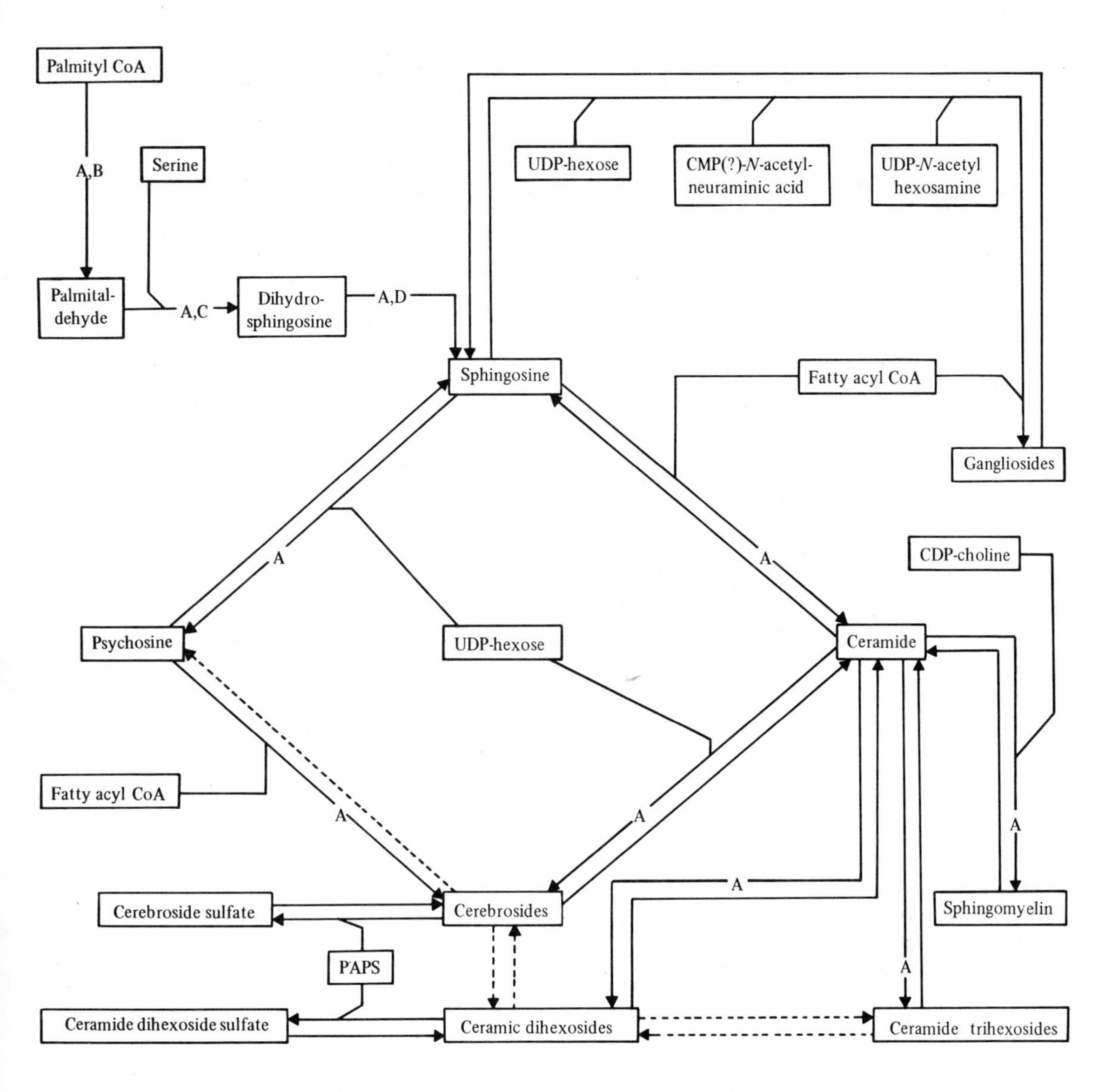

REACTIONS

A: Occurs in microsomes.
B: Requires NADPH.
C: Requires Mn^{2+} and pyridoxal 5-phosphate.
D: Requires a flavoprotein dehydrogenase.

FIGURE 4. Mammalian sphingolipid metabolism. (From Altman, P. L. and Dittmer, D. S., Eds., in *Biological Data Book,* Vol. III, Federation of American Societies for Experimental Biology, Bethesda, Maryland, 1973, 1540. With permission.)

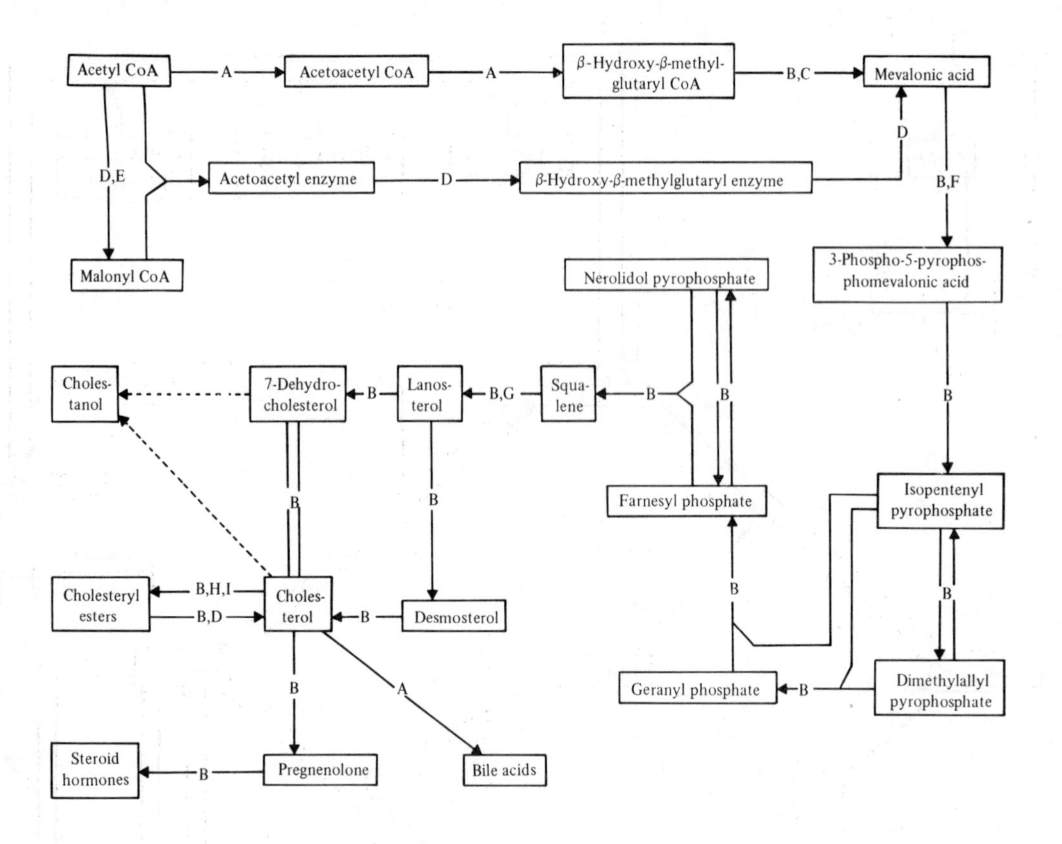

REACTIONS

A: Occurs in mitochondria.
B: Occurs in microsomes.
C: Requires NADPH; not reversible; a target process in control of biosynthesis.
D: Occurs in cytoplasm with soluble enzymes.
E: Requires biotin, ATP, Mg^{2+}, and HCO_3^-; citrate and isocitrate enhance carboxylation of acetyl CoA by complexing with and activating the carboxylase enzyme.
F: Requires ATP and Mn^{2+}.
G: Requires NADPH and O_2.
H: Occurs in blood plasma.
I: Requires lecithin as the acyl donor.

FIGURE 5. Mammalian sterol metabolism. (From Altman, P. L. and Dittmer, D. S., Eds., in *Biological Data Book,* Vol. III, Federation of American Societies for Experimental Biology, Bethesda, Maryland, 1973, 1541. With permission.)

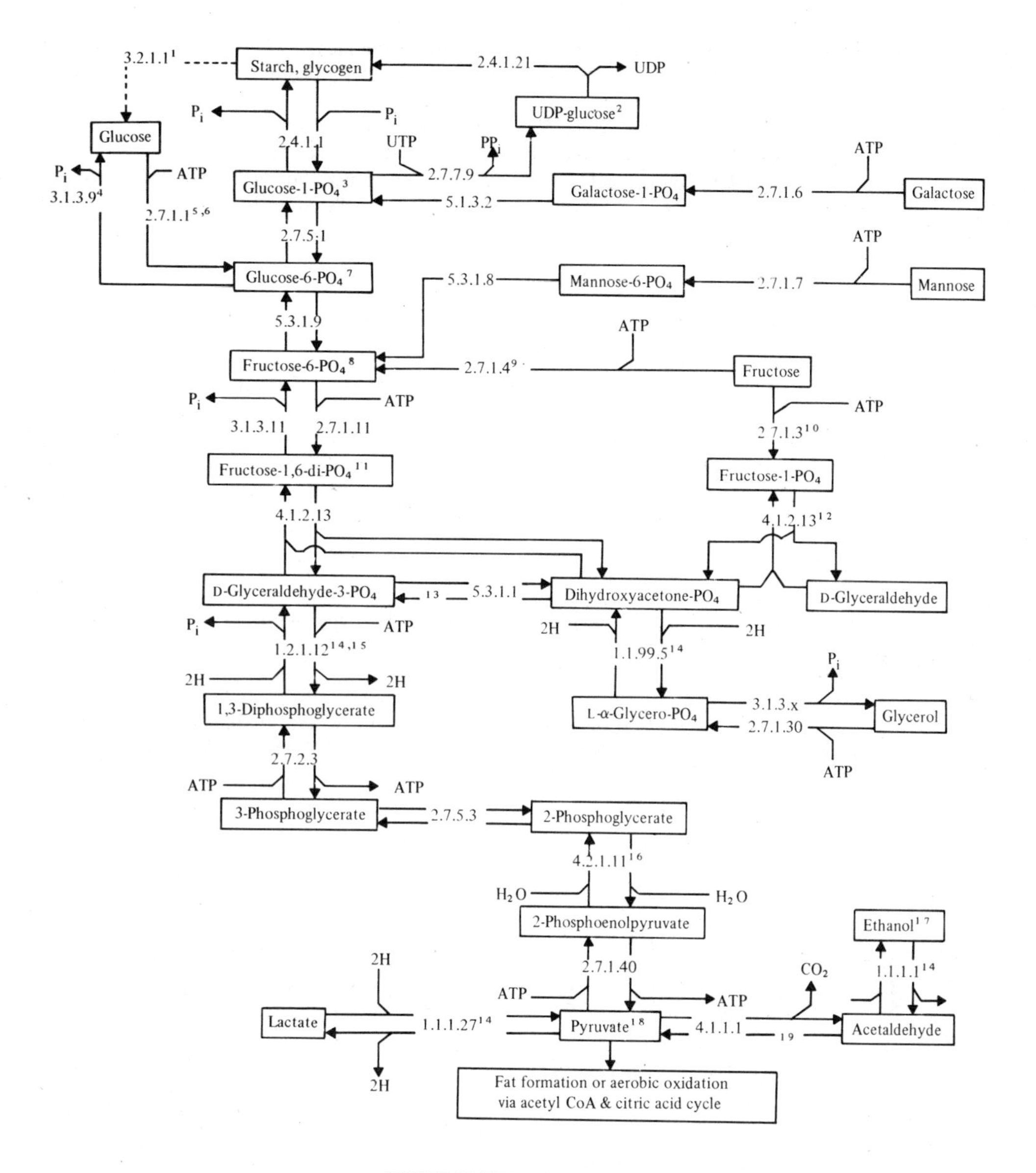

ENZYME KEY [COFACTOR]

1.1.1.1 = alcohol dehydrogenase [NAD$^+$, NADH].
1.1.1.27 = L-lactate dehydrogenase [NAD$^+$, NADH].
1.1.99.5 = glycerolphosphate dehydrogenase [NAD$^+$, NADH].
1.2.1.12 = glyceraldehydephosphate dehydrogenase [NAD$^+$, NADH].
2.4.1.1 = α-glucan phosphorylase [AMP, Mg^{2+}, P$_i$].
2.4.1.21 = UDPglucose-starch glucosyltransferase.
2.7.1.1 = hexokinase [Mg^{2+}].
2.7.1.3 = ketohexokinase [Mg^{2+}].
2.7.1.4 = fructokinase [Mg^{2+}].
2.7.1.6 = galactokinase [Mg^{2+}].
2.7.1.7 = mannokinase [Mg^{2+}].
2.7.1.11 = phosphofructokinase [Mg^{2+}].
2.7.1.30 = glycerol kinase [Mg^{2+}].
2.7.1.40 = pyruvate kinase [K$^+$, Mg^{2+}].
2.7.2.3 = phosphoglycerate kinase [Mg^{2+}].
2.7.5.1 = phosphoglucomutase.
2.7.5.3 = phosphoglyceromutase.
2.7.7.9 = glucose-1-phosphate uridylyltransferase (UDPglucose pyrophosphorylase).
3.1.3.9 = glucose-6-phosphatase.
3.1.3.11 = hexosediphosphatase [Mg^{2+}].
3.1.3.X = phosphoglycerol phosphatase.
3.2.1.1 = α-amylase.
4.1.1.1 = pyruvate decarboxylase [Mg^{2+}].
4.1.2.13 = fructosediphosphate aldolase.
4.2.1.11 = phosphopyruvate hydratase [Mg^{2+}].
5.1.3.2 = UDPglucose epimerase [Mg^{2+}, UDP-glucose].
5.3.1.1 = triosephosphate isomerase.
5.3.1.8 = mannosephosphate isomerase.
5.3.1.9 = glucosephosphate isomerase.

FIGURE 6.

[1] Digestion; glycogen and/or starch are hydrolyzed to glucose in the intestinal lumen.
[2] In plants, ADP glucose is more efficiently utilized for starch synthesis than UDP glucose.
[3] Cori ester.
[4] The reaction, glycogen to glucose-6-PO_4 to blood glucose, takes place in the liver only.
[5] The reaction, glucose to glucose-6-PO_4 to glycogen, takes place in liver, muscle, and other tissues.
[6] Reaction assumed to be inhibited by the growth hormone plus a suprarenal cortical hormone; inhibition by these substances is blocked by insulin, thus favoring conversion of glucose to glucose-6-PO_4.
[7] Robison ester.
[8] Neuberg ester.
[9] In all tissues.
[10] In liver and muscle.
[11] Harden-Young ester.
[12] Liver aldolase also degrades fructose-1-PO_4^-.
[13] This reaction causes each step in the conversion to pyruvate to be doubled quantitatively; thus, 1 mol of glucose gives rise to 2 mol of pyruvate.
[14] NAD^+ acts as acceptor of released hydrogen atoms, becoming NADH in the oxidative direction of this reaction. NADH gives up hydrogen atoms and becomes NAD^+ in the reverse direction. Hydrogen atoms accepted by NAD^+ are passed on in turn to flavoprotein, cytochrome *c*, cytochrome oxidase, and molecular oxygen. If molecular oxygen is not sufficiently available, hydrogen atoms may be passed from NADH to pyruvate to form lactate.
[15] Inhibited by iodoacetate.
[16] Inhibited by fluoride.
[17] End of fermentation in plant tissue.
[18] Pyruvate, followed by conversion to lactate when oxygen supply is deficient (*see* Footnote 14), ends glycolysis in animal tissues. If oxygen is available, pyruvate is oxidized via the citric acid cycle.
[19] Thiamine pyrophosphate required as cofactor in this direction only.

FIGURE 6. Carbohydrate metabolism. (From Altman, P. L. and Dittmer, D. S., Eds., in *Biological Data Book,* Vol. III, Federation of American Societies for Experimental Biology, Bethesda, Maryland, 1973, 1543. With permission.)

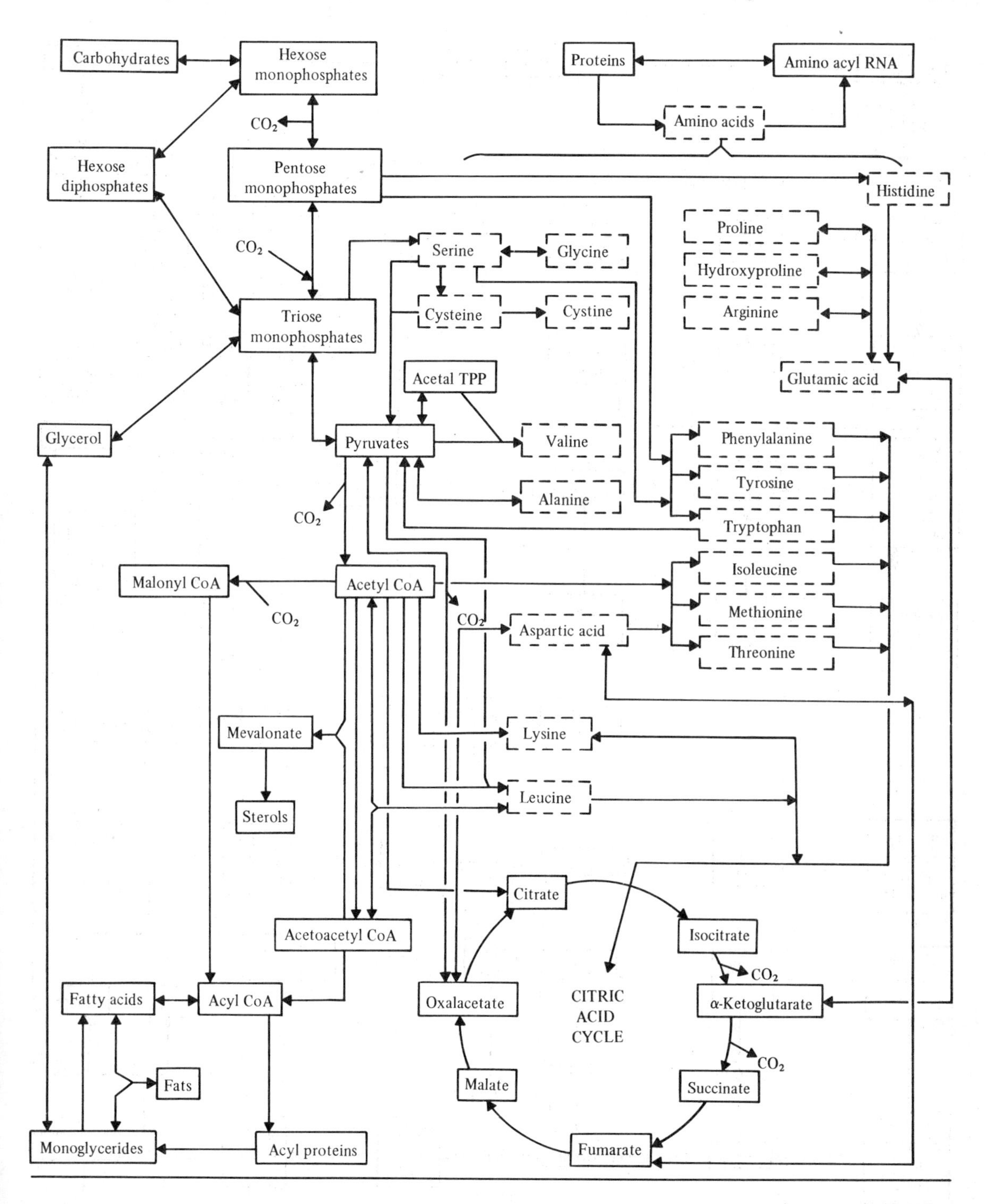

FIGURE 7. Metabolic interrelationship: Lipids, carbohydrates, and proteins. (From Altman, P. L. and Dittmer, D. S., Eds., in *Biological Data Book,* Vol. III, Federation of American Societies for Experimental Biology, Bethesda, Maryland, 1973, 1551. With permission.)

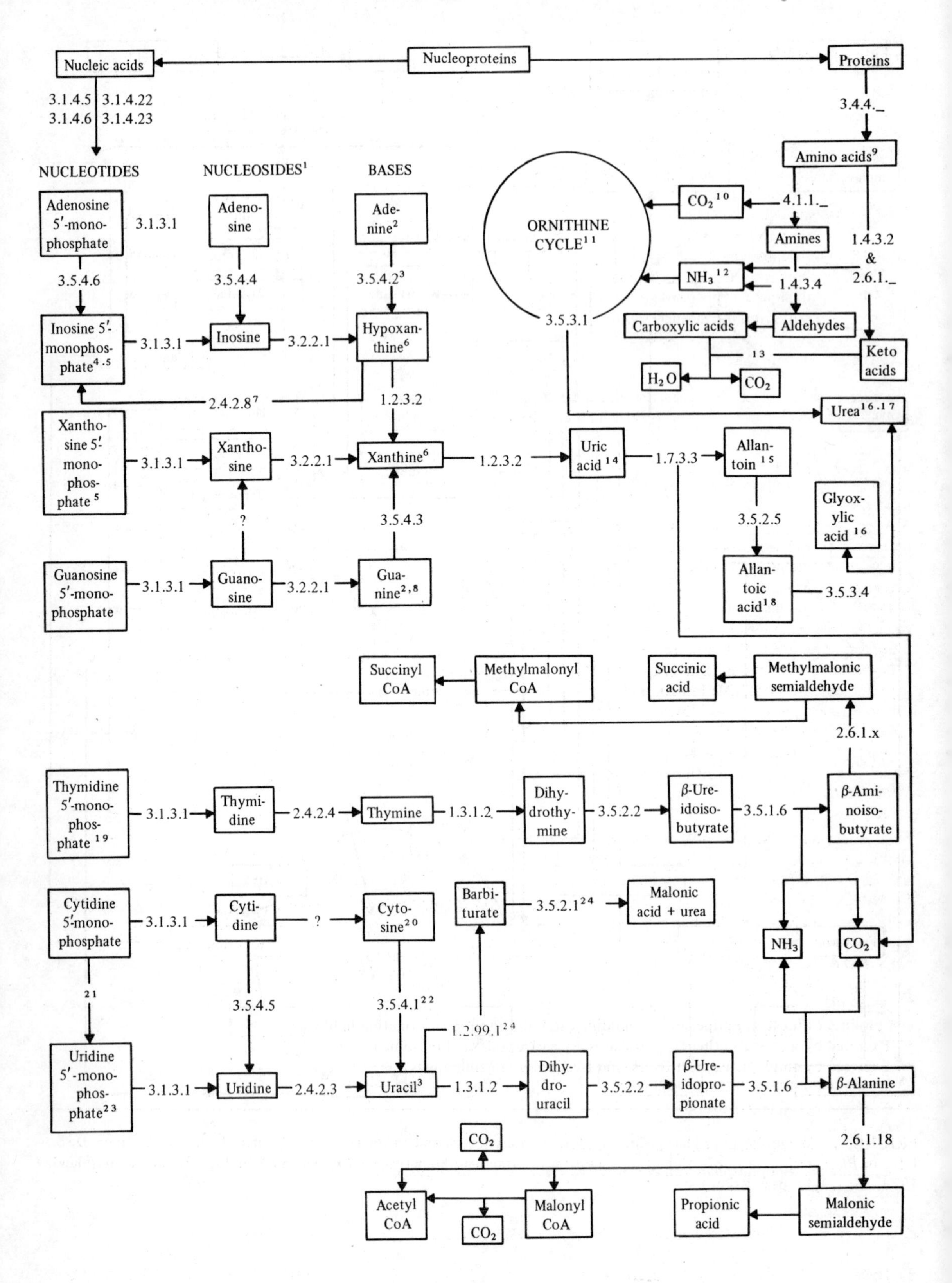
Nucleic acids
Nucleoproteins
Proteins
3.1.4.5 3.1.4.22
3.1.4.6 3.1.4.23
3.4.4._
NUCLEOTIDES
NUCLEOSIDES1
BASES
Amino acids9
ORNITHINE CYCLE11
CO2 10
4.1.1._
Amines
1.4.3.2 & 2.6.1._
NH3 12
1.4.3.4
Adenosine 5'-monophosphate
3.1.3.1
Adenosine
Adenine2
3.5.4.6
3.5.4.4
3.5.4.23
3.5.3.1
Carboxylic acids
Aldehydes
Inosine 5'-monophosphate4,5
3.1.3.1
Inosine
3.2.2.1
Hypoxanthine6
13
Keto acids
H2O
CO2
2.4.2.87
1.2.3.2
Urea16,17
Xanthosine 5'-monophosphate5
3.1.3.1
Xanthosine
3.2.2.1
Xanthine6
1.2.3.2
Uric acid14
1.7.3.3
Allantoin15
?
3.5.4.3
3.5.2.5
Glyoxylic acid16
Guanosine 5'-monophosphate
3.1.3.1
Guanosine
3.2.2.1
Guanine2,8
Allantoic acid18
3.5.3.4
Succinyl CoA
Methylmalonyl CoA
Succinic acid
Methylmalonic semialdehyde
2.6.1.x
Thymidine 5'-monophosphate19
3.1.3.1
Thymidine
2.4.2.4
Thymine
1.3.1.2
Dihydrothymine
3.5.2.2
β-Ureidoisobutyrate
3.5.1.6
β-Aminoisobutyrate
Cytidine 5'-monophosphate
3.1.3.1
Cytidine
?
Cytosine20
Barbiturate
3.5.2.124
Malonic acid + urea
NH3
CO2
21
3.5.4.5
3.5.4.122
1.2.99.124
Uridine 5'-monophosphate23
3.1.3.1
Uridine
2.4.2.3
Uracil3
1.3.1.2
Dihydrouracil
3.5.2.2
β-Ureidopropionate
3.5.1.6
β-Alanine
2.6.1.18
CO2
Acetyl CoA
CO2
Malonyl CoA
Propionic acid
Malonic semialdehyde

FIGURE 8.

1.2.3.2 = xanthine oxidase.
1.2.99.1 = uracil dehydrogenase.
1.3.1.2 = dihydro-uracil dehydrogenase [$NADP^+$].
1.4.3.2 = L-amino-acid oxidase.
1.4.3.4 = monoamine oxidase.
1.7.3.3 = urate oxidase.
2.4.2.3 = uridine phosphorylase.
2.4.2.4 = thymidine phosphorylase
2.4.2.8 = hypoxanthine phosphoribosyltransferase.
2.6.1.– = aminotransferases.
2.6.1.18 = β-alanine aminotransferase.
2.6.1.x = aminotransferase, uncharacterized.
3.1.3.1 = alkaline phosphatase.
3.1.4.5 = deoxyribonuclease I.
3.1.4.6 = deoxyribonuclease II.
3.1.4.22 = ribonuclease I.
3.1.4.23 = ribonuclease II.
3.2.2.1 = nucleosidase.
3.4.4.– = proteases.
3.5.1.6 = β-ureidopropionase.
3.5.2.1 = barbiturase.
3.5.2.2 = dihydropyrimidinase.
3.5.2.5 = allantoinase.
3.5.3.1 = arginase.
3.5.3.4 = allantoicase.
3.5.4.1 = cytosine deaminase.
3.5.4.2 = adenine deaminase.
3.5.4.3 = guanine deaminase.
3.5.4.4 = adenosine deaminase.
3.5.4.5 = cytidine deaminase.
3.5.4.6 = AMP deaminase.
4.1.1.– = amino acid decarboxylases.

[1] 6-Oxypurine nucleotides are split into purines and pentose phosphate by purine nucleoside phosphorylase present in tissues.
[2] Adenine and guanine are the major purines occurring in nucleic acids. Mammals do not require, but can synthesize, purines or pyrimidines from products of protein metabolism.
[3] The route adenine → hypoxanthine is of no importance in animals; adenine deaminase [3.5.4.4] is not found to any extent in mammals.
[4] In the biosynthesis of nucleotides, inosine 5′-monophosphate is the precursor of adenosine 5′-monophosphate and guanosine 5′-monophosphate; in uricotelic species, a portion of the inosine 5′-monophosphate is a precursor of uric acid.
[5] Important in metabolism, but does not occur in nucleic acid.
[6] Excreted to a limited extent by all mammals; primary end product of subjects with xanthinuria, congenital absence of xanthine oxidase [1.2.3.2]. In man, most of xanthine formed is excreted as uric acid. This is the source of $\sim$ 75% of the urinary urate; hypoxanthine, which contributes the other 25%, is primarily a metabolic intermediate.
[7] Recycle found in all normal mammalian tissues; absence resulting from inborn error leads to serious neurological disease.
[8] Excreted by swine and spiders.
[9] For more complete information on amino acid metabolism, *see* Table 5.
[10] May enter into metabolic processes, into the ornithine cycle and be incorporated into and excreted as urea, or be excreted as CO_2.
[11] Urea formation in mammalian liver occurs via the ornithine cycle (Krebs-Henseleit cycle): ornithine → citrulline → arginine succinate → arginine → ornithine. CO_2 and NH_3 enter the cycle via aspartic acid at citrulline; arginine succinate is split into arginine and fumaric acid; arginine is then converted to ornithine with the release of urea.
[12] NH_3, as in the case of CO_2, is also used to synthesize many tissue constituents; hence, it may enter into many metabolic processes, be built into amino acids, be incorporated into urea and excreted, or be excreted as NH_3 across the kidney tubule.
[13] Via the citric acid cycle (prior to entry into this cycle, sulfur-containing amino acids lose their sulfur, usually in the form of sulfate).
[14] Excreted by primates, some reptiles, and some insects as the end product of purine catabolism; excreted by birds as the end product of protein, purine, and pyrimidine catabolism (no urea formation in birds).
[15] Excreted by mammals other than primates, by gastropods, and by some insects.
[16] Excreted by amphibians, most fishes, and freshwater lamellibranchs.
[17] Excreted by mammals as the end product of amino acid metabolism.
[18] Excreted by some teleosts.
[19] Occurs in DNA.
[20] Free cytosine is excreted unchanged in animals.
[21] Cytidine 5′-monophosphate can be converted to uridine 5′-monophosphate.
[22] In yeasts and *Escherichia coli.*
[23] Occurs in RNA.
[24] In *Corynebacterium, Mycobacterium,* and U-i soil bacterium.

FIGURE 8. Pathways of nucleoprotein catabolism. (From Altman, P. L. and Dittmer, D. S., Eds., in *Biological Data Book,* Vol. III, Federation of American Societies for Experimental Biology, Bethesda, Maryland, 1973, 1552. With permission.)

Post-mortem Factors

Post-mortem factors, particularly aging, greatly influence the flavor of cooked meat. The desired flavor is not produced on cooking unaged beef, rather a metallic and astringent flavor results.[84] The enzymes present in raw meat continue to act on meat components during aging and contribute towards the production of flavor precursors. This fact is supported by a number of observations:

Wood[48] identified glucose, glucose-6′-phosphate, fructose-6′-phosphate, and diphosphopyridine nucleotide as reducing substances in raw meat. In an aged muscle, where autolytic processes had been allowed to continue, glucose, fructose, and ribose were the major sugars with traces of sugar phosphates. Apparently during aging, most of the sugar phosphates were enzymatically converted to sugars and inorganic phosphate.

The aging of meat principally involves the action of proteolytic enzymes on muscle proteins, which results in the tenderization of meat. It is commonly accomplished by letting beef cuts hang at temperatures of about 4°C for a period of 1 to 4 weeks. The extent of proteolysis during normal aging is slight but it can be definitely demonstrated. Aging of beef for 2 weeks results in an increase in free amino nitrogen of the nonprotein nitrogen fraction. The contribution of native tissue proteases in meat tenderization has been demonstrated by preparing rabbit and lamb muscles aseptically from freshly killed animals and preserving them under aseptic conditions at 25°C for about 150 days. After 21 days, the electrophoretic pattern of the extracted proteins showed the appearance of several new components that moved faster in the electric field. Microscopic changes were also observed, such as the longitudinal splitting of myofibrils after 50 days.[95]

The effect of post-mortem myotomy on glycolysis and ultimate qualitative characteristics of porcine longissimus and rectus femoris muscles was investigated by Koch et al.[96,97] It was observed that myotomy at or shortly after exsanguination stimulated contractile activity, significantly increased the rate of post-mortem glycolysis, and tended to decrease ultimate qualitative properties. Longissimus muscles incised at the time of exsanguination had lower pH values, glycogen, adenosine triphosphate, creatine phosphate, and higher lactate levels at all post-mortem time periods studied through 2 hr, than muscles not incised until 45 min post-mortem. The effects of myotomy were greater among normal than low quality muscles. The rates of post-mortem pH decline, levels of glycogen, lactate, glucose-6′-phosphate, adenosine triphosphate, and creatine phosphate, as influenced by the post-mortem muscle temperature and myotomy, were investigated. Muscle contraction stimulated by myotomy, particularly in the excised sample, resulted in rapid diminution of adenosine triphosphate and creatine phosphate, which in turn appeared to have contributed to the observed differences in glycolytic rate. An activating mechanism of this nature would not be expected to be as effective on a process that is already rapid as on one that is relatively slow. Thus, muscle contraction resulting from myotomy was more effective in increasing glycolytic rate of normal longissimus muscles than those of lower quality (inherent rapid glycolyzing). Also, since low temperatures (2 to 4°C) are normally used to chill pork carcasses and are known to exert an inhibitory effect upon glycolytic rate, the inhibition would be expected to have a greater effect on the faster process. Thus, the rapid heat dissipation of the rectus femoris muscles decreased glycolytic rate to a greater extent among low quality than normal muscles. Consequently, what appeared to be a discrepancy in response between the longissimus and rectus femoris muscles was in actuality a natural physiological response to differences in experimental procedure.

As has been discussed earlier, in the case of ripened sausages, proteases, lipases, and other enzymes released by micrococci and lactobacilli play a major role in the development of flavor compounds and some flavor precursors.

Nature and Concentration of Flavor Precursors Found in Meat and Their Role in the Development of Meat Flavor

During the last 2 decades, a large number of publications have dealt with the nature of flavor precursors of various types of meat. Recently, Heath[1] reported that a cold water extract of raw beef, after freeze drying and counter current dialysis, showed that only those fractions having a molecular weight of less than 200 produced any beef flavor on heating, and that fractions were composed of 32 amino acids, small peptides, and carbohydrates such as glucose, glucosamine, fructose, and ribose. A partial list of water soluble compounds isolated from meat is given in Table 3.

The literature reports indicate that the mechanisms of meat flavor development include not only Maillard browning reactions and fatty acid oxidation but also inter- and intra-molecular cyclizations as well as numerous reactions which are made possible by the reactivity of ammonia, hydrogen sulfide, mercaptans, and other undefined intermediates, especially at elevated temperatures.[17] It is generally agreed that a major portion of meat volatiles is produced as a result of interaction between sugars and amino compounds. These flavor precursors, in heated and unheated meat samples, have been extensively investigated.[50,51,57] Quantities of individual amino compounds and sugars determined in heated and unheated lyophilized diffusate from beef, lamb, and pork are presented in Tables 6, 7, and 8. The amino compounds found in the meat of three species appear to be quite similar quantitatively as well as qualitatively, except that cysteic acid and ornithine were common only to pork and lamb, and that glutathione was present only in lamb.[51] In all three cases, taurine, anserine-carnosine, and alanine were the major constituents of unheated samples, and losses during heating were relatively large. Decreases in these constituents accounted for 69, 72, and 45% of the total amount during heating of this type of compounds in beef, pork, and lamb, respectively. Other important constituents degraded during heating were glutamic acid, glycine, lysine, serine, cystine, methionine, leucine, isoleucine, and methyl histidine. The loss of essential amino acids ranged from approximately 20% in pork samples to approximately 55% in beef samples. Macy et al.[51] also reported that beef contained largest amount of amino compounds, and pork the least. The precentages of the total amino compounds lost during heating were proportional to initial quantities present. Greater amounts were lost from the beef samples than from the lamb and pork samples.

Macy et al.[50,51] reported the presence of glucose, fructose, ribose, and some unidentified carbohydrate material in beef, lamb, and pork extracts (Tables 6, 7, and 8). The concentrations of these sugars in three species were quite similar; glucose was present in greatest quantities followed by fructose, ribose, and the unknown fraction. Ribose appeared to be the most labile to heating, and fructose the most stable (Tables 6, 7, and 8).

Recently, Jarboe and Mabrouk[61] carried out an extensive investigation of free amino acids, sugars, and organic acids in aqueous beef extracts. These workers confirmed the presence of glucose-6′-phosphate, fructose-6′-phosphate, fructose-1,6′-diphosphate, and adenosine-5′-monophosphate, reported earlier by Wood[48] and others, and provided quantitative data on these compounds. Data of Jarboe and Mabrouk are not in agreement with the earlier work of Macy et al.,[51] possibly due to differences in the state of muscle, and the analytical techniques used.

Batzer et al.[49] in 1962 identified inosinic acid, inorganic phosphate, and a glycoprotein of which the carbohydrate was glucose, as the principal components involved in the development of beef flavor on heating. The principal amino acids in the protein moiety were β-alanine and serine, with smaller quantities of glycine, glutamic acid, α-alanine, leucine and isoleucine. When mixtures of some of these amino acids were added to a solution of glucose, inosine, and inorganic phosphate, meaty flavor developed on heating in fat or water.

Comparatively, little work has been done on the contribution of water soluble components from adipose tissue in the development of meat flavor. The work done by Pepper and Pearson[58] indicates that there are some differences in proteins present in lean muscle and adipose tissue. It was suggested, therefore, that the water soluble fraction from adipose tissue may make a distinct and characteristic contribution to meat flavor above and beyond that related to the lipid portion. The authors further suggest that an additional role of adipose tissue to meat flavor could be derived from the increased number of sulfhydryl groups due to the presence of salt. Since salt is commonly used in cooking, it is possible that it may retard oxidation of these groups, thereby promoting the formation of parent sulfur compounds, which are known to play a major role in the development of cooked meat aroma. An extension of this work was recently reported by Pepper and Pearson.[62] The dialyzable fraction from beef adipose tissue, either treated with salt or untreated and stored under either air or nitrogen, was lyophilized, fractionated by column chromatography, and analyzed by chemical procedures or thin-layer chromatographic techniques. The chromatographic profiles of dialyzable components were unaffected by salt treatments or by storage under air or nitrogen. Amino acid analysis revealed that glycine, glutamic acid, and proline comprised about 46% of the total

TABLE 6

Concentration of Various Amino Compounds and Carbohydrates in Lyophilized Diffusate from Beef before and after Heating

Compound	Concentration mg/100 g tissue	
	Unheated	Heated (boiling water for 1 hr)
Total amino compounds	161.53	71.45
Phosphoserine	0.36	0.29
Glycerophosphoethanolamine	0.02	0.01
Phosphoethanolamine	0.66	0.82
Taurine	9.05	4.02
Urea	0.01	–
Aspartic acid	0.82	0.32
Threonine	1.11	0.74
Serine + asparagine (as serine)	7.53	1.35
Glutamic acid	4.63	2.22
Glycine	2.40	1.32
Alanine	11.28	6.22
Cystine	4.37	–
Valine	2.99	1.47
Methionine	2.01	0.75
Isoleucine	2.04	0.87
Leucine	3.81	2.34
Tyrosine	1.85	0.80
Phenylalanine	1.36	0.97
NH_3 + lysine	6.19	4.11
Histidine	4.10	4.17
Anserine + carnosine (as carnosine)	90.14	38.17
1-Methylhistidine	4.80	0.49
Total carbohydrates	48.66	28.46
Glucose	43.86	25.25
Fructose	3.56	3.21
Ribose	1.09	Trace
Unknown nucleoside (as ribose)	0.15	–

From Macy, R. L., Naumann, D. H., and Bailey, M. E., *J. Food Sci.*, 29, 142, 1964. With permission.

amino acid residues for the untreated samples as compared to 31% for the sample treated with salt. The diffusate from the salt-treated sample contained more arginine, threonine, serine, isoleucine, leucine, tyrosine, and phenylalanine (22%) than the untreated sample (9%). Only traces of sulfhydryls were evident, but positive tests were obtained for aldose and free amino groups. The diffusates from untreated adipose tissue contained about five times as many free amino groups and ten times as many aldoses as that from the salt treatment. This suggests that salt stabilized the adipose tissue from breakdown.

Thin-layer chromatography of beef adipose tissue diffusate confirmed the presence of creatine, creatinine, creatine phosphate, cytosine, uracil, and several fluorescent and phosphate-containing compounds, but purine bases, nucleosides, nucleotides, and lactic acid were absent.

Wasserman and Spinelli[60] have recently published preliminary studies on the components of adipose tissue which may be involved in the development of meat aroma (Table 9). Precursors of the characteristic beef, lamb, and pork aromas were shown to be present in the lipids extracted from adipose tissue, with a chloroform-methanol mixture. Water washing of the extract removed components involved in the formation of these odors. Amino acids and glucose were identified in the wash which produced a nonspecific roast meat aroma on heating to dryness.

TABLE 7

Concentration of Various Amino Compounds and Carbohydrates in Lyophilized Diffusate from Pork before and after Heating

	Concentration mg/100 g tissue	
Compound	Unheated	Heated (boiling water for 1 hr)
Total amino compounds	109.86	88.10
Phosphoserine	0.35	0.23
Glycerophosphoethanolamine	0.02	0.01
Phosphoethanolamine	0.45	1.89
Taurine	12.58	7.94
Aspartic acid	1.37	0.45
Threonine	0.48	0.45
Serine + asparagine (as serine)	2.95	0.53
Glutamic acid	1.95	1.19
Proline	0.64	0.24
Glycine	2.75	1.58
Alanine	4.19	2.80
Cystine	2.11	0.89
Valine	0.30	0.20
Methionine	0.69	1.00
Isoleucine	1.03	0.83
Leucine	1.68	1.00
Tyrosine	0.56	0.37
Phenylalanine	0.51	0.39
Ornithine	Trace	0.56
NH_3 + lysine	4.27	4.06
Histidine	2.55	3.11
Anserine + carnosine (as carnosine)	67.94	58.38
1-Methylhistidine	0.49	Trace
Total carbohydrates	45.90	30.04
Glucose	43.56	28.02
Fructose	2.08	2.02
Ribose	0.20	Trace
Unknown nucleoside (as ribose)	0.06	–

From Macy, R. L., Naumann, D. H., and Bailey, M. E., *J. Food Sci.*, 29, 142, 1964. With permission.

Since adipose tissue is composed of metabolically active cells, it is not surprising that a number of amino acids and at least one sugar component have been identified in tissue extracts. The absence of purine bases, nucleosides, and nucleotides in beef adipose tissue diffusate, reported by Pepper and Pearson,[62] may be due to inadequate detection techniques, as nucleic acids are essential cellular constituents and their metabolic and breakdown products are usually found in living cells.

As reported earlier, Batzer et al.[49] found that inosinic acid, along with glucose and unidentified glycoprotein, was necessary for the development of meaty flavor in beef. Inosinic acid is produced from adenosine 5′-triphosphate post-mortem. Nucleotides are also potential precursors of free ribose and ribose phosphate, which have been implicated in Maillard browning reactions. Macy et al.[56] investigated the effect of cooking on nucleotides, nucleosides, and bases of beef, pork, and lamb. The effect of cooking on beef and pork nucleotides is given in Table 10. These workers reported that inosinic acid was the prominent nucleotide in all three species and it was degraded by heating. Adenylic acid increased during cooking in meat from all three species, possibly due to the hydrolysis of adenosine di- and triphosphate and nucleic

TABLE 8

Concentration of Various Amino Compounds and Carbohydrates in Lyophilized Diffusate from Lamb before and after Heating

	Concentration mg/100 g tissue	
	Unheated	Heated (boiling water for 1 hr)
Total amino compounds	130.01	81.46
Phosphoserine	0.30	0.22
Glycerophosphoethanolamine	0.01	–
Phosphoethanolamine	1.15	1.90
Taurine	26.25	16.47
Urea	Trace	–
Aspartic acid	1.52	1.36
Threonine	3.14	1.36
Serine + asparagine (as serine)	4.74	2.26
Glutamic acid	6.08	2.85
Proline	3.05	0.83
Glycine	4.30	2.25
Alanine	9.18	6.53
Cystine	6.46	5.24
Valine	0.44	0.18
Methionine	2.37	1.53
Isoleucine	2.79	1.87
Leucine	5.62	3.06
Tyrosine	1.98	1.67
Phenylalanine	1.85	1.12
Ornithine	0.95	0.65
NH_3 + lysine	10.51	5.59
Histidine	9.70	7.16
Anserine + carnosine (as carnosine)	25.55	16.28
1-Methylhistidine	2.07	1.08
γ-Amino-*n*-butyric acid	Trace	–
Total carbohydrates	36.42	23.94
Glucose	32.87	21.33
Fructose	2.68	2.61
Ribose	0.52	Trace
Unknown nucleoside (as ribose)	0.35	–

From Macy, R. L., Naumann, D. H., and Bailey, M. E., *J. Food Sci.*, 29, 142, 1964. With permission.

acids. Cytidylic, uridylic, and guanylic acids were present in relatively low concentrations in meat from all three species and changed little during cooking.

It is well known that stiffening of muscle post-mortem is related to decrease in adenosine-5′-triphosphate (ATP) content.[98-103] The formation of adenosine-5′-monophosphate (AMP) results from the removal of two terminal phosphate groups of ATP by the combined catalytic action of ATPase and myokinase. AMP is rapidly deaminated to inosinic acid IMP) by the catalytic action of AMP deaminase. Further degradation of IMP into inosine and hypoxanthine occurs during prolonged storage. The details of these metabolic reactions may be found in Figure 8.

Howard et al.[104] studied nucleotide breakdown in beef tissue and found that with progressive aging the hypoxanthine levels rose to 1.5 to 2.0 μmol/g of meat. Lee and Webster[105] studied the effect of temperature, freezing, frozen storage, thawing, and pH on hypoxanthine production in beef tissue and

TABLE 9

Amino Acid Composition of the Water-soluble Fraction from Chloroform:Methanol Extracts of Adipose Tissue[60]

Compound	Concentration μmol/100 g tissue		
	Pork	Beef	Lamb
Cysteic acid	2.35	1.95	1.50
Taurine	19.40	9.20	34.70
Urea	44.60	39.15	–
Aspartic acid	0.85	3.60	3.25
Asparagine + threonine	10.00	3.85	8.85
Serine	7.00	3.90	8.40
Glutamic acid	4.50	4.90	8.25
Proline	7.05	4.45	10.35
Glycine	16.85	13.20	35.25
Alanine	39.55	25.75	45.40
Valine	5.80	3.90	8.65
Methionine	1.65	0.90	2.30
Isoleucine	5.45	6.55	11.40
Leucine	8.10	4.35	11.30
Tyrosine	4.20	1.65	4.45
Phenylalanine	2.95	2.50	5.80
Ammonia	3.65	2.90	1.15
Lysine	0.75	0.95	1.90
Anserine	0.80	0.50	1.55
Carnosine	5.90	0.85	0.75
Arganine	1.70	0.80	0.11

TABLE 10

Ratios of Sums of Individual Nucleotides and Total Nucleotides of Cooked Relative to those of Raw Beef and Pork[55]

Constituent	Ratio of chemical constituents in cooked and raw meat (raw meat = 1.0)			
	Beef		Pork	
	49°C	71°C	49°C	71°C
Cytidylic acid	0.81	0.99	0.67	0.58
Adenylic acid	1.46	3.38	0.80	1.81
Uridylic acid	0.59	0.75	0.83	0.84
Inosinic acid	0.75	0.61	0.90	0.78
Guanylic acid	0.55	1.41	0.75	0.67
Sum of individual nucleotides	0.77	0.74	0.87	0.83

TABLE 11

Post-mortem Changes in the Concentrations of Nucleotides and Related Compounds in Porcine Longissimus Muscle[59]

Breeds	Post-mortem time (hr)	Concentration μmol/g wet tissue					
		Inosine + hypoxanthine	DPN⁺ + DPNH	5′-IMP	5′-IDP	5′-ADP	5′-ATP
Poland China (n=2)	0	–	0.62	2.11	–	1.09	4.13
	3	1.65	0.10	5.48	0.59	0.38	0.16
	24	2.12	Trace	5.17	0.54	0.63	Trace
	72	2.75	Trace	5.03	0.69	0.54	Trace
	144	2.73	Trace	4.47	0.46	0.36	Trace
Chester White (n=2)	0	–	0.66	0.78	–	0.90	6.04
	3	1.26	0.49	5.67	0.42	0.33	0.38
	24	1.72	0.30	5.44	0.69	0.31	0.21
	72	2.07	0.17	5.20	0.79	0.37	0.23
	144	2.90	0.15	4.59	0.57	0.42	0.11

reported that an increase in temperature and a higher ultimate pH or thawing of frozen samples stimulated the rate of hypoxanthine production.

Tsai et al.[59] investigated the post-mortem metabolism of nucleotides in porcine longissimus muscle. The post-mortem changes in the concentrations of nucleotides and related compounds post-mortem are reported in Table 11. Tsai and co-workers observed that AMP deaminase activity decreased considerably during storage, whereas adenosine deaminase activity remained nearly the same during storage. On the other hand, 5′-nucleosidase activity seemed to increase during the first few hours of storage and then remained constant for several days before declining. The direct consequence of decrease in AMP deaminase activity during storage resulted in the stabilization of IMP concentration within the first few hours of post-mortem. On the other hand, inosine + hypoxanthine concentration in muscle continued to increase for several days, as would be expected from nearly constant adenosine deaminase activity during storage (Table 11).

During the ripening of dry sausage, the concentration of water soluble nitrogenous compounds increases and can reach values up to 25% of the total nitrogen.[106-108] The nature and concentration of these compounds such as amino acids, peptides, nucleotides, and nucleosides determine, to a large extent, the final dry sausage flavor.[109]

Pohja and Niinivaara[110] and Sajber et al.[111] reported that free amino acid production in dry sausage is at least partly due to bacterial protease activity. Although native muscle proteases may continue to hydrolyze proteins in dry sausage, if not inactivated during processing, their contribution in the liberation of free amino acids would be comparatively smaller to bacterial proteases. Cantoni et al.[112] investigated the fate of inosinic acid during dry sausage ripening and reported that inosinic acid was converted to inosine nucleoside and hypoxanthine, respectively, by the action of phosphomonoesterase and nucleosidase. Dierick et al.[113] investigated changes in nonprotein nitrogen compounds during dry sausage ripening. Two batches of sausages were prepared: Batch 1 was a normal sausage preparation;[114] in batch 2 a starter culture, containing lactobacilli and micrococci, and 1% less sugar were added. The concentration of nonprotein nitrogen compounds (Table 12) and free amino acids (Table 13) in the two batches were monitored during ripening. These data show that during the first 3 days of ripening, the rate of free amino acid production is maximal and exceeds the rate of ammonia production and peptide production from proteins. During this period intensive carbohydrate metabolism and bacterial growth also takes place.[114] Subsequent to this, the rate of ammonia production increases but remains lower than amino acid production, whereas the concentration of peptides decreases. The concentration changes for individual amino acids during ripening (Table 13) show that the major amino acids responsible for the increase in total α-amino nitrogen were alanine, leucine, threonine, valine, serine, glycine, and proline. It was also reported

TABLE 12

Concentration of NPN Compounds at Various Stages of Dry Sausage Ripening (mg N/100 g dry matter)

	Batch 1 Stage of ripening (days)				Batch 2 Stage of ripening (days)			
Compound	0–3	3–15	15–36	0–36	0–3	3–15	15–36	0–36
Ammonia	6	28	18	52	2	34	12	48
Free amino acids	47	46	21	114	45	30	72	147
Peptide-nitrogen	34	–43	–7	–16	10	–67	–55	–112
Nucleotide-nitrogen	–1	–20	–1	–22	–16	–5	–2	–23
Nucleoside-nitrogen	8	37	5	50	11	33	14	58

From Dierick, N., Vandekerckhove, P., and Demeyer, D., *J. Food Sci.*, 39, 301, 1974. With permission.

TABLE 13

Concentration of Free Amino Acids at Three Stages of Ripening (mgα-NH_2-N/100 g dry matter)

	Batch 1 Stage of ripening (days)			Batch 2 Stage of ripening (days)		
Free amino acid	0	15	36	0	15	36
Aspartic acid	0.74	1.90	5.30	0	3.79	7.25
Threonine	0.82	3.30	25.20	0	4.95	6.94
Serine	1.73	5.50	9.10	0	7.20	9.85
Glutamic acid	19.00	7.20	5.60	25.40	24.30	18.30
Proline	0	3.40	5.50	0	5.72	6.45
Glycine	3.00	6.15	8.80	0	7.65	14.20
Alanine	10.20	20.90	25.20	3.92	23.10	22.30
Valine	1.44	6.35	8.85	0	7.41	11.95
Methionine	0.56	2.72	3.84	1.60	3.38	5.26
Isoleucine	1.60	3.74	5.45	2.24	4.50	8.10
Leucine	1.06	11.50	13.30	8.30	13.20	17.60
Phenylalanine	0.93	4.50	5.25	2.98	5.00	7.15
Lysine	2.07	4.67	6.35	2.30	3.94	6.46
Histidine	0.73	1.46	0.01	2.77	2.20	0
Tyrosine	0.77	0	0	0	0	0.30
γ-Amino butyric acid	0	2.72	4.07	1.22	7.89	12.50
Ornithine	1.16	0	0	1.98	0	0.83

From Dierick, N., Vandekerckhove, P., and Demeyer, D., *J. Food Sci.*, 39, 301, 1974. With permission.

that during sausage ripening added glutamate disappeared, and at least partly decarboxylated to γ-amino butyric acid. Langner,[115] based on the observation that glutamate is metabolized during dry sausage ripening, questioned its incorporation as a flavor additive in dry sausage. Other amino acids may be decarboxylated during dry sausage ripening, as indicated by the disappearance of histidine, tyrosine, and ornithine. The decarboxylation products of these amino acids are histamine, tyramine, and putrescine, respectively. Data reported by Langner[116] and Reuter and Langner[117] are in contrast with those of Dierick et al.[113] These workers reported an increase in these amino acids. In order to prove the validity of their observation that histidine, tyrosine, and ornithine are indeed decarboxylated during sausage ripening, Dierick et al.[113] determined the concentration of highly basic amines in the ripened samples. Although very small amounts were detected, the concentration of histamine, tyramine, and putrescine was increased at least tenfold, the rate of increase being maximal during the first three days of ripening. Cadaverine, a decarboxylated product of lysine, was also detected in significant quantities. The presence of starter culture in batch 2 produced no striking differences in the nature and concentration of nonprotein nitrogen compounds during ripening, except for a higher final concentration of free amino acids, coupled to a lower peptide-nitrogen and lower increase in threonine concentration (Tables 12 and 13). The lower concentration of amino acids in batch 2 may suggest a higher exopeptidase activity in batch 2.

MECHANISMS OF MEAT FLAVOR DEVELOPMENT

During the heat processing of meat, the types of reactions which appear to be responsible for the development of flavor compounds include:

1. Autoxidation, hydrolysis, dehydration, and decarboxylation of fats giving rise to aldehydes, fatty acids, lactones, and ketones.
2. The nonfatty components may undergo glycoside splitting, oxidation and decomposition to yield volatile and nonvolatile compounds.
3. The complex degradation products of sugars and sugar-amine interactions.

These chemical reactions are discussed in detail in this section. The reactions are divided into subheadings: amino acids, carbohydrates, and lipids, depending on which is the substrate of the reaction.

Amino Acids

The thermal decomposition of amino acids has not been studied at temperatures normally attained in cooking of meat, but several investigations have been made at temperatures of 200°C and above. Based on the energies of activation of these reactions, it is conceivable that some of these reactions would take place at lower (cooking) temperatures.

The thermal decomposition products of alanine, β-alanine, valine, leucine, and isoleucine, which were heated at or above 220°C, were identified by Lien and Nawar (Table 14).[118,119] They have also proposed mechanisms of their formation (Figure 9).

Olefins are produced by the decarboxylation and deamination of aliphatic amino acids; the corresponding paraffins result upon reaction with hydrogen. By this process isobutylene and isobutane are produced from valine, 3-methyl-1-butene and isopentane from leucine, and 2-methyl-1-butene from isoleucine. In addition, the formation of propane and propene from valine, isobutane and isobutylene from leucine, and butane and butene from isoleucine, may be explained by cleavage of carbon-carbon bonds beta to carbonyl group of the amino acid to form a free radical which can then accept or lose a hydrogen atom.

Primary amines are produced by decarboxylation, while the corresponding aldehydes result from both deamination and decarbonylation of the amino acid. The formation of isobutyraldehyde and isobutylamine from leucine cannot be explained by the deamination and decarbonylation reactions.[119] The imines formed from the heat degradation of amino acids probably originate from the reaction of carbonyl compounds with primary amines. Aliphatic ketones react with amines more slowly than aldehydes to form imines which necessitates the use of higher reaction temperatures and longer reaction times than required for the aldehyde.[120] This is probably the reason why Lien and Nawar[118,119] found only aldimines and

TABLE 14

Thermal Decomposition Products from Alanine, β-Alanine, Valine, Leucine, and Isoleucine[118,119]

Alanine	β-Alanine	Valine	Leucine	Isoleucine
Ammonia	Ammonia	Ammonia	Ammonia	Ammonia
Carbon dioxide	Carbon dioxide	Carbon dioxide	Carbon dioxide	Carbon dioxide
Carbon monoxide	Carbon monoxide	Carbon monoxide	Carbon monoxide	Carbon monoxide
Ethane	Ethene	Propane	Isobutane	Butane
Ethene	Propene	Propene	Isobutylene	Butene
Propene	Acetonitrile	Isobutane	Isopentane	Isopentane
2-Butene	Acetone	Isobutylene	3-Methyl-1-butene	2-Methyl-1-butene
Acetaldehyde	Pyridine	Acetone	Acetone	2-Butanone
Ethylamine	3-Picoline	Isobutyraldehyde	Isobutyraldehyde	2-Methylbutyraldehyde
N-Ethylidene-ethylamine	2,4-Lutidine	Isobutylamine	Isovaleraldehyde	2-Methylbutylamine
2-Methyl-5-ethylpyridine	3,5-Lutidine	*N*-Isobutylideneisobutylamine	Isobutylamine	*N*-(2-methylbutylidene)-2-methylbutylamine
N-Ethylpropionamide	2,3,5-Trimethylpyridine	Diisobutylamine	Isoamylamine	bis(2-Methylbutyl)amine
Propionamide	Poly-β-alanine		*N*-Isobutylidene-isoamylamine	
			N-Isoamylidene-isoamylamine	
			Diisoamylamine	

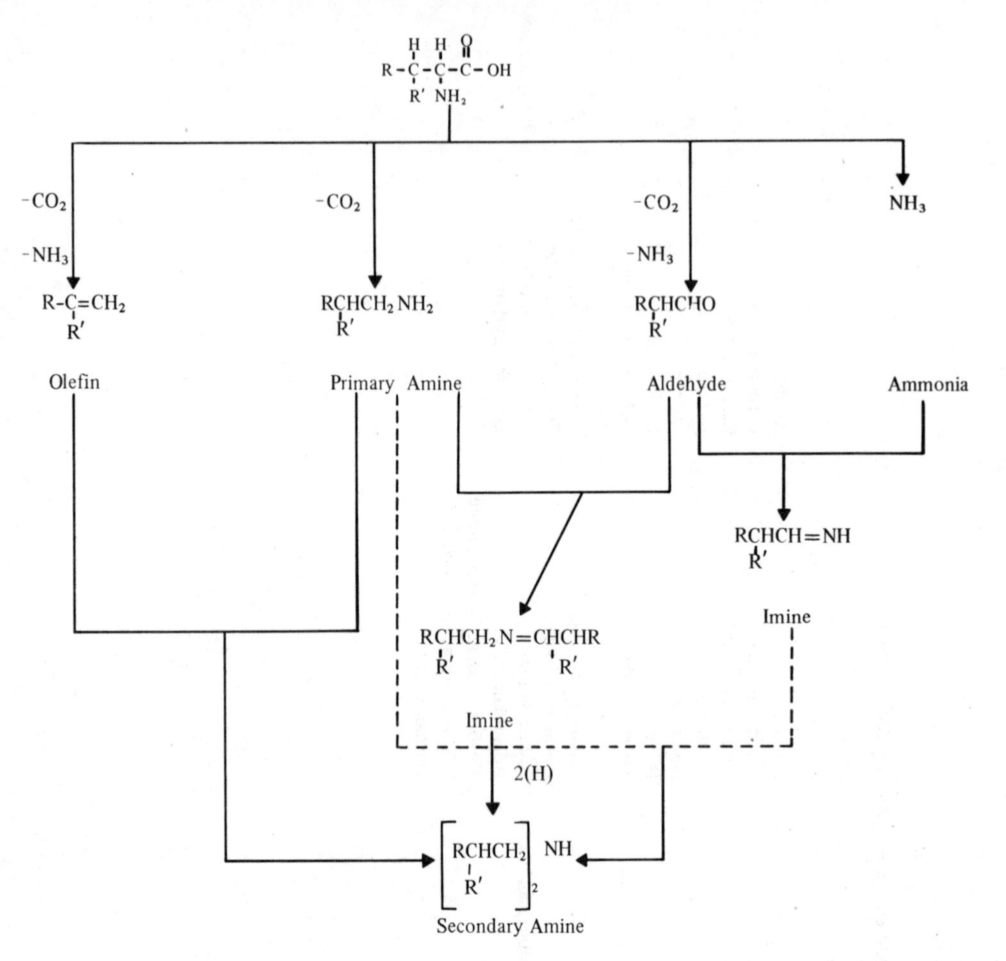

FIGURE 9. Suggested mechanisms for the production of heat degradation products from neutral alimatic amino acids. (see Table 14) (From Lien, Y. C. and Nawar, W. W., *J. Food Sci.*, 39, 911, 1974. With permission.)

not ketimines in their investigation on thermal decomposition products of aliphatic amino acids. Thus the reaction of isobutyraldehyde and isobutylamine, both thermal decomposition products of valine, resulted in the formation of *N*-isobutylidene-isobutylamine. Similarly, *N*-isobutylidene-isoamylamine and *N*-(2-methylbutylidene)2-methylbutylamine were produced from thermal decomposition of leucine and isoleucine, respectively. Aliphatic aldehydes can also react with ammonia to first form simple addition compounds called "aldehyde ammonias." These compounds are unstable and easily decompose to the original constituents, or lose water to form imines.[121] The secondary amines diisobutylamine, diisoamylamine, and bis(2-methylbutyl)-amine result from the addition of hydrogen to the above mentioned imines. An alternative pathway, however, may be the reaction of primary amines with imines to additional products which can be reduced to secondary amines.[122] A third possibility may result from the addition of primary amines to olefins.[123] A characteristic imine, *N*-ethylidene-ethylamine, was detected in the heat breakdown products of alanine. This imine probably resulted from the reaction of acetaldehyde and ethylamine;[119] however, the secondary amine, which was present in the degradation products of valine and leucine, was absent in alanine (Table 14).

The formation of 2-methyl-5-ethylpyridine in the case of alanine may arise from the aldol condensation of acetaldehyde to form crotonaldehyde, followed by dimerization in the presence of ammonia. The most probable reaction mechanism of pyridine synthesis is via Tschitschibabin reactions.[121] This mechanism involves the reaction of aldehyde with ammonia to form imine, followed by aldol-type condensation to produce pyridine derivatives. The absence of pyridine derivatives in pyrolyzed valine, leucine, or isoleucine is probably due to their branched chain structure, which may make the aldol-type condensation difficult.[120] The two amides, propionamide and *N*-ethylpropionamide, also not found in the case of valine, leucine, and isoleucine, were produced in relatively large quantities from alanine. The formation of these two compounds may result from the breakdown of 3,6-dimethyl-2,5-diketopiperazine, which is the dehydration product from two molecules of alanine.

The β-amino acids decompose at lower temperatures than α-amino acids. β-Alanine for example loses its crystalline appearance and develops a yellow color at 200°C, whereas α-amino acids develop yellow color at about 250°C. The decomposition products of β-alanine were also significantly different from those of α-amino acids (Table 14).

Merritt and Robertson[124] listed the major compounds formed on pyrolyzing various common amino acids. Among these compounds benzene, toluene, and ethylbenzene were obtained on pyrolyzing phenylalanine and the corresponding hydroxy-compounds were derived from tyrosine. Imidazole derivatives were formed from histidine.[125] A variety of nitriles have also been reported.[126] Most interesting, however, is Merritt's observation that the compounds formed on pyrolyzing a peptide depended on the sequence of amino acids. Although only dipeptides were used in this study, this presumably holds true for longer peptides and may explain the formation of flavor compounds not obtained on heating model systems of single amino acids.[127]

Carbohydrates

The heat degradation products of carbohydrates, particularly sugars, are important flavoring ingredients, and have been studied extensively. On heating, sugars lose their bound water between 100 to 130°C, without alteration in molecular structure. With increasing temperature, 150 to 180°C, a molecule of water splits from sugars forming anhydrides, and further heating, 190 to 220°C, results in the removal of second water molecule with the formation of furfural from pentoses and hydroxymethylfurfural from hexoses.

Table 15 lists some compounds that have been reported as carbohydrate caramelization and pyrolysis products.[128] Following are likely reaction steps in sugar caramelization: enolization of aldoses with the production of more reactive 2-ketoses; dehydration of ketohexoses, without fission, to 5-(hydroxymethyl)-2-furaldehyde, and dehydration of pentoses similarly to 2-furaldehyde; hydrolytic fission of furaldehydes or intermediates leading to furaldehydes to yield, for example, formic acid and levulinic acid from hexoses, fission of 2-ketoses to yield dihydroxyacetone and glyceraldehyde, glycolaldehyde and four-carbon carbonyl compounds; dehydration of the trioses to yield acetol and pyruvaldehyde; dismutation of biose, trioses, tetroses, and their dehydration products to yield lactic aldehyde, pyruvic aldehyde, lactic acid, glycolic acid, acetaldehyde, acetic acid, formaldehyde, formic acid, acetoin, and diacetyl; self- and cross-condensations of aldehydes and ketones containing active hydrogen; reversion of aldoses and ketoses to di-, tri-, and higher oligosaccharides; dimerization of fructose to fructose anhydrides; cyclodehydration of aldoses to glycosans, and then polymerization; and enolization and dehydration of the synthetic oligosaccharides. Some of these reactions are referred to again under amino acid-carbohydrate reactions.

Amino Acid-Carbohydrate Reactions

Although amino acids and carbohydrates degrade on heating to form flavor compounds, in meat they exist in close proximity and interact with each other to form a variety of compounds.[127] Flavor production by nonenzymatic browning in foods proceeds mainly from reactions of reducing sugars with amines, amino acids, peptides, and proteins.

Aldoses react with primary or secondary amines to give aldosylamines. Under the influence of a suitable catalyst, aldosylamines rearrange to form ketoseamines. This reaction is known as the Amadori rearrangement. Ketoseamines can also be formed directly from aldoses and amines. Ketoseamines are distinguished from aldosylamines by their relatively strong reducing properties and by the formation of sugar degradation products, instead of parent hexoses, on acid hydrolysis. The chemical reactions leading to the formation of flavor compounds, i.e., furaldehydes, acetaldehyde, pyruvic aldehyde, diacetyl, acetic acid, and others, are outlined in Figure 10. These compounds are also produced as caramelization products

TABLE 15

Carbohydrate Caramelization and Dehydration Products[128]

Carbon monoxide	Acetoin
Carbon dioxide	Diacetyl
Formaldehyde	Hydroxydiacetyl
Formic acid	C-Methyl triose-reductone
Acetaldehyde	Levulinic acid
Acetic acid	Furan
Glycolaldehyde	2-Methyl furan
Glyoxal	2,5-dimethyl furan
Glyoxylic acid	Furfuryl alcohol
Lactic aldehyde	2-Furoic acid
Lactic acid	2-Furaldehyde
Acrolein	5-Methyl-2-furaldehyde
Acrylic acid	5-(Hydroxymethyl-2-furaldehyde
Pyruvaldehyde	2-Furyl methyl ketone
Pyruvic acid	2-Furyl hydroxymethyl ketone
Acetone	Isomaltol
Acetol	4-Hydroxy-2,5-dimethyl-3-(2H)-furanone
Dihydroxyacetone	Reductic acid
Glyceraldehyde	1-Methylcyclopentenol(2)-one(3)
Triose-reductone	Maltol
Mesoxalic dialdehyde	

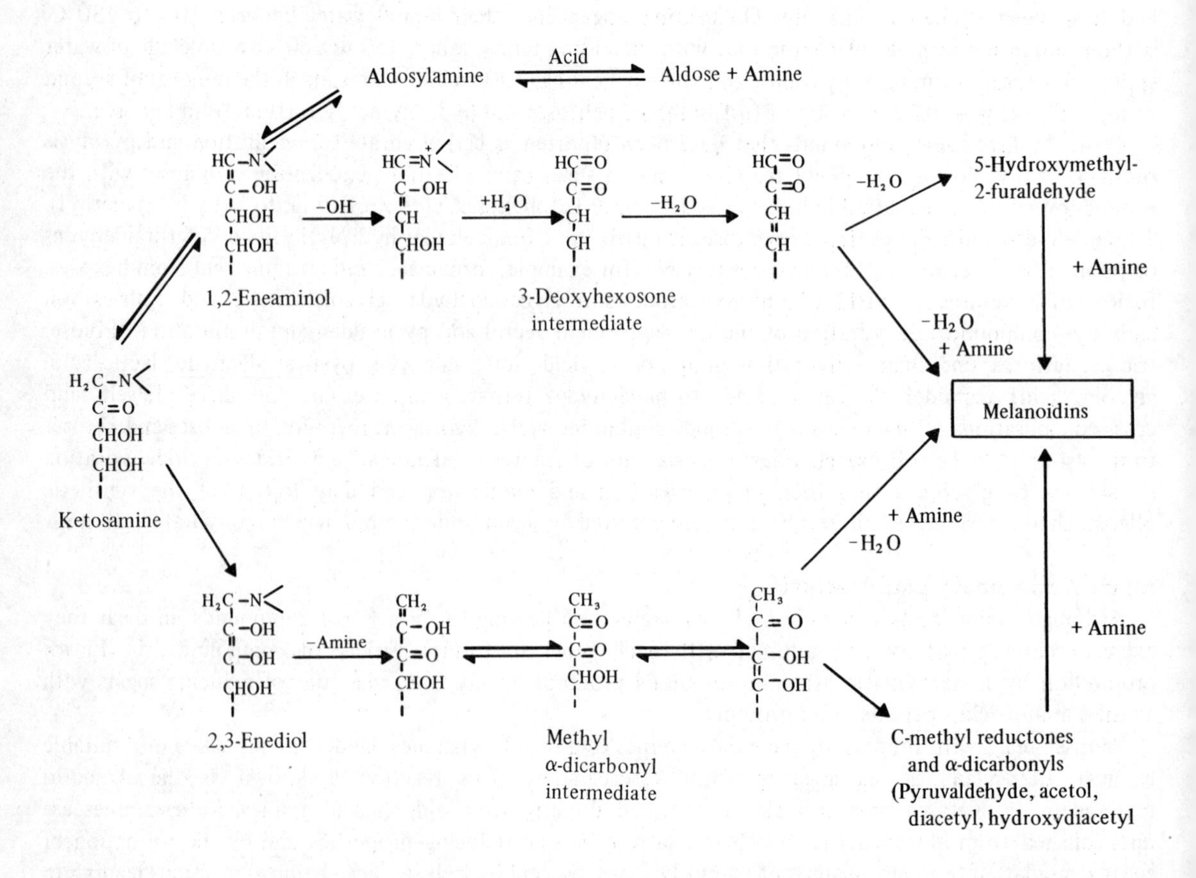

FIGURE 10. Mechanism of the Maillard reactions. (From Hodge, J. E., in *The Chemistry and Physiology of Flavors*, Schultz, H. W., Day, E. A., and Libbey, L. M., Eds., Avi Publishing, Westport, Conn., 1967, 465. With permission.)

of sugars. In the absence of amines it may be assumed that 1,2-enolization of sugars leads to the same transformations shown in Figure 10.[128] The hydroxyl group on C-3 and C-1 would be eliminated (at higher temperatures or under more acidic or alkaline reaction conditions) in the same way shown for glycosylamines and ketoseamines. For example, maltol is formed from maltose by heat alone,[131] and also in the presence of amino compounds;[129,132] but the condensation with amino compounds allows the enolizations and eliminations to take place near neutral pH and at much lower temperature.

The Maillard browning reactions have been reviewed by Reynolds[129,130] and Hodge.[129] Both aldosylamine and ketoseamine reaction schemes provide active compounds for the degradation of α-amino acids to aldehydes and ketones of one less carbon atom by the Strecker degradation mechanism. In this reaction it is required that the amino group must be α, to the carboxyl group and that the carbonyl compound must contain a –C:O–C:O– or a –C:O–(CH:CH)$_n$–C:O grouping. The reaction is illustrated by the following equation:

$$R.CO.CO.R' + R''.CH(NH_2).COOH \longrightarrow R''.CHO + CO_2 + R.CH(NH_2).CO.R$$

If the hydrogen on the α-carbon of the amino acid is substituted, a ketone is produced. For example, acetone is formed from α-aminoisobutyric acid. 2-Furaldehyde produced in Maillard browning reactions participates in the Strecker degradation reaction because it yields a α-diketone and a conjugated dicarbonyl compound by hydrolytic ring opening. Several α-dicarbonyl compounds that are produced by simple heat treatment of carbohydrates are listed in Table 15. Each of these may react amino compounds to produce aldehydes. Furthermore, sugar derived, nonvolatile α-dicarbonyl compounds (Figure 10), and dehydroascorbic acid are likely compounds for inducing the Strecker degradation. The major Strecker degradation aldehydes produced from common amino acids include: formaldehyde, acetaldehyde, propionaldehyde, isobutyraldehyde, isovaleraldehyde, 2-methyl butanal, glycolic aldehyde, lactic aldehyde, methional, benzaldehyde, and several others.[128]

The origins of other important meat volatiles include:

Maillard reaction products of sulfur containing amino acids and sugars. Arroyo and Lillard[133] identified a number of sulfur compounds from methionine, cysteine, and cystine-glucose mixtures including methyl mercaptan, ethyl mercaptan, propyl mercaptan, dimethyl sulfide, dimethyl disulfide, and pentyl mercaptan.

Thialdine has been reported to form from ammonia, hydrogen sulfide, and acetaldehyde.[29] These three compounds have been reported in the headspace of beef broth. Brinkman et al.[29] believe that in view of the ease of thialdine formation from the above mentioned compounds, it is almost certainly present in the broth or the headspace and can be considered as a contributor to beef broth flavor.

Another interesting sulfur compound first identified by Herz[3] in cooked beef is 2,5-dimethyl-1,3,4-trithiolane. This compound has a sulfide odor in concentrated form, but when diluted gives a meaty aroma. It may well be an important contributor to the aroma of boiled beef. Asinger and Thiele[134,135] have described the synthesis of this compound from acetone, hydrogen sulfide, and sulfur in diisobutylamine solvent. Similar conditions could exist locally during boiling of beef, where evolution of both hydrogen sulfide and ammonia occurs. Herz[3] hypothesizes that the sulfur atom could be supplied by active methionine (adenosylmethionine), and acetone could arise from many compounds in beef.

The findings about the origin of thialdine and trithiolane prompted an investigation into the origin of 1-methylthio-ethanethiol. This compound has particularly interesting organoleptic properties; in very dilute aqueous solutions (1 to 5 μg/l) it has a meaty rather than its usual onion-like odor.[29] Scutte and Koenders[136] have proposed a reaction mechanism for the synthesis of this compound in beef broth from known beef components. The proposed system has been substantiated experimentally. 1-Methylthio-ethanethiol is formed when acetaldehyde, methanethiol, and hydrogen sulfide are heated in aqueous solution at pH 6. These immediate precursors are in turn generated under the same conditions from alanine, methionine and cysteine in the presence of Strecker degradation agent such as pyruvaldehyde. The proposed mechanism may be represented as follows:

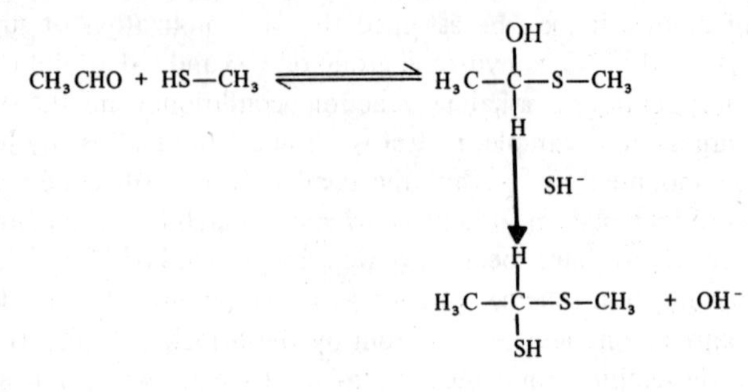

1-Methylthio-ethanethiol

Another important class of compounds identified in cooked meat aroma are pyrazines. Pyrazines are known to arise from the condensation of sugars and amino acids. The necessary precursors, that is, amino acids and sugars, are available in uncooked meat and pyrazine formation should occur readily upon heating. Koehler and Odell[138] have recently investigated the factors affecting the formation of pyrazine compounds in sugar-amine interaction.

Lipids

The role of lipids in the development of the characteristic aroma of meat is a subject of some controversy. The ability of fat to dissolve and retain aroma that originates outside the fat is well known and demonstrated in the case of smoked meat products. Fat trimmed from uncooked meat has little characteristic flavor. It has been reported that rendering of fat trimmed from uncooked meat at low temperature does not produce species specific flavor. However, if rendered at normal cooking temperatures species flavor develops with little or no basic meat character.[98] In the work carried out at Kohnstamm and Co.,[98] rendered fat was steam distilled and distillate solvent extract. Gas-liquid chromatographic studies of meat fat volatiles showed that the characteristic flavor of cooked meat fat depended largely on unsaturated aldehydes. There was considerable qualitative similarity between meat fat volatiles of different species but the quantitative picture showed considerable differences. Fatty acids, methyl ketones and saturated aldehydes were also found and are believed to play a supporting role.[98] Wasserman and Spinelli[138] investigated the role of adipose tissue constituents in the development of species specific flavor. Adipose tissue, which yields the characteristic aroma of the species on heating, was extracted with chloroform-methanol (2:1), and the extract was washed with water. The original extract retained the species specific aroma but in the water-washed residue the characteristic aroma was either missing or greatly reduced. The water soluble material extracted from adipose tissue contained amino acids and at least glucose in the carbohydrate fraction. A typical lean meat aroma developed on heating this fraction.

Lipids may contribute to meat flavor in several ways:

Lipids may act as solvent, trapping aroma components produced elsewhere. Thus, the aroma of water-washed pork lipid extract still had a "piggy" note when heat was first applied.[138] This may be due to 5,α-androst-16-en-3-one, the boar odor identified by Patterson,[85] which is dissolved in the lipids of the animal.

Lipid components or degradation products may react with compounds produced in the lean meat.

Lien and Nawar[139] investigated the interaction of amino acids and triglycerides, using valine and tricaproin as a simple model system, under nonoxidative conditions. The decomposition products identified from valine-tricaproin mixtures heated at 270°C for 1 hr are listed in Table 16. A scheme for the interaction of valine and tricaproin as proposed by Lien and Nawar[139] is shown in Figure 11. Propane, propene, carbon monoxide, and carbon dioxide, which these workers previously identified in the heated triglyceride[141] as well as in heated valine,[118] were also present when both the amino acid and the triglyceride were heated together. Methane, ethane, ethene, butane, butene, pentane, pentene, hexanoic acid, and dicaproin, which were found to form in heated tricaproin, were also present in the heated mixturé. Also, ammonia, acetone, isobutane, isobutylene, and isobutyraldehyde which were produced by

TABLE 16

Decomposition Products Identified from Valine-tricaproin Mixture Heated at 270°C for 1 hr

Carbon monoxide	Isobutylene
Carbon dioxide	Pentane
Ammonia	Pentene
Methane	Acetone
Ethane	Isobutyraldehyde
Ethene	Hexanoic acid
Propane	Caproic nitrile
Propene	Caproic amide
Butane	*N*-Isobutylcaproic amide
Butene	Dicaproin
Isobutane	

From Lien, Y. C. and Nawar, W. W., *J. Food Sci.,* 39, 917, 1974. With permission.

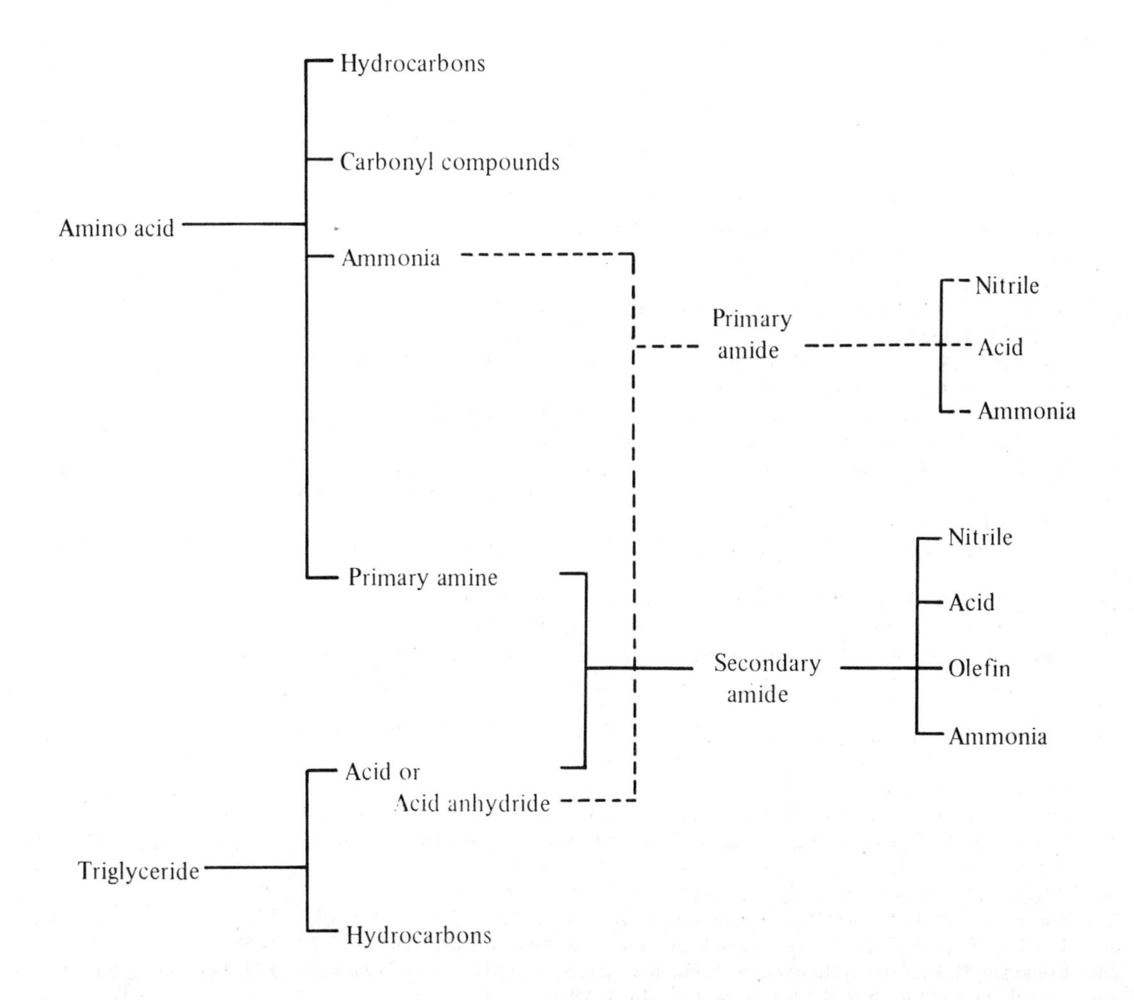

FIGURE 11. Scheme for interaction of amino acid and triglyceride upon heat treatment. (From Lien, Y. C. and Nawar, W. W., *J. Food Sci.,* 39, 917, 1974. With permission.)

heating the amino acid, valine, were individually present when both the amino acid and triglyceride were heated together. On the other hand, acrolein, methyl hexanoate, hexanal, 6-undecanone, 2-oxoheptyl hexanoate, propanediol dicaproate, propenediol dicaproate and oxopropanediol dicaproate (all observed in heated tricaproin), and isobutylamine, *N*-isobutylidene-isobutylamine, and diisobutylamine (all observed in heated valine), could not be detected when valine and tricaproin were heated together. In addition, some compounds not previously observed when the individual substrates were heated have been identified as interaction products formed during heating of the substrate mixture. These are caproic nitrile, caproic amide, and *N*-isobutylcaproic amide. It is evident from the work of Lien and Nawar[118,119,139] that in a simple system containing only one amino acid and one triglyceride, substantial interaction does take place during heating. It is also quite evident that studies on the heat decomposition products of individual components using model systems can aid significantly in unraveling thermal interaction mechanisms in more complex mixtures.

A large number of thermal oxidation products of lipids are formed, in keeping with the size of the molecules, degree of unsaturation, and type of substitution. Thermal oxidation of lipids can occur at 60°C in the presence of a few free radicals, but most of the degradation occurs at 200 to 300°C. Lactones, alcohols, ketones, and lower fatty acids are associated with the thermal oxidation of lipids.

REFERENCES

1. **Heath, H.,** *Flavour Industry,* p. 586, Sept., 1970.
2. **Dwivedi, B. K.,** *CRC Crit. Rev. Food Technol.,* 3(4), 457, 1973.
3. **Herz, K. O.,** Ph.D. thesis, Rutgers – The State University, 1968.
4. **Bender, A. E.,** *Chem. Ind.* (Lond.), 2114, 1961.
5. **Bender, A. E. and Ballance, P. E.,** *J. Sci. Food Agric.,* 12, 683, 1961.
6. **Hornstein, I. and Crowe, P. F.,** *J. Agric. Food Chem.,* 8, 494, 1960.
7. **Hornstein, I., Crowe, P. F., and Sulzbacher, W. L.,** *J. Agric. Food Chem.,* 8, 65, 1960.
8. **Kramlich, W. E. and Pearson, A. M.,** *Food Res.,* 25, 712, 1960.
9. **Merritt, C., Jr.,** in *Food Irradiation,* International Atomic Energy Agency, Vienna, 1966.
10. **Sanderson, A., Pearson, A. M., and Swain, T.,** *Chem. Ind.* (Lond.), 863, 1963.
11. **Self, R., Casey, J. C., and Swain, T.,** *Chem. Ind.* (Lond.), 863, 1963.
12. **Wick, E. L.,** in *Exploration in Future Food Processing Techniques,* MIT Press, Cambridge, Mass., 1963.
13. **Wick, E. L.,** *Food Technol.,* 19, 827, 1965.
14. **Wick, E. L., Murray, E., Mizutani, J., and Koshika, M.,** in *Radiation Preservation of Foods,* ACS Series No. 65, 1967.
15. **Yueh, M. H. and Strong, F. M.,** *J. Agric. Food Chem.,* 8, 491, 1960.
16. **Brennan, M. J. and Bernhard, R. A.,** *Food Technol.,* 18, 742, 1964.
17. **Wilson, R. A., Mussinan, C. J., Katz, I., and Sanderson, A.,** *J. Agric. Food Chem.,* 21, 873, 1973.
18. **Mussinan, C. J., Wilson, R. A., and Katz, I.,** *J. Agric. Food Chem.,* 21, 871, 1973.
19. **Mussinan, C. J. and Walradt, J. P.,** *J. Agric. Food Chem.,* 22, 827, 1974.
20. **Ockerman, H. W., Blumer, T. N., and Craig, H. B.,** *J. Food Sci.,* 29, 123, 1964.
21. **Cross, C. K. and Ziegler, P.,** *J. Food Sci.,* 30, 610, 1965.
22. **Lillard, D. A. and Ayres, J. C.,** *Food Technol.,* 23(2), 251, 1969.
23. **Chang, S. S., Hirai, C., Reddy, B. R., Herz, K. O., and Kato, A.,** *Chem. Ind.* (Lond.), 1639, 1968.
24. **Liebich, H. M., Douglas, D. R., Zlatkis, A., Muggler-Chavan, F., and Donzel, A.,** *J. Agric. Food Chem.,* 20, 96, 1972.
25. **Watanabe, K. and Sato, Y.,** *J. Agric. Food Chem.,* 20, 174, 1972.
26. **Watanabe, K. and Sato, Y.,** *J. Agric. Food Chem.,* 19, 1017, 1971.
27. **Tonsbeef, C. H. T., Copier, H., and Plancken, A. J.,** *J. Agric. Food Chem.,* 19, 1014, 1971.
28. **Tonsbeef, C. H. T., Plancken, A. J., and Weerdhof, T.,** *J. Agric. Food Chem.,* 16, 1016, 1968.
29. **Brinkman, H. W., Copier, H., deLeuw, J. M., and Tjan, S. B.,** *J. Agric. Food Chem.,* 20, 177, 1972.
30. **Herz, K. O. and Chang, S. S.,** *Adv. Food Res.,* 18, 1, 1970.
31. **Alford, J. A., Smith, J. L., and Lilly, H. D.,** *J. Appl. Bacteriol.,* 34, 133, 1971.
32. **Nurmi, E. and Niinivaara, F. P.,** Proc. *10th Eur. Meeting Meat Res. Work,* Roskilde, Denmark, 1964.

33. **Cantoni, C., Molnar, M. R., Renon, P., and Giolitti, G.,** *Die Nahrung,* 11, 341, 1967.
34. **Gervasini, C. and Caserio, G.,** *Wien. Tieraerztl. Monatsschr.,* 56, 418, 1969.
35. **Lubienicki, V. S.,** *Fleischwirtschaft,* 52, 72, 1972.
36. **Wallach, D. P.,** *J. Lipid Res.,* 9, 200, 1968.
37. **Wahlroos, O. and Niinivaara, F. P.,** *Proc. 15th Eur. Meeting Meat Res. Work,* Helsinki, Finland, 1969, 240.
38. **Smith, J. L. and Alford, J. A.,** *J. Food Sci.,* 34, 75, 1969.
39. **Cerise, L., Bracco, U., Horman, I., Sozzi, T., and Wuhrmann, J.,** *Fleischwirtschaft,* 53, 223, 1973.
40. **Demeyer, D., Hoozee, J., and Mesdom, H.,** *J. Food Sci.,* 39, 293, 1974.
41. **Brockerhoff, H.,** *Comp. Biochem. Physiol.,* 19, 1, 1966.
42. **Coretti, K.,** *Fleischwirtschaft,* 45, 21, 1965.
43. **Oterholm, A., Ordal, Z. J., and Witter, L. D.,** *Appl. Microbiol.,* 16, 524, 1968.
44. **De Ketelaere, A., Demeyer, D., Vandekerckhove, P., and Vervaeke, I.,** *J. Food Sci.,* 39, 297, 1974.
45. **Wood, T. and Bender, A. E.,** *Biochem. J.,* 67, 366, 1957.
46. **Bender, A. E., Wood, T., and Palgrave, J. A.,** *J. Sci. Food Agric.,* 9, 812, 1958.
47. **Kazeniac, S. J.,** in *Proc. Flavor Chemistry Symp.,* Campbell Soup Co., Camden, New Jersey, 1961, 37.
48. **Wood, T.,** *J. Sci. Food Agric.,* 12, 61, 1961.
49. **Batzer, O. F., Santoro, A. T., and Landman, W. A.,** *J. Agric. Food Chem.,* 10, 94, 1962.
50. **Macy, R. L., Naumann, D. H., and Bailey, M. E.,** *J. Food Sci.,* 29, 136, 1964.
51. **Macy, R. L., Naumann, D. H., and Bailey, M. E.,** *J. Food Sci.,* 29, 142, 1964.
52. **Koehler, H. H. and Jacobsen, M.,** *J. Agric. Food Chem.,* 15, 707, 1967.
53. **Solms, J.,** *Fleischwirtschaft,* 48, 287, 1968.
54. **Lawrie, R. A.,** *Flavour Industry,* p. 591, Sept. 1970.
55. **Macy, R. L., Jr., Naumann, H. N., and Bailey, M. E.,** *J. Food Sci.,* 35, 78, 1970.
56. **Macy, R. L., Jr., Naumann, H. N., and Bailey, M. E.,** *J. Food Sci.,* 35, 81, 1970.
57. **Macy, R. L., Jr., Naumann, H. N., and Bailey, M. E.,** *J. Food Sci.,* 35, 83, 1970.
58. **Pepper, F. H. and Pearson, A. M.,** *J. Agric. Food Chem.,* 19, 964, 1971.
59. **Tsai, R., Cassens, R. G., Briskey, E. J., and Greaser, M. L.,** *J. Food Sci.,* 37, 612, 1972.
60. **Wasserman, A. E. and Spinelli, A. M.,** *J. Agric. Food Chem.,* 20, 171, 1972.
61. **Jarboe, J. K. and Mabrouk, A. F.,** *J. Agric. Food Chem.,* 22, 787, 1974.
62. **Pepper, F. H. and Pearson, A. M.,** *J. Agric. Food Chem.,* 22, 49, 1974.
63. **Kuninaka, A.,** in *The Chemistry and Physiology of Flavors,* Schultz, H. W., Day, E. A., and Libbey, L. M., Eds., Avi Publishing, Westport, Conn., 1967, 515.
64. **Jones, N. R.,** *J. Agric. Food Chem.,* 17, 712, 1969.
65. **Kirimura, J., Shimizu, A., Kimizuka, A., Ninomiya, T., and Katsuya, N.,** *J. Agric. Food Chem.,* 17, 689, 1969.
66. **Unilever, N. V.,** Netherlands Patent 66,03,250, September 12, 1966.
67. **Unilever, N. V.,** Netherlands Patent 67,00,991, July 22, 1968.
68. **Unilever, N. V.,** Netherlands Patent 67,07,232, November 26, 1968.
69. **Van der Ouweland, G. and Peer, H. G.** (Unilever N. V.), German Patent 1,932,799, January 8, 1970.
70. **Brinkman, H. W. and Van der Heyden, A.** (Lever Brothers Co.), U.S. Patent 3,653,920, April 4, 1972.
71. Kyowa Fermentation Industry Co., Ltd., Netherlands Patent 66,15,345, May 2, 1967.
72. **Kitada, N., Shimazaki, H., and Komato, Y.** (Ajinomoto Co., Inc.), British Patent 1,206,265, September 23, 1970.
73. Ajinomoto Co., Inc., French Patent 1,474,613, March 24, 1967.
74. International Flavors & Fragrances, Inc., Netherlands Patent Appl. 66,09,520, January 9, 1967.
75. **Giacino, C.** (International Flavors & Fragrances, Inc.), South Africa Patent 68,00,518, June 25, 1968.
76. **Thomas, P. D.** (Pfizer, Inc.), U.S. Patent 3,615,698, October 26, 1971.
77. **Broderick, J. J. and Marcus, S. A.** (Kohnstamm, H. and Co., Inc.), U.S. Patent 3,532,515, October 6, 1970.
78. **Barbella, N. G., Tannor, B., and Johnson, T. G.,** *32nd Ann. Proc. Am. Soc. Animal Production,* 1939, 320.
79. **Lowe, B. and Kastelic, J.,** *Iowa Agric. Exp. Stn. Res. Bull.,* 495, 1961.
80. **Crocker, E. C.,** *Food Research,* 13, 179, 1948.
81. **Gruhm, H.,** *Die Nahrung,* 9, 325, 1965.
82. **Bryce-Jones, K., Harries, J. M., Robertson, J., and Akers, J. M.,** *J. Sci. Food Agric.,* 15, 790, 1964.
83. **Huffman, D. L., Palmer, A. Z., Carpenter, J. W., Hargrove, D. D., and Koger, M.,** *J. Anim. Sci.,* 26, 290, 1967.
84. **Weir, C. E.,** *Science of Meat and Meat Products,* Freeman, San Francisco, Calif., 1960, 212.
85. **Patterson, R. L. S.,** *J. Sci. Food Agric.,* 19, 31, 1968.
86. **Patterson, R. L. S.,** *J. Sci. Food Agric.,* 19, 436, 1968.
87. **Lewis, P. K., Jr., Brown, C. J., and Heck, M. C.,** *Food Technol.,* 21, 393, 1967.
88. **Harris, N. D., Strong, D. H., and Sunde, M. L.,** *J. Food Sci.,* 33, 543, 1968.
89. **Johnson, A. R. and Vickery, J. R.,** *J. Sci. Food Agric.,* 15, 695, 1964.
90. **Gunther, H., and Schweiger, A.,** *J. Food Sci.,* 31, 300, 1966.
91. **Herz, K. O.,** M.S. thesis, Rutgers University, 1963.
92. **Bouton, P. E., Howard, A., and Lawrie, R. A.,** *Spec. Rept. Food Invest. Board,* London, No. 66, 1957.
93. **Altman, P. L. and Dittmer, D. S., Eds.,** *Metabolism,* Fed. Am. Soc. Experimental Biol., Bethesda, Md., 1968, 426.

94. **Altman, P. L. and Dittmer, D. S.,** Eds., *Biological Data Book,* Vol. III, Fed. Am. Soc. Experimental Biol., Bethesda, Md., 1973, 1538.
95. **Zender, R., Lataste-Dorolle, C., Collet, R. A., Rowinski, P., and Mouton, R. F.,** *Food Research,* 23, 305, 1958.
96. **Koch, D. E., Markel, R. A., and Purchas, B. J.,** *J. Agric. Food Chem.,* 18, 1073, 1970.
97. **Koch, D. E., Markel, R. A., and Purchas, B. J.,** *J. Agric. Food Chem.,* 18, 1078, 1970.
98. **Frdos, T.,** *Studies Inst. Chem.,* University Szegediensis, 3, 51, 1943.
99. **Bate-Smith, E. C. and Bendall, J. R.,** *J. Physiol.,* 196, 177, 1947.
100. **Bendall, J. R.,** *J. Physiol.,* 114, 71, 1951.
101. **Marsh, B. B.,** *Biochim. Biophys. Acta,* 9, 127, 1952.
102. **Marsh, B. B.,** *J. Sci. Food Agric.,* 5, 70, 1954.
103. **Lawrie, R. A.,** *J. Physiol.,* 12, 275, 1953.
104. **Howard, A., Lee, C. A., and Webster, H. L.,** *Div. Food Preserv. Tech.,* Paper No. 21, C.S.I.R.O., Australia, 1960.
105. **Lee, C. A., and Webster, H. L.,** *Div. Food Preserv. Tech.,* Paper No. 30, C.S.I.R.O., Australia, 1963.
106. **Millet, J. and Henery, M.,** *Proc. 6th Symp. Food Additives,* Madrid, 1960.
107. **Niinivaara, F., Pohja, M., and Komulainen, S. E.,** *Proc. 7th Meeting Eur. Meat Res. Work,* Warsaw, 1961.
108. **Mihalyi, V. and Kormendy, L.,** *Food Technol.,* 21, 108, 1967.
109. **Dahl, O.,** *Fleischwirtschaft,* 50, 806, 1970.
110. **Pohja, M. and Niinivaara, F.,** *Fleischwirtschaft,* 40, 932, 1960.
111. **Sajber, C., Karakas, R., and Mitic, P.,** *Proc. 17th Meeting Eur. Meat Res. Work,* Bristol, England, 1971.
112. **Cantoni, C., Molnar, M. R., Renon, P., and Giolitti, G.,** *Proc. 13th Meeting Eur. Meat Res. Work,* Rotterdam, 1967.
113. **Dierick, N., Vandekerckhove, P., and Demeyer, D.,** *J. Food Sci.,* 39, 301, 1974.
114. **De Ketelaere, A., Demeyer, D., Vandekerckhove, P., and Vervaeke, I.,** *J. Food Sci.,* 39, 297, 1974.
115. **Langner, H.,** *Fleischwirtschaft,* 52, 1299, 1972.
116. **Langner, H.,** *Fleischwirtschaft,* 49, 1475, 1969.
117. **Reuter, G., and Langner, H.,** *Fleischwirtschaft,* 170, 1968.
118. **Lien, Y. C. and Nawar, W. W.,** *J. Food Sci.,* 39, 911, 1974.
119. **Lien, Y. C. and Nawar, W. W.,** *J. Food Sci.,* 39, 914, 1974.
120. **Layer, R. W.,** *Chem. Rev.,* 63, 489, 1962.
121. **Sprung, M. M.,** *Chem. Rev.,* 26, 297, 1940.
122. **Schwoegler, E. J. and Adkins, H.,** *J. Am. Chem. Soc.,* 61, 3499, 1939.
123. **Walter, W., Harke, H. P., and Polchow, R.,** *Z. Naturforsch.,* B22, 931, 1967.
124. **Merritt, C., Jr. and Robertson, D. H.,** *J. Gas Chromatogr.,* 5, 96, 1967.
125. **Schulman, G. P. and Simond, P. G.,** *Chem. Commun.,* 17, 1040, 1968.
126. **Vollmin, J., Kriemler, P., Omura, I., Seibl, J., and Simon, W.,** *Microchem. J.,* 11, 73, 1966.
127. **Wasserman, A. E.,** *J. Agric. Food Chem.,* 20, 737, 1972.
128. **Hodge, J. E.,** in *The Chemistry and Physiology of Flavors,* Schultz, H. W., Day, E. A., and Libbey, L. M., Eds., Avi Publishing, Westport, Conn., 1967, 465.
129. **Reynolds, T. M.,** *Adv. Food Res.,* 12, 1, 1963.
130. **Reynolds, T. M.,** *Adv. Food Res.,* 14, 167, 1965.
131. **Diemar, W. and Hala, H.,** *Z. Lebensm.-Untersuch.-Forsch.,* 110, 161, 1959.
132. **Patton, S.,** *J. Biol. Chem.,* 184, 131, 1950.
133. **Arroyo, P. T. and Lillard, D. A.,** *J. Food Sci.,* 35, 769, 1970.
134. **Asinger, F. and Thiele, M.,** *Angew. Chemie.,* 70, 667, 1958.
135. **Asinger, F. and Thiele, M.,** *Ann. Chem.,* 627, 195, 1959.
136. **Scutte, L. and Koenders, E. B.,** *J. Agric. Food Chem.,* 20, 181, 1972.
137. **Koehler, P. E. and Odell, G. V.,** *J. Agric. Food Chem.,* 18, 895, 1970.
138. **Wasserman, A. E. and Spinelli, A. M.,** *J. Food Sci.,* 35, 328, 1970.
139. **Lien, Y. C. and Nawar, W. W.,** *J. Food Sci.,* 39, 917, 1974.
140. **Lien, Y. C. and Nawar, W. W.,** *J. Am. Oil Chem. Soc.,* 50, 76, 1973.

FISH

Fish was preserved originally to insure adequate supply. Today the preservation of fish is geared to satisfy specific market demands. Various methods of fish preservation are employed: (1) pickling with salt crystals or salt solutions, (2) air or oven drying, (3) smoking, (4) marinating, and (5) sterilization (autoclaving) in tins using precooked fish packed in brine or olive oil.

The method of preservation influences the organoleptic characteristics of the finished product. Cod (Atlantic, *Gadus morhua* L.), salted and dried to about 40% moisture, is marketed under the name *bacalao*. Fish, such as haddock and cod, that is cured without salt by being split and hung in the air to dry is known commercially as stockfish. Herring, cod, salmon, sardine, and eel are usually smoked; eel and herring are also marinated. Boiled sardines, to which olive oil is usually added, are canned much like tuna of the species *Thunnus thynnus* (red) and *Thunnus alalonga* (white tuna). Salmon, mackerel, and other fish suitable for slicing are usually canned after preliminary boiling. The preparation of anchovy paste falls in the category of foodstuffs preserved by salting. Although flavor ingredients are seldom used as flavoring in the above preservation processes, a few spices, such as laurel, are used in marinating and sometimes in smoking fish.

VEGETABLES

Pickled Vegetables

Salt-pickled Vegetables

In general, preservation by covering the vegetable directly with salt (similar to the preservation of meat or fish) is not feasible. Salt-pickling is carried out by pickling the product in salted water. The percentage of salt in the pickling solution is expressed in degrees measured by a special hygrometer or salinometer, which is graduated to show the percentage of saturation of sodium chloride solution. The scale of the salinometer is divided into 100 degrees (1 to 100); one degree corresponds to 0.265% salt in the solution. The solution is saturated when it contains 26.5% pure sodium chloride.

Pickling is effected by using solutions with variable percentages of saturation depending on the final use. The preservative action of salt often is combined with the flavoring action of spices, aromatic herbs, and other flavor ingredients, especially if the pickled products are intended for direct consumption and not for use in subsequent food manufacturing operations. Specific methods of pickling may require the addition of specific flavor ingredients; for example, the addition of orange peels to the solution for home pickling of black olives is a fairly common practice.

An old recipe for the preparation of a flavored solution for generic home pickling of legumes follows:

Pickling Solution for Legumes

Ingredients:

Basil	100 g
Laurel leaves	10 g
Coriander	8 g
Mace	4 g
Juniper, minced	2 g
Salt	1 kg
Water	10 l

Preparation:

1. Mix all ingredients together and bring to a boil.
2. Allow the mixture to cool; strain.
3. Add legumes to pickling solution. The legumes will remain satisfactorily preserved for a few weeks.

Pickling is mainly used for preservation of vegetables in the fresh state. Therefore, unless resulting products with specific flavors are desired, pickled products in general are prepared to exhibit a flavor as close as possible to the original organoleptic characteristics.

Vinegar-pickled Vegetables

The preparation of vinegar-pickled vegetables is somewhat different and requires either the initial boiling of fresh vegetables or a preliminary salt pickling. If the vegetables have been initially salt-pickled, they should be washed thoroughly to remove the salt prior to vinegar pickling. Wine vinegar is highly desirable for pickling, although it may not always be available. If permitted by local regulations, vinegars from various sources and also acetic acid solutions are employed.

Flavoring of vinegar-pickled vegetables, although considered an indispensable operation in all cases, is extremely functional where vinegars other than wine vinegar are employed. A large number of spices, herbs, seeds, fruits, roots, and legumes are used for flavoring vinegar-pickled vegetables. They are used individually or in blends. The flavor ingredients most commonly employed include white and black peppers, garlic, ginger, clove buds, cayenne pepper, coriander, nutmeg, chili, cinnamon, mustard seeds, horseradish, laurel, and paprika. Depending on the application, the whole spice, powder, oleoresins, or essential oils can be used. The flavor ingredients should be dispersed in the vinegar used for pickling.

Chutney

Chutney consists of a salad of comminuted vegetables and fruits in a rather sweet, thick sauce. The choice of vegetables and fruits, which is quite varied, includes onion, carrot, cauliflower, cucumber, beet, raisin, apricot, apple, etc. Chutney sauces contain a variety of products, although they primarily consist of sweet syrup, vinegar, soy, caramel, and several flavor ingredients, usually powdered seeds or fruits from ground spices. Powdered spice also contributes to the thickening of the sauce. Mace, pepper, paprika, and curry are frequently used together with wheat flour and natural gums. (For a further discussion of the flavor ingredients of chutneys, see p. 801.)

The vegetables and fruits are boiled for about 20 minutes together with water and sugar; then spices and salt are added. The mixture is again brought to boiling, then allowed to simmer for about 15 minutes. Vinegar and acetic acid are added, and the mixture is again brought to boiling,; it is allowed to cool and then poured into glass containers together with a thick sauce prepared separately.

Oil-pickled Vegetables

Oil-pickled vegetables, as vinegar-pickled products, must undergo a preliminary cooking. Also vinegar-pickled products may be subsequently pickled in oil. The addition of flavor ingredients to oil-pickled vegetables is effected sometimes by two methods. In the first method flavor ingredients are added during the preliminary cooking of the vegetables; in the second, the flavors are added during preservation. In both cases the flavoring of oil-pickled vegetables is usually less intensive than vinegar-pickled products. Laurel is a flavor commonly employed during preservation; the flavoring consists of whole laurel leaves kept immersed in pickling oil. In addition to preserving vegetables, oil pickling is also frequently used to preserve fish.

Potatoes

Boiled and fried (chips) potatoes or dehydrated potato flakes exhibit different aromas because of the different degradation mechanisms involved as aroma precursors during cooking. Other ingredients, such as fats used for frying, obviously yield different flavor characteristics in the finished products.

Boiled Potatoes

The aroma of boiled potato is believed related to a Strecker's degradation and to the autoxidation of amino acids, sugars, and fats present in the tuber. Following are components reported present in boiled potato flavor:

Components of Boiled Potato Flavor

Alcohols

Ethyl alcohol
Methyl alcohol

Carbonyl Compounds

Acetaldehyde
Acetone
2,4-Decadien-1-al (50%)
2-Decen-1-al
2-Methylbutanal
3-Methylbutanal
2-Methylpropanal
Nonanal
2-Nonen-1-al
Propanal
Propenal
2-Undecen-1-al

Sulfur-containing Compounds

Dimethylsulfide } 90%
Methyl mercaptan }
Dimethylsulfide } 70%
Ethyl mercaptan }

Sulfur-containing Compounds *(continued)*

Diethyl sulfide
Hydrogen sulfide
Methyl ethyl disulfide
Methyl ethyl sulfide
Methyl isopropyl disulfide
Methyl *n*-propyl sulfide
2-Propyl mercaptan
n-Propyl mercaptan

Ketones

Acetone
2-Butanone
2-Heptanone
2-Hexanone
2-Nonanone
2-Octanone
2-Pentanone

Pyrazine Derivatives

Decomposition products of methionine
2,5-Dimethylpiperazine

Fried Potatoes

The flavor of fried potatoes is derived mainly from the autoxidation of the fats and oils used for cooking rather than the action of heat on the potato. Potato aroma is rather unstable and reverts readily from pleasant to unpleasant on aging. Aging causes a decrease of the dienal content with a corresponding increase of carbonyl compounds (following the darkening reaction) together with a characteristic stale odor. The aroma of freshly fried potatoes has been reported caused by the following components:

Aldehydes in Fried Potato Aroma

Heptanal	Octanal
2-Hepten-1-al	2-Octen-1-al
Hexanal	Pentanal
2-Hexen-1-al	

Dehydrated Potato Flakes

The aroma of this product is characterized by the presence of carbonyl compounds derived from the same precursors. The carbonyl compounds are formed as degradation products of saturated and unsaturated fatty acids. These are formed by mechanisms similar to those indicated for boiled potatoes with the added effects of a heat history from the dehydration process. The off-flavors developed during flake manufacture can be controlled in part by using antioxidants.

MISCELLANEOUS PRODUCTS

Tobaccos

The various species of *Nicotiana* (*N. tabacum, N. rustica, N. petuniodes, N. polidiclia*) belong to the family *Solanaceae*. Among these the countless varieties of *N. tabacum* and *N. rustica* represent the most important source for both smoking and sniffing tobaccos. *Nicotiana* plants are native to the Americas; the first seeds were carried to Europe by Cortez in 1519. Because of the widespread *Nicotiana* cultivations throughout the five continents, tobacco varieties are classified according to the geographical name of the growing site—America, Asia, Australia, and Europe. The most important varieties are listed below:

1. American tobaccos: Kentucky (light and heavy)
   Virginia (yellow, sweet, and dark, heavy)
   Maryland (yellow, sweet, aromatic)
   Seed leaf (with silky leaves, sweet)
   Havana (highly aromatic and sweet)
2. Asian tobaccos: Middle East or Turkey
3. Izmirn tobaccos: Sansum and Bafia
4. Australian tobaccos: Sumatra and the Philippines
5. European tobaccos: Produced in the Netherlands, Belgium, Germany, Russia, Hungary, and Italy. They are used in cigarette, cigar, pipe, sniffing, or chewing tobaccos; the latter is steadily declining in use.

Tobacco leaves are treated for their specific end-use. Sniffing tobaccos are made from the most aromatic leaves, which are comminuted, fermented, and flavored to enhance the aroma. Chewing tobaccos, consisting of pressed plaits made from tobacco leaves, are treated with sugar and only slightly flavored. Smoking tobaccos are manufactured using selected qualities most suitable for the specific end-use (e.g., cigarette, pipe, cigar). The dried leaves, free of the woody portion, are used as the outer wrapping for cigars or are finely comminuted for use in pipe and cigarette tobacco blends.

The characteristic aroma of the properly selected tobacco varieties, enhanced and modified by the specific manufacturing process, is considered the most important factor in the finished product because it directs the specific preference of consumers. Natural tobacco flavor may not be extremely palatable because of its constituents and may require additional flavoring. However, the flavoring of tobacco is often misconstrued as a means to mask the natural aroma of less valuable tobacco qualities. Therefore, cigarette manufacturers tend to shy away from claiming additional flavoring in product brands.

Tobacco flavoring may have a twofold function: (1) flavor enhancement and/or improvement, and (2) flavor modification. Flavor enhancement is obtained by the addition of natural tobacco extracts, whereas flavor modification is obtained by the use of "blenders." By extracting the tougher, woody portion of leaves and stems of the best tobacco qualities with water-alcohol or water-ether mixtures, natural tobacco extracts are prepared; these are used to improve the aroma of less valuable tobaccos. Havana tobacco extracts, for instance, are useful in the processing of Puerto Rican tobaccos. Blenders, or finishing flavors, are used to enhance those pleasant flavor notes and to soften the harsh notes. These flavor ingredients are products from various plant sources and make up a substantial portion (up to one-fifth by weight) of the finished product.

During softening of tobacco leaves by conditioning under controlled humidity conditions, the humidified flexible leaves are best suited for the addition of flavors. Specific flavor ingredients, dispersed in liquid aqueous sweetener or solutions (honey, molasses, apple and plum juices, fig tincture, sugar, cocoa, licorice, etc.), are brought to boiling, and the tobacco leaves (previously conditioned with steam) are dipped in the solution. The dip is usually quick and is followed by draining, stretching, drying, and rehumidifying of the leaves.

Each tobacco quality requires specific complementary flavoring and a more or less pronounced sweetening. This at times is carried out by using wine, rum, propylene glycol, alcohol, and glycerol as the dispersing phase for resinoids, essential oils, fluid extracts, and isolates. The most frequently encountered flavor ingredients include storax and opopanax resinoid, Tolu and Peru balsams, absolutes or extracts (melilotus, liatris, tonka bean, vanilla), valerian tincture, orris tincture, and the essential oils of geranium, rose, cinnamon, clove, etc. The essential oils must be terpeneless. Isolates and synthetic ingredients are used at minimum levels and only for marginal flavoring. For example, a few aldehydes enhance the floral headnote. Government regulations (which vary from country to country) provide excellent guidelines as to which products may be used in the formulation of tobacco flavors.

Sniffing tobacco draws its aroma from fermentation of the leaves. Tobacco for cigars is only slightly flavored, cigarette and pipe tobaccos draw the characteristic aroma from selected tobacco qualities (Virginia, Carolina, Burley, Lalaika, and Turkish tobaccos) and from added flavor ingredients and sweeteners.

The formulation of tobacco flavors follows specific criteria entirely different from the flavoring criteria used in food and perfumery. The aroma of the flavor ingredients develops in two stages, *i.e.*, before and after combustion. It is impossible to give fixed guidelines with very few exceptions. Sometimes fruit flavor imitations, such as pear, apple, and peach, may be required to strengthen certain flavor notes. More frequently honey flavor imitations are used. Rum flavor, widely employed in the manufacture of cigars, is usually an imitation prepared by guidelines similar to those used for the preparation of rums for liqueur formulations; these usually contain rum ether.

Chewing Gums

The advent of the chewing gum industry is fairly recent—about one hundred years old. The basic raw material consists of chicle gum (the latex of naseberry or sapodilla) native to Yucatan and Central America. Two basic groups of gum products are manufactured: (1) the so-called chewing gums (in sticks or bonbons) and (2) bubble gums. For the individual products the gum is prepared in standard bases by specialized manufacturers who supply, together with the gum base, recommendations for its use. The gum is employed in amounts varying between 17–30%, depending on the product type and the blend of gums used. Other ingredients present in varying amounts include sucrose (46–59%), glucose (18–23%), water, glycerol, flavor ingredients, and colorants.

Manufacture consists of mixing the ingredients in a special mixer equipped for heating the blend to 60°C; then the mixture is kneaded by passing the plastic dough through cylinders, and finally the mixture is cut into sticks, balls, etc. The gum is heated in the mixer and softened without melting; then glucose, sucrose, water, glycerol, flavor ingredients, and colorants (optional) are added in this order. The homogenous dough is heated to 45°C, and the mass is poured and cut into loaves that are allowed to cool. The loaves are then rolled or stamped, and the manufacturing cycle is terminated by additional operations suitable to prepare the desired end product. The various operations, briefly described above, require special equipment and skill; the discussion of the latter, however, goes beyond the scope of the book.

A number of precautions are required that, in ultimate analysis, characterize the chewing gum production as to smoothness, hardness, plasticity, cohesiveness, lack of tackiness, elasticity, and flavor. The original chewing-gum product was almost flavorless; then it was made sweet and finally aromatic. Today effective flavoring is at the base of most chewing- and bubble-gum production. Therefore, suitable flavor ingredients with specific requirements must be employed. Chewing gum flavor ingredients must be insoluble in water but readily soluble in the gum. This minimizes flavor loss by solubility in saliva and insures the gum acting as a long-lasting flavor carrier. The flavors must also act as plasticizers in the gum and exhibit heat stability throughout the manufacturing cycle. Raw or rectified essential oils are well suited for this application. The most popular chewing-gum flavors include spearmint, peppermint, anise, wintergreen, cinnamon, tutti-frutti, cherry, and banana.

Drug Flavors

The original production of drugs was based on the extraction of herbs. In addition to the "active curative principles," the botanical source contained flavor ingredients that often imparted palatability to the drug extract. By blending various herb extracts, it was often not too difficult to formulate products having sufficiently agreeable odor and flavor for consumption.

With the increasing use of synthetic drugs and drug principles of animal origin, the number of drugs with extremely disagreeable odor and flavor has increased. This, in turn, has greatly increased the need for flavors capable of masking drugs, especially those presented in liquid form for oral consumption. Although the production of drugs today is increasingly oriented toward the manufacture of coated tablets, capsules, injectable specialty products, and suppositories, the formulation of liquid products (syrups and drops) for direct oral consumption together with pills and granulated products still represents a large percentage of drugs requiring suitable flavoring. "Medicines" are often associated with products having exclusively unpleasant odor and flavor. For many years no effort had been made to render more acceptable or palatable the odor or flavor of drugs. Today the trend has reversed; even products for external use, such as pastes, pomades, and unguents, are suitably formulated with agreeable fragrances. Although not all drug principles have disagreeable odor and flavor, some exhibit nauseating aroma characteristics.

The flavoring of solid drugs (tablets, etc.) may be readily attained by hard coating with sugar-based flavors, while the flavoring of liquid products is much more complicated. The simple addition of sugar or other sweeteners to liquids may not only be insufficient but may also produce detrimental effects. The criteria of sweetening the flavor of drugs because of their bitter or salty flavor is absolutely incorrect. The formulation of "medicinal" syrups is perfectly valid as long as certain basic criteria are followed, since the addition of sweeteners to certain products may yield totally unpalatable flavors. On the other hand, flavor modifications may yield unexpected and extremely interesting results. A product with a very bitter taste is a case in point. Theoretically the dispersion of the bitter active principle in the syrup should be sufficient to overcome the bitter flavor; in practice the addition of a few bitter flavor ingredients yields a new bitter note that yields a very palatable flavor sensation with only a minimum addition of sweetener. The addition of fruit flavors and sugar to a drug is not adequate to mask the original unpleasant flavor. For example, sweetening a bitter note and then adding a strawberry, currant, or raspberry flavor may yield totally unacceptable flavor results, whereas shifting the original bitter note to a bitter vermouth note together with the addition of a slight citrus note may yield an excellent solution to the problem. A few exceptions should be mentioned, however, such as drug products for children, who show a marked preference for fruit flavors.

A preliminary, thorough study of the original drug odor and flavor to identify the most suitable modifications for the creation of a new flavor without attempting to transform it into a fruit-flavored product is highly recommended. Following this preliminary investigation and taking into account the expected use levels of the product, the chemical constitution, and the carrier, the task of compounding can be undertaken by selecting the flavor ingredients from broad categories, such as fluid extracts, soluble essential oils, isolates, synthetic products, and concentrated and lyophilized fruit juices. Compounded flavors used for syrups, liqueurs, or pastry should be considered first.

In the choice of flavor ingredients, the chemical composition of the drug must be taken into consideration, since any potential reaction between flavor ingredients and the active drug principles must be avoided. Once any potential reaction between the various ingredients is avoided and the desired flavor results are attained, it is still necessary to control the activity of the drug by suitable analytical means over an extended period of time; certain reactions can develop at an extremely slow rate, and the resulting effects are often unpredictable. As a result of several investigations conducted in France at the Department of Pharmacy of the University of Grenoble, analytical techniques have been developed to monitor both flavor retention and residual pharmacological activity over a period of time in drug preparations. A few flavor houses have developed a series of flavor ingredients specifically suitable for effective drug flavoring, basing their developments on previous experience with typical drug formulations. These flavor ingredients are highly stable and nonreactive toward known drug active principles.

Aside from the preference for fruit flavors by children in general, drug products are commonly flavored with the following notes:

1. Fruit notes: raspberry, strawberry, currant, cherry, peach, banana, plum, orange, lemon, citron, and mandarin
2. Variations of bitter notes: cinchona, rhubarb, gentian, vermouth, and bitter orange
3. Masking flavors with high yields: anise, peppermint, licorice, vanilla, cocoa, and caramel.

Finally, it should be emphasized that in many countries the approval of flavor ingredient for use in food does not include its use for flavoring drug products. For example, current U.S. regulations require that the complete drug formulation, including active principle(s), inert fillers, colorants, and flavor ingredients, be thoroughly tested for pharmacological activity and toxicological safety. The objective is to prevent potential toxicological hazards and/or losses of pharmacological activity resulting from unexpected chemical and physicochemical interactions between the drug principle and other components in the formulation. Similar regulations are followed in West Germany and in Italy, but many other European countries do not regulate the specific use of approved flavor ingredients in drug products.

Animal Feeds

It is impossible to determine exactly whether animals in general or individual animal species in particular experience similar flavor and olfactory sensations to humans. It has been proven, however, that the olfactory sensation is in general much more developed in animals than in humans. A few animal species prefer certain specifically flavored foods, and they clearly show a repulsion for others. Agreeable aromas stimulate similar physiological reactions in animals and humans. It is apparent that flavor ingredients can have a functional use in solving certain animal nutrition problems, for it is well known that palatable food is digested much easier than less appetizing food. This in turn causes feed to be assimilated by animals at a much faster rate than less aromatic feeds. Thus flavors add extra nutritional value to feeds. The efficient consumption of feed has a great economic importance in large-scale farming of domestic animals. The larger the amount of feed ingested and digested by the animal, the higher the yields of meat from poultry and cattle, eggs from poultry, and milk from cows.

The creation of animal feed flavors was developed and conditioned by observations of the alimentary habits of various animals based on species and age. Hogs are fed a sugar-flavored feed, whereas mild flavors are added to the feed of suckling pigs. Cattle generally show a preference for anise- and fenugreek-flavored feed together with sweet, sugar-like flavors. Calves' feed is flavored with milk flavors, whereas the molasses aroma is particularly preferred by more adult animals. Cattle feed often contains preservatives and antioxidants that contribute to the preservation of carotene and vitamin E content in the feed. However, a few preservatives tend to destroy rather than preserve the carotene content; for example, a decrease in carotene was observed in animal feed treated with zinc-bactracin and sodium metabisulfite.

From experiments conducted on dogs, it was observed that the addition of beef-bone flavors favored ingestion of several different foods. Similarly, food flavored with products somewhat undefinable by human taste standards favored the ingestion of feed by chickens, pheasants, etc. Although young animals in general (swine and cattle) show preference for those flavors imitating the flavor of the mother's milk, the type of feed selected and therefore its intrinsic flavor characteristics are also important in the formulation of complementary flavors.

Although it is possible to attain quite satisfactory flavor results using milk flavors, optimum flavoring is achieved with milk flavor formulations specifically tailored for swine or cattle. These flavors must be compounded independently of their specific organoleptic characteristics for optimum masking and blending effects and in combination with the odor and flavor of the ingredients selected for the feed. Flavoring difficulties are greatly increased in special feeds containing anti-stress ingredients, vitamins, and medicines. These require the inclusion of active flavor principles with characteristic notes capable of masking. Pelleted feed requires the use of heat-resistant flavor ingredients for manufacturing purposes. It is important to have available both powdered and liquid flavor ingredients with equivalent aroma yields for both effective surface and mass flavoring of the feed.

The above information may be summarized as follows:

1. Molasses notes are particularly suitable for flavoring cattle feed.
2. Fruit notes in general are preferred by horses.
3. Milk-flavored feed, completed by cocoa notes and similar variations, are particularly attractive to swine.
4. Cocoa flavors, complemented by milk and wintergreen notes, are particularly suitable for flavoring cattle feed.

5. The flavoring of poultry feed is more complex, as experiments conducted in several parts of the world have yielded conflicting results. Together with the mentioned undefinable "human" note, the smoky note appears to have more positive effects.
6. Anise notes are also of interest and can be suitably modified for flavoring rabbit feed.

Feed flavors are being marketed with increasing frequency in powder form (plated over carriers or spray-dried). Such powders very often contain sweeteners in addition to the principle flavor ingredients. The addition of flavors to bulk feed is readily attained by using suitable dispersing agents (surfactants) or by purely mechanical means.

REFERENCES

1. **Fenaroli,** *Riv. Ital. EPPOS,* 44, 256, 263, 1962.
2. **Fenaroli,** *Fr. Ses Parfums,* 7, 285, 289, 1964.
3. **Fenaroli,** Examen des formulations classiques à base d'herbes aromatiques pour les vermouths et les boissons amères, *Parfum, Cosmet. Savons,* 9, 459, 1966.
4. **Berio,** Il gusto di anice dei liquori, *Quintessenza,* (7), 17, Curt Georgi Imes, Milano, Italy.
5. **Fenaroli, Poy and Maroni,** Rhum ed etere di rhum, Part 1, *Riv. Ital. EPPOS,* Sept. 1965.
6. **Fenaroli, Poy and Maroni,** Rhum ed etere di rhum, Part 2, *Riv. Ital. EPPOS,* Feb. 1966.
7. **Haarse and deBrauw,** The analysis of volatile components of Jamaica rum, *J. Food Sci.,* 31, 6, 1966.
8. **Imes (Ed.),** *Essenze Naturali per Rhum Fantasia,* Curt Georgi Imes, Milano, Italy.
9. **Merory,** *Food Flavorings,* 2nd ed., Avi Publishing Co., 1968.
10. **Berio,** I succhi concentrati di frutta, *Quintessenza,* (3), (4), Curt Georgi Imes, Milano, Italy.
11. **Verderio,** L'alterazione dello sciròppo di orzata, *Quintessenza,* (5), Curt Georgi Imes, Milano, Italy.
12. **Hale,** *Chemistry and Technology of Food and Food Products,* Interscience Publishers, 1944.
13. **Insalata,** in *Bottlers' and Glass Packers' Handbook,* McGraw-Hill.
14. **Marion et al.,** *Helv. Chim. Acta,* 50, 1509, 1967.
15. **Staudinger and Reichstein,** British Patents 246,454 and 260,960.
16. **Stoll et al.,** *Helv. Chim. Acta,* 50, 2065, 1967.
17. **Colombo,** Attualitá e futuro degli aromi primordiali, *Riv. Ital. EPPOS,* p. 690, Nov. 1967.
18. **Stoll, Demole, Ferrero and Becker,** *Nature,* 202, 350, 1964.
19. **Reiter and Moller-Madsen,** *J. Dairy Res.,* 30, 419, 1963.
20. **Day and Andersen,** *J. Agric. Food Chem.,* 13, 2, 1965.
21. **Day and Andersen,** *J. Agric. Food Chem.,* 14, 241, 1966.
22. **Day and Andersen,** *J. Dairy Sci.,* 48, 248, 1965.
23. **Day and Andersen,** *J. Dairy Sci.,* 48, 784, 1965.
24. **Shimazono,** *Food Technol.,* 18, 1964.
25. **Bainbridge et al.,** *J. Chem. Soc.,* (101), 2209, 1912.
26. **Mohr,** *Fette Seifen,* 60, 661, 1958.
27. **Weurman et al.,** *J. Food Sci.,* 26, 239, 1961.
28. **von Elzakker et al.,** *Z. Lebensm. Unters. Forsch.,* 115, 222, 1961.
29. **Bailey et al.,** *J. Food Sci.,* 27, 165, 1962.
30. **Quesnel et al.,** *Nature* (London), 199, 605, 1963.
31. **Dietrich et al.,** *Helv. Chim. Acta,* 47, 1581, 1964.
32. **Quesnel,** *J. Sci. Food Agric.,* 16, 596, 1965.
33. **Reymond et al.,** *J. Gas Chromatogr.,* 4, 28, 1966.
34. **Marion,** *Chimia,* 21, 510, 1967.
35. **Marion et al.,** *Helv. Chim. Acta,* 50, 1509, 1967.
36. **Stoll et al.,** *Helv. Chim. Acta,* 50, 2065, 1967.
37. **Flament et al.,** *Helv. Chim. Acta,* 50, 2233, 1967.
38. **Rizzi,** *J. Agric. Food Chem.,* 15, 549, 1967.
39. **Müggler-Chavan et al.,** *Mitt. Geb. Lebensmittelunters. Hyg.,* 58, 466, 1967.
40. **Praag et al.,** *J. Agric. Food Chem.,* 16, 1005, 1968.
41. **van der Wal et al.,** *Recl. Trav. Chim. Pays-Bas,* 87, 238, 1968.
42. **Rohan,** *Food Process. Mark.,* 38, 12, 1969.
43. **van der Wal et al.,** *J. Agric. Food Chem.,* 19, 276, 1971.
44. **Sullivan et al.,** *Quartermaster Res. Eng. Cent. Res. Rep. Anal. Chem. Ser. 12,* 1959.
45. **Zlatkis et al.,** *Food Res.,* 25, 395, 1960.
46. **Radtke et al.,** *Z. Lebensm. Unters. Forsch.,* 119, 293, 1963.
47. **Self et al.,** *Chem. Ind.* (London), p. 863, 1963.
48. **Merritt et al.,** *J. Agric. Food Chem.,* 11, 152, 1963.
49. **Gianturco et al.,** *Tetrahedron,* 19, 2051, 1963.
50. **Gianturco et al.,** *Tetrahedron,* 20, 1763, 1964.
51. **Gianturco et al.,** *Tetrahedron,* 20, 2951, 1964.
52. **Radtke,** Dissertation, Munich, Germany, 1964.
53. **Viani et al.,** *Helv. Chim. Acta,* 48, 1809, 1965.
54. **Reymond et al.,** *J. Gas Chromatogr.,* 4, 28, 1966.
55. **Gianturco et al.,** *Nature,* 210, 1358, 1966.
56. **Heins et al.,** *J. Gas Chromatogr.,* 4, 395, 1966.
57. **Grimmer et al.,** *Dtsch. Lebensm. Rundsch.,* 62, 19, 1966.
58. **Merritt et al.,** *Colloq. Int. Chim. Cafés,* 2, 183, 1966.
59. **Gautschi et al.,** *J. Agric. Food Chem.,* 15, 15, 1967.

60. **Stoll et al.,** *Helv. Chim. Acta,* 50, 628, 694, 1967.
61. **Gianturco,** 4th Symposium on Foods, Oregon State University, Corvallis, 1967, 431.
62. **Bondarovich et al.,** *J. Agric. Food Chem.,* 15, 1093, 1967.
63. **Kung et al.,** *J. Food Sci.,* 32, 455, 1967.
64. **Stoffelsma et al.,** *Recl. Trav. Chim. Pays-Bas,* 87, 241, 1968.
65. **Stoffelsma et al.,** *J. Agric. Food Chem.,* 16, 1000, 1968.
66. **Walter et al.,** *Z. Ernaehrungswiss.,* 9, 123, 1969.
67. **Friedel et al.,** *J. Agric. Food Chem.,* 19, 530, 1971.
68. **Polonia et al.,** *An. Fac. Farm. Porto,* 29, 51, 1969; *Chem. Abstr.,* 75, 74950e, 1971.
69. **Büchi et al.,** *J. Org. Chem.,* 36, 199, 1971.
70. **Maga and Sizer,** *Crit. Rev. Food Technol.,* 4(1), 39, 1973; *J. Agric. Food Chem.,* 21, 22, 1973.
71. **Parliment et al.,** *J. Agric. Food Chem.,* 21, 485, 1973.

BIBLIOGRAPHY

Baked Goods

Büchi, G. and Wüst, H., Synthesis of 2-acetyl-1,4,5,6-tetrahydropyridine, a constituent of bread aroma, *J. Org. Chem.,* 36(4), 609–610, 1971.

De la Guerivière, J. F. and Benoualid, K., Flavoring biscuits, *Fr. Ses Parfums,* 13(69), 260–264, 1970.

Drapron, R., The taste of bread, *Fr. Ses Parfums,* 13(69), 239–243, 1970.

Drapron, R., The taste of bread, *Riv. Ital. EPPOS,* 52(10), 598–604, 1970.

Drapron, R., Bread flavor, *Bull. Anc. Elèves Ec. Fr. Meun.,* (244), 127–132, 1971.

Eldash, A., Precursors of bread flavor. Effect of fermentation and proteolytic activity, *Bakers Dig.,* 45(6), 26–31, 1971.

Hunter, I. R., Walden, M. K., Scherer, J. R., and Lundin, R. E., Preparation and properties of 1,4,5,6-tetrahydro-2-acetopyridine, a cracker odor constituent of bread aroma, *Cereal Chem.,* 46(2), 189–195, 1969.

Johnson, J. A. and Eldash, A. A., Role of non-volatile compounds in bread flavor, *J. Agric. Food Chem.,* 17(4), 740–746, 1969.

Lorenz, K. and Maga, J., Staling of white bread: changes in carbonyl composition and Glc headspace profiles, *J. Agric. Food Chem.,* 20(2), 211–213, 1972.

Markova, J., Honischova, E., and Hampl, J., Aromatic principles of bread and of the intermediates of its preparation, *Brot Gebaeck,* 24(9), 165–166, 1970.

McWilliams, M. and Mackey, A. C., Wheat flavor components, *J. Food Sci.,* 34(6), 493–496, 1969.

Mulders, E. and Dhont, J., Odor of white bread. III. Identification of volatile carbonyl compounds and fatty acids, *Z. Lebensm. Unters. Forsch.,* 150(4), 228–232, 1972.

Mulders, E., DeBrau, T. N., and Van Straten, S., Odor of white bread. II. Identification of components in pentane ether extracts. Pyrazine, lactone, furan, pyrrole, *Z. Lebensm. Unters. Forsch.,* 150(5), 306–310, 1972.

Mulders, E., Maarse, H., and Weurman, C., Odor of white bread. I. Analysis of volatile constituents in the vapor and aqueous extracts, *Z. Lebensm. Unters. Forsch.,* 150(2), 68–74, 1972.

Roiter, I. M., and Borovikova, L. A., Level of aromatic volatile carbonyl compounds at various stages in the production of bread. Bread aroma, aldehyde, crust, fermentation, *Izv. Vyssh. Uchebn. Pishch. Tekhnol.,* (5), 68–70, 1971.

Von Sydow, E. and Anjou, K., The aroma of Rye-Crisp bread, *Lebensm. Wiss. Technol.,* 2(1), 15–18, 1969.

Zyuz'ko, A. et al., Monocarbonyl compounds in bread volatile components. Bread aroma, aldehyde, ketone, *Prikl. Biokhim. Mikrobiol.,* 8(4), 498–504, 1972.

Beverages, Alcoholic

Bourzeix, M., Guitraud, J., and Champagnol, F., Identification of organic acids and determination of their individual amounts in grape juices and wines by chromatography and photodensitometry, *J. Chromatogr.,* 50(1), 83–91, 1970.

Clutton, D. W., The history of gin, *Flavour Ind.,* 3(9), 454–456, 1972.

Coe, G. R., Baldwin, S., and Andreasen, A. A., Colorimetric method for ester content of alcoholic beverages, *J. Assoc. Off. Anal. Chem.,* 54(5), 1225–1230, 1971.

Fenaroli, G., Apéritifs (appetizers), *Fr. Ses Parfums,* 13(69), 227–230, 1970.

Guymon, J. F., Composition of California commercial brandy distillates, *Am. J. Enol. Vitic.,* 21(2), 61–69, 1970.

Jennings, W. G., Wohleb, R., and Lewis, J., Gas-chromatographic analysis of headspace volatiles of alcoholic beverages, *J. Food Sci.,* 37(1), 69–71, 1972.

Kishkovskii, Z. et al., Sugar–amine reaction under distillation conditions. Carbonyl, amine, brandy aroma, *Izv. Vyssh. Uchebn. Zaved. Pishch. Tekhnol.,* (5), 35–39, 1972.

Kozub, G. et al., Volatile substances of sherry, *Sadovod. Vinograd. Vinodel. Mold.,* 27(5), 22–25, 1972.

Lefèvre, P. and Raposo, M., Aromas of wines and brandies. Their formation and their development. Portugese report, *Bull. OIV (Off. Int. Vigne Vin)*, 45(493), 249–258, 1972.

Lichev, V. and Goranov, N., Wine and brandy aromas. Their formation and changes, *Bull. OIV (Off. Int. Vigne Vin)*, 45(494), 317–337, 1972.

Miglio, G., Gas-chromatographic determination of flavor components in brandy, *Branntweinwirtschaft*, 122(2), 29–30, 1972.

Rapp, A., Aromas of wine and brandy. Their formation and evolution, *Bull. OIV (Off. Int. Vigne Vin)*, 45(492), 151–166, 1972.

Schoeneman, R. L., Report on alcoholic beverages, *J. Assoc. Off. Anal. Chem.*, 53(2), 338–341, 1970.

Schoeneman, R. L., Dyer, R. H., and Earl, E. M., Analytical profile of straight bourbon whiskies, *J. Assoc. Off. Anal. Chem.*, 54(6), 1247–1261, 1971.

Tarantola, C., Aromas of wines and brandies. Their formation and development, *Vini Ital.*, 13(72), 211–214, 1971.

Tarantola, C., Aromas of wines and brandies. Their formation and their development. Italian report, *Bull. OIV (Off. Int. Vigne Vin)*, 45(493), 258–266, 1972.

Williams, A., Flavor effects of ethanol in alcoholic beverages, *Flavour Ind.*, 3(12), 604–607, 1972.

Wobben, H. J., Timmer, R., Terheide, R., and DeValois, P. J., Nitrogen compounds in rum and whisky, *J. Food Sci.*, 36, 464–465, 1971.

Beverages, Non-alcoholic

Anon., Soft-drink flavor and the E.E.C. practice, *Soft Drinks Trade J.*, 25, 522–524, 1971.

Beattie, G. G., Soft-drink flavours: their history and characteristics. I. Cola or "Kola" flavours, *Flavour Ind.*, 1(6), 390–394, 1970.

Beattie, G. B. Soft-drink flavours: their history and characteristics. II. Lemonade, *Flavour Ind.*, 1(6), 395–399, 1970.

Beattie, G. B., Soft-drink flavours: their history and characteristics. IV. Orange drinks, *Flavour Ind.*, 1(8), 530–534, 1970.

Beattie, G. B., Soft-drink flavours: their history and characteristics. V. Concentrated orange drinks, *Flavour Ind.*, 1(9), 599–604, 1970.

Beattie, G. B., Soft-drink flavours: their history and characteristics. VI. Ginger beer, *Flavour Ind.*, 1(10), 702–706, 1970.

Beattie, G. B., Soft-drink flavours: their history and characteristics. VII. Lime beverages, *Flavour Ind.*, 1(11), 772–776, 1970.

Beattie, G. B., Soft-drink flavours: their history and characteristics. VIII. Grapefruit drinks, *Flavour Ind.*, 1(12), 836–840, 1970.

Beattie, G. B., Soft-drink flavours: their history and characteristics. IX. Cream soda, *Flavour Ind.*, 2(1), 28–32, 1971.

Beattie, G. B., Soft-drink flavours: their history and characteristics. X. Non-alcoholic grape beverages, *Flavour Ind.*, 2(1), 93–97, 1971.

Blomquist, V. H., Report on flavors and non-alcoholic beverages, *J. Assoc. Off. Anal. Chem.*, 52(2), 259–260, 1969.

Clarke, K. J., Modern trends in flavouring and flavour creation for the soft-drink industry, *Flavour Ind.*, 1(6), 388–389, 1970.

Larry, D., Report on flavors and non-alcoholic beverages, *J. Assoc. Off. Anal. Chem.*, 53(2), 343–344, 1970.

Larry, D., Flavors and non-alcoholic beverages. Flavor analysis, *J. Assoc. Off. Anal. Chem.*, 55(2), 275–276, 1972.

Matsui, M., New flavors for soft drinks. Artificial soft drinks, beverage flavors, *Shokuhin Kogyo*, 14(12), 12–18, 1970.

Melillo, D., Soft-drink beverage flavours, *Perfum. Essent. Oil Rec.*, 60, 108–112, 1969.

Sarzynski, W., Effective use of natural fruit aroma substances in the production of non-alcoholic beverages. Geraniol, mint oil, fruit aroma beverage, *Przem. Ferment. Rolny*, 15(9), 35–40, 1971.

Candy

Cakebread, S., Confectionery ingredients. Flavor. Review, *Confect. Prod.*, 38(4–7), 190–192, 242–246, 314–316, 356–362, 1972.

Editorial, Flavoring pharmaceuticals, *Flavour Ind.*, 3(1), 11–12, 1972.

Fellows, G., Confectionery flavors. Problems and factors affecting taste and choice of ingredients to protect or enhance flavors, *Manuf. Confect.*, 52, 70–78, 1972.

Habersaat, F., Use of Universal aromas in the confectionery industry. Review, *Int. Rev. Sugar Confect.*, 23, 292–294, 1970.

Lee, S. et al., Flavor constituents in various foods. II. Flavor constituents of honey, *Han'guk Sikp'um Kwahakoe Chi*, 3(3), 168–171, 1971.

Olsen, R. D., Ethyl maltol used as flavor potentiator opens doors to exciting tastes in candy, *Candy Ind.*, 135(3), 11, 37, 1970.

Van Eijk, A., The influence of raw materials on the taste of confectionery, *Flavour Ind.*, 3(7), 348–352, 1972.

Coffee

Artem'ev, B. et al., Aromatization of dry coffee extract, *Konservn. Ovoshchesush. Promst.*, 27(5), 18–20, 1972.

Buttery, R. G., Seifert, R. M., Guadagni, D. G., and Ling, L. C., Characterization of volatile pyrazine and pyridine components of potato chips, *J. Agric. Food Chem.*, 19(5), 969–971, 1971.

De Roissart, H. and Johannes, C., Extracts from Plant Products, Especially Coffee (CL. A23F, C11B). Aromatic Volatile Coffee Manufacture, German Patent 2,107,201, September 2nd, 1971.
Feldman, J. R., Ryder, W. S., and Kung, J. T., Importance of non-volatile compounds to the flavor of coffee, *J. Agric. Food Chem.*, 17(4), 733–739, 1969.
Friedel, P. et al., Some constituents of the aroma complex of coffee, *J. Agric. Food Chem.*, 19(3), 530–532, 1971.
Segall, S., Silver, C., and Bacino, S., The effect of reheating upon the organoleptic and analytical properties of beverage coffee, *Food Technol.*, 24, 1242–1246, 1970.
Singh, K. et al., Extraction of coffee oil, *Res. Bull. Panjab Univ. Sci.*, 21 (Pt. 1/2), 49–50, 1970.
Sivetz, M., How acidity affects coffee flavor, *Food Technol.*, 26, 70–77, 1972.
Villanua, L. et al., Analytical determinations of soluble coffee extracts. Aroma component determination, composition, *An. Bromatol.*, 23(3), 259–283, 1971.
Vol'per, I. N. et al., Losses of aromatic substances in the production of soluble coffee. Flavor loss in instant coffee, *Izv. Vyssh. Uchebn. Zaved. Pishch. Teknhol.*, (2), 43–45, 1952.

Dairy Products

Adda, J. and Dumont, J., Different methods for extracting cheese flavors. Review, cheese flavor analysis, *Ind. Aliment. Agric.*, 89(2), 143–145, 1972.
Brewington, C. et al., Conjugated compounds in cow's milk. Glucuronide conjugate, sulfate conjugate, flavor conjugate, *J. Agric. Food Chem.*, 21(1), 38–39, 1973.
Chang, J., Yoshino, U., and Tsugo, T., Cheese ripened mainly with yeast. V. Changes of flavor components in cheese during ripening. *Saccharomyces* cheese ripening, aldehyde, ester, alcohol, ketone, fatty acid, *Nippon Chikusan Gakkai Ho,* 43(10), 561–566, 1972.
Chasov, F. V., Carbonyl compounds in piquant cheese. Cheese flavor, aldehyde, ketone, *Molochn. Promst.*, 32(11), 19-21, 1971.
Dolezalek, J., Hladik, J., and Vys, S. B., Effect of rennet on the ripening, taste and flavor of Emmenthaler cheese, *Sk. Chem. Technol.*, E35, 191–218, 1972.
Douglas, D., Gas-chromatographic mass-spectrometric analysis of volatiles in flavors and biological fluids. Cheese volatiles, urine, *Diss. Abstr. Int. B.*, 33(4), 1432, 1972.
Eopechino, A. A. and Leeder, J. G., Flavor modifications produced in ice-cream mix made with corn syrup. II. CO_2 production associated with the browning reaction, *J. Food Sci.*, 35(4), 398–402, 1970.
Eskamani, A. and Leeder, J. G., Contribution of specific saccharide fractions of corn syrup to the syrup flavor of ice-cream mix, *J. Food Sci.*, 37(3), 328–330, 1972.
Farber, K. T. H. and Weeks, K. J., Some aspects of the chemistry of cheese, *Food Technol. Aust.*, 22, 620–623, 1970.
Ferretti, A., Flanagan, V. P., and Ruth, J. M., Non-enzymatic browning in a lactose-casein model system, *J. Agric. Food Chem.*, 18(1), 13–18, 1970.
Foda, E. A., Role of fat in the favor of Cheddar cheese, *Diss. Abstr. Int. B.*, 32(4), 2265, 1971.
Forss, D., Flavors of dairy fats. Review, lipid flavor, milk fat flavor, *J. Am. Oil Chem. Soc.*, 48(11), 702–710, 1971.
Gordon, D. and Morgan, M., Principal volatile compounds in feed-flavored milk, *J. Dairy Sci.*, 55(7), 751–752, 1972.
Hammond, E. and Seals, R., Oxidized flavor in milk and its simulation. Copper off-flavor, octenone, octanal, *J. Dairy Sci.*, 55(11), 1567–1569, 1972.
Harper, W. J. and Kristoffersen, T., Biochemical aspects of flavor development in Cheddar cheese slurries, *J. Agric. Food Chem.*, 18(4), 564–566, 1970.
Harwalkar, V., Characterization of an astringent flavor fraction from Cheddar cheese. Peptide, *J. Dairy Sci.*, 55(6), 735–741, 1972.
Honkanen, E., Moisio, T., Karvonen, P., and Virtanen, L., On the occurrence of a new lactone compound, *trans*-4-methyl-5-hydroxyhexanoic acid lactone, in milk, *Acta Chem. Scand.*, 22(6), 2041–2043, 1968.
Hosono, A. and Tokita, F., Effect of *Brevibacterium linens* and yeasts on the development of volatile flavor substances in Limburger cheese. I. General properties of yeasts isolated from Limburger cheese and the production of volatile flavor substances by these yeasts. Limburger cheese flavor volatiles, *Nippon Chikusan Gakkai Ho,* 41(3), 131–137, 1970.
Hostettler, H., Appearance, flavor, and texture aspects. Review, milk flavor, physical properties, *Annu. Bull. Int. Dairy Fed.*, (Pt. 5), 6–34, 1972.
Kinsella, J. E., The flavour chemistry of milk lipids, *Chem. Ind.*, (2), 36–42, 1969.
Kohata, Y., Fermentation and flavor, *Hakko Kyokaishi,* 29(10), 479–489, 1971.
Konfortov, A., Production of oil aromatized with butter yeast distillate. Milk diacetyl distillation, rennet, yeast, *Khranit. Promst.*, 21(4), 31–33, 1972.
Kuzdzal-Savoie, S. and Adda, J., The flavor of some types of French cheese, *Fr. Ses Parfums,* 13(69), 252–259, 1970.
Kuzdzal-Savoie, S. and Adda, J., The flavor of some types of French cheese, *Riv. Ital. EPPOS,* 52(12), 714–722, 1970.
Lawrence, R. et al., Cheddar cheese flavor. I. Role of starters and rennets, *N.Z. J. Dairy Sci. Technol.*, 7(2), 32–37, 1972.
Liebich, H. M., Douglas, D. R., Bayer, E., and Zlatkis, A., Volatile flavor components of Cheddar cheese, in *Advances in Chromatography, Proceedings of the 6th International Symposium,* Preston Technical Abstracts Co., 1970, 219–225.
Liebich, H. M., Douglas, D. R., Bayer, E., and Zlatkis, A., Comparative characterization of different types of cheese by gas–liquid chromatography, *J. Chromatogr. Sci.*, 8, 351–354, 1970.

Liebich, H. M., Douglas, D. R., Bayer, E., and Zlatkis, A., The volatile flavor components of Cheddar cheese, *J. Chromatogr. Sci.*, 8, 355–359, 1970.
Lowrie, R. and Lawrence, R., Cheddar cheese flavor. IV. New hypothesis to account for the development of bitterness, *N.Z. J. Dairy Sci. Technol.*, 7(2), 51–53, 1972.
Lueck, H. and Rudd, S., Milk flavored with natural fruit juice, *S. Afr. J. Dairy Technol.*, 4(3), 153–158, 1972.
Manning, D. and Robinson, H., Analysis of volatile substances associated with Cheddar cheese aroma, *J. Dairy Res.*, 40(1), 63–75, 1973.
Martley, F. and Lawrence, R., Cheddar cheese flavor. II. Characteristics of single strain starters associated with good or poor flavor development, *N.Z. J. Dairy Sci. Technol.*, 7(2), 38–44, 1972.
McDaniel, M. R., Sather, L. A., and Lindsay, R. C., Influence of free fatty acids on sweet-cream butter flavor, *J. Food Sci.*, 34(3), 251–254, 1969.
Moreno, V. and Kosikowski, F., Degradation of β-casein by micrococcal-cell-free preparations. Cheese flavor, *J. Dairy Sci.*, 56(1), 33–38, 1973.
Moreno, V. and Kosikowski, F., Peptides, amino acids and amines liberated from β-casein by micrococcal-cell-free preparations. Casein metabolism, cheese flavor, *J. Dairy Sci.*, 56(1), 39–44, 1973.
Motoc, D. and Costin, G., Action of *Penicillium roqueforti* mold. Aroma cheese, *Lucr. Stiint. Inst. Politeh. Galati*, 4, 231–241, 1970.
Neelakantan, S., Shahani, K., and Arnold, R., Lipases and flavor development in some Italian cheese varieties. Fatty acid cheese flavor, provolone, Romano, *Food Prod. Dev.*, 5(7), 52–58, 1971.
Nelson, J. H., Blue-cheese flavor by fermentation, *Food Prod. Dev.*, 4(1), 52–56, 1970.
Ney, K. and Wirotama, I., Flavor of Edelpilzkaese, a German blue mold cheese. Acid, keto acid, carbonyl, ester, amine, alcohol, *Z. Lebensm. Unters. Forsch.*, 149(5), 275–279, 1972.
Ney, K. and Wirotama, I., Unsubstituted aliphatic monocarboxylic acids and volatile aliphatic monoamines in Emmenthal cheese, *Z. Lebensm. Unters. Forsch.*, 149(6), 347–349, 1972.
Nobuhara, A., Syntheses of unsaturated lactones. III. Flavorous nature of some δ-lactones having the double bond at various sites, *Agric. Biol. Chem.*, 33, 1264–1269, 1969.
Palo, V. and Ilkova, H., Direct gas-chromatographic estimation of lower alcohols, acetaldehyde, acetone and diacetyl in milk products, *J. Chromatogr.*, 53(2), 363–367, 1970.
Palo, V. and Krcal, Z., Aroma substances in Olomouc cake cheeses, *Prum. Potravin*, 24(2), 54–57, 1973.
Panouse, J. et al., Cheese flavors. Review, cheese flavor volatiles, *Ind. Aliment. Agric.*, 89(2), 133–140, 1972.
Sandine, W. E. and Elliker, P. R., Microbially induced flavors and fermented foods. Flavor in fermented dairy products, *J. Agric. Food Chem.*, 18(4), 557–562, 1970.
Schwartz, D. P. and Virtanen, A. I., Volatile carbonyl compounds in the milk fat of normally and synthetically fed cows, *Acta Chem. Scand.*, 22(6), 1717–1721, 1968.
Siek, T., Albin, A., Sather, L. A., and Lindsay, R. C., Taste thresholds of butter volatiles in deodorized butter oil medium, *J. Food Sci.*, 34(3), 265–267, 1969.
Singh, S. and Kristoffersen, T., Influence of lactic cultures and curd-nulling acidity on flavor of Cheddar curd slurries, *J. Dairy Sci.*, 54(11), 1589–1594, 1971.
Singh, S. and Kristoffersen, T., Cheese flavor development using direct acidified curd. Lactic acid, glutathione, riboflavine, diacetyl, *J. Dairy Sci.*, 55(6), 744–749, 1972.
Tanaka, H. and Obata, Y., Studies on the formation of the cheese-like flavor, *Agric. Biol. Chem.*, 33, 147–150, 1969.
Urbach, G. et al., Volatile compounds in butter oils. II. Flavor and flavor thresholds of lactones, fatty acids, phenols, indole and skatole in deodorized synthetic butter, *J. Dairy Res.*, 39(1), 35–47, 1972.
Walker, N. J. and Gray, I. K., The glucosinolate of land cress *(Coronopus didymus)* and its enzymic degradation products as precursors of off-flavor in milk – a review, *J. Agric. Food Sci.*, 18(3), 346–352, 1970.

Fruital Flavors

Adriaanse, A. and Klop, W., Simultaneous determination of sugars in jams and fruit juices. I. Determination of glucose using a combined enzymatic-iodometric method, *Lebensm. Wiss. Technol.*, 3(3), 54–56, 1970.
Bayonove, C., Volatile constituents of fruits, grapes and wine, *Ind. Aliment. Agric.*, 88(11), 1559–1567, 1971.
Bertrand, J., Natural fruit flavors. Present processes for industrial recovery, *Fr. Ses Parfums*, 13(69), 234–236, 1970.
Broderick, J. J., Fruit flavor research. The practical flavorist vs the basic researcher, *Food Technol.*, 26, 37–48, 1972.
Büchi, G., Demole, E., and Thomas, A., Syntheses of 2,5-dimethyl-4-hydroxy-2,3-dihydrofuran-3-one (furaneol), a flavor principle of pineapple and strawberry, *J. Org. Chem.*, 38(1), 123–125, 1973.
Chaley, H., Fruits and vegetables. Flavor aroma, plant food, *Food Theor. Appl.*, pp. 251–334, 1972.
Chaveron, H., Chromatographic study of aroma products. Review, pineapple aroma, allyl hexanoate, *Choc. Confiserie Fr.*, (280), 7–20, 1972.
Creveling, R. K. and Jennings, W. C., Volatile components of Bartlett pear: higher-boiling esters, *J. Agric. Food Chem.*, 18(1), 19–24, 1970.
Do, J. Y., Salunkhe, D. K., and Olson, L. E., Isolation identification and comparison of the volatiles of peach fruit as related to harvest maturity and artificial ripening, *J. Food Sci.*, 34(6), 618–621, 1969.
Dupaigne, P., Banana flavor. Review, *Riv. Ital. EPPOS*, 54(4), 357–353, 1972.

Dupaigne, P., Pineapple flavor. Review, recovery of natural flavor, *Riv. Ital. EPPOS*, 54(8), 559–567, 1972.
Flath, R. A. and Forrey, R. R., Volatile components of smooth Cayenne pineapple, *J. Agric. Food Chem.*, 18(2), 306–309, 1970.
Gasco, L., Barrera, R., and de la Cruz, F., Gas-chromatographic investigations of the volatile constituents of fruit aromas, *J. Chromatogr. Sci.*, 7, 228–238, 1969.
Goldstein, J. L. and Wick, E. L., Lipid in ripening banana fruit, *J. Food Sci.*, 34(6), 482–484, 1969.
Houchen, M. et al., Possible precursor of 1-methyl-4-isopropenylbenzene in commercial blackberry flavor essence, *J. Agric. Food Chem.*, 20(1), 170, 1972.
Jennings, W., Volatile compositions of apricots, peaches and Bartlett pears, *Ind. Aliment. Agric.*, 89(2), 121–126, 1972.
Kevei-Pichler, E. and Blalovich, M., Fruit aromas. Fruit aroma extraction, *Elelmiszervizsgalati Kozl.*, 17(3), 125–142, 1971.
Kliewer, W. M., Free amino acids and other nitrogenous substances of table grape varieties, *J. Food Sci.*, 34(3), 274–278, 1969.
Knapp, F. F. and Nicholas, H. J., The sterols and triterpenes of banana pulp, *J. Food Sci.*, 34(6), 584–586, 1969.
Macharoshvili, G., Composition of aromatic substances from apple must, wine, and cider, *Prikl. Biokhim. Mikrobiol.*, 7(5), 566–570, 1971.
Markakis, T. and Amon, A., The presence and origin of 2-pyrrolidone-5-carboxylic acid in processed Concord grape juice and its concentrate, *Food Technol.*, 23, 131–133, 1969.
Myers, M. J., Issenberg, P., and Wick, E. L., Vapor analysis of the production by banana fruit of certain volatile constituents, *J. Food Sci.*, 34(6), 504–509, 1969.
Myers, M. J., Issenberg, P., and Wick, E. L., L-Leucine as a precursor of isoamyl alcohol and isoamyl acetate, volatile constituents of banana fruit discs, *Phytochemistry*, 9, 1693–1700, 1970.
Näf, F., Degen, P., and Ohloff, G., A novel synthesis of 2,6-diolefinic esters: ethyl and methyl *trans*-2,*cis*-6-dodecadienoate, two Bartlett pear constituents, *Helv. Chim. Acta*, 55(1), 82–85, 1972.
Näf-Müller, R. and Willhalm, B., Volatile constituents of pineapple, *Helv. Chim. Acta*, 54(7), 1880–1890, 1971.
Nursten, H. E., Volatile compounds. Aroma of fruits, *Biochem. Fruits Prod.*, (1), 239–268, 1970.
Nursten, H. and Woolfe, M., Examination of the volatile compounds present in cooked Bramley's seedling apples and the changes they undergo on processing. Apple cooking aroma, aldehyde, alcohol, *J. Sci. Food Agric.*, 23(7), 803–822, 1972.
Owades, J. L. and Dono, J. M., Note on cryoscopic determination of grape juice characteristics, *J. Assoc. Off. Anal. Chem.*, 52(3), 651–653, 1969.
Paillard, N., Pitoulis, S., and Mattei, A., Techniques of preparation and analysis of some fruit flavors, *Lebensm. Wiss. Technol.*, 3(6), 107–114, 1970.
Peri, C. and Pompei, C., Extraction and fractionation of phenols from white grapes, *Riv. Vitic. Enol.*, 25(7), 311–320, 1972.
Picard, G., Julien, J., and Riel, R., Study of volatile substances in the juices of four apple varieties, *Can. Inst. Food Technol.*, 4(3), 112–115, 1971.
Pisarnitskii, A., Vereshchagin, P., Macharoshvili, G., and Bogatova, A., Carbonyl compounds and their role in the aroma of fruits and berries, *Prikl. Biokhim. Mikrobiol.*, 6(1), 13–17, 1970.
Ramshan, E. H. and Hardy, P. J., Volatile compounds in dried grapes, *J. Sci. Food Agric.*, 20, 619–621, 1969.
Sevenants, M. R. and Jennings, W. G., Occurrence of 6-pentyl-pyrone in peach essence, *J. Food Sci.*, 36(2), 536, 1971.
Silverstein, R., Biochemistry and physiology of commercially important fruits. 9B. Pineapple. Flavor, review, chromatography, *Biochem. Fruits Prod.*, 2, 325–331, 1971.
Stevens, K. L., Flath, R. A., Lee, A., and Stern, D. J., Volatiles from grapes. Comparison of Grenache juice and Grenache rosé wine, *J. Agric. Food Chem.*, 17(5), 1102–1106, 1969.
Terrier, A. et al., Content of terpenic compounds in *Vitis vinifera* grapes, *C. R. Acad. Sci. Ser. D*, 275(8), 941–944, 1972.
Tressl, R. and Jennings, W., Production of volatile compounds in the ripening banana, *J. Agric. Food Chem.*, 20(2), 189–192, 1972.
Veda, Y. et al., Lipids of fruits and vegetables and their physiological and qualitative role. IV. Changes of volatile fatty acids during maturation of banana fruits. Flavor ripening, *Nippon Shokuhin Kogyo Gakkai Shi*, 17(12), 545–548, 1970.
Wick, E. L. et al., Constituents of banana (*Musa cavendishii* var. *valery*), *J. Agric. Food Chem.*, 17(4), 751–759, 1969.

Meat

Balboni, J. J. and Nawar, W. W., Apparatus for direct collection of volatiles from meat, *J. Agric. Food Chem.*, 18(4), 746, 1970.
Biino, L., Meat aroma in beef. Review, *Atti Soc. Peloritana Sci. Fis. Mat. Nat.*, 16(1/2), 43–51, 1970.
Bratzler, L. J., Spooner, M. E., Weatherspoon, J. B., and Maxey, J. A., Smoke flavor as related to phenol, carbonyl and acid content of bologna, *J. Food Sci.*, 34(2), 146–148, 1969.
Brinkman, H. W., Copier, H., De Leuw, J. J. M., and Tjan, S. B., Components contributing to beef flavor. Analysis of headspace volatiles of beef broth, *J. Agric. Food Chem.*, 20(2), 177–181, 1972.
Cho, I. C. and Bratzler, L. J., Effect of sodium nitrite on flavor of cured pork, *J. Food Sci.*, 35(5), 668–670, 1970.
Dimick, P. S., McNeil, J. H., and Grunden, L. P., Poultry product quality. Carbonyl composition and organoleptic evaluation of mechanically deboned poultry meat, *J. Food Sci.*, 37(4), 544–546, 1972.

Everson, C. W., Danner, W. E., and Hammes, P. A., Bacterial starter cultures in sausage products, *J. Agric. Food Chem.*, 18(4), 570–571, 1970.

Flament, I. and Ohloff, G., Studies on aromas. Part 18(1). On the aroma of roasted beef. I. Pyrazines, *Helv. Chim. Acta*, 54(7), 1911–1913, 1971.

Golovkin, N. and Vasil'ev, A., Variation of aroma compounds during cold storage of meat, *Tr. Vost. Sib. Tekhnol. Inst.*, 3(3), 73–81, 1970.

Gorbatov, V. M. et al., Liquid smokes for use in cured meats, *Food Technol.*, 25, 71–77, 1971.

Gordon, A., Meat and poultry flavour, *Flavour Ind.*, 3(9), 445–453, 1972.

Goryaev, M. et al., Development of a gas-chromatographic method for analyzing volatile aromatic substances in sausage, *Izv. Akad. Nauk Kaz. SSR Ser. Khim.*, 22(6), 43–48, 1972.

Halloran, H., Effect of trimethylamine on flavor of broilers. Fish oil chicken flavor, *Poult. Sci.*, 51(5), 1752–1755, 1972.

Halverson, H., A procedure for isolation and quantitative determination of volatile fatty acids from meat products, *J. Food Sci.*, 37(1), 136–139, 1972.

Harris, N. D. and Lindsay, R. C., Flavor changes in reheated chicken, *J. Food Sci.*, 37(1), 19–22, 1972.

Hornstein, I., Palatability characteristics of meat. Chemistry of meat flavor, in *Science of Meat and Meat Products*, 2nd ed., Price, J. F. and Schweigert, B. S., Eds., W. H. Freeman, 1971, 348–363.

Jones, N. R., Meat and fish flavors. Significance of ribomononucleotides and their metabolites, *J. Agric. Food Chem.*, 17(4), 712–716, 1969.

Kenzhebekov, P., Determination of volatile acids in sausages by gas–liquid chromatography, *Izv. Vyssh. Uchebn. Zaved. Pishch. Tekhnol.*, (4), 36–38, 1972.

Kiratsous, A. S., Meat analogs and flavors, *Cereal Sci. Today*, pp. 147–149, 1969.

Kurkela, R. and Uutela, P., Aroma changes of thiamine glutamic acid solutions during heating. Thiamine glutamate, meat aroma, methionine, *Lebensm. Wiss. Technol.*, 5(2), 43–46, 1972.

Lawrie, R. A., Variation of flavour in meat, *Flavour Ind.*, 1(9), 591–594, 1970.

Liebich, H. M. et al., Volatile components in roasted beef, *J. Agric. Food Chem.*, 20(1), 96–99, 1972.

Lillard, A. and Ayres, J. C., Flavor compounds in country-cured hams, *Food Technol.*, 23, 117–120, 1969.

Lustre, A. O. and Issenberg, P., Phenolic compounds of smoked meat products, *J. Agric. Food Chem.*, 18(6), 1056–1060, 1970.

Macy, R. L., Jr., Naumann, H. D., and Bailey, M. E., Water-soluble flavor and odor precursors of meat. Part 3. Changes in nucleotides, total nucleosides and bases of beef, pork and lamb during heating, *J. Food Sci.*, 35(1), 78–80, 1970.

Macy, R. L., Jr., Naumann, H. D., and Bailey, M. E., Water-soluble flavor and odor precursors of meat. Part 4. Influence of cooking on nucleosides and bases of beef steaks and roasts and their relationship to flavor, aroma and juiciness, *J. Food Sci.*, 35(1), 81–83, 1970.

Macy, R. L., Jr., Naumann, H. D., and Bailey, M. E., Water-soluble flavor and odor precursors of meat. Part 5. Influence of heating on acid-extractable non-nucleotide chemical constituents of beef, lamb and pork, *J. Food Sci.*, 35(1), 83–87, 1970.

Maga, J. A. and Lorenz, K., The effect of flavor enhancers on direct headspace. Gas–liquid chromatography profiles of beef broth, *J. Food Sci.*, 37(6), 963–964, 1972.

Manley, C. H. and Fagerson, I. S., Aroma and taste characteristics of hydrolyzed vegetable protein. Meat aroma, taste, vegetable protein, *Flavour Ind.*, 2(12), 686–690, 1971.

Mussinan, C. and Katz, I., Isolation and identification of some sulfur chemicals present in two model systems approximating cooked meat, *J. Agric. Food Chem.*, 21(1), 43–45, 1973.

Nonaka, M., A new procedure for isolating poultry aroma essence, *Food Technol.*, 25, 45–50, 1971.

Pezacki, W. and Beller, J., Chemical components of meat taste and flavor, *Gospod. Miesna*, 23(8), 6–11, 1971.

Pierson, M. D., Collins-Thompson, D. L., and Ordal, Z. J., Microbiological, sensory and pigment changes of aerobically and anaerobically packaged beef, *Food Technol.*, 24(10), 1171–1175, 1970.

Piotrowski, E. G., Zaika, L. L., and Wasserman, A. E., Studies on aroma of cured ham, *J. Food Sci.*, 35(3), 321–325, 1970.

Pippen, E. L. and Mecchi, E. P., Hydrogen sulfite, a direct and potentially indirect contributor to cooked-chicken aroma *J. Food Sci.*, 34(5), 443, 1969.

Pokorny, J., Aroma of beef and its changes during processing, *Prum. Potravin*, 21(9), 262–263, 1970.

Rozier, J., Flavor of meat and pork products, *Fr. Ses Parfums*, 13(69), 244–251, 1970.

Rozier, J., The flavor of meat and meat products, *Riv. Ital. EPPOS*, 53, 17–26, 1971.

Rozier, J., Meat flavour – a review, *Flavour Ind.*, 2(4), 212–215, 1971.

Schutte, L. and Koenders, E. B., Components contributing to beef flavor. Natural precursors of 1-methylthio-ethanethiol, *J. Agric. Food Chem.*, 20(2), 181–184, 1972.

Thomas, C. P., Dimick, P. S., and McNeil, J. H., Sources of flavor in poultry skin, *Food Technol.* 25, 109–115, 1971.

Tonsbeek, C. H. T., Copier, H., and Plancken, A. J., Components contributing to beef flavor. Isolation of 2-acetyl-2-thiazoline from beef broth, *J. Agric. Food Chem.*, 19(5), 1014–1016, 1971.

Wasserman, A. E., Thermally produced flavor components in the aroma of meat and poultry, *J. Agric. Food Chem.* 20(4), 737–741, 1972.

Wasserman, A. E. and Talley, F., The effect of sodium nitrite on the flavor of frankfurters, *J. Food Sci.*, 37(4), 536–538, 1972.

Watanabe, K. and Sato, Y., Aliphatic γ- and δ-lactones in meat fats, *Agric. Biol. Chem.*, 32, 1318–1324, 1968.
Watanabe, K. and Sato, Y., Lactones in the flavor of heated pork fat, *Agric. Biol. Chem.*, 33, 242–249, 1969.
Watanabe, K. and Sato, Y., Shallow-fried beef. Additional flavor components. Beef flavor, methional, acetylfuran, *J. Agric. Food Chem.*, 20(2), 174–176, 1972.
Watanabe, K. and Sato, Y., Flavor components of shallow-fried beef. Aldehyde, ketone, lactone, pyrazine, *Nippon Chikusan Gakkai Ho*, 43(4), 219–225, 1972.
Watanabe, K. and Sato, Y., Some alkyl-substituted pyrazines and pyridines in the flavor components of shallow-fried beef, *J. Agric. Food Chem.*, 20(2), 1017–1019, 1972.
White, R. H., Howard, J. W., and Barnes, C. J., Determination of polycyclic aromatic hydrocarbons in liquid smoke flavors, *J. Agric. Food Chem.*, 19(1), 143–146, 1971.
Wilson, R. and Katz, I., Review of literature on chicken flavor and report of isolation of several new chicken flavor components from aqueous cooked chicken broth, *J. Agric. Food Chem.*, 20(4), 741–747, 1972.
Zaika, L. L., Meat flavor. Method for rapid preparation of the water-soluble low-molecular-weight fraction of meat tissue extracts, *J. Agric. Food Chem.*, 17(4), 893–895, 1969.
Ziemba, Z. and Malkki, Y., Changes in odour components of canned beef due to processing, *Lebensm. Wiss. Technol.* 4, 118–122, 1971.

Potato Products

Buri, R., Signer, V., and Solms, J., The importance of free amino acids and nucleotides in the flavor of cooked potatoes, *Lebensm. Wiss. Technol.*, 3(3), 63–66, 1970.
Buttery, R., Unusual volatile carbonyl components of potato chips, *J. Agric. Food Chem.*, 21(1), 31–33, 1973.
Buttery, R. G. and Ling, L. C., Characterization of non-basic steam-volatile components of potato chips. Methional, phenylacetaldehyde, *J. Agric. Food Chem.*, 20(3), 698–700, 1972.
Buttery, R. G., Seifert, R. M., Guadagni, D. G., and Ling, L. C., Characterization of volatile pyrazine and pyridine components of potato chips, *J. Agric. Food Chem.*, 19(5), 969–971, 1971.
Buttery, R. G., Seifert, R. M., and Ling, L. C., Characterization of some volatile potato components, *J. Agric. Food Chem.* 18(3), 538–539, 1970.
Guadagni, D. G., Buttery, R. G., Seifert, R. M., and Venstrom, D. W., Flavor enhancement of potato products, *J. Food Sci.*, 36(3), 363–366, 1971.
MacLeod, A. J., The chemistry of vegetable flavour, *Flavour Ind.*, 1(10), 665–672, 1970.
MacLeod, A. J. and MacLeod, G., Flavor volatiles of some cooked vegetables, *J. Food Sci.*, 35(6), 734–738, 1970.
Maruyama, F. T., Identification of dimethyl trisulfide as a major aroma component of cook brassicaceous vegetables, *J. Food Sci.*, 35(5), 540–543, 1970.
Osman-Ismail, F. and Solms, J., Formation of inclusion compounds of potato starch with aromatic substances, *Mitt. Geb. Lebensmittelunters. Hyg.*, 63(1), 88–92, 1972.
Sapers, G. M., Flavor quality in explosion-puffed dehydrated potato. Part 2. Flavor contribution of 2-methylpropanal, 2-methylbutanal and 3-methylbutanal, *J. Food Sci.*, 35(6), 731–733, 1970.
Sapers, G. M. et al., Flavor quality and stability of potato flakes. Volatile components associated with storage changes. Oxidation lipid, benzaldehyde, furfural, *J. Food Sci.*, 37(4), 579–583, 1972.
Sapers, G. M., Osman, S. F., Dooley, J., and Panasiuk, O., Flavor quality in explosion-puffed dehydrated potato. Part 3. Contribution of pyrazines and other compounds to the toasted off-flavor. *J. Food Sci.*, 36(1), 93–95, 1971.
Sapers, G. M., Sullivan, J. F., and Talley, F. B., Flavor quality in explosion-puffed dehydrated potato. Part 1. A gas-chromatographic method for the determination of aldehydes associated with flavor quality, *J. Food Sci.*, 35(6), 728–729, 1970.

Tobacco

Bartle, K. D., Bergstedt, L., Novotny, M., and Widmark, G., Tobacco chemistry. II. Analysis of the gas phase of tobacco smoke by gas chromatography-mass spectrometry, *J. Chromatogr.*, 45(2), 256–263, 1969.
Bell, J. H., Ireland, S., and Spears, A. W., Identification of aromatic ketones in cigarette smoke condensate, *Anal. Chem.*, 41(2), 310–313, 1969.
Cook, C. E. et al., An examination of the hexane extract of flue-cured tobacco involving gel-permeation chromatography *Phytochemistry*, 8, 1025–1033, 1969.
Demole, E. and Berthet, D., Identification of damascenone and damascone in Burley tobacco, *Helv. Chim. Acta*, 54(2), 681–682, 1971.
Demole, E. and Berthet, D., A chemical study of Burley tobacco flavour (*Nicotiana tabacum* L.). I. Volatile to medium-volatile constituents, *Helv. Chim. Acta*, 55(6), 1866–1882, 1972.
Demole, E. and Berthet, D., A chemical study of Burley tobacco flavour (*Nicotiana tabacum* L.). II. Medium-volatile, free-acidic constituents, *Helv. Chim. Acta*, 55(6), 1898–1901, 1972.
Demole, E., Demole, C., and Berthet, D., Chemical study of Burley tobacco flavour (*Nicotiana tabacum* L.). III. Structure, determination and synthesis of 5-(4-methyl-2-furyl)-6-methylheptan-2-one (solanofuran) and 3,4,7-trimethyl-1,6-dioxaspirol(4,5)dec-3-en-2-one (spiroxabovolide), two new flavour components of Burley tobacco, *Helv. Chim. Acta*, 56(1), 265–271, 1973.

Enzell, C. R., Rosengren, A., and Wahlberg, I., Tobacco chemistry. I. The occurrence of C_{11} to C_{19} iso-paraffins and C_{12} to C_{21} anteiso-paraffins in Greek tobacco, *Tobacco,* 169(11), 39–41, 1969.

Fukuzumi, T., Flavor components of oriental tobacco leaves, *Okayama Tabako Shikenjo Hokoku,* (3), 103–134, 1971.

Geisinger, K. R., Jones, T. C., and Schmeltz, I., Further studies of pyrolytic products from some tobacco leaf acids, *Tobacco,* 170(25), 65–66, 1970.

Harke, H. P., Drews, C. J., and Schüller, D., On the occurrence of norbornene derivatives in tobacco smoke condensates, *Tetrahedron Lett.,* (43), 3789–3790, 1970.

Irvine, W. J. and Saxby, M. J., Further volatile phenols of Latakia tobacco leaf, *Phytochemistry,* 8, 2067–2070, 1969.

Kallianos, A., Warfield, A., and Simpson, M., Tobacco Composition Containing 3,5-Disubstituted 2-Hydroxyacetophenones as Tobacco Flavorants (CL. 131 17; A24B), U.S. Patent 3,605,760, September 20th, 1971.

Kaneko, H., Aroma of cigar tobacco. II. Isolation of norambreinolide from cigar tobacco, *Agric. Biol. Chem ,* 35(9), 1461–1462, 1971.

Kaneko, H. and Harada, M., Aroma of cigar tobacco. III. Isolation and synthesis of 3-isopropyl 5-hydroxypentanoic acid lactone, *Agric. Biol. Chem.,* 36(4), 658–662, 1972.

Kaneko, H. and Hoshino, E. K., Isolation from cigar tobacco leaves of tetrahydroactinidiolide (2-hydroxy-2,6,6-trimethylcyclohexylacetic acid γ-lactone), *Agric. Biol. Chem.,* 33(6), 969–970, 1969.

Kaneko, H. and Ijichi, K., Isolation of 2-hydroxy-2,6,6-trimethylcyclohexylidene-1-acetic acid lactone (dihydroactinidiolide) from ether extract of cigar leaves, *Agric. Biol. Chem.,* 32(11), 1337–1340, 1968.

Kaneko, H. and Mita, M., Isolation from cigar tobacco leaves of 2,3-dimethyl-4-hydroxy-2-nonenoic acid lactone, *Agric. Biol. Chem.,* 33(10), 1525–1526, 1969.

Kevanishvili, V. and Oistrapishvili, M., Aroma and tarry substances of tobacco, *Tr. Gruz. Nauchno Issled. Inst. Pishch. Promsti.,* 4, 218–222, 1971.

Kimland, B. et al., Neutral oxygen-containing volatile constituents of Greek tobacco, *Phytochemistry,* 11(1), 309–316, 1972.

Leffingwell, J. C., Young, H. Y., and Bernasek, E., *Tobacco Flavoring for Smoking Products,* R. J. Reynolds Tobacco Co., 1972.

Obi, Y., Muramatsu, M., and Shimada, Y., Evaluation of tobacco quality from pyrolytic aspects. VII. Relationship between chemical constituents of tobacco leaves and gaseous constituents of cigarette smoke, *Agric. Biol. Chem.,* 34(8), 1220–1223, 1970.

Richter, M., Composition of the essential oils of tobacco. I. Preparation and chemical preseparation of essential oils, *Ber. Inst. Tabakforsch. Dresden,* 19, 35–40, 1972.

Roberts, D. L., The structure of a new sesquiterpene isolated from tobacco, *Phytochemistry,* 11, 2077–2080, 1972.

Roberts, D. and Rohde, W., Isolation and identification of flavor components of Burley tobacco, *Tob. Sci.,* 16, 107–112, 1972.

Roeraade, J. and Enzell, C. R., Tobacco chemistry. XIV. Sampling, concentration and examination of tobacco headspace vapors, *J. Agric. Food Chem.,* 20(5), 1035–1039, 1972.

Sakagami, H. and Morishita, I., Spectrophotometric determination of glycyrrhizic acid in the tobacco shreds. Components of licorice root used for tobacco flavoring, *Nippon Nogei Kagaku Kaishi,* 46(9), 443–446, 1972.

Takahara, H. et al., Tobacco extract. Part 7. Aroma components of yellow-leaf tobacco produced in the United States, *Nippon Sembai Kosha Chuo Kenkyusho Kenkyu Hokoku,* 113, 71–75, 1971.

Tamaoki, E. and Noguchi, M., Flavors for Tobacco (CL. A24B, C11B). Carbohydrate Amino Acid Condensate, Japanese Patent 7,109,239, March 9th, 1971.

Webster, C., Reconstituted Tobacco Products of Improved Aroma (CL. A24B). Carotin, German Patent 2,114086, November 4th, 1971.

INDEX

INDEX

A

B

C

D

E

G

H

I

J

K

N

O

P

Q

R

S

W

X

Y

Z